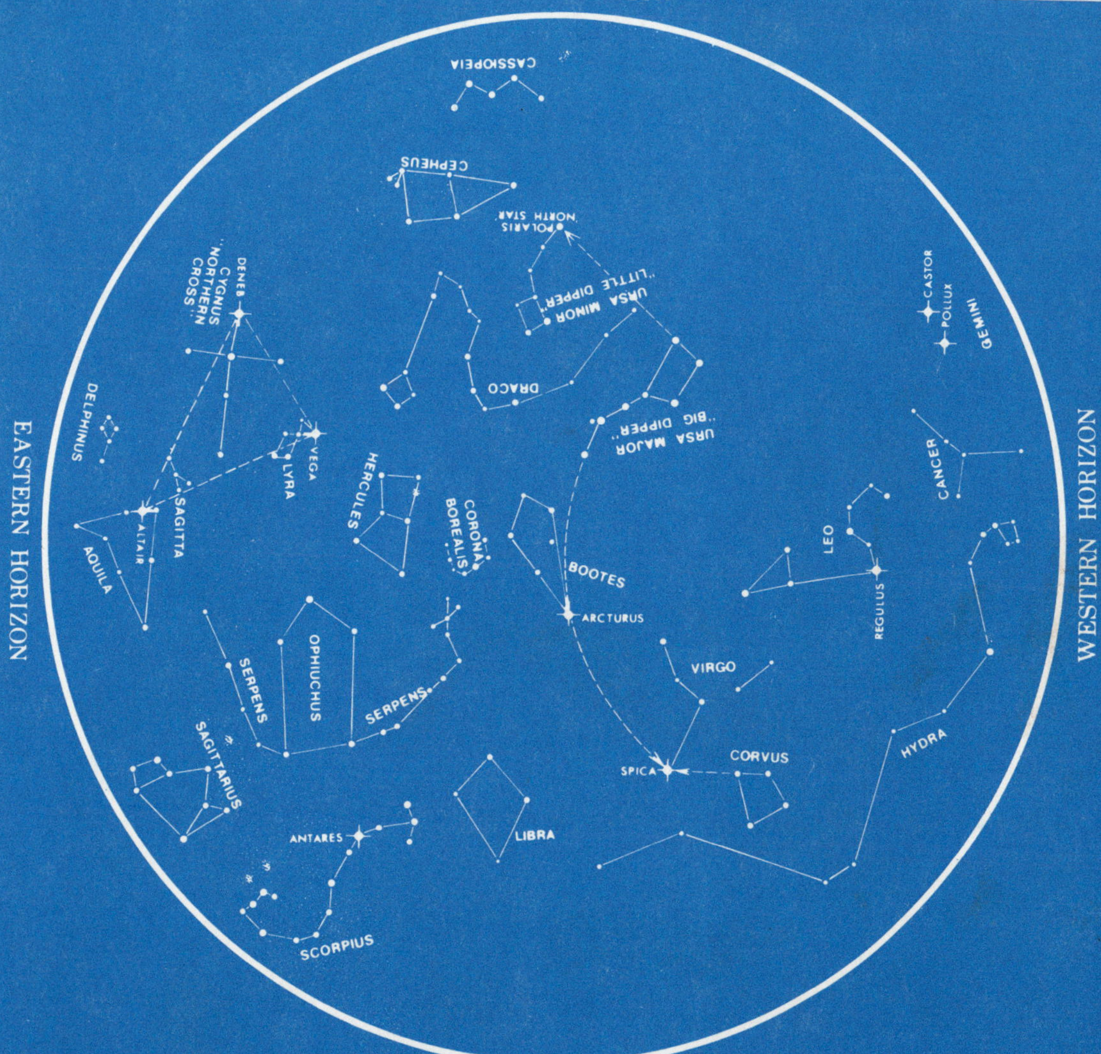

SOUTHERN HORIZON

THE NIGHT SKY IN JUNE

Latitude of chart is 34° N, but it is practical throughout the continental United States.

To use: Hold chart vertically and turn it so the direction you are facing shows at the bottom.

Chart time (Local Standard):
10 p.m. First of month
9 p.m. Middle of month
8 p.m. Last of month

THIRD EDITION

The Dynamic Universe

An Introduction to Astronomy

The great supernova of 1987 (lower right) in the Large Magellanic Cloud. At upper left is the Tarantula nebula, site of active star formation. (Ian Shelton, David Dunlap Observatory, Toronto)

THIRD EDITION

THE DYNAMIC UNIVERSE

AN INTRODUCTION TO ASTRONOMY

Theodore P. Snow
University of Colorado at Boulder

West Publishing Company
St. Paul New York San Francisco Los Angeles

Composition: Parkwood Composition Service
Copyediting: Carol Danielson
Text and cover design: Janet Bollow
Illustrations: Henry Taly Design; House of Graphics
Cover: Supernova 1987A. Photo by Ian Shelton,
David Dunlap Observatory, Toronto.

50 W. Kellogg Boulevard
P.O. Box 64526
St. Paul, MN 55164-1003

Printed in the United States of America

LIBRARY OF CONGRESS CATALOGING-IN-PUBLICATION DATA

Snow, Theodore P. (Theodore Peck)
The dynamic universe.

Bibliography: p.
Includes index.
1. Astronomy. I. Title.
QB43.2.S66 1988 520 87-31599
ISBN 0-314-64212-9

DEDICATION

For Connie

CONTENTS

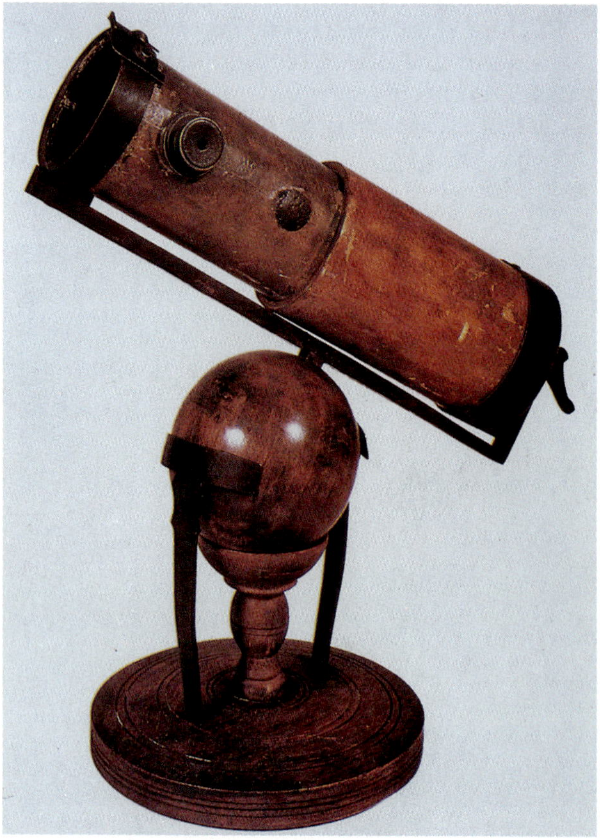

CHAPTER 10

VENUS: A CLOUD-COVERED INFERNO 186

CHAPTER 11

MARS AND THE SEARCH FOR LIFE 202

CHAPTER 12

MERCURY 221

CHAPTER 13

JUPITER: GIANT AMONG GIANTS 231

Chapter 14

Saturn and Its Attendants 250

Chapter 15

The Outer Planets 271

Chapter 16

Space Debris 292

Chapter 17

Adding It Up: Formation of the Solar System 317

SECTION IV

THE STARS 337

CHAPTER 18

THE SUN 339

CHAPTER 19

STELLAR OBSERVATIONS: POSITIONS, MAGNITUDES, AND SPECTRA 363

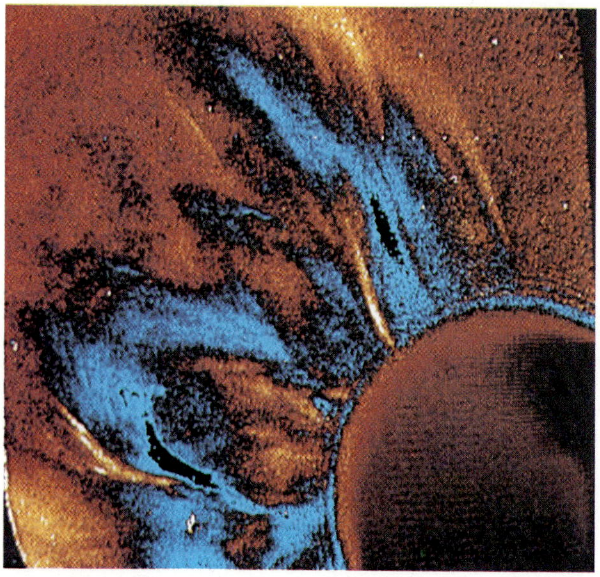

CHAPTER 20

FUNDAMENTAL STELLAR PROPERTIES AND THE H-R DIAGRAM 382

CHAPTER 21

STELLAR STRUCTURE: WHAT MAKES A STAR RUN? 398

CHAPTER 22

STAR CLUSTERS AND OBSERVATIONS OF STELLAR EVOLUTION 415

CHAPTER 23

LIFE STORIES OF STARS 433

SPECIAL INSERT:

THE GREAT SUPERNOVA OF 1987 452

CHAPTER 24

STELLAR REMNANTS 457

SECTION V

THE MILKY WAY 481

CHAPTER 25

STRUCTURE AND ORGANIZATION OF THE GALAXY 483

CHAPTER 26

THE INTERSTELLAR MEDIUM 505

CHAPTER 27

THE FORMATION AND EVOLUTION OF THE GALAXY 526

SECTION VI

EXTRAGALACTIC ASTRONOMY 541

CHAPTER 28

THE NATURE OF THE NEBULAE 543

CHAPTER 29

CLUSTERS AND SUPERCLUSTERS: THE DISTRIBUTION OF GALAXIES 563

CHAPTER 30

UNIVERSAL EXPANSION AND THE COSMIC BACKGROUND 579

CHAPTER 31

PECULIAR GALAXIES, EXPLOSIVE NUCLEI, AND QUASARS 596

CHAPTER 32

COSMOLOGY: PAST, PRESENT, AND FUTURE OF THE UNIVERSE 616

SECTION VII

LIFE IN THE UNIVERSE 639

CHAPTER 33

THE CHANCES OF COMPANIONSHIP 641

APPENDICES

PREFACE

To study astronomy is, in a sense, the most human thing we can do. What distinguishes us from lower creatures, if not our curiosity, our compulsion to explore and discover? And what exemplifies this compulsion better than the study of the universe?

We probe the heavens (and the Earth) by all possible means, and we do it for no other reason than to learn whatever there is to be known. Astronomy has produced many useful byproducts, of course, and could be (and often is) justified solely on that basis. That is not the real reason for astronomy, however.

This textbook represents an attempt by an astronomer to share both the knowledge and the intellectual gratification of our science. There is considerable beauty in the universe for the eye and mind to behold. Just as it is visually stimulating to gaze at a great glowing nebula or a colorful moon, it is pleasing to the intellect to grasp a new understanding of one of the grand themes of the cosmos. It is hoped that the reader of this book will gain by doing both.

This textbook is intended for the student who has not chosen science as his or her major area of study, but who needs an appreciation of science as a vital aspect of preparation for a career. It is as important for such a student to gain some perspective on the general nature of science as it is to learn a great deal of specific information about a particular discipline in the sciences. For that reason, this text stresses the philosophy and outlook of the scientist as well as the knowledge we have gathered about the physical universe we live in.

It is probably as important for the student to understand how we know what we know as it is to understand what we know. In this era of instantaneous communication and universal access to information, we need more than ever to be able to discriminate among competing hypotheses, to be able to judge the reasonableness of ideas that are advanced. This text in astronomy is written with the underlying theme that to know the workings of science is one of the most important tools we have for meeting the challenges of our technological society.

This edition, like its predecessors, covers the entire scope of astronomy, from the most nearby objects to the most distant, from the smallest to the largest. Many questions are left unanswered, because there is so much we do not know, and because it is important for students to be aware of that. Not knowing the answers, but knowing how to pursue them, is the essence of science.

This third edition has some new features, and of course reflects all the exciting new developments in astronomy that have occurred since the second edition was published. Particularly noteworthy are the *Voyager 2* encounter with Uranus, which added in colorful detail a new world to those known to us; the apparition of Halley's comet, which provided us some spectacular images as well as new knowledge of one of the most ancient objects in the solar system; and of course the great supernova of 1987. This last event, the brightest supernova in nearly 400 years, occurred in late spring of 1987, and since then has dominated the entire world of astronomy. The drama of the discovery, the excitement of the search for the progenitor star on old photographs, the many surprises and inconsistencies in the theory of the explosion, and the continuing development of the expanding remnant as we wait to see whether a neutron star has been born out of the violence, all have added up to a marvelous opportunity, not only for scientists, but also for the student seeking to know how science works. We have gone to extraordinary lengths to bring to this edition as complete and current an account of the event as possible. In addition to fitting the supernova story into context in the chapters on stellar evolution, we have inserted a substantial special section on the supernova between chapters 23 and 24, allowing us to update it late in the fall of 1987, just before going to press.

Many other additions to the scientific material have been made, along with some new features. The book is now in full color, allowing the beauty of astronomy to be better appreciated, and also allowing the illustrations to be placed with the text that describes and discusses them, rather than having separate color plates. Many of the drawings have been replaced, and several new ones have been added. All of the end-of-chapter questions have been rewritten, so that now each chapter has a number of review questions and, separately, problems requiring numerical solution. Many tables have been added throughout the text, some containing data and others consisting of summaries of the discussion. The tables are designed to aid student understanding and review, and to provide the instructor with material for various kinds of exercises or problem assignments.

Additional factual data are incorporated into the Appendices, which have been updated for new information (as provided by the *Voyager* planetary encounters, for example). New appendices listing special symbols used in the book and the objects in the famous Messier catalog, a very convenient resource for people wanting to view astronomical objects through binoculars or small telescopes, have been included. In additon, several appendices have been expanded, including those on stellar data, interstellar molecules, major telescopes, and groups and clusters of galaxies.

The arrangement of the text remains traditional, with an introductory section on the background of astronomy, both in history and in basic physics; a section on the solar system, dealing with the planets as individuals before discussing interplanetary bodies and then the formation of the entire system; a section on stars and their lives and deaths; a section on the structure and evolution of our galaxy; a set of chapters on extragalactic astronomy and the universe as a whole; and a final, brief section on the possibilities that life may exist elsewhere. At the beginning of each of these sections is an introduction that leads the student into the material, and at the end of each is a Guest Editorial in which a leading scientist in the field shares his or her thoughts on current problems or controversies and future directions for research.

The book is designed so that the sequence of sections may be easily altered. For example, if it is desired to teach the sections on stars, the galaxy, and the universe before discussing the solar system, one need only skip directly from Chapter 7 to Chapter 18 and then go on to the end before returning to Chapter 8, where the solar system studies begin. One change made in this edition is to move the chapter on the Sun to the beginning of the section on stars, so that skipping or delaying the solar system discussion will not prevent the student form learning about the nearest and best-understood star. The summary chapter on the solar system (Chapter 17) includes enough information on the Sun that the discussion of the system as a whole and its formation is complete as it stands.

The well-received Astronomical Insights have been carried over into this edition, with a substantial number of new ones added. These inserts, placed within the chapters, describe people, discoveries, current controversies or new hypotheses related to the subject matter of the text. They are meant to enhance the students' enjoyment of the material, or add understanding of complex topics, but above all they are designed to increase understanding of the scientific process.

Supplemental materials for this text include an updated version of the *Study Guide* authored by Catharine D. Garmany and myself, along with K. S. Bjorkman and J. O. Bennett (all University of Colorado), and a revised edition of the *Instructor's Manual* by Stephen J. Shawl (University of Kansas), with Bjorkman and Bennett. As before, the *Instructor's Manual* contains helpful discussions of strategies in teaching, provides a large number of exam questions (with answers), and gives complete answers to all the review questions from the main text. The *Study Guide,* intended to help the student get maximum benefit from the text, contains brief chapter summaries, lists of key words and phrases, self-tests, and complete bibliographies of articles on relevent topics, taken from a wide assortment of magazines and journals. In addition to the *Study Guide* and the *Instructor's Manual,* another aid to teaching is offered to large adopters of the text: a set of transparencies for use with overhead projectors, showing a number of useful diagrams and illustrations from the text.

At every step during the preparation of this text, vital assistance was provided by a number of people, whose help is acknowledged with gratitude (with apologies to anyone inadvertantly omitted). The most important guidance and support was provided by my wife, Connie; by the West Publishing Company editor, Denise Simon; and the production editor, Tad Bornhoft. M. M. Allen gathered data for tables and reading lists, and K. S. Bjorkman and J. O. Bennett helped with both the *Study Guide* and *Instructor's Manual.* A very important contribution to the overall quality and accuracy of the book was made by Stephen Shawl, who scrutinized the galley proofs and, as he has during the preparation of earlier editions, made many useful suggestions regarding content as well.

Among my colleagues at the University of Colorado and elsewhere, several have helped by either reviewing sections of the text, providing new figures, or updating data for tables and appendices. Particularly generous in this connection were J. M. Shull, L. Esposito (both University of Colorado), and H. Eichhorn (University of Florida). Others in this category include J. Doggett, J. K. Malville, J. C. Brandt, R. Fesen, and R. Stencel (all University of Colorado); K. Eriksson (Uppsala University, Sweden); U. Fink (University of Arizona); L. Frederick (University of Virginia); R. Dreiser (Yerkes Observatory); W. Golisch (NASA Infrared Telescope Facility); M. Phillips (Cerro Tololo Inter-American Observatory); C. Covault (*Aviation Week*); and D. Malin (Anglo-Australian Telescope).

Reviewers of the revised manuscript were:

Gordon Baird, University of Mississippi
Larry Browning, Marquette University
Heinrich Eichhorn, University of Florida
Laurence Frederick, University of Virginia
Ernest Fusen, Santa Monica Community College
Robert Habershan, Seattle Central Community College
Jeffrey Hayes, Rutgers University
Sally Hoffman, Santa Fe Community College
Rondo Jeffery, Weber State College
Frederick Kleinhans, Indiana University/Purdue University, Indianapolis
Thomas F. Scanlon, Grossmont College
Alex G. Smith, University of Florida
Raymond White, University of Arizona

A special debt is owed to those who wrote the Guest Editorials, for adding their thoughts and visions to my own less elegant discussions. Much of the excitement of astronomy lies in the pursuit of new revelations beyond the scope of current knowledge, and the essays contributed by leaders in this pursuit help immeasurably to impart this excitement to the reader. Those contributing new essays to this edition are John C. Brandt (University of Colorado), J. Roger Angel (University of Arizona), Stanford Woosley (University of California–Santa Cruz), Robert Kirshner (Harvard University), and Margaret Geller (Smithsonian Astrophysical Observatory). In addition, Edwin Krupp (Griffith Observatory) and Frank Drake (University of California–Santa Cruz) have reviewed and permitted the use of their essays written for earlier editions.

For all of these people, and to the students whose responses to my teaching philosophies have also helped to shape this book, I am grateful. With their continued input, I trust that this book will continue to evolve, as does our understanding of the dynamic universe.

SECTION I

The Nighttime Sky and Historical Astronomy

Our study of astronomy begins with an overview of the breadth, relevance, and excitement of the science that encompasses all of the universe. We begin, in Chapter 1, with a broad description of the nighttime sky and of human beings' fascination with its splendor. Next, in Chapter 2, we examine more closely the many phenomena that can be seen and appreciated with only our eyes as observing equipment. This arms us with all the knowledge our ancestors had as they attempted to develop a successful picture of the cosmos and their place in it. In Chapter 2 we go well beyond simply describing the sky; we also uncover the explanations for the observed phenomena. Thus we start with knowledge that took mankind millennia to develop.

We then trace the historical development of astronomy, beginning with the earliest recorded astronomical observations and hypotheses, and following the ofttimes slow growth of science through the ages. The basis of our modern science lies in the Mediterranean region of Europe; therefore we emphasize developments there. We also explore the genesis of astronomy in other parts of the world, where sophisticated knowledge ultimately either led to dead ends or merged with the mainstream of development.

In the fourth chapter we discuss the groundwork for modern science that was laid during the Renaissance, when fresh ideas arose in astronomy, as in all forms of human endeavor. We learn to appreciate the awesome breakthroughs made by giants such as Copernicus, Brahe, Kepler, and Galileo, who led the way toward a correct understanding of the universe and the place of man and his planet in it.

CHAPTER 1

THE ESSENCE OF ASTRONOMY

Telescopes at Kitt Peak Observatory (National Optical Astronomy Observatories).

CHAPTER PREVIEW

The oldest of all sciences is perhaps also the most beautiful. No artificial light show can rival the splendor of the heavens on a clear night, and few intellectual concepts can compare with the beauty of our modern understanding of the cosmos.

Today we study astronomy for a variety of reasons, some technological, some practical, but no one loses sight of the underlying majesty, of the human instinct for intellectual satisfaction. To study astronomy is to ask the grandest questions possible, and to find hints at their answers is to satisfy one of humankind's most deeply ingrained yearnings.

In this text we explore astronomy in the modern context, which is highly technical and sophisticated, but we will endeavor to retain the sense of wonder and beauty that has motivated the science since the beginning. Although some may argue that astronomy has little practical use, we will see that its origins are rooted in very practical requirements for methods of keeping time and maintaining calendars. In this chapter we begin our study by defining astronomy and introducing some simple terminology that will assist us in later chapters.

WHAT IS ASTRONOMY?

Astronomy is the science in which we consider the entire universe as our subject. It is the science in which we derive the properties of celestial objects and from these properties deduce the laws by which the universe operates. It is the science of everything.

Technically, we might say that astronomy is the science of everything except the Earth, or that it is the study of everything beyond the Earth's atmosphere, since the Earth and its atmosphere fall into the purview of other disciplines, such as geophysics or atmospheric science. We will find, however, that the study of astronomy necessarily includes an examination of the properties and the evolution of the Earth and its atmosphere.

In the modern sense, astronomy is probably more aptly called **astrophysics.** Ever since the time of Sir Isaac Newton (the late seventeenth century), the universe has been explored by applying the laws of physics—most of them derived from earthly experiments and observations—to celestial phenomena. Other scientific disciplines enter into our discussions as well: To study the planets, for example, we must know something of geology and geophysics; and to analyze molecules in space, we must understand the principles of chemistry.

Astronomical observations were made and records kept at least as early as the time of the most ancient recorded history (Fig. 1.1), and we believe that the skies were studied and their cyclic motions pondered long before that. In the earliest times, astronomy had a practical motivation: Knowledge of motions in the heavens made it possible to predict and plan for certain significant events, such as the changing of the seasons.

Along with the practical came the whimsical and the spiritual. In ancient times, events in the heavens were thought to exert some influence over the lives of people on Earth, and many early astronomers practiced what today we call astrology. Many a monarch retained the services of an astronomer, not only to foretell the seasons, but also to provide advice on strategies for war, love, politics, and business. In many cultures, religion and astronomy were intimately linked, so astronomers attained the importance that major religious leaders do today.

The rich and diverse Greek civilization that flourished for many centuries before and around the time of Christ developed several sciences—astronomy included—beyond the mystical and the spiritual. The Greeks had many preconceived notions about the nature of the universe, but they also made many rational advances in understanding the heavens, setting the stage for the development of modern science many centuries later. Following the era known as the Dark Ages, astronomy (like many other disciplines) experienced the pangs of rebirth during the Renaissance, becoming a rational and methodical science. During the Renaissance, battles were yet to be fought between religion and science, but the course was set. The work of pioneers such as Copernicus, Galileo,

FIGURE 1.1 AN ANCIENT ASTRONOMICAL SITE. *Stonehenge, a stone monument in England, was built according to astronomical alignments in prehistoric times. Stonehenge is discussed more fully in Astronomical Insight 3.2.* (C. D. McLoughlin)

Kepler, and especially Newton placed astronomy on a firm, physical basis.

Modern astronomy still has a practical aspect, but few who pursue the science do so for that reason. Today we study astronomy primarily for the sake of expanding our knowledge; it is research of the purest sort. Even so, many practical and concrete benefits derive from astronomy; witness, for example, the many technological spin-offs from the space program or the multitude of new physical and chemical processes, applicable here on Earth, that have been discovered through astronomical observation.

No matter how analytical we may be in modern astronomy, we never lose sight of the same basic human feelings that inspired our ancestors. The modern astronomer, who may spend observing time using a large telescope and a variety of complex electronic instruments (Fig. 1.2), still treasures the moments spent outside the telescope building, simply watching the skies with the same tools as the ancients.

FIGURE 1.2 A LARGE MODERN TELESCOPE. *This large reflecting telescope at the Cerro Tololo Inter-American Observatory, located in Chile and operated by the U.S. government, has a diameter of 4 meters, or roughly 13 feet.* (National Optical Astronomy Observatories)

— A TYPICAL NIGHT OUTDOORS —

Let us imagine that we are sitting outdoors on a fine, clear night, and that we are far from city lights and other distractions. This is easy to do and highly recommended; by simply getting out into the countryside, we can see many of the beautiful objects that we will be studying in more detail using this text.

FIGURE 1.3 A CLUSTER OF STARS. *This group of stars is held together by mutual gravitational attraction. This cluster, called the Pleiades, is easily visible to the unaided eye in the autumn and winter.* (Palomar Observatory, California Institute of Technology)

FIGURE 1.4 A GASEOUS NEBULA. *Clouds of gas and tiny dust particles such as this are the birthplaces of stars. This is the Orion Nebula, bright enough to be seen with only a modest telescope or pair of binoculars.* (© 1981 Anglo-Australian Telescope Board)

The most obvious objects in the sky, assuming that the Moon is not in one of its brightest phases, are the stars. They appear in profusion, scattered across the heavens, displaying a wide range of brightnesses and subtle variations in color. They twinkle, giving the appearance of vitality. Here and there we may see concentrations of stars in a cluster (Fig. 1.3), or possibly a dimly glowing gas cloud (Fig. 1.4).

If it is a moonless night, we see a broad, diffuse band of light across the sky. This is the Milky Way, our own galaxy of stars, seen edgewise from an interior position (Fig. 1.5). The Milky Way consists of countless stars intermixed with patchy clouds of interstellar gas and dust.

We may also see a few bright, steady objects that do not appear to twinkle. These are the planets, and we may find as many as five of them to be visible on a given night, distributed along a great arc through the sky. Careful observation over several nights will reveal that the planets are all moving gradually along this arc, changing their positions relative to the fixed stars.

The most prominent object in the nighttime sky is usually the Moon (Fig. 1.6), which shines so brightly when near full that it drowns out the light of all but the brightest stars and planets. The Moon, about one-fourth the diameter of the Earth but some 250,000 miles (or about 60 times the radius of the Earth) away, presents various appearances to us, depending on how much of its sunlit portion we see. The Moon is always to be found somewhere along the same east-west strip of the sky, called the **zodiac,** where the planets travel.

Occasionally we may see brief flashes or trails of light known as **meteors.** These "shooting stars" can be spectacular events, particularly when they arrive with great frequency, as they do during a meteor shower.

Comets are occasional visitors to our sky (Fig. 1.7), as the recent passage of Comet Halley has reminded us all. Perhaps once or twice a year a comet is found that is bright enough to be seen with the unaided eye. These largely gaseous bodies orbit the Sun, as do the Earth and other planets, but in very elongated paths that bring them close enough to the Sun to heat up and glow visibly only for brief periods of days or weeks. Some of the more spectacular, bright comets were interpreted in ancient times as harbingers of catastrophe.

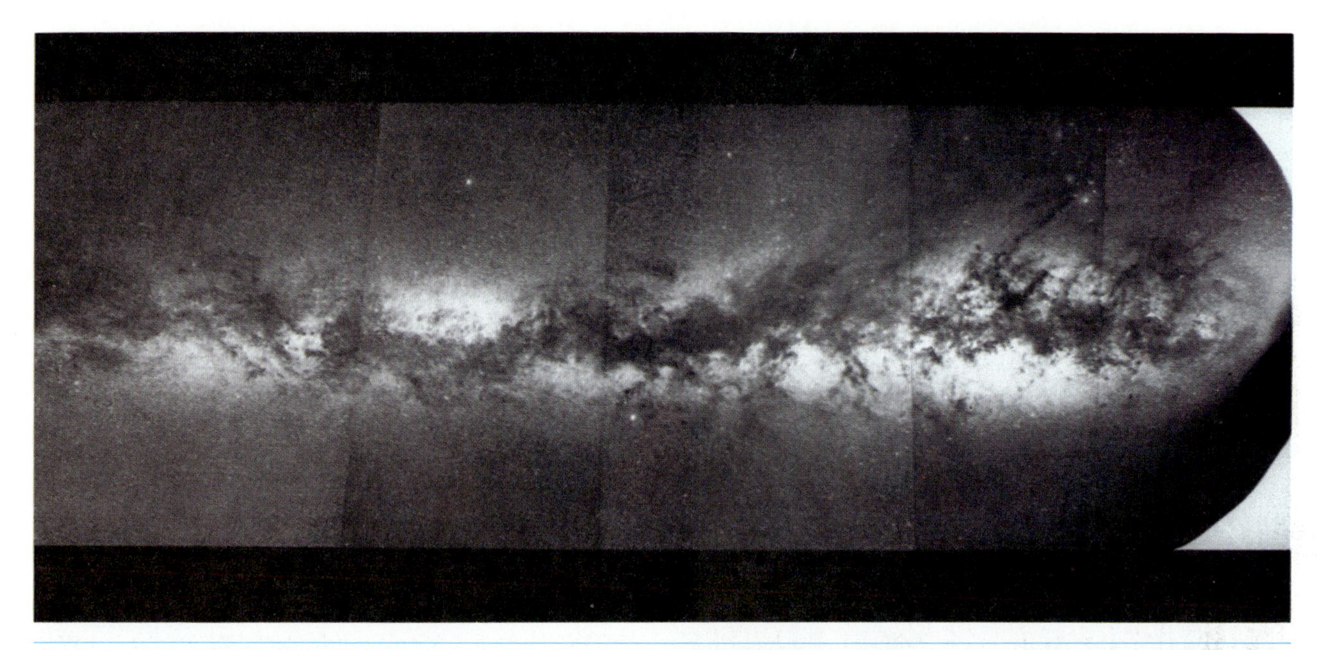

FIGURE 1.5 THE MILKY WAY. *This composite photograph shows the hazy band of light across the sky known as the Milky Way. It is a cross-sectional view of the disk of our galaxy, which contains roughly 100 billion stars.* (Mt. Wilson and Las Campanas Observatories, Carnegie Institution of Washington)

There is a complex pattern of motions in the sky, some of it evident to an alert watcher in an hour or so, other aspects requiring careful observations over hours, days, or weeks. The most obvious motion is the steady rotation of the entire sky; objects rise in the east and set in the west, as a reflection of the earth's rotation. (The terms **rotation** and **revolution** are sometimes interchanged in everyday conversation, but here rotation means the same thing as spin, whereas revolution means orbital motion, such as the motion of the Earth around the Sun or of the Moon around the Earth.) Another motion that can be discerned readily is that of the Moon with respect to the stars. As it orbits the Earth, the Moon moves a distance in the sky equal to its own apparent diameter in just one hour, so it is possible to see its position change with respect to the background stars in a short time. Other cyclical motions, such as those of the planets as they gradually travel along their orbits about the Sun, require more patience and care to be discovered. It is noteworthy, however, that ancient astronomers noticed many of the regular patterns, some of them quite subtle, in the motions of heavenly bodies. That they did so is a testimony to the care and diligence they applied to their studies of the skies.

FIGURE 1.6 THE MOON. *Astronauts on one of the Apollo missions made this photograph from space. The portion at the right in this image is not visible from Earth.* (NASA)

FIGURE 1.7 A COMET. *This is a view of Comet Kohoutek during its 1973 passage through the inner solar system.* (NASA)

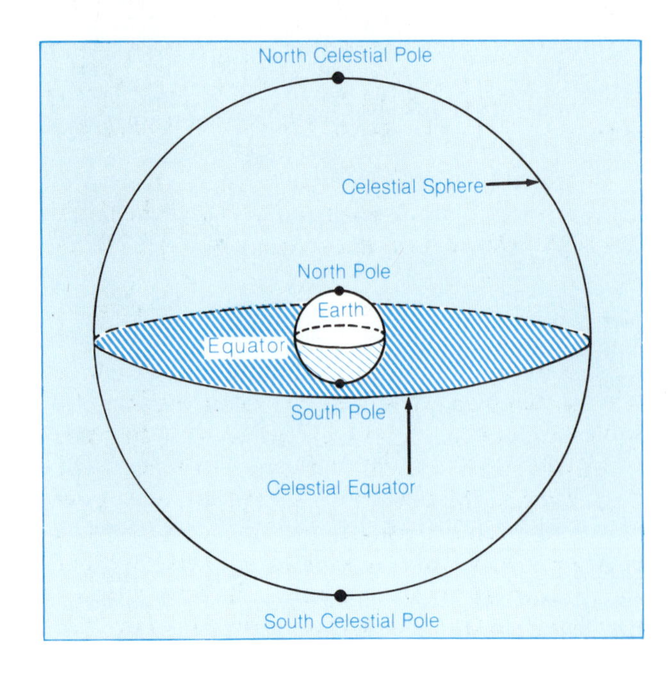

THE VIEW FROM EARTH

When we look at the sky we do not see it in three dimensions, because there are no obvious clues to tell us the distances to the objects we see. Long ago this fact led to the concept of the **celestial sphere** (Fig. 1.8), in which the stars and other objects in the sky are said to lie on the surface of a sphere that is centered on the Earth. Although we no longer think of

FIGURE 1.8 THE CELESTIAL SPHERE. *Because we see the sky in only two dimensions, it is useful and convenient to visualize it as a sphere centered on the Earth, with the stars and other bodies set on the surface of the sphere. We measure positions of objects on the celestial sphere in angular units, because the actual distances are not directly known. (Throughout this text, we will learn about distance measurements in astronomy.)*

ASTRONOMICAL INSIGHT 1.1

THE PHILOSOPHY OF SCIENCE

Science progresses by hypothesis and test, by trial and error. A scientist is willing to revise a theory in the light of new information, rather than adopting a hypothesis and then attempting to ignore or rationalize data that contradict a preconceived notion.

If there is anything that a scientist adopts as an article of faith, it is that no theory is above the possibility of modification—that our knowledge always has room to grow. This willingness to change does not mean that scientific theories are wrong or invalid. A scientific theory should be regarded simply as the best explanation available that is consistent with the known facts. A theory is modified or replaced when more facts become known or when a better explanation is found.

A theory usually starts as a hypothesis, an initial suggestion made to explain some phenomenon that is observed. As new information is gathered, the hypothesis develops into a theory, a framework allowing for prediction and test. Many theories fall by the wayside quickly because new information contradicts them, while others persist and become widely accepted because they are consistent with new data that become available. One feature of any theory is that it leads to specific predictions about what will happen in new situations. Part of the job of testing a new theory is to make predictions from it and then make observations or perform experiments to test those predictions.

One of the most important goals of this text is to impart to the student an understanding of how scientific progress is made—of how theories are developed, tested, and revised. For this reason, many examples of discoveries are described, some in Astronomical Insights such as this one, and others in the main text. These descriptions will help illustrate how scientists develop, test, and modify theories in the light of new information. They will show that even the best scientists make errors, but that the common thread is their willingness to discard incorrect hypotheses when conflicting evidence is found or a better explanation appears. The reader will see that scientists do not always know what to look for, but that our knowledge of the universe grows because scientists seek the best explanation for what is observed, rather than trying to modify what is observed to conform to existing theories. The hope is that these descriptions will help the student develop a sense of how science works, and a better perspective on new theories and controversies that may be encountered beyond the classroom.

this as literal truth, it may still serve as a convenient device for discussing and visualizing the heavens.

Directions and separations of objects on the celestial sphere are measured in angular units, because lacking knowledge of distances to objects, we have no easy way to determine their actual separations in true distance units such as miles or kilometers. Thus we may specify where a star is by saying how many degrees, minutes, and seconds it is from another star or from a reference direction.

We have noted that the stars rise and set with the daily rotation of the Earth. This gives us a natural basis for timekeeping, and our standard units of time are based on the Earth's rotation. The length of the day is equivalent to the rotational period of the Earth (a more specific definition is given in the next chapter).

Anyone who has traveled from one hemisphere to the other may have noticed that the stars that are visible are not the same in the Northern Hemisphere as in the Southern (Fig. 1.9). The portion of the sky that we can see depends on our latitude (our distance, in degrees, north or south of the equator). For those of us living in the Northern Hemisphere, a large region of the southern sky is beyond our view. The constellations that we see vary as we travel north or south, a fact that was well known to early astronomers, who deduced from this (and other) evidence

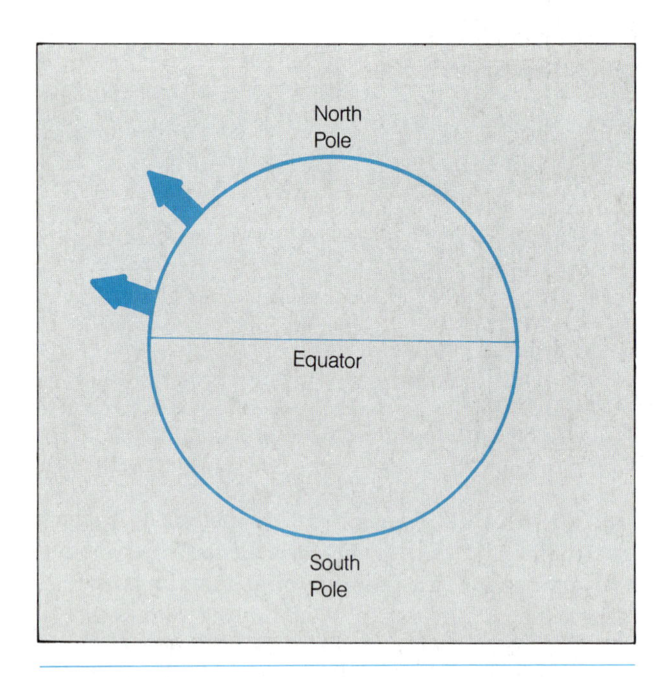

FIGURE 1.9 LATITUDE AND OUR VIEW OF THE SKY. *The portion of the sky that we see depends on where we are with respect to the Earth's equator. The arrows indicate the overhead direction from two different latitudes.*

that the Earth is round. One consequence is that astronomers must have telescopes in both hemispheres in order to study the entire sky.

We usually cannot observe celestial objects during daylight, so our view of the heavens at any particular time is limited to only the half of the sky that is seen at night. Because of the Earth's motion around the Sun, however, the part of the sky that is up during the night gradually changes (Fig. 1.10). Therefore, any part of the sky can be observed at night at some time of the year; we need only wait until the appropriate time to observe a given object.

FROM THE EARTH TO THE UNIVERSE: THE SCALE OF THINGS

We have been describing the appearance of the sky to the unaided eye, which has necessarily limited us to nearby objects such as the Sun, the Moon, the planets, and the stars in our part of the local galaxy. It is interesting now to expand our horizons, and to try to comprehend the scale of the universe beyond this "local" neighborhood. Some of the distance scales we discuss are listed in Table 1.1

Even the nearest star is much farther from the Earth than any solar system object. If we take the average Sun-Earth distance as a unit of measure (we call this the **astronomical unit** or **AU**), the nearest star is nearly 300,000 of these units away from us. The most distant known plant, Pluto, is on the average only about 40 AU from the Sun, so we see that the stars are much more widely dispersed in space than the objects within our solar system.

If we could reduce the scale of the solar system, it might help us to visualize the relative distances. For example, if we let the Earth be the size of a basketball (specifically, let us say that its diameter is

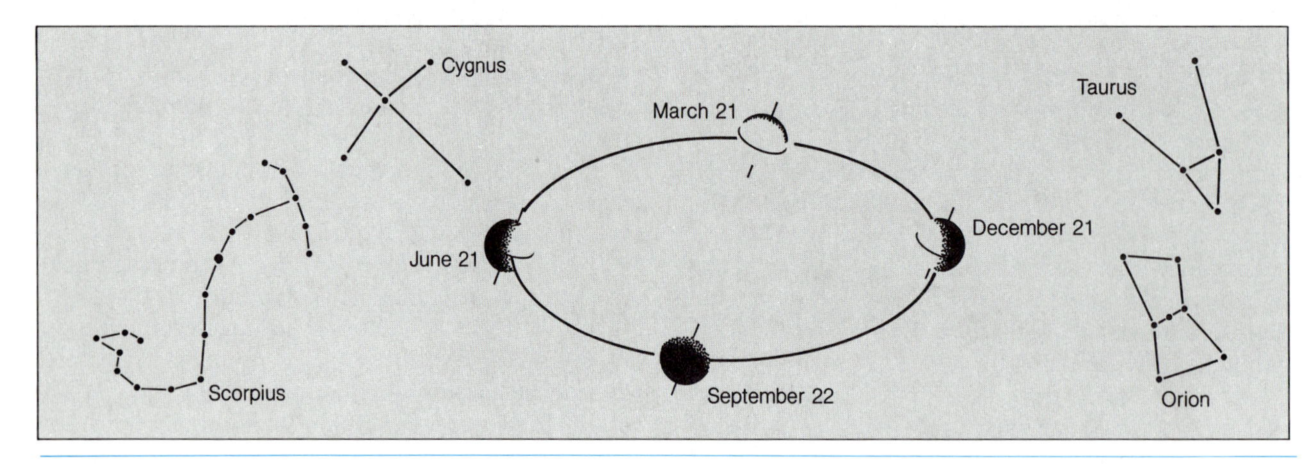

FIGURE 1.10 THE CHANGING VIEW OF THE SKY WITH THE SEASONS. *As the Earth orbits the Sun, the portion of the sky that we can see at night changes. A few prominent constellations are shown here.*

ASTRONOMICAL INSIGHT 1.2

WHAT IS AN ASTRONOMER, ANYWAY?

Throughout this text we will be referring to astronomers, scientists, astrophysicist, and physicists. In a book that endeavors to summarize all that we know about the universe and its contents, we can hardly omit a description of the people who devote their time to developing this knowledge.

There are thousands of people in the United States alone who study astronomy either as a vocation or an avocation. Representative of the latter are the amateur astronomers who, often on their own, but in many cases through local and even nationwide organizations, engage in a variety of astronomical activities. Telescope-making, astrophotography, long-term monitoring of variable stars, public programming, and just plain stargazing are included. If you wish to join such a group, to get advice on buying a telescope, or to learn techniques such as photography of celestial objects, your best bet is to get in touch with an amateur astronomy group. Local clubs exist in most major cities, and regional associations are everywhere. These groups may be difficult to find in the telephone book, but a telescope shop or planetarium is bound to know whom to contact.

The amateur astronomers in the United States outnumber the professionals. There are some 3,500 members of the American Astronomical Society, the principal U.S. professional astronomy organization. These people, as a rule, fall into a limited number of categories: those who work at colleges and universities, where they engage in both teaching and research; those who do research work at government-sponsored institutions, such as the national observatories and federal laboratories like NASA; and those who perform research and related engineering functions in private industry, most often with companies that are involved in aerospace activities.

About 8 percent of the members of the American Astronomical Society are women. This is certainly a low percentage, but a recent study has indicated that there is little discrimination in the hiring of professional astronomers. The small number of women in astronomy thus reflects the low number who choose to enter the field at the graduate level, so it is at that level that improvement must be made.

Most of the funding for research in astronomy, even for those not working directly for federal labs and observatories, comes from the government. A significant function of astronomers on a university faculty is to write proposals, usually to the National Science Foundation or to NASA, for support of the research programs they wish to carry out.

The terms **astronomer** and **astrophysicist** have come to mean pretty much the same thing in modern usage, although historically there was a difference. An astronomer was one who studied the skies, gathering data but doing relatively little interpretation; an astrophysicist was primarily interested in understanding the physical nature of the universe, and therefore carried out comprehensive analyses of astronomical data or did theoretical work, in both cases applying the laws of physics to phenomena in the heavens. Nearly all modern astronomers do astrophysics, however, to varying degrees ranging from the observational astronomer at one end of the spectrum to the pure theorist at the other. The two terms for people engaged in these pursuits are therefore used interchangeably today. Many modern astronomers call upon the fields of engineering (for instrument development), chemistry (in studying planetary and stellar atmospheres and the interstellar medium), geophysics (in probing interior conditions in planets and other solid bodies), and possibly even biology, but always with an underlying foundation in physics.

If you want to become a professional astronomer, you should be aware from the outset that the field is small and that job opportunities are both limited and often subject to the vagaries of federal funding. If you persist, the best course is to study physics, at least through the undergraduate level, and then plan to attend graduate school in an astronomy department or in physics. (This latter option may make you a bit more versatile, and it is never a handicap when entering astronomy later.) On the bright side, demographic studies have shown that there may be a shortage of astronomers for a period beginning in the late 1980s and extending through the rest of this century. Perhaps the timing will be right for you.

1 foot), we can convert the rest of the solar system to the same scale. The Sun would be almost 110 feet in diameter, and the distance from the Sun to the Earth would be nearly 12,000 feet, or more than 2 miles. Pluto would be a tennis ball-size object roughly 90 miles away. The distance from the Sun to Alpha Centauri, the nearest star, would be over 600,000 miles, or more than twice the actual Earth-Moon distance!

Now consider our galaxy, the vast collection of stars to which our Sun belongs (Fig. 1.11). The Milky Way galaxy contains roughly 100 billion stars, arranged in a huge disklike structure having a diameter of about 100,000 light-years. (A **light-year,** the distance light travels in a year at its speed of 186,000 miles per second, is equal to about 5,900 billion miles.) The distance to Alpha Centauri is about 4 light-years, and the most distant stars easily seen with the unaided eye are several hundred light-years away. (The majority of the brightest stars in the sky are actually quite nearby, by galactic standards; see Appendix 8). Light from these stars has been traveling for hundreds of years when it reaches our eyes, and light from the far side of the galaxy takes about 100,000 years to reach us.

The nearest galaxies beyond the limits of the Milky Way are the Magellanic Clouds, two irregularly shaped fuzzy patches of light visible only from the Southern Hemisphere (Fig. 1.12) The Magellanic Clouds lie between 150,000 and 210,000 light-years from the Sun, so we see that they are not very far outside the galaxy. They are considered satellites of the Milky Way, orbiting it in a time of several hundred million years. Light from the Magellanic Clouds takes roughly 200,000 years to reach us. The most distant object visible to the unaided eye is another galaxy similar to the Milky Way, called Andromeda, after

TABLE 1.1 SIZE SCALES IN THE UNIVERSE

Object or phenomenon	Size*
Atomic nucleus	10^{-14}cm
Atom	10^{-8}
Virus	5×10^{-6}
Interstellar dust grain	5×10^{-5}
Bacterium	10^{-4}
Human body cell	5×10^{-3}
Human	2×10^{2}
Planet Earth	1.3×10^{9}
Sun	1.4×10^{11}
Sun-Earth distance	1.5×10^{13}
Distance to nearest star	4.1×10^{18}
Milky Way galaxy	9×10^{22}
Distance to Andromeda galaxy	2×10^{24}
Local group of galaxies	5×10^{24}
Rich cluster of galaxies	10^{25}
Supercluster	10^{26}
The universe	$>10^{28}$

*The sizes listed are meant to be typical, to illustrate the relative scales. For round objects, the diameter is used; for irregular objects, an approximate average dimension is given.

FIGURE 1.11 A GALAXY SIMILAR TO THE MILKY WAY. *We cannot obtain an exterior view of our own galaxy, but it is thought that this one resembles ours. Note that most of the stars lie in a circular disk.* (© 1980 Anglo-Australian Telescope Board)

ASTRONOMICAL INSIGHT 1.3

ASTRONOMY AND ASTROLOGY

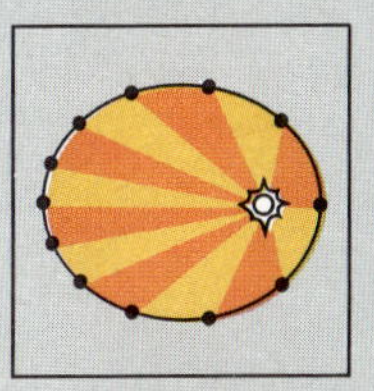

There is an unfortunate tendency in modern society to confuse astrology and astronomy, or, worse yet, to consider one a legitimate alternative to the other. Behind this tendency lies an even more unfortunate misunderstanding of what science is, and the distinction between scientific theory and beliefs adopted on faith.

Astrology, the pseudoscience based on the belief that human lives are influenced by the configurations of heavenly bodies, arose at a primitive stage in human development, when the Earth was thought to be a flat disk under the dome of the heavens. While ancient Greek astronomers did much to raise the study of the heavens to a scientific level, during the subsequent Dark Ages astrology and the governance of human lives by the stars once again became the primary basis for study of the heavens. This changed radically after the Renaissance, when the true nature of the heavens and the motions of astronomical objects were untangled. It is unfortunately true, however, that a segment of the human populace has continued to profess a faith in astrology.

One of the basic lessons people have learned about the universe is that it is easy to make mistakes unless one is careful to be objective and to accept only conclusions that can be verified by repeated observations or experiments, or by making predictions that can be tested. Astrology fails utterly and abysmally to meet these criteria. Serious attempts have been made to test astrological lore by statistical analysis of people born under different signs, with no trace of a correlation ever being seen.

There are many phenomena that still defy the understanding of contemporary science, and many of them deserve more attention than they are getting. But astrology is not one of them. As a phenomenon, it is worthy of study only by the sociological and psychological sciences, for its effects do not exist in the physical universe. It may be interesting party talk to compare astrological signs, but it is well to keep in mind the difference between objective reality and subjective impressions. Failure to do so is failure to understand what science is.

its constellation, which is about 2.3 million light-years away (Fig. 1.13). When we look at the Andromeda galaxy, we are receiving light that has been traveling for more than 2 million years!

Even the distance to the Andromeda galaxy is insignificant compared to the scale of the universe itself. The Milky Way, the Magellanic Clouds, Andromeda, and a number of other galaxies all belong to a concentrated grouping, or cluster, of galaxies. Most of the other galaxies in the universe also belong to clusters (Fig. 1.14), whose diameters can be as large as tens of millions of light-years. Between clusters of galaxies, space is relatively empty (actually, this point is controversial, as we shall see—all we can say for certain is that there are relatively few visible galaxies between clusters).

Clusters of galaxies are themselves grouped into larger conglomerates called **superclusters,** whose size

FIGURE 1.12 THE MAGELLANIC CLOUDS. *These two irregularly shaped, small galaxies lie just outside the Milky Way galaxy and orbit it, taking hundreds of millions of years for each complete orbit.* (Fr. R. E. Royer)

FIGURE 1.13 THE ANDROMEDA GALAXY. *At a distance of over two million light-years, this galaxy is the most distant object visible to the unaided eye. Without a telescope and a time-exposure photograph, the eye sees only an extended, fuzzy patch of light, rather than the detailed view shown here.* (Palomar Observatory, California Institute of Technology)

scales are significant, even on the scale of the universe itself. A supercluster typically may have a diameter measured in the hundreds of millions of light-years. ("Diameter" is probably not a good word; superclusters seem to be sheetlike or filamentary structures, not rounded like clusters of galaxies.) It is difficult to imagine an organized object (or collection of objects) so large that it takes hundreds of millions of years for light to travel across it.

Beyond the size scale of the superclusters, we approach the scale of the universe itself. It is apparent from a variety of lines of evidence that the observable universe has an overall size scale measured in billions of light-years. Light reaching us now from the farthest reaches of the universe has been traveling for many billions of years. This has the fascinating implication that we can see the universe as it was at that long-ago time when the light was emitted, a fact utilized by astronomers who study the origins of the universe.

Considering the sizes and distances of objects in the universe provides us a sobering perspective on ourselves and our tiny planet. It can be quite a revelation to see how much we presume to explain about the universe and how much we think we have learned about the origins, present state, and future evolution of the cosmos and all it contains. It will be wise, however, to keep in mind that there is very much that we do not know.

FIGURE 1.14 A CLUSTER OF GALAXIES. *The faint, fuzzy objects near the center of this photograph are galaxies belonging to a cluster that lies over a billion light-years from the Milky Way. Like our own, each galaxy contains billions of individual stars.* (Palomar Observatory, California Institute of Technology)

4. Because we cannot directly see how far away objects are, we can only measure their positions or separations in angular units. For convenience, we visualize a celestial sphere, centered on the Earth, on which the astronomical bodies lie.
5. The portion of the sky that can be seen depends on the latitude of the observer, and the portion visible at night depends on the time of year.
6. The most distant planet in our solar system is about 40 times farther from the Sun than the Earth is, whereas the nearest star is some 300,000 times, or 4 light-years, farther away. Our galaxy is about 100,000 light-years in diameter, the nearest galaxies are 150,000 to 2.3 million light-years distant, and clusters of galaxies are typically separated by millions to tens of millions of light-years. The size of the observable universe is measured in billions of light-years.
7. The light we receive from a distant object has been traveling toward us for as long as billions of years (in the case of the most distant galaxies), so we can observe the universe as it was long ago.

PERSPECTIVE

The introduction to astronomy provided in this chapter prepares us for a plunge into more detailed discussions. We will begin these in the next chapter by describing the nighttime sky as we see it without telescopes. We will describe in more detail the various bodies and motions observable to the unaided eye, and we will discuss the modern understanding of these phenomena.

SUMMARY

1. Astronomy is the science in which the entire universe is studied. The study of astronomy requires knowledge of several sciences, such as physics, chemistry, geology, and perhaps biology.
2. On a clear, dark night, the unaided eye can see up to five planets, the Moon, countless stars, and the occasional meteor or comet.
3. The Earth's rotation causes all objects in the sky to undergo daily motion, rising in the east and setting in the west.

REVIEW QUESTIONS

Thought Questions

1. Express in your own words the nature of scientific theory, and explain the differences between theory, speculation, and dogma.
2. Discuss how you might test the validity of astrology; that is, describe an experiment or observation that would show whether predictions made by astrologers are correct or incorrect.
3. Why do you think it is best to get away from the city to view the nighttime sky?
4. We have noted that the planets appear to travel across the sky along the same strip. What does this tell us about the orbits of the planets around the Sun?
5. Explain why positions of objects in the sky are measured in angles rather than in units of linear distance such as miles or kilometers.
6. Explain why it is necessary to have observatories in both the Southern and Northern Hemispheres. At what latitude is the largest possible fraction of the sky visible?

*Problems**

1. Express the following numbers in scientific notation (that is, in powers of ten notation): 0.0045; 12,400,000,000; 120; 0.0000000000067; 187,000,000,000,000,000.

2. Perform the following calculations: $1.3 \times 10^4/6 \times 10^{-6}$; $7.48 \times 10^{12} \times 3.18 \times 10^{-8}$; 0.00000000045/1,200,000; 5,780,000,000,000 × 14,000,000,000,000.

3. A degree of angle contains 60 minutes of arc, and a minute of arc contains 60 seconds of arc. How many arcseconds are in a full circle (360°)? The Sun's angular diameter is 30 arcminutes. What is it in degrees and in arcseconds?

4. Suppose the scale of the solar system were changed so that the earth was the size of a marble (with a diameter of 1 centimeter). On this scale, what would be the diameter of the Sun? What would be the Earth-Moon distance? What would be the Sun-Earth distance? How far from the Sun would Pluto be? How far away would the nearest star be? (Note: To answer these questions, you need to know the true sizes and distances involved. Most are given in the text of this chapter, but many can also be found easily in Appendix 7).

*The problems for Chapter 1 are not based directly on mathematical techniques used in the chapter. They are provided as practice exercises for students who need to review the use of scientific notation and the manipulation of large numbers.

5. Using information in Appendix 7, calculate the light-travel time from the Sun to Earth, to Jupiter, and to Pluto. (You will find it easiest to use the speed of light in metric units, which is 300,000 kilometers per second.)

6. If galaxies formed 10 billion years ago, how far away must we look in order to see the universe at a time before any galaxies existed?

ADDITIONAL READINGS

We can further appreciate the essence of astronomy, as well as its beauty, by reading a wide range of books and periodicals. Many bookstores contain volumes of astronomical photographs, as well as numerous books on astronomy written for the layperson.

Periodicals that are particularly well suited to students using this text include *Sky and Telescope*, *Mercury*, and *Astronomy*. *Sky and Telecope* is especially recommended for those wishing to carry out projects such as telescope building or astrophotography; it includes monthly charts showing the positions of the stars and planets as they change throughout the year.

Dean, Geoffrey. 1987. Does astrology need to be true? Part 1, 2. *Skeptical Inquirer* 11 (2):166; 11 (3): 257.

Radner, Dasie, and Michael Radner. 1982. *Science and Unreason*. Belmont, Calif.: Wadsworth, Inc. (discusses the difference between science and pseudoscience)

CHAPTER 2

CYCLES AND SEASONS: MOTIONS IN THE SKY

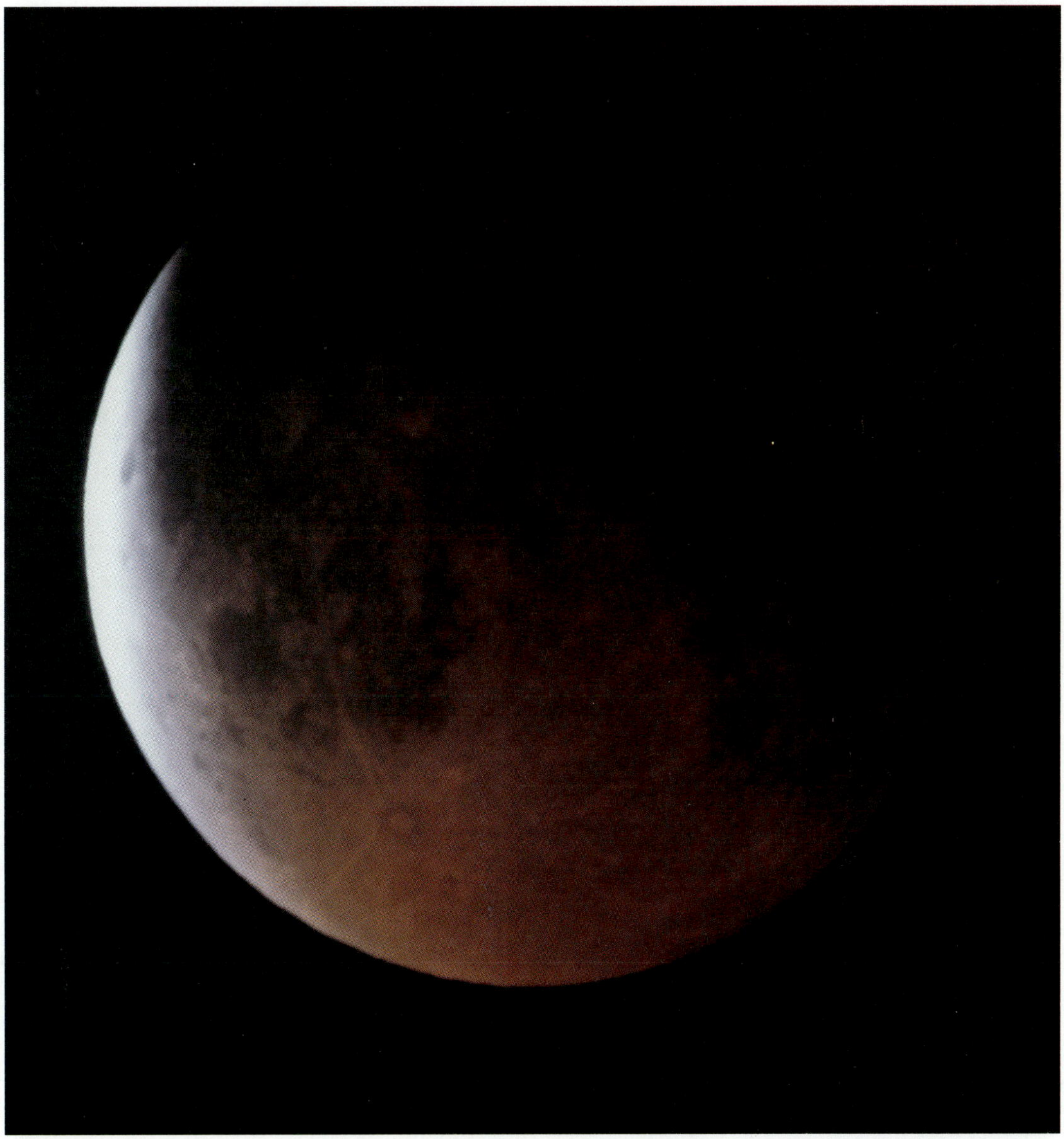

A near-total lunar eclipse. (J. Kloeppel, Sommers Bausch Observatory, University of Colorado).

CHAPTER PREVIEW

We view the heavens from a moving platform. The Earth spins while it travels around the sun in a nearly circular path, and these motions create both daily and yearly cycles of celestial events as seen from the Earth's surface. The other eight planets all behave in similar fashion, and most, including the Earth, have one or more satellites orbiting them. Because we are viewing the skies from a moving vantage point and because all celestial objects have motions of their own, our impression is that the heavens are very complex. It took human beings many centuries of thought, observation, and technological development to arrive at the simple explanation outlined in this paragraph, an explanation that most of us, in these modern times, are exposed to from early childhood.

In this chapter, we will learn about the nighttime sky as it is seen with the unaided eye. We will develop a modern understanding of the objects that can be seen without a telescope and how their simple motions create complex paths through the sky as seen from the Earth. In the process, we will develop an appreciation for the task of the ancient philosophers who strove to comprehend the workings of the universe, for the view of the heavens described here is the sum total of all the evidence available to pretechnological cultures.

TABLE 2.1 PERIODS OF SIGNIFICANT MOTIONS*

Motion	Period
Sidereal day	$23^h56^m4.098^s$
Mean solar day	$24^h00^m00^s$
Tropical year (equinox to equinox)	$365^d5^h48^m45^s$
Sidereal year (fixed stars)	$365^d6^h9^m10^s$
Synodic month	$29^d12^h44^m3^s$
Sidereal month	$27^d7^h43^m11^s$
Mercury: Sidereal period	87.969^d
Synodic period	115.88^d
Venus: Sidereal period	224.701^d
Synodic period	583.92^d
Mars: Sidereal period	1.88089^y
Synodic period	2.1354^y
Jupiter: Sidereal period	11.86223^y
Synodic period	1.0921^y
Saturn: Sidereal period	29.4577^y
Synodic period	1.0352^y
Uranus: Sidereal period	84.0139^y
Synodic period	1.0121^y
Neptune: Sidereal period	164.793^y
Synodic period	1.00615^y
Pluto: Sidereal period	248.5^y
Synodic period	1.0041^y

*Units for lunar and planetary motions are mean solar days or tropical years.

RHYTHMS OF THE COSMOS

The observed motions of celestial bodies result from a combination of rotational and orbital motions, including the spin of the Earth and its annual movement around the Sun, as well as the individual motions of the Moon and the planets. We can best understand the overall machinery of the solar system by examining the individual parts.

Many of the motions in the heavens are cyclical; that is they repeat regularly, in a well-defined period. Table 2.1 lists the periods for several motions in the solar system (unfamiliar terms in the table are explained later in the chapter).

DAILY MOTIONS

The most obvious of the many motions that affect our view of the universe are the daily cycles of all celestial objects caused by the rotation of the Earth. The Earth spins on its axis in about 24 hours, so we, on its surface, see a continuously changing view of the heavens. We see the Sun rise and set, along with the Moon, the planets, and most of the stars. Even though we understand that these daily, or **diurnal,** motions are the result of the Earth's spin, we still refer to them as though the objects in the sky themselves were moving. The rotation of the Earth means that a person standing on the equator covers the entire circumference (about 25,000 miles) in 24 hours, yet our senses give us no feeling of motion.

In science in general, and especially in astronomy, we must always be careful to note that what we observe depends on the frame of reference in which our observations are made. Our view of the sky from a location on the Earth's surface is affected strongly by our reference frame and its own motions. Physicists have learned that there is no absolute reference frame, that in any situation we must define the frame in which we observe and make conclusions. In astronomy we can use the reference frame of the Earth, as we are doing now in discussing our view of the nighttime sky, but we use more general frames in other contexts, such as the framework established by the (approximately) fixed stars, or perhaps the reference frame of distant galaxies.

In the reference frame of the Earth, the Earth's rotation forms the basis for our timekeeping system, since the length of the day is a natural unit of time on which to base our lives—one to which, indeed, nearly all earthly species have adapted. The day is divided into 24 hours, each containing 60 minutes, and each minute consisting of 60 seconds. These divisions are based on the numbering system developed several thousand years ago, largely by the Babylonians (see Chapter 3).

Careful observation shows that, in the reference frame of the Earth, the Sun and the Moon take longer to complete their daily cycles than do the stars. Each has its own motion that carries it in the same direction as the Earth's rotation, opposite to the daily rising and setting of the stars (Fig. 2.1). The Sun and the Moon, therefore, shift to the east with respect to the stars a little each day, so it takes a little longer for each to return to the same position from our point of view. Thus, the Sun rises about 4 minutes later each day, by comparison with the stars, and the Moon rises almost an hour later each day. The planets also move, but so slowly that their diurnal motions are not easily distinguished by eye from those of the stars.

The day that forms the basis for our timekeeping system is based on the **solar day** rather than the **sidereal day** (Fig. 2.2), which is the rotation period of the Earth in the reference frame of the fixed stars. (The term *sidereal* means "with respect to the stars".) Because the Earth's orbital speed is not precisely constant, the length of the solar day varies a little throughout the year. It would be inconvenient to

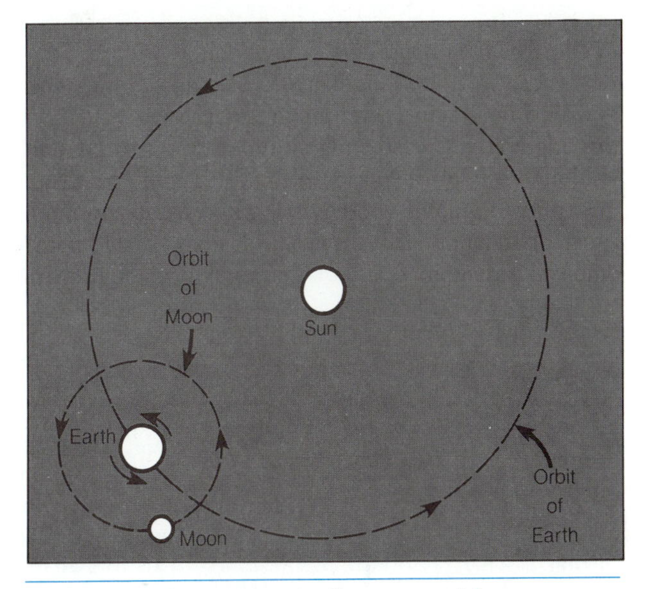

FIGURE 2.1 MOTIONS OF THE EARTH AND MOON. *The Earth and Moon spin and move along their orbits at the same time. The combination of motions affects the apparent path of the Sun and the Moon as seen by an observer on the spinning Earth. (Not to scale.)*

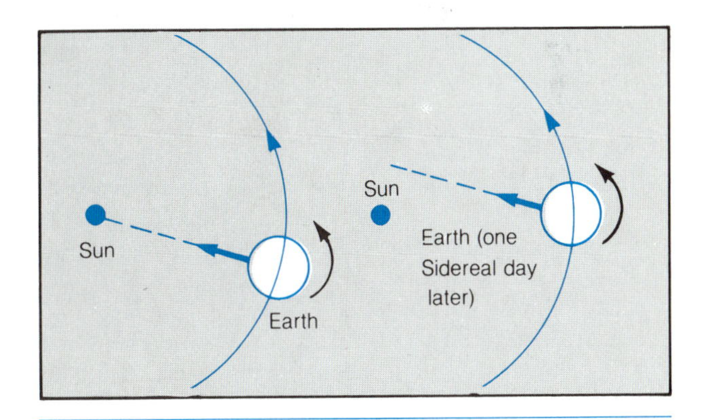

FIGURE 2.2 THE CONTRAST BETWEEN SOLAR AND SIDEREAL DAYS. *The arrow indicates the overhead direction from a fixed point on the Earth. From noon one day (left), it takes one sidereal day for the arrow to point again in the same direction, as seen by a distant observer. Because the Earth has moved, however, it will be about four minutes later when the arrow points directly at the Sun again; hence the solar day is nearly four minutes longer than the sidereal day. (Not to scale.)*

ASTRONOMICAL INSIGHT 2.1

KEEPING IN TOUCH WITH THE STARS

As mentioned in the text, modern timekeeping is done with atomic clocks, which have several advantages over the use of transit telescopes to monitor sidereal time. For example, any physics laboratory can maintain its own atomic clock, and make complicated observations with it, without the need to have a transit telescope. Also, the signals from atomic clocks can be transmitted by radio from laboratory to laboratory, so that experiments requiring precise relative timing can be performed. It must be stressed, however, that the atomic clocks are used only as measures of relative time, the time from some adopted reference time. For this they are very accurate, but they cannot adequately maintain correct time relative to the Earth's rotation, because the Earth's rotation is not perfectly constant.

All methods of timekeeping are still based on the rotation of the Earth, as measured through the observations of stars. Hence, atomic clocks must occasionally be updated to account for small changes that occur in the Earth's rotation rate, when it may speed up or slow down. These changes occur for many reasons, not all of them well understood. Some appear to be random variations in the rotation rate, and others seem to be long-term gradual changes.

One persistent change that always occurs in the same direction is a very gradual slowing of the Earth's rotation rate, probably because of internal frictional forces exerted by the tidal force of the Moon (this is discussed more fully in Chapters 5 and 9, where the effect on the Earth-Moon orbit is described). The length of the day has increased by about 55 seconds since 1900. Given enough time, this slowing of the Earth's rotation rate reduces the number of days in a year, since each day is longer than it used to be. There is fossil evidence that the year once contained about 400 days, instead of the current 365¼.

The short-term random changes in the Earth's rotation rate can cause either increases or decreases in the length of the day. It is thought that some of these changes are caused by the flow of molten material deep in the Earth's core, but the causes of most of them are not known. Some of them are probably related to a slow shrinking of the Earth as its interior gradually cools; these minor effects act to speed up the rotation temporarily.

Over long periods of time, however, the overall trend is toward a decrease in the rotation rate and a lengthening of the day. To keep atomic clocks in synchronization with the Earth's spin, therefore, it is occasionally necessary to add some time to the clocks. Hence, every few years, a "leap second" is added to all official atomic clocks in order to bring their timekeeping into accord with the Earth's rotation, that is, with sidereal time. This is distinct from the leap year, when a full day is added every fourth year. This added day is inserted because the standard calendar has exactly 365 days in a year, whereas the actual length of the year is closer to 365¼ days long. Thus, a full day must be added every 4 years to make up for the one-quarter days that have been dropped. Interestingly, if the length of the day keeps increasing as expected, in a few thousand years the year will contain exactly 365 days, and for a while there will be no need for leap years.

You may occasionally notice small announcements in newspapers about the addition of "leap seconds" to official clocks. These small changes in official time make little practical difference in our daily lives, but now you understand that they are part of a large-scale astronomical process, with important long-term effects.

allow our hour, minute, and second to vary along with it, so the average length of the solar day, called the **mean solar day,** has been adopted as our timekeeping standard. The mean solar day is 3 minutes and 56 seconds longer than the sidereal day.

Precise timekeeping today is measured with atomic clocks, but is still kept synchronized with time based on the Earth's rotation. In order to measure the Earth's rotation period with respect to the stars (that is, the sidereal day), astronomers refer to the **meridian,** the imaginary north-south line passing overhead at an observer's position. As the Earth rotates, each star crosses the meridian once every sidereal day. Special instruments (Fig. 2.3) are used to measure precisely the moment when a star crosses the meridian, thus measuring sidereal time. Allowance must be made in order to determine from this the mean solar time.

Atomic clocks measure intervals of time very accurately, because they are based on the vibration frequencies of certain kinds of atoms, which are very constant. These clocks have several advantages over the measurement of sidereal time based on observations of stars. For example, atomic clocks keep a continuous record of the passage of time, so that they can be referred to whenever needed. Also, identical clocks can be placed in many locations and kept coordinated, allowing many laboratories to precisely measure the time.

We say that atomic clocks measure true physical time, but these clocks are adjusted as needed so that they stay in agreement with sidereal time. Such adjustments are needed because occasionally the Earth's rotation period changes very slightly.

Let us return now to the diurnal motions of celestial objects. We have mentioned that the stars rise and set as the Earth rotates, but in fact not all stars do this. This is because, from a given latitude on the Earth (except right at the equator), it is possible to see an area of the sky, around one of the poles, that is never obscured by the Earth. From a latitude of 30°N, for example, we can see the sky to an angular distance of 30° beyond the North Pole. Therefore, we can always see the whole part of the sky that lies within 30° of the North Pole (but we can never see the sky within 30° of the South Pole), and stars in this part of the sky circle the pole but do not set. The same thing happens around the South Pole for observers in the Southern Hemisphere (Fig. 2.4).

The point about which the stars circle, directly over the Earth's North Pole, is the **north celestial pole,** the point where the earth's rotational pole is projected onto the celestial sphere. Similarly, the projection of the South Pole onto the celestial sphere is the **south celestial pole.** It happens that a bright star, called Polaris, lies very close to the position of the north celestial pole and is therefore almost stationary (in the reference frame of the Earth) throughout the night. The projection of the Earth's equator onto the sky is called, naturally enough, the **celestial equator.** We now will examine the usefulness of the celestial poles and the equator as reference points for measuring star positions.

FIGURE 2.3 A TRANSIT CIRCLE. *Such a device points straight up. It is used to record the times when certain reference stars pass over the meridian and is therefore helpful in measuring the sidereal day.* (U.S. Naval Observatory)

FIGURE 2.4 STAR TRAILS ILLUSTRATING THE EARTH'S ROTATION. *The circular trails are created by stars near the south celestial pole, which completed about half a circle during this all-night exposure.* (© 1980 Anglo-Australian Telescope Board)

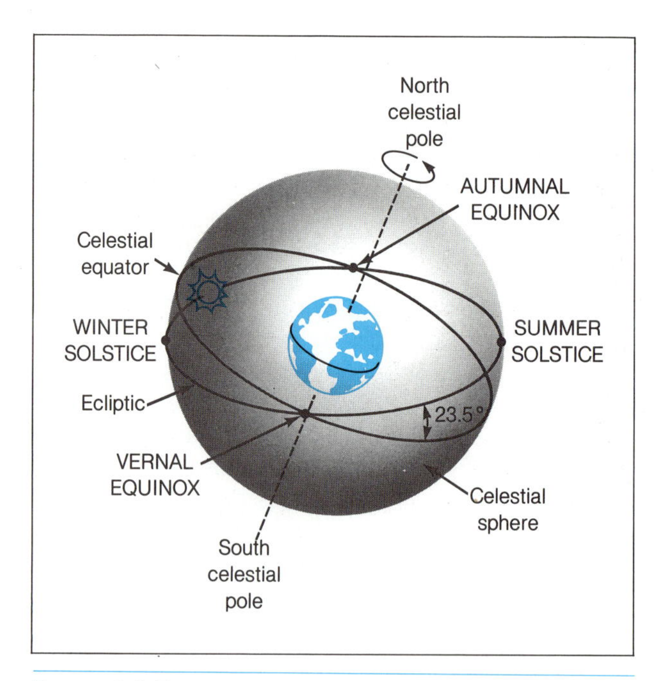

FIGURE 2.5 EQUATORIAL COORDINATES. *Here are illustrated the measurements of star position in declination and right ascension. Declination is the distance (in degrees, minutes, and seconds of angle) of a star north (+) or south (−) of the celestial equator. Right ascension is the distance (in hours, minutes, and seconds of time) of the star to the east of the direction of the vernal equinox, a fixed direction in space.*

Astronomical Coordinate Systems

Our discussion of diurnal motions, latitude, and the changing appearance of the sky as we move over the surface of the Earth has led us to a point where it is convenient to introduce the notion of coordinate systems for the skies. Just as it is useful to be able to describe the position of a place on the surface of the Earth, it is also important for astronomers to be able to record the position of a particular object in the sky so that they can return to it later or communicate to other astronomers where it is. The establishment of a coordinate system naturally requires us to specify the reference frame in which the coordinates are measured. Astronomers use different reference frames for different purposes, but the one most often used is based on the Earth (Fig. 2.5).

In this coordinate system, called the **equatorial coordinate system,** one reference direction is the celestial equator. In order to specify a star's position in this system, two coordinates are needed. One of them is the angular distance of the star north or south of the celestial equator, and is called the **declination** of the star. The declination is analogous to the latitude of a position on the Earth's surface, except that here we speak of angular distance on the sky. North declinations are assigned positive values (such as $+23°14'$ for a star $23°14'$ north of the celestial equator), and south declinations are assigned negative values (the Magellanic Clouds, for example, lie at declinations near $-70°$).

The second coordinate needed to specify an object's position is called **right ascension.** The right ascension measures the angular distance of the object to the east of a fixed direction on the sky, and is therefore analogous to longitude on the Earth's surface. Because the Earth is rotating, however, we cannot measure right ascension from some fixed point defined on the Earth's surface; we must instead specify a reference direction on the sky. This direction is defined by the line of intersection of the Earth's equatorial plane and the plane of its orbit. This is the direction toward the Sun at the time of the Vernal Equinox (defined in the next section). When this coordinate system was established, the line of intersection of these planes pointed toward the constellation Aries, but because of gradual changes in the Earth's orientation (described below) it now points toward Pisces and will soon enter Aquarius; this is why popular astrology refers to the "dawning of the age of Aquarius". Nevertheless, the position toward which this line points is still called the **First Point of Aries.**

Because it takes 24 sidereal hours for a given star to make a complete circuit of the sky as the earth rotates, its position in the east-west direction is measured in units of sidereal hours, minutes, and seconds rather than in angular units. Thus, a star located in the direction of our reference point has a right ascension of $0^h0^m0^s$, whereas a star one-quarter of the way around the sky to the east of that point has a right ascension of $6^h0^m0^s$.

There is one complication in using the equatorial coordinate system. The Earth's rotational axis slowly wobbles (Fig. 2.6) in a motion called **precession,** and this causes the celestial equator to move slowly through the sky. It takes some 26,000 years for the earth to complete one cycle of this motion, which is exactly like the wobbling familiar to anyone who has spun a play top or gyroscope. The motion of our coordinate system is therefore very gradual, causing star positions to shift extremely slowly. Despite the small magnitude of the effect, it was noticed more than 2,000 years ago, and it must be allowed for by modern astronomers when planning observations. Typically, one of the preparations for observing is to "precess" the coordinates of all the target stars (that is, to calculate their current locations in the Earth's reference frame). Catalogs of star positions always specify the date for which they are valid so that astronomers know how much precession to allow for.

Other coordinate systems (that is, other reference frames) sometimes are used in astronomy, although equatorial coordinates are by far the most widely employed, being the most practical for use in observations. Solar system astronomers sometimes use a coordinate system based on the plane of the **ecliptic** (the plane defined by the Earth's orbit around the Sun) and the direction toward the Sun, whereas those studying the Milky Way may use coordinates based on the plane of the galaxy's disk and the direction toward its center. Ecliptic and galactic coordinates are particularly useful in studies of the structure of the solar system or of the galaxy, because an object's position immediately tells the astronomer where it lies with respect to the overall organization of the solar system or galaxy.

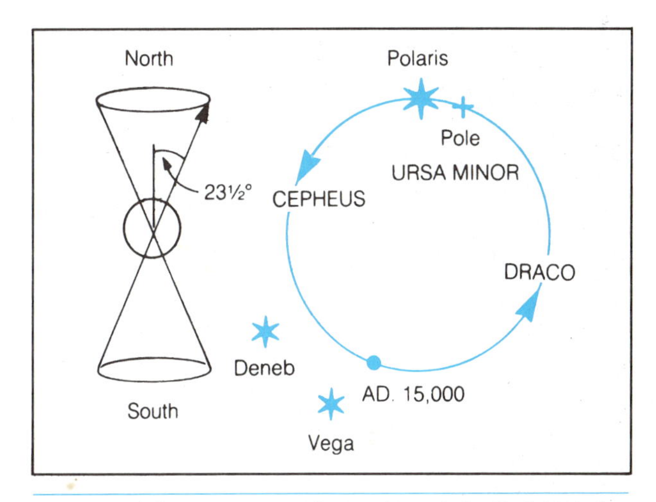

FIGURE 2.6 PRECESSION. *The Earth's axis is tilted 23½ degrees with respect to its orbital plane, and it wobbles on its axis, so that an extension of the axis describes a conical pattern (left) in a time of about 26,000 years. The north celestial pole therefore follows a circular path on the sky (right).*

ANNUAL MOTIONS: THE SEASONS

We turn our attention now to celestial phenomena that are caused by the Earth's motion as it orbits the Sun. One aspect of this, the daily eastward motion of the Sun with respect to the stars, has already been mentioned in connection with the difference between the solar and the sidereal day. It must be emphasized that, in the reference frame of the fixed stars, this is only an apparent motion caused by our changing angle of view as we move with the Earth in its orbit. As the Earth moves about the Sun, it travels in a fixed plane, so the apparent annual path of the Sun through the constellations is the same each year (Fig. 2.7). This apparent path of the Sun, defined by the plane of the Earth's orbital motion about the Sun, is the **ecliptic.** The sequence of constellations through which the Sun passes is called the **zodiac.** There are twelve principal constellations of the zodiac, identified since antiquity and once thought by astronomers to have significance for our daily lives (see the comments on astrology in Chapter 1).

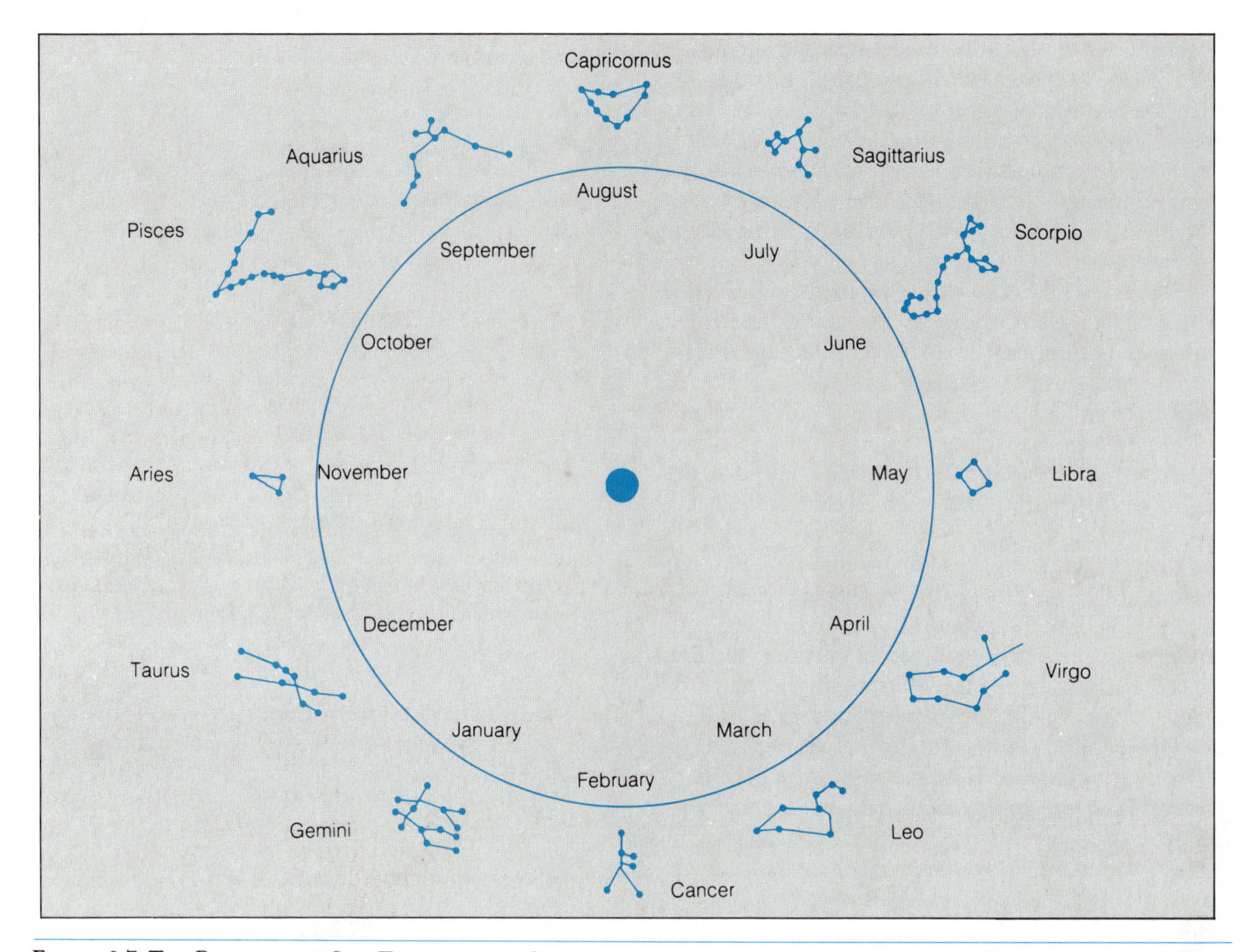

FIGURE 2.7 THE PATH OF THE SUN THROUGH THE CONSTELLATIONS OF THE ZODIAC. *The dates refer to the position of the Earth each month. To see which constellation the Sun is in during a given month, imagine a line drawn from the Earth's position through the Sun; that line will extend to the Sun's constellation. For example, in March the Sun is in Aquarius.*

Because we can see stars only in the nighttime sky, the constellations most easily visible to us at any given time are the ones at least a few hours of right ascension to the east or west of the Sun. This means that most stars, except those near the poles, cannot be observed year-round, as discussed in Chapter 1.

The orbital planes of the other planets and of the Moon are closely aligned with that of the Earth, so the planets and the Moon are always seen near the ecliptic. Hence, most of the major naked-eye objects in the solar system pass through the same sequence of constellations, the zodiac, so it is not surprising that great significance was attached to this sequence by the ancient astronomers.

Besides causing the apparent annual motions of the Sun and planets, the Earth's orbital motion has a second, and far more significant, effect on us: It creates our seasons. The Earth's spin axis is tilted with respect to the perpendicular to its orbital plane, so during the course of a year, portions of the Earth away from the equator are exposed to varying amounts of sunlight. Summer in the Northern Hemisphere occurs when the North Pole is tipped toward the Sun; winter occurs during the opposite part of the Earth's

ASTRONOMICAL INSIGHT 2.2

CELESTIAL NAVIGATION

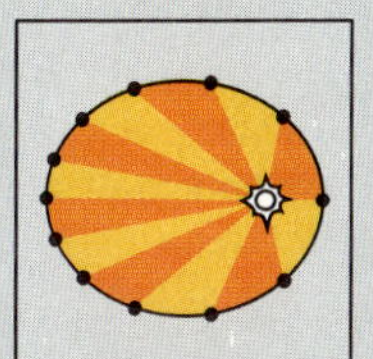

One of the most time-honored and well-known "practical" applications of astronomy is to find one's position on the Earth by observing the stars. Traditionally, this was the only means available to a sailor or someone exploring new territory, whereas today there are many, more modern methods (and less unexplored territory).

The essential method is quite simple, although in practice the technique involves mathematical calculations too complex to describe in detail here. The first requirement is to be able to identify one or more objects in the sky whose coordinates are known (usually from a book of tabulated positions, called an **ephemeris**). Thus, a familiarity with the constellations and the sky in general is required. Ironically, people who have learned celestial navigation are more likely to know the constellations, as well as a number of individual stars and their names, than is the typical professional astronomer.

Once a known star is recognized and its coordinates found in a catalog or ephemeris, the navigator has a reference point in the sky whose position is known. The navigator then measures the angular distance from straight overhead to that star. This measured angular distance is called the **zenith distance** of the star, the angle between the navigator's zenith (directly overhead) and the direction to the star. Given this, as well as the sidereal time, it is possible to locate the spot on the Earth's surface where the star is directly overhead and then draw a circle about that spot (on a globe) showing all the locations where the star would have the measured zenith distance. This does not pinpoint the navigator's location, so the same procedure is carried out using a second star. Now there are two circles of possible locations, and the two intersect at only two points. Unless badly confused, the navigator normally will know which of the two intersection points is the correct location (the two points can be hundreds of miles apart). Hence, measurements of only two stars are usually sufficient, but typically a third star is measured as well to be sure there were no errors. In that case, there are three circles drawn on the map, and all three should intersect in only one point if the measurements have been made correctly.

The measurement of the zenith distance is done by using a device such as a **sextant** that allows angular separations on the sky to be measured. A sextant usually includes a pointer that can be aimed at the star and a means of giving the user a reference to the vertical direction.

In practice, the Sun and the Moon can be used for navigation, although they are both difficult to measure precisely because of their large angular diameters; also, of course, it is difficult to view the Sun directly, because of its brilliance. Furthermore, to use either, it is necessary to have rather complete tables of their positions, which change throughout the year (in the case of the Sun) and throughout the month (in the case of the Moon, whose motion is sufficiently rapid to require tables showing the position hourly). The Sun is often used, however, because the horizon is easily visible during daylight, so that the **altitude,** the angular distance of the Sun above the horizon, can be readily measured in order to find the Sun's zenith distance.

Today, navigation of ships and airplanes is done by various means involving technology that has been developed in the past few decades. The well-traveled areas of the Earth are permeated with navigational radio beams that provide directional information, and modern satellite navigation systems provide very accurate data to ships and planes. In addition, devices called **gyroscopes,** which can maintain a given orientation very accurately over long periods, allow navigators to keep track of their positions over lengthy voyages without any reference to external guideposts such as the sky or a satellite navigation system.

Despite all the modern developments cited here, the art of celestial navigation is not completely lost. Military and commercial airline and ship navigators still learn how to do it and presumably could use the method if all else failed.

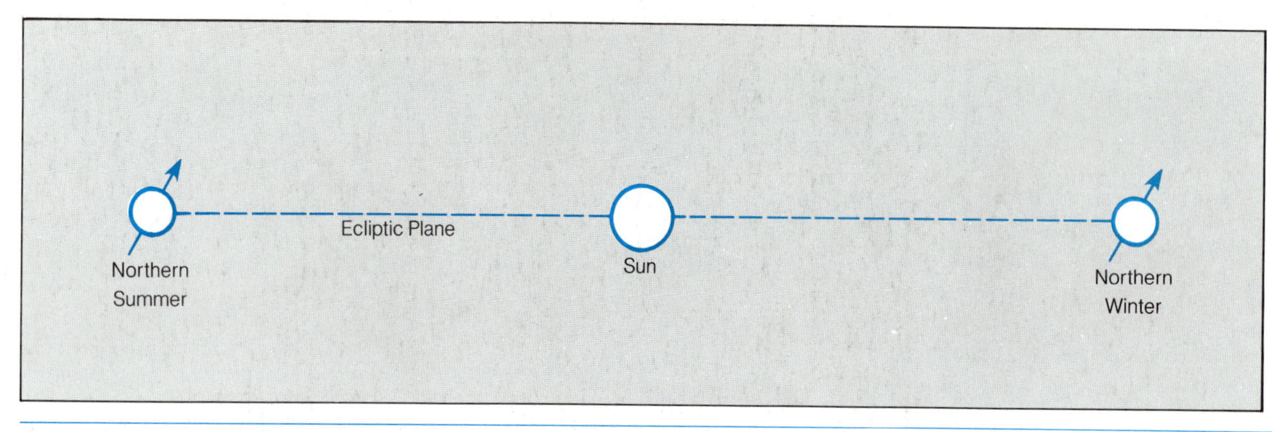

FIGURE 2.8 SEASONS. *The Earth's tilted axis retains its orientation as the Earth orbits the Sun. Thus, at opposite points in the orbit, each hemisphere has winter and summer, depending on whether that hemisphere is tipped toward the Sun or away from it.*

orbit, as the pole, which remains fixed in orientation, is tilted away from the Sun (Fig. 2.8). The tremendous seasonal variations in climate at intermediate latitudes are caused by a combination of two effects: (1) the length of the day varies so that in summer, for example, the Sun has more time to heat the Earth's surface; and (2) the angle at which the Sun's rays strike the ground is more nearly perpendicular in the summer (Fig. 2.9), so the Sun's intensity is much greater, heating the surface more efficiently.

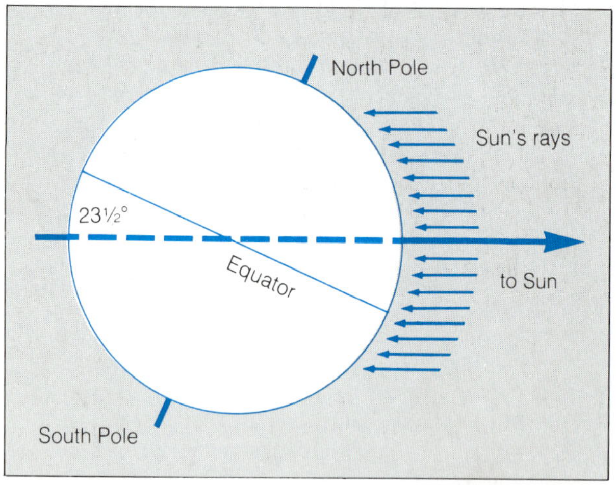

FIGURE 2.9 EFFECT OF THE EARTH'S TILTED AXIS. *The Sun's rays strike the ground at varying angles, depending on the latitude and the time of the year. Therefore, the solar heating, which depends on how directly the rays reach the ground, varies with the season. Here it is summer in the northern hemisphere; the Sun's rays are nearly perpendicular to the Earth's surface at low to moderate northern latitudes, but they strike the surface very obliquely at southern latitudes.*

The Earth's axis is tilted $23\frac{1}{2}^\circ$ from the perpendicular to the orbital plane. Therefore, during the year, the Sun, as seen from the Earth's surface, can appear directly overhead (that is, at the **zenith**) as far north and south of the equator as $23\frac{1}{2}^\circ$, defining a region called the **tropical zone** (Fig. 2.10). For those of us who live outside the tropics, the Sun can never be directly overhead. When the North Pole is tilted most nearly in the direction toward the Sun, an occasion occurring around June 21 and called the **summer solstice,** the Sun passes directly overhead at $23\frac{1}{2}^\circ$N latitude at local noon, but it passes to the south of the zenith for anyone at more northerly latitudes. In Boulder, Colorado, for example, which is right on the fortieth parallel, the Sun on this day reaches a point $40^\circ - 23\frac{1}{2}^\circ = 16\frac{1}{2}^\circ$ south of the zenith, or an altitude of $90^\circ - 16\frac{1}{2}^\circ = 73\frac{1}{2}^\circ$ above the southern horizon (Fig. 2.11).

Traveling farther north, we see that at the time of the summer solstice, sunlight covers the entire north polar region to a latitude as far as $23\frac{1}{2}^\circ$ south of the pole. This defines the **arctic circle** (see Fig. 2.10), and at the time of the solstice the entire circle has daylight for all 24 hours of the earth's rotation. At the pole itself there is constant daylight for 6 months.

During winter in the Northern Hemisphere, the Sun does not rise as high above the southern horizon at local noon as it does during the summer, because the South Pole is now tilted toward the sun. At the **winter solstice,** when the South Pole is pointed most nearly in the direction of the Sun, the Sun's midday

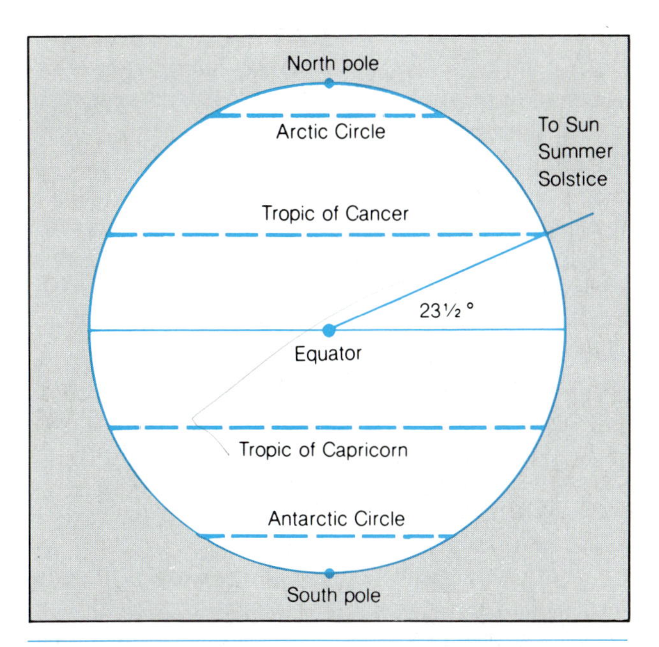

FIGURE 2.10 THE DEFINITION OF LATITUDE ZONES ON THE EARTH. *At summer solstice, the Sun is overhead at 23½ degrees north latitude, its northernmost point. This defines the Tropic of Cancer, the northern limit of the tropical zone. At the same time the entire area within 23½ degrees of the North Pole is in daylight throughout the Earth's rotation, and the boundary of this region is the Arctic Circle. Similarly, the Antarctic Circle receives no sunlight at all during a complete rotation of the Earth. Six months later, the Sun is overhead at the Tropic of Capricorn.*

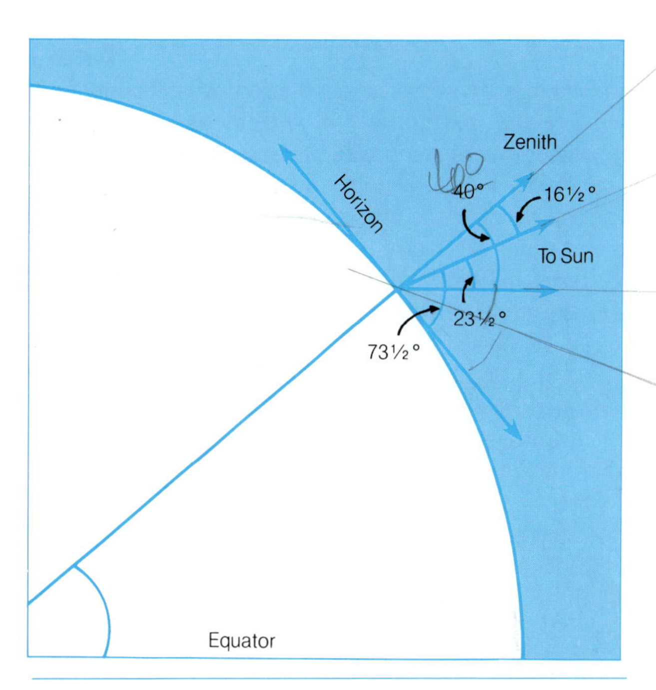

FIGURE 2.11 SUN'S POSITION FROM BOULDER AT THE SUMMER SOLSTICE. *Boulder, Colorado, lies on the fortieth parallel north, so when the Sun is overhead at 23½° N, as it is on the summer solstice, from Boulder it appears to be 16½° south of the overhead position, or 73½° above the southern horizon.*

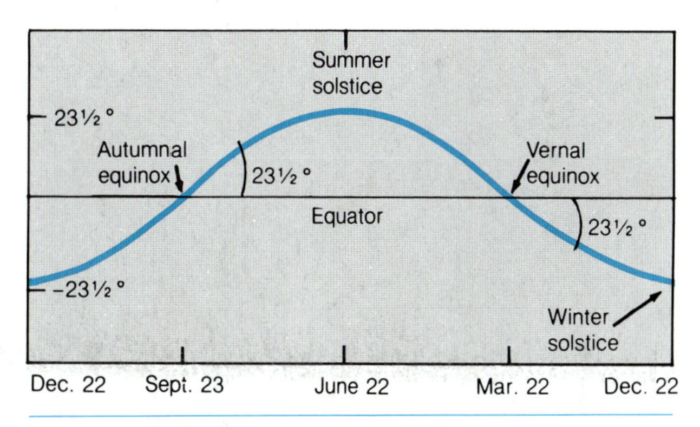

FIGURE 2.12 THE PATH OF THE SUN THROUGH THE SKY. *Because of the Earth's orbital motion and the tilt of its axis, the Sun's annual path through the sky has the shape illustrated here.*

height above the horizon, as viewed from the Northern Hemisphere, is the lowest of the year. Returning to Boulder, we find that on this day, which usually falls near December 21, the sun is $40° + 23½° = 63½°$ south of the zenith, or $90° - 63½° = 26½°$ above the horizon.

If we follow the Sun's motion north and south of the equator throughout the year, we find that it follows a graceful curve as it traverses its range from $+23½°$ (north) declination to $-23½°$ (south) declination (Fig. 2.12). The Sun crosses the equator twice in its yearly excursion, at the times when the Earth's North Pole is pointed in a direction 90° from the Earth-Sun line. At these times, the lengths of day and night in both hemispheres are equal, and these occasions are referred to as the **vernal** (spring) and **autumnal** (fall) **equinoxes.** They occur on about March 21 and September 23, respectively. The direction to the Sun at the time of the vernal equinox coincides exactly with the direction of 0^h right ascension. (Remember, the definition of this direction is that it lies along the line of intersection of the earth's equatorial plane and the orbital plane.)

Calendars

We have referred to the association of astronomy and timekeeping, particularly in ancient times, when astronomers were responsible for predicting major natural events such as changes of season. Today everyone knows when these things will occur and generally keeps track of dates for all purposes by the use of modern calendars (Fig. 2.13), which can be purchased well in advance for any given year. It is interesting to realize, however, that even now our calendars, and hence the annual pattern of our daily lives, are determined by the motions of bodies in the heavens.

A number of calendars were established by various ancient cultures, and all were based to some degree on the length of the year. In several cases it was deemed unlikely or unacceptable that the year should not be evenly divisible into a round number of lunar months or even days, so some of the old calendars adopted lengths of the year that did not agree with the actual period of the earth's orbit of the Sun. All such calendars had to be adjusted every so often or fall badly out of synchronization with the seasons.

The length of the year that is important in calendar making is not the sidereal period of the Earth's orbit, as we might at first expect. The reason is that, because of precession, the seasons will gradually shift with respect to the sidereal year. Consider the length of time from one vernal equinox to the next. As the Earth travels around the Sun, the orientation of its axis does not stay precisely fixed, but instead changes just slightly because of precession. The result is that the next vernal equinox, which occurs when the Earth's axis returns to precisely the same orientation (with respect to the stars) as it had at the previous equinox, takes place just a little sooner than it would have if precession did not occur. The difference between the sidereal year and the **tropical year** (the length of time for a full cycle of seasons to occur) is just 20 minutes. Hence, calendars today are based on the tropical year. Precise values for both sidereal and tropical years are included in Appendix 2.

FIGURE 2.13 MODERN CALENDAR SHOWING THE MOON'S PHASES. (Sierra Club Books/Random House)

Until the time of Julius Caesar, most calendars were based on the assumption that the year included a whole number of days, even though it was well known that this was not the case. The errors that built up were allowed to accumulate until extra days would be inserted in order to bring the calendar back into synchronization with the seasons. Caesar commissioned the development of a new calendar, now called the **Julian calendar,** in which the length of the year was taken to be 365¼ days, only a few minutes in error. Every fourth year an extra day was inserted, a practice that we still follow. The year was divided into 12 months that did not correspond exactly to lunar months, so that at least there would be a whole number of months in a year. Caesar was also responsible for beginning the year on January 1; it had begun in March before his reform, which took place in 46 B.C.

To allow for the few minutes' error that crept into the Julian calendar because the tropical year is not precisely 365 days long, in 1582 Pope Gregory XIII ordered additional reform, with the result that in the modern calendar (called the **Gregorian calendar**), leap year is occasionally not observed. Leap year is ignored (the extra day is not inserted after February 28) on century years not divisible by 400. Thus, there will be a leap year in the year 2000, which is divisible by 400, but not in the years 2100, 2200, and 2300. With this refinement, the Gregorian calendar builds up an error of only 1 day in 3,300 years, a sufficient degree of accuracy for most people's appointment books!

THE MOON AND ITS PHASES

Except for the Sun, the Moon is the brightest object in the sky, and it, too, has its own complex motions as seen from the Earth. The Moon's motion with respect to the stars is much more rapid than that of the Sun, and its therefore more easily noticed. It is not difficult, in fact, to observe this lunar motion during the course of an evening. The moon moves about 12° across the sky every day, or 1° every 2 hours. Its angular diameter is about ½°, which means that the moon moves a distance in the sky about equal to its own diameter every hour.

The plane of the Moon's orbit is closely aligned with that of the Earth's orbit, so the Moon stays near the ecliptic, as seen from the Earth. It takes the Moon a little over 27 days ($27^d7^h43^m11^s.5$) to make one trip around the Earth, in the reference frame of the fixed stars. Because the Earth moves at the same time, however, it appears to us in the Earth's reference frame that the Moon takes longer than 27 days to make one complete circuit; from our point of view, the Moon takes $29^d12^h44^m2^s.8$ to complete its full cycle of phases (Fig. 2.14). The shorter orbital period is called the **sidereal period,** because it is the time required for the Moon to go around the Earth, as seen in a fixed

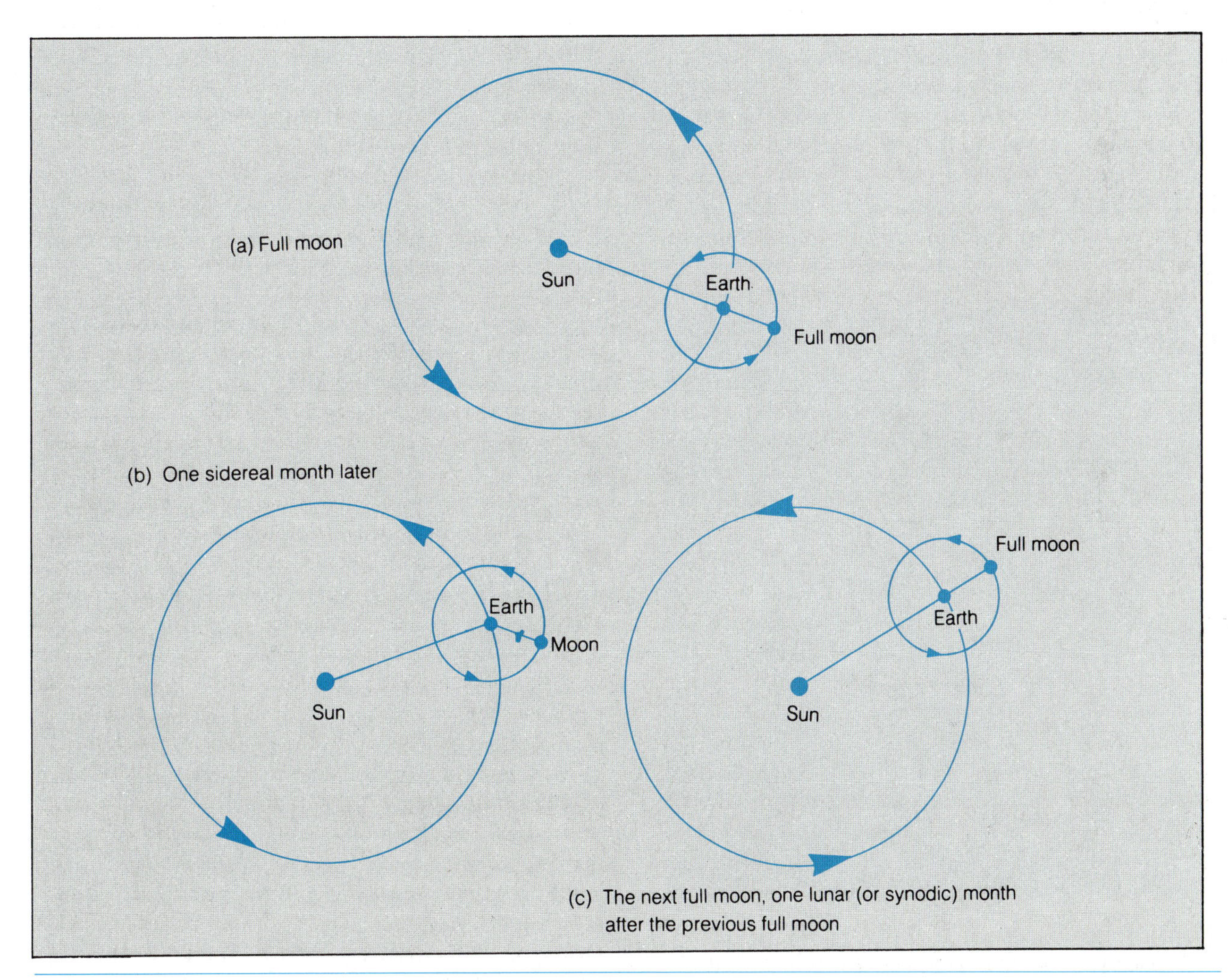

FIGURE 2.14 SIDEREAL AND SYNODIC PERIODS OF THE MOON. *Because the Earth moves in its orbit while the Moon orbits it, the Moon goes through more than one full circle (as seen by an outside observer) to go from one full Moon to the next. Hence the lunar (or synodic) month is about two days longer than the Moon's sidereal period.*

reference frame with respect to the stars. The observed cycle of phases, which is the time required for the Moon to return to a given alignment with respect to the Sun, is called the **synodic period,** or the **lunar month.** The difference is akin to the distinction between the solar and sidereal days discussed in the preceding section. In both cases, it is the Earth's motion about the Sun that lengthens the time it takes to complete a full cycle as we see it.

The Moon always keeps the same side facing the Earth, because its rotation period is equal to its orbital period (Fig. 2.15). It is a common misconception that the Moon is not rotating. It should be clear that the Moon is rotating in the reference frame of the stars, however, for if it were not, we would see all sides of the Moon as it circled the Earth. The fact that the orbital and spin periods are equal is not a coincidence but is due to the tidal forces exerted on the Moon and the Earth by their mutual gravitational pull (this is discussed in Chapters 5 and 9). The phenomenon of matching orbital and spin periods is called **synchronous rotation,** and it is common in the universe, both within the solar system and in double stars.

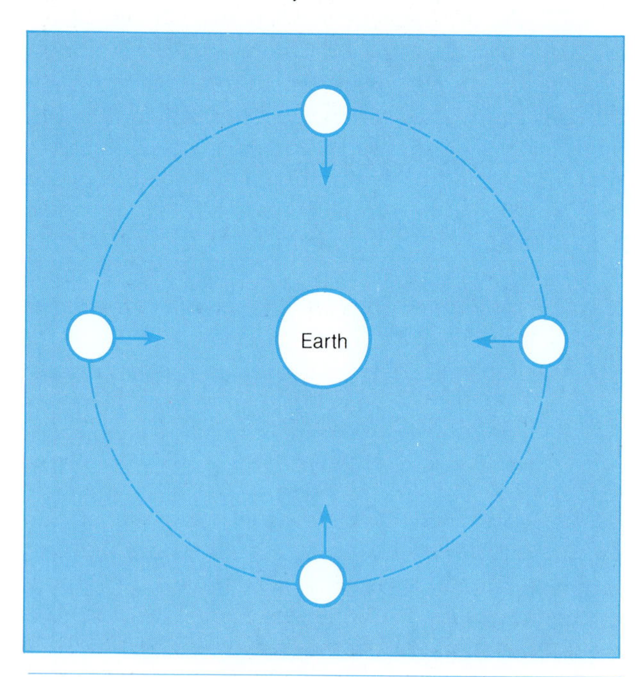

FIGURE 2.15 SYNCHRONOUS ROTATION OF THE MOON. *The arrow, fixed to a specific point on the Moon, illustrates that the Moon spins once during each orbit of the Earth, thereby keeping the same side facing the Earth at all times. (Not to scale.)*

Since the Moon does not emit light of its own but shines instead by reflected sunlight, we easily see only those portions of its surface that are sunlit. As the Moon orbits the Earth and our viewpoint relative to the Sun's direction changes, we see varying fractions of the daylit half of the Moon. This causes the Moon's apparent shape to change drastically during the month, the sequence of shapes being referred to as the **phases** of the Moon (Fig. 2.16). The full cycle of phases is completed during one synodic period, or lunar month, of about 29½ days.

The extremes of the cycle are represented by the **full moon** (occurring when it is directly opposite the Sun, so we see its entire sunlit hemisphere) and the **new moon** (when it is between the Earth and the Sun, with its dark side facing us). The new moon cannot be observed, because it is dark and is so close to the Sun. Hence, our nighttime sky is moonless at the time of the new moon.

Just as we speak of phases of the Moon, which really refer to its apparent shape as seen from the Earth, we also can speak of its **configurations,** which describe its position with respect to the Earth-Sun direction. For example, when a full moon occurs, the Moon is at **opposition** (it is in the direction opposite to that to the Sun), and a new moon occurs at **conjunction** (when the Moon lies in the same direction as the Sun). We can follow the Moon through its phases as we trace its configurations, beginning with the new moon, which takes place at conjunction (refer to Fig. 2.16). During the first week following conjunction, as the Moon moves toward **quadrature** (the position 90° from the Earth-Sun line), it appears to us to have a crescent shape that grows in thickness each night. This is called the **waxing crescent** phase; when the Moon reaches quadrature, in which we see exactly one-half of the sunlit hemisphere, it has reached the phase called **first quarter.** For the next week, as the Moon goes from quadrature toward opposition and we see more and more of the daylit side, its phase is said to be **waxing gibbous.** After the full moon, as it again approaches quadrature, the phase is **waning gibbous,** and this time at quadrature the phase is called **third quarter.** First quarter can be distinguished from third quarter easily by noting the time of night: If the Moon is already up when the Sun sets, it is first quarter, but if it does not rise until midnight, it is third quarter. After third quarter, as the Moon moves closer to the Sun, we see a dimin-

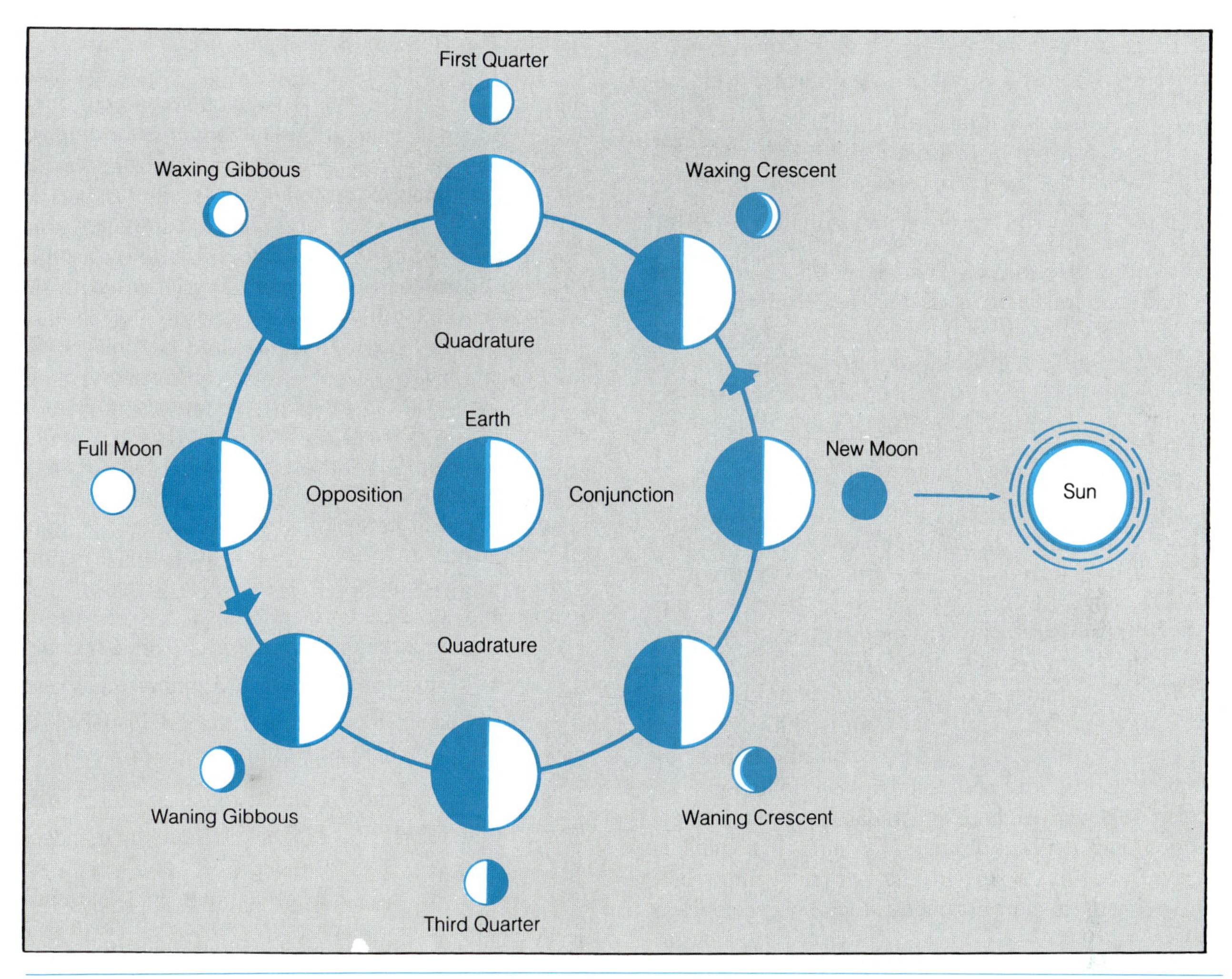

FIGURE 2.16 LUNAR PHASES AND CONFIGURATIONS. *As the Moon orbits the Earth, we see varying portions of its sunlit side. The phases sketched here (outside the circle representing the Moon's orbit) show the Moon as it appears to an observer in the northern hemisphere.*

ishing slice of its sunlit side, and it is in the **waning crescent** phase.

– ECLIPSES OF THE SUN AND MOON –

An eclipse of the Sun or **solar eclipse,** occurs when the Moon passes directly in front of the Sun, as seen from the Earth. A **lunar eclipse,** by contrast, occurs when the Moon passes through the Earth's shadow, so that for a brief period the Moon is not directly illuminated (Fig. 2.17).

From what has been stated so far, it may seem that the Moon should pass directly in front of the Sun on each trip around the Earth, and through the Earth's shadow at each opposition, producing alternating solar and lunar eclipses at 2-week intervals. However, this is not the case; the reason is that the Moon's orbital plane does not lie exactly in the ecliptic but is tilted by about 5°. Therefore, the Moon usually passes just above or below the Sun as it goes through conjunction and similarly misses the Earth's shadow at opposition. The Moon passes through the ecliptic at only two points on each trip around the Earth, where the planes of the Earth's and the Moon's orbits intersect. Because the Moon's orbital plane wobbles slowly, in a precessional motion similar to

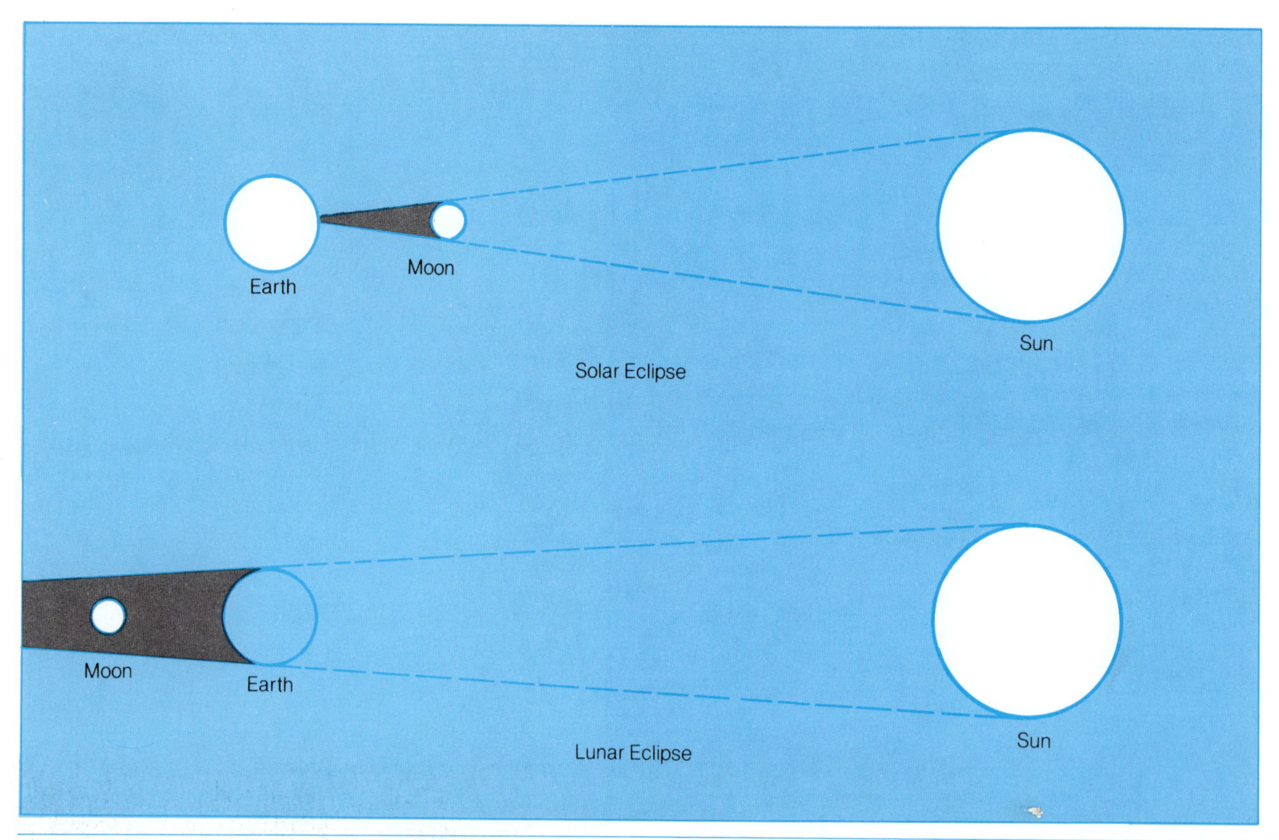

FIGURE 2.17 SOLAR AND LUNAR ECLIPSES. *The relative positions of the Sun, Earth, and Moon are shown at the times of solar (upper) and lunar (lower) eclipses. Only the dark portion of the shadow (the umbra) is shown in each case. (Not to scale.)*

FIGURE 2.18 A TOTAL SOLAR ECLIPSE. *This occurs when the Moon entirely blocks our view of the Sun.* (Fr. R. E. Royer)

that of the Earth's spin axis, the line of intersection with the Earth's orbital plane slowly moves around. The combination of this motion, the Moon's orbital motion, and the movement of the Earth around the Sun creates a cycle of eclipses, with the same pattern recurring every 18 years (see Table 2.2 for a list of future solar eclipses). This cycle of eclipses, called the **saros,** was recognized in antiquity.

It is purely coincidental that the Moon and the Sun have nearly equal angular diameters, so that the Moon neatly blocks out the disk of the Sun during a solar eclipse (Figs. 2.18 and 2.19). The angular diameter of an object is inversely proportional to its distance, meaning that the farther away it is, the smaller it looks. The Sun is much larger than the Moon, but it is also much more distant. The two objects have almost exactly the same angular diameter, because the ratio of the Sun's diameter to that of the Moon just happens to be almost the same as the ratio of the Sun's distance to that of the Moon (Fig. 2.20).

TABLE 2.2 TOTAL SOLAR ECLIPSES OF THE FUTURE

Date of eclipse	Duration (min)	Location
March 29, 1987	0.3	Central Africa
March 18, 1988	4.0	Phillipines, Indonesia
July 22, 1990	2.6	Finland, arctic regions
July 11, 1991	7.1	Hawaii, Central America, Brazil
June 30, 1992	5.4	South Atlantic
November 3, 1994	4.6	South America
October 24, 1995	2.4	South Asia
March 9, 1997	2.8	Siberia, arctic regions
February 26, 1998	4.4	Central America
August 11, 1999	2.6	Central Europe, central Asia
June 21, 2001		South Atlantic, South Africa
December 4, 2002		South Africa, Indian Ocean
March 29, 2006		South Atlantic, Africa, Middle East
August 1, 2008		Greenland, North Atlantic, USSR
July 22, 2009		Indonesia, South Pacific
July 11, 2010		South Pacific
November 13, 2012		South Pacific
March 20, 2015		North Atlantic
August 21, 2017		United States, Atlantic Ocean

If a total solar eclipse occurs at the time when the Moon is farthest from the Earth in its slightly noncircular orbit, it does not quite block all of the Sun's disk, instead leaving an outer ring of the Sun visible. This is called an **annular eclipse.** Because a total (or annular) solar eclipse requires precise alignment of the Sun and the Moon, an eclipse will appear total only along a well-defined, narrow path on the Earth's surface (Fig. 2.21). There is a wider zone outside of that, where the Moon appears to block only a portion of Sun's disk; people in this zone see a partial solar eclipse.

During a lunar eclipse, when the Moon passes through the Earth's shadow, observers everywhere on the nighttime side of the Earth see the same portion of the Moon eclipsed. If the Moon passes through the **umbra** (the dark inner portion of the Earth's shadow), the eclipse is total, as no part of the Moon's surface is exposed to direct sunlight. If the Moon misses being fully immersed in the umbra, it passes through the **penumbra** and undergoes a partial eclipse.

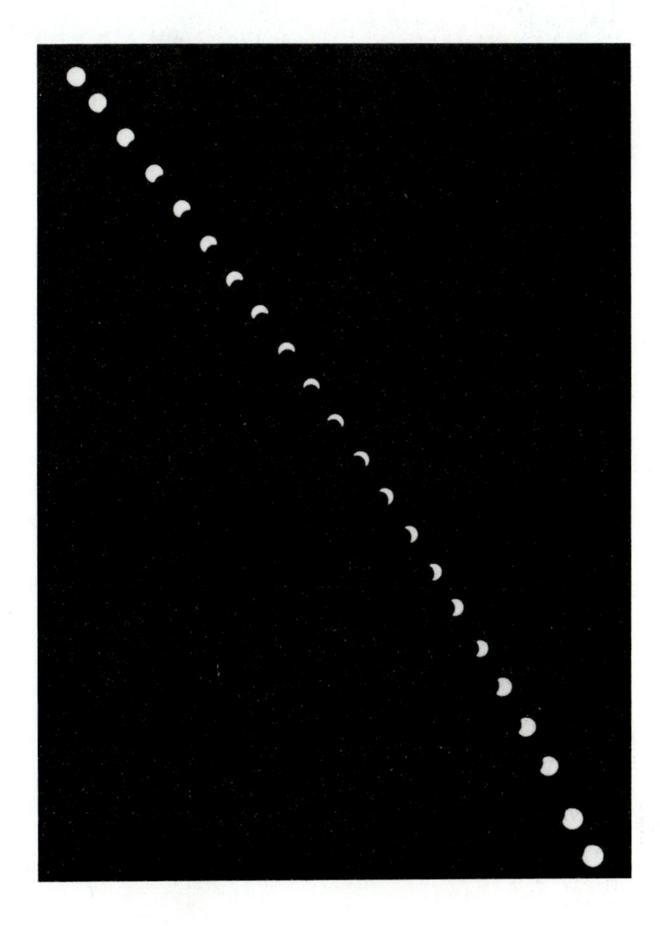

FIGURE 2.19 A SOLAR ECLIPSE SEQUENCE. *This series of photographs illustrates the Moon's progression across the disk of the Sun during a partial solar eclipse.* (NASA)

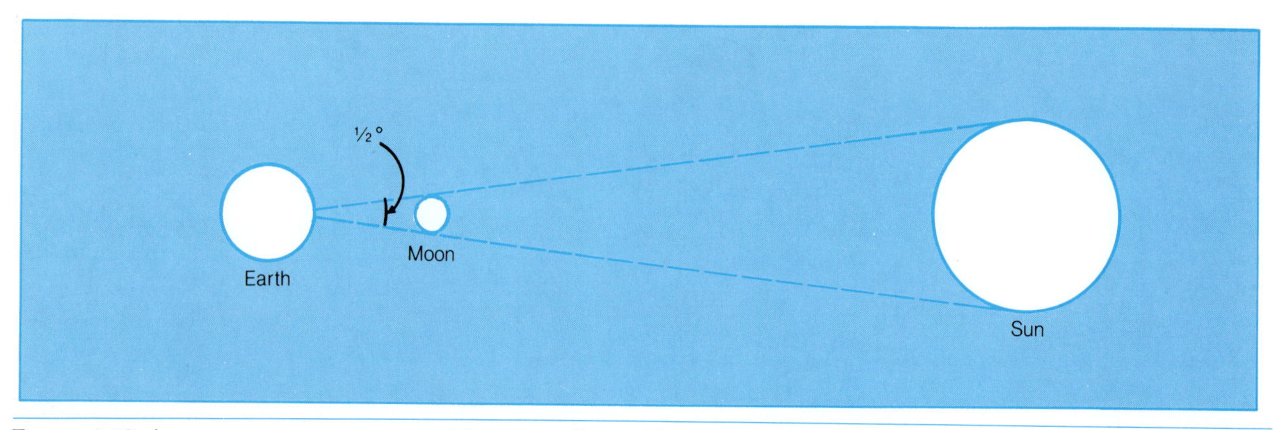

FIGURE 2.20 ANGULAR DIAMETERS OF THE MOON AND SUN. *The actual diameters and the relative distances of the Sun and Moon compensate one another so that the angular diameters are equal. (Not to scale.)*

FIGURE 2.21 THE MOON'S SHADOW DURING A SOLAR ECLIPSE. *This photograph, taken from space, shows the shadow of the Moon on the Earth during a solar eclipse. The eclipse appeared total only to observers on the Earth who were located directly in the center of the shadow's path (that is, in the umbra).* (NASA)

PLANETARY MOTIONS

Just as in the case of the Sun and the Moon, careful observation reveals that the planets, too, move with respect to the background stars. It was this fact that lent the planets their generic name, since **planet** is the Greek word for "wanderer." The planets all orbit the Sun in the same direction as the Earth and, as mentioned earlier, in nearly the same plane, so they appear to move nearly in the ecliptic, through the constellations of the zodiac.

The two planets lying within the orbit of the Earth, Mercury and Venus, are called **inferior planets** and can never appear far from the Sun in our sky (Fig. 2.22). For Mercury, the greatest angular distance from the sun, called the **greatest elongation,** is about 28°, whereas Venus can be seen as far as 47° from the Sun. Like the Moon, the planets have specific configurations, referring to their positions with respect to the Sun-Earth line. An inferior planet is said to be at **inferior conjunction** when it lies between the Earth and the Sun (or in **transit** if the orbit happens to be aligned with that of the Earth so that the planet is seen against the disk of the Sun) and at **superior conjunction** when it is aligned with the Sun but lying on the far side.

The outer planets, or **superior planets,** can be seen in any direction with respect to the Sun, including opposition, when they are in the opposite direction from the Sun (Fig. 2.22). Conjunction for a superior planet can occur only when the planet is aligned with the Sun but on the far side of it, in analogy with a superior conjunction for an inferior planet. **Quadrature** occurs when a superior planet is seen 90° from the direction of the Sun.

Each planet has a sidereal period and a synodic period, the former being the orbital period as seen in the fixed framework of the stars and the latter being the length of time it takes the planet to pass through the full sequence of configurations, as from one conjunction or one opposition to the next (Fig. 2.23). The situation is much like that of two runners on a track. The time it takes the faster runner to overtake the slower one is analogous to the synodic period, whereas the time it takes to simply circle the track corresponds to the sidereal period.

The motion of the Earth has one very important effect on planetary motions. As we go outward from the Sun, each successive planet has a slower speed in its orbit (see the discussion of Kepler's laws of plan-

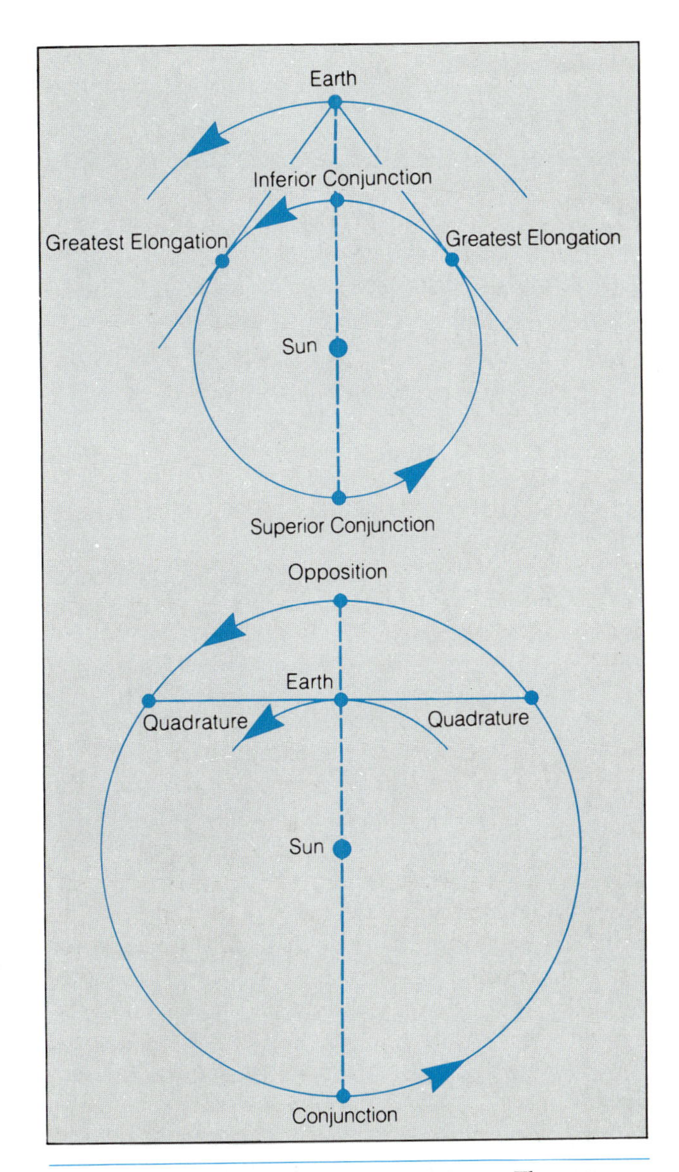

FIGURE 2.22 PLANETARY CONFIGURATIONS. *The upper drawing illustrates the configurations (planetary positions relative to the Sun-Earth line) for an inferior planet; the lower drawing shows the configurations for a superior planet.*

etary motion in Chapter 4). This means that the Earth, moving faster than the superior planets, periodically passes each of them (this occurs once every synodic period). As the Earth overtakes one of the superior planets, there is an interval of time during which our line of sight to that planet sweeps backward with respect to the background stars, making it appear that the planet is moving backward. The same thing hap-

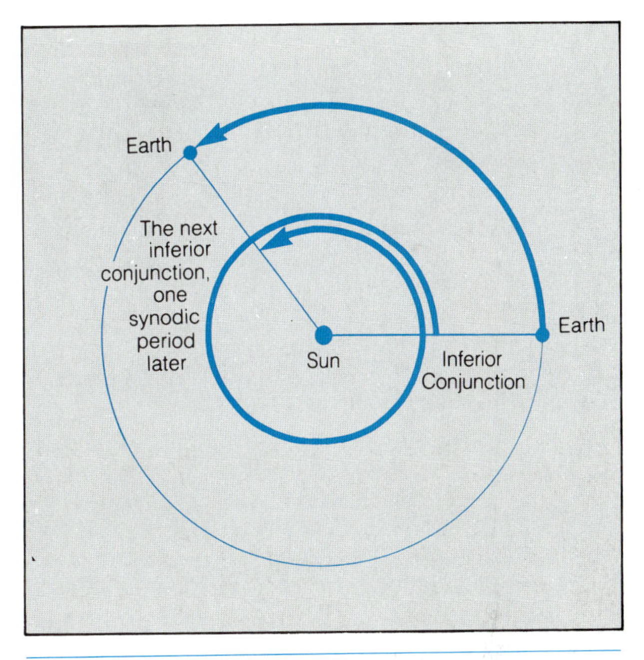

FIGURE 2.23 THE SYNODIC PERIOD OF AN INFERIOR PLANET. *The inner planets travel faster than the Earth in their orbits, and therefore "lap" the Earth, much as a fast runner laps a slower runner on a track. This illustration shows approximately the situation for Mercury, which has a synodic period of about 116 days, or roughly one-third of a year.*

pens when we, in a rapidly moving automobile, pass a slowly moving vehicle on a curved road; for a brief moment the other vehicle appears to move backward with respect to the fixed background. This apparent backward movement, called **retrograde motion,** was thought in ancient times to represent a real motion of the superior planets, rather than being recognized as a reflection of the Earth's motion (Fig. 2.24). This greatly complicated many of the early models of the universe and in fact gave rise to a whole class of theories in which the planets were thought to move in small circles called epicycles, which in turn orbited the Earth (see the discussions of Hipparchus and Ptolemy in Chapter 3).

Even the inferior planets undergo retrograde motion as they overtake the Earth, but this is difficult to observe, since it occurs near the time of inferior conjunction, when they lie near the direction of the Sun.

ASTRONOMICAL INSIGHT 2.3

CHASING A SOLAR ECLIPSE

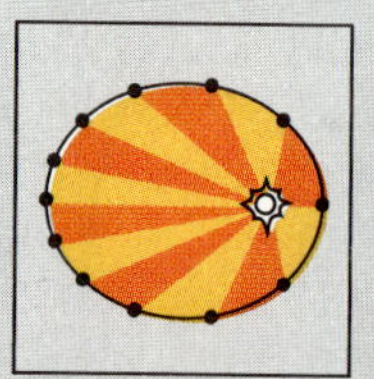

Total eclipses of the Sun are infrequent events, and the path of totality (the region on the Earth where the eclipse appears total) can fall virtually anywhere (see Table 2.2). Total solar eclipses are important to scientists because of the opportunities they provide for study of the Sun, and they can be profoundly moving events to experience. Therefore every total eclipse in modern times has been the cause of several major expeditions of scientists and sightseers, sometimes to quite exotic locales.

Because the angular diameters of the Sun and Moon are so nearly equal, the alignment between them must be very precise in order to produce a total solar eclipse. This means that the eclipse appears total only to observers in a very small region on the Earth. As the Moon moves across the disk of the Sun, this location travels across the Earth's surface; this is why we refer to a "path" of totality. The duration of the eclipse as seen from any single point on the Earth depends on the details of the relative distances of the Moon and the Sun and on the speed of the Moon in its orbit, so the length of time when an eclipse is total varies from one eclipse to another, and even from one point in the path of totality to another. The duration of totality is generally between 3 and 7 minutes.

During this brief time interval, the activity at the location where totality is occurring is frantic, particularly on the part of the scientific expedition members. This is the culmination of months of planning by the solar scientists, as they finally get their chance to collect data on the Sun's outer layers (which are normally invisible due to the overwhelming brightness of the solar disk). Before the eclipse begins there is an air of tension and expectation, as instruments undergo final tests and everyone watches the weather. When totality begins everyone is preoccupied with their observations; some scientists are too busy to look at the eclipse at all.

The sightseers who have come to watch have a better chance than the scientists to enjoy the esthetic qualities of the experience. Those who are busy making photographs, however, may have an experience more like that of the scientists). These people may be friends or relatives of members of the scientific teams, or they may have travelled to the eclipse observation site as part of a tour organized by a travel company. Several such tours usually accompany a total solar eclipse, particularly if the locale is not too difficult to reach and the eclipse will be of substantial duration.

If the weather is clear, a great sense of anticipation builds as time for the eclipse approaches. As *first contact* occurs, when the edge of the Moon first begins to occult the Sun's disk, the partial phase of the eclipse begins. The Moon's silhouette covers more and more of the Sun, and it gradually gets darker at the observing site. As totality nears (usually more than an hour after first contact), the sky becomes much darker. At *second contact,* the beginning of totality, only bits of the Sun, called **Bailey's beads,** are seen as the last rays of sunlight pass through valleys on the edge of the Moon's disk. As the eclipse becomes total, the Sun's **corona** and **chromosphere** become visible. These tenuous outer layers of the Sun, lying above the visible surface, are so dim that normally they cannot be seen directly. The opportunity to observe these layers is one of the main reasons that eclipses are valuable to solar physicists. The corona is an extended, pale white wreath around the Sun, often showing streaks and rays directed outwards. The chromosphere, lying closer to the Sun than the corona, has a distinctly red color, and often shows great jets of gas rising above the solar surface.

During totality at the observing site it is almost as dark as nighttime. Plants and animals behave accordingly: birds become silent, nocturnal animals may emerge from their lairs, and flowers may close up. Quite suddenly this phase ends, as *third contact* is reached, and the Sun begins to emerge from behind the Moon. Soon the world takes on a much more normal appearance to the observers, and the scientists and sightseers begin to relax. The eclipse officially ends at *last contact,* when the Moon's disk completely uncovers the Sun, ending the partial phase of the eclipse. Observers within a broad strip outside of the path of totality see only a partial eclipse which may last for two hours or more.

Even though substantial time and expense is usually required in order to experience a total solar eclipse, those who do it are uniformly enthusiastic and grateful for the opportunity. Perhaps you will have a chance to enjoy the experience on the occasion of one of the future eclipses listed in Table 2.2.

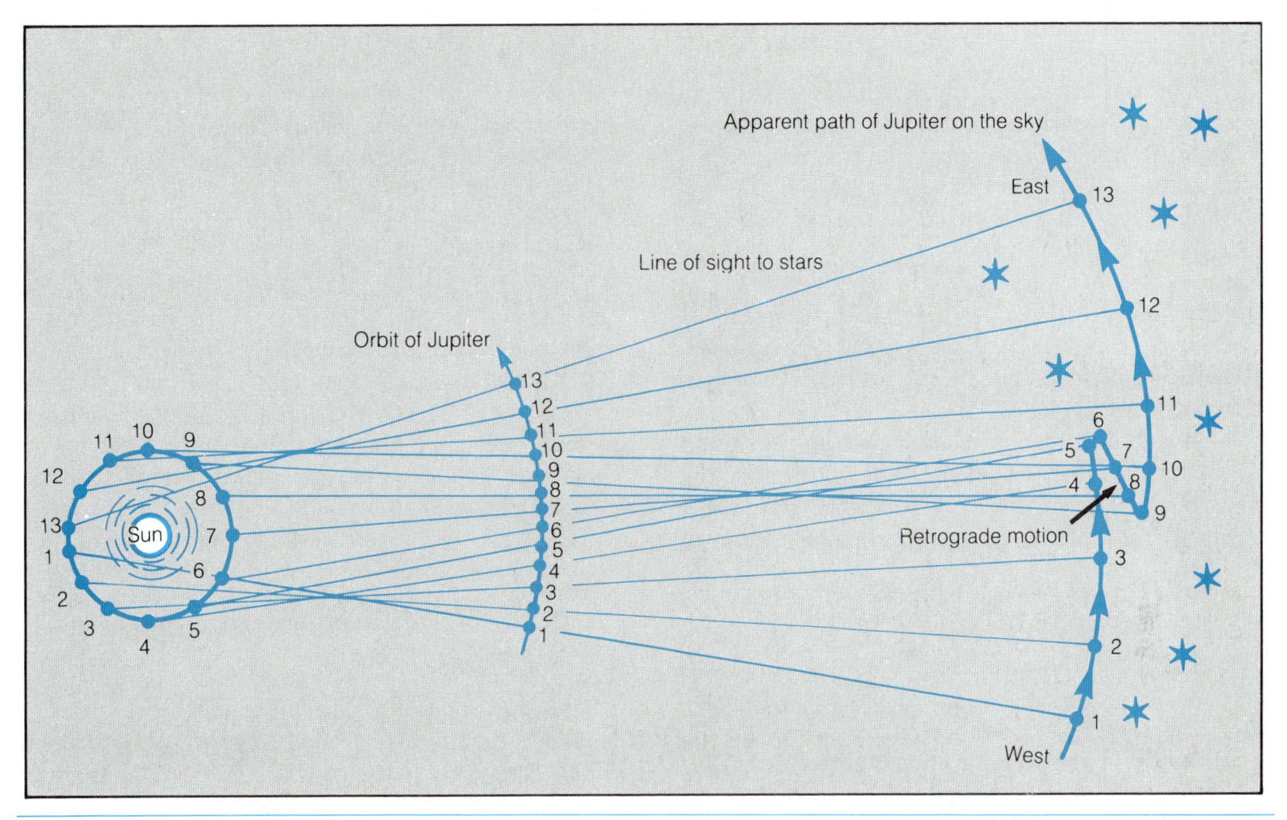

FIGURE 2.24 RETROGRADE MOTION. *As the faster-moving Earth overtakes a superior planet in its orbit, the planet temporarily appears to move backward with respect to the fixed stars. This sketch illustrates the modern explanation of something that took ancient astronomers a long time to understand correctly.*

PERSPECTIVE

We have traced the development of astronomy, beginning with our earliest attempts to understand the world around us. These attempts were based on fear of the unknown and the primitive belief that events in the heavens ruled the inhabitants of the Earth. This kind of astrological outlook emerged independently in many parts of the world, but the Greek philosophers went further in attempting to understand the universe just for the sake of understanding. Political, economic, and religious developments then brought about a prolonged intercession, when no real progress was made. Nearly 900 years after the death of Ptolemy, the first stirrings of intellectual revolution began to be felt.

SUMMARY

1. The Earth's rotation causes diurnal motions: the daily rising and setting of the Sun, the Moon, the stars, and the planets.

2. The solar day is a few minutes longer than the sidereal day because of the Earth's orbital motion about the Sun.

3. The most commonly used astronomical coordinate system, called the equatorial coordinate system, is based on the projection of the Earth's equator onto the celestial sphere (from which declination is measured) and on a fixed direction in the sky (from which right ascension is measured).

4. Precession is a slow drift of the Earth-based equatorial coordinate system with respect to the stars and

is caused by a gradual change in the orientation of the Earth's rotation axis.
5. The orbital motion of the Earth about the Sun causes annual motions, such as the apparent motion of the Sun through the ecliptic, and it is responsible, along with the tilt of the Earth's axis, for our seasons.
6. Modern calendars are based on the tropical year, the length of time from one vernal equinox to the next.
7. The Moon orbits the Earth, while the Earth orbits the Sun, and from the Earth we see the various phases of the Moon as it passes through different configurations.
8. Solar and lunar eclipses occur when the Moon passes directly in front of the Sun or through the Earth's shadow, respectively. The occurrence of these alignments is affected by the tilt and precession of the Moon's orbit.
9. All of the planetary orbits lie nearly in the same plane, so they are always seen along the ecliptic in various configurations with respect to the Sun-Earth line.
10. The planets go through temporary retrograde motion because of the relative speed with which they pass or are passed by the Earth.

REVIEW QUESTIONS

Thought Questions

1. Describe the difference between solar time, sidereal time, and the time kept by atomic clocks.
2. Explain why it is convenient to have the Earth divided into time zones, rather than using local solar time at every location.
3. Can you think of any way to tell, from observations or experiments, that the Earth is moving (spinning and orbiting the Sun)?
4. Explain how our seasons would be different if the Earth's axis were not tilted (that is, if the axis were perpendicular to the Earth's orbital plane). What would the seasons be like if the Earth's axis were tipped 90° instead of 23½°?
5. Use Fig. 2.7 to compare the constellation where the Sun was when you were born with your sign of the zodiac, according to astrology. Explain why they are not the same.
6. Our calendar, like many ancient ones, has the year divided into months that are based on the motions of the Moon. Is it the Moon's sidereal or synodic period that is used in these calendars? Explain.
7. Explain why, on certain nights, the Moon is never visible at all.
8. If a quarter moon is already up when the Sun sets, is it first or third quarter? Explain.
9. Table 2.1 lists sidereal and synodic periods for all the planets. Explain why the synodic periods of the outer planets are not much more than 1 year.
10. Briefly summarize the types of motions that have been described in this chapter. Would all of these motions appear in the nighttime sky to a person on the planet Venus (ignore the fact that Venus is cloud-covered)?

Problems

1. How fast, in miles per hour, does a point on the Earth's equator travel due to the Earth's rotation?
2. The difference between a solar day and a sidereal day is 3^m56^s. Over the course of 1 year, how many days does this difference amount to? Discuss the significance of your answer.
3. What is the right ascension of a star that is directly overhead at midnight at the time of the vernal equinox? What is the right ascension of a star that rises 6 hours after sunset at the time of the vernal equinox?
4. The Mauna Kea Observatory in Hawaii lies at a latitude of roughly 20°N, and the declination of the Magellanic Clouds is about −70°. Can the Magellanic Clouds be observed from Mauna Kea?
5. As noted in the text, precession causes the Sun's position in the zodiac to shift gradually, taking 26,000 years to make a full cycle through all twelve constellations. If the twelve constellations are equal in extent on the zodiac, how long does it take for the Sun to move from one constellation to the next? Discuss how this knowledge could help you determine historically when astrology was founded.
6. Suppose that you live at latitude 30°N. How far above the southern horizon would the Sun appear at noon at the time of the winter solstice? At the summer solstice?
7. What is the declination of the Sun at the time of each solstice and each equinox?

8. Suppose you adopted a calendar in which there were exactly 365 days, instead of the more accurate 365.256 days of the tropical year. How long would it take for an error of one full day to accumulate? How is this related to the leap year that we observe every 4 years?

9. If the Moon's apparent motion is ½° per hour, how long does it take to go all the way around the sky, a full 360°? How is your answer related to the Moon's synodic period; that is, the lunar month?

10. What would the Moon's angular diameter be if its distance from the Earth were doubled? What if the distance were halved instead? What would solar eclipses look like in each case?

ADDITIONAL READINGS

A number of magazines, including *Mercury*, *Sky and Telescope*, *Astronomy*, and the *Griffith Observer*, contain practical information for the sky watcher, such as planetary positions and the seasonal appearance of the constellations. There are also annually published handbooks with similar data. One of the most widely used of these is the *Observer's Handbook*, by Roy L. Bishop (Toronto: Royal Astronomical Society of Canada). Some practical exercises in astronomy can be found in such books as *Astronomy: A Self-Teaching Guide*, by Dinah L. Moché (New York: Wiley, 1981).

It is also possible to find readings on the history of astronomy. Below are listed some books relevant to this chapter, including material on ancient astronomical developments that are not covered here, for example, the astronomies of Asia and the Americas. In addition to locating books such as these, it is also useful to browse through professional journals that cover the history of science or of astronomy. These include *Vistas in Astronomy* and the *Journal for the History of Astronomy*, both of which can be found in most science-oriented libraries.

Heath, T. L. 1969. *Greek astronomy*. New York: AMS Press.

Krupp, E. C. 1987. *In search of ancient astronomies*. Garden City, N.J.: Doubleday.

_____. 1983 *Echoes of the ancient skies*. New York: Harper and Row.

Neugebauer, O. 1969. *The exact sciences in antiquity*. New York: Dover.

CHAPTER 3

Early Astronomy

The nighttime sky at Chaco Canyon, site of Native American astronomical petroglyphs. (J.A. Eddy)

CHAPTER PREVIEW

From the time we first became aware of our environment, we have been preoccupied with the heavens. The speculative mood that we, in these modern times, can conjure up only by disregarding our daily pressures and escaping into the countryside on a clear night to look at the stars must have dominated the nighttimes of the earliest cultures.

For the most part, we can only speculate about the astronomical knowledge of the prehistoric peoples who left no written records. We must assume that our distant ancestors in all parts of the world developed some awareness of astronomy, but we can discuss with real knowledge only those cultures which left records of their achievements. We find that parallel developments occurred in several parts of the world, including Asia, India, the Americas, and of particular importance to us, the Middle East and Mediterranean regions, wherein lie the roots of our modern astronomical science.

To develop the historical sequence leading to the present, we devote this chapter to astronomy in the vicinity of the Mediterranean, the center of the regions where the Babylonian, Egyptian, and Greek civilizations flourished (Fig. 3.1). Later in the chapter, accounts of the early Asian and American astronomies are presented, but largely for historical interest, since these cultures, by accident of geography, did not contribute to the mainstream development.

BABYLONIAN ASTRONOMY

In what is now Iraq, along the valleys of the Tigris and Euphrates rivers, there arose one of the earliest civilizations known to have left written records. The first people in this region were the Sumerians, who occupied the southern portion by about 3000 B.C. and invented a written language, cuneiform script, which was used to record astronomical data on clay tablets (Fig. 3.2).

The entire region was united around 1700 B.C. under the leadership of King Hammurabi, who established his capital at the city of Babylon (not far from the present Baghdad), lending the kingdom the name Babylonia. The clay tablets from that era reveal that the movements of the planets were given great significance. Tables were constructed for astrological purposes, showing that the Babylonians knew a great deal about the various cyclical motions of the heavens. They recognized that the Sun moves through the sky, and they divided the zodiac into twelve parts, probably because the year contains twelve lunar months (although not precisely twelve). The length of the year was known to within 4 minutes' accuracy, but the Babylonian calendar contained twelve 30-day months, creating a discrepancy that was corrected by the addition of an extra month every few years. A numbering system was developed based on powers of 60, much as our own system is based on powers of 10, and it was the Babylonians who bequeathed to us our units of angular measure (360° in a circle corresponding to the 360 days in their year; 60 minutes of arc per degree; and 60 seconds per minute).

The land between the Tigris and the Euphrates, known to the Greeks as Mesopotamia, saw several empires come and go, well into the first centuries A.D. The Assyrians dominated during the era around 600 to 800 B.C., and they preserved the clay tablets containing astronomical data from earlier times. Around 200 B.C., after the center of power had moved west to Greece, a massive amount of data was compiled by the then-occupants of Babylonia, the Chaldeans. The Chaldean tables were sufficiently accurate so that solar, lunar, and planetary motions could be predicted, as could the occurrence of eclipses.

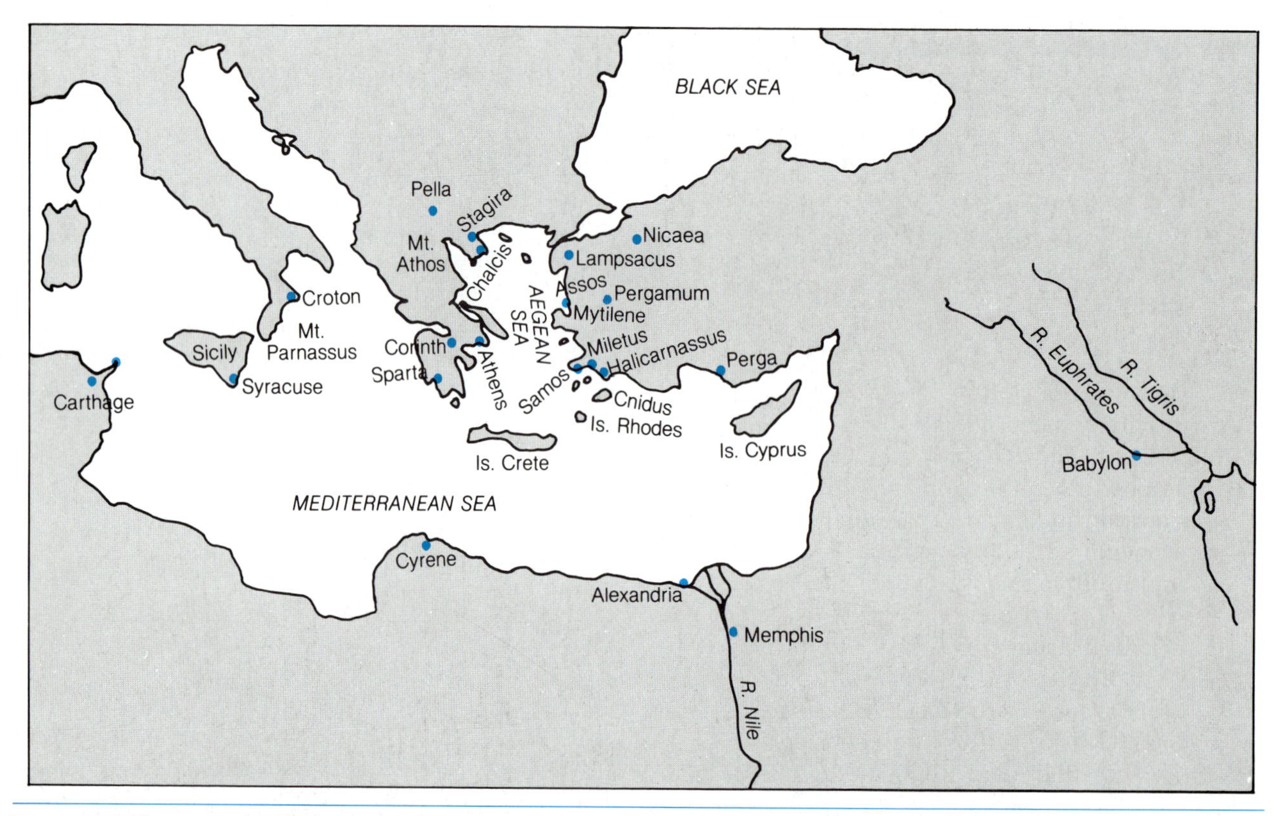

FIGURE 3.1 THE ANCIENT MEDITERRANEAN. *This map shows the locations of many of the sites mentioned in the text. To the right is the region of Babylonia, now Iraq. On the northern shores of the Mediterranean Sea lay most of the Greek empire, in what is now Italy, Greece, and Turkey.* (C. Roman, *The Astronomers*, 1964 [New York: Hill and Wang])

The primary motivation throughout these times was the application of astronomical information to astrology. The Babylonians seem to have speculated very little about the nature of the universe, apparently envisioning the heavens as consisting of a dome supported by the mountains surrounding their region. No evidence has yet been uncovered for any theories of the motions of the heavenly bodies.

Some astronomical development occurred contemporaneously in Egypt, but little was accomplished there beyond the establishment of timekeeping techniques also known to the Babylonians, who used sundials to record local solar time. In Egypt, the main purpose in studying astronomy was neither philosophical nor astrological, but was practical and religious, centered on anticipating the seasons and other events of importance to agriculture and commerce.

As the Babylonian science developed and the region underwent several successive conquests, a separate civilization began to organize itself along the shores of the Mediterranean, a civilization that was to reach the greatest heights of scientific accomplishment in ancient times. The Greek empire that arose there was greatly influenced in its early astronomical development by the Babylonians. Major contributions of the early Greeks are summarized in Table 3.1.

THE EARLY GREEKS AND THE DEVELOPMENT OF SCIENTIFIC THOUGHT

The foundations of the Greek civilization were established some 5,000 years ago in the eastern Mediterranean. There arose a seafaring culture, whose earliest home was the island of Crete, where the legends were born that gave us the names of most of our constellations. This civilization prospered around 1600 B.C. during the reign of King Minos, and it was about this time that the settlement of what is now the Greek mainland began. Much of what we know about this era is gleaned from the writings of Homer in the *Iliad* and the *Odyssey* (Fig. 3.3), written later, probably around 900–800 B.C. Interspersed in these epic poems were the early Greek views of the universe, along with the names of the constellations.

The cosmology of this time consisted of the dome of the heavens, with a disklike earth floating on water. This was similar to the Babylonian view, developed at about the same time, but the Greeks went further, hypothesizing—for what were essentially theoretical reasons—that there was an underworld comparable in scope and complexity to the heavens. Apparently these people had developed a sense of aesthetics and unity in the universe, which shows that their thinking went beyond mere description of the visible skies.

The first formal scientific thought evolved after the time of Homer's writing, and its inception is associated with the philosopher **Thales** (c.624–547 B.C.), who lived in Miletus, on the eastern shore of the Aegean Sea. His principal contribution was the idea that rational inquiry can lead to understanding the universe, going beyond describing it. This general precept was adopted by followers of Thales, who formed the so-called Ionian school of thought, named for that

FIGURE 3.2 A CUNEIFORM CLAY TABLET. *Thousands of tablets such as this one, containing Babylonian records of astronomical data, have been uncovered in what is now Iraq.* (The Granger Collection)

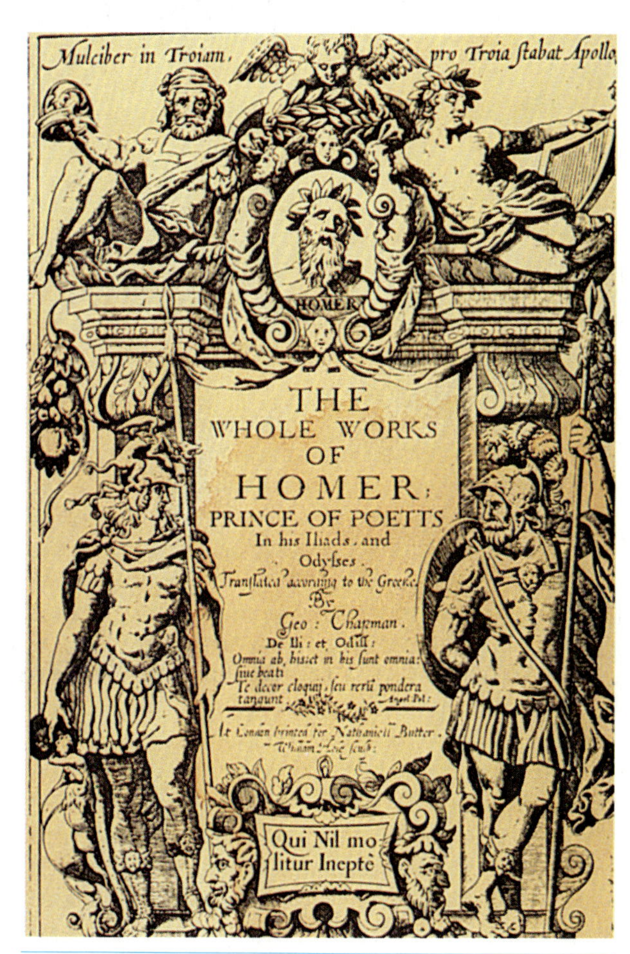

FIGURE 3.3 HOMER'S ODYSSEY AND ILIAD. *In these two epic poems, Homer described the much more ancient legends of astronomical lore from the civilization that had flourished on the isle of Crete.* (The Granger Collection)

TABLE 3.1. NOTABLE GREEK ACHIEVEMENTS

Date	Name	Discovery or achievement
c.900–800 B.C.	Homer	*Iliad* and *Odyssey;* summaries of legends
c.624–547 B.C.	Thales	Rational inquiry leads to knowledge of universe
c.611–546 B.C.	Anaximander	Universal medium; primitive cosmology
c.570–500 B.C.	Pythagoras	Mathematical representation of universe; round earth
c.500–400 B.C.	Philolaus	Earth orbits central fire
c.500–428 B.C.	Anaxagoras	Moon reflects sunlight; correct explanation of eclipses
c.428–347 B.C.	Plato	Material world imperfect; deduce properties of universe by reason
c.408–356 B.C.	Eudoxus	First mathematical cosmology; nested spheres
c.384–322 B.C.	Aristotle	Concept of physical laws; proof that earth is round
c.310–230 B.C.	Aristarchus	Relative sizes, distances of sun and moon; first heliocentric theory
c.273–? B.C.	Eratosthenes	Accurate size of earth
c.265–190 B.C.	Apollonius	Introduction of the epicycle
c.200–100 B.C.	Hipparchus	Many astronomical developments; full mathematical epicyclic cosmology
c. A.D. 100–200	Ptolemy	*Almagest;* elaborate epicyclic model

part of Greece (now Turkey) surrounding Miletus. One of Thales' most singular claims to fame as an astronomer was his supposed prediction of a solar eclipse that took place in 585 B.C., a feat that some historians believe did not actually occur. Much of the Greek knowledge of the motions of the heavens at that time came from the Babylonians, and it is not certain that even they were yet able to predict eclipses accurately.

Thales did create a primitive cosmology, encompassing the idea that all the basic elements of the universe were formed from water, the primeval substance. An associate of his, **Anaximander** (c.611–546 B.C.), altered this view and adopted the idea of a universal medium, of unknown properties, from which all material substance was made. Anaximander also developed the notion, entirely without observational support, that outside the Earth the firmament is filled with flame, contained in rotating tubes with holes that allowed the light within to be seen. Apparently it was his concept that the Sun, Moon, and Stars were openings in the sides of these hoops, so that their rotation created the observed motions of the heavenly bodies.

The Pythagoreans

Pythagoras (c.570–500 B.C.) was born on the island of Samos, off the Aegean coast not far from Miletus. His early life is obscure, but at around the age of forty, following several years of travel and study, he established a school at Croton, on the southern tip of Italy, in which scientific and religious teachings were intermixed. The lasting contribution of this school was the notion that all natural happenings can be described by numbers, laying the foundation for geometry and trigonometry. Pythagoras himself is thought to have been the first to assert that the Earth is round and that all heavenly bodies move in circles, ideas that never lost favor thereafter in ancient times.

Most of what we know about Pythagoras, who himself left no written records, comes from the works of his follower **Philolaus** (c.500–400 B.C.), who added the idea that the Earth moves about a central fire (not the Sun), always hidden on the far side of the Earth. Both he and Pythagoras, for largely philosophical reasons rather than logic, believed the planetary distances to correspond to the lengths of vibrating strings that produced harmonious musical notes. This led to the concept of a celestial harmony that later had considerable influence in philosophical thinking.

There were numerous followers of the Pythagorean school who in various ways refined this view of the universe. One of particular note was **Anaxagoras** (c.500–428 B.C.), who is credited with the realization that the Moon shines by reflected sunlight, rather than by its own power. This awareness, based on observation and deduction rather than personal philosophy, allowed him to correctly attribute solar and lunar eclipses to the passage of the Moon in front of the Sun and through the Earth's shadow, respectively.

Plato and His Followers

On the Greek mainland, a number of democratically ruled city-states arose, of which Athens became preeminent by about the fifth century B.C., dominating artistic and literary affairs. Even after the military defeat in 404 B.C., at the hands of the rival city of Sparta, Athenian culture continued to flourish. It was in this environment that **Plato** (428–347 B.C.; Fig. 3.4) reached maturity, having studied philosophical thought under the tutelage of Socrates.

After some years of traveling, Plato established an academy outside of Athens, in about the year 387 B.C., where he taught his ideas of natural philosophy. His fundamental precept was that what we see of the material world is only an imperfect representation of ideal creation. The corollary of this doctrine was that we can learn more about the universe by reason than by observation, since, after all, observation can give us only an incomplete picture. Hence Plato's ideas of the universe, described in his *Republic,* were based on certain idealized assumptions that he found reasonable. One of the most important was that all motions in the universe are perfectly circular and that all astronomical bodies are spherical. Thus he adopted the Pythagorean view that the Sun, Moon, and planets move in various combinations of circular motion about the Earth. It seems that Plato thought of these objects as affixed to clear, ethereal spheres that rotated. The teachings of Plato, particularly the premise that the mysteries of the cosmos should be solved through reason rather than by direct observation, dominated much of western thinking until the Renaissance, nearly 2,000 years after his time.

FIGURE 3.4 PLATO. *Plato's beliefs that the universe and all celestial objects were perfect and immutable had great influence on later philosophers.* (The Granger Collection)

A pupil of Plato's was **Eudoxus** (c.408–356 B.C.), who made the first attempt to put Plato's view of the universe on a firm mathematical footing; he constructed a series of concentric spheres on which the Sun, Moon, and planets moved in perfect circular motions (Fig. 3.5). Each body required three or four spheres, each with its own rotation rate, to produce the composite motion necessary to duplicate the overall motion. A total of twenty-seven spheres was needed for all the known heavenly bodies. It is not clear whether Eudoxus thought of these spheres as material objects or merely mathematical constructs.

The most renowned student of Plato was **Aristotle** (c.384–322 B.C.) who, after the death of Plato in 347 B.C., established his own small academy across the Aegean Sea at Assos. He then moved to Pella, where he tutored Alexander, the son of King Phillip of Macedonia, leader of the Greek empire. Aristotle later created a school in Athens, where he taught his own view of the universe, one formed from Plato's mold of reason rather than observation.

Aristotle was the first to adopt physical laws and then to show why, in the context of those laws, the universe works as it does. He taught that circular motions are the only natural motions and that the center of the Earth is the center of the universe. He also believed, without any experimental or observational support, that the world is composed of four elements: earth, air, fire, and water. He demonstrated, in the context of his adopted physical laws, that both the universe and the Earth are spherical. He had three ways of proving that the Earth is spherical: (1) only at the surface of a sphere do all falling objects seek the center by falling straight down (it was another premise of his that falling objects are following their natural inclination to reach the center of the universe); (2) the view of the constellations changes as one travels north or south; and (3) during lunar eclipses it can be seen that the shadow of the Earth is curved (Fig. 3.6). By relating his theories to observation in this manner, Aristotle broke with the tradition of Plato to some extent, although, as we have seen, he still approached the problem in the same manner, letting reason rather than observation guide the way.

Aristotle also concluded that the universe is finite in size. This led directly to his belief that the heavenly bodies can follow only circular motions, because otherwise they might encounter the edge of the universe. Another tenet of Aristotle was that the heavenly bodies were made of a fifth fundamental substance, which he called the *aether*.

The Geometric Genius of Aristarchus and Eratosthenes

After Aristotle's period, the center of Greek scientific thought moved across the Mediterranean Sea to Alexandria (the capital city established in 332 B.C. by Aristotle's former pupil, Alexander the Great), near the site of the present-day city of Cairo.

The most prominent astronomer of this era, however, was born on the Mediterranean island of Samos, and it is not well known where he lived and worked,

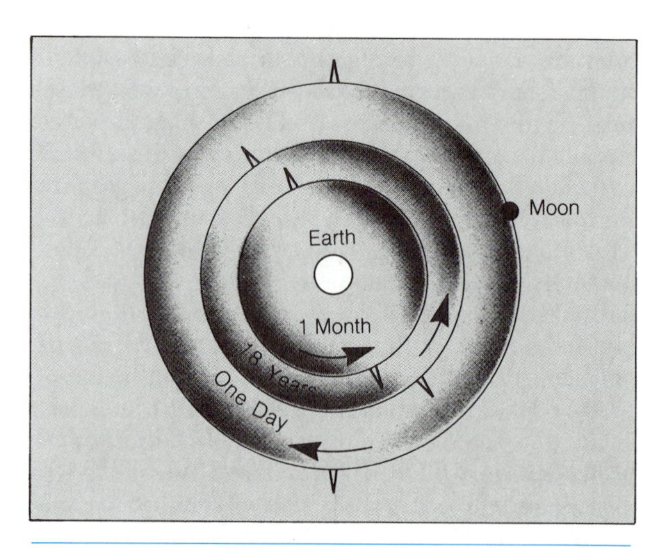

FIGURE 3.5 THE SPHERES OF EUDOXUS. *This shows the relative orientations and directions of rotation for spheres thought to be responsible for the monthly motions of the Moon, as well as its 18-year eclipse cycle.*

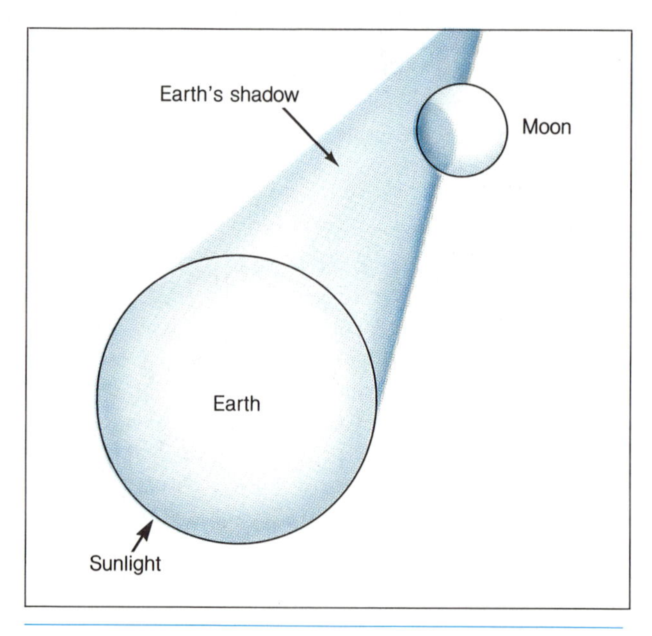

FIGURE 3.6 CURVATURE OF THE EARTH'S SHADOW ON THE MOON. *Only a spherical body can cast a circular shadow for all alignments of the Sun, Moon, and Earth.*

ASTRONOMICAL INSIGHT 3.1

THE MYTHOLOGY OF THE CONSTELLATIONS

As noted in the text, the writings of Homer in the *Iliad* and the *Odyssey*, which appeared around 900–800 B.C., contain descriptions of the constellations and their meanings. These descriptions were based on legends from the Minoan civilization that were already ancient in Homer's time.

Little precise information is available as to when and how the mythology of the constellations arose. However, it has been possible to deduce roughly the era that gave birth to them. Careful examination of the ancient constellations shows them to be distributed symmetrically about a point in the sky that was at the north celestial pole around 2600 B.C. so it is probable that the constellations were invented at about that time. (The pole has since shifted away from the center of these constellations because of precession.) There are few ancient constellations near the southern pole, in regions not visible from the latitude of Crete, a fact that gives independent support to the supposition that the legends arose in the Minoan culture.

It is likely that the constellations were not taken as literally as is usually supposed. Rather than being considered faithful depictions of the people and events with which they are associated, the constellations should more properly be viewed as symbolic representations. They were probably not originally designated on the basis of their imagined resemblance to certain characters; instead, areas of the sky were dedicated in honor of prominent figures of mythology, and the familiar pictures of these figures were then fitted to the patterns of bright stars. This helps account for the lack of obvious resemblance between the star patterns and the figures and events they supposedly represent.

The constellation names were translated from the Greek of Homer to Latin when the Roman Empire rose to dominance, and most of the names with which we are familiar today are the Latin ones. Interestingly, the names of prominent stars went through another transformation, into Arabic, and today's star names are Arabic designations that usually are literal indications of the place these stars hold in their constellations. Betelgeuse, for example, the name of the prominent red star in the shoulder of Orion, is translated as "the armpit of the giant" or "the armpit of the central one." Thus, in today's standard usage, we often adopt Arabic names for stars in constellations with Latin designations, although both translations are based on Greek descriptions of the ancient Minoan legends.

Ancient and nonscientific though they are, the constellations have a significant impact on the nomenclature of modern astronomy. The modern constellations, many of which correspond to ancient ones, refer to very specific regions of the sky with well-defined boundaries that have been agreed on by the international community of astronomers. The "official" constellations are based on those of legend, containing within them the ancient figures of mythology, but their boundaries have been extended so that every part of the sky falls within one constellation or another. A map of the constellations looks a bit like a map of the western United States, where the boundaries are generally straight lines, but the shapes are irregular. As an alternative to the Arabic names for the brightest stars, modern astronomers often use designations based on a star's rank within its constellation, with letters of the Greek alphabet used to indicate specific stars. Thus, Betelgeuse, the brightest star in Orion, is also called α Orionis.

It can be fun and informative to learn the prominent constellations, and this is taught in many general astronomy classes. It is worth keeping in mind, however, that memorizing star patterns on the sky is not the same as scientific inquiry, but should be regarded instead as a convenient method for remembering the locations of objects that are interesting to view.

although it may have been in Alexandria. This was **Aristarchus** (c.310–230 B.C.), the first scientist to adopt the idea that the Sun, not the Earth, is at the center of the universe. This was some 1,700 years before the time of Copernicus, who usually is given credit for this concept. The heliocentric hypothesis of Aristarchus failed to attract many followers at the time because of a lack of concrete evidence that the Earth was in motion, and a general satisfaction with the Aristotelian viewpoint, which had no generally recognized flaws.

It is interesting to consider how Aristarchus reached his revolutionary idea. By geometrical means, he was able to deduce the relative sizes and distances of the Moon and the Sun; and although his results were not very accurate, he did correctly conclude that the Sun is far larger than either the Moon or the Earth. Accepting this, he found it difficult to imagine that the Sun should orbit the lesser Earth, so he adopted the contrary point of view.

A younger contemporary of Aristarchus was **Eratosthenes** (c.273–? B.C.), who worked in Alexandria during his long career of research in geometry and astronomy. His greatest achievement was the determination of the size of the Earth, by a method of great simplicity (Fig. 3.7). It was known that on a certain day in the summer, sunlight penetrated all the way to the bottom of a deep well located at Syene, near present-day Khartoum. On the same day in Alexandria, the Sun's midday position was some 7° south of the zenith, indicating that the distance between the two cities corresponded to 7° of the 360° circle of the Earth's circumference (this assumes that the Sun is sufficiently far away that rays of light from it are essentially parallel when they reach the Earth). Hence the circumference was equal to 360°/7° times the distance between Alexandria and Syene, which was a little less than 5,000 stades, a stadium being a unit of measure equal to about one-tenth of a mile. The circumference of the Earth, as calculated by Eratosthenes, was 252,000 stades, or just under 25,000 miles—within 2 percent of the correct value!

Geocentric Cosmologies: Hipparchus and Ptolemy

By the third century B.C., the need for more precise mathematical models of the universe became apparent, as better observing techniques were developed. A mathematical concept that provided the needed precision while still preserving the precepts of Aristotle was the **epicycle** (Fig. 3.8). It was **Apollonius**

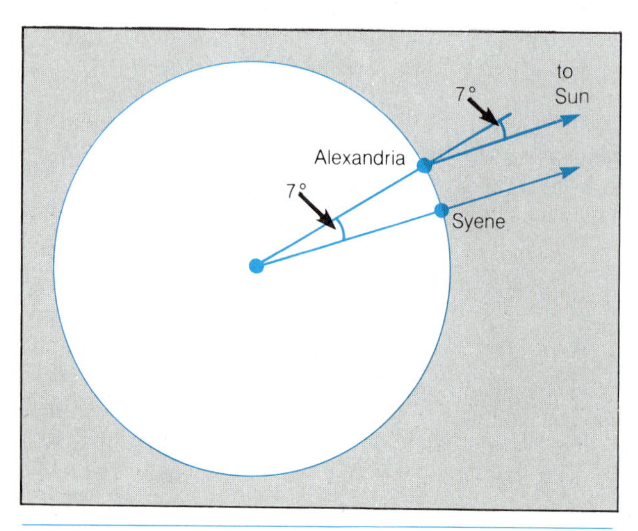

FIGURE 3.7 THE METHOD OF ERATOSTHENES FOR MEASURING THE SIZE OF THE EARTH. *Eratosthenes deduced the portion of a circle that corresponded to the distance between Alexandria and Syene. From this he was able to calculate the circumference of the Earth.*

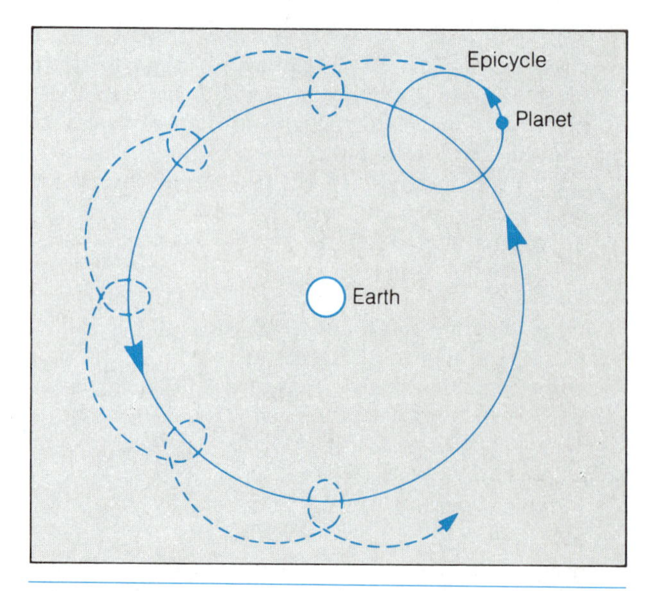

FIGURE 3.8 THE EPICYCLE. *It was realized that planetary motions could be represented by a combination of motions involving an epicycle, which carries a planet as it spins while orbiting the earth.*

(c.265–190 B.C.) who put the epicycle on a firm geometric footing, providing the first mathematically viable alternative to the idea of spheres developed by Eudoxus. An epicycle is a small circle on which a planet moves, whose center in turn orbits the earth following a larger circle called a **deferent.** This construct allows a combination of motions whose speeds and circular radii can be adjusted as needed to provide a close match with the observed motions. The epicycle provided a clear advantage for explaining the retrograde motions of the superior planets, something that confounded Eudoxus' theory that the planets move on rotating spheres.

The epicyclic motions of the celestial bodies were refined further by Hipparchus (Fig. 3.9), who was active during the middle of the second century B.C. (very little is known about his life, even the dates of his birth and death). Hipparchus, most of whose work was done at his observatory on the island of Rhodes, was arguably the greatest astronomer of antiquity. Among his major contributions were: (1) the first use of trigonometry in astronomical work (in fact, he is largely credited with its invention, although many of the concepts were developed earlier); (2) the refinement of instruments for measuring star positions, along with the first known use of a celestial coordinate system, enabling him to compile a catalogue of some 850 stars; (3) the refinement of the methods of Aristarchus for measuring the relative sizes of the Earth, Moon, and Sun; (4) the invention of the stellar magnitude system for estimating star brightnesses, a system still in use today (with minor modifications); and, perhaps most impressively; (5) the discovery of the precession of star positions, which he accomplished by comparing his observations with some that were made 160 years earlier.

FIGURE 3.9 A FANCIFUL RENDITION OF HIPPARCHUS AT HIS OBSERVATORY. *Hipparchus made important discoveries based on his observations and on his advanced methods for analyzing data.* (The Bettmann Archive)

Although Hipparchus did not add much that was new to the then-current models of planetary motions, he did compile an extensive series of observations of the solar motion. This led him to a refinement of the epicycle theory, namely, that the earth was off-center in the large circle on which the sun moved. This idea of an **eccentric,** or off-center, orbit accounted rather accurately for the observed variations of the speed of the sun in its annual motion.

It is worthwhile to consider why Hipparchus, with access to the work of Aristarchus (including the knowledge that the sun is much larger than the earth), did not adopt the heliocentric hypothesis. Apparently he rejected this idea because of one very significant observation: He could not detect any shifting of star positions during the course of the year, shifting that he realized should appear as a result of the changing point of view if the Earth moved about the Sun (Fig. 3.10). The lack of any detectable **stellar parallax,** as such a motion is called, caused many to resist the heliocentric viewpoint for many centuries to come.

After the great work of Hipparchus, almost 300 years passed before any significant new astronomical developments occurred. Claudius Ptolemaeus (Fig. 3.11), or simply **Ptolemy,** lived in the middle of the second century A.D. Attempting to summarize all the world's knowledge of astronomy, he published a treatise now called the *Almagest,* which was based in part on the work of Hipparchus, but which also contained some new developments of his own. The thir-

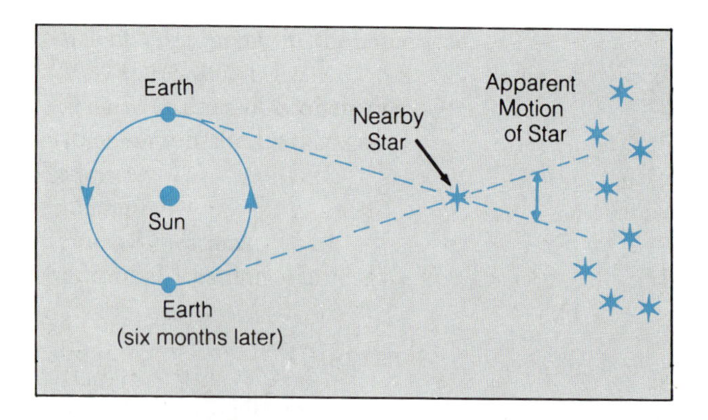

FIGURE 3.10 STELLER PARALLAX. *As the Earth orbits the Sun, our line of sight toward a nearby star varies, causing the star's position (with respect to more distant stars) to change. The change of position is greatly exaggerated in this sketch; in reality, even the largest stellar parallax displacements are too small to have been measured by ancient astronomers.*

FIGURE 3.11 CLAUDIUS PTOLEMY *Ptolemy summarized all of the Greek knowledge of astronomy, including significant contributions of his own, in the* Almagest. (The Granger Collection)

teen books of the *Almagest* range from a summary of the observed motions of the planets to a detailed study of the motions of the Sun and Moon, and from a description of the workings of all the astronomical instruments of the day to a reproduction of Hipparchus's star catalogue. Most important, in his treatise, however, are the detailed models of the planetary motions. Here Ptolemy made his greatest personal contribution, for these models were so accurate in their ability to predict the planetary positions that they were used for the next 1000 years.

To provide this accuracy, Ptolemy had to adopt and extend an idea from Hipparchus, that of the eccentric (Fig. 3.12). To satisfactorily account for all the observed irregularities in a planet's motion, Ptolemy assumed not only that the center of the deferent was off-center from the position of the Earth, but also that the rate of motion of the epicycle as it orbited the Earth was constant as seen from another off-center point, on the opposite side of the Earth from the geometrical center of the deferent. This combi-

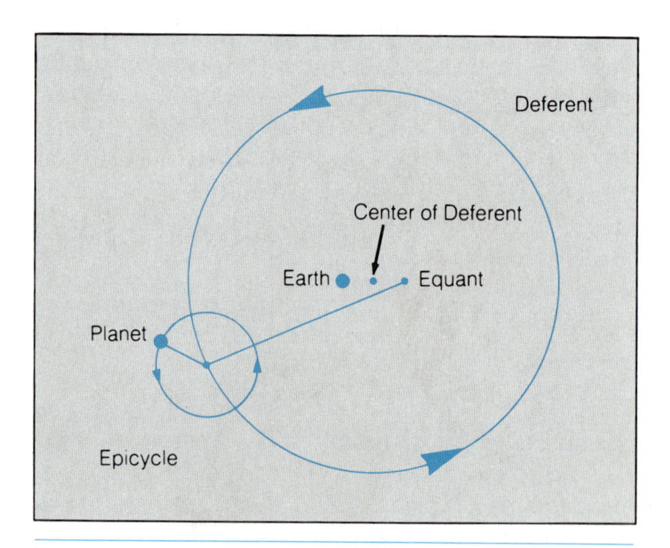

FIGURE 3.12 THE COSMOLOGY OF PTOLEMY. *To account for nonuniformities in the planetary motions, Ptolemy devised a complex epicyclic scheme. The deferent, the large circle on which the epicycle moves, is not centered on the Earth and has a rotation rate that is constant as seen from another displaced point called the equant. This meant that from the Earth, the planet appeared to move faster through the sky when on one side of its deferent than when on the other. Each planet had its own deferent and epicycle, with motion and offsets adjusted to reproduce the observations.*

nation of offsets, with properly chosen dimensions and rotational speeds, allowed good representation of the observed motions while preserving the hypothesis that all the heavenly motions are circular. It is interesting that other doctrines of Aristotle, such as the precise centering of all the planetary orbits on the Earth, were not strictly followed.

It seems from some of Ptolemy's writings that he may not have considered his model of the universe to represent physical reality, but instead may have viewed it as simply a mathematical tool for predicting planetary positions. This is something modern scientists do when confronted with data on phenomena for which the underlying physical principles are not clear, and in this sense, Ptolemy may be considered as one with modern astronomers. Whether Ptolemy believed his model to represent physical reality or not, it was taken literally in the centuries following his time. The notion of a geocentric universe with only circular motions dominated for some fourteen centuries afterward and was expunged only after great turmoil.

The history of Greek astronomy came to an end with the work of Ptolemy, as did most significant astronomical development in Europe and western Asia until the time of Renaissance. We will return to pick up the thread of astronomical knowledge and growth during the Dark Ages, after we digress and consider what was going on in the rest of the world.

Early Astronomy in Asia

Chinese Accomplishments

One of the most ancient recorded astronomies outside the Middle East arose in China. Legends there refer to astronomical activities as far back as 2000 B.C., although the oldest writings are much more recent.

Ancient China had a tumultuous history, with periods of peace, prosperity, and cultural advance occurring throughout numerous centralized dynasties, interspersed with eras of disarray resulting from barbarian invasions and civil unrest. Historical records were sometimes lost during the transitions; thus it is not known just how far in the past the first significant astronomical activity began. Legend refers to a pair of astronomers named Hi and Ho, who, having failed to predict a solar eclipse that occurred in 2137 B.C., were put to death.

We know that timekeeping was well established in China by the fourth century B.C., since records from that era indicate that the length of the year and the lunar month were known with considerable accuracy. The principal roles of astronomers were to keep the time and to officiate over state functions, ensuring that all was done in accordance with the rules of the gods. Astrology played a major role, and the constellations were identified with the emperor and with government organization (Fig. 3.13). The north polar region of the sky, for example, was thought to represent the seat of the emperor, and within a constellation, stars were given royal rank according to their brightness. A catalogue of some 809 stars was constructed in the fourth century B.C. by the astronomer Shih-shen, predating by about 200 years Hipparchus' catalogue.

The prevalent concept of the universe, however, was not nearly as advanced as that of the Greeks. The cosmos was thought to consist simply of the sphere of the sky with its daily rotation, and little

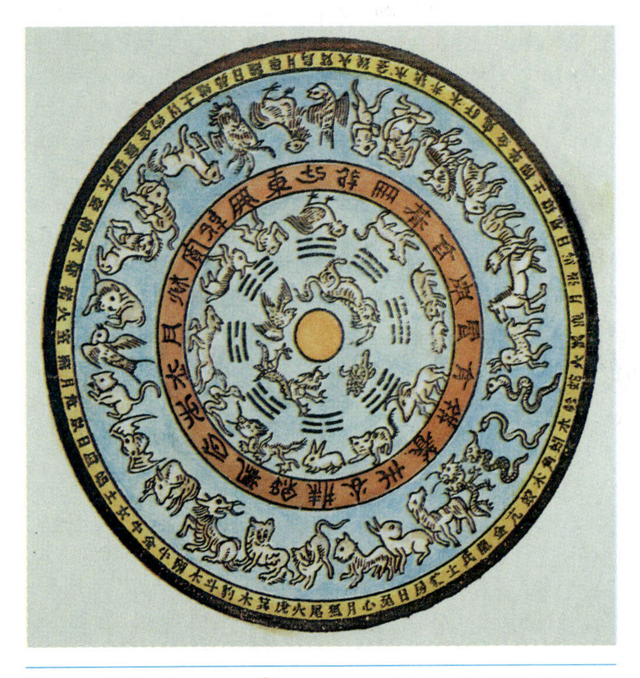

Figure 3.13 Ancient Chinese depiction of the celestial sphere. *Places near the North Pole were assigned to nobility, with lesser beings relegated to lower latitudes.* (The Granger Collection)

significance was attached to the motion of the planets. The poles were considered to be the exalted regions of the sky, whereas the celestial equator and the ecliptic seem to have been given no great importance. Heaven and Earth were thought to be closely related, with events in one domain affecting those in the other; thus, irregularities in the heavens were taken as omens of calamities on Earth.

In later centuries, observations became more accurate, with the employment of simple measuring devices. By the first century A.D., accurate tables of lunar motions had been constructed and eclipses could be predicted. Apparently, precession was known to the Chinese astronomers of this era as well.

Contact with the western world developed during the first few centuries A.D., as a result of Chinese conquests to the west and the eastward diffusion of Greek culture, which reached China by way of India. From this time on, it is difficult to separate independent Chinese developments from those introduced from the West, but the records indicate that numerous important advances were made, particularly in the improvement of measuring devices.

Records of celestial events were kept continuously until medieval times, providing later generations with important data to be found nowhere else. Among the events mentioned in the Chinese records is the appearance of a "guest star" in A.D. 1054, identified in modern times with the supernova explosion that created the Crab Nebula.

Because of China's remote location as seen from Europe, astronomical developments there had little influence elsewhere, except for the maintenance of observing records during the Dark Ages in Europe, when no one else was recording what was seen in the skies.

Hindu Astronomy

In India, as in China, it is evident that some astronomical lore had developed by very early times, but our only knowledge of the oldest accomplishments is based on ancient legends and on the astronomical orientations of some buildings (Fig. 3.14). There are written records of lunar motions dating back to about 1500 B.C., and by 1100 B.C. a calendar had been established, based on the solar year. Apparently all the prominent celestial bodies were named for gods or goddesses and were thought to represent their characters.

FIGURE 3.14 ANCIENT BUILDING IN INDIA *This is one of the largest and finest surviving Hindu temples built along astronomical alignments.* (J. M. Malville)

Astronomy in India flourished during and immediately after the time of the great prince Rama, who was born in the year 961 B.C. During this period, tables of planetary motions were constructed for use in astrology, since the Hindus thought that heavenly occurrences represented earthly affairs. A very complex calendar was adopted that was based on an observed 247-year cycle of solar and lunar motions of great significance. The construction of this calendar, as well as the discovery of precession, show that the Hindu astronomers made use of positional records that must have been maintained over hundreds of years.

Sometime in the first millenium B.C., there was established cultural contact between India and the Middle Eastern civilizations of the Babylonians and the Greeks, and it is difficult to be certain of the direction in which information flowed. Some people claim that the entire basis of the Babylonian and Greek astronomical development was borrowed from India in about the eighth century B.C., whereas others argue that the developments in the two regions were quite independent. In any case, it is clear that by A.D. 500, western influence dominated Hindu astronomy, and that few new developments took place in India after that time.

Ancient Astronomy in the Americas

A modern appreciation of ancient astronomical developments in the New World has been slow in coming. The principal reasons for this were the nearly complete destruction of the writings of the Mayan civilization at the time of the Spanish conquest, and the lack of a written language of the North American natives.

Archeological studies have revealed that in Central and South America sophisticated cultures existed before the time of Columbus. The ancient cities of the Mayan and Incan civilizations, now in ruins, show highly complex and organized structures requiring sound knowledge of astronomical phenomena. It has been discovered that in some cases alignments of principal buildings and avenues were designed to point toward significant astronomical locations, such as the rising point of the sun on the day of its passage through the zenith. A great deal of what we know about astronomy in these ancient cultures is read from the record of their architecture, rather than from their writings.

The center of the Mayan civilization was located in what is now Mexico and several Central American nations to the south, in a region known today as Mesoamerica. Other cultures in this region, which flourished between 500 B.C. and A.D. 1500, were the Olmec, Zapotec, and Aztec.

As in other ancient cultures, the principal motivations behind the Mesoamerican astronomy were timekeeping and astrology. The Mayas developed a very accurate calendar, and measured with great precision the motions of the celestial bodies. A particular significance was associated with the planet Venus, which inspired sacrificial rites and other ceremonies. The motions of Venus were extensively tabulated in a surviving document known as the *Dresden Codex* (Fig. 3.15).

In some locations special astronomical observatories were constructed, usually with many significant alignments made in the directions of important celestial events. The most famous is the Caracol tower at Chichén Itzá, in the Yucatán, a building whose strange architecture defied explanation until the numerous astronomical sightlines it entails were discovered in the twentieth century (Fig. 3.16).

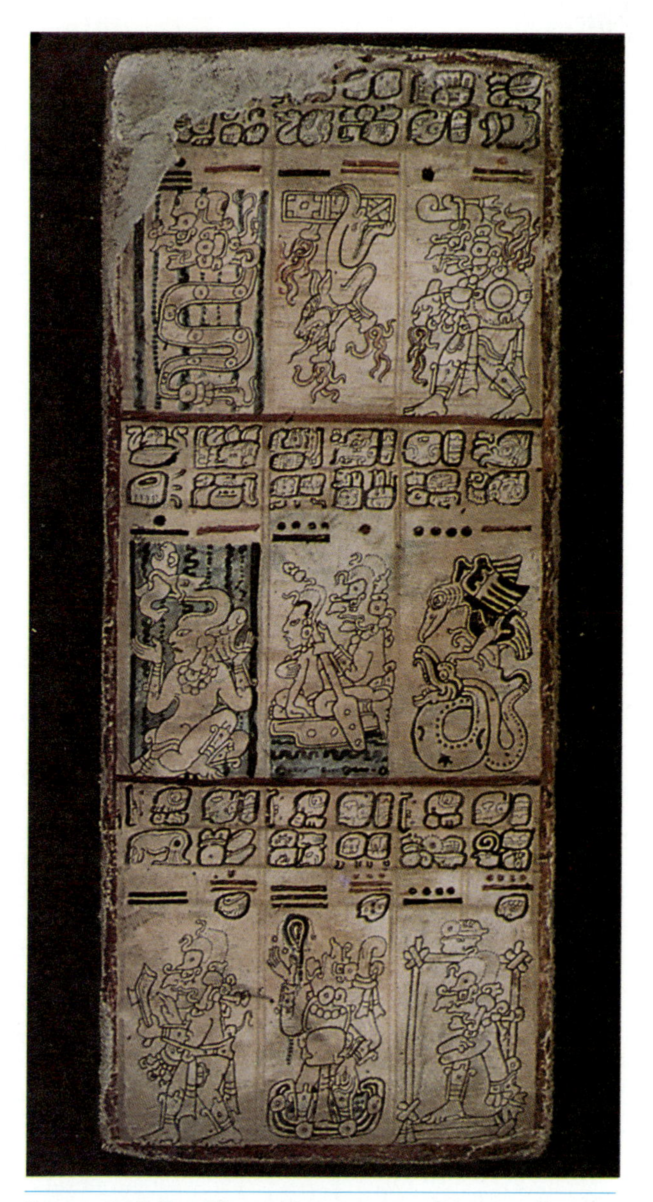

FIGURE 3.15 THE VENUS TABLES IN THE DRESDEN CODEX (MAYA). *These tables show that the Mayans attached particular significance to the planet Venus.* (Historical Pictures Service).

The Mayan cosmology apparently consisted of a layered universe, each level of which contained one type of celestial body. Above the Earth was the domain of the Moon, then the clouds, the stars, the Sun, Venus, comets, and so on. Similarly, there was thought to be an underworld consisting of nine levels, pro-

ASTRONOMICAL INSIGHT 3.2

THE MYSTERIES OF STONEHENGE

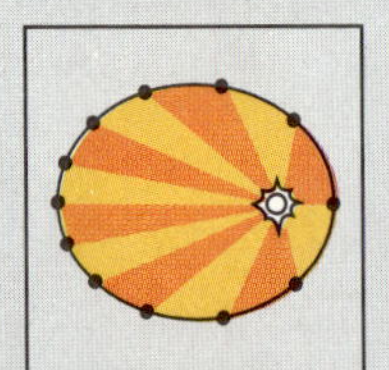

By now we are aware that some form of astronomical influence affected nearly every ancient culture. The timing of major developments seems to have been surprisingly similar in most parts of the world. The earliest records of accurate positional measurements and the establishment of calendars often date from the second millennium B.C., with sufficient sophistication for predictions of eclipses and planetary motions not developing until the last few centuries B.C.

It is most impressive, then, to consider the stone circle monuments of the British Isles, for they predate recorded astronomical achievements in the rest of the world, yet they show that their builders had substantial knowledge of astronomical skills, possibly including the prediction of eclipses.

Of the nine hundred or so of these monuments, by far the best known is Stonehenge, located on Salisbury Plain, about seventy miles west of London. This is a large and complex monument characterized by immense upright stones arranged in various circular patterns, in some instances with horizontal lintels connecting the vertical members.

Little is known about the builders of Stonehenge. The monument was built in three stages, the first commencing about 2800 B.C., and the most recent being completed by 2075 B.C. The sophisticated astronomical principles that guided the construction were incorporated from the beginning, even though it appears that the people inhabiting the region

C. D. McLoughlin

changed. The original Stone Age civilization gave way to a new group, called the Beaker People, who arrived from what is now the Netherlands in approximately 2500 B.C. Subsequent stages were built by this latter group, who apparently embraced the logic that had led to the original construction.

How the stones, which weigh up to fifty tons, were moved to Salisbury Plain is not known. Some came from Wales, possibly by sea, and many came from a region some twenty miles distant. The monument is now in a state of partial ruin and disarray, with some stones missing. Besides the large stones, there are a number of holes that may have once contained wooden posts to mark certain directions. Even today new parts of Stonehenge are occasionally found, in the form of postholes or distant mounds that may have been used to mark significant alignments.

There is no question that Stonehenge was laid out according to astronomical principles. There is substantial controversy, however, about the details. Despite a great deal of publicity in the 1960s over interpretations of Stonehenge as an early computer used to predict eclipses, today there is little agreement as to what is the correct interpretation.

One axis of Stonehenge, delineated by the sight line to the Heelstone from the center of the monument, aligns with the position of sunrise at the summer solstice. However, because of precession the alignment is better now than it was in 2800 B.C., when the first stage was built. (This has led at least one student of Stonehenge to hypothesize that this axis was meant to signify the midwinter rising of the full moon, with which the alignment was better). A circle of fifty-six holes around the center of Stonehenge has also attracted a great deal of attention; some people thought it possible that the holes were used to predict eclipses, although it appears unlikely.

Whatever the correct detailed interpretation of Stonehenge, the fact remains that extensive knowledge of astronomical phenomena was available to its builders as early as 2800 B.C., some two thousand years before other cultures such as the Babylonians are known to have developed such a high level of sophistication. It is unfortunate that we do not know more about the remarkable people who built it.

FIGURE 3.16 THE CARACOL TOWER AT CHICHÉN ITAÁ. *This is one of the most significant of the many astronomically-oriented structures in Mesoamerica.* (J. A. Eddy)

viding symmetry with the heavens, an idea parallel to that of the early Greek cosmologies described by Homer.

In South America, in the Andean territory where modern-day Peru lies, the Incan civilization flourished in the century or so before the Spanish conquest. Here, too, great attention was paid to astrological omens, and much of the astronomical effort was devoted to the construction of accurate timekeeping systems. The Incas had a rather complex calendar that consisted of twelve months, but it was otherwise quite unlike the modern calendar, not even containing the same number of days in a week that we have today. The Incas, like their Mesoamerican neighbors to the north, often laid out their temples and cities in accordance with astronomical reference points, in at least one case using markers set out over an entire valley as a framework for computing dates.

Astronomy practiced by North American natives in ancient times is much more difficult to assess, there being no written language and few records of any kind left, with the exception of scattered rock drawings and artifacts such as mounds, religious symbols, and artwork (Fig. 3.17). It seems evident, however, that the Indians of North America were intimately familiar with the skies and significant events such as the solstices that foretold the changing of the seasons. Among the few records that do exist are illustrations either carved or painted on rocks (Fig. 3.18), found in the Southwest, some of which may depict the appearance of the supernova of A.D. 1054.

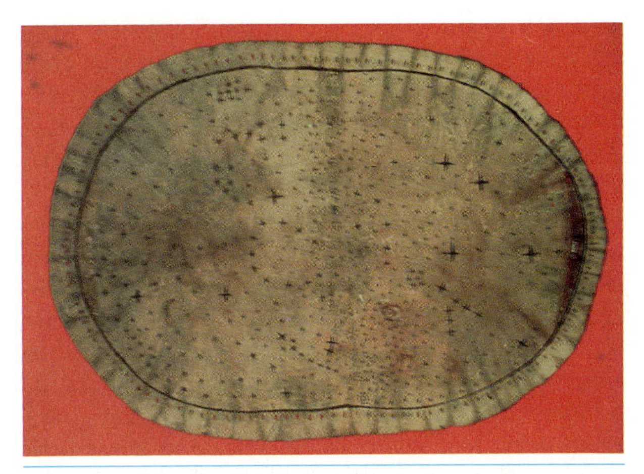

FIGURE 3.17 PAWNEE INDIAN SKY MAP. *This chart, embossed on hide, appears to depict constellations of the northern hemisphere skies.* (From Von Del Chamberlain 1982), *When Stars Came Down to Earth: Cosmology of the Skidi Pawnee Indians of North America* [Ballena Press: Los Altos, Calif.] Skidi Pawnee chart of the heavens, Field Museum of Natural History, photograph by Von Del Chamberlain.)

FIGURE 3.18 NORTH AMERICAN INDIAN PETROGLYPHS. *Drawings such as this depict astronomical objects and events, reflecting substantial knowledge of the skies.* (E. E. Barnard Observatory, photograph by G. Emerson)

This was the same event observed by the Chinese, but except for the possible native-American sighting, it was not mentioned in the records of any other cultures. That the petroglyphs do represent the supernova of 1054 is by no means established, however.

Some structures survive that appear to have astronomical significance, including the sun temple at

Mesa Verde and other ruins in the Southwest that may have been influenced by the Mesoamerican culture to the south. In addition, there are earthen mounds of various descriptions located in many parts of the Midwest, and some seem to represent celestial events or follow astronomically significant alignments.

In the American West there are several circular monuments marked out in rock, called medicine wheels. These were probably left behind by the Plains Indians, and they show astronomical influences to varying degrees. Among the best studied is the Bighorn medicine wheel, located high in the Big Horn Mountains of Wyoming (Fig. 3.19). This is a circular structure some 90 feet in diameter, with rock cairns around its perimeter. Lines traced through these piles of rock coincide with a number of astronomically significant directions, including a primary line pointing in the direction of the rising Sun on the morning of the summer solstice.

Unfortunately, it is impossible to be certain when these monuments were constructed, so we do not know just how long ago our predecessors on this continent developed their understanding of the heavens.

— ASTRONOMY IN THE DARK AGES —

Let us now pick up the main historical thread, with a brief mention of events after the time of Ptolemy.

FIGURE 3.19 THE BIGHORN MEDICINE WHEEL *This prominent and well-studied North American Indian medicine wheel is located in the Big Horn Mountains of Wyoming.* (U.S. Forest Service, provided by J. A. Eddy)

The Christian influence began to be felt throughout what had been the Greek empire, turning people away from the belief that astronomical events dominated their fates, and leading to the dissipation of efforts to improve the understanding of the heavens. The prophet Mohammed was born in the mid-seventh century A.D., and within a century, his followers had conquered most of the Mediterranean and some of southern Europe, and had burned what was left of the library at Alexandria.

The need for accurate calendars caused some revival of interest in astronomy, and the Arabs resurrected a great deal of the Greek knowledge. During their period, our modern numbering system, brought by the Arabs from India, was established, and new mathematical techniques, notably algebra, were developed. Although some new astronomical observations were made by Arab scientists, their principal contribution was to keep alive the Greek knowledge of astronomy and to pass it on to the western European civilization in the twelfth century, through Spain.

In the thirteenth century, Spanish King Alfonso X ordered the construction of new tables of planetary motions, the old tables of Ptolemy finally having begun to show inadequacies. The new tables, derived by a group of astronomers using the epicyclic model of Ptolemy, became known as the Alfonsine Tables, and were used for the next 300 years.

Meanwhile, the philosophical teachings of Aristotle, adhered to by the Islamic astronomers, were adopted by the scientists of the Christian world as well. Thus as the sixteenth century dawned, European astronomy was based almost entirely on the teachings and observations of the ancient Greeks.

PERSPECTIVE

We have traced the development of astronomy, beginning with mankind's earliest attempts to understand the world around him. These attempts were based on fear of the unknown and the primitive belief that events in the heavens ruled the inhabitants of the Earth. This kind of astrological outlook emerged independently in many parts of the world, yet only in the vicinity of the Mediterranean, with the rise of the Greek philosophers, did there exist the intellectual stimulus to understand the universe just for the sake of understanding. Political, economic, and, most of all, religious developments then brought about a prolonged intercession, when no real progress was made. Nearly nine hundred years after the death of Ptolemy, the first stirrings of intellectual revolution began to be felt.

SUMMARY

1. The earliest recorded astronomical data were found in the region known as Babylonia, now Iraq.
2. The first rational inquiry and the earliest cosmological theories arose in the ancient Greek empire, on the shores of the Mediterranean.
3. From 570 B.C., Pythagoras and his followers developed the notion that the universe can be described by numbers, and adopted the belief that the Earth is spherical.
4. Plato, and his pupil Aristotle, stated underlying principles (which could not be tested by observation) that they believed to govern the universe; their principles included the beliefs that the Earth is the center of the universe and that all heavenly bodies are spherical.
5. In the third century B.C., Aristarchus used geometrical arguments to conclude that the Sun, not the Earth, is at the center of the universe.
6. During the same period, Eratosthenes made an accurate measurement of the size of the Earth, by comparing the directions toward the Sun from two widely separated points on the Earth's surface.
7. Hipparchus, in the second century B.C., made precise observations, applied new mathematical techniques to astronomy, compiled a large star catalogue, developed a system for measuring star brightnesses, discovered precession, and developed the epicyclic theory of planetary motion.
8. Ptolemy, in the second century A.D., summarized all astronomical knowledge in the *Almagest,* and used the epicyclic theory to construct tables of planetary motion that were used throughout the Dark Ages.
9. Astronomy in ancient China and India led to significant observational achievements such as the compilation of large star catalogues and the development of accurate calendars, but did not lead to sophisticated cosmological theories.
10. Astronomy in the Americas was highly developed, but few records were left other than the astronomical alignments of buildings and monuments.

REVIEW QUESTIONS

Thought Questions

1. Discuss the relationship between observations of the sky and ancient cultural developments on Earth.
2. Astronomy was studied by ancient cultures for practical and philosophical reasons. Into which of these categories would you put the astronomical interests of the Babylonians? Of the early Greeks? Which reason for studying astronomy do you think is given by most modern astronomers?
3. As far as the records show us, astronomy played some role in the early development of virtually every culture on Earth. Choose a geographical region or society of particular interest to you, and try to find out the extent of astronomical interest and influence in that region or society.
4. What is significant about the early Greek view that there is an underworld, comparable to the heavens?
5. Discuss the relationship between the attitude of Plato toward scientific inquiry and the modern concept of scientific theory, as described in Chapter 1.
6. Even though Aristarchus was correct in asserting that the Sun, not the Earth, was the central body in the solar system, his idea found few followers in his time. Discuss the reasons for this.
7. Discuss the geocentric cosmology of Hipparchus and Ptolemy in the context of our modern view of scientific theory. Did their model conform to the requirement that a theory make predictions that can be tested? How well did it do when tested?
8. In what ways did Ptolemy's epicyclic model obey the "laws of physics" postulated by Aristotle, and in what ways did it violate those laws?
9. Explain in your own words what parallax is, and why nearby stars should undergo parallactic motion as seen from the moving Earth.
10. Why are astronomical developments in the Americas not considered to have been influential in the development of modern astronomy, despite the high level of sophistication reached by early American cultures?

Problems

1. How often did the Babylonians have to add an extra month to their calendar to compensate for their

adoption of a calendar with 12 months of 30 days each?

2. How many arcseconds are there in a degree? How many arcminutes in a full circle?

3. Suppose, in the method of Eratosthenes for measuring the circumference of the Earth, that the distance between Syene and Alexandria was 3,300 km, and the angle to the Sun at Alexandria was 30° on the day that the Sun was overhead at Syene. What would you find for the circumference and diameter of the Earth in this case? How do your answers compare with modern values?

4. Hipparchus discovered precession by comparing star positions that he measured with some that were measured 160 years before, and noticing changes in positions. If his measurements had an uncertainty of 20 arcminutes, how much precession had to occur per year in order for Hipparchus to be able to detect it?

5. Refer to Fig. 3.10, which shows stellar parallax. If Hipparchus could not measure any angle smaller than 20 arcminutes, how close would a nearby star have to be in order to have a parallax motion that Hipparchus could have detected? (Note: You will need to use simple geometry or trigonometry to solve this.)

ADDITIONAL READINGS

Here are some readings on ancient astronomical history. Others may be found by looking up references cited in these articles and books. It is also useful to browse through journals, such as *Vistas in Astronomy* and the *Journal for the History of Astronomy*, both of which can be found in most science-oriented libraries.

Aveni, Anthony F. 1986. Archeoastronomy: Past, present and future. *Sky and Telescope* 72(5):456.

———. 1984. Native American astronomy. *Physics Today* 37(6):24.

Heath, T. L. 1969. *Greek astronomy*. New York: AMS.

Krupp, E. C. 1978. *In search of ancient astronomies*. Garden City, N.J.: Doubleday.

———. 1983. *Echoes of the ancient skies*. New York: Harper and Row.

Neugebauer, O. 1969. *The exact sciences in antiquity*. New York: Dover.

Piini, Ernest W. 1986. Ulugh Beg's forgotten observatory. *Sky and Telescope* 72(6):542.

CHAPTER 4

THE RENAISSANCE

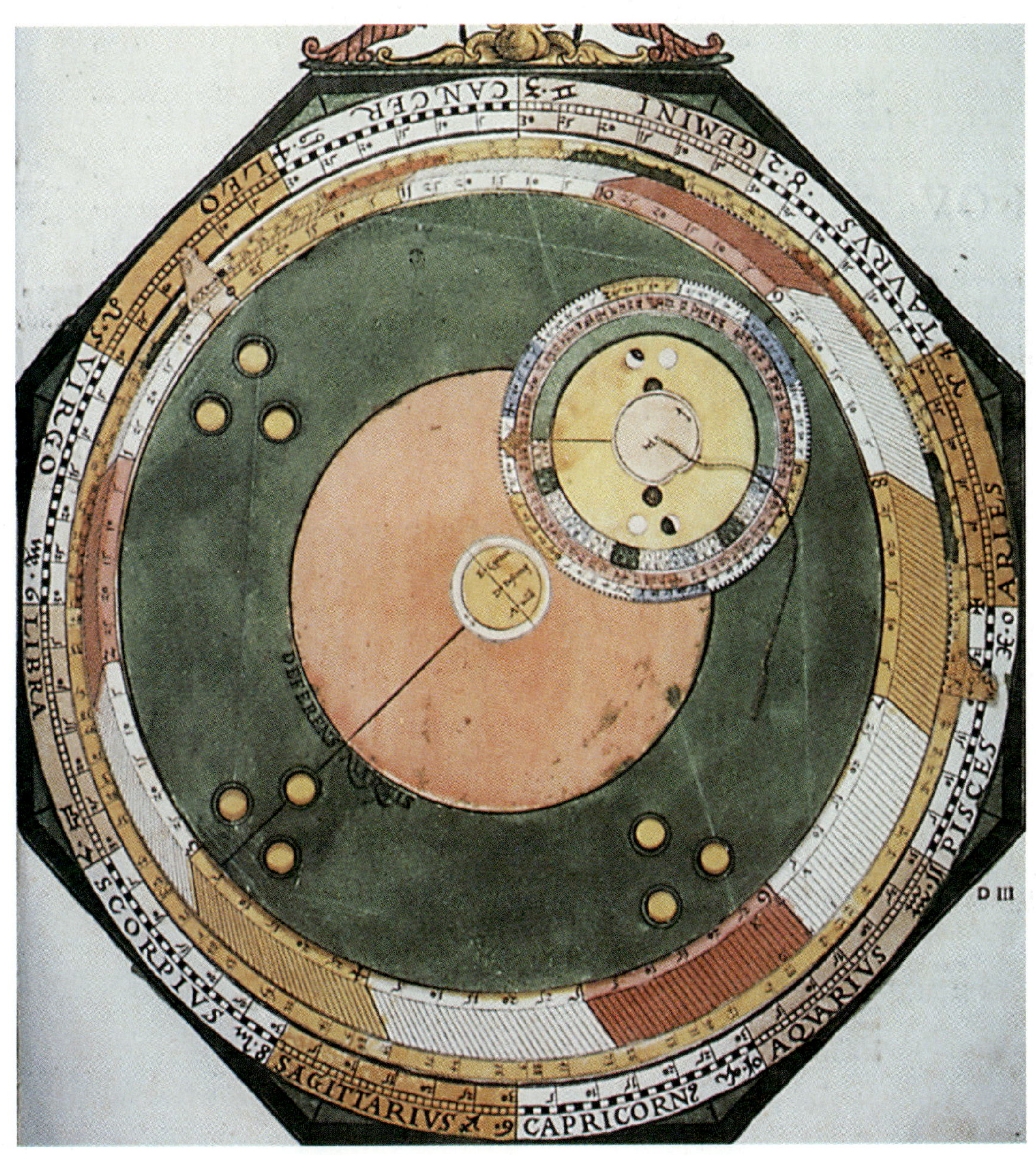

A Renaissance model of the epicyclic cosmology. (© 1981 O. Gingerich, with permission of the Houghton Library, Harvard University)

CHAPTER PREVIEW

The fifteenth century saw the beginnings of a reawakening of intellectual spirit in Europe. Some important scientific studies began at the major universities, increased maritime explorations brought demands for better means of celestial navigation, and the art of printing was discovered, opening the way for widespread dissemination of information.

COPERNICUS: THE HELIOCENTRIC VIEW REVISITED

Some nineteen years before the epic voyage of Columbus, Mikolaj Kopernik (Fig. 4.1) was born in Torun, in the northern part of what is now Poland. As a young man he attended the university in Kracow, where his entrancement with Latin, the universal language of scholars, led him to change his surname to Copernicus. At Kracow he nourished an abiding interest in astronomy, becoming fully acquainted with the Aristotelian view as well as the Ptolemaic model of the planetary motions.

Copernicus continued his academic pursuits in Italy, studying ecclesiastical law at Bologna and Ferrara, and medicine at Padua, before returning to Poland to assist his uncle, the bishop (and therefore chief government administrator) of a region called Ermland. During all this time, he persisted in his studies of astronomy, and it is known that, by 1514, he had developed some doubts about the validity of the accepted system.

His reasons for doing so have been the subject of some uncertainty and misconceptions. It was long assumed that Copernicus was encouraged to adopt the Sun-centered view of the universe because he recognized shortcomings in the geocentric model of Ptolemy. There is, however, no evidence of widespread dissatisfaction with the Ptolemaic system, nor any records indicating that Copernicus himself found serious inaccuracies in it. His reasons for adopting the heliocentric viewpoint were more subtle.

The basis for his conversion was primarily philosophical. The new system, in the mind of Copernicus, presented a pleasing and unifying model of the universe and its motions. Although he was, no doubt, encouraged by the climate of change and cultural revolution that was sweeping Europe with the advent of the Renaissance, he did not adopt the heliocentric hypothesis just to be different, nor did he do it to improve the accuracy or reduce the complexity of the accepted view. The Copernican model was, in fact, no more accurate in predicting planetary motions than was the Ptolemaic system, and it was nearly as complex. Copernicus adhered to the notion of perfect circular motions and was obliged to include small epicycles in order to match the observed planetary

FIGURE 4.1 NICOLAUS COPERNICUS. *The view of Copernicus that the Sun, rather than the Earth, occupied the central place in the cosmos had great influence on later scientists.* (The Bettmann Archive)

orbits. Apparently, one of the most pleasing aspects of the Sun-centered system to Copernicus was the fact that the relative distances of the planets could be deduced (Fig. 4.2), and were found to have a certain regularity, the spacings between planets growing systematically with increasing distance from the Sun. Copernicus was also able to determine the relative speeds of the planets in their orbits, finding that each planet moves more slowly than the next one closer to the Sun.

Copernicus first circulated his ideas informally sometime before 1514, in a manuscript called *Commentariolis,* and it drew increasing attention during the next several years. The Church voiced no opposition, despite the fact that the ideas expressed in the work strongly contradicted commonly accepted Church doctrine. Copernicus was reluctant to publish his findings in a more formal way, for fear of raising controversy, and was continually rechecking his calculations. Eventually a young Protestant scholar known as Rheticus encouraged him to publish his work. Having been persuaded of the validity of the heliocentric theory, Rheticus had published his own summary of the ideas of Copernicus, and then delivered the manuscript written by Copernicus to a Lutheran priest named Osiander, who assumed the responsibility for publishing it.

Fearful of strong religious opposition, Osiander added an unsigned preface in which he stated that the ideas within did not represent physical reality, but should be used only as a mathematical tool for predicting planetary motions. There is no doubt that Copernicus himself believed in the literal truth of his heliocentric model, however. The book was titled *De Revolutionibus Orbium Coelestium (On the Revolution of the Celestial Sphere)*. Copernicus died at just about the time his work was published, so he never knew of its impact upon the world.

The impact was profound, although not immediately so. The many important improvements in understanding the solar system that were embodied in the new theory eventually made a strong enough impression on other scientists to win them to this view. Among the persuasive aspects were the natural ordering of the planets (the relationship between orbital period and distance from the Sun) and the simple explanation for the differing sizes of the retrograde loops that the planets go through. The extent of the backward motion observed for each planet, if it is a

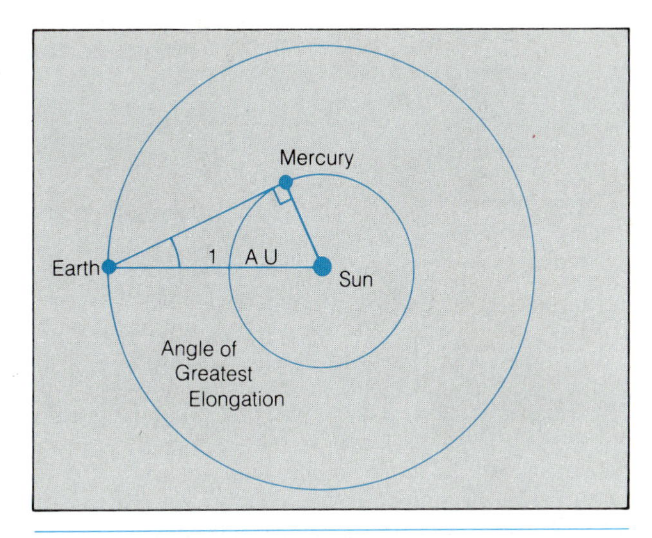

FIGURE 4.2 THE METHOD OF COPERNICUS FOR FINDING RELATIVE PLANETARY DISTANCES FROM THE SUN. *For an inferior planet such as Mercury or Venus, Copernicus knew the angle of greatest elongation, and was therefore able to reconstruct the triangle shown, which provided the Sun-Mercury (or Sun-Venus) distance relative to the Sun-Earth distance (that is, the astronomical unit). For superior planets, similar but slightly more complicated considerations provided the same information.*

result of the Earth's relative motion as Copernicus believed, depends on how far away the planet is. Copernicus also was able to explain the cause of the seasons, and made the daring suggestion that the reason stellar parallax could not be observed was that the stars are simply too far away to have perceptible displacements resulting from the Earth's motion around the Sun.

TYCHO BRAHE: ADVANCED OBSERVATIONS AND A RETURN TO A STATIONARY EARTH

Although Copernicus had contributed little observationally and had not demanded a close match between theory and observation as long as general agreement was found, there was, in fact, a strong need for improved precision in astronomical measurements. The next influential character in the historical sequence filled this need; if he had not done so, further progress

ASTRONOMICAL INSIGHT 4.1

THE ACCURACY OF ANGULAR MEASUREMENTS

We have referred often to measurements of planet and star positions without saying much about how these measurements are made, or about their accuracy. Throughout the history of astronomy the quest for ever more accurate measurements of positions has been constant. In the earliest times, positional measurements, particularly of the Sun, were needed for timekeeping purposes. As scientists and philosophers began to probe the nature of the universe by attempting to deduce the motions of the planets, positional measurements for the Sun and planets became increasingly important. As we learned in Chapter 3, Hipparchus (c. 150 B.C.) was obliged to introduce the notion of the equant into his epicylic model of the cosmos because positional measurements revealed errors in the model. In this chapter we have seen that Tycho Brahe's (c. 1600 A.D.) positional measurements were sufficiently accurate to allow Johannes Kepler to deduce the elliptical shape of planetary orbits, but not accurate enough to reveal stellar parallax.

The accurate measurement of positions in the sky amounts to measuring angles with the best precision possible. The earliest devices for doing this were simple sighting instruments used to determine the height of the Sun above the horizon, often by noting the angle made by the shadow of a thin object such as a peg. This angle could be measured directly, through the use of a quadrant (a device much like the protractor that you used in geometry class), or by measuring the length of the shadow and using trigonometry to deduce the direction toward the Sun. The earliest such techniques produced only rather crude results, with uncertainties of 1° or more. (Keep in mind that by our ordinary standards a degree is a rather small angle; to see this, try to draw a 1° angle, using a protractor.)

By the time of Hipparchus, around 150 B.C., somewhat better accuracies (within 20 arcminutes) were achieved, though the techniques were rather similar. The next major advance came some 1500 years later, when metal-working technology had been developed, and it became possible to engrave angular scales on metal. Tycho in particular pushed this development, building ever larger quadrants. A **quadrant** was a device with a pointer that could pivot so that the pointer could be aimed directly at the target object (see Fig. 4.5). The angle from the horizon to the target could then be read off of a scale engraved on a quarter-circular arc of brass. Tycho's largest quadrant was attached to a fixed wall of his observatory, but others were able to swivel around a vertical post, so that target objects in any direction could be measured. The advantage of large size was that the markings on the metal arc representing degrees and minutes would be farther apart and therefore easier to read accurately. Through the use of such devices (and subsequent averaging of multiple measurements), Tycho's observations produced positional measurements accurate to about one arcminute. This was sufficient to allow Kepler to determine that the orbit of Mars is an ellipse, even though he found other shapes that reproduced the observed motion of Mars to within a few arcminutes.

Within a century following the time of Tycho Brahe, telescope technology made great advances. Measuring angular positions directly from the mounting of the telescope became possible. Most telescopes are mounted so that they can rotate on two axes, one parallel with the Earth's axis, and the other at right angles to the first. Then the position of a star in the usual two coordinates (right ascension and declination; see Chapter 2) corresponds directly to the amount by which each axis of the telescope is turned. By the beginning of the eighteenth century, positions could be measured in this way to an accuracy of better than 10 arcseconds.

In order to make substantial improvements over this, astronomers developed methods for measuring *relative* positions with great accuracy. Now, instead of measuring the position of a planet or star with respect to the Earth's reference frame, astronomers found that angular separations between objects in the sky could be measured more precisely. Selected stars were chosen as standards for reference in the determination of positions of others. (Today these standards are selected and periodically revised by an international panel of astronomers.) At first the rotation axes of the telescope were used for this, as cross-hairs in the eyepiece were set alternately on the target star and on one or more nearby standards, and the amount of rotation was measured (the results of multiple measurements then being averaged). This kind of technique fi-

(continued on next page)

ASTRONOMICAL INSIGHT 4.1

THE ACCURACY OF ANGULAR MEASUREMENTS

(continued from previous page)
nally allowed stellar parallax to be measured, in 1836 (by no less than three independent observers of three different stars!). The largest parallax angle, for the closest star to the Earth, is less than 1 arcsecond, so the precision of positional measurements had indeed improved dramatically from the ancient times. (To envision just how small an angle 1 arcsecond is, try to imagine how big a dime would look as seen from a distance of two miles.)

Soon photographic techniques began to be used in astronomy to measure positions. The measurement consisted of determining the physical distance between star images on a photographic plate where the plate scale (the number of arcseconds per millimeter) is known accurately. This has the advantage that many stars could be measured through the use of only a small amount of telescope time. Modern photographic techniques are capable of measuring separations of stars to an accuracy of 0.01 arcsecond or a little better, and this seems to be about as well as can be done by measuring ordinary stellar images, which themselves are usually about an arcsecond in diameter (because of blurring caused by the Earth's atmosphere).

Nevertheless, astronomers are developing capabilities for even more accurate positional measurements. These generally involve either placing telescopes in space, to avoid the blurring effects of the Earth's atmosphere, or making use of the complex wave properties of light, something we will discuss later (see the discussion of interferometry in Chapter 7). Today accuracies of 0.001 arcsecond appear feasible, and some scientists are beginning to think of even better precision. Throughout the rest of this text, you will learn of the importance of ever better capabilities for measuring angles precisely, for it is one of the everlasting challenges to astronomy.

would have been seriously delayed, as would have the eventual acceptance of the Copernican doctrine.

Tycho Brahe (Fig. 4.3) was born in 1546, some 3 years after the death of Copernicus, in the extreme southern portion of modern Sweden (at the time the region was part of Denmark). Of noble descent, Tycho spent his youth in comfortable surroundings, and was well educated, first at Copenhagen University, then at Leipzig, where he insisted on studying mathematics and astronomy despite his family's wish that he pursue a law career. Interruptions created by war and plague caused Tycho to change universities several times, with most of his formal training eventually taking place at the university in Wittenberg. He developed a strong reputation as an astronomer and as an astrologer (the distinction was still scarcely recognized), and attracted the attention of the Danish

FIGURE 4.3 TYCHO BRAHE. *Tycho's principal contribution, a massive collection of accurate observations of planetary positions, was crucial to subsequent advances in understanding planetary motions.* (The Bettmann Archive)

FIGURE 4.4 DETAIL OF TYCHO'S UNDERGROUND OBSERVATORY AT HVEEN. (Photo Researchers, Inc.)

FIGURE 4.5 THE GREAT MURAL QUADRANT AT HVEEN. *This instrument was used to measure the angular positions of stars and planets with respect to the horizon. Its large size made the angle markings easy to read accurately.* (Photo Researchers, Inc.)

King, Frederick II. In 1575 the king ceded to Tycho the island of Hveen, about 14 miles north of Copenhagen, along with enough servants and financial assistance to allow him to build and maintain his own observatory (Fig. 4.4).

Even before this time, Tycho had shown an acute interest in astronomical instruments, and with the grant to build his observatory, this interest bore fruit. He devised a variety of instruments which, although they really did not encompass any new principles, were capable of more accurate readings than any before his time. The largest was a quadrant built into a wall of the observatory (Fig. 4.5), with a 31½-foot radius. A quadrant is a quarter circle with graduated markings indicating degrees and minutes of arc around its circumference. A sight located at the center allows the observer to read the angular heights of celestial objects above the horizon. The principal improvements of Tycho's instruments over those of his contemporaries were that the angular scales were more precisely marked, and that a new sighting design allowed the scales to be more accurately read.

Tycho made two specific observational discoveries of importance, in addition to amassing a large body of data on planetary motions. One, the observation of a supernova, occurred in 1572, and the other, the observation and analysis of a bright comet, took place in 1577, soon after the observatory became active. By comparing his data with those of astronomers elsewhere, Tycho was able to determine in both cases that the new object was very distant, well beyond the orbit of the Moon (Fig. 4.6). This showed that the heavens were not perfect and immutable. The comet was particularly significant, because Tycho's analysis showed that it passed right through the crystalline spheres of the planets, demonstrating that they could not be the solid material objects supposed to exist by the ancient Greeks. Tycho's analysis of the supernova of 1572 was also quite significant, for it was in recognition of this achievement that King Frederick II granted Tycho the resources to build and operate his observatory. The supernova also provided evidence against the accepted doctrine, which held that the heavens were immutable.

Tycho was unable to accept the heliocentric view, primarily because he could find no evidence that the Earth was moving. He tried and failed to detect stellar parallax, which he supposed he should see with his accurate observations if the Earth really moved. Furthermore, as a strict Protestant, he found it philosophically difficult to accept a moving Earth, when the Scriptures stated that the Earth is fixed at the center of the universe.

On the other hand, he realized that the Copernican system had advantages of mathematical simplicity over the Ptolemaic model, and in the end he was ingenious enough to devise a model that satisfied all of his criteria. He imagined that the Earth was fixed, with the Sun orbiting it, but that all the other planets orbited the Sun (Fig. 4.7). Mathematically, this is equivalent to the Copernican system in terms of accounting for the motions of the planets as seen from the Earth. The idea never received much acceptance, however, and Tycho is remembered primarily for his fine observations.

His observational contribution consisted, in part, of the unprecedented accuracy of his data and, in part of the completeness of his records. Until his time, it had been the general practice of astronomers to record the positions of the planets only at notable points in their travels, such as when a superior planet comes to a halt just before beginning retrograde motion. Tycho made much more systematic observations, recording planetary positions at times other than just the significant turning points in their motions. He also made multiple observations in many instances, allowing the results to be averaged together to improve their accuracy. Tycho himself did not attempt any extensive analysis of his data, but the vast collection of measurements that he gathered over the years contained the information needed to reveal the basis of the planetary motions. The task of unraveling the secrets contained in Tycho's data was left to those who followed him, particularly a young astronomer named Johannes Kepler.

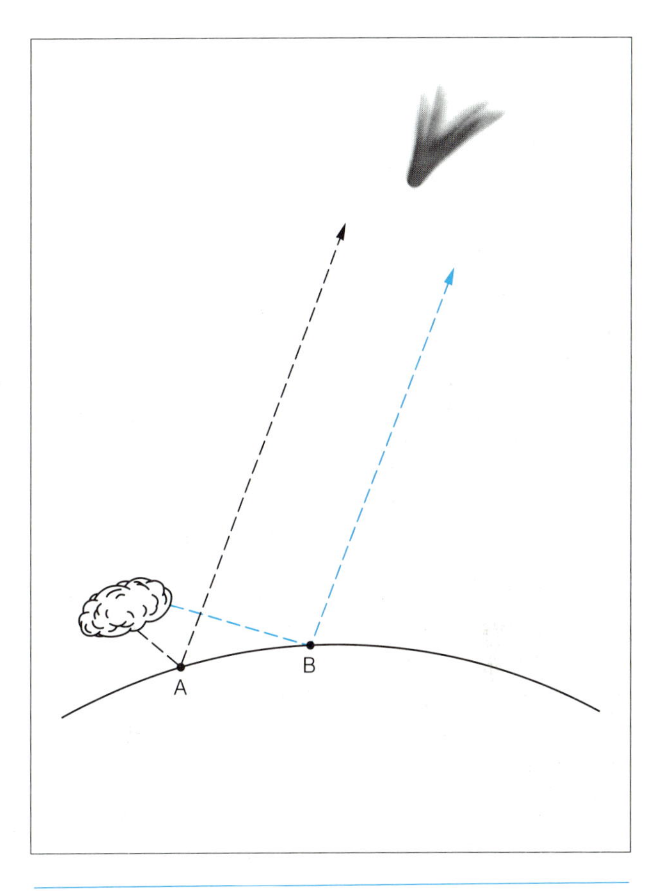

FIGURE 4.6 TYCHO'S ANALYSIS OF THE COMET OF 1577. *Because the direction to the comet as seen from separate locations (A and B) was the same, Tycho showed that the comet was far beyond the Earth's atmosphere, and belonged to the realm of the planets and stars.*

KEPLER AND THE LAWS OF PLANETARY MOTION

Until now it has been possible to follow developments in a straight sequence, but here we must begin to cover events that occurred at nearly the same time, but in different locations. To complete the thread begun with our discussion of Tycho Brahe, we now turn to Johannes Kepler (Fig. 4.8), who worked briefly at the side of Tycho and then spent many years analyzing the great wealth of observational data that Tycho had accumulated. We must keep in mind, however, that during the same period a scientist and philosopher named Galileo Galilei was at work to the south, in Italy, and that the two were aware of each other's accomplishments.

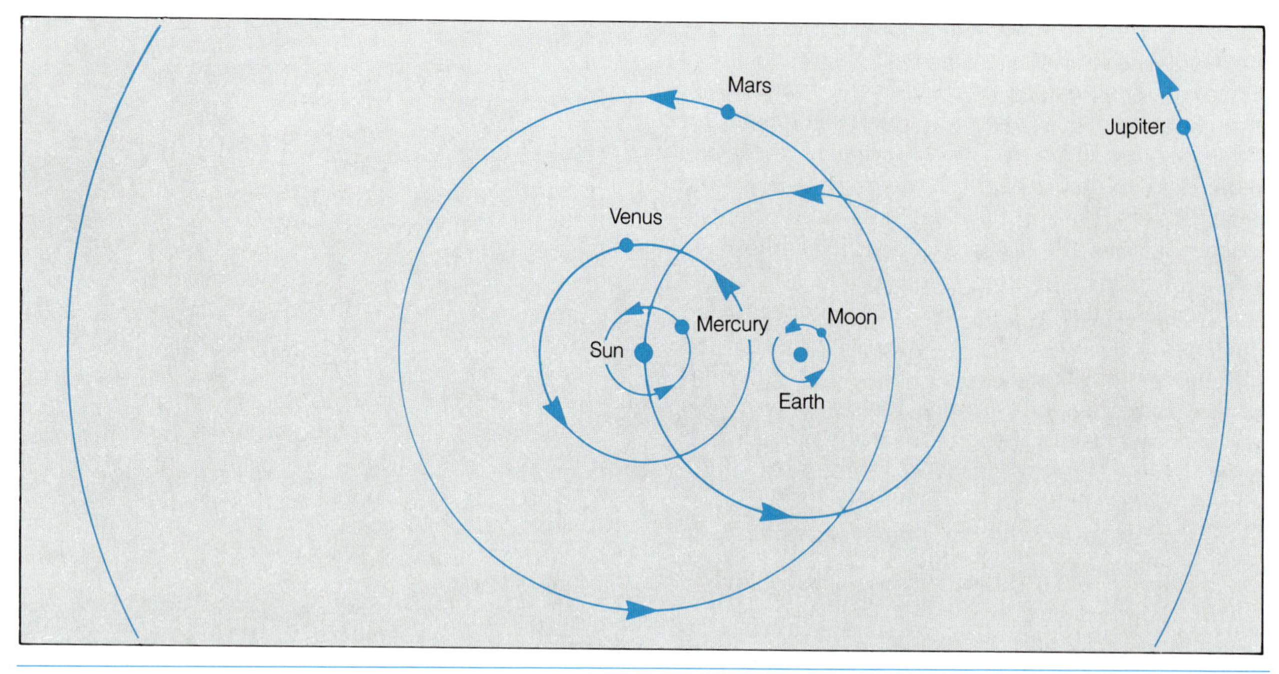

FIGURE 4.7 TYCHO'S MODEL OF THE UNIVERSE. *Tycho held that the Earth was fixed, and the Sun orbited the Earth. The planets in turn orbited the Sun. This system was never worked out in mathematical detail, but it successfully preserved both the advantages of the Copernican system and the spirit of the ancient teachings. It also accounted for the lack of observed stellar parallax, since the earth was fixed.*

Kepler was born in 1571 in Weil, Germany, a year before Tycho's supernova was observed. A sickly youngster, he seemed headed for a career in theology, but he had difficulty conforming to the fundamentalist religious philosopy at the university he attended in Tübingen. While there he encountered a professor of astronomy who inspired in him a strong interest in the Copernican system.

Having completed his education, Kepler accepted a teaching position in the Austrian city of Graz in 1594. While there he published his first scientific work, a book called the *Mysterium Cosmigraphicum*, in which he outlined several arguments favoring the Copernican theory and stated his own view that the distances of the planets from the Sun were determined by a series of regular geometric solids separating the crystalline spheres of the planets (Fig. 4.9). Kepler's reasons for adopting this notion and the Copernican system as well were based on his love of mathematical simplicity and his intuitive feeling that the universe

FIGURE 4.8 JOHANNES KEPLER. *Kepler's fascination with numerical relationships led him to discover the true nature of planetary motions.* (The Bettmann Archive)

was constructed according to a harmonious, unified plan. He devoted his life to seeking out the underlying principles of this universal harmony, and, as we shall see, this led him to a number of profound discoveries. The hypothesis put forth in the *Mysterium Cosmigraphicum* was, of course, not correct, but the book had a major impact anyway, especially because of its eloquent arguments in favor of the Sun-centered universe.

In 1598 a decree banning all Protestants forced Kepler out of Graz and, after first visiting extensively, he accepted a post as chief assistant to Tycho, who had by then moved his observatory to Prague following the death of his sponsor, King Frederick II of Denmark. Tycho died the next year, and Kepler succeeded him as court mathematician to Emperor Rudolph.

Kepler's main mission, owing to both Tycho's wishes and his own, was to develop a refined understanding of the planetary motions and to upgrade the tables used to predict their positions. He set to work first on the planet Mars, for which the data were particularly extensive, and whose motions were among the most difficult to explain in the established Ptolemaic system (and, in fact, in the Copernican system, with its requirement of only circular motions). By a very complex process, Kepler was able to separate the effects of the Earth's motion from those of Mars itself, so that he could map out the path that Mars followed with respect to the Sun.

He soon realized that the Sun must play a central role in making the planets move, and this inspired him to look for a physical link between the Sun and the planets. In this sense he was like a modern scientist and quite unlike other astronomers of his time and before. Whereas they sought only a mathematical device for predicting motions, Kepler believed in a true physical machinery of the solar system, and he sought the underlying principles.

The data on Mars led Kepler eventually to the realization that the Sun could not be at the true center of the planet's orbit. He experimented with various geometric shapes, including epicycles, which he became convinced were contrived and not likely to represent physical reality. At one point, he was forced to reject a system he had devised because of a small error in predicting the position of Mars, an error that would have been acceptable to anyone else who had worked with the planetary motions.

By 1604 Kepler had determined that the orbit of Mars was some kind of oval, and further experimentation revealed that it was fitted precisely by a simple geometric figure called an **ellipse.** This is a closed curve defined by a fixed total distance from two points called **foci** (singular: *focus*), and indeed Kepler found that the Sun was at the precise location of one focus of the ellipse (Fig. 4.10).

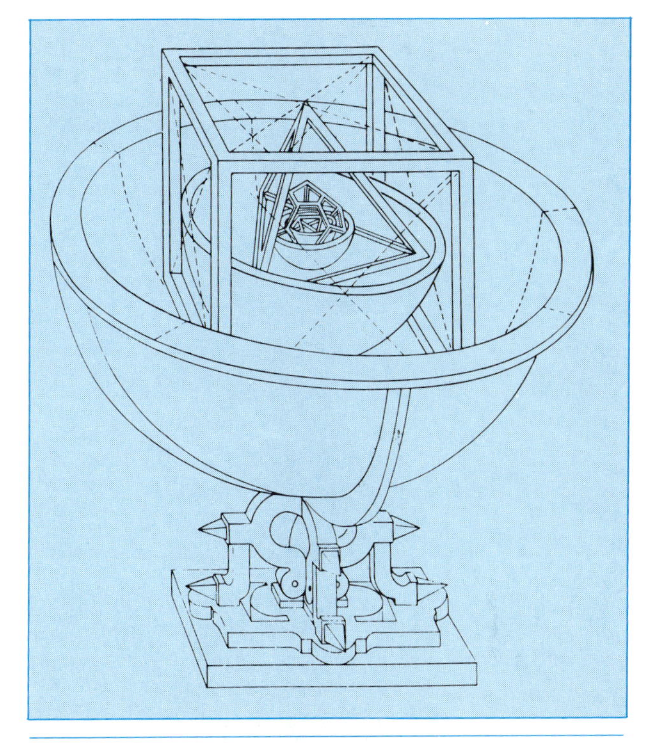

FIGURE 4.9 KEPLER'S NESTED GEOMETRIC SOLIDS. *Kepler expended considerable effort in exploring the possibility that the planetary distances from the Sun were determined by the spacings of nested regular geometric solids (that is, figures whose faces were regular polyhedrons; the number of gaps between possible solids of this sort was equal to the number of known planets). Kepler eventually abandoned this particular scheme, but never relented in his search for regularity in the solar system.* (O. Gingerich)

Further analysis of the motion of Mars revealed a second characteristic, having to do with the fact that the planet moves fastest in its orbit when it is nearest the Sun, and slowest when it is farthest from the Sun. Mathematically, Kepler's second discovery was that a line connecting Mars to the Sun sweeps out equal areas of space in equal intervals of time (see Fig. 4.10).

These two properties of the orbit of Mars were later generalized as Kepler's first two laws of planetary motion, applicable to all the planets. Kepler's

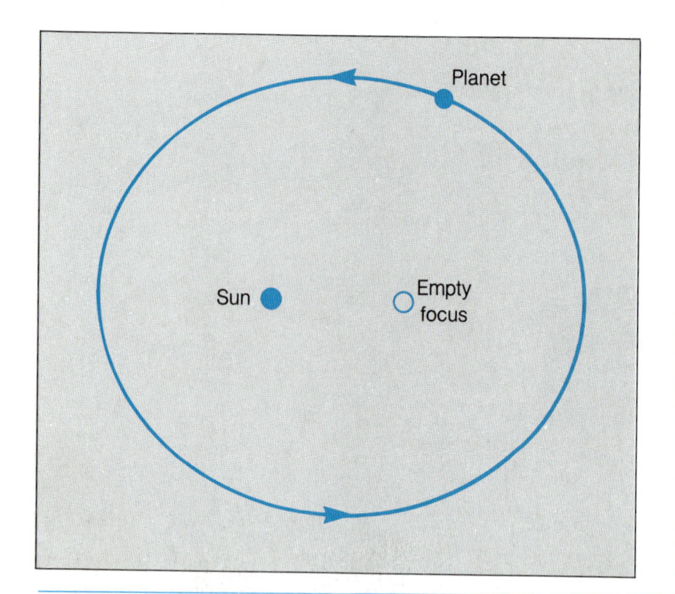

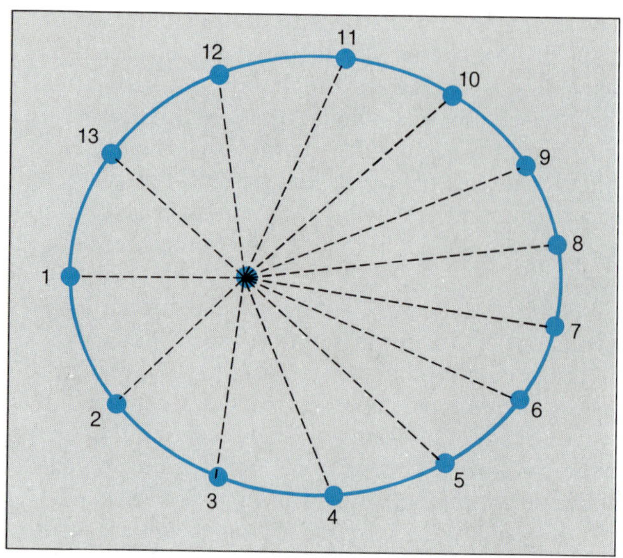

FIGURE 4.10 THE ELLIPSE. *At left is an exaggerated ellipse representing a planetary orbit with the Sun at one focus; at right is a similarly exaggerated ellipse with lines drawn to illustrate Kepler's second law. If the numbers represent the planet's position at equal time intervals, the areas of the triangular segments are equal.*

results on the orbit of Mars were published in 1609 in a book entitled *Astronomica Nova,* or *The New Astronomy: Commentaries on the Motions of Mars.* The book received a great deal of attention.

Kepler's first two laws of planetary motion, in general terms, are: (1) The orbits of the planets are ellipses, with the Sun at one focus; and (2) for each planet, the area swept out in space by a line connecting that planet to the Sun is equal in equal intervals of time.

Most of the planets have orbits that are more nearly circular than is that of Mars, and it is indeed fortunate that Kepler chose to work with the Mars data first, or he might not have been able to discover his first two laws.

In 1612 Emperor Rudolph, who had been supporting Kepler's research, died, and Kepler moved to Linz, in northern Austria. There his work on the planetary motions continued, as he still sought the true nature of the relationship among the planets. In 1619 he published a book entitled *Harmonica Mundi,* or *The Harmony of the World,* in which is reported his discovery of a simple relationship between the orbital periods of the planets and their average distances from the Sun. Now known as Kepler's third law, or simply, the harmonic law, it states that the square of the period of a planet is proportional to the cube of the semimajor axis (which is half of the long axis of an ellipse). In other words, this says that $P^2 = a^3$, where P is the sidereal period of a planet in years, and a is the semimajor axis in terms of the average Sun-Earth distance (that is, in terms of the **astronomical unit** or AU; see Table 4.1).

In another major work called the *Epitome of the Copernican Astronomy,* published in parts in 1618, 1620, and 1621, Kepler presented a summary of the state of astronomy at that time, including Galileo's discoveries. In this book Kepler generalized his laws, explicitly stating that all the planets behaved similarly to Mars, something that had clearly been his belief all along.

By the time the *Epitome* was published, the Roman Catholic church had joined the Lutheran one in taking a very intolerant stance toward what were considered heretical teachings, in contrast with the situation at the time of Copernicus, nearly 100 years before. Kepler's treatise soon found itself in the *Index of Prohibited Books,* along with *De Revolutionibus.*

In 1627 Kepler published his last significant astronomical work, a table of planetary positions based on his laws of motion, which could be used to predict accurately the planetary motions. These tables, which

TABLE 4.1 TESTING KEPLER'S THIRD LAW

Planet	a (AU)	P (years)	a^3	P^2
Mercury	0.387	0.241	0.058	0.058
Venus	0.723	0.615	0.378	0.378
Earth	1.000	1.000	1.000	1.000
Mars	1.524	1.881	3.533	3.538
Jupiter	5.203	11.86	140.85	140.66
Saturn	9.555	29.46	867.98	867.89
Uranus	19.22	84.01	7055.79	7057.68
Neptune	30.11	164.8	27162.32	27159.04
Pluto	39.44	248.5	61349.46	61752.25

he called the *Rudolphine Tables* in honor of his former benefactor, were used for the next several years. The *Rudolphine Tables* represented an improvement in accuracy over any previous tables by a factor nearly of 100, a resounding and remarkable confirmation of the validity of Kepler's laws. In a very real sense, the *Rudolphine Tables* represented Kepler's life's work, since, with their publication, he completed the task set before him when he first went to work for Tycho. Kepler died in 1630, at the age of 59.

GALILEO, EXPERIMENTAL PHYSICS, AND THE TELESCOPE

Very strong contrasts can be drawn between Kepler and his great contemporary, Galileo (Fig. 4.11). Whereas Kepler was fascinated with universal harmony and therefore with the underlying principles on which the universe operates, Galileo was primarily concerned with the nature of physical phenomena and was less devoted to finding fundamental causes. Galileo wanted to know how the laws of nature operated, whereas Kepler sought the reason for the existence of the laws. Galileo's approach was level-headed and rational in the extreme. He used simple experiment and deduction in advancing his perception of the universe. He has frequently been cited as the first truly modern scientist, although others of his time probably deserve a share of that recognition.

Galileo was born in 1564 in Pisa, in northern central Italy. Although his family had intended for him a career in a trade, his obvious intellectual abil-

FIGURE 4.11 GALILEO GALILEI. *From his numerous experiments and observations, Galileo deduced the nature of the cosmos. His methodical attacks on traditional beliefs caused him personal troubles, but greatly influenced contemporary thinking.* (The Granger Collection)

ities were soon recognized, and in 1581 he began formal training at the university in Pisa. There his reluctance to accept dogmatic ideas without question was soon noted, and he earned the nickname "The Wrangler" for his tendency toward debate. Poverty forced him to live at home for a while following his fourth year of study, and for some time he worked at a teaching post in Pisa. In 1592 he moved to the University of Padua, where he was to stay until 1610.

During his early academic years, he carried out numerous experiments in physics, discovering many simple laws, and in the process completely overturning the accepted method of studying science. Whereas a follower of Plato and Aristotle (whose works still dominated in Galileo's time) would proceed by rational thought from standard unproven assumptions, Galileo placed great emphasis on experimental and observational confirmation. When Galileo argued a point, he did so on the basis of what could be demonstrated, although he certainly mixed in a large share of logical deduction. To him, the supreme test of a hypothesis was whether it fit the observed phenomena, not whether the observed phenomena fit a previously adopted hypothesis. In developing this approach, Galileo founded an entirely new basis for scientific inquiry, an achievement in many ways more profound than his contributions to astronomy, which were considerable.

Galileo's discoveries in physics, having to do with the motions of objects, were published in his later years, after his astronomical career had been forcibly ended by Church decree. It was, in fact, an early interest in **mechanics,** the science of the laws of motion, that lured Galileo away from a career in medicine, the subject of his first studies.

Among his most important contributions to mechanics was his elucidation of a principle called **inertia,** the tendency of a moving object to remain in motion or of an object at rest to resist being moved. Galileo was the first to clearly recognize this natural tendency and to understand that an object sliding along a surface stops only because the force of friction acts upon it. He later applied this idea to the motion of the Earth, in explaining how the atmosphere and all the objects on its surface could move along with it.

Galileo also understood the concept of **acceleration,** the rate of change of velocity, and showed that falling objects are subject to a constant acceleration, regardless of their **mass,** the total quantity of matter they contain. Supposedly this point was made in a graphic manner when Galileo is said to have dropped balls of unequal weight from the Leaning Tower of Pisa, demonstrating to a crowd of onlookers that the balls hit the ground at the same time. There is some doubt as to whether, or under what circumstances, this experiment was actually performed, although it is certain that Galileo did carry out comparable experiments, perhaps without attendant ceremony.

Galileo's other contributions to mechanics include the discovery that the oscillation period of a pendulum is approximately independent of how far it swings, which showed that it could be used as a timekeeping device. (Galileo began to build a pendulum-driven clock in his later years, but was forced to abandon the project because of failing eyesight).

Galileo's astronomical discoveries were sufficient to earn him a major place in history, even if he had never done his work in mechanics. It was his astronomical studies, along with his flair for debate and his habit of ridiculing those with whom he disagreed, that made him famous (though not universally loved) in his own time.

His first astronomical contribution was the observation of a supernova in 1604, one also seen by Kepler. Both men showed that its lack of parallax placed it in the distant realm of the stars.

In 1609 Galileo learned of the invention of the telescope, and devised an instrument of his own, which he soon put to use by systematically viewing the heavens with it. Despite the poor quality of the instrument, Galileo made a number of important discoveries almost at once from his observation tower at Padua and reported them in 1610 in a publication called *Sidereus Nuncius (The Starry Messenger)*. Here Galileo showed that the Moon was not a smooth sphere, but instead was covered with craters and mountains (Fig. 4.12). He also reported that the Milky Way, previously thought to be a cloud, actually consisted of countless stars, contrary to the ancient teachings. Most significant of all, he showed that Jupiter was attended by four satellites, whose motions he observed long enough to establish that they orbited the parent planet (Fig. 4.13). All these discoveries, but especially the latter, violated the ancient philosophies of an idealized universe with the Earth at the center.

The satellites of Jupiter showed beyond reasonable question that there were centers of motion other than the Earth.

Once Galileo had begun his astronomical observations, he became fascinated with the structure of the universe, and for a time his work in mechanics was put aside. He spent much of the next several years in pursuit of evidence for the Copernican model (Table 4.2).

Mostly as a result of the reputation Galileo earned with the publication of *Sidereus Nuncius*, he was able to negotiate successfully for the position of court mathematician to the Grand Duke of Tuscany, and

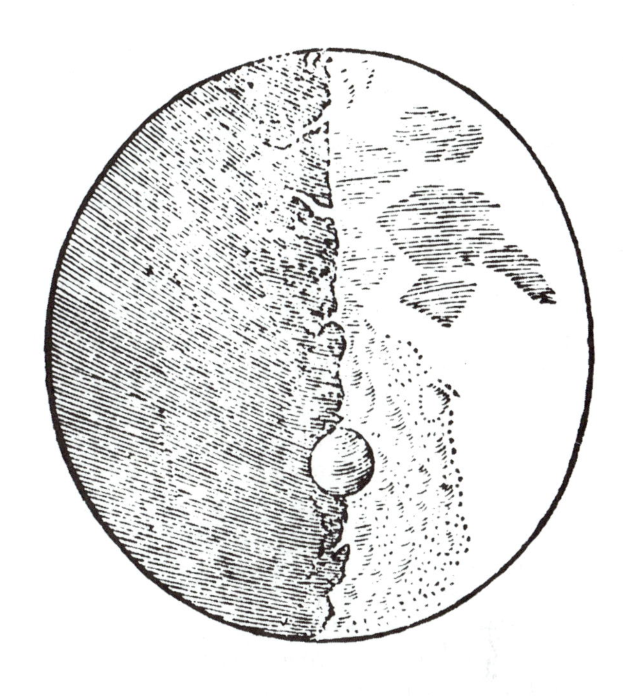

FIGURE 4.12 GALILEO'S SKETCH OF THE MOON. *Galileo's observation of mountains, craters, and "seas" led him to believe that the Moon was not the perfect heavenly body envisioned in the ancient Greek teachings.* (The Granger Collection)

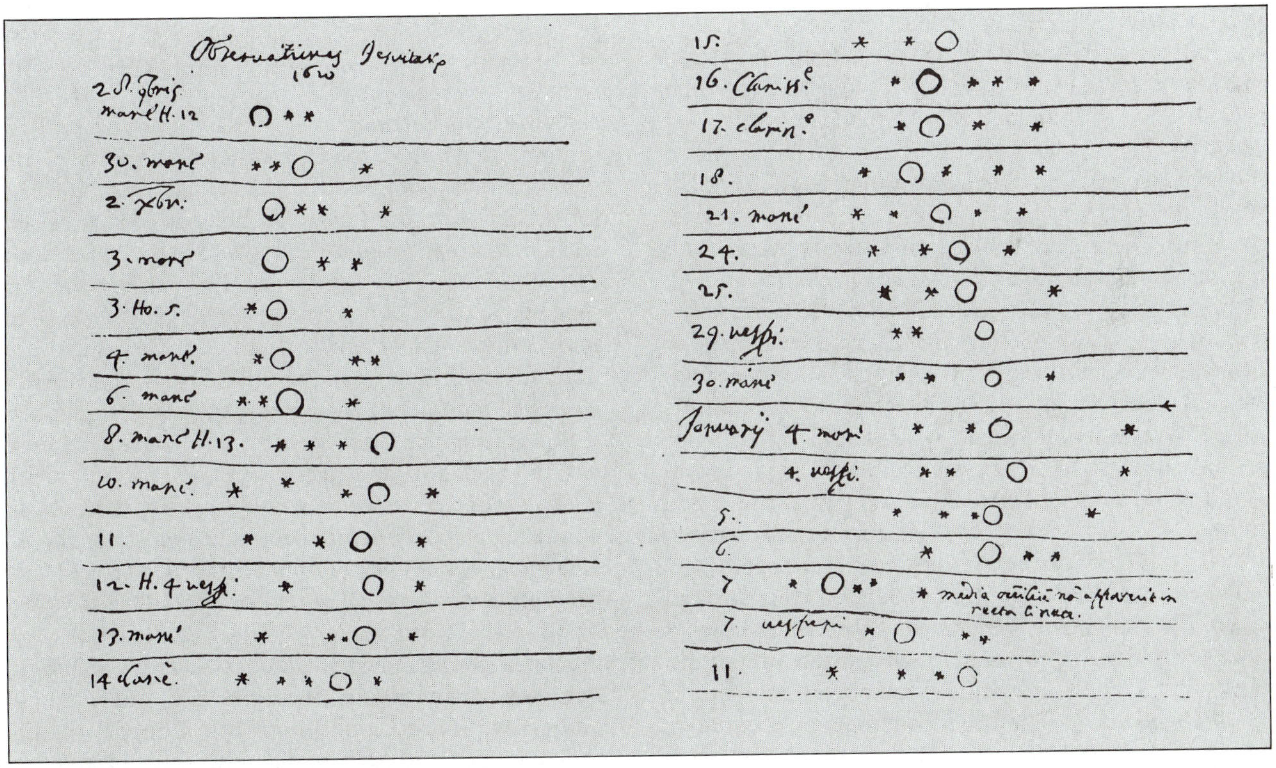

FIGURE 4.13 GALILEO'S SKETCHES OF THE MOONS OF JUPITER. *This series of drawings by Galileo of Jesuit observations is often attributed to the* Starry Messenger, *but in fact was made some years later. Galileo's discovery of moons orbiting Jupiter showed that there are heavenly bodies that do not orbit the Earth.* (Yerkes Observatory)

ASTRONOMICAL INSIGHT 4.2

AN EXCERPT FROM THE *DIALOGUE ON THE TWO CHIEF WORLD SYSTEMS*

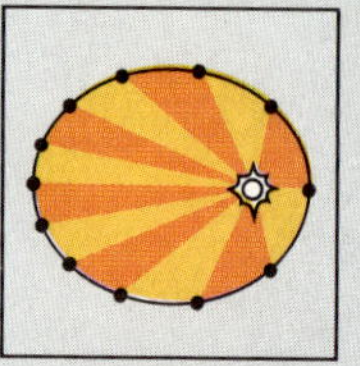

Galileo got into trouble with the Roman Catholic Church when he portrayed the official view of the nature of the universe as not only incorrect, but illogical and even silly. It might help us to see how inflammatory *Dialogue on the Two Chief World Systems* was by reading a small sample from it.

During the course of three days of discussion, the three characters in Galileo's *Dialogue* thoroughly air all the available evidence and arguments bearing on the understanding of mechanics and the structure and nature of the universe. About midway through, a discussion develops that is central to Galileo's principal point: the Sun, not the Earth, is at the center of the universe. In the following excerpt,* we see an example of Salviati's persuasive style, Simplicio's dogged reluctance to give up the ideas of Aristotle, and Sagredo's ready comprehension of Salviati's arguments.

Salviati: *Now if it is true that the center of the world is the same about which the circles of the mundane bodies, that is to say, of the Planets, move, it is most certain that it is not the Earth but the Sun, rather, that is fixed in the center of the World. So that as to this first simple and general apprehension, the middle place belongs to the Sun, and the Earth is as far remote from the center as it is from that same Sun.*

Simplicio: *But from whence do you argue that not the Earth but the Sun is in the center of planetary revolutions?*

Salviati: *I infer the same from the most evident and therefore necessarily conclusive observations, of which the most potent to exclude the Earth from the said center, and to place the Sun therein, are that we see all the planets sometimes nearer and sometimes farther off from the Earth, with so great differences, that, for example, Venus when it is at the farthest is six times more remote than when it is nearest, and Mars rises almost eight times as high at one time as at another. See therefore whether Aristotle was somewhat mistaken in thinking that it was at all times equidistant from us.*

Simplicio: *What in the next place are the tokens that their motions are about the Sun?*

Salviati: *It is shown in the three superior planets, Mars, Jupiter, and Saturn, in that we find them always nearest to the Earth when they are in opposition to the Sun and farthest off when they are towards the conjunction, and this approximation and recession imports thus much, that Mars near at hand appears sixty times greater than when it is remote. As to Venus, in the next place, and to Mercury, we are certain that they revolve about the Sun in that they never move far from it, and in that we see them sometimes above and sometimes below it, as the mutations in the figure of Venus necessarily prove. Touching the Moon, it is certain that it cannot in any way separate itself from the Earth, for the reasons that shall be more distinctly alleged hereafter.*

Sagredo: *I expect that I shall hear more admirable things that depend upon this annual motion of the Earth than were those dependent upon the diurnal revolution.*

In this exchange, Galileo, in the guise of Salviati, advances several arguments based on observations, including the well-known phases of Venus, and the less widely quoted arguments having to do with the varying distances of the planets from the Earth. In this and other passages, Galileo went out of his way to mock the followers of Aristotle, who, according to Salviati in an earlier paragraph, "would deny all the experiences and all the observations in the world, nay, would refuse to see them, that they might not be forced to acknowledge them, and would say that the world stands as Aristotle writes and not as Nature will have it."

The Church did not take the drastic action of excommunicating Galileo for expressing his views in the *Dialogue*, but it did take severe measures, nevertheless. There the matter stood, as scientific and religious attitudes evolved over the subsequent centuries. In a spirit of reconciliation with science and in the interest of clearing up the clouded past, the Roman Catholic Church recently reopened the case of its treatment of Galileo. After a study of the old records, the Church announced in 1983 that Galileo's name had been cleared and that the Church had erred in its treatment of him more than 300 years ago.

*Reprinted from the translation of Galileo's *Dialogue* by G. D. Santillana, by permission of the University of Chicago Press.

ASTRONOMICAL INSIGHT 4.2

AN EXCERPT FROM THE *DIALOGUE ON THE TWO CHIEF WORLD SYSTEMS*

(continued from previous page)

This recent action may seem pointless, since Galileo himself is no longer alive. It was, however, an important statement by a major church that science and religion should tolerate each other—that there is philosophical room for both. In an age when religious attacks on science are occurring anew, this position by an important worldwide religion may have some helpful effect.

in 1610 he moved to Florence, where he was to spend the remainder of his long career. Once established there, Galileo continued his observations and soon added new discoveries to his list. The first of these was the varying aspect of Venus as its position changed with respect to the Sun. Galileo quickly realized that this was analogous to the phases of the Moon—we see from the Earth varying portions of the sunlit side of Venus (Fig. 4.14). This had two important implications: It showed that Venus shines by reflected sunlight rather than by its own power; and it demonstrated that Venus orbits the Sun instead of the Earth, since the apparent size of Venus varies (from being smallest when Venus is fully sunlit to being largest when the dark side of Venus is facing the Earth).

At about the same time, Galileo began to study the dark spots that occasionally appeared on the face of the Sun. These had been independently seen by a number of astronomers, and there was substantial controversy as to what they were. Some thought the spots to be small planets close to the Sun that moved across its face during their orbital motion, but Galileo showed that they must actually be spots on the surface of the Sun. He did this by noting that the spots moved very slowly when near the edge of the Sun's disk, which he attributed to foreshortening as the Sun's rotation carried them toward the Earth on the approaching edge of the Sun and away from the Earth on the receding edge, with significant sideways motion only when they moved across the central portion of the disk (Fig. 4.15). That the Sun should have blemishes on its surface, and rotate as well, were contrary to the established view.

In the years immediately following these discoveries, Galileo began to draw increasingly heavy criticism from the Church, and he made efforts to develop good relations with high-ranking officials in Rome. Nevertheless, in 1616 he was pressured into refuting the Copernican doctrine, and for several years thereafter was relatively quiet on the subject. Except for a well-publicized debate on the nature of comets, Galileo spent most of his time preparing his greatest astronomical treatise, which was finally published, after some difficulties with Church censors, in 1632. To avoid direct violation of his oath not to support the

TABLE 4.2 SOME OF GALILEO'S ARGUMENTS FOR HELIOCENTRIC THEORY

Discovery	Argument
Many faint stars	Difficult to reconcile with idea of stars as points attached to crystalline sphere
Craters on the Moon	Moon is not perfect, immutable heavenly body
Moons of Jupiter	A body other than earth as center of motion
Phases of Venus	Explained only if Venus orbits the Sun, and shines by reflected sunlight
Sunspots	Spots are on solar surface, showing that the Sun is not perfect
Variable planetary sizes	Angular size variations explained by motion of planets around Sun

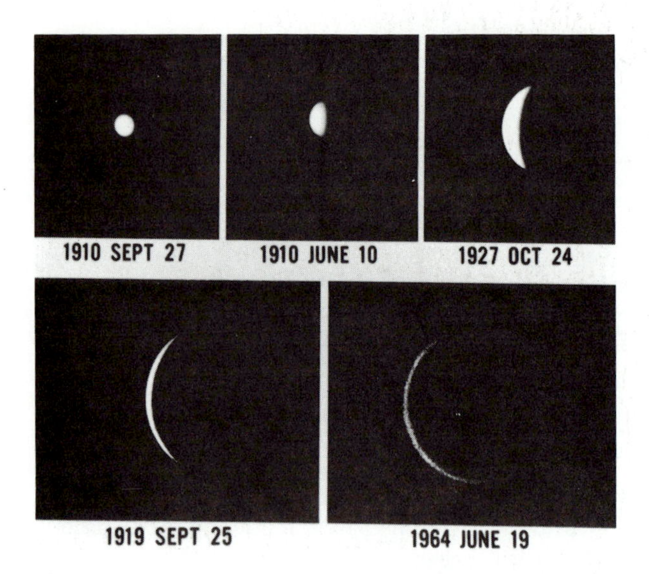

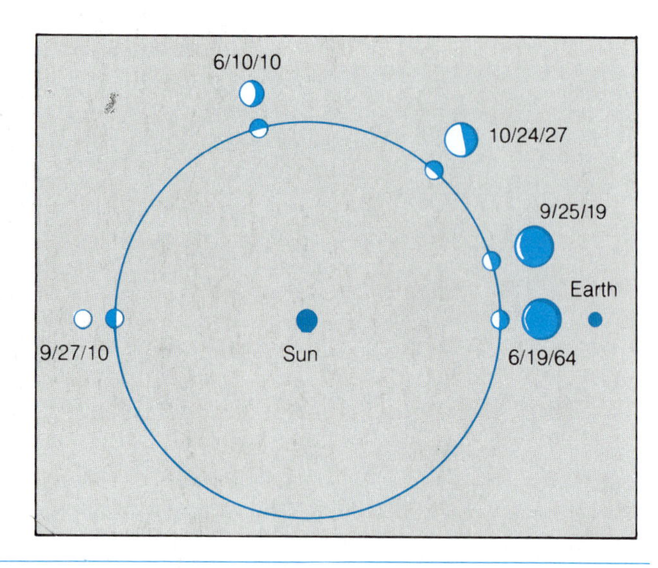

FIGURE 4.14 THE PHASES OF VENUS. *As Venus orbits the sun and its position changes relative to the Sun-Earth line, its phase varies as we see differing portions of its sunlit side. In addition, its apparent size varies because of its varying distance from Earth.* (Photos from NASA)

Copernican heliocentric view, Galileo wrote his book in the form of a dialogue among three characters, one of whom, named Simplicio, represented the official position of the Church; another, Salviati, that of Galileo (although this was, of course, not stated explicitly); and third, Sagredo, who was always quick to understand and agree with Salviati's arguments. In this treatise, called *Dialogue on the Two Chief World Systems,* Galileo, through the character Salviati and at the expense of Simplicio, systematically destroyed many of the traditional astronomical teachings of the Church. The book was published in Italian, rather than in the scholarly Latin, and its contents were therefore accessible to the general populace.

Despite a lengthy preface in which Galileo disavowed any personal belief in the heliocentric doctrine, the Church reacted strongly, and within a few months Galileo was summoned before the Roman Inquisition, while existing copies of the book were heavily censored and further publication was banned. It joined the works of Copernicus and Kepler on the *Index* (where it was to remain until the 1830s), and Galileo himself was sentenced to house arrest, even-

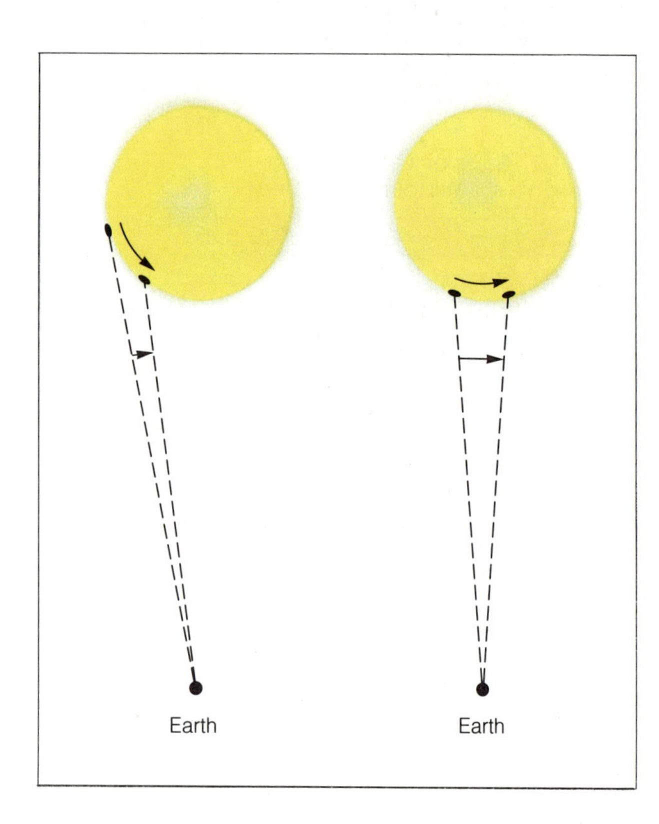

FIGURE 4.15 GALILEO'S ANALYSIS OF SUNSPOTS. *The dashed lines indicate the apparent (angular) motion of a sunspot, as seen from Earth, during two equal time intervals. The variation in observed speed of a sunspot led Galileo to argue that sunspots actually lie on the Sun's surface, and were not shadows of planets passing between the Earth and the Sun.*

tually serving his punishment at his country home near Florence. That is where he spent the remainder of his days.

Before his life came to a close, however, Galileo published one more major work, a summary of his studies of mechanics, based largely on experiments done many years before. The book on mechanics appeared in 1638, and Galileo died in 1642, having suffered blindness the last 4 years of his life.

PERSPECTIVE

We have seen how, in the space of less than 150 years, the science of astronomy advanced from adherence to the 1,500-year-old doctrines of Plato, Aristotle, and Ptolemy to a nearly modern view. Despite Galileo's suppression by the Church, enough questions had been raised through his work and that of Copernicus, Brahe, and Kepler, that serious consideration of the nature of the cosmos thereafter centered on the heliocentric system.

It has been interesting to examine the lives, and especially the frame of mind, of the characters who contributed to this revolution, ranging from Copernicus, whose intuition forced him to reject the established view and adopt an entirely new one; to Tycho, who philosophized little, but contributed much through his observations; to Kepler, whose search for universal harmony led him to the laws of planetary motion; and to Galileo, whose implacable common sense and brilliant deduction put it all on a firm logical footing. The stage was now set for the missing link, the physical cause of the motions in the universe, to be discovered.

SUMMARY

1. Copernicus developed the heliocentric theory because it provided a unifying picture of the universe, not because it was more accurate or simpler than the Earth-centered theory.

2. With his new heliocentric theory, Copernicus was able to calculate the relative distances of the planets from the Sun, and he was able to explain the seasons and the lack of detectable parallax.

3. Tycho Brahe made vast improvements in the quantity and quality of astronomical observations of stars, and especially the planets, accomplishing this by making more precise and complete measurements than his predecessors.

4. Tycho was unable to detect stellar parallax, and for this and philosophical reasons, rejected the heliocentric hypothesis.

5. Kepler sought the underlying harmony among the planets and its physical basis, and in the process he discovered three laws of planetary motion.

6. Kepler's first law states that each planet orbits the Sun in an ellipse, with the Sun at one focus.

7. Kepler's second law states that a planet moves fastest in its orbit when it is closest to the Sun, and slowest when it is farthest away, in such a manner that the area swept out in space by a line connecting the planet to the Sun is equal in equal intervals of time.

8. Kepler's third law relates the periods and semimajor axes of the planets; it states that the square of the period (in years) is equal to the cube of the semimajor axis (in AU).

9. Galileo, by performing experiments, defined several important concepts of physics, including most notably the ideas of inertia and acceleration.

10. Galileo's use of a telescope to observe the heavens led to numerous discoveries that he used to refute the geocentric hypothesis.

11. Despite Church opposition, Galileo brought his ideas before the public in the form of a published fictional dialogue between believers of opposing points of view.

REVIEW QUESTIONS

Thought Questions

1. Summarize the reasons Copernicus adopted the heliocentric view.

2. Describe how the belief of Copernicus conformed with the teachings of Aristotle, and how it violated those teachings.

3. There is a principle adopted by most scientists that the best explanation for something is the simplest one, the one requiring the fewest unprovable assumptions. Copernicus and Hipparchus had different explanations of retrograde motion. Which explanation better satisfies the principle of simplicity? Explain.

4. Summarize the ways in which Tycho Brahe's observations were superior to those made by earlier astronomers.

5. Discuss the philosophical outlook of Johannes Kepler. In what ways was he like the followers of Plato and Aristotle, and in what ways was he like a modern scientist?

6. Why was it so fortunate that Kepler worked first on Mars as he attempted to deduce the nature of planetary motion?

7. Discuss how Kepler's third law, the harmonic law, was the most pleasing to Kepler himself, in view of his philosophical motivations.

8. Summarize the contrasts in style and philosophy between Kepler and Galileo.

9. Make a list of discoveries made by Galileo with the telescope, and explain how each was used by him to support the heliocentric view.

10. Do you think modern scientists are ever subjected to censure and oppression of their views as Galileo was?

Problems

1. Tycho's great quadrant consisted of a quarter circle with a radius of 31.5 feet, covering an angle of 90°. Marks on the quadrant corresponded to degrees and minutes of arc. How far apart were the degree marks? The minute marks?

2. Suppose the true altitude of a certain star above the horizon was 27°14′. Suppose further that Tycho's measurements of the altitude of the star were 27°12′, 27°13′, 27°15′, 27°13′, and 27°16′. How close is the average of these values to the true value? Discuss the comparison between the average of several measurements and the accuracy of individual measurements.

3. Using a board, a couple of thumbtacks, a piece of string, and a pencil, practice drawing ellipses. Notice how the shape of the ellipse varies as the distance between the tacks (foci) is varied. What happens if the two tacks are at the same place?

4. Verify Kepler's third law by comparing P^2 and a^3 for the planets. Use data given in Appendix 7.

5. What would be the period of a planet located 3 AU from the sun? What would be the semimajor axis of a planet whose period was 11.18 years?

6. Galileo cited the changing apparent size of Mars as an argument that Mars orbits the Sun. At its closest approach to Earth (at opposition), Mars is about 0.5 AU away, and at its farthest (conjunction), it is 2.5 AU away. How much larger does it appear from Earth at opposition than at conjunction?

ADDITIONAL READINGS

The references listed here are primarily sources of biographical and historical data on the people discussed in this chapter; many of them contain additional lists of references. Readings on the principles of physics described in this chapter can most easily be found in elementary physics texts, which are available at all levels ranging from completely nonmathematical to any degree of mathematical sophistication desired.

Beer, A., and P. Beer, eds. 1975. *Kepler. Vistas in astronomy,* Vol. 18. New York: Pergamon Press.

_______, and K. A. Strand, eds. 1975. *Copernicus. Vistas in astronomy,* Vol. 17. New York: Pergamon Press.

Cohen, I. B. 1960. Newton in light of recent scholarship. *Isis* 51:489.

_____. 1974. Newton. In *Dictionary of scientific biography.* Vol. 10, ed. C. G. Gillispie. New York: Scribner's.

Drake, S. 1980. Newton's apple and Galileo's dialogue. *Scientific American* 243(2):150.

Draper, J. L. E. 1963. *Tycho Brahe: A picture of scientific inquiry in the sixteenth century.* New York: Dover.

Galileo, G. 1632. *Dialogue on the two chief world systems.* Translated by G. de Santillana. Chicago: University of Chicago Press.

Gingerich, O. 1973. Copernicus and Tycho. *Scientific American* 229(6):86.

______. 1973. Kepler, In *Dictionary of scientific biography,* Vol. 7, ed. C. G. Gillispie. New York: Scribner's.

______. 1983. The Galileo affair. *Scientific American* 247(2):132.

Gingerich, O., ed. 1975. *The nature of scientific discovery.* Washington, D.C. Smithsonian Institution Press.

Kuhn, T. 1957. *The Copernican revolution.* Cambridge, Mass.: Harvard University Press.

Whiteside, D. T. 1962. The expanding world of Newtonian research. *History of Science* 1:15.

GUEST EDITORIAL

SKY SYMBOLS AND CELESTIAL ASTRONOMY IN ANCIENT MEXICO

E. C. Krupp

E. C. Krupp is an astronomer and director of the Griffith Observatory in Los Angeles. Since 1973 Dr. Krupp has been an active participant in the interdisciplinary study of archaeoastronomy. Widely respected as a writer and lecturer, he is the editor and coauthor of In Search of Ancient Astronomies *and* Archaeoastronomy and the Roots of Science, *as well as the author of* Beyond the Blue Horizon *and* The Big Dipper and You *(a children's book). He has visited over seven hundred ancient and prehistoric sites throughout the world, including darkest California, and is especially interested in the celestial component of belief systems. In this essay he not only describes celestial aspects of Mesoamerican cultures but also issues an implicit warning regarding some of the popular pseudoscience of today.*

Once the New World had been populated by Ice Age hunters out of central Asia, the New World Indian civilizations developed in isolation—without significant influence from the Old World's great centers of science. The New World is, then, like an experiment. It provides somewhat different conditions and shows us another way in which cultural development took place. In this sense, access to the ancient New World is like access to another planet. Probes to Venus, Mars, Jupiter, and Saturn have all helped us understand the Earth—and planets in general—a little better. Cross-cultural comparisons shed light on human development, and the study of ancient astonomy allows us to enter, at least marginally the belief systems of the people who preceded us on the planet. The cultural isolation of the New World, then, gives us a chance to see yet another way in which human beings have interacted with the sky.

Astonomy in the New World was embedded in the religious traditions of its civilizations and expressed in ceremony, myth, dynastic ritual, and the calendar. Every 52 years the two primary calendar cycles in use in ancient Mexico arrived again at the same starting point on the same day. One of these day-count cycles comprised 365 days, an interval that clearly approximates the annual cycle of the seasons as the Earth orbits the Sun. The second cycle was completed in 260 days, and the source of this unlikely time span still remains obscure. In any case, any given day would have had a name in the 365-day cycle, as well as a name in the 260-day cycle, and any particular combination of names would have recurred every 18,980 days. This is the lowest common multiple of 260 and 365. It took 73 of the 260-day cycles to complete this round. Of course, the interval is also equal to 52 of the 365-day cycles. The 52-year cycle was observed throughout Mesoamerica, where the same basic calendar was shared by a variety of peoples. Indeed, the 365-day year was known in the Old World, too, by the Egyptians, Mesopotamians, Chinese, and others. No one in the Old World, however, observed a 260-day cycle. This is unique to Mesoamerica.

We know with certainty that the 52-year calendar count was kept by ancient peoples of Mexico, but the evidence for it is not to be found in astronomical texts or in calendar computation guides of the Mesoamericans. Instead, we learn about this 52-year cycle in terms of the ceremonies that accompanied its completion.

The approaching end of a cycle prompted the Aztecs to carry out an important series of rituals culminating in the sacrifice of a human victim on top of Cerro de la Estrella (the "Hill of the Star"), when, on the appointed night, the Pleiades crossed overhead at midnight (the Pleiades are a prominent cluster of stars visible in the evening in late fall). The astronomical component of the ceremony probably was linked with the annual cycle of the seasons. The Pleiades occupy a strategic spot in the sky; they are on the ecliptic. They serve, then, as a seasonal signal, and they were used by just about everyone in antiquity—in both the Old World and the New World—for that purpose. In the Aztec world, the midnight culmination of the Pleiades corresponded with the November

GUEST EDITORIAL

passage of the Sun through the nadir at midnight (the nadir is the direction straight down, through the Earth—the direction exactly opposite the zenith). The Sun was opposite the Pleiades, and the Sun was a primary focus of the Aztec myth, ritual, and religion. Human sacrifices to the Sun were intended by the Aztecs to sustain and energize that brilliant and powerful celestial god, and the Aztecs considered it their sacred duty to provide their god with sustenance. The real message behind the Pleiades' midnight crossing and the sacrificial offering was the rebirth of the Sun from the lowest realm of the underworld.

Every 52 years the years were, in a sense, collected together. The Aztecs called the cermony the Binding of the Years, and as the years were bound, there was a sense that the completion and closure of a round of time could bring time to an end and the destruction of the world. Such apocalyptic interpretations of time are by no means restricted to the ancient Americans. Even as our own calendar bundles up the years toward A.D. 2000, millennial visions accelerate and apocalyptic literature promotes the notion of Doomsday's approach.

For the Aztecs, however, the continuing progress of the Pleiades on the night of possible doom was a signal that another cycle had begun after all. The Sun, reborn in the territory of the dead, would propel time and existence through another 52-year round.

The Aztec empire has fallen, and sacrifices no longer stain the foundations of the temple on the Hill of the Star. However, visitors to Mexico can still encounter signs of the old traditions. At El Tajín, a Huastec antiquity on the Gulf coast, local Totonac Indians perform a traditional ritual, the Flying Pole Dance, largely for the benefit of tourists who come to the site. Four men in cermonial costume dangle by their feet from the top of a 100-foot pole as a fifth man on top plays a flute and a drum. As the four swing around the pole, their ropes lower them closer to the ground. They complete their performance with thirteen circuits of the pole; the number is significant. The number of men (4) times the number of turns (13) is 52. The costumes of the men transform them, in fact, into the symbolic birds that were the "bearers of the years." At El Tajín, long after the conquest of Mexico and the abandonment of the traditional calendars, the Bearers of the Years still fly to the end of the cycle with their burdens of time.

The celestial component of Mesoamerican belief systems shows up in the ancient Aztec ceremony of the Binding of the Years, in the Flying Pole Dance of today's Totonac Indians, and in many other aspects of culture and civilization in Mexico. We can tell that the ritual ball game played at all the great ceremonial centers in Mesoamerica carried a celestial connotation. Most pre-Columbian writings were destroyed in the Spanish conquest of Mexico, but the few surviving hieroglyphic books of the Maya seem to be texts and tables for celestial divination. Inscriptions on the stelae and walls of ruins in the Maya area appear to link the accession of kings—and their deaths—to circumstances in the heavens. The celebrated Bonampak murals, preserved on walls of a temple in the jungles of southern Mexico, depict a ritual battle between two Maya cities, and the battle appears to have been timed to coincide with an appearance of the planet Venus. Elsewhere in Mexico, buildings incorporate astronomical alignments into their architecture and celestial symbols into their design, but most of these structures look more like cermonial monuments than observatories.

Erich von Däniken and other purveyors of ancient astronaut books portray many of the accomplishments of the New World Indian civilizations as products of visits by extraterrestrials. The astronomy of ancient Mexico—expressed in intricate calendars, monumental pyramids, and celestial symbols—is misrepresented, misinterpreted, and explained away in terms of alien intervention. For example, the sarcophagus cover of Lord Pacal, king of Palenque—an evocative Maya site in Chiapas, Mexico—depicts the dead Pacal as falling from the celestial realm into the devouring maw of the underworld. As the source of order and stability on Earth, the king was associated with the Sun, whose orderly, patterned movement orga-

GUEST EDITORIAL

nizes the landscape and the calendar. Surrounded by celestial symbols, Pacal is accompanied into the realm of the dead by the Sun, whose half-skeletal face shows that he, too, is dying. Despite the self-consistency and sense of these symbols, von Däniken insists that the portrait of Pacal illustrates an ancient astronaut blasting off from Mexico at the controls of his own space rocket, or "chariot of the gods."

In fact, ancient astronauts have nothing to do with Lord Pacal, Palenque, or the indigenous traditions of the New World. The ancient astronaut idea is just one more version of an old theme: once the Europeans discovered the great civilizations of the New World, they were never quite able to accept the idea that the native peoples of America were able to develop without the benefit of Old World experience and knowledge. Everyone from the colonists of sunken Atlantis to the Lost Tribes of Israel has been promoted as the inspiration and source of New World civilization. In the space age, ancient astronauts get the credit.

There was, of course, a highly developed practical astronomy in ancient Mexico, but most of the evidence for it is indirect. This fuels the arguments of those who embrace exotic—and incorrect—explanations for New World development. The indirect evidence is, however, ample, and reveals—at least in part—how the Aztecs, the Maya, and their neighbors saw themselves as participants in a dynamic and ordered universe.

SECTION II

THE TOOLS OF ASTRONOMY

Before we can plunge into a full description of the universe and an understanding of it, we must first learn about the methods and tools used by astronomers to gain this understanding. One of the principal goals of this text is to describe *how* we know what we know, going well beyond a simple summary of the phenomena that are observed.

The first of the three chapters in this section is devoted to a description of some of the basic physical principles that underlie the motions and interactions of bodies large and small. We start with a discussion of Isaac Newton, whose intellectual powers unearthed many of these principles, and then we move on to the laws of physics that govern such diverse phenomena as planetary orbits, the motions of molecules in a gas, and the tides on the earth and other bodies.

We would know nothing of the external universe were it not for the light that reaches us from far-away objects, and the next chapter describes the nature of light and the way we decipher its messages. We find that an amazing variety of information can be derived from the spectra of objects like planets and stars. Things once considered forever beyond our grasp are now routinely measured, and in Chapter 6 we learn how this is done.

The final chapter of this section is devoted to the instruments that are used to collect light from afar. Telescopes on the ground and in space are the material monuments to mankind's quet for knowledge of the universe, and in Chapter 7 we will learn the principles of their design and the practical considerations of their construction and use. Having done this, we will, at last, be ready to explore the dynamic universe.

CHAPTER 5

ISAAC NEWTON AND THE LAWS OF MOTION

The law of gravitation acting on the Earth's oceans: low tide at the Bay of Fundy, Nova Scotia, Canada. (© Jeff Apoian, Science Source/Photo Researchers)

CHAPTER PREVIEW

After the publication in 1632 of Galileo's *Dialogue on the Two Chief World Systems,* a 50-year period elapsed during which no major discoveries were made, but substantial progress nevertheless occurred. Telescopic observations were extended to a greater variety of objects, the surface of the Moon was mapped, star catalogues were expanded, and the planets were studied in more detail. Perhaps the greatest astronomer of the period was Christian Huygens (1629–1695), who refined the telescope and was able to detect a moon of Saturn. He also discovered the true nature of the misshapen appearance of the planet (Galileo's crude telescope had only revealed a distortion in the shape of Saturn, whereas Huygens was able to determine that the planet is encircled by a great ring system). The Earth was accurately measured by an assortment of people, and it was even discovered that it is not precisely spherical, but instead bulges at the equator.

THE LIFE OF NEWTON

In the year 1643, a few months after the death of Galileo, Isaac Newton (Fig. 5.1) was born in Woolsthorpe, England, into a climate of inquiry and scientific ferment. Newton's childhood was unremarkable, except that he showed a growing interest in mathematics and science, and at age 18 he entered college at Cambridge, receiving a bachelor's degree in early 1665. He then spent 2 years at his home in Woolsthorpe, largely because the Plague made city living rather dangerous, and it was during this time that he made a remarkable series of discoveries in physics, astronomy, optics, and mathematics, in what surely must have been one of the most intense and productive periods of individual intellectual effort in human history.

Newton had a tendency to exhaust a subject, get bored with it, and go on to new fields, so that nothing of his work was published for some time, during which some of his discoveries were repeated independently by others. Finally, after persistent urging by his friend and fellow astronomer Edmond Halley, in 1687 Newton published a massive work called *Philosophiae Naturalis Principia Mathematica,* now usually referred to as the *Principia* (Fig. 5.2). In this three-volume book, Newton established the science of mechanics (which he viewed as merely background material and relegated it to an introductory section), and applied it to the motions of the Moon and the planets, developing the law of gravitation as well. His work in

FIGURE 5.1 ISAAC NEWTON. *Newton's experiments and brilliant mathematical intuition led him to profound new understandings of physics and mathematics.* (The Granger Collection)

optics was published separately in 1704, although it was probably written much earlier than that.

The *Principia* received great notice, particularly in England. As a result, Newton's later life was a public one, with various government positions which left less and less time for scientific discovery. With the help of younger associates, he did revise the *Principia* on two occasions, in 1713 and in 1726, making some improvements each time. Newton died in 1727, at the age of 84. His impact lives today, in our modern understanding of physics and mathematics. Newton's conclusions on the nature of motions and gravity are still viewed as correct, although it is now realized that there are circumstances in which a more general theory (Einstein's relativity) must be used.

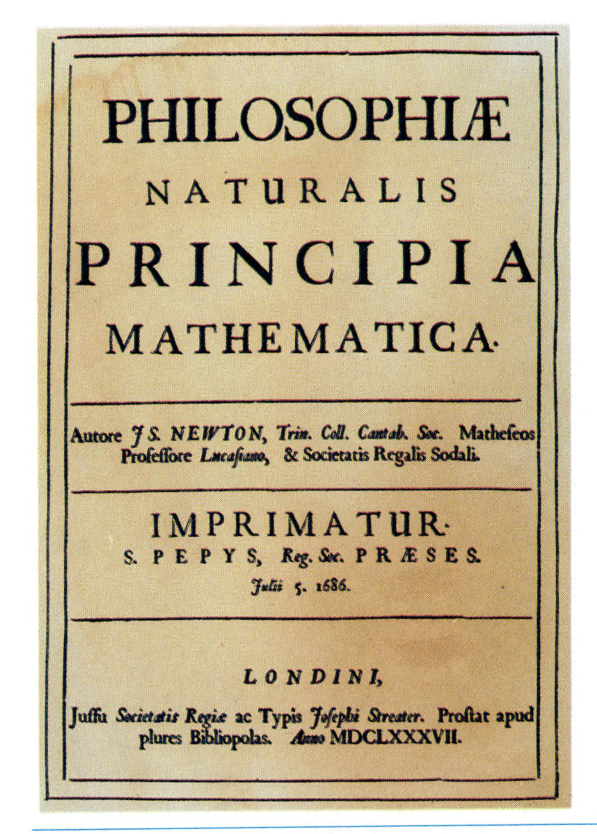

PHILOSOPHIÆ
NATURALIS
PRINCIPIA
MATHEMATICA.

Autore *J S. NEWTON*, *Trin. Coll. Cantab. Soc.* Matheſeos Profeſſore *Lucaſiano*, & Societatis Regalis Sodali.

IMPRIMATUR.
S. PEPYS, *Reg. Soc.* PRÆSES.
Julii 5. 1686.

LONDINI,
Juſſu *Societatis Regiæ* ac Typis *Joſephi Streater*. Proſtat apud plures Bibliopolas. *Anno* MDCLXXXVII.

FIGURE 5.2 THE TITLE PAGE FROM AN EARLY EDITION OF THE PRINCIPIA. *This massive volume is still considered one of the greatest and most influential books ever written.* (The Granger Collection)

NEWTON'S LAWS OF MOTION

Inertia and the First Law

We have already discussed the concept of inertia, first developed by Galileo. Newton expanded the idea, recognizing inertia as a fundamental property of matter, very closely related to another property, the **mass** of an object.

The mass of an object reflects the amount of matter it contains, which in turn determines other properties such as weight and momentum. It is the other properties that are easily observed, not the mass, so mass can be a difficult concept. Mass and weight are particularly easy to confuse, because at the Earth's surface, a given amount of mass will always have the same weight (because the gravitational force is the same everywhere on the Earth). To see the difference, consider an astronaut who goes to the Moon. On the Moon, he weighs only one-sixth as much as he does on Earth, but his mass is the same. The relationship between mass and weight will become clearer in the next section.

Mass is usually measured in units of **grams** or **kilograms.** One gram is the mass of a cubic centimeter of water, and a kilogram, which weighs about 2.2 pounds at sea level, is the mass of 1,000 cubic centimeters (or one **liter**) of water.

Inertia and mass are closely related because the mass of a body determines how much **inertia,** or resistance to change in motion, the body has. The more massive an object is, the more inertia it has, and the more difficult it is to start it moving, to stop it, or to alter its motion once it is moving (Fig. 5.3). For example, imagine two crates of equal size, one containing books, the other pillows. Each has wheels on the bottom, minimizing friction. You may be able to move the box of pillows rather easily, but not the box of books. The box of books contains more mass, and therefore has more inertia, more resistance to being moved. Similarly, a massive object in motion is more difficult to turn or stop than a less massive one. Imagine, for example, the deftness required to bring a large ship to rest at a pier without destroying the pier.

Newton summarized the concept of inertia in what has become known as his first law of motion:

> A body at rest or in a state of uniform motion tends to stay at rest or in uniform motion unless an unbalanced force acts upon it.

FIGURE 5.3 INERTIA. *An object in motion tends to remain in motion.* (H. Roger Viollet)

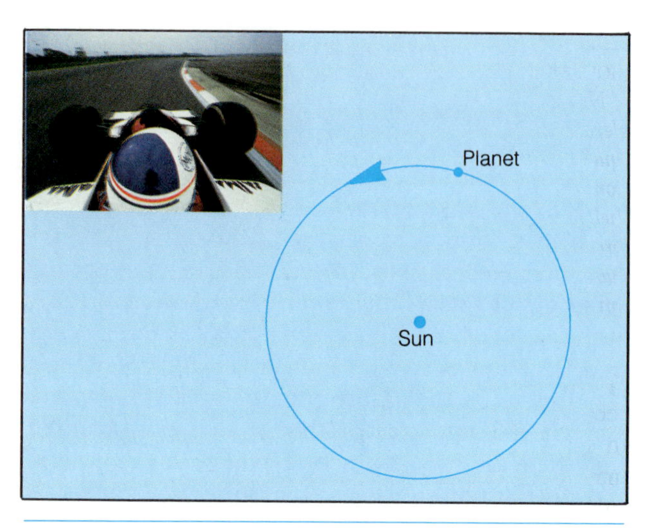

FIGURE 5.4 ACCELERATION. *A car that is speeding up is accelerating, as is a planet that follows a curved path, even if its speed is constant.* Any *change of speed* or *direction is an acceleration.* (Photo Researchers, Inc.)

As we have seen, this had been stated by Galileo, although he did not make it quite so general as Newton, who realized that it applies to the Earth, planets, and all celestial bodies. One very important consequence of the first law of Newton is the implication that a planet should follow a straight line, rather than a curved path around the Sun, unless some force acts upon it. We shall return to this point shortly.

Force and Acceleration

Having stated that a force is required to change an object's state of rest or uniform motion, Newton went on to determine the relationship between force and the change in motion that it produces. To understand this, we must discuss the idea of acceleration.

Acceleration is a general word for any change in the motion of an object. It is acceleration when a moving object is speeded up or slowed down, or when its direction of motion is altered (Fig. 5.4). It is also acceleration when an object at rest is put into motion. A planet orbiting the Sun is undergoing constant acceleration; otherwise it would fly off in a straight line.

Another way of stating Newton's first law is that in order for an object to be accelerated, a force must be applied to it. His second law spells out the relationship among this force, the resultant acceleration, and the mass of the object:

> An object is accelerated when an unbalanced force is applied to it, and the acceleration is proportional to the force and inversely proportional to the mass of the object.

This may be written mathematically as $a = F/m$, where a is the acceleration, F is the force, and m is the mass. More commonly it is written in the alternative form $F = ma$.

We can illustrate the second law as follows: If one object has twice the mass of another and equal forces are applied to the two, the more massive one will be accelerated only half as much. Conversely, if unequal forces are applied to objects of equal mass, the one to which the greater force is applied will be accelerated to a greater speed.

Action and Reaction

Newton's third law of motion is probably more subtle than the first two, although in some circumstances it is quite obvious. It states:

> For every action there is an equal and opposite reaction.

In other words, forces always occur in pairs such that when one object exerts a force on another, the

FIGURE 5.5 ACTION AND REACTION. *Some manifestations of Newton's third law are rather subtle, while others are not. Here a book exerts a force on a table, and the table exerts an equal and opposite force on the book. There is a static situation; there is no motion. When a cannon is fired, however, the shell is accelerated one way and the cannon the other. The force applied to each is the same, but the cannon has more mass than the shell and is therefore accelerated less, as Newton's second law states. (bottom:* U.S. Army)

second object exerts a force on the first that is equal in strength but opposite in direction (Fig. 5.5). This may sound confusing, because the accelerations of the two objects are not necessarily equal, and because in most common situations, other forces such as friction complicate the picture. Furthermore, there are many static situations in which no acceleration of either object occurs.

We can visualize the third law by considering instances in which friction is not important. For example, imagine standing in a small boat and throwing overboard a heavy object, such as an anchor. The boat will move in the opposite direction from the anchor, because the anchor exerts a force on you as you throw it. The "kick" of a gun when it is fired is another example of action and reaction. When a person jumps off the ground by pushing against the Earth, both the person and the Earth are accelerated by the mutual force, but of course the immensely greater mass of the Earth prevents it from being accelerated noticeably.

The third law of motion states the principle on which a rocket works. In this case, hot, expanding gas is allowed to escape through a nozzle, creating a force on the rocket. The gas is accelerated in one direction, and the rocket is accelerated in the opposite direction (Fig. 5.6). Anyone who has inflated a balloon and then let go of it, allowing it to zoom through the air, is familiar with the operating principle of a rocket.

FIGURE 5.6 NEWTON'S THIRD LAW APPLIED TO THE LAUNCH OF A ROCKET. *Hot gases are forced out of a nozzle (or several, as in this case), and in return they exert a force on the rocket that accelerates it. This is the first launch of the space shuttle* Columbia. (NASA)

GRAVITATION AND ORBITS

We return now to a point raised earlier—namely, that the planets would fly off along straight lines if no force were acting upon them. Newton's first law says that this should happen, yet obviously it does not. Newton realized that the planets must be undergoing constant acceleration toward the center of their orbits, that is, towards the Sun (Fig. 5.7). He set out to understand the nature of the force that creates this acceleration.

If you feel confused about the direction of this force, remember what you have just learned about acceleration and inertia: A planet needs no force to keep it moving, but it does requires a force to keep its path curving as it travels around the Sun. What is needed is something to pull the planet inward, toward the Sun. This force can be compared to the tension in the string tied to a rock that you whirl about your head; if you suddenly cut the string, the rock would fly off in whatever direction it happened to be going at the time.

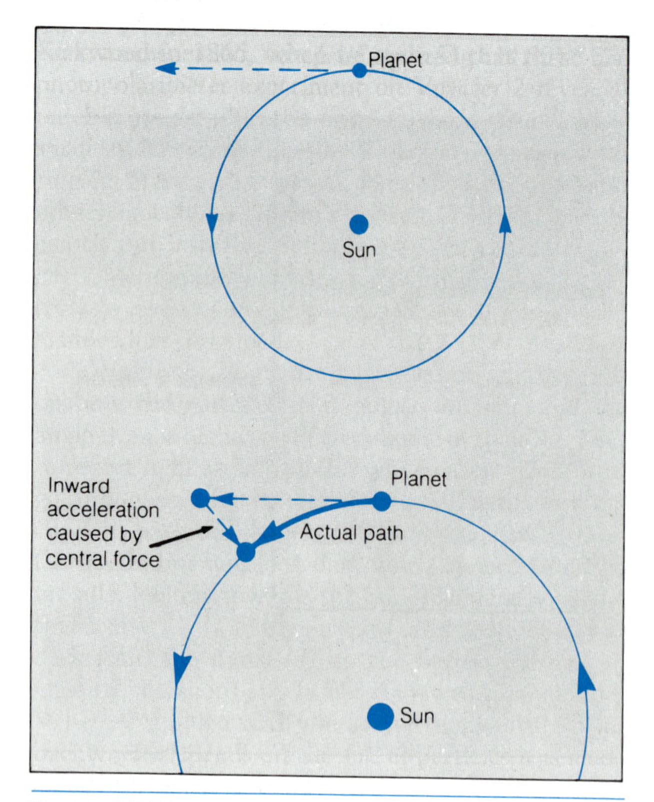

FIGURE 5.7 PLANETARY ORBITS AND CENTRAL FORCE. *A planet would fly off in a straight line if no force attracted it toward the center of its orbit. From diagrams like the one below, Newton was able to determine the amount of acceleration required by an orbiting body to keep it in its orbit.*

What, then, is the string that keeps the planets whirling about the Sun? Newton realized that the Sun itself must be the source of this force, and he made use of Kepler's third law, as well as observations of the Moon's orbit and falling objects at the Earth's surface, to discover its properties. He was led to formulate his law of universal gravitation:

> Any two bodies in the universe are attracted to each other with a force that is proportional to the product of the masses of the two bodies and inversely proportional to the square of the distance between them.

The law of gravitation is one of the fundamental rules by which the universe operates. As we shall see in later chapters, it explains the motions of stars about each other or about the center of the galaxy, the movements of the Moon and planets, and the motions of galaxies about one another. Gravity, in fact, appears to be the dominant factor that will determine the ultimate fate of the universe. Table 5.1 shows the relative gravitational forces on a person on Earth due to various bodies in the universe.

It is useful to consider a few examples illustrating how the law of gravitation is applied. The weight of an object is simply the gravitational force between it and the Earth. If, for example, the diameter of the Earth were suddenly doubled (while its mass remained constant), our weight would decrease by a factor of 4. If the Earth were three times smaller, our weight would be $3^2 = 9$ times greater. If we climb to the top of a high mountain, our weight decreases, but not very much, because even the highest mountains are small compared to the radius of the Earth.

TABLE 5.1 GRAVITATIONAL FORCES ACTING ON A HUMAN STANDING ON EARTH.

Source of force	Relative Strength of Force
Earth	1.0
Moon	3.4×10^{-6}
Sun	8.6×10^{-4}
Venus (at closest approach)	1.9×10^{-8}
Jupiter (at closest approach)	3.3×10^{-8}
Nearest star	1.4×10^{-14}
Milky Way galaxy	2.1×10^{-11}
Virgo cluster of galaxies	10^{-15}

ASTRONOMICAL INSIGHT 5.1

THE FORCES OF NATURE

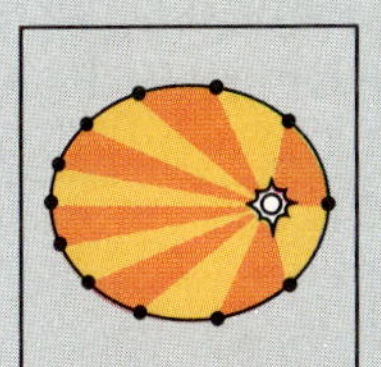

One of the most difficult concepts to grasp is that of a force, particularly when there is no concrete, visible object to exert it. We think of a force as a push or a pull, and we can visualize it easily in certain circumstances, such as when a gardener pushes a wheelbarrow and it moves. It is perhaps less obvious in the case where no movement results, and even less so when no tangible agent exerts the force. In the case of gravity, the force is exerted invisibly and over great distances.

In general terms, a force exerted by a concrete object is referred to as a **mechanical force,** whereas those exerted without any such agent are **field forces.** Gravity is the most familiar force that is created by a field.

There are four basic kinds of field forces in nature, and in fact these forces form the basis for *all* forces, mechanical or field. Gravity was the first of the four to be discovered, although in "discovering" gravity, and in mathematically describing its behavior, Newton did not develop a real fundamental understanding of how or why it works. The reason that gravity was the first to be discovered is simple; no special conditions are required for gravitational forces to be exerted (*all* masses attract each other), and it operates over very great distances. After all, Newton deduced its properties by noting how it controls the motions of planets that are separated from the Sun by as much as a billion kilometers.

The second most obvious type of field force, and the second to be discovered and described mathematically, is the **electromagnetic force.** There is an intimate relationship between electric and magnetic fields. The interaction of these fields with charged particles, first described mathematically by the Scottish physicist James Clerk Maxwell, creates forces. The electromagnetic force is actually much stronger than the force of gravity; the electromagnetic force binding an electron to a proton in the nucleus of an atom is 10^{39} times stronger than the gravitational force between the two particles. The electromagnetic force, like gravity, is inversely proportional to the square of the distance between charged particles. Therefore we might expect this force always to dominate over gravity, as it does on a subatomic scale. But it does not. The reason is that most objects in the universe, composed of vast numbers of atoms containing both electrons and protons, have little or no electrical charge, whereas they always have mass and are therefore subject to gravitational forces. If the planets and the Sun had electrical charges in the same proportion to their masses as the electron and proton do, then electromagnetic forces, rather than gravity, would control their motions.

Besides being immensely stronger, the electromagnetic force also differs from gravity in that it can be either repulsive or attractive, depending on whether the electrical charges are the same or opposite.

The remaining two forces were not discovered until the 1930s. (Maxwell's work on electromagnetics was done in the late 1800s.) Both of these new forces involve interactions at the subatomic level, and it was not until the development of the science of quantum mechanics that they were discovered. One of these is the **strong nuclear force,** which is responsible for holding together the protons and neutrons in the nucleus of an atom, and the other is the **weak nuclear force,** some 10^{-5} times weaker than the strong nuclear force. The stronger nuclear force, in turn, is about 100 times stronger than the electromagnetic force, making it the strongest of all, but it operates only over very small distances. Within an atomic nucleus, where protons with their like electrical charges are held together despite their electromagnetic repulsion for each other, it is the strong nuclear force that acts as the glue that keeps the nucleus from flying apart. The weak nuclear force plays a more subtle role, showing its effects primarily in certain modifications of atomic nuclei during radioactive decay.

From the smallest to the largest scales, these four forces appear to be responsible for all interactions of matter. It is ironic that at the most fundamental level, the mechanism that makes the forces work is not understood. To understand the fundamental nature of forces is one of the principal goals of modern physics. One hope is to develop a mathematical framework that encompasses all of the forces, a framework first sought over 60 years ago by Albert Einstein, and referred to as a **unified field theory.** Progress has been made since Einstein's time, particularly towards unifying the electromagnetic and nuclear forces, but the ultimate goal still eludes us. (This is discussed further in Chapter 32.)

The Earth exerts a gravitational force on an astronaut in orbit, but there is no acceleration relative to the spacecraft, so the astronaut is weightless in the local environment (Fig. 5.8).

The law of gravitation states that the force also depends on the masses of the two objects that are attracting one another. It is easy to imagine that a persons's weight would double if his or her mass doubled; similarly, the force between the Earth and the Moon depends on the masses of these two bodies. If we triple the mass of the Moon, the force is tripled; if we triple the mass of the Moon but decrease the mass of the Earth by a factor of 3 at the same time, the force remains unchanged.

FIGURE 5.8 WEIGHTLESSNESS. *Although astronauts orbiting the earth are subject to the Earth's gravitational force and are "falling around" the Earth along with their spacecraft, they are weightless because they experience no gravitational acceleration relative to their surroundings. This photo shows two astronauts aboard* Skylab, *an orbiting scientific research station.* (NASA)

Although it is useful for the purpose of illustration to work out simple thought examples as we have just done, a more quantitative approach is needed to deal with more realistic situations. As we might have expected, the law of universal gravitation can be written in the form of an equation:

$$F = \frac{Gm_1m_2}{r^2},$$

where F is the strength of the force between two objects whose masses are m_1 and m_2 and whose separation is r. The symbol G represents a constant that is required for the value of F to be expressed in normal units of force. In the system of units used most commonly by astronomers, force is measured in dynes (where 1 dyne is the force exerted by a mass of 1 gram under an acceleration of 1 centimeter per second per second).[1] A person who weighs 150 pounds (or a mass of 67 kilograms) has a weight in this system of 6.7×10^7 dynes, so we see that the dyne is a rather small unit of force. The value of the constant G is 6.67×10^{-8} when the masses are expressed in grams; the separation between them in centimeters; and the force in dynes.

The law of universal gravitation can be used to show something that Galileo had postulated long before Newton's time: namely, that the acceleration due to gravity is independent of the mass of an object. Suppose, in the previous equation, that m_1 represents the mass of the Earth, and m_2 the mass of an object falling at the Earth's surface. In that case, r is the Earth's radius (it can be shown that the Earth acts gravitationally as though all its mass were in a single point at its center, so we say that the separation between the object and the Earth is equal to the Earth's radius).

Now recall that the acceleration on a body is equal to the force applied to it, divided by its mass; hence, the acceleration of the falling object is F/m_2.

[1]While astronomers usually use the **centimeter-gram-second,** or **cgs,** system of units, physicists generally use the **meter-kilogram-second,** or *MK,* system (more commonly known as the **Système Internationale,** or *SI,* system). In this system, force is measured in units of newtons, where 1 newton is defined as the force exerted by a mass of 1 kilogram under an acceleration of 1 meter per second per second. A force of 1 newton is equal to a force of 10^5 dynes.

TABLE 5.2 SURFACE GRAVITIES ON SOLAR SYSTEM BODIES

Body	Surface gravity (g)*
Earth	1.0
Sun	27.9
Moon	0.17
Mercury	0.38
Venus	0.90
Mars	0.38
Ceres (largest asteroid)	0.000167
Jupiter	2.64
Saturn	1.13
Uranus	0.89
Neptune	1.13
Pluto	0.07:

*g = 980 cm/sec².

If we call this acceleration g, because it is due to gravity, then we have

$$g = \frac{F}{m_2} = \frac{Gm_1}{r^2} = \frac{GM}{R^2},$$

since the symbols M and R are generally used for the mass and radius of a planetary body.

At the surface of the Earth, the value of g is 980 centimeters per second per second, meaning that the speed of a falling object increases by 980 centimeters per second for every second of fall. In more familiar units, this corresponds to 32 feet per second per second. To find the surface gravity on some other planet or satellite, the same expression for g can be used, along with the appropriate values for the mass of the planet, M, and its radius, R (see Table 5.2). For example, the Moon has 0.273 times the radius of the Earth, and 0.0123 times the mass. Thus, the acceleration of gravity at the Moon's surface is $0.0123/0.273^2 = 0.165$ that at the surface of the Earth. Therefore, astronauts on the Moon weigh approximately one-sixth as much as they do on the Earth.

ENERGY, ANGULAR MOMENTUM, AND ORBITS: KEPLER'S LAWS REVISITED

Even though Newton made use of Kepler's third law in deriving the law of gravitation, the latter is in fact more fundamental. It was soon possible for Newton to show that all three of Kepler's laws follow directly from Newton's laws of motion and gravitation. Kepler's studies of planetary motions revealed the *result* of the laws of motion and gravitation, whereas Newton found the *cause* of the motions. To appreciate how this was done, we must further discuss some basic physical ideas.

An important concept in understanding not only orbital motions but also many other aspects of astrophysics is **energy.** We can define energy as the ability to do work. Energy can take on many possible forms, such as electrical energy, chemical energy, heat, and others. All forms of energy can be classified as either **kinetic energy,** the energy of motion, or **potential energy,** which is stored energy that must be released (converted to kinetic energy) if it is to do work. A speeding car has kinetic energy because of its motion, whereas a tank of gasoline has potential energy in the form of its chemical reactivity, a tendency to release large amounts of kinetic energy if ignited. Thus a car operates when this potential energy is converted to kinetic energy in its cylinders.

The units used for measuring energy can be expressed in terms of the kinetic energy of specified masses moving at specified speeds. The kinetic energy of a moving object is $\frac{1}{2}mv^2$, where m is the mass of the object and v is its speed. In astronomy the most commonly used unit of energy is the **erg,** a small amount of energy that is equivalent to the kinetic energy of a mass of 2 grams moving at a speed of 1 centimeter per second. (The technical definition is that an erg is the amount of energy required to move a mass of 1 gram a distance of 1 centimeter against a force of 1 dyne.) A larger unit of energy often used by physicists is the **joule.** (A joule is equal to 1 newton-meter per second), the equivalent of a mass of 2 kilograms moving at a speed of 1 meter per second. Thus, 1 joule is equal to 10^7 ergs.

We often speak in terms of **power,** which is simply energy expended per second. The familiar **watt** is 1 joule per second; astronomers, however, tend to use units of ergs per second, which have no special name as a unit. Thus we will speak of the power (or, equivalently, of the **luminosity**) of a star in terms of its energy output in ergs per second.

Using this understanding of energy, we can now discuss orbital motions in a much more general way than we have previously done. Two objects subject to each other's gravitational attraction have kinetic energy resulting from their motions, and potential energy because they each feel a gravitational force. Just as a book on a table has potential energy that can be converted to kinetic energy if it is allowed to fall, an orbiting body also has potential energy by virtue of the gravitational force acting on it.

Newton's laws and the concepts of kinetic and potential energy can be used to show that there are many types of possible orbits when two bodies interact gravitationally. Not all are ellipses, because if one of the objects has too much kinetic energy (exceeding the potential energy caused by the gravity of the other), it will not stay in a closed orbit, but will instead follow an arcing path known as a hyperbola, and will escape after one brief encounter. Some comets have so much kinetic energy that after one trip close to the Sun, they escape forever into space, following hyperbolic paths. A rocket launched from Earth with sufficient velocity will escape into a similar orbit (Fig. 5.9).

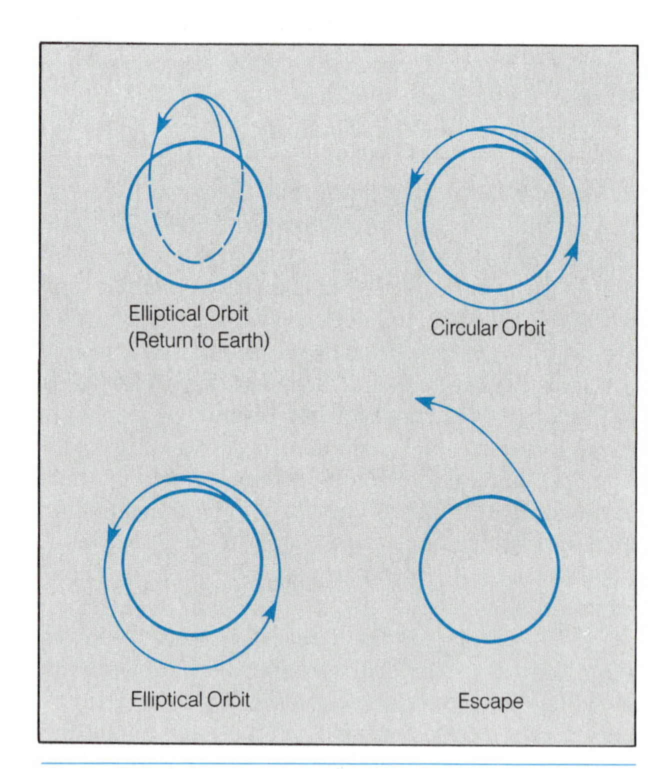

FIGURE 5.9 ORBITAL AND ESCAPE VELOCITIES. *A rocket launched with insufficient speed for circular orbit would tend to orbit the Earth's center in an ellipse, but would intersect the Earth's surface. Given the correct velocity it will follow a circular orbit. A somewhat larger velocity will place it in an elliptical orbit that does not intersect the Earth. If given enough velocity so that its kinetic energy is greater than its gravitational potential energy, however, it will escape entirely.*

If the kinetic energy is less that the potential energy, as it is for all the planets, then the orbit is an ellipse, as Kepler found. It is technically correct to say that a planet and the Sun orbit a common **center of mass** (Fig. 5.10), rather than saying that the planet orbits the Sun. The center of mass is a point in space between the two bodies where their masses are essentially balanced; more specifically, it is the point where the product of mass times distance from this point is equal for the two objects. Since the Sun is so much more massive than any of the planets, the center of mass for most Sun-planet pairs is always very close to the center of the Sun, so the Sun moves very little, and we do not easily see its orbital motion. It is true, however, that the Sun's position changes as it orbits the centers of mass established by its interaction with the planets, especially the most massive ones. In a double-star system, where the two masses are more nearly equal, it is easier to see that both stars orbit a point in space between them. Thus Kep-

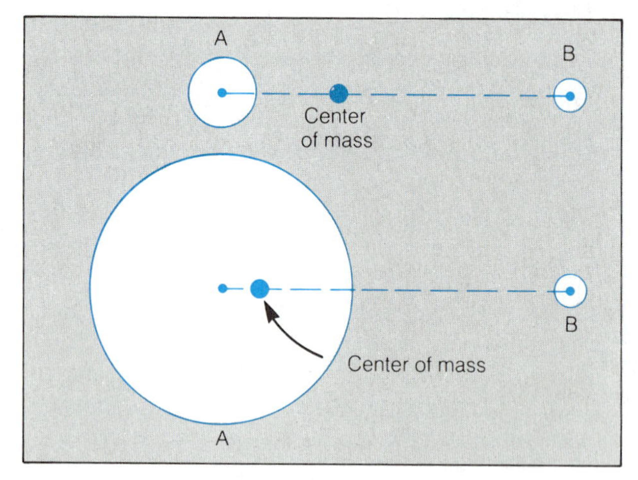

FIGURE 5.10 CENTER OF MASS. *The upper sketch depicts a double star where star A has twice the mass of star B, so the center of mass, about which the two stars orbit, is one-third of the way between the centers of the two stars. The lower sketch shows a case where star A has ten times the mass of star B, so the center of mass is very close to star A. The sun is so much more massive than any of the planets that the center of mass for any sun-planet pair is very near the center of the sun, so the sun's orbital motion is very slight.*

ler's first law as he stated it requires a slight modification:

> Each planet has an elliptical orbit, with the center of mass between it and the Sun at one focus.

The second law also can be restated in terms of Newton's mechanics. Any object that rotates or moves around some center has **angular momentum.** This is dependent on its mass, speed, and distance from the center of motion. In the simple case of an object in circular orbit, the angular momentum is the product mvr, where m is its mass, v its speed, and r its distance from the center of mass.

The total amount of angular momentum in a system is always constant. Because of this, a planet in an elliptical orbit must move faster when it is close to the center than when it is farther away, so that its velocity compensates for the changes in distance (Fig. 5.11). Thus, a planet moves faster in its orbit near **perihelion** (its point of closest approach to the Sun) than at **aphelion** (its point of greatest distance from the Sun). Kepler's second law really stated that

> Angular momentum in a two-body system is constant.

Kepler's third law was also revised by Newton, and this revision is especially important. Newton discovered that the relationship between period and semimajor axis depends on the masses of the two objects. Kepler could not have realized this, because the Sun is much more massive than any of the planets, so that the sum of the masses for any Sun-planet pair is always about the same. Kepler wrote his third law in the form $P^2 = a^3$, where P is the planet's period in years and a is the semimajor axis in astronomical units. Newton realized that the sum of the masses times the square of the period of a pair of bodies in orbit is proportional to the cube of the semimajor axis. In equation form, this says

$$(m_1 + m_2)P^2 = a^3,$$

where m_1 and m_2 represent the masses of the two bodies, for example the Sun and one of the planets. The masses must be expressed in terms of the Sun's mass in this equation. If we use other units, then additional numerical constants are required for the equation to be valid. For example, in the units used by astronomers, where masses are expressed in grams, the period is in units of seconds, and the semimajor axis is expressed in centimeters, the equation becomes

$$(m_1 + m_2)P^2 = \frac{4\pi^2}{G}a^3,$$

where G is the gravitational constant (the equation has the same form in the meter-kilogram-second system of units, but the value of G is different in that system).

The practical importance of Newton's revised version of Kepler's third law is that it becomes possible to use the law to determine the masses of distant objects. In any case, where the period and the semimajor axis of an orbiting object can be observed directly, the equation can be solved for the sum of the masses:

$$m_1 + m_2 = \frac{a^3}{P^2}.$$

For example, consider how Newton could have derived the mass of Saturn. Titan, the largest of Saturn's moons, has an orbital period of 15.945 days = 0.04365 years, and a semimajor axis of 1,222,000 km = 0.00816 AU. Substituting these values for P and a in the expression just cited leads to $m_1 + m_2 = 0.000285$ solar masses; in other words, the sum of the masses of Titan and Saturn is only 0.000285 times the mass of the Sun. Since Titan is very much smaller than Saturn, we can assume that its mass is negligible and

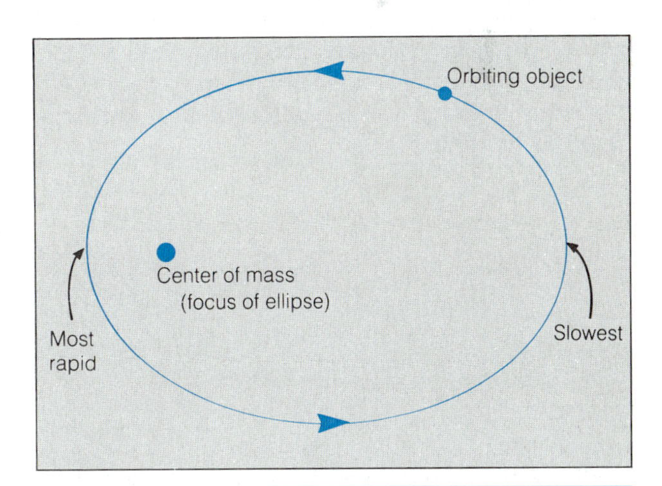

FIGURE 5.11 CONSERVATION OF ANGULAR MOMENTUM. *An object moves in an elliptical orbit with varying speed, and the product of its mass times its velocity times its distance from the center of mass (that is, its angular momentum) is constant. Kepler's second law of planetary motion is a rough statement of this fact.*

that we have found the mass of Saturn. (Compare this with the value in Appendix 7 for the mass of Saturn.)

It is possible to determine the masses of orbiting stars or even orbiting galaxies by the same technique. We will find occasion to use Kepler's third law in this manner throughout our study of astronomy.

ESCAPE VELOCITY AND GAS MOTIONS

The consideration of orbital motions in terms of kinetic and potential energy leads to the concept of an **escape velocity.** If an object in a gravitational field has greater kinetic than potential energy, it will escape the gravitational field entirely. To launch a rocket into space (that is, completely free of the Earth) requires giving it enough upward speed at launch so that its kinetic energy is greater than its potential energy caused by the Earth's gravitational attraction (see Fig. 5.9). It so happens that the speed required to accomplish this is the same for an object of any mass. In equation form it is

$$v_e = \sqrt{2\,GM/R},$$

where v_e is the escape speed, G is the gravitational constant, and M and R are the Earth's mass and radius, respectively. For the Earth this speed is 11.2 kilometers per second, or slightly more than 40,000 kilometers per hour. Escape speeds for all the planets are given in Table 5.3.

TABLE 5.3 ESCAPE SPEEDS

Body	Escape Speed (km/sec)
Earth	11.2
Sun	618
Moon	2.4
Mercury	4.3
Venus	10.3
Mars	5.0
Ceres (largest asteroid)	0.6
Jupiter	60.8
Saturn	36.6
Uranus	21.2
Neptune	23.5
Pluto	1.2

The particles in a gas such as the Earth's atmosphere move all the time, and the speed with which they do so is related to the temperature of the gas (Fig. 5.12). In a hot gas, the atoms and molecules move more rapidly than in a cool gas. This is why a heated gas expands; the individual particles move more rapidly and therefore exert greater pressure on their surroundings. In a strict sense, temperature can be defined in terms of molecular motions (this leads directly to the concept of **absolute zero,** the temperature at which all molecular activity ceases). We can speak of the kinetic energy of individual particles in a gas, or more realistically, the average kinetic energy, which in turn is based on the average particle speed. For any gas, the temperature is proportional to average kinetic energy.

In physics and astronomy, temperatures are usually expressed in terms of the absolute scale, in which zero is absolute zero, and the degrees, equal to one one-hundredth of the difference between the freezing and boiling points of water are called **Kelvins** (K). Water freezes at 273 K and boils at 373 K. Room temperature is about 295 K. For details, see Appendix 4.

When we discuss the planets, we will be concerned with the likelihood of specific gases escaping

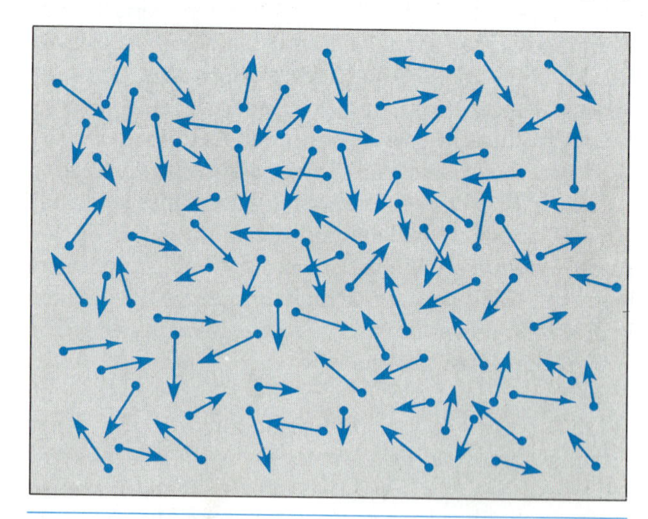

FIGURE 5.12 GAS MOTIONS. *The average kinetic energy of particles in a gas is proportional to the temperature. If the gas particles, either because of high temperature or low mass, exceed a planet's escape velocity, they can escape into space. In any gas, there is a range of particle velocities, and if the average velocity for a particular gas is as much as one-sixth of the escape velocity, it wil eventually escape altogether.*

into space, thus leaving the atmosphere devoid of those gases. We can see from this discussion that the probability that a given gas will escape a planet depends on the temperature of the atmosphere. Recall also that the kinetic energy of a moving particle depends on its mass as well as its speed; this means that in a gas where the particles have a uniform average kinetic energy, the more massive particles, on average, are moving more slowly than the light particles. Therefore, the lightest gases are the most likely to escape; their higher velocities are more likely to exceed the planet's escape velocity. We will find, for example, that some of the planets (including Earth) have lost their light gases such as hydrogen and helium while retaining the heavier ones. We will speak often of this concept in Section III, where we describe the solar system and the properties of the planets.

TIDAL FORCES

A number of important astronomical phenomena can now be understood in terms of gravitational forces. One is tides, which, as we will see, occur in many situations besides our earthly oceans.

We have seen that the gravitational force resulting from a distant body decreases with distance. This means that an object subjected to the gravitational pull of such a body feels a stronger pull on the side nearest that body, and a weaker pull elsewhere. For example, the side of the Earth facing the Moon feels the strongest attraction toward the Moon, and the side opposite the Moon feels the weakest force. The Earth is therefore subjected to a **differential gravitational force,** which tends to stretch it along the line toward the Moon. Of course the Sun exerts a similar stretching force on the Earth, but it is too distant to have as strong an effect as the Moon. (However, its total gravitational force on the Earth is much greater than that of the Moon.) A **tidal force,** as differential gravitational forces are called, depends on how close one body is to the other, because the key is how rapidly the gravitational force drops off over the diameter of the body subject to the tidal force, and it drops off most rapidly at close distances. (The tidal force, in contrast with the total gravitational force, drops off as the cube, not the square, of the distance between two bodies. This is why the Sun, as already mentioned, exerts a greater total gravitational force on the Earth than the Moon does, even though the Moon's tidal force on the Earth is greater. Thus the Moon, not the Sun, dominates tides on the Earth.)

The Earth is more or less a rigid body, so it does not stretch very much as a result of the differential gravitational force of the Moon. The Earth's oceans, however, are fluid, and are influenced strongly by the Moon's differential gravitational force. In Fig. 5.13, consider the bold arrows indicating the strength of the Moon's gravitational force on the Earth on the near side, the center of the Earth, and on the far side. The force on the near side is greater than on the center, so the difference between these forces is a net force toward the Moon, relative to the Earth's center, and this creates a bulge of ocean water on this side. The force on the far side of the Earth is *less* than that on the Earth's center, so the difference between these two forces amounts to a net force *away* from the Moon on the Earth's far side, relative to the center of the Earth, and another bulge of ocean water is created on that side. The *differential* forces are indicated by the thin arrows in Fig. 5.13. Thus, the Earth's oceans have two tidal bulges—one on the side facing the Moon, and one on the opposite side. These two bulges tend to maintain their alignment toward and away from the Moon's direction, so that as the Earth ro-

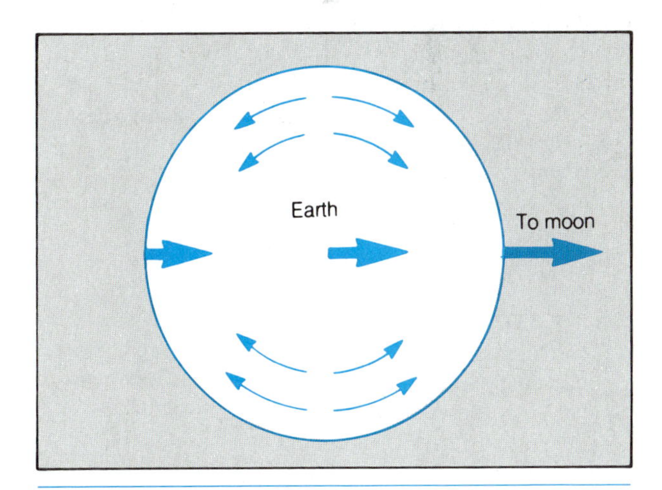

FIGURE 5.13 THE EARTH'S TIDES. *The differential gravitational force caused by the Moon tends to stretch the Earth (large arrows). Seawater at any given point on the Earth is subjected to a combination of vertical and horizontal forces, causing it to flow toward either the side of the Earth facing the Moon or the side opposite it (curved arrows).*

ASTRONOMICAL INSIGHT 5.2

RELATIVITY

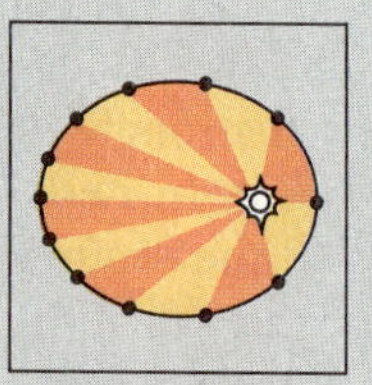

In the text, Newton's work was characterized as being nearly modern in the sense that his laws of motion are still thought today to be adequate representations of the manner in which forces and bodies interact. Although it is true that Newtonian mechanics is sufficient for most practical applications, there are circumstances in which this is not so. By the late 1800s, some difficulties with Newton's laws were becoming recognized, and in the first decades of this century these problems led Albert Einstein to develop his theories of special and general relativity.

The circumstances in which Newton's laws fail are those in which we deal with extremely high velocities (that is, velocities that are a significant fraction of the speed of light), in which case special relativity theory must be used; or when very strong gravitational fields are involved, when general relativity theory is required. Actually, relativity is always a more accurate representation than Newton's laws, but the two theories give virtually identical results except under the circumstances just cited.

Mathematically, both special and general relativity are too complex to be presented here in any detail. It is possible, however, to provide some intuitive description of what is involved. The dilemma that led Einstein to develop special relativity had to do with the speed of light. To see this, first let us consider a case involving lower speeds. Suppose a jet aircraft moving at 1,000 miles per hour fires a missile in the forward direction at a speed relative to the plane of 2,000 miles per hour. The speed of the missile with respect to the ground would then be the sum of the plane's speed and the speed with which the missile left the plane, that is, 3,000 miles per hour. This is the result expected from Newtonian mechanics.

Now imagine a train moving along a track, and suppose that it turns on its headlight. The light leaves the train at the normal speed of light relative to the train, and by analogy with the previous examples, we might expect that the speed of the light with respect to a bystander would be equal to the normal speed of light plus the speed of the train. For firm theoretical and experimental reasons, however, it is thought to be impossible for the normal speed of light to be exceeded. In other words, an observer on the train would actually measure the same speed for the light as would an observer standing by the track, in contradiction to the example of the plane and its missile. In Einstein's time, an actual measurement of this sort had not been made, but to him there was another reason for postulating that the speed of light should be the same for all observers: If this were not so, it would be implied that there was some fundamental difference between one frame of reference and others. There would be a "fundamental" frame in which light would have the speed required by theory, and all other frames would be somehow less fundamental. This would imply that there was a preferred reference frame in the universe, an idea contrary to the lessons learned long before, when it was realized that the Earth was not the center of the universe.

Einstein's solution to the problem involved a surprising postulate: that the rate of passage of time itself depends on the relative speed between observers. Let us think about the train example again. If we believe that the speed of light must be the same for someone on the train as for someone standing by the track, even though the train is moving, then we are forced to accept Einstein's postulate if both observers are to measure the same speed of light. For the two observers to measure the same speed for the light despite their motion with respect to each other, it is necessary for the time interval to be different for the two observers.

If the rate of time depends on the motion of the observers with respect to each other, there are a number of interesting consequences, some of which can be (and have been) tested experimentally. For example, people who travel at high speeds through space will return to earth younger than they would have been if they had stayed home, because time will not pass as rapidly for them. Even the sizes and masses of objects are affected: The length of an object that is moving at high speed will appear smaller to an observer at rest than to someone moving with the object, and its mass will be greater as measured by the person at rest.

Whereas special relativity states that the laws of physics are the same for observers moving at different velocities, general relativity addresses the situation for observers undergoing different accelerations. We have seen that a gravitational field such as the Earth's creates acceleration:

(continued on next page)

ASTRONOMICAL INSIGHT 5.2

RELATIVITY

(continued from previous page)
A falling object increases its speed with every second as it falls. Even without falling, an object is subject to a force because of the acceleration of gravity; we are all familiar with the fact that an object at the surface of the Earth has weight. Now suppose we are somewhere out in space, inside an enclosed spacecraft, and that the spacecraft is accelerating at a rate equivalent to the gravitational acceleration at the Earth's surface. Inside the spacecraft we would "feel" this acceleration, which would create a force (in the direction opposite the spacecraft's motion) exactly identical to the weight we would have on the surface of the Earth. There is no experimental way to tell the difference between the two situations.

Now suppose that the person in the accelerating spaceship fires a gun. The bullet, being "left behind" as the ship continues to increase its speed, will follow a curving path with respect to the spacecraft, falling toward the rear. This is identical to the curving path of a bullet fired at the surface of the Earth; as the bullet travels toward its target, it drops toward the ground because of gravity. Now suppose that the "bullet" is a beam of light. In the spaceship, the beam would curve away from the direction of the ship's acceleration, just as a material bullet would. For the two situations to be fully equivalent, this implies that a beam of light on Earth also would curve downward because of gravity. Thus, Einstein predicted that a gravitational field should bend light, and this was later confirmed experimentally. Since light always follows the shortest possible path between two points, Einstein characterized the effect of a gravitational field as a "bending" of space itself. Hence, the light could travel the shortest path in the gravitational field, but this path would look like a curve to an observer not subjected to the same gravitational field.

Today we speak of **space-time,** indicating that time itself is thought of as a dimension because of the role it plays in governing the measured properties of an object; we also speak of **curved space-time** because of the curvature that occurs in the presence of a gravitational field. Throughout this text, but especially in the later chapters, where we discuss situations involving very high velocities or extreme gravitational fields, these concepts of special and general relativity will come into play.

tates, at any given location on its surface we encounter high tides twice a day, with low tides in between. The separation between high tides is actually a bit longer than 12 hours, because the Moon moves along in its orbit while the Earth rotates.

It is interesting to consider what is happening to the Moon at the same time. It is subjected to a more intense differential gravitational force than the Earth, since the Earth is more massive than the Moon. Even though the Moon is a solid body, its shape is deformed by this force, and it has tidal bulges. The Moon is slightly elongated along the line toward the Earth. This has had drastic effects on the Moon's rotation, causing it to keep one side facing the Earth (Fig. 5.14). The original spin of the Moon is thought to have been much faster than it is today, but tidal forces have slowed the spin until the rotation period and the orbital period became equal. This situation, called **synchronous rotation,** also occurs for many other satellites in the solar system.

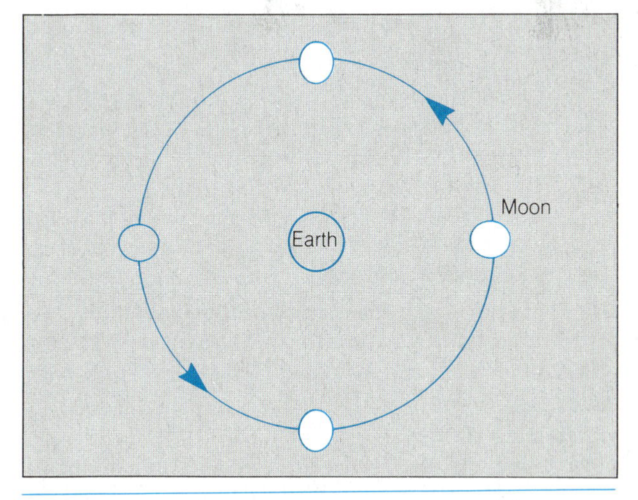

FIGURE 5.14 SYNCHRONOUS ROTATION OF THE MOON. *The Moon is subjected to a tidal force caused by the Earth that is strong enough to slightly deform the shape of the Moon. This sketch shows how the elongated (here greatly exaggerated) Moon keeps one bulge facing the Earth at all times.*

There are many other examples of tidal forces, both in the solar system and outside it. The satellites of the massive outer planets are subjected to severe tidal forces, and there are double-star systems in which the two stars are so close together that they are stretched into elongated shapes. We will discuss these in more detail later, along with galaxies that are affected by tidal forces, sometimes even tearing each other apart.

PERSPECTIVE

We have traced the development of astronomy to the point where we can now describe, in terms of a few simple laws of physics, the motions of the bodies in the solar system. These laws also provide a basis for understanding the motions of more distant astronomical objects. The law of universal gravitation will be invoked time and time again in our study of the solar system and the rest of the universe, because so many important phenomena are explained by it.

Newton's laws of motion and gravitation have stood the test of time rather well. Einstein's theory of general relativity, as we will learn in Chapters 24 and 32, may be viewed as a more complete description of gravity and its interaction with matter. For most situations on the Earth's surface and in space, however, Newton's laws are perfectly adequate.

We are now ready to discuss other laws of physics, particularly those which govern the emission and absorption of light.

SUMMARY

1. Isaac Newton, in the late 1600s, developed the laws of motion, the law of gravitation, and calculus, and made many important contributions to our knowledge of the nature of light and telescopes.
2. Newton's first law states that an object at rest or in a state of uniform motion tends to remain in that state; this is the principle of inertia previously formulated by Galileo.
3. The second law of motion states that when an object is accelerated by a force, the amount of acceleration is equal to the force divided by the mass of the object.
4. Newton's third law states that for every force, there is an equal and opposite force.
5. The law of universal gravitation states that any two objects in the universe attract each other with a force that is proportional to the product of their masses and inversely proportional to the square of the distance between them.
6. Newton's laws, along with the concepts of kinetic and potential energy and angular momentum, can be used to explain orbital motions.
7. Kepler's third law was modified by Newton to show that the relationship between the period and semimajor axis of an orbit is dependent on the sum of the masses of the two objects; this is an important tool for calculating the masses of distant objects.
8. Every body such as a planet or a star has an escape speed, the speed at which an object moving upward has more kinetic than potential energy and will therefore escape into space.
9. The average speed of particles in a gas is a function of the temperature of the gas and the mass of the particles; hence the probability that a given gas will escape a planetary atmosphere depends on these two quantities.
10. Differential gravitational forces are responsible for tides on the Earth and in the interiors of other planets and satellites.

REVIEW QUESTIONS

Thought Questions

1. Summarize the achievements of Newton in physics and astronomy.
2. Contrast the explanation Newton would have given for the fact that a moving object stops unless you keep pushing it with the explanation that Aristotle would have given.
3. Explain, in your own words, the difference between weight and mass.
4. Why does the law of inertia (Newton's first law) imply that there must be a force of attraction (rather than repulsion) between each planet and the Sun?
5. Explain the difference between acceleration and velocity, and between velocity and speed.

6. Explain how a book lying on a table can be stationary, even though it exerts a force on the table and the table exerts a force on it.
7. Explain why an astronaut in an orbiting spacecraft is "weightless." Is there really no force of gravity exerted on the astronaut?
8. Explain the difference between potential and kinetic energy, and between energy and power.
9. Why did Kepler not discover that his third law depends on the sum of the masses of the Sun and each planet?
10. Explain why the Earth's oceans have two high tides per day.

Problems

1. Convert your own sea-level weight into mass, in kilograms.
2. Suppose that you are in orbit building a space station, and you push a crate that has one-third the mass that you have, so that the crate floats off in one direction and you float away in the opposite direction. How does your acceleration compare with that of the crate?
3. Jupiter is about 5 times farther from the Sun than the Earth is, and has about 300 times the mass of the Earth. Compare the gravitational force between Jupiter and the Sun with that between the Earth and the Sun. Perform the same calculation for Saturn, which is about 10 times farther from the Sun than the Earth is, and has about 100 times the mass of the Earth.
4. How much less would you weigh on the summit of Mt. Everest, which is 29,000 feet (8.84 km) above sea level, if you weigh 130 lb at sea level? The radius of the Earth is 6,367 km.
5. What would you weigh on the surface of Titan, the large satellite of Saturn? (Use data from Appendix 7.)
6. Suppose you drop a rock from the edge of a high cliff. How fast is it falling after 3 seconds? Express your answer in feet per second and miles per hour.
7. How many ergs per second are emitted by a 100-watt light bulb?
8. Find the mass of Jupiter, in solar units, using Kepler's third law and data from Appendix 7 on the orbital period and semimajor axis of Callisto. Convert your answer from solar units to grams. How does your answer compare with the value given in Appendix 7?
9. Calculate the escape speed for Mars, and compare this with the escape speed for the Earth.
10. Convert a temperature of 55°F to the centigrade and absolute temperature scales. Convert a temperature of 77 K to the centigrade and Fahrenheit scales.

ADDITIONAL READINGS

Authoritative and thorough information on Newton, as well as many additional references, can be found in the following works.

Christianson, Gale E. 1987. Newton's Principia: A retrospective. *Sky and Telescope* 74(1):8.
Cohen, I. B. 1960. Newton in light of recent scholarship. *Isis* 51:489.
——— 1974. Newton. In *Dictionary of scientific biography*, vol. 10, ed. C. G. Gillispie. New York: Scribner's.
Whiteside, D. T. 1962. The expanding world of Newtonian research. *History of Science* 1:15.

There are many books on conceptual understanding of physics in which more detailed explanations of the principles discussed in this chapter may be found. Many are used as textbooks in introductory physics courses, and should be easy to locate in a bookstore or library. A particularly good example is *Conceptual Physics,* by P. Hewitt 1981 (Little, Brown: Boston).

CHAPTER 6

THE NATURE OF LIGHT

Refraction of light by water droplets. (Photo by Douglas Johnson)

CHAPTER PREVIEW

Some of the tools for unlocking the secrets of the universe became available with the publication of Newton's *Principia* in the 1680s, but others had to wait 200 years or more to be discovered. The laws of motion allow astronomers to understand how the heavenly bodies move, and were of fundamental importance in unraveling the clockwork mechanism of the solar system. To understand the essential nature of a distant object, however, to learn what it is made of and what its physical state is, requires an understanding of light and how it is emitted and absorbed. The only information we can obtain on the nature of a distant object is conveyed by the light from it. Fortunately, an enormous amount of information is there, if we know how to dig it out.

THE ELECTROMAGNETIC SPECTRUM

One characteristic of light is that it acts as a wave. It is possible to think of light as passing through space like ripples on a pond (although, as we will discuss shortly, the picture is actually somewhat more complicated than that). The distance from one wavecrest to the next, called the **wavelength** (Fig. 6.1), distinguishes one color from another. Red light, for example, has a longer wavelength than blue light. It is possible to spread out the colors in order of wavelength, using a prism to obtain the traditional rainbow. Newton was the first to discover that sunlight contains all the colors, and he did so by performing experiments with a prism (Fig. 6.2). Whenever light is spread out by wavelength, the result is called a **spectrum;** more technically, a spectrum is the arrangement of light from an object according to wavelength. The science of analyzing spectra is called **spectroscopy,** and will be discussed at some length later in this chapter.

Let us consider for a moment what lies beyond red at one end of the spectrum or beyond violet at

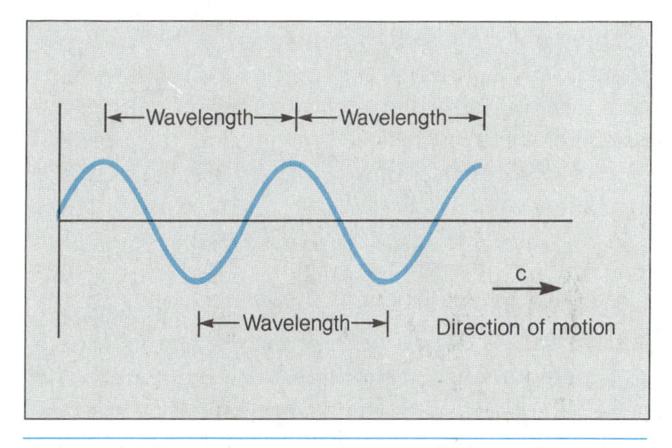

FIGURE 6.1 PROPERTIES OF A WAVE. *Light can be envisioned as a wave moving through space at a constant speed, usually designated* c. *The distance from one wavecrest to the next is the wavelength, often denoted by the Greek letter lambda,* λ. *The frequency* f *is the number of wavecrests to pass a fixed point per second and is related to the wavelength and the speed of light by* $f = c/\lambda$.

FIGURE 6.2 PRISM DISPERSING LIGHT. (Runk/Schoenberger, Grant Heilman Photography)

the other end. By the mid-1800s, experiments had been done to demonstrate that there is invisible radiation from the Sun at both ends of the spectrum. At long wavelengths, beyond red, is **infrared** radiation, and at short wavelengths is **ultraviolet** radiation. The spectrum continues in both directions, virtually without limit. Going toward long wavelengths, after infrared light, are microwave and radio waves; and toward short wavelengths, after ultraviolet, come X-rays and then gamma rays (γ-rays). All these kinds of radiation are just different forms of light, distinguished only by their wavelengths, and together they form the **electromagnetic spectrum** (Fig. 6.3). Electromagnetic radiation is a general term for all forms of light, whether it is visible, X-rays, radio, or anything else.

The reason for the name *electromagnetic radiation* is that the wave motion associated with this radiation consists of alternating electric and magnetic fields (Fig. 6.4). These fields propagate through space (or a medium) as they alternate, with the electric and magnetic fields always lying in planes that are perpendicular to each other. The speed of propagation in a vacuum is always the same (almost exactly 300,000 km/sec, or about 186,000 miles per second), but is slightly slower in a medium such as air or glass. The electric and magnetic fields are said to be *in phase* with each other, meaning that both reach their maximum and minimum values at the same time as the wave propagates, with the relative orientation shown in Fig. 6.4.

The range of wavelengths from one end of the electromagnetic spectrum to the other is immense (see Table 6.1). Visible light has wavelengths ranging from 0.00004 to 0.00007 centimeters (cm). A special unit called the **Ångstrom** (Å) is used to measure light, defined such that 1 cm = 100,000,000 Å, or 1 Å = 0.00000001 cm = 10^{-8} cm. Thus, visible light lies between 4,000 Å and 7,000 Å in wavelength.

A variety of other special units are used by astronomers in referring to electromagnetic radiation of different types, but to reduce confusion, we will use only Ångstroms and centimeters. Even this is perhaps a bit complicated, but the usage is so standard that the student of introductory astronomy should be comfortable with both units.

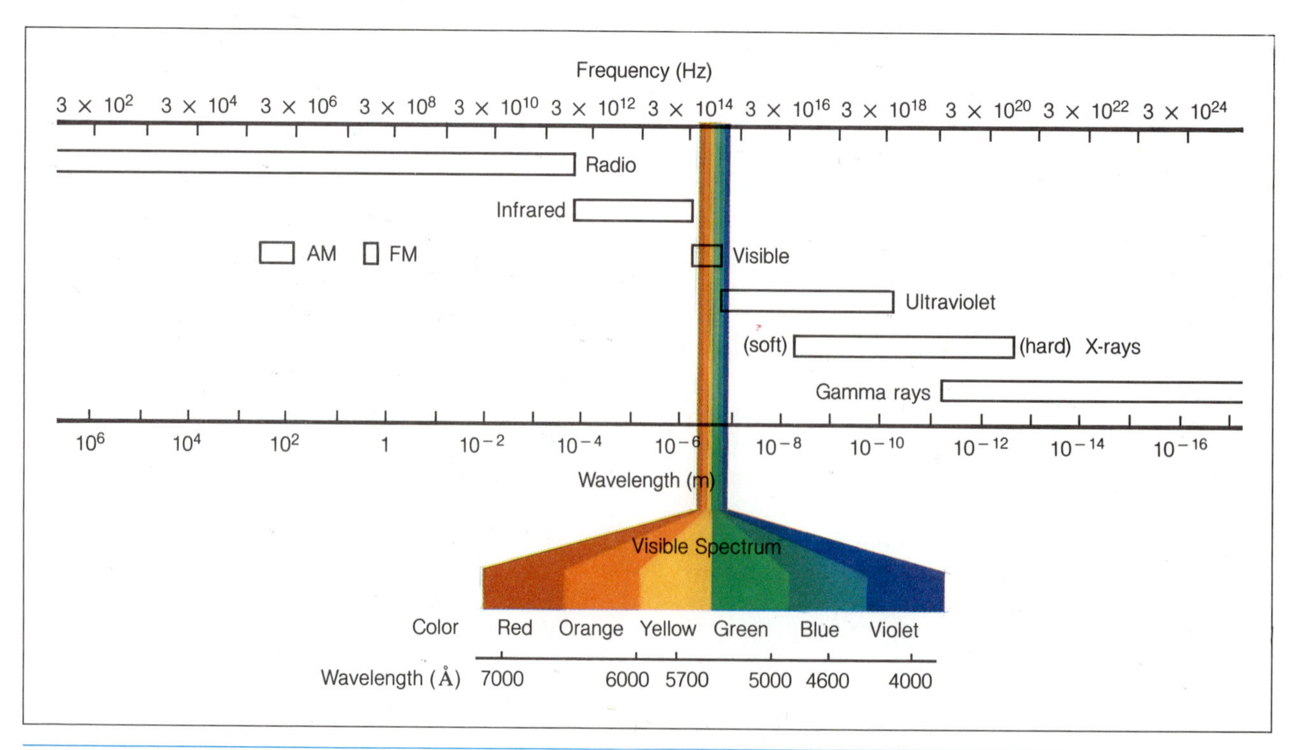

FIGURE 6.3 THE ELECTROMAGNETIC SPECTRUM. *All of the indicated forms of radiation are identical except for wavelength and frequency.*

Infrared light has wavelengths between about 7,000 and a few million Ångstroms; that is, between 7×10^{-5} cm and $2\text{–}3 \times 10^{-2}$ cm. Microwave radiation (which includes radar wavelengths) lies roughly between 0.1 and 50 cm, with no well-defined boundary separating this region from radio, which simply includes all longer wavelengths, up to many meters or even kilometers.

At the other end of the spectrum, ultraviolet light is usually considered to lie between 100 and 4,000 Å, whereas X rays are in the range of 1 to 100 Å, anything shorter than that being considered γ rays.

Thus the entire electromagnetic spectrum has a wavelength range from less than 1 Å to many kilometers; that is, from 10^{-16} to 10^{6} cm. In principle, it extends even farther at both ends, but few physical processes are capable of producing such radiation. The portion of the spectrum to which the human eye responds is extremely limited compared with the full range. The eye undoubtedly has evolved to be sensitive to just this region for a combination of reasons: (1) the Sun emits most strongly in visible wavelengths; (2) the Earth's atmosphere blocks out many wavelengths of radiation, but visible light lies in one of the wavelength regions that can pass through to the ground. Therefore, it was advantageous for creatures dwelling on the Earth's surface to develop the ability to see radiation at these particular wavelengths.

The concept of **frequency** is often used as an alternative to wavelength in characterizing electromagnetic waves. The frequency is the number of waves per second that pass a fixed point, and it is determined by the wavelength and the speed with which the waves move. The speed of light, usually designated c, is constant, and the frequency of light (f) with wavelength λ is $f = c/\lambda$. The standard unit for measuring frequency is the **hertz** (Hz), 1 Hz being equal to 1 wave per second. The frequency of a photon of visible light with a wavelength of 5,000 Å is 6×10^{14} Hz. For a radio photon in the AM part of the spectrum, frequencies are typically around 10^{6} Hz, and at the other extreme, γ-rays have frequencies around 10^{19} Hz.

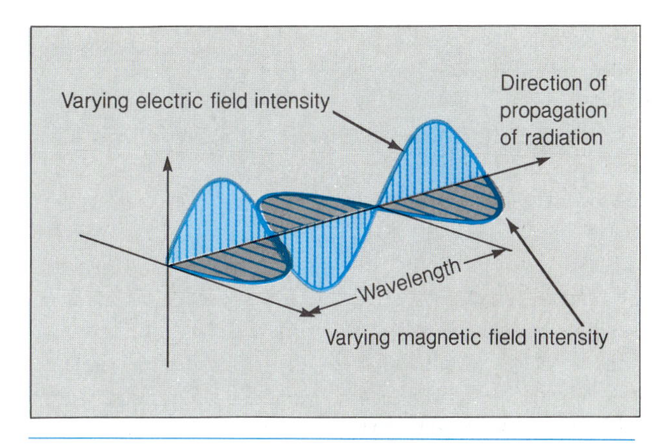

FIGURE 6.4 AN ELECTROMAGNETIC WAVE. *This illustrates how the electric and magnetic fields oscillate in an electromagnetic wave. The directions of the fields and the direction of the wave travel always have the relative orientations shown here.*

TABLE 6.1. WAVELENGTHS AND FREQUENCIES OF ELECTROMAGNETIC RADIATION

Type of Radiation	Wavelength Range (cm)*	Frequency Range (per second)
Gamma rays	$< 10^{-8}$	$> 3 \times 10^{19}$
X-rays	$1 - 200 \times 10^{-8}$	$1.5 \times 10^{16} - 3 \times 10^{18}$
Extreme ultraviolet	$200 - 900 \times 10^{-8}$	$3.3 \times 10^{15} - 1.5 \times 10^{16}$
Ultraviolet	$900 - 4{,}000 \times 10^{-8}$	$7.5 \times 10^{14} - 3.3 \times 10^{15}$
Visible	$4{,}000 - 7{,}000 \times 10^{-8}$	$4.3 \times 10^{14} - 7.5 \times 10^{14}$
Near-infrared	$0.7 - 20 \times 10^{-4}$	$1.5 \times 10^{13} - 4.3 \times 10^{14}$
Far-infrared	$20 - 100 \times 10^{-4}$	$3.0 \times 10^{12} - 1.5 \times 10^{13}$
Radio	> 0.01	$< 3 \times 10^{12}$
(Radar)	(2 – 20)	$(1.5 - 15 \times 10^{9})$
(FM radio)	(250 – 350)	$(85 - 120 \times 10^{6})$
(AM radio)	(18,000 – 38,000)	$(800 - 1{,}600 \times 10^{3})$

*All wavelengths are given in centimeters; recall that 1×10^{-8} cm = 1 Ångstrom.

ASTRONOMICAL INSIGHT 6.1

THE PERCEPTION OF LIGHT

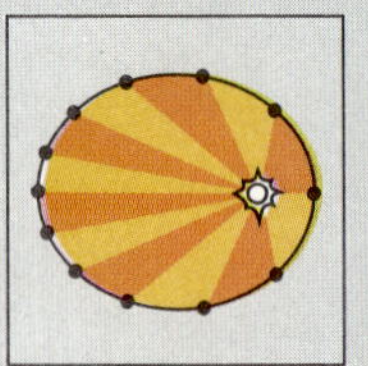

Newton's experiments in optics are commonly thought to have marked the beginning of scientific inquiry into the nature of light, but in fact Newton was preceded in this area by many scientists and philosophers. Much of the early research, and even, to some extent, Newton's work, did not deal with the physical properties of light, but rather was aimed at understanding how the human eye senses light. The ancient Greeks, the earliest philosophers who studied the nature of light, did not recognize the distinction between light and the act of seeing it.

The Greeks studied the principles of vision in much the same way they studied astronomy: They performed very few experiments or even systematic observations, choosing instead to unravel the secrets of the universe by reason and logic. The prevalent concept, described by such notable figures as Hippocrates and Aristotle, was that the eye somehow emitted rays or beams with which it sensed things. This was analogous to the sense of touch, where a hand is extended to feel objects. It was known that objects appeared distorted or bent when viewed under water, but this was attributed to bending of the rays that came from the eye, not to effects on something coming from the object being viewed. It would have been possible for the Greeks to experiment with optics by using natural crystals or eyes removed from dead animals, but no such experimentation was done.

In a definitive summary of all that was known about vision, the Greek scientist Galen, in the second century A.D., wrote of an "animal spirit" that flowed from the brain along the optic nerve to the eye, where it was converted into a "visual spirit" in the retina, a membrane covering the rear interior portion of the eyeball. The lens, in the front of the eyeball, was thought to be responsible for sending out the beam with which the external world was perceived.

In analogy with the astronomical work of Ptolemy, the concepts of the nature of light and vision that were summarized by Galen were accepted as doctrine for some fifteen centuries to come, until the time of the Renaissance. Some of the ancient Greek ideas, such as the notion that the optic nerve was hollow so that the animal spirit could flow along it from the brain, were especially persistent. Even Leonardo da Vinci, who in the late fifteenth and early sixteenth centuries carried out pioneering work in anatomy, accepted much of the old thinking. Da Vinci took a more modern view in that he believed that light rays passing from an object to the eye play an important role, but he was unwilling to entirely give up the idea that other rays emanate outward from the eye.

In the early seventeenth century, none other than Johannes Kepler and the great French philosopher René Descartes developed an understanding of refraction and the formation of an image by a lens, and both correctly viewed vision as the process of image formation on the retina by the lens in the front of the eyeball. Thus, when Christian Huygens and then Newton performed their experiments later in the seventeenth century, it was already well established that seeing was accomplished by sensing something that passes from the object to the eye.

Both men were more concerned with the nature of this radiation than with the properties of the eye, although Newton did discuss the perception of color, which he realized to be entirely a function of the eye. This is an important concept: the rays of light are not colored, nor, technically, is the object being viewed. It is the eye that senses wavelength differences and the brain that interprets them as colors. Light rays consist simply of alternating electric and magnetic fields.

Other advances in understanding the physical properties of light, made by scientists who followed Newton, are described in the text. Research has also been aimed at learning how the human eye perceives light. We now know that the eye has a logarithmic response, meaning that it does not perceive the true range of brightnesses that may be displayed by a series of objects, but instead perceives a lesser range. This point is explained more fully in Chapter 19, in the discussion of the stellar magnitude system.

Entire books have been written on the complex interplay between what the brain perceives and physical reality, and, of course, this interplay is important in astronomy. To a large extent, modern observational astronomy is the science of recognizing and bypassing the limitations or distortions created by our natural light-gathering system.

Particle or Wave?

So far we have discussed light only as a wave, but there are situations in which it acts more like a stream of particles. Newton developed a corpuscular theory of radiation in which he assumed that light consisted of particles. But at the same time other experimenters, including most notably Huygens, performed experiments showing light to have definite wave characteristics.

The fact that light bends around sharp corners, in a process called **diffraction** (which also affects water waves), is evidence for a wave nature, as is the fact that light waves can interfere with each other if their troughs and crests match up in such a way as to cancel each other out. On the other hand, it has been found that light can carry energy only in specific amounts, as though it came in individual packets, and that it can travel in a total vacuum, with no medium to transmit it.

Out of all the seemingly contradictory evidence has developed the concept of the **photon.** A photon is thought of as a particle of light that has a wavelength associated with it. The wavelength and the amount of energy contained in the photon are closely linked; in general terms, the longer the wavelength, the lower the energy. The energy can be expressed mathematically as $E = hc/\lambda$, where h is called the Planck constant, c is the speed of light, and λ is the wavelength. It is important to understand that a photon carries a precise amount of energy, not an arbitrary or random quantity, and that when light strikes a surface, this energy arrives in discrete bundles, like bullets, rather than in a steady stream. When a photon is absorbed, this energy can be converted into other forms, such as heat.

From this discussion, it may be seen that short-wavelength radiation, such as ultraviolet light, X-rays, and γ-rays, has high energy associated with it, whereas longer-wavelength photons have low energies. It is extremely important to remember the relationship between wavelength and energy in the discussions of the emission and absorption of light later in this chapter.

Polarization

The light from a typical source, such as a light bulb or a distant star, consists of vast numbers of photons, each consisting of traveling electric and magnetic fields. Each photon has a characteristic orientation as it travels through space; that is, the planes in which its electric and magnetic fields oscillate remain fixed as the photon travels. Normally, the orientations in a stream of photons from a source are random, but in some circumstances they are not. When light is **polarized,** all the photons tend to have the same alignment of their electric and magnetic planes.

Light can become polarized in many different ways. When light from a source is reflected from a flat surface, usually one orientation is favored, and the reflected light is at least partially polarized. Perfect polarization, in which all the photons have precisely the same orientation, is very rare. Some kinds of stars are embedded in disks of gas that cause the emitted light to be partially polarized. Elongated grains of interstellar dust can partially polarize the light traveling through the galaxy from a star.

It is possible to determine whether the light from a source is polarized by observing it with a polarizing filter. This is a piece of semitransparent glass or plastic that allows only light with a certain orientation to pass through. (Polarizing sunglasses use such material to screen out reflected light, thus reducing glare.) If the filter is rotated, the greatest intensity of light will be seen through it when its orientation is the same as that of the polarized light from the source. Thus, if the brightness of a source is seen to vary when a polarizing filter is rotated, we know that the source is polarized.

Continuous Radiation and Spectral Lines

Newton, in examining the spectrum of light obtained by passing sunlight through a prism, saw only the smooth spread of light over all the colors. Later observers, however, notably W. H. Wollaston and Joseph Fraunhofer, were able to see dark lines across the Sun's spectrum at certain fixed positions (Figs. 6.5 and 6.6). Fraunhofer noted more than 600 of these lines, and in 1817 published a catalogue of many of the stronger ones. (Today these features are still called **Fraunhofer lines.**)

It must be emphasized that the position of each of these lines corresponds to a specific wavelength of light. Remember that the spectrum is a spread of light

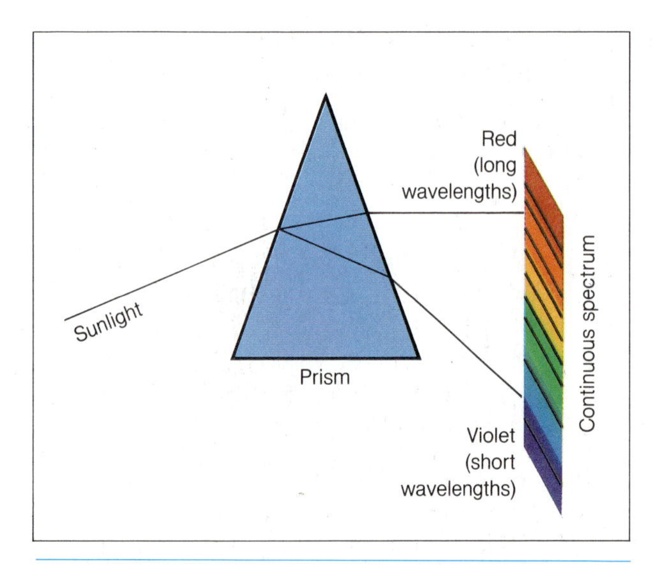

FIGURE 6.5 CONTINUOUS SPECTRUM AND SPECTRAL LINES. *When sunlight is dispersed by a prism, the light forms a smooth rainbow of continuous radiation, gradually merging from one color into the next, with a maximum intensity in the yellow portion of the spectrum. Superimposed on this continuous spectrum are numerous spectral lines, wavelengths where little or no light is emitted by the Sun.*

according to wavelength, and that each position in the spectrum corresponds to a certain wavelength of light. The fact that the lines are dark means that somehow the Sun emits less light at these particular wavelengths than at other wavelengths on either side.

In analyzing the light from a distant object such as a star, an astronomer either may focus attention on these lines, from which certain types of information can be derived, or may ignore the lines and concentrate instead on the overall distribution of the brightness of light as it varies with wavelength. Here we will make a distinction between **continuous radiation,** as the overall distribution is called, and spectral lines (Fig. 6.5).

Continuous Radiation

Stars are glowing objects with a number of properties that can be described by a few simple laws of continuous radiation. Some of these laws were discovered in the mid-1800s through laboratory experiments by people such as Wien, Stefan, Boltzmann, and Rayleigh, and in the 1920s were placed in proper theoretical framework by Bohr and Planck.

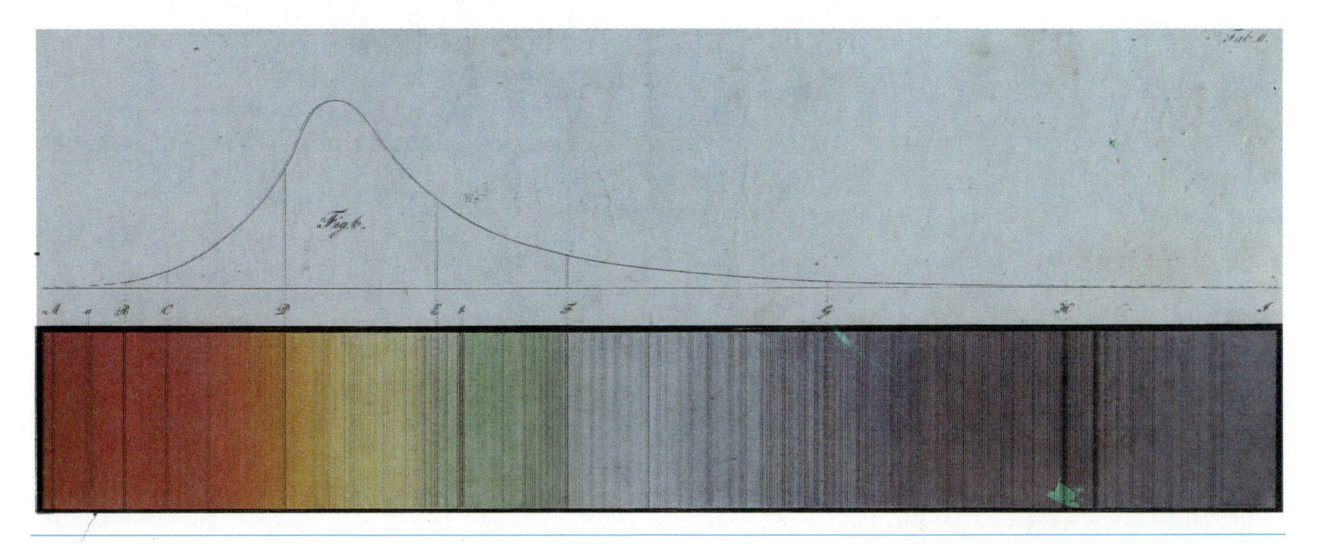

FIGURE 6.6 LINES IN THE SUN'S SPECTRUM. *This is a drawing of the solar spectrum made by Josef Fraunhofer, the first to systematically study the dark lines in the spectrum (now called Fraunhofer lines). The colors illustrate roughly the human-eye response to the wavelengths in the spectrum. Above is a graph showing the relative intensity of sunlight as a function of wavelength.* (Deutsches Museum, Munich)

ASTRONOMICAL INSIGHT 6.2

THE MEASUREMENT OF WAVELENGTHS

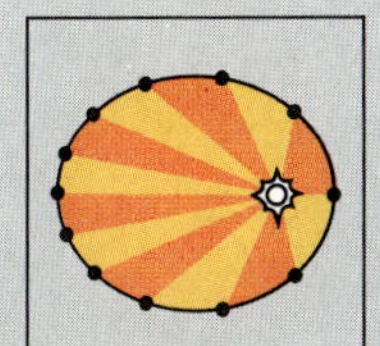

We have seen that visible light has wavelengths that are exceedingly short by any ordinary standards, lying in the range from about 0.00004 to 0.00007 cm. You may have wondered how such small wavelengths are measured. Surprisingly, this was first done nearly 200 years ago, in 1801, in a famous experiment conducted by the English physicist Thomas Young.

Young's experiment relied on the fact that light has the properties of waves. (The experiment actually demonstrated this, and is still considered one of the best arguments for a wave nature of light). When waves run into each other, they **interfere,** meaning that they interact in such a way that the magnitude of the waves is increased if the peaks and valleys match (this is **constructive interference**), or decreased if the peaks of one wave match the valleys of the other (**destructive interference**). Young showed that light waves undergo interference, thus demonstrating their wave nature, and in the process established a method for measuring their wavelengths.

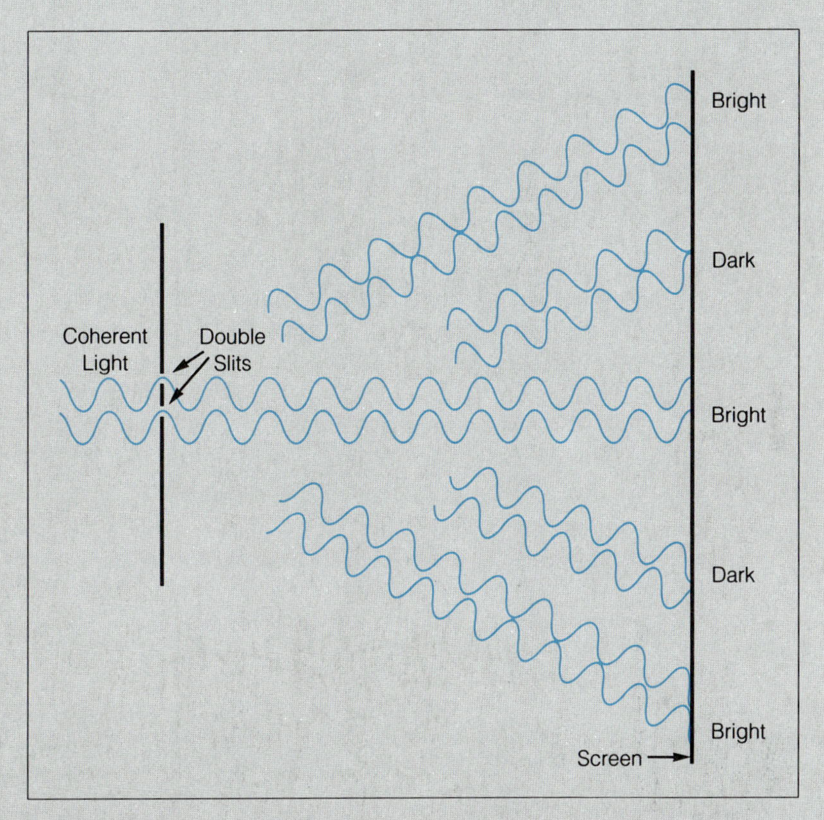

Young allowed light to pass through a pair of closely-spaced parallel slits (see figure), so that a pair of light beams emerged from the slits. What happens next is easiest to see if you imagine that the light all has the same wavelength. We know that the light in the two beams could interfere, depending on whether the troughs and valleys matched or not. If the light emerging from the slits is *in phase,* meaning that the troughs and peaks in the two beams match, then there will be constructive interference as the two beams go straight ahead from the slits. Thus, on a screen placed some distance behind the slits, a bright line is seen directly behind the slits.

Now imagine that instead of two simple beams of light going straight ahead, light actually emerges from the slits in all directions (this is due to another wave property called **diffraction**). Some goes straight ahead, but some emerges in other directions, creating a diffuse glow that covers much of the screen that is placed beyond the slits. We may now view it as though there were many pairs of beams going in many directions as they emerge from the two slits. The pair that travels straight through to the screen interferes constructively as before, combining to create a bright strip on the screen directly behind the slits.

But now consider a pair of beams that emerges at an angle. One of the beams will travel a slightly longer path to the screen than the other,

ASTRONOMICAL INSIGHT 6.2

THE MEASUREMENT OF WAVELENGTHS

(continued from previous page)
because of the separation between the two slits. Thus, the waves in the two beams are shifted with respect to each other, and the amount of shift will determine whether they interfere constructively or destructively. If one beam is shifted by half a wavelength with respect to the other, then the peaks from one will match the troughs of the other, and destructive interference occurs. There is therefore a dark strip at the location on the screen where this pair of beams would have reached it. On the other hand, if one beam is shifted with respect to the other by a distance equal to the wavelength (or a multiple of the wavelength), then the two beams will interfere constructively, creating a bright strip on the screen. The amount of shift between the two beams is related to the angle at which they emerge from the slits; as the angle changes, the shift between the beams changes, first reaching half of the wavelength, then the full wavelength, then one-and-a-half times the wavelength, then twice the wavelength, and so. Thus, interference between the two beams creates a pattern of alternating bright and dark strips on the screen.

The key to measuring the wavelength of the light is that the separation of these bright and dark strips depends on the wavelength of the light. The longer the wavelength, the greater the shift between beams that is required in order to change from constructive to destructive interference, and the greater the separation between adjacent bright and dark strips on the screen. There are details that we have overlooked in this discussion, but you should see that measuring the wavelength of the light in this experiment essentially consists of measuring the separation of the light and dark strips on the screen.

If white light, consisting of all the colors, is used in the experiment, the bright and dark strips become broadened and less clearly separated. This is because different wavelengths of light will create the bright and dark strips at different locations on the screen. Each bright strip will now look like a rainbow because of this, as different colors of light are brightest at different positions. Thus, a pair of slits creates a **spectrum** (actually, a series of spectra), consisting of light that is spread out according to wavelength. Instead of just a pair of slits, a series of many parallel slits can be used, with the same result.

If instead of an array of slits we use a reflecting surface that has many parallel grooves in it, then the reflected light will show the same separation of colors, because of shifts that occur in beams of light reflected from the edges and the bottoms of the grooves. This is how a **diffraction grating** works (see Fig. 7.11).

That an experiment devised as long ago as 1801 is still regarded as a modern tool for analyzing light is quite remarkable. Equally remarkable is that quantities as small as the wavelength of visible light were measured so long before there was any understanding of the processes by which light is emitted and absorbed.

Any object with a temperature above absolute zero emits radiation over a broad range of wavelengths, simply because it has a temperature (Fig. 6.7). Thus, not only stars, but also such commonplace objects as a human body or the wall of a room emit radiation. The continuous radiation produced because of an object's temperature is called **thermal radiation.** (We will discuss only this type of continuous radiation here; in other chapters, such as 13 and 31, we will examine nonthermal sources of radiation.)

For glowing objects such as stars, thermal radiation is emitted over a broad range of wavelengths, with a peak in intensity at some particular wavelength. (Technically, in fact, at least some radiation is emitted at all wavelengths.) For the Sun, this peak is near a wavelength of 5,500 Å in the green portion of the spectrum, but to our eyes the Sun appears yellow because what we see as green corresponds only to a very narrow range of wavelengths. However, the Sun emits almost as strongly over the much broader range of wavelengths that we see as yellow.

A simple relationship between the wavelength of maximum emission and the temperature of an object was discovered in 1893 by W. Wien and is now

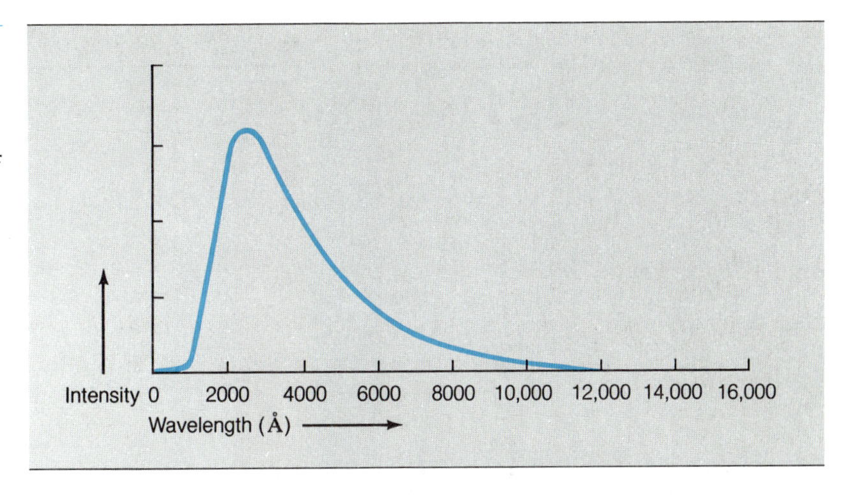

FIGURE 6.7 AN INTENSITY PLOT OF A CONTINUOUS SPECTRUM. *This kind of diagram shows graphically how the brightness of a glowing object varies with wavelength. The curve shown roughly represents a star of surface temperature 10,000 K, whose continuous radiation peaks near 3000Å in the ultraviolet.*

referred to as **Wien's law.** The law states that the wavelength of maximum emission is inversely proportional to the absolute temperature; that is, the hotter an object is, the shorter the wavelength of peak emission (Fig. 6.8). This explains the variety of stellar colors as being a result of a range in stellar temperatures (Fig. 6.9). A hot star emits most of its radiation at relatively short wavelengths and thus appears bluish, whereas a cool star emits most strongly at longer wavelengths and appears reddish. The sun is intermediate in temperature and color.

When speaking of the colors of stars, we must keep in mind that a star emits light over a broad range of wavelengths, so we do not have pure red or pure blue stars. Our eyes receive light of all colors, and stars therefore are all essentially white. Our impression of color arises from the fact that there is a wavelength (given by Wien's law) at which a star emits more strongly than at other wavelengths. It is not as though the star emits only at that wavelength.

Wien's law applies as well to radiation beyond the visible range. The hottest stars actually emit most strongly in ultraviolet wavelengths, and very cool ones emit mostly in the infrared. Indeed, for every object with a temperature above absolute zero, the wavelength of maximum radiation is dictated by the temperature of the object. The human body, as well as all objects in the normal room-temperature environment, radiates in far-infrared wavelengths. Imagine the problems this creates for infrared astronomers, whose telescopes glow at the very wavelengths they are trying to observe! (We will describe the solution to this problem in Chapter 7.)

A second property of glowing objects, known as either **Stefan's law** or the **Stefan-Boltzmann law,** has to do with the total amount of energy emitted per second over all wavelengths and how this total energy is related to the temperature of an object. This law says: The total energy radiated per square centimeter of surface area is proportional to the fourth power of the temperature. This shows that the total energy emitted is very sensitive to the temperature; if we change the temperature a little, we change the energy a lot. If, for example, we double the temperature of an object (such as the electric burner on our stove), we increase the total energy per square centimeter it radiates by $2^4 = 2 \times 2 \times 2 \times 2 = 16$. If one star is 3 times hotter than another, it emits $3^4 = 3 \times 3 \times 3 \times 3 = 81$ times more total energy per square centimeter of surface area.

Notice that we have been careful to express this law in terms of the surface area. The total energy emitted by an object per second, which astronomers call the **luminosity** of an object, depends also on how much surface area it has. If we talk of stars or other spherical objects, which have a surface area of $4\pi R^2$, where R is the radius, we can say that the total energy emitted is proportional to the fourth power of the temperature and to the square of the radius. This can be illustrated by considering two stars, one that is twice as hot but has only half the radius of the other. The hotter star emits $2^4 = 16$ times more energy per square centimeter of surface, but has only $(1/2)^2 = (1/4)$ as much surface area, hence it is $16 \times (1/4) = 4$ times brighter overall. If, on the other hand, this star were twice as hot and 3 times as large in radius

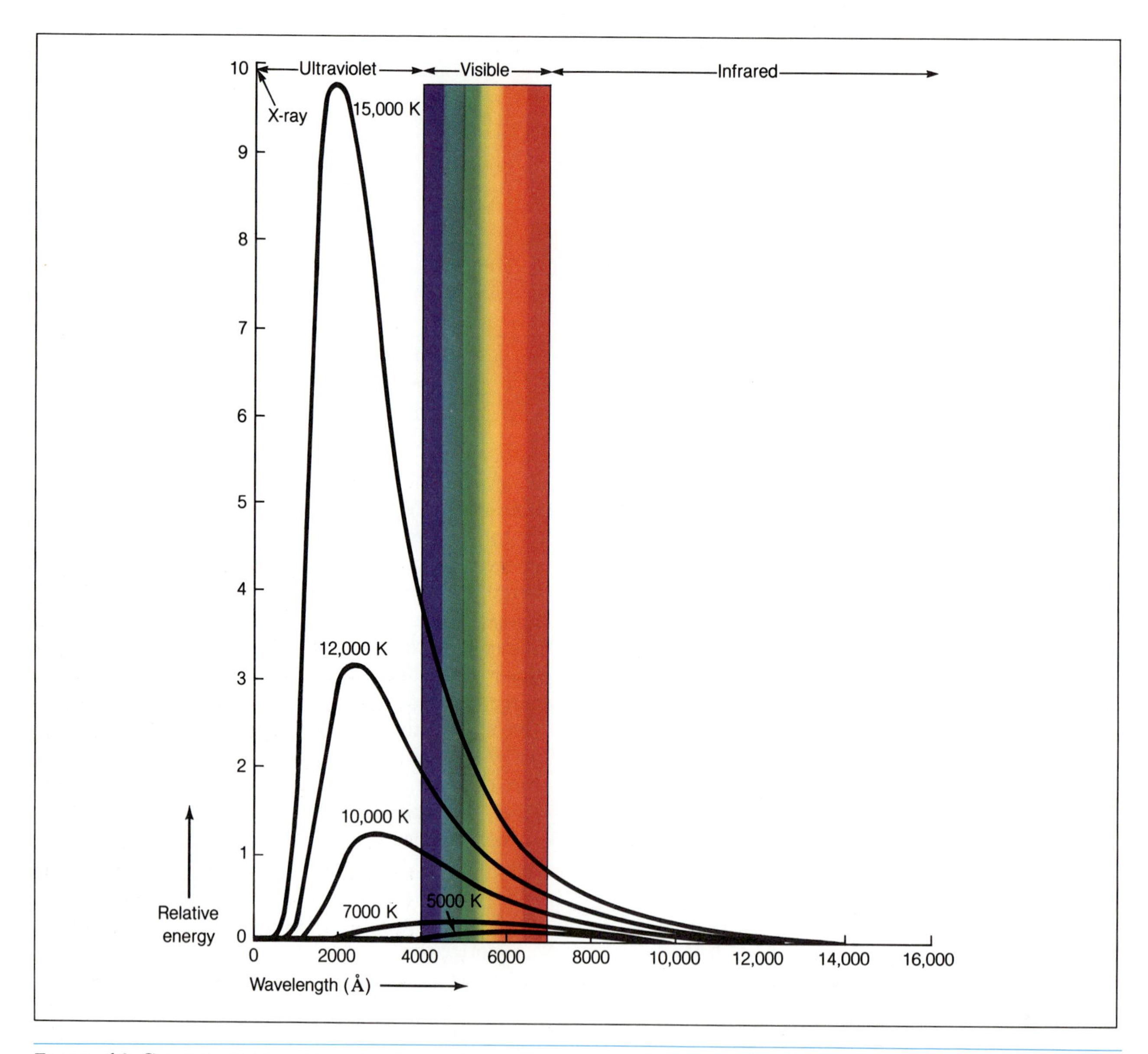

FIGURE 6.8 CONTINUOUS SPECTRA FOR OBJECTS OF DIFFERENT TEMPERATURES. *This diagram illustrates Wien's law, which says that the wavelength of maximum emission is inversely proportional to the temperature (on the absolute scale).*

as the other, it would be $2^4 \times 3^2 = 144$ times more luminous.

Both Wien's law and the Stefan-Boltzmann law were first found experimentally, much in the same manner as Kepler's discovery of the laws of planetary motion. In the case of planetary motions, it remained for Newton to find the underlying reasons for the laws, and he was able to derive them strictly on a theoretical basis. Analogously, Max Planck, the great German physicist who was active early in the twentieth century, found a theoretical understanding of thermal emission and was able to derive Wien's law and the Stefan-Boltzmann law from purely theoretical considerations. The basis of Planck's new understanding was the **quantum** nature of light—the fact, already discussed, that light has a particle nature and carries only discrete, fixed amounts of energy.

Finally, in addition to taking into account the temperature and size of a glowing object, we must consider the effect of its distance. So far we have

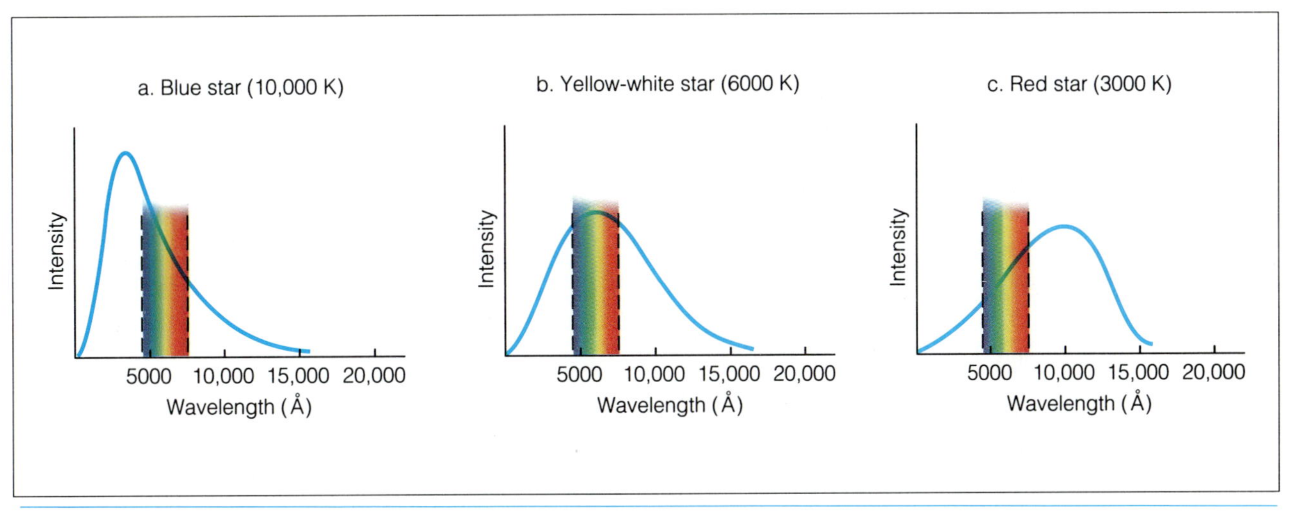

FIGURE 6.9 THE COLORS OF STARS. *The figure shows intensity plots for stars of different temperatures in comparison to the colors perceived by the human eye. This illustrates why stars have a range of colors as seen by the eye.*

discussed the energy as it is emitted at the surface, but not how bright it looks from afar. What we actually observe, of course, is affected by our distance from the object. For a spherical object that emits in all directions, the brightness decreases as the square of the distance, as illustrated in Fig. 6.10 (this should remind you of the law of gravitation.) Thus, if we double our distance from a source of radiation, it will appear $(1/2)^2 = (1/4)$ as bright. If we approach, reducing the distance by a factor of two, it will appear $2^2 = 4$ times brighter. As shown in later discussions of stellar properties, we must know the distances to stars before we can compare other properties having to do with their brightnesses.

— THE ATOM AND SPECTROSCOPY —

We turn our attention now to the spectral lines first noted by Wollaston and Fraunhofer in the spectrum of the Sun. These lines are difficult to see, and are not apparent in the spectrum of light obtained with an ordinary prism. However, with the development of superior techniques for examining the spectrum, their existence became well established (see Fig. 6.6).

In the late 1850s the German scientists R. Bunsen and G. Kirchhoff performed experiments and developed theories that made clear the importance of the Fraunhofer lines. Bunsen observed the spectra of flames created by burning various substances and found

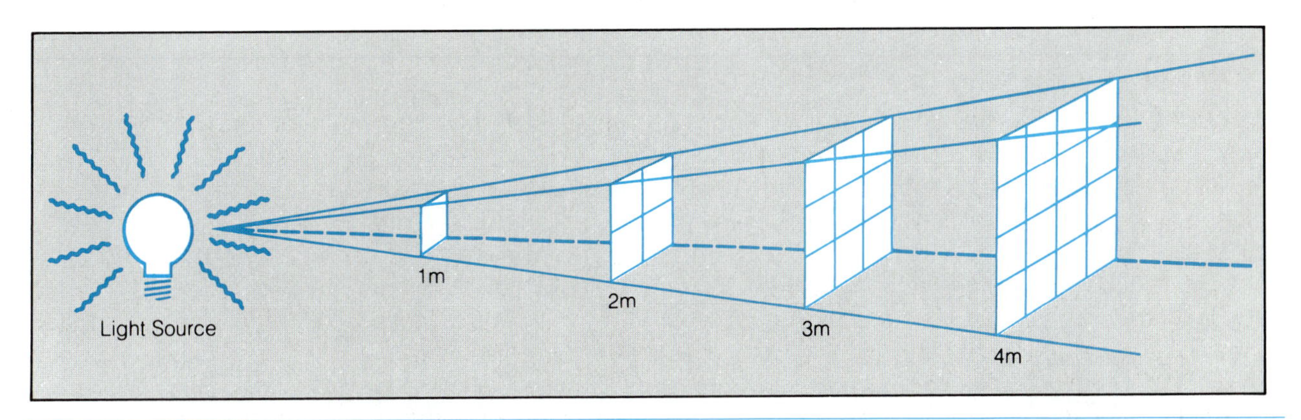

FIGURE 6.10 THE INVERSE SQUARE LAW OF LIGHT PROPAGATION. *This shows how the same total amount of radiation energy must illuminate an ever-increasing area with increasing distance from the light source. The area to be covered increases as the square of the distance; hence, the intensity of light per unit of area decreases as the square of the distance.*

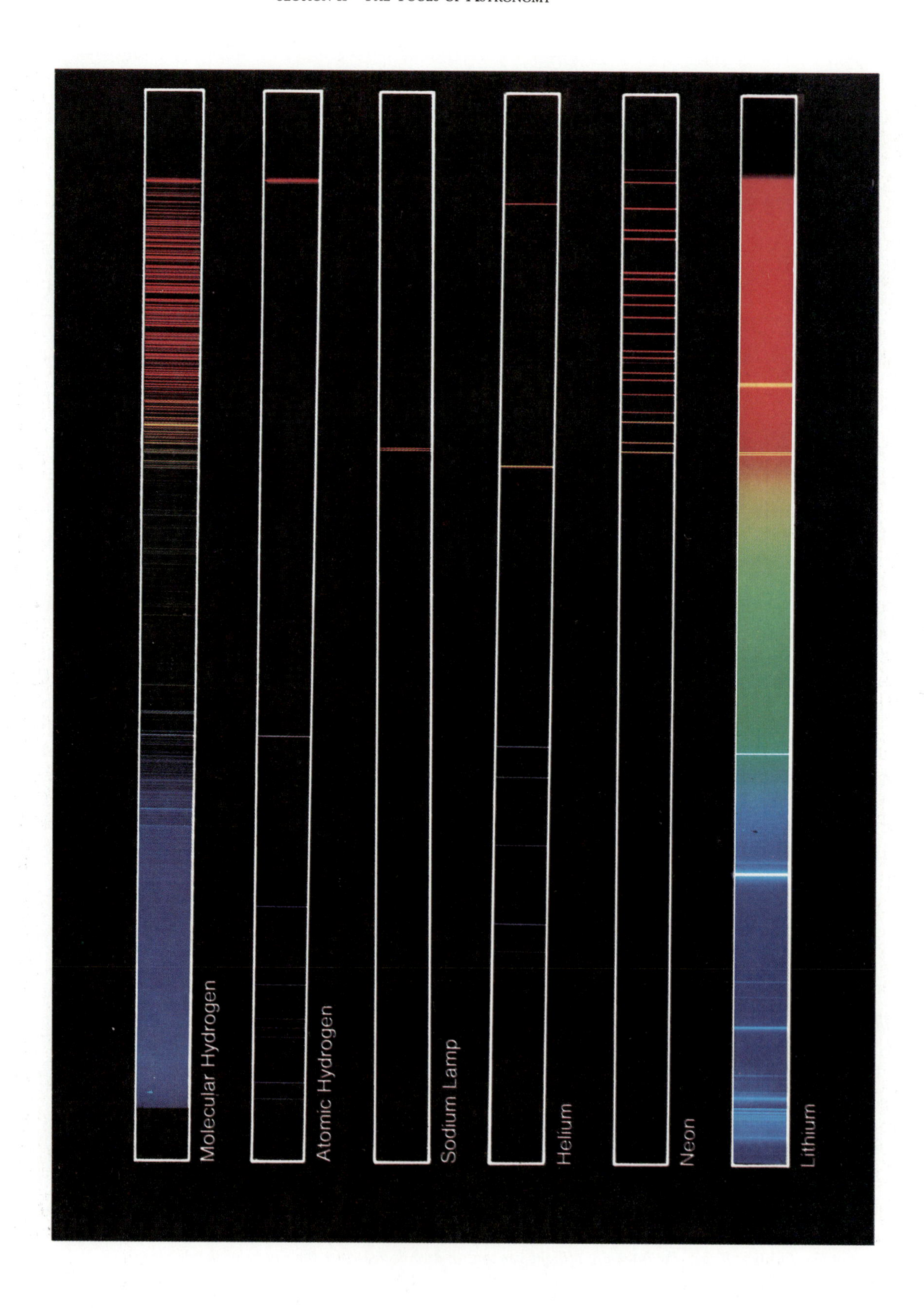
Molecular Hydrogen
Atomic Hydrogen
Sodium Lamp
Helium
Neon
Lithium

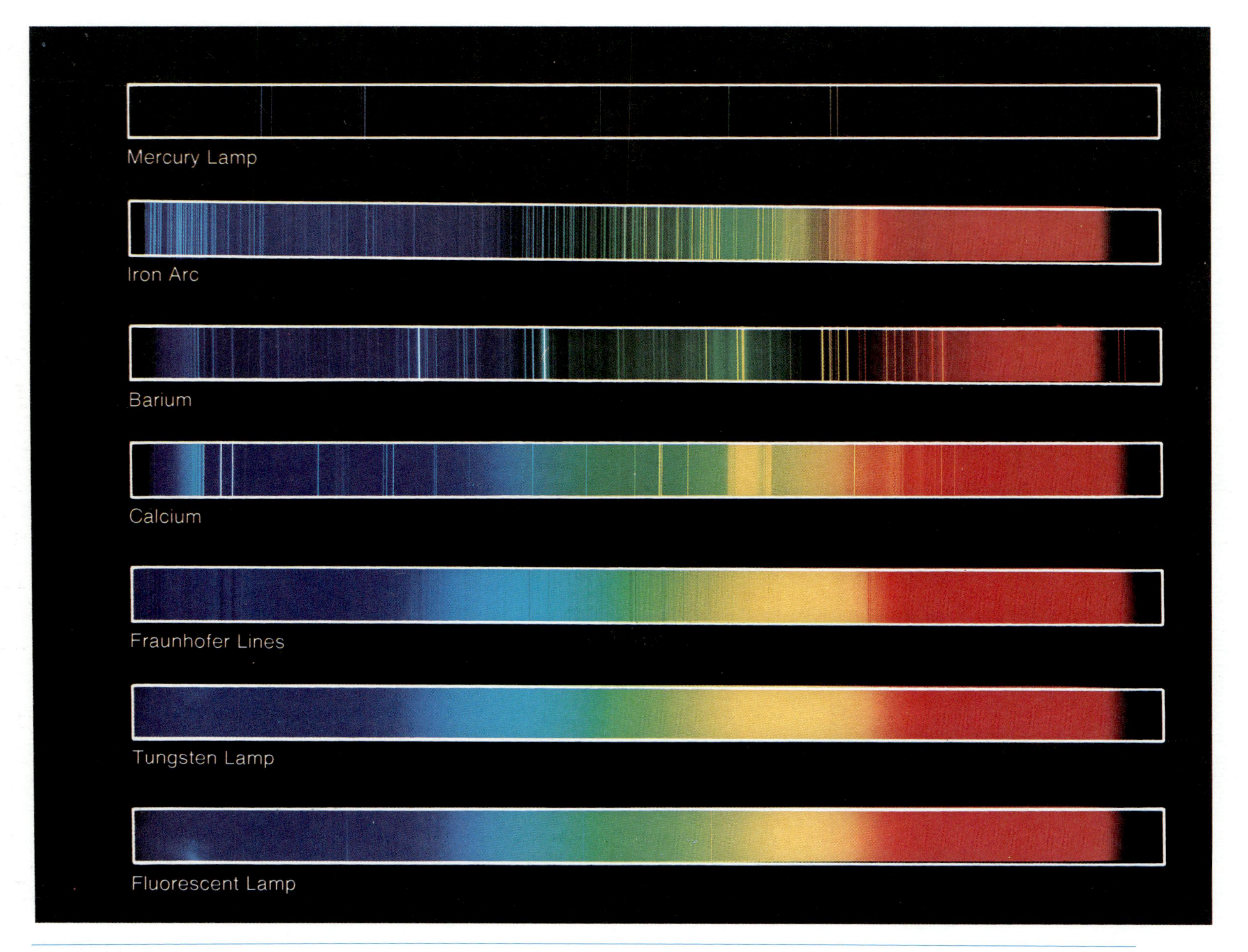

FIGURE 6.11 SPECTRA OF ELEMENTS. *These are emission-line spectra for an assortment of chemical elements. Note that the pattern of lines is distinctly different from one element to the next. Experiments with such spectra, along with observations of dark lines in the Sun's spectrum, led to the realization that is it possible to identify the chemical composition of the Sun and distant stars. (Courtesy of Bausch and Lomb)*

that each chemical element produced light only at specific places in the spectrum (Figs. 6.11 and 6.12). The spectrum of such a flame in this case is dark everywhere except at these specific places, as though the flame were emitting light only at certain wavelengths. The bright lines seen in this situation are called **emission lines** for that reason.

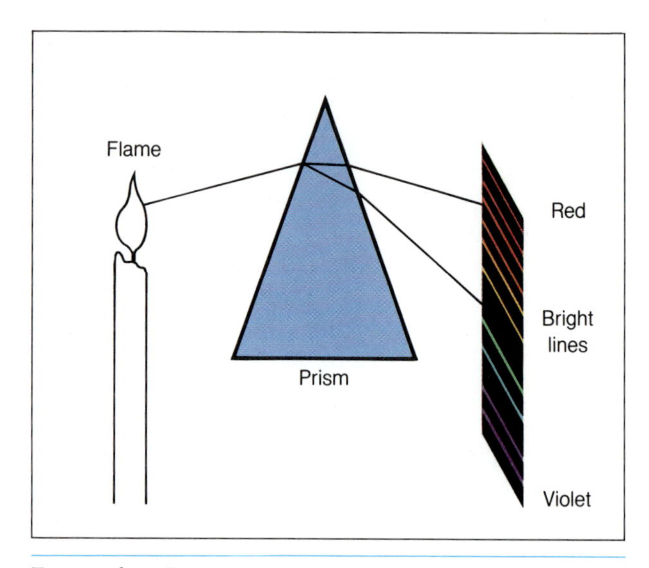

FIGURE 6.12 EMISSION LINES. *The spectrum of a flame is devoid of light except at specific wavelengths where bright emission lines appear.*

It was soon noticed that some of the dark lines seen in the Sun's spectrum coincide exactly in position with some of the bright lines seen by Bunsen in his laboratory experiments. Kirchhoff studied this in detail and was able to show that a number of common elements such as hydrogen, iron, sodium, and magnesium must be present in the Sun because of the coincidence in wavelengths of the lines. This was the first hint that the chemical composition of a distant object could be determined, something previously thought by astronomers to be forever beyond their ability to learn.

Kirchhoff further analyzed the absorption and emission of light and, through a series of experiments, was able to develop a set of rules describing when continuous radiation will be observed, when emission lines will be seen, and when dark **absorption lines** will form instead (Fig. 6.13):

1. A hot, dense gas or a hot solid produces a continuous spectrum (one with no lines).
2. A hot, rarefied gas produces emission (bright) lines.
3. A cool gas in front of a continuous source of light produces absorption (dark) lines.

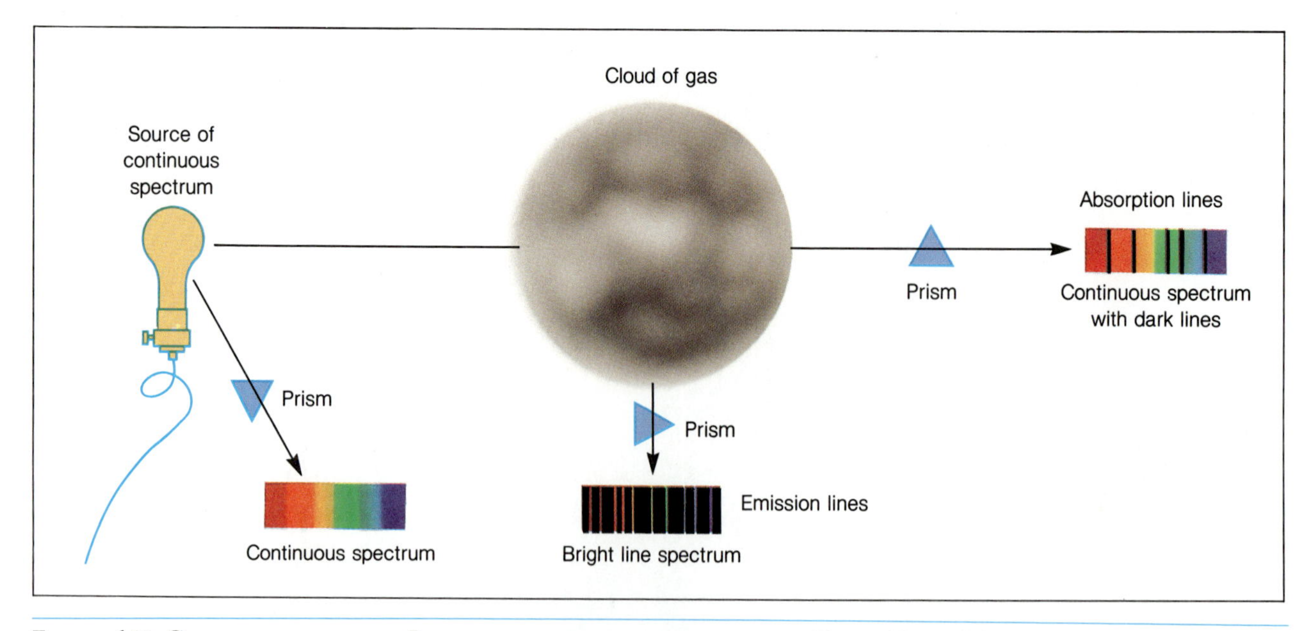

FIGURE 6.13 CONTINUOUS, EMISSION-LINE, AND ABSORPTION-LINE SPECTRA. *The positions of the emission and absorption lines match because the same element emits or absorbs at the same wavelengths. Whether it emits or absorbs depends on the physical conditions, as described in Kirchhoff's laws.*

These rules explain why Bunsen saw only emission lines when he observed the spectra of flames, which are rarefied, hot gases, and they tell us something important about the Sun. The fact that the Sun's spectrum contains absorption lines (Fig. 6.14) implies that the outer layers of the Sun must be cooler than the interior, so that the continuous radiation is formed inside, where it is hot and dense, and absorption lines are formed as this continuous radiation passes out through the outer layers. Thus, a simple understanding of how gases in the laboratory emit and absorb light reveals something about the structure of the Sun.

Energy Levels and Photons

The work of Bunsen and Kirchhoff demonstrated that each element has its own unique spectrum. As we have seen, this discovery had profound implications, for it meant that astronomers could gain information about distant objects previously thought to be forever beyond their grasp. Although astronomers in the later half of the nineteenth century were able to use these astrophysical fingerprints, no real understanding of their nature and cause was developed until the early years of this century.

Spectral lines were studied by a number of scientists, and various regularities in the arrangement of the lines from a given element were noted. Although it was suspected that these regularities reflected some aspect of the structure of atoms, it was not until 1913 that the true relationship was discovered by Niels Bohr. By that time it had been established that an atom consists of a small, dense nucleus surrounded by a cloud of negatively charged particles called electrons. Bohr found that these electrons were responsible for the absorption and emission of light, gaining energy (absorption) or losing it (emission) in the form of photons.

The key to the fixed pattern of spectral lines for each element lies in the fixed pattern of energy levels the electrons can be in. We can visualize an atom as a miniature solar system, with electrons orbiting the nucleus. Each kind of atom (each element) has its own characteristic number of electrons, and in each case the electrons have a certain set of possible orbits. It may be helpful to visualize a ladder, with each rung representing an orbit or energy level (Fig. 6.15). The spacing between the rungs of the ladder for one element, say hydrogen, are different from those of the ladder for any other element. The energy associated with each level increases the higher up the ladder, or the farther from the nucleus, the electron goes.

An electron can absorb a photon of light only if the photon carries the precise amount of energy needed to move the electron to one of the possible discrete levels higher than the one it is in. Imagine an atom in space, with light streaming past it: The atom will let pass every photon that comes by, until it finds one with just the right amount of energy to boost one of its electrons to a higher level. Since the energy of a photon is closely related to its wavelength, this means that the atom will only absorb photons that have specific wavelengths, corresponding to the spacings between its energy levels. Because each kind of atom has its own unique set of energy levels, each can

FIGURE 6.14 SOLAR SPECTRUM. *In this portion of the Sun's spectrum, we see several strong absorption lines, most due to hydrogen atoms in the Sun's outer layers.* (K. Gleason, Sommers Bausch Observatory, University of Colorado)

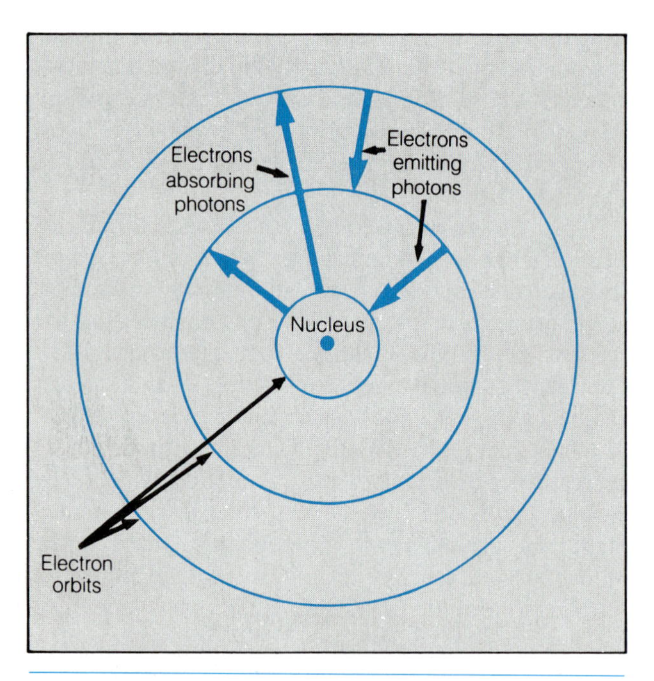

FIGURE 6.15 THE FORMATION OF SPECTRAL LINES.
An electron must be in one of several possible orbits, each representing a different electron energy. If an electron absorbs exactly the amount of energy needed to jump to a higher (more outlying) orbit, it may do so. This is how absorption lines are formed; because the wavelength of the photon absorbed corresponds to the energy difference between the two electron orbits, and that is fixed for any given kind of atom. Conversely, when an electron drops from one orbit to a lower one, it emits a photon whose wavelength corresponds to the energy difference between orbits, which is how emission lines form.

absorb light only in its own unique set of wavelengths. Thus Bohr was able to explain the findings of Bunsen and Kirchhoff.

Emission is the reverse process of absorption. Emission will occur whenever an electron in an upper level drops down to a lower one, and the energy of the photon emitted corresponds, as before, to the energy difference between the two levels. Since the energy levels are fixed for a given element, the emission lines and the absorption lines occur at the same wavelengths. Whether an atom produces emission lines or absorption lines depends on whether the electrons are moving downward or upward.

We can now state Kirchhoff's three rules in a different form:

1. In a hot, dense gas or a hot solid, the atoms crowd together so that their energy levels overlap and all

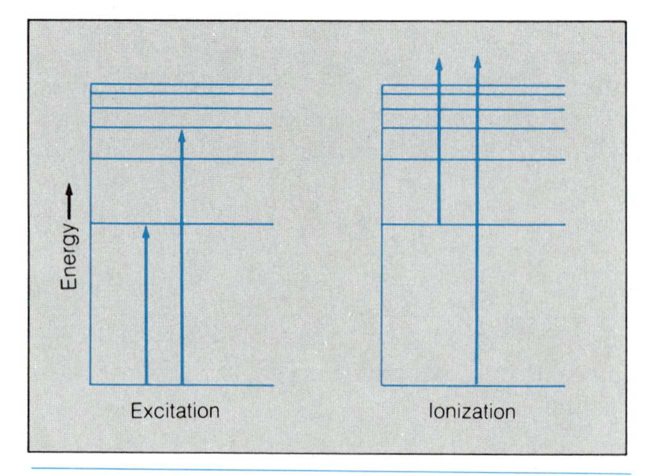

FIGURE 6.16 EXCITATION AND IONIZATION. *In the energy-level diagram at left, electrons are shown jumping up from the lowest level to higher ones. This is excitation, and it can be caused either by the absorption of photons or by collisions between atoms. On the right, electrons are shown gaining sufficient energy to escape altogether, a process called ionization.*

their lines blend together,[1] and we see a continuous spectrum.

2. In a hot, rarefied gas, the electrons tend to be in high energy states and create emission lines as they drop to lower levels.

3. In a cool gas in front of a hot, continuous source of light, the electrons tend to be in low energy levels and absorb radiation from the background continuous source, absorbing photons of specific wavelengths and forming absorption lines.

Ionization and Excitation

We have seen that electrons can move from one level to another by giving up or gaining energy. Even though each atom has many levels, there is always a limit to the range of levels an electron can have. If the electron gains too much energy, more than what is necessary to reach the highest level the atom has, the electron will escape entirely (Fig. 6.16). This process is called **ionization,** and it leaves the atom with one electron short of its usual number. The atom then has a pos-

[1]The production of a continuous spectrum is actually a bit more complicated than this. When free electrons combine with ions, they emit at any wavelength, not just in spectral lines. See the discussion of ionization in the following section.

itive electrical charge and is called an **ion.** It is possible for an atom to lose more than one electron, becoming more highly ionized.

Ionization can occur in two ways. The electron can receive the energy needed for escape either by absorbing a photon with enough energy or by colliding with another atom. In any gas, the individual particles move about randomly, occasionally colliding with each other. Energy may be passed from one atom to another in such a collision, and when it is, the electrons of the atom that gained the energy may move up a few rungs in the ladder of energy levels. If the collision takes place at sufficiently high velocity, one or more of the electrons may gain enough energy to escape. The speed of motion of the atoms depends on the temperature of the gas, so the hotter the gas is, the more ionizations occur. Hence, a hot star has a higher degree of ionization in its outer layers than a cool star does.

The spectrum of an atom changes drastically when it has been ionized, because the arrangement of energy levels is altered, and because different electrons are now available to do the absorbing and emitting of photons. The spectrum of atomic helium, for example, is quite different from that of ionized helium, so the astronomer not only can see that helium is present in the spectrum of a star, but also can tell whether it is ionized. This provides information about the temperature in the outer layers of the star, where the absorption lines are formed. By analyzing the degree of ionization of all the elements seen in the spectrum of a star, the astronomer can determine the gas temperature quite precisely.

Let us now consider an atom that is not ionized, and which collides at relatively low velocity with another atom. Its electrons may gain energy, but if it is not enough to cause ionization, the electrons may land in energy levels that are higher than those they were in before the collision (see Fig. 6.16). Ordinarily an electron will tend to stay in the lowest available energy state, and when it gets into a higher state, the atom is said to be **excited,** or in an **excited state.** When an electron does move to an excited state, it usually will drop back down almost immediately, and it will emit a photon each time it descends from the one level to a lower one. In gas in which collisions are rare, at any given time very few atoms will be in excited states. In a dense gas, however, where collisions occur often, at any moment a certain fraction

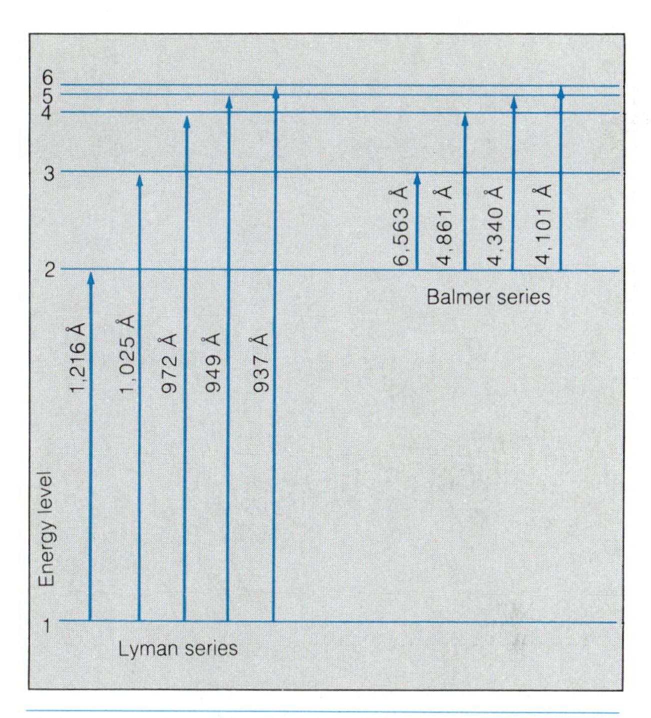

FIGURE 6.17 THE EFFECT OF EXCITATION OF THE SPECTRUM OF HYDROGEN. *As shown here, the absorption lines that can be formed by an atom depend on its degree of excitation. At left are the wavelengths of absorption lines originating in the lowest energy level of hydrogen, while at right are the wavelengths of absorption lines arising from an electron in the first excited level. Note that the Lyman lines are in ultraviolet wavelengths, while the Balmer lines are in the visible portion of the spectrum. This sketch shows only a few of the many energy levels of hydrogen.*

of the atoms will be in excited states. It should be clear that if an electron is in an excited state, the wavelengths at which it can absorb photons are different from those at which the electrons can absorb when it is in the lowest level. The amount of energy required for an electron to move upward in the ladder of energy levels depends on which rung the electron is on to begin with; hence the wavelength of photon that it can absorb depends on which level it is in. Therefore the spectrum of an atom depends on how excited the atom is (Fig. 6.17 and Table 6.2). Observation of which lines appear in the spectrum of a star can tell the astronomer how many electrons are in excited states. This in turn reveals how frequently collisions are taking place in the star's outer layers, so in the end it is possible to infer the density of the gas.

TABLE 6.2 PRINCIPAL LINES IN THE SPECTRUM OF HYDROGEN

Series	Lower Level	Line	Wavelength (Å)
Lyman (ultraviolet)	1	Alpha	1,216
		Beta	1,025
		Gamma	972
		Delta	949
		Epsilon	937
		(Limit)	(912)
Balmer (visible)	2	Alpha	6,563
		Beta	4,861
		Gamma	4,340
		Delta	4,101
		Epsilon	3,970
		(Limit)	(3,646)
Paschen (infrared)	3	Alpha	18,751
		Beta	12,818
		Gamma	10,938
		Delta	10,049
		Epsilon	9,545
		(Limit)	(8,204)
Brackett (infrared)	4	Alpha	40,512
		Beta	26,252
		Gamma	21,656
		Delta	19,445
		Epsilon	18,175
		(Limit)	(14,585)

From careful analysis of a star's spectrum, then, we can determine both the temperature of the outer layers of the star (from the ionization) and the density (from the excitation). Although most of the examples used in the discussion have referred to the Sun or the stars, the same considerations can be applied to the spectra of the planets.

The Doppler Effect

There is something else that spectral lines can tell us about a distant object, in addition to all the physical data we have been discussing. We can also learn how rapidly an object such as a star or planet is moving toward or away from us.

To understand this, we must visualize light from a distant object as though it were a series of concentric waves moving away from the source like ripples on a pond where a pebble has been dropped in (Fig. 6.18).

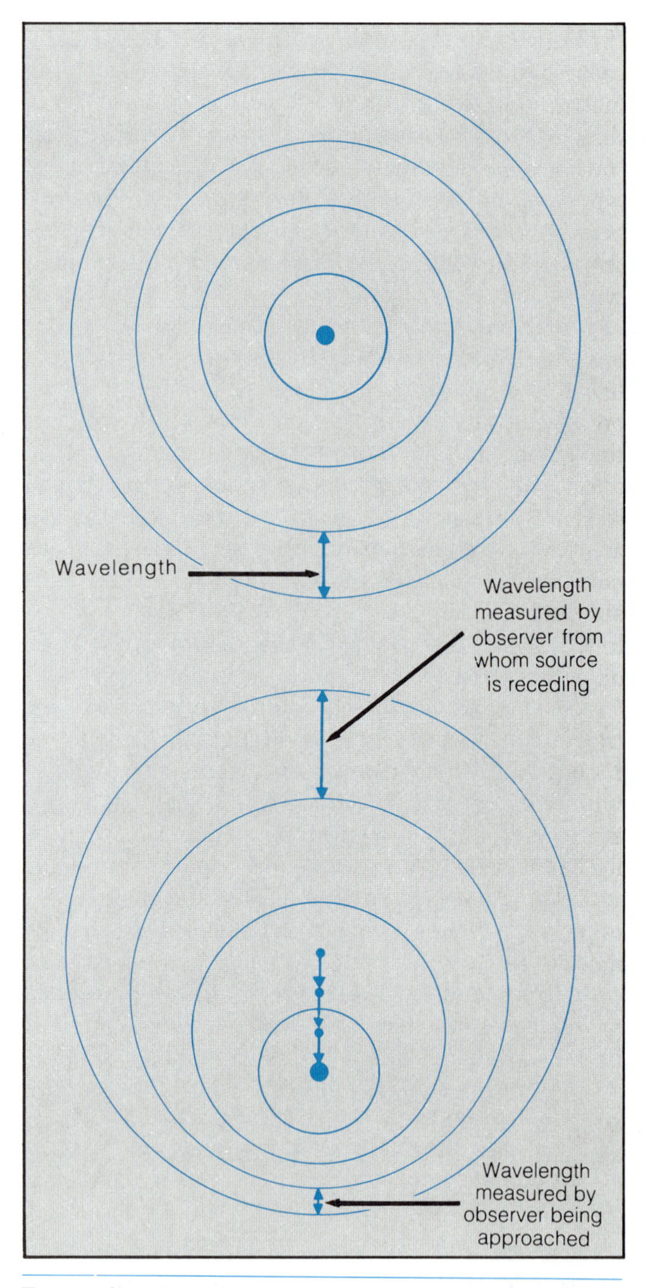

FIGURE 6.18 THE DOPPLER EFFECT. *The light waves from a stationary source (upper) remain at constant separation (that is, constant wavelength) in all directions, whereas those from a moving source get "bunched up" in the forward direction and "stretched out" in the trailing direction (lower). This causes a blueshift or a redshift for an observer who approaches or recedes from a source of light. Note that it does not matter which is moving, source or observer.*

If the source of light is moving, the waves will tend to get bunched up in the direction toward which the motion is occurring, and stretched out in the opposite direction. An observer sitting ahead of the moving source will see waves that are closer together than would be the case if the source were sitting still; that person will observe a shorter wavelength. An observer on the other side, with the source moving away, will see longer wavelengths.

This analogy is oversimplified in some ways, but it does illustrate an effect that really happens with light. When a source of light such as a star is approaching, all the lines in its spectrum are shifted toward wavelengths that are shorter than those observed when the light source is at rest, and if the source is moving away from the observer, all the lines are shifted toward longer wavelength. These two situations are called **blueshifts** (approach) and **redshifts** (recession), respectively, because the spectral lines are shifted toward either the blue or the red end of the spectrum.

A general term for any wavelength shift resulting from relative motion between source and observer is **Doppler shift**, in honor of the Austrian physicist who first explored the properties of such shifts. The effect does not apply just to light waves. Most of us, for example, have noticed the Doppler shift in sound waves when a source of sound passes by. The whistle on an approaching train or the siren on an approaching emergency vehicle will suddenly change to a lower pitch at the moment the train or emergency vehicle passes by, because the wavelength received by the listener suddenly shifts to a longer one.

With the Doppler shift of light, we can determine the speed with which the source of light is approaching or receding, from the formula

$$v = \left(\frac{\Delta\lambda}{\lambda}\right)c,$$

where v is the relative velocity between source and observer, $\Delta\lambda$ is the shift in wavelength (the observed wavelength minus the rest, or laboratory, wavelength of the same line), λ is the laboratory wavelength of the line, and c is the speed of light.

Suppose that we find a star with a spectral line at a wavelength of 5,994 Å. Examination of the pattern of lines in this star's spectrum shows that this is a line of iron, which is measured in the laboratory to have a wavelength of 6,000 Å. Then the speed of the star with respect to the earth is

$$v = \left(\frac{5{,}994\text{Å} - 6{,}000\text{Å}}{6{,}000\text{Å}}\right) \times 300{,}000 \text{ km/sec}$$

$$= \left(\frac{-6\text{Å}}{6{,}000\text{Å}}\right) \times 300{,}000 \text{ km/sec}$$

$$= -0.001 \times 300{,}000 \text{ km/sec}$$

$$= -\ 300 \text{ km/sec}.$$

In this example, we find a negative velocity, because the observed wavelength is smaller than the rest wavelength. This means that the star is coming toward the Earth, and we have a blueshift. If, on the other hand, the observed wavelength of this line had been 6,006 Å, we would have found a velocity of +300 km/sec, and the star would have been moving away from the Earth.

The Doppler shift tells us only about relative motion between the source and the observer. We cannot distinguish whether it is the star or the Earth that is moving, or a combination of the two (which is most likely the case). It is also important to keep in mind that the Doppler shift tells us only about the part of the motion directly toward or away from the Earth, which is called the **radial velocity** (Fig. 6.19). There is no Doppler shift resulting from motion perpendicular to our line of sight. If a star is moving at

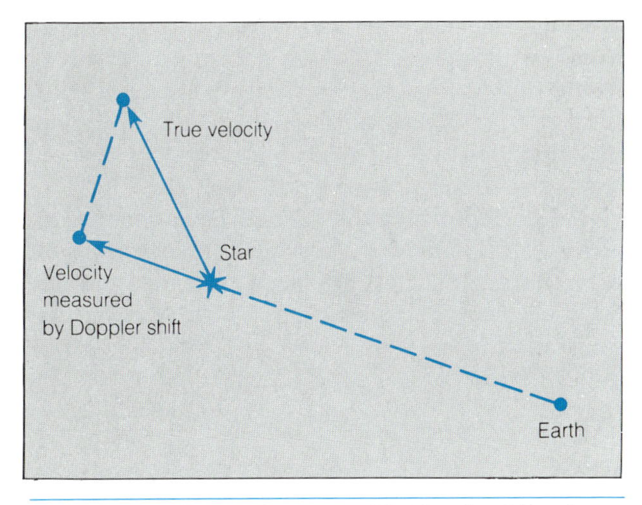

FIGURE 6.19 RADIAL VELOCITY. *The Doppler shift only reflects motion along the line of sight, called the radial velocity. Here a star is moving at an angle to the line of sight, so that from the Doppler shift we can measure only a fraction of its true velocity.*

some intermediate angle with respect to the Earth, as is usually the case, we can determine the part of its velocity that is directed straight toward or away from us (that is, the radial velocity), but we cannot determine its true direction of motion or its speed transverse to our line of sight.

The Doppler shift is another powerful tool for astronomers. From it we learn about the motions not only of stars, but also of planets, galaxies, and any other objects in the universe that have lines in their spectra.

PERSPECTIVE

We now have learned how light is absorbed and emitted, and in doing so have begun to see how astronomers can derive great amounts of information about distant objects. We have seen how the continuous spectrum and overall intensity of light from a star depend on its temperature and size, and how the spectral lines provide information on temperature, density, and motion of a star, as well as chemical composition.

We are almost ready to get on with the job of seeing what is out there, but first we must learn how the light is captured by astronomers so that it can be analyzed.

SUMMARY

1. Visible light is just one part of the electro-magnetic spectrum, which extends from the γ rays to radio wavelengths.
2. Light has properties associated with both waves and particles, and these aspects are combined in the concept of the photon.
3. Astronomers gain information about a source of light, such as a star, from both the continuous radiation and the spectral lines.
4. The continuous radiation provides information on the temperature of a star, through the use of Wien's law. The temperature, luminosity, and radius of a star are related by the Stefan-Boltzmann law. Planck's law is a general one that embodies both Wien's law and the Stefan-Boltzmann law.
5. The observed brightness of a glowing object is inversely proportional to the square of the distance between the object and the observer.
6. Spectral lines are produced by transitions of electrons between energy levels in atoms, molecules, and ions. Absorption occurs when an electron gains energy and jumps to a higher level, and emission occurs when it loses energy and drops to a lower level, the wavelength in both cases corresponding to the energy gained or lost as the electron changes levels.
7. Each chemical element has its own distinct set of spectral-line wavelengths, and therefore the composition of a distant object can be determined from its spectral lines.
8. The ionization and excitation of a gas can be inferred from its spectrum, the former yielding information on the temperature of the gas, and the latter providing data on its temperature and density.
9. Any motion along the line of sight between an observer and a source of light produces shifts in the wavelengths of the observed spectral lines, and measurement of this Doppler effect, as it is called, can be used to determine the relative speed (radial velocity) of source and observer.

REVIEW QUESTIONS

Thought Questions

1. Summarize the various forms of electromagnetic radiation, explaining their differences and similarities. Think of everyday examples of each form of radiation.
2. Explain why the human eye is sensitive to the particular wavelengths of electromagnetic radiation that we can see. What wavelengths do you think beings on a planet circling a much hotter star than the Sun would be able to see?
3. Summarize the ways in which light acts as a stream of particles, and the ways in which it acts as though it consists of waves.
4. Explain how the color of a star is related to its surface temperature.
5. Why are infrared-sensitive cameras useful devices for spotting people hiding in the dark?
6. Explain how the spectral lines formed by a given type of atom are related to the structure of that atom.
7. Explain why an atom can form emission lines under some circumstances and absorption lines under other circumstances.
8. What are the processes that can cause atoms to become ionized?

9. Explain why the Doppler effect tells us only about the component of motion directly along the line of sight between a source of light and an observer.
10. Summarize the information that can be gained about a distant object such as a star or a planet from the analysis of its spectral lines.

Problems

1. What is the wavelength of your favorite FM radio station?
2. How long does it take for a photon of light from the Sun to reach the Earth?
3. A photon in the blue portion of the visible part of the electromagnetic spectrum has a wavelength of 4,500 Å. Express this wavelength in centimeters. What is the energy of this photon in ergs?
4. Calculate the wavelengths of maximum emission for a star of surface temperature 25,000 K, a star of surface temperature 2,500 K, the Sun's corona, whose temperature is 2,000,000 K, and a human body, whose temperature is 310 K.
5. Suppose Mars has a surface temperature 5 times greater than that of Pluto. Which of these two bodies emits more energy per second per square centimeter of surface area, and how much more does it emit?
6. If Mars has twice the diameter of Pluto, compare the relative total energy emitted per second by the two planets.
7. Pluto is approximately 25 times farther from the Sun than Mars is. Compare the intensity of sunlight reaching the two planets.
8. Suppose that in the hydrogen atom the energy difference between two energy states is 4.581×10^{-12} erg. What is the wavelength of a photon emitted when an electron drops from the upper to the lower of these two states? From Fig. 6.17, can you identify which two states these are?
9. The energy needed to ionize a hydrogen atom is 2.18×10^{-11} erg. What wavelength of photon does this correspond to? How fast would an electron have to be moving to have this much kinetic energy? (The electron mass is 9.11×10^{-28} grams; recall the formula for kinetic energy in Chapter 5).
10. Suppose the strong line of hydrogen, whose rest wavelength is 6562.7, is observed in the spectrum of a star and found to have a wavelength of 6563.3 Å. What is the radial velocity of this star?

ADDITIONAL READINGS

The best sources for further explanation of the properties of light and radiation are elementary physics books of the sort used as texts in introductory courses. In addition there are a few nontechnical books on the nature of light, such as the following works.

Bragg, W. 1959. *The universe of light*. New York: Dover.
Rublowsky, J. 1964. *Light*. New York: Basic Books.
Ruechardt, E. 1958. *Light, visible and invisible*. Ann Arbor: University of Michigan Press.

CHAPTER 7

TELESCOPES ON EARTH AND IN SPACE

The Hubble Space Telescope *being assembled.* (NASA)

CHAPTER PREVIEW

The preceding two chapters illustrate the techniques used by astronomers to determine the properties of faraway objects. These techniques, however, require that certain measurements be made, and we have not yet discussed the instruments and methods used to make these measurements. Hence this chapter will outline the principles of telescopes and their uses.

The Need for Telescopes

Although we will discuss a wide variety of telescopes, ranging from those used in space for X-ray and ultraviolet observations, to traditional ground-based telescopes for visible light, to radio antennae, all perform essentially the same task: they gather as much light from a source as possible and bring it to a focus. If we want to observe faint objects, it helps to collect light from as large an area as possible. The human eye has a collecting area only a fraction of a centimeter in diameter, whereas the largest telescopes are several meters in diameter.

The light-gathering power of a telescope depends on the area of its collecting surface. Since the area of a circle is equal to πr^2, where r is the radius, a 2-inch diameter telescope collects four times as much light as a 1-inch telescope, for example. The largest visible-light telescopes are 4 to 6 meters in diameter.

Another basic difference between the eye and the telescope is that the telescope can be equipped to accumulate and record light over a long period of time, through the use of photographic film or an electronic detector, whereas the eye has no capability for storing light. A long-exposure photograph taken through a telescope allows us to obtain pictures of objects that we could not see even when looking through the same telescope.

Both their large size and their capability of making long exposures make it possible for us to use telescopes to detect objects some 100,000,000 times fainter than what we can see with the unaided eye, even under the best conditions.

A third major advantage of large telescopes is that they have superior **resolution,** the ability to discern fine detail. For visible-wavelength telescopes, the Earth's atmosphere creates practical limitations on how fine the resolution can be, but for radio telescopes and optical telescopes in space, the atmosphere is not a problem.

One problem must be overcome in making long exposures: The Earth rotates, and if nothing were done to compensate for this, a star would quickly move out of the field of view. In order to avoid this problem, telescopes are mounted so that they can be moved by a motor in the direction opposite the Earth's rotation, keeping a target object centered in the field of view. The telescope mounting usually allows the telescope to move independently in declination (north-south motion) and right ascension (east-west motion), so the drive mechanism that compensates for the Earth's rotation needs to operate only in the east-west direction. The need to do this, of course, is avoided for telescopes that are not attached to the Earth, such as the various orbiting observatories (discussed later in this chapter).

Refractors and Reflectors

We begin by discussing the principles of telescopes used for visible light, since these have the longest history and because many of the basic ideas can be used for infrared and ultraviolet telescopes as well.

The telescope used by Galileo utilized lenses to bring light to a focus and to magnify the image (Fig. 7.1). This technique was later refined by Huygens and studied by Newton, who realized it has a fundamental shortcoming: The lens acts as a prism, bending different wavelengths of light to different degrees as it passes through the glass, thus making it impossible to

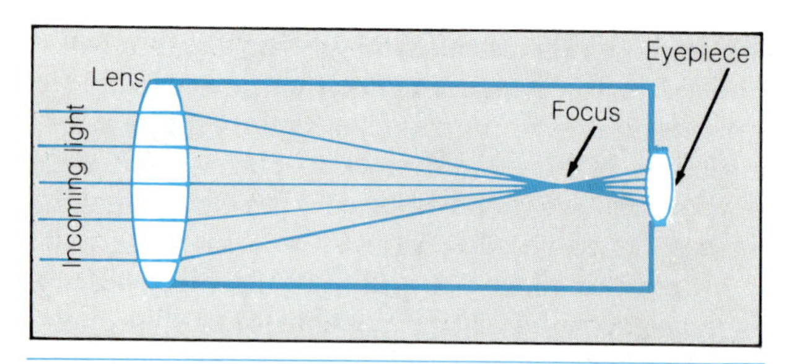

FIGURE 7.1 THE REFRACTING TELESCOPE. *The path of light is bent when it passes through a surface such as that of glass. A properly shaped lens thus can bring parallel rays of light to a focus, where the image of a distant object may be examined. A second lens is used to magnify the image.*

FIGURE 7.2 THE WORLD'S LARGEST REFRACTOR. *This is the 40-inch refracting telescope located at the University of Chicago's Yerkes Observatory in Williams Bay, Wisconsin.* (Yerkes Observatory)

bring all the light of a star to a focus at a single point. The blue light focuses closer to the lens than the longer-wavelength red light: The image of a star gets spread out into a rainbow. This problem, called **chromatic aberration,** was solved much later by the use of lenses made of combinations of different kinds of glass, so that the separation of colors caused by a layer of one kind of glass could be reversed by a layer of another kind of glass.

Telescopes that use a lens to bring light to a focus are called **refractors.** A large lens, called the primary lens, at the end of the telescope tube bends the light rays so that they come together and form an image at a point called the **focus.** The size and curvature of the primary lens determine the focal length, that is, the distance from the primary lens to the focus point when a very distant object is observed. At this point (where the other end of the tube is located) is an eyepiece to magnify the image for the human observer.

The crucial aspect of a telescope is not its ability to magnify images, as is commonly thought, but is instead its ability to collect as much light as possible from a source. This light-gathering power is determined by the size of the principal light-collecting element, in this case the primary lens. This is what determines how sensitive the telescope will be for observing faint objects. The eyepiece does nothing but magnify the image to a scale that is easier to discern; it does not enhance the light-gathering capability of the telescope.

The largest refractors ever made are less than 4 feet in diameter. The biggest is the refractor at the Yerkes Observatory at Williams Bay, Wisconsin, with a primary lens 40-inches in diameter (Fig. 7.2). Another large refractor, with a 36-inch diameter, is located at the Lick Observatory, on Mount Hamilton, east of San Jose, California.

Newton, having discovered (but not overcome the problem of chromatic aberration in refracting telescopes, devised an alternative type of telescope that utilizes a concave mirror to bring the light to a focus (Fig. 7.3, 7.4, and 7.5). This arrangement, called a **reflecting telescope,** has the advantage that all wavelengths of light reflect from a mirror at the same angle, so no separation of colors occurs. A disadvantage, however, is that for us to look at the image created by a concave mirror, we need to block the incoming light, because the image forms in front of the mirror. One solution to this problem was suggested by J. Gregory, who placed a **secondary mirror** in front of the primary to reflect the image back through a hole in the primary to a point behind it, where the observer could look at it without blocking the incoming light. This design is similar to the modern **Cassegrain focus** arrangement (Fig. 7.6) and is used in

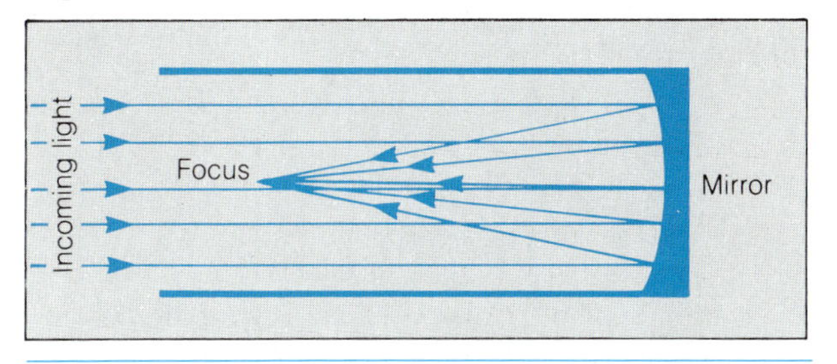

FIGURE 7.3 THE REFLECTING TELESCOPE. *A properly shaped concave mirror can be used instead of a lens to bring light to a focus. Usually an additional mirror is used to reflect the image outside the telescope tube.*

FIGURE 7.4 A REFLECTING TELESCOPE BUILT BY ISAAC NEWTON. *The image is reflected to a focus outside the telescope tube, near the top in this illustration.* (The Granger Collection)

nearly all large, reflecting telescopes. Newton developed a different means of deflecting the image outside the telescope tube; he placed a flat mirror at a 45° angle just before the focal point so that the light passed out through a hole in the side of the tube before coming to a focus. This **Newtonian focus** is used in many small- to moderate-sized reflectors, particularly of the sort often used by amateur astronomers.

Another, more complex, design is sometimes available with large telescopes (Fig. 7.6). Called a **coudé focus,** this arrangement utilizes up to five mirrors (including the primary) to reflect the image out of the telescope tube along one of the main axes of motion to a focus in a separate room, some distance from the telescope. The reason for going to all this trouble is to allow the use of heavy instruments in the analysis of light. In the Cassegrain or Newtonian designs, any equipment needed to record or measure the light must be attached directly to the telescope and must be light enough to move with it. That is not practical if the instrument is very massive. In the coudé design, however, the image does not move, no matter where the telescope is pointed, so large instruments can be fixed in place to analyze it.

Yet another arrangement, possible only with the largest telescopes, is called **prime focus,** (Fig. 7.6). In

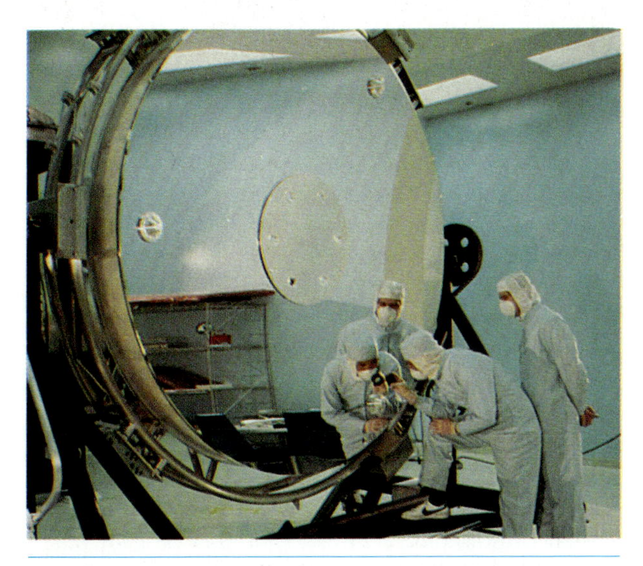

FIGURE 7.5 A LARGE PRIMARY MIRROR. *This photo shows that the proper shape for a telescope is not very highly concave. Only the distorted reflections in this case reveal that the mirror is not flat. This is the 2.4-meter primary mirror for the Space Telescope.* (Perkin Elmer Corporation)

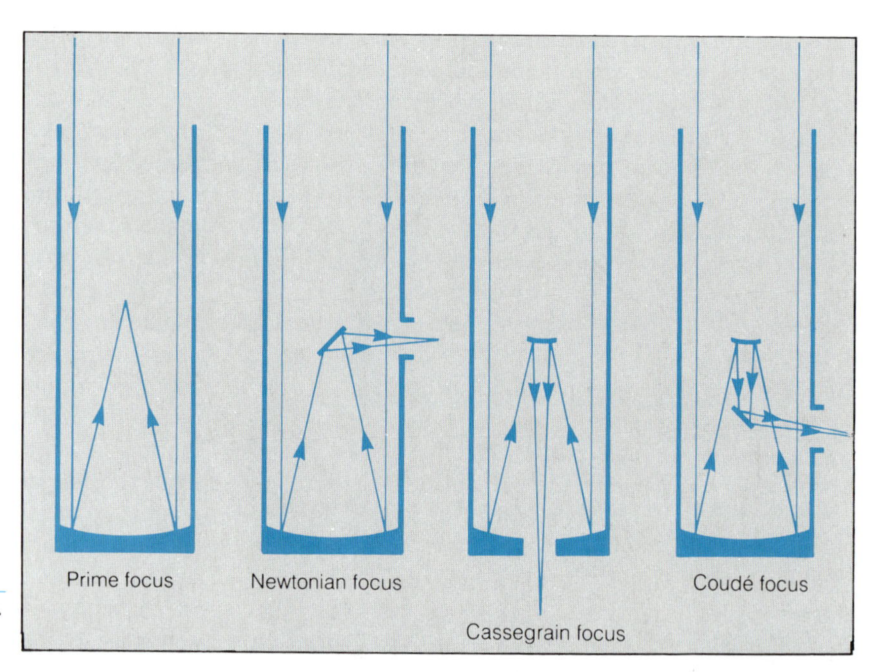

FIGURE 7.6 VARIOUS FOCAL ARRANGEMENTS FOR REFLECTING TELESCOPES.

this case the light is allowed to focus inside the tube, and the astronomer actually sits there, in a small cage, with his equipment for analyzing the light (usually a simple camera in this instance). The advantage of this design is that it reduces the number of reflections to just one, thereby minimizing the loss of light and image quality that inevitably occurs at each reflection. Prime focus is used, therefore, for observations of especially faint objects.

All the focal arrangements used in reflecting telescopes have the disadvantage that some fraction of the incoming light has to be blocked by a secondary mirror or by the observer's cage (if the prime focus is used). This disadvantage is far outweighed by the size advantage offered by the use of mirrors instead of lenses. A lens can be supported only by its edges, since light must be free to pass through. If a lens is large enough to be very heavy, it may sag a little under its own weight, which distorts its light-focusing properties. Even worse, the amount and direction of sag will vary as the telescope is pointed in different directions. A mirror, on the other hand, can be supported from behind and can therefore be kept much more rigid. All the largest telescopes in the world are reflectors for this reason (Fig. 7.7). Another advantage of reflectors is that only one surface has to be shaped to precision, rather than two, as in a lens (or four, in the composite lenses made of two layers of glass to prevent chromatic aberration). Also, because light reflects only from the surface and does not pass through the glass, flaws within the material of the mirror can be tolerated. The biggest telescope for many years was the 200-inch (5-meter) reflector at Mt. Palomar, in southern California (see Fig. 7.7), but a 6-meter telescope was completed in the Soviet Union (Fig. 7.8).

A very promising new kind of telescope was completed in 1979 at Mt. Hopkins, Arizona. This is the Multiple-Mirror Telescope, which consists of six separate 72-inch primary mirrors with secondary mirrors arranged so that each focuses light at the same point (Figs. 7.9 and 7.10) The total collecting area of the six mirrors is equivalent to a single mirror with a 176-inch (4.5-meter) diameter. This system is very complicated, and the difficulties of perfectly aligning the six mirrors are enormous, but the savings in cost and the relative ease with which the smaller mirrors can be constructed make the effort worthwhile.

Auxiliary Instruments

Once the light is brought to a focus, it must then be recorded in some way. The old-fashioned image of an astronomer peering into an eyepiece all night, perhaps taking notes, is replaced in reality by the picture of the astronomer sitting before a rather complex and

FIGURE 7.7 LARGE REFLECTING TELESCOPES. *Clockwise from upper left: The dome of the Canada-France-Hawaii Telescope on Mauna Kea; the 4-meter Mayall Telescope at Kitt Peak National Observatory; the 3.05-meter Shane Telescope at the Lick Observatory; and the 5-meter Hale Telescope on Palomar Mountain. (upper left:* J. G. Timothy; *upper right:* National Optical Astronomy Observatories; *lower right:* Lick Observatory photograph; *lower left:* Palomar Observatory, California Institute of Technology)

impressive control panel, operating sophisticated electronic equipment.

Generally speaking, two kinds of things are done to the light after it is focused: (1) It is often spread out or sorted according to wavelength; and (2) it is then recorded by film or electronic detector.

The first of these steps may be accomplished in a variety of ways, including the use of a prism or other device to spread out the light, or the use of filters to allow only certain wavelengths to pass through. The device that spreads out, or **disperses,** the light according to wavelength and then records the resulting spectrum is called a **spectrograph.** Modern spectrographs usually employ grooved sur-

FIGURE 7.8 THE 6-METER TELESCOPE. *This is the dome building containing the world's largest telescope, located in the Caucasus Mountains of the southern central Soviet Union.* (A. G. D. Phillip)

FIGURE 7.9 THE MULTIPLE-MIRROR TELESCOPE, MT. HOPKINS, ARIZONA. *This observatory is operated by the University of Arizona and the Smithsonian Institution. Six 72-inch mirrors bring light to a common focus, giving the instrument the light-collecting power of a single 4.5-meter telescope.* (Multiple-Mirror Telescope Observatory, University of Arizona and the Harvard-Smithsonian Center for Astrophysics)

faces called **gratings** to disperse the light, rather than prisms. The effect of a grating may be visualized by holding a compact disc that light glances off of it, forming a rainbow of colors (Fig. 7.11).

Besides spectrographs, other common kinds of auxiliary instruments include ordinary cameras, to photograph the field of view (sometimes this is done through a series of filters so that star brightnesses at different wavelengths can be measured); and **photometers,** which use simple photocells like those which control some kinds of automatic doors. A photocell produces an electric current proportional to the brightness of the starlight that hits it, so the brightness is determined by measuring the current.

The final stage of any instrument is the device that actually records the intensity of light. Traditionally, this job has been done with photographic film, and film is still widely used today, but several alternatives have been developed in recent years. These include a variety of electronic instruments that produce pictures like film does, but do so by storing an image electrically so that it can be analyzed easily by computer. Some of these **detectors,** as any device for measuring the intensity of light is called, are very similar to ordinary television cameras, whereas others operate on quite different principles.

The auxiliary instruments used in astronomy range from rather compact, lightweight devices that can be easily mounted directly on the telescope for use at the Cassegrain focus, to bulky, large instruments (usually spectrographs) that can only be used at the coudé focus.

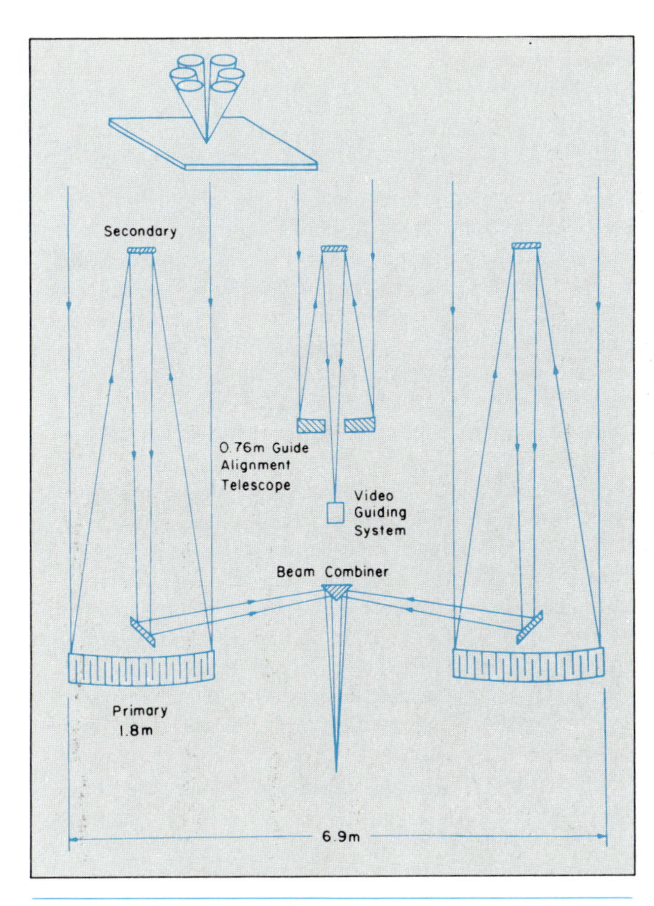

FIGURE 7.10 A SCHEMATIC OF THE MULTIPLE-MIRROR TELESCOPE DESIGN. *This drawing shows how the images from two of the six mirrors are brought to a common focus.* (Multiple-Mirror Telescope Observatory, University of Arizona and the Harvard-Smithsonian Center for Astrophysics)

CHOOSING A SITE: MAJOR OBSERVATORIES

Most major observatories are located in remote regions, for a number of reasons. It is best to minimize absorption of light by the atmosphere, so an observatory should be situated on a high mountain, above as much of the atmosphere as possible. Of course, a site with generally clear weather is imperative, for only radio telescopes can peer through clouds. It is also desirable to build observatories well away from large cities, because the diffuse light from a densely populated area can drown out faint stars. Another consideration is the latitude of the site, which should not be too far north or south of the equator, since that would exclude a large portion of the sky.

FIGURE 7.11 LIGHT DISPERSION BY A GRATING. (Photo Researchers, Inc.)

Other factors that affect the quality of an observing site are the purity of the air, for pollution can reduce the transparency of the atmosphere; and the amount of turbulence ordinarily present in the air above the site, because turbulence creates the twinkling effect known to astronomers as "seeing." This twinkling, which tends to smear out an image, reduces the quality of photographs or spectra. On a good night, the diameter of a star image is less than 1 arcsecond, whereas when the seeing is bad, images may blow up to diameters of 2 arcseconds or more.

The largest observatories in the world (see Appendix 6) generally are located along the western portions of North and South America, in Hawaii, (Fig. 7.12) and in Australia, although the biggest single telescope is in the Caucasus Mountains of the southern central Soviet Union. The U.S. Government, through the National Science Foundation, supports large observatories in North America (Kitt Peak National Observatory, about 50 miles west of Tucson, Arizona; Fig. 7.13) and South America (the Cerro Tololo Inter-American Observatory, on the crest of the Andes, northeast of Santiago, Chile; Fig. 7.14).

TELESCOPES IN SPACE: INFRARED AND ULTRAVIOLET ASTRONOMY

As we saw in our discussion of the emitting properties of glowing objects, the radiation from a star or planet extends beyond the limits detectable by the human eye. It also extends beyond the limits set by the Earth's

ASTRONOMICAL INSIGHT 7.1

A NIGHT AT THE OBSERVATORY

A typical astronomer's night at the observatory will vary quite a bit from one individual to another, and especially from one observatory to another. There is a big difference between sitting in a heated room viewing an array of electronics panels and standing in a cold dome peering into an eyepiece while recording data on a photographic plate, yet both extremes (and various possibilities in between) are part of modern observational astronomy.

Despite apparent differences in practice, however, the principles are very much the same, regardless of where the observations take place, or with what equipment. Here is a brief outline of a typical night at a large observatory, where a modern electronic detector is to be used to record spectra of stars.

By mid-afternoon the astronomer and his or her assistants (an engineer who knows the detector and its workings, and a graduate student) are in the dome, making sure the detector is working properly. To do this, the electronics must be tested in various ways, and test spectra from special calibration lamps must be recorded. Everything in order, they eat supper in the observatory dining hall a couple of hours before sunset, then return to the dome to complete preparations for the night's work. A night assistant (an observatory employee) arrives shortly before dark and is given a summary of the plans for the night. It is the night assistant's job to control the telescope and the dome, and this person is responsible for the safety of the observatory equipment. At large observatories, the telescope is considered so complex and valuable that a specially trained person is always on hand so that the astronomer (who may use a particular telescope only a few nights a year) never handles the controls, except for fine motions required to maintain accurate pointing during the observations.

Before it is completely dark, a number of calibration exposures are taken, using lamps and even the Moon, if it is up. These measurements will help the astronomer analyze the data later, by providing information on the characteristics of the spectrograph and the detector.

Finally, everything is in order, and if things are going smoothly, the telescope will be pointed at the first target star just as the sky gets dark enough to begin work. Observing time on large telescopes is difficult to obtain, and is very valuable; not a moment is to be wasted.

A pattern is quickly established and repeated throughout the night. First, after the night assistant has pointed the telescope at the requested coordinates, the astronomer looks at the field of view to confirm which star is the correct one. (This is often done by consulting a chart prepared ahead of time.) When the star is properly positioned so that its light enters the spectrograph, the observation begins. Depending on the brightness of the star and the efficiency of the spectrograph and the detector, the exposure time may be anywhere from seconds to hours. During this time, the astronomer or the graduate student assistant continually checks to see that the star is still properly positioned, making corrections as needed by pushing buttons on a remote control device.

If the observations are made using an instrument at the coudé focus, the astronomer and the assistants spend the entire night in an interior room in the dome building, emerging only occasionally to view the skies, give instructions to the night assistant, or go to the dining hall for lunch (usually served around midnight).

As dawn approaches and the sky begins to brighten, last-minute strategy is devised to get the most out of the remaining time. Finally, the last exposure is completed, and the night assistant is permitted to close the dome and park the telescope in its rest position (usually pointed straight overhead, to minimize stress on the gears and support system). The astronomer and the assistants make final calibration exposures before shutting down their equipment and trudging off to get some sleep. The new day for them will begin at noon or shortly thereafter.

The scenario just described depicts a typical night at a major modern observatory, such as Kitt Peak, with extensive resources and the largest telescopes. At smaller facilities, such as a typical university observatory, it is more common to find the astronomer working alone, most often using a cassegrain focus instrument, so that he must spend the night out in the dome with the telescope (where it can be very cold in the winter), and refreshing himself with a homemade lunch. The work pattern is similar, however, as is the desire not to waste any telescope time as the research program is carried out.

FIGURE 7.12 THE MAUNA KEA OBSERVATORY. *These are several of the large telescopes on the summit of Mauna Kea, including (from left to right on top of the ridge) the 3.6-meter Canada-France-Hawaii telescope, the University of Hawaii 2.2-meter telescope, and the 3.6-meter United Kingdom infrared telescope. In the foreground are the buildings for two new radio telescopes: a joint Dutch-United Kingdom 15-meter facility (in the large white building), and a 10-meter telescope built by Cal Tech. The planned 10-meter visible-light* Keck Telescope, *to be the largest in the world for several years, will be built just to the left of the 3-meter NASA* Infrared Telescope Facility *at the far left.* (T. P. Snow)

atmosphere, which absorbs all wavelengths shorter than about 3,100 Å, and absorbs in several wavelength regions in the infrared as well. Some portions of the infrared can penetrate to the ground, so infrared astronomy can be done at the major observatories (two of the three large telescopes on Mauna Kea, for example, were built specifically for infrared observations, as was the large telescope at the University of Wyoming). To cover the entire infrared spectrum or any of the ultraviolet (or X-rays), we must, however, place telescopes in orbit above the Earth's atmosphere. (Major space observatories are listed in Table 7.1). This is accomplished through the use of high-altitude planes (Fig. 7.15), balloons (which do not quite get above all the atmosphere), sounding rockets (providing only 5 minutes or so of observing time), or satellites.

Simply putting an infrared telescope above the atmosphere does not solve all the problems. As shown

FIGURE 7.13 A PANORAMA OF TELESCOPES AT KITT PEAK. *This sunset view shows the buildings housing several of the telescopes at Kitt Peak National Observatory near Tucson, Arizona.* (National Optical Astronomy Observatories)

FIGURE 7.14 THE CERRO TOLOLO INTER-AMERICAN OBSERVATORY. *This mountaintop in the Andes, some 200 miles north of Santiago, Chile, is the site of the United States' southern hemisphere national observatory. The largest telescope here is a 4-meter instrument.* (National Optical Astronomy Observatories)

FIGURE 7.15 THE KUIPER AIRBORNE OBSERVATORY. *The aircraft is used for infrared observations at altitudes above much of the water vapor in the atmosphere. Water vapor absorbs infrared light at many wavelengths and interferes with ground-based observations at these wavelengths.* (NASA)

FIGURE 7.16 *IRAS* IN ORBIT. *This is an artist's conception of the* Infrared Astronomical Satellite (IRAS) *as it would appear in orbit. This satellite revolutionized our knowledge of the cold matter in the galaxy.* (NASA)

in the discussion of Wien's law (Chapter 6), a telescope at room temperature glows at infrared wavelengths, which can drown out the infrared light from a target object. We solve this problem, on the ground or in space, by cooling the telescope, or at least the associated instruments, so that the emission is shifted to much longer wavelengths and does not confuse the observations. The cooling is usually done by circulation of liquid nitrogen (which has a temperature of 77 K) through a series of tubes surrounding the instrument. For observations of very long infrared wavelengths, even lower temperatures are required, and are obtained by using liquid helium (4 K). The recent *Infrared Astronomical Satellite (IRAS)* used liquid helium to cool its detectors as it mapped the sky (Fig. 7.16).

Infrared and ultraviolet telescopes work on the same principles as a reflector designed for visible wavelengths, bringing the light to a focus by means of a similar arrangement of mirrors. Radio telescopes operate in essentially the same way, but X-rays and γ-rays require rather different designs.

Ultraviolet astronomy began in the late 1940s with sounding-rocket observations of the Sun, and

TABLE 7.1 MAJOR TELESCOPES IN SPACE

Wavelength Region	Telescope*	Agency	Diameter	Dates
Gamma ray	*(Gamma Ray Observatory: GRO)*	NASA		(1990)
X-ray	*Einstein Observatory*	NASA	0.56 m	1978–1981
	Exosat	ESA		1982–1986
	(Roentgen Satellite: ROSAT)	German-NASA		(1990)
	(Advanced X-ray Astrophysics Telescope Facility: AXAF)	NASA		(1995)
Extreme ultraviolet	*(Extreme Ultraviolet Explorer: EUVE)*	NASA		(1991)
	(Lyman)	NASA-ESA	0.6	(1994)
Ultraviolet	*Copernicus*	NASA	0.9	1972–1980
	International Ultraviolet Explorer (IUE)	NASA-ESA-UK	0.41	1978–
	(Hubble Space Telescope: HST)	NASA-ESA	2.4	(1989)
	(Lyman)	NASA-ESA	0.6	(1994)
Visible	*(Hubble Space Telescope: HST)*	NASA-ESA	2.4	(1989)
	(Hipparcos)	ESA	0.29	(1988)
Infrared	*Infrared Astronomical Satellite (IRAS)*	Dutch-UK-NASA	0.57	1983–1984
	(Shuttle Infrared Telescope Facility: SIRTF)	NASA		(1998)
	(Infrared Space Observatory: ISO)	ESA		
	(Large Depolyable Reflector: LDR)	NASA	10	?
Radio	*(Cosmic Background Explorer: COBE)*	NASA		(1989)
	(Orbital Very Long Baseline Interferometry: Space VLBI)	NASA		(1995)
	(Quasat)	ESA		(1996)

*Instruments in parentheses are planned or under construction, but not yet launched.
†NASA is the U.S. National Aeronautics and Space Administration; ESA is the European Space Agency, a consortium of several countries.
‡Projected launch dates current as of September, 1987, and assume resumed operation of the Space Shuttle by mid-1988.

has progressed steadily ever since, with the development of a series of satellite observatories for ultraviolet spectroscopic observations. Currently one such spacecraft, the *International Ultraviolet Explorer* (Fig. 7.17) is in operation, and the *Space Telescope* (described briefly later in this chapter) will have a variety of ultraviolet capabilities.

FIGURE 7.17 THE INTERNATIONAL ULTRAVIOLET EXPLORER. *This NASA-operated spacecraft has been in operation since 1978, obtaining ultraviolet spectroscopic data for hundreds of astronomers.* (NASA)

THE X-RAY UNIVERSE

Extremely hot gases, according to Wien's law, emit most strongly at very short wavelengths. Temperatures of a million degrees or more produce emission in the X-ray portion of the electromagnetic spectrum, between 1 and 100 Å. There are even spectral lines in this wavelength region, produced by electrons jumping between orbits with very large energy differences.

It is not simple to build an X-ray telescope, however. This is because most materials are not shiny for X-rays; a mirror used in the usual way simply absorbs X-rays rather than reflecting them. It is possible, though, to reflect X-rays if the radiation strikes the reflecting surface at a very oblique angle, and X-ray telescopes can be designed this way (Fig. 7.18). Sometimes several reflections are necessary to bring the X-rays to a focus, because only a small deflection of the radiation can be accomplished at each reflection. The process of dispersing X-rays by wavelength is also more complicated than for visible light, again because most substances absorb X-rays.

Despite the difficulties, a number of rocket and satellite observatories have successfully probed the heavens in X-ray wavelengths, and a wide range of phenomena have been discovered. Sites of violent explosions and other highly energetic regions are the main sources of X-rays, so these spaceborne observations have opened a new universe for astronomers, one whose existence was only hinted at previously.

The most recent X-ray observatories were the U.S. *Einstein Observatory* satellite (Fig. 7.19) and the European *Exosat*, both of which imaged hundreds of astronomical objects in X-ray wavelengths. Currently X-ray astronomy is in a lull between major instruments, but in 1990 the U.S.-German *ROSAT* will be launched, and in the mid-1990s the United States will launch the *Advanced X-ray Astrophysics Facility* (*AXAF*) with a variety of X-ray detectors.

The shortest wavelengths of electromagnetic radiation, the γ-rays, also require special techniques for detection. Here even grazing-incidence reflections will not work, and as yet no instrument has been developed that can bring γ-rays to a precise focus. So far only rather primitive γ-ray telescopes have been flown, but plans are being made for a major satellite mission called the *Gamma Ray Observatory*, which promises to provide important new data on the extremely violent regions in space where nuclear reactions are taking place, such as inside the stars and in supernova explosions.

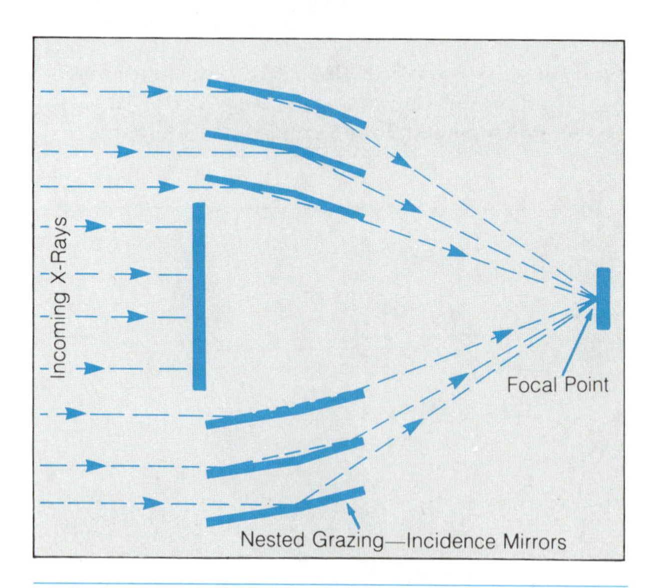

FIGURE 7.18 DESIGN OF AN X-RAY TELESCOPE. *A typical X-ray telescope consists of a set of nested rings, each able to reflect X rays to a focal point by means of grazing-incidence reflections. Several rings are used because each one individually has little effective collecting area.* (W. C. Cash)

RADIO TELESCOPES

We turn now to a discussion of telescopes for another invisible portion of the spectrum, the radio wavelengths. Radio astronomy has a relatively long history, dating back well before World War II. In 1931 Karl Jansky, a scientist with Bell Laboratories, noticed upon testing a radio antenna that the static he was receiving originated in the Milky Way (Fig. 7.20). This was the first indication astronomers had that celestial objects could radiate large quantities of energy outside the visible part of the spectrum, and as such was a tremendously important discovery, although its full significance was not widely recognized at first. Jansky was unable to pursue this discovery much further, but another American, Grote Reber of Wisconsin, became fascinated by Jansky's work and built his own radio antenna. Reber systematically mapped the sky and found many intense sources of radio emission; until the war was over, his work constituted the whole of the science of radio astronomy. British radar technicians during the war found that the Sun is a

FIGURE 7.19 THE EINSTEIN OBSERVATORY. *This X-ray telescope, which operated between 1978 and 1981, made many important discoveries about hot and energetic phenomena in the universe. The inset illustration shows the design of the nested grazing-incidence mirrors that collected X-rays and brought them to a focus.* (Harvard-Smithsonian Center for Astrophysics)

radio emitter. When radio techniques were further refined in the late 1940s, astronomers began to take more interest in the possibilities. One particularly provocative theoretical discovery, the prediction by the Dutch astronomer H. C. van de Hulst that hydrogen atoms in space should emit at a wavelength of 21 centimeters, spurred a great deal of activity, as several research groups worked to develop antennae capable of measuring this emission line. When the 21-centimeter line was detected in 1951, the immense promise

FIGURE 7.20 KARL JANSKY WITH THE RADIO ANTENNA HE USED TO DISCOVER THE FIRST CELESTIAL RADIO SOURCE. *Jansky's discovery of celestial radio emission was pivotal in inspiring the foundation of radio astronomy, although the field was not well established until after World War II.* (Neils Bohr Library)

of radio astronomy as a probe of the galaxy was realized. Now there are several major radio observatories around the world, and they provide insights into such diverse objects as the Sun, exploding galaxies, and molecules deep inside dark interstellar clouds.

The daytime sky is dark in radio wavelengths, so radio observatories need not shut down when the Sun rises, and most are on a 24-hour operating schedule.

A radio telescope works much like a reflecting telescope for visible wavelengths, except that generally everything is on a larger scale. The primary reflecting element need not be a smooth surface; instead it is usually a wire mesh. To radiation with such long wavelengths, a mesh acts like a shiny surface, as long as the wavelength observed is much longer than the spacing between wires in the mesh. At a point centered above the radio antenna, at what would be called the prime focus in a visible-wavelength reflector, is placed a **receiver,** a device that measures the intensity of the radiation. This receiver usually can be tuned so as to screen out unwanted wavelengths, serving in a sense as a filter. The intensity data can then be stored directly in a computer for later analysis.

A radio antenna must be very large in diameter to produce clear pictures of the sky. The resolution of a telescope is usually expressed in terms of how close together two objects can be and still be seen as separate objects, and is determined by the ratio of the wavelength being observed to the diameter of the telescope. Thus if long wavelengths, such as radio emission, are to be observed, the telescope must be very large in order to yield good resolution (that is, sharp images with fine detail). For a visible-wavelength telescope, a diameter of 10 centimeters (about 4 inches) provides a resolution of about 1 arcsecond, but a radio telescope would have to be several kilometers in diameter to achieve comparable resolution. Therefore even the largest radio telescopes (up to around 100 meters in diameter) have resolutions of several minutes of arc, and a radio picture of the sky tends to look fuzzy, with all the fine details smeared out.

A solution to this problem has been found and is being exploited at several radio observatories. By using two or more radio antennae simultaneously, it is possible to simulate the effect of one very large antenna. The farther apart the antennae are, the better the resolution that can be achieved. Whereas a visible-wavelength telescope looking through the Earth's atmosphere typically has a resolution of 1 to 2 seconds of arc (that is, two stars about 1 or 2 arcseconds apart can be distinguished as separate objects), radio telescopes used in combination have managed resolutions of a small fraction of a arcsecond. The techniques, called **interferometry,** is obviously very powerful, but it also is very complex, requiring precise knowledge of when particular radio waves arrive at the separate antennae and sophisticated computer processing to produce images or maps with high resolution (Fig. 7.21). Radio telescopes at various separations ranging from a few hundred feet to many thousands of miles have been employed; imagine the complexity involved in simultaneously using radio telescopes on opposite sides of the Atlantic!

The largest radio observatory in the United States is the National Radio Astronomy Observatory (Fig. 7.22), sponsored by the National Science Foundation, with a 300-foot antenna at Green Bank, West Virginia; a 30-foot dish at Kitt Peak (used mostly for observing interstellar molecules, whose emission lines occur at relatively short radio wavelengths; see Chapter 26); and a new observatory, called the Very Large Array (VLA), near Socorro, New Mexico (Fig. 7.23). The VLA has 27 movable 82-foot antennae arranged in a Y-shaped pattern spanning some 15 miles of the New Mexico desert, and is designed for interferometry, ultimately achieving a resolution of better than 0.1 arcseconds.

Another large U.S. radio observatory, at Arecibo, Puerto Rico, consists of a 1,000-foot dish built into a natural bowl in the mountains (Fig. 7.24). This large diameter provides fairly high resolution and sensitivity, but since the antenna cannot be pointed (it is immovable), observational coverage is somewhat limited in the north-south direction (the Earth's rotation provides east-west coverage).

Major radio observatories outside the United States include the Jodrell Bank Observatory, operated by the University of Manchester in England; the Westerbork Observatory in the Netherlands; the Parkes Observatory in Australia; and the 100-meter dish at Bonn, West Germany.

THE NEXT GENERATION

What kinds of new telescopes will be developed? Astronomers have a flair for dreaming up new and better ways to overcome technical difficulties, so in

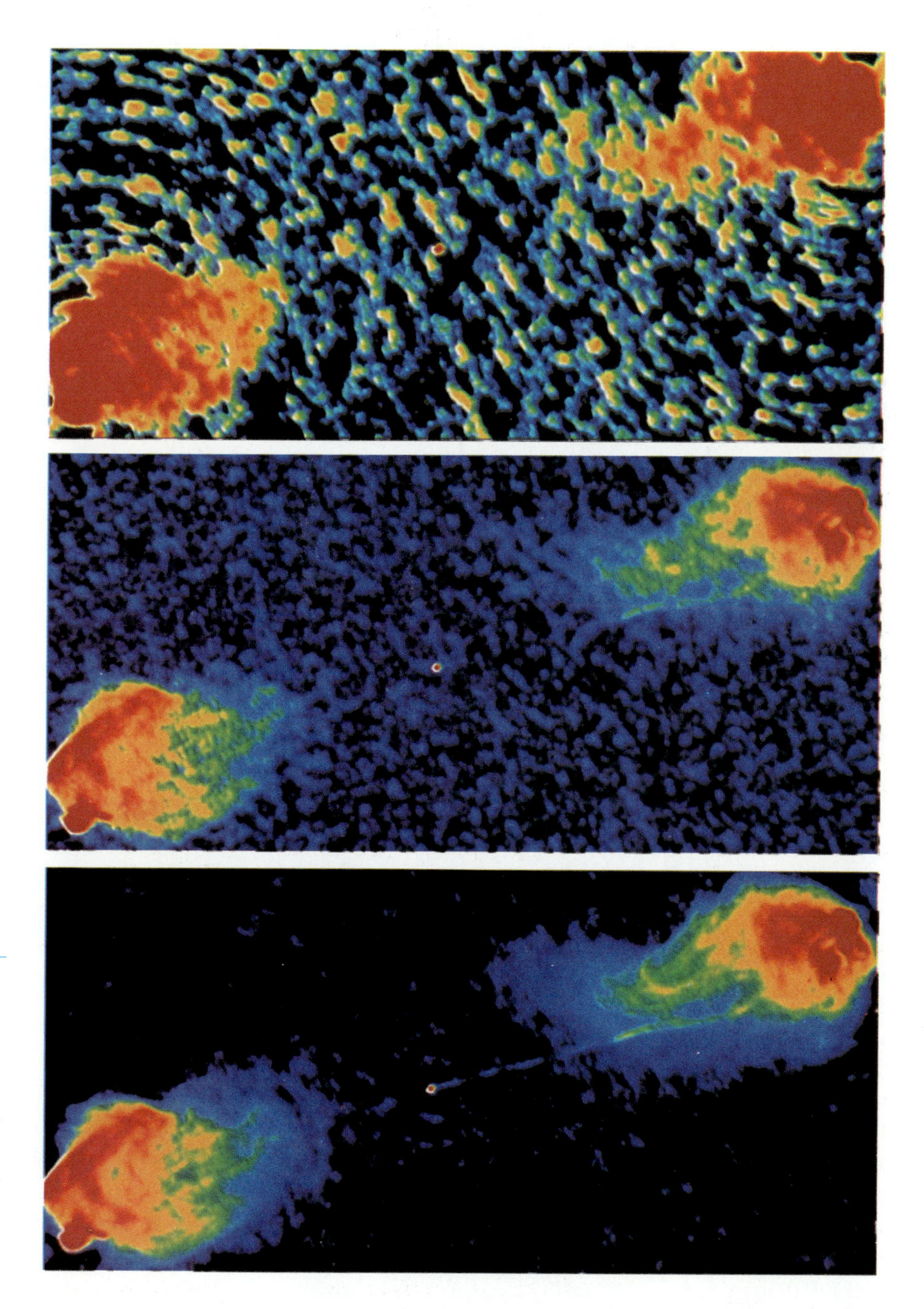

FIGURE 7.21 RADIO INTERFEROMETRY. *A raw interferometric map (top) consists of patterns created by the interference of radio waves received at separate antennae. Complex computer analysis is required in order to produce the final image (bottom). This is the image of a double-lobed radio galaxy (see Chapter 31).* (National Radio Astronomy Observatory)

some cases plans for innovative instruments are already well developed.

There has been widespread discussion of the next generation of telescopes, visible-wavelength instruments with diameters of 10 to 15 meters (remember, the largest telescope today is the Russian 6-meter one). Traditional methods for making telescope mirrors are very difficult for such large sizes, so novel techniques are being developed. One idea, already tested for modest-sized mirrors, is to pour the molten glass into a mold that is rotated so that the surface takes on a concave shape (Fig. 7.25). The mold is kept rotating until the glass hardens, producing a mirror that is already close to the required shape. This vastly reduces the time needed to grind the mirror to its precise shape, and also allows mirrors to be made

FIGURE 7.22 A LARGE RADIO TELESCOPE. *At left, the 300-foot dish at the* U.S. National Radio Astronomy Observatory, *in Green Bank, West Virginia. Despite its size, this telescope can be pointed in any desired direction, just as a visible-light telescope can.* (National Radio Astronomy Observatory)

FIGURE 7.23 THE VERY LARGE ARRAY. *This arrangement of twenty-seven radio dishes spread over a 15-mile wide section of New Mexico desert provides high resolution radio maps through the use of interferometry. The individual telescopes can move along railroad tracks, allowing the separations to be adjusted, depending on the type of data needed.* (National Radio Astronomy Observatory)

FIGURE 7.24 THE GIANT ARECIBO RADIO DISH. *This is the largest single radio antenna in the world. The 1000-foot dish is built into a natural bowl in the mountains of Puerto Rico. Since the telescope cannot be pointed in arbitrary directions, it must rely on the Earth's rotation to scan the sky.* (Arecibo Observatory, part of the National Astronomy and Ionosphere Center operated by Cornell University under contract with the National Science Foundation)

thinner than is possible in the conventional method, where a thick "mirror blank" is cast and then ground to shape. The largest mirrors now being planned for construction by the spin-casting technique will be 8 meters in diameter.

Another method for making large telescopes is to use the multiple-mirror concept, already successfully employed by the Multiple Mirror Telescope (see Figs. 7.9 and 7.10), which uses six 1.8-meter mirrors to duplicate the light-gathering ability of a single 4.5-meter mirror. The National New Technology Telescope (NNTT; Fig. 7.26) and the European Very Large Telescope are the most ambitious projects yet conceived using this technique; each will have four 8-meter mirrors (produced by the spin-casting method for the NNTT), equaling the light-collecting power of a single 16-meter mirror. The site for the NNTT, which is planned as a U.S. National facility, will be Mauna Kea.

The first of the "next generation" telescopes, the 10-meter Keck Telescope (Fig. 7.27), makes use of

FIGURE 7.25 CASTING A MIRROR ON A SPINNING TABLE. *Rotation of the table containing the mirror mold while the glass hardens facilitates construction of large mirrors with concave surfaces for use as telescope primary mirrors. This technique, being developed at the University of Arizona, makes possible relatively low-cost construction of large mirrors. Mirrors as large as 8 meters are being planned using this method.* (University of Arizona)

FIGURE 7.26 THE NATIONAL NEW TECHNOLOGY TELESCOPE. *This is a model of the* NNTT *showing how its four 8-meter mirrors will be mounted so that they move together as the telescope is pointed.* (National Optical Astronomy Observatories)

FIGURE 7.27 THE 10-METER KECK TELESCOPE. *This illustrates the plans for the large telescope being built jointly by Cal Tech and the University of California, to be placed on Mauna Kea.* (Keck Observatory, University of California and the California Institute of Technology)

neither the spin-casting technique nor the multiple-mirror approach. Instead a third type of design is being used, which is a hybrid of the other two. The Keck Telescope will use a segmented mirror; that is, a single 10-meter mirror that consists of a number of hexagonal segments, each of which can be controlled individually to bring its light to a common focal point. This is like the traditional single-mirror design, in that the mirror is a single structure, but like the multiple-mirror concept in that the individual mirror segments act as independent mirrors that can be adjusted individually. The Keck Telescope, a joint project of the University of California and the California Institute of Technology, is already under construction on Mauna Kea and is expected to be in operation by the early 1990s.

In addition to the NNTT, whose funding is not yet secure and whose future is uncertain, and the Keck Telescope, which is under construction, there are plans for several other very large telescopes. The European Southern Observatory, operated by a consortium of European nations (and which already has several telescopes including a 4-meter one in Chile) is planning the Very Large Telescope (mentioned above), making use of four 8-meter mirrors, each of which will be operable as a separate telescope. Several U.S. university groups are developing plans for telescopes in the 5- to 10-meter range as well, so in 10 years there may be as many as half a dozen telescopes in operation that are larger than the current record-holder, the Soviet 6-meter instrument.

In ultraviolet astronomy, scientists are looking forward to the *Hubble Space Telescope,* a 2.4-meter orbiting telescope that will have with it a variety of cameras, photometers, and spectrographs (Fig. 7.28). This facility, currently expected to be launched by the Space Shuttle in 1989, is intended to have a long lifetime, since it will be possible for astronauts to visit it occasionally to make repairs. It is even planned that the *Space Telescope* will be able to return to Earth if necessary to make repairs or refurbish the instruments. In addition to providing opportunities for ultraviolet observations, the *Space Telescope* will have

ASTRONOMICAL INSIGHT 7.2

ASTRONOMY IN THE TROPICS: THE MAUNA KEA OBSERVATORY

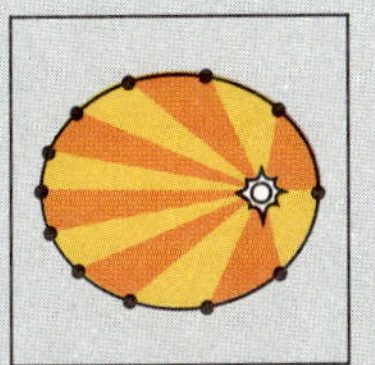

The observatory atop the 13,796-foot dormant volcano, Mauna Kea, on the island of Hawaii, has rapidly reached a position of preeminence in the world of astronomy. Twenty-five years ago, there were no telescopes on the gigantic cinder cone that forms the summit of the mountain, yet today the total light-collecting area of the telescopes there surpasses that of any other observatory in the world, and is continuing to grow. Mauna Kea is becoming recognized as the finest site of all, because of its unique combination of extreme altitude and high atmospheric transmissivity, the stability of the air overhead, its high percentage of clear nights, and its proximity to the equator, providing access to much of the southern sky.

The potential of the Hawaiian Islands as a possible observatory site was recognized about a century ago. When David Kalakaua, king of the Hawaiian empire, stopped in California in the 1880s on his way to visit England, he was given a tour of the Lick Observatory. He was impressed with what he saw, and, realizing that sites in the Islands would satisfy many of the criteria for a good observatory, he later remarked in correspondence that the possibility of stationing a telescope there should be considered. Although a solar observatory eventually was established on the extinct volcano Haleakala on the island of Maui, nothing happened on Mauna Kea until the early 1960s, when a group of scientists at the University of Hawaii won a grant from the National Aeronautics and Space Administration (NASA) to establish an observatory there, beginning with the construction of a 2.2-meter telescope that went into operation in 1970. Before the University of Hawaii astronomers took their bold action, it had been assumed that the nearly 14,000-foot altitude of Mauna Kea and the rugged terrain leading to the summit would make it impossible to build and operate an observatory there. It had generally been thought that any new observatory development would take place on Haleakala, with its 10,000-foot elevation and easier access.

The high altitude of Mauna Kea certainly has an effect on human efficiency, and for that reason construction goes slowly there, and observers must be extra careful not to make careless mistakes. It is also true that the road to the summit is rough and hazardous and that the weather can be harsh; there are many tales of observers stranded at the summit by snowstorms. Nevertheless, Mauna Kea has more than lived up to the expectations of King Kalakaua and the astronomers at the University of Hawaii who established the observatory.

There are now five very large telescopes on Mauna Kea, in addition to the university's 2.2-meter instrument. Two of the new ones are optimized for infrared observations, because the air above Mauna Kea is very dry and therefore relatively free of the infrared obscuration caused by water vapor in the atmosphere. These are the 3.0-meter NASA Infrared Telescope Facility and the 3.8-meter United Kingdom Infrared Telescope. Another recent addition is the 3.6-meter Canada-France-Hawaii Telescope, sponsored jointly by the governments of Canada and France, along with the University of Hawaii (which provides dormitory and food services for all the telescopes on the mountain). The two newest telescopes on Mauna Kea are both radio dishes: the 10-meter California Institute of Technology instrument, and a 15-meter telescope sponsored jointly by the United Kingdom and the Netherlands. Both of these telescopes, designed for observations at wavelengths of less than one millimeter (a virtually unexplored portion of the spectrum), have just gone into operation in early 1987.

The Mauna Kea Observatory has not finished growing. As mentioned in the text, the 10-meter Keck Telescope, a joint effort by the California Institute of Technology and the University of California, is already under construction and will be the world's largest visible-light telescope when it goes into operation in the early 1990s. The National New Technology Telescope, a multiple-mirror instrument with the collecting area of a 16-meter telescope, is being planned for construction on Mauna Kea. Space is being reserved on one of the summit cinder cones for the planned Japanese national telescope. Other large and wonderful instruments will follow in years to come. We can only wonder what King Kalakaua would have thought had he been able to see these developments.

ASTRONOMICAL INSIGHT 7.3

NEW TECHNOLOGY TELESCOPES

Even though ambitious plans for new, unprecedented telescopes are under way, astronomers are already looking forward to even better and larger ones. In today's climate of uncertain funding for large-scale projects, it is not clear which of the grand plans will eventually turn into reality. One of the grandest plans calls for a huge telescope, equivalent in collecting area to a single mirror 16 meters in diameter, called the National New Technology Telescope, or NNTT.

The NNTT is envisioned as a national facility, built with federal funds, unlike the privately-funded large-telescope projects already under way (such as the 10-meter Keck Telescope, described in the text, and two 8-meter telescope projects being undertaken by the University of Arizona with several partner universities). The NNTT has been approved as a high priority of the U.S. astronomical community and has been put into the long-range planning of the National Science Foundation, but it is not known when, or whether, the money (about $125 million) will become available. Ironically, it may happen that the smaller telescopes being built by university groups will divert the interest and funding priorities of the national scientific community enough to delay or kill the NNTT. There are many, however, who think this would be unwise, because none of the current projects will match the capabilities of the NNTT.

After considerable discussion, the design finally chosen for the NNTT is a multiple-mirror design, consisting of four 8-meter mirrors. (It is an interesting exercise for the student to show that these four mirrors have the same collecting area as a single 16-meter mirror.) The design is shown in Figure 7.27.

The four mirrors will be able to act together, bringing light to a single focus point, much like the existing Multiple Mirror Telescope (MMT), but they will have other important capabilities as well. The key special feature, compared with the MMT and the planned telescopes, is that the four mirrors will also be able to operate independently. This will allow, for example, simultaneous observations of a target using different filters or spectrographs, and even more importantly, will allow interferometry to be carried out. The total collecting area of the four mirrors working together will allow observations of very faint objects, while in the interferometry mode, the NNTT will be capable of angular resolutions around 0.005 seconds of arc. (Recall that typically the best resolution achieved from the ground is about 0.1 arcsecond, because of atmospheric effects; interferometry overcomes these effects.) Interferometry will be best done in infrared wavelengths, which will be very useful, for example, for studying the disks of red giant stars. (Angular sizes as small as one-tenth the diameter of Betelgeuse will be resolved, allowing the pattern of flowing gases across the star to be studied.)

No definite time scale has been set for building the NNTT although the site (Mauna Kea) has been chosen. The biggest question today is whether the funding is supported by the U.S. astronomical community. Meanwhile, a consortium of European countries is forging ahead with plans for a telescope of a similar size to be located in South America. If those plans succeed while the NNTT falters, then the U.S. will have relinquished its leadership in large-telescope technology. The next few years will determine the outcome.

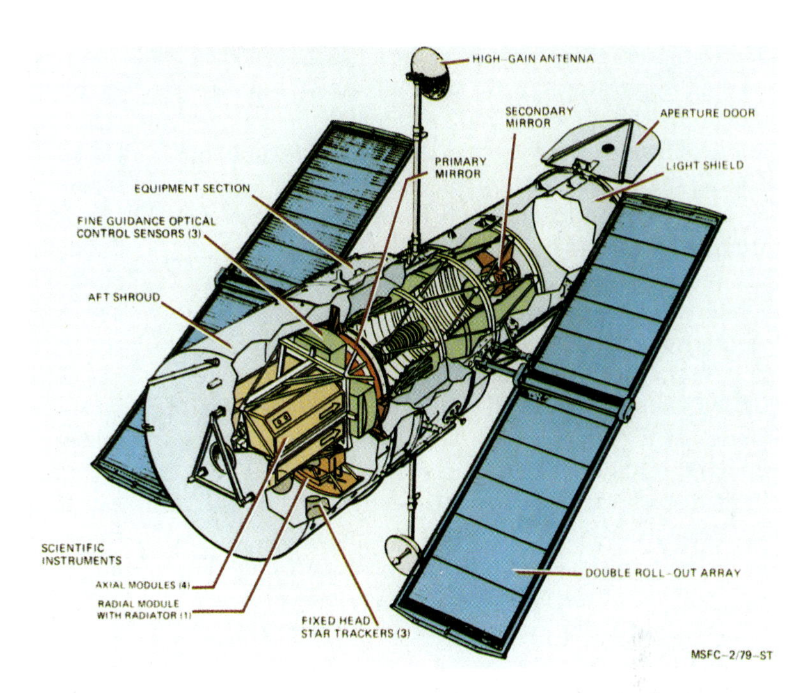

FIGURE 7.28 THE HUBBLE SPACE TELESCOPE. *This is a schematic drawing of the 2.4-meter telescope planned for launch into Earth orbit during 1987. The open end of the telescope is pointed away from us in this view; the near end houses control systems, cameras, and other scientific instruments. The large "wings" are solar panels, which provide electrical power for the spacecraft.* (NASA)

greater resolution and be more sensitive in visible wavelengths than ground-based telescopes, because it will avoid problems introduced by the Earth's atmosphere.

Even farther in the future, we might expect someday to see a very large visible- and ultraviolet-wavelength telescope in space, to capitalize further on the advantages of getting above the atmosphere. In the realm of sheer speculation, some astronomers have begun thinking about placing telescopes on the Moon, where the Earth's atmosphere can be avoided without the size limitations for a satellite in Earth orbit. There has even been some discussion of eventually sending a probe toward Alpha Centauri (the nearest star) that would radio back information on interstellar gas and dust and on the star itself, as seen from close up. Such a mission would take several years, since alpha Centauri is 4 light-years away, so the scientists waiting on Earth will have to be patient.

PERSPECTIVE

We have seen, in broad outline, how astronomers gather the light, in all wavelengths, that carries the information they seek. We have learned that, for many wavelengths, the same basic telescope design will work, although it may be necessary to place the instrument in orbit, above the Earth's atmosphere. To observe other portions of the electromagnetic spectrum, new technology, such as radio telescopes and X-ray and γ-ray instruments, has had to be developed. Today astronomers can look at the universe in nearly any wavelength. We are ready now to see what has been learned.

SUMMARY

1. Telescopes are better than the human eye for astronomical observations because they have larger collecting areas, have superior resolution, and can accumulate light over longer periods of time.

2. Telescopes for visible light use lenses or mirrors to bring light to a focus, but all big telescopes are reflectors, which can be made larger and are free of chromatic aberration.

3. Reflecting telescopes can bring light to a focus in a variety of ways for different purposes.

4. Auxiliary instruments are used to analyze the light once it is brought to a focus, by photographing the image, measuring its brightness, or dispersing the light for spectral analysis.

5. An observatory site must be chosen for clear weather, high atmospheric transmissivity, low turbulence in the air overhead, minimal pollution, remoteness from city lights, and the proper latitude for viewing the desired part of the sky.

6. Infrared and ultraviolet observations can be made through the use of the same optical designs as visible-light telescopes, but must be carried out from above the Earth's atmosphere, which blocks these wavelengths from reaching the ground.

7. Observations in X-ray and γ-ray wavelengths require novel telescope designs and must be carried out from space.

8. Most radio telescopes are similar in design to reflecting telescopes for visible light, but must be built very large in order to provide good resolution.

9. In all wavelengths, but most commonly in the radio portion of the spectrum, several telescopes can be used together to produce high-resolution images by interferometry.

10. Planned future telescopes include the Hubble Space Telescope, several giant reflectors that will be built on the ground, and a variety of radio, X-ray, ultraviolet, infrared, and γ-ray telescopes, many of them to be launched into orbit on the Space Shuttle.

REVIEW QUESTIONS

Thought Questions

1. Summarize the benefits of using a telescope for astronomical observations, compared with using only the naked eye.

2. What are the advantages of reflecting telescopes over refractors?

3. Summarize the various focal arrangements used in reflecting telescopes, and give the advantages of each.

4. Discuss the contrasts between observational astronomy today and what it was like in the time of Tycho Brahe.

5. List the reasons why the summit of Mauna Kea, in Hawaii, is an excellent observatory site.

6. Drawing upon information from Chapter 6, explain why telescopes are needed for observations in wavelengths other than visible light. Try to think of examples of objects that might best be observed in other wavelengths such as the ultraviolet, the infrared, the radio, and the X-ray portions of the spectrum.

7. How is an ultraviolet telescope similar to an ordinary visible-light reflector, and how is it different? How does an infrared telescope compare with a normal visible-light reflector? How does a radio telescope compare?

8. Explain why the design of an X-ray telescope must be very different from that of a visible-light telescope to bring X-rays to a focus.

9. Explain why radio telescopes ordinarily cannot map the sky with as much precision (that is, with resolution as high) as visible-light telescopes can. How do you think ultraviolet telescopes compare in resolving ability with visible-light telescopes?

10. Explain why the *Space Telescope* will have advantages over ground-based telescopes, even for visible-wavelength observations.

Problems

1. The largest refracting telescope has a lens diameter of 40 inches, while the largest reflector has a diameter of 6 meters, or 236 inches. How much greater is the light-collecting power of the 6-meter reflector?

2. Show that the six 72-inch mirrors of the Multiple Mirror Telescope have a total light-collecting area equivalent to that of a single mirror 176 inches in diameter.

3. Suppose the cage in which an astronomer sits to make prime-focus observations with the 200-inch Hale Telescope (on Palomar Mountain) is a cylinder with a diameter of 40 inches. What fraction of the incoming light is blocked by the observer's cage?

4. Suppose you want to observe a planet whose surface temperature is 80 K. At what wavelength does this planet emit most strongly? Compare this wavelength with the wavelength of maximum emission for the telescope detector if it is cooled with liquid nitrogen (77 K) or with liquid helium (4 K). Which coolant is better to use for this observation?

5. The resolving power of a telescope, defined as the smallest separation between two objects that can be seen as separate objects, is proportional to the

wavelength being observed and inversely proportional to the diameter of the telescope. For a telescope of 10 cm diameter, the resolving power for blue light (wavelength 4,000 Å) is 1 arcsecond. What is the resolving power for a telescope of 1-meter diameter, observing at the same wavelength?

6. What is the resolving power for a 1-m infrared telescope, observing at a wavelength of 40,000 Å? What is the resolving power for a police radar detector, a microwave telescope having a diameter of 10 cm, observing at a wavelength of 4 cm?

7. What is the frequency of a 21-cm photon from a hydrogen atom?

8. How large would a radio telescope have to be to obtain a resolving power of 1 arcsecond for observations at a wavelength of 21 cm?

9. The angular diameter of Pluto, when the planet is at opposition, is 0.1 arcseconds. Will the Space Telescope be capable of making out surface markings on Pluto?

10. If an interstellar probe could maintain an average speed of one-half the speed of light, how long would it take to reach alpha Centauri? How long after its launch from Earth would it be until we received the first data from the star?

ADDITIONAL READINGS

Asimov, I. 1975. *Eyes on the Universe*. Boston: Houghton Mifflin.

Bahcall, J. N., and L. Spitzer, Jr. 1982. The space telescope. *Scientific American* 247(1):40.

Bartusiak, Marcia. 1984. Very Large Astronomy (VLA). *Science* 84 5(6):64.

Beatty, J. Kelly. 1985. HST:Astronomy's greatest gambit. *Sky and Telescope* 69(5):409.

diCicco, Dennis. 1986. The Journey of the 200-inch mirror. *Sky and Telescope* 71(4):347.

Evans, David S., and J. Derrall Mulholland. 1986. *Big and bright: A history of the McDonald Observatory*. University of Texas Press.

Field, George B. 1984. The future of space astronomy. *Mercury* 13(4):98.

Gordon, Mark A. 1985. VLBA-A continent-size radio telescope. *Sky and Telescope* 69(6):487.

Habing, H. J., and G. Neugebauer. 1984. The infrared sky. *Scientific American* 251(5):48.

Harrington, S. 1982. Selecting your first telescope. *Mercury* 11(4):106.

Horodyski, Joseph M. 1986. Reach for the stars: The story of Mount Wilson Observatory. *Astronomy* 14(12):6.

Janesick, J., and M. Blouke. 1987. Sky on a chip: The fabulous CCD. *Sky and Telescope* 74(3):238.

King, H. C. 1979. *The history of the telescope*. New York: Dover.

Kirby-Smith, H. T. 1976. *U.S. observatories: a directory and travel guide*. New York: Van Nostrand.

Labeyrie, A. 1982. Stellar interferometry: A widening frontier. *Sky and Telescope* 63(4):334.

Longair, Malcolm. 1985. The scientific challenge of Space Telescope. *Sky and Telescope* 69(4):306.

Meyer-Arendt, J. R. 1972. *Introduction to Classical and Modern Optics*. Englewood Cliffs, N.J.: Prentice Hall.

Miczaika, G. R. 1961. *Tools of the astronomer*. Cambridge, Mass.: Harvard University Press.

Mims, S. S. 1980. Chasing rainbows: The early development of astronomical spectroscopy. *Griffith Observer* 44(8):2.

Overbye, Dennis. 1982. Night flight to the stars. *Discover* 3(8):38. (Kuiper Airborne Observatory)

________. 1984. The secret universe of IRAS. *Discover* 5(1):14.

Physics Today 1982. 35(11): (several review articles on telescopes for various wavelengths).

Shore, Lys Ann. 1987. IUE: Nine years of astronomy. *Astronomy* 15(4):14.

Tucker, Wallace, and Riccardo Giacconi. The birth of X-Ray astronomy. Parts 1, 2. *Mercury* 14(6):178; 15(1):13.

Tucker, Wallace, and Karen Tucker. 1986. *The cosmic inquirers: Modern telescopes and their makers*. Cambridge, Mass.: Harvard University Press.

Verschuur, Gerritt L. 1987. *The invisible universe revealed: The story of radio astronomy*. New York: Springer-Verlag.

GUEST EDITORIAL

TELESCOPES AND THE SEARCH FOR LIFE

J. R. P. Angel

J. R. P. (Roger) Angel is recognized as one of the most innovative and important telescope designers and builders in astronomy today. He has developed novel techniques for constructing thin, highly concave (low focal ratio) mirrors for large telescopes, making feasible a whole new generation of instruments. (The technique of spin-casting large mirrors, a principal Angel innovation, is discussed in Chapter 7). Educated primarily in his native England, Dr. Angel has spent his entire professional career in the United States, initially at Columbia University and, since 1973, at the University of Arizona. His scientific interests originally involved studies of atomic structure, then branched out into a wide array of astronomical problems, and in the past 15 years have included an increasingly large emphasis on the design and development of instrumentation. As a central figure in the development of new large telescopes, he has a role in nearly every major instrument under development in the United States, including the proposed National New Technology Telescope *(Chapter 7). He has won the prestigious Pierce Prize of the American Astronomical Society for his contributions to telescope technology. In addition to his major impact on scientific research and telescope development, Dr. Angel has served the astronomical community in several other ways, as a leader in various advisory committees to the National Science Foundation, to NASA, and to the national Academy of Sciences. In this fascinating article he describes a potential method for detecting planets inhabited by life forms similar to those on Earth.*

Do nearby stars that are like the Sun have planetary systems? If so, do they have planets like the Earth, with abundant liquid water? Have such planets developed life similar to that on Earth? We do not now know the answers to these questions, but we could find them by placing sufficiently powerful telescopes in Earth orbit.

The other planets in the solar system show no obvious signs of life, and their physical conditions are not favorable. If there is or was any life, direct exploration seems to be required to uncover it. But a distant planet that is, like the Earth, totally in the grip of life could readily be identified if we could obtain spectra of its atmosphere.

Spectroscopic analysis can show whether temperature and atmospheric composition favor the formation of life. More significantly, such analysis can show whether the atmosphere has been modified by the presence of life, as has the Earth's. No oxygen was present in the Earth's atmosphere until it was generated from carbon dioxide by living organisms in the process of photosynthesis. As far as we know, free oxygen will not exist except by biological action. Thus the presence of abundant oxygen in the atmosphere of an earthlike planet would constitute strong evidence for life.

Oxygen can be detected because it absorbs light at particular wavelengths, so we might hope to detect oxygen absorption in the light reflected from an earthlike planet. The strongest spectral features, due to ozone (the form of oxygen with three atoms bonded together), lie at ultraviolet wavelengths. These features would be very difficult to detect, because a sunlike star emits relatively little ultraviolet light, so there would be little reflected ultraviolet light from a planet of such a star. The visible-wavelength features due to normal molecular oxygen (O_2) are also difficult candidates for detection, because they are not very strong.

Perhaps the best hope lies in detecting infrared features of oxygen. A planet emits infrared radiation due to its surface temperature (perhaps 300 K for an earthlike planet), and ozone (as well as water vapor and carbon dioxide) in its atmosphere would create absorption features against this thermal emission.

No existing or planned telescopes are powerful enough to detect either the reflected light or the

GUEST EDITORIAL

thermal emission of earthlike planets of even the nearest stars. The *Hubble Space Telescope* and the new generation of ground-based telescopes would just be able to make an optical detection if they were not blinded by the light of the parent star. The planned *Shuttle Infrared Telescope Facility (SIRTF)* space observatory could detect the small amount of radiated heat, but not against the heat of the star and the background of heat emitted by zodiacal dust particles. The brightness ratio of the Sun to the Earth is nearly ten billion in visible light, and ten million in thermal emission. Similar ratios would be expected for any planet supporting earthlike life assuming comparable size to the Earth and distance to the host star. These similarities are needed to obtain the temperature of liquid water, and adequate gravity to preserve the atmosphere.

The huge differences in brightness compounded by the tiny angular separation are what make the detection problem so difficult. Suppose we wish to look not only at the nearest star but at the nearest 100, which lie at distances up to about 20 light-years. About half are in double systems, where conditions may have been quite different to those that led to the formation of the solar system. Most of the remainder are very cool stars which may be too faint and cold to support earthlike planets. But the sample includes nine brighter, single stars that are excellent candidates. Just as in our own solar system, planets warm enough to have liquid water will be relatively close to their star, and will not be nearly as prominent as more distant giant planets. At a distance of 16 light-years, the radius of the Earth's orbit would subtend an angle of ⅕ arcsecond. To detect a planet so close to a bright star, the stellar image must be very sharp. If any light spills out from the image core, it must not fall in the ring at ⅕ arcsecond radius, where the search must be made.

Whenever an image is formed by light passing through an opening such as a telescope aperture, the diffraction and interference of the light coming through different parts of the opening create a central image with circular rings around it. It is possible, by specially treating the metal of a telescope and by making the mirror extraordinarily accurate in smoothness and shape, to broaden the dark gap between the central image and the first bright ring. If this is done well enough, it is possible to broaden the dark gap beyond ⅕ arcsecond, so that the image of an earthlike planet would fall into the gap and be visible. The mirror size needed to do this depends on the wavelength being observed. At the wavelength of a molecular oxygen (O_2) feature in the far-red portion of the visible spectrum (7600Å), the mirror must be 1.5 meters in diameter, and at the wavelength of the infrared feature of ozone (10 microns, or 10,000Å), the mirror diameter must be 20 meters.

The number of photons of 7600Å wavelength received by a 1.5 = meter telescope would be very small, so that a number of such telescopes, and very long exposure times, would be needed to detect an earthlike planet. In the infrared, a greater flux of photons would be received by a 20-meter telescope, and the spectral feature due to ozone is stronger than the one at 7600Å due to O_2. To obtain spectra with a 20-meter infrared telescope good enough for detection of an earthlike planet does not require unreasonably long exposure times. Less than an hour would be needed to detect the ozone signature, if ozone is present in quantities comparable to that in the Earth's atmosphere.

A very dark ring in the diffraction pattern of the star's image will not be obtained unless the imaging comes very close to the theoretical limit of perfection. Thus the distortion introduced by atmospheric turbulence will never be small enough to allow this type of observation from the ground. Even in space, telescopes made for normal diffraction-limited imaging are not good enough. The *Hubble Space Telescope* has the most perfect mirror yet made, but it is about 50 times too rough to detect a planet at visible-light wavelengths. Ripples in its surface spread out starlight into the dark ring. Normally the amount would be negligible, but it would totally swamp the faint light of an earthlike planet.

The spreading effect on infrared waves caused by ripples in the mirror is much less than on visible light, and for the infrared, mirrors of

GUEST EDITORIAL

even slightly inferior quality to the *Space Telescope* would work. NASA has under study a 20-meter infrared space telescope, the *Large Deployable Reflector*. However, the surface accuracy for its proposed diffraction-limited operation at 30 microns wavelength (30,000Å) would be about 100 times worse than needed to detect earthlike planets at 10 microns (10,000Å). Furthermore, it is not presently envisaged that the telescope should be cooled. Cooling to 80 K is needed to detect faint 10-micron emission, to eliminate the thermal background of the telescope itself.

How can telescopes of the same size but of much higher quality than today's billion dollar space instruments be built? If this advance is to be made, we must learn how to get higher accuracy without much increase in cost. The costs of launching and operating satellites will not necessarily be larger because of high accuracy. What we must do, though, is learn how to make very accurate components at acceptable cost, with good ways to ensure that the accuracy can be maintained in the space environment.

For optical detection, mirrors much smoother than the *Hubble Telescope* mirror could be made by better polishing and testing methods. Already there have been significant advances. The ripples characteristic of the short strokes of rigid polishing tools should be eliminated by the method of stressed lap polishing. In this method the mirror is polished by full strokes with a relatively large tool, whose shape is changed as it moves so as always to fit the desired mirror surface. New ways to correct large scale errors are under development. These include glass removal by ion beams. It may also prove possible to control the shape and smoothness of the metallic reflecting surface that is deposited on the glass by computer control of the deposition of the metal layer. This technique could be used for a final touch-up of the surface figure in orbit, when the mirror is under the exact thermal and mechanical conditions of operation.

The large mirror needed for the infrared must be made of segments, as will the 10-meter Keck telescope, now under construction [see Chapter 7]. The challenge of achieving accuracy an order of magnitude higher than in the Keck telescope is not as difficult as might first be expected. In space no wind blows to distort and vibrate the assembly. Also, no weight is pulling in different directions, and the system can be optimized for one fixed operating temperature. A simple three-point movable support is all that is needed for each segment. Precise alignment of the segment array will be facilitated by the light of the bright host star.

Many technical issues need to be worked through beyond simple surface accuracy, but this is not the place for a full discussion. I favor the concept of a large infrared telescope for the search for earthlike planets, because such a telescope would more easily yield a detection of such planets, and would also be a universally powerful tool for astronomy at both visible and infrared wavelengths. However, it will require assembly in orbit, and must rely critically on good thermal design and shielding to allow for operation at 80 K. If instead the search is to be made with a visible-light telescope, the technical challenges are to make the incredibly perfect optics needed to contain the ten billion times brighter star image, and to build up extremely weak signals by very long exposure times.

In conclusion, it is now clear that there are practical possibilities for the detection and spectroscopy of extrasolar earthlike planets. We can look forward to studying the composition of the atmospheres of many planets around nearby stars, assuming the Sun is not unique. It would now be valuable to develop models of atmospheric spectra corresponding to different scenarios for the origin and evolution of life. These are needed to help find the best telescope strategy for discriminating between planets with and without life.

SECTION III

The Solar System

In this portion of the book, we will examine the properties of our system of Sun and planets, with a view toward understanding how this system came to be. We will also learn about the fascinating phenomena that have been revealed through its exploration.

The solar system is complex, and we will see that it is made up of a variety of individual parts. We will also find, however, that there are a number of underlying, general processes at work, so that as we go through the system, examining individual objects, we will be able to make comparisons, to see how these general processes have influenced different situations.

Among the events that we will encounter time and again are tidal forces and other gravitational effects which are responsible for an astonishing assortment of phenomena, such as the spin rates of many of the planets and satellites, the locations of their orbits, and the failure of the asteroids and the ring particles of the outer planets to form large bodies. We also will allude often to the process by which the solar system formed (namely, from the collapse of a rotating interstellar cloud), because on our tour of the universe we will encounter many recognizable artifacts of this process. Another phenomenon that will be mentioned often is the solar wind, a steady stream of subatomic particles from the Sun that pervades the solar system, creating important effects in the outer atmospheres of many of the planets.

As we study the planets, we will make a number of comparisons among them. To establish some of the standards for these comparisons, we will begin with the best-studied and understood of all the planets, the Earth. We will then tour the Moon and the other terrestrial planets, the innermost four (including the Earth), which have similar overall properties. Following this, we will turn our attention to the outer giant planets, starting with Jupiter, the biggest of them all. Again, the first example will serve as the standard for comparison with the others. Finally, we will discuss other bodies in the solar system, which are important leftovers from its formation, before wrapping up the section with a description of the formation and evolution of the solar system.

CHAPTER 8

THE EARTH AS A PLANET

The ultraviolet glow of hydrogen in the Earth's upper atmosphere. (NASA)

CHAPTER PREVIEW

Of all the bodies in the heavens, the Earth (Table 8.1) has been the best studied, although it is equally true that many mysteries remain. Understanding the planet we live on has practical, as well as philosophical, importance. Furthermore, working at such close quarters with the object of study provides us with numerous advantages, including the ability to observe in great detail and over long periods of time, and the possibility of making direct experiments by probing and sampling the surface of the Earth.

TABLE 8.1 EARTH

Orbital semimajor axis: 1.000 AU (149,600,000 km)
Perihelion distance: 0.983 AU
Aphelion distance: 1.017 AU
Orbital period: 365.256 days (1.000 years)
Orbital inclination: 0°0′0″
Rotation period: 23 hr, 56 min, 4.1 sec
Tilt of axis: 23°27′
Diameter: 12,734 km (1.000 $D_{\oplus}$)
Mass: 5.974×10^{27} g (1.000 $M_{\oplus}$)
Density: 5.517 g/cm^3
Surface gravity: 980 cm/sec^2 (1.000 Earth gravity)
Escape velocity: 11.2 km/sec
Surface temperature: 200–300 K
Albedo: 0.31 (average)
Satellites: 1

FIGURE 8.1 THE EARTH AS SEEN FROM SPACE (NASA)

It would be beyond the scope of this text to attempt to summarize all that is known about the Earth. Our discussion will emphasize those aspects that can be compared with what we know of the other planets. We should keep in mind, however, that not only is the Earth much more complex than it is portrayed to be here, but so, undoubtedly, are the other planets. Each is a unique and varied world, probably as diverse as the Earth, except for the influences of life.

From afar, the Earth is a brilliant planet, with an overall bluish color created by the scattering of light in its atmosphere (Fig. 8.1). Much of it is usually cloud covered, and where it is clear, brownish land masses and large regions of water appear. The Earth appears so bright because it has a high **albedo,** the fraction of light that is reflected from it. The Earth's albedo is 0.31, meaning that over 30 percent of the light that strikes it is reflected.

THE ATMOSPHERE

We begin by examining the thin layer of gas that surrounds the solid Earth. The Earth's atmosphere is composed of a variety of gases (Table 8.2), as well as a distribution of suspended particles called **aerosols.** The gases, which are evenly mixed together up to an altitude of about 80 kilometers (km), are primarily nitrogen and oxygen, in the form of the molecules N_2 and O_2. Nearly 80 percent of the gas is nitrogen, about 20 percent is oxygen, and all the other species exist only as traces, although in many cases they are important traces. In addition to this standard mixture of gases, a number of others appear in the atmosphere. Among these, water vapor (H_2O) is important, because of its direct role in supporting life. Others, such as ozone and carbon dioxide, are also essential, but for less direct reasons. Ozone (designated O_3) is a form of oxygen, in which three oxygen atoms are bound together instead of two. Unlike ordinary oxygen (O_2), ozone in the upper atmosphere acts as a shield against ultraviolet light from the sun, which could be harmful to life forms if it penetrated to the ground.

The importance of carbon dioxide (CO_2), which is produced by respiration in animals, by fires, and by many industrial processes, lies in its ability to trap the Sun's radiant energy within the atmosphere, creating overall heating by a process called the **greenhouse effect.** This heating process has progressed to a much greater extent on the planet Venus than on the Earth (and will be discussed more fully in Chapter 10).

The principal atmospheric constituents, nitrogen and oxygen, are both released into the air by life forms on the surface. This process has been important in the evolution of the Earth's atmosphere and leads to the expectation that other planets in the solar system are unlikely to have the same composition (although nitrogen is produced also by processes that do not involve life forms, so we might find it elsewhere). Nitrogen is released into the Earth's atmosphere by the decay of biological material and by emission from volcanic eruptions. Oxygen comes almost exclusively from the photosynthesis process in plants, a process in which carbon dioxide is converted into oxygen with the assistance of radiant energy from the Sun.

TABLE 8.2 COMPOSITION OF EARTH'S ATMOSPHERE

Gas	Symbol	Fraction*
Nitrogen	N_2	0.77
Oxygen	O_2	0.21
Water vapor	H_2O	0.001–0.028
Argon	Ar	0.0093
Carbon dioxide	CO_2	3.3×10^{-4}
Neon	Ne	1.8×10^{-5}
Helium	He	5.2×10^{-6}
Methane	CH_4	1.5×10^{-6}
Krypton	Kr	1.1×10^{-6}
Hydrogen	H_2	5×10^{-7}
Ozone	O_3	4×10^{-7}
Nitrous oxide	N_2O	3×10^{-7}
Carbon monoxide	CO	1.2×10^{-7}
Ammonia	NH_3	1×10^{-8}

*The listed values are fractions by number.

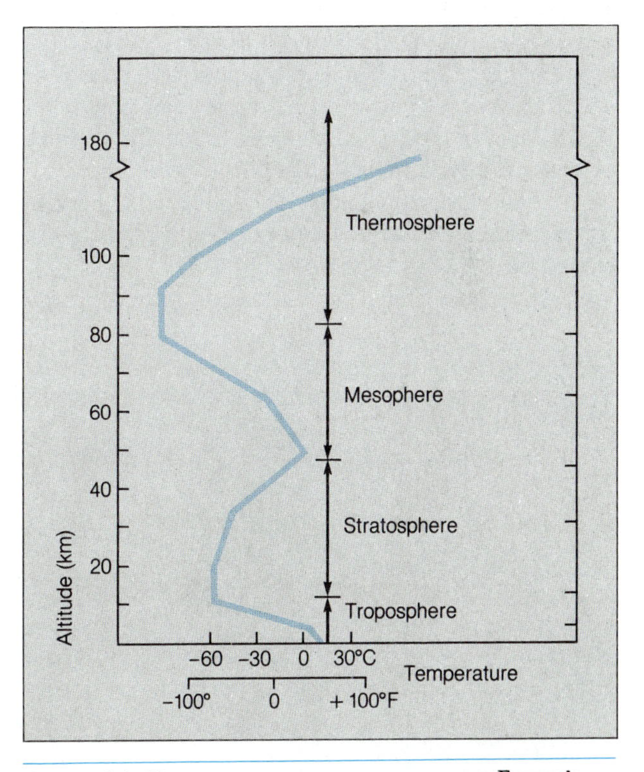

FIGURE 8.2 THE VERTICAL STRUCTURE OF THE EARTH'S ATMOSPHERE. *The distinct zones are defined according to the temperature variation with height, which influences atmospheric motions.*

Heating and Atmospheric Motions

Although there are obvious fluctuations in the temperature of the atmosphere, particularly near the surface, there is a definite, stable temperature structure as a function of height (Fig. 8.2). This structure includes several distinct layers in the atmosphere: the **troposphere,** from the surface to about 10 km in altitude, where the temperature decreases slowly with height and where the phenomenon that we call **weather** occurs; the **stratosphere,** between 10 and 50 km, where the temperature increases with height; the **mesosphere,** extending from 50 to 80 km, where the temperature again decreases; and the **thermosphere,** above 80 km, where the temperature gradually rises with height to a constant value above 200 km. In layers where the temperature decreases with height, there can be substantial vertical motion of the air, whereas in layers where it gets warmer with height, the air is stable, with only lateral motions.

The principal source of heating throughout the atmosphere is the Sun's radiation, but the energy is deposited in different ways at different levels. Much of the heat (that is, infrared radiation) is absorbed by the ground. This heating, which is concentrated at low and middle latitudes on the Earth, is responsible for driving the large-scale motions of the atmosphere, which produce the global wind patterns. Heat is also deposited in the stratosphere where a form of oxygen called **ozone** (O_3) absorbs ultraviolet light from the Sun. This absorption is responsible for the increase of temperature with height in the stratosphere.

There are many scales of motion in the atmosphere, ranging from small gusts a few centimeters in size to the continent-spanning flows thousands of kilometers in scale. Here we will discuss only the large-scale motions, since generally we are able to study only motions of similar scale in the atmospheres of other planets.

The primary influences on the global motions in the Earth's atmosphere are heating from the Sun, as already noted, and the rotation of the Earth. The heating creates regions where the air rises. The air must later cool and fall somewhere else, and a pattern of overturning motions called **convection** is created (Fig. 8.3). Generally, air rises in the tropics and descends at more northerly latitudes, although the situation is more complicated than that (Fig. 8.4). For example, the continents tend to be warmer than the oceans, so that high-pressure regions (characterized by descending air) preferentially lie over water, whereas low-pressure regions (where air rises and cools) tend to be over land.

These tendencies, combined with the rotation of the Earth, create horizontal flows at several levels in the atmosphere. Air that is rising or falling is forced into a rotary pattern by the Earth's rotation. Where air rises in a low-pressure zone, the resultant swirling motion is called a **cyclone** (Fig. 8.5). In the Northern

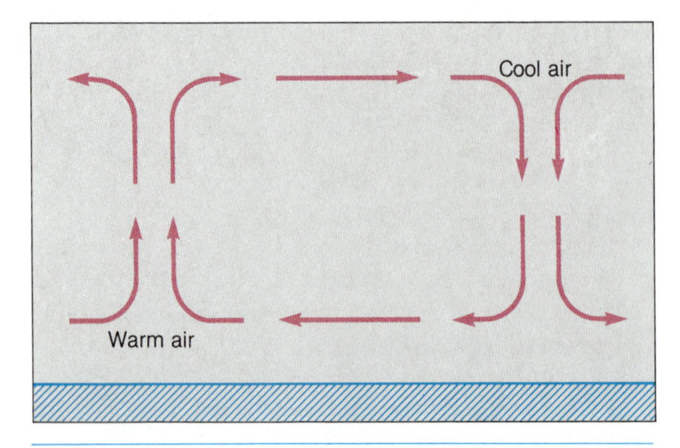

FIGURE 8.3 CONVECTION IN THE EARTH'S ATMOSPHERE. *This is a simplified illustration of the principle of convection, in which warm air rises, cools, and descends.*

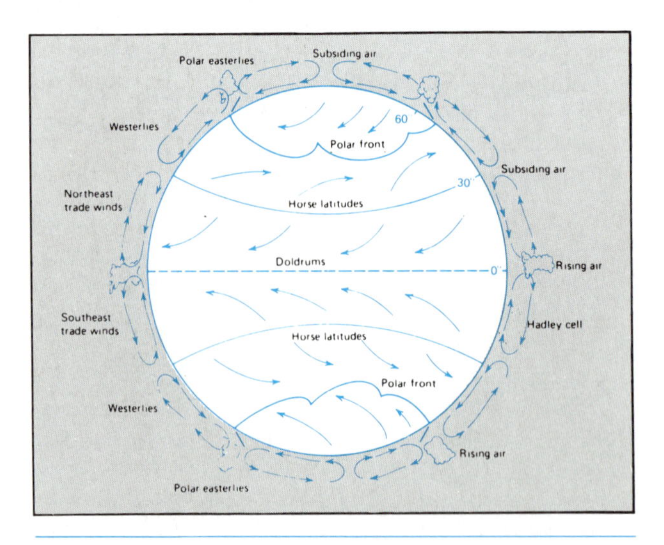

FIGURE 8.4 THE GENERAL CIRCULATION OF THE EARTH'S ATMOSPHERE. *The circulation is made complex by temperature contrasts between seas and land, and by surface features such as mountains.* (From F. K. Lutgens and E. J. Tarbuck 1979, *The Atmosphere: An Introduction to Meteorology,* Fig. 7.3, p. 150 [Englewood Cliffs: Prentice Hall], reprinted with permission of the publisher.)

Hemisphere, the motion of a cyclone is counterclockwise; south of the equator, it is clockwise. Descending air flows, in high-pressure regions, are forced into oppositely directed circular patterns, and are called **anticyclones.** Cyclonic flows sometimes intensify into storms of great strength; hurricanes and monsoons are examples.

The net result of the vertical motions caused by the Sun's heating and the spiral motions caused by the Earth's rotation is the creation of a complex pattern of flows (see Fig. 8.4) with vertical and horizontal components.

Seasonal shifts in the distribution of the Sun's energy input cause changes in the flow patterns, creating our well-known seasonal weather variations. Apparently, other factors can influence the location of the high and low-pressure regions, because there are longer-term fluctuations in the climate whose cause is not well understood. Some of these may be related to cyclic behavior in the Sun (see Chapter 17).

In later chapters, we will examine the flow patterns in the atmospheres of the other planets, and we will see that in many ways they behave like the Earth, except for differences in the amount of solar-energy input and the rate of rotation.

FIGURE 8.5 A STORM SYSTEM ON EARTH. *This is in the Southern Hemisphere, so the circulation about this low-pressure region is clockwise.* (NASA)

THE EARTH'S INTERIOR

At first it may seem surprising that much is known about the deep interior of the Earth. Substantial quantities of data have been collected, however, mostly by indirect means. It is interesting that the same indirect techniques could be applied (through the use of unmanned space probes) to other planets, and in principle it is possible for us to know as much about their interiors as that of our own planet (this would not be true of the outer, gaseous planets, but it does apply to the other terrestrial planets and the Moon).

The primary probes of the Earth's interior are **seismic waves** created in the Earth as a result of major shocks, most commonly earthquakes. These waves take three possible forms, called P, S, and L waves. Studies of wave transmission in solids and liquids have shown that P waves, which are the first to arrive at a site remote from the earthquake location, are **compressional** waves, which means that the oscillating motions occur parallel to the direction of motion of the wave, creating alternating regions of high and low density without any sideways motions (Fig. 8.6). Sound waves are examples of compressional waves. The S waves are **transverse** or **shear,** waves, the vibrations occurring at right angles to the direction of motion (see Fig. 8.6). These waves require that the material they pass through have some rigidity and, unlike the P waves, they cannot be transmitted through a liquid. The L waves travel only along the surface of the Earth and thus do not provide much information on the deep interior.

By measuring both the timing and the intensity of these seismic waves at various locations away from the site of an earthquake, scientists can determine what the Earth's interior is like. The speed of the P waves depends on the density of the material they pass through, and the distribution of the P and S waves reaching remote sites provides data on the location of liquid zones in the interior (Fig. 8.7).

The general picture that has developed from studies is that of a layered Earth (Fig. 8.8 and Table 8.3). At the surface is a crust whose thickness varies from a few kilometers beneath the oceans to perhaps 60 km under the continents (Fig. 8.9). There is a sharp break between the crust and the underlying material that is called the **mantle.** The mantle transmits S waves, so it must be solid; but on the other hand, it undergoes slow, steady, flowing motions in

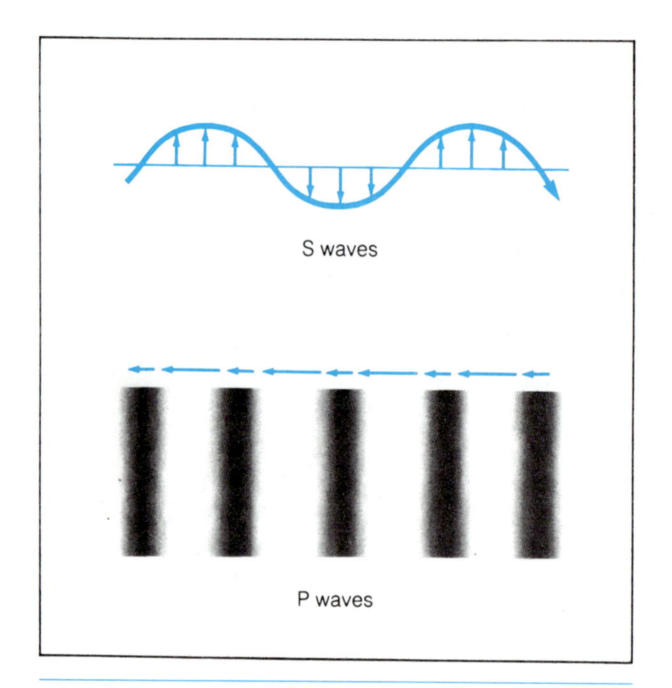

FIGURE 8.6 S AND P WAVES. *An S, or shear, wave (top) consists of alternating motions transverse (perpendicular) to the direction of propagation. Waves in a tight string or on the surface of water are S waves. Compressional, or P, waves (bottom) have no transverse motion but consist of alternating dense and rarefied regions created by motions along the direction of propagation. The arrows as the top represent the relative speeds in and between the dense zones. Sound waves are P waves.*

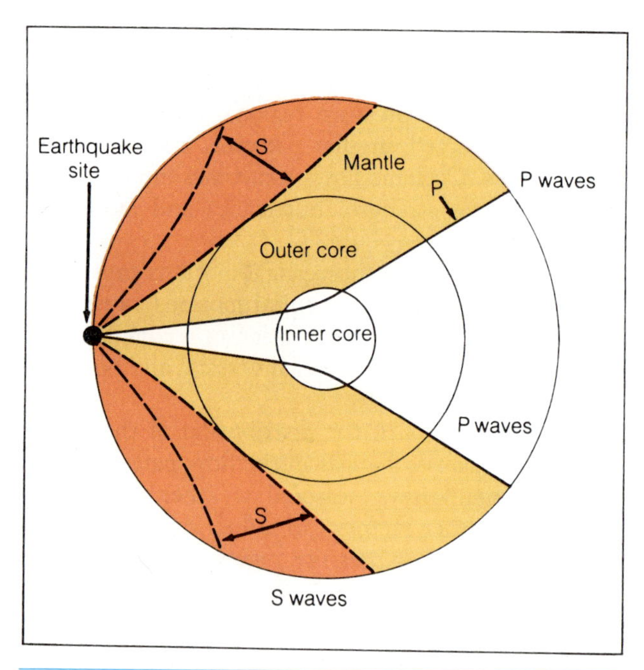

FIGURE 8.7 SEISMIC WAVES IN THE EARTH. *This simplified sketch shows that P waves can pass through the core regions (although their paths may be bent), while S waves cannot. This led to the deduction that the outer core is liquid, because S waves, which require an elastic or solid medium, cannot penetrate liquids.*

its uppermost regions. Perhaps it is best viewed as a plastic material, one that has some rigidity, but which can be deformed, given sufficient time (this is discussed more fully in the next section).

The uppermost part of the mantle and the crust together form a rigid zone called the **lithosphere.** The part of the mantle itself where the fluid motions occur, just below the lithosphere, is called the **asthenosphere.** Below the asthenosphere, there is a more rigid portion of the mantle that extends nearly halfway to the center of the Earth. The lower mantle is called the **mesosphere** (not to be confused with the level in the Earth's atmosphere bearing the same name).

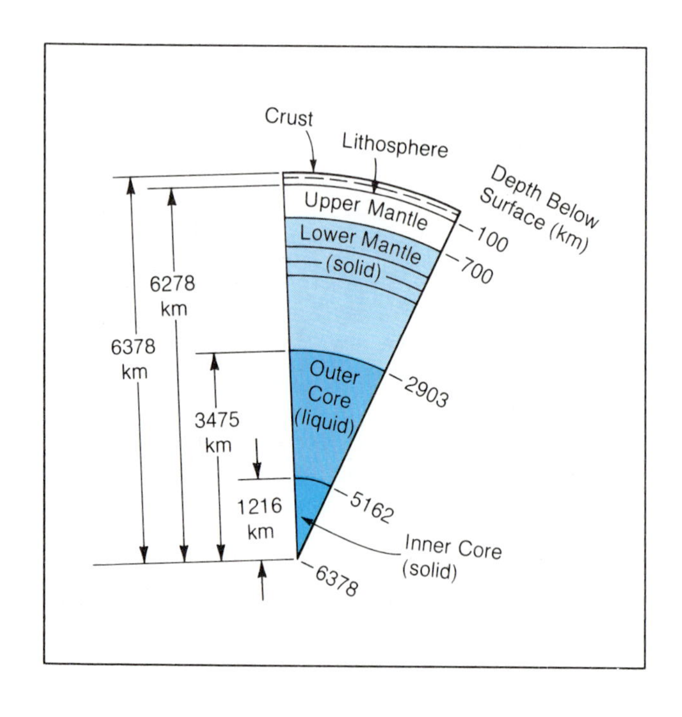

FIGURE 8.8 THE INTERNAL STRUCTURE OF THE EARTH. *This shows the layers that are defined on the basis of contrasts in density or physical state, as deduced from observations of seismic waves.*

TABLE 8.3 EARTH'S INTERIOR

Layer	Depth (km)	Density (g/cm³)	Temperature (K)	Composition
Crust	0–30	2.6–2.9	300–700	Silicates and oxides
Mantle				
Lithosphere	30–70	2.9–3.3	700–1,200	Basalt, silicates, oxides
Asthenosphere	70–1,000	3.9–4.6	1,200–3,000	Basalt, silicates, oxides
Mesosphere	1,000–2,900	4.6–9.7	3,000–4,500	Basalt, silicates, oxides
Core				
Outer core	2,900–5,100	9.7–12.7	4,500–6,000	Molten iron, nickel, cobalt
Inner core	5,100–6,378	12.7–13.0	6,000–6,400	Solid iron, nickel, cobalt

Beneath the mantle is the **core,** consisting of an **outer core** (between 2,900 and 5,100 km in depth) and an **inner core.** The S waves are not transmitted through the outer core, which is therefore thought to be liquid. As a result, the inner core can be probed only with the P waves. Since these waves travel more rapidly through the inner core, its density is thought to be greater than that of the outer core, and the innermost region is thought to be solid.

The overall density of the Earth is 5.5 times that of water (that is, 5.5 grams per cubic centimeter, or 5.5 g/cm³). The rocks that make up the crust have typical densities of less than 3 g/cm³, and the mantle material is thought to have relatively low density also, perhaps 3.5 g/cm³. From these facts, it is deduced that the core must have a density of roughly 13 g/cm³. The most likely substance that could give the core this high density is iron, probably with some nickel and perhaps cobalt.

The high density and probable metallic composition of the core are quite significant, for they show that the Earth has undergone **differentiation,** a sorting out of elements according to their density (Fig. 8.10). This can happen only when the planet is

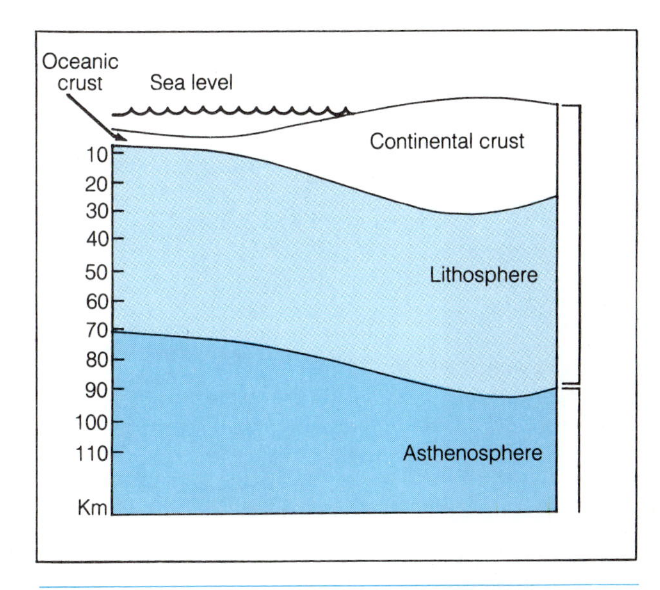

FIGURE 8.9 THE STRUCTURE OF THE EARTH'S OUTER LAYERS. *This illustrates the relative thickness of the crust in continental areas as compared to the seafloor.*

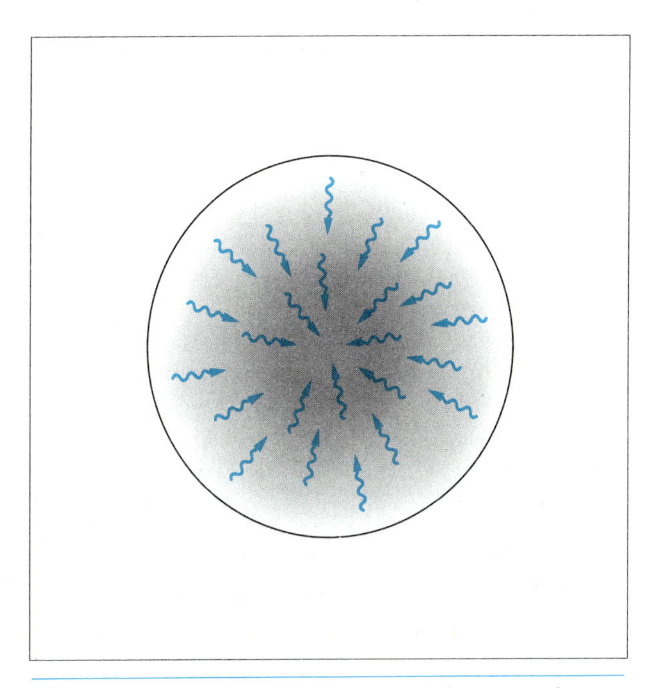

FIGURE 8.10 DIFFERENTIATION. *Heavy elements sank toward the Earth's center, creating the dense core. This process required that the Earth be partially or wholly liquid, and took place very early in the Earth's history.*

in a molten state, so it shows that the Earth was once largely liquid, probably early in its history, soon after it formed. The principal cause of the Earth's molten state most likely is radioactive heating in the interior. Several naturally occurring elements, such as uranium, thorium, and potassium, are radioactive, which means that their nuclei spontaneously emit subatomic particles and, over a long period of time, produce substantial quantities of heat. Another source of heat early in the Earth's history was created by the energy of impact as the smaller bodies that merged to form the Earth came together. Additional heating was caused by the process of differentiation itself, as the heavy elements sank to the center of the Earth, releasing gravitational potential energy. We shall see that differentiation has not occurred in all the bodies in the solar system, and this tells us something significant about their histories.

SURFACE TERRAIN AND TECTONIC ACTIVITY

As in previous sections, here we will deal primarily with large-scale features. One reason for this is that many of the small-scale details of the surface structures of the Earth have been modified by processes, such as flowing water, that do not occur on the other planets (with the exception of Mars; see Chapter 11). Our principal goal here is to describe features of the Earth that can be compared with structures on the other planets.

About three-quarters of the Earth's surface is covered by water. The remainder is covered by a few large continents, whose relative positions have changed steadily over past millennia and are still changing today.

The evidence favoring the idea that continental drift occurs includes the obvious fit between land masses on opposite sides of the Atlantic (Fig. 8.11), the similarities of mineral types and fossils, and the alignment of vestigial magnetic fields in rocks now separated by the ocean that once were together in the same place. In modern times, more direct evidence, such as the detection of seafloor spreading away from undersea ridges and the actual measurement (through the use of sophisticated laser-ranging techniques) of continental motions, has removed any trace of doubt that pieces of the Earth's crust are in motion, and that the continents were once arranged differently (Fig. 8.12).

From all this evidence has arisen a theory of **plate tectonics,** which postulates that the Earth's crust (that is, the lithosphere) is made of a few large, thin, pieces that float on top of the asthenosphere (Fig. 8.13). The flowing motions in the asthenosphere cause these plates to move about constantly and occasionally to crash into each other. The rate of motion is only a few centimeters per year at the most, and the major rearrangements of the continents have taken many millions of years to occur (both the Americas and the European-African system became separated from each other 150 to 200 million years ago).

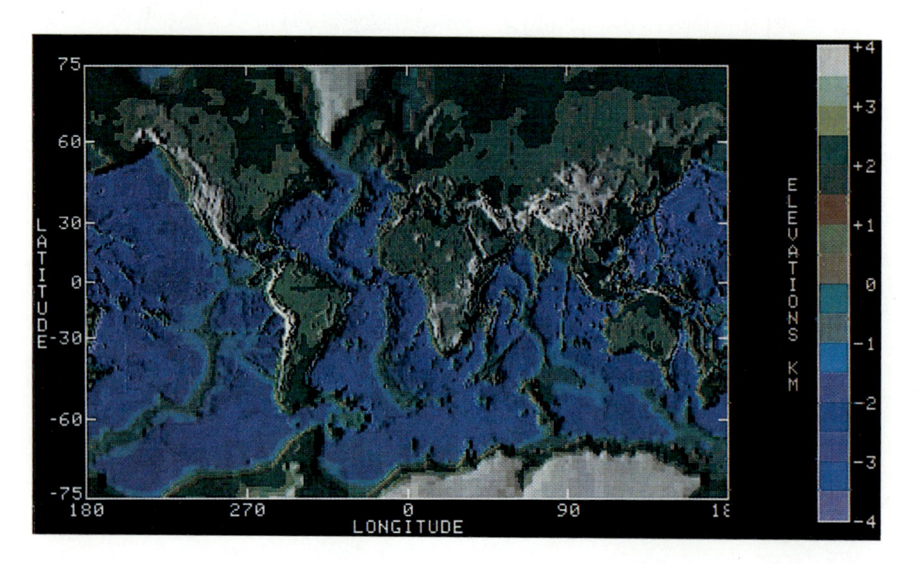

FIGURE 8.11 A RELIEF MAP OF THE EARTH. *This map shows relative elevations over the entire crust of the Earth, including the sea floors.* (M. Kobrick, JPL)

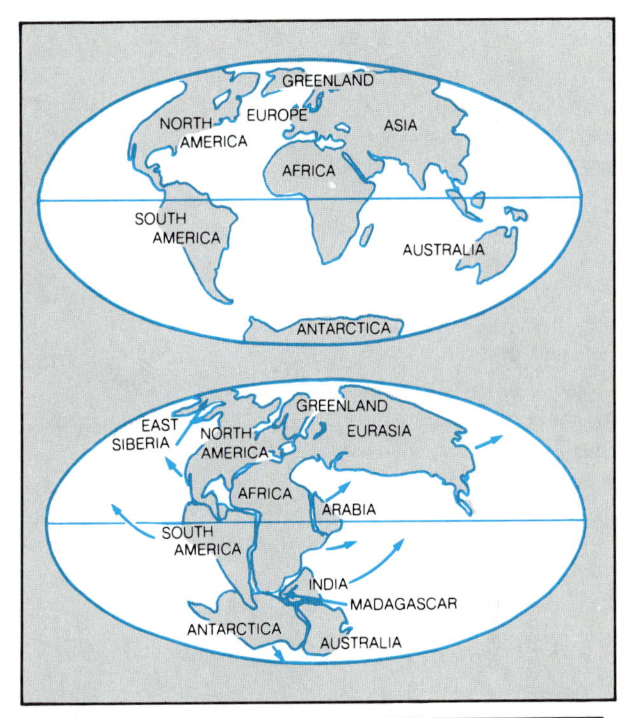

FIGURE 8.12 CONTINENTAL DRIFT. *These maps show the distribution of the continents today and as it was some 200 million years ago.*

The driving force in the shifting of the plates is not well understood. The most widely suspected cause is convection currents in the asthenosphere (Fig. 8.14). This is the same process that occurs in the Earth's atmosphere: Temperature differences between levels cause an overturning motion. If a fluid is hot at the bottom and much cooler at the top, warm material will rise, cool, and descend again, creating a constant churning. The speed of the overturning motions depends largely on the **viscosity** of the fluid, that is, the degree to which it resists flowing freely. The Earth's mantle, as we have already seen, is sufficiently rigid to transmit S waves, and must therefore have a high viscosity as well. Hence, if convection is occurring in the mantle, it is reasonable that the motions should be very slow.

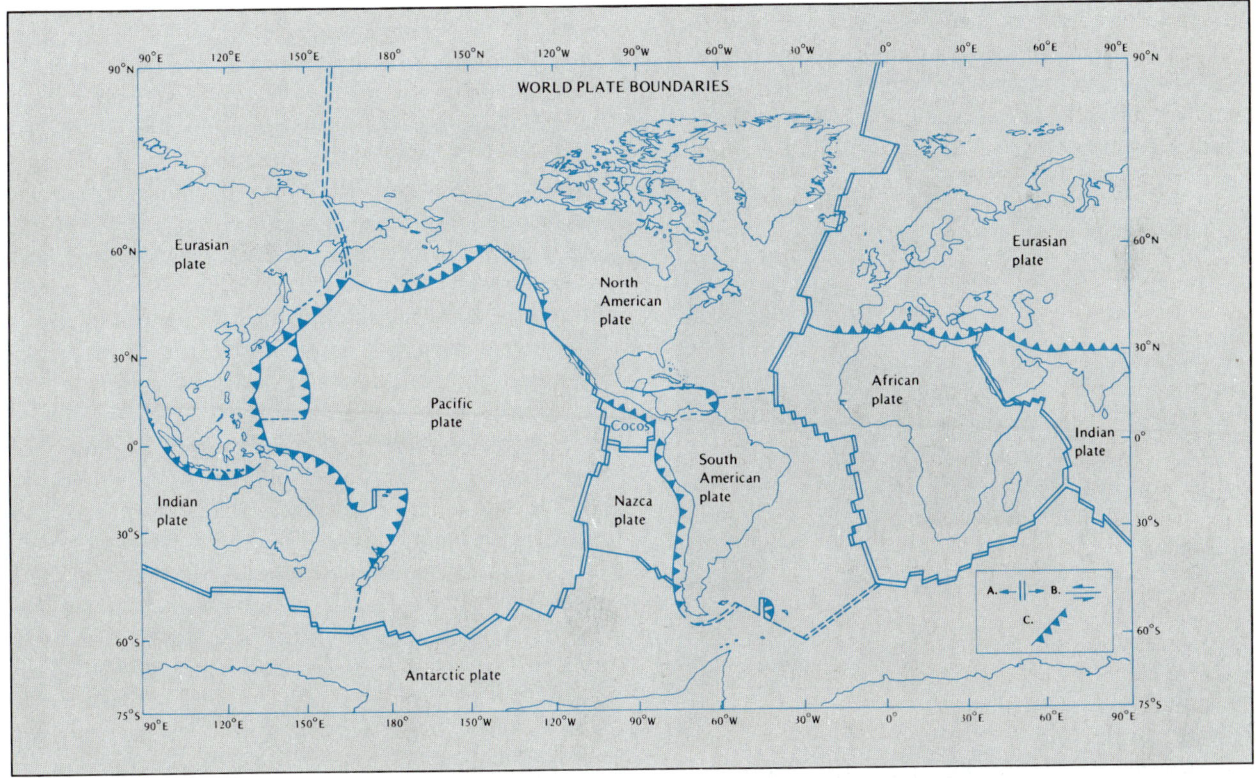

FIGURE 8.13 THE EARTH'S CRUSTAL PLATES. *This map shows the plate boundaries and the direction in which the plates are moving* (From F. S. Sawkins, C. G. Chase, D. C. Darby, and G. Rupp 1978. *The Evolving Earth: A Text in Physical Geology*, Fig. 6.2, p. 163 [New York: Macmillan],© Macmillan. Reprinted with permission of the publisher.)

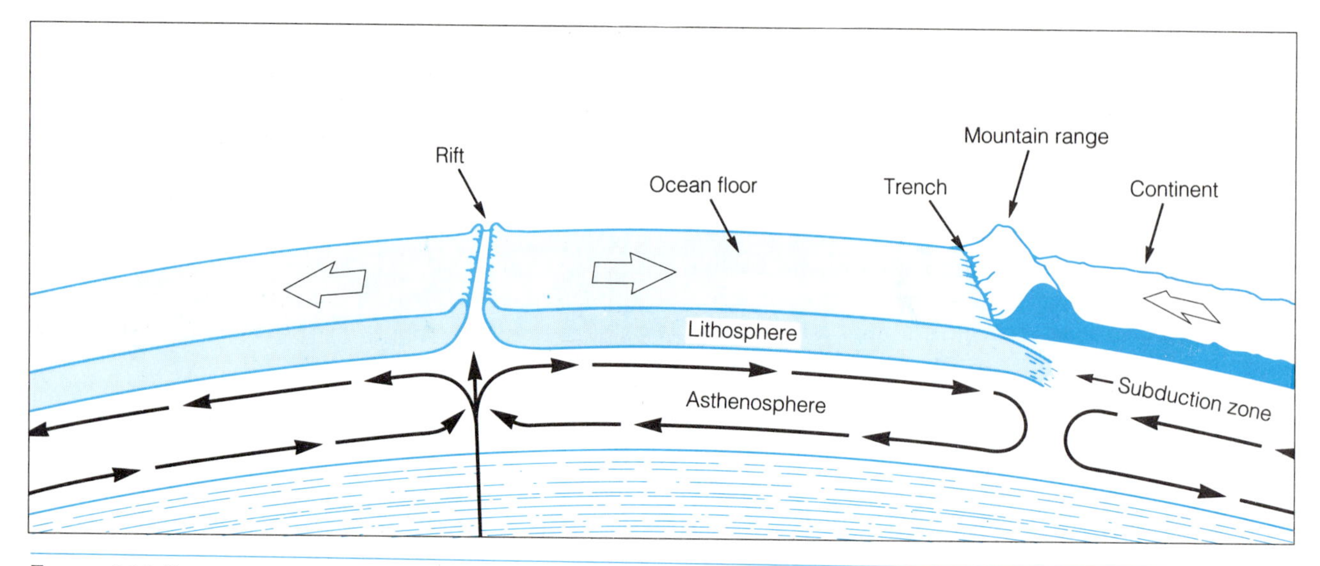

FIGURE 8.14 SCHEMATIC OF THE MECHANISMS OF CONTINENTAL DRIFT. *This sketch shows how continental drift is responsible for uplifted mountain ranges and parallel undersea trenches, where one crustal plate sinks below another (subduction zone), and how a midocean ridge is built up where two plates move away from each other.*

Whether convection is the cause or not, the effects of tectonic activity are becoming well known. When plates collide, one may submerge below the other in a process called **subduction,** which creates an undersea trench, or, particularly if both plates carry continents, the collision may force the uplifting of mountain ranges. The lofty Himalayas are thought to have been created when the Indian plate collided with the Eurasian plate some 50 million years ago. The boundaries where plates either collide or separate are marked by a wide variety of geological activity. Earthquakes are attributed to the sporadic shifting of adjacent plates along fault lines, and volcanic activity is common where material from the mantle can reach the surface, most often along plate boundaries. It has long been recognized, for example, that a great deal of earthquake and volcanic activity is concentrated around the shores of the Pacific, defining the so-called Ring of Fire. Now it is understood that this region represents the boundaries of the Pacific plate.

FIGURE 8.15 VOLCANIC ERUPTION IN HAWAII (Rick Golt, Photo Researchers)

Chains of volcanoes, such as the Hawaiian islands, are also attributed to the action of plate tectonics. In several locations around the world, volcanic hot spots that lie deep and are fixed in place bring molten rock to the surface. As the crust passes over one of these hot spots, volcanoes are formed and then carried away, and a chain of mountains is created as new material keeps coming to the surface over the hot spot. This process is still active, and the Hawaiian islands, for example, are still growing. There is volcanic activity on the southeast part of the largest (and youngest) island, Hawaii (Fig. 8.15). Furthermore, a new island, already given the name Loihi, is rising from the seafloor about 20 km south of Hawaii. Its peak has already risen 80 percent of the way to the surface and has only about 1 km to go, but it will take an estimated 50,000 years to break through.

Plate tectonics, then, accounts for many of the most prominent features of the Earth's surface. Of course, other processes, such as running water, wind erosion, and glaciation, are also important in modifying the face of the Earth.

ROCKS, MINERALS, AND THE AGE OF THE SURFACE

When we examine the surface of the Earth, we see little outward evidence of the large-scale processes. We do not see the continents moving, and we cannot directly perceive the interior. What we do see are rocks. The basic substance of the Earth's crust is rock, although in many areas, soil covers the bedrock. Even in those places, however, gravel, boulders, and rocky stream and lake beds are evident.

Rocks fall into three basic groups according to their origin, called **igneous, sedimentary,** and **metamorphic.** Igneous rocks, formed from volcanic activity, consist of cooled and solidified **magma,** the molten material that flows to the surface during volcanic eruptions. Sedimentary rocks are formed from deposits of gravel and soil that have hardened, usually in layers where old seabeds or coastlines lay. Metamorphic rocks are those which have been altered in structure by heat and pressure created by movements in the Earth's crust.

All three forms can be changed from one to another in a continual recycling process (Fig. 8.16). Igneous and metamorphic rocks can be eroded by wind and water into gravel and soil, later to become sedimentary rocks. Igneous and sedimentary rocks can be transformed into metamorphic rocks. And sedimentary and metamorphic rocks can be melted by pressures in the Earth's crust to become magma and reappear later as igneous rocks.

Because of these evolutionary processes, the surface of the Earth is continually being renewed. Whereas the age of the Earth itself is thought to be some 4.5 billion years, the ages of most surface rocks can be measured in the millions or hundreds of millions of years. The distinction between the age of a planet and the age of its surface is an important one, and it will reappear as we discuss the surfaces of other bodies.

The surface rocks and soil are composed of **minerals,** which are chemical compounds, each consisting of a particular arrangement of atoms. The various kinds of atoms in a mineral are regularly spaced in a set crystal pattern (Fig. 8.17). In most cases, these crystal patterns are confined to very small, microscopic regions that are randomly stuck together in a rock, giving it an overall amorphous appearance.

Some 2,000 minerals have been identified in rocks, but the bulk of the Earth's surface is composed of a

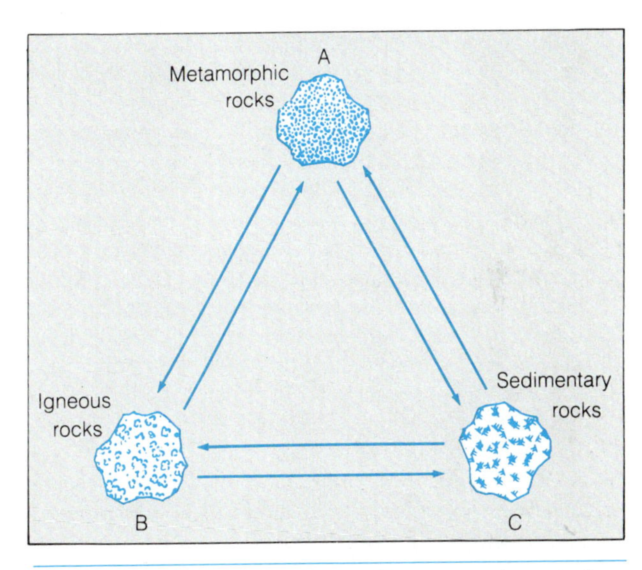

FIGURE 8.16 THE ROCK CYCLE. *This shows schematically how the three basic types of rocks can be converted from one to another.*

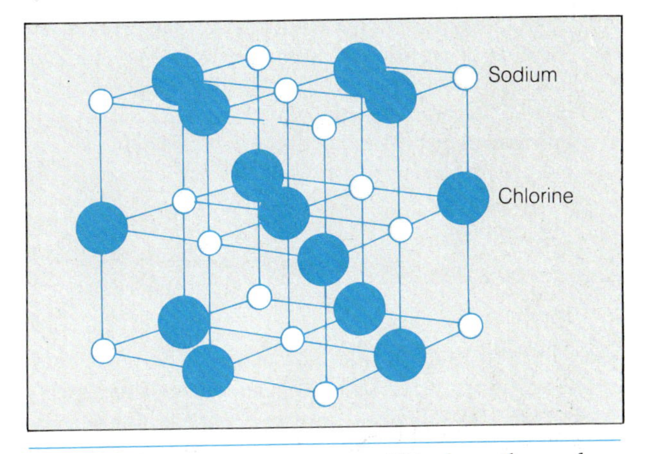

FIGURE 8.17 A TYPICAL CRYSTAL. *This shows the regular arrangement of atoms in the crystalline structure of sodium chloride, common table salt.*

ASTRONOMICAL INSIGHT 8.1

THE AGES OF ROCK

We have spoken of the ages of rocks and of the Earth itself, but we have said little about how these ages are measured. The most direct technique, **radioactive dating,** involves the measurement of relative abundances in rocks of closely related atomic elements.

Recall from our earlier discussions of atomic structure (Chapter 5) that the nucleus of an atom consists of protons and neutrons, and that the number of protons (that is, the **atomic number**) determines the identity of the element. Different **isotopes** of an element have the same number of protons but differing numbers of neutrons. In most elements, the number of protons is not very different from the number of neutrons, but there are exceptions. If the imbalance between protons and neutrons is large enough, the element is unstable, meaning that it has a natural tendency to correct the imbalance by undergoing a spontaneous nuclear reaction. The reactions that occur in this way usually involve the emission of a subatomic particle, and the element is said to be **radioactive.** The emitted particle may be an **alpha particle** (that is, a helium nucleus consisting of two protons and two neutrons) or, as is more often the case, an electron or a **positron,** a tiny particle with the mass of an electron but with a positive electrical charge. When a positron or an electron is emitted, the reaction is called a **beta decay,** and at the same time a proton in the nucleus is converted into a neutron (if a positron is emitted) or a neutron is changed into a proton (if an electron is released). If the number of protons in the nucleus is altered, then this changes the identity of the element. A typical example is the conversion of potassium 40 (^{40}K, with 19 protons and 21 neutrons in its nucleus) into argon 40 (^{40}Ar, with 18 protons and 22 neutrons).

Several elements are known to be radioactive, naturally changing their identity by emitting particles. The rate of change, expressed in terms of **half-life,** is known in most cases. The half-life is the time it takes for half of the original element to be converted. It may be as short as fractions of a second or as long as billions of years. The very slow reactions are the ones that are useful in measuring the ages of rocks.

If we know the relative abundances of the elements that were present in a rock when it formed, then measurement of the ratio of the elements in that rock at the present can tell us how long the radioactive decay has been at work, that is, the age of the rock. In a very old rock, for example, there might be almost no ^{40}K, but a lot of ^{40}Ar. The decay of ^{40}K to ^{40}Ar has a half-life of 1.3 billion years, so from the exact ratio of these two species in a rock, we can infer how many periods of 1.3 billion years have passed since the rock formed.

Other decay processes that are useful in dating rocks include the decay of rubidium 87 (^{87}Rb) into strontium 87 (^{87}Sr), which has a half-life of 47 billion years; and the decays of two different isotopes of uranium to isotopes of lead, ^{235}U to ^{207}Pb and ^{238}U to ^{206}Pb, with half-lives of 700 million years and 4.5 billion years, respectively. (Note that these decays each involve several intermediate steps and that the half-lives given represent the total time for all of those steps.)

Use of these dating techniques has shown that the oldest rocks on the Earth's surface are about 3.5 billion years old and that ages of a few hundred million years are more common. The age of the solar system, and therefore of the Earth itself, is estimated, from isotope ratios in meteorites, to be about 4.5 billion years.

very limited number, about a dozen. The most common, accounting for about 90 percent of the Earth's rocks, are the **silicates,** minerals containing silicon, oxygen, and some metallic elements. Other classes of common minerals include the **oxides,** simple combinations of oxygen and other elements; the **sulfides,** similar combinations of sulfur and other elements; and the **carbonate** and **sulfate** minerals, which contain special combinations of carbon and oxygen or sulfur and oxygen, along with metallic elements. Finally, there are the **halides,** minerals formed by combinations of the haline elements chlorine, iodine, bromine, and fluorine with other elements. Common table salt, sodium chloride, is a halide. Any rock can be classified according to both its type of origin and its mineral content. Two very common examples are basalt and granite, which are both igneous rocks with silicate compositions.

THE MAGNETIC FIELD

The magnetic field of the Earth has guided travelers since antiquity. Certain metallic substances, usually containing iron, are sensitive to the presence of a magnetic field and are used as pointers in compasses; thus, we take advantage of the near alignment of the Earth's magnetic field with the poles (the angle between the magnetic axis and the rotation axis of the Earth is about 11.5°). To see how a compass works, it is convenient to imagine magnetic lines along which iron filings lie when placed near a small bar magnet.

Outer Structure and the Magnetosphere

In cross-sectional view, the Earth's magnetic field is reminiscent of a cut apple, but one that is lopsided, because of the flow of charged particles from the Sun that constantly sweeps past the Earth (see the discussion of the solar wind in Chapter 17). The region enclosed by the field lines is called the **magnetosphere** (Fig. 8.18), and it acts as a shield, preventing the charged particles from reaching the surface.

The past alignment of the magnetic field can be ascertained from studies of certain rocks that contain iron-bearing minerals whose crystalline structure is aligned with the direction of the magnetic field at the time the rocks solidify from a molten state. Traces of the Earth's ancient magnetic activity, called **paleomagnetism,** thus can be deduced from the analysis of magnetic alignments of rocks. Such studies reveal not only that the continents have moved, but also that the magnetic poles of the Earth have moved. Furthermore, the north and south magnetic poles have completely and rather suddenly reversed from time to time, so that the North Pole has moved to the south and vice versa. This flip-flop of the magnetic poles seems to have happened at irregular intervals, typically thousands or hundreds of thousands of years apart.

While a reversal of the poles is taking place, there is, for a short period of time, a much-weakened magnetic field. The magnetosphere is greatly diminished, and charged particles from the solar wind and from **cosmic rays,** highly energetic particles from interstellar space (see Chapter 26), are more likely to penetrate to the ground. These particles can bring about important effects on life forms, including genetic mutations. The sporadic reversals of the Earth's magnetic field may have played a major role in shaping the evolution of life on the surface of our planet.

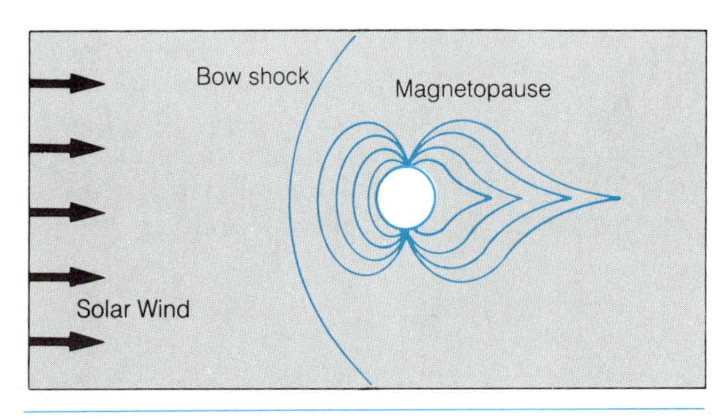

FIGURE 8.18 THE EARTH'S MAGNETIC FIELD STRUCTURE. *This cross-section shows the Earth's field and how its shape is affected by the stream of charged particles from the Sun known as the solar wind.*

The source of the Earth's magnetic field is a mystery, although it is nearly certain that it has to do with the nickel-iron core. A magnetic field is produced by flowing electrical charges, as in the current flowing through a wire wound around a metal rod, which is the basis of an electromagnet. Convection and the Earth's rotation may combine to create systematic flows in the liquid outer core, giving rise to the magnetic field, if the core material carries an electrical charge. This general type of mechanism is called a **magnetic dynamo.** The cause of the reversals of the poles is not well understood but presumably could be related to occasional changes in the direction of the flow of material in the core.

Radiation Belts

We have referred to the magnetosphere and its ability to control the motions of charged particles. When the first U.S. satellite was launched in 1958, zones high above the Earth's surface were discovered to contain intense concentrations of charged particles, primarily protons and electrons. There are several distinct zones containing these particles, and they are now called the **Van Allen belts** (Fig. 8.19), after the physicist who, in the late 1950s, first recognized their existence and deduced their properties. The charged particles, or ions, in these radiation belts are captured from space (primarily from the solar wind), and are forced by the magnetic field to spiral around the lines of force.

Far below the Van Allen belts, the uppermost portion of the atmosphere also contains charged particles, and is called the **ionosphere.** The ionosphere, which extends upward from an altitude of about 60 km, has important effects for us on the surface, particularly in terms of our radio communications. Shortwave radio signals are able to travel around the world because they are reflected back to the ground from the underside of the ionosphere. When the ionosphere is disrupted, as it can be after a particularly violent solar flare, its reflective properties may be destroyed temporarily, interrupting communications. (This problem has been largely overcome by communications satellites, which do not rely on the ionosphere at all.) At the same time, emissions caused by the charged particles may be enhanced, creating displays of the **aurora borealis** (Fig. 8.20), or northern lights (in the Southern Hemisphere they are called the **aurora australis,** or southern lights). The aurorae are always most prominent close to the magnetic poles, because that is where the Earth's magnetic field lines, and the charged particles that are forced to travel along them, come closest to the surface. In middle latitudes, these particles are kept well outside the atmosphere, and aurorae are rarely seen.

— THE EVOLUTION OF THE EARTH —

A great deal has been learned about the general evolution of the Earth. We know it was once largely a molten body with an atmosphere different from that of the present day. Along the way, the atmosphere has undergone major changes, some of them triggered by the development of life on the surface.

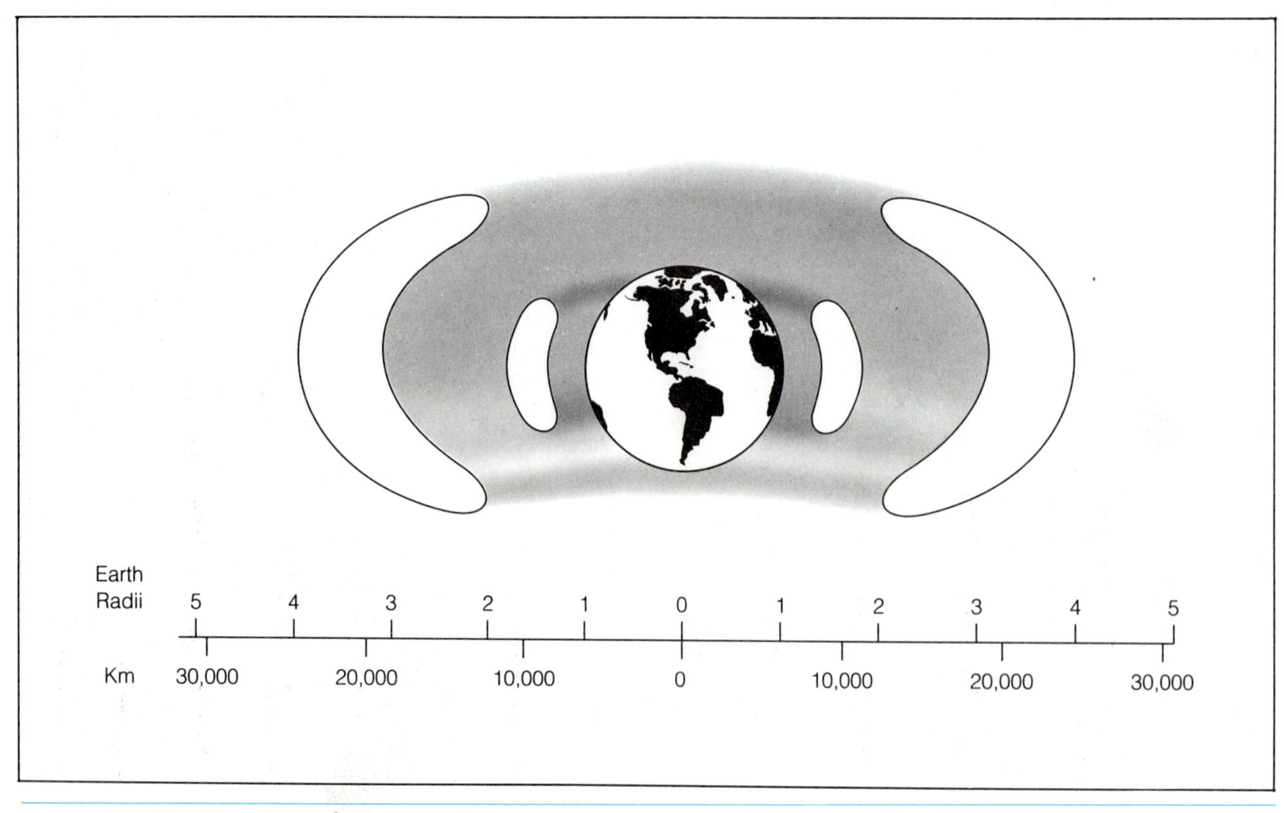

FIGURE 8.19 THE VAN ALLEN BELTS. *This cross-sectional view illustrates the sizes and shapes of the two principal charged-particle belts that girdle the Earth.*

The age of the Earth, estimated from geological evidence, is about 4.5 billion years. The Earth probably formed from the coalescence of a number of **planetesimals,** small bodies that were the first to condense out of the debris that swarmed about the infant Sun in the earliest days of the solar system. As a result of the heat of impact as these bodies merged, and in time, of the heating caused by radioactivity, most or all of the Earth was molten at some point during the first billion years. During this early period, differentiation occurred, with the densest materials (such as iron) sinking to the core. At the same time, **volatile** gases were emitted by the newly formed rocks at the surface. Volatile gases are those which vaporize most easily when heated; they formed the earliest atmosphere of the Earth. The most common of these gases were hydrogen (H_2), ammonia (NH_3), methane (CH_4), and water vapor (H_2O). It is noteworthy that the present-day atmospheres of the cold, outer planets are very similar to that of the primitive Earth. We will explore this point later (for example, in Chapter 18).

The dominant element in the Earth's early atmosphere was hydrogen, but nearly all of it escaped into space by the time the Earth was about 1 billion years old. (This is about when the first simple life forms apparently arose.) Recall from Chapter 4 that individual molecules in a gas whiz around randomly at speeds that depend on both the temperature of the gas and the mass of the particles. At a given temperature, the lightweight molecules move fastest and are therefore most likely to escape the planet's gravitational attraction. This is what happened to the hy-

FIGURE 8.20 THE AURORA BOREALIS. *These colorful displays are triggered by the impact of electrically charged particles from space on the Earth's upper atmosphere.*

ASTRONOMICAL INSIGHT 8.2

TECHNOLOGY AND THE ATMOSPHERE

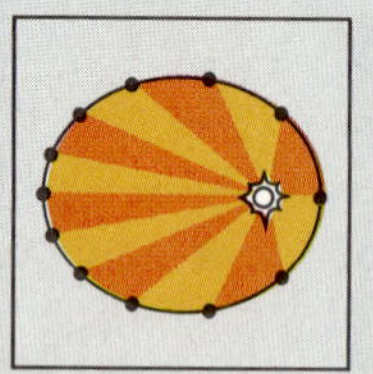

Life forms on the Earth have had a profound impact on the past evolution of the atmosphere, as we have discussed in the text. Much of the nitrogen that dominates the composition of the atmosphere comes from the decay of life forms, and virtually all of the oxygen is the by-product of photosynthesis in plants. Since life first appeared on the Earth, the interplay between biological activity and atmospheric processes has been rich and diverse. Today the interaction continues, with the added impact of a technological society.

Human activities result in many kinds of emissions into the atmosphere that would not occur naturally. These emissions include both gases and tiny particles, or aerosols. The atmosphere is a complex, dynamic entity, and scientists have difficulty predicting how it will react to these emissions. Furthermore, it takes a long time for significant global changes to occur, so it is not easy to see right away what the effects of a change in conditions might be. Despite these difficulties, two gases in particular have been identified as potentially capable of changing the atmosphere enough to alter the Earth's climate. (Many other gases are known or suspected to have other kinds of effects.)

One of these is carbon dioxide (CO_2), a by-product of many industrial processes. Carbon dioxide is an effective absorber of infrared radiation, which means that a high abundance of CO_2 in the atmosphere can trap heat near the Earth's surface, increasing the temperature of the lower atmosphere. In this process, generally known as the **greenhouse effect,** infrared radiation emitted from the surface is not allowed to escape as freely into space as it normally would, because the atmospheric CO_2 absorbs it. The resulting temperature increase in the lower atmosphere could have far-reaching effects. Even a relatively small increase could, over time, result in substantial melting of the ice in arctic regions, in turn raising world-wide sea level enough to inundate coastal areas.

As an illustration of the complexity of atmospheric processes, it is noteworthy that not all scientists agree on what the effects of increased CO_2 would be. Some argue that there would be a global *cooling,* because the CO_2 (along with aerosols associated with it) would block incoming sunlight sufficiently to decrease the heating at the Earth's surface.

An even more controversial technological addition to the atmosphere is created by the so-called **chloro-fluoro-carbons,** complex molecules containing chlorine and other elements. These are commonly used as refrigerants (such as freon) and as propellants in spray cans. In this case, scientists are concerned about complex chemical reactions that may occur in the upper atmosphere involving by-products of the chloro-fluoro-carbons. One of these by-products is chlorine oxide (ClO), which can react with ozone (O_3), destroying the ozone. For the past decade or so, there has been a growing worry that the release of chloro-fluoro-carbons into the atmosphere has slightly decreased the abundance of ozone in the stratosphere (at an altitude of about 16 to 18 km). If so, this could have serious consequences, because the ozone layer is our primary shield against ultraviolet radiation from the Sun, and an increase in the amount of ultraviolet radiation reaching the ground could be quite harmful to life on the Earth.

Recently a deficiency in the quantity of ozone in the atmosphere has been discovered at a particular time and at a particular place. During October (spring in the Southern Hemisphere) the quantity of ozone over the south polar region diminishes. Scientists are not certain that this "ozone hole" is something new; some have suggested that it is a natural phenomenon.

In order to learn more about this south polar ozone deficiency, teams of scientists spent several weeks in Antarctica in late 1986, and will continue to do so for some years. The first of these intensive studies produced detailed measurements of the quantity of ozone and of various molecular species known to react with ozone. Balloons and other devices made measurements at various altitudes in the atmosphere. As a result, we know that the quantity of ozone over the pole has dropped by about 50 percent since the late 1960s. This confirms the theory that the ozone hole is either a new phenomenon, or one that varies—perhaps with the Sun's cycle of activity (see Chapter 18).

The tentative conclusion of some of the scientists involved in the study is that the ozone hole is caused by chloro-fluoro-carbons. This conclusion is based on measurements of other molecular species, on inconsistencies in other hypotheses, and on direct measurements of freon in the

(continued on next page)

ASTRONOMICAL INSIGHT 8.2

TECHNOLOGY AND THE ATMOSPHERE

(continued from previous page)
atmosphere. The abundance of freon has increased by 50 percent over the same time period that the south polar springtime ozone abundance has decreased. All of these arguments are circumstantial and not completely conclusive, but they are consistent with the idea that human technology has begun to erode the Earth's protective ozone shield. There remain many important questions about whether this conclusion is valid, and if so, whether the ozone deficiency will be confined to the south polar region or will spread in time.

The changes in the Earth's atmosphere caused by natural processes related to life forms took millenia to occur, and took place in concert with the development and evolution of those life forms. Now it seems we must take measures to prevent much more rapid changes due to unnatural processes from altering our atmosphere in ways that would make the Earth inhospitable to life.

drogen in the early atmosphere of the Earth (whereas the heavier gases that now dominate the atmosphere are too massive to escape). Another gas that escaped early in the Earth's history is helium, the second most abundant element in the universe, but one so rare on Earth that it was first discovered in the Sun, where spectroscopic observations revealed its presence (hence the name, derived from *helios,* the Greek word for "sun").

The critical reactions that led to the development of life must have taken place before all the hydrogen escaped, because the types of reactions that were probably responsible involve this element. The earliest fossil evidence for primitive life dates back at least 3 billion years, when some hydrogen was still left.

Essentially, no free oxygen was present in the atmosphere until life forms that release this element as a by-product of their metabolic activities had developed. Most plants release oxygen into the atmosphere, and as a supply of this element built up, the opportunity arose for complex animal forms to evolve. Besides being a substance that was necessary for the metabolism of living animals, the oxygen buildup created a reservoir of ozone (O_3) in the upper atmosphere, which in turn began to screen out the harmful ultraviolet rays from the Sun. Once life forms gained a toehold on the continental land masses, the process of converting the atmosphere to its present state began to accelerate. Soon nitrogen from the decomposition of organic matter began to be released into the air in large quantities, and by the time the Earth was perhaps 2 billion years old, the atmosphere had reached approximately the composition it has today.

By this time also the mantle had solidified and the crust had hardened (the oldest known surface rocks are nearly 4 billion years old). The interior has remained warmer than the surface, because heat escapes slowly through the crust, and because radioactive heating of the interior took place over a long period of time and probably is still occurring today.

PERSPECTIVE

We have itemized the Earth's overall properties, from its outermost atmosphere to its core, always keeping in mind the facets of its character that will provide a basis for comparison with what we know about the other planets. We have seen that the Earth is an active world, seemingly placid only because of the long-time scales over which significant changes occur. We are ready now to examine other bodies in the solar system, to see whether they are similarly dynamic. We begin with the Moon, our nearest neighbor.

SUMMARY

1. The Earth, seen from space, is brilliant because of its high albedo.

2. The atmosphere consists of 80 percent nitrogen and about 20 percent oxygen, traces of H_2O and CO_2, and fine suspended particles.
3. The atmosphere is often divided into four vertical zones, based on the variation of temperature with height: the troposphere, stratosphere, mesosphere, and thermosphere.
4. The Sun's heating and the Earth's rotation create the global wind patterns.
5. The Earth's interior, explored with seismic waves, consists of a solid inner core, a liquid outer core, a mantle, and a crust.
6. The crust is broken into tectonic plates that shift around, which accounts for continental drift and most of the major surface features of the Earth.
7. Rocks can be igneous, metamorphic, or sedimentary in origin, and most are silicate minerals by composition.
8. The Earth has a magnetic field, probably created by currents in the molten core, which traps charged particles in zones called radiation belts above the atmosphere.
9. The Earth's evolution from a largely molten planet with a hydrogen-dominated atmosphere to its present state was caused by the loss of lightweight gases into space and by the development of life forms on the surface of the Earth, which helped convert the atmospheric composition to nitrogen and oxygen.

REVIEW QUESTIONS

Thought Questions

1. Summarize the Earth's atmosphere, in terms of chemical composition and physical conditions such as temperature, density, and pressure.
2. Explain how heat from the Sun and rotation of the Earth create the global circulation pattern of the atmosphere. How do the oceans influence the circulation?
3. How do we know anything about the Earth's interior? Do you think the same methods can be applied to other planets?
4. Explain the differences between transverse and compressional waves. Can you think of everyday examples of each kind?
5. If the density of typical surface rocks is 3.5 grams/cm^3, and the Earth's average (overall) density is 5.5 grams/cm^3, what does this tell us about the density in the deep interior? How did this situation arise?
6. Summarize the evidence that tectonic activity, including continental drift, occurs on the Earth.
7. How are the motions in the Earth's atmosphere and those in its interior similar?
8. Explain the distinction between the age of the Earth's surface and the age of the Earth itself.
9. How do you think magnetic field measurements for other planets might be used to give us clues to their internal structure?
10. Summarize the effects of life forms on the evolution of the Earth's atmosphere.

Problems

1. Verify that the Earth's average density is about 5.5 grams/cm^3 by using the figures given in Table 8.1 for the mass and diameter of the Earth. The volume of a sphere is $4\pi r^3/3$, and the density is the mass divided by the volume.
2. From the figures in Table 8.3, calculate the volume of the core (including both inner and outer cores) of the Earth. What fraction of the Earth's total volume is occupied by the core?
3. Assume that all of the Earth outside of the core has a density of 3.5 grams/cm^3, and that the core (inner and outer) has a constant density as well. What must the density of the core be in order to give the Earth its overall density of 5.5 grams/cm^3? (You will need to use your answer to the previous problem.)
4. If North America is approaching Japan at a rate of 3 cm per year, and the present distance between North America and Japan is 3,000 miles, how long will it take for the two to collide?
5. If the half-life of a radioactive element is 5 million years, and a rock sample contains one-half of its original quantity of that element, how old is the rock? How old is it if only one-fourth of the original quantity of the element is left? How old is it if one sixty-fourth is left?
6. We refer to hydrogen and helium as "light" gases. Using data from Appendix 3, compare the masses of hydrogen molecules (H_2) and helium atoms with the masses of the common molecules in today's atmosphere, oxygen (O_2) and nitrogen (N_2). (Hint: Use the atomic weight figures in the Appendix, along with the number of each type of atom in the molecule in question.)

ADDITIONAL READINGS

Battan, L. J. 1979. *Fundamentals of meterorology*. Englewood Cliffs, N.J.: Prentice-Hall.

Ben-Avraham, Z. 1981. The movement of continents. *American Scientist* 69:291.

Burchfiel, B. C. 1983. The continental crust. *Scientific American* 249(3):86.

Carrigan, C. R., and D. Gubbins. 1979. The source of the Earth's magnetic field. *Scientific American* 240(1):118.

Friedman, Herbert. 1986. Sun and Earth. New York: W. H. Freeman.

Ingersoll, A. P. 1983. The atmosphere. *Scientific American* 249(3):114.

Jeanloz, R. 1983. The Earth's core. *Scientific American* 249(3):40.

Leet, L. D., S. Judson, and M. E. Kauffman. 1978. *Physical geology*. 5th ed. Englewood Cliffs, N.J.: Prentice-Hall.

McKenzie, D. P. 1983. The Earth's mantle. *Scientific American* 249(3):50.

Siever, R. 1983. The dynamic Earth. *Scientific American* 249(3):30.

——— 1975. The Earth. *Scientific American* 233(3): 82.

——— 1974. The steady state of the Earth's crust, atmosphere, and oceans. *Scientific American* 230(6):72.

Taubes, Gary. 1983. The curious case of the tektites. Discover 4(6):74.

Toon, O. B., and S. Olson. 1985. The warm Earth. *Science 85* 6(8):50.

Wahr. J. 1986. The Earth's inconstant rotation. *Sky and Telescope* 71(6):545.

Wilson, J. T., ed. 1972. *Continents adrift*. San Francisco: W. H. Freeman. (A collection of readings from *Scientific American*)

CHAPTER 9

THE MOON

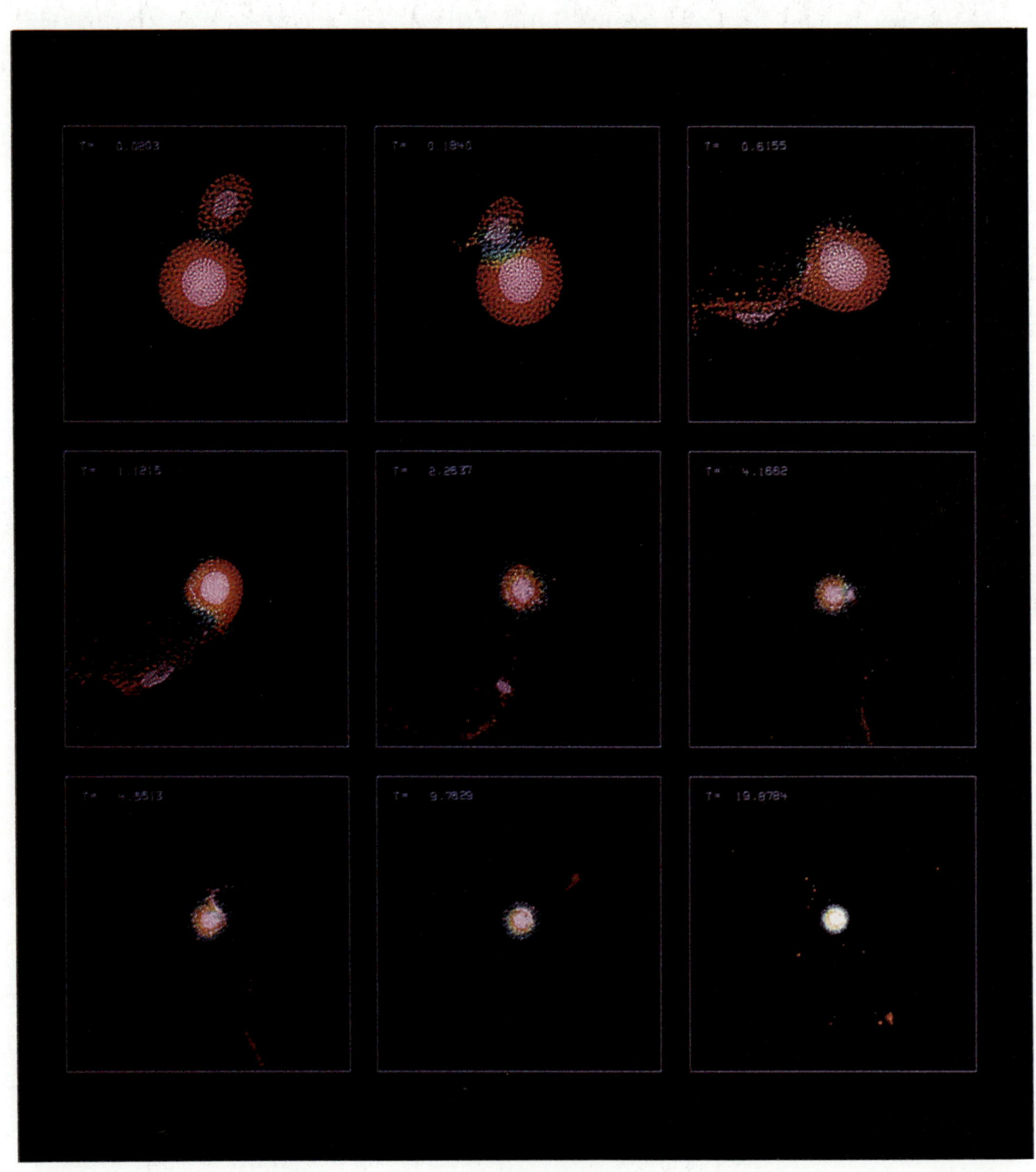

A computer simulation of the Moon's formation due to the impact of a large planetesimal on the young Earth. (W. Benz and W. Slattery, Los Alamos National Laboratories)

CHAPTER PREVIEW

Our closest neighbor in the solar system is the Earth's satellite, the Moon (Table 9.1). The fact that the Moon orbits the Earth was correctly recognized in the earliest days of civilization, when scientists first began to speculate on the nature of the heavenly bodies. In more modern times, the Moon and its motions have helped astronomers understand the workings of the laws of physics, particularly gravity, that govern orbital motions; and concepts derived from studies of the lunar orbit have been applied to many kinds of faraway objects. Thus the Moon served as a sort of laboratory for earthly scientists even before experiments could be performed on its surface.

We have already discussed the Moon's phases and other celestial phenomena associated with its motions, such as eclipses. In this chapter, we will concentrate instead on the nature of the Moon itself.

TABLE 9.1 MOON

Mean distance from Earth: 384,401 km ($60.4R_{\oplus}$)
Closest approach: 363,297 km
Greatest distance: 405,505 km
Orbital sidereal period: $27^d\ 7^h\ 43^m\ 12^s$
Synodic period (lunar month): $29^d\ 12^h\ 44^m\ 3^s$
Orbital inclination: 5°8′43″

Rotation period: 27.32 days
Tilt of axis: 6°41′ (with respect to orbital plane)

Diameter: 3,476 km ($0.273\ D_{\oplus}$)
Mass: 7.35×10^{25} ($0.0123\ M_{\oplus}$)
Density: 3.34 g/cm^3
Surface gravity: 0.165 Earth gravity
Escape velocity: 2.4 km/sec

Surface temperature: 400 K (day side); 100 K (dark side)
Albedo: 0.07

GENERAL PROPERTIES

Seen from afar, the Moon is a very impressive sight (Figs. 9.1 and 9.2) except in comparison with the Earth, which far outshines it (Fig. 9.3). The Moon's albedo is only 0.07, compared with the Earth's albedo of 0.31. In the first photographs that showed both objects together, taken by the *Voyager* spacecraft, the image of the Moon had to be artificially enhanced to make it show up (Fig. 9.4). Our normal view of the Moon from the Earth's surface shows it to be a brightly lit object, especially impressive when full. The Moon is pockmarked over its entire ½° angular diameter with a variety of surface features, the most prominent of which are the light and dark areas. The latter, thought by Renaissance astronomers to be bodies of water, are called **maria** (singular: *mare*), after the Latin word for "seas." Closer examination of the Moon with primitive telescopes also revealed numerous circular features called **craters,** named after similar-looking structures on volcanoes.

The pattern of surface markings on the face of the Moon is quite distinctive, and because it never

FIGURE 9.1 THE FULL MOON. *This is a high-quality photograph taken through a telescope on the earth.* (Mt. Wilson and Las Campanas Observatories, Carnegie Institution of Washington)

FIGURE 9.2 THE MOON AS SEEN FROM SPACE. *Here is a view obtained by one of the* Apollo *missions, showing portions of the near (left) and far (right) sides. On the left horizon is Mare Crisium, and in left center are Mare Marginis (upper) and Mare Smythii (lower). No maria are seen in the right-hand half of this image, there are almost none on the entire lunar far side.* (NASA)

changes, it was recognized long ago that the Moon always keeps the same face toward the Earth. The far side cannot be observed from the Earth, but it has been thoroughly mapped by spacecraft (see Fig. 9.2). The synchronous rotation of the Moon (described in Chapter 5) is attributed to tidal forces exerted on the Moon by the Earth. (Later in this chapter, the history of this situation will be discussed, as we learn of the Moon's formation and evolution.)

With no trace of an atmosphere (because the escape velocity is so low that all the volatile gases were able to escape long ago), erosion caused by weathering does not occur on the Moon, so craters, mountain ranges, and other features retain a jagged appearance indefinitely. Other types of features seen

FIGURE 9.3 EARTHRISE. *This photograph, taken from lunar orbit, dramatically contrasts the stark, sterile Moon to the hospitable environment of the Earth.* (NASA)

FIGURE 9.4 THE EARTH AND THE MOON FROM DEEP SPACE. *This photograph, obtained by one of the* Voyager *spacecraft, shows the contrast in albedo of the Earth and the Moon (enhanced here).* (NASA)

on the moon include **rays,** light-colored streaks emanating from some of the large craters, and **rilles,** winding valleys that resemble earthly canyons.

The Moon's mass is only about 1.2 percent of the mass of the Earth, and its radius is a little more than one-fourth the Earth's radius. The Moon's density, 3.34 grams/cm^3, is below that of the Earth; it more closely resembles the density of ordinary rock. This indicates that the Moon probably does not have a large, dense core and that it, therefore, most likely did not undergo strong differentiation. Infrared measurements showed, even before the space program allowed direct measurements to be taken, that the lunar surface is subject to extreme temperatures, ranging from 100 K (-279°F) during the 2 week night to 400 K (261°F) during the day.

Of course, all the long-range studies of the Moon were almost instantly antiquated by the development of the space program, which has featured extensive exploration of the Moon, first by unmanned robots, then by men (Fig. 9.5). The Moon was the first celestial body to be probed by spacecraft sent from Earth, and it is still the only one to be visited by manned missions. The Soviet Union had the first successes,

FIGURE 9.5 MAN ON THE MOON. *The* Apollo *missions, six of which included successful manned landings on the moon, represent humankind's only attempt so far to visit another world.* (NASA)

obtaining photographs of the far side in 1959, and the United States launched spacecraft that began to carry out many complex and sophisticated observations and experiments in the 1960s. The *Ranger* spacecraft were designed to simply crash-land on the Moon, taking photographs on the way. Following the first success of this program in 1964, the United States launched several *Surveyor* and *Lunar Orbiter* missions, all of which were spectacularly successful. Five *Surveyors* landed on the Moon between 1966 and 1968, with the primary goal of analyzing surface conditions for future manned landings, and five *Lunar Orbiter* missions circled the Moon during the same period, mapping it photographically, again in preparation for the manned missions to come.

The historic first manned landing occurred on July 20, 1969. This was the *Apollo 11* mission and was followed by five more manned landings (Fig. 9.6), the last being *Apollo 17*, which took place in late 1972. Each mission incorporated a number of scientific experiments. Some involved observations of the Sun and other celestial bodies from the airless Moon, but most were devoted to the study of the Moon itself.

The following sections describe our current understanding of the Moon, based to a very large extent on the information gleaned from the *Apollo* program.

FIGURE 9.6 THE LUNAR ROVER. *The later* Apollo *missions used these vehicles to travel over the moon's surface, allowing the astronauts to explore widely in the vicinity of the landing sites.* (NASA)

A Battle-Worn Surface

Viewed on any size scale, from the largest to the smallest, the Moon's surface is irregular, marked throughout by a variety of features. We have already mentioned the maria, the large, relatively smooth, dark areas (Fig. 9.7). The maria appear darker than their surroundings because they have a relatively low albedo (that is, they reflect less sunlight). Despite their smooth appearance in comparison with the more chaotic terrain seen elsewhere on the Moon, the maria are marked here and there by craters.

Much of the rest of the lunar surface is covered by rough, mountainous terrain called **highlands**. Even though the maria dominate the near side of the Moon, the highland regions actually cover most of its surface. The *Lunar Orbiter* missions found that, oddly enough, there are few maria on the far side.

FIGURE 9.7 THE LUNAR "SEAS." *Here is a broad vista encompassing portions of three maria: Mare Crisium (foreground); Mare Tranquilitatis (beyond Mare Crisium); and Mare Serenitatis (on the horizon at upper right). These relatively smooth areas are younger than most of the lunar surface, having been formed by lava flows after much of the cratering had already occurred.* (NASA)

FIGURE 9.8 CRATERS ON THE MOON. *This fine Earth-based photograph reveals the details of a heavily cratered region on the lunar surface.* (Palomar Observatory, California Institute of Technology)

FIGURE 9.9 A LARGE CRATER. *This crater is named Eratosthenes, after the ancient Greek scientist who made the first accurate measurement of the size of the Earth (Chapter 3). The crater rim and central peak, both indicative of impact craters, are clearly visible, as is a quantity of ejecta surrounding the crater.* (NASA)

FIGURE 9.10 CRATERS WITHIN CRATERS. *Craters often overlap or lie within each other on the Moon's surface. Relative ages of the craters can be estimated from the sequence of impacts that formed them.* (NASA)

Craters are seen literally everywhere (Fig. 9.8). They range in diameter from hundreds of kilometers (Fig. 9.9) to microscopic pits that can be seen only under intense magnification. In some regions, the craters are so densely packed together that they overlap. As noted earlier in this chapter, the lack of erosion on the Moon means that craters can survive for billions of years, and there is plenty of time for younger craters to form within the older ones (Fig. 9.10). The fact that relatively few craters are seen in the maria indicates that the surface in these regions has been transformed in relatively recent times, after most of the cratering had already occurred.

Most of the craters on the Moon are **impact craters,** formed by collisions of interplanetary rocks and debris with the lunar surface, rather than by volcanic eruptions. The particular shapes of the craters, the central peaks in some of them (see Fig. 9.9), and the trails left by ejecta, or cast-off material created

ASTRONOMICAL INSIGHT 9.1

TRAVEL TO THE MOON

In the nineteenth century, Jules Verne wrote of a journey to the Moon in a spaceship that was fired from a gigantic cannon. In the twentieth century, people have traveled to the Moon in a somewhat more sophisticated manner, but the basic principles are not much different from those of Verne's original speculation.

As we learned in our discussions of orbits and energy in Chapter 5, a certain minimum escape velocity is needed in order to leave the Earth's gravitational field. Verne evidently realized that this velocity was quite high, but he imagined that a giant cannon could do the job, whereas the task was actually accomplished by rocket engines. Furthermore, because of the enormous energies required and because of the need to slow the descent to the Moon's surface (a detail that Verne overlooked), the actual manned journeys to the Moon involved a very complex series of rocket firings and maneuvers.

Some of the early unmanned probes sent to the Moon in the mid-1960s were simply sent out from the Earth on escape trajectories that intersected the Moon's orbit. These, including most notably the U.S. *Ranger* series, crashed into the lunar surface, taking photographs or other scientific measurements on the way. The rockets that got these probes to the Moon always consisted of two or more distinct stages. It was (and remains) difficult to construct a single rocket engine capable of producing the necessary acceleration from rest on the Earth's surface to escape velocity, so the job was done by a series of rockets, the second stage taking over after the first had burned out, and so on.

The giant rockets that transported humans to the Moon for the six successful *Apollo* landings (as well as a few preliminary *Apollo* missions that were not carried all the way to the point of landing on the Moon) consisted of three stages. The complete vehicle that lifted off from the Earth's surface was some 360 feet tall, but most of its bulk was shed on the trip, as the first stages used their fuel and were discarded. Two stages were required to launch the third stage—the capsule containing human passengers—into Earth orbit. As in the case of most Earth-orbiting spacecraft, the *Apollo* launches were always made in the direction of the Earth's spin (that is, toward the east) so that some of the horizontal velocity needed to achieve orbit was provided by the rotation of the Earth. This explains why the principal U.S. launch site is on the East Coast: so that the launch trajectories can carry the rockets over water rather than inhabited land.

Getting away from the Earth is only part of the difficulty for a vehicle intended to land softly on the lunar surface. As a spacecraft approaches the Moon (or any other body), it is accelerated by the lunar gravitational field and would indeed crash violently if not somehow slowed first. This problem was solved in the *Apollo* missions by having the spacecraft enter lunar orbit instead of going directly to the Moon's surface. Even so, rocket motors had to be used to brake the approach velocity; otherwise, the craft would have simply curved around the moon and escaped into deep space. Once the *Apollo* capsule reached lunar orbit, a small landing vehicle (with two of the three astronauts on board) made the descent to the Moon's surface, firing rockets first to slow the vehicle so that it dropped out of lunar orbit and then firing them again to soften the impact upon landing.

Because of the relative weakness of the lunar gravitational field (and the subsequently lower escape velocity), this same craft with its modest rocket was then able to launch itself into lunar orbit, where it rejoined the larger lunar orbiter vehicle (and the third astronaut) for the return journey to the Earth. This sequence of events was certainly more complicated than that envisioned by Jules Verne, but this is only because it proved more difficult than he imagined to create the necessary launch speed and to land softly once the Moon was reached.

It is not certain what the future holds for travel to the Moon, but it is clear that the technical difficulties will be the same, and the solutions to them similar to those employed in the *Apollo* program. The Space Shuttle, for example, although not capable of going to the Moon, uses multiple stages to reach Earth orbit and would need additional boosters to travel to higher orbits or to escape the Earth altogether. Most of the energy required for going to the Moon is that needed to reach Earth orbit, so it is likely that future lunar missions will involve relatively modest spacecraft that are carried to Earth orbit (perhaps by the Shuttle itself or its technological descendants) and then launched to the Moon. It has even been suggested that future spacecraft designed to travel beyond Earth orbit will be constructed in space from components carried there individually by shuttles from Earth.

FIGURE 9.11 A CRATER WITH EJECTA. *This crater on the lunar far side is a good example of a case where material ejected by the impact has created rays of light-colored ejecta. Close examination of such features often reveals secondary craters, formed by the impacts of debris blasted out of the lunar surface by the primary impact.* (NASA)

by the impacts (Fig. 9.11), all suggest that the craters were formed in this manner. The rays seen stretching away from some craters are strings of smaller craters and debris formed by the ejecta from the larger, central crater.

Some of the lunar mountain ranges reach heights greater than any on Earth. They are more jagged than earthly mountains, again because there is no erosion, and they lack the prominent drainage features usually found in terrestrial ranges.

The rilles are rather interesting features, resembling dry riverbeds (Fig. 9.12). Apparently they were formed by flowing liquid, but rather than water, lava was responsible. In some cases there are even lava tubes that have partially collapsed, leaving trails of sinkholes. The rilles and, as we shall see, the maria as well, indicate that the Moon has undergone stages when large portions of its surface were molten.

FIGURE 9.12 RILLES. *Here is shown the intersection of two rilles, river-like valleys on the Moon that are thought to have formed as a result of lava flows early in the Moon's history.* (NASA)

ROCKS AND MINERALS ON THE MOON

The *Apollo* astronauts found a surface that in many areas is strewn with loose rock, ranging in size from pebbles to boulders as big as a house (Figs. 9.13 and 9.14). The rocks generally show sharp edges (owing to the lack of erosion), and occasional cracks and fractures. In most cases the large boulders appear to have been ejected from nearby craters, and are therefore thought to represent material originally from beneath the surface.

Almost 800 pounds of small lunar rocks were brought to Earth by the *Apollo* astronauts. This gave scientists the opportunity to study the lunar surface characteristics in detail and to compare them with Earth rocks and soils (Fig. 9.15). Thus, extensive chemical analyses, as well as close-up observation of rocks in place on the Moon, were possible. At present, the rock samples from the Moon are housed in many scientific laboratories around the world, where analysis continues. Specimens have also found their way into museums, and in at least one case (the Smithsonian Institution's Air and Space Museum in Washington, D.C.), a lunar rock can be touched by visitors.

The lunar soil, called the **regolith,** consists of loosely packed rock fragments and small glassy minerals probably created by the heat of meteor impacts (Fig. 9.16). In addition to the loose soil, there are three morphological types of surface specimens. The most common are the **breccias,** which consist of small

FIGURE 9.13 A FIELD OF BOULDERS. *Rocks in a wide variety of sizes are strewn over much of the Moon's surface. Most have been blasted out of the surface by impacts.* (NASA)

FIGURE 9.14 A LARGE BOULDER. *Rocks on the lunar surface range in size from tiny pebbles to massive objects like this.* (NASA)

rock fragments cemented together and resembling chunks of concrete. There are similar kinds of rock on Earth, except that those on Earth are formed in stream beds where water plays a role in shaping them. The lunar breccias contain jagged, sharp rock fragments, and they were probably fused together by pressure created in meteoroid impacts. The other common morphological types of lunar rocks are two basalts with small openings permeating their interiors (basalt is a silicate rock common on Earth).

All lunar rocks are pitted, on the side that is exposed to space, with tiny **micrometeorite craters.** These are formed by the impact of tiny bits of interplanetary material no bigger than grains of dust. From the degree of pitting on a rock, it is possible to deduce how long it was exposed to space, and how often it may have been turned over by impacts.

The simple classification of lunar rocks according to shape does not reveal much of significance in terms of the composition and history of the Moon itself. A more meaningful analysis of samples brought to Earth was done, in which their chemical makeup and mineral structure were examined. Several important discoveries were made. First, the relative abundances of some atomic isotopes were found to be very similar to those of earthly rocks. Of particular importance are the oxygen isotopes, which are known to vary in relative abundance throughout the solar system. The fact that these ratios are the same on the Earth and the Moon indicates that the two bodies must have formed out of very similar material, probably in the same part of the solar system. This has some impact on theories of the origin of the Moon, as discussed later.

A second discovery made from the chemical analysis of Moon rocks is that they contain lower abundances of the volatile elements than do earthly rocks. This probably implies that the Moon's volatile elements escaped sometime before the Moon formed.

The ages of lunar rocks could also be estimated by using the radioactive-dating techniques described in Chapter 8. The lunar rock and soil samples from highland regions were found to be very old by earthly standards, as old as 3.5 to 4 billion years. Rocks from the maria are not quite so old, but they still date back several billion years. In addition, a correlation was found between the ages of the rocks and the types of minerals they contain. This provides detailed information on the history of the Moon, because the conditions needed to form the different minerals are known,

FIGURE 9.15 MOON ROCKS. *One of the thousands of lunar samples brought back to Earth by the* Apollo *astronauts, the rock at left is an example of a breccia. At right is a thin slice of a Moon rock, illuminated by polarized light, which causes different crystal structures to appear as different colors.* (NASA)

and the correlation with age reveals when the various required conditions existed.

— CLUES TO INTERIOR CONDITIONS —

Some of the experiments carried out on the Moon by the *Apollo* astronauts were aimed at revealing the interior conditions. As in the case of the Earth, the best way to do this is to monitor seismic waves caused by quakes on the Moon. Therefore, the astronauts carried with them devices for sensing vibrations in the lunar crust and, because it was not known whether natural moonquakes occurred frequently, they also brought along devices for thumping the surface to make it vibrate. They were even able to take advantage of quakes caused by the impact of artificial objects on the surface, such as a discarded fuel tank that crashed into the Moon. It turned out that natural moonquakes do occur, although not with great violence. The seismic measurements continued after the *Apollo* landings, with data radioed to Earth by instruments left in place on the lunar surface.

The measurements showed that the regolith, or surface soil, is typically about 10 meters thick and is supported underneath by a thicker layer of loose rubble. The crust is 50 to 100 kilometers thick at the *Apollo* sites and may be somewhat thicker on the far side, where there are no maria (Fig. 9.17). Beneath the crust is a mantle, consisting of a well-defined lithosphere, which is rigid, and beneath it an asthenosphere, which is semiliquid. The innermost 500 kilometers consist of a relatively dense core, but not as dense as that of the Earth.

FIGURE 9.16 THE LUNAR SOIL. *Here is the bootprint of an astronaut, showing how the fine lunar soil can be compacted. In the absence of erosion, this bootprint may last for millions of years.* (NASA)

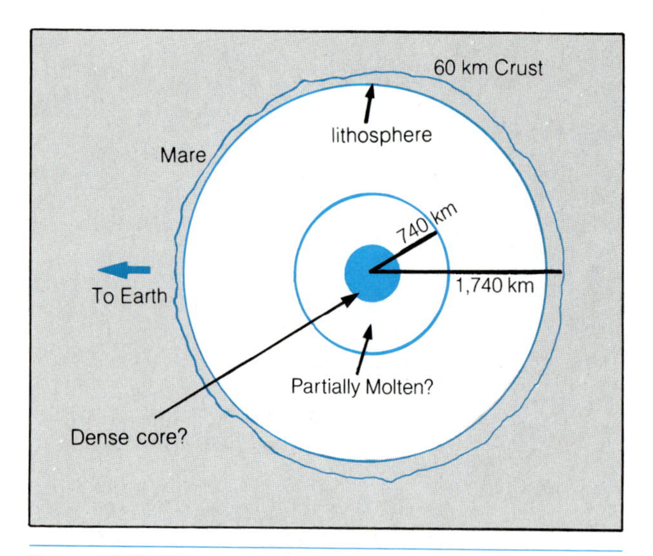

FIGURE 9.17 THE INTERNAL STRUCTURE OF THE MOON. *This cross-section illustrates the interior zones inferred from seismological studies. The existence of a dense core is not certain. Note that the maria lie almost exclusively on the side facing the Earth, where the lunar crust is relatively thin.*

The seismic measurements indicate no truly molten zones in the Moon at the present time, although temperature sensors on the surface discovered a substantial heat flow from the interior. This is probably caused by radioactive minerals below the surface.

There is no detectable overall magnetic field, further evidence against a molten core. The chief distinction among the internal zones in the lunar interior is density, with the densest material closest to the center. Thus, moderate differentiation has occurred in the Moon, which implies that it was once molten.

There is no trace of present-day lunar tectonic activity, perhaps the greatest single departure from the geology of the Earth. The force that has had the greatest influence in shaping the face of the Earth has no role today on the Moon. Instead, the lunar surface is entirely the result of the way the Moon formed (which may have involved some tectonic activity in the early stages) and the manner in which it has been altered by lava flows and by the incessant bombardment of rocks and debris from space.

FORMATION AND HISTORY

The moon's beginnings are somewhat obscure to us, despite the close-up examination that has been afforded by the *Apollo* program. We know a great deal about the present-day Moon, and in trying to ascertain how the Moon formed, we must seek the hypothesis that explains all the observed properties and develop from that a theory that can be confirmed by prediction and test. It is important to realize that, while scientists are beginning to agree on a preferred hypothesis, none proposed so far has been fully developed into a quantitative theory.

Traditionally there have been three types of hypotheses regarding the origin of the Moon: (1) that the Moon and the Earth formed together, as two bodies, out of the same material, (2) that the Moon consists of material that split off from the Earth some time after its formation, and (3) that the Moon formed elsewhere in the solar system and was later captured by the Earth's gravitational field. None of these three ideas has received universal acceptance, either because of contradictions with observation or because they invoke special events that seem very unlikely to have occurred. Very recently, a fourth hypothesis, that the Moon formed as a result of a gigantic impact on the newly forming Earth, has been proposed, and appears to explain what is known of the Moon better than the others. Let us examine some of the relevent evidence, and attempt to see why this latest idea may be better than the others.

The Moon has major chemical similarities with the Earth, but also some important differences. Certain forms of oxygen and other elements, for example, are very similar in the two bodies, which offers support for the theory of simultaneous formation. On the other hand, the Moon has a lower overall iron content and a higher abundance in general of the so-called **refractory elements,** those which are not easily vaporized. This contrast, the low abundance of easily vaporized volatile elements on the Moon, and evidence found in the radioactive isotopes and apparent melting history of lunar surface rocks, indicate that the entire surface layer of the Moon, down to a depth of some 300 kilometers, was molten for a while during the first few million years after the Moon's formation. These facts argue that the Moon somehow was subjected to much more heating early in its history than was the Earth.

The so-called **coeval formation** or **binary accretion** hypothesis, in which the Moon and the Earth are assumed to have formed as two bodies when planetesimals coalesced in the early solar system, has the

ASTRONOMICAL INSIGHT 9.2

MAPPING THE MOON[1]

All the major topographic features on the Moon, and many minor ones, have names. These names are used in maps of the Moon and are recognized by the international community of astronomers. It is interesting to examine how they were assigned.

Galileo was the first to have a chance to name lunar features, because he was the first to look at the Moon through a telescope that enabled him to see them. He is responsible for the terms *maria* (the dark areas that he thought were seas) and *terrae* (the highlands). Following Galileo's lead, as telescope technology improved and more and more people examined the Moon, a number of early lunar cartographers chose names for prominent craters, mountain ranges, and other regions. Some of the names that were assigned, such as those devised by the court astronomer to the king of Spain (who named features after Spanish nobility), were doomed to be forgotten. Others, such as the suggestion by a German astronomer that lunar mountain ranges be named after terrestrial features, are still used today.

The most influential of the lunar cartographers of medieval times was the Italian priest Joannes Riccioli, who, with his pupil Francesco Grimaldi, assigned names to the maria after human moods or experiences, and named craters after famous scientists. Thus we have names for maria such as Mare Imbrium (Sea of Showers), Mare Tranquilitatis (Sea of Tranquility, where the first manned moon landing occurred), and Mare Serenitatis (Sea of Serenity); and craters named Tycho, Hipparchus, and Archimedes. There were political overtones to Riccioli and Grimaldi's nomenclature; they published their map in 1651, at a time when the heliocentric hypothesis was still not something that people embraced publicly. With this in mind, Riccioli and Grimaldi gave geocentrists like Ptolemy and Tycho very large, prominent craters, but assigned Galileo only a very small one, and placed the name of Copernicus on a crater in the Sea of Storms (Mare Nubium, which can also be translated as Sea of Clouds).

Other names were added during the eighteenth and nineteenth centuries, but until 1921, no formal international agreement was made to use a specific, uniform naming system. In that year, the newly formed International Astronomical Union designated a committee to oversee and standardize lunar nomenclature. Things were uneventful thereafter, until 1959, when the Soviets obtained the first photographs of the far side of the moon, and promptly named numerous features after prominent Russians. Many of these names (but not all) were later approved by the International Astronomical Union. Perhaps the most controversial was the Sea of Moscow, which reportedly was finally approved (amid laughter) when the Soviet representative declared, in response to criticism that it was inconsistent with tradition to name a mare after a city, that Moscow is a state of mind, just as tranquility and serenity are.

More recently, numerous additional features have been mapped and named, as the sophistication of lunar probes has improved, and especially since manned exploration of the Moon has occurred. The Apollo astronauts assigned rather colloquial names to features at each landing site, and because of all the attendant publicity, these names became well known before there was time for the International Astronomical Union to even consider them for approval. Most of them were based on characteristics of the features themselves, such as a terraced crater that was called Bench, and one with a bright rim that was dubbed Halo. Groups of craters were named for their pattern, such as a pair called Doublet and a group named Snowman, which included individual craters with names such as Head. Some of the terrain mapped by the Apollo astronauts received names of prominent scientists, in keeping with the tradition established by Riccioli and Grimaldi more than 300 years earlier. Many of the names invented by the Apollo astronauts were eventually approved by the International Astronomical Union, as were a dozen craters named for American and Soviet astronauts and cosmonauts.

Just about the time the lunar cartography was getting resolved (after the years of confusion brought on by the lunar exploration program), unmanned probes began to obtain images of Mercury, Venus, and Mars, and the whole problem of naming extraterrestrial landmarks arose again. Except for Mars, whose large-scale light and dark regions had

(continued on next page)

[1]Based on information from F. El-Baz, 1979, "Naming Moon's Features Created 'Oceans of Storms'", *Smithsonian,* 9(10):96.

ASTRONOMICAL INSIGHT 9.2

MAPPING THE MOON

(continued from previous page)
been named on the basis of telescopic observations from Earth, no tradition existed for naming features on other planets. After considerable discussion, the International Astronomical Union adopted rules for doing so. New features on Mars will be named after villages on Earth. Venus will bear the names of prominent women and radio and radar scientists. Mercury will have features named after people famous for pursuits other than science. And features of the outer planets (or, as we shall see, the satellites thereof) will have names assigned from mythology. Perhaps it will thus be possible to map the rest of the solar system with a minimum of controversy.

advantage that it explains the general chemical similarities between Earth and Moon rocks. It has difficulty, however, in explaining why Moon rocks have a lower content of volatile elements. Another, perhaps more serious, problem is explaining why two separate bodies would have formed, rather than just one. It is very difficult to see how the prelunar material could remain in orbit about the Earth instead of settling in and becoming part of the Earth. (Another way of saying this is that it is difficult to see how the Earth-Moon system could have had enough angular momentum to form two bodies in orbit, rather than a single body.)

The fission hypothesis also easily explains the general similarities between Earth and Moon rocks. The lower volatile content of Moon rocks can also be understood if the material that split off of the Earth to form the Moon came from the Earth's mantle, which has a similarly low volatile content to Moon rocks. The major problem with this idea has always been that it is very difficult to find a reason for the early Earth to split, forming the Moon. The most widely cited explanation is that the young Earth was spinning so rapidly that rotational forces caused some of the outermost material of the Earth to be flung outward, forming a disk around the Earth that later coalesced to form the Moon. The difficulty is that there is no known way to make the early Earth spin rapidly enough for this to occur. (It is calculated that the length of the day would have had to be about 2 hours!) Recently, it has been shown that even then fission would not occur unless the Earth were completely liquid, with no internal friction (that is, viscosity), and this is thought to be very unrealistic.

The capture hypothesis, which assumes that the Moon formed independently of the Earth and then became trapped into orbit about the Earth, has several problems. One is that it seems probable that the detailed compositions of Moon and Earth rocks would differ more than they do, because the early Moon and the early Earth would have been subjected to different environments. A far more serious objection arises from the difficulty in finding a way to have the two bodies join together in mutual orbit, rather than just passing by each other when they happened to meet. Unless something happened to slow their relative speed of approach, the two would have too much kinetic energy to become trapped into orbit, if they started out as free bodies. Various suggestions have been made for overcoming this problem, such as assuming that the incoming Moon was slowed by a cloud of debris orbiting the Earth, or that the Moon was broken apart by tidal forces and its remains went into Earth orbit where they coalesced to re-form the Moon, but none of these events is considered very likely to have occurred.

The most recent hypothesis, that the Moon formed as the result of a giant impact on the young Earth, seems to fit all the known facts, and so far has not encountered any serious objections. (It remains to be seen whether this means it is really better than the others, or just that it is so new that objections are yet to be found.) Detailed mathematical models for the formation of planetesimals in the early solar system show that as these bodies built up, there was a time when nearly all the preplanetary material was in the form of large planetesimals. Thus, there was apparently an era when the solar system had many bodies that were almost of planetary size. This makes it probable that there were occasional collisions between objects as large as the Moon or a planet. The suggestion regarding the Moon's origin is that the young Earth, more or less already as large as it is today, collided with a huge planetesimal, as large as Mars or even larger. The impact would have occurred at low relative speed, because the Earth and the other body would have been orbiting the Sun in closely matched orbits, but the energy of impact would still have been sufficient to vaporize a huge quantity of material, some of which then would have gone into orbit about the earth. In this picture, the Moon then formed from the coalescence of the debris that orbited the Earth. The low volatile content of Moon rocks would be explained by the vaporization that occurred upon impact, and the large amount of angular momentum represented by the Moon's orbital motion would have been contributed by the energy of the impact.

It is ironic that we know so little for certain about the formation of our nearest neighbor in space. Perhaps this is a lesson for us; the more detailed information we have, the more difficult it is to find a theory that adequately accounts for all the known facts. In any event, there is general agreement on the evolution of the Moon once it had formed.

Apparently the Moon was largely molten very early in its history, and then a rigid crust formed as it cooled. The formation of this crust was probably delayed because of heating due to impacts of meteorites, and for a while the surface was molten, with large quantities of gases escaping. The Moon's low escape velocity allowed all gases to dissipate into space, and thus the Moon has no atmosphere. There was little differentiation, because the Moon's low mass did not provide sufficient gravitational force to bring about significant separation of the heavy elements. After a time, the crust cooled and hardened, but the interior remained molten longer, probably because of radioactive heating.

The next major stage in the Moon's life involved extensive volcanic activity. Probably triggered by large impacts from space that cracked the young Moon's crust, molten material from the interior flowed over large areas of the surface, particularly in the lowlands. Thus the maria were formed. It appears possible that the extensive maria on the near side of the Moon (the only significant maria the Moon has) were created as the result of a single tremendous impact. This also caused the near side of the moon to have a somewhat higher surface density than the far side, and this helped the Earth's tidal forces alter the Moon's spin to bring about synchronous rotation. The Moon's original rotation rate was apparently much faster than it is today, but the Earth's tidal force has acted to slow the rotation, and now the denser side of the Moon is pointed permanently toward the Earth. The internal friction that had been acting to slow the Moon's rotation has been eliminated. As we will learn in later chapters, synchronous rotation such as this is the rule for all satellites that are relatively close to their parent planets.

It is known that the Earth and Moon were once much closer together than they are now and that they gradually separated as the Moon's rotation was slowed by tidal forces. Because the total amount of angular momentum in a system must remain constant, the loss of angular momentum as the Moon's spin slowed was compensated by the Moon's growing distance from the Earth, which increased the angular momentum of its orbital motion.

Throughout the Moon's early history, it was continually bombarded by the rocky material still floating around the young solar system, as was the Earth. On our planet, erosion and tectonic processes have erased most evidence of cratering, but on the Moon, once the crust cooled, craters that were formed by the impacts of material hitting the surface were able to survive. The heavily cratered regions of the Moon we see today were shaped at that time, within the first billion years after the Moon formed. When the lava flows that created the maria occurred, they obliterated craters in the lowlands, so today there are relatively few craters in the maria. Those which are seen there were formed by meteoroids that hit the Moon after the lava flows cooled.

After the maria were formed, some 3 billion years ago, little else happened to modify the Moon's structure. The rate of cratering decreased as the interplanetary debris either escaped the solar system or got swept up by the planets and other large bodies, so only a moderate amount of catering has occurred since the maria formed. The Moon's interior has gradually cooled, eventually reaching its present nonmolten state.

PERSPECTIVE

The Moon is the best-studied object in the solar system, other than the Earth. The fact that its fate is so closely intertwined with that of the Earth enhances our interest in it, and it is important to note the similarities and differences between the two bodies. They are alike in age and composition, except for the lower quantities of volatile elements on the Moon. The lunar interior is geologically very quiet, whereas the Earth is still a dynamic, evolving body.

We are now ready to take a larger step out into the solar system, to visit and explore Earth's near twin, the planet Venus.

SUMMARY

1. The general nature of the Moon (its surface topography and its physical conditions), was ascertained from Earth-based observations.
2. Much more detailed information has come from the Soviet and U.S. space programs, which have sent unmanned and manned probes to the Moon.
3. The lunar surface consists of relatively smooth areas called maria, as well as highland mountainous regions, and is marked everywhere by impact craters.
4. The lunar soil is called the regolith, and all the rocks on the surface are igneous (most are silicates), with low abundances of volatile gases.
5. Seismic data show that the Moon has a crust 50 to 100 kilometers thick, a mantle, and a core extending about 500 kilometers from the center.
6. The Moon has no present-day tectonic activity and no magnetic field, indicating there is probably no liquid core.
7. The hypothesis that currently appears to have the fewest contradictions with observations suggests that the Moon formed as the result of a gigantic impact on the young Earth by a very planetesimal. The energy of the collision would have vaporized a large quantity of material, some of which would have gone into Earth orbit, where it coalesced to form the Moon. Other hypothesis, each of which has serious problems, are that the Moon formed together with the Earth; that it formed as the result of splitting of the young Earth; or that it formed elsewhere in the solar system and then was captured by the Earth's gravitational field.
8. The Moon's evolution consisted of a molten state, followed by hardening of the crust and subsequent large-scale lava flows, which created the maria. Since that time (about 1 billion years after the Moon's formation), the Moon has been geologically quiet.

REVIEW QUESTIONS

Thought Questions

1. The Moon's surface is characterized by mountainous regions, large, relatively smooth lowland areas (the maria), and craters everywhere. In what ways is the Moon's surface similar to that of the Earth, and in what ways is it different?
2. What is the evidence that the craters on the Moon are formed by impacts, rather than by volcanic eruptions?
3. Summarize the similarities and contrasts between lunar rocks and those found on the Earth's surface.
4. Why does the Moon have much more extreme fluctuations in surface temperature than does the Earth?
5. Why does the Earth not have as many craters as the Moon?
6. Why are there relatively few craters in the maria?
7. Compare the Moon's interior structure with that of the Earth.
8. Even though the craters on the Moon are thought to have been formed by impacts, there is evidence for volcanic activity in the past. What is the evidence?
9. Summarize the competing theories on the formation of the Moon, giving the strong points and weak points of each.
10. Discuss the role of angular momentum in the formation of the Moon and in the subsequent evolution of its spin and its orbit.

Problems

1. If the Moon and the Earth had equal diameters, so that they would appear the same size to an outside observer, the Earth would appear brighter because of its higher albedo. How much brighter would the Earth appear to be? Now take into account the fact that the Earth's diameter is approximately four times greater than that of the Moon, and estimate how much brighter the Earth appears to an outside observer.

2. Using information given in Chapter 5, calculate the surface gravity on the Moon, compared to that on the Earth. If you dropped a rock from the edge of a high cliff on the Moon, how fast would it be falling after 3 seconds?

3. If impacts by objects from space occurred at a uniform rate over the entire history of the Moon, and the maria have only one-fourth as many craters (per square kilometer) as the rest of the Moon, how much younger than the rest of the lunar surface would the maria be? Is this consistent with what the text says about the ages of the maria and of the Moon itself? What does this imply about the rate of impacts over the Moon's history?

4. What fraction of the Moon's volume is occupied by its core, if the core has a radius of 500 km?

ADDITIONAL READINGS

Anderson, D. L. 1974. The interior of the Moon. *Physics Today* 27(3):44.

Beatty, J. Kelly. 1986. The making of a better moon. *Sky and Telescope* 72(6):558.

Brownlee, Shannon. 1985. The whacky theory of the Moon's birth. *Discover* 6(3):65. (the Moon's origin attributed to a giant impact)

Burgess, E., and J. E. Oberg. 1976. Science on the Moon. *Astronomy* 4:684.

Cadogan, P. 1983. The Moon's origin. *Mercury* 12(2):34.

French, B. M. 1977. What's new on the Moon? *Sky and Telescope* 53:164.

Goldreich, P. 1972. Tides and the Earth-Moon system. *Scientific American* 226(4):42.

Goles G. G. 1971. A review of the Apollo project. *American Scientist,* 59:326.

Hammond, A. L. 1973. Lunar science: analyzing the Apollo legacy. *Science* 179:1313.

Kelly, I. W. 1986. The Moon was full and nothing happened. *Skeptical Inquirer* 10(2):129.

Register, Bridget Mintz. 1985. The fate of the moon rocks. *Astronomy* 13(12):14.

Sanduleak, N. 1985. The Moon is acquitted of murder in Cleveland. *Skeptical Inquirer* 9(3):236.

Taylor, S. R. 1975. *Lunar science: a post-Apollo view.* Elmsford, N.Y.: Pergamon Press.

Wood, J. A. 1975. The Moon. *Scientific American* 233(3)92.

CHAPTER 10

VENUS: A CLOUD-COVERED INFERNO

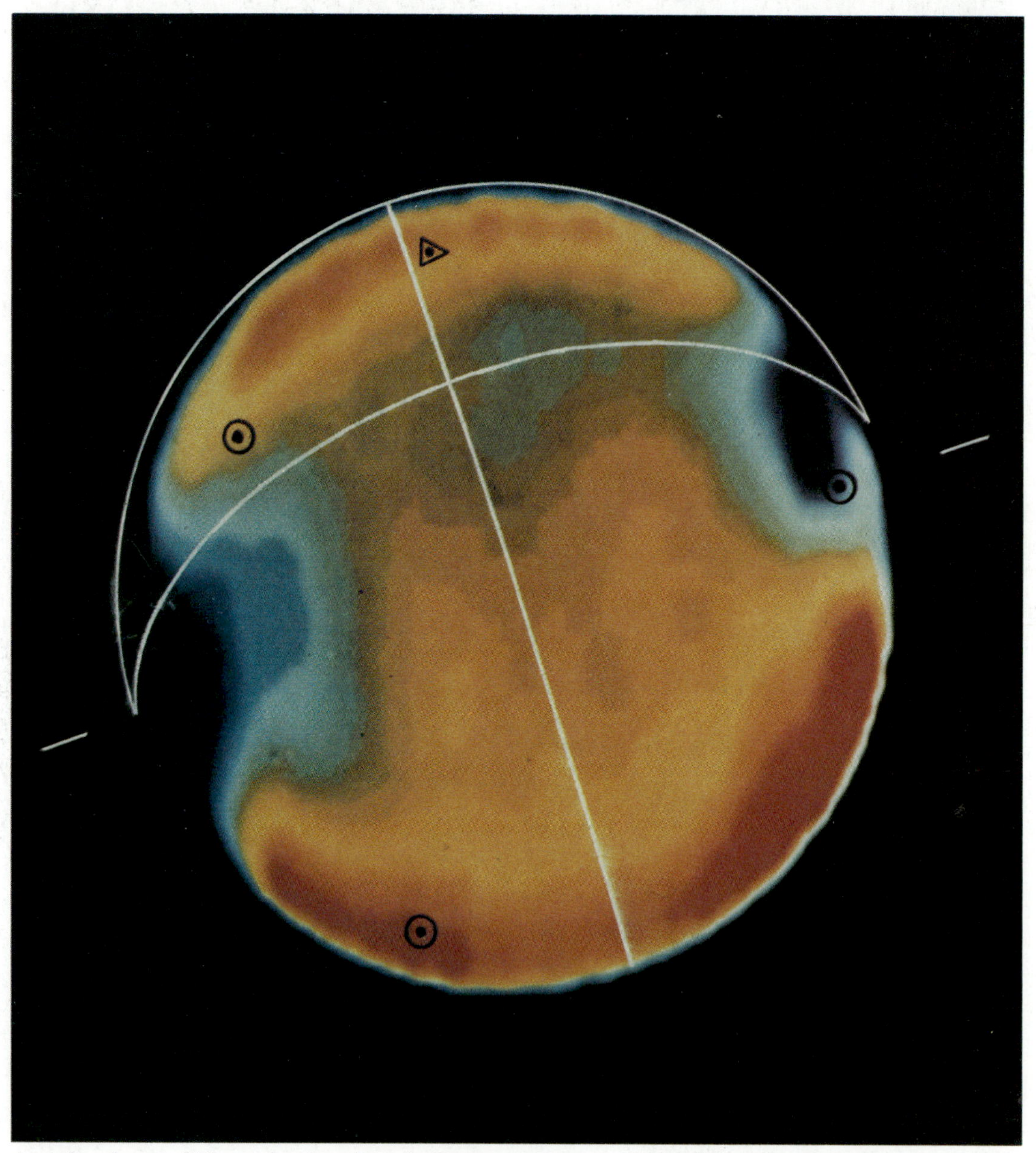

An infrared map of Venus, showing where the Pioneer Venus *probes entered the atmosphere.* (NASA)

CHAPTER PREVIEW

Of all the planets, Venus (Table 10.1 and Fig. 10.1) is the most prominent in our nighttime sky. It has played a role in the astronomical lore since antiquity. Because Venus is an inferior planet, it never appears far from the sun, as seen from the Earth (its greatest elongation is 47°). Thus it is visible to us either early in the evening, shortly after sunset, or in the wee hours just before dawn. In ancient times, the morning and evening stars were thought to be two separate planets, named Phosphorous and Hesperus, but by the sixth century B.C. the two had been identified as a single planet alternately appearing on either side of the Sun.

Bright as it is, Venus has become a natural subject for careful scrutiny by astronomers throughout history, and it has played an important role in some major developments. For example, the fact that it never wanders far from the Sun prompted early speculation that perhaps it orbits the Sun rather than the Earth, and Galileo's observation of the phases of Venus was an important bit of evidence favoring the heliocentric theory.

In modern times, a great deal of scientific effort has gone into studies of Venus, because its proximity to Earth, both in location and in general characteristics, means that Venus may teach us valuable lessons about our own planet.

TABLE 10.1 VENUS

Orbital semimajor axis: 0.723 AU (108,200,000 km)
Perihelion distance: 0.718 AU
Aphelion distance: 0.728 AU
Orbital period: 224.7 days (0.615 years)
Orbital inclination: 3°23′40″
Rotation period: 243 days (retrograde)
Tilt of axis: 3°
Diameter: 12,110 km (0.951 $D_{\oplus}$)
Mass: 4.87×10^{27} grams (0.815 $M_{\oplus}$)
Density: 5.3 g/cm^3
Surface gravity: 0.90 Earth gravity
Escape velocity: 10.4 km/sec
Surface temperature: 750 K
Albedo: 0.75
Satellites: none

OBSERVATIONS FROM NEAR AND FAR

A substantial body of knowledge about Venus was developed, particularly within the last two decades, from remote observation. Its size and orbital parameters, as well as some data on the composition of its upper atmosphere and an estimate of its mass, all were derived from Earth-based observations, as were some surprising discoveries about its rotation and its surface temperature.

FIGURE 10.1 THE CRESCENT VENUS. *This photograph, taken from Earth, shows Venus when it is near inferior conjunction; thus we see only a sliver of its sunlit side.* (NASA/JPL)

Venus is brilliant because it is shrouded in clouds that efficiently reflect sunlight. Its albedo is 0.76. The same clouds that create the brilliance also hide the surface from telescopic view, and for this reason special techniques have been required to penetrate the clouds.

Although it was once popular to imagine that Venus is a pleasantly tropical planet, with lots of water and, no doubt, plant and animal life, what was discovered from early radio observations was vastly different. Radio waves penetrated the cloud cover, and the message they brought to Earth is that the surface of Venus is very hot, about 750 K (477°C, or about 900°F). This discovery made it obvious that, despite outward appearances of similarity, Venus and the Earth had followed different paths. No longer could Venus be viewed as a likely abode for life.

The reason for the high temperature was quickly ascertained (Fig. 10.2). Spectroscopic measurements dating back to the 1930s showed that carbon dioxide (CO_2) is a major constituent of the atmosphere. This molecule has many spectral lines in the infrared part of the spectrum, so that a CO_2 atmosphere is effectively opaque to infrared light.

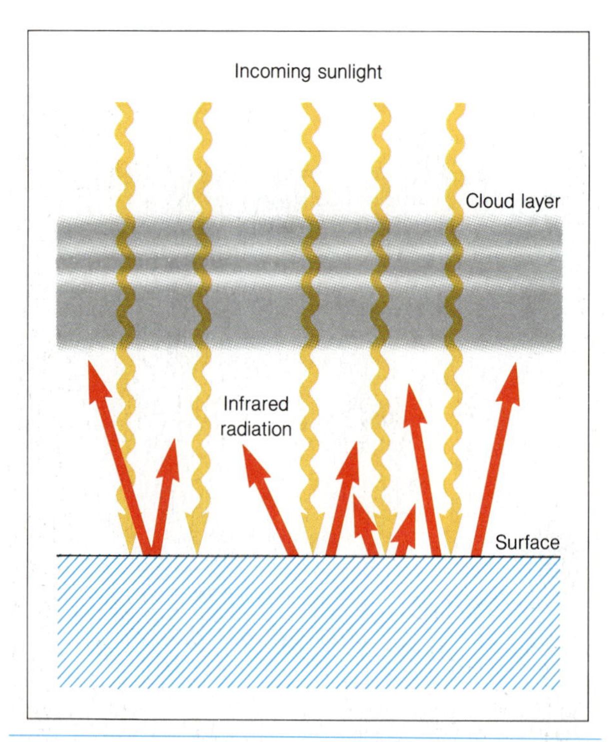

FIGURE 10.2 THE GREENHOUSE EFFECT. *Visible light from the sun reaches the surface of Venus and heats it, which causes the ground to emit infrared radiation. The carbon dioxide in the atmosphere efficiently absorbs infrared radiation, so heat is trapped near the surface.*

Visible light from the Sun can penetrate the clouds of Venus from above and is absorbed at the surface, heating it. When the ground responds by emitting infrared radiation, as it must according to Wien's law (Chapter 6), this radiant energy cannot rapidly escape because of absorption by the CO_2, and heat is trapped near the surface. This heating mechanism is called the **greenhouse effect,** because it is similar to what happens in a greenhouse when light enters through the glass and heats the interior.

A second surprising discovery was made from Earth, again through the use of radio (specifically radar) wavelengths. The first radar measurements of the surface were made in 1961, and the Doppler effect was used to determine the rate of rotation. Contrary to expectations, the planet was found to rotate very slowly, and even more astonishingly, to do so in the backward direction, the direction opposite that of the other orbital and rotational motions in the solar system. This is called **retrograde rotation.**

The spin of Venus takes 243 Earth days, so slow that its day is actually longer than its year (the orbital period is 225 days). Here we must distinguish between the sidereal day, the 243 days just mentioned, and the solar day, the apparent time it takes for the Sun to go through one daily cycle, as seen from a point on the surface of Venus. The length of the solar day on Venus is 116.8 days.

The reason for the unusual rotation of Venus is not understood. Because it is so strongly at odds with the spins of the other planets, it seems likely that something unusual happened either during the formation of Venus or at some later time.

Both the Soviet Union and the United States focused on Venus as a logical and interesting target for exploration by spacecraft, and a number of missions have been flown there by both nations (Table 10.2). Spacecraft can be launched toward Venus most easily when the relative positions of the Earth and Venus are favorable. The United States launched *Mariner 2* toward Venus on such an occasion in 1962, and this spacecraft flew within 35,000 kilometers of the cloud tops, making various measurements. The next opportunity came in 1965–1966, when both the United States and the Soviet Union

TABLE 10.2 PROBES TO VENUS

Spacecraft	Launch	Encounter	Nature of Encounter
Mariner 2	Aug. 27, 1962	Dec. 14, 1962	Flyby, 34,833 km
Venera 4	June 12, 1967	Oct. 18, 1967	Hard landing, dark side
Mariner 5	June 14, 1967	Oct. 19, 1967	Flyby, 4,100 km
Venera 5	Jan. 5, 1969	May 16, 1969	Hard landing, dark side
Venera 6	Jan. 10, 1969	May 17, 1969	Hard landing, dark side
Venera 7	Aug. 17, 1970	Dec. 15, 1970	Soft landing, dark side
Venera 8	March 27, 1972	July 22, 1972	Soft landing, day side
Mariner 10	Nov. 3, 1973	Feb. 5, 1974	Flyby, 5,700 km
Venera 9	June 18, 1975	Oct. 22, 1975	Orbiter plus soft lander
Venera 10	June 14, 1975	Oct. 25, 1975	Orbiter plus soft lander
Pioneer Venus 1	May 20, 1978	Dec. 4, 1978	Orbiter
Pioneer Venus 2	Aug. 8, 1978	Dec. 9, 1978	Four hard landers, both sides
Venera 11	Sept. 9, 1978	Dec. 25, 1978	Flyby (25,000 km), soft lander
Venera 12	Sept. 14, 1978	Dec. 21, 1978	Flyby (25,000 km), soft lander
Venera 13	Oct. 30, 1981	March 1, 1982	Flyby, soft lander ?
Venera 14	Nov. 4, 1981	March 5, 1982	Flyby, soft lander?
*Venera 15**	? (1984)	? (1985)	Details unknown
*Venera 16**	? (1984)	? (1985)	Details unknown

*The *Venera 15* and *Venera 16* spacecraft were renamed *Vega 1* and *Vega 2* after encountering Venus, when they went on to fly by Comet Halley in March, 1986.

took advantage by launching Venus probes. The U.S. *Mariner 5* mission was another flyby, but the Soviets chose instead to have their *Venera 4* probe descend into the atmosphere of Venus, and it relayed back to Earth important information on conditions there, before it failed on its way down to the surface. *Veneras 5* and 6, launched in 1969, also entered the atmosphere of Venus, as did *Veneras* 7 and 8 in 1970 and 1972.

The U.S. *Mariner 10* mission in 1973–1974 flew by Venus (and later Mercury), obtaining detailed data on the structure of the clouds and their motions. This and earlier *Mariner* spacecraft showed that Venus has no magnetic field.

The ultraviolet images obtained by the early *Mariner* flyby missions showed structure in the clouds (Fig. 10.3). Continued observations of this structure revealed that the cloud tops are moving rapidly, circling the planet every 4 days at speeds of about 100 meters per second (360 kilometers per hour). This atmospheric circulation occurs in the retrograde direction, the same direction as the rotation of the planet. The temperature of the cloud tops was derived from infrared measurements, which led to an estimate of 240 K (−27°F).

FIGURE 10.3 A CLOSE-UP PORTRAIT OF VENUS. *This is a full-disk ultraviolet image of Venus obtained by the* Mariner *10 spacecraft; it shows structure in the clouds.* (NASA)

ASTRONOMICAL INSIGHT 10.1

INTERPLANETARY RADAR

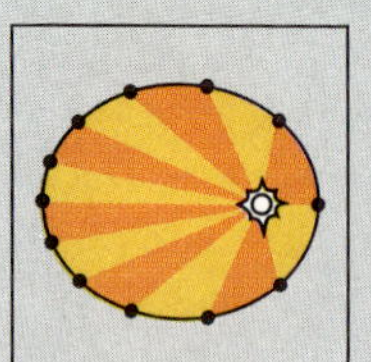

Many of us are familiar with radar in its conventional uses, such as the detection of other aircraft by an airliner, or the measurement of automobile speeds by a police officer. Radar also has very important scientific uses, including the study of planetary properties.

Radar was developed during World War II, when emission from the Sun in short radio wavelengths was discovered and its potential for defense purposes was recognized. The term "radar" refers to radio radiation in specific wavelengths (a few centimeters) in the microwave spectrum. Radiation at these wavelengths reflects efficiently from various kinds of surfaces (particularly metallic ones). Radar always travels at the speed of light, so the distance to an object can be found precisely by measuring how long it takes for the echo to be received after radiation is sent toward the object. This is very useful in astronomy, as are other capabilities that the use of radar provides scientists.

Several of the world's large radio telescopes are equipped to make radar measurements. They can send powerful bursts of radio radiation out into space and can detect the return signal, even if it is very weak. In this way the precise distances to the Moon and the nearby planets, as well as some asteroids (see Chapter 16) are measured. This is the most accurate technique available for making these measurements. As we shall see in later chapters, our knowledge of distances within the solar system determines how well we know other distance scales in the universe (in particular, see the discussion of the "distance pyramid" in Chapter 30). Thus, radar astronomy is very important to astronomers in all areas.

In addition to measuring distances, radar techniques reveal surface features on other planets, and can tell us how they are moving as well. The efficiency of a surface for reflecting radar waves is influenced by the composition of the surface and by its texture. This means, for example, that mountainous areas and smooth lowland regions on a planet will reflect radar with different efficiencies, so that the two kinds of surface can be distinguished from one another by radar measurements. This kind of analysis has been particularly useful for studies of Venus, since its surface cannot be seen in visible light because of the global cloud cover. Radar observations of Venus, made from the Earth, have revealed the locations of mountains, plains, and several large circular features that may be craters. Close-up radar observations of Venus, made from the *Pioneer-Venus* orbiter, have produced detailed relief maps of the planet (see Fig. 10.9).

To measure the motion of a distant object using radar, scientists make use of the Doppler effect (Chapter 6). The Doppler effect is the shift in wavelength (and frequency) of a wave due to relative motion between source and observer. If the source of waves is approaching the observer, the observed wavelength is shorter than if there were no relative motion; if source and observer recede from each other, the observed wavelength is increased. In radar measurements, the Doppler shift is caused by the motion of the object that reflects the radar waves. A police radar device

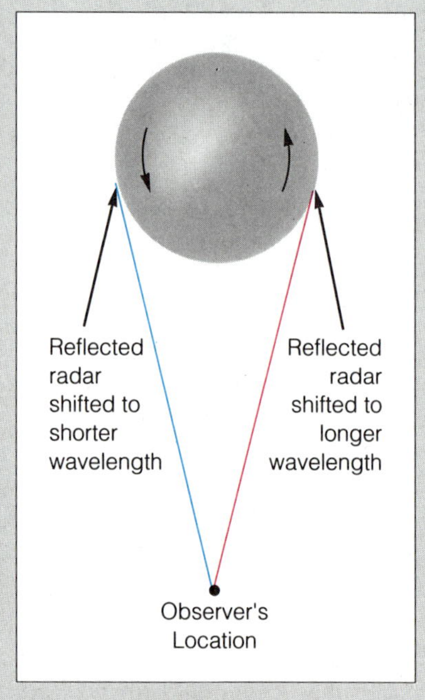

measures the speed of an oncoming car by detecting the Doppler shift caused by the relative motion of the car. In similar fashion, radar provides astronomers with a method for measuring speeds of nearby bodies in space.

Of particular interest to astronomers has been the measurement of rotational speeds of the planets. In order to measure how fast a body like a planet is spinning, scientists measure the speed of approach of one edge of the planetary disk, and the speed of recession of the other edge (see drawing). This provides a measurement of the rotational speed, and from this speed and the radius of the planet, the rotational period can be derived. It was measurements like this that first revealed the unusually slow, backward rotation of Venus. As we will see in Chapter 12, similar measurements produced a major

(continued on next page)

ASTRONOMICAL INSIGHT 10.1

INTERPLANETARY RADAR

(continued from previous page)
surprise when the rotational speed of Mercury was measured.

Today many of the basic measurements of planetary distances and rotational speeds that are of interest have been made, so this aspect of radar astronomy is less utilized than it once was. There are still active programs to measure surface details of planetary objects, however, and studies of asteroids are being carried out as well. Plans are being made for a new Venus orbiter, called the *Venus Radar Mapper,* whose primary job will be to make more detailed surface relief maps than previously possible. It is really quite remarkable how widespread the applications of radar technology are, both on Earth and in space.

The most recent probes sent to Venus have provided spectacular new information on conditions beneath the clouds. Whereas the earlier Venera instruments had failed (as a result of the intense heat and pressure) before reaching the surface, the *Venera 9* and *10* missions (in the mid- 1970s), *Veneras 11* and *12* (1978), and most recently, *Veneras 13* and *14* (1982), all succeeded in landing and operating on the surface. *Veneras 9, 10, 13,* and *14* carried cameras and sent back photographs of the ground surrounding the landing sites. These images showed that the surface is brightly lit, with enough sunlight penetrating the clouds to actually cast shadows, and that the ground is strewn with rocks, many of them with surprisingly sharp edges. This indicates that, despite the hostile conditions, there is very little erosion acting on the surface of Venus, or that some geological process renews the surface from time to time.

TABLE 10.3 COMPOSITION OF THE ATMOSPHERE OF VENUS

Gas	Symbol	Fraction*
Carbon dioxide	CO_2	0.96
Nitrogen	N_2	0.035
Sulfur dioxide	SO_2	1.5×10^{-4}
Water vapor	H_2O	1×10^{-4}
Argon	Ar	7×10^{-5}
Carbon monoxide	CO	4×10^{-5}
Neon	Ne	5×10^{-6}
Hydrogen chloride	HCl	4×10^{-7}
Hydrogen fluoride	HF	1×10^{-8}

*Values are fractions by number.

The most recent U.S. probe to Venus was called *Pioneer Venus,* and consisted of two separate spacecraft, an orbiter and a package of probes designed to penetrate the atmosphere. As of late 1987, the orbiter was still in operation, gathering data on the cloud structure, composition, and motions, as well as on conditions in the upper atmosphere, above the clouds. The five small probes that descended into the atmosphere when the spacecraft arrived at Venus in late 1978 gathered data on physical conditions such as temperature and pressure, and data on chemical composition of the atmosphere as a function of altitude. The *Pioneer Venus* probes reached the ground safely and continued to operate there for a while (up to 1 hour).

THE ATMOSPHERE OF VENUS

We can summarize what has been learned about the atmosphere of Venus by discussing three distinct aspects: its composition, vertical structure, and motions.

The most important ingredient in the atmosphere of Venus, already identified from Earth-based measurements, is carbon dioxide (Table 10.3). The *Venera* and *Pioneer Venus* probes showed that CO_2 dominates all through the atmosphere to the ground and makes up about 96 percent of the atmosphere. It is interesting that the overall quantity of CO_2 on the Earth is about the same as that on Venus, except that on Earth the CO_2 is mostly in carbonate rocks or dissolved in seawater, and not in the atmosphere.

The next most abundant gas in the atmosphere of Venus is nitrogen (in the form of N_2), which accounts for most of the remaining 4 percent or so. In addition, traces of water vapor and other species such as oxygen, argon, neon, and sulfur (in the form of compounds such as sulfur dioxide and sulfuric acid) exist there. The sulfuric acid (H_2SO_4) apparently condenses out of the lower clouds, forming droplets like rain. These circumstances, along with the high temperature and pressure at the surface, make it difficult for spacecraft to operate on Venus.

THE STRUCTURE OF THE ATMOSPHERE

We have already seen that the atmospheric conditions on Venus vary quite a lot from the top of the clouds to the surface. The temperature increases from 240 K at cloud tops to 750 K at the ground, and the pressure at the bottom of the atmosphere reaches the rather incredible value of 90 Earth atmospheres, or more than 1,300 pounds per square inch, equivalent to the pressure about 3,000 feet below the surface of the ocean. The temperature decreases smoothly from the surface to the top of the clouds.

The clouds in the atmosphere of Venus are composed of small droplets of sulfuric acid and some other, as yet unidentified, kinds of particles. The H_2SO_4 droplets are about 2 millionths of a meter (about 0.0002 cm) in diameter.

There are three distinct zones of clouds at separate altitudes (Fig. 10.4). The uppermost layer, the one visible from afar, lies some 60 km above the surface of Venus and averages about 10 km in thickness. The middle cloud layer, about 6 km thick, is suspended at an altitude of roughly 53 km, whereas the lowest layer lies at 48 km and is relatively thin, extending only 2 km or so vertically. Below the lower cloud is clear atmosphere, all the way to the ground. Above the uppermost clouds is a haze layer, consisting of much smaller particles than the droplets that form the clouds, but of unknown composition.

Clouds form in the atmosphere wherever the combination of temperature, pressure, and relative abundance of H_2SO_4 corresponds to the conditions required for this gas to condense. This is exactly analogous to the formation of clouds in the Earth's atmosphere, except that on the Earth, the gas that condenses is water vapor rather than sulfuric acid. It happens that the proper conditions for condensation occur at three different levels in the atmosphere of Venus.

The clouds also contain quantities of the gas sulfur dioxide (SO_2), which is an efficient absorber of ultraviolet light. We see from ultraviolet photographs (Figs. 10.3 and 10.5) that there is considerable structure in the clouds, because the SO_2 abundance varies with depth: therefore, in places where the winds stir up regions with quantities of SO_2, or where the cloud structure allows us to see to greater depths, the clouds look dark (see Fig. 10.3). Thus, ultraviolet photographs show some contrast in the clouds, and are useful for mapping the motions of the upper regions (Fig. 10.6).

The atmosphere seems to be less changeable than that of the Earth, where cloud structures come and go with changes in the weather. The relative stability of Venus is probably a result of a combination of factors, including the greater density and pressure of the atmosphere, and the slow rotation of the planet, which minimizes circulation patterns driven by rotation.

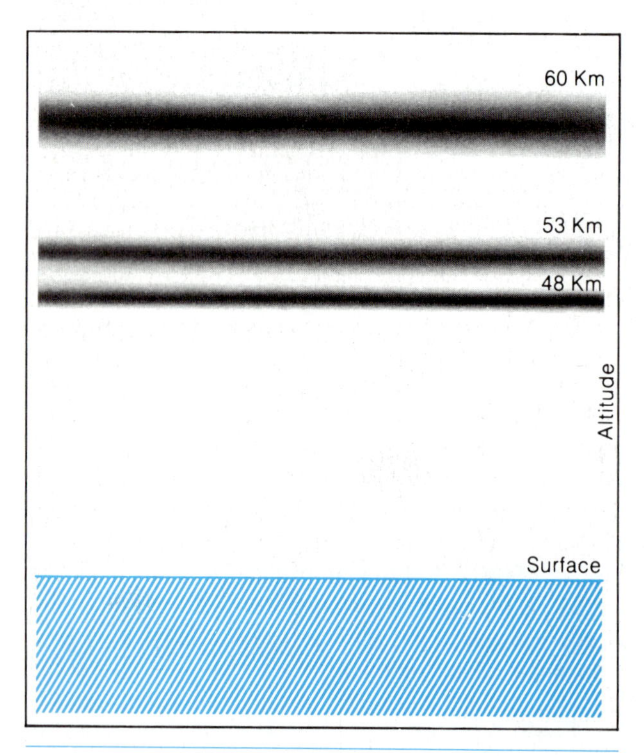

FIGURE 10.4 THE CLOUDS OF VENUS. *The sulfuric acid clouds are separated into three distinct layers. These layers occur at altitudes where the combination of temperature and pressure causes sulfuric acid to condense.*

Another important distinction between Earth and Venus is that there is no magnetic field around Venus. There is no magnetosphere and no protective shield of magnetic field lines to prevent charged particles from entering the atmosphere. This creates a complex electrical environment at the top of the atmosphere, including lightning discharges, which probably affects the chemical processes there as well.

Atmospheric Motions

We have seen that the upper atmosphere of Venus has a general circulation pattern at the cloud tops, moving in the same direction as the planet's rotation (Fig. 10.6). The velocity is high at the cloud tops but diminishes to nearly zero at the surface.

There are complex vertical motions in addition to these horizontal ones. Because of solar heating near the equator, there is a rising current there that spreads out in all directions, descending when it reaches cooler regions near the poles and on the dark side of the planet. This pattern is similar to the circulation cells in the Earth's atmosphere, except that on Venus there is insufficient rotation of the planet to break up the flow into numerous small cells. On Venus it is more of a global circulation pattern, with warm gas rising in the location called the **subsolar point,** where the Sun is directly overhead, and descending in a very broad region away from that point (Fig. 10.7). Besides minimizing rotational forces that might otherwise break up this simple flow pattern into smaller cells, the slow spin of Venus also means that the subsolar point moves very slowly as the planet rotates. Thus, the heating from the Sun lingers, enhancing the flow pattern.

Some of the gas that rises at the subsolar point flows all the way around to the dark side of Venus before cooling and descending. As it does so, free atoms combine to re-form molecules, which then emit light as they cool further. This creates aurorae on the dark side of Venus, a phenomenon that was suspected from Earth-based observations, but which was not confirmed until the flight of *Pioneer Venus* (Fig. 10.8).

FIGURE 10.5 AN ULTRAVIOLET IMAGE FROM PIONEER VENUS. *The dark regions seen here are places where atmospheric sulfur dioxide, which absorbs ultraviolet light, is exposed.* (NASA)

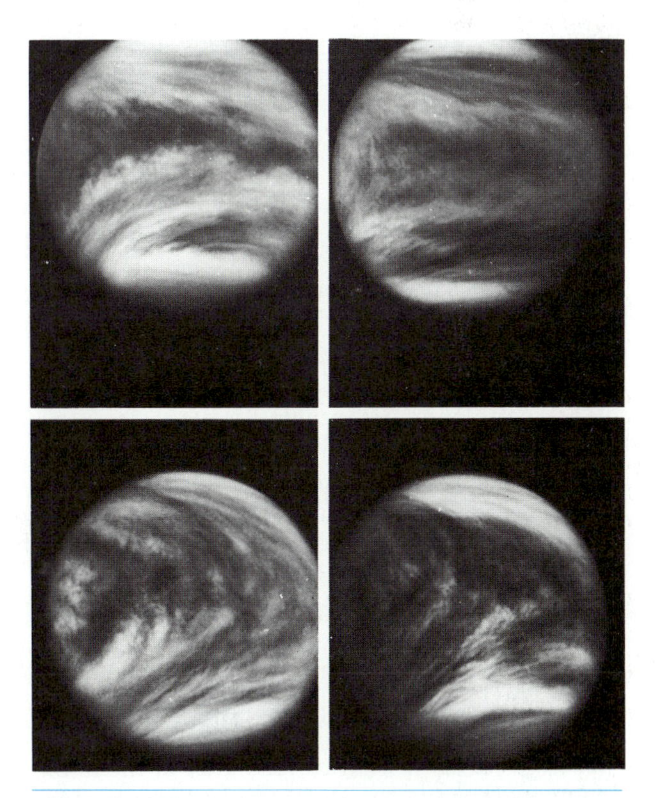

FIGURE 10.6 CIRCULATION OF THE ATMOSPHERE OF VENUS. *These four ultraviolet views show two rotations of the planetary cloud cover. The upper two are separated by one day; the lower two were obtained about a week after the first pair and are also separated by a day. The atmospheric motion is from right to left in these images.* (NASA)

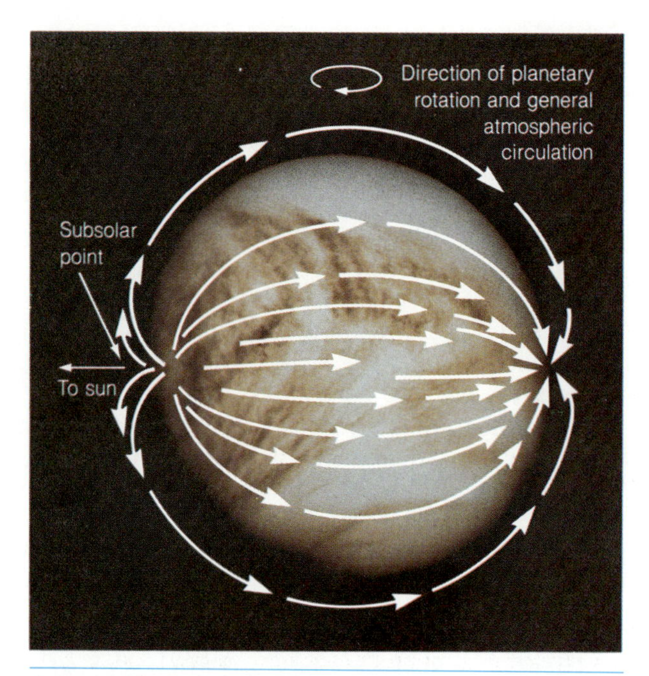

FIGURE 10.7 HIGH-ALTITUDE CIRCULATION. *Well above the clouds of Venus, which are rotating from right to left in this sketch, gases heated at the subsolar point circulate around the planet toward the dark side, where they cool and descend.* (Photo from NASA)

FIGURE 10.8 AIRGLOW ON THE NIGHT SIDE OF VENUS. *Atoms produced by the disruption of molecules on the sunlit side circulate around to the dark side, where they cool and combine once again into molecules. The newly formed molecules have excess energy, which is released in the form of light, creating a glow in the upper atmosphere on the night side.* (LASP, University of Colorado, sponsored by NASA)

These aurorae are quite distinct from those in the Earth's upper atmosphere, which are created by charged particles from space.

In and between clouds, additional vertical circulation patterns exist, depending on where the Sun's heat energy is deposited. This, in turn, depends on the cloud composition; we have already seen that SO_2 is particularly effective as an absorber of solar energy, so there is extra heating in levels where SO_2 is abundant. There appear to be at least four flow patterns at different levels involving circulation between the equatorial and polar regions.

THE SURFACE OF VENUS

Radar experiments have provided most of the information available on the surface of Venus. Observations made from Earth had insufficient resolution to reveal much in the way of surface features, although there were some indications of highlands and lowlands. The radar instrument on the *Pioneer Venus* orbiter, however, is capable of making fairly detailed measurements and can detect features as small as 30 kilometers across. As a result, global relief maps of Venus have been constructed, showing many interesting features (Fig. 10.9).

The surface has been classified into three general types of terrain: (1) rolling plains, covering about 65 percent of the planet; (2) highlands, covering about 8 percent; and (3) lowlands, occupying the remaining 27 percent.

The rolling plains are characterized by many craters and circular basins, some of them apparently lava-filled, resembling small-scale lunar maria. There is uncertainty about whether the craters are all caused by meteor impacts, or whether some may be volcanic in origin.

The highlands are concentrated in three major regions. Two of these, called Ishtar Terra (Fig. 10.10) and Aphrodite Terra, are comparable in size to the continents of Africa and Australia on the Earth, whereas the third, Beta Regio, is much smaller. The

ASTRONOMICAL INSIGHT 10.2

VOLCANOES ON VENUS

Until the *Voyager* spacecraft flew past Jupiter in the late 1970s, the Earth was the only body known to have current volcanic activity. Gigantic volcanoes had been found on Mars in the early 1970s, but they are apparently extinct. *Voyager* found volcanoes in the act of erupting on Io, one of the moons of Jupiter, making that satellite the first nonterrestrial body known to have current volcanic activity. Now Venus has become the second, if a recent suggestion proves to be correct.

No space probe or other device has actually observed an eruption in progress on Venus, but a variety of evidence leads to the conclusion that eruptions have occurred recently. As we see in this chapter, there are surface features on Venus that appear to be volcanoes. It has been argued that the number of these now present implies that they are still forming from time to time. Other data, including observations of lightning storms and determinations of atmospheric chemical abundances, lend support to the idea that volcanoes may erupt occasionally on Venus.

The strongest argument, however, comes from recently reported variations in the quantity of sulfur dioxide (SO_2) in the atmosphere of Venus. This molecule was detected in abundance by the *Pioneer Venus* orbiter when it arrived at Venus in 1978, but its abundance declined since then as the orbiter has continued to make measurements. Furthermore, ultraviolet measurements made by other instruments before 1978 failed to reveal any SO_2, even though it should have shown up if it had been as abundant as it was later found to be by *Pioneer Venus.* Apparently there was a sudden increase in the quantity of atmospheric SO_2 in 1978 or shortly before. This molecule is one of the major products of volcanic venting, so it has been suggested that a large-scale eruption took place on Venus in the late 1970s.

Other evidence supporting this idea comes from observations, also by *Pioneer Venus,* of a haze layer above the clouds of Venus. This haze has been attributed to tiny particles suspended in the atmosphere, and such particles are another common product of volcanic eruptions. It is known that no such haze existed in the 1960s, when observations were made from Earth that would have revealed its presence if it had been there.

Based on the quantity of SO_2 and haze apparently injected into the atmosphere, estimates have been made of the magnitude of the proposed eruption. These estimates suggest that the energy released was comparable to that of the famous Krakatoa explosion that occurred on Earth (near Java) in 1883. This was a very violent event, devastating an entire island, producing widespread tidal waves, creating sound waves heard thousands of miles away, and putting enough dust into the atmosphere to affect the passage of sunlight through it for several years. It may prove possible within the next few years to test the hypothesis that Venus is volcanically active. Of course, one possible form of verification will have been found if a new influx of SO_2 is discovered by the still-operating *Pioneer Venus* or by some other ultraviolet instrument. The typical time between eruptions major enough to be detected in this indirect way is not known, so there is no way of guessing whether a new rise in the SO_2 abundance will occur soon.

Another test of the theory will occur in the 1990s, when the *Venus Radar Mapper (VRM),* with its high-resolution radar imaging instrument, is launched. This device should send back data on the surface structure sufficiently detailed to tell us whether there are volcanoes that are either now erupting or have recently done so. It will be interesting to see what is found.

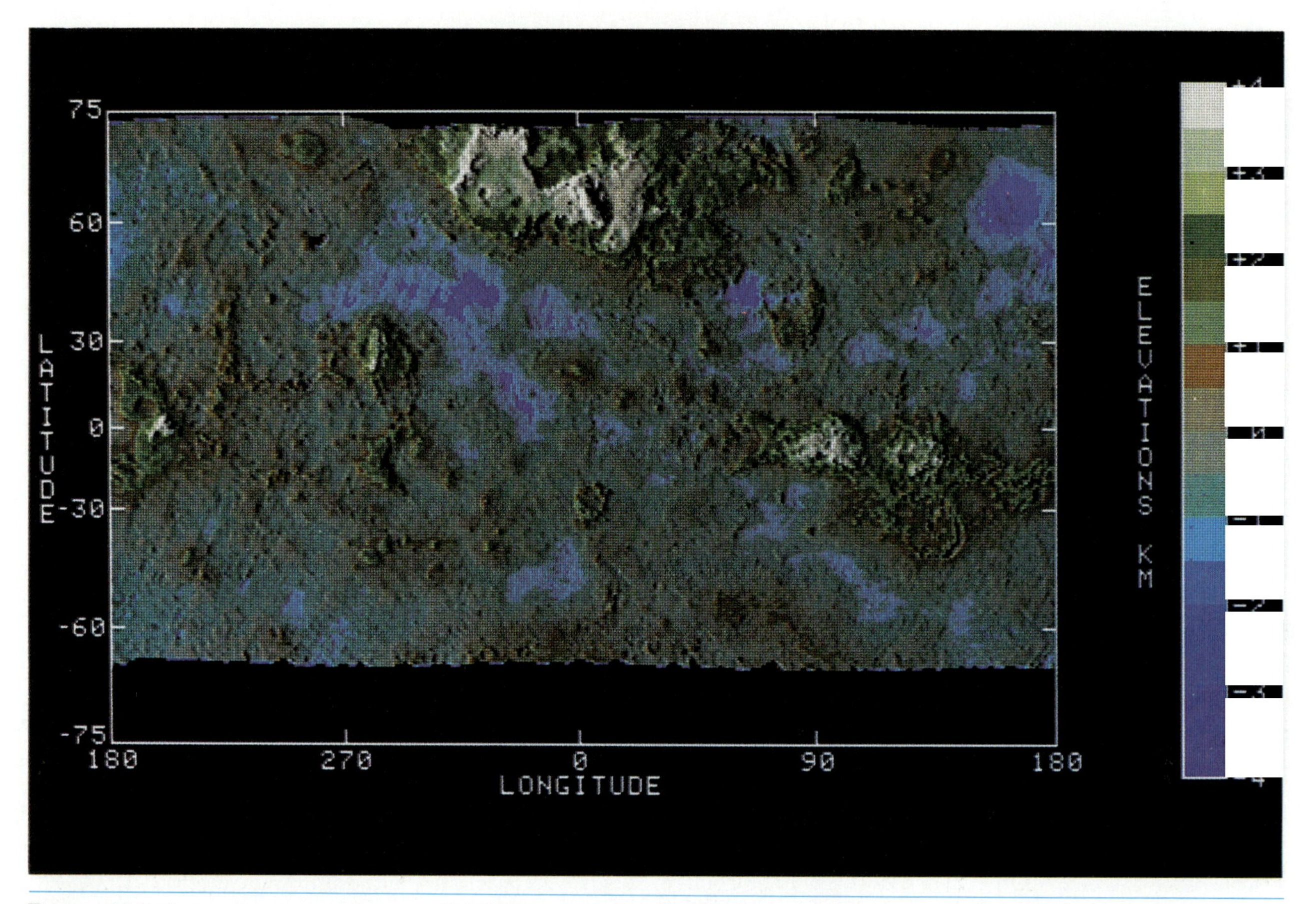

FIGURE 10.9 A RELIEF MAP OF VENUS. *This image, constructed on the basis of* Pioneer Venus *radar maps of the surface, shows the three major highland areas: Ishtar Terra at the top, Aphrodite Terra at right center, and Beta Regio at left center.* (NASA/JPL)

maximum elevation above the mean surface level is greater than that of Mt. Everest above sea level, but the many canyons are much shallower than the deepest undersea trenches on Earth. There are many adjacent parallel valley and trench systems in the highlands of Venus, forming systems up to 9,000 kilometers in length.

Given the similarities in size and average density of Venus and the Earth, it is interesting to consider whether the internal structure and processes are also similar. One point of comparison is tectonic activity. Venus apparently does not have such active continental drift as the Earth; there is not a planetwide system of plate boundaries and ridges and trenches, comparable to that on the Earth. The reasons for this difference between the Earth and Venus are not clear.

There is evidence, however, in the form of the valley and trench systems in the highlands, that tectonic activity has played some role in the formation of these regions. The highlands consist in part of large volcanic mountains that built up as a result of prolonged lava-flow episodes. The smallest of the three highland areas, Beta Regio, appears to consist entirely of a few large volcanoes. Apparently there is volcanic activity on Venus currently. This was deduced from indirect evidence, primarily an atmospheric SO_2 content that has been found to vary with time. Since this gas is emitted by volcanoes, the variation in its abundance is thought to be due to volcanic eruptions.

The combination of the valley and ridge systems and the large volcanoes indicates that there are convection currents in the upper mantle, and that there

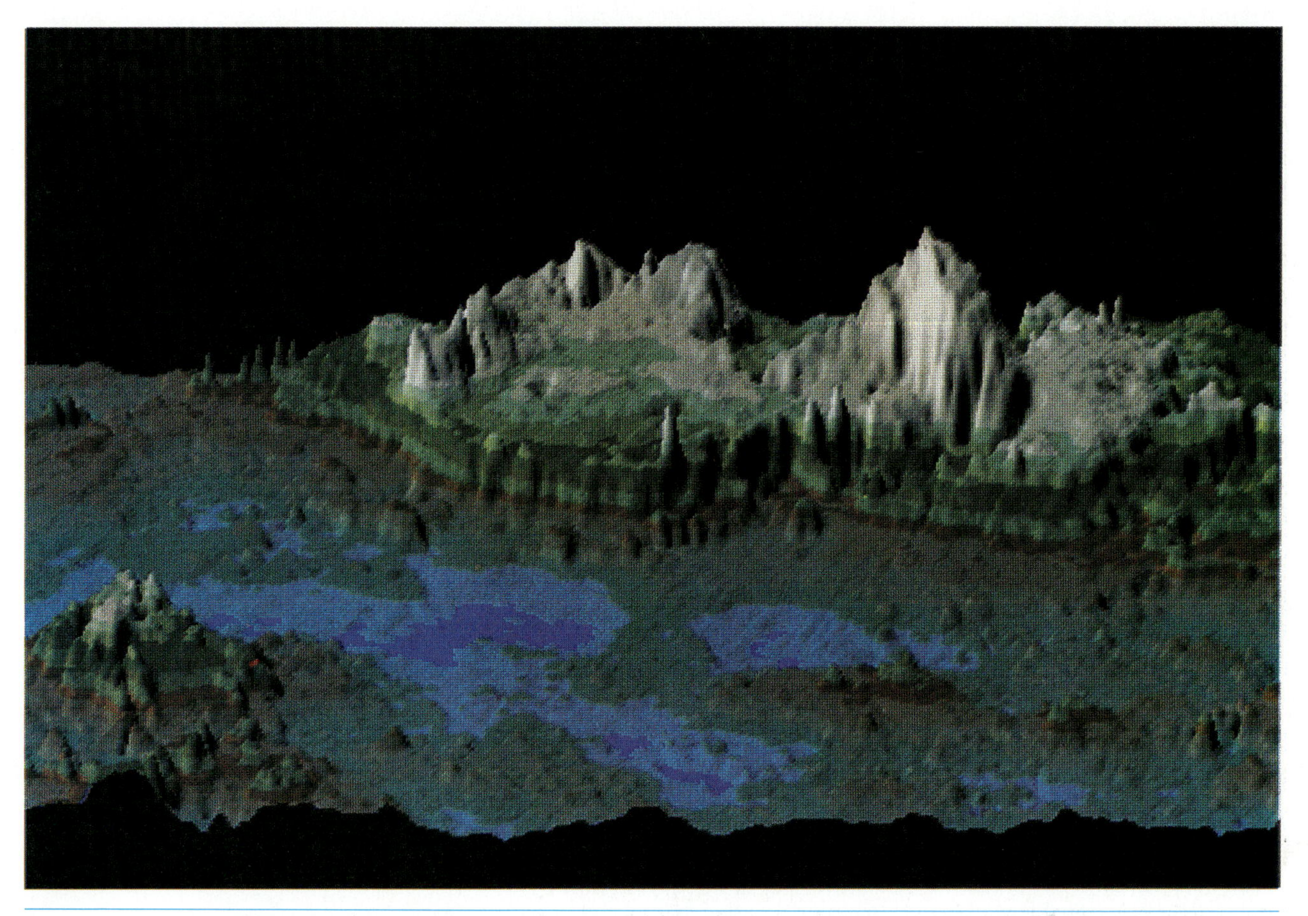

FIGURE 10.10 A PORTION OF ISHTAR TERRA. *This computer reconstruction illustrates (on an exaggerated vertical scale) the vertical relief of this highland region, and includes the largest mountain on Venus, Maxwell Montes, more than 11 kilometers above the mean surface level.* (NASA/JPL)

are volcanic hot spots where lava flowed upward for long periods. Thus Venus's internal activity and structure may be much like that of the Earth, except its crust has not broken up into plates that move about.

The reason for the lack of continental drift on Venus is not known, although there are some interesting speculations. One suggestion is that because Venus has such a high surface temperature, its crust is warmer and therefore more buoyant than that of the Earth. On both planets, the highlands (or continents) consist of old material that essentially floats, in equilibrium with the denser material of the lowlands (or seafloors). On Earth, the seafloor material is as dense as the underlying asthenosphere and therefore is easily forced below continental margins in the subduction zones that produce the great undersea trenches. On Venus, the lowland material is less dense than the underlying mantle and therefore cannot easily be forced downward. The result is that convection currents inside Venus have not succeeded in creating crustal plates that can move about by slipping over one another, because subduction cannot take place.

Some information about surface rocks was derived from the photographs and other measurements obtained by the *Venera* landers (Figs. 10.11 and 10.12).

FIGURE 10.11 SURFACE ROCKS ON VENUS. *These views are from the first Soviet* Venera *landers. They reveal that sunlight penetrates strongly to the surface and that the surface rocks have sharp edges, which indicates a lack of erosion.* (TASS from SOVFOTO)

As we saw earlier, the rocks have sharp edges, indicating a lack of erosion, something that was a surprise until scientists realized that the winds at the surface of Venus are almost nonexistent.

Measurements of surface rocks show a basaltic composition, similar to the predominant crustal rocks on the Earth and the Moon. Basalt, as noted in Chapter 8, is an igneous silicate rock with a moderate density. This, combined with the fact that the average density of Venus is comparable to that of the Earth (5.3 grams/cm compared with 5.5 for the Earth), implies that Venus is differentiated, with a large, dense core.

Other than this very indirect evidence, little is known about the interior of Venus. The fact that its internal structure is similar to that of the Earth might make us suspect that Venus has a magnetic field, but as we have seen, it does not. This may be explained by the slow rotation of the planet, which could be insufficient to create a magnetic dynamo in its core.

THE EARTH AND VENUS: SO NEAR AND YET SO DIFFERENT

We are now equipped with enough information to discuss the history of Venus and to develop some understanding of what might be responsible for the extreme differences between conditions there and on the Earth. This is important, for we need to know how precarious our situation is, so that we know how to avoid turning the Earth onto the path that Venus has followed.

Venus is thought to have formed from rocky debris orbiting the infant Sun in a great disk, just as the Earth did. The early evolution also must have been similar, with Venus undergoing a molten period when its dense elements sank to the center, leaving a lighter crust composed largely of carbonates and silicates. Before the crust cooled, volatile gases escaped from the surface, forming a primitive atmo-

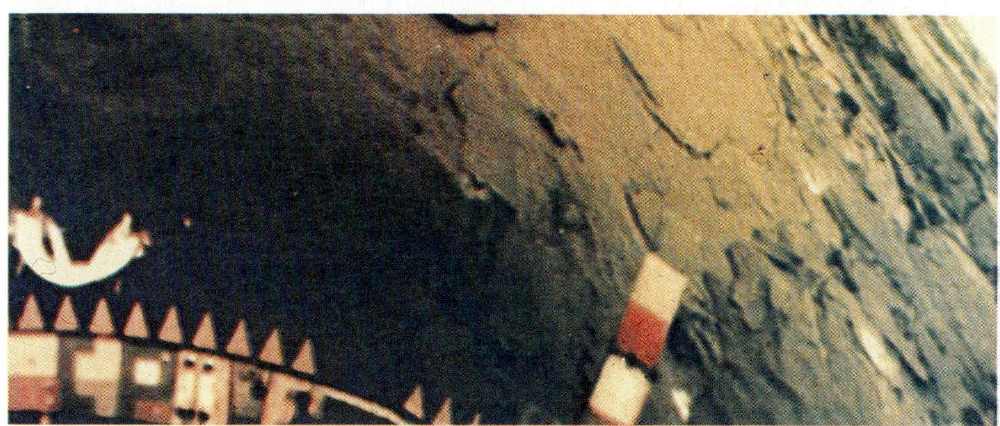

FIGURE 10.12 COLOR OF THE SURFACE OF VENUS. *These photographs, obtained by the Soviet* Venera 13 *and* 14 *landers, reveal the yellowish color of the surface of Venus.* (TASS from SOVFOTO)

sphere of hydrogen compounds much like the earliest atmosphere on Earth.

At this point, a major contrast developed between the Earth and Venus. As oxygen escaped from the rocks on the Earth and combined with hydrogen to form H_2O, much of it could persist in the liquid state, and the Earth had oceans from that time onward. On Venus, however, which is closer to the Sun, liquid water could not exist, but instead stayed in the atmosphere as water vapor. Venus is 0.72 AU from the Sun, so the intensity of sunlight at its surface is $1/.72^2 = 1/.52 = 1.92$ times greater than at the surface of the Earth (recall the inverse-square law, discussed in Chapter 6). This is not a very great difference, but it was extremely significant.

Once the atmosphere of Venus became contaminated with water vapor, the greenhouse effect went into operation, further heating the surface. The temperature got sufficiently high to bake the carbon dioxide out of the carbonate rocks, and with no oceans to dissolve it and return it to rock, this gas soon dominated the atmosphere. On the Earth, the CO_2 that was outgassed was largely dissolved in seawater and then deposited in rocks, where it still exists. The development of life on Earth also had a major effect that was absent on Venus, gradually altering the Earth's atmospheric composition to one dominated by oxygen and nitrogen, which do not create a strong greenhouse effect.

The increasing CO_2 content in the atmosphere of Venus enhanced the heating caused by the greenhouse effect, leading to the extreme surface conditions known today. Eventually the water vapor in the atmosphere was dissociated by the high temperatures, releasing the hydrogen atoms so that they escaped into space.

Meanwhile the surface of Venus started to follow an evolution much like that of the Earth. Convection apparently developed soon after the interior differentiated, giving rise to the tectonic activity that has shaped the highland masses, although on Venus, unlike on the Earth, these continental areas did not begin to move about. Volcanic activity has been a continuing feature of the environment, however, just as it has on Earth. In a strictly geological sense, Venus may almost be a twin of the Earth.

PERSPECTIVE

The study of Venus has been very instructive, not only for what it tells us of the planet itself, but also for the lessons learned about the Earth. We must not forget that, except for a rather small difference in the intensity of sunlight at an early stage in its history, the Earth might have ended up with a crushing, hot CO_2 atmosphere like that on Venus.

We turn our attention next to another of the terrestrial planets, the next closest to Earth. In examining Mars we will be able to draw other parallels and contrasts with Earth.

SUMMARY

1. Observations of Venus from Earth, particularly radio data, revealed that it is extremely hot at the surface, and rotates slowly in the retrograde direction.
2. Soviet and U.S. spacecraft have flown by Venus and probed its atmosphere. Several Soviet probes have landed on the surface and sent back information on conditions there.
3. The atmosphere of Venus consists mostly of carbon dioxide, with a bit of nitrogen and traces of sulfur dioxide and sulfuric acid.
4. The atmospheric temperature increases from 240 K at the cloud tops to 750 K at the surface, where the pressure is 90 times that at sea level on the Earth. The heating is caused by the greenhouse effect.
5. The clouds, composed of droplets of sulfuric acid, form three distinct layers between 48 and 60 km in altitude. The atmosphere is clear below the clouds.
6. At the cloud tops, high winds circle the planet in four days, whereas it is calm at the surface. Above the clouds there are global convection currents.
7. The surface of Venus, consisting of rolling plains, highlands, and lowlands, shows evidence of limited tectonic activity but no continental drift. There is evidence of current volcanic activity.
8. Venus evolved in a manner very different from the Earth, primarily because it is so close to the Sun that liquid water could not exist on its surface. The lack of water allowed CO_2 to stay in the atmosphere and discouraged the development of life.

REVIEW QUESTIONS

Thought Questions

1. Summarize the composition of the atmosphere of Venus, and compare it with the composition of Earth's atmosphere.
2. How does the role of sulfur dioxide (SO_2) in the atmosphere of Venus compare with the role of ozone in the Earth's atmosphere?
3. Why does Venus have so much carbon dioxide (CO_2) in its atmosphere, while the Earth has so little?
4. Why could the mass of Venus not be deduced from Earth-based observations in the same manner as the masses of most of the other planets?
5. How are the clouds in the atmosphere of Venus like clouds in the Earth's atmosphere, and how are they different?
6. Would you expect Venus to have charged-particle belts surrounding it, comparable to the Earth's Van Allen belts? Explain.
7. Summarize the evidence that Venus has undergone tectonic activity, and the extent of that activity compared to that on the Earth.
8. What is the evidence that Venus may have current volcanic activity?
9. Explain why Venus has never had liquid oceans, and the effect this has had on the evolution of the planet's atmosphere.
10. Why are scientists concerned about increasing levels of carbon dioxide in the Earth's atmosphere?

Problems

1. What is the wavelength of maximum emission from the surface of Venus, whose temperature is 750 K?
2. Suppose a radar echo from one edge of Venus has its wavelength shifted by $\Delta\lambda/\lambda = 6 \times 10^{-9}$. What

is the rotational velocity (that is, the speed of rotation at a point on the surface) of Venus? If the radius of the planet is 6,000 km, use the rotational speed you just calculated to find the rotation period of Venus.

3. If Venus and the Earth are the same size, how much brighter than the Earth is Venus, given that Venus has a higher albedo (0.76 compared to 0.31) and receives more intense sunlight (almost twice as much; the figure is given in the text)?

4. Sea level atmospheric pressure on the Earth is 14.7 lb/in^2, and on Venus it is about 1,300 lb/in^2. How many times higher is the pressure on Venus?

ADDITIONAL READINGS

Beatty, J. Kelly. 1985. A radar tour of Venus. *Sky and Telescope* 69(2):507.

———. 1985. A Soviet space odyssey. *Sky and Telescope* 70(4):310 (Vega at Venus).

———, B. O'Leary and A. Chaikin, eds. 1981. *The new solar system*. Cambridge, England: Cambridge University Press.

Head, J. W., C. A. Wood, and T. A. Mutch. 1977. Geologic evolution of the terrestrial planets. *American Scientist* 65:21.

Pettingill, G. H., D. B. Campbell, and H. Masursky. 1980. The surface of Venus. *Scientific American* 243(2):54.

Schubert, G., and C. Covey. 1981. The Atmosphere of Venus. *Scientific American* 245(1):66.

Weaver, K. F. 1975. Mariner unveils Venus and Mercury. *National Geographic* 147:848.

Young, A., and L. Young. 1975. Venus. *Scientific American* 233(3):70.

CHAPTER 11

MARS AND THE SEARCH FOR LIFE

Martian plains on a frosty morning. (NASA)

CHAPTER PREVIEW

The planet Mars (Table 11.1), while not as brilliant as Venus, is nevertheless a prominent object in the heavens (Fig. 11.1). The fourth planet from the Sun and our next nearest neighbor after Venus, Mars has played an important role in the development of astronomy, lending itself first to mythological interpretations, and later providing the basis of Kepler's discoveries on the motions of the planets. In recent times, Mars again became a center of considerable human speculation, as people mused over the possibility that life might exist there. The most recent chapter in the story, the exploration of Mars by unmanned spacecraft, has dispelled such notions but has not diminished our fascination.

TABLE 11.1 MARS

Orbital semimajor axis: 1.524 AU (227,900,000 km)
Perihelion distance: 1.381 AU
Aphelion distance: 1.667 AU
Orbital period: 1.881 years (687.0 days)
Orbital inclination: 1°51′0″
Rotation period: $24^h\ 37^m\ 22.6^s$
Tilt of axis: 23°59′
Diameter: 6,762 km (0.531 $D_\oplus$
Mass: 6.39×10^{26} g (0.107 $M_\oplus$)
Density: 3.96 g/cm^3
Surface gravity: 0.38 Earth gravity
Escape velocity: 5.0 km/sec
Surface temperature: 130–290 K
Albedo: 0.15 (average value)
Satellites: 2

OBSERVATIONS AND GENERAL PROPERTIES

From antiquity, Mars has attracted special attention because of its distinct and unusual reddish color, apparent to even the unaided eye. The Greeks named this planet *Ares,* after their god of war, and the Roman name *Mars* has the same connotations.

The first telescopic observations of the red planet were made by Galileo, who noted that its disk varies in angular diameter as its position with respect to the Sun varies. About 50 years later, Huygens became the first to sketch surface features as he saw them on the planet's disk, something that later led to one of the more fascinating scientific debates since medieval times. Successive generations of astronomers attempted to study the surface markings, a very sub-

FIGURE 11.1 THE EARTH-BASED VIEW OF MARS. *The vague surface markings seen here inspired some early astronomers to suggest the existence of an extensive Martian civilization.* (Lick Observatory photograph)

FIGURE 11.2 THE MARTIAN "CANALS." *This sketch by Giovanni Schiaparelli shows the linear features eventually interpreted to be artificial channels in which water flowed.* (Historical Pictures Service, Chicago)

jective business, especially since there was no photographic film with which to record the image. Photographs are not necessarily superior to the human eye in this case, anyway, because variations in the turbulence of the Earth's atmosphere may allow momentary clear glimpses, but these few instances of high clarity will be smeared out in a photograph obtained over a long exposure time.

On the planet's surface, Huygens noted a dark patch, which was named Syrtis Major. Additional details were seen later, and by the 1870s observers had named a number of features, using Latin words related to bodies of water, which they thought the features to be. At about this time, Father Secchi, the Roman Catholic priest who later was to play a pioneering role in the development of stellar spectrum analysis, saw what he believed to be linear features on the surface of Mars, and he referred to them as *canali,* an Italian word referring generally to natural channels of water. Giovanni Schiaparelli, who was then the director of an observatory in Milan, imagined that he saw a network of these features on the surface of Mars, and in 1877, when Mars was at opposition, he produced a detailed sketch mapping them (Fig. 11.2). Soon other observers around the world were seeing the canali, and in due course the name began to be interpreted as meaning canals, artificially constructed channels for transporting water.

At this point a wealthy Boston aristocrat name Percival Lowell entered the story. He fantasized that an extensive and well-developed Martian civilization had built the canals for irrigation, in response to the growing dryness of the planetary climate. Lowell (Fig. 11.3) published a book full of these speculations in 1895, which created a groundswell of public interest in the notion. Despite skepticism on the part of some astronomers who either could not see the canals or realized how unreasonable it was that Martian canals should be visible from Earth even if they really existed, the idea of a Martian civilization became fixed in the public mind. To this day, a wide variety of science fiction novels and movies (starting with H. G. Wells's *The War of the Worlds*) have been inspired by these beliefs.

Lowell probably invested more personal energy and effort in the Martian civilization than anyone

FIGURE 11.3 PERVICAL LOWELL *A wealthy enthusiast of astronomy, Lowell became fascinated with the notion of a Martian civilization.* (Lowell Observatory photograph)

else. He dedicated the rest of his life to the study of Mars, establishing an observatory at Flagstaff, Arizona, for the purpose. The Lowell Observatory has since grown to become a major research facility, still supported partially by bequests from the Lowell family. Although his contributions to modern knowledge of Mars were ultimately minimal, Lowell did carry on observations of other planets as well, with a bit more scientific detachment. Some valuable advances were made at the Lowell Observatory, most notably the discovery of Pluto, the ninth planet (see Chapter 15).

OBSERVATIONS FROM NEAR AND FAR

As long ago as 1666, the rotation period of Mars was known to be slightly more than 24 hours, and in the late 1700s the British astronomer William Herschel had deduced the tilt of the Martian rotation axis; it is very similar to the Earth's tilt. In 1877, the two Martian moons were discovered, which allowed, through the use of Kepler's third law, accurate measurement of the planet's mass.

Spectral analysis of the light reflected to Earth from Mars started in the first decades of the twentieth century. Photographs taken through red and blue filters revealed a haze in the Martian atmosphere that scatters blue light. The first spectroscopic measurements in the 1930s revealed an absence of water vapor and oxygen, of particular interest in view of the prevailing idea that life might exist on Mars. More recently, absorption lines of carbon dioxide were identified in the spectrum of Mars.

Seasonal changes in the coloration of certain regions on Mars (Fig. 11.4), well away from the polar caps, were once interpreted as being due to foliation of green plants, which were thought to undergo the same sort of annual cycle as earthly plants. By the 1960s, good photographs had revealed evidence of large-scale dust storms in the Martian atmosphere, and this inspired American astronomer Carl Sagan

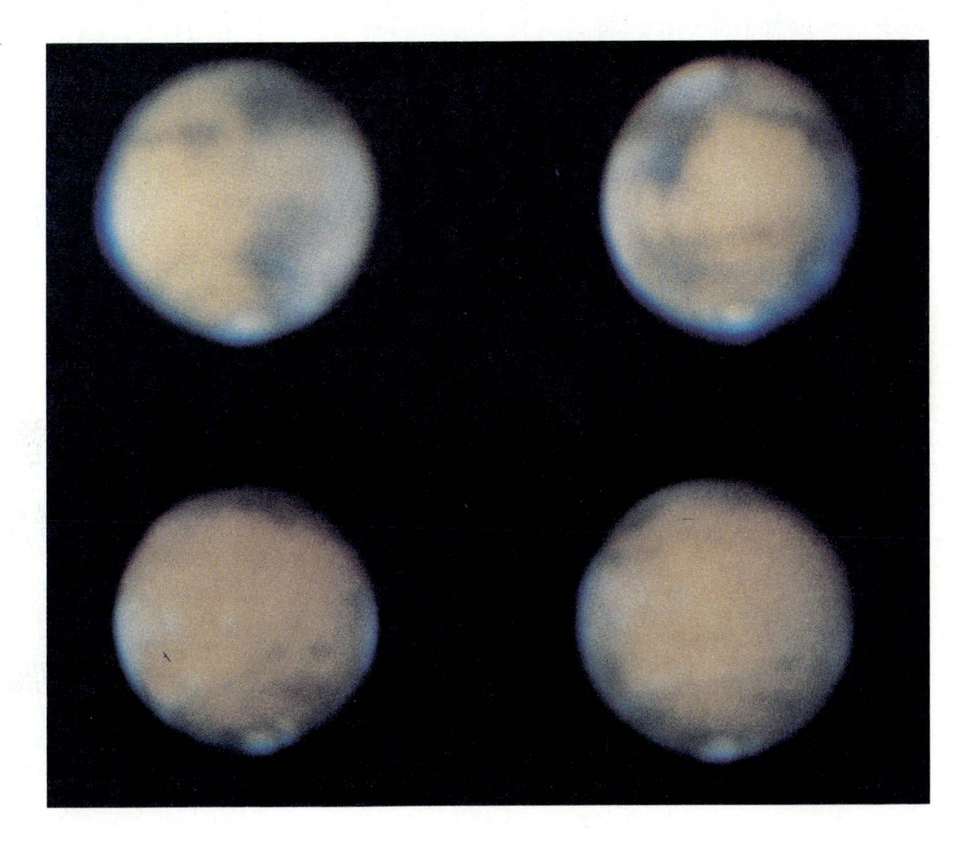

FIGURE 11.4 VARIATIONS IN THE MARTIAN SURFACE MARKINGS. *This series of photos shows the changes in the polar ice caps and surface markings that occur with the seasons on Mars.* (Mt. Wilson and Las Campanas Observatories, Carnegie Institution of Washington)

TABLE 11.2 PROBES TO MARS

Spacecraft	Launch	Encounter	Type of Encounter
Mariner 4	Nov. 28, 1964	July 14, 1985	Flyby, 10,000 km
Mariner 6	Feb. 24, 1969	July 30, 1969	Flyby, 3,200 km
Mariner 7	March 27, 1969	Aug. 4, 1969	Flyby, 3,200 km
Mariner 9	May 30, 1971	Nov. 13, 1971	Orbiter
Viking 1	Aug. 20, 1975	June 16, 1976	Orbiter plus lander
Viking 2	Sept. 9, 1975	Aug. 7, 1976	Orbiter plus lander

to explain the seasonal color variations in terms of seasonal shifts in the Martian wind patterns, alternately covering and uncovering some areas of the planetary surface. This explanation satisfactorily accounts for most of the observed variations, although some remain unexplained.

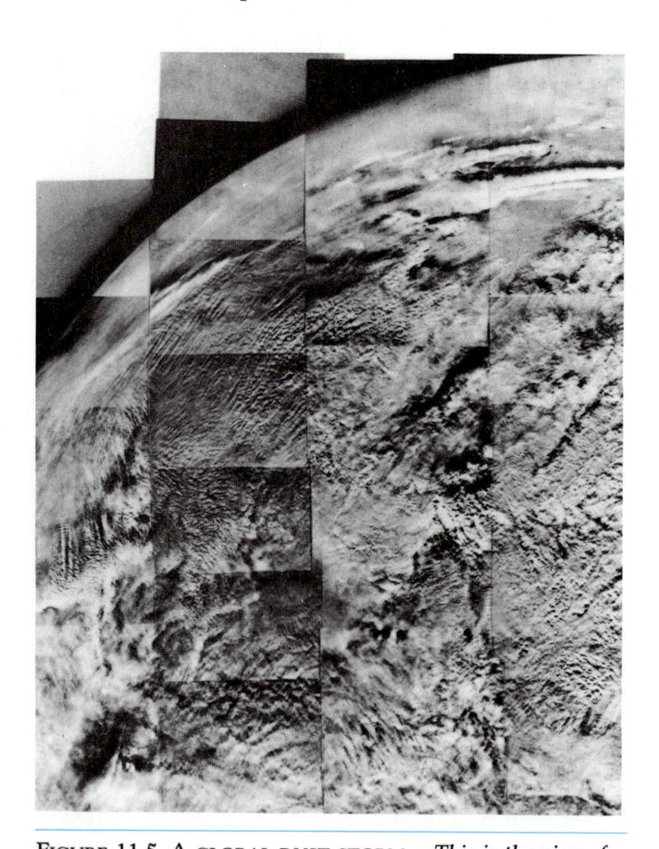

FIGURE 11.5 A GLOBAL DUST STORM. *This is the view of Mars that confronted the* Mariner 9 *spacecraft for the first several weeks after it went into orbit around the planet.* (NASA)

Infrared observations of Mars yielded data on its surface temperature, found to vary between 145 K (about −198°F) and 300 K (81°F). At high noon in the summer near the equator of Mars, the temperature is comfortable, by earthly standards.

Given our profound interest in Mars and the history of speculation over the possibility that life exists there, it is not surprising that the red planet became an early target for unmanned interplanetary probes (Table 11.2). The first successful mission to Mars was *Mariner 4,* launched in 1964, and this was followed by other *Mariner* missions, most notably *Mariner 9,* which orbited the red planet and surveyed it photographically. This task was stymied for several months by a global dust storm, which completely obscured the surface of Mars (Fig. 11.5). The storm finally abated, however, allowing *Mariner 9* to make many profound discoveries, including several gigantic volcanoes and a canyon that dwarfs anything on the Earth.

The crowning achievement to date of the Martian exploration program came some 4 years later, with the launch of the two *Viking* spacecraft in the late summer of 1975. Each consisted of an orbiter, destined to circle Mars as *Mariner 9* had done, and a lander, designed to descend to the surface and land intact (Fig. 11.6). The orbiters and the landers each carried a complex assortment of instruments designed for a variety of measurements. The orbiters improved on the quality of the older *Mariner 9* photographic survey of the entire planet (Fig. 11.7), and the landers were able to carry out many sophisticated experiments, including, most notably, panoramic photographs of the Martian landscape and sensitive tests for the presence of microscopic life forms in the soil. The results are discussed later in this chapter.

– THE MARTIAN ATMOSPHERE AND – SEASONAL VARIATIONS

Mars has a very thin atmosphere compared to that of the Earth, with a surface pressure of 0.006 times the sea-level pressure on earth. The Martian air is about 95 percent CO_2 and is similar to Venus in this regard (Table 11.3). Besides CO_2, other species include nitrogen, which accounts for 2.7 percent; argon, which constitutes another 1.6 percent; and other trace gases, which together total less than 1 percent. Water vapor is present in variable quantities, reaching levels of nearly one-tenth of 1 percent in the Martian summer. This is occasionally sufficient to form clouds of ice particles.

The Martian polar caps have been somewhat controversial, because of the difficulty of determining whether they are made of water ice or frozen CO_2 (dry ice). Based on temperature measurements, the conclusion was finally reached that the surface cap material is frozen CO_2, but that some water ice is also present, lying beneath the CO_2. The seasonal variations in the polar caps are due to the accumulation of CO_2 ice during the Martian winter, followed by evaporation in the summer. The northern and southern caps experience different seasonal variations, due to differences in the intensity of the seasonal changes in the weather (described below). Whereas the northern cap annually loses all of its CO_2, exposing the water ice cap beneath, in most southern summers the south polar cap retains enough CO_2 to keep the underlying water ice protected. Thus, the southern cap varies relatively little, compared to the seasonal variations in the northern cap.

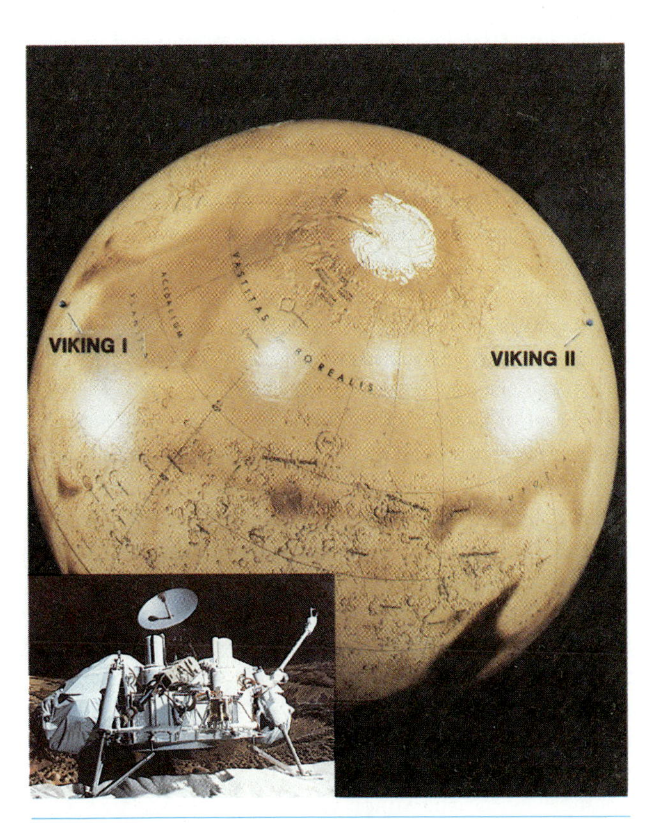

FIGURE 11.6 THE VIKING LANDING SITES. *This globe of Mars indicates the locations of the landing sites for the two* Viking *spacecraft. One of the landers is depicted in the inset.* (NASA)

FIGURE 11.7 A MARTIAN PHOTOMOSAIC. *This is a very detailed view of much of the Maritan surface, made by merging several* Viking *images. Valles Marineris crosses the planet left-to-right near the center, and at right are three of the giant volcanoes.* (U.S. Geological Survey)

TABLE 11.3 COMPOSITION OF THE MARTIAN ATMOSPHERE

Gas	Symbol	Fraction*
Carbon dioxide	CO_2	0.95
Nitrogen	N_2	0.027
Argon	Ar	0.016
Oxygen	O_2	0.0013
Carbon monoxide	CO	7×10^{-4}
Water vapor	H_2O	3×10^{-4}
Neon	Ne	2.5×10^{-6}
Krypton	Kr	3×10^{-7}
Ozone	O_3	1×10^{-7}
Xenon	Xe	8×10^{-8}

*Values are fractions by number.

The general question of water on Mars is complex, and for a long time was the basis of a serious scientific mystery, quite apart from the older speculations about canals. Enough water should have been released from surface rocks early in the history of the planet to cover its entire surface to a depth of tens to thousands of meters, so the lack of water on the planet was difficult to explain. There was probably no energy source on Mars to dissociate H_2O molecules, as happened on Venus, but on the other hand, the atmospheric pressure is too low to allow liquid water to exist on the surface. The atmosphere, which gets rather cold at night, cannot support much water in the form of vapor, and the amount in the ice caps is much too small to explain the discrepancy. The answer had to be sought elsewhere. We will discuss current ideas about the fate of the Martian water supply in the next section.

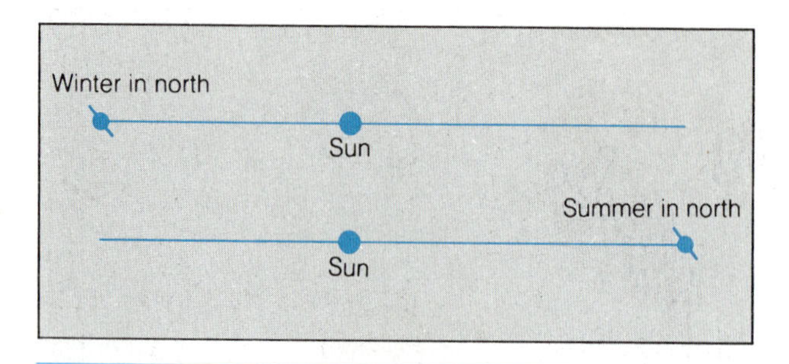

FIGURE 11.8 THE MARTIAN SEASONS. *This exaggerated view shows that in the northern hemisphere both summer and winter are moderated by the varying distance of the planet from the Sun, whereas in the southern hemisphere both seasons are enhanced. The extreme temperature fluctuations in the southern hemisphere give rise to the winds that cause the seasonal dust storms.*

The existence of clouds in the atmosphere of Mars was recognized long ago, from Earth-based observations. While some are due to blowing dust, many are white clouds formed of ice crystals, primarily H_2O, although some CO_2 clouds also exist. The clouds are found in lowlands in the early Martian morning and in regions where winds push the air up mountain slopes, forcing the H_2O vapor to cool and condense. In late summer in each hemisphere, extensive cloud systems form over the polar caps.

Winds on Mars are always present, often with substantial velocities (although the thinness of the atmosphere reduces the impact of the winds on surface features). The global flow patterns are created by the daily variations in solar heating at the surface and are strongly modified by terrain such as the large mountains that exist on Mars. The prevailing winds tend to move from west to east, but with a veer either towards or away from the poles, depending on the season. Typical wind velocities are 35 to 50 km/hr.

The fact that Mars has seasons was recognized long ago, when variations in the sizes of the polar ice caps were noticed. The cause of the seasons is a bit more complex than on Earth, however, because of the elliptical shape of the orbit of Mars. Earth's orbit is so nearly circular that the intensity of sunlight hitting its surface is almost constant year-round (in fact, the Earth is closest to the Sun in early January, and the slight extra solar imput surely does not moderate the winter weather in the northern hemisphere). The orbit of Mars is more elongated, as we have already learned, with the result that at its closest approach to the Sun, Mars is about 17 percent nearer than when it is farthest away, on the other side of its orbit (Fig. 11.8). Therefore, the Sun's intensity on the Martian surface is $1/.83^2 = 1.45$ times greater at closest approach. This is enough to have a pronounced effect on the climate. Mars approaches the Sun most closely during winter in the northern hemisphere, and is farthest away during the northern summer; hence the shape of the orbit tends to moderate both seasons in the north. By the same token, the southern seasons are enhanced by the orbital shape, so that this hemisphere has much more extreme variations from summer to winter. The southern polar cap grows larger than the northern one in its winter, but it diminishes to a smaller remnant in summer. As noted, however, only the southern cap retains sufficient CO_2 ice during the summer to keep the underlying water ice covered, possibly because the CO_2 layer grows much thicker during the harsh southern winter.

The most striking seasonal variations on Mars are the widespread dust storms, which have been observed for several decades to occur at the time when Mars is closest to the Sun, during the southern summer. The extra solar heating in the south creates winds that can begin to lift fine particles into the air, where they can stay suspended for lengthy periods. Some of the dust is raised by small-scale circular winds known on earth as **dust devils,** which have been detected in photographs from the *Viking* orbiters. Once the seasonal winds begin, the dust storm grows to cover at least the southern hemisphere, and occasionally the entire planet, and it lasts for weeks. As we noted earlier, one result of these annual storms is seasonal variations in the dust cover on the ground in certain regions, creating the well-known seasonal color variations seen from Earth.

WATER, TECTONICS, AND THE MARTIAN SURFACE

The surface of Mars (Fig. 11.9) can be classified into two types of areas: plains, which dominate much of the northern hemisphere; and rough, cratered terrain, which covers much of the south. The plains are low-elevation regions covered with lava flows, and altogether they spread over some 40 percent of the planet. The remainder is occupied by the rougher cratered land, much of which consists of highlands which are older than the lowlands. One huge uplifted region, called Tharsis, covers nearly one-quarter of the entire Martian surface, and averages some 6 km elevation above the mean surface level. The formation of this gigantic bulge on Mars is not well understood, but most suggestions invoke tectonic activity driven by convection in the mantle that occurred very early in the planet's history.

A major valley system (Fig. 11.10), extending some 4,500 km in length, was named Valles Marineris in honor of *Mariner 9,* whose photographs revealed its gargantuan proportions. This valley would dwarf the Grand Canyon, which is 5 to 10 times smaller in every dimension. Valles Marineris is 700 km across at its widest, and 7 km deep at its deepest. It probably was formed when the Tharsis highland was uplifted, creating extensive cracks in the crust.

Another major feature of the surface is created by the astounding volcanoes that exist on Mars, primarily in three locations. These are the most conspicuous features on the planet, except for the polar caps. The grand champion is Olympus Mons (Fig. 11.11), a monster with a height of 27 km (nearly 90,000 feet) above the mean surface level, and a base

FIGURE 11.9 A RELIEF MAP OF MARS. *The highest elevations are shown in white in this view; note the massive Tharsis plateau (left), and the huge volcanoes.* (M. Kobrick, JPL)

FIGURE 11.10 VALLES MARINERIS. *A tremendous valley as long as the breadth of North America is put into perspective in the left-hand global mosaic, which shows the valley at center right. The dark spot above is the large volcano Ascreaus Mons, and the white patch near the bottom is the Argyre Basin, covered with frost. The image at right shows detail in a portion of Valles Marineris.* (NASA)

FIGURE 11.11 OLYMPUS MONS. *This is the largest mountain known to exist in the entire solar system. Its base is comparable in size to the state of Colorado, and its height is about three times that of Mount Everest.* (NASA)

some 700 km across (Mt. Everest, on Earth, rises "only" 8.8 km or 29,000 feet). Olympus Mons is in the northern hemisphere, not far from the Tharsis plateau. The other great volcanic structures are formed by chains of peaks stretching across vast areas of the Martian plains.

The largest of the giant volcanoes apparently formed because of a lack of continental drift at the time and location of their formation. If they had formed on moving plates in the Martian crust, they should have moved away from the subsurface hot spots that created them before they got so big. On the other hand, some of the smaller volcanoes, particularly those on the Tharsis plateau, are aligned much like the Hawaiian islands, suggesting that there was some drifting when they formed.

The entire question of Martian tectonic activity is unsettled. There is no well-developed system of crustal plates, and no evidence of current continental drift, although evidently there has been some in the past. It has generally been believed that the volcanoes are quite old (perhaps 3 billion years), but some have suggested that they could be as young as 500 million years, implying that tectonic activity took place as recently as that. The Tharsis plateau and the nearby Valles Marineris formed about 4 billion years ago, indicating that Mars was geologically active at that early stage in its history. All of the age estimates are based on crater counts and are therefore quite uncertain. It would be helpful, indeed, to be able to perform radioactive dating measurements on rocks from various parts of the Martian surface.

The present-day lack of tectonic activity is consistent with the absence of any detected magnetic field, since most theories of magnetism invoke a fluid core, something else that Mars may not have.

Close examination of the surface of Mars reveals some interesting details. There are craters in most regions, created by impacts, as on the Moon. Some regions appear completely disorganized, resembling loosely piled rocks and rubble, as though landslides had occurred there. In some locations, often adjacent to this so-called **chaotic terrain,** there are winding valleys and flood plains that appear to have been formed by flowing water (Figs. 11.12, 11.13, and 11.14). These features have created a great deal of excitement because, as we learned earlier, the lack of water on Mars has been a major puzzle. All of the channels are quite old, having formed around the time when the Tharsis plateau was uplifted.

The fact that some of the channels are adjacent to chaotic regions suggests that water has been present on Mars in the form of ice underneath the surface, probably in the form of **permafrost.** One theory of the formation of the channels is that, early in the planet's history, underground heating caused the occasional sudden melting of a pocket of permafrost, creating a large reservoir of water. This water quickly ran off, carving a channel in the surface, and leaving behind empty caverns or porous soil in the subsurface volume that it evacuated. The soil there collapsed, creating the kind of jumbled, disorganized appearance associated with the chaotic terrain. This hypothesis seems satisfactory for the situations in which a channel emanates from a chaotic region, but there are channel systems not associated with chaotic terrain, so some other explanation may be needed. One possibility is that Mars had substantial quantities of sur-

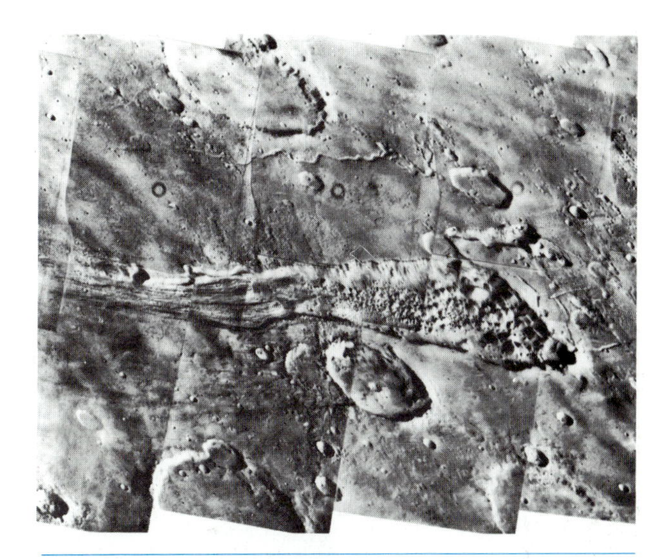

FIGURE 11.12 CHAOTIC TERRAIN. *This* Viking *orbiter photo shows a region that apparently subsided when permafrost below the surface suddenly melted and flowed away (toward the left).* (NASA)

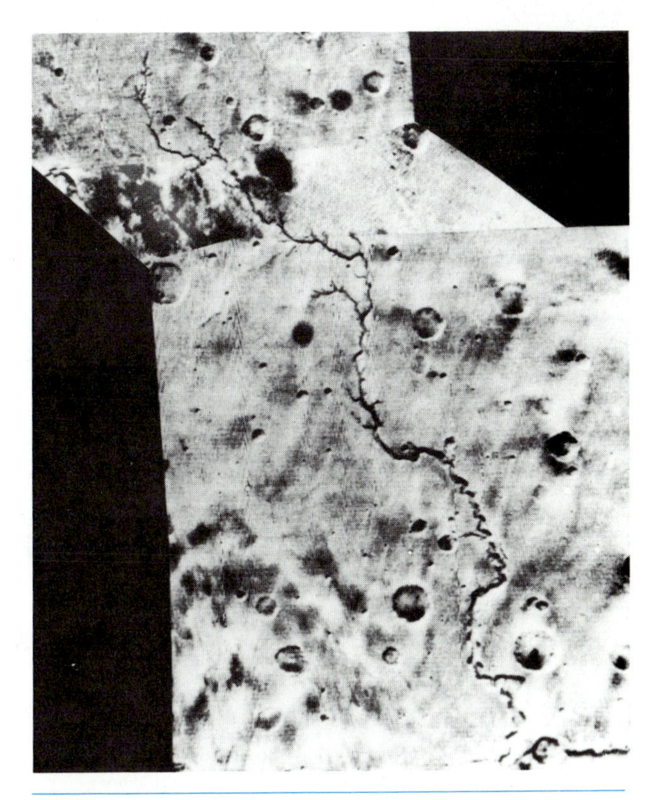

FIGURE 11.13 AN ANCIENT RIVER CHANNEL ON MARS. *This is one of the most striking examples of evidence for flowing water in the Martian past.* (NASA)

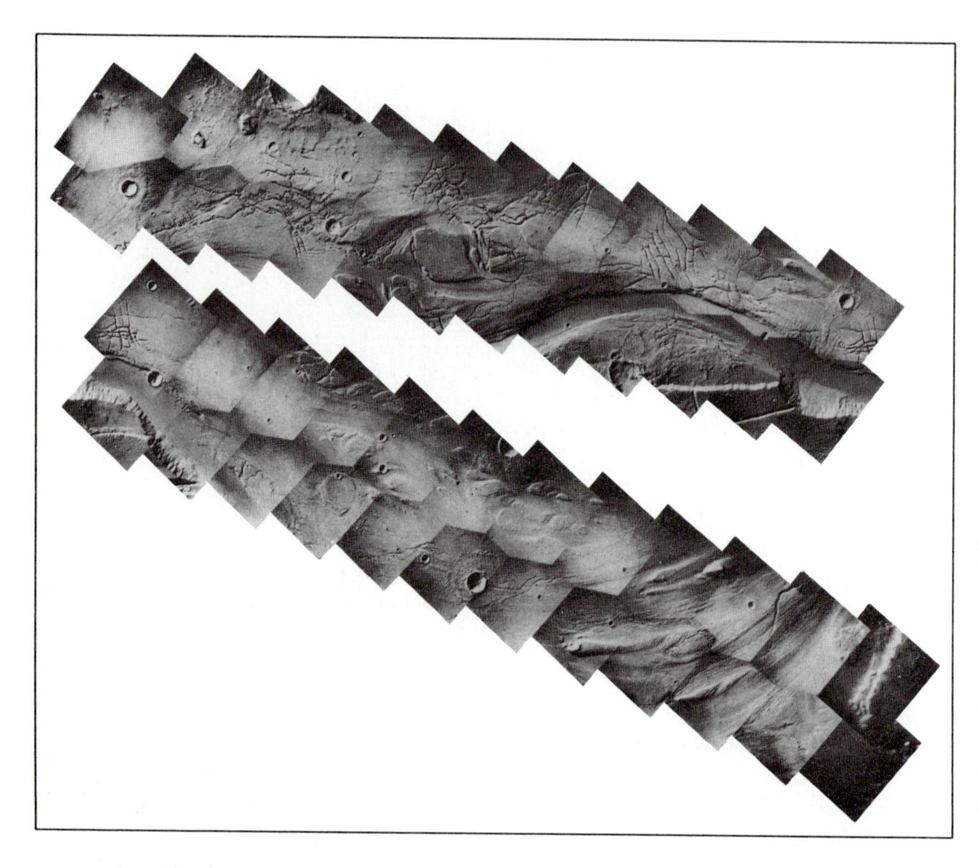

FIGURE 11.14 A FLOOD PLAIN. *Here is clear evidence that water once inundated this region of the Martian surface. The surface patterns indicate that the prominent craters in this photo already existed when the water flowed. The two mosaics show parts of Kasei Valles.* (NASA)

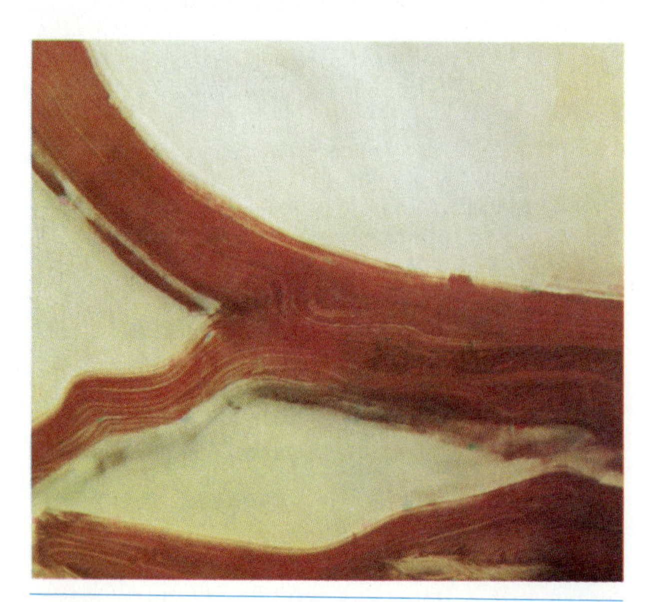

FIGURE 11.15 LAMINATED TERRAIN. *This photo shows some of the curious layered surface features seen near the edges of the polar ice caps.* (NASA)

face ice at times in the past, and that melting of this ice created the channels (see Astronomical Insight 11.1). Another suggestion is that Mars once had a sufficiently thick and warm atmosphere that liquid water could have persisted on its surface; that is, Mars may once have had seas or extensive lakes. In this view, long-term changes in the atmosphere reduced its pressure, so that eventually the water evaporated and escaped or froze out near the poles.

Another entire class of surface features on Mars are those created by winds. We have already learned about the seasonal dust storms and the variations in the surface coloration that they cause, but there are other features created by the winds, including light and dark streaks and the **laminated terrain** (Fig. 11.15), curious areas where the surface has apparently been deposited in layers and then eroded at the edges. The stair-step structures that result are especially prominent around the edges of the polar caps and may be formed of thin layers of dust and ice that are precipitated by the annual storms.

ASTRONOMICAL INSIGHT 11.1

ICE AGES ON MARS AND THE EARTH

It is mentioned in the text that one possible explanation for the dry river channels on Mars is that there were significant deposits of ice on the surface of the planet at times in the past. The suggestion has been made that Mars once had regular ice ages, and that they were caused by the same forces that created the ice ages on the Earth.

The planet Jupiter, next in sequence from the Sun after Mars, is by far the most massive of all the planets. As we will see in upcoming chapters, gravitational effects of Jupiter have played major roles in the history of the solar system, particularly in the influence of Jupiter's tidal forces on the asteroid belt. It has recently been shown that these same tidal forces, acting on Mars and on the Earth, may be responsible for long-term variations in the climates of those two planets.

The orbits of Jupiter, Mars, and the Earth are not precisely in the same plane. Furthermore, the orbit of Mars is a somewhat more elongated ellipse than is typical of most of the planets. Due to a combination of these facts, there are times when Jupiter and Mars come closer together than normally is the case; these times occur roughly every 100,000 years. When the two are at closest approach, the tidal force exerted on Mars by Jupiter is the strongest. At these times, the force exerted on Mars can cause its tilt to change, by as much as 15° in either direction from the normal tilt of about 25°. When the tilt of Mars is increased, then its seasonal variations in climate are even more extreme than they are today.

The effect of Jupiter's tidal force on Mars was even more important in the early history of Mars, before the Tharsis plateau was uplifted. The influence of tidal forces on a planet's tilt depends on the shape of the planet, and the shape of Mars was changed significantly when Tharsis was created. During the early times, when the tilt of Mars varied by more than it does today, the polar ice caps were completely melted every 100,000 years or so, because of the extreme variations in the seasons. Much of the water that was released from the caps was deposited in the form of ice sheets in the equatorial zones. Mars had ice ages.

When Tharsis was uplifted, Jupiter's effect on the tilt of Mars was reduced, and the ice ages stopped. All of the dry river channels that mark the surface of Mars today were formed *before* the formation of the Tharsis plateau, which suggests that they were produced by the melting and runoff of ice during the ice ages. Hence, the runoff channels and flood plains on Mars may be the remnant of an ancient time when the planet formed sheets of surface ice around its midsection every 100,000 years.

Jupiter is so massive that it has exerted a similar influence on the Earth, even though the distance is much greater. At somewhat irregular intervals, the tidal force on the Earth due to Jupiter is maximized, and the tilt of the Earth's axis is changed. In this case, the effect is small; the tilt is altered only by about 1° in either direction from the normal 23.5°. Nevertheless, calculations of the timing and the magnitude of the effect suggest that these occasional small variations in the Earth's tilt, and the changes in the Sun's heating that they cause, are sufficient to trigger the great ice ages known to have afflicted the Earth every few thousand years.

It is interesting to think that both Mars and the Earth have undergone such dramatic events as ice ages because of a common phenomenon. The role of tidal forces in general, and of Jupiter in particular, will be a recurrent theme throughout the rest of this section of the text. Meanwhile, it is important to keep in mind that the scenario described here is not the only hypothesis regarding the origin of water-formed features on the surface of Mars, and that there is considerable uncertainty about the role of water on Mars.

There are also sand-covered regions where the surface has been shaped into dunes, just like those seen on earthly deserts and beaches (Fig. 11.16). The sand in the dune regions is distinct from the much finer dust that is suspended in the air during the major storms. A vast field of dunes girdles the planet just outside the north polar cap, creating both a dark belt visible from afar and a mystery for scientists, who can offer no plausible explanation of its origin.

A great deal more will be learned about the surface of Mars, and the role of water on the red planet, when the U.S. *Mars Observer* spacecraft reaches Mars (in 1992, according to the current schedule). This orbiter will be equipped with a variety of instruments designed to probe the composition of Mars' atmosphere and its changing water vapor content, and to perform sophisticated analyses of the nature and composition of the surface. It will operate long enough to monitor the seasonal variations and should clear up many of the uncertainties pointed out in this text.

The Soviets are planning an extensive campaign to explore Mars, with the first of several missions to be launched in July, 1988. This mission, consisting of two spacecraft, will closely examine the Martian moons. The next Soviet mission, to be launched in 1992, will orbit Mars for close-up studies, and a third possible mission may land robots on Mars in order to probe the soil to depths as great as 100 feet. Both the U.S. and the Soviet Union have also begun to study possible missions to bring samples of Martian soil and rocks back to the Earth. Mars will be the focus of some exciting exploration in years to come.

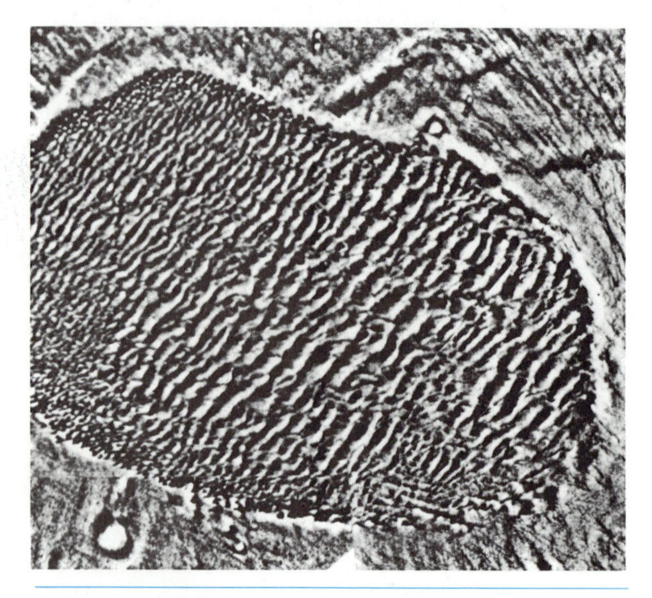

FIGURE 11.16 MARTIAN SAND DUNES. *These dunes are testimonials to the incessant action of winds on Mars.* (NASA)

SURFACE ROCKS AND THE INTERIOR OF MARS

The *Viking* landers were able to carry out some close-up analyses of Martian soil and surface rocks. Both landing sites were in the plains regions of the northern hemisphere, and at both places the landscape consisted of dusty soil, with low ridges, a dense scattering of rocks and boulders, and craters here and there (Figs. 11.17 and 11.18). The rocks lying around were probably ejected by the impacts that formed nearby craters.

The chemical composition of the surface minerals was measured (Fig. 11.19), and striking differences were found between Mars and the Earth. The soil of Mars appears to have formed from the breakdown of igneous rocks such as basalt, but it contains unusually high concentrations of silicon and iron. The iron, in the form of iron oxide (i.e., rust) is responsible for the reddish color.

The high abundance of iron is quite significant, for it indicates that Mars is not highly differentiated. Apparently the crust was not sufficiently molten for long enough to allow the heavy elements to sink to

FIGURE 11.17 A GROUND-LEVEL PANORAMA. *This is the first photograph made by the* Viking 1 *lander. It shows a rock-strewn plain extending in all directions.* (NASA)

the center of the planet, and Mars does not have a large iron-nickel core. There probably is a dense core (measurements of the gravitational field show evidence for this), but it must be considerably smaller than the cores of the other terrestrial planets. Ironically, even though Mars has a high surface content of iron, its *total* iron content is smaller than that of the other terrestrial planets. The explanation for this lies in the history of element condensation during the formation of the solar system, something we discuss in Chapter 17.

These data lead to a picture of an early Mars that, due to its low mass or its distance from the Sun, was not fully molten for long and which formed a very thick crust early in its history. Some tectonic activity has occurred, but there have not been extensive recent motions of the crust. Because of this, massive volcanoes formed, some of them building up to tremendous heights because of the lack of continental drift to carry them away from the hot spots that created them.

PROSPECTING FOR LIFE: THE *VIKING* EXPERIMENTS

The most widely publicized aspect of the dual *Viking* missions was the attempt made by the robot landers to detect evidence of life forms on Mars. This was done in several ways, the first and most straightforward being simply to look with the television cam-

FIGURE 11.18 DRIFTING DUST ON MARS. *In these two photos of the same landscape, there is more fine dust cover on the right than on the left.* (NASA)

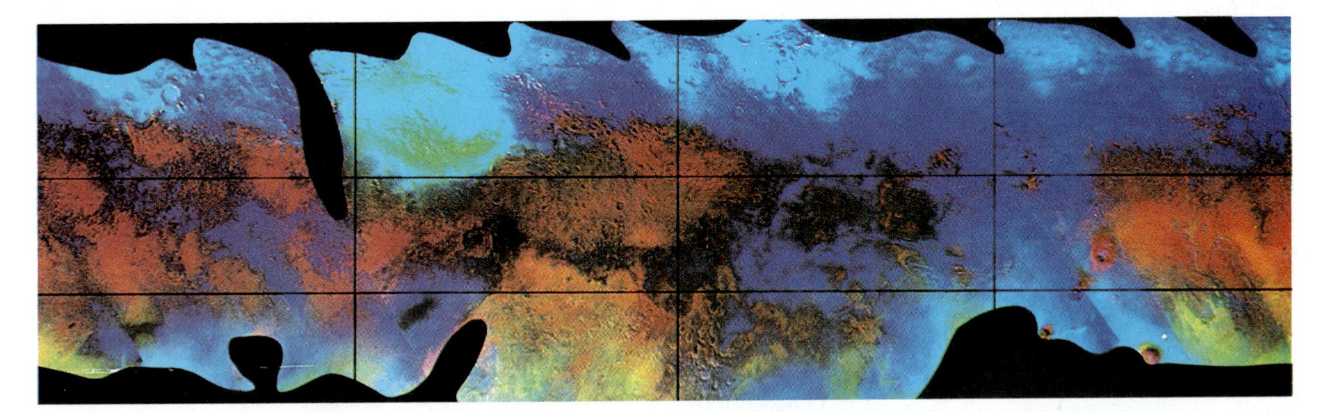

FIGURE 11.19 A GEOCHEMICAL MAP OF MARS. *This map, reconstructed from* Viking *orbiter data, uses colors to indicate predominant minerals in an equatorial strip between 30° N and 60° S latitude on Mars. The red colors indicate iron oxides, and the blues are less oxidized basalts. Turquoise indicates fogs and frosts, and oranges, browns, and yellows show where sand or dust covers the surface.* (NASA)

eras for any large plants or animals. None were seen, and more sophisticated tests were tried.

The tests for life were done by three different experiments on board the landers, each of which used soil scooped up by a mechanical arm (Fig. 11.20) and deposited in containers for analysis. The basic idea of all three was similar: to look for signs of metabolic activity in the sample. All living organisms on Earth, even microscopic ones, alter their environment in some way just by existing. Usually the effects involve chemical changes as the organism derives sustenance from its surroundings and ejects waste material.

Two of the three, the *pyrolytic release experiment* and the *gas-exchange experiment,* showed no evidence for possible life forms. The third, however, the *labeled release experiment,* created much excitement among scientists because its initial results duplicated the expected effects of active life forms. In this experiment, a sample of Martian soil was placed in a container, and a nutrient solution was added, in the expectation that organisms in the soil would metabolize the nutrient and release waste gases into the chamber. To make such released gases detectable, a small quantity of radioactive carbon (^{14}C, each atom consisting of the usual six protons but eight neutrons instead of six) was added to the nutrient. If metabolic activity took place in the soil sample, then some ^{14}C should appear in gases emitted by the sample, and that is just what happened. As a check, the same experiment was performed on other samples that had been heated so that any life forms present would have been killed, and indeed no tracer gases were emitted from those samples. The combination of activity in the normal samples and the lack of activity in the sterilized examples was consistent with the presence of life forms in the Martian soil.

The only flaws in the picture were that the other experiments failed to show positive results, and that the speed with which the reactions occurred in the labeled release experiment was much higher than expected for metabolic activity. Gases containing ^{14}C were released more quickly and in greater quantity than would have been possible for any earthly microorganisms. Furthermore, and perhaps most telling, a given sample would only react positively *once,* even though nutrient was added to some samples several times. Real life forms should be capable of eating more than one meal. Eventually it was concluded that the observed activity must have been due to an unexpected type of rapid chemical reaction between the

FIGURE 11.20 SAMPLING THE MARTIAN SOIL. *Here the scoop on* Viking 2 *is shown as it digs up a sample of Martian soil for one of its life-detection experiments.* (NASA)

ASTRONOMICAL INSIGHT 11.2

PRIMORDIAL LIFE ON MARS?

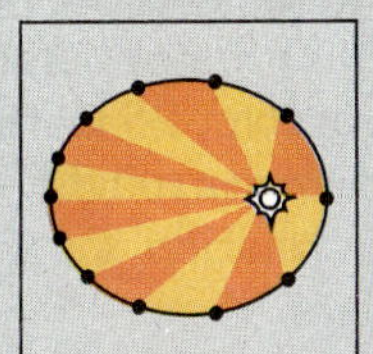

The *Viking* landers failed to find any conclusive evidence for the existence of living forms in the Martian soil. Some scientists have argued that the experimental techniques employed were not sensitive enough, and that microscopic life could still be present on Mars. Many others are more pessimistic, pointing out that all life forms we know of need liquid water, and that water does not exist in this form on Mars.

Even if Mars has no life today, some scientists have suggested that life may once have thrived on the red planet. We believe that life on Earth began through complex chemical reactions that took place in the early oceans (this is discussed in more detail in Chapter 33). The oldest fossil evidence for life on Earth is found in rocks in western Australia that are about 3.5 billion years old. Thus, life began within the first billion years after the formation of the Earth.

As discussed in the text, we believe that Mars originally contained as much water and carbon dioxide (in proportion to its mass) as Venus and the Earth. Today we do not know where all of this water and carbon dioxide is, but we believe some of the water is stored below the surface in the form of ice. Additional quantities of water and carbon dioxide may have escaped into space, possibly as the result of major impacts due to asteroids crashing into Mars. (The asteroids are rocky fragments that orbit the Sun, most of them in the space between Mars and Jupiter). The ancient river channels on Mars testify to the fact that liquid water has existed there, although there is some question as to whether this was a temporary phenomenon or a long-term condition.

If Mars really did have an early period when the atmosphere was dense, then the planet would very likely have been much warmer than it is today, because of greenhouse heating due to the high carbon dioxide abundance. The elevated atmospheric pressure, combined with the warmer temperatures, would have allowed lakes or oceans to exist. In short, conditions much like those on the early Earth might have prevailed. If so, there is no reason to rule out the possibility that primitive life forms may have developed. The search for traces of such life forms is one of the important reasons that scientists are making plans for new missions to Mars.

In order to seek evidence of fossil life forms on Mars, soil samples must be scrutinized in various ways, so the task demands a landing on the surface of the planet. The most comprehensive testing could be done by using sophisticated equipment, so the idea of returning samples of the soil to the Earth has been proposed. Even more daring is the proposal that a manned mission go to Mars and conduct searches there. People in vehicles equipped for roving the surface could explore areas where standing water once might have been, and could probe depths where rocks with fossil traces might have been undisturbed.

Both the U.S. and the Soviet Union are making plans for future missions to Mars. At the moment, the only definite U.S. mission is the *Mars Observer* (scheduled for a 1992 launch from Earth), an orbiter which will tell us little about possible evidence for primordial life. The Soviets are planning three types of missions, two of which involve making probes of the Martian surface. One will insert shafts with scientific instruments deep into the soil, to measure conditions below the surface; the other will be a sample return mission. Scientists in both nations have begun to discuss the possibility of a joint manned mission, so perhaps in our lifetimes we will see a definitive search for ancient life on Mars. If evidence for such life is found, this will tell us that the Earth is not the only place where life started, as profound and far-reaching a discovery as any ever made.

soil and the nutrient, probably involving oxides in the soil whose chemical properties were altered by heating, so that the reaction did not occur in the sterilized samples.

A further test for evidence of life was carried out by a device called a **mass spectrometer,** which is capable of analyzing a sample to determine the types of molecules it contains. The mass spectrometers aboard the *Viking* landers found no evidence in Martian soil samples of **organic molecules,** those containing certain combinations of carbon atoms that are always found in plant or animal matter on Earth. Hence, no doubt with some reluctance, the *Viking* scientists concluded that no evidence for life on Mars had been found.

The lack of evidence for life on Mars is not the same as evidence for lack of life. After all, these experiments could test samples at only two localities on a whole planet, and furthermore, the tests were predicated on the assumption that Martian life forms, if they exist, would be in some way similar to those on Earth. Now that a great deal is known about the chemical properties of the Martian soil, new experiments probably could be devised that would be relatively free of the confusion created by the nonbiological activity detected in the Viking experiments. Perhaps one day soon, further attempts to find living organisms on Mars will be made.

THE MARTIAN MOONS

The two satellites of Mars, discovered telescopically in 1877, are rather insignificant compared to the Earth's moon, or to the major satellites of the outer planets (Table 11.4). Both are very near to the parent planet and therefore have relatively short periods: Phobos (Fig. 11.21), the inner moon, circles Mars in just 7 hours, 39 minutes; while Deimos (Fig. 11.22) has a period of 30 hours, 18 minutes. Because Phobos orbits the planet in a time less than a Martian day, to an observer on the planet's surface this satellite would cross the sky twice daily, in the west-to-east direction.

Close-up photos of Phobos and Deimos reveal that each is a small, irregularly shaped chunk of rock, pitted with craters, and, in the case of Phobos, covered by linear grooves. Both satellites are somewhat elongated, Phobos having a length of some 28 km and a width of 20 km, while Deimos measures roughly 16 by 10 km. Both are in synchronous rotation, keeping one end permanently pointed toward Mars.

A striking and unusual characteristic of the surface of Phobos is that it is covered, over most of its area, with parallel grooves some 100 to 200 meters wide and 10 to 20 meters deep. These features are most prominent in the vicinity of a very large crater on Phobos and are least evident on the opposite side. A number of ideas for their origin have been suggested, but the one that appears to fit the data best is the hypothesis that they were created by the same impact that carved out the major crater. It has been suggested that this impact fractured Phobos along planes where its material was weakest, and that heat from the impact caused some melting of subsurface rock, which then partially filled the fractures. The complete lack of similar grooves on Deimos, in this view, is due to the fact that this satellite has never suffered such a major impact, as shown by its lack of large craters.

Clearly Phobos and Deimos were not formed in the same way as the Earth's moon. The Moon has a much larger mass relative to its parent planet and evidently went through some geological evolution, whereas Phobos and Deimos are tiny and show no signs of having been acted on by the kinds of forces, such as vulcanism, that have forged the surface of the Moon. The origin of the two Martian moons is not well understood, but many have pointed out the general similarities between them and typical asteroids.

TABLE 11.4 SATELLITES OF MARS

No.	Name	Distance (R_M*)	Period (days)	Diameter (km)	Mass (g)	Albedo
1	Phobos	2.77	0.3189	27 × 21.6 × 18.8	9.6×10^{18}	0.06
2	Deimos	6.93	1.2624	15 × 12.2 × 11	1.9×10^{18}	0.07

*The distances given are orbital semimajor axes, in units of the radius of Mars, which is R_M = 3,384 km.

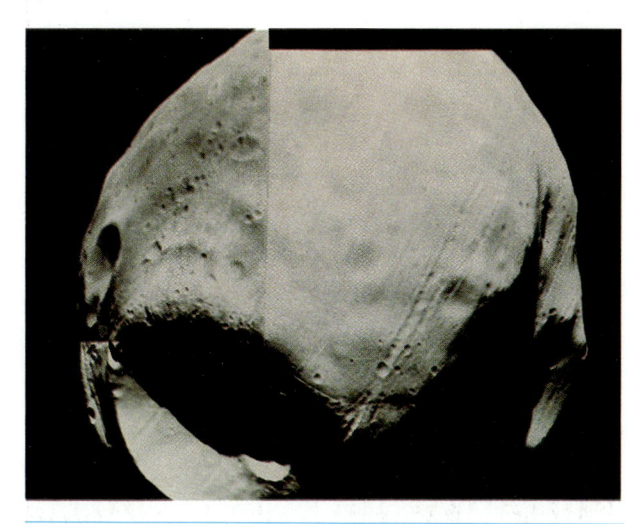

FIGURE 11.21 THE MARTIAN MOON PHOBOS. *This* Viking *mosaic shows two interesting features: the parallel grooves that mark the surface of Phobos, and the major impact crater whose formation may have caused the stresses that created the grooves.* (NASA)

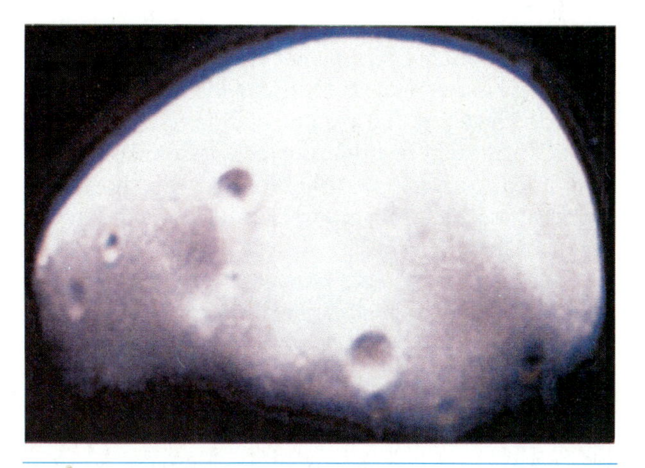

FIGURE 11.22 DEIMOS. *The more distant of the two moons of Mars, Deimos has a relatively featureless surface.* (NASA)

PERSPECTIVE

The mysterious red planet is still mysterious, but in new ways. Most of the age-old questions related to its changing appearance, the possibility of civilizations thriving on its surface, and its distinctive color have been answered through close-up examination. What we have found is a planet like the Earth, except one not as fully developed, in the geological sense. Tectonic activity was arrested in the planet's youth, and the low temperatures and low escape velocity combined to leave Mars with only a thin atmosphere of CO_2, its water hidden away underground.

We are almost finished with our tour of the terrestrial planets, except for Mercury, the innermost, which is discussed in the next chapter.

SUMMARY

1. Telescopic observations of Mars show a thin carbon dioxide atmosphere, polar caps that vary in size with the Martian seasons, and a very low abundance of water vapor.
2. The Martian atmosphere has almost the same composition as that of Venus, but has only 0.6 percent of Earth's sea-level pressure.
3. Seasonal variations on Mars are strongly influenced by the relatively elongated shape of its orbit.
4. The Martian surface consists of plains and rough highlands, with evidence for ancient tectonic activity but no continental drift, similar to Venus.
5. The Martian water is stored in the permanent ice caps, and possibly beneath the surface, in the form of permafrost.
6. Surface rocks have a high iron content, and Mars has only a moderate average density, both facts indicating that the planet has undergone little differentiation.
7. The Viking experiments found no evidence for life forms in the Martian soil.
8. Mars has two tiny satellites, resembling asteroids in size and shape.

REVIEW QUESTIONS

Thought Questions

1. Before close-up observations of Mars were made by the *Mariner* and *Viking* probes, what were the best arguments against the idea that the dark markings on Mars were canal systems?
2. Compare the atmospheric composition of Mars with that of Venus. How do the physical conditions (pressure and temperature) compare?
3. The composition of the atmosphere of Mars (and other planets) can be determined by measuring the

spectrum of sunlight reflected from the planet. Gases in the atmosphere of the planet create absorption lines that allow scientists on Earth to identify the gases. How do you think the absorption lines created by the planet's atmosphere are distinguished from absorption lines in the Sun's spectrum? How do you think they are distinguished from absorption lines created by the Earth's atmosphere?

4. Summarize the role of water on Mars; that is, where is it and how does its distribution vary with the seasons?

5. Explain why the seasonal variations on Mars are so different in the northern and southern hemispheres of the planet.

6. Compare the atmospheric circulation and winds on Mars with those on the Earth.

7. Discuss the evidence for tectonic activity on Mars, and whether it is thought to be occurring today.

8. Why is Mars red?

9. In the *Viking* searches for life on Mars, what assumptions were made about the nature of life forms that might be present there?

10. Mars has little or no magnetic field. Discuss the effect of this on possible Martian life forms, and on humans who might someday visit there.

Problems

1. From Fig. 11.2, estimate how wide some of the dark features are, relative to the diameter of Mars. Knowing that the planet's diameter is 6,768 km, calculate how wide these dark features must be. Is your answer reasonable if the features are canals?

2. To illustrate why Earth's seasonal variations are not affected by the varying distance of Earth from the Sun, calculate the relative intensity of sunlight on the Earth when it is closest to the Sun, compared to when it is farthest. At closest approach, the Earth is 0.983 AU from the Sun, and at its farthest, it is 1.017 AU away. Compare this variation with that on Mars (figures are given in the text).

3. Calculate the mass of Mars from data on the period and semimajor axis of one of its satellites, using Kepler's third law. How does your answer compare with the value given in Table 11.1?

ADDITIONAL READINGS

Arvidson, R. E., A. B. Binder, and K. L. Jones. 1978. The surface of Mars. *Scientific American* 238(3):76.

Beatty, J. K., B. O'Leary, and A. Chaikin, eds. 1981. *The new solar system.* Cambridge, Eng.: Cambridge University Press.

Carr, M. S. 1983. The surface of Mars: A post-*Viking* view. *Mercury* 12(1):2.

———. 1976. The volcanoes of Mars. *Scientific American* 234(1):32.

Cordell, B. M. 1986. Mars, Earth and ice. *Sky and Telescope* 71(1):17.

Horowitz, N. H. 1977. The search for life on Mars. *Scientific American* 237(5):52.

———. 1986. *To utopia and back: The search for life in the solar system.* New York: W. H. Freeman.

Murray, B. C. 1973. Mars from Mariner 9. *Scientific American* 228(1):48.

Leovy, C. B. 1977. The atmosphere of Mars. *Scientific American* 237(1):34.

Pollack, J. B. 1975. Mars. *Scientific American* 233(3):106.

Veverka, J. 1977. Phobos and Deimos. *Scientific American* 236(2):30.

Young, R. S. 1976. Viking on Mars: A preliminary survey. *American Scientist* 64:620.

CHAPTER 12

MERCURY

Mercury as seen from space. (NASA)

CHAPTER PREVIEW

Mercury (Table 12.1) is the closest of the planets to the Sun, and the most difficult of the four terrestrial planets to observe from Earth (Fig. 12.1). Its greatest elongation is only 28° (Fig. 12.2), so it is always so near the Sun in our sky that it can be viewed only at dawn or dusk, rising or setting at most barely 2 hours before or after the Sun.

As in the case of Venus, Mercury for a long time was thought of as two distinct planets, given the names Mercury and Apollo.

THE EARTH-BASED VIEW

Few people ever see Mercury, despite the fact that it is rather bright (a little brighter than Sirius, the brightest star in the sky). The difficulty of observing it has made Mercury the most enigmatic of the terrestrial planets, with little of its nature observable by Earth-based telescopes. The best photographs are taken during daylight so that Mercury can be viewed well above the distortion caused by Earth's atmosphere near the horizon, but are degraded by the high degree of atmospheric turbulence or poor "seeing" that prevails during the day due to the Sun's heating of the atmosphere. These photographs show little except a fuzzy disk with a hint of surface markings possibly resembling the lunar maria.

The orbit of Mercury is unusually eccentric; that is, its elliptical shape is more highly elongated than those of all the other planets except distant Pluto, and the Sun is quite far off center in Mercury's orbit. Mercury is over 50 percent father away from the Sun at its greatest distance than when it is closest. As we will see, this variation in distance from the Sun has had some important consequences. The orbit is also tilted by the unusually large angle of 7° with respect to the ecliptic. At an average distance from the Sun of 0.387 AU, Mercury has an orbital period of 88 days, the shortest of all the planets.

Mercury is small, as planets go, with a diameter only slightly larger than that of the Moon. At least three satellites in the solar system, Ganymede of Jupiter, Titan of Saturn, and Triton of Neptune, are bigger. The mass of Mercury has been determined from its gravitational influence on nearby objects such as comets, and especially the asteroid Icarus, which passed nearby in 1968. More accurate measurements were later possible, when the U.S. spacecraft *Mariner 10* flew by Mercury. Mercury's mass is about 5.6 percent of the Earth's mass, and its density is 5.4 grams/cm^3, very similar to that of the Earth, implying that Mercury probably has a large, dense core. No trace of an atmosphere was detected from Earth-based observations.

Orbital Eccentricity and Rotational Resonance

As we have seen, observations of Mercury are difficult to make, and no surface features on the planet can be seen distinctly. Careful observations of the fuzzy markings that do appear led some observers in

TABLE 12.1 MERCURY

Orbital semimajor axis: 0.387 AU (57,900,000 km)
Perihelion distance: 0.467 AU
Aphelion distance: 0.308 AU
Orbital period: 87.97 days (0.241 years)
Orbital inclination: 7°0′15″
Rotation period: 58.65 days
Tilt of axis: 28°
Diameter: 4,864 km (0.382 $D_\oplus$)
Mass: 3.33×10^{26} g (0.0558 $M_\oplus$)
Density: 5.50 g/cm^3
Surface gravity: 0.38 Earth gravity
Escape velocity: 4.3 km/sec
Surface temperature: 700 K (day side); 100 K (dark side)
Albedo: 0.06
Satellites: None

the past to the conclusion that Mercury was in synchronous rotation, always keeping the same side facing the Sun. As close to the Sun as Mercury is, this was not a surprising idea, for the tidal forces acting on it are immense.

Radio observations of Mercury done in the early 1960s allowed its surface temperature to be calculated, through the use of Wien's law (see Chapter 6). The daylit side was hot, as expected, with a temperature of about 700 K. Because observers thought that the other side never saw the light of day, they expected to find very low temperatures there. It was a surprise, therefore, when a temperature of about 100 K was deduced for the dark side. This is certainly cold, but not as cold as it would have been if this side of the planet never faced the Sun.

Radar observations soon provided the solution to this mystery. The Doppler shift of radar signals reflected off of Mercury showed that the planet's rotation period is shorter than its 88-day orbital period. Mercury is not in synchronous rotation, but instead rotates once every 59 days, so that all portions of its surface are exposed to sunlight some of the time.

But why 59 days? This was a bit of a puzzle in itself, because most planets spin much more rapidly. It was noticed that 59, or, more precisely, 58.65 days, is precisely two-thirds of the orbital period of 88 (actually 87.97) days. Thus, Mercury rotates exactly three times for every two trips around the Sun (Fig. 12.3).

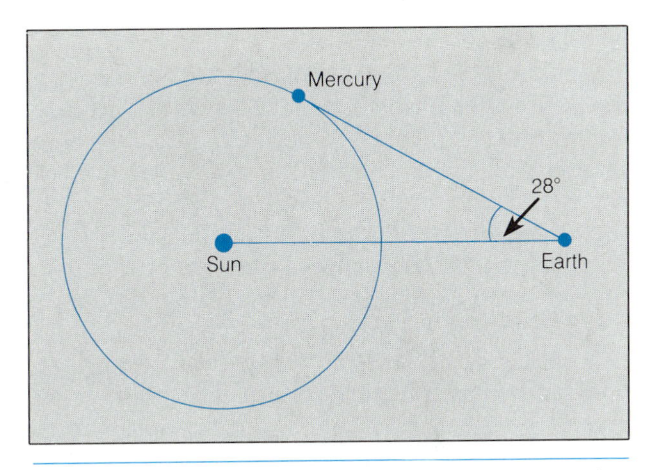

FIGURE 12.2 MERCURY'S GREATEST ELONGATION. *The farthest the planet ever gets from the Sun is 28 degrees, which means that it always rises and sets within two hours of the Sun.*

FIGURE 12.1 MERCURY AS SEEN FROM THE EARTH. *The planet is never far from the Sun and is therefore difficult to observe.* (NASA)

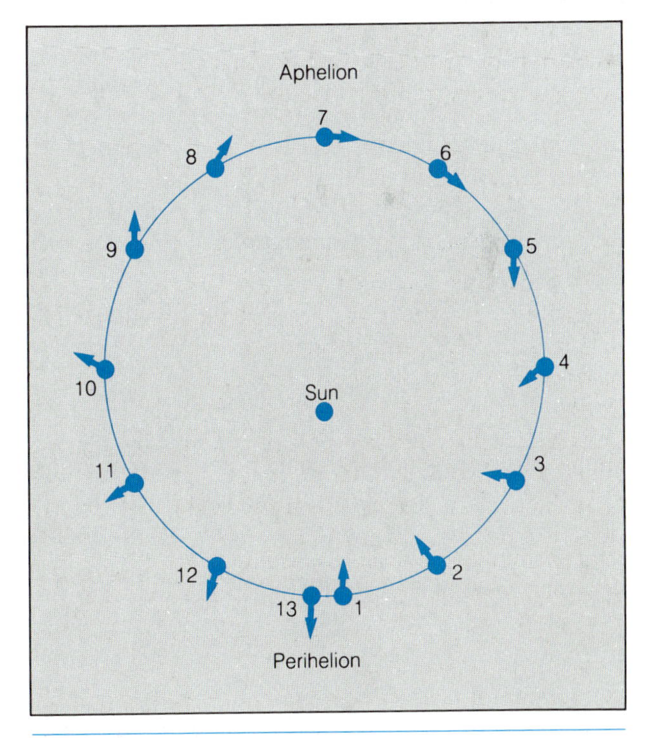

FIGURE 12.3 SPIN-ORBIT COUPLING. *This sketch shows how Mercury spins one-and-a-half times while completing one orbit around the sun. At perihelion it always has the heavy side facing directly toward or away from the Sun.*

This is no mere coincidence; clearly the gravitational tidal forces exerted by the Sun are at work. The key is what happens at the point where Mercury is closest to the Sun, called **perihelion,** for it is here that the tidal forces are the strongest. If Mercury has one side denser than the other, as the Moon does, then the tidal forces exerted by the Sun will act to ensure that this side is aligned with the Sun's direction when Mercury is at perihelion. One way to accomplish this, of course, could be synchronous rotation, so that this side would always point toward the Sun, but because Mercury's orbit is so elongated, the urge to keep its heavy side facing the Sun is most significant at perihelion.

The planet spins one-and-a-half times during each orbit; thus the heavy side either faces the Sun or exactly the opposite direction at each perihelion passage. In either case, the tidal force on Mercury is balanced, so that there is no tendency to alter the rotation further. Apparently the planet once rotated much faster, but whenever it passed through perihelion with the heavy side pointed in some random direction, tidal forces exerted a tug that tended to change its spin. This went on until the rotation slowed to the present rate, so that the tidal forces were always balanced when Mercury was at perihelion. If the orbit had not been so eccentric, or if Mercury had been symmetric instead of having a relatively dense side, the tidal forces would have been more uniform throughout, and Mercury might today be in synchronous rotation.

The 3:2 relationship between Mercury's orbital and spin periods is an example of **spin-orbit coupling,** a general term applied to any situation where the spin of a body has been modified by gravitational forces so that a special relationship between the orbital and the spin periods is maintained. Synchronous rotation is the most common example, and we will find several cases of it as we explore the rest of the solar system.

One remaining question regarding Mercury and its spin is why the planet should be asymmetric, with one side denser than the other. We will find a hint of a possible answer in a later section.

The combination of orbital and rotation speeds on Mercury produces a very unusual occurrence for anyone who might visit this planet. At closest approach to the Sun, when Mercury is moving most rapidly in its orbit, the orbital speed is actually greater than the rotation speed at its surface. For a short time (lasting a few hours) the Sun would appear to turn around and move backward across the sky, from west to east. Imagine the difficult time our ancestors would have had explaining this retrograde motion of the Sun if a similar phenomenon had occurred on the Earth!

RENDEZVOUS WITH *MARINER*

Mercury has not drawn the concentrated attention of the U.S. space program that Venus and Mars have, but nevertheless it was observed at close range by one probe, *Mariner 10.* This spacecraft, launched in late 1973, flew by Venus in February of 1974 and then set its course for Mercury. The gravitational pull of Venus was used to modify *Mariner 10*'s path so that this could be done. The spacecraft arrived at Mercury in late March, 1974.

An extraordinary opportunity presented itself when *Mariner 10* reached Mercury and continued into solar orbit. It proved possible to adjust its orbit so that the period was 176 days, exactly twice that of Mercury. This meant that the spacecraft made one orbit about the Sun for every two of Mercury, and the two therefore encountered each other repeatedly (Fig. 12.4). There was sufficient power to operate the instruments aboard *Mariner 10* on each of its next two meetings with Mercury, so there were actually three close encounters during which observations were made. The approaches to the planet were very close on the first and third encounters, when *Mariner 10* passed only a few hundred kilometers above the surface, and much further away on the second flyby, when the spacecraft passed by some 50,000 kilometers away.

There was one drawback to the *Mariner 10* orbit. Although it did allow ample opportunity for close-up observation of Mercury, only one side of the planet could be viewed. Because the spacecraft visited Mercury every other orbit, the same side of the planet was facing the Sun each time, since Mercury rotates three times in every two orbits around the Sun. If *Mariner 10* could have been placed in an orbit with an 88- or 264-day period, it could have viewed opposite sides of Mercury each time the two met. As it

ASTRONOMICAL INSIGHT 12.1

VISITING THE INFERIOR PLANETS

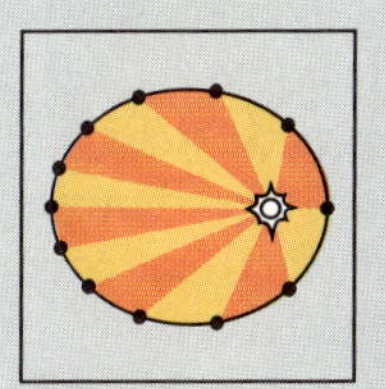

In Chapter 5 there was some discussion of the mechanics of placing a satellite in the Earth's orbit, but nothing on how to aim a spacecraft that is designed to visit another planet. The principles involved are particularly simple when an inferior planet is to be the target.

The concept of energy was discussed in Chapter 5, where we pointed out that in a system of two orbiting bodies, there is kinetic energy due to the motions of the bodies, and potential energy due to their gravitational attraction for each other. For a planet orbiting the Sun, the greater the total energy (the sum of kinetic plus potential), the larger the orbit. Thus for an object to go from the orbit of the Earth to that of an inner planet, it must lose energy.

A spacecraft sitting on the launchpad is moving with the Earth in its orbit, at a speed of 29 km/sec. To make the rocket fall into a path that intercepts the orbit of an inner planet, we must diminish this speed. Therefore, we launch the rocket backward with respect to the Earth's orbital motion, thereby lowering its velocity with respect to the Sun, and decreasing its orbital energy. If the launch speed is properly chosen, the spacecraft will fall into an elliptical orbit that just meets the orbit of the target planet. In order to reach Mercury from the Earth, the rocket must be launched backward with a speed relative to Earth at 7.3 km/sec.

There are, of course, a few additional considerations. For one thing, the rocket has to be launched with enough speed to escape the Earth's gravity. The escape speed for the Earth is 11.2 km/sec, so in fact we have to launch our Mercury probe with a speed in excess of this value, which is more than the speed needed to attain our trajectory to Mercury. The launch speed must therefore be calculated to take the Earth's gravity into account, so that the rocket escapes the Earth, but in the process is slowed just the right amount to give it the proper course. The gravitational pull of the target planet must also be taken into account, for this speeds the spacecraft up as it approaches.

Another important consideration is timing; the target planet must be in the right spot in its orbit at just the moment when the spacecraft arrives. It is for this reason that the term **launch window** is used. The Earth and the target planet must be in specific relative positions at the time of the launch. The interval between launch windows for a given planet is simply its synodic period. For a probe to Mercury, the travel time is about 106 days, so Mercury actually makes a little more than one complete orbit while the probe is on its way.

The gravitational pull of the target planet can be used to good advantage in modifying the orbit of a spacecraft that flies by. If the probe is aimed properly, its trajectory may be altered as it goes by in just the right way to send it on to some other target. This technique was used with Mariner 10, first to send it from Venus to Mercury, and then to modify its orbit again so that it returned to Mercury repeatedly thereafter.

Sending spacecraft to the outer planets is a bit more difficult, because the spacecraft has to have more energy than what it gets from the Earth's motion. The launch is therefore made in the forward direction, so that the rocket has the speed of the Earth in its orbit plus its own launch speed with respect to the Earth.

Ironically, the seemingly simple tasks of launching a probe to the Sun is one of the most complex. The most straightforward solution would be to launch the spacecraft backward from the Earth with a speed of 29 km/sec, entirely canceling out the Earth's orbital speed, so that the probe would then fall straight in toward the Sun. This is a prohibitively high launch speed, however, so in practice we will probe the Sun by first sending the spacecraft out around Jupiter. The spacecraft will fly by the massive planet in such a way (backward with respect to Jupiter's orbital motion) that its own orbital energy is reduced; the craft can then fall into the center of the solar system. A mission to explore the outer layers of the Sun in this manner is currently in the planning stages.

was, though, only one side of the planet was charted photographically (Fig. 12.5); close-up examination of the other side will have to wait for some future mission.

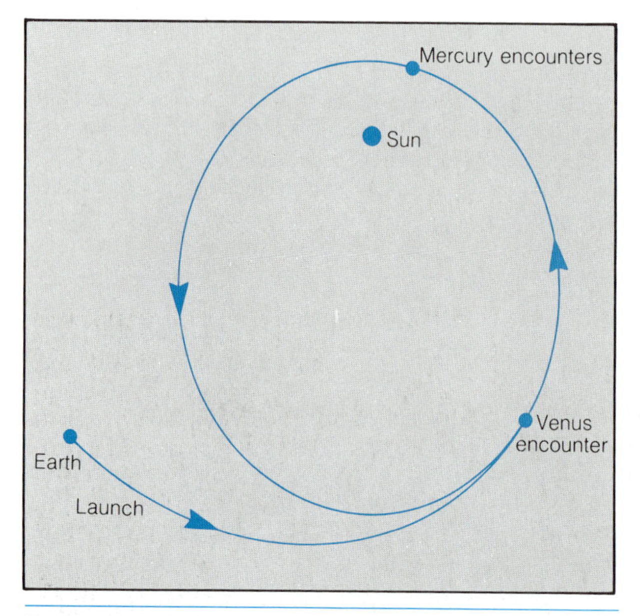

FIGURE 12.4 THE TRAJECTORY OF MARINER 10. *The spacecraft flew by Venus and then encountered Mercury repeatedly because its orbital period was adjusted to equal with twice the orbital period of Mercury.*

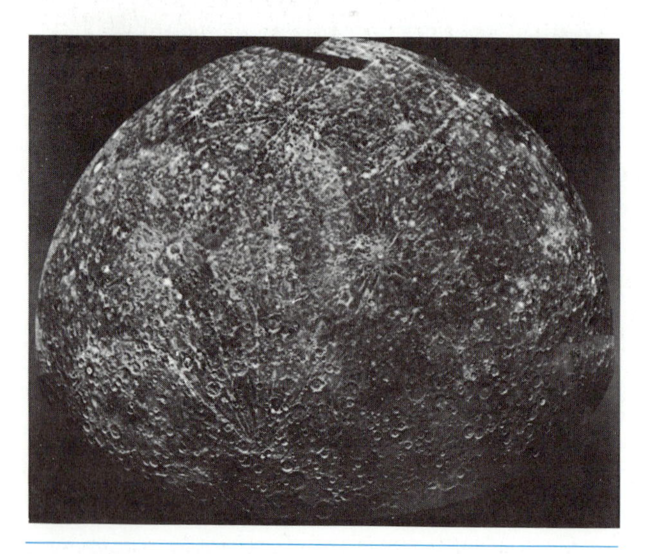

FIGURE 12.5 MERCURY AS SEEN FROM SPACE. *This* Mariner 10 *view is a mosaic of several images.* (NASA)

MAGNETIC FIELD AND INTERNAL CONDITIONS

The relatively high density of Mercury, known long before the *Mariner 10* encounters, indicated the likelihood of a dense core inside this planet, similar to that of the Earth and Venus. This might have led us to believe that Mercury had a magnetic field, except for the fact that the slow rotation was thought insufficient to produce the internal electrical currents needed to activate a magnetic dynamo.

Thus it was a bit of a surprise when *Mariner 10* detected a magnetic field around Mercury. Its strength is not great, about 1 percent as strong as that of the Earth, but nevertheless it is there. Like the Earth, Mercury has a magnetosphere whose shape is modified by the charged particles flowing past from the Sun.

Scientists concluded that Mercury's core must be relatively large (Fig. 12.6), in order to produce a magnetic field despite the slow rotation of the planet. The core apparently extends about three-fourths of the way out from the center to the surface, whereas

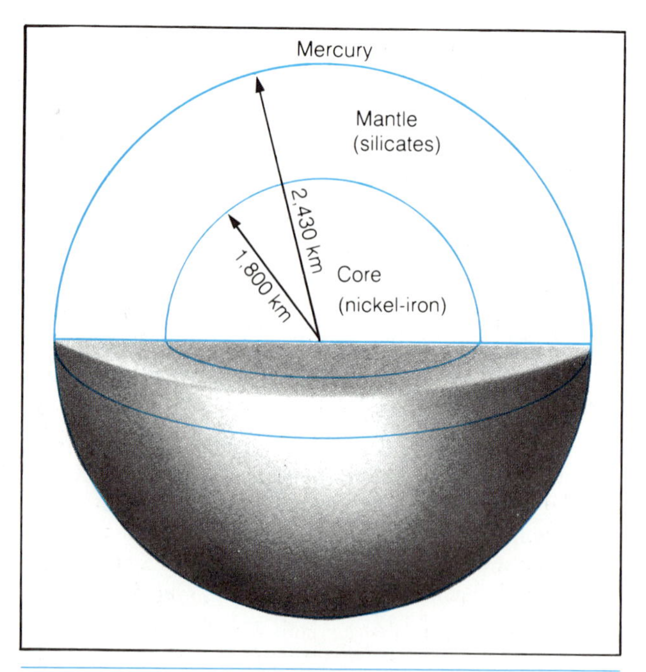

FIGURE 12.6 MERCURY'S INTERNAL STRUCTURE. *The presence of a magnetic field, along with the planet's relatively high density, implies the presence of a large core.*

the Earth's core is contained within the inner half of its radius.

The implication of such a large core is not only that Mercury is differentiated, but that it contains a relatively high overall abundance of heavy elements to begin with. This says something about its formation, namely, that a greater portion of its volatile, lightweight gases escaped than was the case for the other terrestrial planets. Thus the temperature in the vicinity of Mercury must have been relatively high early in the history of the solar system, higher than at the places where Venus, the Earth, and Mars formed. As in the other cases where well-developed cores exist, Mercury must have been fully molten, staying in that state long enough for differentiation to occur.

Not much is known about the mantle of Mercury, which occupies the outer 25 percent or so of its radius. The density there is lower than in the core, but its other properties are unknown. The photographs taken by *Mariner 10* show no obvious evidence of tectonic or volcanic activity, although we must keep in mind that half of the surface has yet to be examined.

SURFACE FEATURES AND GEOLOGICAL HISTORY

The surface of Mercury, at first glance, strongly resembles the lunar landscape (Fig. 12.7). It even looks quite similar on second glance, the differences being revealed only under rather detailed scrutiny.

Perhaps the most obvious distinction is found in the prominent and extensive cliffs, referred to as **scarps** (Fig. 12.8), which exist in many locations and form a network that girdles the planet. These are quite unlike anything seen on the other terrestrial planets or the Moon. Their appearance suggests that Mercury shrank a little after its crust hardened, causing it to shrivel and crack.

Another less obvious difference between Mercury and the Moon is that on Mercury the craters tend to be more widely separated, with more smooth space between them (Fig. 12.9). Apparently, when impacts occur on Mercury and form craters, the ejecta do not travel as far as on the Moon, so there are less extensive rays or other ejecta debris around the large craters (Fig. 12.10). The reason for this is clear: Mercury has a higher surface gravity than the Moon, by about a factor of 3, so that the rubble blasted out of the surface by an impact does not travel as far before falling back down. The craters on Mercury also tend to have a flatter appearance than those on the Moon, for the same reason. There are extensive smooth areas on Mercury which were apparently formed by lava flows, much like the lunar maria.

One very large impact crater was seen by *Mariner 10,* just at the dividing line between daylight and darkness (Fig. 12.11). The basin, formed by the tremendous crash of a massive body hitting the surface, resembles a giant bull's-eye, with concentric rings around it, and it extends over a diameter of some

FIGURE 12.7 MERCURY OR THE MOON? *This* Mariner 10 *view is a mosaic of several images. The planet bears a strong resemblance to the moon, although there are significant contrasts, particularly in internal structure.* (NASA/JPL)

1,400 kilometers. The gigantic crater, called Caloris Planitia, happens to lie on the side of Mercury that is facing the Sun at perhilion every other orbit. (The name Caloris was given to the crater for this reason, because this position is the hottest place on Mercury every other time it passes close to the Sun.) We noted earlier that Mercury is apparently lopsided, with an asymmetric distribution of mass; the position of this

FIGURE 12.8 A SCARP SYSTEM ON MERCURY. *The lengthy series of cliffs extending along the upper left is one of many on the planet's surface.* (NASA/JPL)

FIGURE 12.9 CRATERS ON MERCURY. *Because Mercury has a greater surface gravity than the moon, impact craters have lower rims and are shallower, and ejecta do not travel as far.* (NASA/JPL)

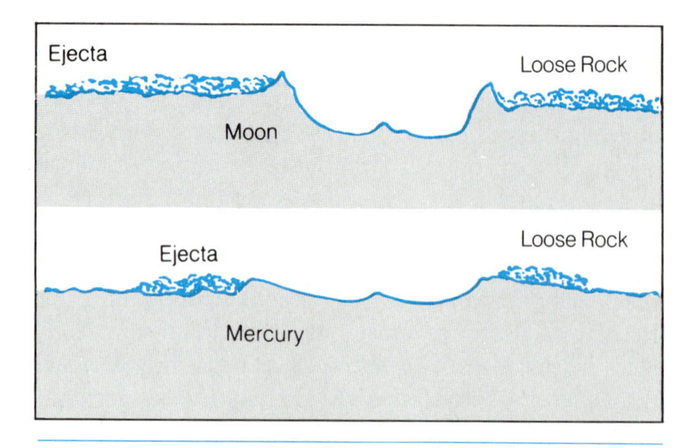

FIGURE 12.10 CRATER FORMATION ON THE MOON AND ON MERCURY. *The Moon's lower surface gravity allows higher rim walls to remain, and allows ejecta to travel farther from the point of impact than on Mercury.*

FIGURE 12.11 CALORIS PLANITIA. *This photo shows half of the immense impact basin known as Caloris Planitia. This region is directly facing the sun at perihelion on every other orbit.* (NASA/JPL)

huge impact basis suggests that perhaps the object that crashed into Mercury was responsible for making the planet denser on one side.

Directly opposite Caloris Planitia, on the far side of Mercury, is a curiously wavy region, so strange and unprecedented in appearance that is called the **weird terrain.** The rippled appearance there may have been created by seismic waves that raced around the planet when the impact occurred that formed Caloris Planitia. When the waves met on the far side of the planet, they distorted the surface there, resulting in the odd terrain seen today. Similar, but less prominent, features are seen on the Moon, at positions opposite some of the largest impact craters. On alternate orbits of Mercury about the Sun, when Caloris Planitia is on the dark side, the weird terrain is at the point directly beneath the Sun, and temporarily holds the dubious distinction of being the hottest place on the planet.

Infrared measurements made by *Mariner 10* indicate that there is a fine layer of dust on Mercury, much like that on the Moon. The surface rocks themselves are probably also similar to Earth and Moon rocks, except perhaps for a lower abundance of volatile elements.

Despite the combination of relatively low surface gravity and high temperature, which allowed nearly all the gases present on Mercury to escape long ago, a trace of an atmosphere does appear to exist. *Mariner 10* detected very small quantities of a number of gases, which are thought to originate in the solar wind, the steady stream of particles that sweeps past Mercury.

All these data indicate that Mercury's formation and evolution must have been quite similar to that of the Moon. Mercury and the Moon went through comparable stages of accretion and melting at the beginning, followed by hardening of the crust along with major lava flows (perhaps triggered by large meteorite impacts), and succeeded finally by a long period of cooling and decreased cratering.

We are not sure just when the tremendous impact that created Caloris Planitia took place, nor how long ago tidal forces succeeded in locking the planet's rotation into its 3:2 coupling with the orbital period. We can surmise that there must have been substantial internal heating before then, as tidal forces exerted by the Sun created internal friction. Perhaps there is still some excess internal heating caused by tidal forces, which helps to maintain the large molten core needed to create the planet's magnetic field.

PERSPECTIVE

In Mercury, we have found a world that is more like the Moon than a planet in some ways. Yet in others is quite similar to the Earth. It is Mercury's surface features that are similar to the Moon's (and in this sense the resemblance is superficial), whereas it is Mercury's interior, with its large dense core and magnetic field, that is apparently more like the interior of the Earth. In discussing Mercury, we have seen a rather unique example of spin-orbit coupling, which emphasizes the importance of tidal forces.

Having completed our examination of the terrestrial planets, we turn our gaze outward, bypassing the asteroid belt for now, to focus on the outer planets. We will begin our exploration of this part of the solar system with Jupiter, the king, or at least the prince (after the Sun) of the solar system.

SUMMARY

1. Mercury is so close to the Sun that it is difficult to observe from Earth, and only very general properties have been derived from telescopic observations.
2. Radar measurements showed that Mercury rotates exactly one-and-a-half times for each orbit around the Sun, in a form of spin-orbit coupling that was created by a combination of the Sun's tidal force on Mercury and the highly elliptical orbit of the planet.
3. The *Mariner 10* space probe, which made three close flybys of Mercury, provided the most detailed information on the planetary properties.
4. Mercury has a magnetic field and a high density, indicating that it has a large, dense core, probably partially molten.
5. The surface of Mercury is nearly identical in general appearance to that of the Moon, except for a planetwide system of scarps, and minor differences in the heights of impact craters and the distances traveled by ejecta.

6. The great impact basin Caloris Planitia is located at the subsolar point at perihelion on alternate orbits of the Sun, and may be the site of the mass imbalance that caused Mercury's spin to fall into resonance with its orbital period.

REVIEW QUESTIONS

Thought Questions

1. What is unusual about Mercury's orbit, compared to the orbits of the other planets discussed so far?
2. Explain how the shape of Mercury's orbit has led to the unusual spin-orbit coupling of the planet's rotation with its orbital period.
3. What do you think the rotation period of Mercury would be if the planet had a circular orbit?
4. Explain why *Mariner 10* was able to photograph only one side of Mercury, despite the fact that it visited the planet three separate times.
5. What could you say about the internal structure of Mercury, based only on the knowledge that the planet's overall density is relatively high?
6. Discuss the similarities and contrasts between the surface of Mercury and that of the Moon.
7. From what you know of the sizes and albedos of Mercury and the Moon, do you think the Moon would be visible from Earth if it orbited Mercury?
8. Why does Mercury not have a significant atmosphere?

Problems

1. If Mercury is 50 percent farther from the Sun at aphelion than at perihelion, how much more intense is sunlight on its surface at perihelion?
2. At greatest elongation, Mercury is 28° from the Sun, as seen from Earth. What is the longest possible time between sunset and the time when Mercury sets, or between the time when Mercury rises and sunrise? Discuss how this affects observations of Mercury made from the Earth.
3. *Mariner 10* was put into a solar orbit having twice the orbital period of Mercury. What was the semimajor axis of the orbit of *Mariner 10*?
4. Using information from Chapter 5, calculate the surface gravity on Mercury and compare it to the surface gravity on the Moon.
5. Calculate and compare the escape velocities of Mercury and the Moon.

ADDITIONAL READINGS

Beatty, J. K., B. O'Leary, and A. Chaikin, eds. 1981. *The new solar system.* Cambridge, England: Cambridge University Press.

Hartmann, W. K. 1976. The significance of the planet Mercury. *Sky and Telescope* 51(5):307.

Head, J. W., C. A. Wood, and T. A. Mutch. 1977.. Geologic evolution of the terrestrial planets. *American Scientist* 65:21.

Murray, B. C. 1975. Mercury. *Scientific American* 233(3):58.

Weaver, K. F. 1975. Mariner unveils Venus and Mercury. *National Geographic* 147:848.

CHAPTER 13

JUPITER: GIANT AMONG GIANTS

An infrared view of Jupiter. (© 1983 Anglo-Australian Telescope Board)

CHAPTER PREVIEW

The largest planet in our solar system, Jupiter (Table 13.1) is completely unlike any of the inner four. In discussing the terrestrial planets, we could draw comparisons with the Earth, but our modest home is not in the same league with Jupiter, and we will find that almost no aspect of this gaseous giant can be compared with the Earth. Entirely new standards must be adopted here, and Jupiter, the closest of the giants to Earth, and the best studied, will serve as the model for our later discussions of the others.

The Jovian system is very complex, with its many fascinating satellites, its ring, the extensive and intense radiation belts, and the fantastic appearance of the planet itself. Like Venus and Mars, Jupiter has played a role in the historical development of astronomy, particularly with Galileo's discovery of the four large moons, whose existence he cited as an argument for centers of motion other than the Earth.

TABLE 13.1 JUPITER

Orbital semimajor axis: 5.203 AU (778,300,000 km)
Perihelion distance: 4.951 AU
Aphelion distance: 5.455 AU
Orbital period: 11.86 years (4.333 days)
Orbital inclination: 1°18′17″
Rotation period: $9^h\ 55^m\ 30^s$
Tilt of axis: 3°5′
Mean diameter: 138,450 km (10.79 $D_\oplus$)
Polar diameter: 134,240 km
Equational diameter: 142,800 km
Mass: 1.901 × 10^{30} (318.1 $M_\oplus$)
Density: 1.33 g/cm^3
Surface gravity: 2.64 Earth gravities
Escape velocity: 60 km/sec
Surface temperature: 130 K (cloud tops)
Albedo: 0.34
Satellites: 16

Jupiter truly is a giant planet, containing almost three times the mass of all the other planets together. It orbits at an average distance of 5.2 AU from the Sun, so that it is never closer to the Earth than 4.2 AU. Its mass (318 times that of the Earth) is spread over such a large volume that the average density is only 1.3 grams/cm^3, barely greater than that of water.

OBSERVATION AND EXPLORATION

Even a small telescope reveals colorful structure in the Jovian atmosphere, with reddish brown **belts** and light-colored **zones** encircling the planet (Fig. 13.1). The Great Red Spot, a giant reddish oval in the southern hemisphere that dwarfs the Earth in size, has apparently been a persistent feature for at least 300 years. The thick atmosphere prevents any hope of our seeing a solid surface underneath; as we will see, modern theories hold that there is no well-defined solid surface, anyway. Spectroscopy of Jupiter's at-

FIGURE 13.1 THE VIEW OF JUPITER FROM EARTH. *This is a good photograph from an Earth-based telescope. One moon is visible: the dark spot seen against the planet's disk.* (NASA)

mosphere reveals compounds of the general type that are thought to have been present in the primitive atmospheres of the terrestrial planets; i.e. hydrogen-bearing molecules such as methane (CH_4), ammonia (NH_3), and hydrogen itself (H_2), as well as helium. This composition is rather significant, for, as we shall discuss, it means that the evolution of the Jovian atmosphere was arrested at a very early stage. Jupiter is certainly a cold planet by our earthly standards, with a temperature at the cloud tops of about 130 K.

Jupiter's rotation is very rapid, despite the large size of the planet. The rotation period is difficult to pinpoint, because the planet spins differentially, meaning that it goes around more quickly at the equator than near the poles. This can only happen in a nonrigid object. The Jovian day is just under 10 hours long, which means that the cloud tops near the equator must travel at a speed close to 45,000 km/hr. This rapid spin has flattened the planet a bit, giving it an oblate shape, so that the diameter through the equator is noticeably larger than the diameter through the poles (see Figs. 13.1 and 13.2).

Jupiter was discovered quite accidentally to be a source of radio emission, and further observations have shown a variety of activity in the radio portion of the spectrum, some of it understood, and some not. A general radio glow from the planet, at a wavelength of a few centimeters, is simply the thermal radiation expected according to Wien's law (although the peak emission occurs at shorter wavelengths, in the infrared).

In addition to the thermal radiation, Jupiter emits bursts of radio static much like the emission associated with lightning in the Earth's atmosphere, and it is now known that there is lightning on Jupiter. A form of radio emission was also detected that is caused by charged particles moving through a magnetic field. This **synchrotron radiation** showed that Jupiter has both a strong magnetic field and intense radiation belts surrounding the planet, analogous to the Van Allen belts on Earth.

The intensity and spectrum of the infrared radiation from Jupiter imply that the giant planet is producing some energy in its interior. The total amount of energy being emitted by the planet is about 1.68 times greater than the amount it receives from the Sun. The probable explanation for this is discussed later.

Observations from Earth also revealed that Io, the innermost of the four Galilean satellites, has strange properties, including the fact that it modulates the Jovian radio emission as it passes in front of the planet, and the fact that Io is accompanied by a cloud of atomic and ionized gas that streams along behind it in its orbital path around Jupiter.

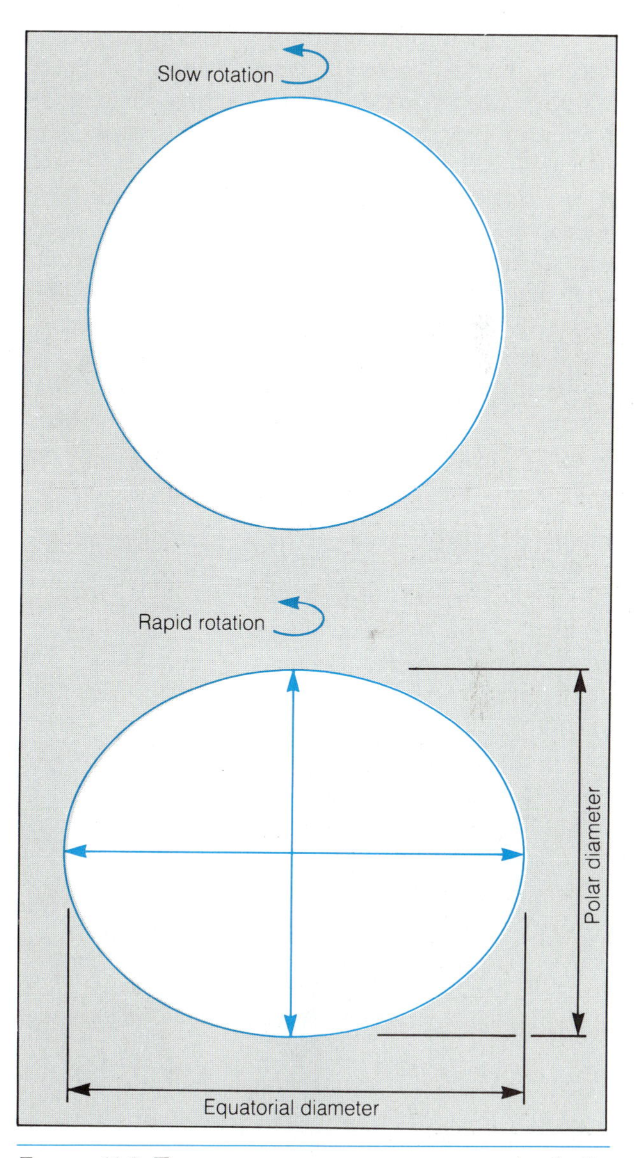

FIGURE 13.2 THE EFFECT OF RAPID ROTATION. *An elastic or fluid body is distorted out of a spherical shape by rotation. Jupiter is "flattened" by its rotation so much that its equatorial diameter is 6 percent larger than its polar diameter.*

TABLE 13.2 PROBES TO JUPITER

Spacecraft	Launch	Encounter	Type of Encounter
Pioneer 10	March 2, 1972	Dec. 3, 1973	Flyby, 131,000 km
Pioneer 11	April 6, 1973	Dec. 2, 1974	Flyby, 46,400 km
Voyager 1	Sept. 5, 1977	March 5, 1979	Flyby, 280,000 km
Voyager 2	Aug. 20, 1977	July 9, 1979	Flyby, 960,000 km

Like the terrestrial planets, Jupiter has been a target of the U.S. space program (Table 13.2). Sending probes to the outer planets requires some modifications to the technology that has been so successful in reaching the inner planets. The spacecraft must endure a passage through the asteroid belt, must have their own internal power source (small nuclear reactors have been used) because they travel too far from the Sun to use solar panels, and must be able to communicate with the Earth over great distances.

These challenges were successfully met by the *Pioneer 10* and *11* missions (an image of Jupiter from *Pioneer 11* is shown in Fig. 13.3), and more recently by the *Voyager 1* and *2* spacecraft (see Fig. 13.4). *Pioneers 10* and *11* reached Jupiter in late 1973 and late 1974, and carried out an assortment of studies of the planet's atmosphere, its magnetic field and radiation belts, and its satellites. The *Voyager* spacecraft flew past Jupiter in March and July of 1979, and carried out similar types of research, as well as sending back some 30,000 sensational images of Jupiter and its moons, obtained with their fine high-resolution cameras. Perhaps more than any other facet of the planetary exploration program, these pictures of Jupiter (and later of Saturn) have caught the public imagination, and today copies are seen everywhere.

Pioneer 11, as well as both *Voyager* spacecraft, went on to visit Saturn, using the gravitational pull

FIGURE 13.3 A CLOSE-UP VIEW OF JUPITER. *This image of Jupiter was obtained by the* Pioneer *spacecraft from a distance of just over a million kilometers. While far superior to any earth-based photos, this image was soon to be surpassed by those from the* Voyager *missions.* (NASA)

FIGURE 13.4 A VOYAGER IMAGE OF JUPITER. *This photo was obtained by* Voyager *when it was still millions of kilometers from Jupiter. Already a vast amount of detail is evident in the atmospheric structure.* (NASA)

ASTRONOMICAL INSIGHT 13.1

CLOSE ENCOUNTERS OF A DISTANT KIND

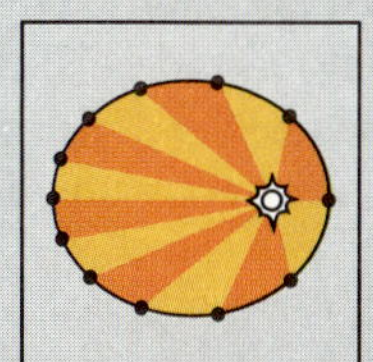

The process of discovery in astronomy is usually a methodical, gradual one. Typically an astronomer gathers data for a few nights at an observatory, then spends days or weeks analyzing the data before determining the results. Things are far different, however, on the occasion of a planetary encounter by a space probe such as *Voyager* or its predecessors.

The preparation for a planetary encounter is a combination of literally *years* of design and planning, and months or years of waiting for the probe's arrival at the target planet once it has been launched. The full scientific analysis after the encounter is thorough and methodical, and the results go through the normal scientific evaluation before being published months later in technical journals. What makes an encounter unique in astronomy, however, is that brief period of time when the data are arriving from the spacecraft. During that exciting time, the team of scientists responsible for the probe find themselves trying to explain the data and images as they are coming in, as the public and the news media literally look over their shoulders.

Probably the best example of the pressures and excitement of planetary encounters are provided by the two *Voyager* spacecraft, which were launched in 1976 and arrived at Jupiter in 1979 and Saturn in 1980, after which *Voyager 1* took a course out of the solar system, while *Voyager 2* went on to Uranus, which it reached in early 1986, and Neptune, which it will fly by in 1989. The *Voyager* program, as all other NASA planetary exploration missions, was managed by the Jet Propulsion Laboratory (usually called simply JPL), located in Pasadena, California, and operated under NASA funding by a contractual agreement with the California Institute of Technology. The spacecraft itself and the scientific instruments were built by various contractors (primarily groups of university scientists, in the case of the instruments) and assembled and tested at JPL. The launches took place at NASA's Kennedy Space Flight Center at Cape Canaveral, Florida, but the in-flight mission operations and the encounter activities were all managed from JPL, through a network of deep-space radio antennae scattered around the world.

During the so-called cruise phases of the *Voyager* craft—the lengthy periods between encounters—operations were quite routine, and most of the scientists spent their time at their home institutions, carrying on normal duties. The excitement and the pace increased each time an encounter was imminent, however, with more and more time spent at JPL, making detailed plans. After all the months of waiting, the encounters themselves took place very rapidly, with the spacecraft passing by the target planet in a matter of hours. Therefore, the sequence of observations had to be carefully planned and programmed into the on-board computers ahead of time. A further complication was presented by the fact that radio commands to the spacecraft took substantial amounts of time to reach them; the light-travel time from Earth to Jupiter, for example, was about 45 minutes when the *Voyager* encounters took place. Thus, it was impossible to relay commands to the probes and get an instantaneous response. During the weeks and days leading up to the encounter, team scientists were busy checking the command sequence and obtaining preliminary data. Among the most important tasks was refining positional data on target objects such as satellites, so that the cameras would succeed in getting the desired images. This was especially crucial for the *Voyager 2* encounter with Uranus, because the spacecraft went by the planet very quickly, and because the plane of the satellite orbits and of the rings is tipped perpendicular to the ecliptic, so that *Voyager* came in perpendicular to this plane, and had to busily swivel around, getting all of its images in a very short time. The position of each satellite to be observed during the Uranus encounter had to be known very precisely, because even a small error would have caused the cameras to miss the target. Furthermore, the cameras had to be programmed to swivel during exposures, to avoid blurring the images due to the rapid motion of the spacecraft as it flew past. All of this was done very successfully, despite the fact that radio communications during the encounter took nearly 3 hours one way!

(continued on next page)

ASTRONOMICAL INSIGHT 13.1

CLOSE ENCOUNTERS OF A DISTANT KIND

(continued from previous page)

During the days or hours of an encounter, all of the team scientists gather at JPL to examine the data as they come in. Much of the time, members of the news media watch along with the scientists as the images are displayed on a large screen in an auditorium; there are few other examples in all of science where the public literally sees that data at the same time as the scientists. Add the fact that every encounter revealed major surprises and mysteries, and imagine the combination of pressure and excitement the scientists experienced as they were called on to explain what they were seeing, almost as they saw it for the first time. This gave the public a rare but distorted opportunity to see how scientists make hypotheses and refine them.

Future close encounters with distant objects such as planets, comets, and asteroids are already being planned. The next major event will be *Voyager 2*'s encounter with Neptune in August of 1989, when again things will happen very quickly and communications will be even slower than they were at Uranus. Other upcoming planetary probes include the *Galileo* mission to Jupiter, which will place a probe in orbit around the planet, allowing for somewhat more leisurely examination, and a mission called *Comet Rendezvous and Asteroid Flyby*, or *CRAF*, which will travel along with a comet for several months as it approaches the Sun. The schedule for *Galileo* is currently uncertain; it was to have been launched in the spring of 1986, but major delays have been caused by the shuttle *Challenger* disaster. The *CRAF* mission, if approved for funding in 1987, could be launched in 1991 for a rendezvous with Comet Wild 2 in 1995, encountering an asteroid on the way to the comet. Other plans for future missions include a new Mars orbiter and a Venus radar mapper. We can look forward with anticipation to more exciting times at JPL as each of these missions unfolds.

of Jupiter to correct their courses and to provide a boost along the way. Through much of the 1980s, all of the outer planets are on the same side of the Sun, and this led to the idea of a "grand tour." It was realized that a space probe could visit all the outer planets, using each along the way to provide a gravitational boost towards the next. While the full tour is not being tried, *Voyager 2* is traveling most of it. This spacecraft, having flown by Saturn in August of 1981, encountered Uranus in January of 1986, and will visit Neptune in August of 1989.

ATMOSPHERE IN MOTION

Let us now return to Jupiter itself, and consider the structure and especially the complex motions in its atmosphere. The upper portion of the atmosphere is made of hydrogen and hydrogen compounds, as already mentioned. Helium is also present in large quantities. The presence of these gases represents a major contrast with the terrestrial planets, which long ago lost to space all of their original hydrogen and helium. The overall chemical composition of Jupiter (Table 13.3) is essentially identical to that of the Sun, except that on Jupiter it is cold enough for most of the material to be in molecular form, whereas the Sun is so hot that few molecules can exist, and most atoms are ionized. It is noteworthy that the principal species in the Jovian atmosphere, CH_4, NH_3, H_2, and helium,

TABLE 13.3 COMPOSITION OF JUPITER'S ATMOSPHERE

Gas	Symbol	Fraction*
Hydrogen	H_2	0.90
Helium	He	0.10
Water vapor	H_2O	1×10^{-6}
Methane	CH_4	7×10^{-4}
Ammonia	NH_3	2×10^{-4}

*The values listed are the fraction of the total number of molecules in the atmosphere. Note that hydrogen and helium themselves total 100 percent; this is not precisely true but reflects the dominance of these two species.

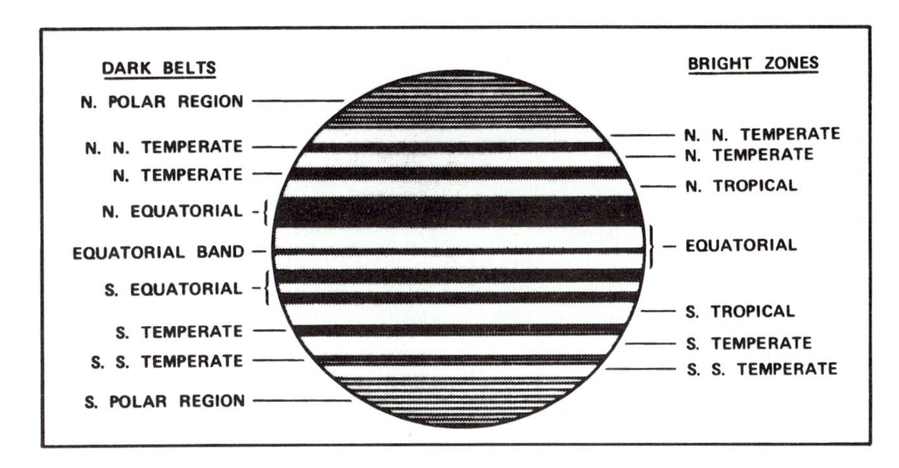

FIGURE 13.5 BELTS AND ZONES ON JUPITER. *This schematic drawing names the major features that give Jupiter its banded appearance.* (NASA)

are the ones thought to be present in the Earth's atmosphere at the time life forms first appeared.

Heavy elements, such as metals, are not observed in the Jovian atmosphere, but are probably present, having sunk to the core of the planet. Jupiter, being fluid nearly all the way to the center, is highly differentiated.

The bright and dark colors in the upper atmosphere are probably caused by slight differences in molecular composition, which in turn are created by subtle differences in temperature and pressure. The dark belts are regions of low pressure at the cloud tops, where the atmospheric gas is descending, while the light-colored zones are high pressure regions where the gas is rising. The banded appearance of the planet (Fig. 13.5) is created by the rapid rotation, which stretches what would otherwise be circular flow patterns into elongated strips that circle the planet. Just as on the Earth, air that is ascending forms a cyclonic flow, and it rotates in a clockwise direction in the southern hemisphere, and counterclockwise in the north (Figs. 13.6 and 13.7). Where cool gas is de-

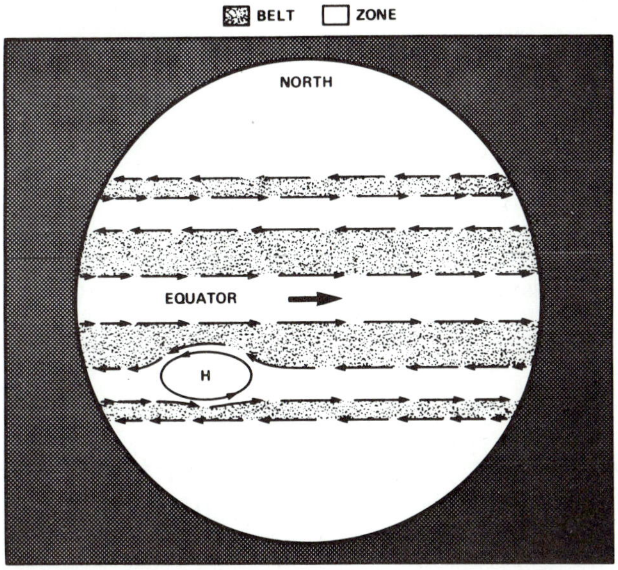

FIGURE 13.6 CIRCULATION OF THE JOVIAN ATMOSPHERE. *This cross-section indicates how the horizontal circulation pattern is related to vertical convective motions in the outer layers of the atmosphere. This is analogous to curculation in the earth's atmosphere, except that the rapid rotation of Jupiter stretches rotary patterns into belts and zones.* (NASA)

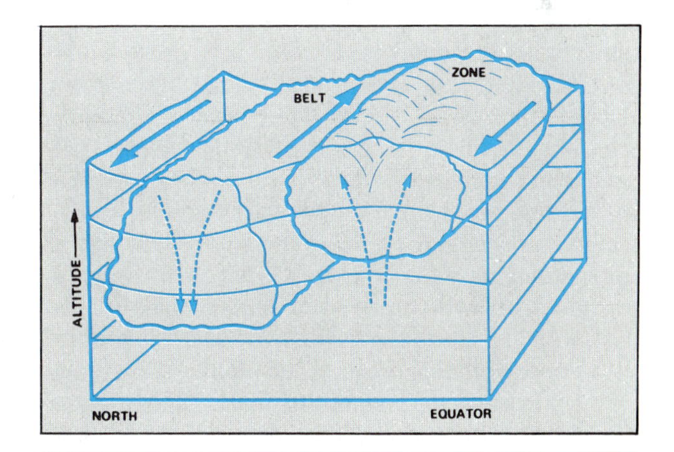

FIGURE 13.7 VERCIAL MOTIONS IN THE JOVIAN ATMOSPHERE. *This cross-section indicates how the horizontal circulation pattern is related to vertical convective motions in the outer layers of the atmosphere. This is analogous to circulation in the earth's atmosphere, except that the rapid rotation of Jupiter stretches rotary patterns into belts and zones.* (NASA)

FIGURE 13.8 THE GREAT RED SPOT. *This is a* Voyager *image of the rotating column of rising gas that has been present on Jupiter for at least 300 years.* (NASA)

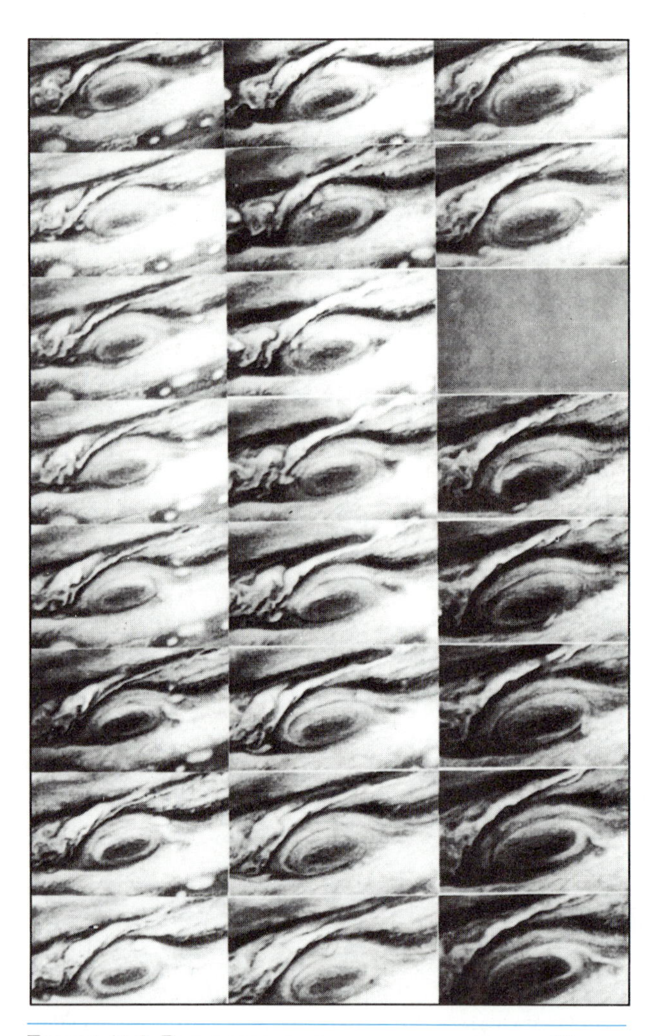

FIGURE 13.9 ROTATION OF THE GREAT RED SPOT. *The sequence starts at the upper left and goes down each column. Each frame is separated from the ones above and below it by about twenty hours. Note the two white spots initially in the upper left, and trace how they enter the spot, move around it counterclockwise, and then move out of view at the lower left.* (NASA)

scending, the anticyclonic flow, as it is then called, is in the opposite direction.

Besides the regular pattern of belts and zones, Jupiter's atmosphere displays a wide variety of spots and other features, the Great Red Spot (Figs. 13.8 and 13.9) being the largest and most prominent. There are also a number of smaller spots, both dark and light in color. Just like the belts and zones, these are cyclonic (if light colored) or anticyclonic (if dark). Thus, the Great Red Spot may be characterized as a storm system, although it is a very long-lived one, having been present for at least 300 years. The Great Red Spot is not well understood, having properties that are unusual, even for Jupiter. One surprising fact is that the central portion of it consists of gas that is rising, reaching much greater heights than the belts and zones normally do, yet it is dark colored, in contrast with the rising gas in the light-colored zones.

Very detailed photographs of Jupiter's cloud tops reveal a fantastic picture of turbulent, chaotic flows where the belts, zones, and spots interact (Fig. 13.10). Movie sequences reconstructed from images obtained from the *Voyager* spacecraft as they approached Jupiter show an incredible complexity of motions all happening at once, as in some gigantic machinery of gears and wheels.

INTERNAL STRUCTURE AND EXCESS RADIATION

No direct probes of the interior of Jupiter have been made, so what we know has been surmised largely from theoretical calculations. This may seem like an exceedingly uncertain basis for any detailed ideas about interior conditions, and of course there is room for

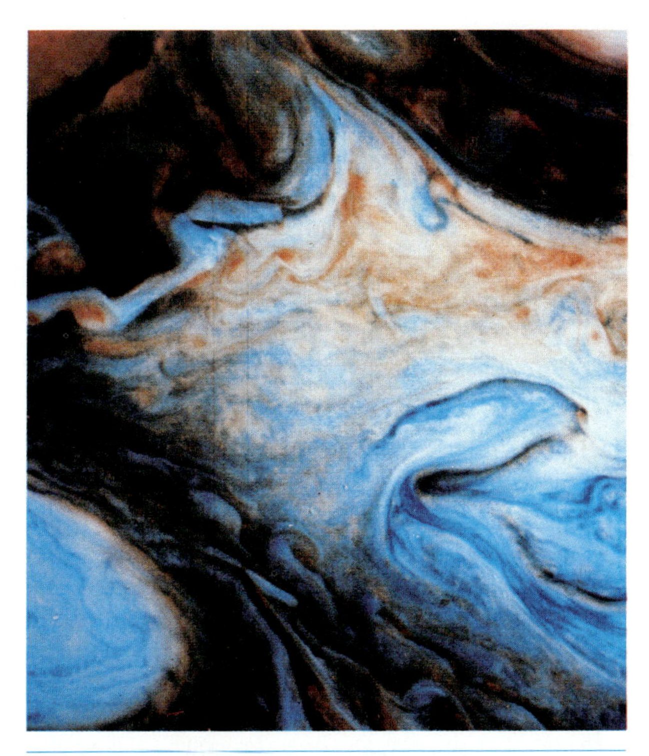

FIGURE 13.10 TURBULENT MOTIONS IN THE JOVIAN ATMOSPHERE. *In this* Voyager *image, colors are altered to enhance contrast and reveal complex details.* (NASA)

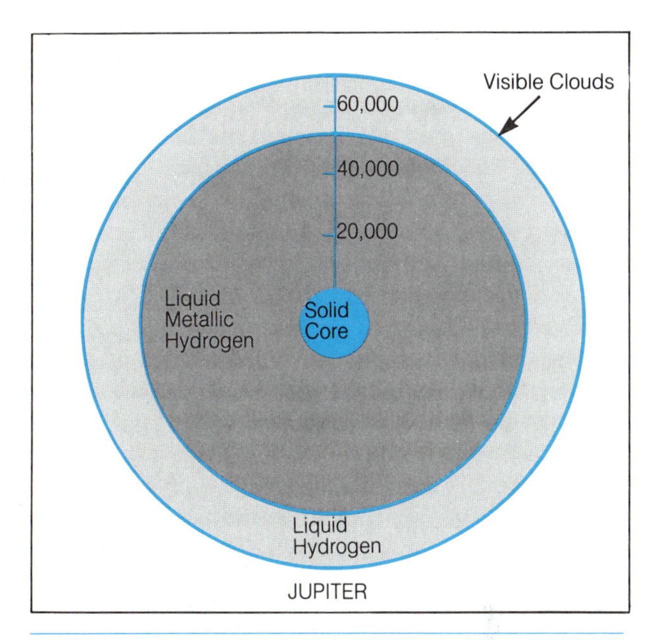

FIGURE 13.11 INTERNAL STRUCTURE OF JUPITER. *This cutaway illustrates the rough model of the Jovian interior described in the text.*

error, but in fact the laws of physics dictate rather explicitly what the internal structure must be like. The general procedure for calculating a theoretical model of Jupiter is very much like that described in Chapter 21 for determining the internal properties of a star. A set of equations describing the relationships among various properties such as pressure and temperature, as well as processes such as heat flow, is solved, starting with the known size, mass, composition, and surface characteristics of the planet.

While different assumptions made by people working in this field lead to somewhat different results, the major features come out the same (Fig. 13.11). Below a relatively thin layer of clouds (perhaps 1,000 km thick) is a zone consisting of hydrogen in liquid form, extending more than one-third of the way towards the center. The pressure in this region is as great as 3 million times sea-level pressure on Earth, and the temperature is as high as 10,000 K or more. An even thicker zone below the liquid hydrogen has such extreme pressure that hydrogen is forced into a state called **liquid metallic hydrogen**, which has a somewhat rigid structure like certain metals. Liquid metallic hydrogen cannot be produced on Earth, because there is no known way to create the necessary pressure, but the simplicity of hydrogen creates confidence in the theoretical calculations that predict its behavior under these extreme conditions. The central core of Jupiter, below the liquid metallic hydrogen, is probably solid rocky material of very high density, containing all the heavy elements of the entire planet, which sank to the center as the planet differentiated. In the core, the temperature is expected to be as great as 30,000 K. The mass of the core is probably in the range of 10 to 20 times the mass of the Earth, indicating that as much as 6 percent of the planet's mass is contained in only 0.5 percent of its volume.

Except possibly for the boundary of this rocky central zone, Jupiter has no solid surface, instead merging gradually from gas to liquid to metallic liquid as we go inward from the outside.

As noted earlier, the planetary radio emission shows the presence of an internal heat source, and this was included in the theoretical calculations just described. The source of this internal heat is difficult to understand, however, because Jupiter is not sufficiently massive to have nuclear reactions taking place in its core, as in a star. For a while, it was thought that Jupiter might be shrinking gradually, converting gravitational energy into heat in the process, but this now appears to be ruled out. The most likely explanation, based on calculations of the heat capacity and compressibility of the interior, is that the interior is still hot from the time of the planet's formation, some 4.5 billion years ago. Such a prolonged retention of heat is made possible by the characteristics of the outer layers, which do not efficiently transmit heat from the interior to be lost to space. The atmospheric circulation, already described, is caused in part by heating of the lower atmosphere from below.

THE MAGNETIC FIELD AND RADIATION BELTS

It has been known for some time that Jupiter has a magnetic field, as mentioned earlier. This was originally deduced from observations of radio emission, and was later verified by the instruments on board the *Pioneer* and *Voyager* probes.

The Jovian field at the planet's cloud tops is about 10 times the strength of the Earth's magnetic field, and is oriented in the opposite direction, with the north magnetic pole pointing south, and vice versa. This should not be considered too surprising, since we know that the Earth's magnetic poles have reversed themselves from time to time. Perhaps the Jovian field does the same, and has its poles aligned with those of the Earth some of the time.

The magnetic axis of Jupiter does not exactly parallel the rotation axis, but is instead tilted by 11° (Fig. 13.12). This is also similar to the Earth, but on Jupiter it has some important effects on the distribution of charged particles in the radiation belts, as we shall discuss shortly.

The origin of the magnetic field is thought to be roughly similar to that of the Earth's field; that is, electrical currents created by flows inside the planet create a magnetic dynamo. The structure of the field indicates that the dynamo operates not far below the

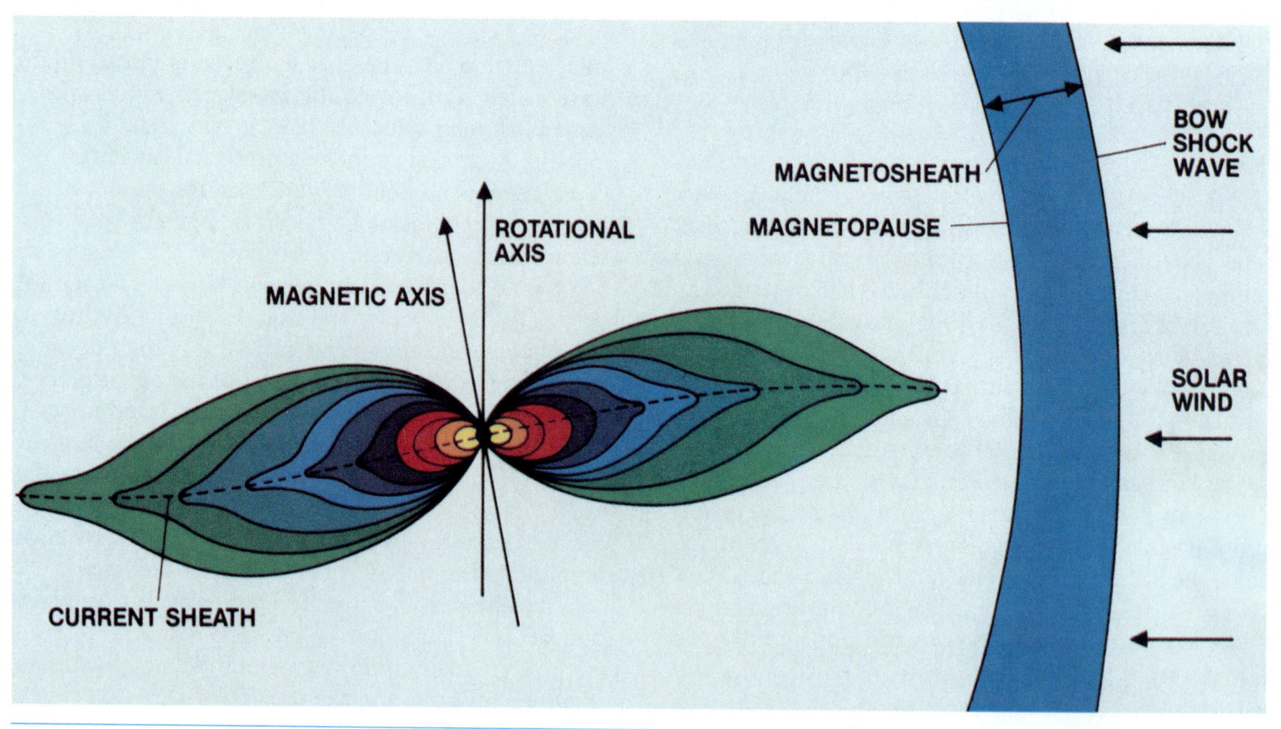

FIGURE 13.12 THE JOVIAN MAGNETOSPHERE AND RADIATION BELTS. *The shape of the magnetosphere is influenced by the solar wind and by the rapid rotation of Jupiter. The result is a sheet of ionized gas that is closely confined to the equatorial plane but wobbles as the planet's off-axis magnetic field rotates.* (NASA/JPL)

surface of the planet, probably in the liquid hydrogen zone.

Within the magnetosphere of Jupiter are very intense radiation belts (see Fig. 13.12), with much higher densities of charged particles than in the Van Allen belts of Earth. The passages of the *Pioneer* and *Voyager* spacecraft through these zones was rather perilous, since electronic circuits can be damaged by an environment full of speeding electrons and ions. Some data transmissions were garbled or lost, but on the whole the craft survived intact and were able to continue functioning normally.

One mystery concerning the Jovian radiation belts was where the great quantities of charged particles come from. A number of possibilities have been considered, but most were not feasible, and gradually it was realized that the answer had to do with the inner Jovian satellites. The five closest to the planet all orbit within the magnetosphere, where they sweep up charged particles as they move along their paths. The discovery, from Earth-based observations, that Io is accompanied by a cloud of glowing gas, led to the speculation that perhaps some of the satellites emit particles that enrich the radiation belts. As we shall see in the next section, the *Voyager* missions were soon to provide spectacular confirmation that just such a process is occuring.

THE SATELLITES OF JUPITER

Jupiter has sixteen known satellites (Table 13.4), ranging from Metis (also designated J-16), closest to Jupiter, to faraway Sinope, the outermost. Most of the sixteen are small, icy objects, with diameters of less than 100 km. The outermost four have retrograde orbits that are quite eccentric, and inclined by several degrees to the orbital plane of the more regular, inner satellites. The fact that these four vary from the norm may suggest that their origin is different from the others, which probably formed from interplanetary debris swirling around the infant Jupiter at the time the solar system coalesced.

Four of the Jovian satellites, the fifth through the eighth in order from Jupiter, are much larger than the rest, all being comparable to or larger than the Earth's Moon (Fig. 13.13). These are the four spotted by Galileo with his primitive telescope, and they are easily bright enough to be seen with the unaided eye, except that the brilliance of nearby Jupiter drowns them out.

The *Pioneer* and *Voyager* spacecraft were sent along trajectories aimed at providing close-up views of these Galilean satellites. Of particular interest was Io, the innermost of the four, because of its mysterious influence on radio emissions from Jupiter and because

TABLE 13.4 SATELLITES OF JUPITER

No.	Name	Distance (R_J)	Period (days)	Diameter (km)†	Mass (g)	Density (g/cm³)
16	Metis	1.79	0.295	20:	9.5×10^{19}	
15	Adrastea	1.81	0.298	20 × 20 × 15	1.9×10^{19}	
5	Amalthea	2.54	0.498	270 × 170 × 150	7.2×10^{21}	
14	Thebe	3.11	0.675	110 × ? × 90	7.6×10^{20}	
1	Io	5.91	1.769	3,630	8.92×10^{25}	3.53
2	Europa	9.40	3.551	3,138	4.87×10^{25}	3.03
3	Ganymede	15.0	7.155	5,262	1.49×10^{26}	1.93
4	Callisto	26.3	16.689	4,800	1.08×10^{26}	1.70
13	Leda	156	238.7	16:	5.7×10^{18}	
6	Himalia	161	250.6	186:	9.5×10^{21}	
10	Lysithea	164	259.2	36:	7.6×10^{19}	
7	Elara	165	259.7	76:	7.6×10^{20}	
12	Ananke	297	631	30:	3.8×10^{19}	
11	Carme	317	692	40:	9.5×10^{19}	
8	Pasiphae	329	735	50:	1.9×10^{20}	
9	Sinope	332	758	36:	7.6×10^{19}	

*The symbol R_J stands for the equatorial radius of Jupiter, which is 71,400 km.
†A colon (:) following a value for diameter indicates an uncertain value, and no density has been calculated for these cases (or for satellites having irregular shapes).

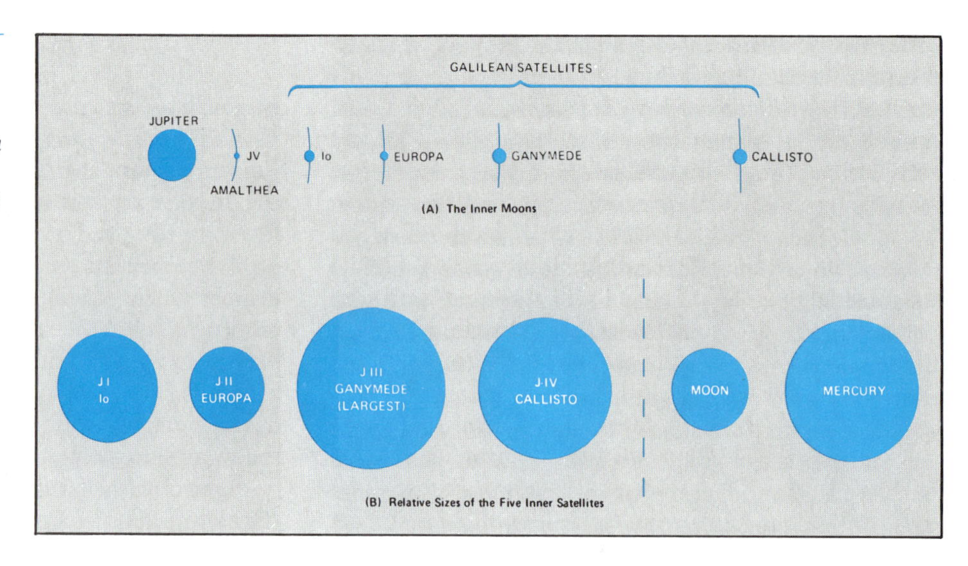

FIGURE 13.13 THE MAJOR JOVIAN SATELLITES. *These profiles indicate the relative distances of the major satellites from Jupiter and their relative sizes. The Earth's Moon and the planet Mercury are included for comparison* (NASA)

of the glowing cloud of gas that surrounds it.

The sizes of the Galilean satellites, as well as their masses, have been most accurately determined by the *Pioneer* and *Voyager* probes. While each of the four has its own distinctive character, the Galilean satellites fit a pattern, with systematic trends from one to the next. Their densities decrease steadily with distance from Jupiter, Io having the greatest density and Callisto, the outermost of the four, the least. Their albedos also decrease in the same manner. These two facts, along with measurements of the surface temperatures, indicate that the outer satellites contain large quantities of water ice, which can be very dark when seen from afar, especially if it contains some gravel and grit. The inner Galilean satellites contain less ice and greater fractions of rock. The surfaces of these bodies also show marked contrasts with one another. The color photographs taken by the *Voyager* probes showed tremendous complexity, revealing the Galilean satellites to be wondrous worlds, each with its own special effects.

Callisto, the farthest from Jupiter, is covered entirely with craters (Fig. 13.14) and has the appearance of a celestial target range that has been bombarded for ages with rocks and missiles from space. The impact points look white, in contrast to the overall dark coloration of the surface. This contrast is due partly to the removal by the impacts of the weathered, old surface layer of the ice, and partly to the fact that when ice is fractured, multifaceted cracks and fissures are created, which reflect and scatter sunlight. At one point on the surface of Callisto is a tremendous circular feature resembling a giant bull's-eye, where a very large chunk of material from space left its mark.

Ganymede, next in from Callisto, is the second largest satellite in the solar system. Until recently, Titan, the largest of the satellites orbiting Saturn, was

FIGURE 13.14 CALLISTO. *The outermost of the Galilean satellites, Callisto has the oldest surface.* (NASA)

thought to be bigger, but *Voyager* measurements showed that the solid body hidden within the dense atmosphere of Titan is actually smaller than Ganymede. The surface of Ganymede is dark but is marked with linear, light-colored streaks (Figs. 13.15 and 13.16) that show evidence of tectonic activity in the satellite's past, with discontinuities and fractures where the crustal plates have shifted. Numerous impact craters are seen, though they are not as dense as on Callisto, indicating that some geological process has eradicated the oldest craters on Ganymede.

The next satellite in sequence is Europa, whose surface is reminiscent of the old sketches of canals on Mars (Fig. 13.17). Europa is covered by random linear features and resembles nothing so much as a cloudy crystal ball that has been cracked throughout. The dark lines appear to be rifts in the rocky crust where water has oozed to the surface and frozen, leaving a smooth surface, but one with strong color contrasts. Recent theoretical studies show that Europa may have liquid water underneath the surface ice. Thus, this moon of Jupiter may be the one place in the solar

FIGURE 13.16 FEATURES ON GANYMEDE. *This close-up of the surface shows a complex pattern of grooves and ridges, which were probably created by deformations of an icy crust. Some impact craters are also visible.* (NASA)

FIGURE 13.15 GANYMEDE. *The surface of this moon shows evidence of less geological activity than the inner two Galilean satellites, but it still has banded segments indicative of surface motions and stresses.* (NASA)

FIGURE 13.17 EUROPA. *This is the second of the Galilean satellites in progression outward from Jupiter. Its linear features and relative lack of craters indicate that this is a tectonically active moon, with a young surface.* (NASA)

ASTRONOMICAL INSIGHT 13.2

THE MANY ROLES OF JUPITER'S GRAVITY

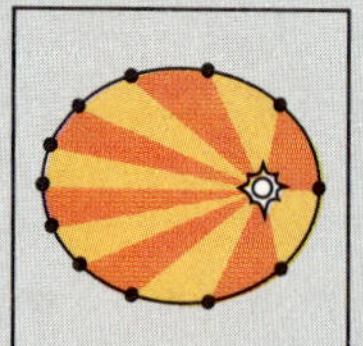

The motions of the planets are dominated by the force of gravity due to the Sun, which contains more than 99 percent of all the mass in the solar system. Most of the rest of the mass resides in Jupiter, so it might be expected that this large planet should exert some influence on other bodies. Even so, you might be surprised at the great variety of effects that are attributed to the gravitational influence of Jupiter.

One effect, discussed in this chapter, is the incessant volcanic activity on Io, the innermost of the major satellites of Jupiter. Io is subjected to tremendous tidal forces due to its proximity to the parent planet, and the resulting internal stresses caused by this (and by the frequent alignments with Europa) keep Io hot and molten inside, triggering frequent volcanic eruptions. Similar but less severe heating occurs in Europa as well, causing some scientists to believe that this moon may have large quantities of liquid water below its surface.

Between Jupiter and Mars a vast swarm of small, rocky bodies called **asteroids** or **minor planets** orbit the Sun. These bodies are thought to be left over from the time of the formation of the solar system. Some are probably fragments of larger objects that were destroyed by collisions, and others may be in their original shape, having undergone little change since their formation. It is believed that all of the planets formed from the coalescence of bodies like the asteroids, so one might ask why the asteroids have not merged to form a planet. The answer, apparently, is that tidal forces due to Jupiter have prevented such a planet from forming. As the asteroids overtake Jupiter in their orbits around the Sun, Jupiter's gravity causes them to depart from the perfectly elliptical paths dictated by Kepler's laws. These **perturbations** prevent the asteroids from coalescing, and furthermore have caused collisions between asteroids, occasionally violent enough to break them up. The **meteorites** that reach the Earth are thought to be remnants of asteroids that have been destroyed by collisions induced by Jupiter's gravity. Meteorites and asteroids are described in more detail in Chapter 16.

In addition to being responsible for their existence, Jupiter influences the asteroids in another way. There are gaps in the asteroid belt, discovered over 100 years ago by Daniel Kirkwood. These gaps occur at distances from the Sun where orbiting particles would have periods equal to one-half, one-third, one-fourth, and other simple fractions of the orbital period of Jupiter. Asteroids that may have once existed in these locations would have regularly been aligned with Jupiter, and over time, the gravitational influence of Jupiter apparently has altered the orbits of these asteroids. The result is gaps in the asteroid belt at the locations where objects would have orbital periods that were simple fractions of Jupiter's period. These clear zones, known as **Kirkwood's gaps,** are discussed more fully in Chapter 16.

There are a few asteroids whose orbits are governed by Jupiter's gravity in another way. It was discovered in the nineteenth century by the French scientist and mathematician J. L. Lagrange that when two bodies such as Jupiter and the Sun are in mutual orbit, there are points in the orbit of the less massive body (Jupiter in this case) in which additional small bodies may be trapped. These points, called the L-4 and L-5 points, lie 60° ahead and 60° behind Jupiter. At these places the combined gravitational effects of Jupiter and the Sun create stable positions where small bodies can orbit the Sun. Following the mathematical prediction that this could happen, several asteroids were found to be occupying both points in the orbit of Jupiter. These groups of asteroids are called the **Trojan asteroids,** and will be discussed further in Chapter 16.

There is one additional effect of Jupiter's gravity worth mentioning here. This one is very speculative, and by no means widely accepted as definitely occurring. It has recently been suggested that the great ice ages on the Earth, as well as long-term climatic changes on Mars, have been induced by slow alterations of the axial tilts of these planets that were caused by the gravitational effects of Jupiter. Tidal forces exerted by Jupiter may be responsible for changes in the axial tilt of Mars, which in turn may have caused the early climate on Mars to be changed radically. There is evidence that the climate of Mars has experienced periods of relatively warm conditions alternating with very cold epochs, when ice sheets covered much of the surface. The same is true of the Earth, which is known to have undergone several ice ages, some as recently as only a few thousand years ago.

It is interesting and perhaps ironic that modern scientists are finding evidence that the planets, or at least Jupiter, do affect occurrences on the Earth, although the causes and the effects are far from what is normally predicted by astrologers.

system other than the Earth where oceans exist. Like Ganymede, Europa shows evidence of tectonic activity, particularly in this pattern of linear fractures. The surface of Europa shows very few craters, implying that some process has removed the traces of most of the impacts that occurred in the past.

We finally reach Io, the source of great mystery. Seen close-up, Io is a brilliantly colorful object with a wonderfully variegated red-orange-yellow surface, marked here and there by black smudges and circular features resembling volcanic craters (Fig. 13.18). White frosty patches are seen in places. No impact craters are found, indicating that Io has a very young surface, constantly being reprocessed.

The explanation for all these fantastic features was discovered by *Voyager 1,* in a photograph taken as the spacecraft left Io behind, looking back towards it with the sunlight passing the edge of the satellite. There, illuminated by the Sun's rays, was a plume of gas streaming out of Io's interior, in a volcanic eruption (Fig. 13.19). This was the first detection of active volcanism on any body other than the Earth, and soon nearly 10 active regions were found on Io's surface (Fig. 13.20).

Apparently Io is undergoing incessant volcanic activity, and the surface is continually being coated with deposits ejected from its interior. The dominant colors are due to sulfur compounds, which can range from black through red to yellow, depending largely on temperature. Perhaps even more than Venus, Io resembles the classical picture of Hell, with its tortured landscape and sulfurous fumes.

Tidal Forces and the Volcanoes on Io

We have seen that the four Galilean satellites show some clear-cut trends in properties as we progress from one to the next, with density and surface albedo decreasing outwards while the ice content increases, and at the same time the ages of the surfaces increase steadily as we go outwards.

When all of this was discovered by the *Voyager* probes, scientists did not have to wait long for an explanation, for it had already been published. The

FIGURE 13.18 IO. *The innermost of the four Galilean satellites, Io has long been a source of mystery. On the full-disk image at left we see a number of volcano vents (dark markings) amid a surface of sulfur compounds. No impace craters are visible, indicating that this is a very young surface. At right is an image showing Io against the backdrop of Jupiter's atmosphere.* (NASA)

FIGURE 13.19 AN ERUPTION ON IO. *This image of an eruption from a volcanic vent with a plume of ejected gas dramatically illustrates Io's present state of dynamic activity.* (NASA)

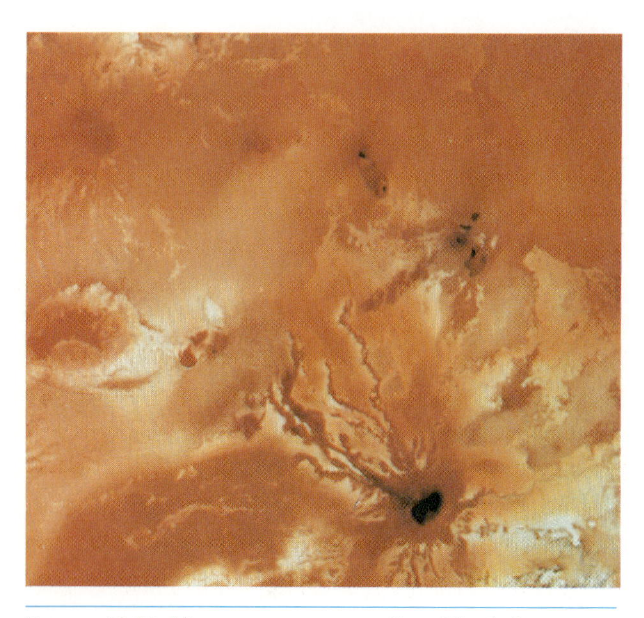

FIGURE 13.20 VOLCANIC VENTS ON IO. *The darkest material seen here indicates the locations of recently active volcanoes.* (NASA)

presence of volcanoes on Io had actually been predicted, from a consideration of the effects on the satellite of the tremendous gravitational force of Jupiter and the gravitational forces on Io due to the other Jovian satellites. As close to Jupiter as Io is, it is subject to very strong tidal forces. One result is that Io, as well as the other Galilean satellites and Amalthea (Fig. 13.21), are in synchronous rotation, always keeping the same side facing the parent planet, just as the Earth's Moon does. This fact alone does not account for the volcanic activity on Io, however, because once synchronous rotation was established, there would be no further internal stress unless other forces acted on Io as well. It happens that the next Galilean satellite, Europa, has almost exactly twice the orbital period of Io, and therefore the two are frequently lined up together, a phenomenon called **orbital resonance.** At these times, extra tidal stress is exerted on Io by Europa, and the periodic squeezing of Io that results creates internal stress and heating. This apparently keeps the interior of Io fully molten, accounting for all the volcanic activity. The heating

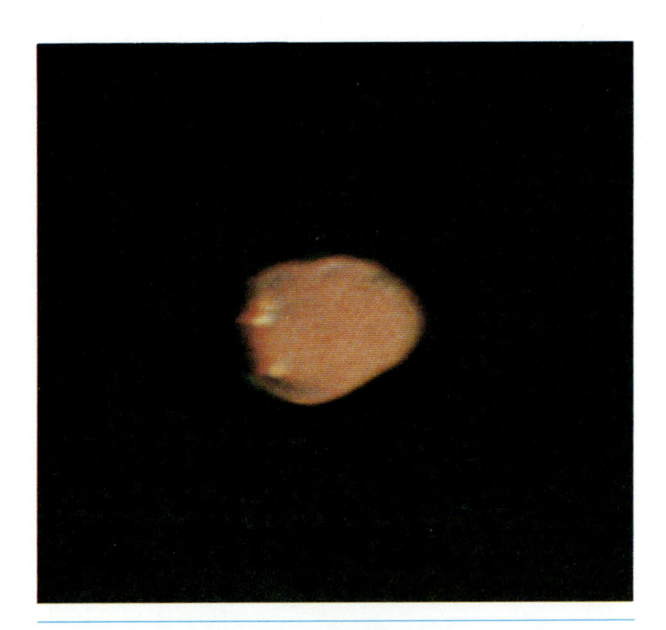

FIGURE 13.21 AMALTHEA. *Orbiting Jupiter very close to the planet, this tiny, potato-shaped moon always keeps the same end pointed toward Jupiter.* (NASA)

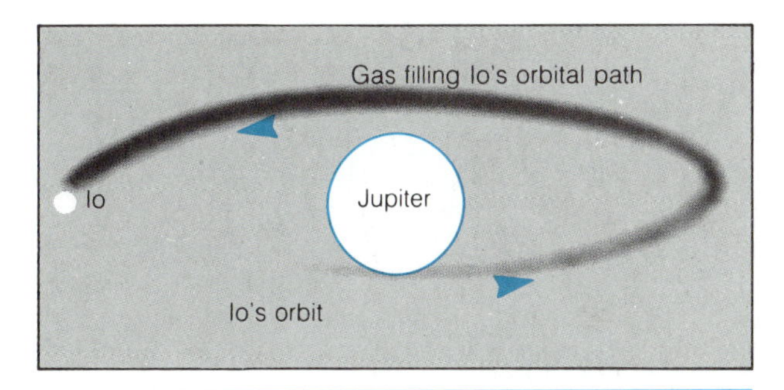

FIGURE 13.22 THE IO TORUS. *This sketch illustrates the geometry of the ring of gas that fills the orbit of Io.*

and continual eruptions also explain why Io has such a high density compared to the other Galilean satellites, since any water or other volatiles that might have been present initially have long ago been vaporized and expelled from the satellite.

Similar processes have affected the other three large satellites, but to a decreasing degree for each one in sequence going outwards from Jupiter. Thus Europa, Ganymede, and Callisto, in that order, show fewer signs of volcanic and tectonic activity, have older, less processed surfaces, and contain higher abundances of ice and volatiles.

The eruptions on Io help explain another puzzle that we have already mentioned. Io is attended by a cloud of gas that fills its orbital path in a gigantic doughnut shape called a *torus* (Fig. 13.22). The discovery of volcanoes on Io offered an immediate explanation of the source of this cloud. Recall that Jupiter rotates very rapidly, taking only about 10 hours to spin once. This is much shorter than the orbital period of Io, so that as Jupiter rotates, its magnetic field sweeps past Io at high speed. Charged particles from the volcanoes are captured by the magnetic field and are carried along with it, quickly being spread out along the full length of Io's orbital path. The particles become more highly ionized in the acceleration process and eventually become part of the Jovian radiation belts. Since Io lies in Jupiter's equatorial plane, the torus is thickened by the wobbling of the magnetic equator of Jupiter, due in turn to the 11° misalignment of the magnetic and rotation axes.

The immediate vicinity of Io is filled with charged particles, accounting for the electrical properties that modify radio emission from Jupiter as Io plows through the radiation belts.

The *Voyager* missions were not yet finished making major discoveries about the complex and fascinating Jovian system. A bright streak seen in some of the photographs turned out to be a ring around the planet, very close in, well within the orbit of Io and even inside of Amalthea's distance from Jupiter (Figs. 13.23 and 13.24). The ring is very thin and is not

FIGURE 13.23 THE JOVIAN RING. *This* Voyager 1 *image revealed the first evidence of a ring girdling Jupiter. Here a segment of the ring is seen at the right.* (NASA)

FIGURE 13.24 THE GEOMETRY OF THE JOVIAN RING. *Here the ring has been drawn in, showing its size relative to Jupiter. The ring radius is about 1.8 times the radius of the planet; it is estimated to be no more than 30 kilometers thick.* (NASA)

bright enough to be seen from Earth (although its presence was suspected from *Pioneer* observations). There are indications of some structure, consisting of a relatively bright outer rim and a diffuse disk extending inwards towards the planet's atmosphere. Its discovery made Jupiter the third of the four giant planets found to have at least one ring. Whatever the origin of planetary rings, no longer could they be viewed as special, singular features, but rather must be a natural consequence of the formation of the giant planets.

PERSPECTIVE

The planet Jupiter, with its satellites, ring, and radiation belts, is the most complex and varied member of the solar system. It is like a universe all its own, with a distinctive and unique family of celestial bodies. While it is modified by the Sun, which provides some heat energy as well as solar wind particles to deform and feed the Jovian magnetosphere, Jupiter would probably exist in very much the same condition with no help from the Sun. It is interesting to speculate whether the universe may contain many objects like Jupiter, just short of being stars in their own right.

Our journey through the realm of the giant planets now continues, with our sights set next on Saturn, another fascinating world that has recently been studied by space probes.

SUMMARY

1. Jupiter is gigantic, with 318 times the Earth's mass, but it is so large that its average density is only one-fourth that of the Earth.
2. Jupiter is totally unlike the terrestrial planets, having no solid surface, a thick atmosphere of hydrogen compounds, and rapid rotation.
3. Jupiter's atmosphere has a pattern of belts and zones, which are planet-girdling wind systems created by a combination of convection and the rapid rotation of the planet.
4. Internally, Jupiter has a small, probably solid core formed by differentiation, a thick zone of liquid me-

tallic hydrogen outside the core, an even thicker layer of liquid hydrogen above that, and a thin layer of clouds at the top.
5. Jupiter's strong magnetic field and its rapid rotation combine to create and maintain intense radiation belts.
6. Jupiter has 16 satellites, of which the four Galilean moons are by far the largest.
7. The Galilean satellites show a range of properties indicating a decreasing degree of tectonic and geological activity from the innermost (Io) to the outermost (Callisto).
8. Io, the innermost Galilean moon, has volcanic activity and is accompanied in its orbit by a cloud of gas that forms a torus around Jupiter, whose size and orientation is controlled by Jupiter's magnetic field.
9. Jupiter has a thin, dim ring.

Review Questions

Thought Questions

1. Describe the differences in overall composition between Jupiter and the terrestrial planets.
2. What is the significance of the fact that Jupiter is a source of radio emission?
3. Discuss the similarities and contrasts between the atmospheric circulation in Jupiter's atmosphere and that in the Earth's atmosphere.
4. In what ways is the circulation in the Great Red Spot unusual, even for Jupiter?
5. How is anything at all known about the interior of Jupiter?
6. What is thought to be the source of the excess heat in the interior of Jupiter?
7. Compare Jupiter's magnetic field structure and radiation belts with those of the Earth.
8. Discuss the evidence for geological activity on the four Galilean satellites.
9. Explain why Io has volcanic activity. What is the role of Europa in this explanation?
10. Describe the structure and the origin of the Io torus.

Problems

1. Sum up the masses of the other eight planets, and compare the total with the mass of Jupiter alone.
2. If the radius of Jupiter's solid core is 12,000 km and the mass of the core is fifteen Earth masses, what is the average density of the core? Compare your answer with the typical densities of the terrestrial planets and their cores.
3. Show that Kepler's third law is valid for the satellites of Jupiter, by computing P^2 and a^3 for several of them, and showing that P^2 is proportional to a^3 (that is, that the ratio of P^2 to a^3 is constant). Can you think of a system of units in which P^2 will equal a^3 for the satellites of Jupiter?
4. Io has the highest density among the Galilean satellites, largely because it has little or no ice in its interior, while the others have retained ice in varying proportions. What would the density of Io be if its volume were doubled by adding ice to it? (Ice has a density of 1 gm/cm^3.) Compare your answer with the densities of the other Galilean satellites.

Additional Readings

Allen, D. A. 1983. Infrared views of the giant planets. *Sky and Telescope* 65(2):110.
Beatty, J. K. 1979. The far-out worlds of Voyager 1. *Sky and Telescope* 57:423 (Part 1); 57:516 (Part 2).
Beatty, J. K., B. O'Leary, and A. Chaikin, eds. 1981. *The new solar system.* Cambridge, England: Cambridge University Press.
Cruikshank, D. P., and D. Morrison. 1976. The Galilean satellites of Jupiter. *Scientific American* 234(5):108.
Elliot, J., and R. Kerr 1985. How Jupiter's ring was discovered. *Mercury* 14(6):162.
Ingersoll, A. P. 1981. The meteorology of Jupiter. *Scientific American* 245(6):90.
Johnson, T. V., and L. A. Soderblom. 1983. Io. *Scientific American* 249(6):60.
Kauffman, William. 1984. Jupiter: Lord of the planets. *Mercury* 13(6):168.
Morrison, David. 1985. The enigma called Io. *Sky and Telescope* 69(3):198.
Morrison, D., and J. Samz. 1980. *Voyage to Jupiter.* NASA Special Publication SP-439. Washington, D.C.: NASA.
Soderblom, L. A. 1980. The Galilean moons of Jupiter. *Scientific American* 242(1):68.
Squyres, S. W. 1983. Ganymede and Callisto. *American Scientist* 71(1):56.
Washburn, Mark. 1983. *Distant Encounters: Voyagers 1 and 2 at Jupiter and Saturn. New York: Harcourt, Brace and Javanovich.*
Wolfe, R. E. 1975. *Jupiter. Scientific American* 233(3):118.

CHAPTER 14

SATURN AND ITS ATTENDANTS

The satellite Dione seen against Saturn's atmosphere. (NASA)

CHAPTER PREVIEW

Perhaps the most awe-inspiring sight in the solar system, one that can be appreciated with even a modest telescope, is presented by the planet Saturn (Fig. 14.1 and Table 14.1). With its banded atmosphere and delicate ring system (Fig. 14.2), this planet, more than any other celestial object, has come to represent the realm of outer space in the imagination of mankind. Saturn is unusual in almost every respect. Besides the fantastic ring structure, it has a lower overall density than any other planet, it has an assortment of mysterious moons, and one of its satellites, Titan, is one of only two nonplanetary objects in the solar system known to have an atmosphere.

For a long time, Saturn, at its distance of nearly 10 AU from the Sun, was the outermost planet known. Its long orbital period, close to 30 years, means that its synodic period is not much longer than a year, and we get annual opportunities to view it in our nighttime skies.

In this chapter we will discover Saturn's nature, both in the distant view and as it is seen from nearby, for space probes have visited it, too. Many comparisons will be made with the other giant planet we have studied, and some similarities with Jupiter will be noted, along with numerous contrasts.

TABLE 14.1 SATURN

Saturn
Orbital semimajor axis: 9.555 AU (1,427,000,000 km)
Perihelion distance: 9.023 AU
Aphelion distance: 10.086 AU
Orbital period: 29.46 years (10,759 days)
Orbital inclination: 2° 29′33″
Rotation period: $10^h\ 39^m.4$
Tilt of axis: 26° 44′
Mean diameter: 113,570 km (8.91 $D_\oplus$)
Polar diameter: 110,700 km
Equatorial diameter: 120,660 km
Mass: 5.69×10^{29} (95.12 $M_\oplus$)
Density: 0.68 g/cm^3
Surface gravity: 1.13 Earth gravities
Escape velocity: 36 km/sec
Surface temperature: 95 K (at cloud tops)
Albedo: 0.34
Satellites: 17 (five additional satellites are suspected)

FIGURE 14.1 SATURN AS SEEN FROM EARTH. *This photo was taken under good conditions, and shows some structure in the ring system.* (Link Observatory photograph)

FIGURE 14.2 A SKETCH OF SATURN AS SEEN FROM EARTH. *This drawing, done by E. E. Barnard in 1898, shows the rings tilted to the greatest possible extent.* (Yerkes Observatory)

GENERAL PROPERTIES

As in many previous cases, it was Galileo who made the first important discoveries about Saturn, using his telescope to observe it in 1610. He saw a brilliant yellowish object that appeared strongly elongated in shape. He later referred to Saturn as the planet with "ears."

Some 50 years after Galileo's observations, Huygens, with better data, was able to deduce the ringlike structure of the extensions on the planet, and thus explained the shape noted by Galileo. Extended observations showed the rings to be so thin that when viewed edge-on, they are nearly impossible to see. Saturn's rotation axis is tilted with respect to its orbital plane by 27°, and the rings, which lie in its equatorial plane, are therefore tilted through a range 27° above and 27° below the orbital plane as Saturn goes around the Sun.

Huygens also discovered Titan, the largest of Saturn's satellites, and shortly thereafter the French astronomer G. D. Cassini found four more satellites and a prominent gap in the rings (called the Cassini division to this day).

The mass of Saturn was deduced from the satellite orbits, using Kepler's third law, and this led to the realization that the planet's density is very low, only about 0.7 grams/cm^3. This is about half the density of the Sun, and little more than half that of Jupiter. The density of Saturn is actually less than that of water.

Careful study of photographs showed that the rings, as seen from the Earth, appear to be three in number, with two inner ones (one of which is brighter than the other), and a third that lies outside of the Cassini division. From observations made when the rings are viewed edge-on, it was determined that they can be no more than about 10 km thick. It was also discovered that when they are seen nearly face-on, background stars and the disk of Saturn itself are visible through them, as though they were made of sheer fabric. This supported a suggestion made in the late 1800s that the rings actually consist of innumerable small chunks of debris, each orbiting Saturn in accordance with Kepler's third law. Spectroscopic measurements, and later radar observations, used the Doppler effect to determine the speeds of the particles, and confirmed this idea by directly showing that they travel at orbital speeds consistent with Kepler's laws. The radar data also revealed that at least some of the particles are chunks of material some centimeters or meters in size.

The planet itself rotates very rapidly, with a period of about 10 hours, similar to that of Jupiter, and resulting in a similarly oblate shape. The rotation is also differential, with a shorter period at the equator than at the poles.

When spectroscopic measurements of the atmosphere were first carried out, familiar chemical compounds such as methane (CH_4), molecular hydrogen (H_2), and some ammonia (NH_3) were found. The lower temperature of Saturn compared to Jupiter caused some differences, however, in that much of the ammonia on Saturn has apparently crystallized and precipitated out of the atmosphere in the form of snow, and a few more complex molecular species are present on Saturn. No doubt the overall composition of Saturn is the same as that of Jupiter, since both formed out of the same material originally, and neither has lost any significant fraction of its gases to space. (However, Saturn has a lower helium content in its clouds, probably because some of the helium has sunk to lower levels in the planet's interior; this is discussed below.) With their large masses and low temperatures (about 130 K at the Jovian cloud tops, and 95 K in the upper atmosphere of Saturn), the combination of high escape velocity and low average particle speed has prevented any large quantity of particles from escaping.

Like Jupiter, Saturn is a source of radio emission, with a combination of different origins. Synchrotron emission signifies the presence of radiation belts, where the planetary magnetic field has captured an extensive cloud of electrons and ions, and thermal radiation shows that Saturn also emits excess energy from its interior.

Another similarity between Saturn and Jupiter is the attention each has received from the U.S. space program. Three of the four probes sent to Jupiter so far have also visited Saturn, using the gravitational pull of Jupiter to assist and guide them on their way. Only *Pioneer 10* did not do this, having taken a path through the Jovian system that was unsuited for such a maneuver.

Pioneer 11, with its charged particle monitors, radio antennae, and camera, uncovered several new details about Saturn and its attendant rings and satellites. At least one new moon was discovered, and

ASTRONOMICAL INSIGHT 14.1

THE ELUSIVE RINGS OF SATURN*

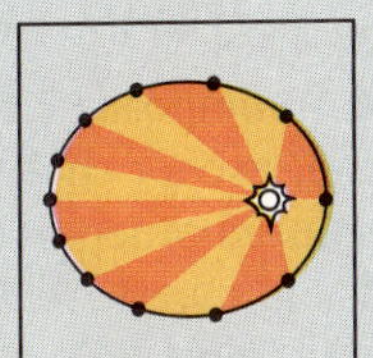

In the text we said little about the discovery of the ring system encircling Saturn, except that Galileo was perplexed by the appearance of the planet, and that, some years after his death, the problem was solved by Christian Huygens. The story is actually much more complex and interesting than is implied by that simple description. The mystery of Saturn's appearance, and the extensive efforts devoted to solving it, was the leading problem in astronomy from the 1630s to the end of the seventeenth century.

Galileo first observed Saturn through a telescope in 1610, and immediately saw that its appearance was strange. He believed that he saw a spherical planet attended on either side by smaller spherical bodies, and he concluded that these were two satellites of Saturn. He made repeated observations thereafter, but to his surprise, he saw no changes in the positions of the suspected satellites, at least not for a while. Frustrated by the lack of change, Galileo stopped observing Saturn regularly and was therefore shocked when, in 1612, he found it to be a simple sphere, with no trace of the peculiar extensions or satellites seen earlier (the rings were edge-on to the Earth at that time). Galileo resumed frequent observations, and saw the shape of Saturn change steadily. In 1616, he saw an elliptical overall shape, with dark spots near each end, and this is what gave him the impression of ears (the dark spots were the small gaps between the rings and the planet, where the blackness of space can be seen in the background).

Galileo exchanged correspondence with other observers, and soon Saturn was being watched by a number of astronomers, all of whom were puzzled by its unorthodox behavior. When the planet regained a simple spherical appearance in 1642, interest was further heightened, and by this time Saturn was the central problem in astronomical research and discussion. After 1642, many sketches of the planet were published by a number of observers, and soon a variety of suggestions were made as to the cause of its variable and strange appearance.

It is interesting that the resolution of most of the telescopes used was actually good enough to provide some detail, and the explanation of the changing appearance in terms of a ring circling Saturn quite possibly could have been developed years before it actually was put forth by Huygens. The problem was that the idea of a ring was completely alien to the usual astronomical phenomena, and nobody made the mental leap to this idea. Meanwhile, the data on the question continued to accumulate.

In early 1655, soon before the ring was to disappear again as it presented itself edge-on to the Earth-based observer, Huygens had his first look at Saturn. His telescope, contrary to popular belief, was really little better than those of many of his contemporaries, and he was able to see no more detail than they. Furthermore, he made his first observation at a time when the rings were so nearly edge-on that he had no chance of seeing the true nature of Saturn's peculiar shape. Nevertheless, on the basis of his own limited data and especially his knowledge of the appearance seen by others, Huygens deduced that all the observations could be explained by a flat ring that circled the planet in its equatorial plane (the inclination of the equator of Saturn had been deduced from the orbital motions of the satellite, Titan, which had been discovered by Huygens during his observations of 1655).

A number of alternative views were proposed at about the same time, and it took years for Huygens' solution to become generally accepted. It is interesting that it was essentially a theoretical solution, rather than something that he saw simply by looking through a good telescope, as is commonly assumed.

*Based on information from Van Helden, A. 1974, Saturn and his anses, *Journal Hist. Astronomy,* 5(2):105; and Annulo Cingitur: the solution to the problem of Saturn, *Journal Hist. Astronomy,* 5(3):155.

an additional ring was found, lying outside of those previously known or suspected. The magnetic field of Saturn was directly measured for the first time and was found to be somewhat stronger than that of the Earth, with the north and south magnetic poles upside down compared to Earth (as in the case of Jupiter), and the magnetic axis well aligned with the rotation axis, unlike both Jupiter and the Earth. The radiation belts are largely confined to the equatorial plane by the rapid rotation of Saturn, again mimicking Jupiter.

Hardly more than a year after *Pioneer 11* encountered Saturn in September, 1979, *Voyager 1* arrived, and a great wealth of additional information quickly followed *Pioneer's* discoveries. The high-resolution color cameras on *Voyager* provided new revelations as sweeping in their impact as their previous portraits of Jupiter. *Voyager 2* reached Saturn in August of 1981, following a different trajectory and viewing it from a different perspective.

As in the case of Jupiter, the *Voyager* missions have added breadth, color, and a certain amount of fantasy to our picture of a newly explored planetary system (Fig. 14.3). Now we can discuss in some detail the weather patterns on Saturn, the composition and structure of its atmosphere and interior, and the nature of its dazzling array of rings and satellites.

FIGURE 14.3 A VOYAGER PORTRAIT. *This image of Saturn was obtained from several million kilometers away, and it shows much more detail than can be seen in the best earth-based photos.* (NASA)

ATMOSPHERE AND INTERIOR: JUPITER'S LITTLE BROTHER?

To the faraway observer, the atmospheres of Jupiter and Saturn look similar in overall color and banded appearance. Although these aspects are generally similar, there are also many contrasts (Table 14.2).

As noted earlier, the chemical compositions of the two atmospheres are much alike, except that the colder temperature on Saturn has caused much of the ammonia to precipitate out and has allowed the formation of somewhat more complex molecules. An example that was detected from Earth-based spectroscopic data is ethane (C_2H_6), and other derivatives of methane are probably present.

Whereas Jupiter has a colorful and distinct contrast between belts and zones, and turbulent, swirling winds that show up with great clarity, Saturn has a muted overall appearance. The belts and zones are not so distinct, and the dark and light spots that can be seen do not contrast strongly with their surroundings.

Observations show that Saturn's cloud layer is much thicker than that of Jupiter (Fig. 14.4), although it is lacking the high haze layer that characterizes the larger planet. The lower surface gravity of Saturn allows the cloud layer to persist to much greater heights, with the result that the structure of the lower atmosphere is much more heavily obscured than on Jupiter. Hence, Saturn does not have the striking banded appearance of Jupiter, and the atmospheric flow patterns are correspondingly more difficult to discern.

Enough can be seen through the haze on Saturn to show that the atmospheric flow patterns, while similar in some ways to those on Jupiter, also display significant differences. As before, rising gas forms cyclonic flows, and descending cool gas creates anticyclones, in each case stretched by the rapid planetary rotation into elongated strips circling the planet. Like Jupiter, Saturn has many spots and oval flow patterns (Fig. 14.5).

The speed of the flow patterns on Saturn is much greater than on Jupiter. At the equator, for example, where Jupiter has wind velocities of about 100 meters/sec (360 km/hr), Saturn has winds moving as rapidly as 400 to 500 meters/sec (up to 1,800 km/hr). The belt and zone structure on Saturn extends closer

TABLE 14.2 COMPARISON OF SATURN AND JUPITER

Phenomenon	Comparison
Mass	Jupiter is 3.34 times more massive
Diameter	Jupiter is 1.2 times larger
Density	Jupiter is almost twice as dense
Atmosphere	
Structure	Saturn has deeper clouds, muting the contrast of the belts and zones
Composition	Nearly the same, except that ammonia has precipitated out of Saturn's atmosphere, and Saturn has more complex molecules, in both cases because of Saturn's lower temperature
Circulation	Wind speeds are greater on Saturn; belts and zones extend closer to poles on Saturn
Internal structure	Very similar, but Jupiter has higher internal density and more extended liquid metallic hydrogen zone
Internal heat	Both have excess internal heat, but apparently for different reasons: Jupiter's is thought to be simply left over from its formation; Saturn's is thought due to continuing differentiation as helium separates and sinks
Seasons	Jupiter has none; Saturn has significant seasons because of its almost 27° tilt
Magnetic field	Both strong and of similar origin, although Saturn's is less extensive and may form deeper in the interior
Rings	Saturn has extensive, complex system of rings; Jupiter has just one thin, dim ring

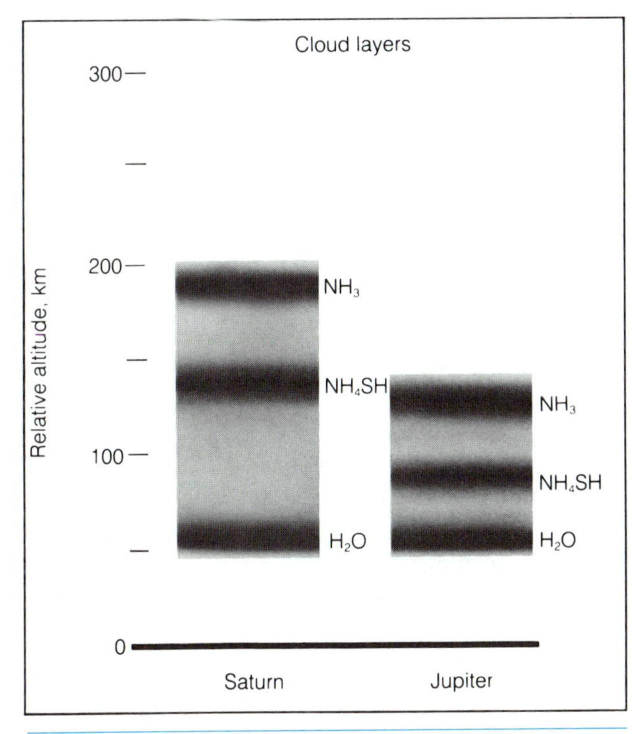

FIGURE 14.4 THE CONTRASTING CLOUD THICKNESS ON JUPITER AND SATURN. *The deeper cloud zone on Saturn is largely responsible for the relative lack of contrast in its visible surface features.* (Data from Beatty, J. K., B. O'Leary, and A. Chaikin (eds.) 1981, The New Solar System (Cambridge, England: Cambridge University Press)

FIGURE 14.5 CIRCULATION IN THE ATMOSPHERE OF SATURN. *Even though the contrast is not as vivid on Saturn as on Jupiter, the ringed planet does have complex atmospheric motions, driven by a combination of convection and rapid planetary rotation.* (NASA)

to the poles than on Jupiter (Fig. 14.6), while the light and dark strips are broader in the north and narrower in the south. These contrasts between Saturn and Jupiter may be caused by the fact that Saturn, with its axial tilt of nearly 27°, has seasons and therefore variations in the Sun's intensity, whereas Jupiter, whose axis is nearly perpendicular to the ecliptic (its tilt is only a little over 1°), does not. To directly observe the effects of Saturn's seasons will take a long time, in view of its nearly 30-year orbital period.

FIGURE 14.6 BANDS IN THE ATMOSPHERE OF SATURN. *These are* Voyager 1 (upper) *and* Voyager 2 (lower) *images, showing atmospheric bands. Comparison of the images shows some differences in contrast, primarily because the viewing angles were different.* (NASA)

There are contrasts in the internal structures of the two planets as well. First and foremost, the density inside Saturn is much lower, because its much smaller mass is spread out over nearly the same volume as Jupiter. The solid core inside Saturn is probably not as dense as the core of Jupiter and may contain some ice (Fig. 14.7). Outside this core, the internal pressure, while not so great as inside Jupiter, is thought to be sufficient to form liquid metallic hydrogen, which may inhabit a zone extending about halfway out to the surface. Above this is thought to be a thick layer of liquid hydrogen and helium, topped off by the layer of clouds that is visible from afar.

Like Jupiter, Saturn also radiates substantially more energy than it receives from the sun, by a factor of 1.78 in this case. The explanation offered for Jupiter does not work as well here; Saturn, with its lower mass, could not have retained enough primordial heat from its formation to still be as warm internally as the excess radio emission implies. Some additional source of heat must be present, and its nature is not clear. One suggestion is that differentiation is still going on in the metallic and liquid hydrogen zones, as the heavier helium atoms gradually sink towards the center of the planet, releasing

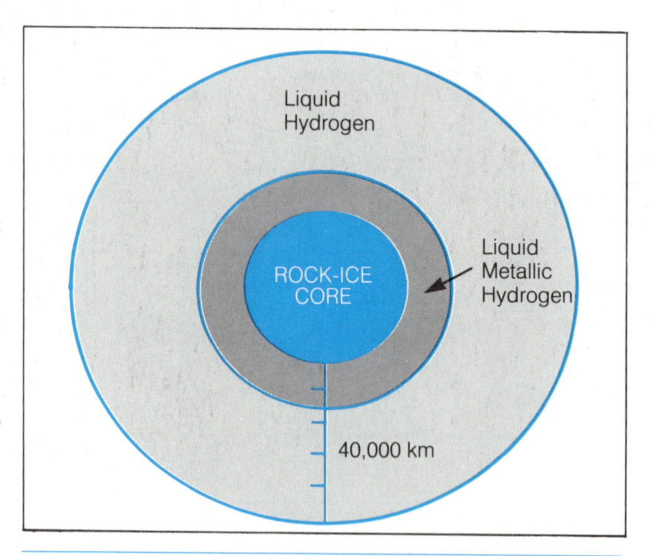

FIGURE 14.7 THE INTERNAL STRUCTURE OF SATURN. *This cross-sectional view should be compared with the structure of Jupiter shown in Fig. 13.11.*

gravitational energy as they do so. The likelihood that this process is really at work inside Saturn is enhanced by the observation that Saturn has relatively less helium in its outer layers than does Jupiter, consistent with the idea that Saturn's helium has sunk, while Jupiter's has not. The probable reason that this has not occurred inside Jupiter is that the interior of the larger planet is warmer and more turbulent, preventing the helium from quietly sinking towards the center.

The magnetic field of Saturn is, no doubt, formed by a dynamo in the interior. The structure of Saturn's magnetic field suggests that the depth at which the field forms is greater than in the case of Jupiter. The zone of trapped ions and electrons around Saturn is still quite large, and most of the satellites and all of the rings are within this zone, sweeping up charged particles as they move along their orbits. Because of variations in the flow of ions from the Sun, the magnetosphere of Saturn fluctuates in size, with the result that Titan, the largest satellite, is sometimes within the magnetosphere and sometimes outside of it.

Filling the orbit of Titan, and extending quite a bit farther inward and outward, is a cloud of hydrogen atoms resembling in shape the Io torus that girdles Jupiter. In this case, however, the gas is not ionized, so its structure is not controlled by the planetary magnetic field.

THE SATELLITES OF SATURN

Before the *Pioneer* and *Voyager* encounters, Saturn was thought to have at least 10 satellites, some of them small and rather difficult to see from Earth. (An eleventh satellite was suspected, having been reported in 1978, but was not considered confirmed until *Pioneer 11* also detected it.) The first of Saturn's moons to be discovered was Titan, noted by Huygens in 1655. Four more were found by Cassini in the latter part of the seventeenth century, and in the next 300 years an additional five were seen. The innermost was thought for a time to be the satellite Janus, first reported in 1966, but the suspected eleventh satellite, simply called S-11, was actually closer in orbiting very near the outer portion of the rings visible from Earth.

The *Pioneer 11* and *Voyager 1* and *2* encounters have clarified the situation quite a bit. *Pioneer 11* confirmed the presence of S-11 (now called Epimetheus; Fig. 14.8). In addition, four new satellites were discovered by the *Voyager* missions and two from Earth-based observations, so the current total is sev-

FIGURE 14.8 SOME OF THE SMALL SATELLITES OF SATURN. *Taken at various ranges, these images do not reflect the true relative sizes. The upper middle and lower left moons are the co-orbital satellites (Epimetheus and Janus), and the two at lower right (Prometheus and Pandora) are the pair that "shepherd" the F ring (discussed later in this chapter). The upper left satellite (Helene) is the one that shares the orbit of Dione, and at upper right and lower left center are the pair (Telesto and Calypso) that share the orbit of Tethys.* (NASA)

enteen, with three more suspected on the basis of *Voyager* data (Table 14.3). The confounding former eleventh satellite turned out to be sharing the same orbit with another satellite (Janus), something previously never seen. Both of these moons are elongated in shape and orbit with their long axes pointed towards Saturn, in synchronous rotation. Two of the new satellites, also very small, were found orbiting in the same path as Dione, one of the larger satellites. These moons, designated Helene and No Name II (because its existence is only suspected), are held in position about 60° ahead of Dione in its orbit by the combined gravitational forces of Saturn and Dione. Two small satellites, Telesto and Calypso, share an orbit with Tethys, another of the relatively large moons.

Another pair of small satellites, Prometheus and Pandora, were found to be orbiting at nearly the same distance from Saturn, with a thin ring, discovered by *Pioneer 11,* between them. As we shall see in the next section, these satellites probably are helping to maintain this ring by their gravitational forces.

The innermost of the new satellites discovered by the *Voyager* probes, named Atlas, actually orbits well inside the three rings that lie farthest out, although it is still just outside the broad, bright ring that is the outermost one visible from Earth. A small moon (No Name I) is suspected to exist between the larger satellites Tethys and Dione, and there may be two more (No Name II and No Name III) between Dione and Rhea. Very recently, radio observations have led to the possible detection of two more satellites, so that the total number may be as high as 22.

The major satellites, those easily observed from Earth, turned out to be as varied in their individual qualities as the Galilean satellites of Jupiter. The biggest of them all, Titan (Fig. 14.9), was especially important for *Voyager 1* to examine closely, since it was already known to have an atmosphere, making it nearly unique in that respect among all the satellites of the solar system. Recall that, until *Voyager 1* revealed the great depth of Titan's atmosphere, this satellite was thought to be larger than Ganymede, whereas in fact

TABLE 14.3 SATELLITES OF SATURN

No.	Name	Distance(R_S)*	Period	Diameter (km)	Mass (g)	Density (g/cm^3)†	Albedo
15	Atlas	2.282	$0^d.602$	40 × 20			0.9
16	Prometheus	2.311	0.613	140 × 100 × 80			0.6
17	Pandora	2.349	0.629	110 × 90 × 80			0.9
11	Epimetheus	2.510	0.694	140 × 120 × 100			0.8
10	Janus	2.510	0.695	220 × 200 × 160			0.8
1	Mimas	3.075	0.942	392	4.5×10^{22}	1.43	0.5
2	Enceladus	3.947	1.370	500	7.4×10^{22}	1.13	1.0
13	Telesto	4.885	1.888	34 × 28 × 26			0.5
14	Calypso	4.885	1.888	34 × 22 × 22			0.6
3	Tethys	4.885	1.888	1,060	7.4×10^{23}	1.19	0.9
	No Name I	5.47:	?	15–20			
4	Dione	6.256	2.737	1,120	1.05×10^{24}	1.43	0.7
12	Helene	6.267	2.737	36 × 32 × 30			0.7
	No Name II	6.30:	?	15–20			
	No Name III	7.79:	?	15–20			
5	Rhea	8.737	4.518	1,530	2.50×10^{24}	1.33	0.7
6	Titan	20.26	15.945	5,150	1.35×10^{26}	1.89	0.21
7	Hyperion	24.55	21.277	410 × 260 × 220	1.71×10^{22}		0.3
8	Iapetus	59.03	79.330	1,460	1.88×10^{24}	1.15	0.5/0.05
9	Phoebe	214.7	550.5	220			0.06

*The symbol R_S stands for the equatorial radius of Saturn, taken here to be equal to 60,330 km. A colon (:) following a value for distance indicates an uncertain value.

†No density value is given when the diameter is uncertain or the satellite shape is irregular.

Figure 14.9 Titan. *The largest of Saturn's moons. Titan is almost unique among all the satellites in the solar system in having an atmosphere.* (NASA)

Ganymede has the honor of being the largest satellite in the solar system.

It was hoped that the vertical structure of Titan's atmosphere (Fig. 14.10) could be probed and the surface below examined, but it turned out that the atmosphere is so thick that it is impossible to see to any great depth. In addition, there is an obscuring haze layer at the top of the atmosphere that makes direct observations more difficult. Nevertheless, radio and infrared sensors could penetrate deeply, providing data which, along with spectroscopy of the upper layers, were sufficient to allow some rather detailed estimates of the surface conditions to be made.

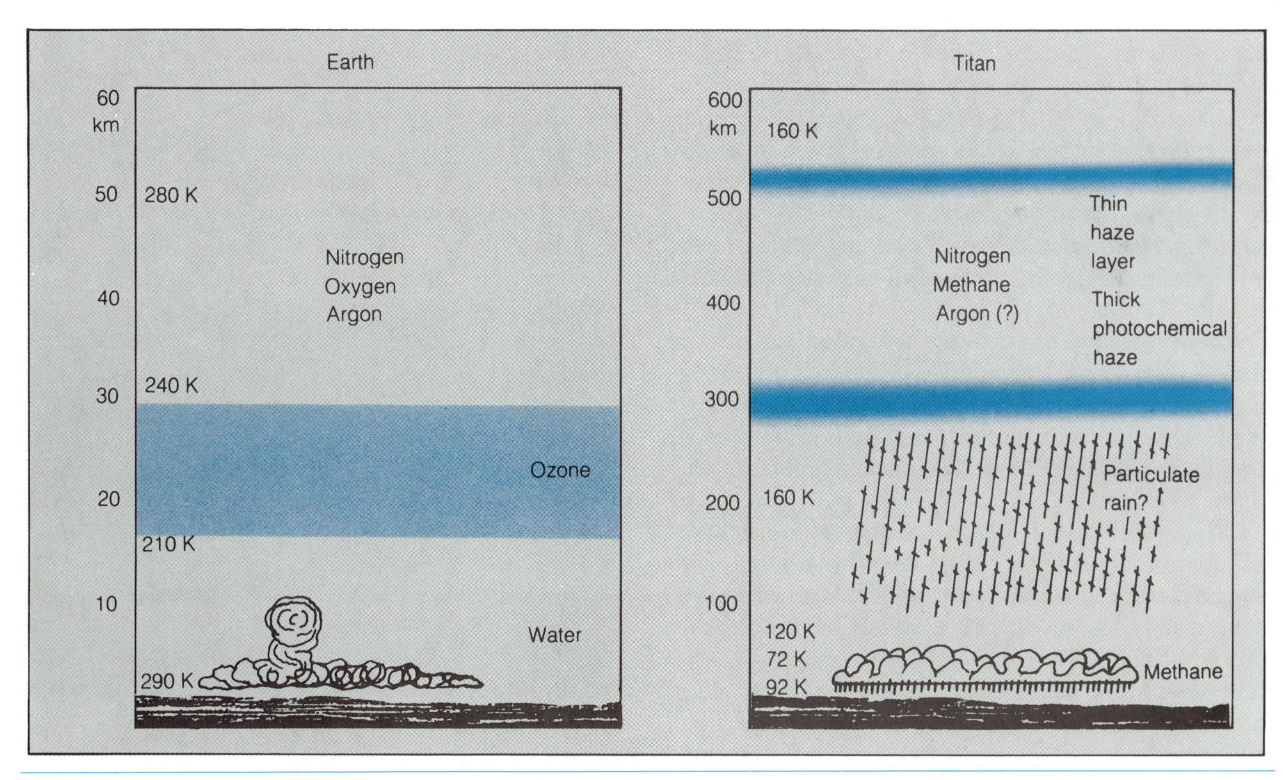

Figure 14.10 Titan's atmosphere. *This diagram compares conditions in the atmospheres of Titan and the Earth. Note that the vertical scales are not the same; Titan's atmosphere is much deeper.* (Data from Beatty, J. K., B. O'Leary, and A. Chaikin (eds.) 1981, *The New Solar System* [Cambridge, England: Cambridge, University Press])

First, regarding the solid body of Titan itself, it was ascertained that its density is just under 2 grams/cm^3, corresponding to an internal composition of about half rock and half ice. This is very similar to Ganymede, the largest Jovian satellite.

We can only speculate as to what the surface of Titan might be like. There is methane in the atmosphere, and radio data showed that the surface temperature is about 90 K, very near the value where methane can exist in any of the three forms: liquid, solid, or vapor. Thus, there may be liquid methane on the surface, along with methane ice formations such as glaciers, as well as gaseous methane in the atmosphere. In short, methane on Titan may play very much the same role as does water on the Earth. If it does, there is probably substantial erosion, so that old impact craters and tectonic formations, if they exist, are continually being smoothed over.

Besides Earth, only Mars, Venus, and now Titan, are known to have any significant abundance of nitrogen in their atmospheres. The origin of the nitrogen on Titan may be volcanic, since it is a volatile gas that can be emitted from certain kinds of rocks when they are heated. Therefore, there may also be volcanic features on the surface of this satellite.

The atmospheric pressure is about 50 percent greater at the surface of Titan than at sea level on Earth, and Titan's atmosphere extends to five times the height of Earth's atmosphere above the surface. It is not well understood why Titan should have been able to retain such a thick atmosphere, when no other satellite (except Triton) has any atmosphere at all. Certainly the low temperature and large mass have helped, but in these regards Titan does not appear to have any great advantage over Ganymede, for example. Perhaps the atmosphere of Titan is being continually replenished by some ongoing process, such as volcanic activity.

The satellites of Saturn are generally classified into four groups: the tiny inner moons that orbit near and between the rings; Titan, which is in a class by itself; distant Phoebe, a weird misfit of a satellite that also is without peer; and the seven remaining satellites, which are intermediate in size and similar in many respects. We have already discussed the first two of these categories, and we turn our attention now to the intermediate-sized moons.

These seven, named Mimas, Enceladus, Tethys, Dione, Rhea, Iapetus, and Hyperion, are actually quite varied in some respects. They have diameters ranging from about 400 to 1,500 km (whereas Titan's diameter is 5,150 km). Masses for most of them were determined from their gravitational influences on the trajectory of the *Pioneer* and *Voyager* spacecraft, and from these measurements the densities were estimated. In all cases the densities are rather low, indicating a high ice content, and a few of these moons may be nearly pure ice, with densities near 1 gram/cm^3.

Among the intermediate-sized satellites, a variety of surface features were seen in the *Voyager* photographs, including heavy cratering in many cases, and also including some peculiar structures. On Dione (Fig. 14.11) were found sinuous, branching valley systems and white, wispy patterns that look deceptively like the high-altitude cirrus clouds of Earth. These may be frozen gases that crystallized immediately upon being emitted from surface rocks or volcanoes, falling to the ground in cloudlike arrangements. Linear trenches were found on a few of the satellites, including Dione, Tethys (Fig. 14.12), and Rhea (Fig. 14.13), possibly the result of fractures caused by massive impacts. The innermost of this group of satellites, Mimas (Fig. 14.14), is marred by a tremendous impact crater that is fully one-fourth as large as the diameter of the satellite itself.

Mysteries abound. Iapetus (Fig. 14.15), one of the larger members of this group, has a bright and a dark side, the one some ten times more reflective than

FIGURE 14.11 DIONE. *This intermediate-sized moon of Saturn has wispy, light-colored features on its surface that may be icy deposits of material that escaped from the interior and crystallized. This side of Dione, which is in synchronous rotation, always trails as the satellite orbits Saturn; the leading side is more heavily cratered.* (NASA)

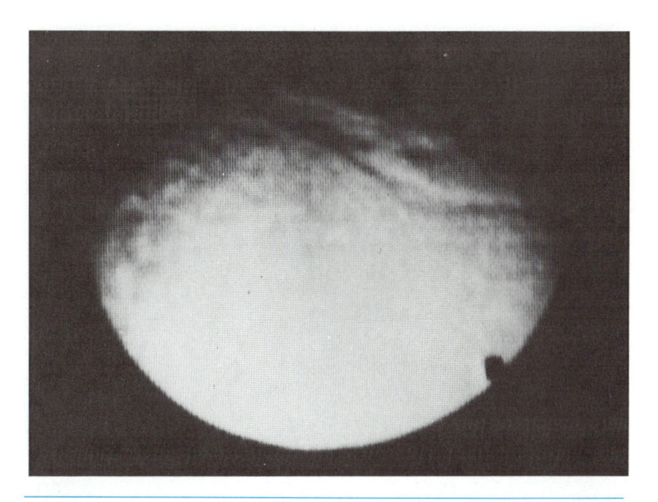

FIGURE 14.12 TETHYS. *Another of the intermediate-sized moons, Tethys is characterized by large linear trenches, an example of which is seen here* (NASA).

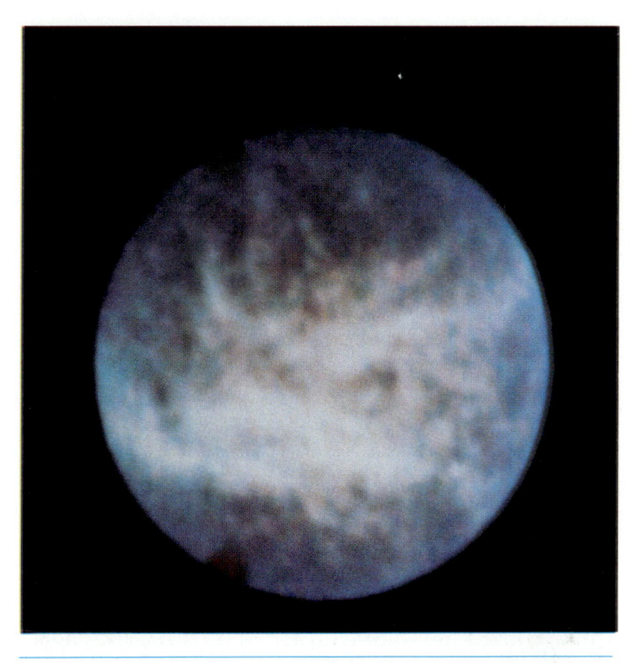

FIGURE 14.13 RHEA. *One of the seven intermediate-sized moons, Rhea shows evidence of cratering and light-colored areas that are probably composed of ice.* (NASA)

the other. *Voyager 2* images indicate that the dark side of Iapetus may be covered by some sort of deposit, perhaps resembling soot. Enceladus (Fig. 14.16) is the shiniest object in the solar system, with an albedo

FIGURE 14.14 MIMAS. *This satellite shows many impact craters, including one very large one that has about one-fourth the diameter of the satellite itself. This moon is probably composed chiefly of ice, as are several of the other satellites of Saturn.* (NASA)

FIGURE 14.15 IAPETUS. *This satellite has very unusual bright and dark areas. It appears that the dark material may be some sort of deposit.* (NASA)

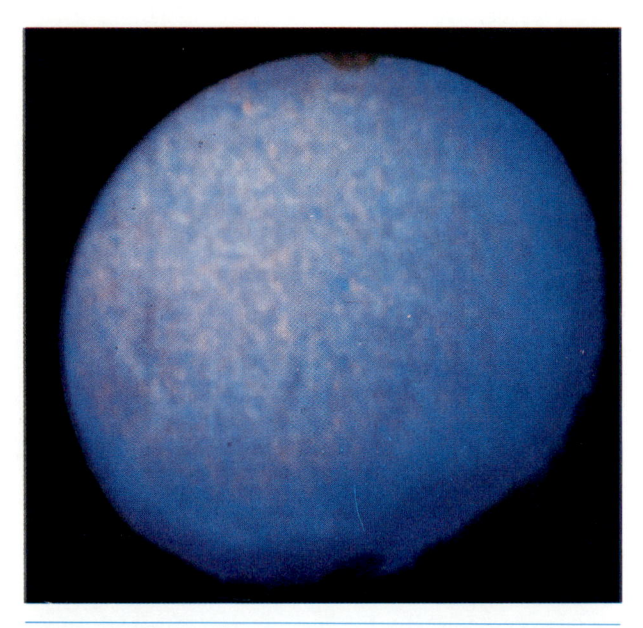

FIGURE 14.16 ENCELADUS. *This is the shiniest object in the solar system, with an albedo of approximately 1, just like a mirror. The surface may have been coated with deposits from internal melting and outgassing.* (NASA)

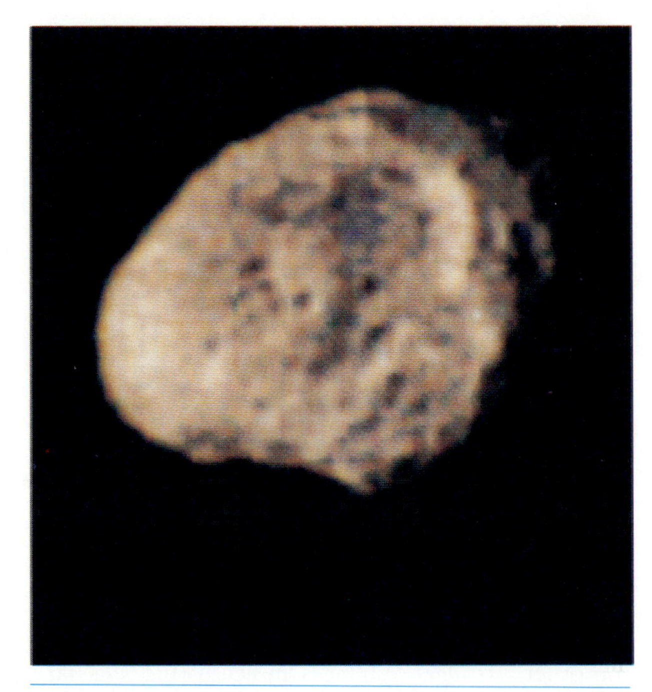

FIGURE 14.17 HYPERION. *This satellite has a very unusual shape and is marked by several impact craters.* (NASA)

near 1, meaning that it reflects just about all the light that hits it. The orbit of this satellite is coupled to that of Dione, with Enceladus orbiting Saturn once for every two orbits of Dione, so that the two are frequently lined up with Saturn. This may create sufficient tidal friction inside Enceladus to make it volcanically active, much like Io. In this case, the eruptions may consist of ejections of water vapor, so that the surface of Enceladus may be continually recoated by ice. Another hypothesis is that Enceladus is being continuously coated by ice particles from the tenuous E ring, within which its orbit lies. The satellite Hyperion (Fig. 14.17) was found to be strangely asymmetric in shape despite its moderately large size, perhaps indicating that it was never molten.

Perhaps the strangest of all the satellites of Saturn is Phoebe, nearly four times farther out than all the rest, and orbiting in the retrograde direction, with the plane of the orbit tilted nearly perpendicular to the others. This oddball among oddballs, which has a dark surface, may be a stray asteroid that was captured by Saturn's gravitational field and forced into orbit.

THE RINGS

We turn our attention now to the stunning ring system of Saturn, probably the most spectacular sight of all (Fig. 14.18). The *Voyager* data proved these structures to be even more complex and fascinating than had been imagined.

The rings consist of countless individual particles, ranging in diameter from a fraction of a centimeter in some portions of the system to a few meters in others. Each particle has its own orbit, with its period determined by its distance from Saturn according to Kepler's third law. Thus, the rings do not rotate like rigid hoops around the planet, in which case all the particles would have the same period (meaning that the outer ones would have to travel fastest). Instead, they are fluid structures, moving slowest in the outermost portions.

The three distinct rings visible from the Earth had been labeled the A, B, and C rings, in order from the outer one inwards (Table 14.4). The possibility of a faint ring inside the C ring was raised some time ago, and this D ring was actually found, as were new

E, F, and G rings outside of the A ring (the letters are assigned in the order of discovery).

Even before the full complexity of the ring system was known, some idea of how the rings formed had been developed. The rings are all very close to Saturn, and this led to the suspicion that tidal forces may have had something to do with their formation. There is a certain distance from any massive body, called the **Roche limit,** inside which the tidal forces pulling an object of a given size apart are greater than the gravitational force holding it together. The exact location of the Roche limit depends on the mass and size of the orbiting object. We have discussed the fact that the Moon has a tidal bulge on it due to the Earth's gravitational pull; to envision what the Roche limit means, imagine that the bulge becomes so severe that the Moon is actually pulled apart and broken into pieces.

The rings of Saturn are all within the Roche limit for a satellite of any significant size. Therefore, it may be that there once was a satellite that wandered too close to Saturn and was broken apart, leaving behind the debris that now forms the rings. More likely, the rings are made of particles that have been orbiting Saturn since its formation but that were so close to it that tidal forces prevented their ever merging together to form a satellite.

Another possibility, recently proposed but gaining some acceptance, is that the rings consist of debris from former moons that were destroyed by major impacts with large meteoroids (see Chapter 16). Such impacts are more probable near a massive planet than elsewhere, because the gravitational field of the planet attracts by-passing objects. Furthermore, there is evidence that the rings of Saturn (and of Uranus) are not permanent features, but may be rather young.

Regardless of the origin of the rings, the many mechanisms that govern their structure are the same. Most of these mechanisms involve gravitational effects of Saturn's many satellites.

Gravitational forces are responsible for the major gap in the ring system of Saturn, although the details are not entirely clear. The Cassini division, between the A and B rings, lies at just the right distance from Saturn that any particle orbiting there would be in orbital resonance with Mimas, the innermost of the larger satellites. By this we mean that the orbital period of such a particle would be precisely half of the period of Mimas, so that every two trips around, the particle and Mimas would line up on the same side of Saturn. These frequent alignments subject particles in this region to extra-strong gravitational forces. The result is that particles are cleared out of this region by the small gravitational tugs exerted on them at each alignment with Mimas. Calculations of this process are not entirely satisfactory, however, because it appears to take too long to clear out the

FIGURE 14.18 THE RINGS OF SATURN. *This* Voyager *image reveals some of the fantastically complex structure of the ring system.* (NASA)

TABLE 14.4 THE RINGS OF SATURN

Feature	Distance (R_S)*
D ring inner edge	1.11
C ring inner edge	1.233
B ring inner edge	1.524
B ring outer edge	1.946
A ring inner edge	2.021
A ring gap center	2.212
A ring outer edge	2.265
F ring center	2.326
G ring center	2.8
E ring inner edge	3:
E ring outer edge	8:

*R_S stands for the equatorial radius of Saturn, taken here to be equal to 60,330 km.50 colon (:) following a value for distance indicates an uncertain value.

observed gaps. It is possible that collisions between particles are enhanced by the alignments, so that the particles are deflected more efficiently than would be the case without collisions. There is little doubt that orbital resonance with Mimas is responsible for the gap, despite the present uncertainty in some of the details. Other satellites, or other fractions of the orbital period of Mimas (such as the location where an orbiting particle would have exactly one-third the period of Mimas, for example) could create other gaps in the rings, and indeed such gaps have been found. Orbital resonances with satellites of Saturn led to the expectation that the rings might have complex structure, even before *Pioneer* and *Voyager* data were available.

Despite this expectation, when the first *Voyager* photos of the rings were obtained, scientists were amazed by what they saw. None of the rings had a smooth appearance when viewed from nearby, but instead the prominent ones visible to us from Earth are each composed of a number of thin rings that came to be called "ringlets" (see Fig. 14.18). The photopolarimeter experiment on *Voyager 2* revealed much more detail in the rings, showing structures as small as 100 meters in size. There are thousands of ringlets in the entire system, some appearing dark and others light. Even Cassini's division is not completely empty, but instead is inhabited by several dark ringlets. There apparently is a narrow gap as well that truly is empty, probably created by Mimas in the manner just described.

Whether a ring will appear dark or light depends on how the particles in it reflect sunlight, and the angle from which it is viewed. Some of the rings tend to reflect light straight back, while others allow it to go through in a forward direction. A ring that reflects sunlight back will appear bright when viewed from the Earth, but will look dark when viewed from the far side, looking towards the Sun. Conversely, a ring that scatters light in the forward direction will appear dark from the Earth (as do the narrow ringlets in Cassini's division) and bright if viewed from behind. Whether a given ring will scatter light forwards or backwards depends on the size of particles it is made of, and also on their shape and composition.

From considerations of their light-scattering properties, it has been possible to deduce the average sizes of the particles in the various rings of Saturn. The outermost ones tend to be made of very tiny particles, a few ten-thousandths of a centimeter in diameter, while the inner rings, in contrast, seem to consist mostly of chunks of material a few meters across. The composition of all the particles is probably mostly ice, and their shapes are irregular. The total mass of all the ring material together is very small, only about a millionth of the mass of our Moon, so if the ring particles could have formed into a satellite of Saturn, it would have been a rather small one.

The complex pattern of rings and ringlets seen by the *Voyager* cameras required a more sophisticated explanation than the orbital resonance hypothesis described above, although that process is undoubtedly at work. Apparently at least three other mechanisms exist for shaping the rings of Saturn.

One of these is also caused by gravitational effects of Saturn's moons. The clue to this was presented by the orbit of the two small satellites, discovered by *Voyager 1,* which orbit just inside and just outside the F ring (Fig. 14.19). The gravitational interaction between the ring particles and these satellites acts to keep the particles confined to orbits between them (Fig. 14.20). The moon just outside the ring, according to Kepler's laws, moves a little more slowly than the ring particles. Whenever a ring

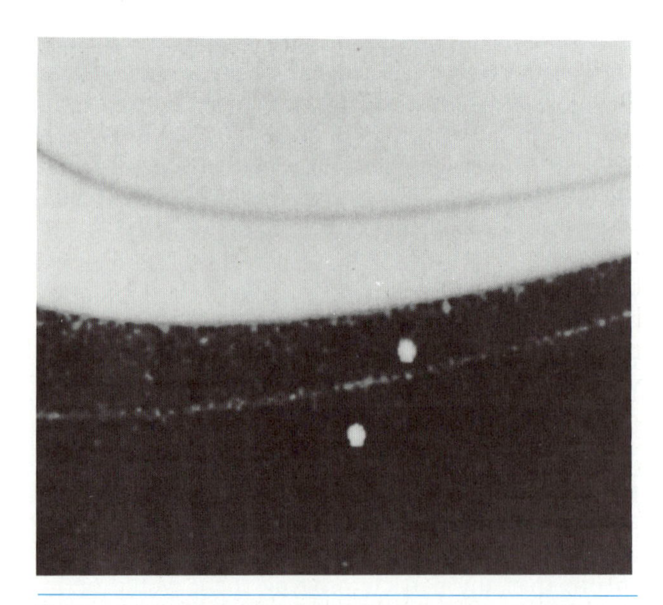

FIGURE 14.19 THE SHEPHERD SATELLITES AND THE F RING. *The two small satellites (Pandora and Prometheus) that maintain the F ring through gravitational interaction are both visible in this image.* (NASA)

particle overtakes this satellite at a close distance, the particle loses energy due to the gravitational tug of the satellite, and drops to a slightly lower orbit. On the other hand, the satellite on the inside of the ring slowly overtakes the ring particles, and when it passes close by one, the particle gains energy and moves out to a higher orbit. In this way the two satellites ensure that the ring particles stay in orbit between them, because any particle that strays too close to either one will have its orbit altered in just such a way that it moves back into the ring. For this reason these two satellites are called "shepherd" satellites, because they behave like sheep dogs, keeping their flock in line by nipping at their heels.

Voyager 2 data showed that another mechanism, suspected before *Voyager* missions, was also influencing the structure of the rings. Mathematical calculations show that, in any flattened disk system such as Saturn's rings, small gravitational forces can create a spiral wave pattern, something like the grooves on a phonograph record. The waves take the form of alternating dense and rarefied regions, and the entire spiral pattern rotates about the center of the system. Similar spiral density waves have long been thought responsible for the spiral arm structure of galaxies like our own, and so the *Voyager* data have shown that the rings of Saturn can be considered a small-scale model of a spiral galaxy.

It was predicted from calculations that the spacing between spiral density waves should increase steadily with distance from the center. The photopolarimeter data from *Voyager 2* (Fig. 14.21) revealed that the expected increase in spacing between ringlets occurs in several portions of the ring system, confirming that at least some of the ring structure is created by spiral density waves. The ringlets that are formed by this mechanism are not separate, circular structures, but instead are part of a continuous, tightly wound spiral pattern that slowly rotates. Individual ring particles, by contrast, follow elliptical orbits, and alternately pass through the ringlets and the spaces between. (It is important here to distinguish between the motion of the wave and the motions of the particles, which do not move with it; a reasonable analogy would be a cork bobbing on water waves, but not moving along with them). The reason there are more particles in the ringlets than between them at any given moment is that they move more slowly as they pass through the density waves, and tend to

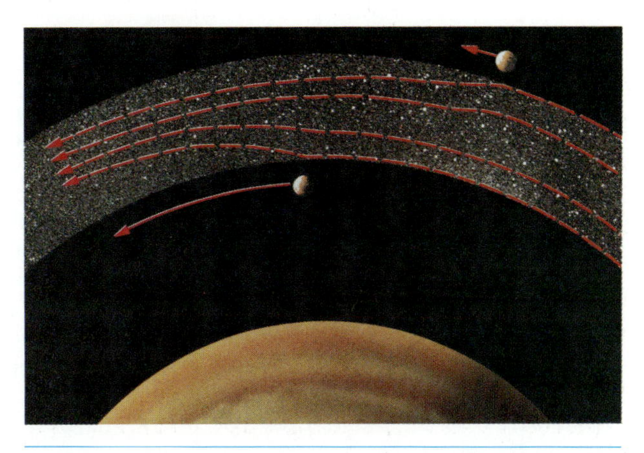

FIGURE 14.20 THE ACTION OF SHEPHERD SATELLITES. *This illustrates how a pair of small moons can keep particles trapped in orbit between them. The inner shepherd satellite overtakes the particles, accelerating them if they fall inward, thus moving them out. Similarly, the outer shepherd slows particles that move too far out, so that they fall back in.* (NASA/JPL)

FIGURE 14.21 VERY FINE STRUCTURE IN THE RINGS. *This is a synthetic image of a small section of the ring system, reconstructed from stellar occultation data obtained by the photopolarimeter experiment on* Voyager 2. *The finest details visible here are only about 100 meters in size, whereas the best* Voyager *camera images have a resolution of about 10 kilometers. Data like these have established that spiral density waves play a role in shaping the rings.* (NASA)

ASTRONOMICAL INSIGHT 14.2

RESOLVING THE RINGS

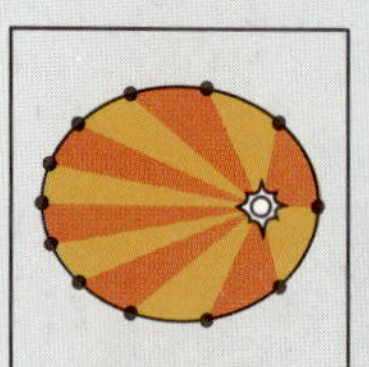

The *Voyager* missions to Saturn have produced a wealth of new information on the rings, primarily because of the high resolution of the images that they have sent back to Earth. Recall, from our discussions in Chapter 7, that resolution is the ability to make out details. In our past discussion, we emphasized that resolution can be enhanced by the use of large telescopes, by placing observatories above the Earth's turbulent atmosphere, or by the use of special interferometry techniques. There is another way to achieve high resolution, however: getting close to the subject of the observations. This is what the *Voyager* spacecraft did.

The cameras aboard *Voyagers 1* and 2 yielded images of Saturn's rings with a maximum resolution of about 10 km; that is, details as small as 10 km in size could be seen. This was quite an improvement over the best photographs taken from Earth, which have a resolution about 4,000 km. The *Space Telescope,* when it is placed in Earth orbit in the late 1980s, will resolve details in the ring system about ten times smaller than this, roughly 400 km, still far inferior to *Voyager* images.

There was an instrument on the *Voyager* spacecraft that provided much better resolution than even the cameras. This was the photopolarimeter, a device for measuring the intensity of light, as well as its polarization. The photopolarimeter on *Voyager 1* failed before that spacecraft reached Saturn, but the one on *Voyager* 2 was able to carry out an observation that has provided by far the most detailed information yet on the ring system.

The observation was rather simple: As *Voyager* 2 approached Saturn, the photopolarimeter measured the brightness of a star (Delta Scorpii) that happened to lie behind the rings, from the point of view of the spacecraft. As *Voyager 2* swept by the planet with the photopolarimeter trained on the star, the spacecraft motion caused the ring system to move across in front of the star. As it did so, the ringlets and gaps caused the brightness of the star to fluctuate. When a dense ringlet passed in front of the star, its brightness temporarily dimmed; when a gap passed in front, the star appeared brighter. This kind of observation is known as a stellar occultation.

The photopolarimeter was able to record the changes in the star's brightness very rapidly. This, combined with the rate at which the rings appeared to move across in front of the star, determined how fine the details were that could be detected in the ring structure. As it turned out, the resolution of this experiment was about 100 meters, a factor of 100 better than the best images obtained by the *Voyager* cameras! This meant that ringlets or gaps as small as the length of a football field could be detected.

Of course, the photopolarimeter could only measure the ring structure along a single cross section; it could not provide a full picture of the entire system. Its image of the rings was one dimensional, but it nevertheless led to profound new understanding, particularly of the spiral density waves described in the text.

The data produced by the photopolarimeter were originally in the form of plots showing the variations in the star's intensity as a function of time (see Fig. 14.21). Dips in the intensity corresponded to the passage of ringlets in front of the star. This information was then translated into a plot of ring density as a function of distance from Saturn. The data were processed a step further so that things were easier to visualize; synthetic images of the rings, simulating those taken by a camera, were produced (it was assumed that the rings were circular, and different colors were used to represent varying degrees of ring density; see Fig. 14.21). In this way, false-color images of the rings were produced, somewhat similar to the actual photographs obtained by the cameras, but showing details a factor of 100 smaller.

The *Voyager* 2 photopolarimeter was able to use the same stellar occulation technique to analyze fine structure in the rings of Uranus, during the encounter in January, 1986. The results are discussed in the next chapter.

congregate there, just like cars in a traffic jam that build up at a point where the flow is slowed down.

The gravitational disturbances that create the density waves in portions of Saturn's ring system are caused by satellites. The strongest disturbances occur at resonance points where the particles have orbital periods that are a simple fraction of the periods of satellites orbiting at greater distances. These are also the locations of gaps in the rings in some cases, as explained earlier, so we sometimes find that a gap lies just at the inner edge of a series of ringlets that are created and maintained by a spiral density wave.

Some of the rings have asymmetric shapes (Figs. 14.22 and 14.23) that cannot be maintained by any simple gravitational forces due to Saturn or its moons. For example, one of the inner ringlets is not circular, but is instead flattened into a slightly oval shape (and does not seem to be part of a spiral density wave) and the F ring, the one that is held in place by the pair of shepherd satellites, has a very peculiar braided structure, as though three rings had been interwoven. No good explanation has been found for either kind of behavior. Comparison of *Voyager 1* and *Voyager 2* photographs showed that the irregular structure of the F ring varies with time.

The fact that noncircular and irregular rings are not stable shapes has contributed to the idea, already mentioned, that the rings may not be permanent features. It is known that major impacts on moons can and do occur (recall the huge crater on Mimas), therefore it is not unreasonable to suppose that occasionally an impact is forceful enough to shatter a moon, creating a new ring or rings from the debris. Thus, the rings of Saturn and the other giant planets may be young features, rather than having been present since the beginning of the solar system.

FIGURE 14.22 ASYMMETRIES IN THE RINGS. *This composite shows that the rings are not perfectly circular; here the thin ring within the dark gap is thinner on one side of the planet than on the other and slightly displaced.* (NASA)

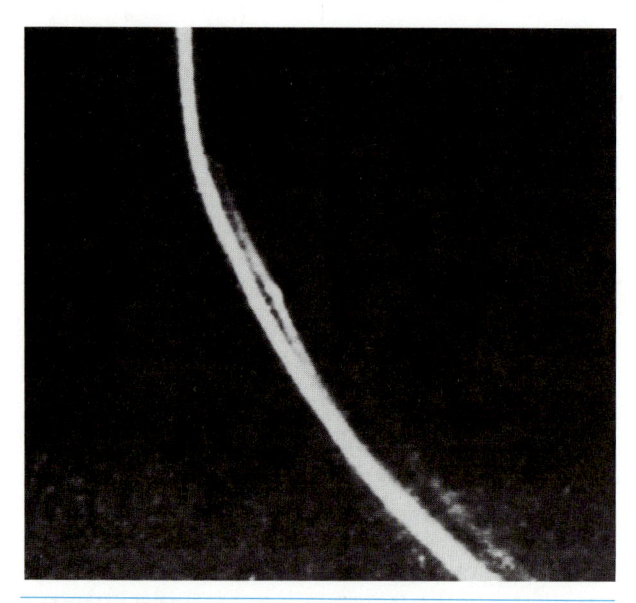

FIGURE 14.23 THE F RING. *Here we see the unusual braided appearance of this thin, outlying ring. The cause of the asymmetric shape is not well understood.* (NASA)

Some of the asymmetric structure in the rings may be created by electrical, rather than gravitational, forces. Electrons from the belts colliding with ring particles can cause the buildup of negative charges on the particles. Once these particles have electrical charges, they are subject to electromagnetic forces due to the planet's magnetic field. If a charged ring particle is sufficiently small, its motion can actually be controlled by electromagnetic forces, rather than gravitational ones. It is therefore possible that Saturn's magnetic field may exert some influence over the smaller particles in the ring system, distorting the shapes of some of the rings. The magnetic field rotates with Saturn at a rate much faster than the particles move in their orbits, and the magnetosphere itself varies in size and intensity with fluctuations in the stream of charged particles coming from the sun, so it is possible to imagine nonsymmetric forces acting to shape the rings.

It is almost certain that electromagnetic forces are responsible for the distribution of particles in one particular phenomenon associated with the rings. A haze layer of very fine particles is suspended above and below the plane of the rings, apparently held there by electric forces. These suspended particles were first noticed because they create spokelike structures pointing away from Saturn (Fig. 14.24), that move around the planet. The origin of these dark, asymmetric features was quite a mystery until it was

FIGURE 14.24 THE DARK SPOKES. *At the left the spokes appear bright because they scatter light in the forward direction, and they were photographed looking toward the sun. At right they are viewed looking away from the sun; so they appear dark.* (NASA)

realized that the particles there are so tiny that they easily could acquire enough electrical charge to have their motions governed by the electromagnetic field.

A great deal of work lies ahead before astronomers fully understand all the influences acting on the myriad particles and rings orbiting Saturn. Meanwhile we can be content to marvel at their wondrous appearance, and to speculate about new mysteries yet to be revealed.

PERSPECTIVE

Our visit to the ringed planet has taught us much about how complex nature can be. Every planet, even every satellite, that we have discussed has been revealed as a distinct individual, and Saturn is perhaps the most unique of the lot.

While many parallels could be drawn with Jupiter, in almost every respect, from the internal structure to the atmospheric motions, from the satellites to the rings, Saturn has shown us strong contrasts. This planet may be a cousin of Jupiter, but it is certainly not an identical twin.

We are nearing the completion of our tour of the planets, for only three of the nine remain, and two of them (Neptune and Pluto) have not been visited by space probes and therefore remain enigmatic. All three are discussed in the next chapter.

SUMMARY

1. Saturn's atmosphere and internal structure are different from Jupiter, in the lack of contrast in the global wind patterns, the fact that Saturn has seasonal variations, the greater velocities of its winds, and its different source of excess internal heat.
2. Saturn has seventeen known satellites, and five more are suspected. The known satellites include the giant Titan, seven intermediate moons, and a large number of tiny, irregular satellites.
3. Titan has a thick atmosphere dominated by nitrogen, but it also includes methane, which may exist in gaseous, liquid, and frozen states on the surface.
4. The intermediate moons all have very low densities, indicating that they have icy compositions, and some show evidence of resurfacing, by frozen internal gases that were released, or by debris from space.
5. The complex structure of Saturn's ring system is created by a combination of effects: orbital resonances with the satellites; spiral density waves; shepherding by pairs of satellites; and electromagnetic forces acting on charged particles. The rings may not be permanent features, but instead may be relatively young.

REVIEW QUESTIONS

Thought Questions

1. Compare the composition of the atmospheres of Jupiter and Saturn with that of primitive Earth's atmosphere.
2. How are the atmospheric circulations of Jupiter and Saturn similar, and how do they differ?
3. How often are the rings of Saturn seen edge-on from the Earth?
4. How is the internal structure of Saturn thought to differ from that of Jupiter? Are the sources of excess internal heat thought to be the same? Explain.
5. Compare the extent to which Jupiter and Saturn are differentiated with the degree of differentiation in the terrestrial planets.
6. How is methane on Titan comparable to water on Earth?
7. Summarize the properties of the seven intermediate-sized moons of Saturn. Is there any evidence for recent geological activity on any of them?
8. Explain the probable origin of the rings of Saturn.
9. Describe the probable cause of the Cassini division in the rings, and explain how it is similar to the cause of volcanic activity on Io, the innermost Galilean satellite of Jupiter.
10. Summarize all the processes that are thought to affect the structure of the ring system of Saturn.

Problems

1. Under the best observing conditions, astronomers on Earth can make out details as small as one-twentieth of the angular diameter of Saturn. How many kilometers thick can the rings be without being visible when viewed edge-on?
2. Given the rotation period and diameter of Saturn as listed in Table 14.1, calculate how rapidly a point on the equator is moving as the planet spins. Compare

your answer with the wind speed given in the text for atmospheric circulation near the equator.

3. Calculate and compare the wavelengths of maximum emission for Jupiter and Saturn, from their cloud-top temperatures (130 and 95 K, respectively).

4. Saturn emits about 1.8 times as much energy as it receives from the Sun, while Jupiter emits 1.7 times as much as it receives. Which actually emits more excess energy? (Note: You need to take into account the difference in the amount of solar energy they receive at their different distances from the Sun.)

5. Calculate the surface gravity and the escape velocity on Titan, and compare with that of the Earth. Do your values explain how Titan can have a thicker and denser atmosphere than Earth?

6. Calculate the distance from Saturn at which a ring particle should have one-third the orbital period of Mimas. Is there a gap between rings at this distance? Similarly, does Titan create a gap at the distance where ring particles would have one-half of its orbital period?

ADDITIONAL READINGS

Allen, D. A. 1983. Infrared views of the giant planets. *Sky and Telescope* 65(2):110.

Beatty, J. K., B. O'Leary, and A. Chaikin, eds. 1981. *The new solar system* Cambridge, Eng.: Cambridge University Press.

Esposito, Larry. 1987. The changing shape of planetary rings. *Astronomy* 15(9):6.

Hunten, D. M. 1975. Saturn. *Scientific American* 233(3):130.

Ingersoll, A. P. 1981. Jupiter and Saturn. *Scientific American* 145(6):90.

Morrison, D. 1981. The new Saturn system. *Mercury* 10(6):162.

———. 1982. *Voyages to Saturn.* NASA Special Publication SP-451. Washington, D.C.: NASA.

Overbye, Dennis. 1981. Lord of the rings: Saturn. *Discover* 2(1):24.

Pollack, J. B., and J. N. Cuzzi. 1981. Rings in the solar system. *Scientific American* 145(5):104.

Soderblom, L. A., and T. V. Johnson. 1982. The moons of Saturn. *Scientific American* 246(1):100.

Washburn, Mark. 1983. *Distant Encounters: Voyagers 1 and 2 at Jupiter and Saturn.* New York: Harcourt, Brace and Jovanovich.

CHAPTER 15

THE OUTER PLANETS

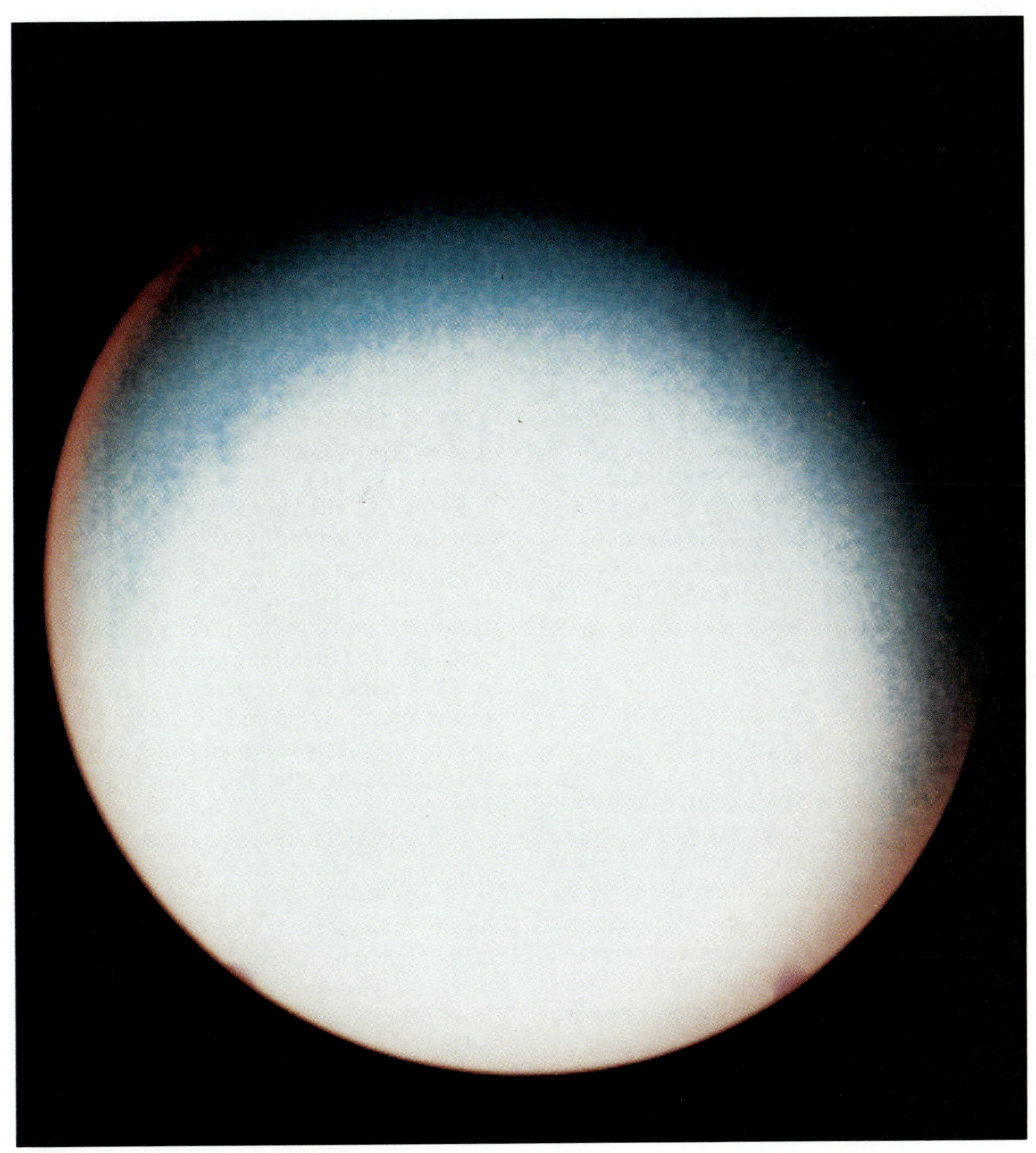

A close-up view of Uranus obtained by Voyager 2. (NASA)

CHAPTER PREVIEW

TABLE 15.1 URANUS

Orbital semimajor axis: 19.22 AU (2,875,000,000 km)
Perihelion distance: 18.31 AU
Aphelion distance: 20.13 AU
Orbital period: 84.01 years (30,685 days)
Orbital inclination: 0°46′23″
Rotation period: $17^h\ 14^m$
Tilt of axis: 97°55′
Mean diameter: 51,670 km (4.05 $D_\oplus$)
Polar diameter: 51,060 km
Equatorial diameter: 52,290 km
Mass: 8.69×10^{28} (14.54 $M_\oplus$)
Density: 1.20 g/cm^3
Surface gravity: 0.89 Earth gravities
Escape velocity: 21.2 km/sec
Surface temperature: 52 K (minimum cloud temperature)
Albedo: 0.34
Satellites: 15

The three planets that we have not yet discussed, orbiting in the distant reaches of the solar system far from the warmth of the Sun, are the least understood. These three—Uranus, Neptune, and Pluto—all lie well beyond the orbit of Saturn and were not known to the ancient astronomers. Until very recently, they were also little understood by modern astronomers. Now one of them, Uranus, has been visited by the *Voyager 2* spacecraft, and we have detailed information on this planet and its rings and satellites, much as we do for Jupiter and Saturn.

URANUS

The seventh planet from the Sun, Uranus (Table 15.1) is marginally bright enough to be seen with the naked eye and was actually included in a number of star charts compiled from telescopic observations in the seventeenth and eighteenth centuries. At its distance of nearly 20 AU, its period is so long (about 84 years) and its motion so slow that for quite a while it was not noticed that its position constantly changes with respect to the background stars.

Discovery and Observations from Earth

The planet was finally discovered in 1781, when the English astronomer William Herschel found an object during a routine sky survey that seemed unusual for a star. It had a disklike appearance, rather than being a single, twinkling point of light, and it appeared blue-green. Herschel at first suspected that he had found a comet, but after observing it for an extended time and then calculating its orbit, he found it to be moving about the Sun in a nearly circular path, about twice as far out as Saturn. This established that Herschel had found a planet, for no comet is visible at such a great distance from the Sun, nor would a comet move in a circular orbit. Herschel initially named the new planet after King George III of England, but the name was never widely accepted, and the recognized name became Uranus, after the Greek god representing the heavens.

Some information was derived from Earth-based telescopic observations of Uranus. For example, its mass was estimated using Kepler's third law (Uranus had five known satellites until the recent *Voyager* encounter, which revealed 10 more), and its radius was estimated from its angular diameter measured from photographs made by high-altitude balloon experiments (Fig. 15.1) and from knowledge of its distance from Earth. Spectroscopic observations revealed some information on the composition of the atmosphere, and infrared data provided estimates of its surface temperature and indications of surface markings (Fig. 15.2), which are very difficult to see from Earth. A system of relatively dim, thin rings was

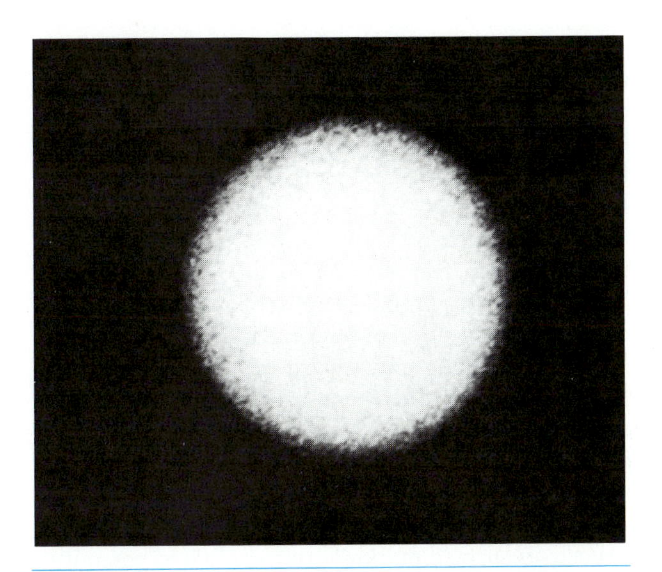

FIGURE 15.1 URANUS. *This exceptionally fine photograph was obtained from a high-altitude balloon, which was able to avoid much of the blurring effect of the Earth's atmosphere. No surface markings are evident, except for some darkening at the edges of the disk.* (Project Stratoscope, Princeton University, supported by the National Science Foundation)

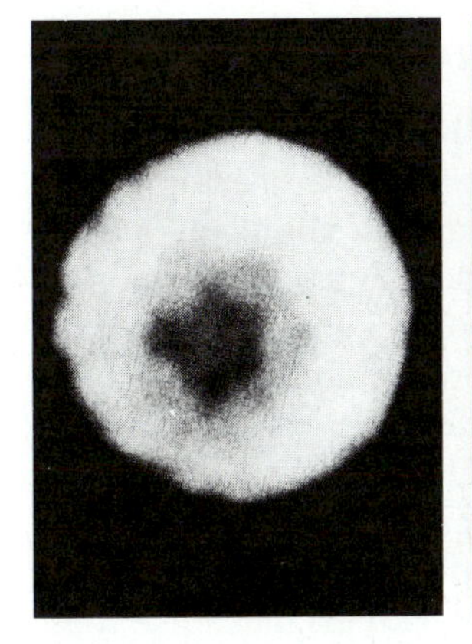

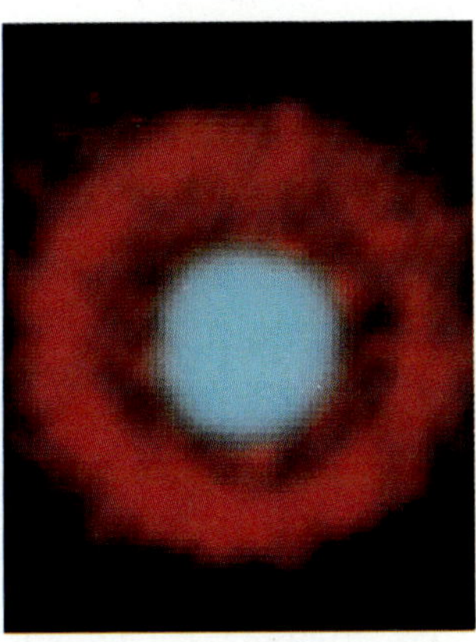

FIGURE 15.2 INFRARED VIEWS OF URANUS. *The image at left, obtained through a filter matching a wavelength at which methane absorbs light, shows zonal structure in the planet's atmosphere. At right is an image in which colors have been used to indicate relative temperatures. The blue disk of the planet is surrounded by red for the rings, which are colder.* (NASA, left photo courtesy of H. Reitsema)

discovered in 1977, when observations of a star that was about to pass behind the planet showed a series of occultations due to the rings. Five major satellites, including two found by Herschel in 1787, were discovered from Earth-based observations, although more were suspected because of the structure in the ring system. (Recall the role of Saturn's moons in governing the structure of its rings.)

One of the most striking peculiarities about Uranus, known from spectroscopic observations, is that its rotation axis is tipped over very far. The pole that we normally think of as the north pole (where rotation is counterclockwise as seen from above) actually points slightly (8°) *below* the plane of the planet's orbit. We say, therefore, that the inclination of the planet is 98°, which technically means that the rotation of the planet is retrograde, since when viewed from above the ecliptic plane, the rotation is backward. Seasons on Uranus, of course, are quite extreme (Fig. 15.3). Presently, the "south" pole of the planet is pointed almost directly toward the Sun, and this was the orientation of the planet when *Voyager 2* visited in January of 1986.

The extreme tilt of Uranus stands as a anomaly in the solar system and is not easily explained by theories of planetary formation (see Chapter 17). The fact that all five of the known satellites, as well as the recently discovered rings of Uranus, orbit in the equatorial plane of the planet (that is, the plane of the satellite and ring orbits is nearly perpendicular to the ecliptic) suggests that whatever disturbed the planet away from the usual orientation must have happened at some time before the satellites had formed, perhaps when they were still part of a disk of orbiting debris.

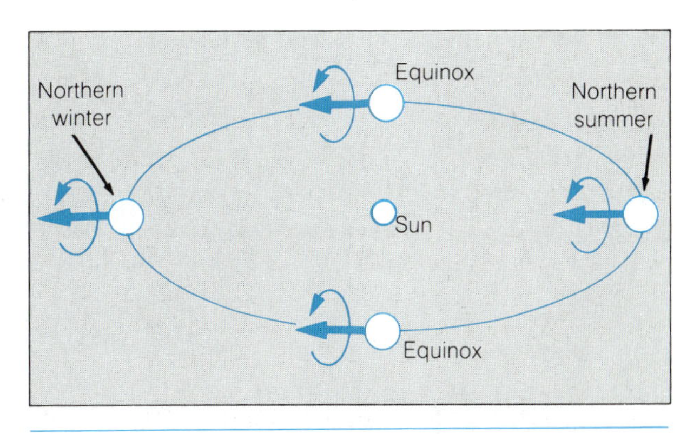

FIGURE 15.3 SEASONS ON URANUS. *The unusual tilt of Uranus creates bizarre seasonal effects. The solid arrow in each case indicates the direction of the planet's north pole, and the curved arrow, the direction of its rotation.*

Atmosphere and Interior

The atmosphere of Uranus has a composition similar to that of Jupiter and of Saturn: Hydrogen molecules dominate, helium is the next most abundant species, and methane (CH_4) is common. Ammonia is not seen in large quantities, presumably because it has condensed into solid form and precipitated out. The blue-green color of the planet is caused by methane, which absorbs red light, so that light reflected from the upper layers of atmosphere is deficient in red.

Uranus has no source of excess internal heat, yet it has atmospheric circulation patterns much like those of Jupiter and Saturn (although far less vivid). This indicates that, even at the remote distance of Uranus from the Sun, sunlight can provide enough heat to cause convection, which, along with the planet's rapid rotation (its period is 17.24 hours, as derived from *Voyager* measurements of radio emission), creates belts and zones that girdle the planet. One major surprise was the discovery that the direction of flow in the atmosphere is just backward from what had been expected on the basis of theoretical calculations. Theory indicates that when the poles of a rotating planet are warmer than the equator, which should be the case on Uranus, the winds near the equator should be slower than the rotation, meaning that they should flow backward with respect to the body of the planet. Instead, the equatorial winds on Uranus flow faster than the planet's rotation, so they are in the forward direction with respect to Uranus. This unexpected wind pattern is probably caused by an equally unexpected distribution of temperatures over the planet: Rather than being much warmer at the poles than at lower latitudes, the temperature was found to be quite uniform over most of the planet, with the coldest zones at intermediate latitudes (about 30° north and south of the equator).

The *Voyager* images of Uranus (Fig. 15.4) showed almost no surface markings, except when computer-processing techniques were used to enhance contrast (Fig. 15.5). Then the zonal flow patterns were revealed. Also seen were occasional white clouds near the equator, apparently due to rising plumes of methane ice crystals which are quickly spread out by the planetary rotation (Fig. 15.6).

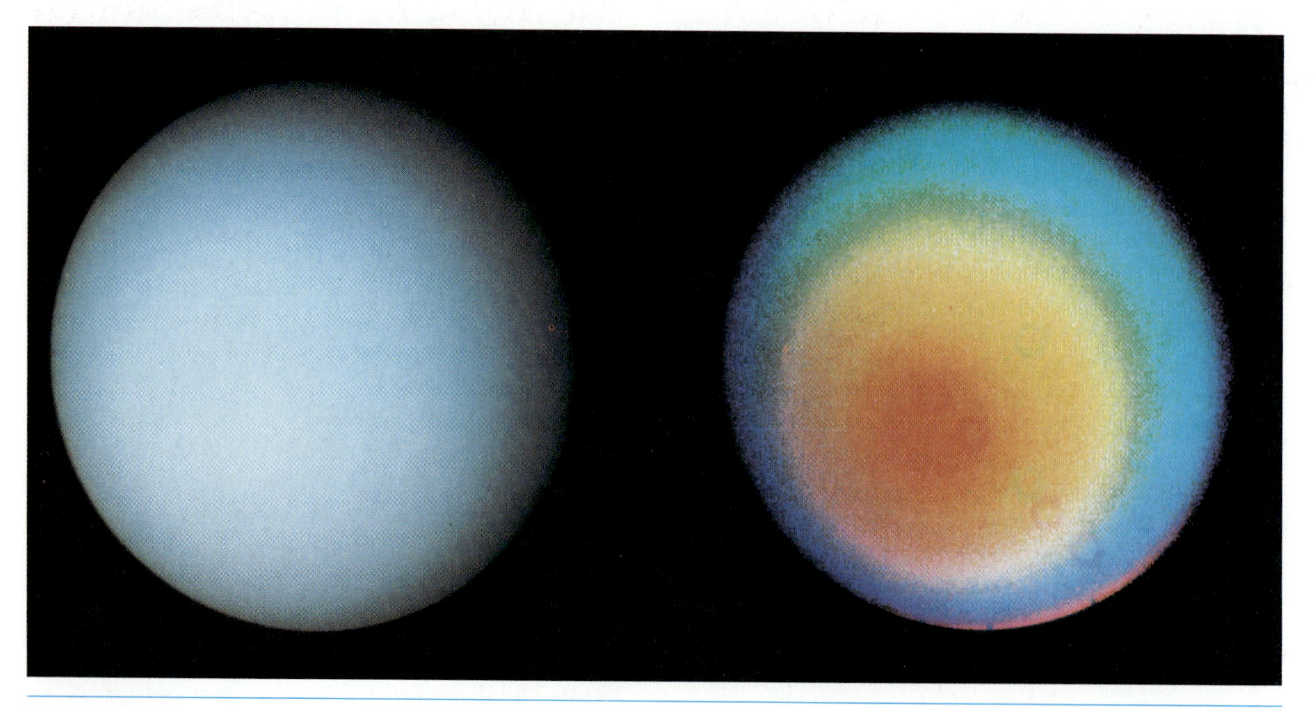

FIGURE 15.4 URANUS AS SEEN BY VOYAGER. *The image at left shows how featureless the planet appears when viewed in ordinary visible light, even when seen close-up. At right, color enhancement has been used to increase the contrast and show bands.* (NASA/JPL)

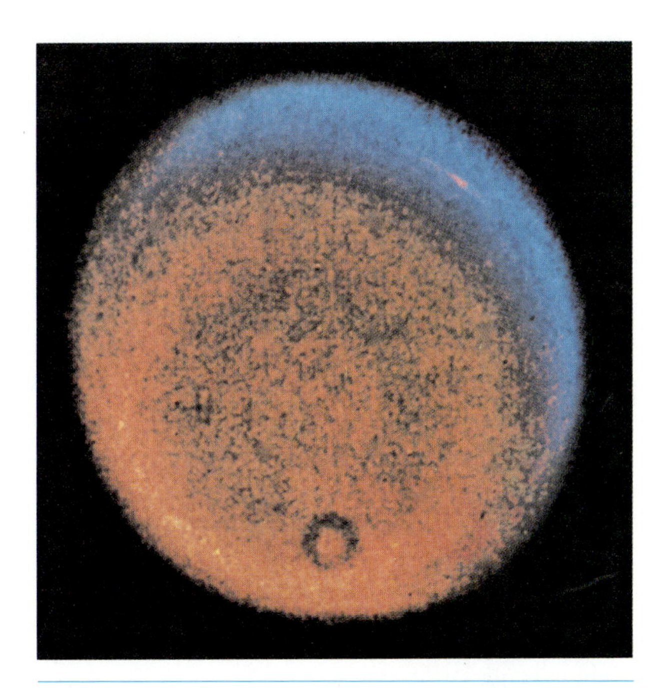

FIGURE 15.5 ZONAL FLOWS ON URANUS. *This computer-enhanced image shows the belts and zones on Uranus, which are the equivalent of those on Jupiter and Saturn except that the contrast is very low. Despite the lack of internal heating on Uranus, the flow patterns in its atmosphere are very similar to those of the larger giant planets.* (NASA/JPL)

A cloud of hydrogen atoms surrounds Uranus, produced by the breakup (dissociation) of hydrogen molecules (H_2) due to collisions with rapidly moving electrons in the charged particle belts that surround the planet.

The magnetic field of Uranus was not detected until *Voyager*'s approach, although the presence of a field had been suspected and was no surprise. What was a surprise, however, was its orientation: The magnetic axis is tipped some 60° with respect to the rotation axis. Thus, the magnetic axis is roughly perpendicular to the plane of the planet's orbit, so that the magnetosphere, the volume surrounding the planet where charged particles are trapped by the magnetic field, has much the same appearance as for any other planet, with an extended tail pointing away from the Sun because of the solar wind (Fig. 15.7). As the planet rotates on its highly tipped axis, this tail rotates about.

The magnetic field of Uranus is thought to arise in an interior level where water may exist, and may

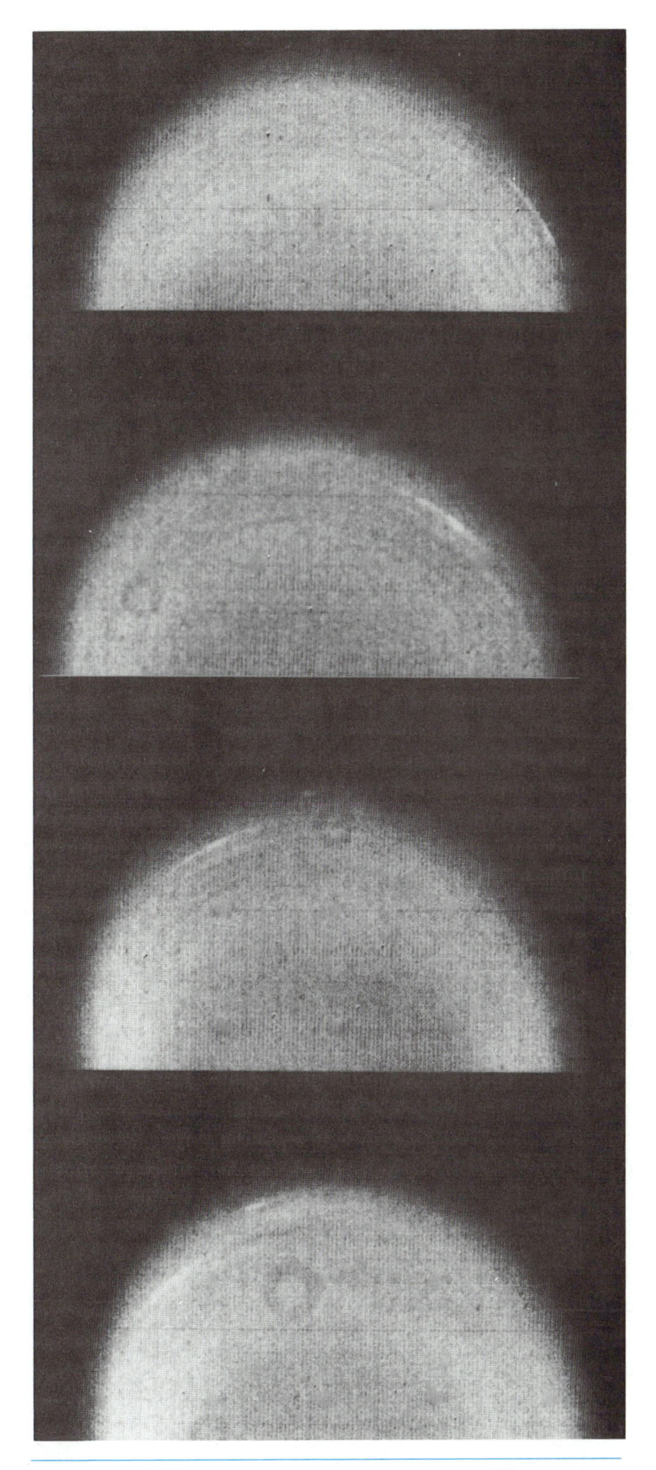

FIGURE 15.6 CLOUDS IN THE ATMOSPHERE OF URANUS. *The white streak seen here is a cloud of methane ice crystals. Such clouds occasionally rise through the atmosphere near the equator and are then carried around the planet by the winds, as seen in this sequence.* (NASA/JPL)

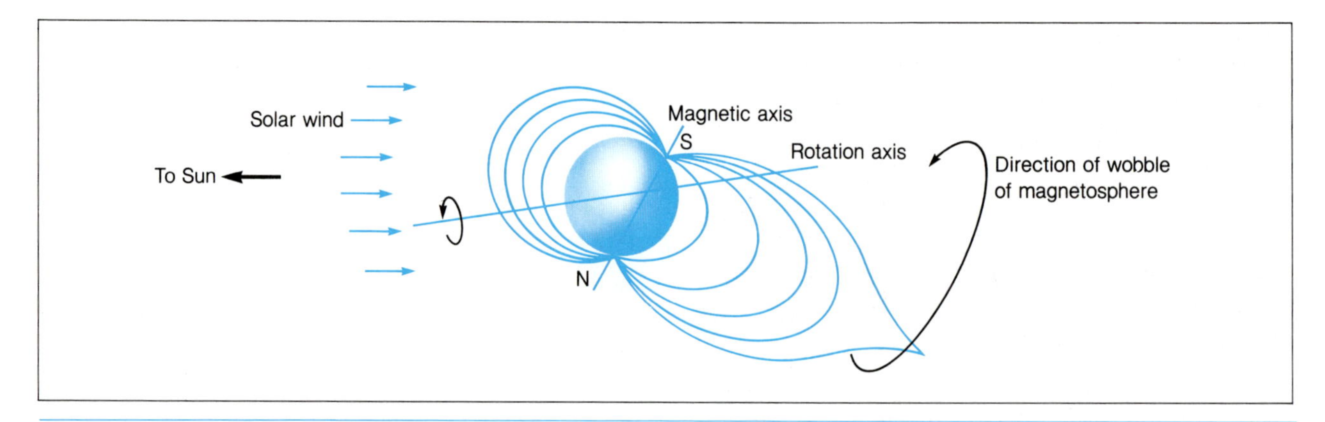

FIGURE 15.7 THE MAGNETOSPHERE OF URANUS. *Due to the combination of the high tilt of the rotation axis of Uranus and the misalignment between its rotation and magnetic axes, the magnetosphere of Uranus resembles that of one of the planets whose rotation axis is more normal. The tail of the Uranian magnetosphere rotates as the planet spins around its highly tilted axis.*

be sufficiently compressed to conduct electricity. Little is known about the internal structure of the planet, although now that detailed data are available on the magnetic field, the mass distribution (as deduced from gravitational effects on *Voyager 2*), and the atmospheric properties, detailed interior models will soon be developed. It is thought that there is a rocky core, probably about the size and mass of the Earth, outside of which is a liquid hydrogen zone, perhaps overlain by a water zone, and, in the lower clouds, layers of methane ice crystals (along with other ices such as ammonia and water). There is probably no liquid metallic hydrogen zone, because the pressure inside Uranus is not sufficient.

Satellites and Rings

Uranus has fifteen known satellites (Table 15.2), five of which were detected from Earth (Figs. 15.8 and 15.9). All fifteen are in synchronous rotation due to tidal forces from Uranus. The ten new satellites discovered by *Voyager* are all small objects, some of which

TABLE 15.2 SATELLITES OF URANUS

Name	Distance (R_U)*	Period (Days)	Diameter (km)	Mass (g)	Density (g/cm³)	Albedo
1986U7	1.901	0.33	40			0.1
1986U8	2.058	0.38	50			0.1
1986U9	2.264	0.43	50			0.1
1986U3	2.364	0.46	60			0.1
1986U6	2.398	0.48	60			0.1
1986U2	2.471	0.49	80			0.1
1986U10	2.528	0.51	80			0.1
1986U4	2.674	0.56	60			0.1
1986U5	2.880	0.62	60			0.1
1986U1	3.289	0.76	170			0.07
Miranda	4.949	1.413	484	7.5×10^{22}	1.26	0.34
Ariel	7.305	2.520	1,330	1.4×10^{24}	1.14	0.40
Umbriel	10.186	4.144	1,110	1.3×10^{24}	1.82	0.19
Titania	16.672	8.706	1,600	3.5×10^{24}	1.60	0.28
Oberon	22.318	13.463	1,630	2.9×10^{24}	1.28	0.24

*The distances are orbital semimajor axis, given in units of the equatorial radius of Uranus, which has the value R_U = 26,145 km.

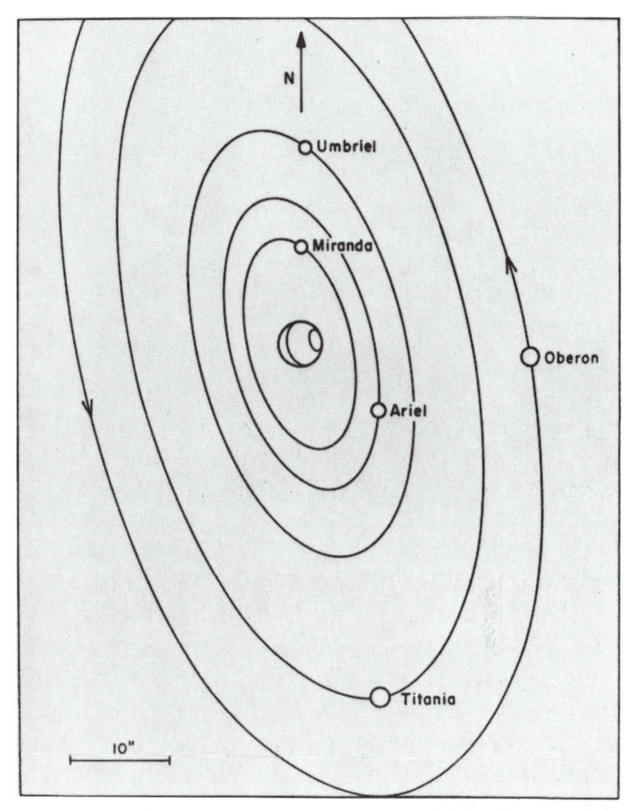

FIGURE 15.8 URANUS AND ITS SATELLITES. *This is an Earth-based photo, showing all five of the large moons that are identified in the sketch.* (NASA)

are within the system of rings. These satellites, as well as the five large ones, govern the structure of the ring system, just as the many moons of Saturn influence its rings. There are shepherd satellites, for example (Fig. 15.10), and cases of orbital resonances which create gaps in the rings.

Each of the five major satellites was observed in some detail by *Voyager*. The outermost of the five, Oberon (Fig. 15.11), has a heavily cratered surface and shows little sign of recent geological activity, yet it has a huge mountain and scarp systems indicative of past episodes of activity. Next in is Titania (Fig. 15.12), which shows evidence of having a younger surface. It appears that somehow Titania has undergone a more recent episode of cratering from impacts than Oberon, and Titania has an extensive series of cliffs that may be the result of expansion due to the freezing of water ice in the interior. The middle one of the five major moons, Umbriel (Fig. 15.13), shows

FIGURE 15.9 THE MAJOR MOONS AS SEEN BY VOYAGER. *This is a* Voyager 2 *image, showing the planet and the positions of the major satellites during approach.* (NASA/JPL)

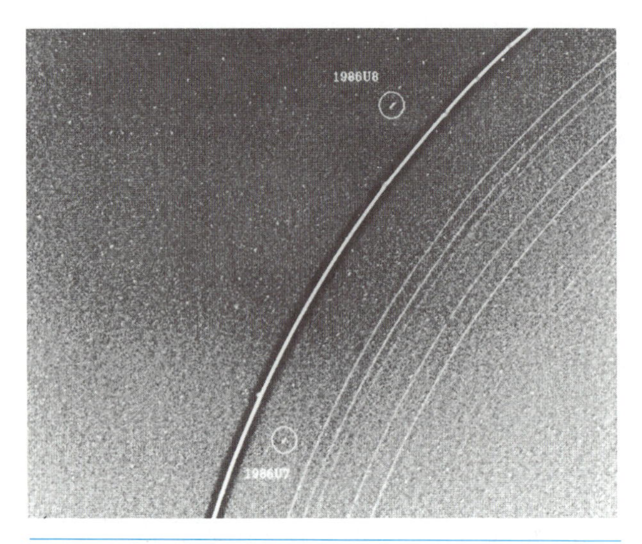

FIGURE 15.10 SHEPHERD SATELLITES. *Here we see two small moons on either side of a thin ring around Uranus. The gravitational effects of these moons keep the ring confined.* (NASA/JPL)

FIGURE 15.11 OBERON. *This is the outermost of the five major satellites of Uranus.* (NASA/JPL)

FIGURE 15.12 TITANIA. *This is the next major satellite inward from Oberon. Titania shows signs, such as more recent cratering, of having a younger surface than Oberon and Umbriel, the next moon inward.* (NASA/JPL)

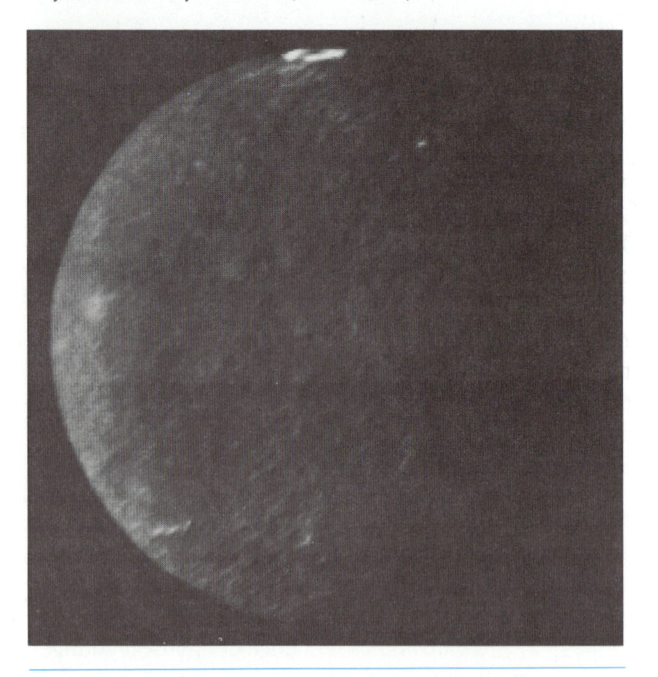

FIGURE 15.13 UMBRIEL. *The third of the five major moons of Uranus, Umbriel is also the darkest, its surface apparently having been coated by some sort of low-reflectivity deposit. The many craters show that the surface is old and has not undergone major geological activity in recent times.* (NASA/JPL)

FIGURE 15.14 ARIEL. *This satellite shows signs of geological activity. Its bright surface and evidence of recent ice flows indicate probable internal motions and outgassing.* (NASA/JPL)

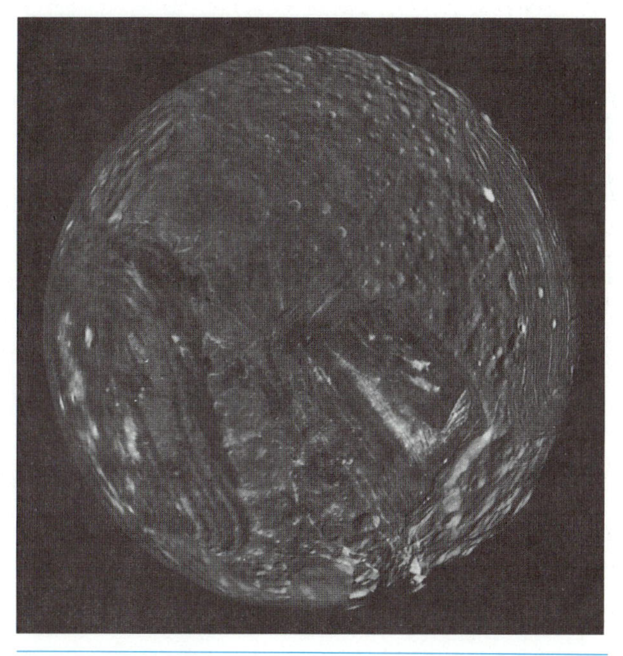

FIGURE 15.15. MIRANDA. *The innermost of the five large moons of Uranus, Miranda shows a surprising degree of recent geological activity. Tidal stresses are an inadequate explanation for the youth of the surface of this satellite, and more exotic explanations are needed.* (NASA/JPL)

a surface equally cratered and apparently as old as that of Oberon. Umbriel is the darkest of the major satellites, with an albedo of only 0.19, which indicates that its surface may have been coated with dark dust or debris at some time in the past. The next satellite, Ariel (Fig. 15.14), is the brightest of the Uranian moons, evidence of a young surface coated by ices that have probably escaped from the interior. Ariel also has extensive faults, and there is evidence of icy flows on the surface.

The strangest of the large satellites is Miranda, the smallest and innermost (Fig. 15.15). Miranda has several remarkable features on its surface that indicate recent or current geological activity, more than can be easily explained by tidal stresses due to Uranus and the other moons. There are grooved regions, resembling similar markings on Ganymede (Jupiter's largest satellite), and at least one more or less rectangular feature jokingly called the Circus Maximus, because of its resemblance to a Roman racetrack (Fig. 15.16). This feature is raised in the center and consists of terraces that drop off sharply toward the outside, and can only be explained as due to uplifting of Miranda's crust. The cause of this kind of uplifting is unknown, but one suggestion is that a large mass of ice, which is more buoyant than rock, forced its way to the surface from the interior. Normally, the lower-density ice would have been on the surface in the first place, so how the ice got into the interior is a mystery. A rather bizarre suggestion has been made: that Miranda was once broken apart by a collision with an asteroidlike body, and then coalesced again into a satellite, with some of the original surface material ending up in the interior. In any case, Miranda provided one of the major surprises of the *Voyager* encounter with Uranus and will take some imaginative thinking to be explained.

The rings of Uranus (Fig. 15.17) are thinner and dimmer than those of Saturn, perhaps resembling the single ring of Jupiter more than those of Saturn. The ring particles have low reflectivities, one reason for the difficulty of observing them from Earth. Apparently, this is because some of these rings are made of relatively large particles, which must be very dark.

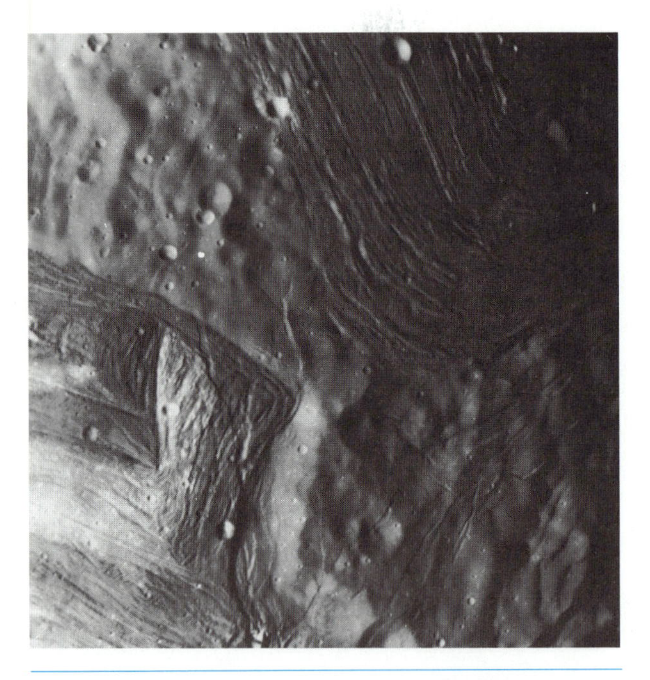

FIGURE 15.16 THE "CIRCUS MAXIMUS." *This unusual feature (right) on Miranda consists of an elevated inner portion, and sharply sloping terraces around it, resembling a Roman racetrack. The cause of this feature is unknown, but it would require some source of uplift in the interior. At lower left is a similar, chevron-shaped feature.* (NASA/JPL)

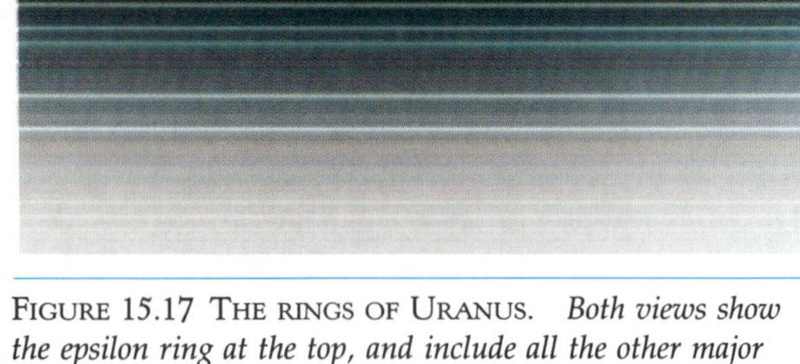

FIGURE 15.17 THE RINGS OF URANUS. *Both views show the epsilon ring at the top, and include all the other major rings inward toward the planet. The color-enhanced image brings out greater detail.* (NASA/JPL)

Also, large particles do not reflect light well, particularly in the backward direction; recall the discussion of scattering by ring particles in Chapter 14. The typical particles in the rings of Uranus are roughly 1 meter or more in diameter, whereas many of Saturn's rings consist of particles a million times smaller. The lack of small particles in the Uranian rings may be due to the extended cloud of hydrogen atoms that surrounds the planet, which could create sufficient drag on small particles to force them to fall into the planet. However, *Voyager 2* discovered very dark, dusty rings around Uranus—how these rings of small particles can survive for long in the presence of the hydrogen cloud is not known.

Another unusual characteristic of the rings around Uranus is that several of them are asymmetric; a particular ring may be wider in some places than in others, or it may not be perfectly circular, or it may consist only of an arc that does not completely encircle the planet. This suggests that the rings are not stable, permanent features, but instead that they vary with time and may look quite different the next time a space probe visits.

We have already mentioned, in our discussion of Saturn's rings, the recently developed theory that planetary ring systems are transitory phenomena and not permanent features. The *Voyager* observations originally stimulated this idea, because of the difficulty in understanding how the rings could last for a long time in the shapes they have. Not only does it appear that the dust rings have a very short lifetime,

TABLE 15.3 THE RINGS OF URANUS

Ring	Distance (R_U)*	Width (km)
1986U2R	1.41–1.51	2,500
6 ring	1.60	1–3
5 ring	1.62	2–3
4 ring	1.63	2
Alpha ring	1.71	8–11
Beta ring	1.75	7–11
Eta ring	1.81	2
Gamma ring	1.82	1–4
Delta ring	1.85	3–9
1986U1R	1.91	1–2
Epsilon ring	1.96	22–93

*Distances are in units of the equatorial radius of Uranus, which is R_U = 26,145 km.

TABLE 15.4 NEPTUNE

Orbital semimajor axis: 30.11 AU (4,504,300,000 km)
Perihelion distance: 29.85 AU
Aphelion distance: 30.37 AU
Orbital period: 164.79 years (60,188 days)
Orbital inclination: 1°46′22″

Rotation period: 18^h 12^m
Tilt of axis: 28°48′

Mean diameter: 49,930 km (3.91 $D_\oplus$)
Polar diameter: 49,410 km
Equatorial diameter: 50,450 km
Mass: 1.03×10^{29} (17.2 $M_\oplus$)
Density: 1.58 g/cm^3
Surface gravity: 1.13 Earth gravities
Escape velocity: 23.5 km/sec

Surface temperature: 50 K
Albedo: 0.29

Satellites: 2

but the several narrow rings, with relatively sharp edges, will gradually broaden and disperse unless some mechanism—such as shepherd satellites—act to confine them. As we have noted, some shepherd satellites were found, but not enough to explain all the ring structures.

If the rings are actually young and short-lived, then it may seem like a remarkable coincidence that at the present time all four of the giant planets happen to have rings. Because such coincidences are unlikely, it seems probable that there is some mechanism for occasionally forming new rings. The best current theory, one that is gaining acceptance, is that rings are created by the destruction of satellites in collisions. This idea was mentioned in connection with Saturn's rings (see Chapter 14), and is supported by ample evidence of major collisions of rocky objects from space with existing satellites. We have seen, for example, that Uranus' moon Miranda may have been broken apart and re-formed in its past, and Saturn's moon Mimas bears the scar (a huge crater) of a collision that must have nearly destroyed it. Such collisions are more likely with the moons of the giant planets than in other locations of the solar system because the immense gravitational fields of these planets help attract the colliding bodies. (This effect is called gravitational focusing.)

We will see in the next section that Neptune provides even more evidence in favor of the idea that planetary rings are temporary structures.

NEPTUNE

Perhaps more than any other two planets, Uranus and Neptune may be regarded as twins. Neptune, the eighth planet (Table 15.4), orbits at a distance of some 30 AU from the Sun, taking 165 years to make one circuit. Its size is very similar to that of Uranus, but its mass is somewhat greater, giving it a density of 1.7 grams/cm^3, the greatest of the gaseous giants.

Since Neptune is even farther from the Sun than Uranus, less is known about it, and this disparity will not be eliminated until 1989, when the well-traveled *Voyager 2* spacecraft will fly by Neptune.

The Discovery: A Victory for Newton

Neptune is a rather dim object, too faint to be seen with the unaided eye, and even too inconspicuous to have been included on star charts made before the mid-1800s. Rather than being found by accident, as Uranus was, the existence of Neptune was first deduced indirectly.

When Uranus was found by Herschel in 1781 and its orbit subsequently computed, astronomers were naturally eager to see whether its path conformed perfectly to the predictions of Newton's laws of mo-

ASTRONOMICAL INSIGHT 15.1

SCIENCE, POLITICS, AND THE DISCOVERY OF NEPTUNE

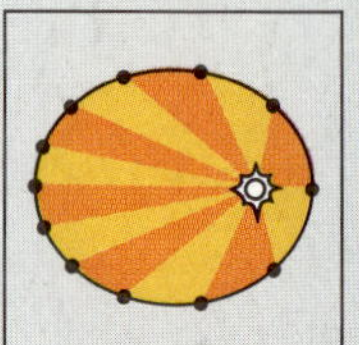

Many times, throughout the history of science, a great discovery was just waiting to be made, as soon as the necessary technology or preliminary information had been developed. If Galileo had not turned his telescope to the heavens, surely someone else would have done so very soon, and would have found the wonders that he discovered. If Newton had not developed his laws of mechanics when he did, they would have become apparent to some other student of the motions of the Moon and planets.

The discovery of Neptune was such a case. The irregularities in the orbit of Uranus were well known by the mid-1800s, and the mathematical techniques needed to calculate from them the position of the eighth planet were also standard. It was only a matter of someone recognizing the possibility and carrying out the computation. As it happened, two people did so, unbeknownst to each other, and the result was one of the more fascinating disputes in the history of science.

In England, a young college graduate named John C. Adams made the prediction in 1845 that an eighth planet should be at a certain location in the constellation Aquarius. Being young and unknown, Adams had difficulty convincing the established astronomers of England that an effort should be made to find the predicted planet. A contributing factor was the development of a minor tiff with the Astronomer Royal, with the result that the latter simply ignored the affair.

Within a few months, a French scientist named Urbain Leverrier carried out calculations that led him to the same conclusion, although he was totally unaware of Adams' work, which had not been published. Leverrier's work did appear in scientific journals, however, and this prompted the Astronomer Royal in England to take the idea of an eighth planet a bit more seriously, and a search was begun.

The English astronomers were hindered by the lack of good star charts for the appropriate region of the sky, and substantial preliminary work was necessary before an earnest search could begin. Meanwhile, Leverrier contacted an acquaintance at the observatory in Berlin, who had excellent charts of Aquarius, and discovered Neptune on the very first night that he looked for it. This was in 1846.

England and France were international rivals, and years of contentious arguments followed over who should be credited with the prediction of Neptune's existence. There never was an amicable settlement of this question; it just gradually faded from the forefront of French-English affairs. Today, the heat of the debate having cooled with the passage of time, Adams and Leverrier are generally credited equally with the discovery.

---

tion and gravitation. Since Uranus had been included on star charts dating back some 90 years before Herschel's discovery, it was possible to compare its observed and computed positions over a long period. Much to the dismay of the scientists who did this, discrepancies were found. The mismatches between the expected and observed positions of the planet were not very large, amounting to 2 minutes of arc by 1840, but it was considered very serious, for this was too large an error by far to be allowed by the laws of motion. Something was amiss, and people began to question the validity of Newton's work. Eventually it was suggested that the discrepancies were due to the gravitational influence of an eighth planet, and the planet was soon discovered within 1° of the predicted location. The name Neptune was taken from that of the Roman god of the seas.

Properties of Neptune

Neptune is difficult to examine in any great detail, even more so than Uranus. The best photographs show little more than a spot of light, and only rather large-scale features can be detected (and some have been). The fine resolution of the *Space Telescope* will allow smaller details to be seen on Neptune, if they exist. The angular diameter of Neptune, as seen from the Earth at opposition, is 2.3 seconds of arc, and the *Space Telescope* will be capable of discerning features as small as about one-tenth of an arc second in size. Of course, this instrument will also provide better views of Uranus than any that are presently available, but presumably *Voyager 2* will be sending back close-ups of that planet by about the time the *Space Telescope* is launched.

For now, we surmise that Neptune closely resembles Uranus, since the two are so similar in all respects that can be measured from Earth. The diameter of Neptune, like that of Uranus, was measured from a stellar occultation, and the mass was deduced from the application of Kepler's third law to the orbits of its satellites.

The rotation period of Neptune has proven difficult to measure, but recent observations indicate that it is 18.2 hours, and that its rotation axis is tilted by 29°, similar to the tilts of several of the other planets.

The atmosphere may be even colder than that of Uranus, with a temperature of perhaps 50 K. Spectroscopy reveals the presence of hydrogen and methane, and again it is suspected that ammonia has precipitated out. It is likely that the internal structure is similar to that of Uranus, with the probable exception that Neptune has a larger core, since it is known that its overall density is greater.

Surprising as it may seem, some information has been found on atmospheric motions on Neptune. With the use of special filters that match the wavelengths of methane gas absorption lines, it is possible to infer, from brightness variations measured through these filters (Fig. 15.18), the flow pattern of methane on Neptune. The information is not very detailed, but there are indications of a pattern of zonal winds like those on Jupiter and Saturn, with velocity differences between zones of about 110 km/sec. This tends to confirm our natural expectation that there are strong similarities among all the giant, gaseous planets.

Satellites in Chaos, and a Possible Ring System

Neptune has two known satellites (Table 15.5). One of which, Triton, noted in the same year that Neptune itself was discovered, is quite large, but not as big as the largest moons of Jupiter and Saturn. Recent spectroscopic observations have shown that Triton has a gaseous atmosphere containing methane, making this

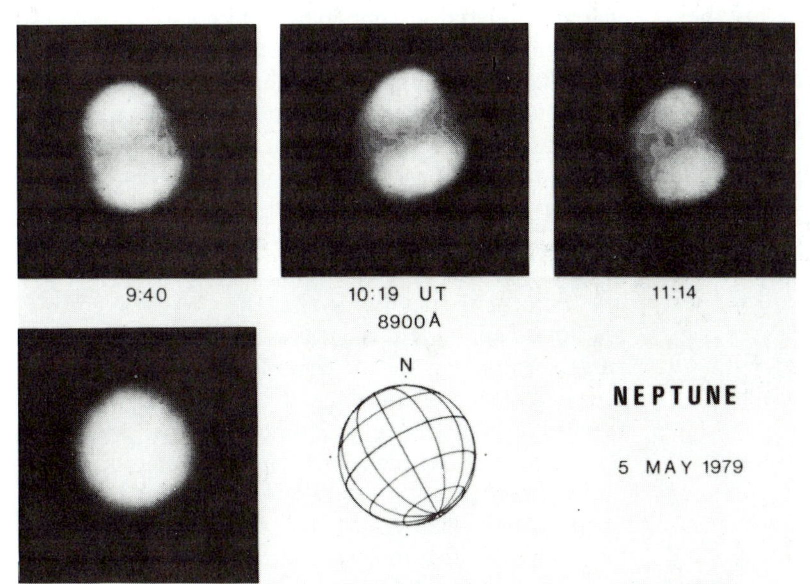

FIGURE 15.18 NEPTUNE. *This series of images was obtained in infrared light. The three upper images were made at a wavelength where methane absorbs light, so the dark areas are regions of methane concentration. The bright areas north and south of the equator are thought to have ice crystals and haze above the methane.* (NASA)

TABLE 15.5 SATELLITES OF NEPTUNE

No.	Name	Distance (R_N)*	Period (days)	Diameter (km)	Mass (g)
1	Triton	14.046	5.877	3,800	1.3×10^{26}
2	Nereid	218.5	360.2	300	2.1×10^{22}

*Distances are semimajor axis in units of the equatorial radius of Neptune, which is R_N = 25,225 km.

the second satellite (along with Titan) known to have an atmosphere. The other satellite of Neptune, Nereid, is very small, with a diameter of roughly 500 km (compared to Triton's diameter of 3,800 km), and was not detected until 1949.

Neptune has a fairly normal rotation, but the motions of its two satellites are extremely unusual (Fig. 15.19). Triton orbits the planet backwards, taking about 6 days to complete each retrograde orbit. Furthermore, the orbit is tilted by an unusually large angle, about 20°, with respect to the planet's equatorial plane.

The orbit of Nereid is perhaps even more bizarre. It goes around in the normal direction (opposite the motion of Triton), and its orbital plane is inclined 28° to Neptunes's equatorial plane. Nereid is very distant from Neptune, and had a very long period for a satellite, about one Earth year. The strangest aspect of its orbit, however, is its shape, which is very eccentric. Its distance from Neptune varies by a factor of 5 between its closest approach and the farthest reaches of its orbit. Newton's laws of motion allow for the possibility of such highly flattened orbits, and as we shall see, the comets follow similar or even more elongated paths around the Sun, but no planet nor any other satellite in the solar system has such an orbit.

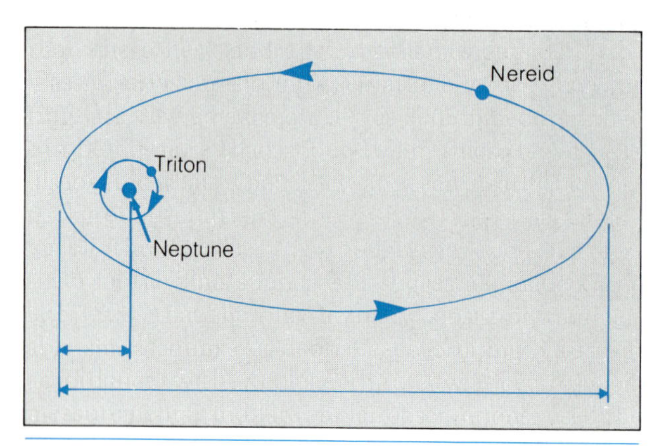

FIGURE 15.19 THE SATELLITES OF NEPTUNE. *This scale drawing illustrates the unusual orbital characteristics of the two moons, Triton and Nereid.*

The strange motions of Neptune's two satellites cannot be explained as any sort of natural consequence of the formation of the planet, but must instead indicate that these satellites were captured, or that some immense gravitational disturbance disrupted the system at some time in the past, after Neptune and its satellites had formed. Perhaps a near-collision with a massive body caused the upheaval. Some interesting speculations on this will be discussed in the next section, for it has been suggested that Pluto, the ninth planet, may have been a character in the drama.

It was recently found that Neptune may have a ring system, which would make the presence of rings a feature of all four of the giants. On two separate occasions, when Neptune was due to undergo a stellar occultation, the star's brightness dipped just before the planet itself blocked the star completely. This is analogous to the situation when the rings of Uranus were discovered, except that in both cases involving Neptune, there was no corresponding dip in the star's intensity as it emerged from behind the planet on the other side. Analysis of the geometry of the occultations showed, however, that the path of the star with respect to Neptune was such that it might have missed passing through the equatorial plane, where a ring system is most likely to lie, on its way out from the planet.

Another possibility is that the rings of Neptune are not complete, but instead consist of arclike structures that only partially encircle the planet, as was found for some of the rings of Uranus. Therefore, it is possible that the star was occulted by a ring system on one side of the planet, but missed being occulted on the other side (Fig. 15.20). This arc-structure theory has now been confirmed. Apparently there are between two and four such arcs. Partial rings such as these are thought to be very unstable, therefore it is believed that Neptune's rings provide additional

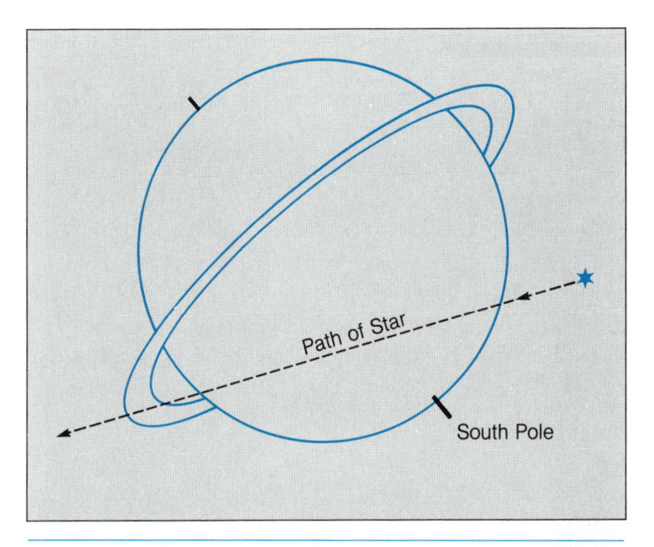

FIGURE 15.20 RINGS OF NEPTUNE? *This sketch roughly depicts the characteristics of the ring or rings of Neptune, whose existence was tentatively inferred from two recent stellar occultation observations.* (E. F. Guinan, C. C. Harris, and F. P. Maloney, Villanova University)

evidence for the conclusion, discussed in the previous section, that planetary rings are temporary features.

PLUTO

The ninth and last of the known planets is Pluto (Table 15.6), orbiting at an average distance of nearly 40 AU from the Sun. From that perspective, the Sun would appear as a very bright star but surely would not produce much warmth. This planet is the most obscure of them all, and we will raise many questions about its nature.

The Discovery: An Inspired Accident

Following the triumph of Neptune's discovery, continued analysis of the orbits of both Uranus and Neptune seemed to indicate further, rather minor, discrepancies in their motions. The natural inclination of astronomers, given the recent experience with Neptune, was to speculate that yet another planet, beyond the orbit of Neptune, might be responsible. This was an example of a common human tendency to expect to find additional examples of fascinating new discoveries, and in this case the expectation was stronger than the evidence. In any case, several independent calculations, including most notably one carried out in 1915 by Percival Lowell (of Martian canal fame), led to predictions of the location of the supposed ninth planet, and the search was on.

TABLE 15.6 PLUTO

Orbital semimajor axis: 39.44 AU (5,900,000,000 km)
Perihelion distance: 29.58 AU
Aphelion distance: 49.30 AU
Orbital period: 248.5 years (90,700 days)
Orbital inclination: 17°10′12″
Rotation period: 6.39 days
Tilt of axis: 112°
Diameter: 2,200–2,300 km (0.17–0.18 $D_{\oplus}$)
Mass: 1.4 × 10^{25} (0.002 $M_{\oplus}$)
Density: 1.4–2.0 g/cm^3
Surface gravity: 0.06 Earth gravity
Escape velocity: 1.2 km/sec
Surface temperature: 40 K
Albedo: 0.4
Satellites: 1

Lowell, in particular, set out to find this new planet, and he spent the last years of his life searching photographic plates for it. The process was made difficult by the fact that the predicted position was in a portion of the Milky Way that is crowded with stars, so that a search for a dim object that slowly changed position among all the fixed stars was tedious and difficult. Lowell was unsuccessful.

In 1930, the American astronomer Clyde Tombaugh (Fig. 15.21), having taken up the search at the Lowell Observatory some time after Lowell himself had died, found a tiny dot of light that displayed the expected motion (Fig. 15.22), and he found it only 7° away from the position that Lowell had predicted. The new member of the Sun's family was given the name Pluto, after the Roman god of the underworld (the name also honors Percival Lowell, since its first two letters are Lowell's initials).

Subsequent analysis of Pluto showed it to be far too small and undermassive to have created any noticeable disturbance in the paths of Uranus and Neptune. Reexamination of the data that Lowell and others had used in predicting the existence of a ninth planet showed, furthermore, that there was no evidence for such a planet. The minor discrepancies be-

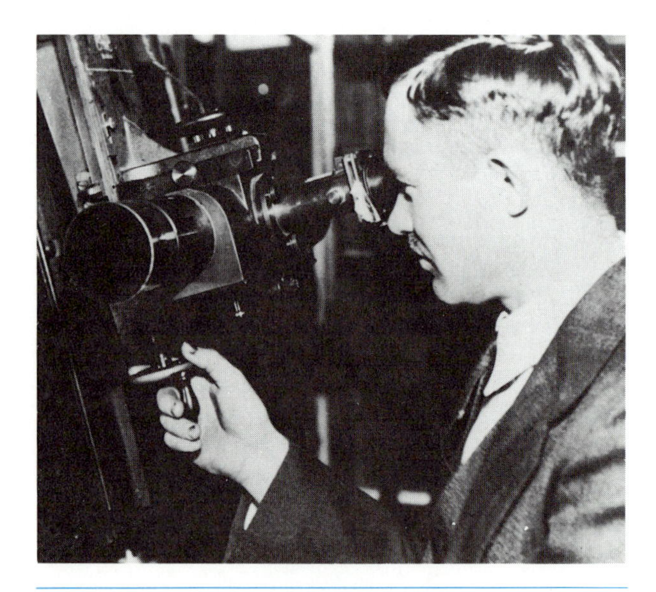

FIGURE 15.21 CLYDE TOMBAUGH AT THE BLINK COMPARATOR. *This photo shows the discoverer of Pluto peering into the eyepiece of the device he used to find the planet on photographic plates. The blink comparator allows two images of the same field of stars to be examined alternately (with the use of a mirror that flips between two positions) so that moving objects stand out.* (Lowell Observatory photograph)

FIGURE 15.22 THE DISCOVERY OF PLUTO. *These are portions of the original photographs on which the ninth planet was discovered in 1930. The position of the planet in each is indicated by an arrow.* (Lowell Observatory photograph)

tween calculated and observed positions were all within the uncertainty of the measurements. Apparently Lowell and the others who made the calculations wanted to see evidence for a ninth planet, and so they found it, whereas dispassionate scientific analysis would not have shown that it was there. It is a remarkable coincidence, then, that Pluto was found not far from the position Lowell had predicted. Even without that prediction Pluto would probably have been discovered by Tombaugh anyway, for he was in the process of searching the entire ecliptic plane.

Planetary Misfit

The little we know about Pluto shows it to be a nonconformist. Throughout our discussions of the other planets, certain systematic regularities have shown up, and Pluto violates many of them.

First, its orbit is highly irregular, compared to those of the other planets. It is both noncircular and tilted with respect to the ecliptic. At its greatest distance from the sun, Pluto is nearly 70 percent farther away than at its closest, when it actually moves closer to the Sun than Neptune for a time. (In fact, Pluto is temporarily the eighth planet from the Sun right now, having moved inside Neptune's orbit in 1978, not to reemerge until 1998.) Pluto's orbit is tilted by 17° to the ecliptic, whereas the tilts of all the other planetary orbits are 7° or less.

In physical characteristics, Pluto also seems not to fit in with the other planets. The other outer members of the solar system are giant planets, with thick atmospheres. While it has been difficult to pinpoint the mass of Pluto, it has long been clear that this planet does not resemble the other outer planets. For a long time, the only information on the mass of Pluto came back from the lack of any significant effect on the motion of Neptune, which implied that Pluto's mass had to be less than about 10 percent of that of the Earth. The diameter has been difficult to determine, the most recent results from interferometric measurements (Chapter 7), indicating a value near 3,000 km. Very recently, new techniques, discussed below, have revised this value.

Spectroscopy has shown that Pluto has methane ice and possibly methane gas as well, so there is a thin atmosphere and perhaps also a frost of methane ice on the surface. Most molecular compounds formed of common elements freeze out into solid form at the low temperature (perhaps 40 K) thought to characterize Pluto.

Brightness variations appear to indicate that Pluto has some surface markings and is rotating with a period of 6.39 days. The orientation of the rotation axis has been difficult to determine, but new information indicates a large tilt, probably about 112° (with respect to the perpendicular to the planet's orbital plane, which is itself tilted 17° with respect to the ecliptic).

The small size and apparent low mass of Pluto, combined with its unusual orbit, led to the speculation that perhaps it was once a satellite of one of the giant planets, and that it somehow escaped to follow its own path around the Sun. Recall the disorder of the satellites of Neptune; it has been shown that if Pluto had once been a satellite of Neptune, a three-way collision among Pluto, Triton, and Nereid could have altered the orbits of the latter two and allowed Pluto to escape into its present orbit around the Sun. This would have required an extremely special set of circumstances, with the three meeting at just the right speed and direction, but this does not mean it could not have happened. After all, something unusual must have occurred to create the motions of Triton and Nereid. Thus it seemed that a whole set of anomalous motions in the outer solar system could be explained by a single catastrophic event.

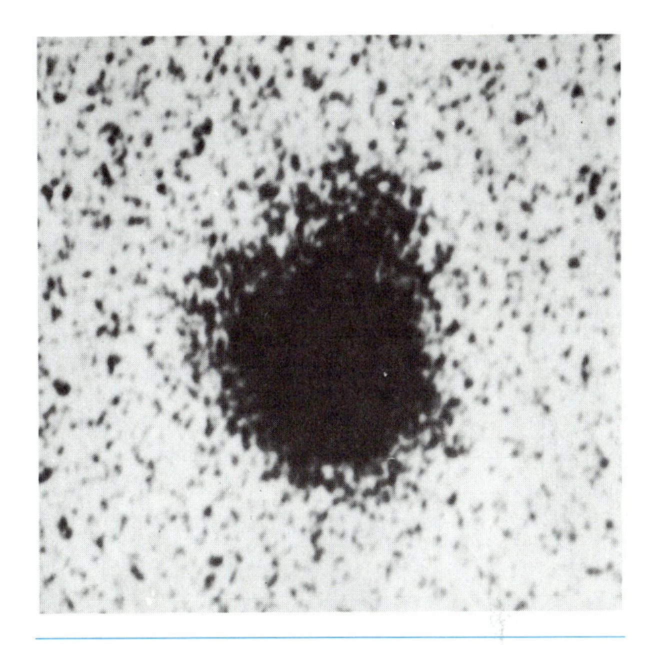

FIGURE 15.23 THE DISCOVERY OF CHARON. *This is the photograph on which Pluto's satellite, the bulge at upper right, was discovered.* (U.S Naval Observatory)

This theory has recently suffered a setback, however. Careful examination of photographs of the planet have shown that Pluto has an extension that moves back and forth from one side to the other (Fig. 15.23). Apparently this extension is a satellite very close to the planet, orbiting Pluto with a period of just over 6 days. To create the distorted appearance of Pluto, the satellite must be relatively large and close to it. Interferometric observations have since confirmed the reality of Pluto's satellite, and have shown that it is separated from the planet by about 19,000 km. The satellite was given the name Charon.

The idea that Pluto should have a satellite of its own does not square with the theory that Pluto itself is an escaped satellite of one of the giant planets. We have to drop that idea, unless Charon split off from Pluto in the cataclysmic process that caused its escape from Neptune, a very unlikely scenario.

The orbital period of Charon is 6.39 days, precisely equal to the rotation period of Pluto. Thus, this

ASTRONOMICAL INSIGHT 15.2

THE SEARCH FOR PLANET X

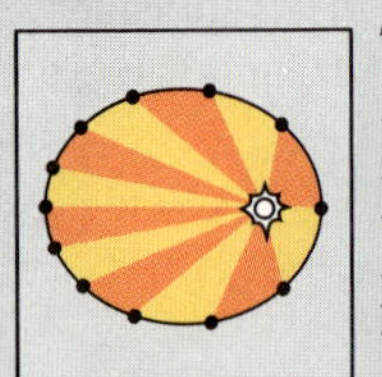

The discovery of Pluto was by no means the end of the search for new planets. Quite to the contrary, this event sharpened the interest in continuing the search, and it was carried on for an extended period.

The technique by which Pluto was discovered was tedious but efficient, in that it was thorough. Observers accomplished the task by comparing photographs taken at different times (usually a few days apart), to see whether any of the objects in the field of view moved. This would ordinarily be very difficult to do, particularly in areas of the sky that are crowded with stars, but was made easier with the aid of a special instrument called a **blink comparator.** This is a microscope for viewing two photographic plates simultaneously, with a small mirror that flips back and forth, providing, in rapid succession, alternating views of the two plates. If the plates are mounted so that the images of fixed stars are perfectly aligned, then an object that moved between the times of the two exposures will appear to blink on and off at two separate locations, while all the fixed objects are steady. Examining a region of the sky using this technique, therefore, consists of mounting in the blink comparator two plates taken at different times, then systematically scanning them with the movable eyepiece so that each object could be checked, to see if it is blinking. The entire job could take days or weeks for each pair of plates.

The blink comparator technique, while time-consuming, has the capability of allowing thorough searches of large regions of the sky. The search for a tenth planet, often referred to as Planet X, continued at Lowell Observatory for 13 years after the discovery of Pluto. Clyde Tombaugh, who had discovered Pluto, was the principal worker in the extended search. Nothing was found.

During and since that time, occasional reports of evidence for a tenth planet have surfaced, but always the evidence has been indirect. Reanalysis of the motion of Neptune has convinced some astronomers that there really may be discrepancies attributable to gravitational perturbations caused by another planet. Other researchers have pointed to the motions of certain comets as indicative of gravitational influence exerted by Planet X. Most of these predictions have suggested a distance from the Sun of 50 to 100 AU. The motion of Halley's comet, a bright and very famous object that visits the inner solar system every 76 years, has been analyzed by several scientists, who find that its arrival near the Sun is usually several days later than it should be, according to the laws of motion. This has been attributed by some to the gravitational influence of a tenth planet, but no such object could be found at the predicted position. One of the predictions based on the motion of Halley's comet was rather spectacular: The inferred Planet X had a mass three times that of Saturn, and an orbit inclined by 120° to the ecliptic!

Modern information on comets has explained the discrepancy in the motion of Halley's comet another way. As it nears the Sun and heats up, volatile gases vaporize and stream outwards from the heated side, acting like a rocket exhaust to slow the comet's motion (this process is described in more detail in Chapter 16).

The routine search for new objects continues today, with occasional success, but no new planets have been found. There was a brief flurry of excitement in 1977, when a Sun-orbiting body was discovered whose elliptical path carries it almost as far from the Sun as Uranus, and as close in as Saturn. It is clearly not a planet, but the exact nature of Chiron, as it is called, is uncertain. It may be an asteroid in a highly unusual orbit, or an old cometary nucleus that never approaches the Sun closely enough to develop a luminous tail (see Chapter 16). Perhaps eventually Planet X will be found, but if so, it will undoubtedly be a dim, cold, distant object, much too faint to be seen without a large telescope.

is a case in which both the satellite and the parent planet are locked in synchronous rotation. This is the only case in the solar system of mutually synchronous rotation (although it happens commonly in closely spaced double stars), and it is not clear how it occurred within the 4.5 billion year age of the solar system. To an observer on the surface of Pluto, Charon would appear fixed in position overhead, while the background stars streamed past because of the planet's rotation.

Charon is unusually large compared to the planet it orbits. No doubt this has helped create sufficiently strong tidal forces to produce Pluto's synchronous rotation. Pluto might be accurately described as a double planet.

The orbital plane of Charon is tilted about 68° with respect to the plane of Pluto's orbit about the Sun, and Charon orbits in the retrograde direction (Fig. 15.24). The tentative conclusion that Pluto's rotation axis is tipped 112° with respect to the perpendicular to its orbital plane is based on the assumption that Charon orbits in the plane of Pluto's equator.

The discovery of Charon gave astronomers a chance to determine the mass of the planet, using Kepler's third law. It was found that Pluto's mass is

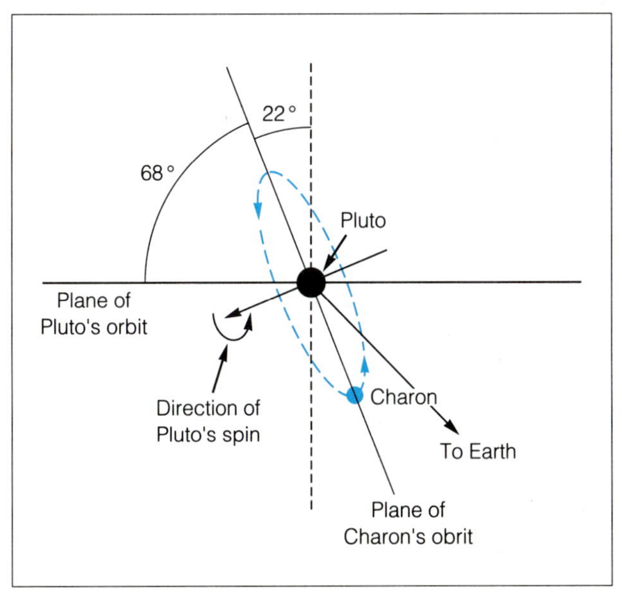

FIGURE 15.24 CHARON'S ORBIT. *This shows the orientation of Pluto's rotation axis and Charon's orbit, with respect to the plane of Pluto's orbit about the Sun (which is tilted 17° with respect to the plane of the Earth's orbit).*

only about 0.002 of the Earth's mass. The diameter of Pluto has recently been accurately measured as well, again with help from Charon. It happens that the orbital plane of Charon has just became aligned with the direction toward Earth, so that Charon eclipses Pluto each orbit. The duration of the eclipses allows astronomers to determine the diameter of Pluto, which is found to be 2,200–2,300 km. This, combined with the mass of Pluto, yields a density for the outermost planet of about 2 grams/cm^3. Evidently, Pluto is not a rocky planet, as had previously been assumed, but rather is made largely of ices, much like the satellites of Jupiter and Saturn.

PERSPECTIVE

We have thoroughly digested all of the available information on the planets. We certainly have many unanswered questions about them, but on the other hand, a number of systematic trends have revealed themselves to us. Two of the outer three planets fit the pattern established by the others, being gaseous giants with many apparent similarities to Jupiter and Saturn. The last planet, Pluto, is mysterious, and remains an anomaly to be reckoned with.

Nearly all the pieces of the puzzle are now in place. We need only examine the nonplanetary bodies roving through the solar system to see what they can tell us of its origins, and then we will address the formation of the system. We turn our attention now to the minor bodies, the asteroids, comets, and meteoroids.

SUMMARY

1. Uranus has been examined by the *Voyager 2* space probe and has been found similar to the other giant planets in that it has a banded atmosphere and an internal structure that is highly differentiated, with a solid core and layers of liquid and gaseous light elements above the core.
2. The spin axis of Uranus, as well as the orbital plane of its fifteen satellites, is tilted 98°, creating strange seasonal effects and an unusual magnetosphere.
3. Uranus has fifteen satellites and at least nine thin, dim rings, whose structure is governed by the satellites in the same ways as the rings of Saturn.

4. The major satellites of Uranus show varying degrees of geological activity; the innermost of these moons, Miranda, is surprisingly active geologically, with many very young surface features.
5. Neptune was discovered because of its gravitational effects on Uranus, a discovery that vindicated the laws of motion and gravitation developed by Newton.
6. Neptune's two satellites have unusual orbital properties, possibly created by a major gravitational disturbance at some time in the past.
7. Pluto, the ninth planet, was discovered fortuitously close to the position predicted by Percival Lowell on the basis of suspected irregularities in the orbit of Neptune.
8. Pluto is unusual in that its orbit is highly eccentric and tilted, and it much smaller and probably different in structure than the other outer planets.
9. Pluto has a relatively large satellite that provides data on the planet's mass and density, showing that Pluto is probably composed largely of ice.

REVIEW QUESTIONS

Thought Questions

1. Summarize what has been learned about Uranus and Neptune from Earth-based observations, and discuss the advances made when a probe such a *Voyager 2* is able to make close-up observations.
2. How long is a day at the north pole of Uranus? What is the length of day at the equator when the north pole is pointed 90° away from the Sun?
3. Summarize the major differences between Uranus and Neptune, and between Jupiter and Saturn.
4. The innermost major satellite of Jupiter, Io, is undergoing current geological activity. The innermost major moon of Uranus, Miranda, apparently has undergone geological activity recently. Why is such activity readily explained for Io, but quite unexpected and unexplained for Miranda?
5. How are the rings of Uranus similar to those of Saturn, and in what ways are they different?
6. Contrast the means by which Uranus, Neptune, and Pluto were discovered.
7. Discuss the kinds of information *Voyager 2* might gather on Neptune when the encounter occurs in August, 1989.
8. How are the satellites and possible rings of Neptune different from those of the other major planets?
9. Summarize the many ways in which Pluto is unlike any other planet in the solar system.
10. Explain why Pluto has been forced into synchronous rotation with Charon, while no other planet is in synchronous rotation with its satellite.

Problems

1. Suppose that when Uranus occulted the light from a background star, it took 1 hour, 40 minutes to pass in front of the star. The apparent motion of Uranus as seen from the Earth is 0.0005 arcseconds per second of time. Calculate the angular diameter of Uranus.
2. When Uranus is situated directly between the Sun and Neptune, the gravitational disturbance on its orbit caused by Neptune is maximized. At this time, Neptune is one-half as far from Uranus as the Sun is. The mass of the Sun is about 20,000 times greater than the mass of Neptune. Calculate the fraction by which the Sun's attraction for Uranus is reduced because of Neptune's gravitational force on Uranus.
3. Use Kepler's third law to calculate the semimajor axis of Nereid's orbit around Neptune. (Note: This will be easiest if you use the form of the law derived by Newton, including the sum of the masses.) How does your answer compare with the typical sizes of the orbits of other satellites of the major planets?
4. By how much does the intensity of sunlight on Pluto vary during the planet's orbit around the Sun (its distance from the Sun varies from about 30 Au to about 50 AU)? How does the Sun's intensity on Pluto at perihelion compare with the intensity of sunlight on the Earth?

ADDITIONAL READINGS

Allen, D. A. 1983. Infrared views of the giant planets. *Sky and Telescope* 65(2):110.
Beatty, J. Kelly. 1986. A place called Uranus. *Sky and Telescope* 71(4):333.
———. 1985. Pluto and Charon: The dance begins. *Sky and Telescope* 69(6):501.
———. 1987. Pluto and Charon: The dance goes on. *Sky and Telescope* 74(3):248.
———. 1986. Voyager's triumph. *Sky and Telescope* 72(4):336.
———, B. O'Leary, and A. Chaikin, eds. 1981. *The new solar system.* Cambridge, Eng.: Cambridge University Press.

Chaikin, Andrew. 1986. Voyager among the ice worlds. *Sky and Telescope* 71(4):338.

Croswell, Ken. 1986. Pluto: Enigma on the edge of the solar system. *Astronomy* 14(7):6.

Elliott, J. L., E. Dunham, and R. L. Mills. 1977. The discovery of the rings of Uranus. *Sky and Telescope* 53(6):412.

Harrington, R. S., and B. J. Harrington. 1980. Pluto: Still an enigma after 50 years. *Sky and Telescope* 59(6):452.

Hunten, D. M. 1975. The outer planets. *Scientific American* 233(3):130.

Ingersoll, Andrew P. 1987. Uranus. *Scientific American* 256(1):38.

Johnson, Torrence V. 1987. The moons of Uranus. *Scientific American* 256(4):48.

Mullholland, Derrall. 1982. The ice planet: Pluto. *Science 82* 3(10):64.

Overbye, Dennis. 1986. Voyager was on target again: At Uranus. *Discover* 7(4):70.

Tombaugh, Clyde W. 1986. The discovery of Pluto. *Mercury* 15(3):66.

———. 1979. The search for the ninth planet. *Mercury* 8(1):4.

CHAPTER 16

SPACE DEBRIS

Halley's comet in 1985. (© 1985 Anglo-Australian Telescope Board)

CHAPTER PREVIEW

The planets are the dominant objects among the inhabitants of the solar system (except, of course, for the Sun), but they are not entirely alone as they follow their regular paths through space. The abundant craters on planetary and satellite surfaces have shown us that there must have been a time when interplanetary rocks and gravel were more abundant than they are today, covering any exposed surface with impact craters. Today the rate of cratering is much lower than it once was, but some vestiges of the space debris that caused it still remain, orbiting the Sun and occasionally becoming obvious to us as they pass near the Earth or enter its atmosphere.

There are at least four distinct forms of interplanetary matter, some of which are closely related. In this chapter we will discuss the **asteroids** (more commonly, the **minor planets**), myriad rocky chunks up to several hundred kilometers in diameter that orbit the Sun between Mars and Jupiter; the **comets,** whose ephemeral and striking appearances have caused us to pause and marvel since ancient times; **meteors,** the bright streaks often visible in our nighttime skies, and the objects that create them; and the **interplanetary dust,** a collection of very fine particles that fill the void between the planets.

The Minor Planets

As we have already seen, some of the early students of planetary motions were concerned with the distances of the planets from the Sun. Kepler spent enormous amounts of time and energy seeking a mathematical relationship that would describe the distances of the planets in terms of geometrical solids. What he finally did discover was something different, a relationship between the distances and orbital periods. Kepler's third law can be used to predict the period of a planet, given its distance from the Sun; or its distance, if its period is known. But it did not provide any underlying basis for explaining why there are planets only at certain distances from the Sun.

In 1766, a German astronomer named J. D. Titius found a simple mathematical relationship that seemed to accomplish what Kepler had set out to do. Titius discovered that if we start with the sequence of numbers 0, 3, 6, 12, 24, 48, and 96 (obtained by doubling each one in order), then add 4 to each and divide by 10, we end up with the numbers 0.4, 0.7, 1.0, 1.6, 2.8, 5.2, and 10.0, corresponding closely to the observed planetary distances in astronomical units. A few years after Titius found this numerological device, it was popularized by another German astronomer, Johann Bode, and eventually became known as Bode's law, or the Titius-Bode law.

The sequence of numbers dictated by Bode's law includes one, 2.8 AU, where no planet was known to exist. After the discovery of Uranus in 1781, and the recognition that its distance fits the sequence (the next number is $(192 + 4)/10 = 19.6$, and the semimajor axis of Uranus's orbit is 19.2 AU), a great deal of interest arose in the idea that there ought to be a planet at 2.8 AU, since Bode's law was doing so well in predicting the positions of the others.

A deliberate search for such a planet was begun in 1800, but the discovery of the sought-after object came accidentally, when an Italian astronomer named G. Piazzi noted a new object on the night of January 1, 1801, and within weeks found from its motion that it was probably a solar system body.

When the orbit of the new object was calculated, its semimajor axis turned out to be 2.77 AU, placing the object in solar orbit between Mars and Jupiter, where a prediction had been made that a new planet might be found. The new fifth planet was named Ceres.

A little over a year after the discovery of Ceres, a second object was found orbiting the Sun at approximately the same distance, and was named Pallas. Because of their faintness, both Ceres and Pallas were obviously very small bodies, and were not respectable

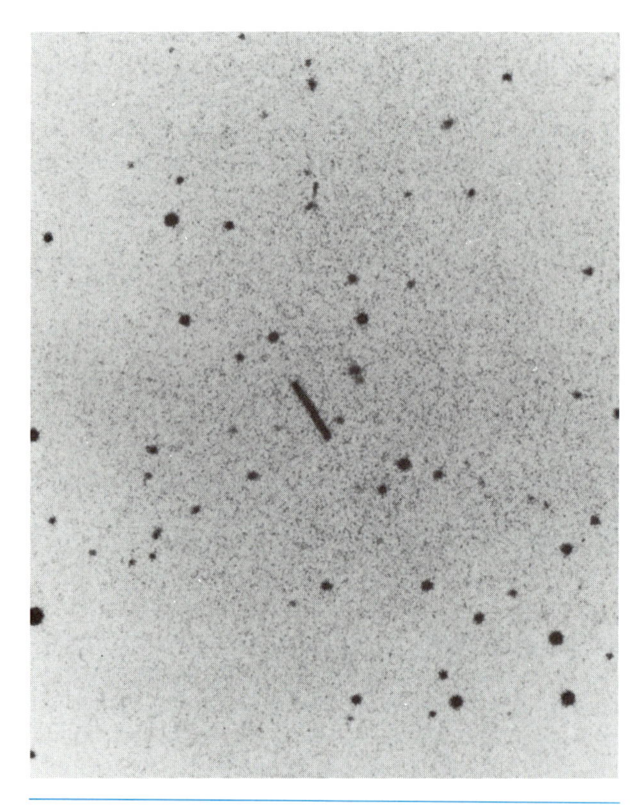

FIGURE 16.1 ASTEROID MOTION. *The elongated image in this photo is the trail made by an asteroid that moved relative to the fixed stars during the twenty-minute exposure. Most new asteroids are discovered on photos of this type, although the first few were spotted visually.* (Eleanor F. Helin, Palomar Observatory)

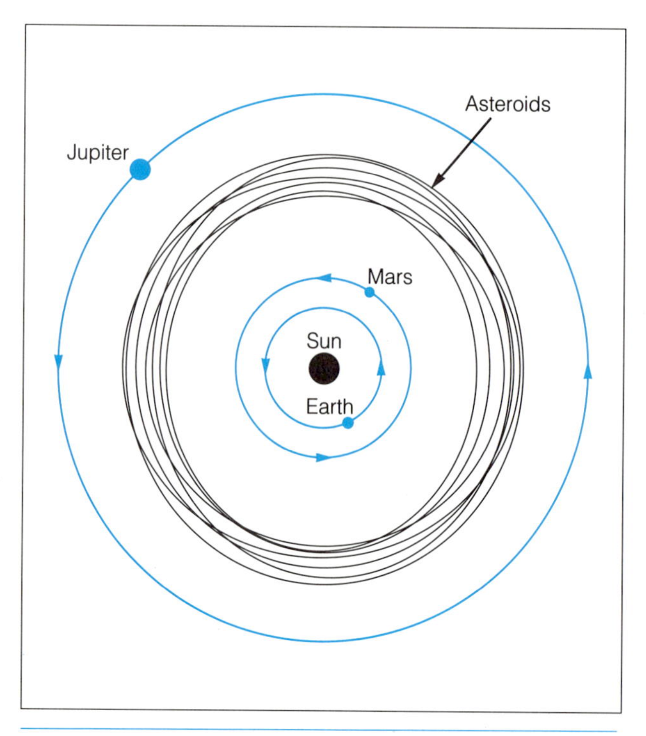

FIGURE 16.2 ASTEROID ORBITS. *This shows typical orbital paths for the majority of asteroids, which orbit the Sun between Mars and Jupiter.*

planets. By 1807, two more of these **asteroids,** as they were then being called, had been found and were designated Juno and Vesta. A fifth, Astrea, was discovered in 1845, and in the next decades, vast numbers of these objects began to turn up. The efficiency of finding them was improved greatly when photographic techniques began to be used. A minor planet or asteroid will leave a trail on a long-exposure photograph, because of its orbital motion (Fig. 16.1).

Today the number of known asteroids is in the thousands, with about 2,000 of them sufficiently well observed to have had their orbits calculated (Fig. 16.2) and logged in catalogs. The total number is probably much more, perhaps 100,000.

It is possible to deduce some of the properties of the asteroids from a variety of observational evidence. Measurements of their brightnesses can lead to size estimates; spectroscopy of light reflected from their surfaces provides information on their chemical makeup; and direct analysis of meteorites that may be remnants of asteroids adds data on their internal properties.

The largest asteroids were the first to have their diameters estimated, simply by calculating what size they had to be to reflect the amount of light observed. Much more recently, infrared brightness measurements of both large and small asteroids have provided information on their sizes, since at temperatures of a few hundred degrees they glow at infrared wavelengths. The Stefan-Boltzmann Law (Chapter 6) then can be used to determine the total surface area from the intensity of the emission.

A few direct measurements of asteroid sizes have also been possible, in cases in which these objects have passed sufficiently close to Earth that their angular sizes could be directly determined. In recent times, the angular diameters of some asteroids have been successfully measured by the technique of interferometry.

The result of using these assorted techniques is that asteroids are known to have a wide range of diameters (Table 16.1), a few as large as several hundred kilometers. Ceres, the largest, is about 1,000 km in diameter, and the majority are rather small, with diameters of 100 or 200 km or less. The largest are apparently spherical, whereas the smaller ones often are jagged, irregular chunks of rock (Fig. 16.3), varying in brightness as they tumble through space, reflecting light with different efficiencies on different sides. A few asteroids are binary, consisting of two chunks orbiting each other as they circle the Sun. The total mass of all the asteroids together is small by planetary standards, amounting to only an estimated 0.04 percent of the mass of the Earth.

The compositions derived from spectroscopic analyses are highly varied. The principal ingredients range from metallic compounds to nearly metal-free

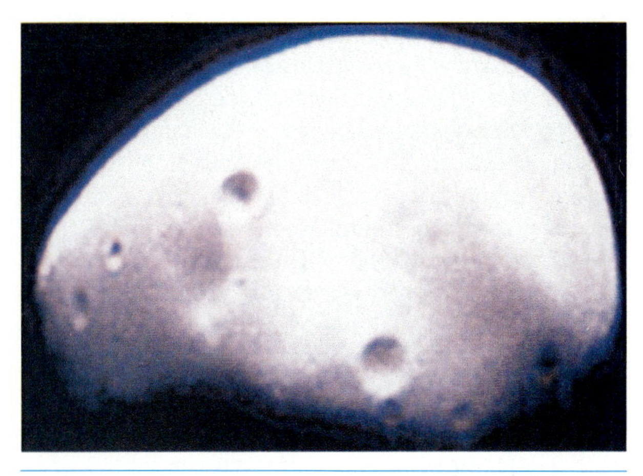

FIGURE 16.3 A PORTRAIT OF AN ASTEROID? *Many asteroids probably bear a general resemblence to this irregularly shaped object, which is Deimos, one of the tiny Martian moons.* (NASA)

TABLE 16.1 SELECTED ASTEROIDS

No.	Name	Year of Discovery	Diameter (km)	Mass (g)	Period (years)	Distance (AU)*
1	Ceres	1801	760	1×10^{24}	4.60	2.766
2	Pallas	1802	480	3×10^{23}	4.61	2.768
3	Juno	1804	200	2×10^{22}	4.36	2.668
4	Vesta	1807	480	2×10^{23}	3.63	2.362
6	Hebe	1847	220	2×10^{22}	3.78	2.426
7	Iris	1847	200	2×10^{22}	3.68	2.386
10	Hygiea	1849	320	6×10^{22}	5.59	3.151
15	Eunomia	1851	280	4×10^{22}	4.30	2.643
16	Psyche	1852	280	4×10^{22}	5.00	2.923
51	Nemausa	1858	80	9×10^{20}	3.64	2.366
511	Davida	1903	260	3×10^{22}	5.67	3.190

*Periods are orbital periods; distances are orbital semimajor axes.

ones, from carbon-dominated to silicon- bearing minerals, and there are several classes whose composition is not known. Among the biggest asteroids in the main belt between Mars and Jupiter, the majority (about three-quarters) are carbonaceous, meaning that they contain carbon in complex molecular forms. Most of the rest are composed chiefly of silicon-bearing compounds, and about 5 percent are metal-rich, the primary metals being nickel-iron mixtures.

KIRKWOOD'S GAPS: ORBITAL RESONANCES REVISITED

As increasing numbers of asteroids were discovered and cataloged throughout the nineteenth century, calculations of their orbits showed remarkable gaps at certain distances from the Sun. One such gap is at 3.28 AU, and another is at 2.50 AU.

The explanation for these was offered by Daniel Kirkwood in 1866, when he realized that these distances correspond to orbital periods that are simple fractions of the period of Jupiter (Fig. 16.4). The giant planet, at its distance of 5.2 AU from the Sun, takes 11.86 years to make a trip around it, whereas an asteroid at 3.28 AU, if one existed there, would have a period exactly half as long, 5.93 years. Thus this asteroid and Jupiter would be lined up in the same way frequently, every time Jupiter made one orbit. The asteroid would therefore be subjected to regular tugs by Jupiter's gravity, and apparently this has prevented asteroids from staying in such orbits, at this and other distances where the orbital periods would result in regular alignments. The gap at 2.50 AU corresponds to orbits with a period one-third that of Jupiter. These orbital resonances are exactly analogous to the ones created by the moons of Saturn, which are responsible for some of the gaps in the ring system of that planet (see Chapter 14). Since Jupiter is the most massive and the nearest of the outer planets to the asteroid belt, it has by far the greatest effect in producing gaps, but in principle the other planets could do the same thing.

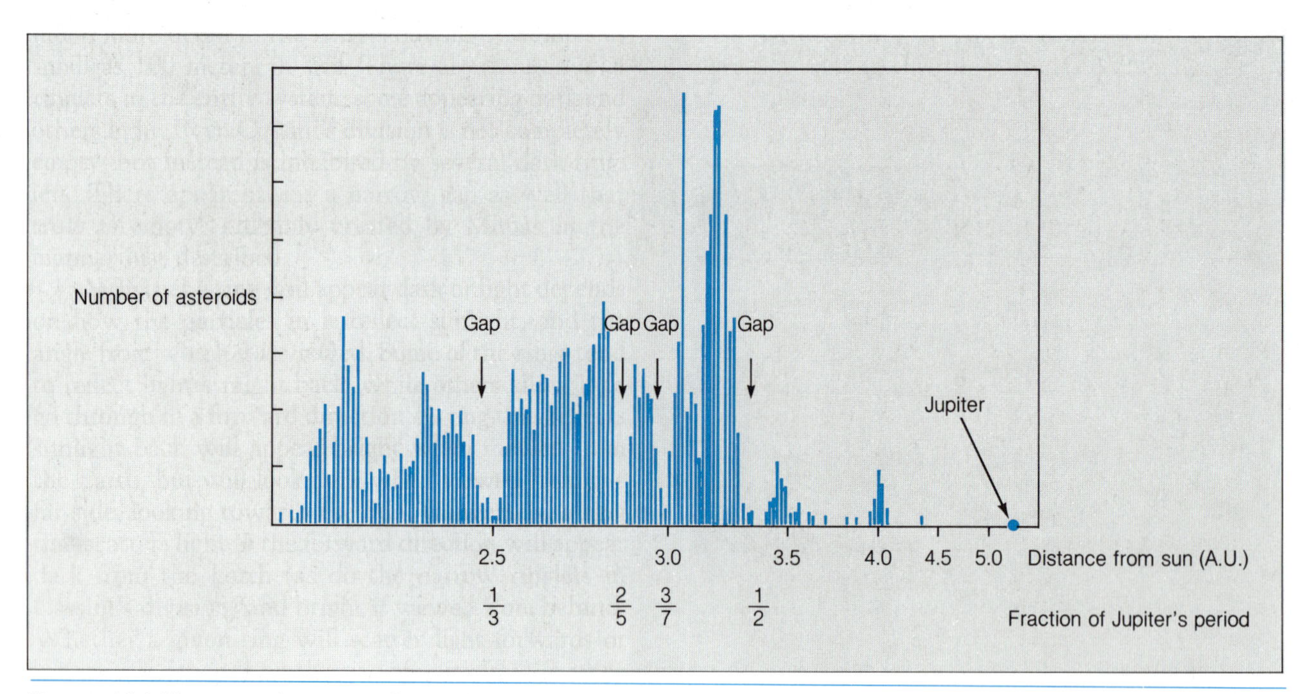

FIGURE 16.4 KIRKWOOD'S GAPS. *This graph shows the distribution of asteroids relative to the oribts of the Earth, Mars, and Jupiter. Gaps appear at the distances from the Sun where asteroids would have exactly one-half, one-third, or other simple fractions of the orbital period of Jupiter.*

— THE ORIGIN OF THE ASTEROIDS —

The most natural explanation of the asteroids seemed for a long time to be the breakup of a former planet, creating a swarm of fragments that continued to orbit the Sun. This argument was weakened when the total mass of the asteroids was estimated and found to be much less than that of any ordinary planet. An additional, rather strong argument against the planetary remnant hypothesis can be cited: There is no known reasonable way to break a planet apart once it has formed, whereas it is easy to understand why material orbiting the Sun at the position of the asteroid belt could never have combined into a planet in the first place. It became simpler to accept the idea that the debris never was part of a planet, rather than to find a way to have it first form one and then break apart.

The invisible hand of Jupiter's gravity is invoked again. If, as it is suspected, the planets formed from a swarm of debris orbiting the young Sun in a disk, no planet could have formed in a location where the pieces of debris could not stick together. Calculations show that once Jupiter had formed, its immense gravitational force would have stirred up the material near its orbit, so that collisions between particles would occur at speeds too great to allow them to stick together. It is as though Jupiter wielded a giant spatula, stirring up the debris near it and keeping it spread out as a loose collection of rocky fragments. Even today, Jupiter is still at work, keeping the asteroids mixed up, occasionally causing collisions between them that can sometimes break them up.

The study of meteorites, some of which probably originated as pieces of asteroids that broke apart in collisions, gives us a chance to examine material from the early solar system. It is interesting to note that some of the asteroids apparently have undergone differentiation, developing nickel-iron cores, because some meteorites are almost pure chunks of this material. There must have been sufficient heat inside some of the asteroids to create a molten state, allowing the heavy materials to segregate from the rest.

In contrast with the nickel-iron asteroids, some of the other types, particularly the carbonaceous ones, apparently have undergone almost no heating, since they contain high quantities of volatile elements that would have been easily cooked out.

— COMETS: FATEFUL MESSENGERS —

Among the most spectacular of all the celestial sights are the comets. With their brightly glowing heads and long, streaming tails, along with their infrequent and often unpredictable appearances, these objects have sparked the imagination (and often the fears) of people through the ages.

In antiquity, when astrological omens were taken very seriously, great import was attached to the occasion of a cometary appearance (Fig. 16.5). Ancient descriptions of comets are numerous, and in many

FIGURE 16.5 CALAMITY ON EARTH ASSOCIATED WITH THE PASSAGE OF COMETS. *This drawing is from a seventeenth century book describing the universe.* (The Granger Collection)

cases these objects were thought to be associated with catastrophe and suffering.

Included among the teachings of Aristotle was the notion that comets were phenomena in the Earth's atmosphere. There was no good evidence for this idea; apparently Aristotle adopted it because he believed that any phenomena that were changeable were associated with the Earth, whereas the heavens were perfect and immutable. In any case, this idea was accepted for centuries to come. Tycho Brahe in 1577 was able to prove that comets were too distant to be associated with the Earth's atmosphere, because they do not exhibit any parallax when viewed from different positions on the Earth. If a comet were really located only a few kilometers or even a few hundred kilometers above the surface of the Earth, its position as seen from the Earth would change from one location to another. Tycho was able to show that this was not the case, and that therefore comets belonged to the realm of space.

Halley, Oort, and Cometary Orbits

A major advance in the understanding of comets was made by a contemporary and friend of Newton, Edmond Halley. Aware of the power of Newton's laws of motion and gravitation, Halley reviewed the records of cometary appearances and noted one outstanding regularity. Particularly bright comets seen in 1531, 1607, and 1682 seemed to have similar properties, and Halley suggested that in fact all three were appearances of the same comet, orbiting the Sun with a 76-year period.

Calculations showed, using Kepler's third law, that for a period of 76 years, this object must have a semimajor axis of nearly 18 AU. Halley realized, therefore, that in order to appear as dominant in our skies as the comet does, it must have a highly elongated orbit (Fig. 16.6), so that it comes close to the Sun at times, even though its average distance is well beyond the orbit of Saturn, nearly as far out as Ura-

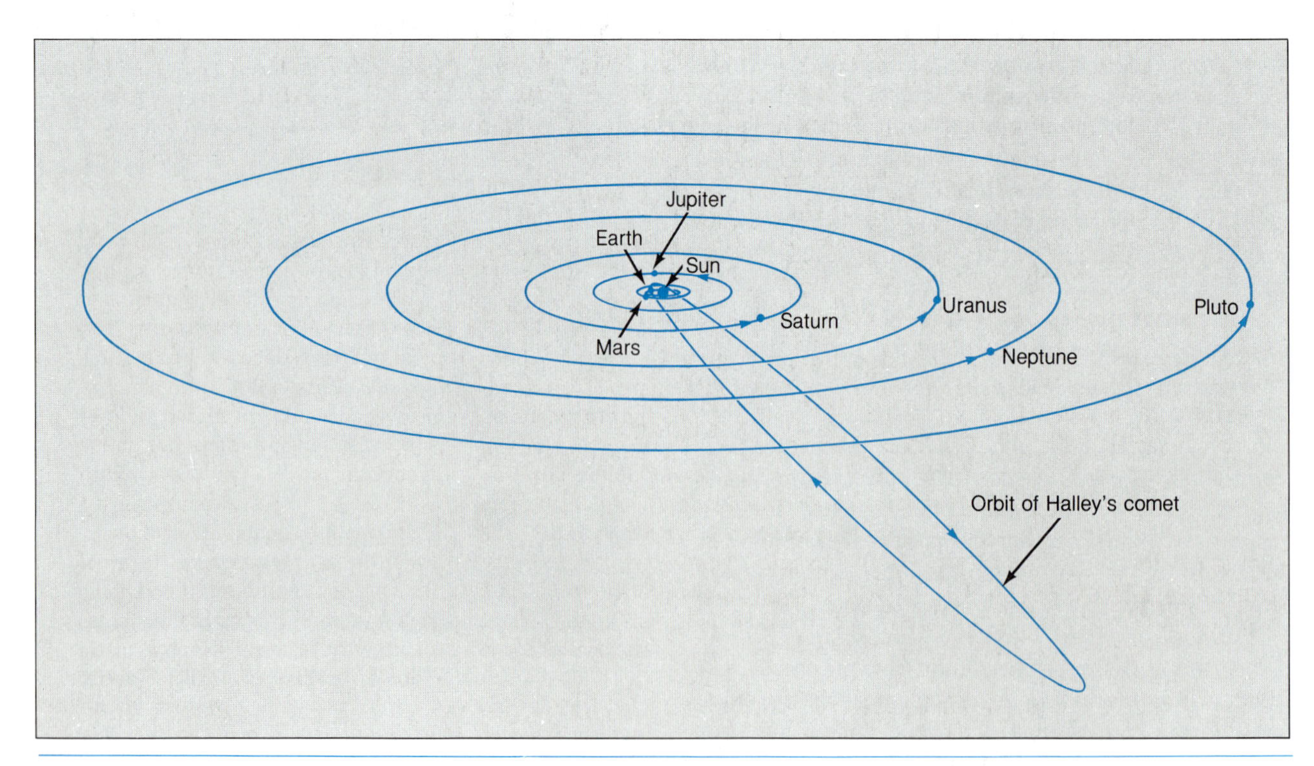

FIGURE 16.6 A COMETARY ORBIT. *This is a rough scale drawing of the orbit of Halley's comet. Most comets actually have much more highly elongated orbits than this, and correspondingly longer periods.*

ASTRONOMICAL INSIGHT 16.1

ENCOUNTERS WITH HALLEY

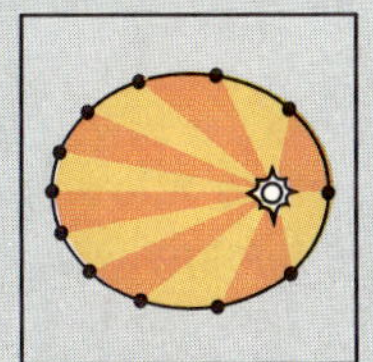

Even though the 1986 visit of Halley's comet to the inner solar system did not provide very good views from Earth (particularly from the Northern Hemisphere), scientists had an extraordinary opportunity to observe the comet from close range, using space probes. No less than five spacecraft were sent from Earth to rendezvous with the comet, all within a few days of each other in March, 1986, when Halley crossed the plane of the Earth's orbit after passing behind the Sun. In addition to these five missions, two other spacecraft, originally launched for other purposes, observed the comet from locations in space far from the Earth.

The five missions designed to encounter Halley's comet included the European Space Agency's *Giotto* mission, two Soviet *Vega* spacecraft, and two Japanese probes, called *Sakigake* and *Suisei.* This collection of spacecraft carried an assortment of instruments ranging from magnetic field detectors to spectrographs to particle detectors to cameras for obtaining images of the comet. *Giotto* passed closest to the nucleus of Halley, zipping by only about 600 km from it, while the *Vega* and the Japanese pairs passed by at greater distances. Both *Vegas* and *Giotto* suffered damage caused by impacts of dust particles from the comet; *Giotto* was sent spinning by an impact just seconds before its closest approach to the nucleus, and communications were lost temporarily. Nevertheless, it succeeded in obtaining unprecedented images of the nucleus.

The two existing spacecraft that observed Halley were the *Pioneer Venus* probe, which has been orbiting Venus since 1978, and the *International Cometary Explorer* (or *ICE*), a probe originally sent into space to measure interplanetary particles and magnetic fields. The *ICE* craft actually carried out the first space rendezvous with a comet, passing through the tail of comet Giacobini-Zinner in September, 1985, before making more remote observations of Halley in early April, 1986. The *Pioneer Venus* spacecraft, able to view the comet from a rather different vantage point than Earth, observed ultraviolet emission from the extended halo of hydrogen gas that surrounds the comet. Since this hydrogen is a by-product of water, the major constituent of the nucleus, measurements of the hydrogen led directly to estimates of the rate at which the nucleus was losing mass as the Sun's heat caused water ice to vaporize from the nucleus.

As expected, the combination of all the space encounters and observations of Halley greatly extended our knowledge of comets. In general, the results supported the so-called "dirty snowball theory," in which a comet is viewed as a loose conglomerate of rock and gravel held together by ice that vaporizes as the comet passes through the inner solar system. There were, however, some surprises along with a great amount of refinement of this picture; the major discoveries are described in the text.

The encounters with Halley, besides all of their scientific yield, also prompted a remarkable degree of international cooperation. The Soviet *Vega* spacecraft, which reached the comet before *Giotto,* provided essential information about the exact position of the nucleus, so that *Giotto* could be steered precisely to its close encounter. On the ground, a massive international comet-observing program known as International Halley Watch was carried out over several months and involved free and rapid exchange of information among scientists from many countries.

A remarkable future comet encounter mission is now being planned in the United States. A spacecraft called the *Comet Rendezvous and Asteroid Flyby* (or simply *CRAF*) will, if the plan is funded, fly by an asteroid and then match orbits with Comet Wild 2 in 1995, and stay with it in its solar orbit for at least 3 years thereafter, observing the nucleus from very close range (30 km) as it approaches the Sun and develops its coma and tail. This will provide even more detailed information on the life and times of a comet than the recent Comet Halley spectacular.

nus. Such an eccentric orbit, as a very elongated ellipse is called, had not previously been observed, even though Newton's laws clearly allowed the possibility.

Since Halley's time, searches of ancient reports of comets have revealed that Halley's comet has been making regular appearances for many centuries. The earliest records are provided by the ancient Chinese astronomers, who apparently observed its every appearance for well over 1,000 years, possibly beginning as early as the fifth century B.C.

The most recent visit of Halley's comet was in 1985 and 1986, when it did not pass very close to the Earth (Fig. 16.7), and the best views were from the Southern Hemisphere (Fig. 16.8). A much more spectacular appearance occurred on the comet's previous visit in 1910 (Fig. 16.9), when the Earth actually passed through the rarefied gases of its tail.

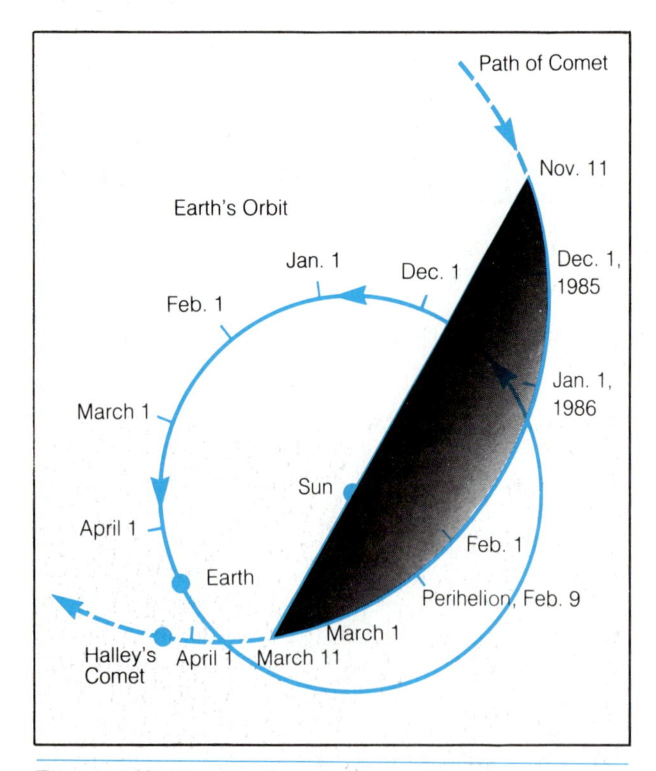

FIGURE 16.7 THE PATH OF HALLEY'S COMET IN 1986. *This graph shows the path of Halley's comet and the relative position of the Earth during the recent appearance. Note that the comet was on the far side of the Sun when it passed closest to it.*

Other spectacular comets have been seen (Fig. 16.10), and a number have rivaled Halley's comet in brightness. Traditionally, a comet is named after its discoverer, and there are astronomers around the world who spend long hours peering at the nighttime sky through telescopes, looking for a piece of immortality.

When a new comet is discovered, a few observations of its position are sufficient to allow computation of its orbit. The results of many years of comet-watching have shown that there are many comets whose orbits are so incredibly stretched out that their periods are measured in thousands or even millions of years. These comets, for all practical purposes, are seen only once. They return thereafter to the void of space well beyond the orbit of Pluto, there to spend millenia before visiting the inner solar system again.

The orbits of comets, particularly these so-called long-period ones, are randomly oriented. These comets do not show any preference for orbits lying in the plane of the ecliptic, in strong contrast with the planets, and about half go around the Sun in the retrograde direction.

Consideration of these orbital characteristics, especially the large orbital sizes, led Dutch astronomer Jan Oort to suggest that all comets originate in a cloud of objects that surrounds the solar system. The **Oort cloud,** as it is now called, is envisioned to be a spherical shell with a radius of 50,000 to 150,000 AU, extending, therefore, a significant fraction of the distance to the nearest star, which is almost 300,000 AU from the Sun.

Occasionally, a piece of debris from the Oort cloud is disturbed from its normal path, either by a collision with another object or perhaps by the gravitational tug of a nearby star, and it begins to fall inwards towards the Sun. Left to its own devices, a comet falling in from the Oort cloud would follow a highly elongated orbit with a period of millions of years, appearing to us as one of the long-period comets when it made its brief, incandescent passage near the Sun. In many cases, a comet is not left to its own devices, however. Instead it runs afoul of the gravitational pull of one of the giant planets, most often Jupiter. When this happens, the comet may be speeded up, so that it will escape the solar system entirely after it loops around the Sun. Or it may be slowed

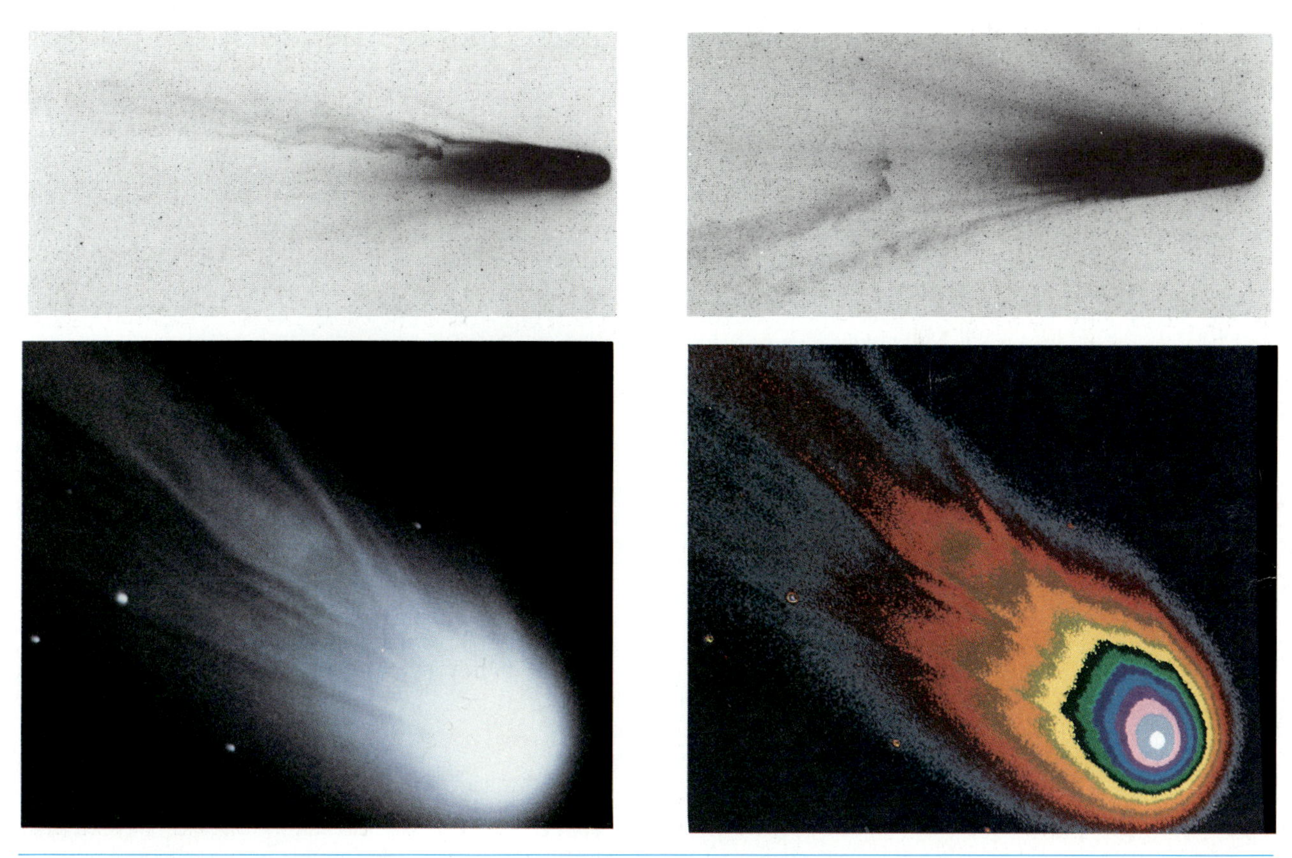

FIGURE 16.8 HALLEY'S COMET IN 1986. *Here are a few of the fine photographs made from Earth during the apparition of Halley's comet in 1985 and 1986. The two negative images on top, obtained only days apart, show changes in the detailed structure of the tail, including a disconnection event, as part of the ion tail broke free of the comet (top right). The two images at the bottom were both made from the same photograph, the color-coded version being made to enhance slight brightness variations. (top:* ©1986 Royal Observatory,Edinburgh; *bottom:* U. Fink, University of Arizona)

down, dropping into a smaller orbit with a shorter period, so that it becomes one of the numerous comets that are seen to reappear frequently (Table 16.2).

By invoking the existence of the Oort cloud, we can account for all the observed cometary orbits, and this concept is widely accepted today. We still must consider how the Oort cloud itself came into being, however. The best guess is that it consists of material left over from the formation of the solar system, and therefore it is thought that the comets are made of very old matter, representing the primordial stuff from which the planets and the Sun were created. The formation of the Oort cloud will be discussed at greater length in Chapter 17, where the formation and evolution of the solar system as a whole are described.

FIGURE 16.9 HALLEY'S COMET AS IT APPEARED IN 1910. (Palomar Observatory, California Institute of Technology)

TABLE 16.2 SELECTED PERIODIC COMETS

Comet*	Period (years)	Semimajor Axis (AU)	Year of Next Appearance
Encke	3.30	2.21	1987.6
Temple (2)	5.26	3.0	1988.6
Schwassmann-Wassermann (1)	6.52	3.50	1987.8
Wirtanen	6.65	3.55	1987.9
Reinmuth (2)	6.72	3.6	1987.8
Finlay	6.88	3.6	1988.2
Borrelly	7.00	3.67	1988.5
Whipple	7.44	3.80	1993.1
Oterma	7.89	3.96	1990.0
Schaumasse	8.18	4.05	1993.0
Wolf	8.42	4.15	1992.9
Comas Sola	8.58	4.19	1987.0
Vaisala	10.5	4.79	1991.9
Schwassmann-Wassermann (2)	16.1	6.4	1998.6
Neujmin (1)	17.9	6.8	2002.7
Crommelin	27.9	9.2	2012.6
Olbers	69	16.8	2025
Pons-Brooks	71	17.2	2025
Halley	76.1	17.2	2062.5

*A number in parentheses following the name of a comet indicates cases where a single observer has discovered more than one comet.

FIGURE 16.10 ANOTHER BRIGHT COMET. *This is Comet West, which appeared in 1970 and was one of the brightest comets in recent years.* (Fr. R. E. Royer)

THE ANATOMY OF A COMET

A remarkably accurate picture of the nature of a comet was developed many years ago by American astronomer Fred Whipple. Whipple's model, sometimes called the "dirty snowball," envisions that for most of its life, a comet is just a frozen chunk of icy material, probably consisting of small particles like gravel or larger boulders, embedded in frozen gases. This picture was developed on the basis of observations of many comets over the years, and its basic features were verified by the close-up examination of Halley's comet in early 1986, when several space probes were able to make observations from close range (Fig. 16.11). Of course, a great many new details were discovered as well.

As a comet passes through the outer reaches of its orbit, far from the Sun, it does not glow, has no tail, and is not visible from Earth. As it approaches the Sun, however, it begins to warm up as it absorbs sunlight, and the added heat causes volatile gases to

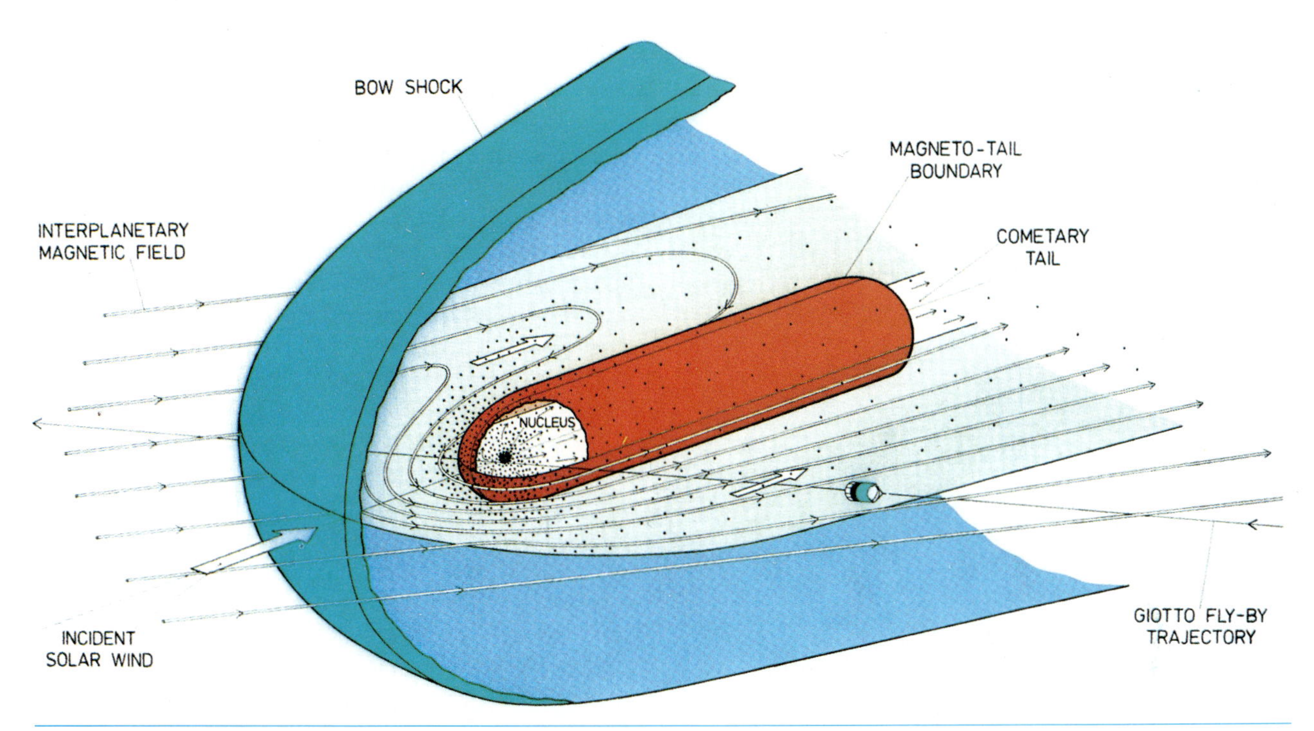

Figure 16.11 Probes to Comet Halley. *This shows the path taken by the European Space Agency's* Giotto *spacecraft, which made the closest approach to the nucleus of the several probes that encountered the comet. The other probes flew along similar paths, crossing ahead of the nucleus, but not as close to it.* (European Space Agency)

escape. A spherical cloud of glowing gas called the **coma** develops around the solid **nucleus** (Fig. 16.12). The principal gases found in the coma of Halley's comet (Table 16.3) are H_2O (about 80 percent by number), carbon dioxide (CO_2; about 3.5 percent), and carbon monoxide (CO; most of the remaining 16.5 percent). Other slightly more complex species, such as ammonia (NH_3) and methane (CH_4) are probably present also, along with hydrogen molecules (H_2). The observed species glow by a process called **fluorescence.** The molecules absorb ultraviolet light from the Sun, causing them to be excited to high energy levels, and then they emit visible light as they return to low energy states. A cloud of hydrogen atoms, resulting from the breakup of the molecules in the coma, extends out to great distances from the nucleus. The visible coma may be as large as 100,000 km in diameter, while the halo of hydrogen atoms may extend as much as ten times farther from the nucleus, as demonstrated by ultraviolet observations of Hal-

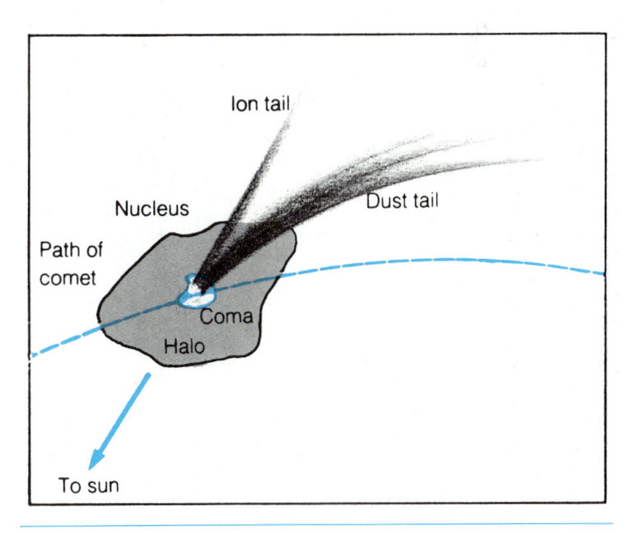

Figure 16.12 Anatomy of a comet. *This sketch illustrates the principal features of a comet, though not necessarily to correct scale.*

TABLE 16.3 DETECTED CONSTITUENTS OF COMETS

Coma		Ion Tail	
Atoms	*Molecules*	*Ions*	*Molecular Ions*
Hydrogen (H)	H_2O	C^+	CO^+
Oxygen (O)	CO_2	Ca^+	CO_2^+
Sulfur (S)	C_2	O^+	H_2O^+
Carbon (C)	C_3	S^+	OH^+
Sodium (Na)	CH	Fe^+	CH^+
Iron (Fe)	CN	Na^+	CN^+
Potassium (K)	CO		N_2^+
Calcium (Ca)	CS		H_3O^+
Vanadium (V)	NH		CS_2^+
Chromium (Cr)	OH		S_2^+
Manganese (Mn)	NH_2		CS^+
Cobalt (Co)	NCN		
Nickel (Ni)	CH_3CN		
Copper (Cu)	S_2		
	HCO		
	NH_3		

ley's comet in 1986 (Fig. 16.13). The solid nucleus is relatively tiny, having a diameter of a few kilometers. The nucleus of Halley was found to be irregularly shaped, with a long axis of about 15 km and a width ranging up to about 10 km (Fig. 16.14). The nucleus was found to have a coating of very dark material (albedo only 0.04), probably a matrix of carbonaceous dust left behind by the evacuation of ices that evaporated previously. Hence, the dirty snowball idea has been modified to incorporate a complete outer coating of dirt, along with the traditional picture of mixed dirt and ice in the interior of the nucleus.

Apparently the gas escaping from the nucleus of a comet is released in jets that can be so powerful that they alter the orbital motion of the comet. It has been known for years that comets can be slowed in this way as they approach the Sun, but the close encounters with Halley's comet provided dramatic confirmation, the images revealing spectacular fountains of gas and dust erupting from beneath the crust of the nucleus (Fig. 16.15). Apparently, these gas jets

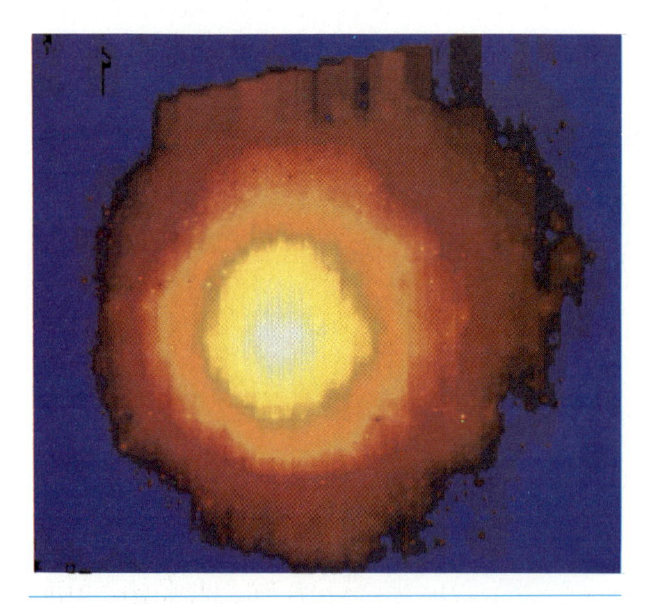

FIGURE 16.13 THE HALO OF HALLEY'S COMET. *This is an ultraviolet image, made by the* Pioneer Venus *spacecraft from Venus orbit, showing the extent of the hydrogen cloud surrounding the comet's nucleus. The hydrogen is a by-product of water vapor that is ejected from the nucleus and then dissociated by the Sun's ultraviolet radiation. The halo was about 500,000 kilometers, or roughtly ⅓ AU in diameter, when this photo was taken.* (LASP, University of Colorado, sponsored by NASA)

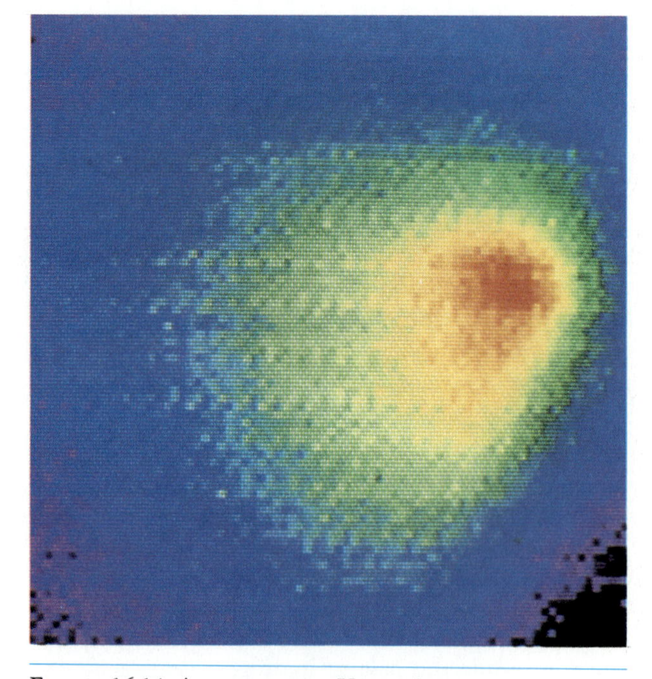

FIGURE 16.14 A CLOSE-UP OF HALLEY'S COMET. *This false-color image was obtained by a Soviet* Vega *spacecraft. The brightest area (in red) represents the nucleus of the comet.* (TASS from SOVFOTO)

are created by heat that is absorbed by the dark crust and conducted into the icy interior, where the heat causes ices to evaporate in a process called **sublimation,** in which solid ice is converted directly into gas. The released gases expand and create outward pressure, which is relieved as the gases break through the crust, forming the jets that are observed. As the nucleus rotates, jets are active only on the sunlit side, so each individual jet turns on and off as it rotates toward and then away from the Sun. Surprisingly, examination of photos of Halley's comet in 1910 indicate that the jets occurred in the same locations then as they did in 1986. Apparently, the jets occur at weak points in the crust of the comet, creating craters that persist over the years between encounters with the Sun's warming radiation.

As a comet nears the Sun, the solar radiation and the solar wind force some of the gas from the coma to flow away from the Sun, forming the tail, which in some cases is as long as 1 AU. Often there are two distinct tails (Fig. 16.16), one formed of gas from the coma, usually containing molecules that have been ionized, such as CO^+, N_2^+, CO_2^+, and CH^+: and another formed of tiny, solid particles released from the ice of the nucleus. The **ion tail,** the one formed of ionized gases, is shaped by the solar wind, and therefore points almost exactly straight away from the Sun at all times. It has been observed many times, for Halley and other comets, that all or part of the ion tail can detach itself and dissipate, then to be replaced by a new tail (see Fig. 16.8). The reasons for this behavior are not well understood.

The other tail, called the **dust tail,** usually takes on a curving shape, as the dust particles are pushed away from the Sun by the force of the light they absorb. This **radiation pressure** is not strong enough to force the dust particles into perfectly straight paths away from the Sun, so they follow curved trajectories that are a combination of their orbital motion and the outward push caused by sunlight.

One of the major discoveries made by the probes that encountered Halley's comet and Comet Giacobini-Zinner (see Astronomical Insight 16.2) was that the ionized gases surrounding a comet create a

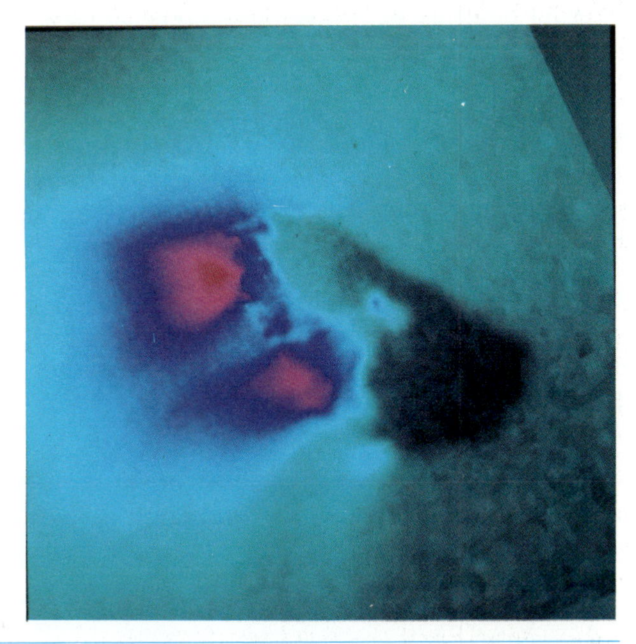

Figure 16.15 The nucleus of Comet Halley. *These are two versions of the same composite of some 60 images obtained by the* Giotto *spacecraft as it flew past the nucleus of the comet. Detailed surface markings in the crust of the nucleus can be seen, as well as the bright jets of warm dust that were being ejected toward the Sun (to the left).* (Max Planck Institute for Aeronomie, courtesy of A. Delamere, Ball Aerospace Corporation)

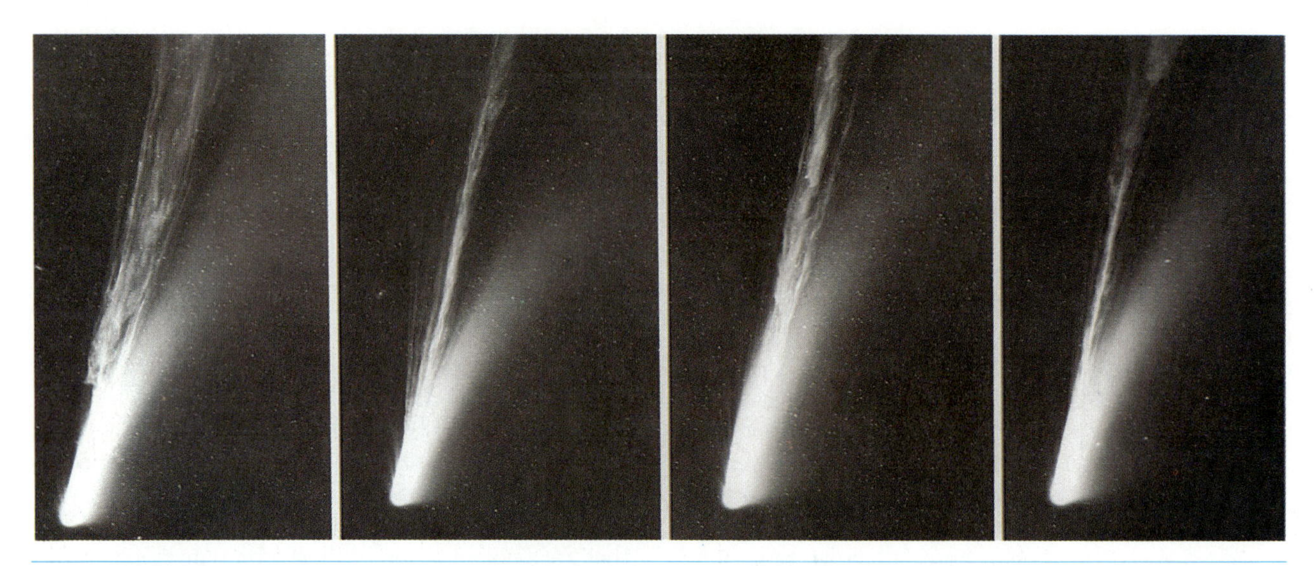

FIGURE 16.16 TWO TAILS. *These photos of Comet Mrkos (1957) show distinctly the two types of tails that characterize many comets. The ion tail points straight up in these views, while the dust tail curves to the right.* (Palomar Observatory, California Institute of Technology)

trapped magnetic field. A "bow shock," analogous to a boundary of a planet's magnetosphere, builds up on the sunward side of a comet, where the solar wind encounters the trapped magnetic field and flows around it. In the direction away from the Sun, the magnetic field that is wrapped around the comet forms a sheet of trapped charged particles called a **current sheet.**

The gases that escape from the nucleus of a comet as it approaches the Sun are highly volatile and would not be present in the nucleus if it had ever undergone any significant heating. This tells us that comets must have formed and lived their entire lives in a very cold environment, probably never getting as warm as even 100 K before falling into orbits that bring them close to the Sun. If a comet is so easily vaporized, then once it has begun to follow a path that regularly brings it close to the Sun, its days are numbered. It may take many round trips, but eventually it will dissipate all of its volatile gases, leaving behind nothing but rocky debris. It has been estimated that Halley's comet was losing some 50 tons per second of water ice when it passed nearest the Sun in early 1986! (Despite this high mass-loss rate, the comet should last at least another 100,000 years before the nucleus is entirely dissipated. There is some speculation that a comet as large as Halley may eventually build up a crust thick enough to halt further sublimation of its interior ices. If this is so, Halley's comet may be immortal, but it will eventually stop developing a coma and tail on its visits to the inner solar system.) Several times, a comet has failed to reappear on schedule, but has been replaced by a few pieces or perhaps a swarm of fragments (Fig. 16.17). In time, the remains of a dead comet will be dispersed all along the orbital path, so

FIGURE 16.17 THE BREAKUP OF COMET WEST. *This dramatic sequence shows the nucleus of Comet West fragmenting into four pieces.* (New Mexico State University)

that each time the Earth passes through this region, it encounters a vast number of tiny bits of gravel and dust, and we experience a meteor shower.

METEORS AND METEORITES

Occasionally one of the countless pieces of debris floating through the solar system enters the Earth's atmosphere, creating a momentary light display as it evaporates in a flash of heat created by the friction of its passage through the air. The streak that is seen in the sky is called a **meteor** (Fig. 16.18). Most of us are familiar with this phenomenon, commonly called a shooting star, since it is often possible to see one in just a few minutes of sky-gazing on a clear night. On rare occasions an especially brilliant meteor is seen, possibly persisting for several seconds, and these spectacular events are called **fireballs** or **bolides.**

The piece of solid material that causes a meteor is called a **meteoroid.** Most are very small, amounting to nothing more than tiny grains of dust or perhaps fine gravel. A few, however, are larger solid chunks, which are responsible for the bright fireballs.

Occasionally one of the larger meteoroids survives the arduous trip through the atmosphere and reaches the ground intact. Such an object is called a **meteorite** (Fig. 16.19), and examples can be found in museums around the world. Meteorites have been the subject of intense scrutiny, for until the past 18 years or so, they were the only samples of extraterrestrial material scientists could get their hands on.

Historically, the notion that rocks could fall from the sky was scoffed at by scientists, until a meteorite that was seen to fall near a French village in 1803 was found and examined just after it fell. Such falls are rare, but they are observed occasionally, and have even been known to cause damage (but so far, few injuries).

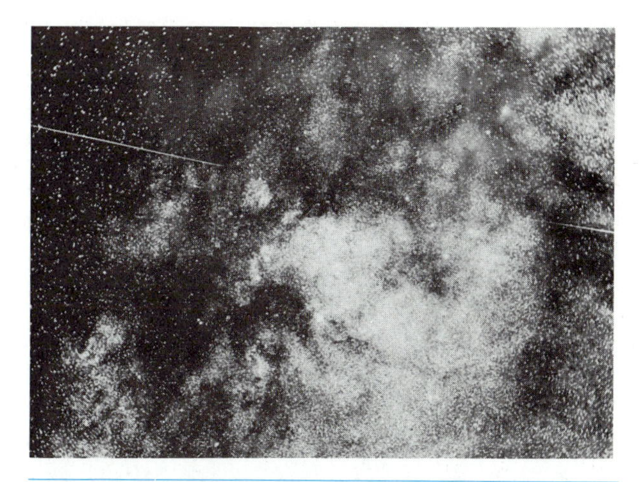

FIGURE 16.18 TWO BRIGHT METEORS. *The streaks of light in this photo are created by a tiny particle entering the Earth's atmosphere from space.* (Yerkes Observatory)

FIGURE 16.19 A METEORITE. *This is a stony meteorite; the black coloring is due to heating as the object passed through the atmosphere. The light-colored spots are breaks in the fusion crust where interior material is exposed.* (Griffith Observatory, Ronald A. Oriti Collection)

Primordial Leftovers

Meteorites are old, much older than most surface rocks on the Earth. This fact enhances the interest of scientists in studying them, for in doing so they get a glimpse into the history of the solar system.

Meteorites generally can be grouped into three classes: the stony meteorites, which comprise about 93 percent of all meteorite falls; the iron meteorites, accounting for about 6 percent; and the stony-iron meteorites, which are the rarest. These relative abundances were determined indirectly, because the different types of meteorites are not equally easy to find on the ground. The majority of all meteorites found are the iron ones, which, as we have just seen, comprise only a small fraction of those that fall. The stony meteorites look so much like ordinary rocks that they are usually difficult to pick out, and some are burned up on their way through the atmosphere. A particularly productive place to search for meteorites is

Antarctica, where a thick layer of ice conceals the native rock. Meteorites that fall there are relatively easy to find, and there is little chance to confuse them with Earth rocks.

The stony meteorites are mostly of a type called **chondrites,** so named because they contain small spherical inclusions called **chondrules** (Fig. 16.20). These are mineral deposits formed by rapid cooling, most likely at an early time in the history of the solar system, when the first solid material was condensing. A few of the stony meteorites are **carbonaceous chondrites** (Fig. 16.21), thought to be almost completely unprocessed since the solar system was formed, and therefore representative of the original stuff of which the planets were made. The primordial nature of the carbonaceous chondrites, like the carbonaceous asteroids mentioned in an earlier section, is deduced from their highly volatile contents, which indicates that they were never exposed to much heat. One particularly fascinating aspect of these meteorites is that in at least one case, complex organic molecules called **amino acids** were found inside a carbonaceous chondrite meteorite, showing that some of the ingredients for the development of life were apparently available even before the Earth formed.

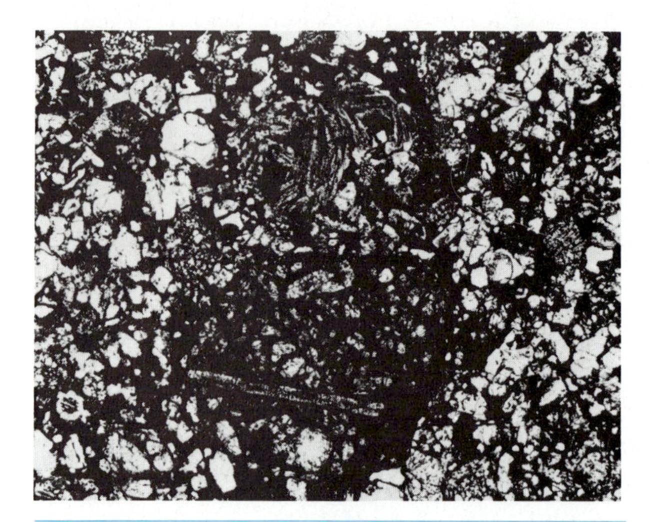

FIGURE 16.20 CROSS-SECTION OF A CHONDRITE. *This shows the many chondrules (light patches) embedded within the structure of this type of stony meteorite.* (Griffith Observatory, Ronald A. Oriti Collection)

The iron meteorites have varying nickel contents, and sometimes show an internal crystalline structure (Fig. 16.22) that indicates a rather slow cooling process in their early histories. This has important implications for their origin, as we shall see.

FIGURE 16.21 A CARBONACEOUS CHONDRITE. *This example, not of the type in which amino acids have been found, shows a large chondrule (the light-colored spot, upper center), which is about 5 millimeters in diameter.* (Griffith Observatory, Ronald A. Oriti Collection)

FIGURE 16.22 CROSS-SECTION OF A NICKEL-IRON METEORITE. *This example shows the characteristic crystalline structure indicative of a slow cooling process from a previous molten state. Such meteorites are thought once to have been parts of larger bodies that differentiated.* (Griffith Observatory, Ronald A. Oriti Collection)

ASTRONOMICAL INSIGHT 16.2

THE IMPACT OF IMPACTS

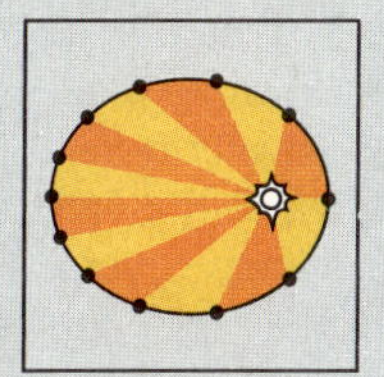

It is well established that impacts by objects from space have occurred throughout the solar system, because every exposed surface has craters. Recently scientists have found evidence that impacts may have played a far more important role in the shaping of solar system bodies than just to create craters. Impacts may have modified biological and geological processes on Earth, might have caused the formation of our Moon, may explain various anomalies in the motions of other planets, and might provide evidence for a companion star orbiting the Sun. All of these conclusions are unproven, but the amount of attention they are receiving is testimony to their appeal among scientists in many fields.

Mathematical analyses of the process by which planetesimals formed and built up by coalescence during the early history of the solar system shows that there was a time when most of the preplanetary material was in the form of very large planetesimals several thousand kilometers in diameter. These orbited the Sun in a disk, so that planetesimals near each other traveled in the same direction with only relatively small velocity differences between near neighbors. Planets formed when these large planetesimals collided and merged. Even with low relative speeds, vast amounts of heat were released upon impact, making much of the material molten and allowing the two bodies to merge into a single spherical one. In most cases, the planet that resulted had its rotation axis roughly perpendicular to the plane of the solar system disk, and its rotation in the prograde direction. It is easy to see, however, that the tilt of the axis and the direction of spin could have been modified by the direction of impact of the major planetesimals that merged to form the planet. We see that the slow retrograde spin of Venus, and the large tilt (98°) of Uranus could have resulted from the off-center impacts of large planetesimals. It was also suggested (as discussed in Chapter 9) that the Earth's Moon might have formed as the result of the impact of a Mars-size planetesimal during the latter stages of formation of the Earth.

While the very massive impacts of planetesimals occurred only during the very early stages of solar system formation, we know that impacts by much smaller bodies still take place in modern times. The Barringer crater in Arizona is only about 25,000 years old, a major impact occurred in Siberia in 1908, and we find traces of craters scattered over the Earth (these are described in the text). In the past few years, it has been suggested that impacts of large meteorites or cometary nuclei on the Earth have been responsible for major modifications in the climate, which in turn may have been linked to mass extinctions of life forms on the Earth. Geological evidence for impacts (consisting of thin layers of debris deposited over large areas of the Earth) coincide in age with times of major extinctions. The most widely discussed event seems to have occurred at just about the time (65 million years ago) when the dinosaurs became extinct. This suggestion is highly controversial, because it appears that the extinctions were not as sudden as the impact scenario supposes, but there is strong evidence that such an impact occurred. A location for the impact (in the Indian Ocean) has even been suggested.

Very recently it was hypothesized that impacts have been responsible for the sporadic reversals in the Earth's magnetic field (described in Chapter 8). There is a pattern of very striking coincidences between the times of major impacts in the past and the times when reversals of the Earth's magnetic field occurred. The impacts are dated by the deposition of thin layers of debris, and the magnetic field reversals are dated by measurements of the magnetic orientation of rocks which show what the field direction was when the rocks formed. If indeed impacts somehow triggered magnetic field reversals, the actual mechanism is not well understood. It has been suggested that changes in climate caused by dust in the atmosphere could have trapped more of the world's water in the form of polar ice, thus changing the mass distribution of the Earth enough to affect its rotation. The change in rotation of the Earth, it is argued, could disrupt the internal currents that give rise to the magnetic field, causing the field to vanish temporarily. When the field reestablished itself later, its orientation would have an even chance of being the opposite of what it had been before the impact.

Some scientists have suggested that there is a regular pattern of major impacts in the Earth's geological record, as though the bodies that caused the impacts had a higher

(continued on next page)

ASTRONOMICAL INSIGHT 16.2

THE IMPACT OF IMPACTS

(continued from previous page)
chance of striking the Earth at specific time intervals. Some fascinating explanations for this periodic behavior have been offered. The most well known of these is the suggestion that the Sun has a companion star orbiting it, with a period of some 26 million years. This star, dubbed Nemesis, would regularly pass near the Oort cloud and disturb the cometary bodies there, so that some of them would fall in toward the inner solar system, thus increasing the likelihood of impacts on the Earth and the other planets. Many astronomers do not believe that there is a regular pattern to the impacts of the past, however, and no evidence for a companion star has yet been found, despite intensive searches.

To follow the development of ideas about the role of impacts in solar system history is a fascinating lesson in how science operates. All of the elements are present: observational data (unusual rotations of some planets, layers of debris on Earth, the fact that extinctions occurred, evidence for magnetic field reversals); uncertainties in the data (mostly involving the timing of events in the past); and hypotheses. What is lacking so far are theories sufficiently well developed to allow prediction and test. It will be interesting to keep watch on developments over the next few years as some of these novel ideas are refined, and others fall by the wayside.

Dead Comets and Fractured Asteroids

The origins of the meteoroids that enter the Earth's atmosphere can be inferred from what we know of the properties of meteorites and of the asteroids and comets.

As we have seen, most meteors are caused by relatively tiny particles that do not survive their flaming entry into the Earth's atmosphere. During **meteor showers,** when meteors can be seen as frequently as once per second, all seem to be of this type. As noted earlier, these showers are associated with the remains of comets that have disintegrated (Table 16.4), leaving behind a scattering of gravel and dust. Therefore, it is thought that the most common meteors, those created by small, fragile meteoroids, are a result of cometary debris.

The larger chunks that reach the ground as meteorites may have a different origin. It is likely that there are occasional collisions among the asteroids, sometimes sufficiently violent to destroy them, dispersing the rubble that is left over throughout the solar system. Most meteorites are probably fragments of asteroids. The iron meteorites apparently originated in asteroids that had undergone differentiation, while the stony ones either came from the outer portions of differentiated asteroids, or from smaller bodies that never underwent differentiation at all. The chondrites probably fall into the latter category, since the chondrules reflect a rapid cooling that would have characterized very small bodies. This is why the chondrites are thought to be the most primitive of the meteorites, having undergone no processing in the interiors of large bodies.

From our studies of the other planets and satellites, we know that there was a time long ago when frequent impacts occurred, forming most of the craters seen today. Certainly the Earth was not immune, and no doubt was also subjected to heavy bombardment.

TABLE 16.4 MAJOR METEOR SHOWERS

Shower	Approximate Date	Associated Comet
Quandrantid	January 3	—
Lyrid	April 21	Comet 1861 I
Eta Aquarid	May 4	Halley's Comet
Delta Aquarid	July 30	—
Perseid	August 11	Comet 1862 III
Draconid	October 9	Comet Giacobini-Zinner
Orionid	October 20	Halley's Comet
Taurid	October 31	Comet Encke
Andromedid	November 14	Comet Biela
Leonid	November 16	Comet 1866 I
Geminid	December 13	—

FIGURE 16.23 IMPACT CRATERS ON THE EARTH. *This map shows the locations of major craters thought to have been created by impacts of massive objects.* (Griffith Observatory)

The difference, of course, is that the Earth has an atmosphere, along with flowing water and glaciation, all of which combine to erase old craters in time. A few traces are still seen, however; there is a very large basin under the Antarctic ice that is probably an ancient impact crater, and a portion of Hudson's Bay in Canada shows a circular shape thought to have a similar origin. A number of other suspected ancient impact craters have been found throughout the world (Fig. 16.23 and Table 16.5).

TABLE 16.5 SOME KNOWN OR SUSPECTED IMPACT CRATERS ON THE EARTH

Location	Diameter (km)	Location	Diameter (km)
Amirante Basin, Indian Ocean	300	Deep Bay, Saskatchewan	13.7
Sudbury, Ontario	140	Deep Bay, Saskatchewan	12
Vredefort, Orange Free State, South Africa	100	Lake Bosumtwi, Ashanti, Ghana	10.5
Manicouagan	100	Chassenon Structure, Haut-Vienne, France	10
Sierra Madera, Texas	100	Wolf Creek, Western Australia	8.5
Charlevoix Structure, Quebec	46	Brent, Ontario	3.8
Clearwater Lake West, Quebec	32	Chubb (New Quebec), Quebec	3.2
Mistastin Lake, Labrador	28	Steinem, Swabia, Germany	2.5
Gosses Bluff, Northern Territory, Australia	25	Henbury, Northern Territory, Australia	2.2
Clearwater Lake East, Quebec	22	Boxhole, Central Australia	1.8
Haughton, Northwest Territories	20	Barringer Crater, Winslow, Arizona	1.2
Wells Creek, Tennessee	14		

FIGURE 16.24 METEOR CRATER NEAR WINSLOW, ARIZONA. *The impact that created this crater occurred about 25,000 years ago.* (Meteor Crater Enterprises)

Although the frequency of impacts has decreased, major impacts still occur on rare occasions. The Barringer crater (Fig. 16.24), near Winslow, Arizona, was formed only about 25,000 years ago, for example, and the possibility exists that other large bodies could still hit the Earth. Given a long enough time, it is almost inevitable.

MICROSCOPIC PARTICLES: INTERPLANETARY DUST AND THE INTERSTELLAR WIND

The space between the planets plays host to some very tiny particles, in addition to the larger ones we have just described. There is a general population of small solid particles, perhaps a millionth of a meter in diameter, called interplanetary dust grains. There is also a very tenuous stream of gas particles flowing through the solar system from interstellar space.

The presence of the dust has been known for some time from two celestial phenomena, both of which can be observed with the unaided eye, though only with difficulty. The dust particles scatter sunlight, so that under the proper conditions a diffuse glow can be seen where the light from the Sun hits the dust. This is analogous to seeing the beam of a searchlight stretching skyward; you see only the beam where there are small particles (either dust or water vapor) that scatter its light, so that some of it reaches your eye.

One of the phenomena created by the interplanry dust is the **zodiacal light** (Fig. 16.25), a faintly illuminated belt of hazy light that can be seen stretching across the sky (along the ecliptic) on clear, dark nights, just after sunset or before sunrise. The second observable phenomenon created by the dust is a small bright spot seen on the ecliptic in the direction opposite the Sun. This diffuse spot, called the **gegenschein** (Fig. 16.26), is created by sunlight that is reflected straight back by the interplanetary dust, which is concentrated in the plane of the ecliptic. This is analogous to seeing a bright spot on a cloud bank or low-lying mist when you look at it with the Sun directly behind you; the bright spot is just the reflected image of the Sun and is the counterpart of the gegenschein.

The *IRAS* satellite (see Chapter 7), which mapped the sky in infrared wavelengths, observed interplanetary dust. The grains of dust are cold, and emit light at infrared wavelengths, so these observations reveal much more information about the dust than observations at other wavelengths. The *IRAS* sky maps show that the zodiac glows at infrared wavelengths due to the concentration of dust in the ecliptic plane. In addition, the infrared maps revealed streaks of dust following cometary orbits, and a huge spiralling ring of dust between the orbits of Mars and Jupiter (Fig. 16.27). This band of dust, which extends above and below the plane at the ecliptic, is thought to have been created by a collision between asteroids or between a comet and an asteroid.

It has been possible to collect interplanetary dust particles for direct examination (Fig. 16.28). This is done most commonly by the use of high-altitude balloons, but recently a surprising new technique was developed in which the dust particles are found in sludge scooped off the ocean floor. The Earth is constantly being pelted with dust particles (which add about 8 tons per day to its mass!), and those that fall into the oceans can lie undisturbed on the seabed for long times. Studies of the grains show that they are probably of cometary origin, having been dispersed throughout space from the dead nuclei of old comets.

As we will learn in Chapter 26, the space between stars in our galaxy is permeated by a rarefied gas medium. In the Sun's vicinity, the average density of this gas is far below that of any artificially created vacuum; it amounts to only about 0.1 particle per cubic centimeter (i.e., there is one atom, on average, in every volume of 10 cm^3, corresponding to a cube about 1 inch on each side). Because of the motion of the Sun, the interstellar gas streams through the solar system with a velocity of about 20 km/sec. The presence of this ghostly breeze, consisting mostly of hydrogen and helium atoms and ions, was discovered in the early 1970s, when observations made from satellites revealed very faint ultraviolet emission from the hydrogen and helium atoms in the gas. The interstellar wind, tenuous as it is, has very little effect on the other components of the solar system but is nevertheless studied with some interest for what it may tell us about the interstellar medium.

FIGURE 16.25 THE ZODIACAL LIGHT. *The diffuse glow of reflected sunlight from dust grains in the ecliptic plane is clearly visible in this photograph made from the summit of Mauna Kea.* (W. Golisch)

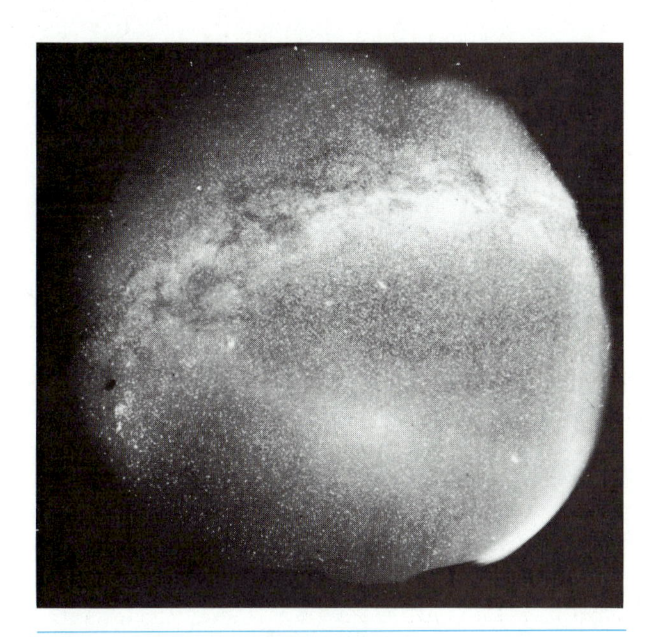

FIGURE 16.26 THE GEGENSCHEIN. *This photo shows the Milky Way stretching across the upper portion and a diffuse concentration of light at lower center, which is the gegenschein. It is created by light reflected directly back to Earth from interplanetary dust in the direction opposite the Sun.* (Photo by S. Suyama, courtesy of J. Weinberg)

FIGURE 16.27 DUST BANDS IN THE OUTER SOLAR SYSTEM. *The* IRAS *satellite's infrared sensors detected emission from this spiral-shaped tail of dust, which may have been created by a collision between an asteroid and a comet. At lower left is the Milky Way, as seen in the infrared by* IRAS. (NASA)

FIGURE 16.28 AN INTERPLANETARY DUST GRAIN. *This is a microscopic view of a tiny particle from interplanetary space. The amorphous structure is highly variable from one grain to another.* (NASA photograph, courtesy of D. E. Brownlee)

PERSPECTIVE

The interplanetary wanderers discussed in this chapter have given us insight into the history of the solar system and have told us much about its present state as well. We have found two primary origins of the various objects: comets and asteroids. The former account for most of the meteors and for the interplanetary dust, and the latter are responsible for the meteorites, including the massive bodies that formed the major impact craters in the solar system.

In the next chapter we consider the overall properties of the solar system as we discuss theories of its formation and evolution.

SUMMARY

1. Thousands of minor planets have been discovered and cataloged, displaying a variety of sizes (up to 1,000 km in diameter) and compositions (ranging from metallic ones to rocky minerals).
2. Gaps in the asteroid belt are created by orbital resonances with Jupiter, whose gravitational influence was probably also responsible for preventing the asteroids from coalescing into a planet in the first place.
3. Comets are small, icy objects that develop their characteristic comae and tails only when in the inner part of the solar system.
4. Comets apparently originate in a cloud of debris very far from the Sun, occasionally falling in, either to bypass the Sun and return to the distant reaches of the solar system for millenia, or to be perturbed by the gravitational influence of one of the planets, becoming periodic comets.
5. When near the Sun, a comet ejects gases that glow by fluorescence. Periodic comets eventually lose all of their icy substance in this process and disintegrate into swarms of rocky debris.
6. A comet may have two tails, one created by ionized gas, and the other made of fine dust particles.
7. A meteor is a flash of light created by a meteoroid entering the Earth's atmosphere from space, and a meteorite is the solid remnant that reaches the ground in some cases.
8. Meteorites are either stony, stony-iron, or iron in composition, and are very old, providing information on the early solar system.
9. Most meteors are created by fine debris from comets, but most meteorites are fragments of asteroids.
10. Interplanetary space is permeated by fine dust particles and by an interstellar wind of hydrogen and helium atoms from the space between the stars.

REVIEW QUESTIONS

Thought Questions

1. How is the asteroid belt similar to the rings of Saturn?
2. Why is the study of asteroids, comets, and meteorites of particular interest to scientists concerned with the history of the solar system?
3. Discuss how Bode's law stands in comparison with scientific theories; that is, how is it like a theory and how is it not?
4. Summarize the role of Jupiter in influencing asteroids, meteorites, and comets.
5. How do the motions of comets differ from those of the planets?
6. Discuss how the "dirty snowball" model of comets compared with what was found from the detailed close-up observations of Comet Halley in 1986.
7. Summarize the life story of a periodic comet.
8. Why are carbonaceous chondrites especially important clues to the early history of the solar system?
9. Would an observer on the surface of Mercury see meteors in the nighttime sky? Might he or she find meteorites on the ground there?
10. Summarize the evidence for the existence of interplanetary dust in the solar system.

Problems

1. What is the orbital period of an asteroid whose semimajor axis is 2.8 AU?
2. At what distance from the Sun would an asteroid have one-tenth the orbital period of Jupiter? Is there a gap in the asteroid belt at this distance?
3. Use Wien's law and the Stefan-Boltzman law to find the surface temperature and diameter of an asteroid whose wavelength of maximum emission is 0.003 cm and whose luminosity is 2×10^{20} erg/sec.
4. Suppose a comet in the Oort cloud has a semimajor axis of 100,000 AU. The nearest star like the Sun is Alpha Centauri, 4 light-years (about 300,000 AU from the Sun. Compare the gravitational force on the comet due to the Sun with that due to Alpha Centauri, when the comet lies in the same direction from the Sun as Alpha Centauri.

ADDITIONAL READINGS

Anatomy of a comet. 1987. *Sky and Telescope* 73(3).

Beatty, J. Kelly. 1986. An inside look at Halley's Comet. *Sky and Telescope* 71(5):438.

Cassidy, W. A., and L. A. Rancitelli. 1982. Antarctic meteorites. *American Scientist* 70(2):156.

Chapman, C. R. 1975. The nature of asteroids. *Scientific American* 23(91):24.

Chapman, R. P., and J. C. Brandt. 1985. An introduction to comets and their origins. *Mercury* 14(1):2.

Gehrels, Tom. 1985. Asteroids and comets. *Physics Today* 38(2):32.

Gingerich, Owen. 1986. Newton, Halley and the comet. *Sky and Telescope* 71(3):230.

Greenstein, George. 1985. Heavenly fire: Tunguska. *Science 85* 6(6):70.

Hartmann, W. K. 1975. The smaller bodies of the solar system. *Scientific American* 233(3):142.

Larson, Stephan, and David H. Levy. 1987. Observing Comet Halley's near nucleus features. *Astronomy* 15(5):90.

Morrison, D. 1976. Asteroids. *Astronomy,* June 1976, p. 6.

Ronan, C., and M. Mohs. 1986. Scientist of the year: Edmund Halley. *Discover* 7(1):52.

Sagan, Carl, and Ann Druyan. 1985. *Comet.* New York: Random House.

Tatum, J. B. 1982. Halley's comet in 1986. *Mercury* 11(4):126.

Van Allen, J. A. 1975. Interplanetary particles and fields. *Scientific American* 233(3):160.

Wagner, Jeffrey K. 1984. The sources of meteorites. *Astronomy* 12(2):6.

Whipple, Fred L. 1986. Flying sandbanks or dirty snowballs: Discovering the nature of comets. *Mercury* 15(1):2.

———. 1974. The nature of comets. *Scientific American* 230(2):49.

CHAPTER 17

ADDING IT UP: FORMATION OF THE SOLAR SYSTEM

An artist's conception of the dust around Vega. (NASA)

CHAPTER PREVIEW

We have collected a considerable quantity of information about the solar system and its contents. Along the way, we have uncovered various systematic trends, such as the similarities in internal structure and composition of the terrestrial planets and the comparisons among the outer planets, but so far we have said very little about how all of this arose. In discussing the histories of the individual planets, we have referred to some of the important processes thought to have been involved, such as the condensation of the first solid material from a cloud of gas surrounding the young Sun, but it remains for us to put together a coherent story of the formation and evolution of the entire system.

We have so far said little about the Sun, the central object in the solar system. For the purposes of this chapter, we need to know something about the Sun, although most of the discussion of it is deferred until the next chapter, which opens the section of this text on stars and their properties. For our present discussion, we need to keep in mind that the Sun is by far the most massive body in the solar system, having a mass over 300,000 times that of the Earth. It is composed primarily of hydrogen (about 70 percent, by mass) and helium (about 27 percent), with the other elements in trace amounts, carbon, nitrogen, and oxygen being the most abundant. (Table 18.2 shows the composition of the Sun in detail). The Sun is gaseous throughout, with a surface temperature of just under 6,000 K and a central temperature of roughly 10 million degrees. It has hotter layers above the surface; of particular interest for this chapter is the **corona,** an extended zone as hot as 1 to 2 million degrees, and the **solar wind,** a stream of charged particles (electrons and protons) flowing outward from the Sun throughout the solar system.

Before we describe the modern theory of solar system formation, we will review the overall properties of the system that must be accounted for, and then we will have a look at the historical developments and alternative suggestions that scientists have considered over the years.

A SUMMARY OF THE EVIDENCE

Several facts must be explained by any successful theory regarding the formation of the solar system. In a way, this is much like a mystery story, in which the detective (the scientist seeking the correct explanation) has certain clues that reveal isolated parts of the story, from which past events must be reconstructed.

One category of clues in our mystery has to do with the orbital and spin motions of the planets and satellites in the solar system (Figs. 17.1 and 17.2; Table 17.1). First and most obvious, the planetary orbits lie in a common plane, all are nearly circular, and all go around the Sun in the same direction. Pluto violates these rules to an extent, because its orbit is both elongated and tilted (by 17°) with respect to the ecliptic, but even this misfit goes around the Sun in the same direction as the other planets. The spins of nearly all the planets are in this same direction, as are the orbital and spin motions of the vast majority of the satellites in the solar system. The direction of these motions, counterclockwise when viewed from above the North Pole, is said to be **direct,** or **prograde,** motion.

Another thing we must understand in a successful theory is why the planets formed where they did, and not elsewhere. A final clue having to do with the motions of the planets is that nearly all of them have small **obliquities,** meaning that their equatorial planes are nearly aligned with their orbital planes. Even the Sun has its axis of rotation nearly perpendicular to its orbital plane, the ecliptic. Only Uranus, with its 98° tilt, does not fit the pattern.

A second general type of clue in the mystery is concerned with the natures of the individual planets, and the systematic trends from planet to planet that we have discussed (Table 17.2). First, we must consider the planetary compositions, seeking an explanation for why the inner planets—the terrestrial ones—

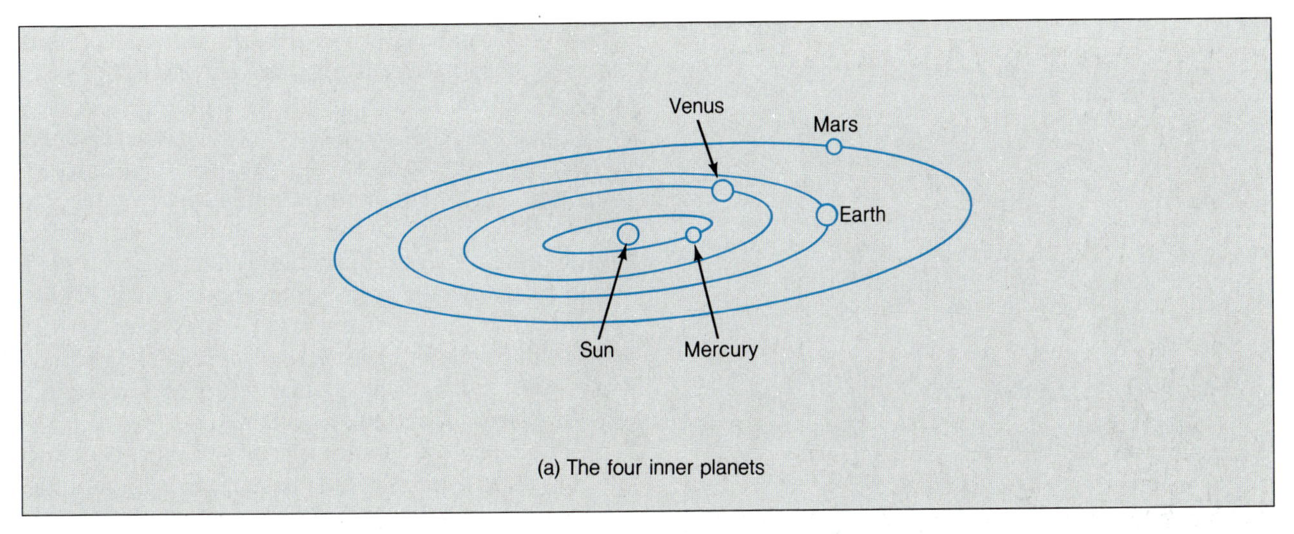

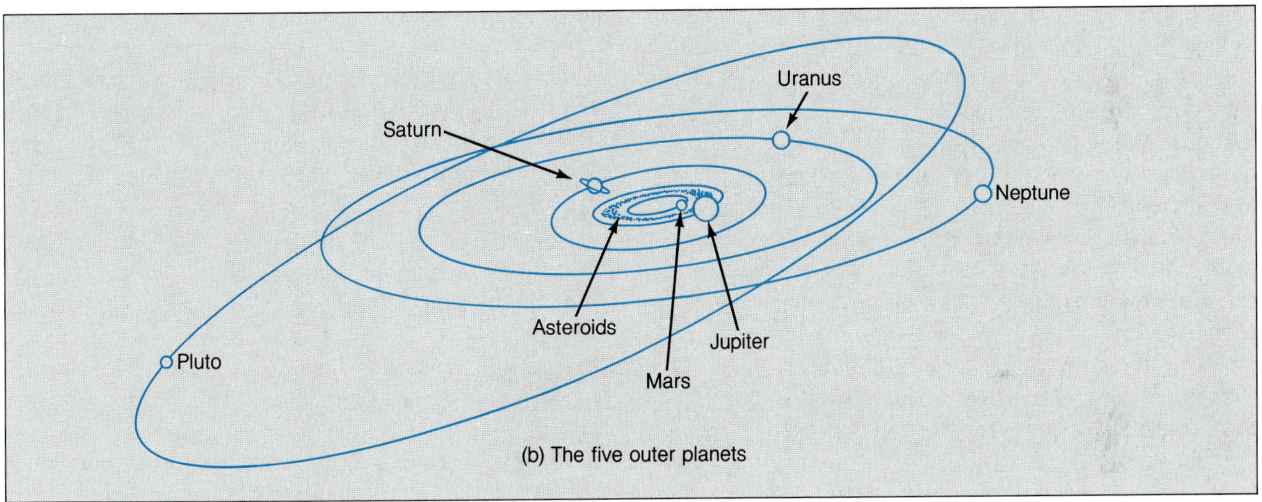

FIGURE 17.1 THE COALIGNMENT OF PLANETARY ORBITS. *All but one of the nine planets have nearly circular orbits in a common parallel plane. The exception is Pluto, whose orbit is tilted 17 degrees with respect to the ecliptic, and is sufficiently eccentric (elongated) that Pluto is actually closer to the Sun than Neptune at times.*

have high densities, indicating that they are made mostly of rock and metallic elements; whereas the outer planets have low densities, and consist primarily of lightweight gases such as hydrogen and helium, as well as ices composed of these and similarly volatile gases. Within each of the two groups of planets, there are relatively minor differences in chemical composition, such as the various isotope ratios we have discussed, which must also be accounted for.

TABLE 17.1 PROPERTIES OF PLANETARY ORBITS

Progression of planetary distances from the Sun.
Planetary and satellite orbits lie in a common plane.
Nearly all orbital and spin motions are in the same direction.
Spin axes of most planets and satellites are nearly perpendicular to the ecliptic.
Planetary and satellite orbits are nearly circular.

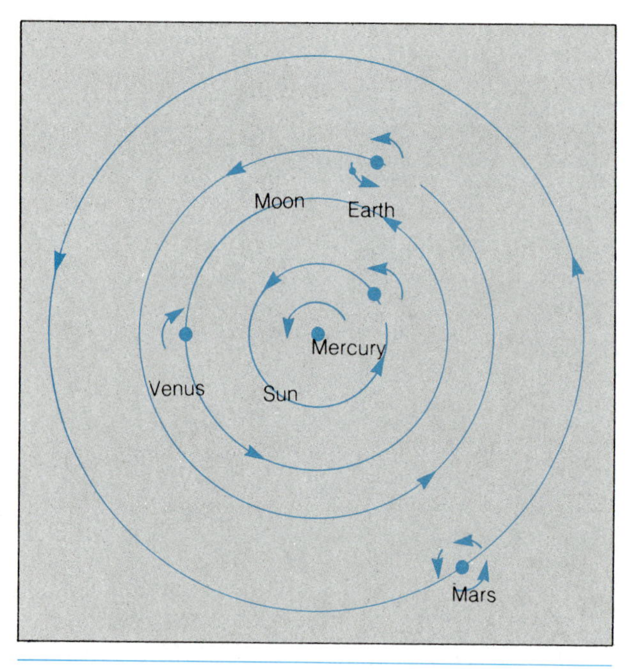

FIGURE 17.2 PROGRADE MOTIONS IN THE INNER SOLAR SYSTEM. *The rotations and orbital motions of the planets are generally in the same direction. This sketch of the orbits of the inner four planets shows that Venus has retrograde rotation but that the other three planets and their satellites obey the rule. The outer planets (except for Pluto) and nearly all their satellites do also.*

The Sun rotates in a time of about 25 days, which posed a major mystery for a long time. According to many early theories, the Sun should be spinning much faster. The difficulty arises in theories that suggest that the solar system formed from the collapse of a cloud of gas and dust; these theories, as we shall soon see, are the most successful ones. The laws of physics tell us that the angular momentum of an object (see Chapter 5) must remain constant. This means that, if a spinning object shrinks in size, it must spin faster to compensate, thereby maintaining constant angular momentum. Thus the Sun, thought to have formed at the center of a collapsing cloud, should have a very rapid rotation rate, rather than the leisurely 25-day period it has. As we will see, the explanation for this was a long time in coming, and for a while astronomers were sidetracked in their effort to understand the formation of the solar system.

A final category of information that bears on the origin of the solar system is the distribution and nature of the various kinds of interplanetary objects that were described in Chapter 16. We learned there that asteroids, comets, meteroids, and interplanetary dust inhabit the space between the planets, and we found that the motions and compositions of these objects reflect conditions very early in the history of the solar system. Somehow this information must fit into our overall picture.

Let us now take a look at some of the ideas that have been developed historically, in the face of all this evidence.

TABLE 17.2 PHYSICAL PROPERTIES OF THE PLANETS

There are two major types of planets: terrestrial and gaseous giant.
Giant planets are much larger and more massive than the terrestrial planets but have lower densities.
Terrestrial planets have low abundances of volatile elements; giant planets have plentiful volatiles.
Giant planets have more satellites than terrestrial planets.
Giant planets tend to have ring systems; terrestrial planets do not.
Giant planets rotate more rapidly than do terrestrial planets.

CATASTROPHE OR EVOLUTION?

In discussing theories of the formation of the solar system, we will confine ourselves to those that were suggested after the time of Copernicus, when it was established that the planets orbit the Sun. Several primitive ideas that were based on the geocentric notion were outlined in Chapter 3.

The first serious theory based on the heliocentric view of the solar system was developed by the French mathematician and philosopher René Descartes, who in 1644 advanced the idea that the solar system formed from a gigantic whirlpool, or vortex, in a universal fluid, with the planets and their satellites forming from smaller eddies. This hypothesis was rather crude, without any clearly specified idea of the nature of the cosmic substance from which the Sun and planets arose, but it did account for the fact that all the orbital motions are in the same direction.

The hypothesis of Descartes was the first of a general type known as **evolutionary theories,** theories in which the formation of the solar system occurred as a natural by-product of the sequence of events that produced the Sun. In an evolutionary theory, no special circumstances are needed to create the planets, other than the fact that the Sun formed. This leads to a natural expectation that other planets may be orbiting other stars.

Further elaboration on the idea of Descartes was developed by the German philosopher Immanuel Kant, who in 1755 used the principles of the recently discovered Newtonian mechanics to argue that a rotating gas cloud would flatten into a disk as it contracted (Fig. 17.3). Kant's theory was called the **nebular hypothesis,** because it invoked formation of the Sun and planets from an interstellar cloud, or nebula. In 1796, Pierre Simon de Laplace, a French mathematician, added to this general idea the notion that as the spinning cloud flattened into a disk, concentric rings of material broke off because of rotational forces, so that at one point the early solar system would have looked very much like the planet Saturn with its rings (see Fig.17.3). Each ring was then supposed to have condensed into a planet. Soon the problem of accounting for the slow solar rotation was recognized, and further development of the evolutionary theories was stymied.

Meanwhile, an alternative class of theories, called **catastrophic theories,** had been suggested. These concepts of the formation of the planetary system envision a solitary Sun to begin with, but then invoke some singular, cataclysmic event that disrupted the Sun and formed the planets. The first of these ideas was that of another Frenchman, Georges Louis de Buffon, who in 1745 suggested that a massive body (which he referred to as a comet, although it was much larger than any cometary nucleus) passed so near the Sun that its gravitational pull forced material out of it (Fig. 17.4), this gas then condensing to form the planets.

Buffon's idea was largely ignored until the beginning of the twentieth century, when the difficulty with the Sun's slow spin forced scientists to consider alternatives to the evolutionary theories. By 1905, the English astronomers T. C. Chamberlain and F. R. Moulton had suggested an elaboration of Buffon's original idea, in which another star was the object that passed near the Sun, creating tidal forces that caused matter from inside the Sun to stream out, thereafter condensing into planets.

FIGURE 17.3 THE HYPOTHESIS OF KANT AND LAPLACE. *Descartes's simple vision of a vortex was refined by Kant, who realized that rotation should cause a collapsing cloud to take a disklike shape; and by Laplace, who hypothesized that a rotating disk would form detached rings that could then condense into planets.*

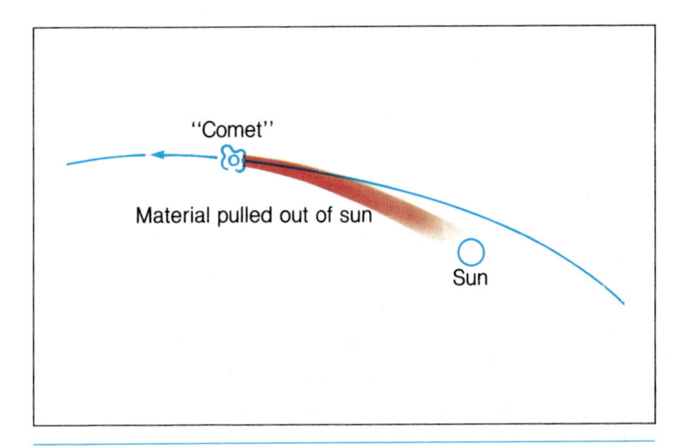

FIGURE 17.4 THE CATASTROPHE HYPOTHESIS. *According to this theory, a passing body would gravitationally draw from the Sun the material that would subsequently form planets.*

Interestingly, in modern times, the idea of mass transfer from one star to another in close double-star systems has been discovered to be very important under certain circumstances (see Chapters 21 and 23), so the idea of gravitational forces extracting matter from the Sun is not out of the question. Further refinements to the bypassing-star hypothesis were soon provided by the British scientists H. Jeffreys and J. H. Jeans.

A disturbing problem with the catastrophic theories was the low probability that the Sun could have collided with another star, given the great distances between stars in our part of the galaxy. Of course, this did not rule out the possibility, for even a low-probability event can occur, but it was sufficiently worrisome to lead to the development of an alternative catastrophic theory, one in which the Sun started out as a member of a triple system but was then freed from the gravitational bonds of the other two when the three nearly collided. In the process of expelling the Sun from the triple-star system, the gravity of the other stars tore some of the solar material out into a tail, which later condensed to form the planets. (More recently, a similar three-body interaction has been suggested to explain the strange orbits of Neptune's two satellites and the expulsion of Pluto into its own solar orbit; see Chapter 15.) This theory, developed by H. N. Russell, R. A. Lyttleton, and F. Hoyle, replaced the difficulty of forming the Sun and the planets together with one of forming the Sun and two other stars together. The advantage was that the slow rotation of the Sun did not have to be produced during its formation but could have developed later, when it nearly collided with the other stars and was ejected.

By the late 1930s the catastrophic theories were becoming quite unwieldy, not only because of the special circumstances that had to be assumed, but also because of several fundamental problems. For example, calculations showed that, even if some material really were pulled out of the Sun's interior, it would be so hot that it would expand and dissipate into space, rather than condensing to form planets. Another difficulty was how to pull material out of the Sun, giving the extracted matter sufficient angular momentum for the planets, but without giving it enough speed to escape entirely.

A more modern, but quite serious, objection to the catastrophic theories has to do with the abundance of the hydrogen isotope deuterium. Deuterium is a form of hydrogen whose nucleus contains a proton and a neutron, in contrast with the normal form of hydrogen whose nucleus consists solely of a proton. It happens that deuterium is destroyed in nuclear reactions that will inevitably occur if the gas containing deuterium is subjected to high enough temperatures, such as would be the case inside a star. Even in the Sun's outer layers, the temperatures are high enough to destroy deuterium. Hence, if the solar system were formed of material pulled out of the Sun, we would expect to find no deuterium in the planets and interplanetary bodies. To the contrary, deuterium is found to be a normal constituent of the solar system, which means that it must have been present in the primoridal material from which the system formed and that this material was never part of the Sun. (In Chapter 32 we discuss the origin of deuterium in the universe.) This emphatically rules out all catastrophic theories that suggest the planets formed from gas pulled out of the Sun or any other star (except possibly a very cool one).

In the 1940s a new type of theory was proposed by the Soviet astronomer O. Y. Schmidt and was developed in some detail by H. Alfvén, the Swedish astrophysicist who was later to win a Nobel Prize for his work on the behavior of ionized gas in the solar magnetic field. The idea was that once the Sun had formed, its gravitational field trapped material from the surrounding interstellar medium, which then formed into planets. This general idea, called the **accretion theory** of solar system formation, never gained much popularity, because the problems with the evolutionary theory began to be solved at about the time the accretion theory was being developed. Most research efforts after this time centered on the evolutionary theory.

Progress was being made in the 1940s on the general question of how a contracting cloud would flatten into a disk and then break up into eddies that could form planets. The German physicist C. F. von Weiszäcker showed that the disk would tend to rotate differentially, meaning that the inner parts would orbit the center faster than the outer regions, and that this would cause the disk to break up into eddies (Fig. 17.5). The work of von Weiszäcker even showed how the relative sizes of the eddies would vary with distance from the center, accounting for both the relative distances of the planets from the Sun and for their varying sizes. More detailed analyses of the sta-

FIGURE 17.5 VON WEISZÄCKER'S THEORY OF A TURBULENT DISK. *This sketch shows how a rotating disk would form eddies, according to the theory of C. F. von Weizsäcker.*

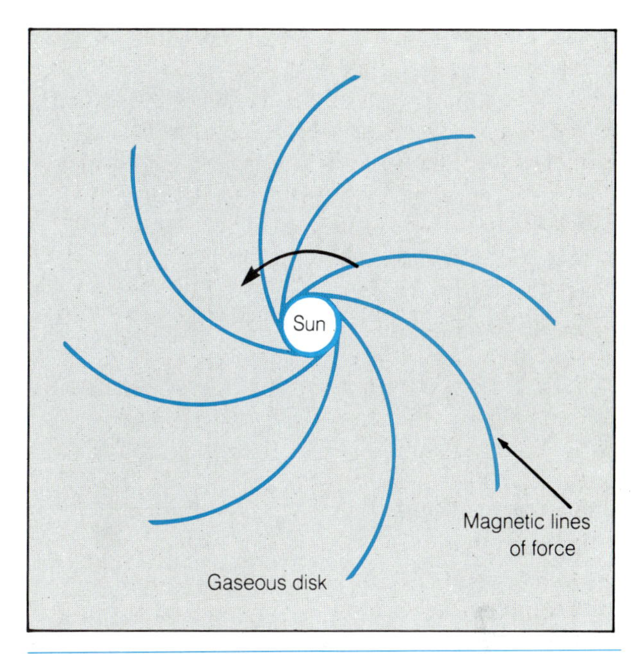

FIGURE 17.6 MAGNETIC BRAKING. *The young, rapidly spinning Sun has the magnetic field structure shown, the field lines rotating with the Sun. The early solar system was permeated with large quantities of debris and gas, some of which was ionized and therefore subject to electromagnetic forces. The Sun's magnetic field exerted a force on the surrounding gas, which in turn created drag that slowed the Sun's rotation.*

bility of rotating disks were later carried out by many others, who showed that, rather than forming regular eddies, as envisioned by von Weiszäcker, a rotating disk would become clumpy, forming localized regions in which the density was high enough to allow material to fall together gravitationally. This process would lead naturally to the breakup of the disk into a number of small, solid bodies. These in turn could merge later to form the planets.

The breakthrough in understanding the Sun's slow rotation came in the early 1960s, with the discovery of the solar wind. Because the Sun continuously expels a stream of charged particles, interplanetary space is filled with them, and the Sun's magnetic field tries to pull them along with it as the Sun rotates (Fig. 17.6). This creates a constant drag on the Sun (imagine trying to spin a pinwheel under water). This drag, given long enough to act, slows the Sun's spin. If the Sun was born with a rapid rotation, this **magnetic braking** process could easily have slowed it to the present rate in the billions of years that have passed. Most likely, the density of charged particles in space around the Sun was much greater in the early days of the solar system, so the drag created as the Sun's magnetic field tugged on them was probably greater than it is now, and the braking could have taken place in a relatively short time.

With the understanding of the magnetic braking process, most of the major pieces had fallen into place, and all that remained was further refinement of the theory, accounting for the details. There are still many unanswered questions, but the general idea of how the solar system formed is probably correct, as described in the next section.

A MODERN SCENARIO

Table 17.3 summarizes the current theory of solar system formation. The first step was the collapse of an interstellar cloud, one that must have been rotating before it began to fall in on itself. Our galaxy has a large quantity of interstellar material in it that con-

TABLE 17.3 TIMETABLE FOR SOLAR SYSTEM FORMATION

Approximate Time	Event
0	Collapse of interstellar cloud; disk formation
400,000 years	Central condensation becomes hot enough for pressure to balance gravity; matter still falling in
1 million years	Second collapse stage is followed by new nearly stable phase, with heating by slow gravitational contraction; central object is now called the protosun; infalling gas and dust shrouds protosun from exterior view
1–100 million years	T Tauri wind
100 million years	Start of nuclear reactions in core
0.1–1 billion years	Magnetic braking slows Sun's rotation; major cratering occurs throughout solar system

tains both gas and small dust grains (about the size of the interplanetary grains described in Chapter 16). The interstellar medium is not uniformly distributed throughout space, but instead is concentrated here and there in large amorphous regions called interstellar clouds, some of them sufficiently dense to block out starlight, appearing as dark regions in photographs of the sky (Fig. 17.7). Even in these areas of relatively high concentration, the material is so rarefied that a quantity of mass equal to that of the Sun is spread over a volume up to 10 or more light-years across.

The composition of the interstellar material is apparently quite uniform, consisting of about the same mixture of elements as the Sun. Hence the Sun and other stars are born with a standard composition of about 73 percent hydrogen (by mass), about 25 percent helium, and just traces of all the other elements. The planets must also have formed out of material with this composition, yet, as we have seen, their present makeup is quite different from this, especially in the case of the terrestrial planets.

Somehow the extended, tenuous gas in interstellar space had to become concentrated in a very small volume on its way to becoming a star. Gravitational forces were responsible for pulling the cloud together, but there must have been an initial push or a chance condensation to get the process started. (One possibility is that a shock wave from a violent event such as a supernova explosion caused the presolar cloud to begin its collapse; see Chapter 22 for details). Once the cloud was falling in on itself, the rest of the process leading to the formation of the Sun was inevitable, dictated by physical laws

A combination of theory and observation of the formation of other stars tells us how the collapse proceeded. The innermost portion of the cloud fell in on itself very quickly, leaving much of the outer material still suspended about the center. The rotation of the cloud sped up as its size diminished, and if the cloud had a magnetic field to begin with (as most do), the field was intensified in the central part as a result of the condensation.

The core of the cloud began to heat up from the energy of impact as the material fell in. Eventually it began to glow, first at infrared wavelengths, and finally, after a prolonged period of gradual shrinking, at visible wavelengths. Nuclear reactions began in the center when the temperature and pressure were sufficiently high, and the Sun began its long lifetime as a star, powered by these reactions.

The steps in the formation of the Sun described up to this point are fairly well understood and have been observed to be taking place today in many cloudy regions of the galaxy. In a number of stellar nurseries, infrared sources are found embedded inside dark interstellar clouds, indicating that newly formed stars are hidden there, still heating up. The details of planet formation are a little sketchier, however, because we have no way to observe the process as it occurs. All that we know about it has had to be inferred from observations of the end product: the present-day solar system.

Figure 17.7 Regions of star formation. *At left is the Orion nebula, in the "sword" of the constellation Orion, and at right is the region near the M supergiant Antares (bright object at right center). In both regions, newly-formed stars are known from infrared observations to lie within the dark, dusty clouds seen here.* (*left:* ©1985 Anglo-Australian Telescope Board; *right:* ©1979 Royal Observatory, Edinburgh)

As the central part of the interstellar cloud collapsed to form the Sun, the outer portions were forced into a disk shape by rotation, just as Kant had argued. At this stage there was an embryonic sun surrounded by a flattened, rotating cloud called the **solar nebula** (Fig. 17.8). The inner portions of the nebula were hot, but the outer regions were quite cold.

Throughout the solar nebula, the first solid particles began to form, probably by the growth of the interstellar grains that were mixed in with the gas. For every element or compound, there is a combination of temperature and pressure at which it "freezes out" of the gaseous form, in direct analogy with the formation of frost on a cold night on Earth. The **volatile elements** require a very low temperature in order to condense, so these materials tended to stay in gaseous form in the inner portions of the solar nebula but condensed to form ices in the outer portions. The elements that condense easily, even at high temperatures, are called **refractory elements,** and these were the ones that formed the first solid material in the warm inner portions of the solar nebula. This material therefore consisted of rocky debris containing only low abundances of the volatile species.

In due course, rather substantial objects built up that resembled asteroids in size and composition, and these are referred to as **planetesimals.** Mathematical analysis of the process by which particles orbiting the young Sun collided and built up their sizes shows that the growth of large planetesimals happened rather quickly (taking place in only a few thousand years), and that most of the material in the disk would have gone into large planetesimals. Thus a stage was reached in which collisions between large bodies were likely.

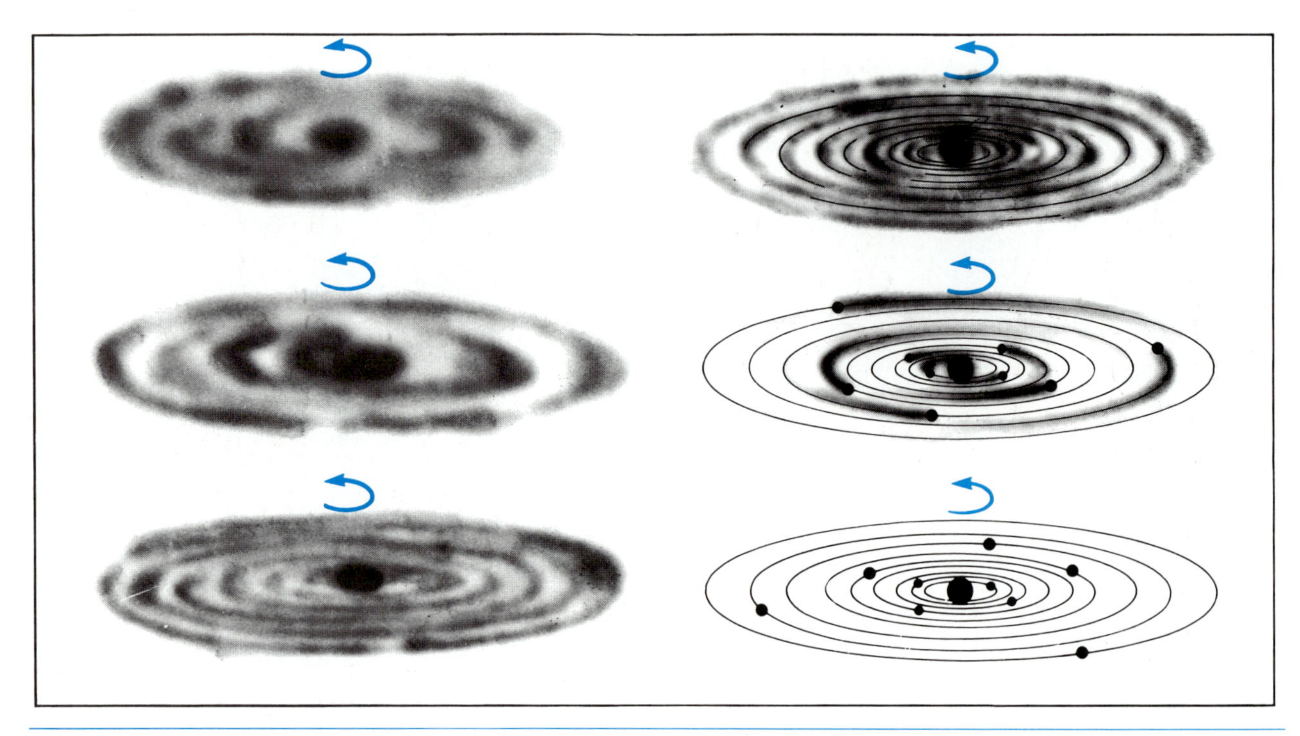

FIGURE 17.8 STEPS IN THE SUN'S FORMATION. *An interstellar cloud, initially very extended and rotating very slowly, collapses under its own gravitation. This happens most quickly at the center. As the collapse occurs, the internal temperature rises and the rotation rate increases. Eventually the central condensation becomes hot and dense enough to be a star, with nuclear reactions in its core.*

Because the planetesimals were all orbiting the Sun in a disk, their relative speeds would have been rather small, so that when collisions occurred, the colliding bodies could stick together. Even at low speeds (perhaps 10 km/sec, typically), enough energy would be released in a collision between large planetesimals to partially vaporize or melt both objects. This would help them merge into a single spherical body and would also aid the process of differentiation, which requires that a body be molten. In most cases, the planet that was formed in this way rotated in the same direction as the overall rotation of the disk, but it is possible that an off-center collision could alter the direction of spin. This may have happened during the formation of Venus and Uranus, accounting for the slow retrograde rotation of Venus and the large tilt of Uranus. As we learned in Chapter 9, many scientists now think that the Earth's Moon formed as the result of a grazing collision between the young Earth and a Mars-sized planetesimal.

The scenario just described apparently applies to the terrestrial planets, which therefore seem to have formed in two stages: (1) the condensation of refractory elements, leading to the development of planetesimals; and (2) the accretion of the planetesimals to form planets. The low quantity of volatile elements that characterizes the terrestrial planets was already established when the planets formed, and then was exaggerated by the release of volatile gases that occurred during their early histories, when they underwent molten periods.

There is more uncertainty about the sequence of events that led to the formation of the outer planets. These planets contain a much higher proportion of volatile gases, which would naturally be expected because of the lower temperatures in the outer portions of the solar nebula. Thus, planetesimals that formed there would have contained higher relative abundances of gases like hydrogen and helium, and so would the planets that formed later through the coalescence of these planetesimals.

The most recent theory of the formation of the giant planets suggests that their cores formed from the coalescence of planetesimals, just as the terrestrial

ASTRONOMICAL INSIGHT 17.1

THE EXPLOSIVE HISTORY OF SOLAR SYSTEM ELEMENTS

Even according to the evolutionary theory that is now widely accepted, the development of the solar system required explosive events. The evolution of the universe itself may be viewed as the result of a series of explosions, including the initial one from which the universe has been expanding ever since. A variety of astronomical evidence (discussed thoroughly in Chapter 32) shows that the universe began some 10 to 15 billion years ago in a single point, and, at some time in the past, began to expand outward. The initial temperature and density were so extreme that the birth of the universe must have been an explosive event, commonly called the big bang.

For a brief time during the early stages of the universal expansion, conditions were suitable for nuclear fusion reactions to occur, and the primordial substance was converted from a soup of subatomic particles to one of recognizable elements, primarily hydrogen and helium, with almost no trace of heavier species. Eventually, as the universe continued to expand, galaxies and individual stars formed. Just as nuclear reactions in the Sun form a heavier element (helium) from a light one (hydrogen), similar reactions take place inside all stars, gradually enriching the universal content of heavy elements. Stars more massive than the Sun undergo more reaction stages, going from the production of helium to that of carbon, and from carbon to other elements. When these stars die, often explosively, their newly formed heavy elements are returned to interstellar space and are available for inclusion in new stars that form later. Thus, during the 5 to 10 billion years that passed between the time of the big bang to the time of the solar system formation, the chemical makeup of our galaxy gradually changed, so that roughly 2 percent of the total mass was in the form of elements heavier than helium.

Since this roughly represents the composition of the Sun and the primordial material in the solar system, it might seem that there is no more to the story of the origin of the elements. There are compelling reasons to believe, however, that the formation of the solar system was influenced by at least two local explosive events. Certain atomic isotopes that are present in solar system material are the result of explosive nuclear reactions that must have occurred relatively recently, compared to the time that has passed since the big bang. These isotopes, one a form of aluminum (with twenty-six instead of the normal twenty-seven neutrons and protons) and the other a form of plutonium (with 244 protons and neutrons instead of 242) are radioactive, with half-lives short enough that they could not have been present when the solar system formed unless they had been created relatively recently in explosive events. The plutonium isotope in question probably formed in a stellar explosion called a supernova that occurred a few hundred million years before the solar system formed, whereas the aluminum must have been created in a nearby supernova only a million years or so before the solar system was born.

The second of these inferred explosions may have played a crucial role in triggering the formation of our Sun and planetary system. We have said that the solar system formed from an interstellar cloud that collapsed. What we have not discussed, however, is what caused this collapse to occur. Normally, an interstellar cloud is quite stable as an interstellar cloud, and it will not fall in on itself unless something happens to compress it. Various mechanisms can bring this about, and one of them is a shock wave from an explosive event such as a supernova. Thus, in a singular event sometimes called the "bing bang," our solar system may have been born as the direct consequence of the death of a nearby star. We derive from this event not only some minor anomalies in the abundance of nuclear isotopes, but also the very existence of the solar system. It is interesting to note that, even in an evolutionary origin as discussed in this chapter, the solar system has been strongly influenced by what can only be described as catastrophic events.

planets themselves formed. In the outer solar system, however, where the temperatures were much lower than in the inner portions of the system, the solid objects that formed from the coalescence of planetesimals were able to gravitationally trap gas from large volumes in their vicinities, and these planets therefore accreted extended, gaseous atmospheres. Tidal forces from the Sun played an important role in this scenario; for the inner planets, these forces helped prevent the accretion of gases from the surrounding nebula, but in the outer solar system, farther from the Sun, the tidal forces were weaker and did not prevent the trapping of gases by the giant planets.

This viewpoint is prompted in part by the extensive ring and satellite systems of the outer planets, which resemble miniature solar nebulae. The suggestion is that these planets, in accreting gas from their surroundings, formed disks much like the solar nebula itself. Gravitational and rotational effects would have caused the material falling in to form a disk in the equatorial plane of each planet, and then instabilities in the disk would have caused eddies to form, leading to the coalescence of satellites (Fig. 17.9). Ring systems formed wherever the disks extended closer to the parent planet than the Roche limit, where tidal forces prevented the formation of large satellites. The rapid rotation rates of the giant planets are explained in this view by the fact that angular momentum conservation would force them to spin more rapidly as the disks contracted (and by the fact that no magnetic braking process would have acted, in contrast with the early Sun).

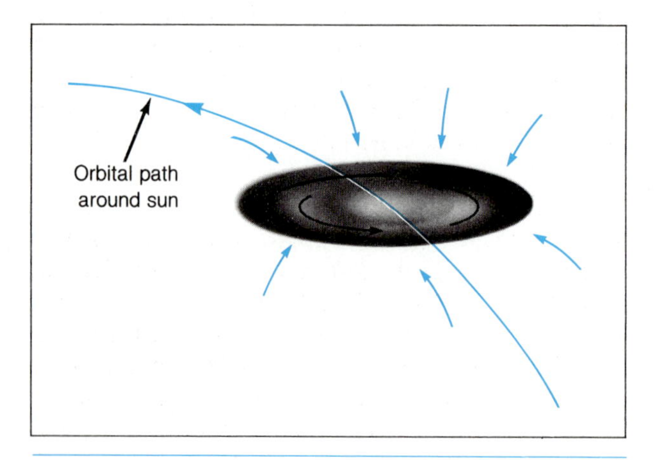

FIGURE 17.9 FORMATION OF A GIANT OUTER PLANET. *Here a rotating disk has formed around a condensation in the outer solar system. Lumps in the disk grow to become satellites, except in the innermost portion where tidal forces prevent this, and a ring system forms instead. The entire system of planet, rings, and moons orbits the Sun.*

It is interesting that Uranus, despite its large (98°) tilt, has its satellites and rings lying in its equatorial plane. This indicates that this planet accumulated its disk of icy debris after the event occurred that caused its highly tilted orientation, and this is consistent with the picture of outer planet formation just described. In this view, material trapped by Uranus would be forced into a disk in the equatorial plane of the planet, regardless of the orientation of that plane. Hence the tilt of Uranus must have been established before it trapped the material that was to form its satellites and rings.

The asteroids probably formed as planetesimals, similar to those that eventually created the terrestrial planets, but they were prevented from coalescing into a planet by the gravitational effects of Jupiter, as we noted in Chapter 16. The comets probably formed farther out, though not at the distance of the Oort cloud, where they now reside. At such a great distance (50,000 to 150,000 AU) from the center of the solar nebula, it is doubtful that any condensation could have occurred, because the density would have been too low. Therefore the comets are believed to have condensed at intermediate distances, probably near the orbits of Uranus and Neptune, and then were forced out to their present great distances by the gravitational effects of the major planets. Both Jupiter and Saturn, lying closer to the Sun, would have forced the cometary bodies farther out by speeding them up a little each time they passed by. Eventually, most of the comets retreated to a great distance, forming the Oort cloud.

Once the planets and satellites were formed, the solar system was nearly in its present state, except that the solar nebula had not totally dissipated. A lot of gas and dust were still swirling around the Sun, along with numerous planetesimals that had not yet accreted onto planets. The Sun's magnetic field, pulling at ionized gas in the nebula, was very effective during this time in slowing the solar rotation through the magnetic braking process that we have already described. Meanwhile, the remaining planetesimals were floating about, now and then crashing into the planets and their satellites. Most of the cratering in the solar system occurred during this period, in the first billion years.

The leftover gas in the solar nebula was dispersed rather early, when the infant Sun underwent a period of violent activity, developing a strong wind which swept the gas and tiny dust particles out into space (Fig. 17.10). This left behind only the larger solid bodies, the newly formed planets and the planetesimals. This wind phase has been observed in other newly formed objects called **T Tauri** stars and is apparently a natural stage in the development of a star like the Sun.

Once the majority of the remaining interplanetary debris was eliminated, either by the T Tauri wind or by accretion onto planetary bodies, the solar system looked much like it does today. Each planet went its own way, either continuing to evolve geologically, as most of the terrestrial planets have, or remaining perpetually frozen in its original condition, as the outer planets apparently have.

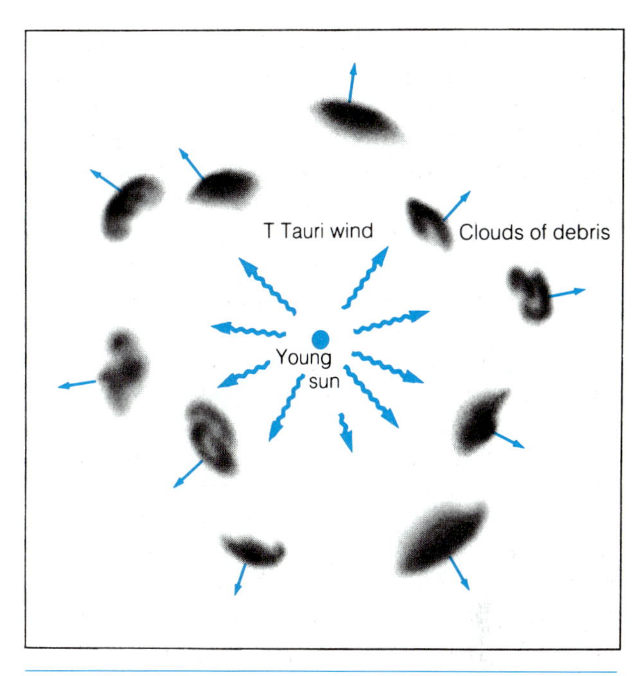

FIGURE 17.10 THE T TAURI PHASE. *A star like the Sun is thought to go through a phase very early in its lifetime during which it violently ejects material in a high-velocity wind. This wind sweeps away matter left over from the formation process.*

ARE THERE OTHER SOLAR SYSTEMS?

As far as we know, the entire process by which our planetary system formed was a by-product of the Sun's formation, requiring no singular, catastrophic event. We might expect, therefore, that the same processes have occurred countless times as other stars were born, and therefore there may be many planetary systems in the galaxy. Even if we rule out double- and multiple-star systems, in which perhaps all the material was either ejected or included in the stars, a vast number of candidates remain.

Detecting planets orbiting distant stars is not easy. For a long time astronomers suspected one particular star of having a planet. The gravitational pull of a very massive planet would cause its parent star to wobble a little in position as the planet orbited it. A nearby star called Barnard's star was thought for a long time to be wobbling in such a way, but a recent reexamination of the data have shown that there is no clear-cut indication of such a motion, and at the moment the case for the existence of a massive planet circling Barnard's star is not well established.

Another technique is to search carefully for variations in the velocities of stars suspected of having planets. Here the Doppler shift is used to measure the speed of the star relative to Earth (see Chapter 6). Small, regular changes in this velocity could indicate that the star is orbiting a small companion. Very recently, a study using this technique has revealed several nearby stars with dim companions, some of which may be massive planets. Planets as small as the Earth are much more difficult to detect in this way, because they would cause only minute variations in the velocity of their parent star.

Infrared observation shows some promise as a technique for detecting other planetary systems. This was illustrated by the *Infrared Astronomical Satellite* (*IRAS*), which mapped most of the sky in far-infrared wavelengths and found several stars that are surrounded by disks containing tiny solid particles of the sort thought to be the precursors of planetesimals (Fig. 17.11). Other kinds of new technology may also provide new information in this area. The *Space Telescope,* for example, will orbit above the Earth's atmosphere and will therefore be free of the distortion and fuzziness that plagues ground-based telescopes. The *Space Telescope* is expected to allow astronomers to see objects comparable to Jupiter, if they orbit nearby

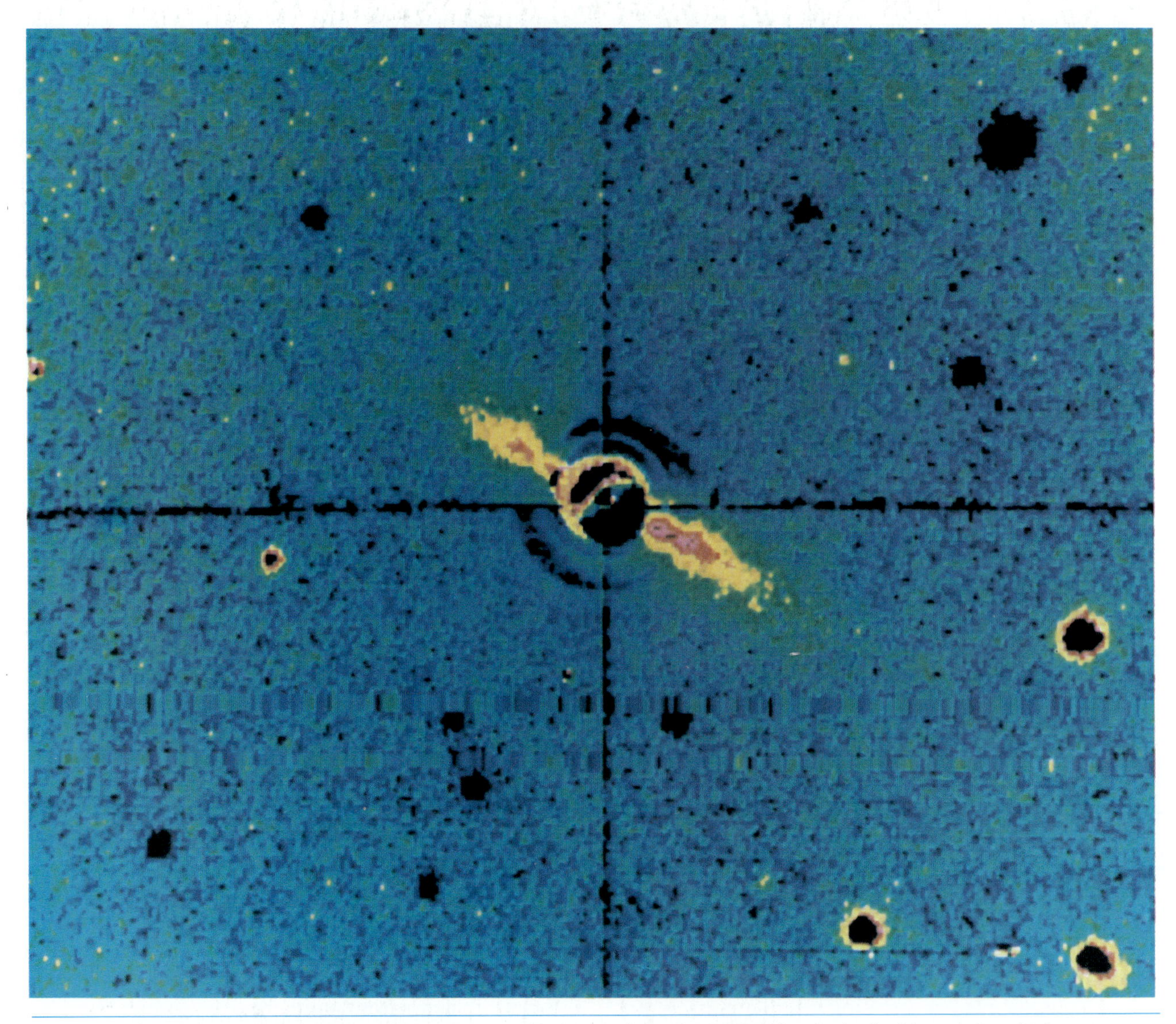

FIGURE 17.11 A STAR WITH A PREPLANETARY DISK. *This image of the star β Pictoris, obtained with an infrared telescope on the ground, shows a disk of solid particles edge-on, much like the solar nebula. Infrared data have revealed a number of stars with surrounding material, supporting the suggestion that planetary systems may form around many stars.* (NASA/JPL, courtesy of B. A. Smith)

stars. There is little hope of directly detecting smaller objects such as the terrestrial planets, however.

The fine resolution of the *Space Telescope* images will also allow much more precise measurements of stellar motions than have been achieved from the ground, so that tiny oscillations in the positions of stars will be more easily detected. Again, this will probably not allow the detection of small planets like the Earth, but could locate objects as large as Jupiter. The use of interferometry at ground-based observatories may also provide new hope of detecting small oscillations in star positions indicative of the presence of planets. Astronomers are now waiting eagerly for the chance to use these new instruments and techniques in a search for planets in other solar systems. It will be fascinating to learn the results in the coming years.

PERSPECTIVE

In this chapter we have outlined a natural, evolutionary process that seems to account for all of the observed properties of the solar system. The orbital characteristics of the planets, their compositions, the Sun's slow rotation, and the nature and motions of the interplanetary material have all been explained. Many details are yet to be worked out, but the overall scenario is probably correct in essence.

With this, we have completed our examination of the solar system, and we are ready to move on. The next logical step is to explore the realm of the stars.

SUMMARY

1. The facts that must be explained by a successful theory of the formation of the solar system include the systematics of the orbital and spin motions of the planets and satellites, the contrasts between the terrestrial and giant planets, the slow rotation of the Sun, and the existence and properties of the interplanetary bodies.
2. There are two general classes of theories: catastrophic theories and evolutionary theories.
3. Catastrophic theories require some singular event, such as a near-collision between stars, to distort the Sun by tidal forces, pulling out matter that can then condense to form the planets.
4. Evolutionary theories postulate that the planets formed as a natural by-product of the formation of the Sun. These theories, although they are simpler, were not accepted for a while because of their failure to explain the slow spin of the Sun.
5. When the solar wind and its creation of magnetic braking was discovered in the 1960s, the evolutionary theory became most widely accepted.
6. In the modern theory, the planets formed from condensations in the solar nebula, the flattened disk of gas and dust that formed around the young Sun.
7. The terrestrial planets formed from the coalescence of planetesimals, solid, asteroidlike objects that condensed from the hot inner portions of the solar nebula.
8. The giant planets are thought to have formed when solid cores coalesced from planetesimals, then trapped surrounding gas, which formed their atmospheres. The extensive ring and satellite systems of these planets formed from disks of gas and solid particles in their equatorial planes.
9. The evolutionary theory leads to the notion that many other planetary systems should exist around other stars, but so far none has been detected because of the difficulty of observing planets at interstellar distances.

REVIEW QUESTIONS

Thought Questions

1. Summarize the overall pattern of orbital and spin motions in the solar system that must be explained by any successful theory of solar system formation. Describe any exceptions to the normal pattern.
2. Describe the general contrasts between the terrestrial planets and the giant outer planets.
3. The acceptance of evolutionary models of solar system formation was delayed considerably because of the difficulty in explaining why the Sun's rotation is so slow. Explain why this was difficult. Discuss the parallels between this, the difficulty early astronomers had in accepting the heliocentric model, and the long delay in the general acceptance that continental drift occurs.
4. What are the chief arguments against the catastrophic theories of solar system formation?
5. Itemize the effects of the solar wind on the planets.
6. If the solar system formed from an interstellar cloud of uniform chemical composition, explain how the terrestrial planets developed compositions that are so different from that of the Sun.
7. Based on the information in this chapter regarding the formation of the Sun, how do you think astronomers can observe other stars in the process of formation?
8. Summarize the contrasts between the formation of the terrestrial planets and the formation of the giant gaseous planets.

9. Despite the arguments against an overall catastrophic model for the formation of the solar system, many have suggested that catastrophic events have helped to modify the system. Describe these catastrophic events.
10. Discuss the methods that can be used to search for planets orbiting other stars. Which method do you think is most sensitive, that is, most likely to succeed?

Problems

1. Using data from Appendices 2 and 7, calculate the total fraction of the mass of the solar system that is contained in the Sun. How does your answer fit into the model for solar system formation described in the text?
2. To illustrate how much of the volatile material in the inner solar system was not captured by the terrestrial planets, calculate what the mass of the Earth would be if it had the other elements in the same proportion to iron as the Sun has. Assume that the mass of the Earth's iron core is 7×10^{26} g, then calculate the mass of other elements needed so that they have the same relative abundance as in the Sun. (Hints: You can ignore all elements but hydrogen and helium in this calculation, and you should use the last column of Table 18.3 for the relative abundances by mass.) Compare your answer with the masses of the giant planets.
3. Suppose the Oort cloud defines the outer boundary of the interstellar cloud that was to collapse and form the solar system. Calculate what the density of this cloud would have been; that is, what is the density of a sphere having the mass of the Sun and a radius of 100,000 AU? If the average atom in this cloud had a mass of 2×10^{-24} g, to how many atoms per cubic centimeter does your answer correspond?
4. Suppose the early Sun had a temperature of 1,000 K at some time during its collapse and heating phase. At what wavelength would it have emitted most of its energy?
5. How much dimmer would Jupiter appear if it orbited the nearest star, 4.3 light-years, or 270,000 AU away? (Compare its brightness as seen from 270,000 AU with its brightness when it is closest to Earth, 4.2 AU away.) In view of your answer, discuss the feasibility of directly observing a planet like Jupiter orbiting another star.

ADDITIONAL READINGS

Beaty, J. K., B. O'Leary, and A. Chaikin, eds. 1981. *The new solar system*. Cambridge, Eng.: Cambridge University Press.
Cameron, A. G. W. 1975. The origin and evolution of the solar system. *Scientific American* 233(3):32.
Falk, S. W., D. N. Schramm. 1979. Did the solar system start with a bang? *Sky and Telescope* 58(1):18.
Friedman, Herbert. 1986. *Sun and Earth*. New York: W. H. Freeman.
Head, J. W., C. A. Wood, and T. A. Mutch. 1977. Geologic evolution of the terrestrial planets. *American Scientist* 65:21.
Nicholson, Iain. 1982. *The Sun*. New York: Rand-McNally and Co.
Noyes, Robert W. 1982. *The Sun, our star*. Cambridge, Mass.: Harvard University Press.
Overbye, Dennis. 1982. John Eddy: Solar detective, *Discover* 3(8):68.
Reeves, H. 1977. The origin of the solar system. *Mercury* 7(2):7.
Sagan, C. 1975. The solar system. *Scientific American* 233(3):23.
Schramm, D. N. and R. N. Clayton. 1978. Did a supernova trigger the formation of the solar system? *Scientific American* 237(4):98.
Wetherill, G. W. 1981. The formation of the earth from planetesimals. *Scientific American* 244(6):162.
Williams, George E. 1986. The solar cycle in precambrian times. *Scientific American* 255(2):88.

GUEST EDITORIAL

THE *HUBBLE SPACE TELESCOPE* AND SOLAR SYSTEM EXPLORATION

John C. Brandt

Dr. Brandt has been involved with space astronomy for much of his career. After receiving a doctorate from the University of Chicago in 1960, he spent a few years each as a faculty member at the University of California—Berkeley and as a staff astronomer at Kitt Peak National Observatory, before joining NASA in 1967. At that time his research interests lay in work on the recently-discovered solar wind, and at NASA he became head of the Solar Physics Branch at the Goddard Space Flight Center. As NASA evolved, Dr. Brandt's post became Chief of the Laboratory for Astronomy and Solar Physics, one of NASA's principal scientific research centers in astronomy. During his tenure at NASA, Dr. Brandt was an Adjunct Professor of Astronomy at the University of Maryland, where he taught graduate and undergraduate astronomy courses. He has authored several books (including at least one text for introductory astronomy classes), and nearly 300 scientific papers. His research interests have broadened, and today he is concentrating on comets, having been a leading United States figure in the recent International Halley Watch *and in space-based observations of Halley's comet. In 1986 Dr. Brandt left NASA and joined the University of Colorado, freeing himself from administrative duties so that he could concentrate on research. He is the Principal Investigator on the* Goddard High Resolution Spectrograph, *one of the major scientific instruments*

to be launched with the Hubble Space Telescope, *and in this article he shares some of his thoughts about the impact of the* Space Telescope *on solar system research.*

The soon-to-be-launched *Hubble Space Telescope (HST)* has a well-advertised potential for major impact on astronomy beyond the solar system. From the *HST* we can expect pioneering studies of stars, the interstellar gas and dust, and galaxies. Moreover, cosmological questions—questions about the origin and ultimate fate of our universe that come to mind as we view the vast night sky or a brilliant sunset—will surely be studied as well.

Somewhat less publicized are the capabilities of the *HST* with regard to solar system science. It is this subject that I would like to explore here. But first, a basic question: Why study the solar system?

Besides issues of pure science that primarily interest specialists, and the obvious fact that our planetary system is the only one available for examination at close hand, study of the solar system addresses some interrelated themes of very general interest. In a broad sense, the solar system is our environment and we would like to understand our relationship to and dependence on that environment. For example, would extensive use of chlorofluorocarbons in spray cans change our climate to resemble that of too-cold Mars, where water freezes, or that of too-hot Venus, where water vaporizes? Or, perhaps if we understood the chemistry and transport of chlorofluorocarbons we could prevent such a dramatic change in our climate. [The issue of atmospheric changes caused by chlorofluorocarbons is discussed in Astronomical Insight 8.2.]

Why are the atmospheres and climates of the planets so different? Is it solely their distance from the Sun? Or are these differences a legacy of the solar system's formation? Why are the compositions of the outer planets (Jupiter and beyond) so different from those of the inner planets (Mars and closer in)? Most of these questions touch on the subject of comparative planetology, and some bear on the origin of life. Are there conditions and chemical compounds sufficiently complex to have allowed the development of life? These are fascinating questions that the capabilities of the *HST* will en-

GUEST EDITORIAL

able us to explore. What then are these capabilities?

The *HST* is a complete observatory designed for operation in orbit for 15 years. The primary mirror is over 94 inches (2.4 meters) in diameter. While not as large as the largest ground-based mirrors, it will be the largest ever orbited. When this observatory is operational, it will provide observations free of atmospheric turbulence (which blurs images), free of airglow (which hides faint objects), and free of atmospheric absorption in the ultraviolet and infrared regions (which in some cases completely prohibits observations from the ground). The useful range of the *HST* is 1,150 to 10,000Å, although not all instruments cover the entire range. The power of the telescope can be illustrated by its resolution of 0.1 arcsecond. In other words, an object can be seen as a detail at a distance 2 million times its size.

Five first-generation scientific instruments will fly on the *HST*. The *Wide-Field and Planetary Camera,* which covers the entire wavelength region of the *HST,* can provide, in its planetary mode, full-disk images of the planet Jupiter. These images are obtained with short exposures at almost the resolution limit of the telescope. Such images are comparable in quality to *Voyager* images of Jupiter taken just a few days before closest approach. The *Faint Object Camera,* which covers the ultraviolet and visual wavelength regions, is designed to fully exploit the resolving power of the telescope system. It does so with reduced field of view.

The two spectrographs to be carried aboard the *HST* differ primarily in their ability to divide light into small increments by wavelength. The measure of this ability is the spectral resolving power, defined as the wavelength of the observation divided by the size of the resolved increment of wavelength. One of the spectrographs, the *Goddard High Resolution Spectrograph,* has principal resolving powers of 20,000 and 100,000, the latter being almost unprecedented in space astronomy. (These resolving powers mean, for example, that at a wavelength of 1200Å, details in the spectrum as small as 0.06Å and 0.0012Å, respectively, could be seen.) This instrument operates only in ultraviolet wavelengths and on targets brighter than those within the capabilities of the *Faint Object Spectrograph,* whose principal resolving powers are 100 and 1,000. The *Faint Object Spectrograph* operates at ultraviolet and visual wavelengths well into the red.

The fifth instrument, the *High Speed Photometer,* is designed to measure the brightnesses of objects over most of *HST*'s spectral range with very high time resolution. Time variations shorter than one second are usually impossible to detect from the ground because of variations in the Earth's atmosphere. This instrument should measure them down to 0.00001 second.

Finally, the system that is used to accurately point the *HST* makes possible scientific investigations in the field of *astrometry,* the precise measurement of position. The positions, of stars for example, typically should be 10 times more accurate than traditional ground-based methods.

The second-generation instruments for *HST* are already being prepared. The farthest along is a replacement wide-field and planetary camera similar to that of the first generation. Later, an instrument designed for operation in the near infrared and an imaging spectrograph will be available. These are designed to be placed by astronauts into the *HST* in space, along with other maintenance items such as batteries.

Clearly, an observatory with the capabilities just described should have tremendous impact on solar-system science through the years. For example, the imaging systems can check for double asteroids and provide a continuous, detailed record of weather on other planets. The spectrographs can search for exotic molecules and monitor changes in chemical composition.

But what about the grand schemes and cosmological considerations mentioned earlier? Particularly relevent to these questions is a specific observation to be made with the *Goddard High Resolution Spec-*

GUEST EDITORIAL

trograph on board the *HST*—an observation related to comets.

The role of comets in the formation of the solar system has long been a topic of interest and speculation. The "new" comets that we now see in the inner solar system come from the Oort cloud, a cosmic deepfreeze containing an essentially spherical distribution of comets located on average some 50,000 AU from the Sun. This cloud is probably replenished from an inner belt much closer to the Sun. But where were the comets formed? Perhaps the ratio of deuterium (heavy hydrogen) to normal hydrogen could pin down the location. Chemical reactions proceed at different rates for these two forms of hydrogen, and hence the ratio gives clues to the physical conditions in their region of formation (by ice condensation). Some idea of the range of values is as follows.

In the interstellar gas, deuterium is one part in 100,000. In Uranus, deuterium is one part in 10,000. In meteorites, deuterium can be as high as one part in 1,000. In some dense interstellar clouds, deuterium can reach one part in 100 or higher. Obviously, this ratio is a sensitive indicator of physical conditions. If we have a good measurement and a sound basis for interpretation, we have valuable information on the origin of comets. I had planned to measure the fundamental line of deuterium at 1215.34Å (near the fundamental line of normal hydrogen at 1215.67Å) to obtain this ratio in Halley's comet. Unfortunately, Halley's Comet came before the launch of the *HST*, but I plan to carry out the measurement on another bright comet using the *Goddard High Resolution Spectrograph* aboard the *HST*. On the positive side, an *in situ* measurement was in fact made on Halley's comet by the European Space Agency's *Giotto* spacecraft, but the value of the deuterium-to-hydrogen ratio obtained is somewhat uncertain.

However, the *in situ* measurements of Halley's comet have served to sharpen our interest in comets and the origin of life. Our knowledge of comets has soared because of the increased scientific activity associated with the apparition of Halley's comet and the direct exploration of comets Halley and Giacobini-Zinner by six spacecraft. Because some moderately complex molecules were previously detected in comets (such as hydrogen cyanide and methyl cyanide), there was hope that complex molecules might be directly detected. Our most optimistic dreams appear to have come true. Recent analysis of data from the *Giotto* spacecraft shows regular mass peaks at about 50 atomic mass units, which suggests the presence of chain molecules enriched in carbon, oxygen, and hydrogen. The data are consistent with molecules of the type

$$(\ldots - CH_2 - O - CH_2 - O \ldots).$$

These are polymerized (meaning large molecules formed from repeated small units) formaldehyde or polyoxmethylene. Curiously, formaldehyde (H_2CO) has long been a feature of models of cometary nuclei. These polymers explain the measurements in the comet's atmosphere (coma) and, if bonded to grains on the surface of the nucleus, help to explain the very dark surface.

Excitement builds if we combine these results with the results for the deuterium-to-hydrogen ratio for water ice in Comet Halley. These values are consistent with the view that the Earth received its volatiles (such as water, methane, and noble gases) from a veneer of volatile-rich materials of cometary origin. If this is true, the comets may have supplied the original complex molecules necessary for the origin of life. These recent developments simply heighten our interest in comets as the Rosetta stones of the solar nebula because the polymers must have been incorporated into the comets at the time of their formation. Thus, an understanding of the chemical composition of comets is fundamental to our consideration of the origin of life and other philosophical questions.

What is the role of the *HST* in this area? It is too soon to give many specifics beyond our hope of obtaining improved, direct measurements of the deuterium-to-hydrogen ratio in several comets. But a broad role for the *HST* in this and other solar system scientific areas is clear. Information about the solar system comes to us in three fundamental

GUEST EDITORIAL

ways: (1) *Direct exploration:* Instruments are sent and measurements are made. These are critical, but they are expensive and produce only snapshots. (2) *Laboratory work:* Meteorites (from asteroids and Mars) and returned samples (Moon rocks, for example) are available for detailed analysis. This area is limited by the samples available. (3) *Remote sensing:* Examples of remote sensing are ground-based observations and observations that will be made from the *HST*. This approach provides new results as observing techniques are improved (that is, as spatial or spectral resolution is increased and new wavelength bands become available). Remote sensing is the only feasible approach when observations over a long period of time are required, such as the observation of long-term climatic change. We fully expect the *HST* to make monumental contributions in the remote sensing area, but it should be noted that the combined use of all three approaches yields the maximum power. For the study of comets over the next few years, new observations will come only from the remote sensing approach.

Finally, we should always expect the unexpected from the *HST* as applied to the solar system. Our theories are quite useful for planning purposes but often are not good enough for accurate predictions. We are clearly in for some exciting surprises.

SECTION IV

THE STARS

This section is devoted to the stars and their properties. In its seven chapters we learn about how stars are observed, how their fundamental properties are deduced from observations, and how stars work. We see what the processes are that control the structure of stars, and we learn how stars live, evolve, and die. We discover that stars are dynamic, changing entities, and only seem fixed and immutable because the time scales on which they evolve are vastly longer than the human lifetime.

The first of the seven chapters describes our own Sun, its properties and structure, and how it compares to other stars in the universe. The next chapter describes the three basic types of stellar observations; measurements of positions, brightnesses, and spectra. The emphasis is on raw information that is gathered from observation, whereas discussion of how this information is used to derive the more basic stellar parameters is reserved for the next chapter. There we systematically consider several stellar properties, discuss how each is determined, and summarize the typical values that are found.

Chapter 21 then initiates the discussion of how stars function and evolve by describing the physical processes that occur in stars, and how these processes govern the stellar parameters previously discussed. Here we draw upon knowledge of our own Sun, whose well-observed properties provide vital information about other stars as well.

Chapter 22 illustrates how astronomers can learn about stellar evolution from observations, particularly of star clusters. We will find that despite the immensity of the time scales involved, it is possible to deduce the life stories of stars.

The final two chapters, 23 and 24 describe the evolution of stars; from their births to their deaths. In Chapter 23 we have to rely on the results of theoretical calculations, but we are guided by observations discussed earlier. In Chapter 24 we discuss the remnants of stars that have

SECTION IV

finished their lives, and here we discover some of the most bizarre and exciting objects in the universe.

Between chapters 23 and 24 a special insert discusses the excitement created by the Great Supernova of 1987, the brightest supernova seen in nearly four hundred years. We will learn what scientists are discovering through their study of this spectacular occurrence.

Having studied the stars, we will then be ready to consider large groupings such as our own galaxy, which contains many billions of stars, and whose own evolution is governed in part by that of the individual stars within it.

CHAPTER 18

THE SUN

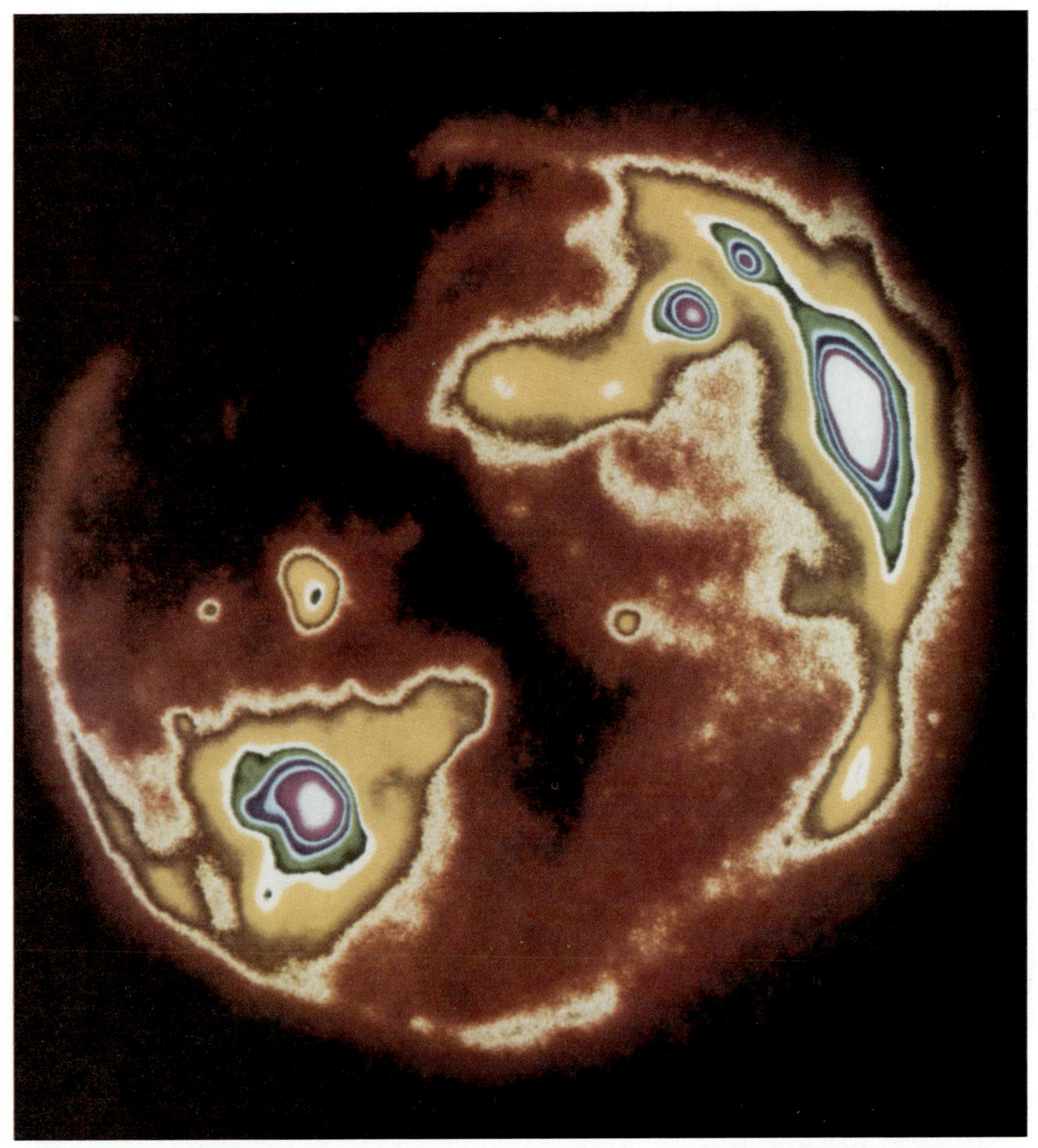

An X-ray image of the Sun, showing bright active regions and dark coronal holes. (NASA/Marshall Space Flight Center)

CHAPTER PREVIEW

Our star, the Sun, is rather ordinary by galactic standards. It has modest mass and size, and there are stars as much as a few hundred times larger in diameter and a million times more luminous. Its temperature is also moderate, as stars go. In many respects the Sun is entirely a run-of-the-mill entity.

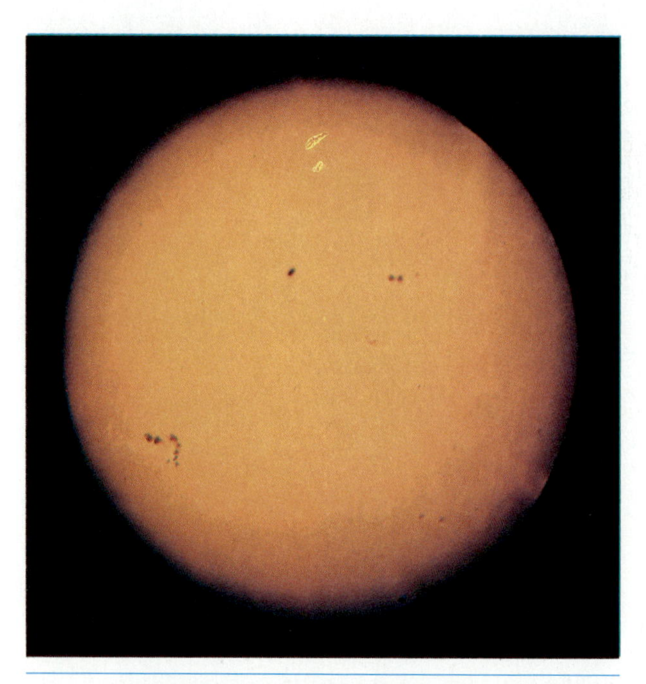

FIGURE 18.1 THE SUN. *This is a visible-light photograph of the Sun at a time when only a few sunspots were present. The overall yellow color reflects the surface temperature of almost 6000 K. The spots and the darkening of the disk near the edges are discussed in the text.* (National Optical Astronomy Observatories)

Because the Sun is an ordinary star, and because we have far more detailed knowledge of it than of other stars, it is useful to begin our discussion of stars with a chapter on the Sun. In subsequent chapters, we will see how the information on the Sun helps us to understand the nature of other stars. Conversely, we will find also that studies of other stars help us to learn more about the nature of the Sun.

BASIC PROPERTIES AND INTERNAL STRUCTURE

Perhaps the most obvious attribute of the Sun is its brightness; it emits vast quantities of light. Its **luminosity,** the amount of energy emitted per second, is about 4×10^{33} ergs per second (for comparison, recall that a 100-watt light bulb emits 10^9 erg/sec). The intensity of sunlight reaching the Earth (above the atmosphere) is about 1.4×10^6 erg/cm²/sec. This quantity is known as the **solar constant,** although it varies slightly according to activity in the Sun's outer layers (discussed later in this chapter).

The Sun is a ball of hot gas (Fig. 18.1). It dwarfs any of the planets in mass and radius (Table 18.1). Its density, on average, is 1.41 g/cm³, not much more than that of water, but its center is so highly compressed that the density there is about ten times greater than that of lead. The interior is gaseous rather than solid, because the temperature is very high, around 10 million degrees at the center, diminishing to just under 6,000 K at the surface. At these temperatures, the gas is partially ionized in the outer layers of the Sun, and is completely ionized in the core, all electrons having been stripped free of their parent atoms.

The Sun is held together by gravity. All of its constituent atoms and ions attract each other, and

TABLE 18.1 THE SUN

Diameter: 1,391,980 km (109.3 $D_\oplus$)
Mass: 1.99×10^{33} g (332,943 $M_\oplus$)
Density: 1.409 g/cm³
Surface gravity: 27.9 Earth gravities
Escape velocity: 618 km/sec
Luminosity: 3.83×10^{33} erg/sec
Surface Temperature: 6,500 K (deepest visible layer)
Rotation period: 25.04 days (at equator)

the net effect is for the solar substance to be held in a spherical shape. Gas that is hot exerts pressure on its surroundings, and this pressure, pushing outward, balances the force of gravity, which pulls the matter inwards. This balance is called **hydrostatic equilibrium,** with gravity and pressure equaling each other everywhere. The deeper we go into the Sun, the greater the weight of the overlying layers, and the more the gas is compressed. The higher the pressure, the greater the temperature required to maintain the pressure, so we find that the pressure and temperature both increase as we approach the center.

The composition of the Sun (Table 18.2) is the same as that of the primordial solar system and of most other stars: About 73 percent of its mass is hydrogen, 25 percent is helium, the rest is made up of traces of other elements. As we have already seen, this is similar to the chemical makeup of the outer planets and would represent the terrestrial planets as well, except that they have lost most of their volatile elements such as hydrogen and helium. It is apparent that all the components of the solar system formed together, from the same material.

The ultimate source of all the Sun's energy is located in its core, within the innermost 25 percent or so of its radius (Fig. 18.2). Here nuclear reactions create heat and photons of light at γ-ray wavelengths. It is this radiation that eventually reaches the surface and escapes into space, but it is a laborious journey. Each photon is absorbed and reemitted many times along the way (Fig. 18.3), gradually losing energy. In the process, most γ-ray photons become visible light photons, and the energy they lose heats the surroundings. Because a photon travels only a short distance before it is absorbed, and because when it is reemitted, it is in a random direction, progress towards the surface is very slow. It takes an individual photon as long as a million years to migrate from the center of the Sun to the surface, even though the light travel time if it were unimpeded would be only 2 seconds. If the Sun's energy source were suddenly turned off, we would not be aware of it for a million years!

Throughout most of the solar interior, the gas is quiescent, without any major large-scale flows or currents. The energy from the core is transported by the radiation wending its slow pace outwards, except in the layers near the surface, where convection occurs, and heat is transported by the overturning motions of the gas. As we shall see, the bubbling, boiling action in the outer portions of the Sun creates a wide variety of dynamic phenomena on the surface.

The Sun rotates, and it does so differentially. This is apparent from observations of its surface features, which reveal that, like Jupiter and Saturn, the Sun goes around faster at the equator than near the poles. The rotation period is 26.5 days at the equator, 28 days at middle latitudes, and even longer near the poles. As we will see, the differential rotation probably plays an important role in governing variations in the solar magnetic field, which in turn have a lot

TABLE 18.2 THE COMPOSITION OF THE SUN'S PHOTOSPHERE

Element	Symbol	Number of Atoms*	Fraction of Mass
Hydrogen	H	1.0000	0.735
Helium	He	0.0851	0.248
Lithium	Li	1.55×10^{-9}	7.85×10^{-9}
Beryllium	Be	1.41×10^{-11}	9.27×10^{-9}
Boron	B	2.00×10^{-10}	1.58×10^{-9}
Carbon	C	0.000372	0.00326
Nitrogen	N	0.000115	0.00118
Oxygen	O	0.000676	0.00788
Fluorine	F	3.63×10^{-8}	5.03×10^{-7}
Neon	Ne	3.72×10^{-5}	0.000547
Sodium	Na	1.74×10^{-6}	2.92×10^{-5}
Magnesium	Mg	3.47×10^{-5}	0.000615
Aluminum	Al	2.51×10^{-6}	4.94×10^{-5}
Silicon	Si	3.55×10^{-5}	0.000727
Phosphorus	P	3.16×10^{-7}	7.14×10^{-6}
Sulfur	S	1.62×10^{-5}	0.000379
Chlorine	Cl	2.00×10^{-7}	5.17×10^{-6}
Argon	Ar	4.47×10^{-6}	0.000130
Potassium	K	1.12×10^{-7}	3.19×10^{-6}
Calcium	Ca	2.14×10^{-6}	6.26×10^{-5}
Scandium	Sc	1.17×10^{-9}	3.84×10^{-8}
Titanium	Ti	5.50×10^{-8}	1.92×10^{-6}
Vanadium	V	1.26×10^{-8}	4.68×10^{-7}
Chromium	Cr	5.01×10^{-7}	1.90×10^{-5}
Manganese	Mn	2.51×10^{-7}	1.01×10^{-5}
Iron	Fe	3.98×10^{-5}	0.00162
Cobalt	Co	3.16×10^{-8}	1.36×10^{-6}
Nickel	Ni	1.91×10^{-6}	8.18×10^{-5}
Copper	Cu	2.82×10^{-8}	1.31×10^{-6}
Zinc	Zn	2.63×10^{-8}	1.25×10^{-6}
(all others combined)		(less than 10^{-8} of total)	

*The numbers given are relative to the number of hydrogen atoms.

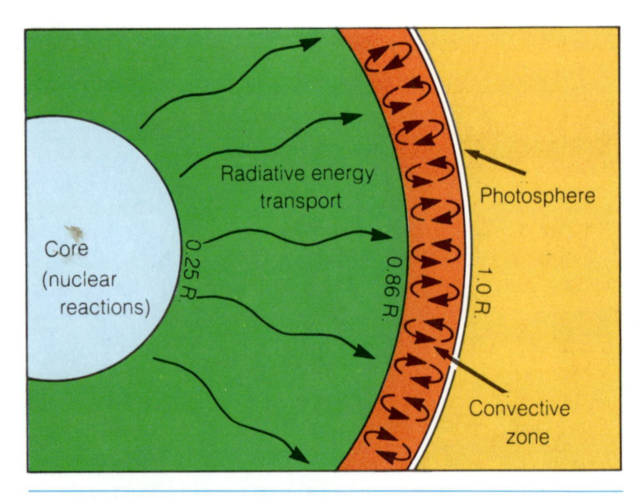

FIGURE 18.2 THE INTERNAL STRUCTURE OF THE SUN. *This drawing shows the relative extent of the major zones within the Sun, except that the depth of the photosphere is greatly exaggerated.*

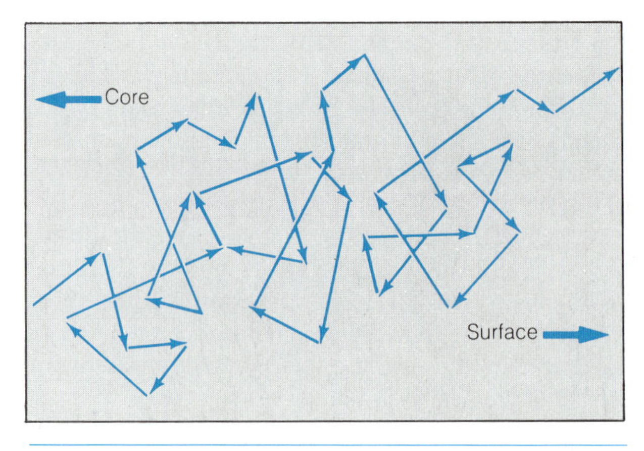

FIGURE 18.3 RANDOM WALK. *A photon is continually being absorbed and re-emitted as it travels through the Sun's interior, each time it is re-emitted, it is in a random direction. Thus its progress from the core, where it is created, to the surface is very slow.*

to do with behavior of the most prominent surface features, the sunspots.

There is evidence that the core of the Sun rotates a little more rapidly than the surface. The only clue for this is found in the presence of very subtle oscillations on the solar surface, which may be wave motions affected by the more rapid spin of the interior. The spin of the solar core is probably a direct result of the collapse and accelerated rotation of the interstellar cloud from which the Sun formed. (The origin of the Sun's rotation was discussed in Chapter 17.)

NUCLEAR REACTIONS

One of the biggest mysteries in astronomy in the early decades of this century was presented by the Sun. The problem was how to account for the tremendous amount of energy it radiates, particularly perplexing in view of geological evidence that the Sun has been able to produce this energy for at least 4 or 5 billion years. Two early ideas, that the Sun is simply still hot from its formation, or that it is gradually contracting, releasing stored gravitational energy, were both ruled out, since neither mechanism could possibly supply the energy needed to run the Sun for a long enough time.

The first hint at the solution came in the first decade of the 1900s, when Albert Einstein developed his theory of special relativity. He showed that matter and energy are equivalent, according to the famous formula $E = mc^2$, where E is the energy released in the conversion, m stands for the mass that is converted, and c for the speed of light. If mass could be converted into energy, enormous amounts of energy could be produced. Physicists began to contemplate the possibility that somehow this energy was being released inside the Sun and stars.

By the 1920s, following the pioneering work on atomic structure by Max Planck, Niels Bohr, and others, the concept of nuclear reactions began to emerge. These are transformations like chemical reactions, except that in this case it is the subatomic particles in the nuclei of atoms that react with each other. There can be **fusion** reactions, in which nuclei merge to create a larger nucleus, representing a new chemical element; and there are **fission** reactions, in which a single nucleus, usually of a heavy element with a large number of protons and neutrons, splits into two or more smaller nuclei. In either type of reaction, energy is released as some of the matter is converted according to Einstein's formula. In the 1920s, many physicists explored these possibilities, and by the 1930s Hans Bethe had suggested a specific reaction sequence that might operate in the Sun's core.

Bethe envisioned a fusion reaction in which four hydrogen nuclei (each consisting of only a single pro-

ASTRONOMICAL INSIGHT 18.1

THE SOLAR NEUTRINO MYSTERY

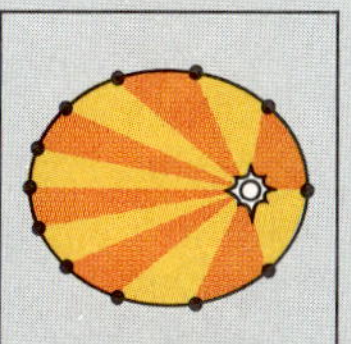

Nearly all that we know about the reactions in the Sun's core is based on theoretical calculations, and of course astronomers and physicists have been interested in verifying the results of these calculations. We think that the theory is basically correct, because it neatly accounts for the Sun's observed energy output, but nevertheless a more direct confirmation has been sought.

One way to do this was suggested by the early work of Enrico Fermi. Fermi postulated the existence of a tiny subatomic particle called the **neutrino,** which, his calculations showed, ought to be released at certain stages in nuclear reactions. This is a strange little particle, having no mass and no electrical charge, but which can carry away small amounts of energy, thus balancing the equations describing the reactions. Because of its ephemeral properties, a neutrino hardly interacts with anything at all (it can interact with other matter only through the weak nuclear force, see Astronomical Insight 5.2). Thus, neutrinos produced in the Sun's core ought to escape directly into space at the speed of light, in sharp contrast with the photons from the core, which, as we have seen, take a million years or more to get out.

A possible test of the reactions going on inside the Sun, therefore, would be to measure the rate at which neutrinos are coming out. Unfortunately, the same properties that allow them to escape the Sun so easily also make them very difficult to catch and count. There is an indirect technique, however, based on the fact that certain nuclear reactions can be triggered by the impact of a neutrino.

A chlorine atom, for example, can be converted into a radioactive form of argon when it encounters a neutrino. Accordingly, an experiment was set up a few years ago, and is still in operation, in which a huge tank containing a chlorine compound is monitored to see how many of its atoms are being converted into radioactive argon atoms, which in turn indicates how many neutrinos are passing through the tank.

Much to everyone's surprise, neutrinos have not been detected coming from the Sun in the expected numbers, and at present no satisfactory explanation has been found. Clearly there is something not yet understood about nuclear reactions in the Sun, but it is not clear whether the problem is in our understanding of nuclear reactions or in our understanding of the internal properties of the Sun. Unfortunately, the neutrinos that can react with chlorine are produced by a minor offshoot of the main energy-producing reactions in the Sun, so it is possible that the main reaction sequence occurs as expected, but that the rate at which this other reaction occurs is slower than thought.

Ideally, we would like to detect the neutrinos produced directly in the proton-proton chain. These neutrinos are even harder to catch, however, and so far no one has managed to build a detector for them. Plans are under way to do so, using an element called gallium with which the solar neutrinos can react. To build a gallium neutrino telescope will require several years' worth of the world gallium supply, so this project will be lengthy and expensive. Nevertheless, groups in the Soviet Union and in Israel, each with many international collaborators, are at work on the task. We can hope that within a few years a more direct test of solar nuclear reactions will be possible.

Until then, scientists will continue the quest to explain the perplexing results already in hand. One recent suggestion invokes another kind of elementary particle whose existence is not as well established as that of the neutrino. A family of so-called Weakly Interacting Massive Particles (WIMPs) has been postulated to exist. These particles, relatively massive by subatomic standards but capable of interacting with other matter only through the weak nuclear force, could be produced in solar nuclear reactions. If so, they are capable of carrying away enough energy to account for the low observed rate of neutrino emission.

Interestingly, one of the theoretical consequences of having WIMPs carry away energy from the Sun's core is that the core temperature would be slightly cooler than it is thought to be in present models of interior conditions. A lower core temperature would help explain another mystery having to do with the Sun's internal pulsations (see Astronomical Insight 18.2). The frequency of vibration due to internal pulsations depends on the speed of

(continued on next page)

ASTRONOMICAL INSIGHT 18.1

THE SOLAR NEUTRINO MYSTERY

(continued from previous page)
sound in the Sun's core, and the speed of sound depends on the temperature. A lower temperature would bring theoretical and observed vibration frequencies into agreement, so some of the scientists who study the Sun's pulsations are attracted to the WIMP hypothesis. WIMPs also have the potential to solve far-reaching questions about the structure of our galaxy and the distribution of mass in the universe, as we discuss in later chapters.

Questions regarding WIMPs and neutrino properties await technological advances in order to be answered. One such advance would be the advent of gallium detectors for neutrinos; another would be the development of particle accelerators capable of producing the hypothetical WIMPs, so that their existence could be verified and their properties tested. The proposed **Superconducting Super Collider,** now being planned by the United States government, will be capable of this. Perhaps within a few years the solar neutrino mystery will be solved. Until then, it remains one of the outstanding problems in modern astrophysics.

ton) combine to form a helium nucleus, made up of two protons and two neutrons. The reaction actually occurs in several steps (Table 18.3): (1) two protons combine to form **deuterium,** a type of hydrogen that has a proton and a neutron in its nucleus (one of the two protons undergoing the reaction converts itself into a neutron by emitting a positively charged particle called a **positron**) and another particle called a **neutrino,** which has very unusual properties described in Astronomical Insight 18.1; (2) the deuterium combines with another proton to create an isotope of helium (^{3}He) consisting of two protons and one neutron; and (3) two of these ^{3}He nuclei combine, forming an ordinary helium nucleus (^{4}He, with two protons and two neutrons in the nucleus) and releasing two protons. At each step in this sequence, heat energy is imparted to the surroundings in the form of kinetic energy of the particles that are produced, and in step (2) a photon of γ-ray light is emitted as well.

The net result of this reaction, which is called the **proton-proton chain,** is that four hydrogen nuclei (protons) combine to create one helium nucleus. The end product has slightly less mass than the ingredients, 0.007 of the original amount having been converted into energy. We will learn in Chapters 21 and 23 that other kinds of nuclear reactions take place in some stars, but the vast majority produce their energy by the proton-proton chain.

It is easy to see that the proton-proton chain can produce enough energy to keep the Sun shining for billions of years. The total mass of the Sun is about 2×10^{33} grams. Only the innermost portion of this, perhaps 10 percent, undergoes nuclear reactions, so the Sun started out with about 2×10^{32} grams of mass available for nuclear reactions. If 0.007 of this, or 1.4×10^{30} grams, is converted into energy, then to find the total energy the Sun can produce in its lifetime, we multiply this mass times c^2, finding a total energy of $E = (1.4 \times 10^{30}\text{ g})(3 \times 10^{10}\text{ cm/s}^2)^2 = 1.3 \times 10^{51}$ erg. The rate at which the Sun is losing energy is its luminosity, which is about 4×10^{33} ergs per second. At this rate, the Sun can last for $(1.3 \times 10^{51}\text{ ergs})/(4 \times 10^{33}\text{ erg/s}) = 3.2 \times 10^{17}$ seconds, or just about 10 billion years. Geological evidence shows that the solar system is now about 4.5 billion years old, so we can expect the Sun to keep shining for another 5 billion years or more. (In Chapter 23, we will see what happens to stars like the Sun when the nuclear fuel, hydrogen, runs out.)

Nuclear fusion reactions can take place only under conditions of extreme pressure and temperature, because of the electrical forces that normally would keep atomic nuclei from ever getting close enough together to react. Nuclei, which have positive charges because all of their electrons have escaped, must collide at extremely high speeds in order to overcome the repulsion caused by their like electrical charges.

TABLE 18.3 STRUCTURE OF THE SUN

Zone	Radius* (R)	(km)	Temperature (K)	Density (g/cm³)	Mass Fraction (M)*
Interior	0.00	0	1.6×10^7	160	0.000
	0.04	28,000	1.5×10^7	141	0.008
	0.1	70,000	1.3×10^7	89	0.07
	0.2	139,000	9.5×10^6	41	0.35
	0.3	209,000	6.7×10^6	13	0.64
	0.4	278,000	4.8×10^6	3.6	0.85
	0.5	348,000	3.4×10^6	1.0	0.94
	0.6	418,000	2.2×10^6	0.35	0.982
	0.7	487,000	1.2×10^6	0.08	0.994
	0.8	557,000	7.0×10^5	0.018	0.999
	0.9	627,000	3.1×10^5	0.0020	1.000
	0.95	661,000	1.6×10^5	0.0004	1.000
	0.99	689,000	5.2×10^4	0.00005	1.000
	0.995	692,000	3.1×10^4	0.00002	1.000
	0.999	695,300	1.4×10^4	0.0000001	1.000
Photosphere	1.000	695,990	6.4×10^3	3.5×10^{-7}	1.000
	1.000 +	280	4.6×10^3	4.5×10^{-8}	1.000
Chromosphere	1.000 +	320	4.6×10^3	3.1×10^{-8}	1.000
	1.001 +	560	4.1×10^3	3.6×10^{-9}	1.000
(Transition)	1.002 +	1,900	8.0×10^3	3.4×10^{-13}	1.000
	1.003 +	2,400	4.7×10^5	4.8×10^{-15}	1.000
Corona	1.003 +	2,400	5.0×10^5	1.7×10^{-15}	1.000
	1.2 +	140,000	1.2×10^6	8.5×10^{-17}	1.000
	1.5 +	348,000	1.7×10^6	1.4×10^{-17}	1.000
	2.0 +	696,000	1.8×10^6	3.4×10^{-18}	1.000

*The radii are expressed in fractions of the Sun's radius R, as well as in kilometers. Above 695,990 km, the kilometer values are heights above the Sun's "surface," or lower boundary of the photosphere. The mass fractions are expressed in units of the Sun's mass M., which has the value 1.989×10^{33} g. (Data from C. W. Allen, 1973, *Astrophysical Quantities*, London: Athlone Press.)

The speed of particles in a gas is governed by the temperature, and only in the very center of the Sun and other stars is it hot enough (around 10 million degrees) to allow the nuclei to collide fast enough to fuse. This is why only the innermost portion of the Sun can ever undergo reactions. The high pressure in the Sun's core causes nuclei to be crowded together very densely, and this means that collisions will occur very frequently, another requirement if a high reaction rate is to occur.

STRUCTURE OF THE SOLAR ATMOSPHERE

Observations of the Sun's appearance when viewed in different wavelengths of light make it clear that the outer layers are divided into several distinct zones (Fig. 18.4). The "surface" of the Sun that we see in visible wavelengths is the **photosphere,** with a temperature ranging between 4,000 and 6,500 K. When we view the Sun at the wavelength of the strong line of hydrogen at 6,563 Å, we see the **chromosphere,** a layer above the photosphere whose temperature is 6,000 to 10,000 K. Outside of that is the very hot, rarefied **corona,** best observed at X-ray wavelengths, and whose temperature is 1 to 2 million degrees. Between the chromosphere and the corona is a thin region, called the **transition zone,** where the temperature rapidly rises. Overall, the tenuous gas within and above the photosphere is referred to as the solar atmosphere. If we consider the temperature throughout this region, we find that it decreases outward through the photosphere, reaching a minimum value

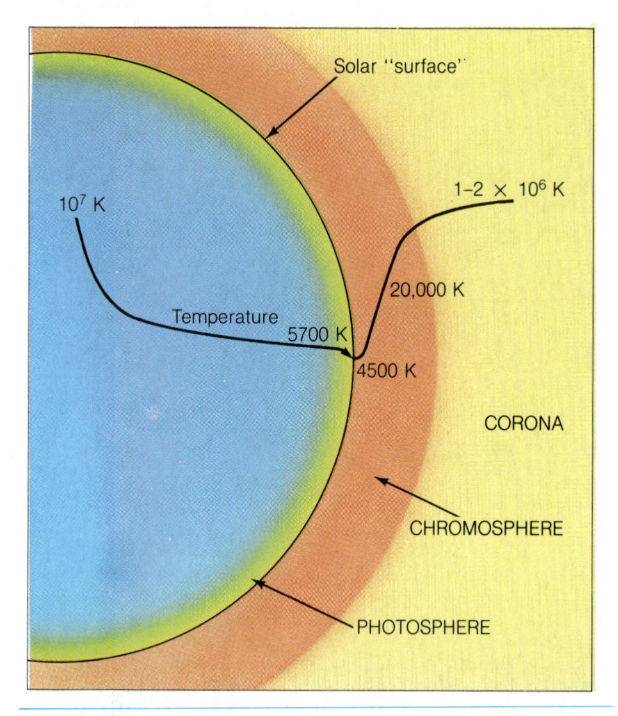

FIGURE 18.4 THE STRUCTURE OF THE SUN'S OUTER LAYERS. *This diagram shows the relative temperatures of the convective zone, the photosphere, the chromosphere, and the corona.*

of about 4,000 K. From there the trend reverses itself, and the temperature begins to rise as we go farther out. The chromosphere, immediately above the temperature minimum, is perhaps 2,000 km thick. Above there the temperature rises very steeply within a few hundred kilometers, to the coronal value of over a million degrees. Clearly, something is adding extra heat at these levels; shortly we will consider where this heat comes from.

First let us discuss the photosphere, the "surface" of the Sun as we look at it in visible light. This is the level we see because it is here that the density becomes great enough for the gas to be opaque, so that we can see no farther into the interior. It is in the photosphere that the Sun's absorption lines (Fig. 18.5) are formed, as the atoms in this relatively cool layer absorb continuous radiation coming from the hot interior.

A photograph of the photosphere reveals a cellular appearance called **granulation** (Fig. 18.6). Bright regions, thought to represent areas where convection in the Sun's outer layers causes hot gas to rise, are bordered by dark zones where cooler gas is apparently descending back into the interior. Recent detailed studies of granules and their changes with time have shown their behavior to be more complex than pre-

FIGURE 18.5 FRAUNHOFER LINES. *This is a portion of the Sun's spectrum, showing some of the dark lines first noticed by Wollaston and catalogued by Fraunhofer. The two lines near the far left in the upper segment are due to ionized calcium. The center segment shows several lines of the Balmer series of hydrogen, and the strongest Balmer line, H-alpha, is seen in the red portion of the lower segment.* (K. Gleason, Sommers Bausch Observatory, University of Colorado)

viously thought: Granules can change appearance rapidly and can disintegrate quickly and disappear.

The temperature of the photosphere, roughly 6,000 K, is measured from the use of Wien's law and from the degree of ionization in the gas there. The density, roughly 10^{17} particles per cubic centimeter in the lower photosphere, was found from the degree of excitation, as described in Chapter 6. This is lower

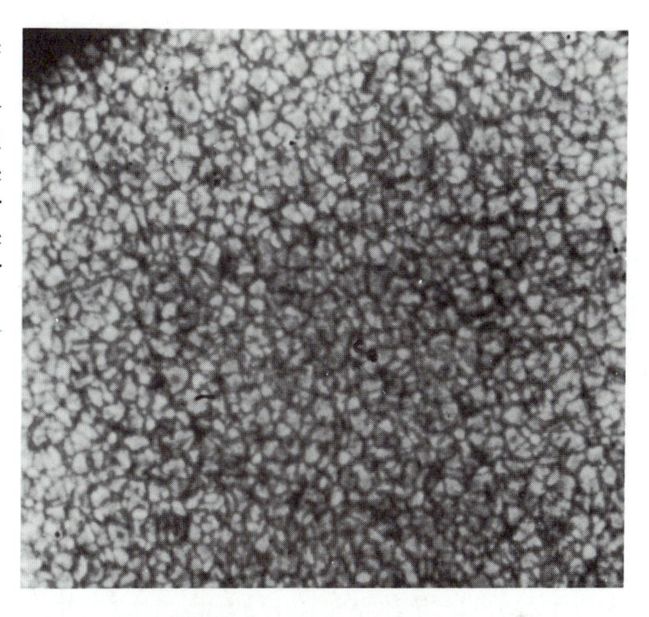

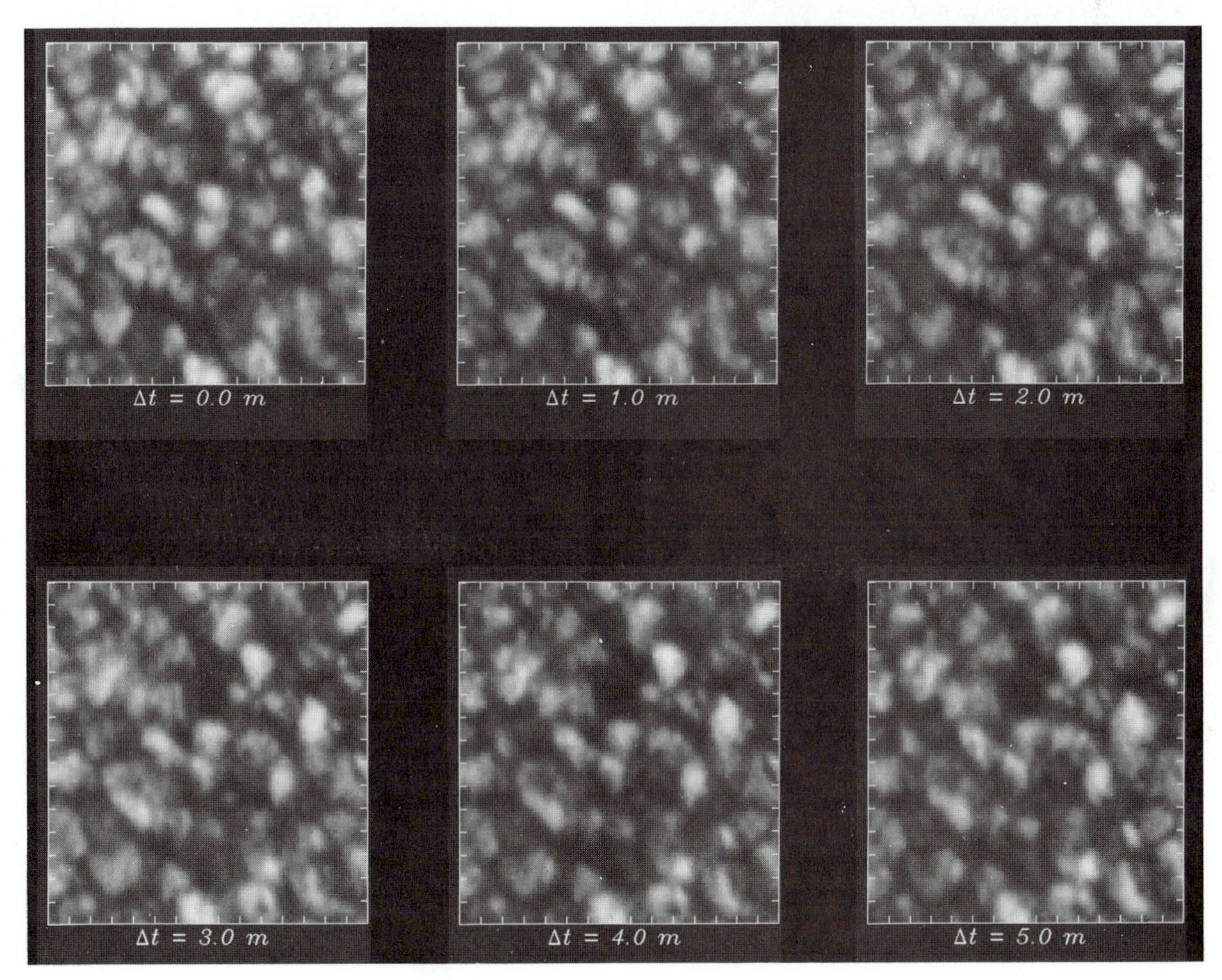

FIGURE 18.6 SOLAR GRANULATION. *The upper photograph, obtained by a high-altitude balloon above much of the atmosphere's blurring effect, distinctly shows the granulation of the photosphere, which is due to convective motions in the Sun's outer layers. The six images in the lower figure, obtained from space, show how granules change from minute to minute (top:* Project Stratoscope, Princeton University, sponsored by the National Science Foundation; *bottom:* Courtesy of A. Title, Lockheed Aerospace Corporation)

ASTRONOMICAL INSIGHT 18.2

MEASURING THE SUN'S PULSE

In the early 1960s, measurements of spectral lines formed in the solar photosphere revealed small Doppler shifts that alternated between blue- and redshifts. This seemed to indicate that the Sun's surface was rising and falling, with a period of about 5 minutes. The reported Doppler shifts were quite small (less than one one-hundredth of an angstrom), near the limit of detectability, and the measurements were correspondingly difficult. Today the existence of this and other periodic oscillations of the Sun's surface is the basis for an entire research field, sometimes called solar seismology.

Oscillations of the solar surface can be likened to a musical instrument with a resonant cavity. When waves of any kind are created inside a cavity with reflecting walls, certain wavelengths or frequencies persist, while all others die out. The reason is that waves reflected from the walls interfere with incoming waves, the valleys and crests either adding together or canceling each other out. In an enclosed space, the wavelengths that add together constructively are determined by the dimensions of the enclosure; such wavelengths are characteristic of the particular enclosure and correspond to its so-called resonant frequency. In a musical wind instrument, this frequency determines the pitch of the tone that is emitted. The musician can control the pitch by altering the dimensions of the internal cavity (by opening and closing valves in most cases).

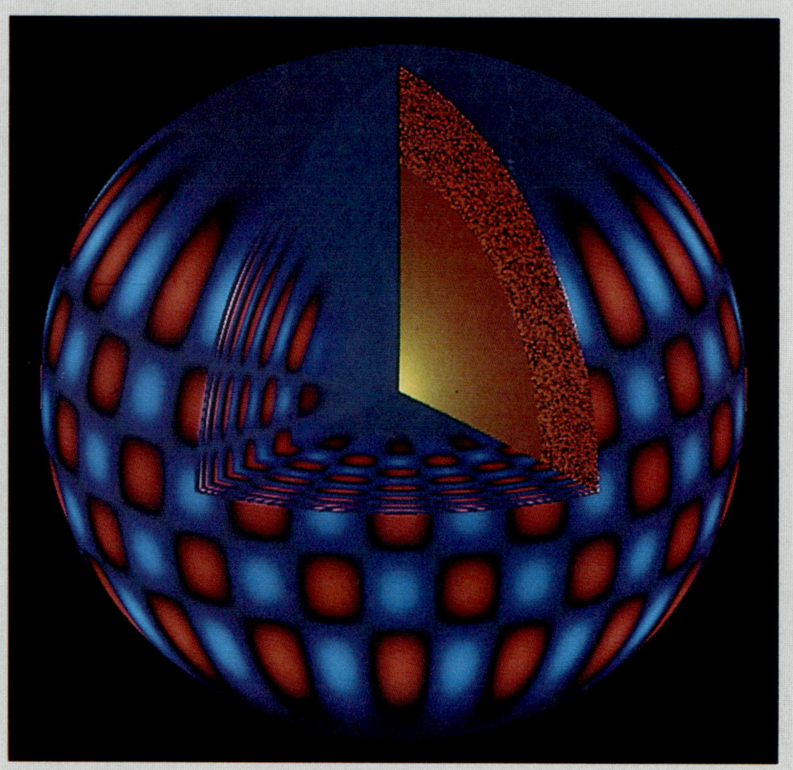

(National Optical Astronomy Observatories)

Inside the Sun, there are no walls, but physical conditions that vary with depth can create barriers as real as walls for certain types of waves. It has been deduced that sound waves (i.e., compressional waves) can persist in the Sun's convective zone, and are confined above and below by the manner in which the Sun's temperature varies with depth. Sound waves in a certain range of wavelength are reflected back as they travel farther in or farther out, and this creates a resonant cavity in the outermost portion of the Sun for sound waves whose period of pulsation is about 5 minutes. These waves create a standing pattern of crests and valleys on the solar surface, and at any particular point on the surface, the gas rises and falls with this period.

This phenomenon is quite interesting for its own sake but is worthy of careful study for another rea-

ASTRONOMICAL INSIGHT 18.2

MEASURING THE SUN'S PULSE

son as well: The properties of the waves or oscillations of the solar surface can reveal information on conditions in the solar interior, where the waves arise. For example, study of the frequency range of the waves can yield information on exactly how the temperature varies with depth inside the Sun, since it is this variation that determines the location of the lower boundary of the resonant cavity. Perhaps even more interestingly, analysis of the surface oscillations can provide information on the internal rotation of the Sun. This is possible because waves traveling over the solar surface in opposite directions are either spread out or squeezed closer together, depending on whether they move in the direction of the Sun's rotation or in the opposite direction. This means that waves going in opposite directions have different frequencies, and the difference in frequency depends on the speed of rotation of the Sun in the layer where the waves arise. In effect, it is possible to infer the Sun's internal rotational velocity at the lower boundary of the resonant cavity in which the waves oscillate.

The determination of internal rotational speeds using observations of surface oscillations is very complex and requires very sophisticated treatment of extensive quantities of data. At present, it has not been possible to carry this effort very far, but it has been determined from the 5-minute oscillations that the Sun is rotating more rapidly near the bottom of the convective zone than at the surface, leading to speculations that the rotation near the core may be even more rapid. If so, this could have important implications for the Sun's overall structure and evolution, and a great deal of importance is being attached to further study of this phenomenon.

While the 5-minute oscillation of the Sun is now well established and understood, other reported oscillations are not. There is a 3-minute oscillation that may be caused by a similar resonant cavity at the level of the chromosphere, and apparently there are oscillations lasting much longer that are driven by waves in very deep resonant cavities, near the solar core. These controversial oscillations, having periods of 40 and 160 minutes, have the potential of revealing conditions such as the rotation rate very deep inside the Sun, and their study is therefore of great importance.

For a variety of reasons, the detection of long-period oscillations is very difficult. It is important, for example, to observe the Sun continuously over many cycles of the oscillations, which means it is necessary to observe without interruption for several days. This may sound impossible, because of the Sun's daily rising and setting, but in fact it can be done from one of the Earth's poles during local summer. Therefore, solar studies have been done from the scientific research station at the South Pole. During one of these studies, involving six days of continuous observations of the Sun, the 160-minute oscillation was convincingly detected.

Another approach to making continuous observations of the oscillations is to establish a worldwide network of observers, so that the Sun can be observed, with similar instruments, throughout the Earth's 24-hour rotation period. One such network, called the Global Oscillation Network Group (GONG), is being established by the U.S. National Solar Observatory. GONG, which will consist of several identical observing stations located at sites around the world, is expected to provide data on solar oscillations for years to come.

Beyond simply detecting the existence of the long-period oscillations, further refinements are needed if they are to be exploited as tools for probing the Sun's deep interior. To observe them in enough detail to do this requires entirely new technology, and at present there are several research groups at work developing the necessary instruments. The problem is to find a way to observe very small Doppler shifts in spectral lines at many positions on the Sun's surface, and to make these measurements repeatedly at very closely spaced intervals of time. The velocities involved are very small, only a meter per second or less, so the Doppler shift in a typical visible-wavelength line is only a few millionths of an angstrom. To accurately measure such small shifts is a formidable task, but one that probably will be successfully completed within a few years. When this has been accomplished, a wealth of new information on the internal properties of our local star will be mined.

FIGURE 18.7 THE CHROMOSPHERE. *Parts of the chromosphere is seen as a thin red region at the edge of the solar disk in this eclipse photograph.* (Photograph by Patrick Wiggens, Hansen Planetarium)

than the density of the Earth's atmosphere, which is about 10^{19} particles per cubic centimeter at sea level.

Most of what we know about the Sun's composition is based on the analysis of the solar absorption lines, so strictly speaking, the derived abundances represent only the photosphere. We have no reason to expect strong differences in composition at other levels, however, except for the core, where a significant amount of the original hydrogen has been converted into helium by the proton-proton reaction.

The photosphere near the edge of the Sun's disk looks darker than in the central portions. This effect, called **limb darkening,** is caused by the fact that we are looking obliquely at the photosphere when we look near the edge of the disk. We therefore do not see as deeply into the Sun there as we do when looking near the center of the disk. Therefore, at the limb, the gas we are seeing is cooler than at disk center, and radiates less.

The chromosphere lies immediately above the temperature minimum. The fact that this region forms emission lines tells us, according to Kirchhoff's laws, that the chromosphere is made of hot, rarefied gas, hotter than the photosphere behind it. When viewed through a special filter that allows only light at the wavelength of the hydrogen emission line at 6,563 Å to pass through (Fig. 18.7), the chromosphere has a distinctive cellular appearance referred to as **supergranulation,** similar to the photospheric granulation, but with cells some 30,000 km across instead of about 1,000 km. There is also fine-scale structure in the chromosphere, in the form of spikes of glowing gas called **spicules** (Fig. 18.8). These come and go, probably at the whim of the magnetic forces that seem to control their motions.

The outermost layer of the Sun's atmosphere is the corona, which extends a considerable distance above the photosphere and chromosphere. The corona is irregular in form, patchy near the Sun's surface, but with radial streaks at great heights suggestive of outflow from the Sun (Fig. 18.9). The density of the coronal gas is very low, only about 10^9 particles per cubic centimeter.

As we have already seen, the corona is very hot, containing highly ionized gas. The source of the energy that heats the corona to such extreme temperatures is not well understood, although a general picture has emerged, as we will see.

FIGURE 18.8 SPICULES. *This photograph shows the transient features in the chromosphere known as spicules. These spikes of glowing gas, which are apparently shaped by the Sun's magnetic field, come and go irregularly.* (Sacremento Peak Observatory, National Optical Astronomy Observatories)

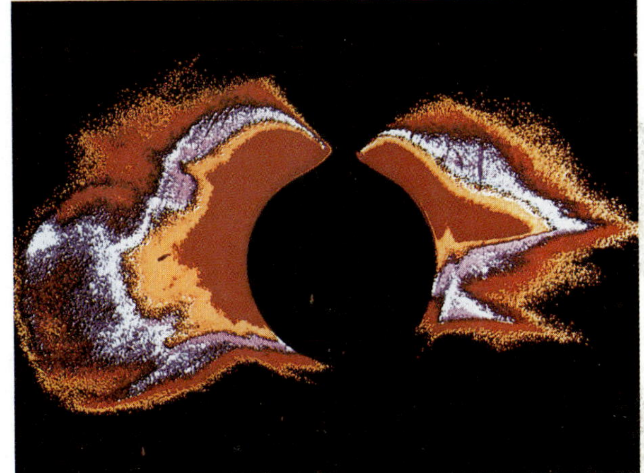

FIGURE 18.9 THE CORONA. *The upper photograph, obtained during a total solar eclipse, shows the type of structure commonly seen in the corona. There are giant looplike features and an overall appearance of outward streaming. Bits of more intense light from the chromosphere are seen around the edges of the Moon's occulting disk. Visible at left is the planet Venus. The lower image, obtained from space using a disk to block out the Sun itself, illustrates brightness gradations with false colors.* (*top:* High Altitude Observatory, National Center for Atmospheric Research, sponsored by the National Science Foundation; *bottom:* NASA)

X-ray observations reveal that the corona is not uniform, but instead has a patchy structure (Fig. 18.10). There are large regions that appear dark in an X-ray photograph of the Sun, where the gas density is even lower than in the rest of the corona. These regions are called **coronal holes** and, as we shall see, are probably created and maintained by the Sun's magnetic field. The coronal holes, as well as the overall shape of the corona, vary with time (Fig. 18.11), showing that the corona in general is a dynamic region. Another impressive sign of this is presented by the **prominences** (Figs. 18.12 and 18.13), great geysers of hot gas that spurt upward from the surface of the Sun, taking on an arc-shaped appearance. These are usually associated with sunspots, and both phenomena are linked to the solar activity cycle (to be discussed shortly).

The hot outer layers of the Sun have provided astronomers with a second major mystery regarding the solar energy budget. In contrast with the mystery of the Sun's internal energy source, the mechanism for heating the chromosphere and corona has not yet been entirely deduced. There is a great deal of energy

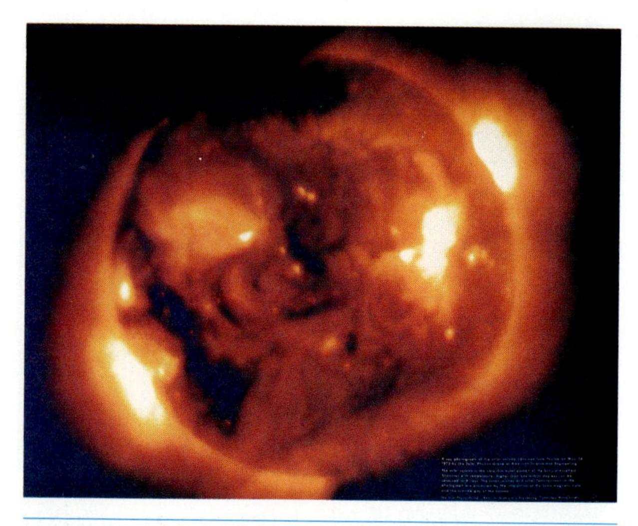

FIGURE 18.10 AN X-RAY PORTRAIT OF THE CORONA. *This image, obtained by* Skylab *astronauts, shows the Sun in X-ray light, which reveals only very hot, dense regions. The bright regions are places where the corona is especially dense, and the dark regions, known as coronal holes, are places where it is much more rarefied. The structure seen here changes with time.* (NASA)

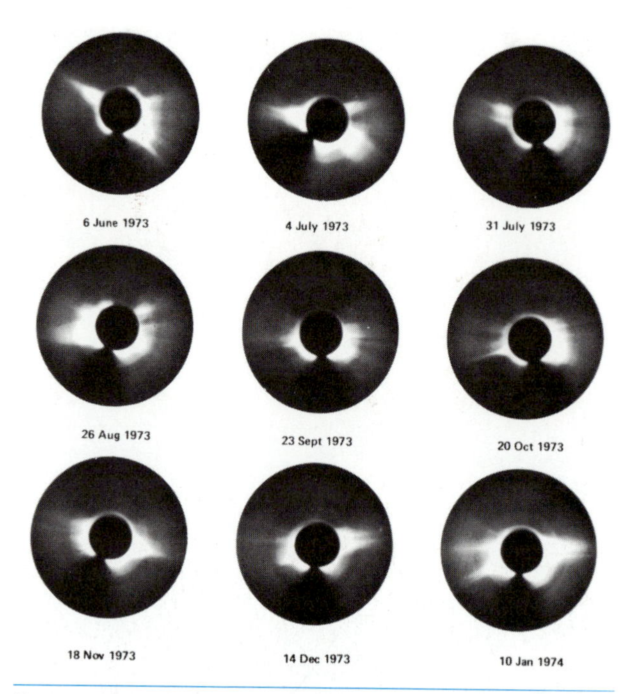

FIGURE 18.11 CHANGES IN THE CORONA. *This series of photographs was obtained using a special shutter device that blocks out the solar disk and allows the corona to be observed without a total solar eclipse. Here we see that the structure of the corona changes quite markedly over a period of a few months.* (High Altitude Observatory, National Center for Atmospheric Research, sponsored by the National Science Foundation)

available in the form of gas motions in the convection zone just beneath the solar surface, and it is generally accepted that the heating of the corona somehow comes from this energy. It is not clear how the energy is transported to such high levels, however, although there are several ideas, each invoking some form of waves. Waves of any kind carry energy, and it has been suggested that either sound waves or magnetic waves of some kind are the agents that transfer energy from the convection zone to the corona. There is certainly enough energy in the convective motions in and below the photosphere; the problem is one of understanding how this energy is transported into the higher levels.

Recent satellite observations in ultraviolet and X-ray wavelengths have shown that stars similar to the Sun in general type also have chromospheric and coronal zones, so if we can understand how the Sun operates, we will also gain a deeper understanding of how other stars work. Similarly, observations of other stars, particularly of the relationship of chromospheric and coronal activity to stellar properties such as age and rotation, can help us learn more about the Sun.

THE SOLAR WIND

The long, streaming tail of a comet always points away from the Sun, regardless of the direction of motion of the comet. The significance of this was fully realized in the late 1950s, when the first U.S. satellites revealed the presence of the Earth's radiation belts and the fact that they are shaped in part by a steady flow of charged particles from the Sun. The solar wind reaches a speed near the Earth of 300 to 400 km/sec. It is this flow of charged particles that forces cometary tails always to point away from the Sun.

The existence of the solar wind is evidently a natural by-product of the same heating mechanisms that produce the hot corona of the Sun. It was originally thought that particles in this high- temperature region move about with such great velocities that a steady trickle escapes the Sun's gravity, flowing out-

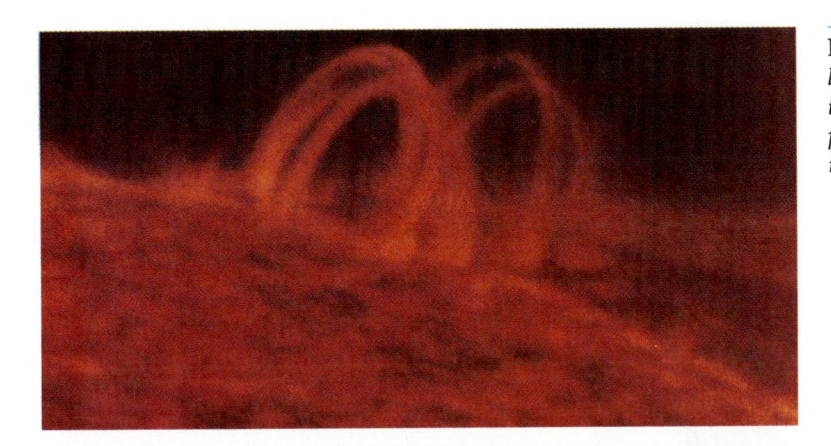

FIGURE 18.12 A PROMINENCE. *Here the looplike structures thought to be governed by the Sun's magnetic field are readily seen. This photograph was obtained by* Skylab *astronauts using an ultraviolet filter.* (NASA)

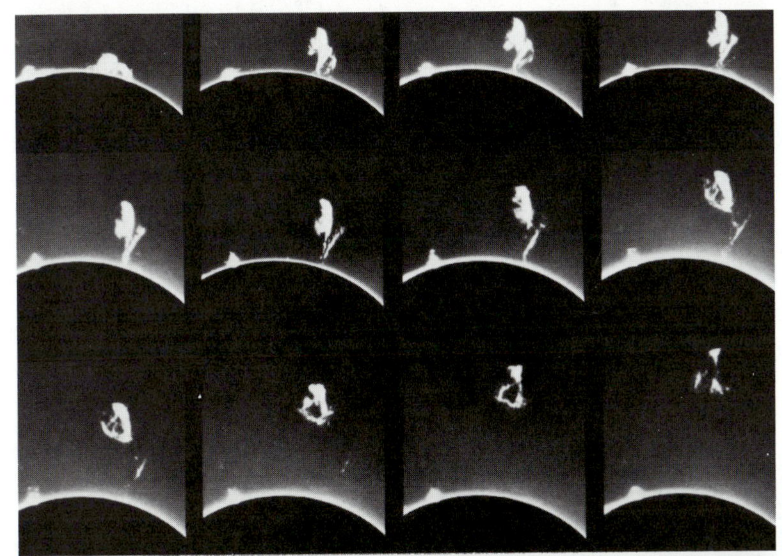

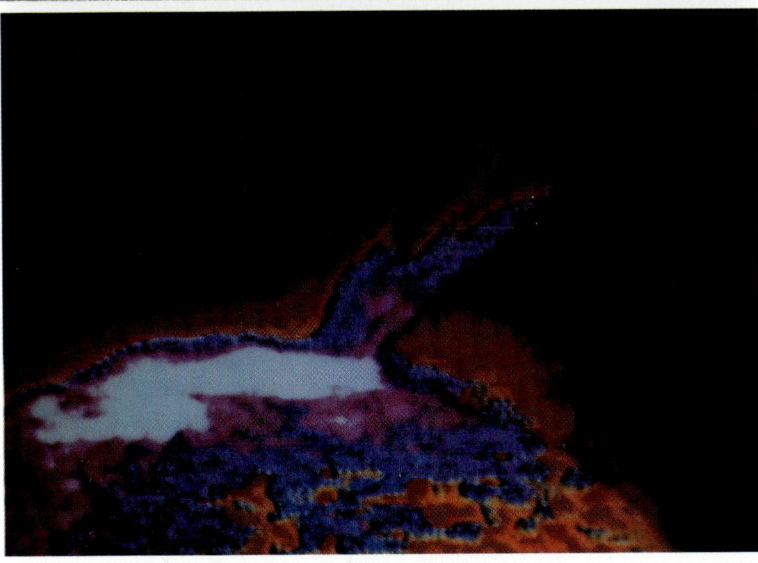

FIGURE 18.13 AN ERUPTIVE PROMINENCE. *The striking sequence at top shows an outburst of ionized gas from a prominence seen on the Sun's limb. The photographs were obtained by using a special device to block out the Sun's disk. At bottom is an ultraviolet view of another eruptive prominence, showing the extent reached in less than two hours.* (*top:* High Altitude Observatory, National Center for Atmospheric Research, sponsored by the National Science Foundation; *bottom:* NASA)

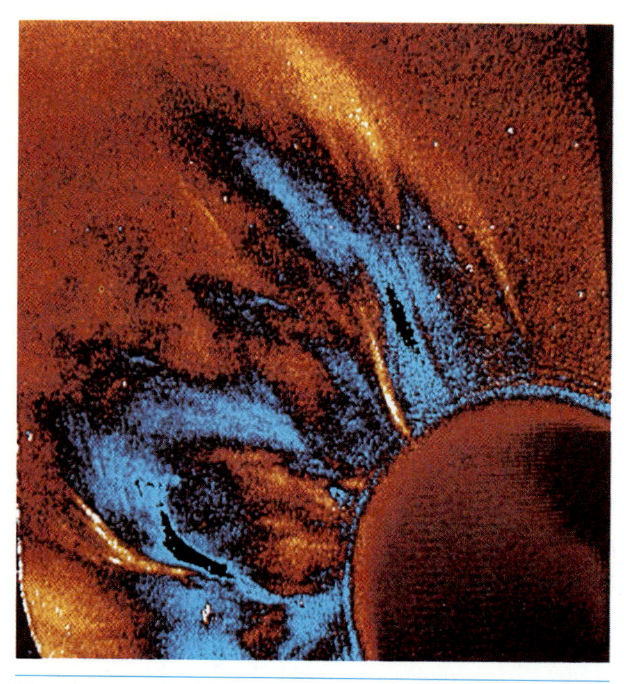

FIGURE 18.14 THE SOLAR WIND. *This far-ultraviolet image was obtained by the* Solar Maximum Mission *in Earth orbit. A special color separation technique was used to show the outward flow of gas.* (Data from the *Solar Maximum Mission,* NASA and the High Altitude Observatory National Center for Atmospheric Research, sponsored by the National Science Foundation)

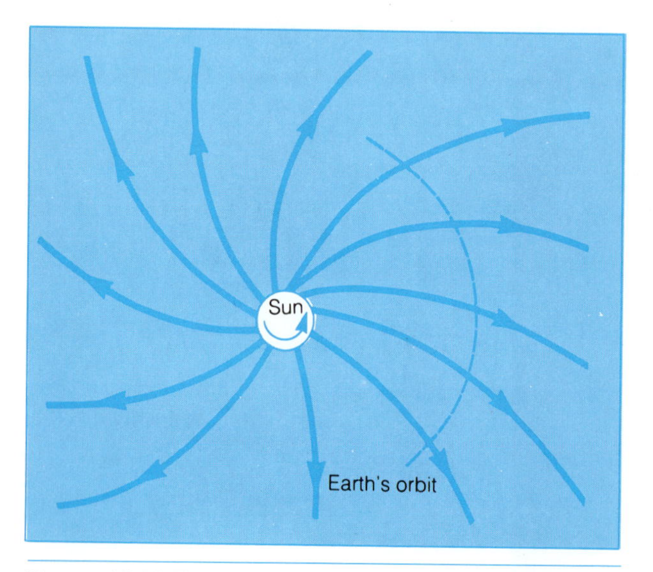

FIGURE 18.15 THE SOLAR WIND. *This schematic diagram illustrates how ionized gas from the Sun spirals outward through the solar system in a steady stream. The solar magnetic field creates sectors of variable density in the wind.*

ward into space. Subsequently, however, X-ray observations of the Sun have shown that the situation is not that simple. It is the solar magnetic field that governs the outward flow of charged particles. The coronal holes, mentioned earlier (see Fig. 18.10), are regions where the magnetic field lines open out into space. Charged particles such as electrons and protons, constrained by electromagnetic forces to follow the magnetic field lines, therefore escape into space only from the coronal holes. The speed of the solar wind is relatively low close to the Sun, but accelerates outwards, quickly reaching the speed of 300 to 400 km/sec already mentioned, after which it is nearly constant. The wind nearly reaches its maximum speed by the time it passes the Earth's orbit, and beyond there it flows steadily outward, apparently persisting at least to a distance beyond the orbit of Neptune. (The *Pioneer 10* spacecraft, on its way out of the solar system, still detected solar wind particles as it crossed the orbit of Neptune in mid-1983, and it has not yet found a limit to the solar wind.) At some point in the outer solar system, the wind is thought to come to an abrupt halt where it runs into an invisible and tenuous wall of matter swept up from the interstellar medium that surrounds the Sun.

Most of the direct information we have on the solar wind comes from satellite and space probe observations (Fig. 18.14), because the Earth's magnetosphere shields us from the wind particles. Solar wind monitors are placed on board the majority of spacecraft sent to the planets. One striking discovery has been the fact that the wind is not uniform in density, but instead seems to flow outwards from the Sun in sectors, as though it originates only from certain areas on the Sun's surface. This is explained by the X-ray data already mentioned, which indicate that the wind emanates only from the coronal holes. Because the base of the wind is rotating with the Sun, the wind sweeps out through space in a great curve, similar to the trajectory of water from a rotating lawn sprinkler (Fig. 18.15).

Occasional explosive activity occurring on the Sun's surface releases unusual quantities of charged particles, and some three or four days later, when this burst of ions reaches the Earth's orbit, we experience disturbances in the ionosphere that can interrupt short-wave radio communications and cause auroral displays. These **magnetic storms,** as they are often called, are outward manifestations of a much more complex overall interaction between the Sun and the Earth.

At the end of the next section, we will discuss this interaction, generally called the solar-terrestrial relation.

SUNSPOTS, SOLAR ACTIVITY AND THE MAGNETIC FIELD

The dark spots on the Sun's disk were observed more than 300 years ago. They were cited by Galileo as evidence that the Sun is not a perfect, unchanging celestial object, but rather has occasional flaws. Observations of the spots, which individually may last for months, have, over the centuries since, revealed some very systematic behavior. The number of spots varies, reaching a peak every 11 years, and during the interval between peak numbers of spots, their locations on the Sun change steadily from middle latitudes to a concentration near the equator. At the beginning of a cycle, when the sunspot number is maximized, most of them appear in activity bands about 30° north or south of the solar equator. During the next 11 years, the spots tend to lie ever closer to the equator, and by the end of the cycle they are nearly on it (Fig. 18.16). By this time, the first spots of the next cycle may already be forming at middle latitudes. A plot of sunspot locations during a cycle that clearly shows this effect is called a "butterfly diagram" because of the shape the pattern of spots makes (Fig. 18.17).

A hint at the origin of the spots was found when their magnetic properties were first measured. This is accomplished by applying spectroscopic analysis to the light from the spots, taking advantage of the fact that the energy levels in certain atoms are distorted by the presence of a magnetic field. This in turn causes

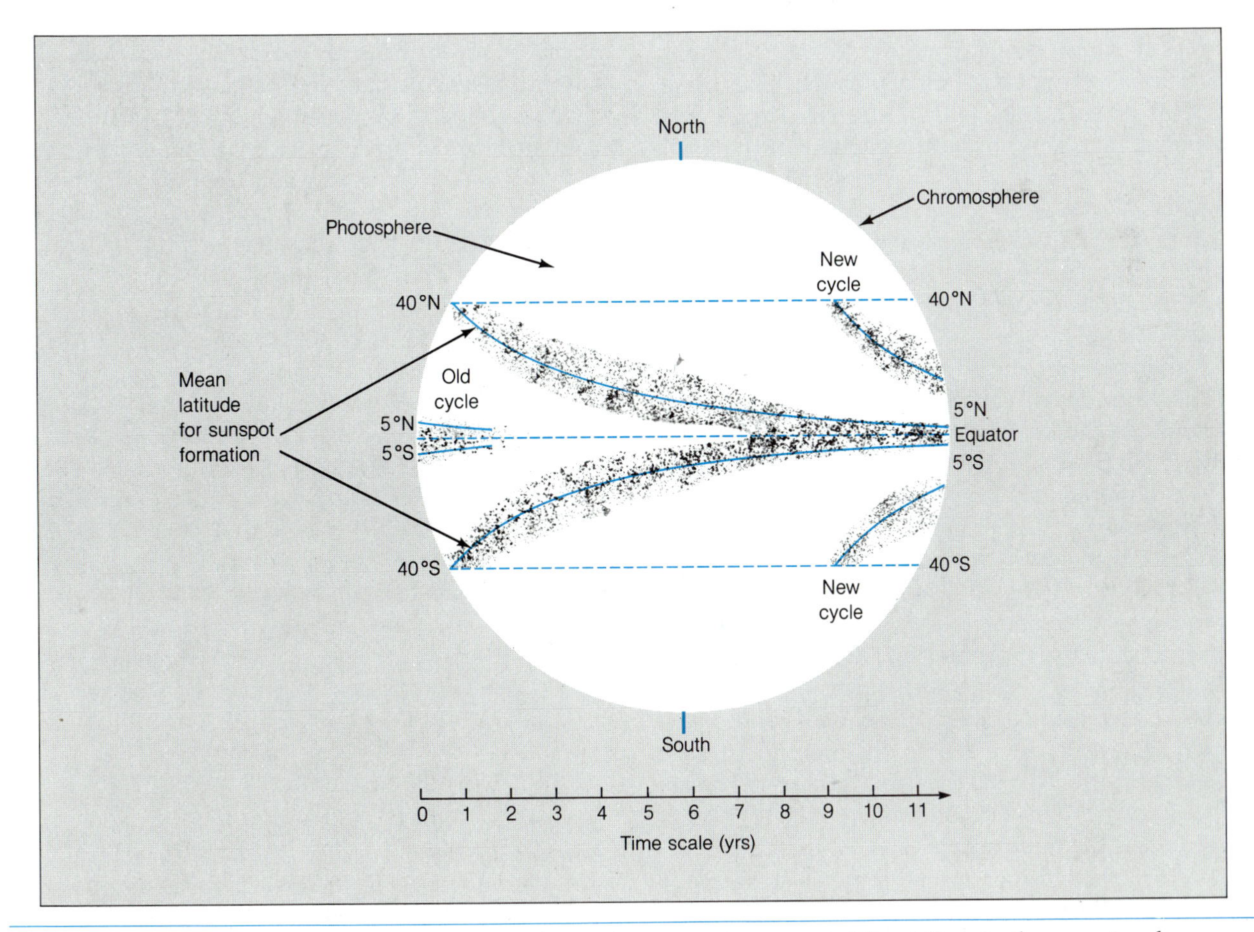

FIGURE 18.16 SUNSPOT LOCATIONS. *This drawing shows where sunspots appear at different times in the sunspot cycle.*

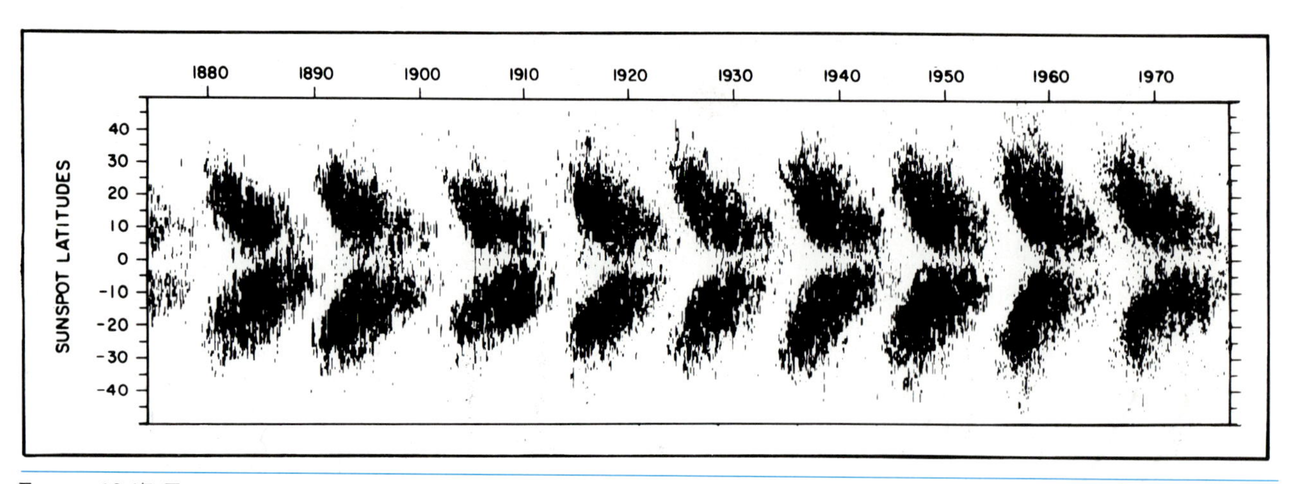

FIGURE 18.17 THE BUTTERFLY DIAGRAM. *This is a plot of the latitudes of observed sunspots through several cycles of solar activity. During each cycle, the spots gradually shift their favored locations closer and closer to the solar equator* (Prepared by the Royal Greenwich Observatory and reproduced with the permission of the Science and Engineering Research Council, courtesy of J. A. Eddy)

the spectral lines formed by those levels to be split into two or more distinct, closely spaced lines. The degree of line-splitting, which is referred to as the **Zeeman effect,** depends on the strength of the magnetic field, so the field can be measured from afar simply by analyzing the spectral lines to see how widely split they are. This technique works for distant stars as well as for the Sun (except that for distant stars we cannot measure different portions of the disk separately).

When the first measurements of the Sun's field were made in the first decade of this century, the field was found to be especially intense in the sunspots, about 1,000 times stronger than in the surrounding gas. It is now thought that the intense magnetic field associated with a sunspot inhibits convection, so that heated gas from below cannot rise up to the surface. Hence the spot is cooler than its surroundings and therefore is not as bright (Fig. 18.18). The typical temperature in a spot is about 4,000 K, compared to the roughly 6,000 K temperature of the photosphere. Using Stefan's law, we see that the intensity of light emitted in a spot compared to the surroundings is $(4{,}000/6{,}000)^4 = 0.2$; i.e. the brightness of the solar surface within a sunspot is only about one-fifth of the brightness in the photosphere outside.

When the sunspot magnetic fields were measured (Fig. 18.19), they were found to act like either north or south magnetic poles; i.e. each spot has a specific magnetic direction associated with it. Furthermore, pairs of spots often appear together, the two members of a pair usually having opposite magnetic polarities. During a given 11-year cycle, in every sunspot pair in the same hemisphere the magnetic polarities always have the same orientation. For ex-

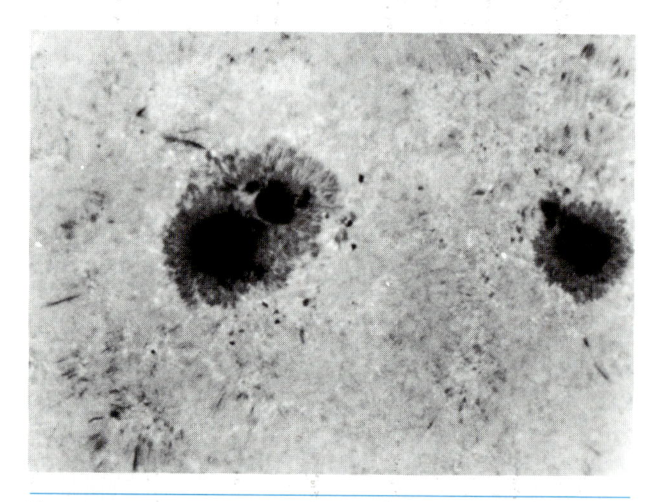

FIGURE 18.18 SUNSPOTS. *This is a telescopic view of a group of spots, showing their detailed structure. They appear dark only in comparison with their much hotter surroundings.* (Mt. Wilson and Las Campanas Observatories, Carnegie Institution of Washington)

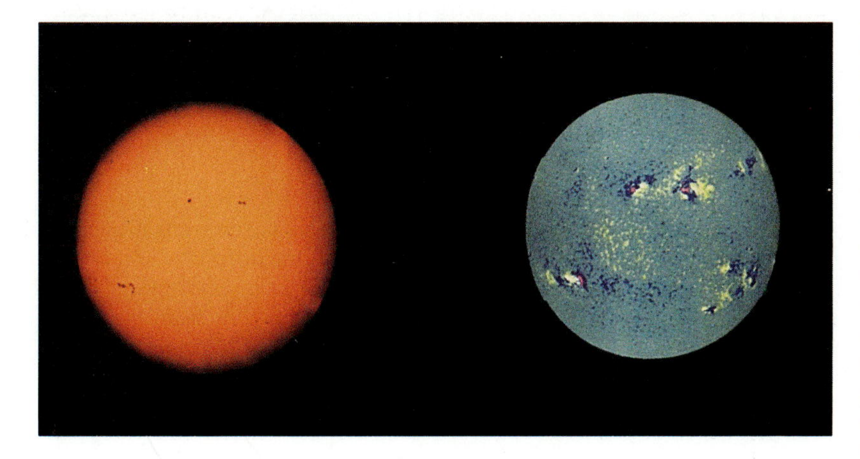

FIGURE 18.19 MAGNETIC FIELDS ON THE SUN. *The* magnetogram, *or map of magnetic field strength (right), was made through the measurement of the splitting of spectral lines by magnetic fields (the Zeeman effect). Here colors (dark blue and yellow) are used to indicate regions of opposing magnetic polarity. Note that the polarities in the southern hemisphere are reversed with respect to those in the north. At left is a visible-light photograph of the Sun taken at the same time, so that the correspondence between the surface magnetic field and the visible sunspots can easily be seen.* (National Optical Astronomy Observatories)

ample, during one 11-year cycle, the spot to the east in each pair will generally be the one with north magnetic polarity, and the one to the west has south magnetic polarity. During the next cycle, all the pairs will be reversed, with the south magnetic spot to the east, and the north polar spot to the west. In the other hemisphere, these relative orientations are exactly the opposite. Between cycles, when this arrangement is reversing itself, the Sun's overall magnetic field also reverses, with the solar magnetic poles exchanging places. It is actually 22 years before the Sun's magnetic field and sunspot patterns repeat themselves, so the solar magnetic cycle is truly 22 years long.

Sunspot groups, often called **solar active regions** (Fig. 18.20), are the scenes of the most violent

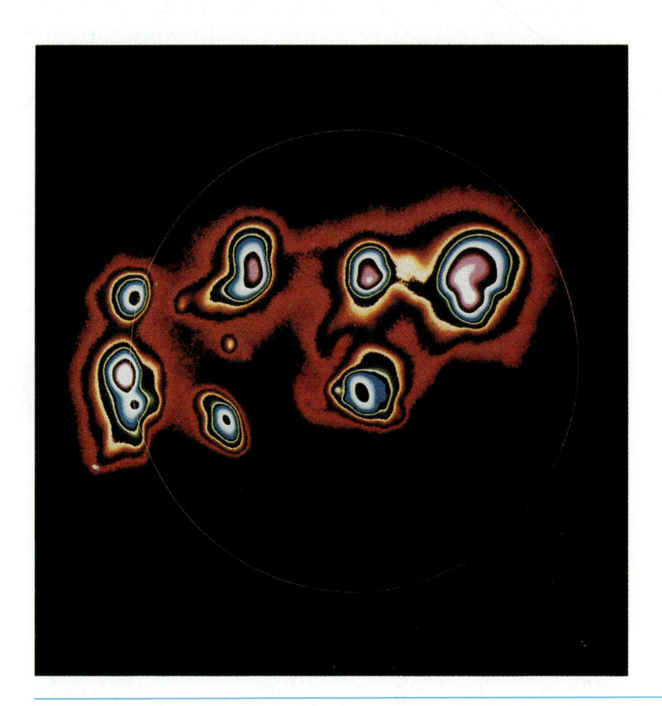

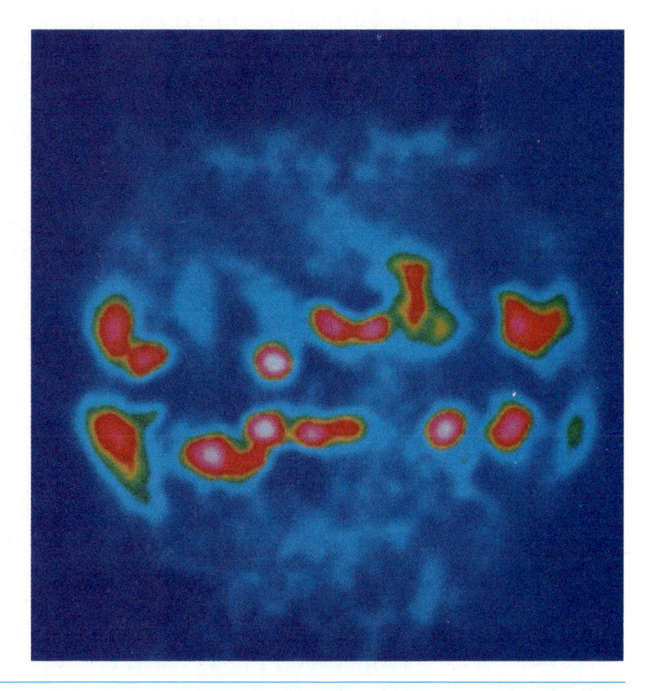

FIGURE 18.20 SOLAR ACTIVE REGIONS. *At left is an X-ray image and at right is a radio map. In both, bright colors indicate the solar active regions, which emit strongly in X-ray and radio wavelengths.* (*left:* NASA; *right:* National Radio Astronomy Observatories)

ASTRONOMICAL INSIGHT 18.3

SOLAR ACTIVITY AND THE EARTH'S CLIMATE

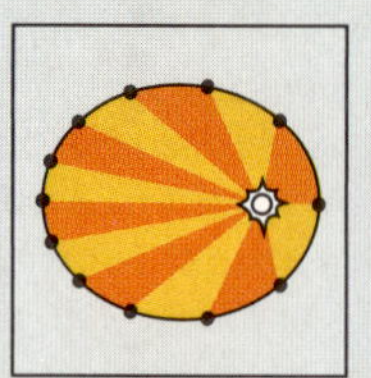

In the text a possible connection between the Sun's activity cycle and weather conditions on the Earth is mentioned. This has been a subject of uncertainty and some controversy—some scientists have claimed to find correlations between the two, and others have disputed these findings. So many factors influence weather that it is a very difficult task to isolate any one factor to see what its effect is. Furthermore, no clear identification has been made of a mechanism that would allow the Sun's activity to modify weather on the Earth. Scientists are often skeptical of claimed phenomena if no likely cause is known.

Recently two lines of evidence have been found supporting the suggestion that the Earth's climate indeed varies with solar activity. One of these involves measurements of the energy of radiation that reaches the Earth from the Sun. The amount of energy reaching the Earth is measured in units of ergs/cm^2/sec and is called the **solar constant.** Historically this measurement has been made at the Earth's surface. Corrections had to be made for energy absorbed by the Earth's atmosphere, and this made precise measurements difficult. For the past decade, however, there have been instruments in space for measuring the solar constant, and much more accurate data have been made available.

It has been discovered that the solar constant varies; that is, the amount of energy reaching the Earth changes from time to time. There is a reduction in the solar constant when sunspot groups move across the side of the Sun that faces Earth. By a mechanism not yet understood, the sunspots block some of the Sun's radiation, and this causes a slight reduction in the energy reaching the Earth. Such variations could be responsible for variations in the Earth's weather, so now scientists have a possible cause for 11-year climatic variations (although it is still not universally accepted that these variations occur).

Another discovery made from the space-based measurements of the solar constant is perhaps more important. During the nearly 10 years that these measurements have been made, there has been a steady decline in the solar constant. It has been calculated that if this decline continues for 100 years, the reduction in solar energy reaching the Earth would be sufficient to create an extended period of very cold weather. Such a period occurred around the time of the Maunder Minimum, an interval in the late 1600s when the Sun stopped undergoing its regular 11-year variations in sunspots (see text). That period, called the "Little Ice Age," was a time of great hardship in many parts of the world. Scientists now wonder whether it may have been caused by a long-term reduction in the solar constant, and some fear that we may be currently headed for another Little Ice Age. On the other hand, the solar constant measurements cover such a short time that no one can say whether the decline will continue long enough to have drastic effects on our climate.

Information on past solar activity and its influence on the Earth's climate recently has come from an entirely different source. A rock formation in western Australia has a regular pattern of dark layers in it, and the pattern of these layers apparently corresponds to the solar cycle. Evidently when this rock formed on the floor of an ancient lake some 670 million years ago, annual deposits of silt created the layers. The thickness of each annual layer was determined by the amount of silt released by the melting of glacial ice (the Earth was wrapped in a deep ice age at the time). The amount of ice that melted was, in turn, determined by the Sun's radiant energy. Thus, it is thought, the thickness of each layer in this rock formation is an indication of the Sun's energy output during the year the layer was deposited.

If this interpretation is correct (and some scientists have doubts), then the rocks contain invaluable information on the Sun's activity cycles over some 19,000 years, or about 1600 11-year cycles. This provides scientists an unprecedented opportunity to determine whether there are longer-term cycles in the Sun's behavior that might account for such phenomena as the Maunder Minimum and large-scale variations in the Earth's climate, such as the variations in the solar constant might cause. Analysis of the rock indeed shows that long-term cycles do occur, but that the basic 11-year pattern was present as long ago as 670 million years.

It is even possible to make predictions about the solar activity cycle in future years, based on the assumption that the Sun behaves today as it did 670 million years ago. One prediction is that the next 9 or 10 cycles should be less intense than the past several ones, and that the average length of the cycle should increase to 13 or 14 years. Perhaps these changes will begin to be observed in our lifetimes.

forms of solar activity, the **solar flares.** These are gigantic outbursts of charged particles, as well as visible, ultraviolet, and X-ray emission, created when extremely hot gas spouts upwards from the surface of the Sun (Fig. 18.21). Flares are most common during sunspot maximum, when the greatest number of spots are seen on the solar surface. Close examination of flare events shows that the trajectory of the ejected gas is shaped by the magnetic lines of force emanating from the spot where the flare occurred. Charged particles flow outwards from a flare, some of them escaping into the solar wind. If the flare occurs where the part of the wind arises that hits the Earth, then the Earth is bathed by an extra dosage of solar wind particles some three days later, creating effects on radio communications that have already been described. The extra quantity of charged particles entering the Earth's upper atmosphere can also be responsible for unusually widespread and brilliant displays of aurorae. Apparently flares occur when twisted magnetic field lines suddenly reorganize themselves, releasing heat energy and allowing huge bursts of charged particles to escape into space.

The combination of all these bits of data on sunspots, magnetic fields, and solar activity cycles has led to the development of a complex theoretical picture, one that successfully accounts for many of the observed phenomena. This theory envisions ropes or tubes of magnetic field lines inside the Sun, connecting its north and south magnetic poles. At some places these tubes become kinked, and loops break through the surface, creating pairs of sunspots with opposing magnetic polarities where they emerge and reenter. Early in the sunspot cycle, the magnetic tubes break through the surface at middle latitudes, but later, as the Sun's magnetic field is moving towards reversal of the poles, they do so near the equator. This accounts for the latitude dependence of the spots during a cycle, as shown by the butterfly diagram. When the solar magnetic field reverses itself every 11 years, so do these magnetic ropes, so that when the new cycle begins, the sunspot pairs have their polarities reversed compared with the pairs of the previous cycle.

Records of sunspot counts maintained over centuries have allowed astronomers to study the long-term behavior of the Sun. It has been found that the 22-year cycle is not perfectly repeatable, but that there have been longer-term variations in solar activity (Fig. 18.22). Most striking was a prolonged period in the late 1600s when the cycle seemed to stop, with no marked periods of sunspot maxima. This epoch of reduced sunspot activity has been called the **Maunder Minimum,** after its discoverer, E. W. Maunder (also the originator of the butterfly diagram). Only time will tell whether the Maunder Minimum was part of some much longer-period cycle in the Sun's behavior.

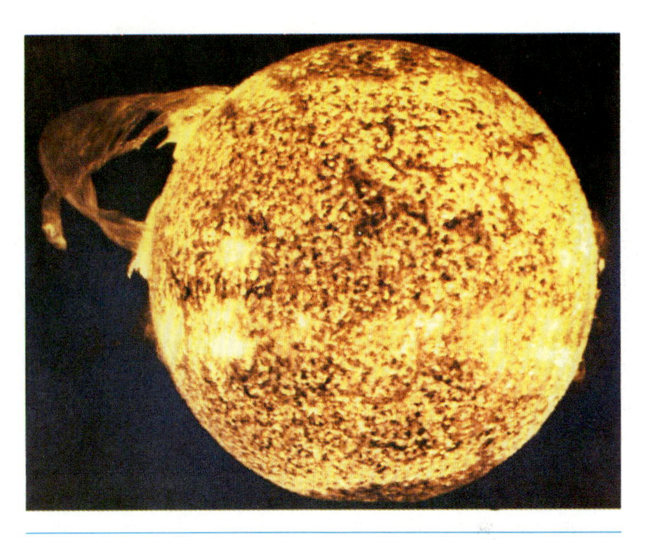

FIGURE 18.21 A MAJOR FLARE. *The gigantic looplike structure in this ultraviolet photograph obtained from* Skylab *is one of the most energetic flares ever observed. Supergranulation in the chromosphere is easily seen over most of the disk.* (NASA)

The origin of the Sun's magnetic field, and its periodic pole reversals, is probably a dynamo, similar in nature to those thought to be at work in the interiors of the planets that have fields. Unlike the terrestrial planets, however, the Sun is not rigid, and consequently differential rotation may play a role in creating the instability that causes the dynamo to reverse itself regularly, every 11 years. From the solar behavior, we might speculate that the magnetic fields of Jupiter and Saturn, which also rotate differentially, may also reverse themselves from time to time.

It is known from X-ray and ultraviolet observations of other stars that many stars, particularly relatively cool ones like the Sun, not only have chromospheres and coronae, but also have activity cycles. From this, we deduce that these stars must also have magnetic fields that reverse themselves periodically. Statistical studies of large numbers of stars help reveal

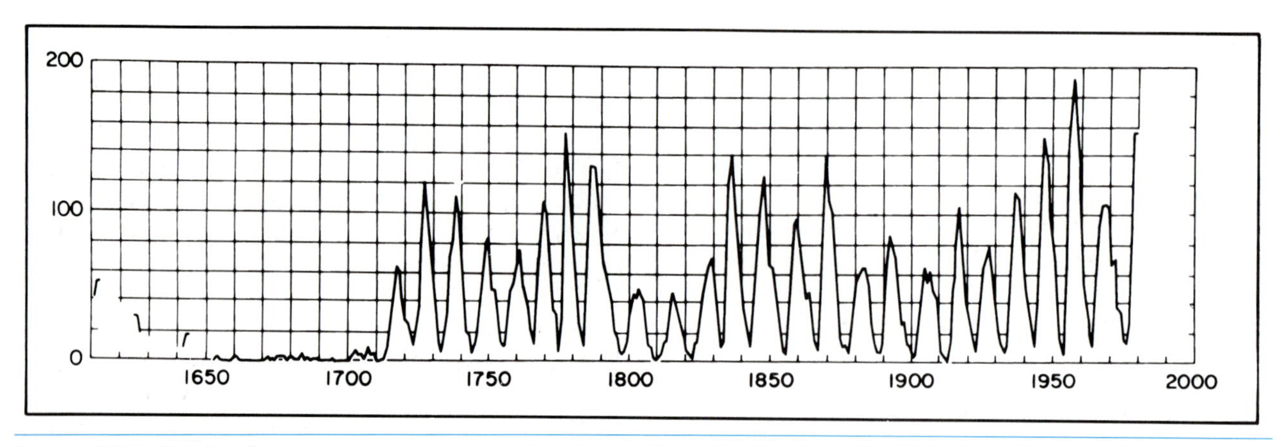

FIGURE 18.22 THE SUN'S LONG-TERM ACTIVITY. *This diagram illustrates the relative level of activity (in terms of sunspot numbers) over three centuries. It is clear that the level varies, including a span of about fifty years (1650–1700), known as the Maunder minimum, when there was little activity. There is evidence that the Sun has long-term cycles that modulate the well-known twenty-two-year period.* (J. A. Eddy)

how their activity cycles depend on such phenomena as rotation and stellar age. For example, we know (as described in Chapter 17) that the Sun's rotation has slowed since the time of its formation. From studies of other stars, we find that this is a general occurrence, that younger stars tend to rotate more rapidly than older ones. The degree of chromospheric and coronal activity also diminishes with age, suggesting that rotation of a star is responsible for its activity cycle. Thus, the Sun probably had more intense activity when it was younger than it does now. This is just one example of the kinds of things we can learn by combining information on the Sun and on other stars like it.

The solar activity cycle may have some indirect effects on the climate of the Earth. Eleven-year patterns have been reported in the occurrence of droughts, and it seems possible that such patterns are related to the solar cycle, although it is not known how. During the time of the Maunder Minimum, the Earth's climate was in chaos, with terrible droughts in many areas and particularly severe winters in Europe and in North America.

There are other aspects of the relationship between solar activity and the Earth's atmosphere. As the solar wind fluctuates in intensity, the Earth's magnetosphere varies in extent. We have already seen that solar flares, which occur most often during sunspot maximum, create important effects on the ionosphere. The chemistry of the upper atmosphere may also be strongly influenced by variations in the solar ultraviolet emission that are linked in turn to the solar activity cycle. The entire question of solar-terrestrial relations is an important one that merits and is receiving further attention.

PERSPECTIVE

Our Sun, the source of nearly all our energy, is a very complex body. As stars go, it is apparently normal in all respects, so we imagine that other stars are just as complex, even though we cannot observe them in such detail.

We have explored the Sun, both in its deep interior and in the outer layers that can be observed directly. We know that nuclear fusion is the source of all the energy, and that the size and shape of the Sun are controlled by the balance between gravity and pressure throughout the interior. Somehow heat is transported above the surface, keeping the chromosphere and the corona hotter than the photosphere. Perhaps most intriguing of all is the solar activity cycle and its relationship to the Sun's complex magnetic field.

We are now prepared to study stars in general, using the Sun as an example to help us understand some of what we learn about other stars.

SUMMARY

1. The Sun is an ordinary star, one of billions in the galaxy.
2. The Sun is a spherical, gaseous object whose temperature and density increase toward the center. The Sun emits light from its surface, whose ultimate source is nuclear reactions in its core.
3. Energy is produced in the core by nuclear fusion and is slowly transported outward by radiation, except near the surface, where energy is transported by convection.
4. The nuclear reaction that powers the Sun is the proton-proton chain, in which hydrogen is fused into helium.
5. The outer layers of the Sun govern the nature of the light it emits. These layers consist of the photosphere, with a temperature of about 6,000 K; the chromosphere, where the temperature ranges between 6,000 and 10,000 K; and the corona, where the temperature is over 1,000,000 K.
6. The photosphere, which is the visible surface, creates absorption lines, while the hotter chromosphere and corona create emission lines.
7. The excess heat in the outer layers is somehow transported there from the convective zone just below the surface, but the mechanism for transporting the heat is not well known.
8. The Sun emits a steady outward flow of ionized gas called the solar wind. The wind originates from coronal holes and is controlled by the solar magnetic field.
9. The sunspots occur in 11-year cycles that are part of the 22-year cycle of the Sun's magnetic field reversals. The spots are regions of intense magnetic fields where magnetic field lines from the solar interior break through the surface.
10. Studies of other stars show that those similar to the Sun in temperature also have chromospheres, coronae, and activity cycles.
11. The solar activity cycle may have important effects on the Earth's climate.

REVIEW QUESTIONS

Thought Questions

1. Explain in your own words why hydrostatic equilibrium causes the Sun to have a spherical shape. Do you think the rotation of the Sun might affect its shape?
2. Why do nuclear reactions occur only in the central core of the Sun?
3. Explain why the use of special filters to isolate the wavelengths of certain spectral lines allows astronomers to separately examine different layers of the Sun.
4. Why are the granules in the Sun's photosphere bright, and the descending gas that surrounds them relatively dark?
5. Explain, in the context of Kirchhoff's rules (Chapter 6), why the chromosphere forms emission lines while the photosphere forms absorption lines.
6. Describe the evidence for the existence of the solar wind.
7. Discuss the relationship among the Sun's magnetic field, coronal holes, and the solar wind.
8. Why do we say that the solar activity cycle is 22 years long, even though the sunspot maxima occur every 11 years?
9. How would the appearance of sunspots be altered if they had weaker magnetic fields than their surroundings? Would the explanation of them given in the text have to be altered if this were the case?
10. Describe a research program that you could undertake to determine the effects of solar variations on the Earth's climate.

Problems

1. Suppose that an X-ray photon released at the Sun's center has a frequency of $f = 1 \times 10^{20}$ Hz. If half of its energy is converted into heat and the other half into visible photons with frequency $f = 5 \times 10^{14}$ Hz

as the energy propagates outward to the Sun's surface, how many visible photons are produced, to be radiated away into space?

2. How would the lifetime of the Sun be altered if its luminosity were ten times greater than it is and its mass were twice as large?

3. Calculate the wavelengths of maximum emission for the solar photosphere (temperature 6,500 K), the chromosphere (20,000 K), and the corona (2×10^6 K). How would each of these layers be best observed?

4. If the average speed of the solar wind between the Sun and the Earth is 250 km/sec, how long does it take for a particle to reach the Earth once it is emitted by the Sun? How is your answer related to the time it takes for a solar flare to begin to affect radio communications on the Earth?

5. Suppose the temperature in a sunspot were one-half the temperature in the surrounding photosphere. How much less intense would the radiation be from the sunspot than from the photosphere?

ADDITIONAL READINGS

Eddy, J. A. 1977. The case of the missing neutrinos. *Scientific American* 236(5):80.

Gibbon, E. G. 1973. *The quiet sun.* NASA Publication NAS 1.21. Washington, D.C.: NASA.

Gough, D. 1976. The shivering sun opens its heart. *New Scientist* 70:590.

Newkirk, G., and K. Frazier. 1982. The solar cycle. *Physics Today* 35(4):25.

Parker, E. N. 1975. The sun. *Scientific American* 233(3):42.

Waldrop, Mitchell. 1985. First sightings: Solar systems elsewhere. *Science 85* 6(5):26.

Walker, A. B. C., Jr. 1982. A golden age for solar physics. *Physics Today* 35(11):61.

Wallenhorst, S. G. 1982. Sunspot numbers and solar cycles. *Sky and Telescope* 64(3):234.

Wilson, O. C., A. H. Vaughan, and D. Mihalas. 1981. The activity cycle of stars. *Scientific American* 244(2):104.

Wolfson, R. 1983. The active solar corona. *Scientific American* 148(2):104.

CHAPTER 19

STELLAR OBSERVATIONS: POSITIONS, MAGNITUDES, AND SPECTRA

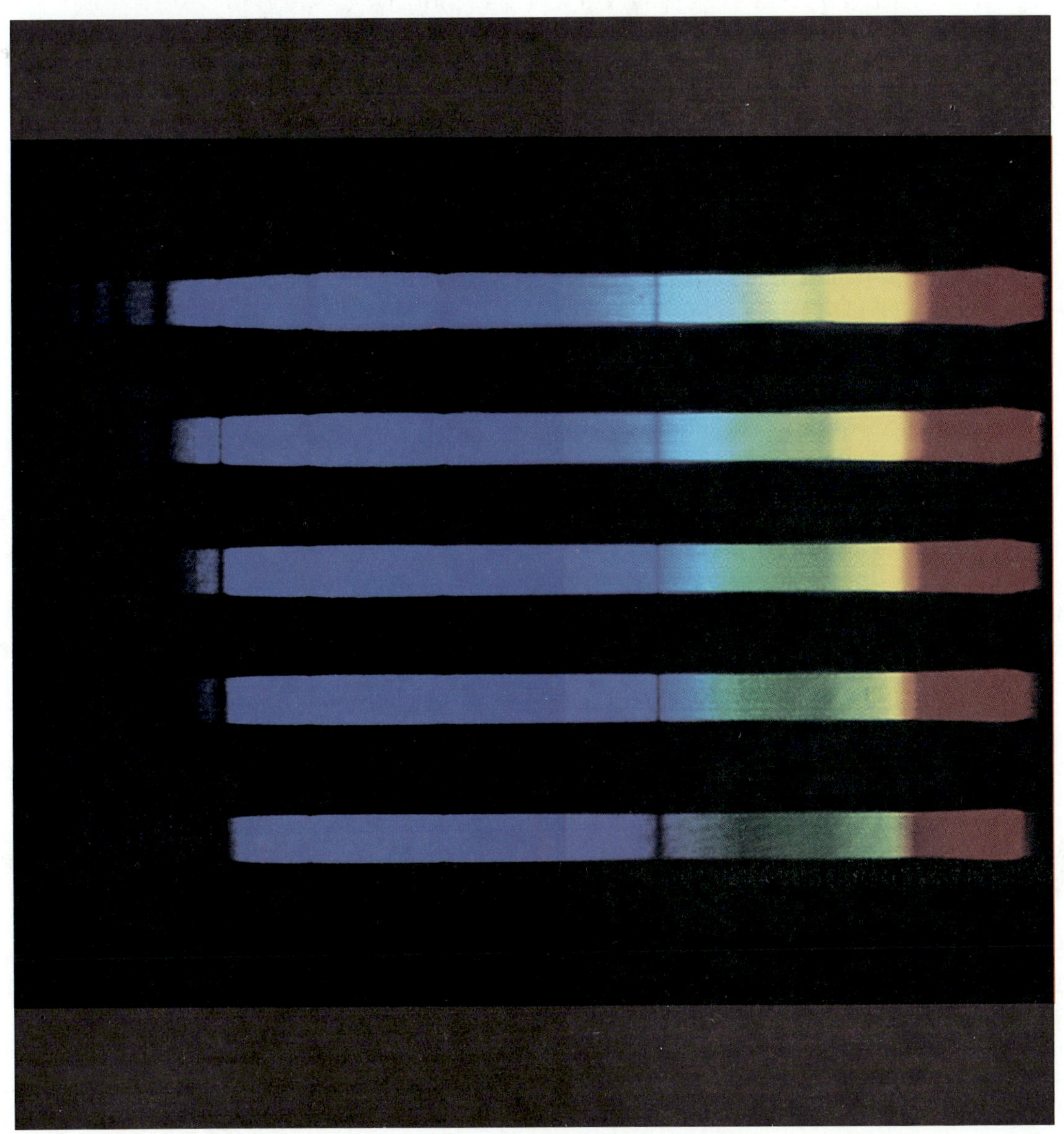

Spectra of the bright star Vega. (National Optical Astronomy Observatories)

CHAPTER PREVIEW

POSITIONAL ASTRONOMY

For many centuries, positional measurements were practically the only kind of observation made, as we learned in Chapter 3. Although some heed was paid to stellar brightnesses and especially to changes that occurred in the heavens, it was the positions of the stars and planets that were systematically observed and recorded. This was an inexact science until the first sextants (Fig. 19.1) and quadrants were available to provide accurate measurements of angular distances. In modern times, highly precise positions can be measured.

One of the principal techniques used today in **astrometry,** the science of measuring star positions, is to photograph the sky and to carefully measure the positions of the star images on the photographic plates. If a number of photographs of a given portion of the sky are taken and measured separately, the results can be averaged to produce a more accurate determination than is possible by measuring a single photograph. Today astronomers can measure relative stellar positions to precision of about 0.01 second of arc (recall that a second of arc is one-sixtieth of a minute of arc, and that an arcminute is one-sixtieth of a degree; there are 1,296,000 arcseconds in a full circle). When the *Space Telescope* is in operation in the late 1980s, even more accurate measurements will be possible, because the fuzziness of star images caused by the Earth's atmosphere will be eliminated.

All stars in the galaxy have individual motions (Fig. 19.2), consisting in part of the systematic rotation of the galaxy as a whole, and in part of random

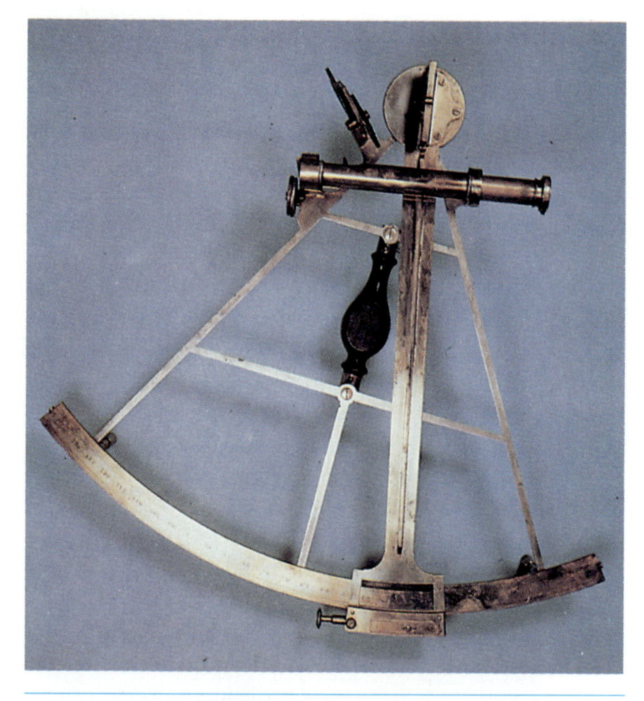

FIGURE 19.1 A SEXTANT. *Instruments of this type were used for centuries for the measure of star positions relative to each other. In modern times, sextants are used as tools for navigation.* (The Granger Collection)

motions of individual stars. The situation could be likened to a traveling swarm of bees: The individual bees move randomly within the swarm, and the group as a whole has an overall motion as well. The apparent speed of a star (measured in units of changing angle with time) depends on how fast the star is actually moving and on how far away it is. If a star is too far from the Sun, even the most careful measurements will fail to reveal any motion. The apparent gradual changes in a star's position are called **proper motions** and are expressed in terms of arcseconds per year.

Astrometric measurements reveal another kind of motion, one referred to often in Chapters 3 and 4. From ancient times through the Renaissance, astronomers were perplexed by the fact that no stellar parallax could be measured, even with the measurement precision available to Tycho Brahe. This failure to detect a parallax led many, including Brahe, to reject the notion that the Earth was orbiting the Sun, since it seemed obvious that if it were, the stars would

FIGURE 19.2 PROPER MOTIONS. *Each star has its own motion on the sky. We call the angular movement of a star its proper motion. Proper motions are small (at most a few arcseconds per year), and require separate photographs taken years apart to be measured.*

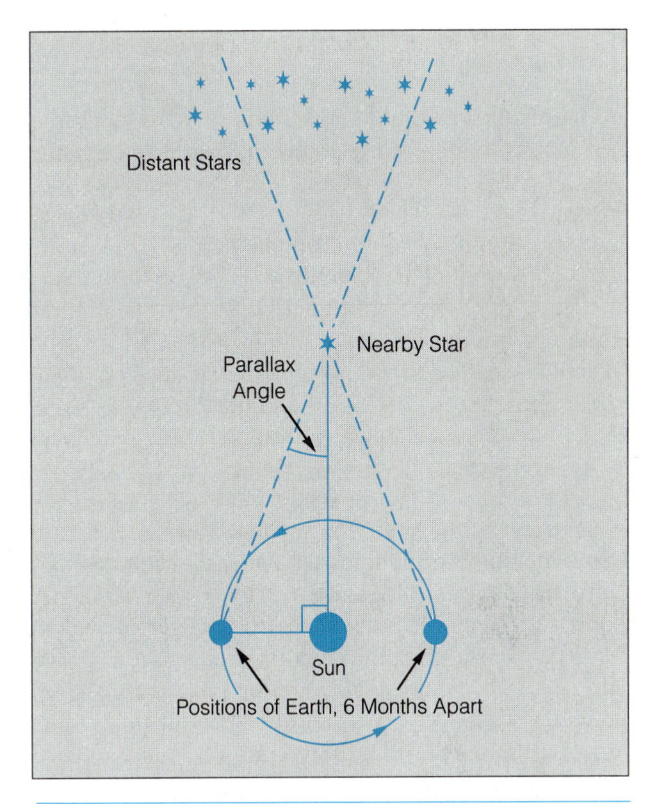

FIGURE 19.3 STELLAR PARALLAX. *As the Earth orbits the Sun, our line of sight to a nearby star varies in direction, so that the nearby star appears to move back and forth with respect to the more distant background stars. The parallax angle* p *is defined as one-half of the total angular motion. If* p *is one arcsecond, the distance to the star is 206,265 AU, or one parsec.*

appear to shift back and forth as a reflection of our motion. Eventually, the heliocentric theory won out, but stellar parallax was still undetected.

Finally, in 1838, parallaxes were detected independently by F. W. Bessel, F. G. W. Struve, and T. Henderson, each observing a different nearby star. This was clearly a case of a great discovery awaiting the development of suitable techniques for making the measurement. The reason stellar parallaxes had defied earlier observers became obvious: Even for the closest stars, the maximum shift in position was just over one second of arc! The stars are simply much farther away than the ancient astronomers had dreamed possible. The annual parallax motion of the star alpha Centauri, the nearest to the Sun, is comparable to the angular diameter of a dime as seen from a distance of about 2 miles, a very small angle indeed.

The successful detection of stellar parallax led to a direct means of determining distances to stars (Fig. 19.3). The amount of shift in a star's apparent position resulting from our own motion about the Sun depends on how distant the star is; the closer it is, the bigger the shift. The parallax p is defined as one-half the total angular shift of a star during the course of a year. This angle, once it is measured, tells us the distance to a star, in a unit of measure called the **parsec.** A star whose parallax is 1 arcsecond is, by definition, 1 parsec away (the word parsec, in fact, is a contraction of *parallax-second,* meaning a star whose parallax is 1 arcsecond). In mathematical terms, the distance to a star whose parallax is p is $d = 1/p$. Thus, if a star has a parallax of 0.4 arcseconds, it is $1/0.4 = 2.5$ parsecs away. The closest star has a parallax of less than 1 arcsecond; we conclude, therefore, that even this star is more than 1 parsec away.

In more familiar terms, a parsec is equal to 3.26 light-years, or 3.08×10^{18} centimeters, or 206,265

ASTRONOMICAL INSIGHT 19.1

STAR NAMES AND CATALOGS

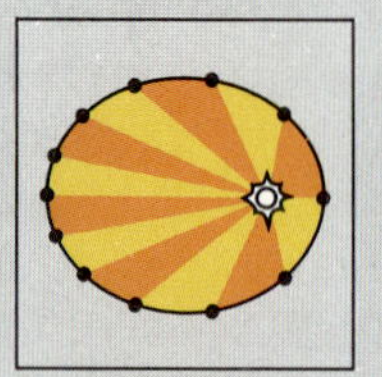

From the earliest times when astronomers systematically measured star positions, they began to compile lists of these positions in catalogs. Hipparchus developed an extensive catalog, as did early astronomers in China and other parts of the world, more than 2,000 years ago (see Chapter 3).

Listing stars in a catalog of positional measurements requires some kind of system for naming or numbering the stars, and a variety of such systems have been employed. The ancient Greeks designated stars by the constellation and the brightness rank within the constellation; for example, the brightest star in Orion is α Orionis, the next brightest is β Orionis, the next is γ Orionis, and so on. (The constellation names are spelled in their Latin versions, since this system was perpetuated by the Romans and the Catholic Church after the time of the Greeks). Today many stars are still referred to by their constellation rankings, using the Greek alphabet to designate the rank.

After the time of Ptolemy, when Western astronomy went into decline for some 1,300 years, Arab astronomers, occupying northern Africa and southern Europe, carried on astronomical traditions. These people assigned proper names to many of the brightest stars, and these names, such as Betelgeuse (α Orionis), Rigel (β Orionis), and Bellatrix (γ Orionis), are also still in use.

With the advent of the telescope, when many new stars were discovered that were too faint to be seen with the naked eye, catalogs rapidly outgrew the old naming systems. Generally, each catalog now assigned numbers to the stars, usually in a sequence related to their positions. Thus, modern catalogs, such as the *Henry Draper Catalogue*, the *Boss General Catalogue*, the *Yale Bright Star Catalogue*, and the *Smithsonian Astrophysical Observatory Catalog*, list stars by increasing right ascension (i.e., from west to east in the sky, in the order in which the stars pass overhead at night). In some cases, separate listings are made for different zones of declination (i.e., for different east-west strips of sky, separated in the north-south direction). Because each of these catalogs includes a different particular set of stars, each has its own numbering system, and often no cross-reference to other catalogs is provided. Hence a given star may have a constellation-ranked name, an Arabic name, and numbers in a variety of catalogs, but only by matching up the coordinates is it possible to be sure you are referring to the same star in different catalogs. Actually, things are not quite so bad as that, since some modern catalogs do provide cross-references to others. Nevertheless, one of the principal tasks of a would-be astronomer is to become familiar with catalogs and their uses.

astronomical units (AU). The use of parallax measurements to determine distances is a very powerful technique and is the only direct means astronomers have for measuring how far away stars are. Unfortunately, parallaxes are large enough to be measured only for stars rather near us in the galaxy. The smallest parallax that can be measured is barely less than 0.01 arcsecond, and accurate measurements are possible only for parallaxes larger than this. Thus, reliable distance measurements can be made only out to $d = 1/0.01 = 100\ pc$ or so. The galaxy, on the other hand, is more than 30,000 parsecs (100,000 light-years) in diameter! Clearly, other distance-determination methods are needed if we are to probe the entire galaxy. One very powerful method will be described in the next chapter.

STELLAR BRIGHTNESSES

The second general type of stellar observation is the measurement of the brightnesses of stars. This was first attempted in a systematic way by Hipparchus, who, more than 2,000 years ago, established a system of brightness rankings that is still with us today. Hipparchus ranked the stars in categories called **magnitudes,** from first magnitude (the brightest stars) to sixth (the faintest visible to the unaided eye). In his

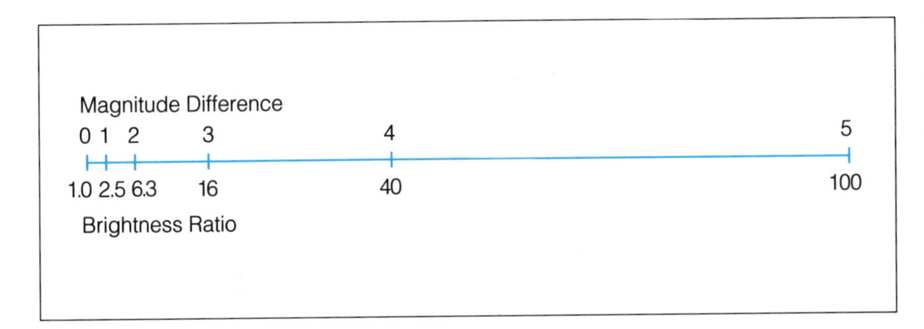

FIGURE 19.4 STELLAR MAGNITUDES. *Magnitude differences are indicated above the line, below the line are shown the corresponding brightness ratios.*

catalogue of stars, and in all since then, these magnitudes are listed along with the star positions.

By the mid-1800s, astronomers were beginning to make precise measurements of stellar brightnesses, rather than estimates made by the naked eye. It was discovered that what the eye perceives as a fixed *difference* in brightness from one magnitude to the next actually corresponds to a fixed brightness *ratio*. Measurements showed that a first-magnitude star is about 2.5 times brighter than a second-magnitude star, a second-magnitude star is 2.5 times brighter than a third-magnitude star, and so on (Fig. 19.4). The ratio between a first-magnitude star and a sixth-magnitude star was found to be nearly 100. In 1850, the system was formalized by the adoption of this ratio as exactly 100; thus, the ratio corresponding to a 1-magnitude difference is the fifth root of 100, or $100^{1/5} = 2.512$.

Thus a precise, quantitative magnitude system was developed, and with the worldwide adoption of a few standard stars used for calibration, all astronomers could record magnitudes in a comparable way. To do so no longer depended on rough estimates made by naked eye. Today photoelectric devices are used to make the measurements (Fig. 19.5). These are devices that produce an electric current when light strikes them; they are used in many familiar applications, such as door openers in modern buildings. The amount of electrical current produced is determined by the intensity of light, so an astronomer need only measure the current to determine the brightness of a star.

Let us work out a few examples to gain familiarity with the magnitude system. If a first-magnitude star is 2.512 times brighter than a second-magnitude star, and a second-magnitude star is 2.512 times brighter than a third-magnitude star, then the first-magnitude star is $2.512 \times 2.512 = 6.3$ times brighter than the third-magnitude star. Similarly, a second-magnitude star is $2.512 \times 2.512 \times 2.512 = 15.9$ times brighter than a fifth-magnitude star. All we need to remember is that a difference of one magnitude corresponds to a brightness ratio of 2.512. It is sometimes confusing that the brighter a star is, the smaller the magnitude is.

Once magnitude could be measured precisely, it was found that stars have a continuous range of brightnesses, and do not fall neatly into the various magnitude rankings. Therefore, fractional magnitudes must be used. Deneb, for example, has a magnitude of 1.26 in the modern system, although it was classified simply as a first-magnitude star in the old days. Each of the former categories is found to include a range of stellar brightnesses. This is especially true for the first-magnitude stars, some of which turned out to be as much as 2 magnitudes brighter than others. To measure these especially bright stars in the

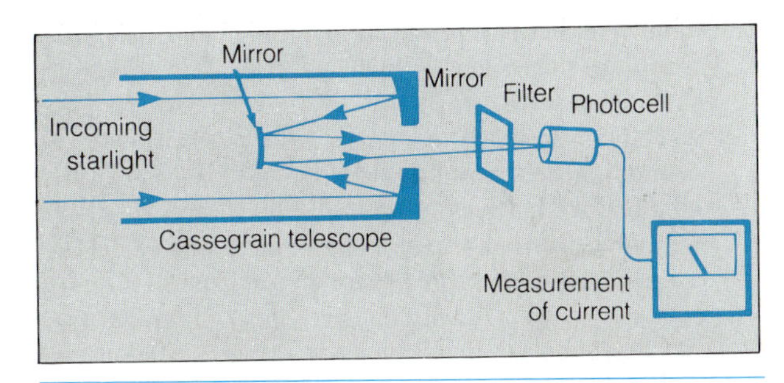

FIGURE 19.5 MEASURING STELLAR MAGNITUDES. *This is a schematic illustration of photometry, the measurement of stellar brightnesses. Light from a star is focused (often through a filter that screens out all but a specific range of wavelengths) onto a photocell, a device that produces an electric current in proportion to the intensity of light. The current is measured and, by comparison with standard stars, converted into a magnitude.*

modern system requires the adoption of magnitudes smaller than 1. Sirius, for example, which is the brightest star in the sky, has a magnitude of −1.42. By using negative magnitudes for very bright objects, astronomers can extend the system to include such objects as the Moon, whose magnitude when full is about −12, and the Sun, whose magnitude is about −26 (thus the Sun is 25 magnitudes, or a factor of $2.512^{25} = 10^{10}$, brighter than Sirius). A simple mathematical method for dealing with any magnitudes, even fractional or negative ones, is described in Appendix 10. Table 19.1 lists the magnitudes of a variety of objects.

Of course, some stars are too faint to be seen by the human eye, so the magnitude scale must extend beyond sixth magnitude. With moderately large telescopes, astronomers can measure stars as faint as fifteenth magnitude, and with long-exposure photographs taken with the largest telescopes, they can measure stars as faint as twenty-fifth magnitude. Such a star is nineteen magnitudes, or a factor of $2.512^{19} = 4 \times 10^7$, fainter than the faintest star visible to the unaided eye. The Space Telescope is expected to detect stars as faint as twenty-eighth magnitude, which is a factor of $2.512^3 = 15.9$ fainter yet.

TABLE 19.1 APPARENT MAGNITUDES OF FAMILIAR OBJECTS

Object	Apparent Magnitude
Sun	−26.74
100-W bulb at 100 ft	−13.70
Moon (full)	−12.73
Venus (greatest elongation)	−4.22
Jupiter (opposition)	−2.6
Mars (opposition)	−2.02
Sirius (brightest star)	−1.45
Mercury (greatest elongation)	−0.2
alpha Centauri (nearest star)	−0.1
Large Magellanic Cloud	+0.1
Saturn (opposition)	+0.7
Small Magellanic Cloud	+2.4
Andromeda galaxy (farthest visible object)	+3.5
Brightest globular cluster (47 Tucanae)	+4.0
Orion nebula	+4
Uranus (opposition)	+5.5
Faintest object visible to naked eye	+6.0
Neptune (opposition)	+7.9
Crab nebula	+8.6
3C273 (brightest quasar)	+12.80
Pluto (opposition)	+14.9

The stellar magnitudes we have discussed so far all refer to visible light. As we learned in Chapter 6, however, stars emit light over a much broader wavelength band than the eye can see. We also learned that the wavelength at which a star emits most strongly depends on its temperature. Therefore, by measuring a star's brightness at two different wavelengths, astronomers can learn something about its temperature. Filters that allow only certain wavelengths of light to pass through are used for this purpose. Typically, a star's brightness is measured through two such filters, one that passes yellow light, and one that allows blue light to pass, resulting in the measurement of V (for visual, or yellow) and B (for blue) magnitudes. Since a hot star emits more light in blue wavelengths than in yellow, its B magnitude is smaller than its V magnitude. For a cool star, the situation is reversed, and the B magnitude is larger than V (remember that the magnitude scale is backward in the sense that a smaller magnitude corresponds to greater brightness).

The difference between the B and V magnitudes is called the **color index** (Fig. 19.6). The exact value of this index is closely related to the surface temperature of a star, so stellar temperatures can be estimated simply by measuring the V and B magnitudes. A very hot star might have a color index $B - V = -0.3$, whereas a very cool one might typically have $B - V = +1.2$.

One additional type of magnitude should be mentioned, although it is a very difficult one to measure directly. This is the **bolometric magnitude,** which includes all the light emitted by a star at all wavelengths (Fig. 19.7). To determine a bolometric magnitude requires ultraviolet and infrared observations, as well as visible, and can best be done with the use of telescopes in space. This is particularly true for hot stars, which emit a large fraction of their light in ultraviolet wavelengths. For cool stars, which emit little ultraviolet light but much infrared radiation, bolometric magnitudes can be measured from the ground with fair accuracy. The bolometric magnitude of a star is always smaller than the visual magnitude, because more light is included when all wavelengths are considered than when only the visual wavelengths are measured. We will see in Chapter 20 how bolometric magnitudes are related to other properties of stars, particularly the luminosity.

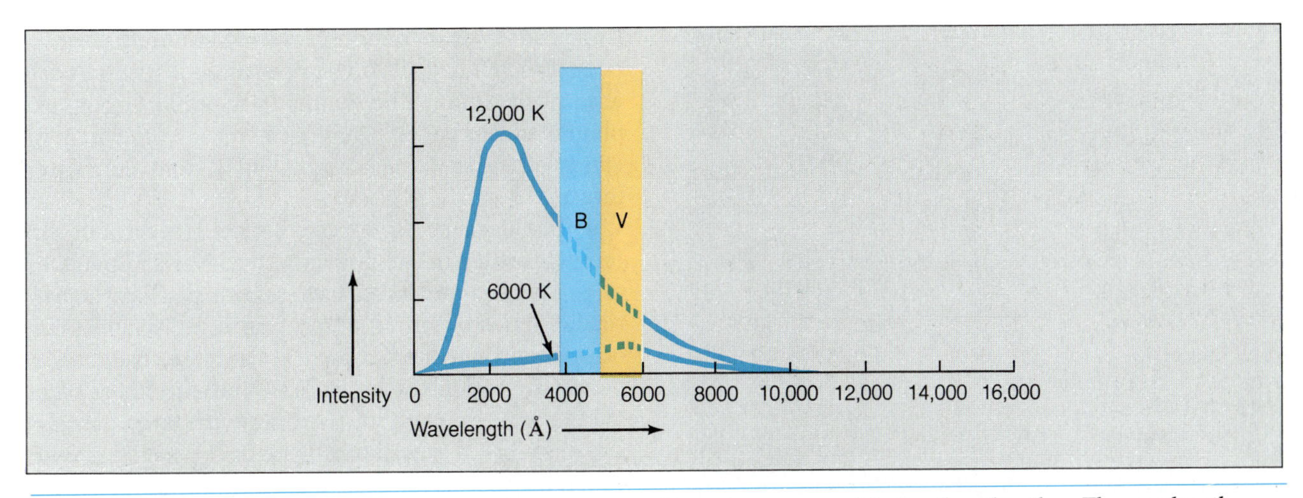

Figure 19.6 The color index. *Here are continuous spectra of two stars, one much hotter than the other. The wavelength ranges over which the blue* (B) *and visual* (V) *magnitudes are measured are indicated. We see that the hot star is brighter in the* B *region than in the* V *region of the spectrum; therefore, its* B *magnitude is* smaller *than its* V *magnitude, and it has a* negative B − V *color index. The opposite is true for the cool star.*

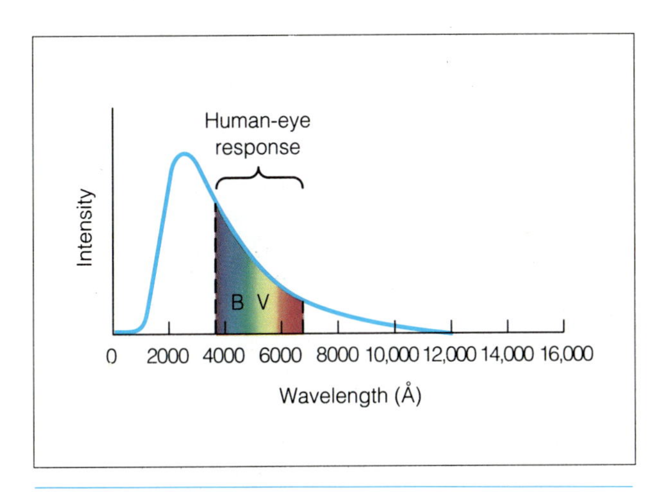

Figure 19.7 Bolometric magnitude. *The bolometric magnitude includes all the light emitted by a star at all wavelengths.*

The Appearance of Stellar Spectra

Observations of stellar spectra began in the mid-1800s, well before scientists understood the physics of the atom and how it forms spectral lines. A great deal was learned about the appearance of the spectra of stars, even though the full significance was not appreciated until much later. For most of the nineteenth century, stellar spectra were examined through the use of a **spectroscope,** which is simply an eyepiece focused on the light emerging from a prism, which in turn is placed at the focus of a telescope. The first stellar spectra to be recorded, in the 1870s, were exposed on photographic plates, a technique that is still widely used. Spectra of many stars can be photographed at the same time when a thin prism is placed in front of a telescope (Fig. 19.8). As we learned in Chapter 7, however, modern astronomers are turning more and more to new electronic detectors to record the spectra of stars.

Regardless of how it is done, the basic principle is the same: The light is dispersed according to wavelength, and its intensity is recorded at each wavelength. The result is called a **spectrogram** if recorded photographically (Fig. 19.9), or simply a spectrum if recorded by other means.

Normal Stars and Spectral Classification

Among the first astronomers to systematically examine spectra of a large number of stars was the Jesuit priest Angelo Secchi, who in the 1860s cataloged hundreds of spectra, using a spectroscope. He found the appearance of the spectra to vary considerably

FIGURE 19.8 STELLAR SPECTRA. *This photograph was made through a telescope with a thin prism in the light beam, so that each stellar image was dispersed into a spectrum. It is possible to measure and classify the spectra of large numbers of stars using this technique.* (University of Michigan Observatories)

from star to star, although they were consistent in one respect: They all showed continuous spectra with absorption lines. The work of Kirchhoff soon explained this to be the result of the relatively cool outer layers of a star, which absorb light from the hotter interior.

The primary differences between spectra of different stars lay in the differing strengths and patterns of the absorption lines. One star might have a particular series of strong absorption lines, whereas another might not have these lines at all, or they might be much weaker. Since Bunsen and Kirchhoff were able to identify many of the lines with known chemical elements, it was naturally assumed that stars with different patterns of lines were made of different elements. As we will see, this was not quite correct.

By the end of the nineteenth century, others had taken up and extended the work of Secchi; they used photographic plates to record the spectra, which allowed more detailed analyses (see Fig. 19.8). Most notable were a group of astronomers at Harvard University, led by Edward C. Pickering. A sequence of spectral categories or classes was established, based on the different types of spectra that were being found, and Pickering's group set out to classify the spectra of all the stars in the sky brighter than ninth magnitude, some 200,000 in all. This immense task was eventually carried out by Annie J. Cannon (Fig. 19.10), who today is regarded as the founder of modern spec-

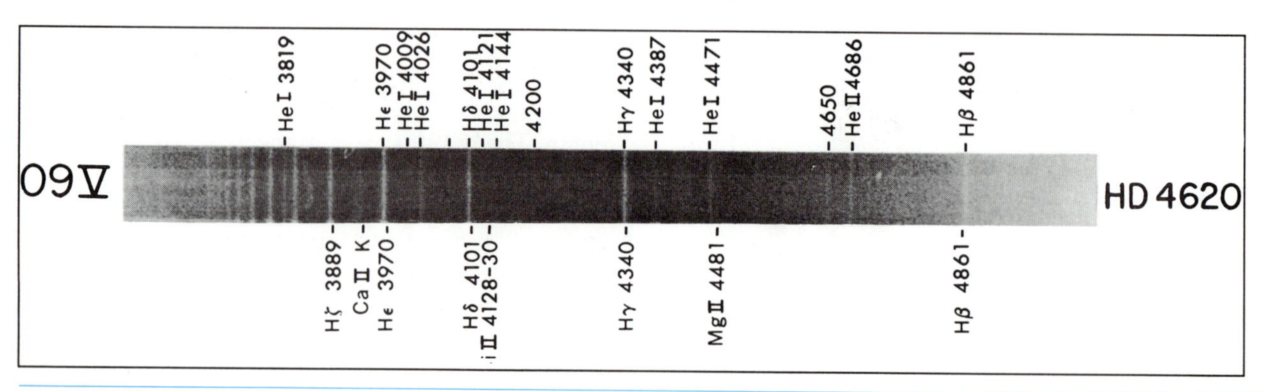

FIGURE 19.9 A STELLAR SPECTRUM. *This photograph of a stellar spectrum shows the manner in which they are usually displayed, as negative prints. Hence light features are absorption lines, wavelengths at which little or no light is emitted by the star.* (From Abt, H., W. W. Mogan, and R. Tapscott. 1968. *An Atlas of Low-Dispersion Grating Stellar Spectra* (Tucson-Kitt Peak National Observatory)

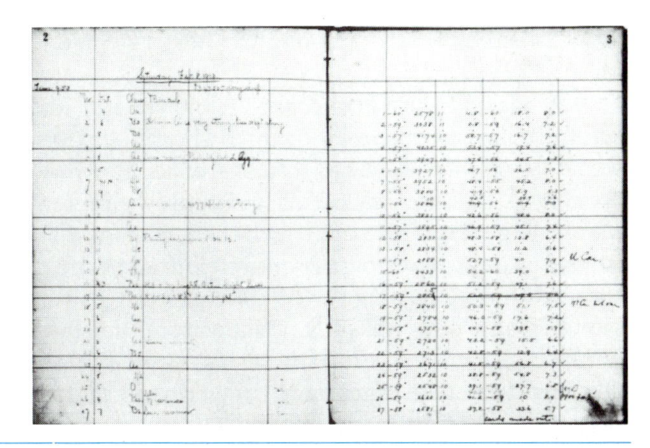

Figure 19.10 Annie J. Cannon. *A member of the Harvard College Observatory for almost fifty years, Cannon classified the spectra of several hundred thousand stars. At right is shown a page from one of her notebooks. Today she is recognized as the founder of modern spectral classification.* (Harvard College Observatory)

tral classification (and as a pioneer of woman's leadership in astronomy).

Stars were placed in categories according to which lines were prominent in their spectra, and the categories were simply labeled with letters of the alphabet. The sequence began with A stars, which showed the strongest lines of hydrogen, the simplest element. Other classes followed, based on the relative strengths of the lines of hydrogen and other elements, as each class was defined by reason of similar appearance and then assigned a letter of the alphabet. The spectral types were found, however, to represent a smooth sequence if arranged in the order O, B, A, F, G, K, M, (Fig. 19.11), although at that time the reason for this sequence was unknown. The Harvard classifications were collected in a massive compilation called the *Henry Draper Catalogue,* and today stars are commonly referred to by their HD numbers. (The collection was named after an astronomer whose estate sponsored its publication. Draper had been the first, in 1872, to photograph a stellar spectrum.)

Through the work of Bohr, who showed how spectral lines are formed by atoms, and the work of later researchers such as Sir Arthur Eddington, Henry Norris Russell, and Sir James Jeans, it was later deduced that stars of different spectral classes do not differ significantly in chemical composition; they are all made of basically the same ingredients. What does

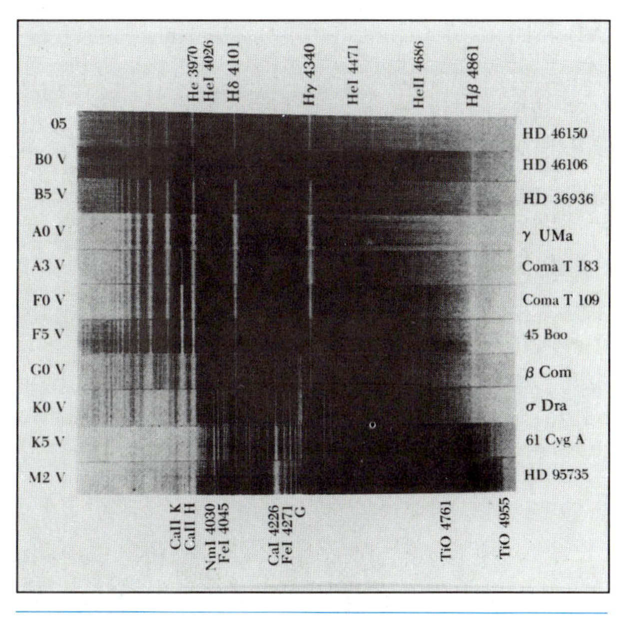

Figure 19.11 A comparison of spectra. *Here we see several spectra representing different spectral classes. A few major absorption lines are identified. These are arranged in order of decreasing stellar temperature (top to bottom).* (Mt. Wilson and Las Campanas Observatories, Carnegie Institution of Washington)

differ from one spectral type to the next is temperature.

We learned in Chapter 6 that the state of ionization of a gas determines which spectral lines it forms. When an atom has lost one of its electrons, the ion, as it is now called, has an entirely different set of energy levels through which the remaining electrons can jump, so it has an entirely different set of spectral lines. The number of electrons lost by the atoms in a gas depends on its temperature; thus, the hottest stars have spectra of highly ionized gases, whereas the coolest stars show little or no ionization (Fig. 19.12). The difference in ionization from one temperature to another is responsible for the differences in the appearance of the spectra from one class to the next. Hence, the sequence of spectral types is really a sequence of stellar temperatures. Refinements in the classification guidelines, along with improvements in the quality of spectrograms, have led to the creation of subclasses, so that we have B0, B1, B2 stars, and so on. Each major division has up to ten subdivisions. The Sun, in this modern system, is a G2 star.

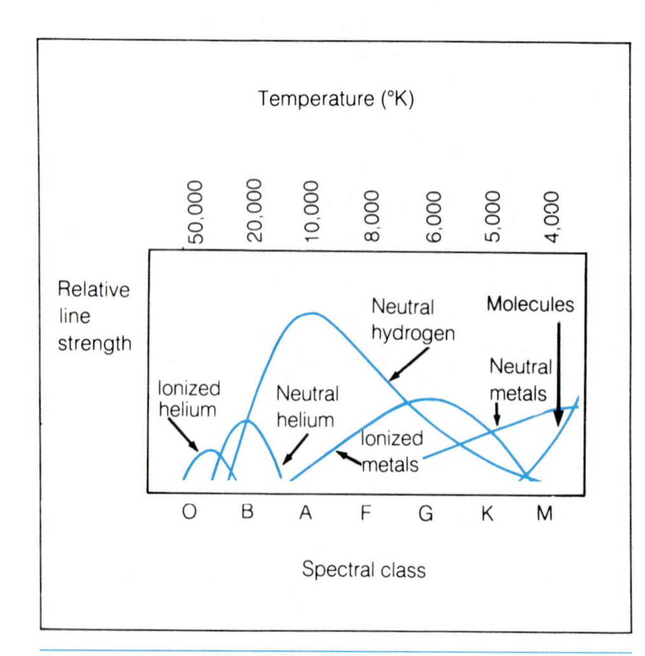

FIGURE 19.12 IONIZATION FOR STARS OF VARIOUS SPECTRAL TYPES. *This diagram shows which ions appear prominently in the spectra of stars of different classes. Note that the degree of ionization evident for the hot stars (left) is much greater than for the cool ones (right).*

Peculiar Spectra

Not all stars fall into one of the standard spectral types, and astronomers generally refer to any such misfits as "peculiar" stars. The fact is that these oddballs may be rare, but in many cases they probably represent normal stages in the evolution of stars.

Most of the peculiar stars are so designated because of unusual absorption lines in their spectra, unusual relative strengths of the lines, or emission lines. These peculiarities, in most instances, do represent abnormal chemical abundances in the outer layers of the star, contrary to our statement that most stars have about the same composition. Actually, the peculiar stars may have essentially normal element abundances, but for complex reasons, have undergone a kind of differentiation or sifting process that resulted in overabundances of certain elements at the surface. In other cases, elements produced by nuclear reactions inside the star (see Chapter 21) are brought to the surface by internal currents, thereby altering the surface composition.

Some stars are referred to as peculiar because they have emission lines in their spectra (Fig. 19.13). As we have seen, in a normal star the cool outer layers form absorption lines, so how do we under-

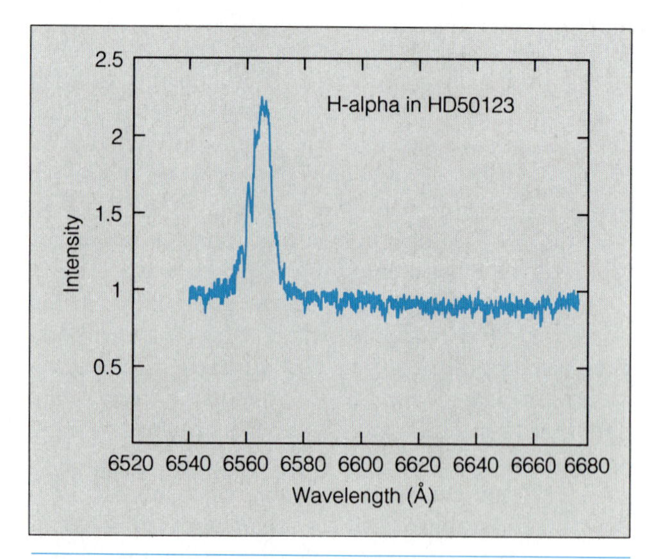

FIGURE 19.13 A STELLAR EMISSION LINE. *This graph shows an emission line due to hydrogen in a star belonging to a peculiar class of stars known to have hot gas surrounding them.* (T. P. Snow; data obtained at the Canada-France-Hawaii Telescope)

ASTRONOMICAL INSIGHT 19.2

STELLAR SPECTROSCOPY AND THE HARVARD WOMEN

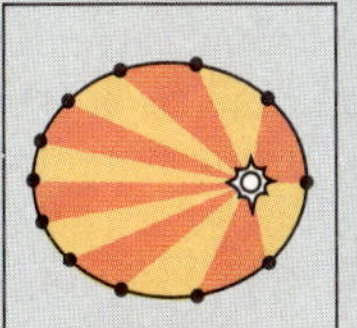

There is some irony in the fact that Harvard University, long a stronghold of the all-male tradition in American colleges, was also the institution that nurtured some of the nation's first leading women astronomers. Today, women are still underrepresented among professional astronomers, but were even more so at the turn of the century when the foundations of modern stellar spectroscopy were developed at the Harvard College Observatory.

In 1877, when Edward C. Pickering became director of the Observatory, Secchi's work (described in the text), based on visual inspection of stellar spectra through a spectroscope, was the only attempt that had been made to classify stars according to the appearance of their spectra. Henry Draper, an American amateur astronomer, had in 1872 become the first to photograph the spectrum of a star. Upon Draper's death in 1882, his widow endowed a new department of stellar spectroscopy at Harvard. Pickering, as director, hired among his assistants a number of women, several of whom went to work on the problem of classifying spectra of stars.

An innovative technique was used to photograph spectra of large numbers of stars. A thin prism was placed in front of the telescope, so that each star image on the photographic plate at the focus was stretched, in one direction, into a spectrum. If color film had been used (it was not), each stellar image would have looked like a tiny rainbow. Pickering and his group refined this **objective prism** technique to the point where all stars in a field of view as faint as ninth or tenth magnitude would appear as spectra well-enough exposed for classification. Thus, it became possible to amass stellar spectra in vast quantities.

The problem of sorting out and classifying the spectra fell to Pickering's associates. One of them, Williamina Fleming, published the first *Draper Catalogue of Stellar Spectra* in 1890. In this catalog, some 10,351 stars in the Northern Hemisphere were assigned spectral classes A through N, in a simple elaboration of a rudimentary classification scheme adopted earlier by Secchi. A number of Fleming's classes were later dropped.

While the first catalog was being prepared, a niece of Henry Draper, Antonia Maury, joined the staff and set to work on the analysis of spectra of bright stars. For these objects, spectra could be photographed using thicker prisms (actually, a series of two or three thin prisms), so that the spectra were more widely spread out according to wavelength, allowing greater detail to be seen. Antonia Maury concluded, on the basis of these high- quality spectra, that stars should be grouped into three distinct sequences, rather than just one. She had found that some stars had unusually narrow spectral lines, others were rather broad, and a third group was in between. It was later found that her sequence *c*, the thin-lined stars, are giant stars (the lower atmospheric pressure in these stars causes the lines to be less broadened than in main sequence stars). Maury's discovery helped Ejnar Hertzsprung confirm the distinction he had recently found between giant and main-sequence stars, and he thought her work to be of fundamental importance. Unfortunately, Maury's separate sequences were not adopted in the further work on classification at Harvard, which was thereafter based on a single sequence.

The most important of the Harvard workers in spectral classification arrived on the scene in 1896. Her name was Annie J. Cannon, and she gradually modified the classification system to the present sequence, finding that the arrangement O, B, A, F, G, K, M was a logical ordering, with smooth transitions from one type to the next (at this time, there was still no knowledge of the fact that this was a temperature sequence). Cannon was also able to distinguish the gradations so finely that she established the ten subclasses for each major division in use today. In 1901 she published a catalog of classifications for 1,122 stars and then embarked on her major task: the classification of more than 200,000 stars whose spectra appeared on survey plates covering both hemispheres. By this time she was so skilled that she could reliably classify a star in a few moments, and most of the job was done in a 4-year period between 1911 and 1914. The resulting *Henry Draper Catalogue* appeared in nine volumes of the *Annals of the Harvard College Observatory*, the final one being published in 1924. Pickering died in 1919, be-

(continued on next page)

ASTRONOMICAL INSIGHT 19.2

STELLAR SPECTROSCOPY AND THE HARVARD WOMEN

(continued from previous page)
fore the catalog was complete, and was succeeded as director by Harlow Shapley.

Annie Cannon continued her work, later publishing a major *Extension* of the catalog, along with a number of other specialized catalogs. She died in 1941, and the American Astronomical Society subsequently established the Annie J. Cannon prize for outstanding research by women in astronomy.

The successes of the Harvard women were not confined to stellar spectroscopy. Another major area of interest to Pickering and later to Shapley was the study of variable stars, and Henrietta Leavitt, who joined the group as a volunteer in 1894, played a leading role in this area. Following some early work in the establishment of standard stars for magnitude determinations, Leavitt by 1905 was at work identifying variables from comparisons of photographic plates (she was eventually to discover more than 2,000 of them). In the process, she examined the Magellanic Clouds for variables, and noticed that their periods of pulsation correlated with their average brightnesses. This was a discovery of profound importance, for the period-luminosity relationship for variable stars was to become an essential tool for establishing both the galactic distance scale and the intergalactic scale (see discussions in Chapters 25 and 28).

The Harvard women, including those discussed here and several whose names are now obscure, were a remarkable group, responsible for a number of major advances in the science of astronomy at a time when its basis in physics was just becoming clear. In this day of awakening recognition of the proper role of women in all areas, it is fitting to consider and appreciate the pioneering work done by this group.

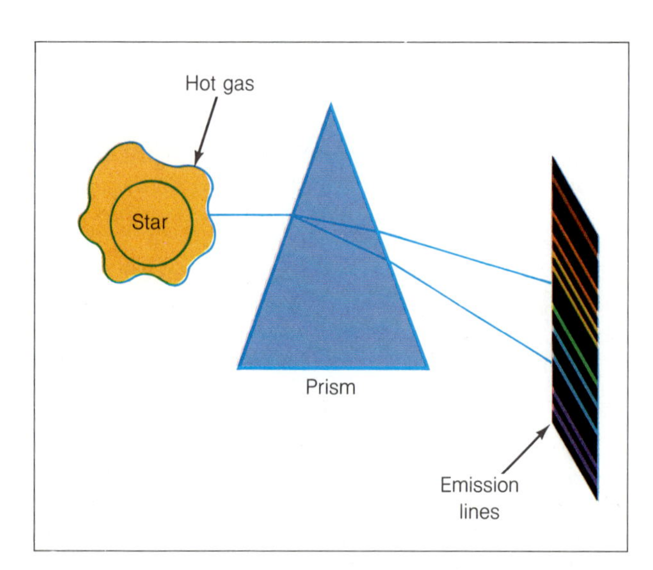

FIGURE 19.14 STELLAR EMISSION LINES. *Very few stars have emission lines in their visible-wavelength spectra (but many do in ultraviolet wavelengths; see Chapter 21). When such lines are present, it usually signifies the existence of hot gas above the surface. According to Kirchhoff's second law, a rarefied, hot gas produces emission lines.*

stand these unusual cases? According to Kirchhoff's second law, a hot, rarefied gas forms emission lines, and this is how the emission lines in the spectra of some stars are formed. In these stars, hot gas extends above the surface, and it is in this region that the emission lines are produced (Fig. 19.14). In most cases, astronomers do not yet understand why this extra gas is present, although they have found some indications of its source. Many of the hottest and most luminous stars, for example, have steady "winds" of outflowing material, and these winds maintain a cloud of gas around the star, creating emission lines.

Other stars with gas swirling around them are found in double-star systems (see "Binary Stars" section), where one star may lose matter to the other. Such systems have received much attention recently because they often involve exotic remnants of stellar deaths (these are discussed in Chapter 24).

BINARY STARS

About half of the stars in the sky are members of double or **binary,** systems, in which the two stars

TABLE 19.2 TYPES OF BINARY SYSTEMS

Type of System	Observational Characteristic
Visual binary	Both stars visible through telescope (usually requires photograph); change of relative position confirms orbital motion
Astrometric binary	Positional measurements reveal orbital motion about center of mass
Spectrum binary	Spectrum shows lines from two distinct spectral classes, revealing presence of two stars
Spectroscopic binary	Spectral lines undergo periodic Doppler shifts, revealing orbital motion. If two sets of spectral lines are seen shifting back and forth (which occurs only if the two stars are comparable in magnitude), it is a double-lined spectroscopic binary; if only one set of lines is seen (in the more common situation where one star is much brighter than the other), it is a single-lined spectroscopic binary.
Eclipsing binary	Apparent magnitude varies periodically because of eclipses, as stars alternately pass in front of each other

orbit each other regularly. All types of stars can be found in binaries, and their orbits come in many sizes and shapes. In some systems the two stars are so close together that they literally are touching, and in others they are so far apart that it takes hundreds or thousands of years for them to complete one revolution.

Each of the three types of observations can be used to detect binary systems (Table 19.2). Positional measurements, of course, tell us when two stars are very near each other in the sky. Sometimes this occurs by chance, when a nearby star happens to lie nearly in front of one that is in the background; this is called an **optical double** and is not a true binary system, since the two stars do not orbit each other (Fig. 19.15).

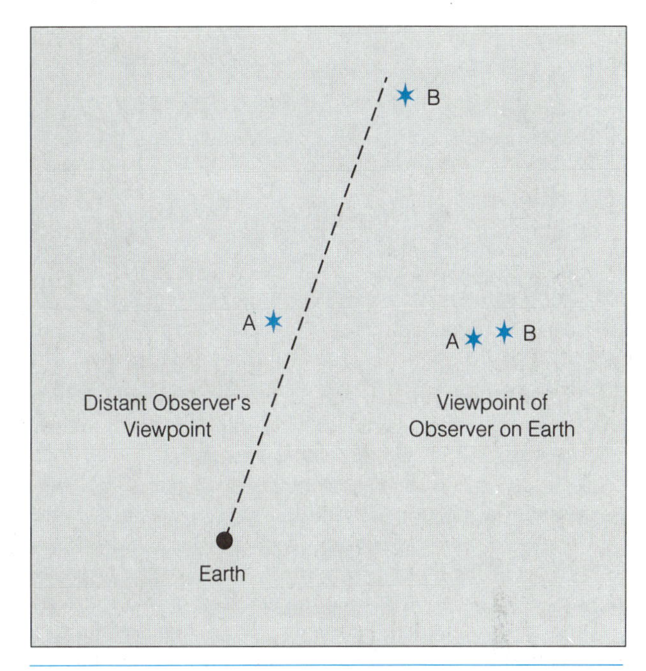

FIGURE 19.15 AN OPTICAL DOUBLE. *At left we see the line of sight from the Earth passing close to two stars (A and B) that are far apart but in nearly the same direction from Earth. At right we see that these two stars appear very close together on the sky. This is not a true binary system.*

When a pair of stars is seen close together, and measurements show that they are in motion about one another, this system is called a **visual binary.** Accurate positional measurements are needed to reveal the orbital motion, because the two stars are very close together in the sky, and because they appear to move very slowly about each other. The distance between the two stars must be many AUs in order for both stars to appear separately (as seen from the Earth), and therefore the orbital period is many years.

It is possible for binary systems to be recognized even when only one of the two stars can be seen. If positional measurements over a lengthy period of time reveal a wobbling motion of a star as it moves through space (Fig. 19.16), it may be inferred that the star is orbiting an unseen companion. Such systems, detected because of variations in position, are called **astrometric binaries.**

Simple brightness measurements can also tell us when a star which may appear to be single is actually

FIGURE 19.16 AN ASTRONOMIC BINARY. *Careful observation of the motion of a star across the sky in some cases reveals a curved path such as the one shown here. Such a motion is caused by the presence of an unseen companion star, so that the visible star orbits the center of mass of the binary system as it moves across the sky.*

part of a binary system. If our line of sight happens to lie close to the plane of the orbit, so that the two stars alternately pass in front of each other as we view the system, then the observed brightness will decrease each time one star is in front of the other (Fig. 19.17). This is called an **eclipsing binary.**

Finally, spectroscopic measurements can also be used to recognize binary systems, again even in the case in which a star appears single. If two stars are actually present, and if they have nearly equal brightnesses, spectral lines from both will appear in the spectrum. If we find a star with line patterns representing two different spectral classes, we can be sure that two different stars are present; such a system is called a **spectrum binary.** Even if it is not clear that two different spectra are present (that is, if the two stars have similar spectral classes, or if one is too faint to contribute noticeably to the spectrum), the Doppler effect may reveal the fact that the star is double. Because the stars are orbiting each other, they move back and forth along our line of sight (as long as we are not viewing the system face-on), and this motion produces alternating blueshifts and redshifts of the spectral lines (Fig. 19.18). Binaries in which these periodic Doppler shifts are seen are called **spectroscopic binaries,** and they are more common than spectrum binaries.

A given binary system may fall into more than one of these categories. For example, a relatively nearby system seen edge-on could be a visual binary, if both

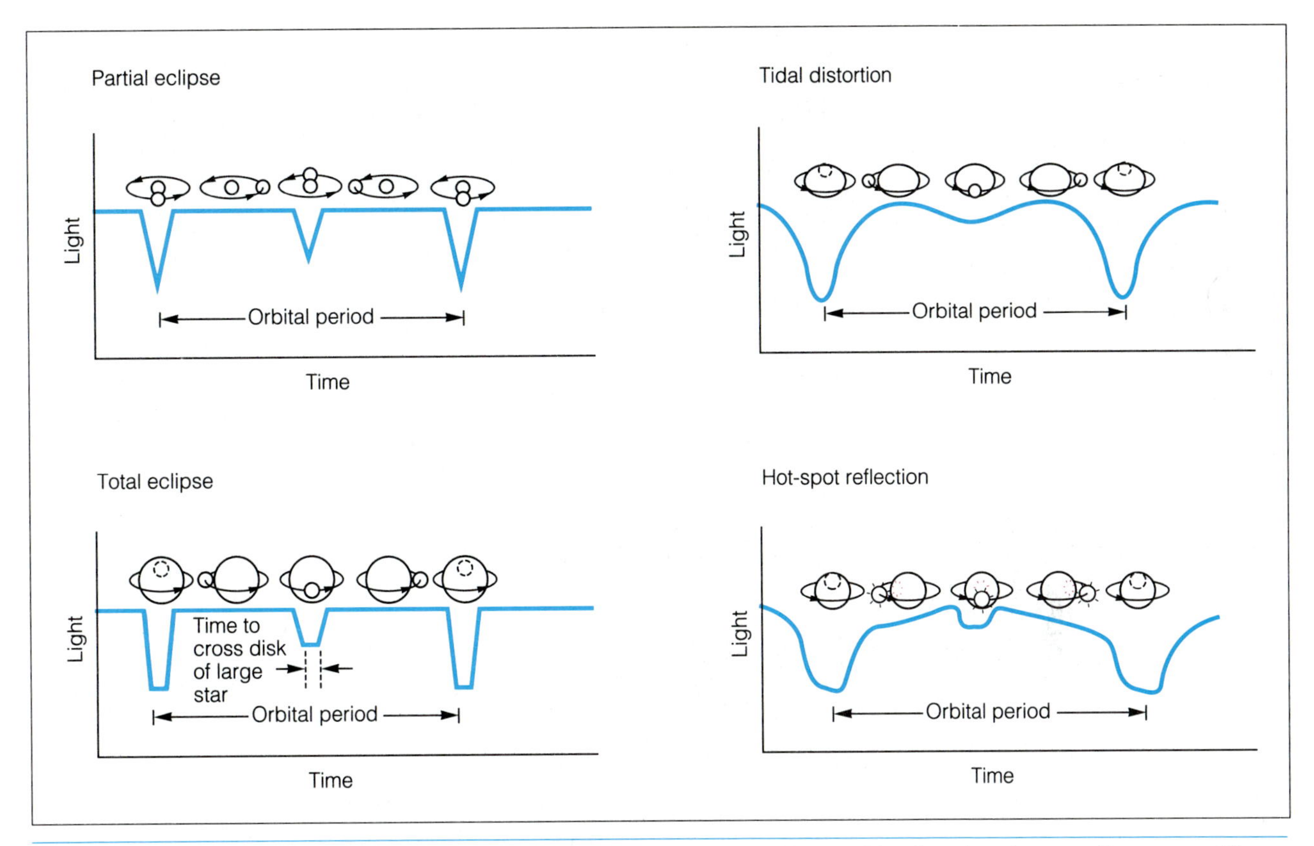

FIGURE 19.17 ECLIPSING BINARIES. *In each sketch we see a pair of stars in binary orbit oriented so that an eclipse occurs. The light curves illustrate how the brightness of the system varies as the eclipses occur.*

stars can be seen in the telescope; it may also be an eclipsing binary, if the two stars alternately pass in front of each other; and it almost surely will be a spectroscopic binary, since the motion back and forth along our line of sight is maximized when we view the orbit edge-on.

Binary stars are important, not just because there are so many of them, but also because they provide a direct means of determining many basic stellar properties, as we will learn in the next chapter.

VARIABLE STARS

Although we tend to think of stars as eternal, steady beacons of light in the nighttime sky, close examination reveals that some of them are not so steady. There are stars that pulsate regularly, varying in brightness as they do so. A few of the brightest of these have been known since ancient times, but most have been found by modern astronomers who made systematic searches for them, usually by comparing photographs of the same region of sky, taken at different times. Such a comparison will reveal any stars that change dramatically in brightness from one photograph to the next. (It is quite possible that most or all stars oscillate in some manner, but usually the pulsations do not cause noticeable changes in brightness, as in the case of our own Sun; see Astronomical Insight 18.2.)

Like peculiar stars, variable stars have been classified according to their observed properties, primarily the period of pulsation but also the shape of the curve representing the variation in brightness (called a **light curve**). Most of these classes are named after the prototype, which is usually the first (and often the brightest) star discovered in that class. Examples in-

FIGURE 19.18 SPECTROSCOPIC BINARIES. *Here are spectrograms of spectroscopic binaries, showing the periodic Doppler shifts due to orbital motion. The sequence of three spectra above shows a single-lined spectroscopic binary, in which lines of only one star are seen to shift back and forth. Below are spectra of a double-lined spectroscopic binary, in which lines of both stars are visible when the two are moving in opposite directions along the line of sight. The emission-line spectra at top and bottom of each set of stellar spectra are used to provide a reference for the wavelength scale.* (Yerkes Observatory)

clude the **δ Cephei** (or **Cepheid**) stars, with periods of 1 to 100 days; the **RR Lyrae** stars, with pulsation times of less than 1 day; and the **Mira,** or long-period, variables, which can take up to 2 years to complete one cycle.

A pulsating variable star physically expands and contracts as it goes through its cycle (Fig. 19.19). This is known because the spectrum shows a periodic Doppler shift as the star's outer layers rise and fall in concert with the variations in brightness. The surface temperature also varies, and so does the spectral class. The brightness may change by as much as one magnitude (or much more, especially in the case of the long-period variables), or as little as a few hundredths of a magnitude.

As with most peculiar stars, the pulsating variables are normal stars passing through a particularly unstable point in their evolutions. A star has a natural vibration frequency, just like a bell, so that if it is disturbed at all, it will pulsate at this frequency for a while before dying down to a steady state again. For certain stars, however, a kind of resonance is established, and once the vibrations start, they continue. These are the pulsating variables.

Some of these stars have certain uniform properties, such as average luminosities, that allow them to be used as distance indicators; these types of variables will be discussed in Chapter 25. In other chapters (particularly Chapter 24), we will discuss stars that vary in a highly irregular fashion rather than pulsating at a steady rate. Many of these irregular and eruptive variables are stars that are in late stages of their lifetimes. Thus, we will describe them along with other types of dying stars.

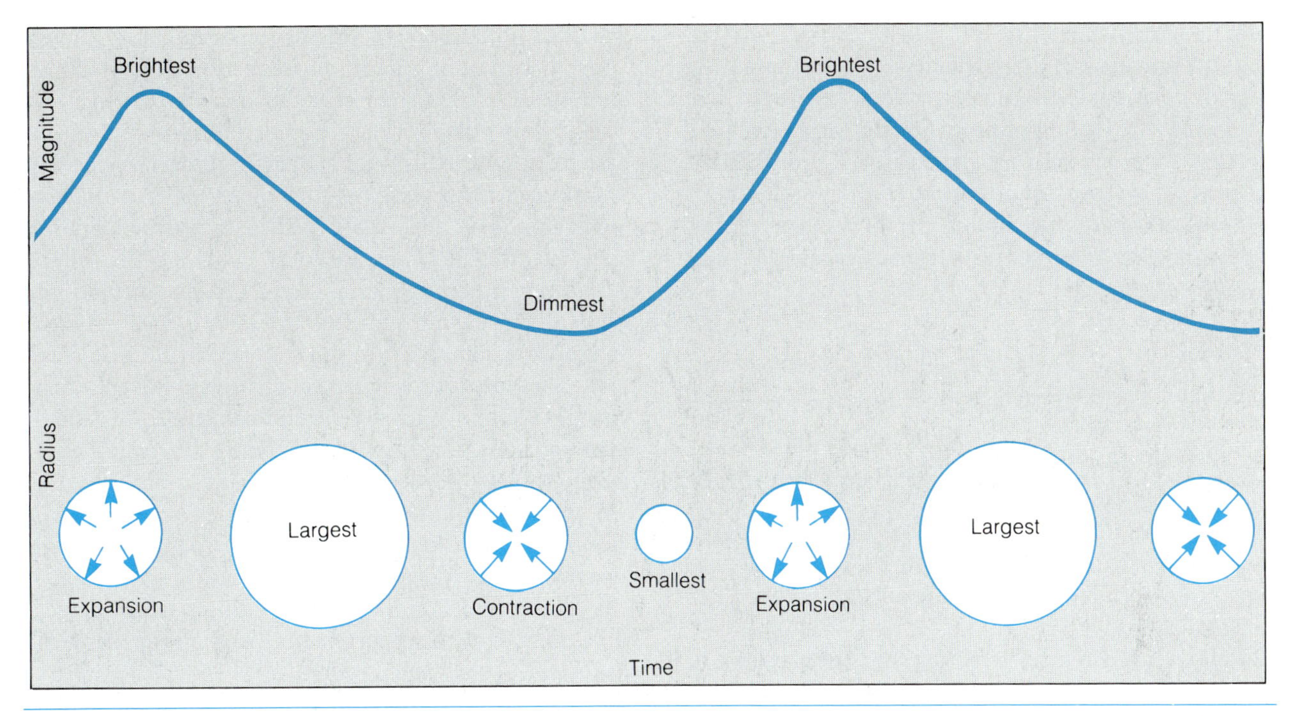

FIGURE 19.19 A PULSATING VARIABLE STAR. *The curved line shows how the brightness (in magnitude units) varies as the star expands and contracts. The sequence of sketches illustrates how the expansion and contraction phases are related in sequence to the variations in brightness. The surface temperature also varies.*

PERSPECTIVE

In this chapter we have learned the basics of stellar observations, by discussing the three fundamental types of observations and the characteristics of stars revealed by each. The stellar properties we have discussed—namely, positions, brightnesses, and spectra—represent the superficial information that comes directly from the observations. We are now ready to see how much more can be learned about stars by combining these directly observed properties with our knowledge of the laws of physics.

SUMMARY

1. All that we know about stars is derived from the light we receive from them.
2. There are three basic types of stellar observation: measurements of position, brightness, and spectra.
3. Positional astronomy (that is, astrometry) has developed techniques capable of measuring position to an accuracy of about 0.01 arcseconds, which is sufficient to detect proper motions, binary motions in some cases, and stellar parallaxes for stars up to 100 parsecs away.
4. Astronomers generally carry out stellar brightness measurements by using the stellar magnitude system, in which a difference of 1 magnitude corresponds to a brightness ratio of 2.512.
5. Magnitudes can be measured at different wavelengths, allowing the determination of color indices, or over all wavelengths, resulting in the determination of bolometric magnitudes.
6. Stellar spectra contain patterns of absorption lines that depend primarily on the surface temperatures of stars, and which therefore can be used to assign stars to spectral classes that represent a sequence of temperatures.
7. Peculiar stars are those that do not conform to the usual spectral types, most often because they have unusual surface compositions, but sometimes because

they have emission lines.
8. Astronomers can detect binary-star systems on the basis of positional variations of one or both stars (astrometric binaries), brightness variations (eclipsing binaries), or composite or periodically Doppler-shifted spectral lines (spectrum or spectroscopic binaries).
9. Some stars are variable, in brightness or spectral appearance, in many cases because of physical pulsations.

REVIEW QUESTIONS

Thought Questions

1. Compare the techniques used today for positional measurements in astronomy with those used by the ancient Greeks. How much more accurate are today's techniques?
2. How can the proper motion and the parallax of a star be distinguished from one another?
3. Does the Sun have annual parallax motion, as seen from the Earth? Explain.
4. Briefly discuss the role of stellar parallax in the development of the heliocentric theory.
5. Describe how the ancient stellar magnitude system has evolved into the modern one. What has been retained, and what has changed?
6. Explain in your own words why the magnitude is smaller, the brighter a star is. Why do extremely bright objects have negative magnitudes?
7. Why is the difference between visible and bolometric magnitudes largest for stars whose temperatures are very different from that of the Sun? Why is the bolometric correction always negative?
8. Explain, in terms of what you learned about ionization in Chapter 6, why stars of different temperature have different absorption lines in their spectra.
9. What does the presence of emission lines in the spectrum of a star tell you about that star? Explain.
10. Summarize the different types of binary stars and how they are observed.

Problems

1. Convert the following distances from parsecs to light- years: 100 *pc*; 1.33 *pc* (the distance to the nearest star); 8,000 *pc* (the distance to the center of our galaxy); 46,000 *pc* (the distance to the large Magellanic Cloud, one of the closest external galaxies to our own); 700,000 *pc* (the distance to the nearest galaxy like our own).
2. What is the distance (in parsecs) to the following stars: (a) a star whose parallax is $p = 0''.25$; (b) a star whose parallax is $p = 0''.04$; (c) a star whose parallax is $p = 0''.005$?
3. How much fainter is a twelfth-magnitude star than an eleventh-magnitude star? How does a fourth-magnitude star compare with a fifth-magnitude star?
4. Compute the brightness ratio for: (a) a fourth-magnitude star and a seventh-magnitude star; (b) a star with magnitude $m = 17$ and a star with $m = 22$; (c) a star of $m = 6$ and a star of $m = -1$.
5. How much brighter is the Sun ($m = -26$) than the full Moon ($m = -12$)? How does the full Moon compare with Venus at its brightest ($m = -4$)?
6. How much fainter are the faintest stars detectable to modern telescopes ($m = 26$) than the brightest stars in the sky ($m = -1$)?
7. Use Wien's law to find the temperature where the wavelength of maximum emission moves from the V band to the B band (assume that the wavelength boundary between the bands occurs at 5,000 Å). Discuss your result in relation to the color index $B-V$.
8. Suppose a star has a bolometric correction of -2. If its visual magnitude is 13, what is its bolometric magnitude? What fraction of its total energy is emitted in the visual band?
9. In a spectroscopic binary system a spectral line of one of the stars is observed to undergo period Doppler shifts. The rest wavelength of the line is 4570.90 Å, and it shifts periodically from an observed wavelength of 4570.44 Å to 4571.36 Å. What is the orbital velocity of this star with respect to its companion? (Assume that the orbit is circular.)
10. In an eclipsing binary system, the duration of the eclipse of star B by star A is 100 hours, and the velocity of star A is 50 km/sec with respect to star B. What is the diameter of star B?

ADDITIONAL READINGS

Probably the best sources of additional information on stars and their observational properties are general astronomy textbooks, of which there are a number readily available in nearly any library. A few more specific resources are listed here as well.

Aller, L. H. 1971. *Atoms, stars, and nebulae.* Cambridge, Ma.: Harvard University Press.

Beyer, Steven L. 1986. *The star guide.* Boston, Mass.: Little, Brown, Inc.

Kaler, James B. 1987. The B stars: Beacons of the sky. *Sky and Telescope* 74(2):174.

______. 1986. Cousins of our sun: The G stars. *Sky and Telescope* 72(5):450.

______. 1986. "The K stars: Orange giants and dwarfs. *Sky and Telescope* 72(2):130.

______. 1986 M stars: Supergiants to dwarfs. *Sky and Telescope* 71(5):450.

______. 1986. The origins of the spectral sequence. *Sky and Telescope* 71(2):129.

______. 1987. The temperature F stars. *Sky and Telescope* 73(2):131.

______. 1987. White Sirian stars: Class A. *Sky and Telescope* 73(5):491.

McAlister, H. A. 1977. Binary star speckle interferometry. *Sky and Telescope* 53(5):346.

Mihalas, D. 1973 interpreting early type stellar spectra. *Sky and Telescope* 46(2):79.

Rubin, Vera. 1986. Women's work: Women in modern astronomy. *Science 86* 7(6):58.

Strand, K. A., ed. 1965. *Basic astronomical data.* Chicago: University of Chicago Press.

Struve, O., and V. Zeberge 1962. *Astronomy of the twentieth century.* New York: Crowell Collier and Macmillan.

Upgren, A. 1980. New parallaxes for old: a coming improvement in the distance scale of the universe. *Mercury* 9 (6):143.

Welther, Barbara. 1984. Annie Jump Cannon: Classifier of the stars. *Mercury* (1):28.

P2. a) $d = \frac{1}{p}$

$= \frac{1}{.25} = 4$ pc

b) $d = 25$ pc

c) 200 pc

CHAPTER 20

FUNDAMENTAL STELLAR PROPERTIES AND THE H-R DIAGRAM

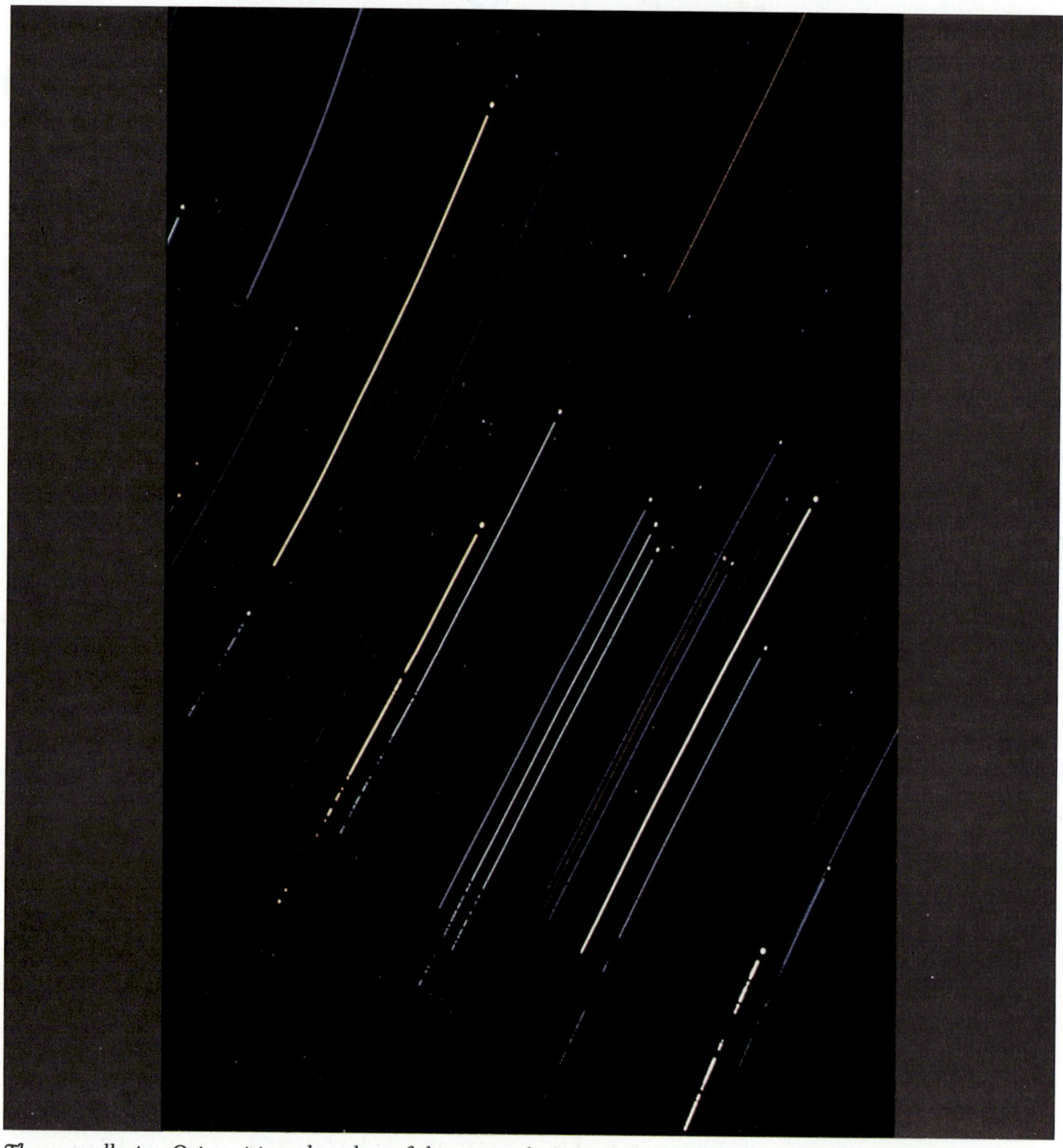

The constellation Orion rising, the colors of the stars indicating their temperatures. (National Optical Astronomy Observatories)

CHAPTER PREVIEW

Before being able to understand how stars work, astronomers needed to determine their physical properties and how they are related to each other. A number of fundamental quantities characterize a star, and in this chapter we will see how the observations described previously can be used to determine those quantities. These include the luminosities, the temperatures, and radii, and, above all, the masses of stars. In this chapter a new, very powerful distance-determination method will also be described.

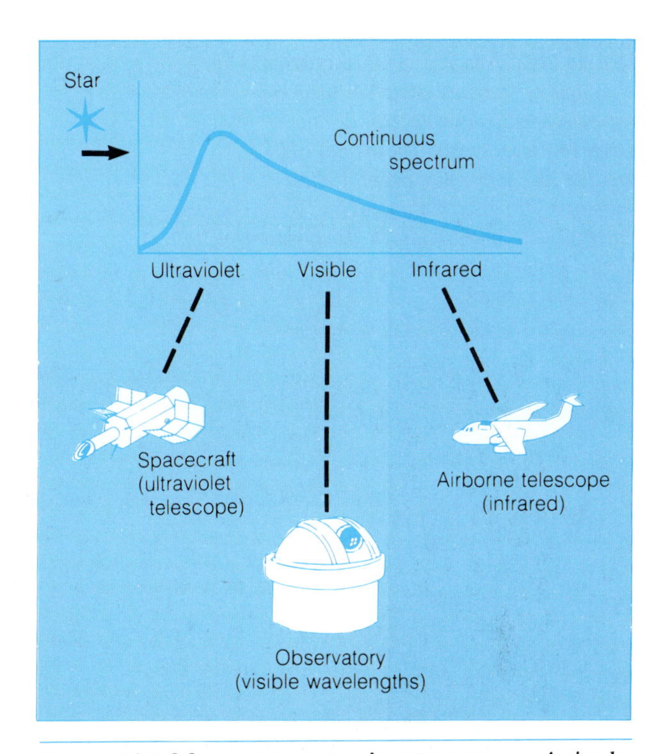

FIGURE 20.1 MEASURING A STAR'S LUMINOSITY. *A simple concept in principle, the determination of a stellar luminosity can be complex. The star's brightness must be measured at all wavelengths where it emits light (that is, its bolometric magnitude must be measured), and then allowance must be made for its distance, in order to calculate how much energy the star is emitting from its surface.*

ABSOLUTE MAGNITUDES AND STELLAR LUMINOSITIES

The luminosity of a star is the total amount of energy it emits from its surface, in all wavelengths; it is usually measured in units of ergs per second (see Chapter 5). If the distance to a star is known, it seems an easy task to measure the apparent brightness and then allow for the distance by taking into account the inverse-square law of light propagation. It is, however, a very painstaking observational problem to measure the intensity of light from a star in absolute units, especially when it must be done over all wavelengths, including the ultraviolet and infrared (Fig. 20.1). Because of the complexities associated with such measurements, astronomers have done this only for a relatively small number of stars, and they have determined the luminosities of others by making comparisons with these standard stars.

The luminosity of a star cannot be determined unless its distance is known, because the brightness we measure directly is affected by distance (recall our discussion of the inverse square law in Chapter 6). Partly for this reason, partly because of familiarity with the magnitude system, and partly because it is easier to make comparative (rather than absolute) measurements, we speak in terms of how bright stars would look if they were all the same distance from us. To do this, we define the **absolute magnitude,** which is the magnitude a star would have if it were seen at a standard distance of 10 parsecs (Fig. 20.2). This distance was chosen arbitrarily, and is used by all astronomers. Thus a comparison of absolute magnitudes reveals differences in luminosities, because all the effects of distance have been canceled out. Table 20.1 lists absolute magnitudes for a variety of objects.

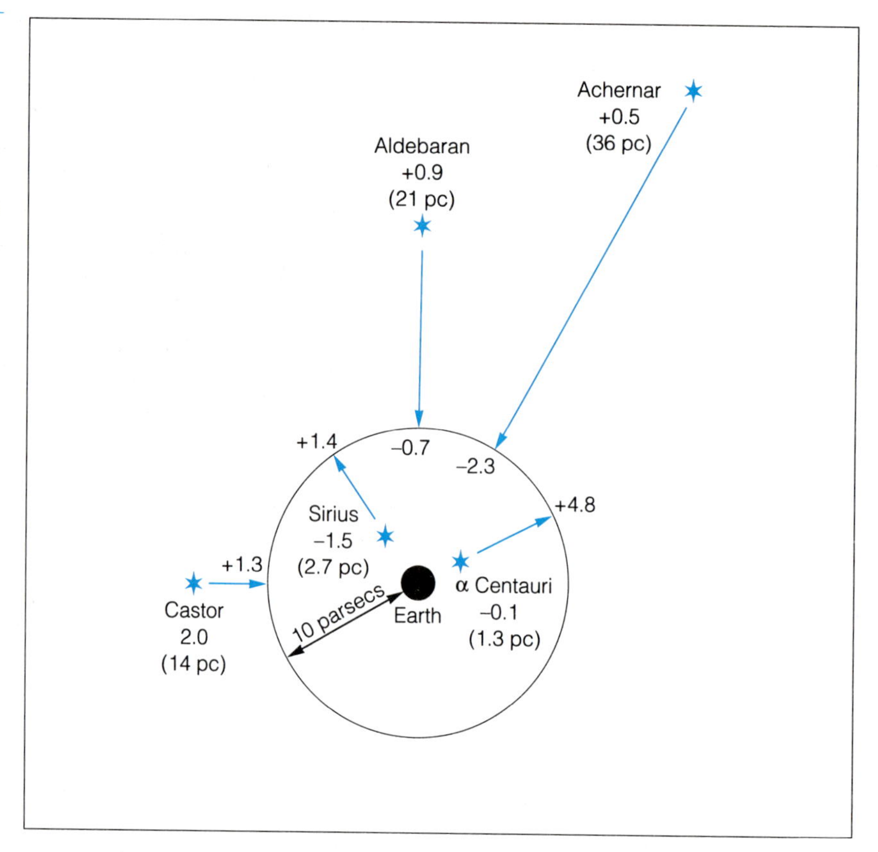

FIGURE 20.2 ABSOLUTE MAGNITUDES. *A sample of stars are shown at their correct relative distances (actual distances given in parentheses). The arrows indicate the imaginary movement of these stars to a 10 parsec distance from the Earth. The numbers indicate the apparent (actual distance) and absolute (10 pc distance) magnitudes.*

To understand how absolute magnitudes are determined, consider a specific example. Suppose a certain star is 100 parsecs away and has an **apparent magnitude** (the observed magnitude) of 7.3. To determine the absolute magnitude, we must find out what this star's magnitude would be if it were only 10 parsecs away. The inverse-square law tells us that since the star would be a factor of 10 closer, it would appear a factor of $10^2 = 100$ brighter. Thus, if it were only 10 parsecs away rather than 100, this star would be a factor of 100 brighter, corresponding to exactly 5 magnitudes brighter. Therefore, its absolute magnitude is $7.3 - 5 = 2.3$.

It is always possible to derive the absolute magnitude from the measured apparent magnitude and the distance, by following a similar line of thought. Of course, the answer generally will not involve simple factors of 100, as did the example just cited, but the logic is the same. If we are doing these conversions mentally, it helps to remember that a factor of 10 in distance always corresponds to a difference of 5 magnitudes. Thus, if in our example the star had been 1,000 parsecs away rather than 100, then to move it to a distance of 10 parsecs would have required bringing it a factor of $100 = 10^2$ closer, so its absolute magnitude would have been $5 + 5 = 10$ magnitudes brighter than its apparent magnitude. Similarly, a star 10,000 parsecs away has an absolute magnitude 15 magnitudes brighter (smaller) than its apparent magnitude.

It should be clear that if we somehow knew the absolute magnitude of a star and measured its apparent magnitude, we could work these arguments backward to determine the distance to the star. This is a very important point, and we will discuss it later.

Let us now return to the question of luminosities. By determining the absolute magnitudes of stars, we can compare their luminosities, since the distance effect has been removed. Thus if one star's absolute magnitude is 5 magnitudes smaller than another's,

TABLE 20.1 VISUAL ABSOLUTE MAGNITUDES OF SELECTED OBJECTS

Object	Absolute Magnitude
Typical bright quasar	−28
Brightest galaxies	−25
Milky Way galaxy	−20.5
Andromeda galaxy	−21.1
Type I supernova (maximum brightness)	−18.8
Large Magellanic Cloud	−18.7
Type II supernova (maximum brightness)	−17
Small Megallanic Cloud	−16.7
Typical globular cluster	−8
The most luminous stars	−8
Typical nova outburst	−7.7
Vega (bright star in summer sky)	+0.5
Sirius	+1.41
alpha Centauri (nearest star)	+4.35
Sun	+4.83
Venus (greatest elongation)	+28.2
Full moon	+31.8
100-W bulb	+66.3

then we know that its luminosity is a factor of 100 greater. To actually express the luminosity in terms of ergs per second, we make a comparison with one of the standard stars for which a direct measurement of the luminosity has been made. Strictly speaking, since the luminosity refers to energy emitted over all wavelengths, we must use bolometric absolute magnitudes in order to determine luminosities (recall that a bolometric magnitude is simply a magnitude measured over all wavelengths).

What astronomers find when they determine stellar luminosities is that the values from star to star can vary over an incredible range. There are stars with luminosities as small as 10^{-4} that of the Sun to as great as 10^6 that of the Sun, a range of 10 billion from the faintest to the most luminous! The luminosity is by far the most highly variable parameter for stars; the others that we will discuss cover only ranges of a factor of 100 or so from one extreme to the other.

SURFACE TEMPERATURES OF STARS

We have already learned something about how the temperatures of stars are determined, and we need not add much to that. In Chapter 19 we saw that the color of a star depends on its surface temperature, as does the spectral type, so that by observing either, we can deduce the temperature. Recall that the color index, the difference between the blue (*B*) and visual (*V*) magnitudes, is a measured quantity that indicates temperature. A negative value of $B-V$ means that the star is brighter in blue than in visual light, and therefore is a hot star. A large positive value indicates a cool star. A specific correlation between color index and surface temperature has been established and is used to determine temperatures from observed values of $B-V$.

More refined estimates of temperature can be made from a detailed analysis of the degree of ionization, which is done by measuring the strengths of spectral lines formed by different ions. This is basically the same as simply estimating the temperature from the spectral type, since in either case the point is that the strength of spectral lines of various ions depends on how abundant those ions are, which in turn depends on the temperature.

The temperature referred to here has been referred to as the surface temperature, although stars do not have solid surfaces. We are really referring to the outermost layers of gas, where the absorption lines form. This region is called the **photosphere** of a star, and it actually has some depth (although it is very thin compared with the radius of the star).

Stellar temperatures range from about 2,000 K for the coolest M stars to 50,000 K or more for the hottest O stars. The temperature of the Sun, a G2 star, is a little less than 6,000 K.

THE HERTZSPRUNG-RUSSELL DIAGRAM

We have seen that temperature and spectral type are closely related, and we have learned how astronomers deduce the luminosities of stars. In the first decade

FIGURE 20.3 HENRY NORRIS RUSSEL. *One of the leading astrophysicists of the era when a physical understanding of stars was first emerging, Russell made many major contributions in a variety of areas.* (Princeton University)

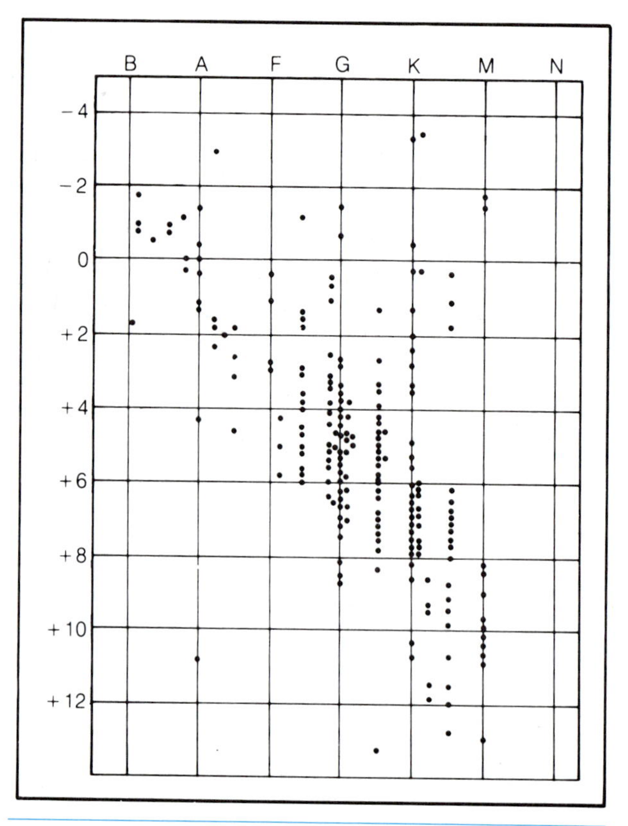

FIGURE 20.4 THE FIRST H-R DIAGRAM. *This plot showing absolute magnitude versus spectral class was constructed in 1913 by Henry Norris Russell.* (Estate of Henry Norris Russell, reprinted with permission)

of this century, the Danish astronomer Ejnar Hertzsprung and, independently, the American Henry Norris Russell (Fig. 20.3), began to consider how luminosity and spectral type might be related to each other. Each gathered data on stars whose luminosities (or absolute magnitudes) were known, and found a close link between spectral type (that is, surface temperature) and absolute magnitude (or luminosity). This relationship is best seen in the diagram constructed by Russell in 1913 (Fig. 20.4), now called the **Hertzsprung-Russell,** or **H-R, diagram.** In this plot of absolute magnitude (on the vertical scale) versus spectral class (on the horizontal axis), stars fall into narrowly defined regions rather than being randomly distributed. A star of a given spectral class cannot have just any absolute magnitude, and vice versa.

Most stars fall into a diagonal strip running from the upper left (high temperature, high luminosity) to the lower right (low temperature, low luminosity) (Fig. 20.5). This strip has been given the name the **main sequence.** A few stars are not in this sequence, but instead appear in the upper right (low temperature, high luminosity). Since the spectra of these stars indicate that they are relatively cool, their high luminosities cannot be a result of the fact that they have greater temperatures than the main-sequence stars of the same type. The only way one star can be much more luminous than another of the same temperature is if it has a much greater surface area; recall that the two stars will emit the same amount of energy per square centimeter of surface per second. Hertzsprung and Russell independently realized that these extra-luminous stars located above the main sequence must be much larger than those on the main sequence, and they named these stars **giants** and **supergiants.**

The distinction among giants, supergiants, and main-sequence stars (known as **dwarfs**) has been incorporated into the spectral classification system used by modern astronomers. A **luminosity class** has been added to the spectral type with which we are already familiar. The luminosity classes, designated by Roman numerals following the spectral type, are I for supergiants (further subdivided into classes Ia and Ib), II for extreme giants, III for giants, IV for stars just a

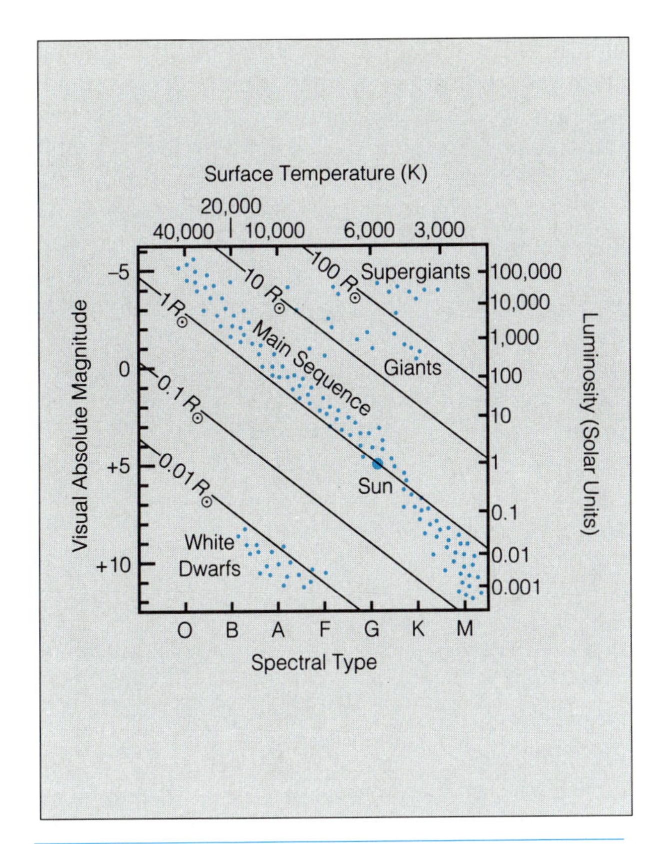

FIGURE 20.5 A MODERN H-R DIAGRAM. *This diagram gives alternative units for both axes, with spectral type (bottom) or surface temperature (top) on the horizontal axis, and visual absolute magnitude (left) or luminosity (right) on the vertical axis. Note that a bolometric correction has been applied in order to compare visual absolute magnitudes to luminosities.*

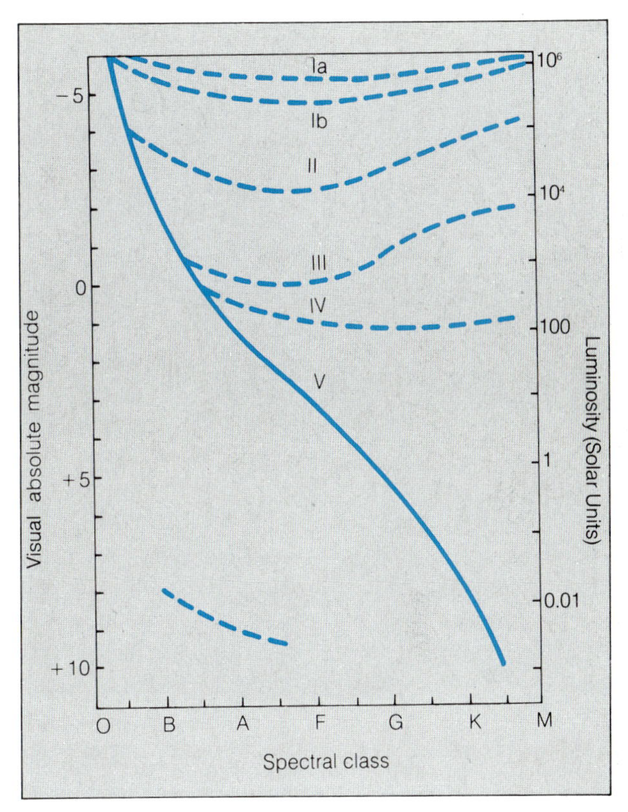

FIGURE 20.6 LUMINOSITY CLASSES. *This H-R diagram shows the locations of stars of the luminosity classes described in the text. A complete spectral classification for a star usually includes a luminosity class designation, if it has been determined.*

bit above the main sequence, and V for main-sequence stars, or dwarfs (see Fig. 20.6). Thus the complete spectral classification for the bright summertime star Vega, for example, is A0V, meaning that it is an A0 main-sequence star. The red supergiant in the shoulder of Orion, Betelegeuse, has the full classification M2Iab (it is intermediate between luminosity classes Ia and Ib). It is usually possible to assign a star to the proper luminosity class from examination of subtle details of its spectrum.

Another group of stars has become known (mostly since the time Russell first plotted the H-R diagram), that do not fall into any of the standard luminosity classes, but which instead appear in the lower left (high-temperature, low-luminosity) corner of the diagram. Since these stars are hot but not very luminous, they must be very small, and they have been given the name **white dwarfs.** These objects have some very bizarre properties, which will be discussed in Chapter 24.

The H-R diagram is a fundamental tool for understanding stars and the relationships among their various properties, as well as how they evolve. We will be referring to it continually throughout this section of the book.

Spectroscopic Parallax

The H-R diagram can be used to find distances to stars, even stars that are very far away. The idea is really very simple: If we know how bright a star is intrinsically (that is, how much energy it is actually

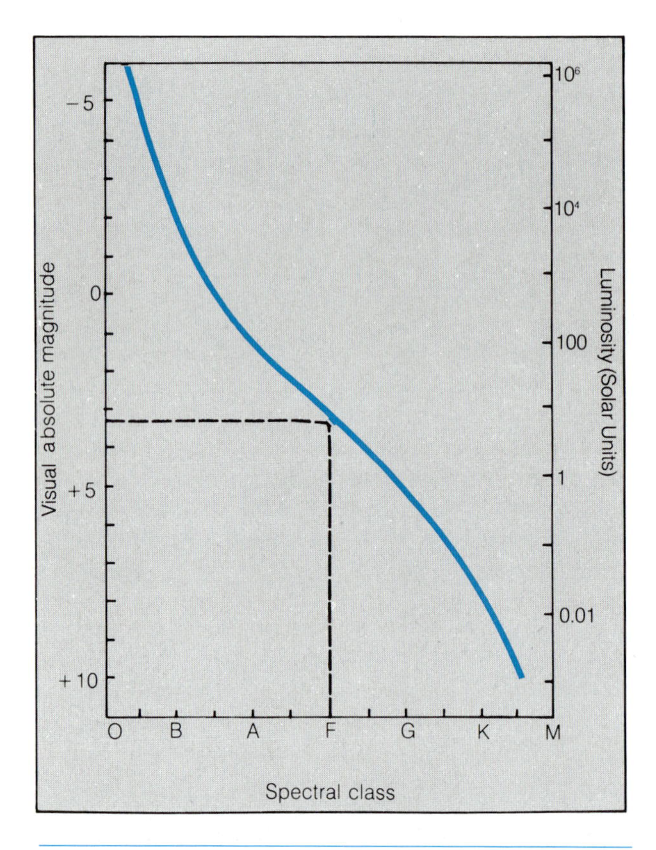

FIGURE 20.7 SPECTROSCOPIC PARALLAX. *Knowledge of a star's spectral class (including the luminosity class) allows the absolute magnitude, and hence the distance, to be determined. This figure shows how the absolute magnitude (+3.3) of an F0 main sequence star is read from the H-R diagram. The distance can then be found by comparing the absolute and apparent magnitudes of the star (as explained in the text and Appendix 10).*

emitting from its surface), and we measure how bright it appears to be, we can determine how far away it is, because we know that the difference between its intrinsic brightness and its observed brightness is caused by the distance. The only problem lies in knowing the intrinsic brightness of the star (its luminosity), and this is where the H-R diagram comes in.

Once we determine the spectral class of a star, we can place it on the H-R diagram (as long as we are sure we know the luminosity class, so we know whether it is on the main sequence or is a giant or supergiant). After we have placed it on the diagram, we simply read off of the vertical axis the absolute magnitude of the star, which is a measure of its luminosity (Fig. 20.7). A comparison of the absolute magnitude with the apparent magnitude, therefore, amounts to the same thing as a comparison of the intrinsic and apparent brightnesses of the star, and from such a comparison the distance can be found.

Consider these examples. If the difference $m - M$ (that is, the apparent minus the absolute magnitude), which is called the **distance modulus,** is 5, then the star appears 5 magnitudes, or a factor of 100, fainter than it would at the standard distance of 10 parsecs. A factor of 100 in brightness is created by a factor of 10 change in distance, so this star must be ten times farther away than it would be if it were at 10 parsecs distance; therefore, it is $10 \times 10 = 100$ parsecs away. Similarly, a star whose distance modulus $m - M$ is 10 is 1,000 parsecs away. If $m - M = 15$, then the distance is 10,000 parsecs. It should be obvious that if $m = M$ (that is, $m - M = 0$), then the distance to the star must be 10 parsecs, because this is the distance that defines the absolute magnitude. As a rule of thumb, it helps to remember that for every 5 magnitudes of difference between the apparent and absolute magnitudes, the distance increases by a factor of 10.

This method is very powerful, because it can be used for very large distances. All that is needed is to be able to place a star on the H-R diagram so that its absolute magnitude can be determined, and to measure its apparent magnitude.* Because this distance-determination technique requires that the spectrum of a star be classified so that it can be placed on the H-R diagram, it is called the **spectroscopic parallax** method (the word parallax is used by astronomers as a general word for distances, even though, technically speaking, no parallax is measured in this case).

It should be noted that any time it is possible to determine the absolute magnitude of an object, even if it is by some means other than placing it on the H-R diagram, its distance can be found by comparing the absolute and apparent magnitudes. Any object whose absolute magnitude can be determined so that it can be used as a distance indicator is called a **stan-**

*There is an additional bit of information needed: The interstellar medium contains a haze of fine dust particles that tends to make stars appear dimmer than they otherwise would, and allowance must be made for this effect. The interstellar dust is discussed in Chapter 26, and the method for correcting distance measurements is given in Appendix 10.

ASTRONOMICAL INSIGHT 20.1

FUNDAMENTAL PROPERTIES AND INTERFEROMETRY

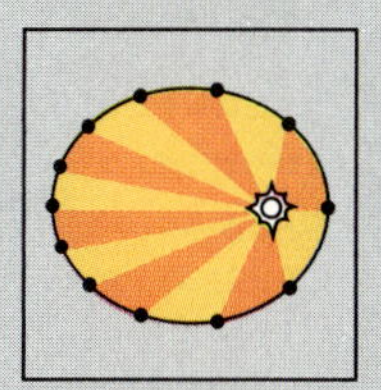

In order for astronomers to gain any understanding of the physical processes that govern stars, they must have reliable information on the properties of the stars. We cannot hope to understand the relationships among basic properties such as luminosity, mass, diameter, and temperature until we can accurately measure these properties. This chapter is devoted to descriptions of how astronomers measure basic stellar parameters, emphasizing traditional techniques. Today major advances are being made using new methods, particularly **interferometry.**

Interferometry, discussed briefly in Chapter 7, involves using the wave nature of light to improve resolution, the ability to detect and measure fine detail. Interferometry takes advantage of the manner in which light waves interfere with each other, and the fact that the interference depends sensitively on the relative pathlengths followed by separate beams of light. Separation of beams can be accomplished in at least two different ways: separate telescopes can be used, so that precise timing of the arrival of light waves provides precise directional information; or the Earth's atmosphere can be allowed to do the separation. As light from a small source such as a star passes through the atmosphere, density fluctuations in the air cause the light to be split into separate bundles which can interfere with each other. Analysis of the interference pattern can then lead to a reconstruction of the source image.

The basic principles of interferometry have been understood for several decades, but application has been very difficult because of the precision required, both in instruments and in measurements. The precision requirements vary with observed wavelength, and thus it has been in radio astronomy where the first widespread applications have occurred. Today, however, there is intense activity in visible-light interferometry, made possible by modern technology. Advances in computers have been particularly important, allowing the necessary high-speed capacity for analyzing interference patterns.

The principal stellar property that can be measured with interferometry is the diameter. What is actually measured is the angular diameter; to convert this to a physical diameter requires knowing the distance to the star. Angular diameters for stars within a few hundred parsecs of the Earth can be as small as 0.001 arcsecond (that is, a *milliarcsecond*) or less. Recall that there are 60 arcseconds in an arcminute, 60 arcminutes in a degree, and 360 degrees in a full circle; multiplying these together shows that there are 1,296,000 arcseconds in a full circle. Thus, an arcsecond is a very small angle (about the angular size of a dime seen at a distance of two miles), yet modern interferometry allows angular diameters one thousand times smaller to be measured (try to imagine the angular size of a dime seen from a distance of 2000 miles!).

Interferometry also leads to better knowledge of stellar masses, because the improvement in resolution offered by interferometry allows better determination of binary star orbits. As we learn in this chapter, the masses of stars are measured by the application of Kepler's third law to orbits of stars in binary systems. It is rarely possible for astronomers using ordinary telescopes to directly measure the separation of the two stars, but interferometry allows this separation to be measured for many more systems. Thus, interferometry allows astronomers to measure stellar masses more accurately and for more stars, a great improvement in our knowledge of basic stellar properties.

Other stellar parameters can be determined from the ones that are directly measured using interferometry. If the diameter of a star is measured and the luminosity is known, then the surface temperature can be derived from the Stefan-Boltzmann equation, which relates surface temperature, luminosity, and radius (see Chapter 6). These properties, together with the mass of a star (also determined with the help of interferometry), are required in order to construct meaningful models of interior conditions (as described in the next chapter).

In addition to improving our knowledge of basic properties of stars, interferometry can also tell us much about other interesting aspects. For example, interferometric measurements may be the best method we have for detecting planets orbiting

(continued on next page)

ASTRONOMICAL INSIGHT 20.1

FUNDAMENTAL PROPERTIES AND INTERFEROMETRY

(continued from previous page)
other stars. (Here infrared wavelengths are particularly useful, because planets are cool and emit mostly infrared radiation.) Interferometry can also tell us a great deal about star formation and young stars still surrounded by gas and dust. Closer to home, interferometry allows solar system astronomers to measure planetary and satellite diameters, and to determine fundamental properties of interplanetary bodies such as asteroids.

Several new interferometry programs are being started or are underway all over the world. Most involve the use of multiple telescopes, usually mounted on tracks so that their separations can be changed (this allows greater accuracy and a greater range of angular diameters that can be measured). These projects are identical in concept to the *Very Large Array*, the huge radio interferometry observatory in New Mexico (Chapter 7). Other new projects are designed to do **speckle interferometry,** the technique where the image of a star is reconstructed after being broken up by passage of starlight through the Earth's atmosphere. There are even plans being made to put interferometric instruments into space, so that they can start with images that are free of the distortions caused by the Earth's atmosphere. If this can be done, then it may become possible to measure angular diameters as small as one-millionth of an arcsecond (that is, a *microarcsecond*). This would allow direct measures of a stellar diameters for stars the size of the Sun up to distances of nearly 10,000 parsecs, equivalent to the distance from the Sun to the center of our galaxy. This will truly broaden the horizons of astronomers studying not only the properties of stars, but also the galaxy.

dard candle. We will see in later sections how important these objects are in exploring the size scale both of the galaxy and of the universe.

STELLAR DIAMETERS

We have already seen that a star's position in the H-R diagram depends partly on its size, since the luminosity is related to the total surface area. If two stars have the same surface temperature (and therefore the same spectral type), but one is more luminous than the other, we know that it must also be larger. The Stefan-Boltzmann law (see Chapter 6) specifically relates luminosity, temperature, and radius; use of this law allows the radius to be determined if the other two quantities are known.

Eclipsing binaries provide another means of determining stellar radius, one that is independent of other properties. Recall that these are binary systems in which the two stars alternately pass in front of each other, as we view the orbit nearly edge-on. The eclipsing binary is very likely also to be a spectroscopic binary, so that the speeds of the two stars in the orbits can be measured from the Doppler effect. We therefore know how fast the stars are moving, and from the duration of the eclipse we know how long it takes one to pass in front of the other; the simple formula "distance equals speed times time" thus gives us the diameter of the star that is being eclipsed (Fig. 20.8). Even if no information on the orbital velocity is available, the relative diameters of the two stars can be deduced from the relative durations of the alternating eclipses.

Eclipsing binaries provide the most direct means of measuring stellar sizes, but unfortunately, there are not many of them. In most cases the radii are estimated from the luminosity and temperature, as just described. In a few cases of nearby stars, stellar radii have been measured directly by use of interferometry, which overcomes the blurring effects of the Earth's atmosphere. Only relatively nearby, large stars can be measured this way, however (Fig. 20.9).

Stars on the main sequence do not vary greatly in radius, ranging from perhaps 0.1 times the Sun's radius for the *M* stars at the lower right-hand end to

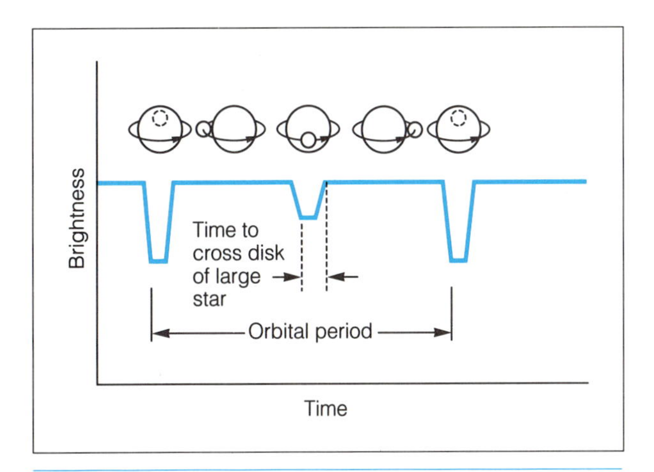

FIGURE 20.8 MEASURING STELLAR DIAMETERS IN AN ECLIPSING BINARY. *The duration of each eclipse is combined with knowledge of the orbital speeds of the stars (from the Doppler effect) to yield the diameters of the stars.*

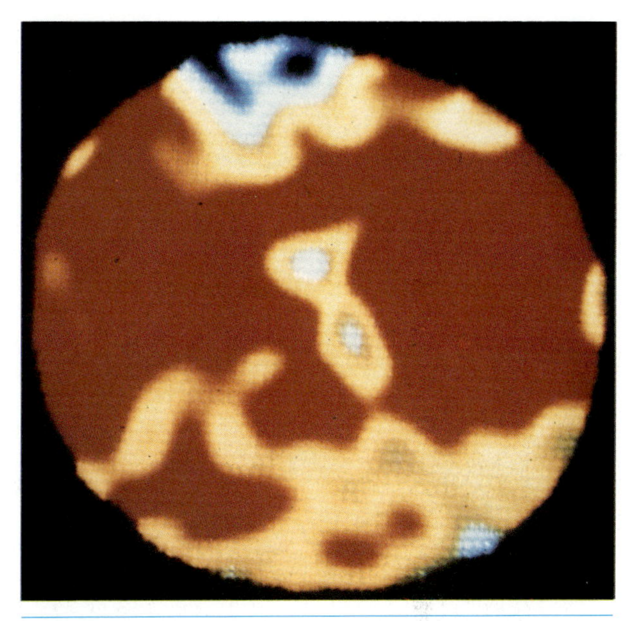

FIGURE 20.9 SURFACE FEATURES ON BETELGEUSE. *The large areas of contrasting color on this image of the red supergiant Betelgeuse (α Orionis), obtained by use of interferometry, probably represent convective cells on the star's surface.* (National Optical Astronomy Observatories)

10 or 20 solar radii at the upper left. Of course, large variations in size occur as we go away from the main sequence, either toward the giants and supergiants, which may be more than 100 times the size of the Sun, or toward the white dwarfs, which are as small as 0.01 times the size of the Sun.

STELLAR ROTATION

To measure how rapidly a star is spinning is a fairly straightforward procedure, but there is one serious complication. Part of the light that reaches us from a spinning star is emitted from the portion of its surface that is moving toward us, and part is emitted from the portion that is moving away. Thus, because of the Doppler effect, spectral lines formed in the approaching portion are blueshifted, and those formed in the receding portion are redshifted. When we measure the spectrum of a star, the spectral lines are broadened, being made up of light from all parts of the surface. The more rapid the rotation, the greater the Doppler shift, and the broader the spectral lines (Fig. 20.10). Thus, in principle, a star's velocity of rotation can be determined from measurements of the width of its spectral lines.

The complication is that we do not know the orientation of the star's rotation axis. If we view a star from straight above its pole, there will be no Doppler shift or line broadening, no matter how fast it is spinning, because there will be no motion along our line of sight (recall that the Doppler effect occurs only for motions directed straight toward or away from the observer). If we view a star in the plane of its equator, on the other hand, then the Doppler shift does cause its spectral lines to be broadened, and we can make a direct measurement of its rotational speed. Unfortunately, we rarely know whether we are viewing a star pole-on, equator-on, or, as is more likely, at some random angle in between. The result is that we usually cannot precisely determine the true rotational speed of a particular star; instead, by sampling large numbers of stars, we can make general statements about the range of stellar rotation speeds that exist.

Many stars rotate slowly, with surface speeds of less than 10 km/sec (the Sun's rotational speed at the equator is about 2 km/sec), and rotation periods of at least several days or a few weeks. There are exceptions, particularly among the hot stars (which, as we will see, are young objects that may still be spinning rapidly as a result of their formation). In these cases,

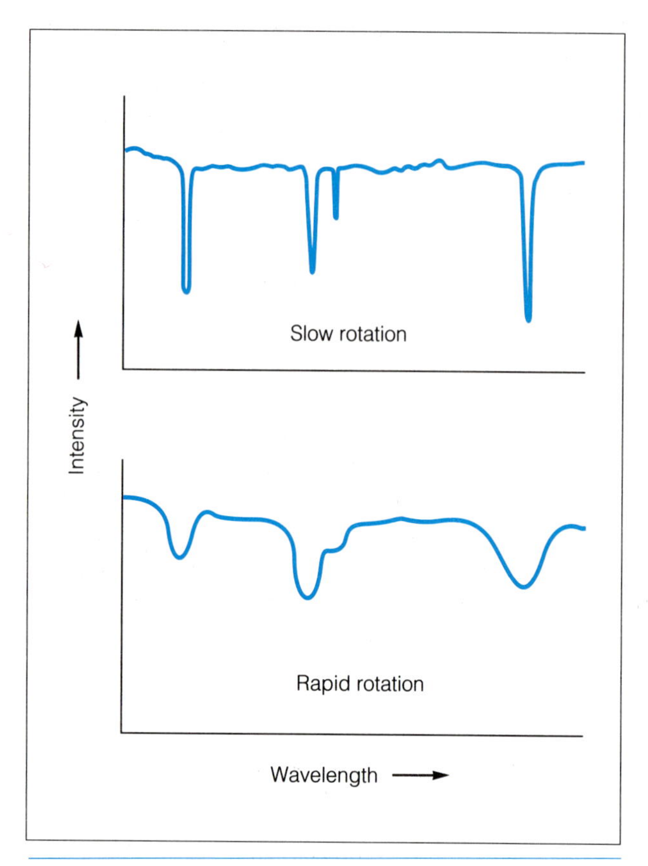

FIGURE 20.10 THE EFFECT OF STELLAR ROTATION. *The Doppler effect causes spectral lines in a rapidly rotating star (bottom) to be broadened, because one edge of the star's disk is rapidly approaching us while the other edge is receding. Little broadening occurs in a slowly rotating star (top).*

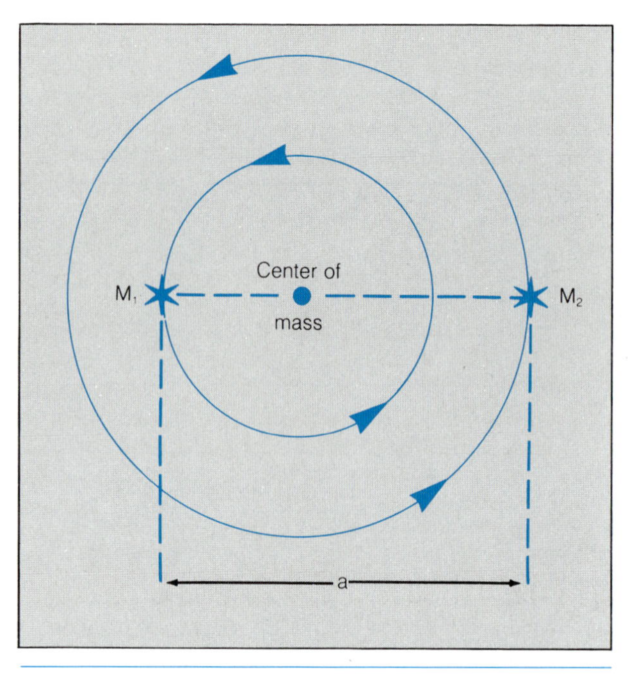

FIGURE 20.11 BINARY STAR ORBITS. *This figure illustrates the terms used in Kepler's third law. Two stars of masses* m_1 *and* m_2 *(*m_1 *larger than* m_2 *in this case) orbit a common center of mass, each making one full orbit in period* P. *The semimajor axis* a *that appears in Kepler's third law is actually the sum of the semimajor axes of the two individual orbits about the center of mass; this sum corresponds to the average distance between the two stars.*

rotational velocities up to 450 km/sec are found, corresponding to rotation periods as short as 1 or 2 days.

BINARY STARS AND STELLAR MASSES

The mass of a star is the most important of all its fundamental properties, for it is the mass that governs most of the others. (This point will be discussed at some length in the next chapter.) Unfortunately, there is no direct way to see how much mass a star has. The only way to measure a star's mass is by observing its gravitational effect on other objects, and this is possible only in binary star systems, in which the two stars hold each other in orbit by their gravitational fields. The fact that binary systems are common provides us with many opportunities to determine masses by analyzing binary orbits.

The basic idea is rather simple, although the application may be quite complex, depending on the type of binary system. Kepler's third law is used, in the form derived by Newton. Remember that if the period P is measured in years, the average separation of the two stars (the semimajor axis a) is measured in astronomical units, and the masses m_1 and m_2 in units of solar masses, then Kepler's third law is

$$(m_1 + m_2)P^2 = a^3$$

We need only observe the period and the semimajor axis in order to solve for the sum of the two masses. Careful observation of the sizes of the individual orbits (actually, of the relative distances of the two stars from the center of mass; see Fig. 20.11) also yields the ratio of the masses; when both the sum and the

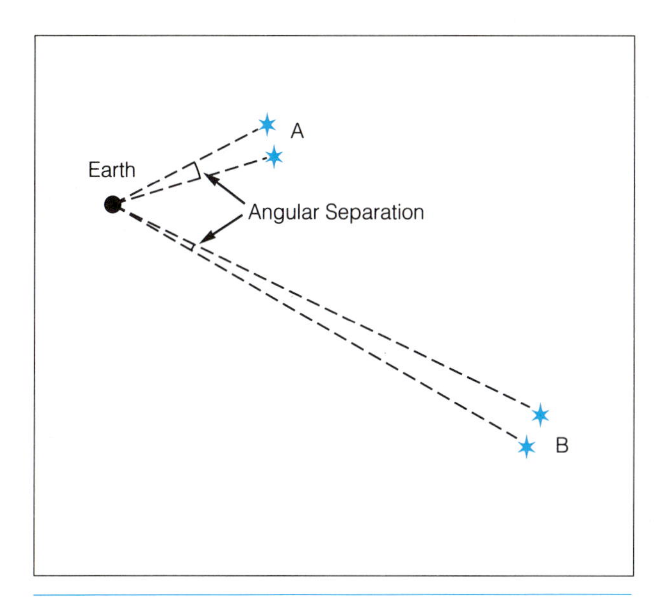

FIGURE 20.12 THE EFFECT OF DISTANCE ON MEASUREMENTS OF BINARY STAR ORBITS. *Here two binary systems have the same actual separation, but from Earth the nearer system (A) appears more widely separated than the more distant system (B). Therefore the distance to a binary system must be known in order to analyze the stellar orbits.*

ratio are known, it is simple to solve for the individual masses.

Complications arise when some of the needed observational data are difficult to obtain. The period is almost always easy to measure with some precision, but not so the semimajor axis. One main problem is that the apparent size of the orbit is affected by our distance from the binary (Fig. 20.12), and so the distance must be well known if a is to be accurately determined. Another problem is that the orbital plane is inclined at a random, unknown angle to our line of sight, so that the apparent size of the orbit is foreshortened by an unknown amount. In some cases it is possible to unravel these confusing effects by carefully analyzing the observations, and in other cases it is not. Even so, some information about stellar masses can be gained, but usually only in terms of broad ranges of possible values, rather than precise answers.

The masses of stars vary along the main sequence from the least massive stars in the lower right to the most massive in the upper left. The M stars on the main sequence have masses as low as 0.05 solar masses, whereas the O stars reach values as great as 60 solar masses. It is likely that stars occasionally form with even greater masses (perhaps up to 100 solar masses), but as we will see in the next chapter, such massive stars have very short lifetimes, so it is rare to find one.

The giants, supergiants, and white dwarfs have masses comparable to those of main-sequence stars. Hence, their obvious differences from main-sequence stars in other properties such as luminosity and radius have to be the result of something other than extreme or unusual mass. (This is discussed in the next chapter.)

For main-sequence stars, there is a smooth progression of all stellar properties from one end to the other (Table 20.2). The mass and the radius vary by similar factors, whereas the luminosity changes much more rapidly along the sequence. The mass of an O star is perhaps 100 times that of an M star, whereas the luminosity is greater by a factor of 10^9 or more.

STELLAR COMPOSITION

The next property we should consider is the chemical makeup of a star. We have already learned that this does not vary much from one star to another; that the major observed differences among the spectral types are a result of temperature effects. It is important, however, to be able to measure the elemental abundances in stars, both to see what they are made of and to understand the variations that exist.

We learned previously that each chemical element has its own set of spectral lines. It would therefore seem simple to determine the composition of a star by seeing which elements are represented in its spectrum. This is complicated, however, by the fact that the temperature and density, which control the ionization and excitation, have a very profound, dominant influence on the spectrum.

To measure chemical abundances, therefore, requires a more subtle analysis. It is necessary to know the temperature and density fairly well to begin with, so that these effects may be taken into account, and then to analyze the strengths of the spectral lines to see what the relative abundances are. This process is carried out most accurately by using complex computer programs that can calculate a simulated spec-

TABLE 20.2 PROPERTIES OF MAIN SEQUENCE STARS*

Spectral Type	Mass ($M_\odot$)	Temperature	$B-V$	Luminosity ($L_\odot$)	M_V	BC	Radius ($R_\odot$)
O5V	40	40,000	−0.35	5×10^5	−5.8	−4.0	18
B0V	18	28,000	−0.31	2×10^4	−4.1	−2.8	7.4
B5V	6.5	15,500	−0.16	800	−1.1	−1.5	3.8
A0V	3.2	9,900	0.00	80	+0.7	−0.4	2.5
A5V	2.1	8,500	+0.13	20	+2.0	−0.12	1.7
F0V	1.7	7,400	+0.27	6	+2.6	−0.06	1.4
F5V	1.3	6,580	+0.42	2.5	+3.4	0.00	1.2
G0V	1.1	6,030	+0.58	1.3	+4.4	−0.03	1.1
G5V	0.9	5,520	+0.70	0.8	+5.1	−0.07	0.9
K0V	0.8	4,900	+0.89	0.4	+5.9	−0.19	0.8
K5V	0.7	4,130	+1.18	0.2	+7.3	−0.60	0.7
M0V	0.5	3,480	+1.45	0.03	+9.0	−1.19	0.6
M5V	0.2	2,800	+1.63	0.008	+11.8	−2.3	0.3

*Masses are in units of $M_\odot$, the solar mass (whose value is 1.99×10^{33} gm); luminosities are in units of $L_\odot$, the solar luminosity (having a value of 3.83×10^{33} erg/sec), and radii are in units of $R_\odot$, the solar radius (whose value is 6.96×10^{10} cm). The temperatures are **effective temperatures,** which are a measure of surface temperature. The heading $B-V$ stands for the color index, M_V stands for visual absolute magnitude, and BC stands for bolometric correction.

trum for comparison with the observed spectrum (Fig. 20.13). In the calculation of the theoretical spectrum, the temperature, density, and chemical composition are all varied until the best match with the observed spectrum is found. This is a costly and time-consuming process, and is not done in such detail for many stars.

In general terms, the result of stellar composition studies shows that most stars consist almost entirely of hydrogen and helium, with everything else constituting, at most, a small percentage of the total mass. We have already learned that the Sun and indeed the entire solar system were formed out of material with similar composition; we now find this to be a nearly universal effect. This provides an important clue regarding how the universe itself formed, as we shall see in Section V.

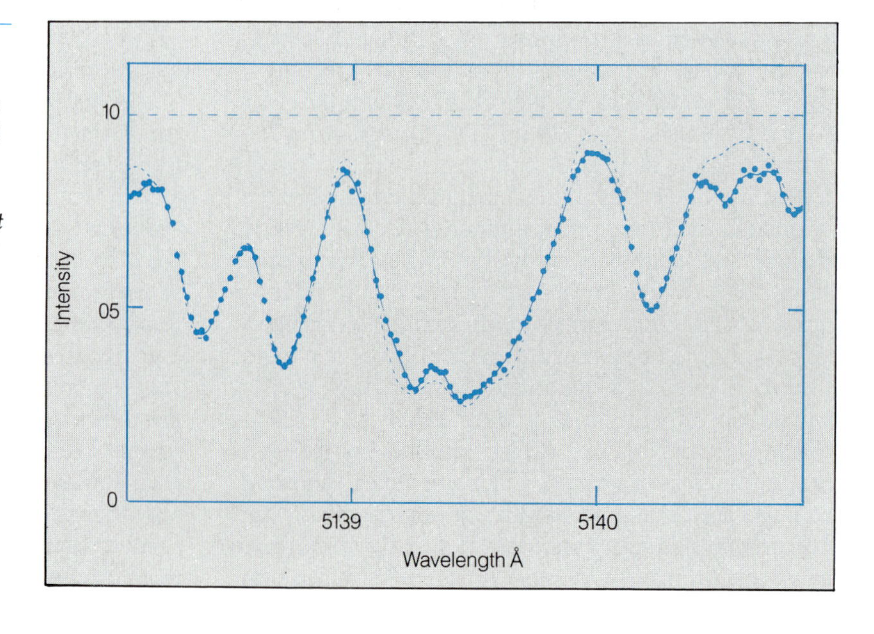

FIGURE 20.13 DETERMINATION OF STELLAR COMPOSITION. *This diagram illustrates a modern technique for measuring the chemical composition of a star. A theoretical spectrum (solid line) is compared with the observed spectrum (dotted and dashed line). The abundances of elements assumed present are varied in the computed spectrum until a good match is achieved.* (D. L. Lambert)

MAGNETIC FIELD

An often overlooked, but undeniably fundamental, property of a star is its magnetic field. We expect that stars should have fields; after all, the Sun and many of the planets do, and we suspect that magnetism in these bodies originates in fluid motions in their interiors. There is every reason to expect the same processes to be at work in other stars.

To determine whether a distant star has a magnetic field, and furthermore to measure its strength, is very difficult. It was found some years ago that some spectral lines of certain elements are split into two or more closely spaced lines in the presence of a magnetic field (Fig. 20.14). In a process called **Zeeman splitting,** electromagnetic forces split certain energy levels of an atom, so that an electron can be in one or two or more different levels, whereas there would only be one if there were no magnetic field. Thus a spectral line involving one of these split levels is itself split into two lines because the electron that absorbs a photon of light may be in either level. The amount of splitting is determined by the strength of the magnetic field, so in principle it is possible to determine the field strength by measuring how widely the lines are split.

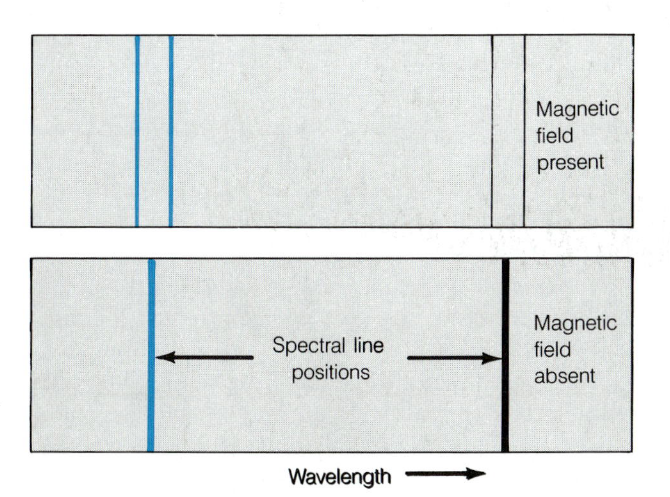

FIGURE 20.14 THE MEASUREMENT OF A MAGNETIC FIELD. *The presence of a magnetic field splits certain absorption lines of some elements, in a process called Zeeman splitting. The amount of separation is a measure of the strength of the magnetic field.*

In practice, such measurements are very difficult, however, for a number of reasons. The splitting of spectral lines is normally very small, so very precise measurements are needed. Furthermore, stellar rotation or gas motions in the outer layers can cause the spectral lines to be broadened by the Doppler effect, so that the separate lines are merged into one, and their separation cannot be measured. Because of these difficulties, we do not have much information on magnetic-field strengths in typical stars. The fields have been measured in a number of stars having especially strong ones, and values up to hundreds or thousands of times greater than the Sun's magnetic-field strength have been found.

The lack of more complete data on magnetic-field strengths in normal stars is frustrating, because it is likely that the fields play important roles in stellar formation and evolution.

PERSPECTIVE

We now know how all the basic properties of stars are derived from observational data. We can categorize stars, classify them, and describe them in any way we wish. Now we are ready to see how they work and why the various quantities are related the way they are and not in any other way.

SUMMARY

1. We determine stellar luminosity through knowledge of the distance to a star and the star's apparent magnitude. The absolute magnitude, a measure of luminosity, is the magnitude a star would have if it were seen from a distance of 10 parsecs.
2. Stellar temperatures can be inferred from the $B-V$ color index, estimated from the spectral class, or determined from the degree of ionization in the star's outer layers.
3. The Hertzsprung-Russell diagram shows that the luminosities and temperatures of stars are closely related, and that stars which do not fall on the main sequence are either larger (as with red giants) or smaller (white dwarfs) than those on the main sequence.
4. We can measure the distance to a star by first determining the star's spectral class, then using the

H-R diagram to infer its absolute magnitude, and finally comparing the absolute magnitude with its apparent magnitude to yield the distance. This technique is called spectroscopic parallax.
5. We can determine stellar diameters directly in eclipsing binaries through knowledge of the orbital speed and the duration of the eclipses.
6. The rotational speeds and periods of stars are determined from the Doppler-shifting of the spectral lines, which causes the lines to be broadened. However, these speeds are almost always ambiguous because of the unknown orientation of the rotation axis.
7. We can derive stellar masses in binary systems through the use of Kepler's third law, when the period and the orbital semimajor axis are known from observations (which is possible only if the distance is known).
8. The chemical composition of a star can be deduced from the strengths of the absorption lines in its spectrum, and is most commonly and accurately done by comparison with theoretically computed spectra.
9. The magnetic-field strengths of stars are probably important in their structure and evolution, and in some cases can be inferred from the Zeeman splitting of spectral lines. However, these strengths are not easily measured for most stars.

REVIEW QUESTIONS

Thought Questions

1. Why is it difficult to directly measure a star's brightness at all wavelengths?
2. Explain the relationship between the luminosity of a star and its absolute magnitude. Discuss both absolute visual magnitudes and absolute bolometric magnitudes.
3. Summarize all the methods that have been discussed so far for determining the surface temperatures of stars.
4. Explain the significance of the fact that stars in an H-R diagram (representing luminosity versus surface temperature) are not scattered randomly, but instead fall into well-defined regions.
5. Explain why a G2 star that lies above the main sequence must have a larger diameter than a G2 star that lies on the main sequence.
6. Explain the spectroscopic parallax method for finding the distances to stars. Why do you think the word "parallax" is used?
7. Discuss the role of standard candles in determining distances to astronomical objects.
8. Summarize the methods by which stellar diameters are measured.
9. Could Kepler's third law, in the form discovered by Kepler himself, have been used to determine the masses of stars? Explain.
10. Describe how the chemical composition of a star is determined. Explain how part of the method depends on observations, and part of it depends on theoretical calculations.

Problems

1. Determine the absolute magnitude of the following stars: (a) a star 1,000 *pc* from the Sun with apparent magnitude $m = 12.3$; (b) a star 10,000 *pc* from the Sun with $m = 12.3$; (c) a star 10 *pc* away with $m = 6.8$; (d) a star 1 pc away with $m = 2.1$.
2. If a star has absolute bolometric magnitude $M_{bol} = 1.5$, and another star has $M_{bol} = 4.5$, which is more luminous, and by how much?
3. If one star is four times hotter than another, and has twice as large a radius, which is more luminous, and by how much?
4. Using the spectroscopic parallax method, find the distances to the following stars: (a) a star with apparent magnitude $m = 7.2$ and absolute magnitude $M = -2.2$; (b) a star with $m = 18.6$ and $M = 8.6$; (c) a star with $m = 9.0$ and $M = 9.0$; and (d) a star with $m = 5.8$ and $M = -4.2$.
5. What is the rotational speed of a star whose 4500.9 Å line is broadened by the Doppler effect to a width of 0.90 Å?
6. What is the sum of the masses in a binary system whose orbital period is 1 year and whose semimajor axis is 2 AU? What are the individual masses of the two stars if the center of mass between them is one-fourth of the way from the more massive star (star A) to the less massive one (star B)?

Additional Readings

Abt, H. A. 1977. The companions of sun-like stars. *Scientific American* 236(4):96.

De Vorkin, D. 1978. Steps towards the Hertzsprung-Russell diagram. *Physics Today* 31(3):32.

Evans, D. S., T. G. Barnes, and C. H. Lacy. 1979. Measuring diameters of stars. *Sky and Telescope* 58(2):130.

Getts, Judy. 1983. Decoding the Hertzsprung-Russell diagram. *Astronomy* 11(10):16.

Philip, A. G. D., and L. C. Green. 1978. The H-R diagram as an astronomical tool. *Sky and Telescope* 55(5):395.

CHAPTER 21

STELLAR STRUCTURE: WHAT MAKES A STAR RUN?

A computer simulation of stellar pulsations. (European Southern Observatory, D. Baade)

CHAPTER PREVIEW

We have learned how astronomers determine from observations the physical characteristics of stars. From this we have learned a great deal about their surface properties, but very little about what goes on inside them, or how they evolve. To understand these secrets, it is necessary to apply the laws of physics and calculate theoretically the internal structure and evolution, requiring that the calculations reproduce the observable properties. If this is done successfully, then we can have some confidence that the calculations also accurately describe the internal conditions. We have already seen what can be learned, in our discussions of the Sun in Chapter 18. In this chapter, we will rely on our previous discussions, for it is possible to apply the same physical principles to other stars that were useful for the Sun.

WHAT IS A STAR, ANYWAY?

As we found for the Sun, a star is a spherical ball of hot gas. The outer layers are partially ionized, and the interior, where the temperature and pressure are much higher, are fully ionized, so that the gas there consists of bare atomic nuclei and free electrons. Continuous radiation is generated at the core of a star, and from there photons of light make their way slowly out to the surface, where they are emitted into space from a layer called the **photosphere,** just as in the Sun. Absorption lines are formed in the photosphere, so that the spectra of normal stars are absorption-line spectra, just as the Sun's is. The interior regions of a star are very dense because of gravitational compression, but they are still gaseous because of the high temperature. Temperatures range from 10 to 100 million degrees absolute in the cores of stars.

Like the Sun, most stars are made primarily of hydrogen, although, as we will see, this changes gradually over a star's lifetime because of nuclear reactions. (Recall that in the Sun's core, hydrogen is gradually being converted into helium.)

The internal structure of a star, like that of the Sun, is governed by the balance between gravity and pressure that we call **hydrostatic equilibrium.** The gas particles all exert gravitational forces on each other, so that the entire star is held together by its own gravity. This force is always directed towards the center; thus, the star is forced into a symmetric, spherical shape. The fact that gas is compressible explains how gravity creates a state of high density inside.

A gas that is compressed heats up, causing it to exert greater pressure on its surroundings. Thus, a star's interior is very hot and the internal pressure is high. If it were not for this pressure, gravity would cause a star to keep shrinking. A balance is struck between gravity, which is always trying to squeeze a star inward and pressure, which pushes outward (Fig. 21.1). This balance, which can be stable for long periods of time only if there is a source of energy inside the star, plays a dominant role in determining the internal structure. Whenever the balance is disturbed, as it is during some phases of a star's lifetime, the internal structure changes (this is discussed in the next chapter).

It is the mass of a star that causes a star to reach a certain state of internal compression and no other. The amount of gravitational force is set by the total mass, and this force in turn determines how much pressure is needed to balance gravity. The star will be compressed until this pressure is reached, and then it will become stable. The pressure required to balance gravity dictates the temperature inside the star. Hence a star's mass determines the internal density and temperature as well as the overall size of the star, since this is a function of the mass and the density.

Luminosity is also governed largely by the mass. The luminosity of a star is simply the amount of energy per second generated inside it that eventually reaches the surface and escapes into space, and it is determined by the temperature in the interior. As we have just seen, the temperature is set primarily by the

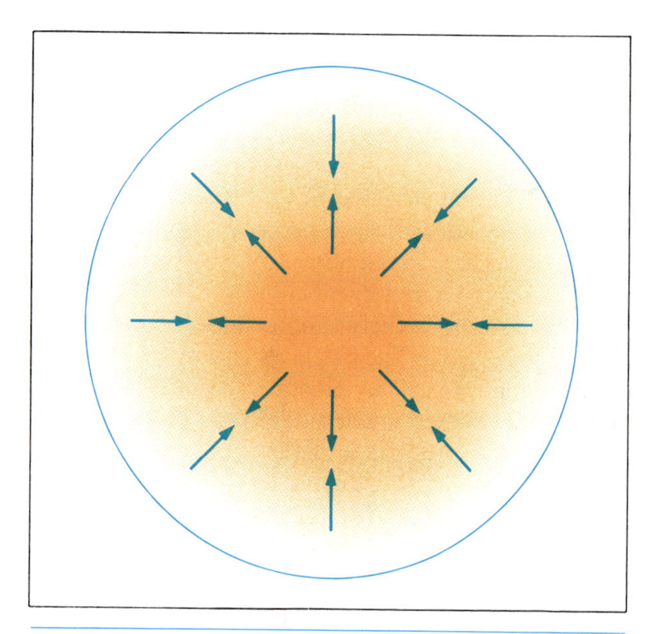

FIGURE 21.1 HYDROSTATIC EQUILIBRIUM. *This cutaway sketch of a star shows its spherical shape. The arrows represent the balanced inward forces of gravity and outward pressure, and the shading indicates that the density increases greatly toward the center. Equilibrium is reached when the core becomes sufficiently hot to attain the pressure necessary to counterbalance gravity.*

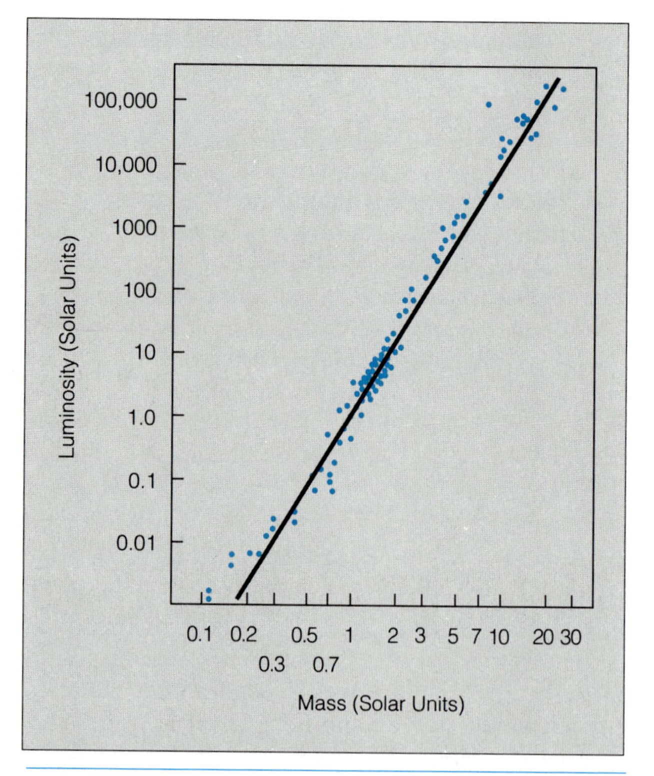

FIGURE 21.2 THE MASS-LUMINOSITY RELATION. *This diagram, based on observed luminosities and masses for stars in the main sequence, shows how well the two quantities correlate with each other.*

mass. The dependence of luminosity on mass is demonstrated by making a graph of luminosity versus mass (Fig. 21.2). This is the **mass-luminosity relation,** and it was discovered observationally before there was a full theoretical understanding of it.

Thus, virtually all the observed properties of stars depend on their masses (Fig. 21.3). This explains why the main sequence represents a smooth run of masses increasing from the lower right to the upper left in the H-R diagram: The luminosity and temperature, which define a star's place in the diagram, both depend on mass.

Of course, there are stars that do not fall on the main sequence, yet their masses are not different from those main-sequence stars. This tells us that there must be something other than mass that can influence a star's properties. This other parameter is the chemical composition. As we shall see, this varies as a star ages. The supergiants, giants, and white dwarfs have different chemical makeups than main-sequence stars. In the next section we will see how these differences arise.

The amount of internal compression, and hence the temperature and luminosity of a star, depend on the average mass of the individual nuclei in the star's core. The heavier the particles that make up the gas, the more tightly they are compressed by gravity, the hotter it gets, and the greater the luminosity. When a star is formed, it consists mostly of hydrogen, but as it ages its core material is converted to helium at first, and later possibly to other, even heavier elements. In the process, the core heats up and the star becomes more luminous.

The fact that all properties of a star depend on just its mass and composition was recognized several decades ago and is usually referred to as the **Russell-Vogt Theorem,** after the astrophysicists who first stated it. In modern times, it is known that other influences, such as magnetic fields and rotation, must play roles in governing a star's properties as well. Just what these roles are is not yet well understood, however.

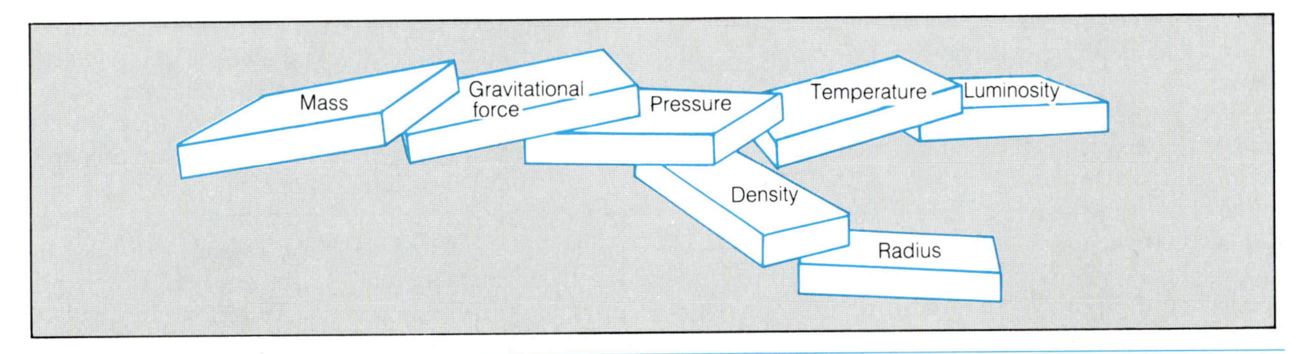

FIGURE 21.3 THE IMPORTANCE OF MASS. *A star's mass is the single quantity that governs all its other properties for a given composition. The sequence shown here is the same in general, but the details vary for different chemical compositions.*

NUCLEAR REACTIONS AND ENERGY TRANSPORT

We have seen that the core of a star must be very hot, in order to maintain the pressure required to counterbalance gravity. Since energy flows outward from the core, there must be some source of heat in the interior. If there were not, then the star would gradually shrink.

Nuclear fusion reactions provide the only source of heat capable of maintaining the required temperature over a sufficiently long period of time. A particular type of reaction was described in Chapter 18, where we discussed the Sun's interior. Recall that in atomic fusion reactions, nuclei of light elements such as hydrogen combine to form nuclei of heavier elements. In the process, a small fraction of the mass is converted into energy according to Einstein's famous equation $E = mc^2$. It is this energy, in the form of heat and radiation, that maintains the internal pressure in a star.

Fusion reactions can occur only under conditions of extremely high temperature and density. Protons all carry positive electrical charges and, therefore, repel each other. The force (called the **strong nuclear force**) that holds the protons and neutrons together in a nucleus and which causes fusion to occur, can overcome this repulsive electric force only over very short distances. Therefore the nuclei must collide at very high velocities (Fig. 21.4) in order to combine, so that they can get close enough together despite their electromagnetic repulsion for each other. The high temperature of a stellar core imparts high speeds to the nuclei, so that they collide with great energy, and the high density means that collisions will be frequent. Even so, only occasionally do two nuclei combine in a fusion reaction; a single particle may typically collide and bounce around inside a star for millions or even billions of years before it reacts with another. There are so many particles, however, that reactions are continually occurring despite this.

The amount of energy released in a single reaction between two particles is small, about 10^{-5} erg. An erg (defined in Chapter 5) is itself a small quantity; a 100-watt light bulb radiates 10^9 ergs per second. Hence, a light bulb would require 10^{14} reactions per second to keep glowing, if nuclear reactions were its energy source. A star like the Sun emits more than

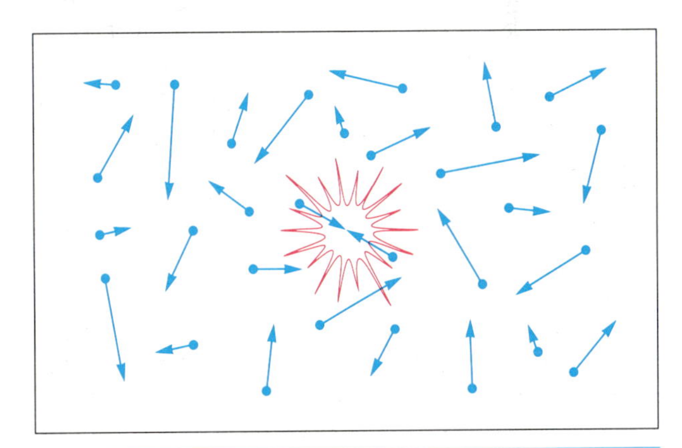

FIGURE 21.4 NUCLEAR FUSION. *The core of a star is a sea of rapidly moving atomic nuclei. Occasionally a pair of these nuclei, most of which are simple protons (hydrogen nuclei), collides with sufficient velocity to merge and form a new kind of nucleus (deuterium in this case, composed of one proton and one neutron). Energy is released in the process.*

10^{33} ergs per second, so a tremendously large number of reactions must be occurring in its interior at all times.

In Chapter 18 we learned that the reaction that powers the Sun is the proton-proton chain, in which hydrogen is converted into helium, with 0.007 of the initial mass being converted into energy. The net result of this reaction is the combination of four hydrogen nuclei into one helium nucleus; two of the protons must be converted into neutrons in the process. There is another reaction sequence, called the **CNO cycle** (see Appendix 11), that has the same net effect, but operates primarily in stars hotter and more massive than the Sun. *All* stars on the main sequence produce their energy by one of the two processes that convert hydrogen into helium in their cores; the specific reaction that is dominant depends on the internal temperature, which in turn depends on the star's mass. The point on the main sequence that divides proton-proton chain stars from CNO stars corresponds roughly to spectral class F0; stars above that point produce most of their energy by the CNO cycle, and those below that point do so by the proton-proton chain.

Once it is produced in the core, energy must somehow make its way outward to the stellar surface. For most stars, two different energy-transport mechanisms are at work at different levels (Fig. 21.5). One of these is **radiative transport,** meaning that the energy is carried outward by photons of light. In the core, where it is very hot, the photons are primarily γ-rays, but as they slowly move outward, being continually absorbed and reemitted on the way, they are gradually converted to longer wavelengths. When the light emerges from the stellar surface, it is primarily in the visible wavelength region (or the ultraviolet or infrared, if the star is very hot or very cool). Radiative transport is a very slow process, as we learned in the case of the Sun; there it takes about a million years for a single photon to make its way from the core to the surface.

The second means of energy transport inside stars is **convection,** a process already discussed briefly in Chapter 18. Convection is an overturning of the gas, as heated material rises and cooled material sinks. The same process causes warm air to rise towards the ceiling of a room and is responsible for the overturning of water in a pot that is being heated from the bottom. When conditions are right for convection to occur, it is much more rapid and efficient than radiative transport. Energy travels outward much faster in con-

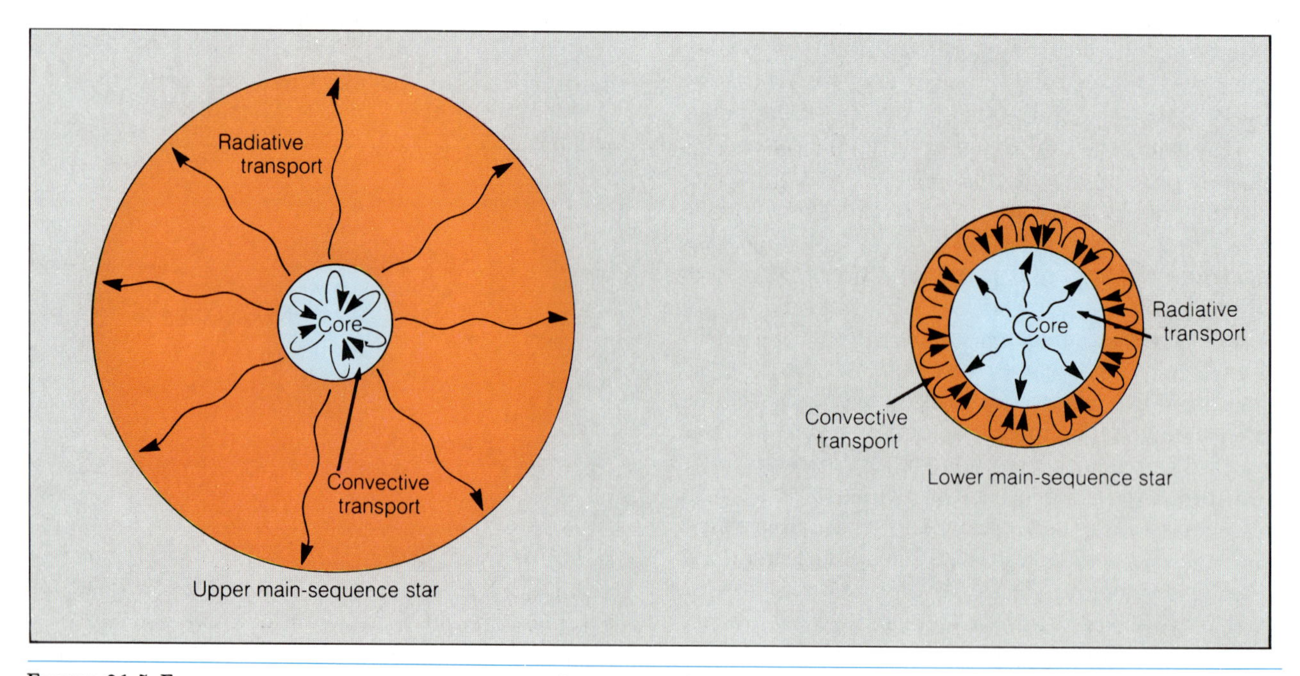

FIGURE 21.5 ENERGY TRANSPORT INSIDE OF STARS. *In a star on the upper portion of the main sequence, energy is transported by convection in the inner zone and radiation in the outer regions. The opposite is true of lower main sequence stars.*

vective zones than in zones where radiation is the chief means of energy transport.

Whether radiative transport or convection will be the dominant energy-transport mechanism in a star depends on the temperature structure (specifically, on how rapidly the temperature decreases with distance from the center). Model calculations show that most stars have both a convective zone and a radiative region. In stars like the Sun (i.e., for stars on the lower half of the main sequence, of spectral types F, G, K, and M), convection occurs in the outer layers, while radiative transport is the principal means of energy transport in the interior. For stars on the upper main sequence, the situation is reversed: There is convection in the central core but not in the rest of the star, where radiative transport is responsible for conveying the energy to the surface.

STELLAR LIFE EXPECTANCIES

The lifetime of a star is measured by the amount of energy it can produce in nuclear reactions and the rate at which the energy is radiated away into space (Fig. 21.6). When all the available nuclear fuel has been used up, the star undergoes major changes in its structure and properties, and it leaves the main sequence (this will be discussed in detail in the next chapter). In Chapter 18 we found that the Sun will have a life expectancy of about 10 billion years, assuming that 10 percent of its mass undergoes reactions, and that 0.007 of that mass is converted into energy. For other stars in which the CNO cycle is the dominant reaction, the calculation is made the same way, and the fraction of the initial mass converted into energy is the same. The calculation involves determining the total amount of energy available to the star during its lifetime (derived from the formulae $E = mc^2$), and dividing this by the luminosity, which is the rate at which energy is expended.

We can apply similar calculations to other stars (see Table 21.1). A star near the top of the main sequence, for example, may have 50 times the mass of the Sun, but at the same time it uses its energy as much as 10^6 times more rapidly. (Recall how rapidly the luminosity of a star varies with its mass, as discussed in the last chapter.) These numbers lead to an estimated lifetime that is $50/10^6 = 5 \times 10^{-5}$ times that of the Sun. This star will last only half a million years! (Actually, in such a massive star, more than 10 percent of the total mass can undergo reactions, and the lifetime is accordingly longer than this simple estimate; it is a few million years.) By astronomical standards, such massive stars exist for only an instant before using up all their fuel and dying. This is one reason why stars of this type are very rare; only a few may be around at any given moment, regardless of how many may have formed in the past (as it

FIGURE 21.6 STELLAR LIFETIMES. *Even though the more massive star (top) has a greater supply of nuclear fuel, it loses energy so much more rapidly than the lower-mass star (bottom), that its lifetime is far shorter.*

TABLE 21.1 MAIN-SEQUENCE LIFETIMES

Spectral Type	Hydrogen-Burning Lifetime (years)
O5V	5×10^5
B0V	9×10^5
B5V	8×10^7
A0V	4×10^8
A5V	1×10^9
F0V	3×10^9
F5V	5×10^9
G0V	9×10^9
G5V	1×10^{10}
K0V	2×10^{10}
K5V	4×10^{10}
M0V	2×10^{11}
M5V	3×10^{11}

happens, very massive stars also do not form very often, another reason for their rarity).

Now let us consider a lower main-sequence star, for example an M star with mass 0.05 solar masses, and luminosity 10^{-4} times that of the Sun. Its lifetime is $0.05/10^{-4} = 500$ times *greater* than the Sun's lifetime, or about 5×10^{12} years. This is a very long time, greater than the age of the universe. No star ever formed with such a low mass has yet had time to use up all of its fuel; all such stars born are still on the main sequence. For this reason, the vast majority of all stars in existence today are low-mass stars (it is also true that these stars form in greater numbers, adding further to their dominance in numbers).

Heavy Element Enrichment

Clearly the nuclear reactions have an effect on the chemical composition of a star, because they change one element into another. We have stressed that all stars consist of about the same mixture of hydrogen and helium, with a trace amount of other elements, but now we find that, in the core of a star, hydrogen is gradually converted into helium. While this may not immediately affect the surface composition, which is what astronomers can measure directly from spectral analysis, it does change the internal composition. It is this change that causes a star to evolve, because the core density is altered as hydrogen nuclei combine into the heavier helium nuclei, and because when the hydrogen runs out, the star must make major structural adjustments as its primary source of internal pressure disappears.

These adjustments are discussed in Chapter 23; for now, let us consider only the change in abundances of the elements. There are other nuclear reactions that can take place in stars, after the hydrogen-burning stage has ended, if the core temperature reaches sufficiently high levels. One such reaction is the conversion of helium into carbon, by a sequence called the **triple-alpha reaction** (see Appendix 11; helium nuclei are called alpha-particles, and in this reaction, three of these combine to form one carbon nucleus). When the triple-alpha reaction takes place, the core of a star changes its composition from helium to carbon.

Many additional reactions can occur, if in later stages the core temperature goes even higher (Fig. 21.7). These reactions produce ever-heavier products, so that as a star goes through successive stages of nuclear burning, its internal composition changes from what had originally been predominantly hydrogen to elements as heavy as iron (which has twenty-six protons and thirty neutrons in its nucleus).

These reactions require ever-higher temperatures, because as the particles become more massive, it takes more energy to keep them moving fast enough to react. Furthermore, the heavier nuclei have greater electrical charges and therefore stronger repulsive forces tending to keep them apart. For both of these reasons, reactions involving the fusion of heavy nuclei require higher temperatures than those in which lightweight nuclei such as hydrogen undergo fusion.

It is the most massive stars that go through the greatest number of reaction stages and, as we have just seen, these stars live only a short time. In later chapters we will discuss the role played by these massive stars in the enrichment of chemical abundances in the galaxy.

Stellar Chromospheres and Coronae

Let us now consider other processes that may affect a star's structure and evolution. Some very important ones take place near the surface, rather than deep inside.

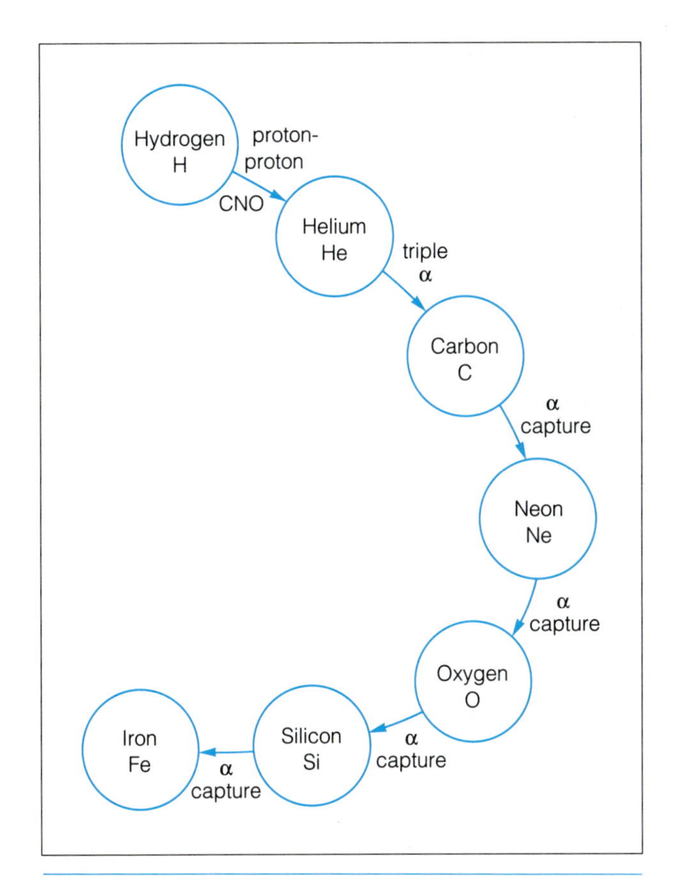

FIGURE 21.7 THE ENRICHMENT OF HEAVY ELEMENTS. *As a star ages, each time one nuclear fuel is exhausted another ignites if the temperature in the core rises sufficiently (which depends on the mass of the star). The net result is an enrichment of heavy elements.*

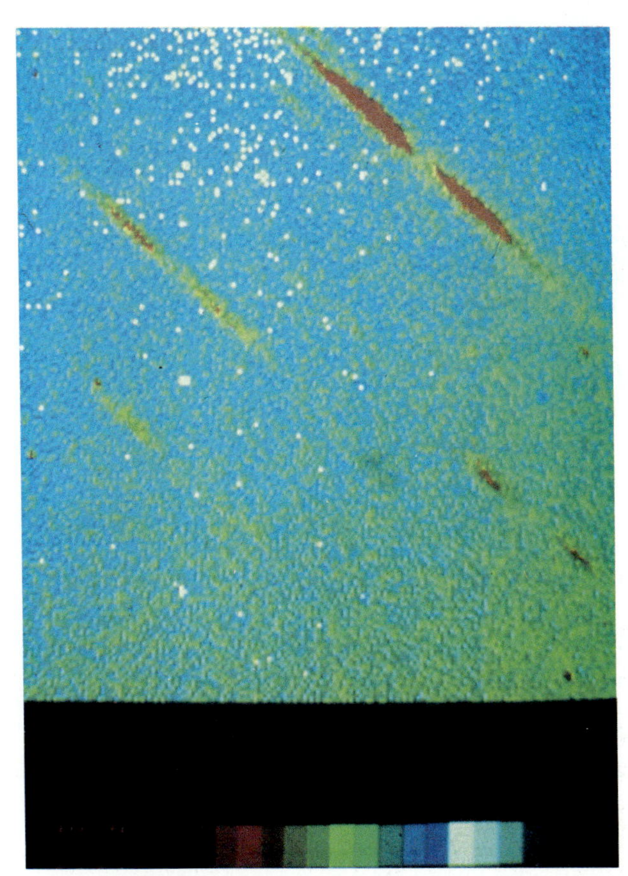

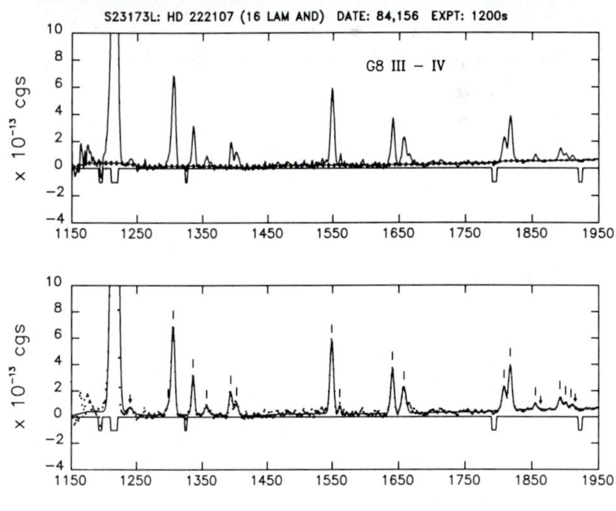

FIGURE 21.8 ULTRAVIOLET EMISSION LINES IN A COOL STAR. *At top is a spectrum obtained by the* International Ultraviolet Explorer, *showing, in several segments of ultraviolet spectrum, emission lines (red features). At bottom is a plot showing the intensity of these emission lines (in a different star). (top:* NASA; *bottom:* J. Bennett)

As we have seen, in most stars convection occurs at some level inside. For stars on the upper portion of the main sequence, convection takes place near the center and has no directly observable consequences. For stars on the lower portion of the main sequence, however, where convection occurs in the outer layers, there are important effects. Stars of spectral types F, G, K, and M, where surface convection is thought to occur, also are found to have chromospheres and coronae. (Recall from Chapter 18 that chromospheres and coronae produce ultraviolet emission lines and X-ray emission, and these signatures are generally found in *all* cool stars: Figs. 21.8 and 21.9.)

The presence of chromospheres and coronae seems to be linked with the presence of convection in the

ASTRONOMICAL INSIGHT 21.1

STELLAR PULSATIONS AND MASS LOSS

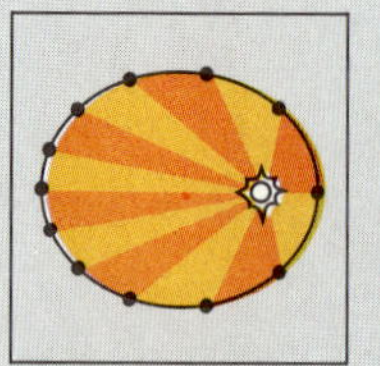

It is noted in the text that stars in the upper portion of the H-R diagram all lose mass through stellar winds, but that the nature of the winds in the hot stars is quite different from that in the cool stars. The hot stars have very rapid winds, with velocities up to several thousand kilometers per second, whereas the cool giants and supergiants have slower winds, with speeds of perhaps 10 to 20 km/sec. The rate at which mass is lost can be comparable, however, because the cool-star winds are much denser. It has generally been assumed that, since the winds in the two types of stars are so different, the causes of the winds must also be different. Recent evidence suggests, to the contrary, that the basic mechanisms may actually be very similar.

We have learned a little about pulsating variable stars, such as the Cepheid variables and the RR Lyrae stars. These are stars that physically expand and contract, growing brighter and dimmer in the process. Other kinds of pulsating variable stars have also been known to exist, such as the so-called long-period variables, red supergiants that take as long as 1 or 2 years to go through a single cycle of expansion and contraction. It happens that these stars are the ones in the upper right-hand portion of the H-R diagram that have the strongest stellar winds, suggesting a connection between the pulsations and the loss of mass. Specifically, it has been suggested that every time such a star expands to its largest size and then begins to contract, some of the gas from its outer layers continues to drift outward, because the surface gravity of such an extended star is very low. It is known from infrared observations that the escaping matter can condense to form tiny, solid grains of material. Photons of light from the star (which is very luminous, and can provide many photons) then push the grains outward, creating a steady flow. The outward force exerted on the grains by photons from the star is called **radiation pressure.**

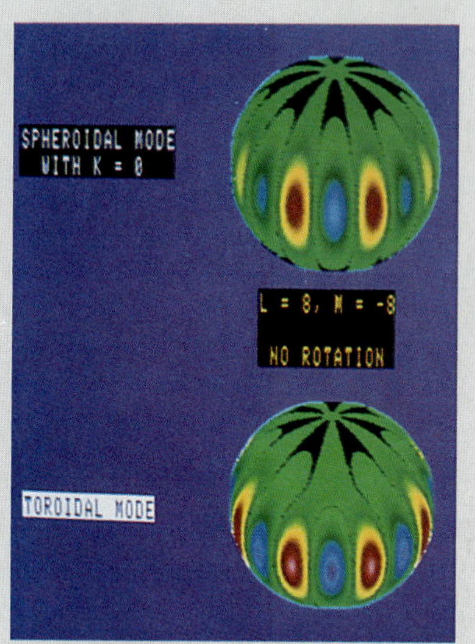

European Southern Observatory, D. Baade

The idea that radiation pressure plays a role in creating the winds from cool stars is not new; what has not been well understood is the mechanism for getting the outflow started, and it now appears that stellar pulsations may provide the answer. Similarly, it has been known for some time that the very high velocities reached by winds from hot luminous stars are due to radiation pressure acting on gas atoms and ions, but there has been great difficulty in discovering how the outflow gets started in the first place, so that radiation pressure can then take over and provide the acceleration. Now it has been found that many hot stars pulsate in a very subtle way, and that these pulsations may provide the missing mechanism for initiating hot star winds, just as the larger pulsations of cool stars may do for their winds.

A star can pulsate **nonradially,** meaning that it does not simply expand and contract as a whole, but instead that its surface consists of separate regions that alternately rise and fall. A very simple example would be a star that alternately is elongated (a bit like a cigar) and squashed. Much more complex "modes" of pulsation are possible, however, with the result that the surface of a star may have many zones that rise and fall over small distances, but with high frequency. The causes of such pulsations are complex, having to do with internal rotation (see Astronomical Insight 18.2, "Measuring the Sun's Pulse").

It has been suggested very recently that the small velocities in rising regions on a star's surface may be sufficient to allow radiation pressure to act on that gas, accelerating it to very high velocities and creating the observed winds. The general phenomenon has been dubbed "mass-tossing," and has striking overall similarities to the mechanism described here for the winds in cool stars. Some details are quite different, such as the type of stellar pulsation, the velocity, and the types of particles that are pushed by radiation pressure, but the general mechanism is surprisingly similar. Detailed theoretical and observational studies of stellar pulsations and their possible relationship to stellar winds are among the most active areas of research in stellar structure today.

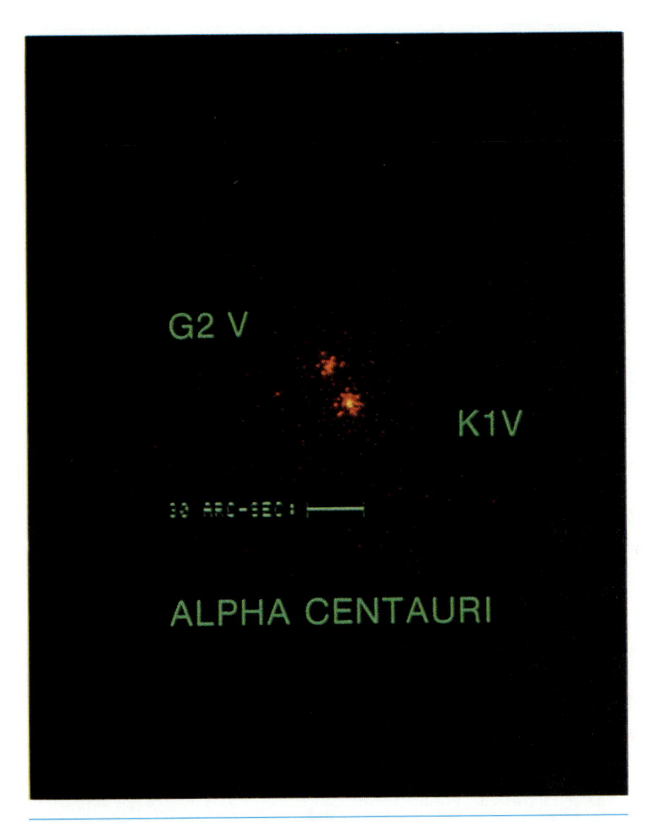

FIGURE 21.9 AN X-RAY IMAGE OF COOL STARS. *This image, obtained by the* Einstein Observatory's *X-ray telescope, shows our nearest neighbor,* α *Centauri. This is a multiple-star system, and here we see that two of the stars, a G star and a K star, emit X-rays from their coronae.* (Harvard-Smithsonian Center for Astrophysics)

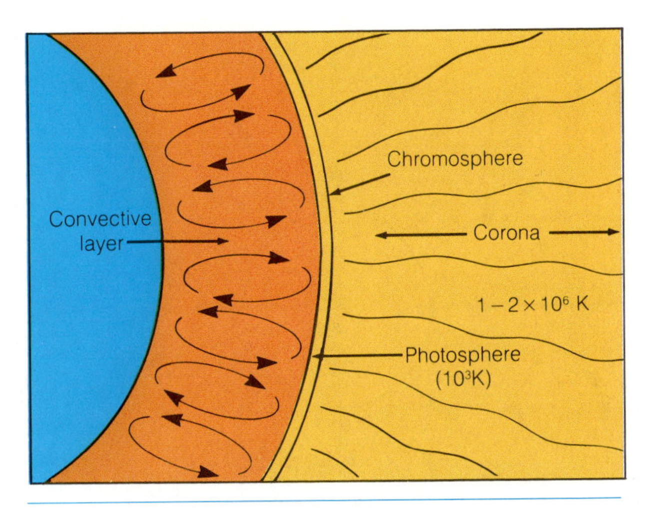

FIGURE 21.10 A STELLAR CHROMOSPHERE AND CORONA. *Stars in the lower portion of the main sequence all have chromospheres and coronae, analogous to the Sun (see Chapter 18 for more details). The photosphere is the region in which the star's continuous spectrum and absorption lines are formed; the chromosphere is a thin, somewhat hotter region just above; and the corona is a very hot, extended region outside of that. The source of heat for the chromosphere and corona is probably related to convective motions in the star's outer layers.*

outer layers (Fig. 21.10). Stars on the lower portion of the main sequence generally have both phenomena. Somehow, the kinetic energy of the turbulent motions in the convective layer is transported into higher zones, where it causes heating. As we learned in our discussion of the Sun, the precise mechanism for converting the energy of convection into heat is not understood.

What of the red giants and supergiants? These stars can have surface temperatures comparable to the lower main-sequence stars, but obviously their internal structure is rather different, since they are so much larger. These stars also have convection in their outer layers, however, and indeed they also have chromospheres, although it is not certain that the extremely high temperatures characteristic of coronae are present.

STELLAR WINDS AND MASS LOSS

So far we have said nothing about the possibility that the hot stars might have chromospheres or coronae. Some of the upper main-sequence stars are known to have emission lines in their spectra, indicating that they have some hot gas above their surfaces. Visible-wavelength emission lines are found in only a few extremely hot or luminous stars, however.

In the late 1960s, with the first observations of ultraviolet spectra, it was immediately found that many hot stars (nearly all the O stars and many of the hotter B stars) have ultraviolet emission lines. Furthermore, there are absorption features that show enormous Doppler shifts, indicating that the stars are ejecting material at speeds as great as 3,000 kilometers per second! The most extreme **stellar winds,** as these outflows are called (Fig. 21.11), occur in the O stars, but most cool giants and supergiants also have winds, usually with lower speeds.

The cause of the winds is not known, although there are indications of how the high velocities are reached. Once gas begins to move outward, light from

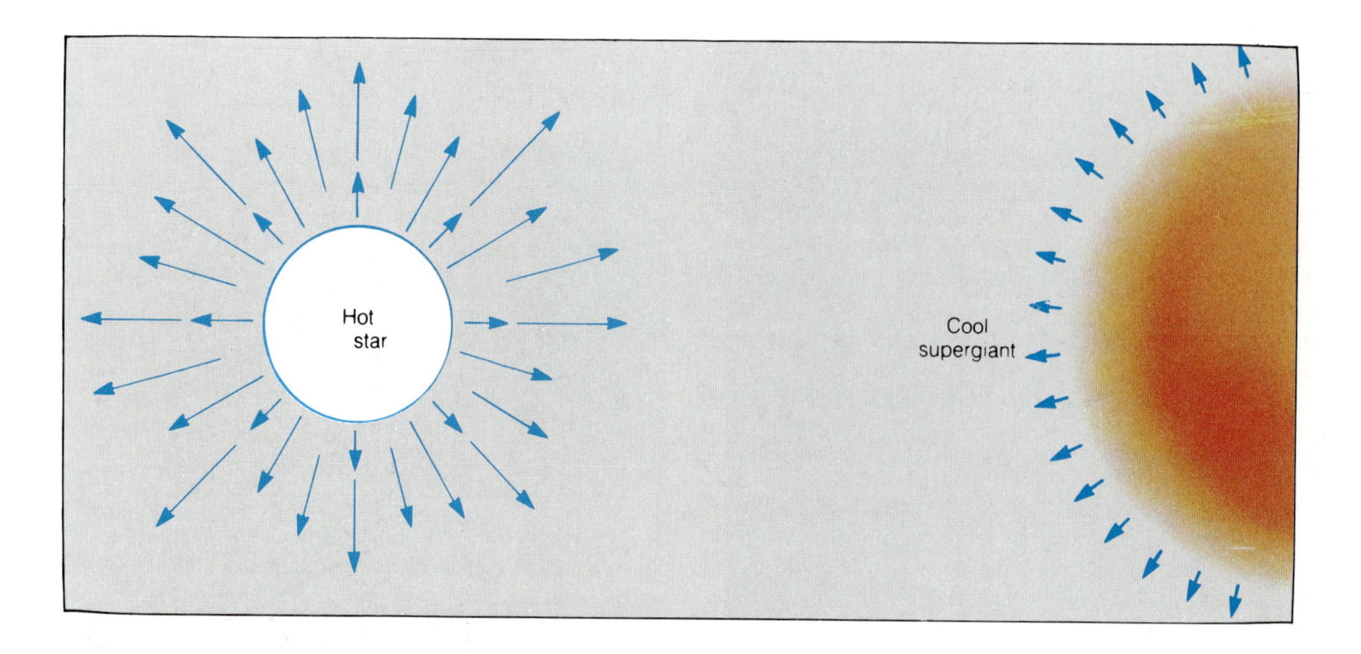

FIGURE 21.11 STELLAR WINDS. *In the drawing at top a luminous hot star (left) ejects gas at a very high velocity; the lengths of the arrows indicate that the gas accelerates as it moves away from the star. A luminous cool star (especially a K or M supergiant) is so extended in size that the surface gravity is very low, and material drifts away at relatively low speeds (right). In both types of stars, radiation pressure probably helps accelerate the gas outward. At bottom is a photograph of a hot star with an extensive bubble of gas around it, created by a stellar wind.* (© 1984 Anglo-Australian Telescope Board)

the star can exert sufficient force to accelerate it to high speeds. This force, called **radiation pressure,** is very weak (you certainly can't feel the breeze from a light bulb, for example), but O and B stars are so luminous that strong acceleration of the wind is possible. For the most luminous of these stars, radiation pressure may be sufficiently strong to initiate the winds, but for the slightly less luminous O and B stars, calculations show that radiation pressure is not capable of creating the initial outflow but can accelerate the gas to high speed once the flow begins. Some other mechanism must start the outflow first. It is not known what this mechanism is, but recently it has been suggested that subtle pulsations (see Astronomical Insight 21.1) might be responsible.

Very high degrees of ionization are observed in the winds from O and B stars. Recently X-ray emission from stars with winds has been discovered, and it is possible that the X-rays are responsible for ionizing the gas (X-ray photons have high energies and cause ionization when they are absorbed by atoms). The X-rays must be produced in some very hot region, either gas in the wind that is heated by turbulence, or a hot coronal region in the star's outer layers. It would be a surprise to theorists if these hot stars have coronae, because the O and B stars are not thought to have convection in the outer layers. Perhaps this is incorrect, and they do have convection zones near their surfaces, or perhaps some other process is creating the coronae in these stars.

Whatever the cause of the winds, they have important consequences. Analysis of the ultraviolet emission lines shows that, in some cases, stars are losing matter at such a great rate that a large fraction of the initial mass may be lost during their lifetimes. An O star might lose as much as one solar mass every 100,000 years. If such a star begins life with twenty or thirty solar masses and lives a million years, it could lose ten solar masses, a significant fraction of what it had to begin with. This has important effects on how such a star evolves, as we shall see.

The supergiants in the upper right of the H-R diagram also lose mass through stellar winds (Fig. 21.12), but these winds have a distinctly different nature. The red supergiants are so large that their surface gravity is very low, and the gas in the outer layers is not tightly bonded to the star. Radiation pressure can easily push the gas outward, at the relatively low speed of 10 or 20 kilometers per second. The amount of mass lost can be just as great as in the hot stars, however, because these low-velocity winds are much denser than the high-speed winds from the O and B stars.

Because the red supergiants are relatively cool objects, the gas in their outer layers is not ionized, and even molecular species form there. In addition, small solid particles called dust grains can condense, so that the star becomes shrouded in a cloud. In extreme cases the dust cloud becomes so thick that little or no visible light escapes to the outside. The dust grains become heated, however, and emit infrared radiation, so the star can still be detected with the use of an infrared telescope.

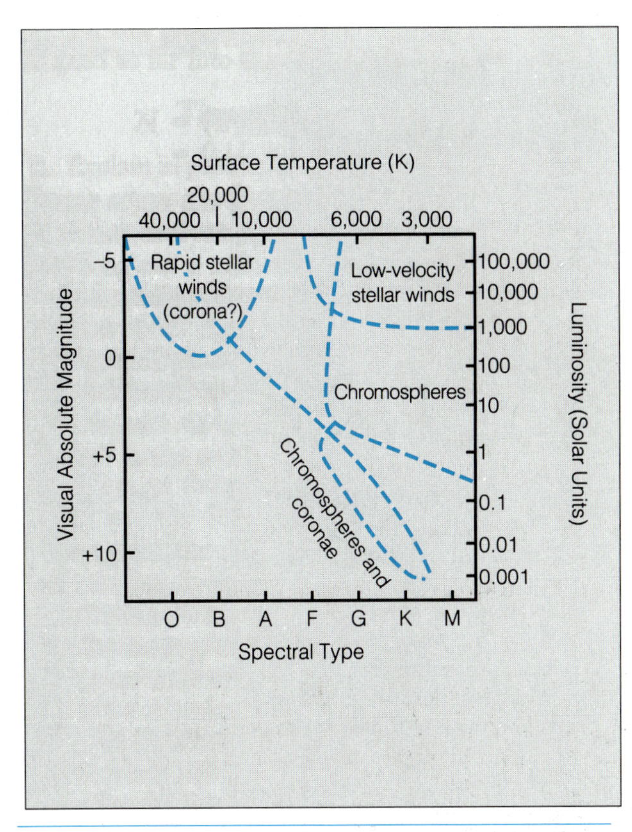

FIGURE 21.12 CHROMOSPHERES, CORONAE, AND STELLAR WINDS IN THE H-R DIAGRAM. *Here we see that most stars of type F or cooler have chromospheres, although coronae may be confined to the main sequence. Very luminious stars, hot and cool alike, have sufficiently strong winds to result in significant loss of mass.*

Some of the dust that forms in the outer layers of red supergiants escapes into the surrounding void, and interstellar space gets contaminated in this way with a kind of interstellar haze (the interstellar dust is discussed at greater length in Chapter 26).

MASS EXCHANGE IN BINARY SYSTEMS

All of the processes discussed so far can occur in single stars or in stars that are part of binary systems. There are additional factors that come into play in certain kinds of binaries, but which do not affect single stars.

We have seen that stars can lose matter into space. If such a star has a companion, then the com-

ASTRONOMICAL INSIGHT 21.2

STELLAR MODELS AND REALITY

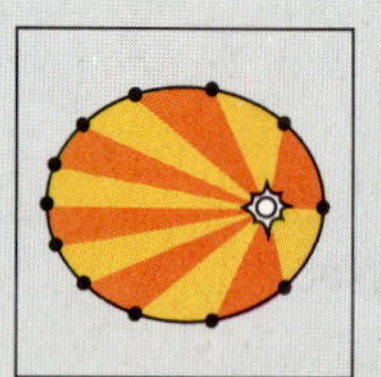

In this chapter we discuss the physical processes that are believed to govern the internal conditions and external properties of stars. We will find that, despite the remoteness of stars and the fact that all of our observational information is derived only through the analysis of light emitted from their surfaces, we know quite a bit about their insides. It may seem surprising that we could have any confidence at all in our theories of stellar interiors and atmospheres, but the fact that we do derives from our faith that the same laws of physics apply to distant stars just as they do on Earth. Given that, stars are actually rather simple objects, in comparison with, for example, a solid planet or a biological system.

Stars are made entirely of gas, and this is the major reason for the simplicity of describing them theoretically. Gases obey certain well-defined laws, which are verified through experiment and theory. Add the fact that stars are basically symmetric objects and stable (not changing with time), and we can invoke additional physical laws in the form of equations for the forces acting on gas particles in a star and for the production and transport of energy. Within the context of the adopted physical laws, there is only one solution to the set of equations that describe a star's structure and properties. This is a very powerful statement, for it implies that all we need to do is be sure to include the correct laws of physics in our calculations, and we will get the correct solution.

Aside from the fact that our known laws of physics may be incomplete, the major difficulty in calculating theoretical models of stars lies in the fact that some of the processes occurring in stars are quite complex when examined in sufficient detail. For example, it is easy to treat the transport of energy from the core to the surface of a star when the energy is carried solely in the form of radiation, but very difficult when the energy is transported by flowing streams of heated gas, in the process we call convection. The calculations are simple when a star is spherically symmetric, but they become more complex when the star is rotating rapidly, so that its shape becomes somewhat flattened. Magnetic fields in stars are probably very complex, if our Sun is any example, and theory has not yet fully treated even the Sun's magnetic field. These are cases in which we know the laws of physics well enough, but their application is made complex because of the vast amount of detail that must be incorporated into our calculations.

Despite the complications, stellar model calculations are quite good. We know this because it is possible to achieve very good agreement between observed quantities (such as stellar luminosity, radius, surface temperature, and others) and those calculated from theory. The first test of any theoretical model for a star is to see whether the model successfully reproduces the observed surface conditions. If it does, then we have confidence that the model must also represent the interior fairly well.

Today's stellar structure theorists are busily refining their models, not so much because they are incorrect, but because it is becoming increasingly possible to add details to the calculations. Just as new telescope technology increases the ability of astronomers to observe astronomical objects, new advances in computer technology increase the capability of theorists to represent reality in ever greater detail. The first stellar models were calculated with slide rules or slow, mechanical desk calculators, and it took weeks or months to calculate a single model. In the past two or three decades, model complexity has increased in proportion to computer capacity and speed, and it is now possible to calculate many models (for different stages in the evolution of a star, for example) in hours. Today we are on the brink of a new generation of "supercomputers," machines with vastly greater speed and capacity for massive calculations, and no one will benefit from this more than the stellar structure theorist. As the capabilities of the computers increase, so will the ability of astrophysicists to represent reality in their calculations.

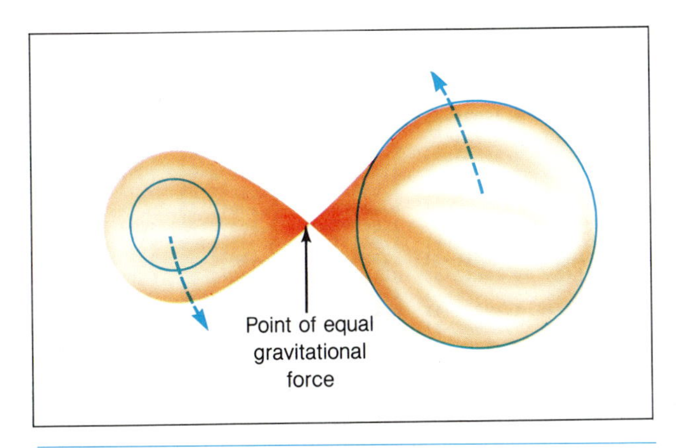

FIGURE 21.13 MASS TRANSFER IN A BINARY SYSTEM. *The more massive star in a binary will be the first to evolve and swell up as a red giant. In the process, its outer layers may come sufficiently close to the companion star to be pulled to it gravitationally. As we will see in later chapters, such a transfer of mass can radically affect both stars.*

panion may catch some of the cast-off material and gain mass in the process (Fig. 21.13). In some cases the companion actually helps its neighbor to lose matter. There is a point between the two stars where their gravitational forces just balance, and a particle of gas that reaches that point can fall either way, into either star. If one of the stars is swelling up on its way to becoming a red giant, when its outer layers reach this balance point, gas will begin to flow over onto the other star. It is actually possible for a pair of stars with unequal initial masses to reverse themselves, so that the one that started out with most of the mass ends up with the least.

This mass-exchange phenomenon has been observed in several binary systems. These systems are characterized by flowing streams of gas swirling around the two stars, creating emission lines with Doppler shifts indicating the motions of the gas. In some of these systems, the star that is receiving mass has finished its evolution and has become a stellar remnant such as a white dwarf, neutron star, or a black hole (see Chapter 24), and the gas falling in creates spectacular effects ranging from nova explosions to intense X-ray emission.

No matter how it happens, the loss or gain of mass by a star will have important effects on how it evolves. We have already learned that the mass of a star determines all of its other properties; hence it follows that the other properties will vary if the mass does.

STELLAR MODELS: TYING IT ALL TOGETHER

Once astrophysicists have understood all the phenomena that govern a star's internal structure and its outward appearance, they must combine their knowledge of all these processes into a theoretical calculation (Fig. 21.14) that will tell them what goes on inside the star, and what will happen to it in time. When the calculations correctly reproduce the observable quantities, it is assumed that they also correctly illustrate the nonobservable aspects, such as the star's interior conditions. We can have reasonable confidence in such calculations, because it can be demonstrated that the resulting solutions are unique; i.e., no other solution to the mathematical equations can exist. The most difficult question is whether we know all the correct equations and physical parameters to begin with.

The calculation of a model star consists of simultaneously solving a set of basic mathematical equations describing the physical processes, and solving them repeatedly for different depths inside the star. In this way, the calculations may proceed from the observed surface conditions to the center, or from assumed central conditions to the surface, with adjustments made until the calculations match the observed properties of the star. To carry out such a calculation requires a substantial amount of time on a large computer.

To see how a star evolves, it is necessary to compute the model many times over, each time taking into account changes produced in the star that took place in the previous step in the calculation. For example, a certain amount of hydrogen is converted into helium at each step, and this change in the chemical composition of the core must be taken into account in the next step. The conversion of hydrogen into helium uses up the hydrogen fuel, so that the star will eventually exhaust its supply, and it causes the density of the core to increase, because helium nuclei are more massive than hydrogen nuclei. These things cause the star's overall structure to change, and the

FINAL MODEL

POINT	RADIUS	DENSITY	TEMP	PRESSURE	ENERGY GEN		LUMIN	OPACITY	DELAD	DELRAD	BETA
1	0.	2.8526E+00	3.5455E+07	1.7651E+16	1.9682E+05	1	0.	3.4602E-01	2.8868E-01	3.0013E+00	7.7421E-0
2	5.4736E+08	2.8526E+00	3.5455E+07	1.7651E+16	1.9680E+05	1	3.0569E+32	3.4602E-01	2.8868E-01	3.0013E+00	7.7421E-0
3	7.3124E+08	2.8525E+00	3.5455E+07	1.7650E+16	1.9679E+05	1	9.1397E+32	3.4602E-01	2.8868E-01	3.0011E+00	7.7421E-0
4	8.8736E+08	2.8525E+00	3.5455E+07	1.7650E+16	1.9677E+05	1	1.6376E+33	3.4602E-01	2.8868E-01	3.0009E+00	7.7421E-0
5	1.0365E+09	2.8525E+00	3.5455E+07	1.7650E+16	1.9675E+05	1	2.6288E+33	3.4602E-01	2.8868E-01	3.0008E+00	7.7421E-0
6	1.1927E+09	2.8524E+00	3.5454E+07	1.7649E+16	1.9673E+05	1	3.9864E+33	3.4602E-01	2.8868E-01	3.0006E+00	7.7421E-0
7	1.3540E+09	2.8524E+00	3.5456E+07	1.7649E+16	1.9671E+05	1	5.6460E+33	3.4602E-01	2.8869E-01	3.0004E+00	7.7421E-0
8	1.5283E+09	2.8523E+00	3.5458E+07	1.7648E+16	1.9668E+05	1	8.3930E+33	3.4602E-01	2.8869E-01	3.0001E+00	7.7421E-0
9	1.7159E+09	2.8523E+00	3.5457E+07	1.7648E+16	1.9664E+05	1	1.1881E+34	3.4602E-01	2.8869E-01	2.9998E+00	7.7421E-0
10	1.9206E+09	2.8522E+00	3.5457E+07	1.7647E+16	1.9659E+05	1	1.6659E+34	3.4602E-01	2.8869E-01	2.9994E+00	7.7421E-0
11	2.1449E+09	2.8521E+00	3.5456E+07	1.7646E+16	1.9654E+05	1	2.3202E+34	3.4602E-01	2.8869E-01	2.9989E+00	7.7422E-0
12	2.3910E+09	2.8519E+00	3.5455E+07	1.7645E+16	1.9647E+05	1	3.2162E+34	3.4602E-01	2.8869E-01	2.9983E+00	7.7422E-0
13	2.6639E+09	2.8518E+00	3.5454E+07	1.7643E+16	1.9639E+05	1	4.4432E+34	3.4602E-01	2.8869E-01	2.9976E+00	7.7422E-0
14	2.9647E+09	2.8515E+00	3.5453E+07	1.7641E+16	1.9628E+05	1	6.1231E+34	3.4602E-01	2.8869E-01	2.9967E+00	7.7423E-0
15	3.2977E+09	2.8513E+00	3.5452E+07	1.7639E+16	1.9615E+05	1	8.4230E+34	3.4602E-01	2.8869E-01	2.9956E+00	7.7423E-0
16	3.6666E+09	2.8510E+00	3.5450E+07	1.7636E+16	1.9600E+05	1	1.1571E+35	3.4602E-01	2.8869E-01	2.9942E+00	7.7424E-0
17	4.0755E+09	2.8506E+00	3.5448E+07	1.7632E+16	1.9580E+05	1	1.5880E+35	3.4602E-01	2.8869E-01	2.9925E+00	7.7424E-0
18	4.5291E+09	2.8501E+00	[illegible]	1.7628E+16	[illegible]	1	2.1776E+35	3.4602E-01	2.8869E-01	[illegible]	[illegible]
19	5.0324E+09	2.8496E+00	3.5443E+07	1.7623E+16	1.9527E+05	1	2.9841E+35	3.4602E-01	2.8870E-01	2.9878E+00	7.7426E-0
20	5.5910E+09	2.8488E+00	3.5439E+07	1.7616E+16	1.9491E+05	1	4.0871E+35	3.4602E-01	2.8870E-01	2.9847E+00	7.7427E-0
21	6.2112E+09	2.8480E+00	3.5434E+07	1.7608E+16	1.9446E+05	1	5.5949E+35	3.4602E-01	2.8870E-01	2.9808E+00	7.7429E-0
22	6.8994E+09	2.8469E+00	3.5428E+07	1.7598E+16	1.9391E+05	1	7.6580E+35	3.4602E-01	2.8870E-01	2.9760E+00	7.7431E-0
23	7.6646E+09	2.8455E+00	3.5421E+07	1.7585E+16	1.9324E+05	1	1.0468E+36	3.4602E-01	2.8871E-01	2.9701E+00	7.7434E-0
24	8.5141E+09	[illegible]	3.5412E+07	1.7570E+16	[illegible]	1	1.4307E+36	3.4603E-01	2.8872E-01	[illegible]	7.7437E-0

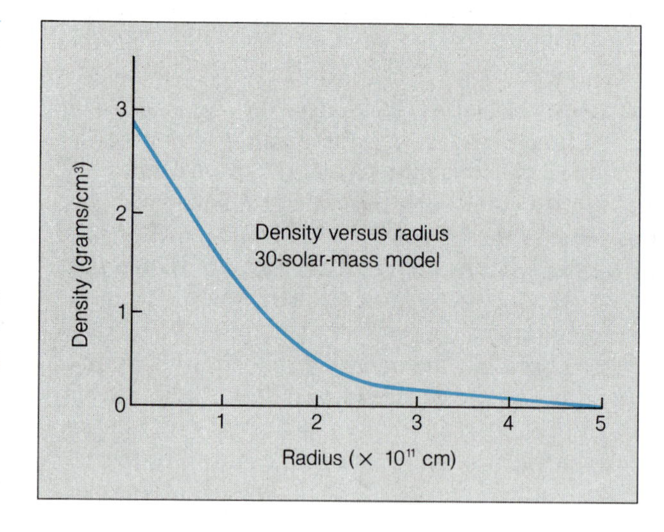

FIGURE 21.14 BUILDING A MODEL STAR. *Modern astrophysicists reconstruct what is happening inside of stars with the use of complex computer programs that simultaneously solve large numbers of equations describing the physical conditions. Often the result is plotted, as a convenient means of showing the implications of these calculations.* (C. Hansen, University of Colorado)

calculations keep track, allowing the evolution to be traced. In the next chapters we will discuss the evolution of stars, and how it is determined from a combination of theory and observation.

PERSPECTIVE

We now are acquainted with all the physical processes that affect a star; we know what a star is and what makes it run. We have seen that mass plays a dominant role in dictating all the other properties, although the composition, which changes with time, is also important. We are prepared to see what the life story of a star is—how it is born, how it lives, and how it ends its days.

SUMMARY

1. A star is gaseous throughout, with temperature and density increasing towards the center.
2. The balance between gravity and pressure within a star, called hydrostatic equilibrium, governs the internal structure.
3. The mass of a star, along with its chemical composition, governs its central temperature and pressure (through hydrostatic equilibrium), its luminosity, its radius, and its internal structure.
4. Nuclear fusion reactions occur in the core of a star, where the temperature and density are high enough to allow nuclei to collide with sufficient velocity and frequency.
5. In most stars (those on the main sequence), hydrogen is converted into helium in the reactions in the core.
6. The energy inside a star is transported by convection or by radiation. On the lower main sequence (including the Sun), radiative transport dominates in

the inner parts of the star, while convection operates in the outer layers. On the upper main sequence, convection is dominant in the core, while radiative transport dominates in the outer layers.

7. The lifetime of a star depends on its mass and how rapidly it uses up its nuclear fuel; it decreases dramatically from the lower main sequence to the upper main sequence.

8. Nuclear reactions in stars, and the recycling of matter between stars and interstellar space, result in a gradual increase in the abundance of heavy elements in the universe.

9. Stars in the cool half of the H-R diagram have chromospheres and coronae, in analogy with the Sun.

10. Stars in the hot portion of the H-R diagram, and very luminous cool supergiants as well, have stellar winds that can cause the loss of large fractions of the initial stellar masses. The loss of mass can have significant effects on how the stars evolve.

11. In certain binary star systems, matter is transferred from one star to the other, with important effects on the structure and evolution of both.

12. To determine the interior conditions in a star, astronomers must compute them from known laws of physics and observations of the surface conditions.

REVIEW QUESTIONS

Thought Questions

1. How do we know that a star must be in a state of balance between gravity and pressure? What would happen if this balance were disturbed?

2. Explain in your own words why mass is a star's most fundamental property; that is, explain how the mass of a star is the most important factor in determining its other properties.

3. Why do nuclear reactions in stars occur only in the innermost region, that is, in the core?

4. Summarize the ways in which energy produced in a star's core is transported outward, to the surface. Point out ways in which the transport varies with the type of star.

5. Explain why the lifetime of a star is *shorter,* the more massive the star.

6. As a star ages and undergoes successive nuclear reaction stages, a higher temperature is required to start each new reaction. Why?

7. Compare the stellar winds from O stars with the solar wind.

8. Summarize the differences between an upper main-sequence star (an O or B star) and one of the lower main sequence (K or M), in structure, energy generation and transport, and outer layers.

9. If mass is exchanged between two stars in a binary system, how do you think it might affect the lifetimes of the two stars?

10. Discuss stellar model calculations. Why do we think they tell us anything realistic about conditions inside stars, where we cannot observe? How does a mathematical model of a star stand as a scientific theory; that is, does it have the attributes of a theory?

Problems

1. If a single proton-proton chain reaction produces only 10^{-5} ergs, how many reactions per second must occur in order to supply the Sun with the 4×10^{33} erg/sec that it radiates? How many reactions (CNO cycle, which produces about the same amount of energy per reaction) must occur per second in an O star whose luminosity is 10^{38} erg/sec?

2. Estimate the hydrogen-burning lifetime of a star a) of fifteen solar masses and luminosity 10,000 times that of the Sun, b) and a star of 0.2 solar masses and luminosity 0.008 that of the Sun

3. When they hydrogen in a star's core has all been converted to helium, there are only one-fourth as many nuclei in the core as there were before the reactions. If the average distance between nuclei is the same, by how much has the core's density increased?

4. Suppose the speed of the wind from an O supergiant star is 2,000 km/sec. What is the Doppler-shifted wavelength of the three-times ionized carbon line, whose rest wavelength is 1548.8 Å (in the ultraviolet)?

5. If a star has an initial mass of twenty solar masses and a lifetime of 5 million years, how much of its mass will it lose if the rate of mass loss is one solar mass every 500,000 years?

6. Suppose the Sun becomes a red supergiant star with a radius of 200 times its present radius. By how much would its surface gravity and escape velocity be decreased? Discuss the importance of your answer in view of the stellar winds from red supergiants.

ADDITIONAL READINGS

Aller, L. H. 1971. *Atoms, stars, and nebulae.* Cambridge, Mass.: Harvard University Press.

Boss, A. P. 1985. Collapse and formation of stars. *Scientific American* 252(1):40.

Cox, A. N., and J. P. Cox. 1967. Cepheid pulsations. *Sky and Telescope,* May, p. 278.

Giampapa, Mark S. 1987. The solar-stellar connection. *Sky and Telescope* 74(2):142.

Graham, J. 1984. A new star becomes visible in Chile. *Mercury* XIII(2):60.

Hoyle, F. 1975. *Astronomy and cosmology: A modern course.* San Francisco: W. H. Freeman.

Jones, B., Lin, D., and R. Hanson. 1984. T Tauri and TIRC: A star system in formation. *Mercury* XIII(3):71.

Perry, J. R. 1975. Pulsating stars. *Scientific American* 232(6):66.

Shu, F. 1982. *The physical universe.* San Francisco: W. H. Freeman.

Smith, E. P., and K. C. Jacobs. 1973. *Introductory astronomy and astrophysics.* Philadelphia: Saunders.

Snow, T. P. 1981. Dieting for stars, or how to lose 10^{25} grams per day (and feel better, too!). *Griffith Observer* 45(5):2.

Weymann, R. J. 1978. Stellar winds. *Scientific American* 239(2):34.

CHAPTER 22

STAR CLUSTERS AND OBSERVATIONS OF STELLAR EVOLUTION

The globular cluster 47 Tucanae. (National Optical Astronomy Observatories)

CHAPTER PREVIEW

How can we watch a star evolve? After all, even the most short-lived ones are around for hundreds of thousands of years, and human studies of astronomy date back only three or four millennia. Occasionally we catch a star in the act of change, as it makes a transition from one stage in its evolution to another, but clearly we cannot hope to see a star through its whole lifetime, watching it form, live, and die.

We can, however, piece together a story of stellar evolution by examining many stars of different ages, and deducing from this the sequence of events that occurs in a single star. An apt analogy is the deduction of the life story of a tree from the examination of a forest, where seeds, young saplings, mature trees, and dead logs are all found. For stars, matters are complicated by the fact that the manner of evolution depends on the mass of the star, whereas for a group of trees we expect all to have about the same properties. As we shall see, however, stars, particularly groups of them, provide certain clues for the detective trying to reconstruct their life stories.

PROPERTIES OF CLUSTERS

Within our galaxy, not all stars are uniformly distributed, but instead many are located in concentrated regions called clusters. A few of these are sufficiently prominent to be visible to the unaided eye, and others can be viewed with a small telescope or a good pair of binoculars (Table 22.1). The Pleiades (Fig. 22.1), a bright group to the west of Orion, is perhaps the best-known cluster for Northern Hemisphere observers, being easily visible in the evening sky throughout the late fall and early winter. Other

TABLE 22.1 SELECTED CLUSTERS OF STARS

Cluster	Type	m_V	Distance (pc)	No. Stars	Diameter (pc)	Age (years)
47 Tuc	Globular	4.0	5,100	10^4	10	1.5×10^{10}
M103	Open	6.9	2,300	30	5	1.6×10^7
h Persei	Open	4.1	2,250	300	16	1.0×10^7
χ Persei	Open	4.3	2,400	240	14	1.0×10^7
M34	Open	5.6	440	60	30	1.3×10^8
Perseus	Open	2.2	167	80	12	1.0×10^7
Pleiades	Open	1.3	127	120	4	5.0×10^7
Hyades	Open	0.6	42	100	5	6.3×10^8
S Mon	OB	4.3	740	60	6	6.3×10^6
τ CMa	Open	3.9	1,500	30	3	5.0×10^6
Praesepe	Open	3.7	159	100	4	4.0×10^8
o Velorum	Open	2.6	157	15	2	2.5×10^7
M67	Open	6.5	830	80	4	4.0×10^9
θ Carinae	Open	1.7	155	25	3	1.3×10^7
Coma	Open	2.8	80	40	7	5.0×10^8
ω Cen	Globular	3.6	5,000	10^5	20	1.5×10^{10}
M13	Globular	5.9	7,700	10^5	11	1.5×10^{10}
M22	Globular	5.1	3,000	10^5	9	1.5×10^{10}
M15	Globular	6.4	14,000	10^6	11	1.5×10^{10}

FIGURE 22.1 THE PLEIADES. *This relatively young galactic cluster is a prominent object in the Northern Hemisphere during the late fall and winter*. (© 1980 Anglo-Australian Telescope Board)

clusters easy to find with a small telescope are the Hyades, the double cluster in Perseus, and M13 (Fig. 22.2), a fantastic, spherical collection of hundreds of thousands of stars. It is apparent that the stars in a cluster are gravitationally bound together.

There are several distinct types of star clusters. Those containing a modest number of stars (up to a few hundred), found in the disk of the galaxy, are referred to as **galactic,** or **open** clusters (Figs. 22.3 and 22.4). The Pleiades (see Fig. 22.1) and the Hyades are examples. Loose groupings of hot, luminous stars are also found in the plane of the galaxy, and are

FIGURE 22.2 THE GLOBULAR CLUSTER M13. (U.S. Naval Observatory)

FIGURE 22.3 THE OPEN CLUSTER N6C3293. (© 1977 Anglo-Australian Telescope Board)

FIGURE 22.4 THE OPEN CLUSTER NGC4755. (National Optical Astronomy Observatories)

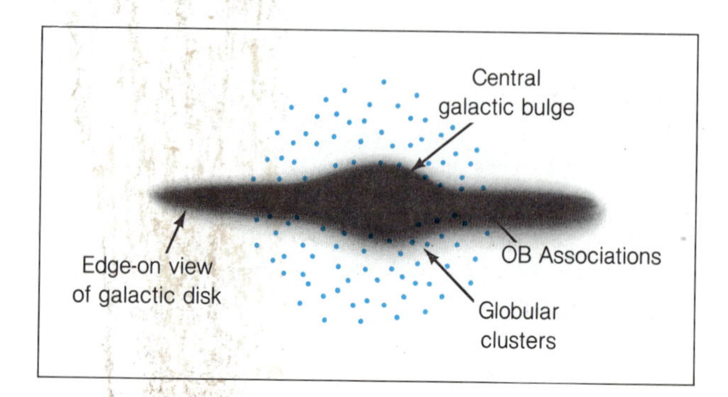

FIGURE 22.5 THE LOCATIONS OF CLUSTERS IN THE GALAXY. *The open or galactic clusters are found in the disk of the Milky Way, as are the OB associations, which are always located in spiral arms (see Chapter 25). The globular clusters, to the contrary, inhabit a large spherical volume, and are not confined to the plane of the disk.*

called **OB associations,** being dominated by stars of spectral types O and B. These clusters, in contrast with the other types mentioned here, are probably not gravitationally bound together, but appear grouped simply because the stars have recently formed together. The giant **globular clusters,** such as M13 (Fig. 22.2), are found in a spherical volume about the galactic center and are not confined to the plane of the galaxy (Fig. 22.5). In later chapters, especially Chapters 25 and 27, we will discuss how the various types of clusters fit into the story of the formation and evolution of the galaxy. For now we will exploit certain properties of the clusters that help us see how stars evolve.

There are two important assumptions astronomers make about a cluster of stars: (1) the stars in a cluster are all at the same distance from us; and (2) they formed together so that all are the same age and have the same composition initially.

The Assumption of a Common Distance

The first of these assumptions is, strictly speaking, only approximately true. Clearly, a cluster of stars has some spatial extent, so that the stars on the near side are closer to us than those on the far side. The diameter of a cluster, however, is rarely more than a few parsecs and is usually small compared with the distance from the Earth to the cluster. Therefore, for all practical purposes, the stars in a cluster are all at the same distance from us. This means that any observed differences from star to star within a cluster must be real differences and cannot be a false effect created by differing distances from us. Brightness differences, for example, have to be the result of variations in the luminosities of stars within the cluster.

Because distance effects are eliminated when we compare stars within a cluster, it is possible to plot an H-R diagram for a cluster without first determining the absolute magnitudes of the stars. We simply plot apparent magnitude against spectral type, or, more commonly, against the color index $B - V$, which is easier to measure for a large number of stars (Fig. 22.6). To do this requires only the determination of the visual (V) and blue (B) magnitudes of the cluster stars, a rather straightforward observational procedure, and then the construction of a plot of V versus $B - V$. The result, called a **color-magnitude** diagram, looks just like an H-R diagram, except that the vertical scale is an apparent-magnitude scale. This is because the differences in magnitude of the cluster stars are the same, whether we use apparent or absolute magnitudes.

One advantage of plotting cluster H-R diagrams is that these diagrams can be used to determine the distances to clusters (Fig. 22.7). A comparison of a cluster H-R diagram with a standard one (that is, one with an absolute-magnitude scale) allows us to see what the absolute-magnitude scale for the cluster diagram should be. This in turn tells us the difference between the absolute and apparent magnitudes for the cluster, and from this difference we deduce the distance. This procedure, known as **main-sequence fitting,** depends only on the assumption that the cluster has a main sequence that is identical to that of the standard H-R diagram, a safe assumption as long as the chemical composition of the cluster is not unusual.

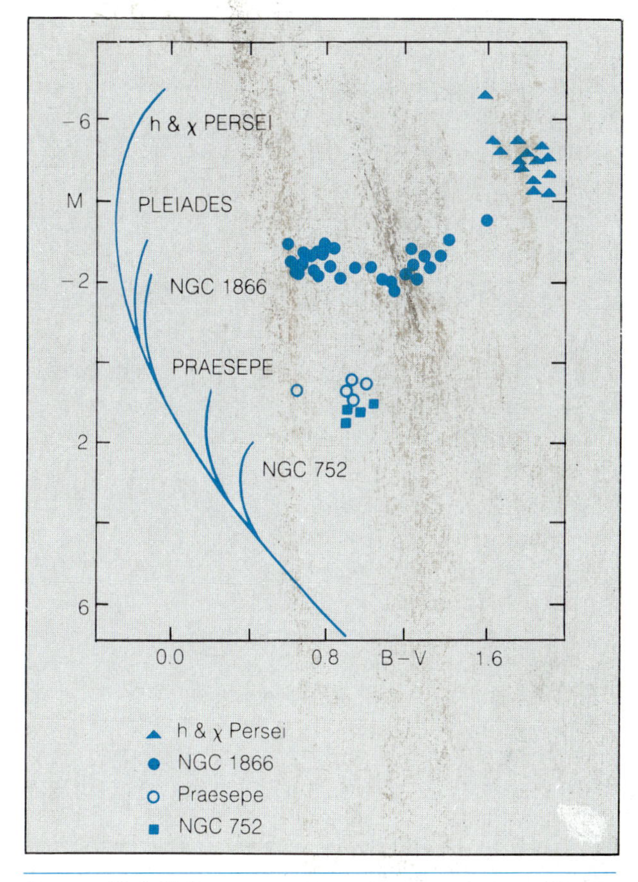

FIGURE 22.6 H-R DIAGRAMS FOR STAR CLUSTERS. *Here, plotted on the same axes, are H-R diagrams (using color index rather than spectral type) of several clusters. Different symbols are used in the right-hand portion to distinguish among the giant stars in different clusters.*

The Common Age Assumption

The second major assumption made about clusters, that all the stars are the same age, is very important in studying stellar evolution. The basic premise is that the stars formed together, out of a common cloud of interstellar material, and did not just happen to come together. The basis for this assumption is the extremely low probability that a number of stars could ever encounter each other and form a cluster by chance, as a result of random stellar motions. A close encounter between stars is very rare indeed, and furthermore, even when one does occur, it is very unlikely to result in gravitational bonding. Binary systems as well as clusters must form as binaries or clusters. A corollary is that all stars in a cluster must begin with the same chemical composition, since they form from a common source of material.

The knowledge that all stars in a cluster have the same age and initial composition gives astronomers tremendous leverage in deducing the evolutionary histories of stars. The only major parameter that differs from one star to another within a cluster is mass, so the observed differences between stars in a cluster give us clues as to how the properties of stars depend on mass. Comparison of H-R diagrams of clusters of different ages then gives us a picture of how these different masses evolve.

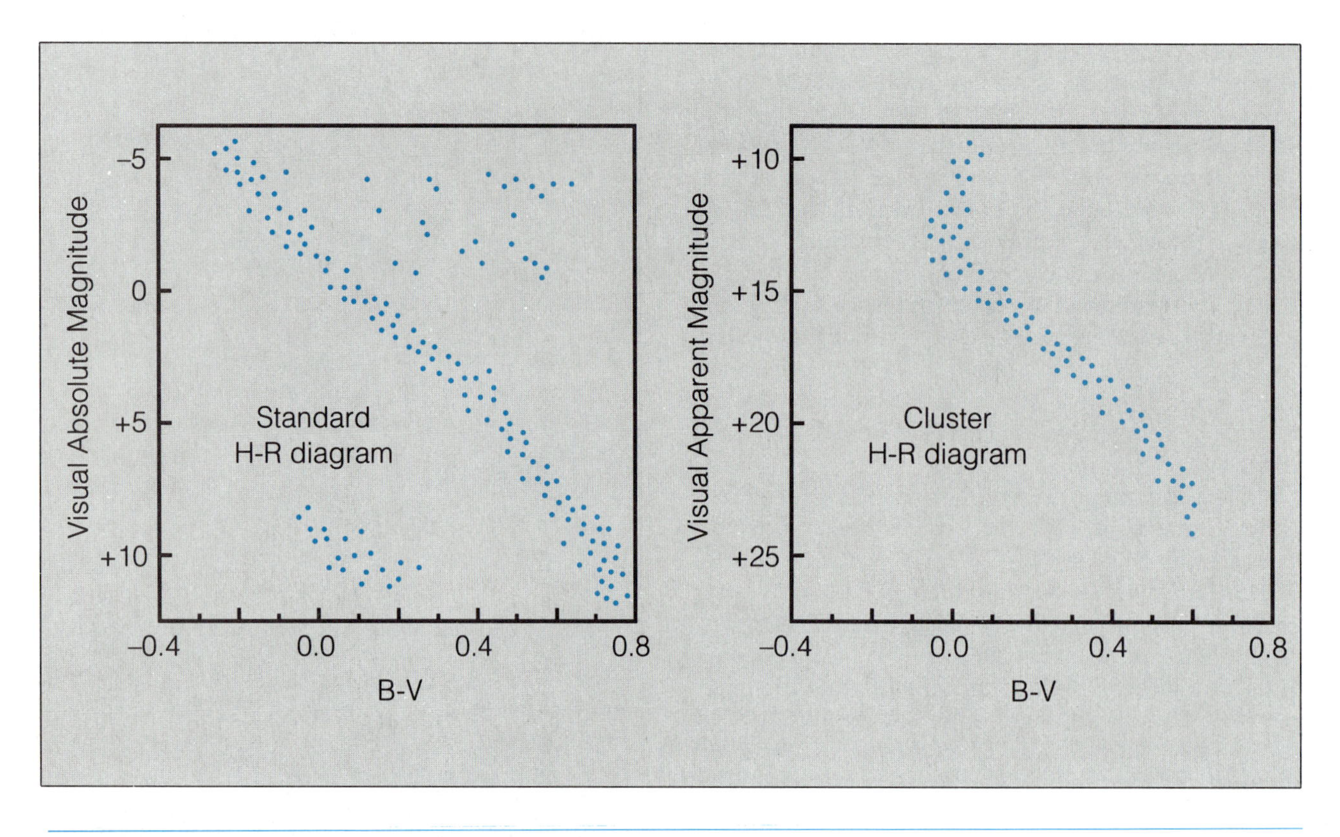

FIGURE 22.7 MAIN SEQUENCE FITTING. *This is a standard H-R diagram, with a cluster H-R diagram next to it using apparent magnitude on the vertical axis. A comparison of the two diagrams leads to an estimate of the distance to the cluster.*

— THE MAIN-SEQUENCE TURNOFF — AND CLUSTER AGES

Striking differences are found between H-R diagrams of different clusters. The Pleiades, for example, have an almost complete main sequence, but no red giants. By contrast, a globular cluster has no stars on the upper portion of the main sequence, but it has a large number of red giants. A wide variety of intermediate cases can be found.

From considerations discussed in the last chapter, we expect massive stars to evolve most rapidly. These are the stars that start out near the top of the main sequence, and they have such tremendous luminosities that they use up their nuclear fuel rapidly. Tying this together with the observed differences between cluster H-R diagrams leads to the conclusion that the upper main-sequence stars must evolve into red giants when they have used up all their hydrogen fuel. In a very young cluster, even the most massive stars might still be on the main sequence, whereas in an older cluster, these massive stars will have had time to use up their core hydrogen and become red giants. Hence, the Pleiades are relatively young, whereas a globular cluster is very old.

As a cluster ages, more and more of its stars use up their hydrogen and leave the main sequence. The upper main sequence gradually erodes away as its stars move to the right in the H-R diagram, toward the red-giant region (Figs. 22.8 and 22.9). This process creates a definite cutoff point on the main sequence, above which there are no stars. The location of this cutoff depends entirely on the age of the cluster. The position of this **main sequence turnoff,** as it is com-

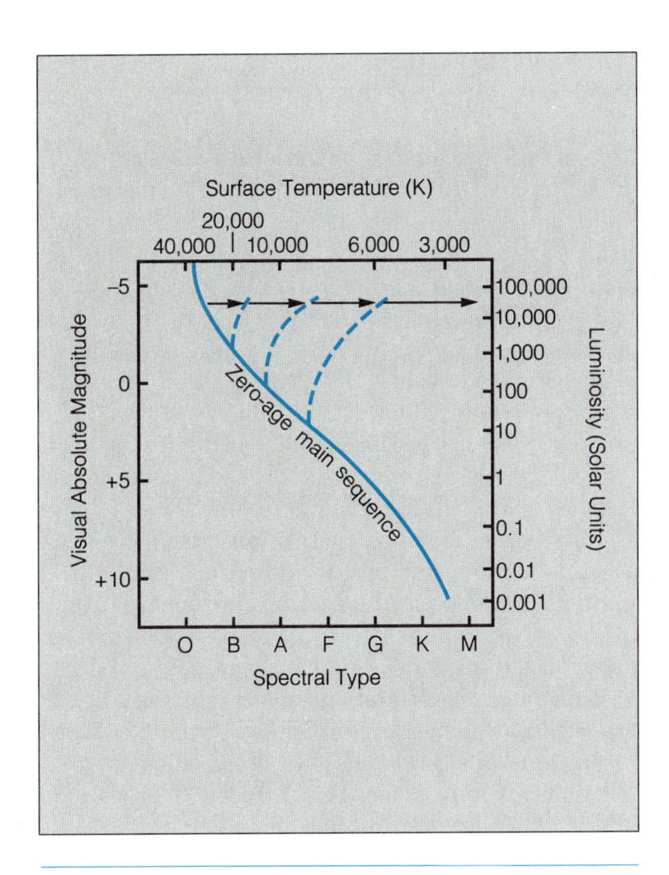

FIGURE 22.8 THE EVOLUTION OF THE MAIN SEQUENCE. *As a cluster ages, the stars on the upper main sequence move to the right. The farther down the main sequence this has happened, the older the cluster.*

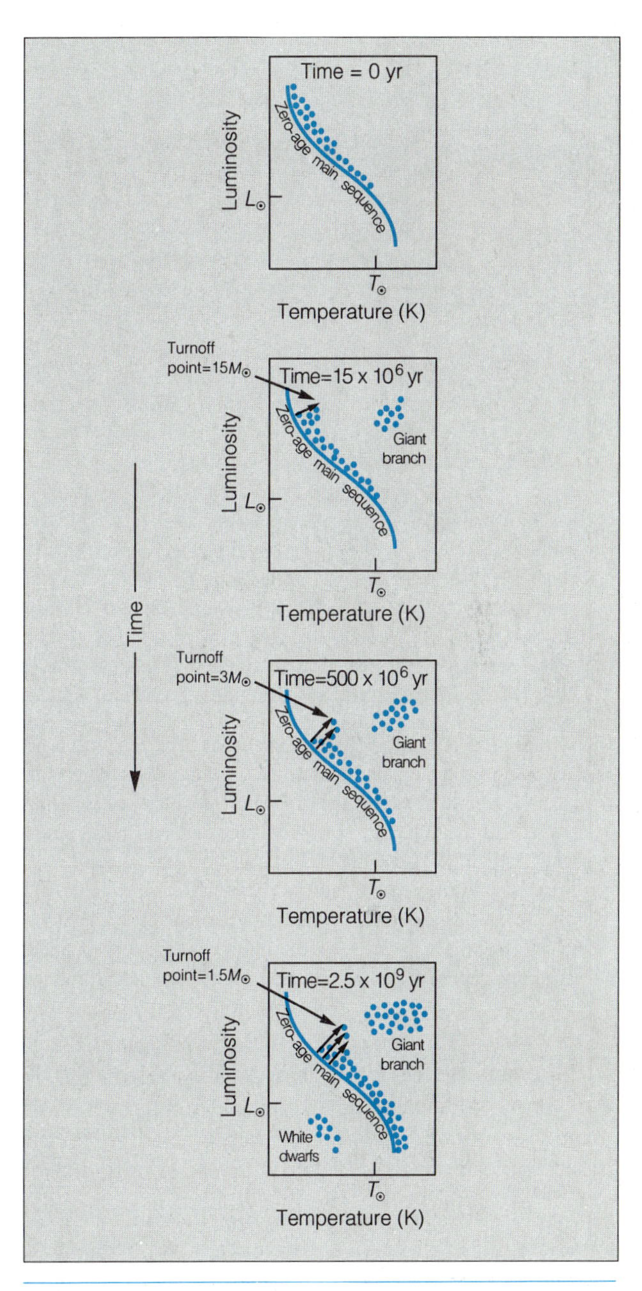

FIGURE 22.9 EVOLUTION OF THE H-R DIAGRAM FOR A STAR CLUSTER. *This shows how the H-R diagram for a cluster of stars changes as the cluster ages and the most massive stars evolve to become red giants.*

monly called, indicates the age of a cluster. If, for example, we find a cluster whose main sequence stops at the position of stars like the Sun (that is, it has no stars above G2, or $B-V = 0.6$, on the main sequence), then we know that all stars more massive than one solar mass have burned their hydrogen and evolved toward the red-giant region. In the last chapter, we learned that the Sun's lifetime for hydrogen burning is about 10 billion years. We conclude, therefore, that such a cluster is just about 10 billion years old. If it were younger, its main sequence would extend farther up, and if it were older, even the Sun-like stars would have had time to evolve into red giants, and the main-sequence turnoff would be farther down.

ASTRONOMICAL INSIGHT 22.1

STAR TRACKS

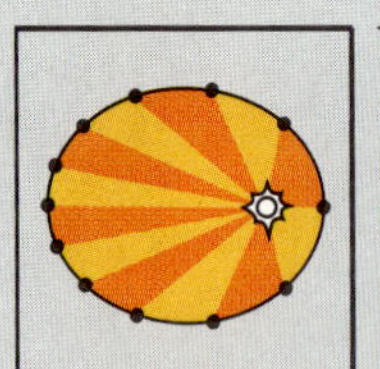

We have seen that as a star evolves, its position in the H-R diagram changes. The correct deduction of how a star's position changes was not easily made, however. During the years following the development of the H-R diagram, great intellectual effort was concentrated on discovering how stars evolve, but because the necessary information was not yet complete, many of the first ideas were wrong. This resulted in an assortment of views on the direction of motion in the H-R diagram, and evolutionary tracks thought to be made by stars during their lifetimes changed directions a few times.

By the second decade of this century it was well established that the red giants formed a distinct grouping of stars in the H-R diagram, above the main sequence. Nothing was known of the true energy source that powers stars, and little was known of stars' chemical composition. It was thought that stars glowed by virtue of energy released in a slow gravitational contraction, which caused the stars to shrink and heat, at least in their early stages. It was also suspected that within the main sequence, the hot stars were youngest and the cool stars the oldest. In 1913 Henry Norris Russell, in an address to the American Astronomical Society, put these ideas together when he suggested a comprehensive theory of stellar evolution. He proposed that a star begins life as a red giant, and for a while it contracts and heats, moving across the top of the H-R diagram from right to left. Eventually the temperature gets so hot inside the star that the gas there becomes highly ionized (now we know this is true from the start). Because ionized gas (it was supposed incorrectly) cannot compress and heat further, the star begins to cool while still contracting, and it moves down the main sequence from upper left to lower right. The evolutionary track of a star was therefore thought to resemble a backward figure seven, and the main sequence was thought to be just an age sequence.

In the decade following Russell's hypothesis, several astrophysicists, including most notably Sir Arthur Eddington and Sir James Jeans, developed theoretical models of stellar structure. These models, which of necessity incorporated many approximations, appeared to be consistent with the evolutionary scenario of Russell. But by the early 1920s, discrepancies began to arise. The chief problem was that there appeared to be no reason for the ionized gas in a star's interior to become incompressible, and therefore no reasonable explanation for why a contracting star should stop heating up and start cooling when its interior became ionized. The new theoretical work, along with developing evidence for a close relation between a star's mass and its luminosity, led to the first suggestion (again by Russell) that the main sequence represented a mass sequence, where stars are stable and spend a long time with no movement in the H-R diagram. Russell suggested that the difference between the main-sequence stars and the red giants was that the two groups had different internal energy sources, although his ideas of what those sources might be were rather vague. He still thought that the red giants were young stars, and that they moved down to the main sequence as their first energy source was exhausted and the second took over. Thus, in 1923 (when this was hypothesized), the evolutionary tracks attributed to stars were much like those known today, but with the motion in the exact opposite direction.

The proper sequence of events in a star's lifetime finally began to be sorted out in the late 1920s and early 1930s, with Russell once again leading the way. Two major clues were developed: (1) cluster H-R diagrams (studied by Hertzsprung, among others) showed that as a cluster ages, it loses stars from its upper main sequence and gains red giants; and (2) the Russell-Vogt theorem was developed, which demonstrated that a star's properties are all determined specifically by its mass and composition. It finally became clear that a star moves from the main sequence to the red-giant region, and that this movement must be caused by a change in its internal composition. It was even suggested that the change of composition involves the conversion of hydrogen into helium, although it was not until the late 1930s that the exact nature of this process was understood.

The study of star tracks in the H-R diagram demonstrates how plausible, but incorrect, theories can be deduced from incomplete information, yet it also shows how patient and careful collection of data will eventually lead to the correct picture. The incorrect hypotheses developed along the way were important, for they helped inspire the continued efforts to solve the problem.

YOUNG ASSOCIATIONS AND STELLAR INFANCY

Earlier in this chapter we mentioned OB associations, loosely bound groups of O and B stars. We now know that these are very young clusters, probably the youngest in the galaxy, because they contain stars on the extreme upper portion of the main sequence. In many cases, the stars are moving away from the center and escaping the association, so within a few million years no concentration of stars will remain.

If the OB associations are the youngest ensembles of stars in the galaxy, we might expect to find star formation actually occurring in or near them. Some young clusters show remnants of the process of star birth, in the form of gas and dust surrounding the stars (Figs. 22.10, 22.11, and 22.12). The Pleiades provide a good example of this; photographs show bluish clouds of dust, shining by the reflected light of the stars in the cluster (see Fig. 22.1). In other cases, not only these **reflection nebulae,** but also hot, glowing gas clouds called **HII regions** may be found. The notation *HII* refers to hydrogen in its ionized form; these hot gas clouds are called HII regions because they glow most strongly in emission lines of hydrogen (these lines form when protons and electrons combine to form hydrogen atoms, which can occur only in a region that is ionized). HII regions appear red because hydrogen has a very strong emission line at a wavelength of 6,563 Å in the red portion of the spectrum.

Since several young clusters are permeated by gas and dust left over from star formation, it is natural to expect that in even younger, newly forming clus-

FIGURE 22.10 REGIONS OF YOUNG STARS AND NEBULOSITY. *Here are two regions populated by recently-formed stars and their associate interstellar gas and dust.* (© 1981, 1984 Anglo-Australian Telescope Board)

FIGURE 22.11 A VERY YOUNG CLUSTER WITH NEARBY DUST CLOUD. *The cluster, NGC6520, contains several hot, young blue stars. The nearby dust cloud is so dense that it blocks out the light from the stars behind it.* (© 1980 Anglo-Australian Telescope Board)

FIGURE 22.12 A DARK CLOUD. *This is the well recognized Horsehead nebula. Many stars in the process of formation have been detected, by infrared technique, within the dark and dusty regions seen here.* (© 1984 Anglo-Australian Telescope Board)

ters, the gas and dust would be much denser, so dense, in fact, that the embryonic stars might be hidden from the view of observers. Hence the search for stars in the process of forming takes us into dense interstellar clouds.

How do we probe these dark and dusty regions? For a couple of reasons, this is best done by observing in infrared wavelengths. One reason is that infrared radiation penetrates much father through clouds of gas and dust than does visible light. Another is that the dust around a newborn star is hot, and it glows at infrared wavelengths. Therefore, to see infant stars, we must look for infrared sources buried in dark clouds (Figs. 22.13 and 22.14).

We have found many regions in our galaxy in which star formation is apparently taking place. One of the best observed of these is in the sword of Orion, in the Orion nebula, a great, glowing HII region. A very young cluster of stars is located there, and infrared observations reveal a number of infrared sources embedded within the dark cloud associated with the cluster (see Fig. 22.14).

In other, similar regions, faint variable stars called **T Tauri** stars are found. These appear to be newborn

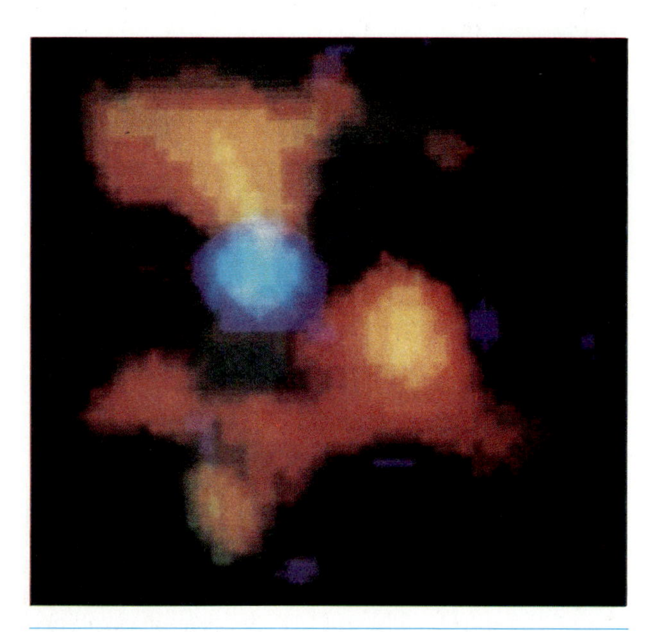

FIGURE 22.13 AN EMBEDDED INFRARED SOURCE. *In this infrared photograph of a region within the Orion nebula, blue indicates the location of an intense, relatively warm source that is thought to be a newborn star.* (R. D. Gehrz, J. Hackwell, and G. Grasdalen, University of Wyoming)

FIGURE 22.14 THE ORION NEBULA. *At top is a normal photograph, showing the familiar nebula. Below is an infrared image of the same region. Here little of the glowing gas can be seen, but the young, hot stars that are embedded within the outer portions are clearly visible. Note that the Trapezium, a close group of four brilliant young stars, is visible in both images. (top:* U.S. Naval Observatory; *bottom:* © 1984 Anglo-Australian Telescope Board)

stars just in the process of shedding the excess gas and dust from which they formed; in fact, spectroscopic measurements indicate that some material may still be falling into these stars, as though the collapse of the cloud from which they are forming is not yet complete. We learned in Chapter 17 that the Sun, in its evolution, probably went through an early phase as a T Tauri star.

STAR FORMATION

The process of star formation deduced from the types of observations just described begins with the gravitational collapse of an interstellar cloud (Fig. 22.15). It is not certain what causes a cloud to begin to fall in on itself, but once it starts, gravity takes care of the rest (see Astronomical Insight 22.2). Calculations show that the innermost portion collapses most quickly, while the outer parts of the cloud are still slowly

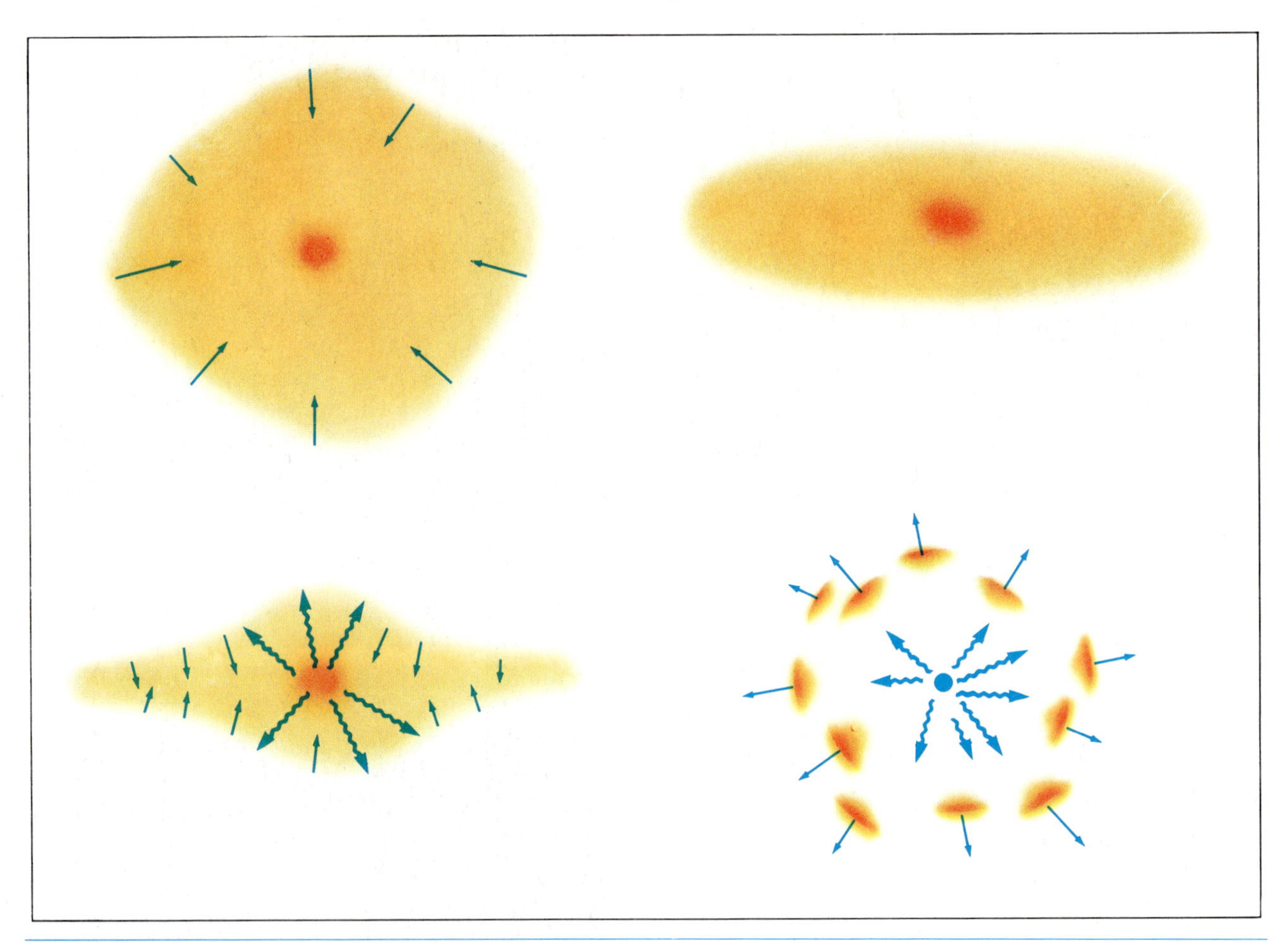

FIGURE 22.15 STEPS IN STAR FORMATION. *A rotating interstellar cloud collapses, most rapidly at the center. At first infrared radiation escapes easily and carries away heat, but eventually the central condensation becomes sufficiently dense that this radiation is unable to escape, and the core becomes hot. After a lengthy period of slow contraction, nuclear reactions begin in the protostar. At some stage, in a process not well understood, the young star develops a strong wind that helps clear away remaining debris.*

picking up speed. The temperature in the core builds up as the density increases, but for a long time the heat escapes as the dust grains in the interior radiate infrared light. Eventually, however, the cloud becomes opaque in the center, and radiation no longer can escape directly into space. After this, the heat builds up much more rapidly, and the resulting pressure causes the collapse to slow to a very gradual shrinking. Material still falling in crashes into the dense core, creating a violent shock front at its surface. The luminosity of the core, which by this time may have a temperature of 1,000 degrees or more, depends on its mass. If the mass is large, that is, several times the mass of the Sun, then the luminosity may be sufficient to blow away the remaining material by radiation pressure. For a lower-mass **protostar,** as the central dense object is called, material continues to fall in. Eventually the density of this material becomes low enough that the protostar can be seen through it. Regardless of mass, a point is reached at which a starlike object becomes visible to an observer; this stage may be identified with the T Tauri stars mentioned earlier. If the obscuring matter is swept away violently, the object may appear to brighten up dra-

ASTRONOMICAL INSIGHT 22.2

TRIGGERS OF STAR FORMATION

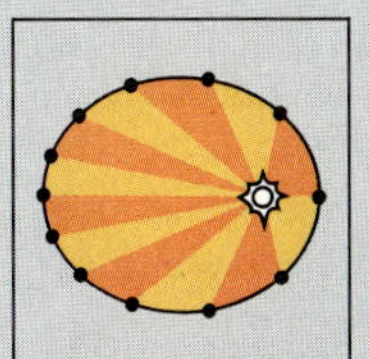

An interstellar cloud can collapse under its own gravity only if its density exceeds a certain value for its size, such that the inward gravitational force overcomes the ordinary gas pressure as a result of random motions of the cloud particles. There is a specific relationship among density, temperature, and cloud size called the **Jeans criterion** (named after Sir James Jeans, the British astrophysicist who derived it in the early decades of the twentieth century), which predicts the mass of cloud that can collapse under a given set of physical conditions. This criterion states that, as long as there are normal interstellar cloud densities (several thousand to perhaps a million atoms per cubic centimeter in dark clouds), only very massive clouds can collapse gravitationally. This poses the dilemma of explaining how stars with masses like that of the Sun or smaller can form.

One answer to this question has been to assume that as a massive cloud collapses, it breaks up into smaller fragments that can then individually collapse into stars. This is possible because as the overall density increases during collapse, the Jeans criterion allows smaller and smaller masses to condense individually. Furthermore, this picture provides a nice explanation of how a cluster of stars may form, starting with a very massive cloud that creates fragments as it collapses, the fragments in turn forming individual stars.

But what of individual, isolated stars? There is no evidence, for example, that the Sun was ever a member of a cluster. Evidently it is possible for small-mass clouds to collapse and form stars, contrary to the Jeans criterion.

A hint at the answer to this problem is found in regions of active star formation, such as the Orion nebula. Here observations show that the interstellar material is chaotic; it is stirred up by violent shock fronts, probably themselves triggered by supernova explosions or strong stellar winds. The entire interstellar medium is in a constant state of turmoil (see Chapter 26), so that a cloud of gas and dust is likely to be ravaged sooner or later by a passing shock front. Theoretical calculations show that in some circumstances this event will cause the cloud to compress enough to make it collapse, even though the cloud's initial density may not have been high enough to satisfy the Jeans criterion.

Supernova explosions occur in massive stars when all the nuclear fuel in the core is depleted (see Chapter 23 for a more complete discussion). These are the same stars that have very short lifetimes. Hence, when a cluster of stars forms, its most massive members will evolve and explode as supernovae rather quickly. The shock waves from these explosions can then trigger the collapse of interstellar material remaining from the original formation of the cluster, creating a new generation of stars. The most massive of these new stars will also evolve quickly, themselves dying in supernova explosions and creating shock waves that can give birth to yet another generation of stars. A sort of chain reaction of star formation is set off, gradually consuming an entire interstellar cloud complex.

There is some evidence that the Sun's formation was associated with a nearby supernova explosion. Certain atomic isotopes found in meteorites show evidence of having formed out of an isolated interstellar cloud that was compressed by shocks created in a supernova outburst. This was not part of a chain-reaction process like that just described, presumably because the Sun's predecessor cloud was not in a dense region of gas and dust.

ASTRONOMICAL INSIGHT 22.3

HYPERACTIVE YOUNG STARS

For many years a class of peculiar stars called **T Tauri** stars has been recognized and studied. These stars, named after the brightest and first-identified example, are characterized by cool-star spectra (most are G stars), and by prominent emission lines. The T Tauri objects are often associated with dense clouds of interstellar matter, and are therefore generally best observed at infrared wavelengths, which penetrate the interstellar dust far better than visible light.

The presence of emission lines has been taken as an indicator that these stars have very active chromospheres, a supposition borne out by recent ultraviolet observations. Whereas emission lines in visible wavelengths do not provide much information on the temperature of the gas, there are many good diagnostics of hot gas in ultraviolet wavelengths. Data obtained primarily with the *International Ultraviolet Explorer* (see Chapter 7) have revealed a number of bright ultraviolet emission lines indicative of a very hot gas. Many T Tauri stars have been found by the *Einstein Observatory* satellite to emit X-rays as well. There is no doubt that the T Tauri stars have chromospheres, probably heated by energy transported from the outer convective zones of the stars to higher levels. This is analogous to the Sun's chromosphere, except that the amount of energy involved in the T Tauri chromospheric heating is apparently greater than in the Sun.

B. J. Bok

There are indications, in the Doppler shifts that sometimes are seen in the emission and absorption lines in T Tauri star spectra, that gas is flowing outwards at high velocities. Careful study of the Doppler shifts has shown that the situation is not simple, however. Several cases have been found of T Tauri stars apparently *receiving* gas that is falling in, and certain stars have even shown both infall and outflow at the same time. Probably a more accurate picture would be to assume that, rather than having an overall wind flowing one way or the other, a typical T Tauri star has gas simultaneously rising and falling, in analogy with the solar prominences whose motions are governed by the Sun's magnetic field.

There are other indirect indications that magnetic fields play important roles in T Tauri stars. From observations of bright patches called **Herbig-Haro objects** that are sometimes seen amid the nebulosity near T Tauri stars, it is inferred that gas is ejected at great velocities from the stars and is beamed along very specific directions. The Herbig-Haro objects appear to be "hot spots" where the high-velocity beams strike surrounding material (see photograph). Very recent radio observations have directly revealed a linear jet or beam of hot gas emanating from one suspected T Tauri star, helping to clarify this interpretation of the Herbig-Haro objects. The best mechanism for explaining these beams of gas is to invoke a magnetic field, with outflowing ionized gas from the poles of the star moving along the field lines. Recently these **bipolar flows** from young stars have begun to receive a great deal of attention because they may be a standard phenomenon in star formation, rather than isolated occurrences.

The general consensus on T Tauri stars is that they are very young objects on their way to becoming stars. This conclusion is based on their locations within dense regions of interstellar material, on their

ASTRONOMICAL INSIGHT 22.3

HYPERACTIVE YOUNG STARS

apparent instability (the spectra and the gas motions they reveal are often variable with time), and on comparisons of their positions in the H-R diagram with theoretical calculations of the properties of newly forming stars. The "T Tauri wind," a phase thought to be undergone by young stars in general, is usually invoked to explain how the early solar system was cleared of leftover debris from the formation of the planets (see Chapter 17).

Very recently, the picture of T Tauri stars as young suns was supported by a spectacular discovery. Using a technique called **speckle interferometry** (see Astronomical Insight 20.1) in infrared wavelengths, astronomers found that the prototype star, T Tauri itself, is a binary system. Further analysis of the properties of the dim companion have established that it is not a star at all but is instead a planet, some five to twenty times as massive as Jupiter. This is a striking confirmation of the evolutionary theory of solar system formation (discussed in Chapter 17) and lends credence to the general supposition that planetary systems may be quite common. It is fitting that this discovery should be associated with the "standard" young star T Tauri, although not surprising, since this is the brightest and best-studied example of its class.

matically in a very short time. In a few cases, a previously rather faint star was found to have brightened suddenly by several magnitudes.

In any case, the protostar continues to shrink slowly, growing hotter in its core. The shrinking stops when the temperature becomes high enough for nuclear reactions to begin to occur in the center. This is a major landmark in the process of star formation, for once the reactions have started, the star is on its way to becoming stable (that is, it shrinks no further), and it lies on the main sequence, where it will spend most of its life converting hydrogen into helium in its core.

It is interesting to trace on the H-R diagram the path the star takes as it forms (Fig. 22.16). When it is a cold, dark cloud, its position is far to the right

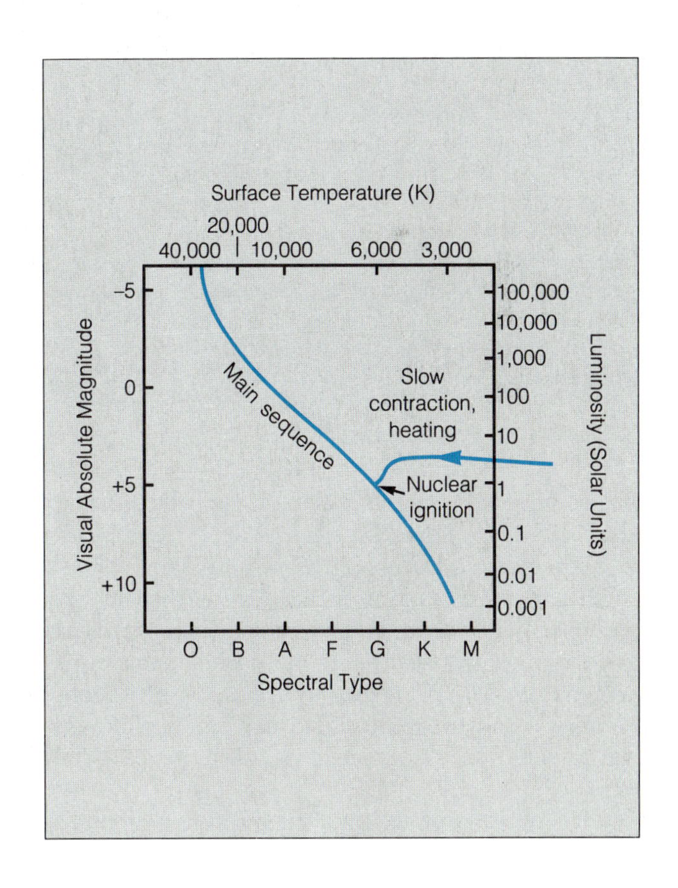

FIGURE 22.16 THE PATH OF A NEWLY FORMING STAR ON THE H-R DIAGRAM. *As a protostar heats up, it eventually becomes hot and luminous enough to appear in the extreme lower right-hand corner of the H-R diagram. When the core becomes opaque, the rapid collapse slows to a gradual shrinking, and the protostar moves across to the left as it heats. Eventually nuclear reactions begin, and the star is on the main sequence.*

TABLE 22.2 STEPS IN STAR FORMATION (ONE SOLAR MASS)

1. Gravitational collapse of interstellar cloud: Collapse proceeds most rapidly at the center, where the density grows most quickly. The cloud core remains cool because heat escapes in the form of infrared radiation.

2. Slow contraction of core: Once the core density is high enough so that the cloud becomes opaque to infrared radiation, heat is trapped, and the temperature and pressure in the core rise. The central portion of the cloud contracts very slowly for a while.

3. Infall: Material from the outer portions of the cloud continues to fall in, creating heating and shock waves. When the hydrogen molecules in the core begin to break up because of the additional heating, the breakup of molecules takes up heat from the gas, reducing the pressure, so that a new collapse phase occurs.

4. Final contraction: After the second collapse, the internal pressure again rises to balance gravity, and the central condensation enters a new phase of very slow contraction. At this point, the protostar is still surrounded by gas and dust, and is usually observable only as an infrared source.

5. Nuclear ignition: The protostar continues to heat gradually as the contraction goes on. Eventually, the core becomes hot enough to start nuclear fusion reactions, and the contraction stops because there is an internal source of energy to provide pressure that balances gravity fully. The star is now on the main sequence. Observationally, it may still be associated with gas and dust and may still undergo a T Tauri phase.

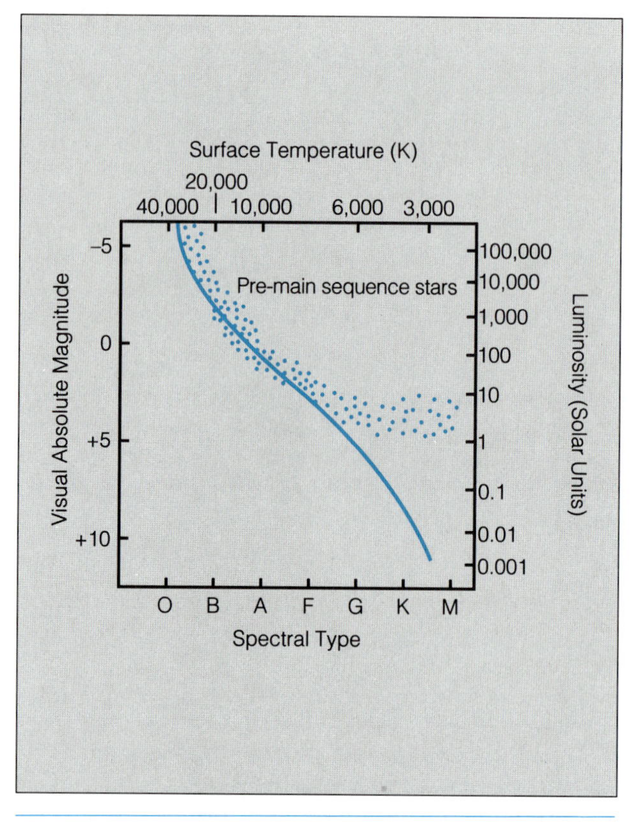

FIGURE 22.17 PRE-MAIN SEQUENCE STARS. *H-R diagrams of very young clusters often show stars that have not yet reached the main sequence, but are still moving toward it. At the same time, some of the massive, short-lived stars at the top of the main sequence may have already evolved away from it, having used up their hydrogen fuel.*

and down, well off the scales of temperature and luminosity appropriate for stars. As the cloud heats up in the core and begins to glow in infrared wavelengths, it moves up and to the left, eventually attaining sufficient luminosity to fall onto the standard H-R diagram, but still off to the right of the main sequence. When it reaches the phase of slow contraction, it moves gradually to the left and down, finally reaching the main sequence when the reactions begin.

The entire process, from the beginning of cloud collapse to the point at which the newly formed star is on the main sequence (Table 22.2), takes many millions or even billions of years for the least massive stars, and only a few hundred thousand years for the most massive. In a cluster with stars of various masses, the most massive stars may form, live, and die before the least massive ones even reach the main sequence. The H-R diagram for such a cluster shows stars located to the right of the main sequence at the lower end (Fig. 22.17).

PERSPECTIVE

The observational evidence tells us that stars must form out of interstellar material, and that they spend most of their lives on the main sequence. H-R diagrams of clusters show that, upon completion of the main-sequence lifetime, a star moves up and to the right, becoming a red giant. By combining these observational deductions with the results of stellar model calculations, we can deduce the detailed life stories of individual stars.

SUMMARY

1. Steps in stellar evolution are deduced from observations of stars in various stages, from newly formed ones to those in the late stages of their lives.
2. Clusters are groups of stars that formed together out of the same material.
3. The relative brightnesses of stars in a cluster reflect real differences in luminosity, because all the stars in a cluster are essentially at the same distance from us.
4. The distances to clusters can be determined by plotting a color-magnitude diagram and comparing the location of the main sequence with that on a standard H-R diagram.
5. The fact that all stars in a cluster have the same age and initial composition means that the observed differences from star to star within the cluster are entirely a result of differences in stellar mass. This allows us to determine how stars of different mass evolve.
6. As stars evolve, they move off of the main sequence, becoming cooler and more luminous, and move into the red-giant region of the H-R diagram.
7. The age of a cluster can be inferred from the location of its main-sequence turnoff point.
8. In associations of young, hot stars, interstellar material is often left over from the process of star formation.
9. Stars in the process of forming are often embedded within dense clouds and are best observed in infrared wavelengths.
10. A star forms from the gravitational collapse of an interstellar cloud, which condenses most rapidly at the center. The new star begins to heat up only after the gas in the core of the cloud becomes opaque, and the resulting protostar continues to slowly shrink until its interior becomes hot enough for nuclear reactions to begin.

REVIEW QUESTIONS

Thought Questions

1. Summarize the importance of star clusters in the study of stellar evolution.
2. Describe the various types of star clusters mentioned in the text. Include an indication of where they are likely to be found in the galaxy.
3. If a cluster is observed to have O stars in it, how do we know that it must be a young cluster? Is it possible for the same cluster to also contain M stars?
4. Suppose two stars in a cluster have identical $B-V$ color indices and apparent magnitudes. How do their masses and other properties compare? Could you make the same assumption for a pair of stars not in the same cluster, but which also have identical color indices and apparent magnitudes?
5. Describe the main sequence fitting technique for finding distances to star clusters. On what assumptions about the stars in a cluster does this method depend?
6. Explain, in your own words, how the main sequence turnoff for a cluster H-R diagram is related to the age of the cluster.
7. Explain why infrared observations are an important method for observing star formation.
8. Summarize the sequence of events thought to take place as a star forms.
9. Is a young star in hydrostatic equilibrium when it is in the slow gravitational contraction stage? Explain.
10. What determines the location on the main sequence where a star will lie after it has formed?

Problems

1. Comparison of the main sequence of a cluster color-magnitude diagram with a standard H-R diagram shows that the apparent magnitudes of the stars on the cluster diagram main sequence are 5 magnitudes larger than the absolute magnitudes of the corresponding stars on the standard H-R diagram. How far away is the cluster?
2. Using what you learned about stellar lifetimes and about the properties of main-sequence stars (Table 20.2), calculate the age of a cluster whose brightest main-sequence stars belong to spectral type F5. (Hint:

Use Table 20.2 to find the mass and luminosity of an F5V star, then calculate its lifetime). What is the $B-V$ color index for the main sequence turnoff for this cluster (again, use Table 20.2)?

3. When a star leaves the main sequence, it becomes a red giant, having a larger radius and lower surface temperature. Suppose the Sun becomes a red giant, with a radius twenty times greater and a surface temperature one-half its present value. By how much does its luminosity increase? How many magnitudes brighter (in bolometric magnitudes) is it?

ADDITIONAL READINGS

Allen, David A. 1987. Star formation and the IRAS galaxies, *Sky and Telescope* 73(4):372.

Bok, B. J. 1972. The birth of stars. *Scientific American* 227(2):48

———. 1981. The early phases of star formation. *Sky and Telescope* 227(2):48.

Boss, Alan P. 1985. Collapse and formation of stars. *Scientific American* 252(1):40.

Cohen, M. 1975. Star formation and early evolution. *Mercury* 4(5):10.

Gehrz, R. D., D. C. Black, and P. M. Solomon. 1984. The formation of stellar systems from interstellar molecular clouds. *Science* 224(4651):823.

Graham, John. 1984. A new star becomes visible in Chile. *Mercury* 13(2):60.

Herbst, W., and G. E. Assousa. 1979. Supernovas and star formation. *Scientific American* 241(2):138.

Jones, B., D. Lin, and R. Hanson. 1984. T Tauri and TIRC: A star system in formation. *Mercury* 13(3):71.

Lada, Charles J. 1986. A star in the making. *Sky and Telescope* 72(4):334.

Lada, C. J. 1982. Energetic outflows from young stars. *Scientific American* 247(1):82.

Loren, R. B., and F. J. Vrba. 1979. Starmaking with colliding molecular clouds. *Sky and Telescope* 57(6):521.

Scoville, N., and J. S. Young. 1984. Molecular clouds, star formation, and galactic structure. *Scientific American* 250(4):42.

Spitzer, L. 1983. Interstellar matter and the birth and death of stars. *Mercury* 12(5):142.

Waldrop, M. Mitchell. 1983. Stellar nurseries. *Science 83* 4(4):40.

Werner, M. W., E. E. Becklin, and G. Neugebauer. 1977. Infrared studies of star formation. *Science* 197:723.

Zeilik, M. 1978. The birth of massive stars. *Scientific American* 238(4):110.

CHAPTER 23

Life Stories of Stars

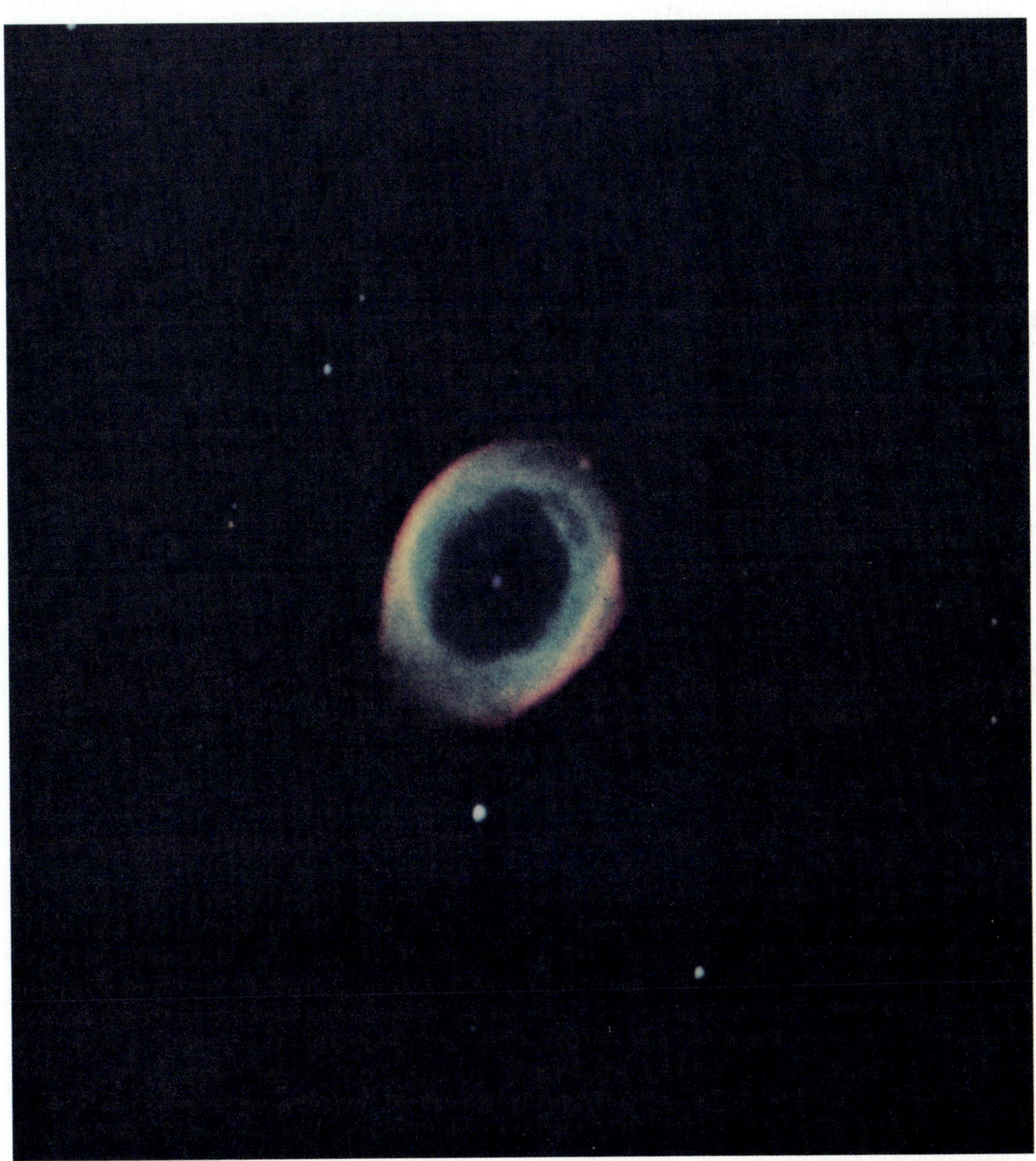

The Ring nebula, a planetary nebula. (Lick Observatory photograph)

CHAPTER PREVIEW

Combining our knowledge of stellar structure with observations showing how stars evolve, we can develop a picture of the story of an individual star, tracing it from its birth to its death. Because so much depends on the mass of a star, we will discuss three separate examples: a star like the Sun, a star with a mass a few times that of the Sun, and a very massive star. In the process we will see many important differences among the three cases.

We have already discussed the formation of stars, finding that the process is rather similar for all masses, except for the time required. We will not reiterate this discussion here; we will begin our story for each star on the main sequence.

STARS LIKE THE SUN

A star with the mass of the Sun, when it arrives on the main sequence, is similar to the Sun in all its properties. Thus it is a G2 star, with a surface temperature around 6,000 K and a luminosity of about 10^{33} erg/sec. Its composition initially is about 73 percent hydrogen by mass, and roughly 25 percent helium, with only about 2 percent of the mass composed of other elements. It has convection in the outer layers, and it has a chromosphere and a corona. The star must change with time, because its core composition changes. Table 23.1 summarizes the stages in the life of a star like the Sun.

TABLE 23.1 EVOLUTION OF A 1-$M_\odot$ STAR

1. *Hydrogen burning:* During the 10^{10} years the star spends on the main sequence, hydrogen is converted to helium through the proton-proton chain. The luminosity gradually increases as the core becomes denser and hotter.

2. *Development of degenerate core and evolution to red giant:* When hydrogen in the core is used up, the core contracts until it is degenerate. A hydrogen shell source outside the core still undergoes reactions, causing the outer layers to expand and cool. The star becomes a red giant.

3. *Helium flash:* The core eventually gets hot enough to ignite the triple-alpha reaction, in which helium is converted to carbon. Because the core is degenerate and cannot expand and cool to counteract the new source of heat, the reaction rapidly takes place throughout the core. So much heat energy is added that degeneracy is quickly destroyed, and the reactions become stable.

4. *Stable helium burning:* Once reactions are taking place in the core again, the major energy production is there, and the shell source becomes less important. The outer layers contract, and the surface heats up; the star moves back to the left on the H-R diagram.

5. *Second red giant stage:* When the core helium is used up, the energy again comes from a shell (an inner shell, where helium is converted to carbon, and an outer shell, where hydrogen is still converted to helium). The outer layers expand again, and the star becomes a red giant for the second time.

6. *Mass loss:* Through a stellar wind and through pulsational instabilities, the star loses its outer layers, exposing the hot core. The core continues shrinking but never again is hot enough for nuclear reactions. The star becomes a planetary nebula periodically as the outer layers are shed.

7. *White dwarf:* The core again becomes degenerate as it shrinks. Eventually the outer layers are gone altogether, leaving only the hot, degenerate core, which is a white dwarf.

In the core, the proton-proton chain converts hydrogen into helium. As we saw in Chapter 20, this process lasts some 10 billion years for a star of 1 solar mass. During this time, the core gradually shrinks as the hydrogen nuclei are replaced by a smaller number of the heavier helium nuclei, and the internal tem-

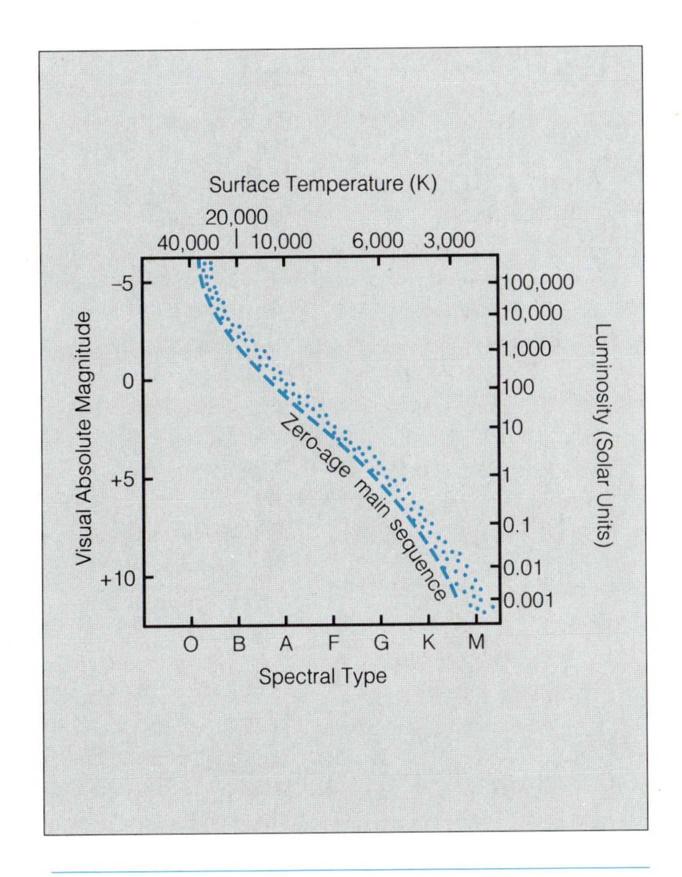

FIGURE 23.1 THE WIDTH OF THE MAIN SEQUENCE. *As stars on the main sequence gradually convert hydrogen into helium in their interiors, the cores shrink a little, and get hotter. This increases the luminosity, and the stars gradually move upward on the H-R diagram. As a result, the main sequence, consisting of stars of a variety of ages, is not a narrow strip, but has some breadth.*

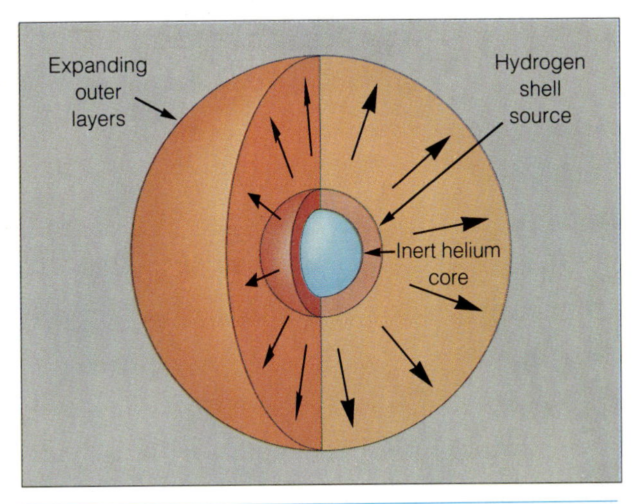

FIGURE 23.2 THE HYDROGEN SHELL SOURCE. *After the hydrogen in the core of a star is completely used up, the core, now composed of helium, continues to shrink and heat. This causes a layer of gas outside the core to reach the temperature required for nuclear reactions. A shell source, in which hydrogen is converted into helium, is ignited. The star's outer layers expand and cool, and the star becomes a red giant.*

perature rises. This causes an increase in luminosity, and the star gradually moves up in the H-R diagram. The main sequence, therefore, is not a perfectly narrow strip but has some breadth, since stars on it slowly move upward as their luminosities increase (Fig. 23.1). The starting point, the lower edge of the main sequence, is called the **zero-age main sequence (ZAMS),** because this is where newly formed stars are found.

When the hydrogen in the core is gone, reactions there cease. By this time, however, the temperature in the zone just outside has nearly reached the point at which reactions can take place, and with a little more shrinking and heating of the core, the proton-proton chain begins again in a spherical shell surrounding the core (Fig. 23.2). At this point, the star has an inert helium core, and it is producing all its energy in the hydrogen-burning shell, which steadily moves outward inside the star, eating its way through the available hydrogen.

The core continues to shrink and heat. This heating enhances the nuclear reactions in the shell, causing it to produce more and more energy. As the source of energy is heated and moves toward the surface, the outer layers of the star are forced to expand, and as this occurs, the surface cools. The luminosity of the star increases because of its increased surface area, so it moves upward on the H-R diagram (Fig. 23.3). It also moves to the right, because the surface temperature is decreasing. The star rapidly becomes a red giant, reaching a size of 10 to 100 times its main-sequence radius (Fig. 23.4). The outer layers are constantly overturning as convection occurs to a great depth (Fig. 23.5).

The helium core becomes extremely dense. At first the temperature is not high enough for any new nuclear reactions to start there. (The triple-alpha reaction, in which helium nuclei combine to form car-

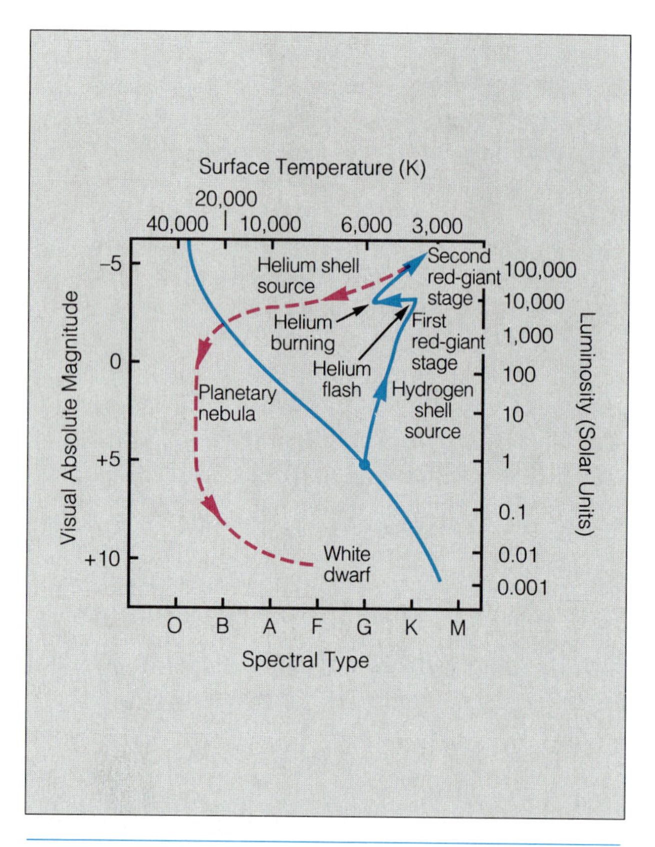

FIGURE 23.3 EVOLUTIONARY TRACK OF A STAR LIKE THE SUN. *This H-R diagram illustrates the path a star of one solar mass is thought to follow as it completes its evolution. The dashed portion, following the second red giant stage, is less certain than the earlier stages.*

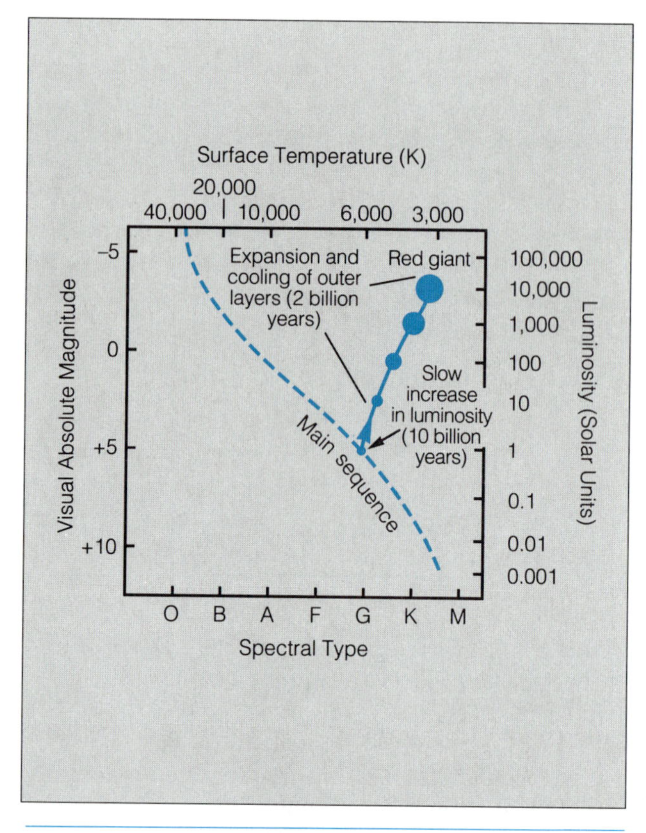

FIGURE 23.4 THE DEVELOPMENT OF A RED GIANT. *As a star's outer layers expand because of the shell hydrogen source, they cool. At the same time the surface area increases, raising the star's luminosity. The star therefore moves up and to the right on the H-R diagram.*

bon, requires a temperature of about 100 million degrees.) As the core continues to shrink, the matter there takes on a very strange form. A new kind of pressure gradually takes over from the ordinary gas pressure that has been supporting the core. Electrons have a property that prevents them from being squeezed too close together, and this creates the new pressure. The gas in the stellar core contains many free electrons, and eventually their resistance to being compressed becomes the dominant pressure that supports the core against further collapse. When this happens, the gas is said to be **degenerate.**

A degenerate gas has many unusual properties. One of them is that the pressure no longer depends on the temperature, and vice versa. If the gas is heated further, it will not expand to compensate, as an ordinary gas would. As we will soon learn, there are important consequences if nuclear reactions start in the degenerate region of a star.

We now have a red giant star, with highly expanded outer layers. Near the center is a spherical shell in which hydrogen is burning in nuclear reactions, and inside of this is a degenerate core containing helium nuclei. The core is small, but it may contain as much as a third of the star's total mass.

The Helium Flash

During the red giant phase, the helium core continues to be heated by the reactions going on around it. Eventually the temperature becomes sufficiently high

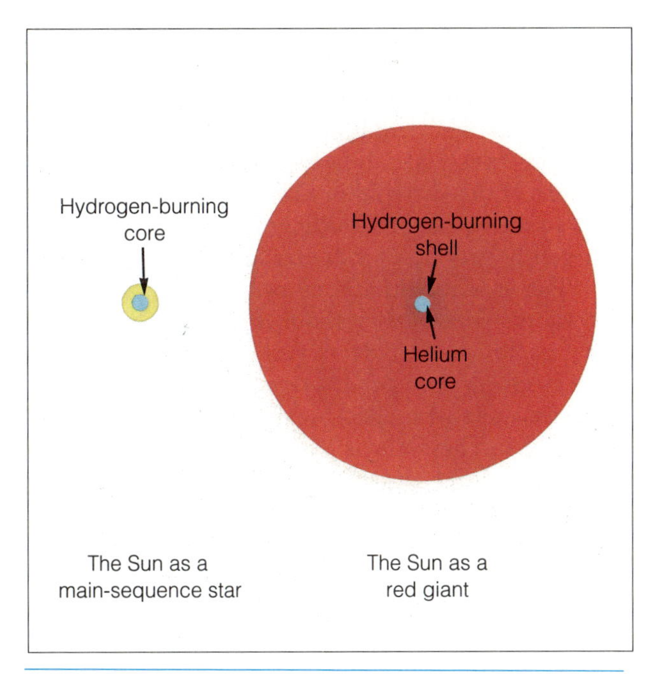

FIGURE 23.5 STRUCTURE OF A RED GIANT. *When a star like the Sun becomes a red giant its diameter grows much larger, but most of its mass remains concentrated in a small inner region. Energy is supplied by nuclear reactions in the hydrogen shell source.*

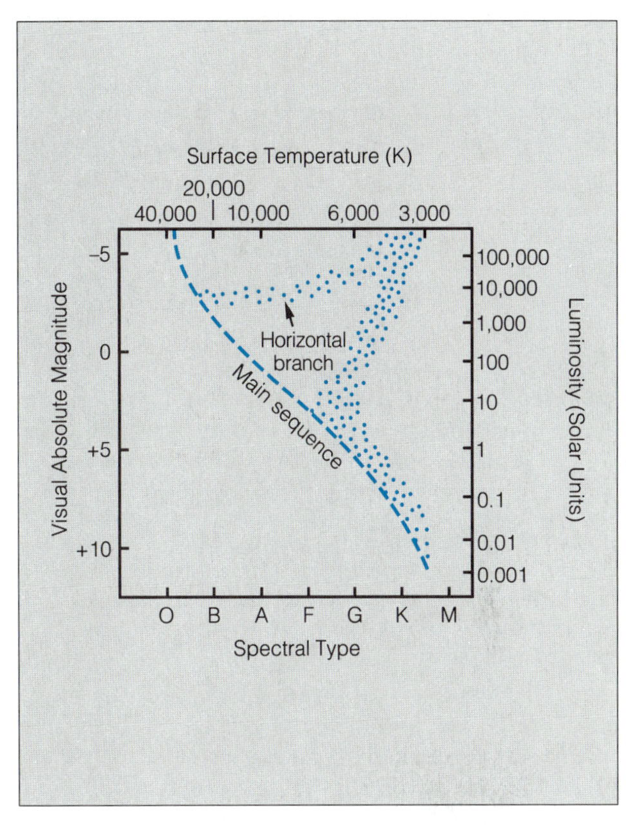

FIGURE 23.6 THE H-R DIAGRAM FOR A GLOBULAR CLUSTER. *These star clusters are very old, and all the upper main sequence stars hve evolved to later stages. Here we see a number of stars in or approaching the red giant stage, and a number on the horizontal branch, where they move following the helium flash. Only very old clusters have a horizontal branch. (Besides their great ages, globular clusters also have a relatively low content of heavy elements, and this affects how the stars evolve. This point is discussed more fully in Chapter 27.)*

for the triple-alpha reaction to begin. When it does, there are spectacular consequences.

Ordinarily, when a gas is heated, it expands, and this limits how hot it can get, because an expanding gas tends to cool. In a degenerate gas, however, no such expansion occurs, because pressure does not depend on temperature in the usual way. Hence, when the reactions begin in the degenerate core of a red giant, the temperature goes up quickly, and the core retains the same density and pressure. The increased temperature speeds up the reactions, producing more heat, in turn accelerating the reactions even further. There is a rapid snowball effect, and in an instant (literally seconds), a large fraction of the core is consumed in the reaction. This spontaneous runaway reaction is called the **helium flash.** Although it creates dramatic consequences for the star's interior, calculations show that there is little or no immediate effect visible to an observer. The overall direction of the star's evolution is changed, however, which would be apparent after thousands of years.

The helium flash quickly disrupts the core, destroying its degeneracy. The core then returns to a more normal state in which temperature and pressure are linked, and the triple-alpha reaction continues, now in a more stable, steady fashion. The outer layers of the star begin to retract as the star reverts to a more uniform internal structure, and as they do so, they become hotter. The star moves back to the left on the H-R diagram (see Fig. 23.3). Old clusters, particularly globular clusters, are found to have a number of stars in this stage of evolution, forming a sequence called the **horizontal branch** (Fig. 23.6). This is seen

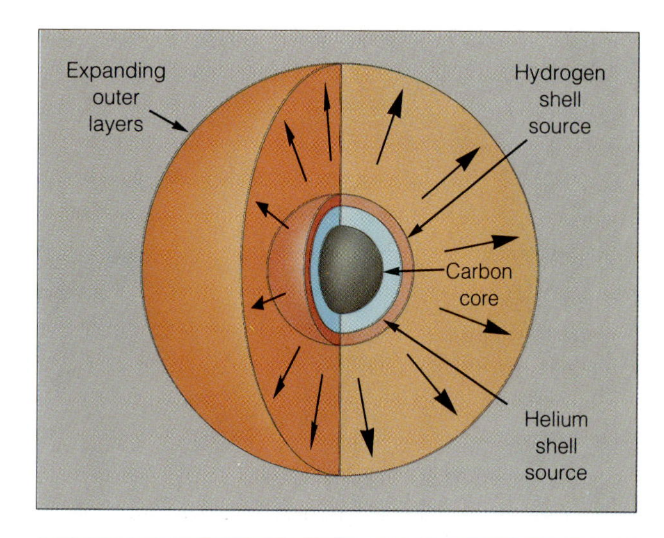

FIGURE 23.7 INTERNAL STRUCTURE LATE IN A STAR'S LIFE. *The core of this star has undergone hydrogen-burning and helium-burning, and is composed of carbon. Hydrogen and helium shell sources are still undergoing reactions and producing energy.*

only in clusters old enough for stars as small as 1 solar mass to have evolved this far; as we learned in previous chapters, this requires an age of 10 billion years or more.

In due course, the helium in the core becomes exhausted, leaving an inner core of carbon (Fig. 23.7). Helium still burns in a shell around the core; there could even be an active hydrogen-burning shell farther out in the star at the same time. The star expands and cools, becoming a red giant for the second time. This second red giant stage is short-lived, however, lasting perhaps 1 million years. A degenerate core again develops, but there is not another dramatic flare-up in its interior. In fact, no more nuclear reactions ever occur in the core of this low-mass star, because the temperature never reaches the extremely high levels needed to cause heavy elements like carbon to react.

Planetary Nebula to White Dwarf

The star in its second red giant stage may be viewed as having two distinct zones: the relatively dense interior, consisting of a carbon core inside a helium-burning shell which in turn is surrounded by a hydrogen-burning shell; and the very extended, diffuse outer layers. The inner portion may contain up to 70 percent of the mass but occupies only a small fraction of the star's volume.

As the nuclear fuel in the interior runs out, the core shrinks. The details of what happens to the outer layers are not clear, but apparently the star ejects portions of this material in a series of minor outbursts that could be likened to blowing smoke rings. The star gently gets rid of its outer layers.

The evidence that this occurs is primarily observational, since a successful theory of how it happens has not yet been developed. A number of objects have been observed that seem to consist of a hot, compact star surrounded by a shell of expanding gas (Fig. 23.8). These shells often have a bluish green color because of the emission lines by which they glow, and through a telescope or on a photograph they bear some resemblance to the planets Uranus and Neptune. Because of this resemblance, they are called **planetary nebulae.** Some, such as the Dumbbell Nebula (see Fig. 23.8) or the Ring Nebula (Fig. 23.9), are well-known objects visible with a small telescope.

The star that remains in the center of a planetary nebula lies far to the left in the H-R diagram (see Fig. 23.3), having a surface temperature that may be as high as 100,000 K or more, far hotter than any main-sequence star. Evidently, when the shell is being ejected, the star's surface temperature increases as hot inner gas becomes exposed on the surface. In essence, such a star is a naked core, left behind when the outer layers are cast off.

Some nuclear reactions may still be going on in a shell inside the star, but they do not last much longer, since the fuel in the core is depleted. The density, already very high, increases as the stellar remnant condenses under the force of gravity. A larger and larger fraction of the star's interior becomes degenerate. As the star shrinks, it stays hot, but its luminosity decreases because of its diminishing surface area, and it moves downward on the H-R diagram (see Fig. 23.3). Eventually the shrinking stops, and the star becomes stable again, but now it is a very small object, so dim that it lies well below the main sequence, in the lower left-hand corner of the H-R diagram. It is called a **white dwarf.**

A white dwarf is a bizarre object in many ways. It is made of degenerate matter whose peculiar properties govern its internal structure. It has a mass as great as that of the Sun (some white dwarfs are even

FIGURE 23.8 PLANETARY NEBULAE. *At top is the Dumbbell Nebula; at bottom, the Helix Nebula. In both, the blue-green color is produced by emission lines of twice-ionized oxygen. The central stars, which are responsible for ejecting the gas of the nebulae, are visible at the center of each. (top:* Lick Observatory photograph; *bottom:* © 1979 Anglo-Australian Telescope Board.)

slightly more massive), yet it is approximately the size of the Earth. This means that its density is incredibly high, roughly a million times that of water. A cubic centimeter of white-dwarf material would weigh a ton at the surface of the Earth!

A white dwarf does not do much. It has no more nuclear fuel, so it does not evolve further. It just sits there, radiating away the heat energy still contained inside and slowly cooling off. A white dwarf can be stirred into action under certain circumstances (as described in the next chapter), but in most cases, this is the end of the line for a star like the Sun.

FIGURE 23.9 THE RING NEBULA. *This is one of the best-known and most beautiful planetary nebulae, the Ring nebula in the constellation Lyra.* (Lick Observatory photograph)

THE MIDDLEWEIGHTS

Let us now consider a star of significantly greater mass than the Sun, say 5 to 10 solar masses. For this star, much of the evolution is similar to that of the 1-solar-mass star, although there are some major differences (Table 23.2). One of the contrasts, of course, is the time scale, because this star will have perhaps 100 times the solar luminosity and will use up its fuel accordingly faster. Let us quickly escort this star through its development, starting again on the main sequence. The path in the H-R diagram is shown in Figure 23.10.

Our middleweight star lands on the middle to upper portion of the zero-age main sequence when it has completed its formation, and nuclear reactions begin. Its spectral type is B8 if its mass is 5 solar masses, and B2 if it is 10 solar masses. The nuclear reaction in the core is the CNO cycle, producing helium from hydrogen. There is no convection in the outer layers, but there is in the core. The surface temperature is roughly between 12,000 and 20,000 K, depending on the mass.

TABLE 23.2 EVOLUTION OF A 5-$M_{\odot}$ STAR

1. *Hydrogen burning:* While on the main sequence, the star converts hydrogen to helium in its core by the CNO cycle. The time on the main sequence is much shorter than for a 1-$M_{\odot}$ star—about 100 million years instead of 10 billion.

2. *First red giant stage:* When the core hydrogen is gone, a shell source powers the star, causing the outer layers to expand and cool as the star becomes a red giant. The core does not become degenerate because of its high temperature, and there is no helium flash.

3. *Subsequent red giant stages:* Each time a nuclear fuel source in the core is used up, the shell source outside the core becomes the dominant source of energy, and the star becomes a red giant. The core then contracts and heats until a new fuel ignites, and the star shrinks again, moving back to the left on the H-R diagram. A star may undergo several red giant loops in this way, progressively burning helium, carbon, oxygen, neon, and other elements in the sequence of alpha-capture reactions.

4. *Mass loss:* During each red giant or supergiant phase, mass is lost through stellar winds. Eventually the loss of mass and the requirement for ever-higher temperatures to start new reaction phases combine to halt all nuclear reactions.

5. *Final stage:* The form of remnant left depends on how much mass remains after all the red giant stages. If the final mass is under 1.4 $M_{\odot}$, the star will end as a white dwarf whose composition is determined by the number of nuclear reaction stages that took place. If the final mass is between 1.4 and 2 or 3 $M_{\odot}$, then the remnant will be a neutron star. If the final mass is over the neutron star limit, then a black hole results. It is thought that most stars with initial masses below 8 $M_{\odot}$ lose enough to become either white dwarfs or neutron stars.

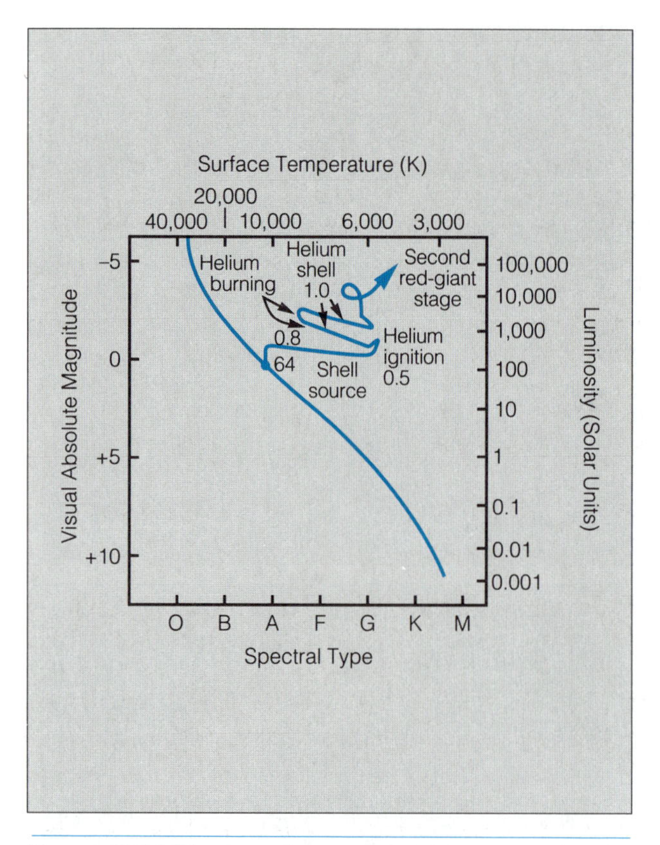

FIGURE 23.10 THE EVOLUTIONARY PATH OF A SOLAR-MASS STAR. *This star goes through a more complicated series of stages than a less massive star. There are probably additional red giant stages following those indicated here, but calculations of the evolution have not yet been carried that far. The numbers indicate times in units of 10^6 years.*

Multiple Red Giant Loops

As in the case of the Sun-like star, the more massive star on the main sequence gradually brightens as its core becomes more compressed. The hydrogen in its core is gone (after about 100 million years) before a shell outside the core is ignited, and for a brief period the star simply contracts and heats, with no reactions taking place. During this time, it turns sharply to the left in the H-R diagram. After only 1 or 2 million years, however, hydrogen begins burning again, in a shell outside the core, and the star reverses itself and heads for the red giant region. Within a few hundred thousand years, it reaches the upper right-hand portion of the H-R diagram. In this case, the core is not degenerate, because the temperature is high enough to prevent a collapse to the necessary density, so no helium flash occurs. When the core is hot enough, the triple-alpha reaction quietly begins, and the star contracts and moves to the left in the H-R diagram. This star does not move far to the left, however, and never really leaves the red giant region before it develops a new reaction shell outside the core, and expands and cools once again, reversing itself and heading back to the right.

What happens next is not clear, but a sketchy picture is emerging. The core temperature continues to increase, and it is possible for new nuclear reactions to begin. These reactions, called **alpha-capture reactions,** start with carbon and helium nuclei (remember, a helium nucleus consisting of two protons and two neutrons is called an alpha particle). A carbon nucleus can capture an alpha particle, forming an oxygen nucleus. This in turn can capture another alpha particle, forming neon, and so on. If the core temperature reaches high enough levels, a complex sequence of reactions can occur, gradually building up a supply of heavy elements in the star's core. Each time one fuel source is exhausted and shell burning begins, the star expands and moves upward and to the right in the H-R diagram. Each time a new reaction starts in the core, the star settles down and moves to the left. Thus, a middleweight star can execute a number of left-to-right loops in the red giant region of the H-R diagram. The number of loops depends on the mass. The situation is very complicated, however, and theoretical calculations do not give definitive answers.

One thing that complicates matters is that extremely luminous red giants lose mass through strong, dense winds (as described in Chapter 21). The rate of mass loss can be sufficient to significantly alter the mass of the star in a few hundred thousand years. If a star's mass is reduced, its evolution is modified. The central density and temperature are lower, which in turn reduces the number of reaction stages the star goes through. At some difficult-to-determine point, all reactions cease. Then the red giant contracts and moves down and to the left in the H-R diagram, possibly passing through a phase as a planetary nebula.

Death: To Go Out Quietly or with a Bang?

Whether or not these middleweight stars end up as white dwarfs is uncertain. It has been shown theoretically, and confirmed from observations, that a white dwarf cannot exist if the star's mass is greater than a certain limit, about 1.4 times the mass of the Sun. If a mass is any larger, it creates sufficient gravitational force to overcome the pressure of the degenerate electron gas in the core, and the star collapses further. As we will see, this collapse leads to a violent ending in some stars. Stellar wind and planetary-nebula ejection might reduce the mass far enough, however, so it is possible that many stars that start out in the middleweight division end up as white dwarfs. Current theoretical models and stellar wind observations indicate that stars initially having up to 8 solar masses can lose enough mass to end up as white dwarfs.

The white dwarfs formed from the middleweight stars are different from those formed by lower-mass stars, however. Recall that for a star of 1 solar mass, the most advanced nuclear reaction that occurs is the triple-alpha process, converting helium into carbon. The 5- to 10-solar-mass star goes well beyond this point, however, so heavier elements are built up in its core. The white dwarf that results, then, may have an interior made of oxygen, neon, or even heavier elements (Fig. 23.11), whereas one produced by a lower-mass star consists of carbon or perhaps only helium.

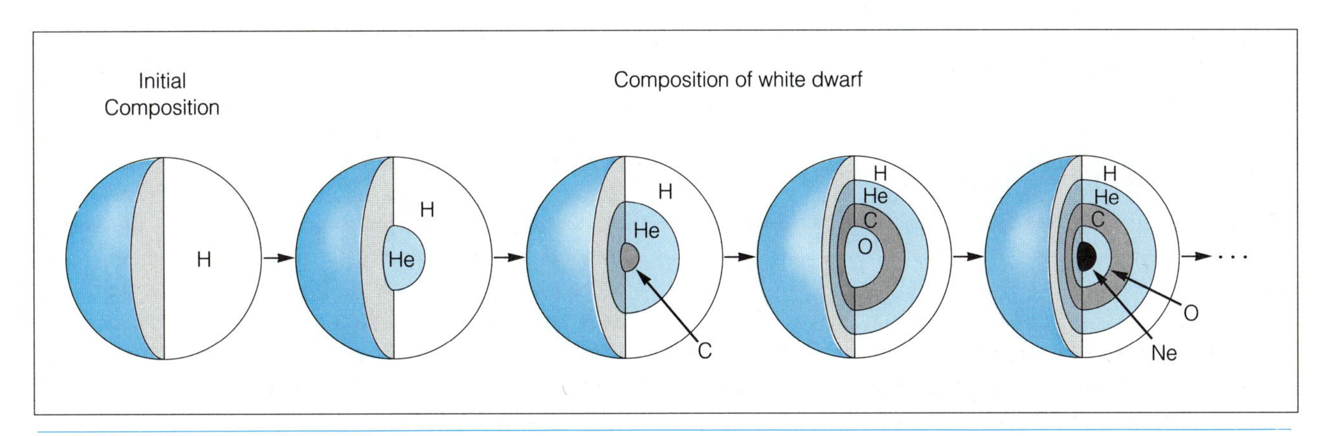

FIGURE 23.11 WHITE-DWARF COMPOSITION. *The more nuclear-burning stages a star goes through, the greater the variety of elements produced in its interior. The white dwarf that results, therefore, has a composition that depends on the number of nuclear-burning stages, and this, in turn, depends on the mass of the star.*

TABLE 23.3 HISTORICAL NAKED-EYE SUPERNOVAE

Date	Peak Magnitude	Duration	Type	Comment*
185 A.D.	−8	20 months	I?	Remnant: RCW86
393	−1	8 months	?	Remnant: unidentified
1006	−9	Several years	I	Remnant: PKS 14 59-41
1054	−5	22 months	II	Remnant: Crab nebula
1181	0	6 months	?	Remnant: 3C58
1572	−4	18 months	I	Tycho's supernova
1604	−2.5	12 months	I	Kepler's supernova
1680	+6?	—	II?	Remnant: Cassiopeia A
1885	+6	A few days	I pec.	In Andromeda galaxy
1987	+2.9	9 months + ?	II	In Large Magellanic Cloud

*Several of the comments given here are catalog numbers of the remnants of the listed supernovae. Most of the catalogs are lists of radio sources, and each has its own particular designation system.

What of the relatively massive stars that do not shed enough material to become white dwarfs? These stars contract beyond the white dwarf stage and therefore exceed the stupendous densities achieved by the white dwarfs. There seem to be two possible end results: The star may contract to another kind of degenerate state, where it can stabilize; or it may collapse indefinitely, never again reaching stability (this possibility is discussed in the next section). In either case, the collapse may be violent, with the infalling matter creating an outburst called a **supernova** explosion.

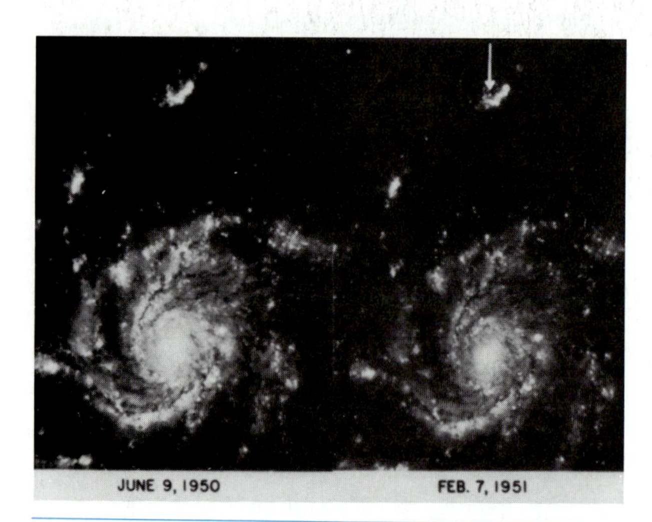

FIGURE 23.12 A SUPERNOVA IN A DISTANT GALAXY. *Before the distinction between novae and supernovae became clear, these flare-ups (arrow) contributed to the controversy over the nature of the nebulae. Now that these occurrences in galaxies are known to be supernova explosions, comparisons of their apparent and assumed absolute magnitudes provide distance estimates.*

Only a few supernovae nearby enough to be seen without a telescope have been recorded (Table 23.3). Many more have been observed in modern times, in distant galaxies (Fig. 23.12), where the incredible brilliance of supernovae can be measured. The spectra of supernovae fall into at least two distinct classes: Those showing no lines of hydrogen are designated **Type I supernovae,** while those having hydrogen lines are **Type II supernovae.** There are distinctions in the light variations of the two types as well (Fig. 23.13).

Current theoretical models indicate that a star in the mass range between 8 and 12 solar masses late in its lifetime (that is, after mass loss has occurred) will form a **neutron star.** This is a degenerate stellar remnant consisting of neutrons, which, like electrons, cannot be forced closer together than a certain limit. The pressure created by this resistance to being squeezed together balances gravity, preventing the star from further collapse. The density of a neutron star is far greater than that of a white dwarf, reaching values of more than 10^{14} grams/cm^3 (equivalent to the density of an atomic nucleus). Theoretical models of neutron stars indicate that they cannot exceed 2 or 3 solar masses, because for higher masses, even degenerate neutron gas pressure cannot prevent further

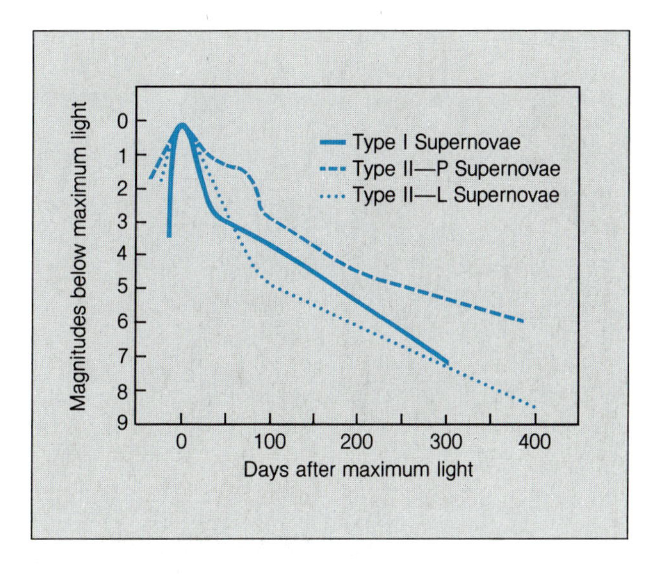

FIGURE 23.13 SUPERNOVA LIGHT CURVES. *The two types of supernovae are readily distinguished by their spectra and their light curves, which show marked differences. Type I supernovae usually have no hydrogen lines in their spectra and are thought to be the explosions of carbon white dwarfs that have accreted matter. Type II supernovae have hydrogen lines in their spectra and are the explosions that occur in massive stars when all nuclear fuel is gone. As seen here, the peak brightness is slightly different, and the rate of decline quite different for the two types, which were distinguished historically, before the mechanism for either had been understood.* (J. Doggett and D. R. Branch)

collapse. Thus, when a star in the mass range 8 to 12 solar masses becomes a neutron star, it loses 5 to 9 solar masses in the process. The mass that does not become part of the neutron star is ejected in a supernova explosion.

Apparently the explosion occurs because the core of the star, during the late phases of nuclear reactions, becomes degenerate. The core may be composed of oxygen, magnesium, neon, or silicon, depending on how many reaction stages have taken place (Fig. 23.14). In any case, when the core becomes hot enough for new reactions to occur after it has become degenerate, the new reactions are very rapid and very violent because of the degeneracy (this is analogous to the helium flash discussed earlier in this chapter). The

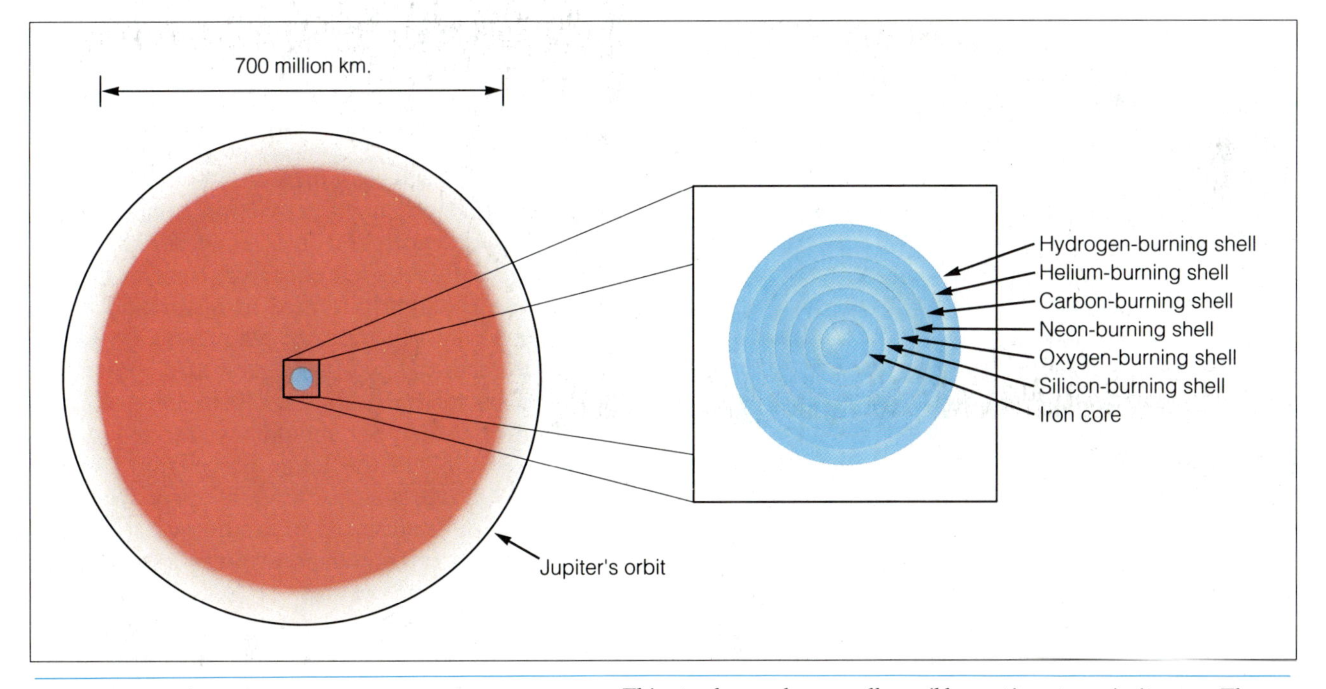

FIGURE 23.14 A MASSIVE STAR AT THE END OF ITS LIFE. *This star has undergone all possible reaction stages in its core. The core consists of multiple layers of different composition, including six active shell sources and an inert iron core. This star is about to undergo core collapse and supernova explosion.*

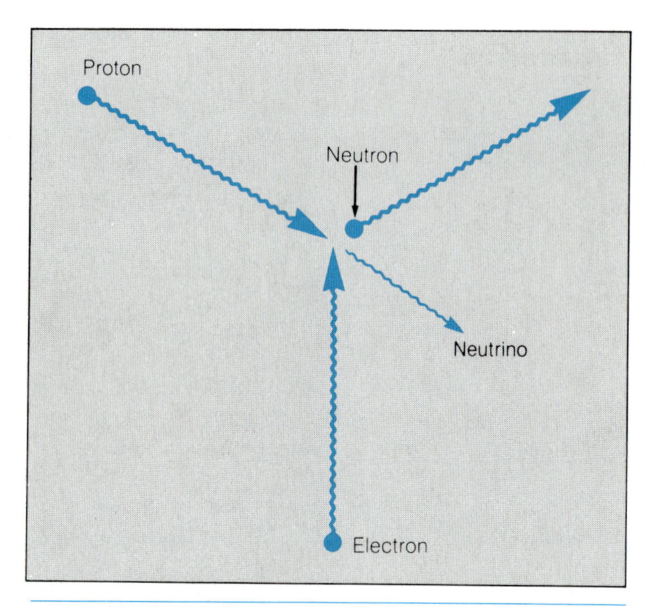

FIGURE 23.15 INVERSE BETA DECAY. *In this process, which only occurs at a significant rate under conditions of very high pressure and temperature, a proton and an electron are forced together, forming a neutron and a neutrino. Energy is absorbed in the process, rather than being released.*

reactions quickly consume all the available fuel in the core, and the star collapses rapidly. During the collapse the immense core pressure forces protons and electrons together, forming neutrons (Fig. 23.15) and thus a neutron star. Once this happens, the rest of the matter falling in rebounds violently, resulting in a supernova explosion. The explosion is aided by the pressure of outward-flowing neutrinos formed in the nuclear reactions in the core. Recall from our discussions in Chapter 18 that neutrinos are elementary particles formed in nuclear reactions, having no mass but able to carry away energy. They interact very little with ordinary matter, but so many are produced in the core collapse that they exert a very strong pressure force on the outer layers of the star, helping to make them expand outward explosively. Thus the outer material of the star is ejected, leaving behind a neutron star.

The energy released in a supernova explosion can be as much as 100 times the total energy a star like the Sun releases in its entire 10-billion-year lifetime. Only a fraction of the supernova's energy is released in the form of visible light, but even that fraction is enough to make the supernova as bright for a while as an entire galaxy (see Fig. 23.12). The major part of the energy of the explosion is released in the form of neutrinos, and some is converted into kinetic energy of the expanding gas. The burst of neutrinos from the 1987 supernova in the Large Magellanic Cloud was detected on Earth, and the measurements indicated that about 99 percent of the total energy of the explosion was released in the form of neutrinos.

The supernova process just described is one of two mechanisms believed by astronomers to occur. The other type of supernova involves the explosion of a white dwarf that has gained enough mass to push it over the 1.4 solar mass limit, and will be discussed at greater length in the next chapter. The white dwarf explosion is usually identified observationally as a Type I supernova, whereas the explosion of a middleweight star at the end of its nuclear burning lifetime, as described in this section, is usually a Type II supernova. (The distinction is based on whether there is hydrogen in the outer layers of the exploding star; if there is, it is a Type II supernova, if not, it is Type I.)

The neutron star that is formed by the explosion of a middleweight star has properties even more fantastic than a white dwarf. Its diameter is about 10 kilometers, yet it contains more than 1.4 solar masses, and its density far exceeds that of the white dwarf. A cubic centimeter of neutron-star material would weigh a billion tons at the Earth's surface, greater than the total mass of all living humans!

Observations have confirmed that neutron stars can form during supernova explosions, because neutron stars have been found in the center of the expanding cloud of debris left over from such an event. One of these apparently formed in historical times, when the supernova that created the present-day Crab Nebula occurred; this was the "guest star" observed by Chinese astronomers in A.D. 1054. Observation and theory tell us that one should have formed in the supernova of 1987 in the Large Magellanic Cloud, but we will have to wait for a while to see whether this is the case, because the expanding gas cloud is still too thick for astronomers to probe the interior for evidence of a neutron star.

Neutron stars are exceedingly dim objects, very difficult to see directly, and were for a long time thought to be unobservable. The ways in which they have been observed will be described in the next chapter. For now, we have finished our life story of

ASTRONOMICAL INSIGHT 23.1

THE MYSTERIES OF ALGOL

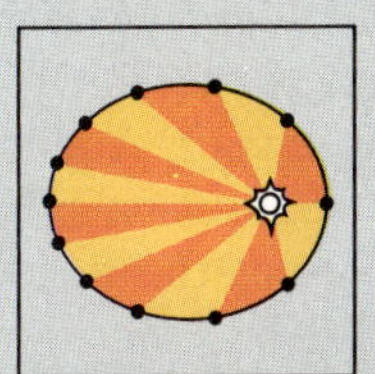

The bright double star known as Algol (β Persei) has played a leading role in the discovery of two important astronomical phenomena. Both of these phenomena, the existence of binary stars and the process of mass transfer in close binaries, went unrecognized for some time because they seemed too radical for most astronomers to accept.

The first report that Algol's brightness varies came as long ago as 1667, when the Italian astronomer Geminiano Montanari noted that the star occasionally appeared fainter than normal. At that time a few variable stars were known, but no explanation for them had been found. It was not until 1782 that anyone rediscovered the variability of Algol; in that year the English astronomer John Goodricke found that the light variations were periodic, with a brief decrease in brightness occurring every 69 hours. Goodricke made the then-novel suggestion that the dimmings could be due to the presence of a companion star that orbited Algol, eclipsing it each time it passed in front. At that time there was controversy over earlier suggestions that some stars appeared to be double, and the prevailing view (including that of the eminent English astronomer William Herschel) was one of skepticism.

In independent studies of Goodricke's idea, the Swiss astronomer Daniel Huber and an obscure English clergyman named William Sewall soon analyzed the light variations of Algol to deduce crudely the properties of the suspected companion. The writings of both men, published originally in 1787 and 1791, were ignored by their contemporaries and then lost, not being rediscovered until well into the twentieth century. Meanwhile, the notion that stars could be double gradually became accepted, as additional Algol-type stars were found, and observations of visual binaries became common. Interestingly, it was William Herschel who, in 1803, conclusively proved the existence of double stars when he was able to actually measure orbital motion in a visual binary system. Finally, in 1889, spectroscopic observations of Algol itself revealed the orbital motion of the bright star about its companion, establishing that stars can be double even though they present a single-point image when viewed in a telescope or when photographed.

Much more recently, Algol played a similar leading role in the discovery of mass transfer in binary systems. During the twentieth century, Algol has been widely observed, using photometric and spectroscopic techniques, and a paradox was revealed as an understanding of stellar evolution arose. Analysis of the data on Algol using Kepler's third law and the shape of the light curve (a graph showing the brightness of the star versus time) showed that the visible star is a B main-sequence star having a mass of 3.7 times that of the Sun, whereas the fainter star is a G giant star having a mass of only 0.81 times the Sun's mass. From observations of cluster H-R diagrams and stellar evolution theory, it was known that more massive stars evolve faster than less massive ones. Thus the fact that the *less* massive member of the Algol system was already a red giant, while the *more* massive star still on the main sequence presented a major mystery, which became known as the Algol paradox.

The solution to this puzzle was found only in the past 20 years or so, as ultraviolet observations have revealed massive stellar winds from some kinds of stars, and other evidence (mostly from telescopes in space) for mass transfer between stars in close binary systems has been found. Evidently the G star in the Algol system originally contained more mass and was once an upper main-sequence star, whereas its companion started out on the lower main sequence. As the first star, being more massive, became a red giant, it swelled up to such a large size that matter began to flow across to the companion. The first star became less massive and fainter than the companion, which in turn became hotter and brighter as its mass grew. Today the original situation is reversed, with the star that started as the less massive one now being the more massive. At some future time, there may be another reversal, as the B star becomes a red giant and swells up, spilling mass back onto its companion.

The existence of binary stars and the process of mass transfer in close binaries were both profound discoveries in astronomy. It is ironic that hints of both were provided by Algol, but in both cases the evidence was found long before astronomers were ready to understand and accept the implications.

the middleweight star, and we are ready to see how the most massive stars evolve.

THE HEAVYWEIGHTS

The most massive stars form in basically the same manner as their less massive relatives, but they do so more quickly. All phases of their evolution are speeded up. As we learned in the previous chapter, the massive stars in a cluster can go through the entire evolution from birth to death before the less massive ones even reach the main sequence. The evolutionary track of a massive star is shown in the H-R diagram in Fig. 23.16.

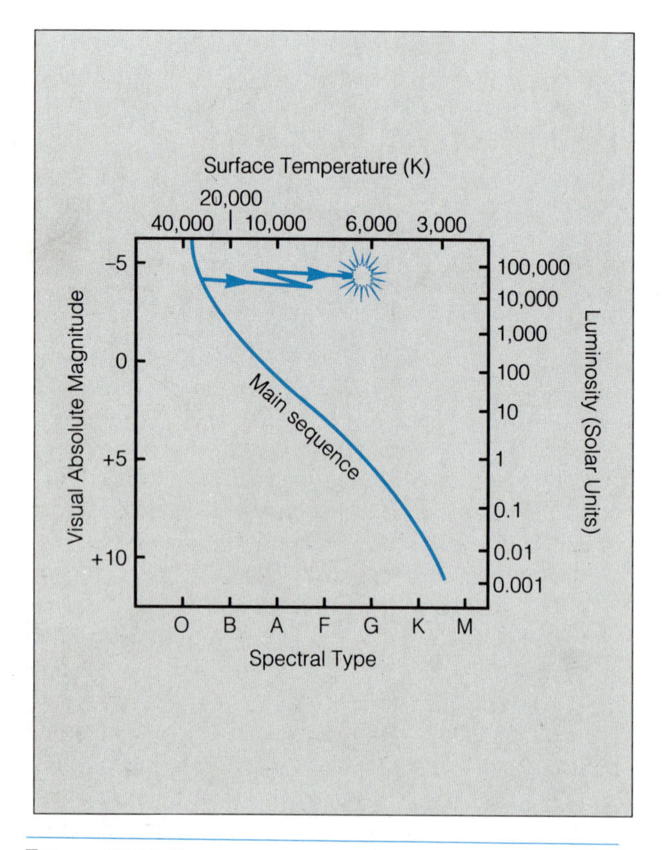

FIGURE 23.16 EVOLUTION OF AN O STAR. *The evolutionary track of a very massive star is quite simple: it moves to the right (not up, because mass loss by a stellar wind counteracts the effects of expansion), briefly moves to the left during helium burning, then goes back to the right. It may go back and forth (rapidly) a few times, but within a total lifetime of about one million years, it exhausts all nuclear fuel and explodes in a supernova.*

Stars containing twenty or more solar masses start their main-sequence lifetimes as O stars. They have surface temperatures of 30,000 K or more and luminosities of 10^5 to 10^6 times that of the Sun. The dominant nuclear reaction during the main-sequence phase is the CNO cycle. There is convection in the cores of these stars, but probably not at the surface, even though there is some excess heating there.

As in the other cases, these stars undergo gradual core contraction as the hydrogen there is converted into helium. By contrast with the less massive stars, however, the O stars do not move upward in the H-R diagram; instead, they move straight across to the right. The reason is that O stars steadily lose matter through high-velocity winds (described in Chapter 21), so that the tendency to increase their core temperature and hence the luminosity is offset by the decreased mass of the overlying layers.

The details of the later stages of evolution for these massive stars are not very clear. Most likely, several stages of nuclear burning take place as the star goes through numerous loops in the red giant region of the H-R diagram, leading ultimately to the formation of iron in the core. At the same time, the wind may rip away so much of the star's substance that the outer portion of the core is exposed, creating a "peculiar" object called a **Wolf-Rayet star** (Fig. 23.17). These are hot stars with very strong winds that have excess quantities of carbon (a result of the triple-alpha reaction) or nitrogen and oxygen (the products of the CNO cycle) at their surfaces. When the outer layers have been stripped off, we get a direct view of the innards of the star, where material that has recently undergone nuclear reactions is seen. The convection that is taking place in the core helps to bring this processed matter up to the surface.

When iron has been formed in the core of a star (see Fig. 23.14), no further nuclear reactions can support the star against the inward force of gravity. Indeed, any additional reactions that could occur would help gravity, because reactions involving elements heavier than iron are all **endothermic,** meaning that more energy is required to create the reaction than is produced. Hence if these reactions start, they actually cool the star's interior, thereby reducing the pressure

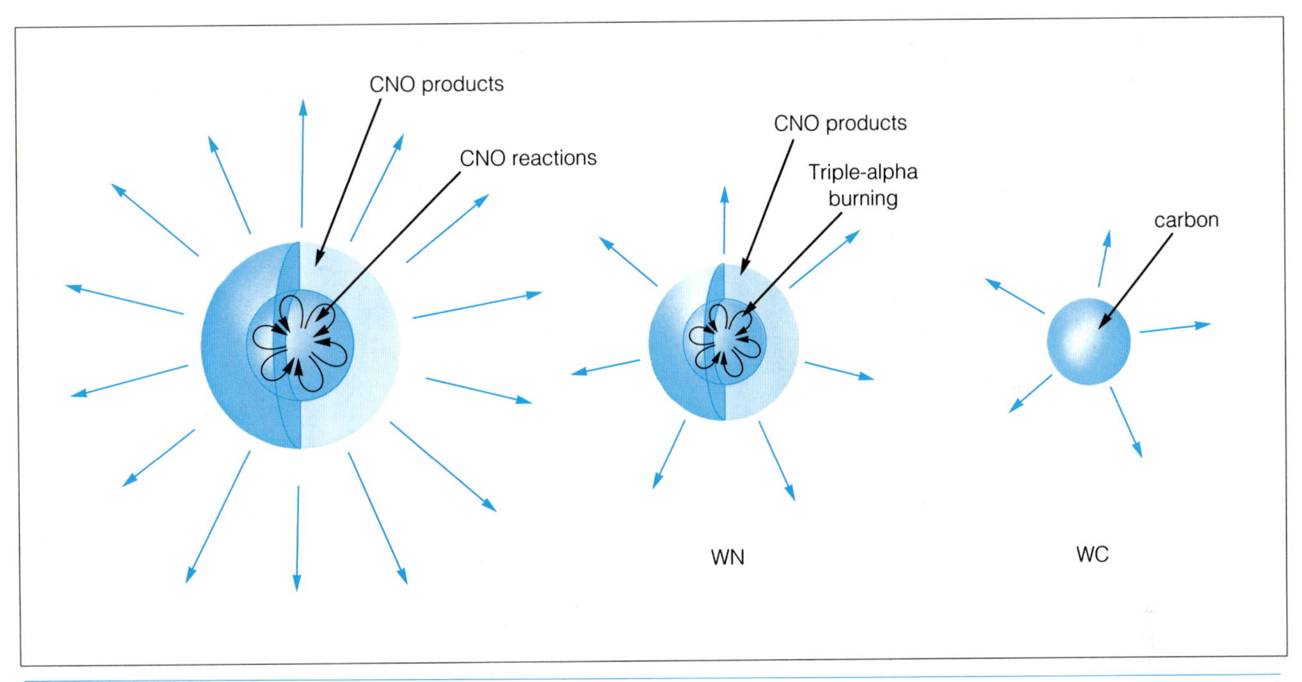

FIGURE 23.17 FORMATION OF A WOLF-RAYET STAR. *A massive star loses its outer layers through a stellar wind; meanwhile its core is undergoing CNO cycle reactions, gradually being transformed into helium. When the outer layers have been blown away, the exposed inner region of the star, which contains enhanced quantities of nitrogen and oxygen as by-products of the CNO cycle, is a type of Wolf-Rayet star known as a WN star. If the star has progressed to helium burning by the triple-alpha process, it is a WC star with an overabundance of carbon.*

and allowing gravity to dominate further. Current theoretical models indicate that stars having more than 12 solar masses in their cores will proceed all the way to iron formation, after which they become unstable because no new reactions can occur.

It is not known whether the collapse of a very massive star results in a supernova explosion or not. If it does, the details may be different than in the supernova process described above, for stars in the 8- to 12-solar-mass range. This is because it is thought that the core of a more massive star does not become degenerate; recall that core degeneracy is the ultimate trigger for the supernova process just described. The fact that the core of a very massive star does not become degenerate is what allows it to proceed all the way to iron production without blowing up first.

When a massive star has reached the stage where it has an iron core, the reactions stop, and the star begins to collapse, having no means of support (Fig. 23.18). As the collapse begins, the compression of the core causes endothermic reactions to start, and suddenly there is an effective vacuum at the center, as the heat there goes into the reactions. The star now collapses very quickly, in free-fall, and it implodes. It is as though the rug were suddenly pulled out from under the outer layers, and they come crashing down into the center of the star. Theory and observation do not yet tell us what happens next. It is thought possible that rebound from the core may occur, as the core forms a neutron star and the outer layers explode in a supernova. On the other hand, many think it more likely that no rebound occurs, and the core simply continues to collapse, with no supernova taking place.

When the collapse does not stop at the neutron star stage and there is no supernova, what remains of the original star? The star simply continues to collapse. Nothing can ever stop it. The resulting object,

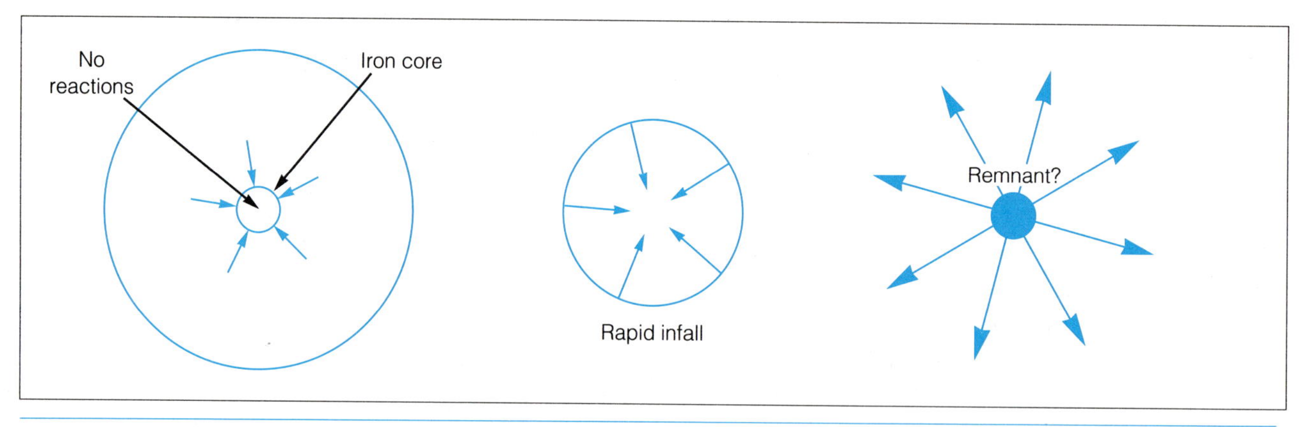

FIGURE 23.18 A SUPERNOVA EXPLOSION. *When the last possible reaction stage is completed, the star's outer layers collapse inward because the source of internal pressure is gone. The collapse is so violent that the star "bounces back," and explodes as a supernova. The explosion is actually caused by the outward pressure of neutrinos created in rapid nuclear reactions.*

first envisioned mathematically nearly 70 years ago, is called a **black hole,** and today there is observational evidence that such things exist.

Although a black hole cannot be seen because its gravitational field is so strong that even light is trapped and cannot get out, there are indirect means by which its presence may be detected. These means, and the information about black holes that can be obtained from observations, are described in the next chapter. For now, we return to our discussion of stellar evolution.

— EVOLUTION IN BINARY SYSTEMS —

It was pointed out in Chapter 21 that some special effects may occur in certain binary systems. If the two stars are sufficiently far apart, say, at least several astronomical units, then most likely each evolves on its own, as though the companion were not there. If, however, they are very close together, then significant interaction takes place which can drastically alter the normal course of events.

Observations of binary-star systems in which the less massive star appeared to be more evolved than its companion provided the first evidence that normal stellar evolution can be modified by interaction between the two stars. Ordinarily the more massive star would be more evolved, since all stages of evolution proceed more quickly for a more massive star. Systems were found, for example, containing a red giant and a more massive main-sequence star. The red giant had apparently gone all the way through its hydrogen-burning phase and expanded as shell reactions started, but the main-sequence star was still undergoing hydrogen burning, despite its greater mass. It was subsequently discovered that, because stars at various stages can eject matter, mass can be exchanged between the two stars (particularly if it is a close binary system), thus altering their rates of evolution.

As we have seen, stars can have strong winds at different stages in their lives. Very massive stars have high-speed winds from the start, whereas less massive stars have low-speed, dense winds when they have evolved to the red giant or supergiant stage. Even without a wind, a red giant or supergiant can transfer matter to its companion when it is sufficiently swollen so that its outer layers reach the point of equal gravitational force between the two stars (see Chapter 21 for more information of this).

Regardless of how it happens, once mass begins to transfer from one star to the other, the evolution of one or both is modified (Fig. 23.19). The one that gains mass speeds up its evolution, whereas the one that loses it may slow down. This can explain how a binary system can contain an evolved low-mass star and a more massive main-sequence star: The low-mass red giant had more mass to begin with, so it has evolved farther, but in its later stages has transferred enough material to the other star to make the com-

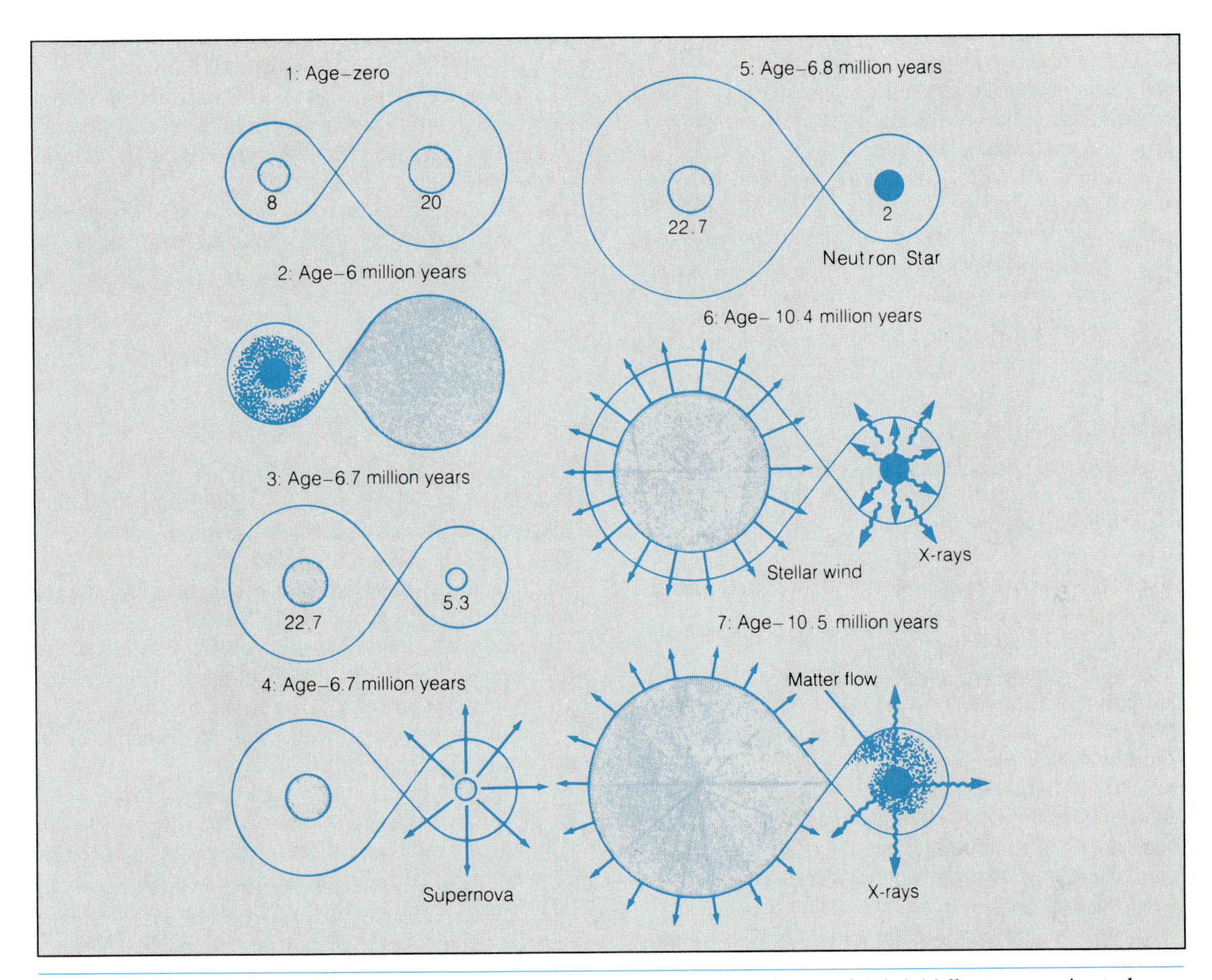

FIGURE 23.19 MASS EXCHANGE AND EVOLUTION IN A CLOSE BINARY SYSTEM. *The star that is initially more massive evolves faster, and as a red giant it deposits material on the companion. This speeds up the evolution of the second star, so that it expands and returns matter to the first star. If the first one is by this time a compact object, such as a neutron star, the consequences of this stage of mass transfer can be violent (see discussion in Chapter 24).*

panion the more massive one. The most spectacular observational effects occur when one star has gone all the way through its evolution and has become a white dwarf, neutron star, or black hole, and then it gains new material from the companion star. Special effects, often including X-ray emission, are then observed, and it is under these circumstances that stellar remnants are most easily detected. This will be discussed at some length in the next chapter.

PERSPECTIVE

We have now seen how stars are born, live their lives, and die, and we know how the sequence of events depends on mass. The least massive stars, those with much less than 1 solar mass, have not been discussed much, simply because they evolve so slowly that none has yet had time, since the universe began, to complete its main-sequence lifetime. The evolution of these stars has had little effect on the rest of the galaxy, whereas the much more rare heavyweight stars are profoundly important, because it is only through their rapid evolution that the galaxy has been enriched with heavy elements.

Before we finish discussing stars and turn our attention to larger-scale objects in the universe, we must consider the properties of the stellar remnants left behind as stars die.

SUMMARY

1. A star with the mass of the Sun spends about 10^{10} years on the main sequence, producing its energy by nuclear reactions that convert hydrogen to helium in its core.
2. When the core hydrogen is gone, the reactions stop there but continue in a spherical shell outside the core. The outer layers expand and cool, and the star becomes a red giant.
3. The core continues to shrink and heat until it becomes degenerate. When it eventually gets hot enough, a helium flash occurs, after which the star is powered by the triple-alpha reaction, converting helium into carbon.
4. Following the helium-burning phase, the star may undergo instabilities that cause the ejection of one or more planetary nebulae.
5. The star eventually becomes degenerate throughout and ends its evolution as a white dwarf.
6. A more massive star, in the 5- to 10-solar-mass range, undergoes many of the same evolutionary steps as the 1-solar-mass star, but it does so much more quickly.
7. The moderate-mass star may go through several red giant phases, as successive nuclear reaction stages produce ever-heavier elements in the stellar core.
8. This star will become a white dwarf only if it loses enough mass, through stellar wind or other processes, to bring its mass below the 1.4-solar-mass limit for white dwarfs.
9. If the star is too massive at the end of its evolution to become a white dwarf, it may explode in a supernova, leaving a neutron star.
10. The most massive stars also undergo the same evolutionary steps as the less massive stars, but they do so even more quickly.
11. There will be several red giant stages, producing elements as heavy as iron in the stellar core.
12. When all possible reaction stages are completed, the star will implode and may explode in a supernova, or perhaps more likely will continue to collapse, forming a black hole.
13. In close binary systems, mass can be transferred from one star to the other, speeding the evolution of the one that gains mass.

REVIEW QUESTIONS

Thought Questions

1. Explain why the main sequence is somewhat broadened rather than being a thin strip in the H-R diagram. Where do you think the Sun lies within the broad strip of the main sequence?
2. We have learned that when helium-burning reactions start in a star like the Sun, they begin in an instantaneous "helium flash." Why is there not a "hydrogen flash" earlier in a star's lifetime, when it forms and nuclear reactions begin for the first time?
3. Explain why a star like the Sun becomes a red giant when its hydrogen fuel is gone.
4. White dwarfs have a small range of masses; that is, most have masses that are very similar. Despite this, white dwarfs can have a wide variety of different compositions. Explain how this can be so.
5. Summarize the contrasts between the evolution of a 1-solar-mass star and a star of 5 solar masses.
6. When a star undergoes successive nuclear reaction stages, each stage is shorter than the previous one. Explain why. (*Hint:* You need to recall what you have learned about the temperatures required for successive reactions, and the dependence of luminosity on temperature.)
7. Discuss the role of stellar winds in the evolution of massive stars.
8. Compare and contrast the evolution of a 5-solar-mass star with that of a star of 15 solar masses.
9. The relative abundances of the elements in the galaxy (Appendix 3) show that the abundances generally decline with increasing atomic weight, with hydrogen the most abundant, helium next, and so on. Contrary to this trend, iron is relatively abundant, compared to other elements near it in atomic weight.

Based on what you have learned about stellar evolution, explain why this is so.
10. Discuss the effects of mass exchange in binary systems on the evolution of the two stars.

Problems

1. A star like the Sun may become as large as 20 times its original radius when it is a red giant, and as small as 0.01 times its original radius when it is a white dwarf. Compute its density, surface gravity, and escape velocity for the red giant and white dwarf stages, relative to their values when the star is on the main sequence. Assume that the mass remains constant throughout.
2. If a star loses one-tenth of a solar mass each time it ejects a planetary nebula, how many times would a star of initial mass of 2.4 solar masses have to eject planetary nebulae in order to become a white dwarf?
3. What is the wavelength of maximum emission for the central star of a planetary nebula whose surface temperature is 100,000 K? What kind of telescope would be best for observing this object?
4. If a neutron star has a radius that is 1/500 times that of a white dwarf of half the mass of the neutron star, compare the average densities of the two.
5. A supernova may have an absolute magnitude as small (bright) as -19. How much more luminous is such a supernova than the Sun, whose absolute magnitude is $+5$?

ADDITIONAL READINGS

Balick, Bruce. 1987. The shaping of planetary nebulae. *Sky and Telescope* 73(2):125.

Bethe, Hans A. 1985. How a supernova explodes. *Scientific American* 252(5):60.

Flannery, B. P. 1977. Stellar evolution in double stars. *American Scientist* 65:737.

Greenstein, George. 1985. Neutron stars and the discovery of pulsars. Part 1, 2. *Mercury* 14(2):34, 14(3):66.

Kaler, J. 1981. Planetary nebulae and stellar evolution. *Mercury* 10(4):114.

Kawaler, S. D., and D. E. Winget. 1987. White dwarfs: Fossil stars. *Sky and Telescope* 74(2):132.

Murdin, Paul, and Lesley Murdin. 1985. *Supernova.* Cambridge Mass.: Cambridge University Press.

Page, T., and L. W. Page, eds. 1968. *The evolution of stars.* New York: Macmillan.

Schaefer, Bradley E. 1985. Gamma-ray bursters. *Scientific American* 252(2):52.

Seeds, M. A. 1979. Stellar evolution. *Astronomy* 7(2):6.

Seward, Fredrick. 1986. Neutron stars in supernova remnants. *Sky and Telescope* 71(1):6.

Shklovskii, I. S. 1978. *Stars: their birth, life, and death.* San Francisco: W. H. Freeman.

Sweigart, A. V. 1976. The evolution of red giant stars. *Physics Today* 29(1):25.

Wyckoff, 1979. Red giants: the inside scoop. *Mercury* 8(1):7.

These articles are all related to Supernova 1987A:

Helfland, David. 1987. Bang: The supernova of 1987. *Physics Today* 40(8):24.

Schorn, Ronald A., 1987. Neutrinos from hell. *Sky and Telescope* 73(5):477.

———. 1987. SN1987A: Watching and waiting. *Sky and Telescope* 74(1):14.

———. 1987. A supernova in our backyard. *Sky and Telescope* 73(4):382.

———. 1987. Supernova 1987A's fading glory. *Sky and Telescope* 74(3):258.

———. 1987. Supernova shines on. *Sky and Telescope* 73(5):470.

———. 1987. A surprising supernova. *Sky and Telescope* 73(6):582.

Time. March 23, 1987. 129(12):60.

SPECIAL INSERT

THE GREAT SUPERNOVA OF 1987

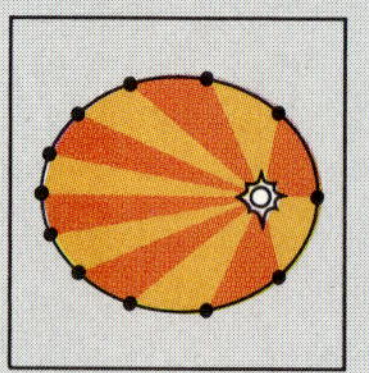

One of the most spectacular and scientifically bountiful events in the history of astronomy was first observed on February 23, 1987. A supernova bright enough to be easily seen with the naked eye was discovered in the Large Magellanic Cloud, providing astronomers the first opportunity in nearly four hundred years to study a supernova so bright. Ian Shelton, a Canadian astronomer, working at Canada's southern hemisphere observatory at La Silla, Chile, noticed the new object on a photographic plate as he was developing it. He rushed outside and was able to see with his own eyes what the photograph had told him: The Large Magellanic Cloud contained a new star, at that time just brighter than fifth magnitude. Later that night, other astronomers in New Zealand and Australia independently discovered the supernova (Fig. 1). While some now refer to it as Supernova Shelton, most astronomers use the more mundane name Supernova 1987A, which indicates that this was the first supernova detected in 1987 (typically a dozen or more are found each year in faraway galaxies).

Astronomers around the world were galvanized by the discovery as the news traveled through the International Astronomical Union's telegraph service, and by telephone and computer networks. NASA's *International Ultraviolet Explorer* (*IUE*; see Chapter 7) was making ultraviolet spectroscopic observations within 14 hours, and ground-based observers everywhere in the southern hemisphere trained their telescopes on it. Radio telescopes also observed the supernova within hours of the discovery. Unfortunately, the event occurred at a time when space astronomy was in a lull between major missions (the *Space Telescope*, for example, still sits on the ground because of delays in the space shuttle program), so it was not possible to obtain all the data that would have been desirable. A Japanese X-ray telescope called *Ginga* was launched on February 24, 1987, however, and it was soon able to make X-ray observations. The Soviet space station *Mir* also observed it in X-rays. NASA moved quickly to mobilize a number of sounding-rocket and high-altitude aircraft and balloon experiments as well, so there will be substantial coverage of the supernova in wavelengths that do not reach the ground. Perhaps the most spectacular observation of all, however, did not concern electromagnetic radiation—it was the detection of neutrinos from the explosion (to be discussed below).

Spectroscopic data quickly

FIGURE 1 THE REGION OF THE SUPERNOVA BEFORE AND AFTER THE EXPLOSION. *At left is a photograph made before the supernova occurred; the photograph at right was made afterwards. At this point the supernova was a fourth-magnitude object, on its way to its eventual peak brightness of magnitude 2.9. The large size of its image compared to normal stars is due to blurring on the photographic emulsion caused by the brilliance of the supernova; it was actually a point source, no bigger than the images of the other stars.* (©1987 Anglo-Australian Telescope Board)

SPECIAL INSERT

THE GREAT SUPERNOVA OF 1987

showed that the supernova was of Type II, because hydrogen lines were seen in its spectrum. The expansion velocity implied by the Doppler shifts of these lines was nearly 20,000 km/sec. Recall that standard models assume that Type II supernovae result from the core collapse of massive stars that still have normal (that is, hydrogen-rich) gas in their outer layers. As we will see, however, Supernova 1987A has several characteristics that do not fit the standard interpretation. (there is, in fact, a growing consensus that the classification scheme needs revision, because apparently there are more than just two combinations of explosion mechanism and stellar composition.)

Of the many unique opportunities the supernova is providing astronomers, one of the most important is the chance to analyze the star that blew up. Several photographic plates of that region exist in observatory files around the world, so it was possible to examine them to determine what kind of star the progenitor was. This examination led to some initial confusion, because careful study showed that there were no fewer than three stars very close to the position of the explosion (Fig. 2), and for a week or two identification of the star that blew up was uncertain. Finally, a combination of extensive measurements of many plates and ultraviolet data from the *IUE* satellite established beyond reasonable doubt that the progenitor was a twelfth-magnitude blue supergiant (spectral type B3I), listed in catalogs as Sk-69 202. This discovery was a major surprise for many astronomers, because until then it had been thought that only red supergiants exploded as supernovae of Type II.

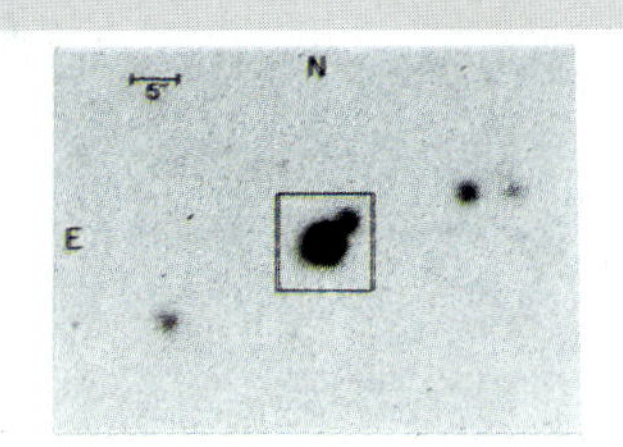

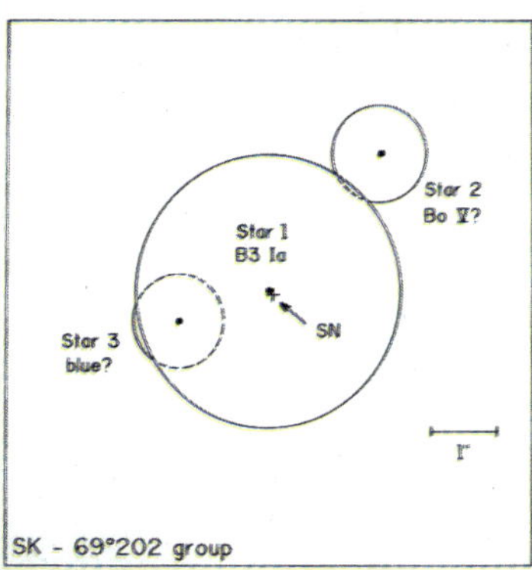

FIGURE 2 TRACKING DOWN THE PROGENITOR. *This photograph and diagram illustrate the relative positions of the three stars seen in pre-supernova photographs to lie near the position of the outburst. The circles represent the uncertainties of the measured positions of each of the three stars. Eventually Star 1 (Sk-69 202, a B3I supergiant) was determined to be the one that exploded. Note the scale; all three stars lay within just a couple arcseconds of each other.* (M. Phillips, Cerro Tololo Inter-American Observatory)

The "unusual" nature of the progenitor, once it was established, helped explain some other peculiar features of the explosion. For example, the supernova never reached the luminosity normally expected, and it reached its peak brightness (just brighter than third magnitude) on May 20, some 85 days (an unusually long time) after its discovery. Radio data, on the other hand, showed a rapid decline almost from the beginning, whereas other Type II supernovae typically do not reach their peak radio emission until hundreds of days after the outburst. The *Ginga* and *Mir* satellites initially showed no X rays coming from Supernova 1987A, although this may not be unusual (there is very little information on X-ray emission in the early stages of other supernovae). Now, however, X rays from the supernova have begun to be seen, showing that some of the high-energy radiation from inside the expanding gas cloud is beginning to leak out.

All of these peculiarities fit well with supernova models involving core collapse in a massive star that is not as large as a red supergiant. The small size of the blue supergiant Sk-69 202(compared to a red supergiant) accounts for the low initial luminosity, since relatively little surface area is available to radiate light. Another very important measure of the size of the star came from the detection of neutrinos, which escape almost directly from the core as the star collapses. The observed light is emitted later, when the shock front caused by the collapse reaches the star's surface. Thus, the neutrinos reached the Earth before the supernova was visible, but the short (4-hour) delay (equivalent to the time it took for the shock to travel from

(continued on next page)

SPECIAL INSERT

THE GREAT SUPERNOVA OF 1987

(continued from previous page)
the star's core to its surface) indicates that the star could not have been as large as a red supergiant. The best fit to all the observations suggests that Sk-69 202 had an initial mass of about 19 solar masses and a diameter roughly 40 times the diameter of the Sun.

As already mentioned, it was a surprise to most astronomers that such a star should explode, although some model calculations had suggested that it could happen. It is still not clear what stage of evolution the star was in: Was it just leaving the main sequence on its way to becoming a red supergiant, or had it already been a red supergiant at least once and then moved back to the left on the H-R diagram as its core contracted and it shed its outer layers in a stellar wind? As yet there is no consensus on the answer to this question, although most supernova theorists seem to be leaning toward the latter point of view. In either case, the explosion itself is thought to have been triggered by the formation of an iron core that then collapsed violently and rebounded, the standard mechanism described in the text. A complicating factor in all these explanations is that the Large Magellanic Cloud has lower overall abundances of heavy elements than "normal" spiral galaxies like our own and most others where Type II supernovae are observed, and lower abundances of heavy elements can alter the evolution of a star. We can be certain that the occurrence of Supernova 1987A will stimulate new efforts to model the evolution of massive stars.

Since Supernova 1987A failed (by a factor of about 15) to reach the magnitude expected for Type II supernova, many astronomers at first assumed that this was a freak event and not any kind of common occurrence. Further analysis showed, however, that many supernovae of comparable brightness could occur in other galaxies without being detected, because they would not stand out so well. Therefore, supernovae like 1987A might actually occur fairly frequently (there have been a small number of other "peculiar" supernovae with some similarities to 1987A).

The detection of neutrinos from Supernova 1987A was a spectacular event for astronomy and for physics. A burst of neutrinos was detected simultaneously by two separate experiments, one in Japan and one in Ohio. This was the first time that extraterrestrial neutrinos had been detected (except for those from the Sun—see Chapter 18), and this event is regarded by some as establishing a new branch of astronomy. As already explained, the timing of the arrival of the neutrinos provided valuable information on the size of the star. However, the neutrinos represent something far more important in terms of understanding supernovae: The energy the neutrinos carried away from the explosion was greater than the supernova's energy in all other forms by a factor of roughly 100. The visible light emitted over the first several months amounts to some 10^{49} ergs of energy; the kinetic energy of the expanding gas cloud entails perhaps a few times 10^{51} ergs; and the energy released in neutrinos is about 3 or 4 times 10^{53} ergs. These numbers are thought to be typical of Type II supernovae, according to model calculations, but Supernova 1987A provided the first confirmation of the important role of neutrinos. What we see of a supernova, as brilliant as it is, is only a small fraction of the energy produced. Furthermore, even though Supernova 1987A was dimmer than many other supernovae in visible wavelengths of light, its total energy output was comparable to the brightest of supernovae.

Physicists who study elementary particles are as excited as the astronomers by the detection of neutrinos from Supernova 1987A. Detailed analysis of the energy distribution of the neutrinos that were detected, along with the exact timing of their arrival, provides information on their basic properties. As mentioned elsewhere in this text, there has been some debate on the question of neutrino mass. The standard theories of neutrinos say that they are massless, yet controversial experimental data, along with a host of astronomical problems that would be solved, suggest that neutrinos may have a very small mass. The data from the supernova neutrinos are not as complete as would have been desirable (fewer than 20 neutrinos were detected) but the information that was obtained still allows something to be learned about neutrino masses. Preliminary analysis indicates that the mass must be zero or very small, though not necessarily too small to solve the astronomical problems al-

SPECIAL INSERT

THE GREAT SUPERNOVA OF 1987

luded to here (such as the total mass content of the universe and the low observed neutrino emission from the Sun).

The light emitted by the supernova as it continues to shine brightly arises from at least three distinct sources of energy. The most important energy source during the early days of the supernova was heating by shock waves created in the rebound following the initial collapse of the star's core. This heating produced most of the supernova's luminosity at first. A second energy source is radiation from a neutron star, the remnant of the star that exploded. So far there has been no direct confirmation that a neutron star formed, but it seems very likely because neutron star formation is required in the successful models of the explosion and especially because it is the only known way to produce the observed quantity of neutrinos. Another probable source of energy is the decay of radioactive elements produced in the explosion. The most abundant radioactive isotope expected is a form of nickel (^{56}Ni), which quickly decays to a form of radioactive cobalt (^{56}Co) and then more slowly to normal iron (^{56}Fe). Calculations show that an initial mass of about 0.075 solar masses of ^{56}Ni would provide the observed energy from Supernova 1987A. As this isotope decays to iron, the emitted energy decreases in a particular manner (called exponential decay) that should be clearly recognizable in the shape of the supernova's light curve (a graph showing how the brightness changes with time; see Fig. 3). For the first few months the light curve did not follow this shape (probably because other sources of energy dominated), but recent observations show that it is now doing so, with precisely the rate of decline expected from the decay of ^{56}Co to ^{56}Fe.

The expanding shell of gas from the supernova was so thick for the first several months that there was no chance of seeing into the interior, of actually detecting a neutron star or measuring other properties of the material inside. It is expected that this situation will gradually change as the shell expands and becomes more tenuous. There must be a massive flux of gamma rays and X rays trying to make their way out through it, and eventually we may expect this radiation to begin to leak through. Theoretical calculations show that one of the first signs of this thinning will be the detection of X rays, and

(continued on next page)

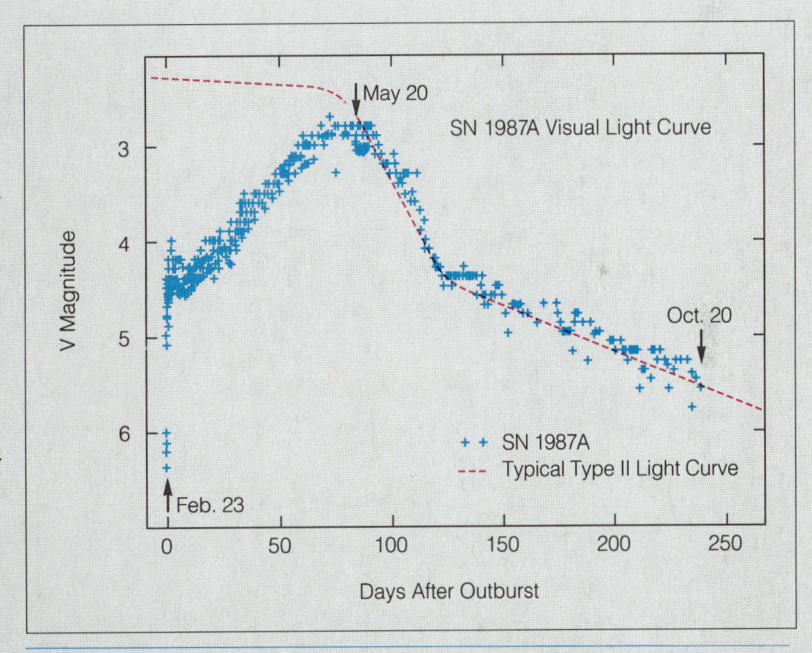

FIGURE 3 THE LIGHT CURVE OF SUPERNOVA 1987A. *This graph shows the variation in magnitude of Supernova 1987A since its discovery up to late Fall 1987. The early part of the light curve is very unusual compared to other known supernovae, but once this one reached its peak it began to behave very much like the light curves of other Type II supernovae. The reasons for the difference in the early period is that the star that exploded was smaller than is thought typical, so that it had less surface area to emit light. The later part of the light curve is consistent with energy production by the decay of radioactive nickel (^{56}Ni).* (J. Doggett and R. Fesen, University of Colorado)

SPECIAL INSERT

THE GREAT SUPERNOVA OF 1987

(continued from previous page)

the very recent detection of X rays by the *Ginga* satellite is therefore an important confirmation of this aspect of the theory. (These are X rays from the radioactive decay; those from the neutron star are not expected to begin escaping for several more months.) Another sign will be the formation of strong ultraviolet emission lines in the outer regions of the expanding cloud, as X rays begin to reach these regions and excite the gas. For this reason, the *IUE* satellite is also monitoring the supernova, and it may have begun to detect some of the expected emission lines. Several gamma ray, X-ray, and ultraviolet telescopes will be launched (by sounding rocket or high-altitude balloon) in Australia in coming months to detect the high-energy emission. (In a spectacular recent development, gamma-ray emission lines formed by the radioactive decay of ^{56}Co have been detected, virtually proving that, as expected, the supernova is now powered by this energy source.) Parallel infrared observations have apparently recently detected the hoped-for "echo" as the light from the supernova explosion reaches outlaying interstellar clouds in the vicinity. Such clouds are expected to be present as a result of the earlier mass-losing phases of the progenitor star, particularly if the star had gone through one or more red-supergiant stages. Infrared observations will also provide information on solid dust particles that could form by condensation in supernova outbursts; such dust permeates interstellar space (see Chapter 26), and it will be helpful to assess how much of it comes from supernovae.

FIGURE 4 THE MYSTERY COMPANION. *The small spot next to the much larger image of the supernova is the mysterious companion object first detected in late March.* P. Nisenson, Harvard-Smithsonian Center for Astrophysics)

While much of the behavior of Supernova 1987A is now reasonably well understood, there are still some mysteries. The most outstanding of these is the presence of a new companion to the supernova—there is glowing spot (Fig. 4) displaced by only 0.06 arcseconds from the supernova itself that was not there before the explosion. When it was detected early in the summer of 1987, this mystery spot was very bright (about 10 percent as bright as the supernova itself), but repeat observations later in the summer failed to confirm its presence. At that time the Large Magellanic Cloud was very low on the horizon, so that the observations had to be made through a lot of Earth atmosphere, and this limits the effectiveness of the interferometric measurements originally used to detect the companion (see Chapter 7). A better opportunity to detect it was to occur in November, 1987, and at the time of this writing astronomers were eagerly waiting to see what would be found. It is not known whether the mystery spot is a physical blob of gas that was ejected from the supernova (which would require a speed of nearly 40 percent of the speed of light), or whether it is material that was already present in interstellar space and was excited to glow because of radiation (either light or a beam or jet of rapid electrons) from the supernova. Perhaps this mystery will have been solved by the time this text is published.

The occurrence of Supernova 1987A has already had major impact on virtually every field of astronomy. The technical journals are filled with articles on theory or observations of the supernova, and we can expect this to continue for months or years. Eventually, as the expanding shell evolves into a normal supernova remnant, it is expected to become one of the brightest radio and X-ray sources in the sky, as the ionized gas disperses and emits energy by synchrotron and other nonthermal processes. In due course, perhaps in months or years, the gaseous remnant may become thin enough to allow astronomers to detect the neutron star, particularly if it is a pulsar (discussed in the next chapter). For decades or centuries the gaseous remnant will be a strong source of visible and ultraviolet emission lines, for centuries it will remain an intense X-ray source, and for millennia it will be a strong radio emitter. Many generations of astronomers will take part in the study of this supernova, and many future astronomy classes will learn about it.

CHAPTER 24

STELLAR REMNANTS

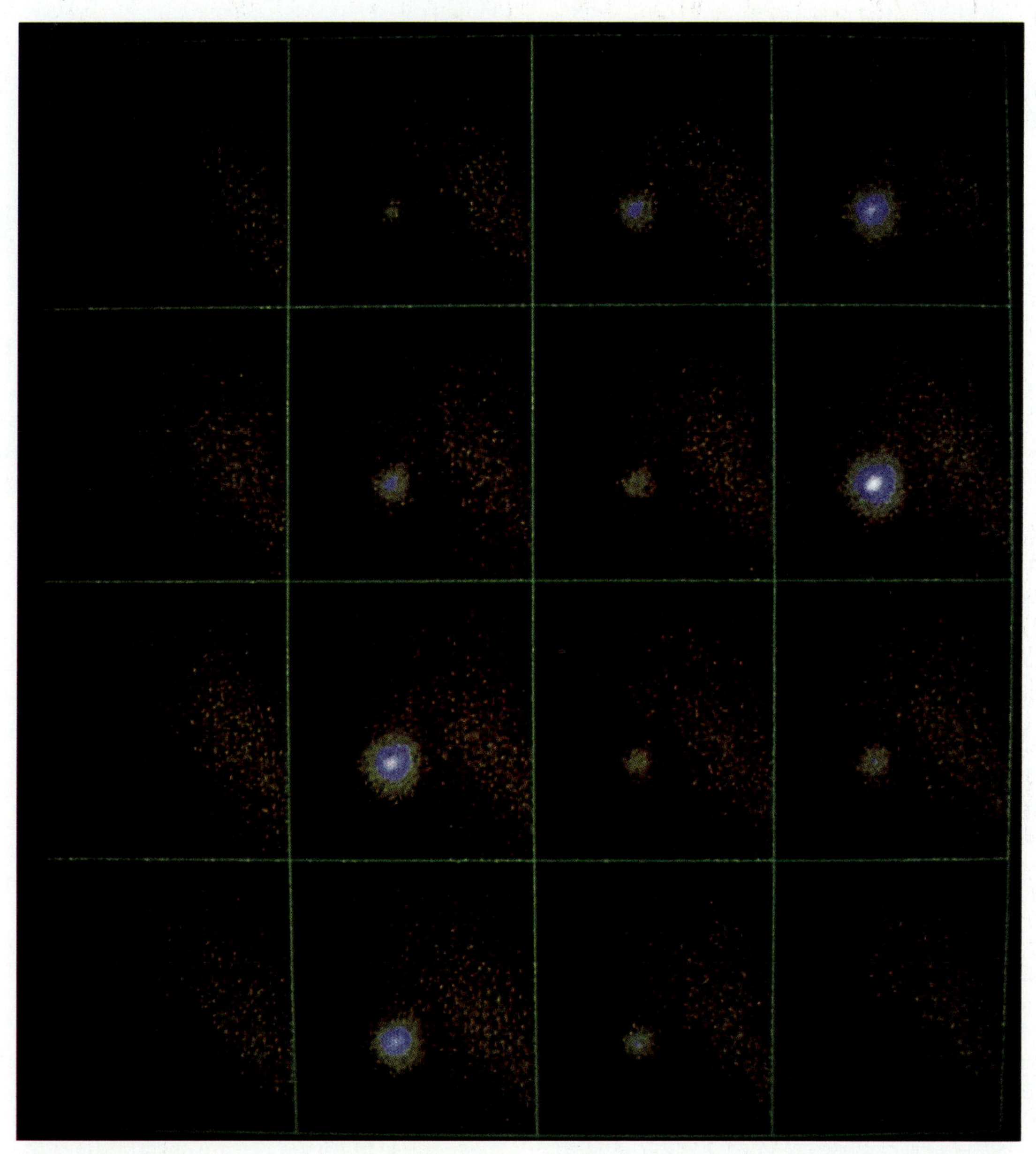

X-ray pulses from the Crab pulsar. (Harvard-Smithsonian Center for Astrophysics, courtesy F. R. Harnden)

CHAPTER PREVIEW

FIGURE 24.1 SIRIUS B. *The dim companion to Sirius was the first white dwarf to be discovered; analysis of its mass, temperature, and luminosity led to the realization that it is very small and dense.* (Palomar Observatory, California Institute of Technology)

Our story of stellar evolution is almost complete. We have seen how the properties of stars are measured, and we have learned how they form, live, and die. The only missing link is what is left of a star after its life cycle is completed.

As we saw in the previous chapter, the form of remnant that remains at the end of a star's life depends on the mass the star had when it finally ran out of nuclear fuel. Three possibilities were mentioned: white dwarf, neutron star, and black hole. In this chapter each of these three bizarre objects is discussed, with emphasis on the observational properties.

– WHITE DWARFS, BLACK DWARFS –

Table 24.1 summarizes the properties of a typical white dwarf. An isolated white dwarf is not a very exciting object, from an observational point of view. It is rather dim (Fig. 24.1), and its spectrum is nearly featureless, except for a few very broad absorption lines created by the limited range of chemical elements it contains (Fig. 24.2). The great width of the lines is caused by the immense pressure in the star's outer atmosphere; pressure tends to smear out the atomic energy levels, because of collisions between atoms. In effect, the energy levels become broadened, and this results in broadened spectral lines when transitions of electrons occur between the levels.

TABLE 24.1 PROPERTIES OF A TYPICAL WHITE DWARF

Mass: 1.0 $M_\odot$
Surface temperature: 10,000 K
Diameter: 0.0015 $D_\odot$ (1.0 $D_\oplus$)
Density: 5 × 10^5 g/cm^3
Surface gravity: 1.3 × 10^5 g
Luminosity: 2 × 10^{31} erg/sec (0.005 $L_\odot$)
Visual absolute magnitude: 11

Additional broadening or shifting of spectral lines may occur because of magnetic effects. If the star had a magnetic field before its contraction to the white dwarf state, that field is not only preserved but is intensified in the process of contraction. Thus a white dwarf may have a very strong magnetic field, and as we have seen previously (in Chapters 18 and 20), a magnetic field causes certain spectral lines to be split into two or more parts. In a white dwarf, where the lines are already very broad, this splitting (called the

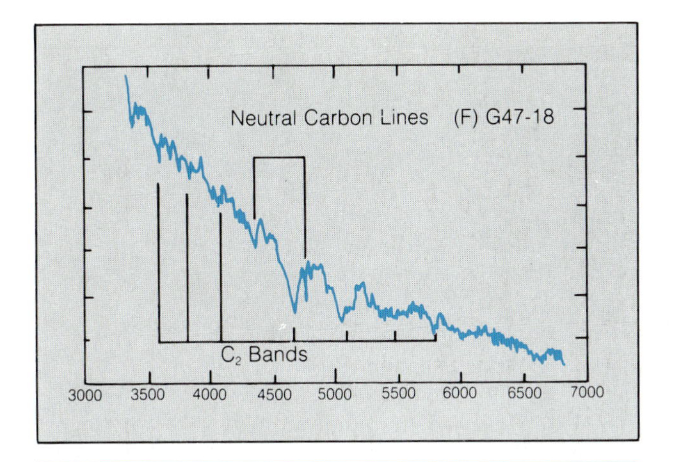

FIGURE 24.2 A WHITE DWARF SPECTRUM. *Here we see that white dwarf spectra are without many strong features, and that the spectral lines tend to be rather broad. This white dwarf has a large abundance of carbon, as a result of the triple-alpha reaction. (G. A. Wegner)*

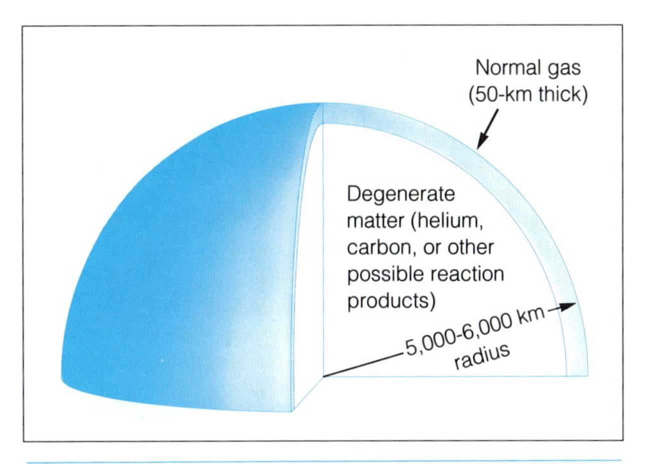

FIGURE 24.3 THE INTERNAL STRUCTURE OF A WHITE DWARF. *The star's interior is degenerate and would cool very rapidly, except that the outer layer of normal gas acts as a very effective insulator, trapping radiation so that heat escapes only very slowly.*

Zeeman effect) usually just makes the lines appear even broader.

The lines in the spectrum of a white dwarf are also shifted, always toward longer wavelengths. This is not caused by motion of the star; if it were, it would be telling us that somehow all the white dwarfs in the sky are receding from us. This is instead a different type of redshift, one caused by gravitational forces. One of the predictions of Einstein's theory of general relativity is that photons of light should be affected by gravitational fields. White dwarfs, with their immensely strong surface gravities, provide a confirmation of this prediction. In later sections, we will see even more extreme examples of these **gravitational redshifts.**

Left to its own devices, a white dwarf will simply cool off, eventually becoming so cold that it no longer emits visible light. At this point, it cannot be seen, and simply persists as a burnt-out cinder that may be referred to as a black dwarf.

The cooling process takes a long time, however; several billion years may go by before the star becomes cold and dark. This may seem surprising, since the white dwarf has no source of energy, staying hot only as long as it can retain the heat that it contained when it was formed. The outer skin of the white dwarf acts like a very efficient thermal blanket, however, since it consists of gas that is nearly opaque. Radiation cannot penetrate it easily, and because there is no other way of transporting heat away from the interior into space, it takes a long time for the heat to filter out.

The thermal blanket created by the outer atmosphere is very thin; most of the volume of the star is filled with degenerate electron gas (Fig. 24.3), as discussed in the last chapter. The atmosphere, consisting of ordinary gas, may be only about 50 km thick (remember, the white dwarf itself is about the size of the Earth, with a radius of some 5,000 to 6,000 km). The composition of the interior may be primarily helium, if the star never progressed beyond the hydrogen-burning stage; it may be carbon, if the triple-alpha reaction was as far as the star got in its nuclear evolution before dying; or it may be some heavier element, if the original star was sufficiently massive to have undergone several reaction stages. In any case, however, the mass of the white dwarf must be less than the limit of 1.4 solar masses.

White Dwarfs, Novae, and Supernovae

Ordinarily, a white dwarf cools off and is never heard from again. There are circumstances, however, in which it can be resurrected briefly for another role in the cosmic drama. It has recently become apparent

ASTRONOMICAL INSIGHT 24.1

THE STORY OF SIRIUS B

Sirius, also known as α Canis Majoris, is the brightest star visible in the heavens. In 1844 it was discovered to be a binary, because of its wobbling motion as it moves through space (recall the discussion of astrometric binaries in Chapter 19). Its period is about 50 years, and the orbital size about 20 astronomical units; analysis of these data using Kepler's third law leads to the conclusion that the companion must have a mass near that of the Sun. At the distance of Sirius (only 2.7 parsecs), a star like the Sun should be easily visible, yet the companion to Sirius defied detection for some time.

Part of the difficulty lay in the overwhelming brightness of Sirius itself, whose light would tend to drown out that of a lesser, very close star. Astronomers were so intrigued by the mysterious nature of the companion, however, that intensive efforts were made to find it, and finally, in 1863, it was seen with an exceptionally fine, newly developed telescope.

It was immediately confirmed that this star was unusually dim for its mass, having an apparent magnitude of 8.7 (the apparent magnitude of the primary star is −1.4) and an absolute magnitude of 11.5, nearly 7 magnitudes fainter than the Sun. Spectra obtained with great difficulty (again because of the brightness of the companion) showed that this object, called Sirius B, has a high temperature, similar to that of an O star.

When Henry Norris Russell constructed his first H-R diagram in 1913, he included Sirius B, finding it to lie all by itself in the lower left-hand corner. The only way for the star to have such a low luminosity while its temperature was so high was for it to have a very small radius, and it was deduced at that time that Sirius B was about the size of the Earth. The incredible density was estimated immediately, and it was realized that a totally new form of matter was involved. The term white dwarf was coined to describe Sirius B and other stars of this type, which were discovered soon in other binary systems.

The physics of this new kind of matter took some time to be understood, partly because the theory of relativity had to be applied (the electrons in a degenerate gas move at speeds near that of light). The Indian astrophysicist S. Chandrasekhar was the first to solve the problem of the behavior of this kind of gas, and by the 1930s he was able to construct realistic models of the structure of a white dwarf. From this work came the 1.4 solar mass limit we have referred to in the text.

Modern observational techniques offer the possibility of detecting white dwarfs more easily. For example, even though Sirius B is about 10 magnitudes (or a factor of 10,000) fainter than Sirius A in visible wavelengths, its greater temperature makes it relatively brighter in the ultraviolet. There is even a wavelength, about 1,100 Å, below which the white dwarf is actually brighter than its companion. Hence Sirius B and other white dwarfs have been detected with ultraviolet telescopes, and with the development of more sophisticated space instruments, it will become possible to study the properties of many white dwarfs in this way.

that white dwarfs are central characters in producing spectacular events.

We discussed in the previous chapter the evolution of stars in close binary systems, where mass can transfer from one star to the other. Consider such a system, in which the star that was originally more massive has gone far enough in its evolution to become a white dwarf. Now the companion begins to transfer matter to the white dwarf (Fig. 24.4), probably because the companion has expanded to become a red giant. The white dwarf may respond to the addition of the new material in a variety of ways, depending on its composition and on the rate at which mass is added. All of the possible responses lead, sooner or later, to an explosive outburst, or to many repeated outbursts, because of the degenerate nature of the white dwarf.

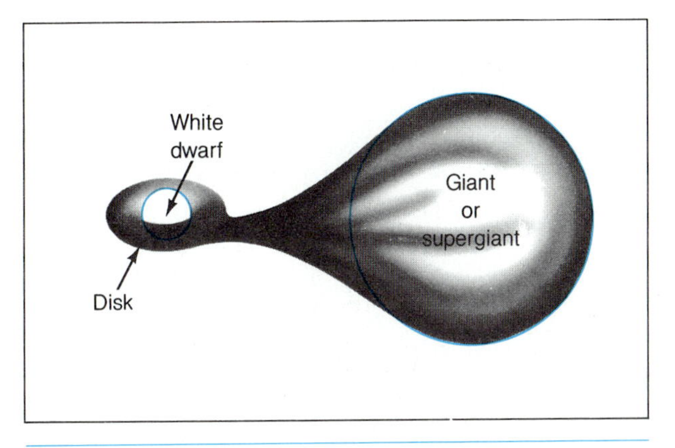

FIGURE 24.4 MASS TRANSFER TO A WHITE DWARF. *Matter is transferred onto a white dwarf by a red giant companion. Theoretical calculations show that the new material will orbit the white dwarf in a disk but will eventually become unstable and fall down onto the white dwarf. When it does so, nuclear reactions flare up, creating a nova outburst.*

If the transferred gas reaches the white dwarf with a high velocity, so that heating occurs because of the energy of impact on the white dwarf's surface, there may be an instant nuclear explosion, enhanced by the fact that the degenerate dwarf material cannot expand and cool when heated (recall the discussion of the helium flash in the previous chapter). In other cases, mass may accrete onto the white dwarf for some time, perhaps decades or centuries, before the new material becomes hot enough to undergo nuclear reactions. The intensity of the outburst depends on how much matter is involved. If enough material is consumed in this explosion, the formerly very dim white dwarf may become more luminous than any ordinary star (Fig. 24.5). This is a **nova.** Its luminosity may be 20,000 to 600,000 times that of the Sun (corresponding to an absolute magnitude in the range −6 to −9). A star that becomes a nova may do so repeatedly, usually with separations of decades between outbursts.

In some binary systems, rather small amounts of matter reach the white dwarf at regular intervals, creating relatively minor flare-ups. A general name for such systems is **cataclysmic variables,** and there are a number of specific types, each named after a particular prototype. The frequency of the outbursts can be as high as several times a day, but typically it is once every few days.

If material trickles onto the white dwarf at a slow rate, so that not much heating is caused by the impact, then the white dwarf can gradually gain mass, without the infalling material causing any explosions. In this way, a white dwarf can approach the 1.4 solar mass limit, beyond which degenerate electron gas pressure can no longer support it. As a white dwarf gains mass, it becomes *smaller;* this is another peculiar property of degenerate matter. Thus, the white dwarf increases in density. What happens as the white dwarf approaches the mass limit depends on its composition; it may, more or less quietly, suddenly contract, becoming either a neutron star or a black hole. If its composition consists of carbon, however, theory shows that nuclear reactions will begin instead, and the degeneracy will cause these reactions to very rapidly consume much of the white dwarf's mass, creating an immense explosion. This is another mechanism for producing a supernova explosion.

In the previous chapter we discussed supernovae caused by the collapse and rebound of a massive star at the end of its nuclear lifetime; now we see that supernovae can also be caused by the addition of mass to a white dwarf. Supernovae created in these two different ways are generally similar in their maximum brightness and in the length of time required to become dim again. But they are rather different in many details, such as the types of spectral lines observed, and the shape of the **light curve,** a graph showing the brightness variations with time (see Fig. 23.13).

FIGURE 24.5 NOVA CYGNI 1975. *This was one of the brightest novae in recent years. The photograph at left shows the region without the nova; at right is the same region near the time of peak brightness, when the apparent magnitude of the object was near +1.* (E. E. Barnard Observatory, photograph by G. Emerson)

Supernovae caused by the explosion of a massive star are usually Type II supernovae, because hydrogen lines appear in the spectrum, while those produced in the explosion of a white dwarf are Type I supernovae, having no hydrogen features.

White dwarfs, then, are most prominent in binary systems in which mass transfer from a normal companion takes place. We will see later in this chapter that other forms of stellar remnants are also best observed in these circumstances.

TABLE 24.2 PROMINENT SUPERNOVA REMNANTS

Remnant	Age (yr)	Distance (pc)
Vela X	10,000	1,600
Cygnus Loop	20,000	2,500
Lupus	975	4,000
IC 443	60,000	5,000
Crab Nebula	934	6,500
Puppis A	4,000	7,200
Cassiopeia A	200	10,000
Tycho's Supernova	400	9,800
Kepler's Supernova	370	20,000

SUPERNOVA REMNANTS

We have learned that stars can explode as supernovae in two different ways, in some cases leaving a stellar remnant (a neutron star or black hole). In all cases there will be another form of matter left over—an expanding gaseous cloud called a **supernova remnant.**

Several supernova remnants are known to astronomers (Table 24.2). A few, like the prominent Crab Nebula (Fig. 24.6), are detectable in visible light, but many are most easily observed at radio wavelengths. This is partly because visible light is affected by the interstellar dust that can hide our view of a remnant, whereas radio waves are largely unaffected. But it is primarily because, as a supernova remnant ages, its emission in visible light fades out long before it stops emitting radio waves. Thus there is an ex-

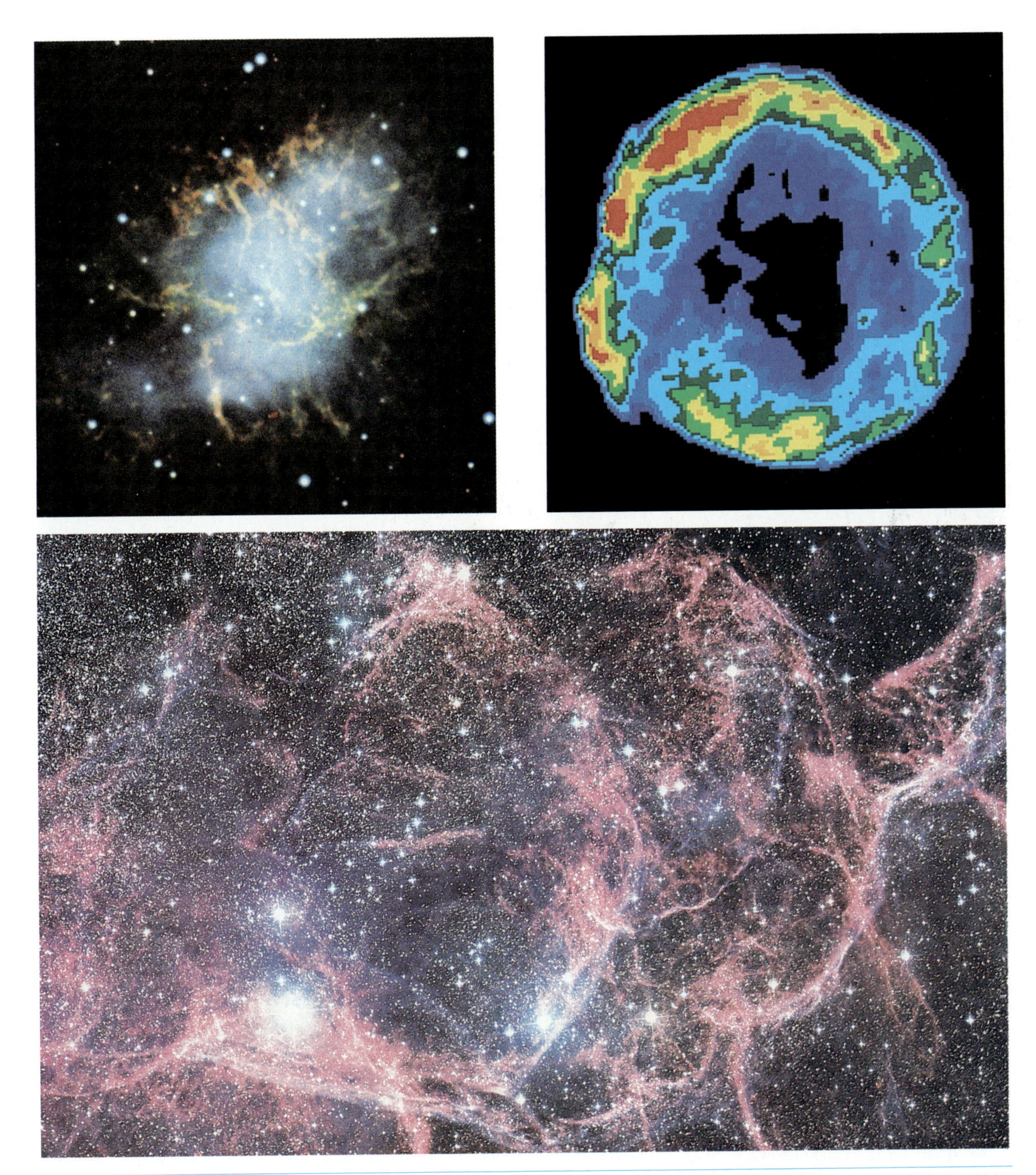

FIGURE 24.6 SUPERNOVA REMNANTS. *At upper left is the Crab Nebula, the remnant of a star that was seen to explode in 1054 A.D. Below is a photograph of part of the Veil Nebula, showing wispy, filamentary structure. At upper right is a radio image of the remnant of Tycho's supernova. (Top left:* Lick Observatory photograph; *top right:* National Radio Astronomy Observatory; *bottom:* © 1980 Anglo-Australian Telescope Board)

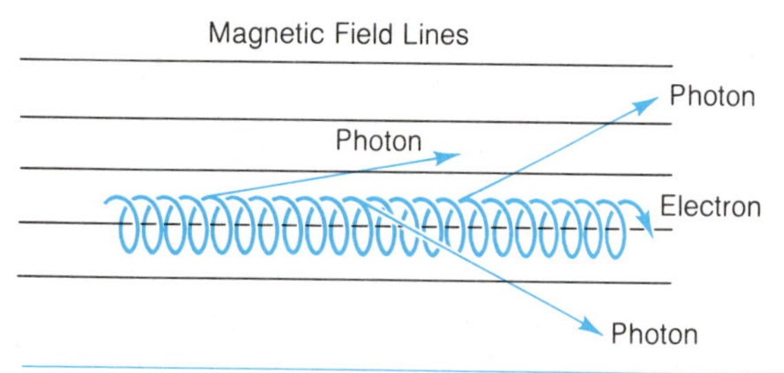

FIGURE 24.7 THE SYNCHROTRON PROCESS. *Rapidly moving electrons emit photons as they spiral along magnetic field lines. The radiation that results is polarized, and its spectrum is continuous but lacks the peaked shape of a thermal spectrum. The electrons must be moving very rapidly (at speeds near that of light), so whenever synchrotron radiation is detected, a source of large quantities of energy must be present.*

tended time period (10,000 to 20,000 years) when the remnant emits radio radiation and little else. The radio emission is not a result of the temperature of the gas, as in thermal emission, but is instead produced by a different mechanism, called **synchrotron emission.**

Synchrotron emission, given its name by the particle accelerators in which it is produced on Earth, occurs when electrons move rapidly through a magnetic field. The electrons must have speeds near that of light; as they travel through the magnetic field, they are forced to move in a spiraling path, and they emit photons as they do so (Fig. 24.7). The emission occurs over a very broad range of wavelengths, including some visible, ultraviolet, and even X-ray radiation (Fig. 24.8), but these other wavelengths are

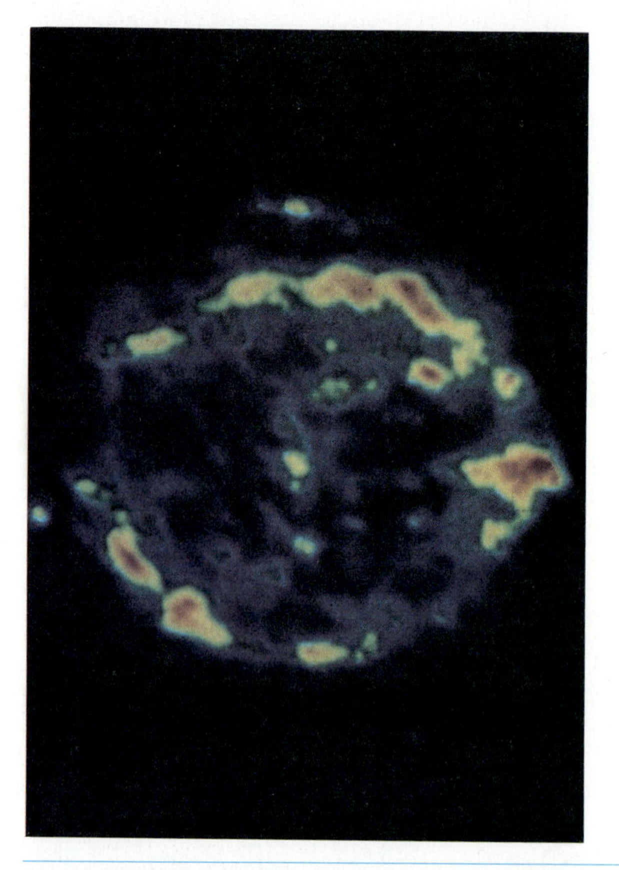

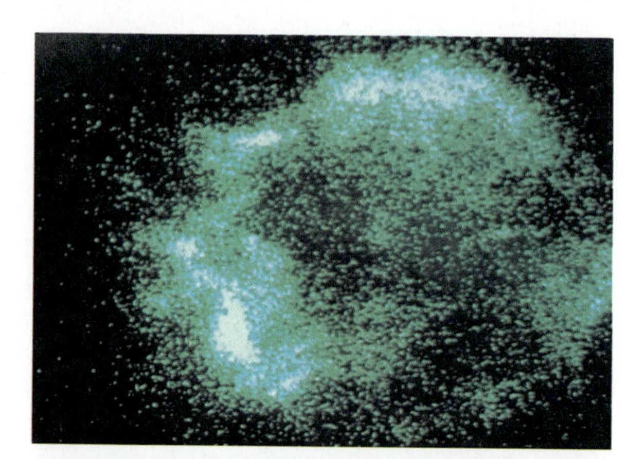

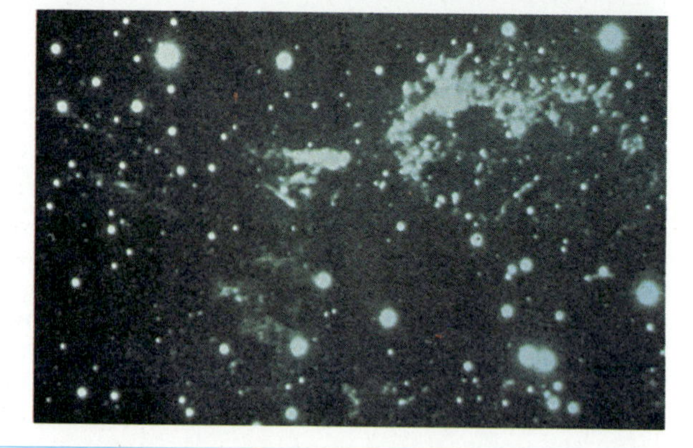

FIGURE 24.8 EMISSION FROM A SUPERNOVA REMNANT. *This shows three different images of the same supernova remnant Cassiopeia A. At left is a radio message, at upper right an X-ray image, and at lower right a visible-light image. The remnant emits at all these wavelengths by the synchrotron process; it is less obvious in visible light because there are many other, brighter sources in the field of view.* (*left:* National Radio Astronomy Observatory; *top right:* Harvard-Smithsonian Center for Astrophysics; *bottom right: K. Kamper and S. van den Bergh)*

ASTRONOMICAL INSIGHT 24.2

RECONSTRUCTING AN EXPLODED STAR*

Supernova explosions are rare events. We seldom get to observe one that is nearby. (Recall that the one in 1987 in the Large Magellanic Cloud was the first in nearly 400 years bright enough to be visible to the unaided eye.) When we find them in distant galaxies, we usually know nothing about the star before it exploded. Most of what we know about which kinds of stars blow up is based on theoretical understandings of nuclear reactions and stellar structure. Our knowledge would be more complete if we knew more about the progenitors of supernovae. It has been possible to reconstruct some information in the case of one recent supernova in addition to the great supernova of 1987.

The most recent supernova explosion known to have occurred in our galaxy took place some 300 years ago. The explosion itself was not noted in historical records, which implies that it may not have reached as great a brightness as those usually seen in distant galaxies. We know about the 300-year old event because the remnant of expanding gas that was left behind is very easily observed today. Known as Cassiopeia A, it is one of the brightest objects in the radio sky, and is observed in visible and X-ray wavelengths as well. Images of this remnant are shown in Fig. 24.8.

Scientists believe that as a massive star ages and undergoes many successive nuclear reaction stages, it develops a layered internal structure, often likened to an onion. At the core is the material currently undergoing reactions, outside of the core is a layer of whatever element was produced by the previous reaction stage, and so on. When such a star proceeds all the way to iron formation in the core and explodes, the layered structure should be preserved in the expanding supernova remnant. The surface material will fly outward most rapidly, the next layer a little slower, and so on. Thus it may be possible, by measuring the properties of debris at different distances from a supernova explosion, to reconstruct the internal structure of the progenitor. This has recently been done successfully for the star that formed Cassiopeia A.

Very careful visible-light observations have revealed fragmentary wisps of gas far outside the main part of the supernova remnant. According to the scenario just described, these filaments of gas, which have been flying away at higher speed than the rest of the remnant, must be fragments of the original surface of the star. This provides astronomers an opportunity to compare the surface material of the exploded star with observed surface properties of known types of stars, to see what kind of star the progenitor was.

Spectroscopic analysis has shown that the gas in these outlying fragments contains large quantities of nitrogen. This leads to the conclusion that the progenitor star was a nitrogen-rich **Wolf-Rayet star,** a hot star of high mass but only moderate diameter. These stars are observed here and there in the galaxy, and are believed to be the remains of large, massive stars whose outer layers have been stripped away by stellar winds. (See the discussions of Wolf-Rayet stars in Chapter 23 and of stellar winds in Chapter 21.) This discovery was made before the supernova of 1987, so the progenitor of Cassiopeia A was the first to be identified.

It is not surprising that a Wolf-Rayet star should explode when its nuclear lifetime is over. These stars, despite having lost a fraction of their original mass due to stellar winds, still have masses 20 or more times that of the Sun. Furthermore, the supernova that formed the Cassiopeia A remnant was apparently not as luminous as many that are observed in other galaxies, and this can be explained if the progenitor had only a moderately large radius. (A Wolf-Rayet star, while much larger than the Sun, is still considerably smaller than a massive supergiant or a hot upper main sequence star.) Possibly many supernovae occur in relatively small stars, and go unobserved due to their modest luminosities (modest only by comparison with the very brilliant supernovae that are seen in the most faraway galaxies).

Unfortunately, the technique used to reconstruct the progenitor of the Cassiopeia A supernova remnant cannot be widely applied to other remnants, because in time (a few hundred years) the gaseous fragments cool to the point where they no longer can be found and analyzed spectroscopically. Thus, the number of instances where we know what kind of star blew up will remain very small.

*Based on information provided by Dr. Robert Fesen, whose work is described here.

often not as easily detected as radio, because the radio emission is the strongest and most persistent.

Supernova remnants that are detected in visible wavelengths usually glow in the light of several strong emission lines, most notably the bright line of atomic hydrogen at a wavelength of 6,563 Å, in the red portion of the spectrum. These lines often show large Doppler shifts, indicating that the gas is still moving rapidly as the entire remnant expands outwards from the site of the explosion that give it birth. Often a filamentary structure is seen, suggestive of turbulence, but probably modified in shape by magnetic fields.

Remnants are visible today at the locations of several famous historical supernova explosions. The Crab nebula is the most prominent, having been created in the supernova observed by Chinese astronomers in A.D. 1054. Other supernovae seen by Tycho Brahe (in 1572) and by Kepler and Galileo (1604) also left detectable remnants, although neither is as bright in visible wavelengths as the Crab. Apparently a supernova remnant can persist for 10,000 years or more before becoming too dissipated to be recognizable any longer. One of the best-studied remnants, Cassiopeia A (Fig. 24.8), is apparently only about 200 years old, but the supernova that created it was quite dim, and not widely observed. This supernova may have been obscured by interstellar dust, or it may have been an unusually dim one intrinsically.

The energy of a supernova explosion is immense, comparable to the total amount of radiant energy the Sun will emit over its entire lifetime. Some of this energy takes the form of mass motions as the remnant expands. This kinetic energy heats the expanding gas to very high temperatures. The entire process has a profound effect on the interstellar gas and dust that permeates the galaxy, as we shall see in Chapter 26.

NEUTRON STARS

We have referred to a neutron star as a stellar remnant composed entirely of neutrons that are in a degenerate state, similar to that of the electrons in a white dwarf. The pressure created by the degenerate neutron gas is greater than that of a degenerate electron gas, so a star slightly too massive to become a white dwarf can be supported by the neutrons. Again, however, there is a limit on the mass one of these objects can have. This limit, which depends on other factors such as the rate of rotation, is between 2 and 3 solar masses. Thus, there is a class of stars, whose masses at the end of their nuclear lifetimes are between 1.4 and about 2 or 3 solar masses, that become neutron stars.

The structure of a neutron star (Table 24.3 and Fig. 24.9) is even more extreme than that of a white dwarf. All of the mass is compressed into an even smaller volume (now the radius is about 10 km), and the gravitational field at the surface is immensely strong.

TABLE 24.3 PROPERTIES OF A TYPICAL NEUTRON STAR

Mass: 1.5 $M_{\odot}$
Radius: 10 km
Density: 10^{14} g/cm^3
Surface gravity: 7×10^9 g
Magnetic field: 10^{12} Earth's field
Rotation period: 0.001–100 sec

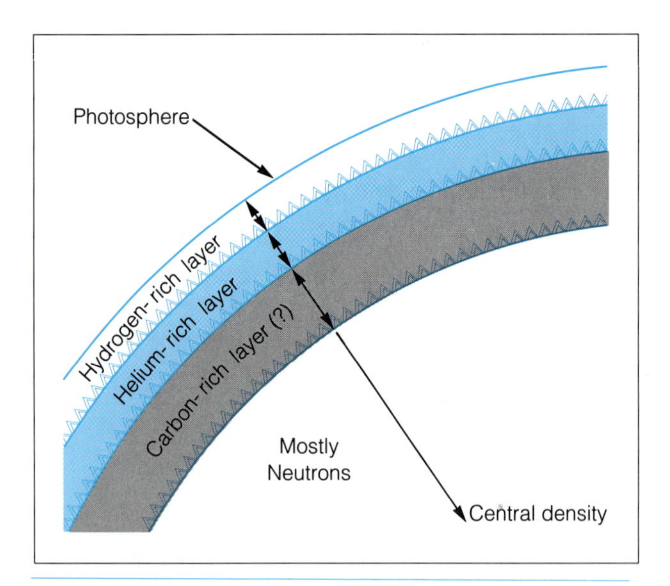

FIGURE 24.9 THE STRUCTURE OF A NEUTRON STAR. *This diagram shows that the outer regions of a neutron star may consist of thin layers of various elements that were produced by nuclear reactions during the star's lifetime. These outer layers were thought to have a rigid crystalline structure because of the intense gravitational field of the neutron star.*

The layer of normal gas that constitutes its atmosphere is only centimeters thick, and beneath it may be zones of different chemical composition resulting from previous shell-burning episodes, each zone only a few meters thick at the most. Inside these surface layers is the incredibly dense neutron gas core, which takes the form of a crystalline lattice. The temperature throughout is very high, but because the surface area is so small, a neutron star is very dim indeed. It may have a magnetic field that is also much more intense than even that of a white dwarf, resulting from the compression of the star's original field.

Pulsars: Cosmic Clocks

The properties of neutron stars were predicted theoretically several decades ago, but until 1967 it was not expected that they could be observed, because of their low luminosities. In that year, however, an accidental discovery by a radio astronomer established a new class of objects called **pulsars,** which are now believed to be neutron stars. As it turned out, pulsars are only one of two distinct ways in which neutron stars can be detected; the other is described in the next section.

Pulsars are radio sources that flash on and off very regularly, and with a high frequency (Fig. 24.10). One of the most rapid of them, the one located in the Crab Nebula, repeats itself 30 times a second, and at least a few others are now known to flash even more rapidly than the Crab pulsar; one of these pulses at the incredible frequency of 885 times per second! The slowest known pulsars have cycle times of more than a minute, on the other hand. In every case, the pulsar is only "on" for a small fraction of each cycle.

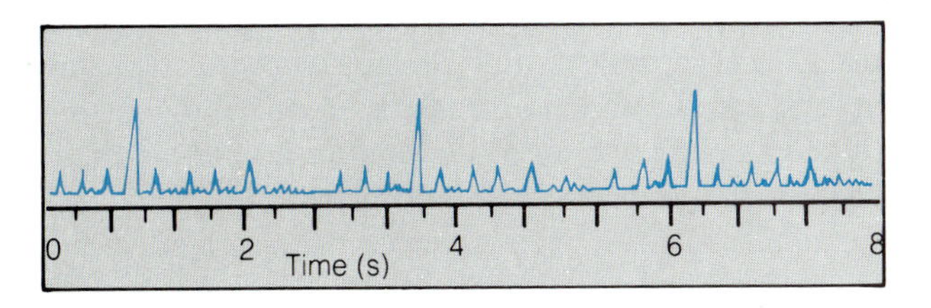

FIGURE 24.10 THE RADIO EMISSION FROM A PULSAR. *Here, plotted on a time scale, is the radio intensity from a pulsar. The radio emission is weak or nonexistent except for a very brief flash once each cycle (in some cases there is a weaker flash between each adjacent pair of strong flashes).*

When pulsars were discovered, there was a great deal of excitement and speculation, including the suggestion that they were beacons operated by an alien civilization. Once the initial shock of discovery wore off, however, a number of more natural explanations were offered by astronomers who sought to establish the identity of the pulsars. It was well known that a variety of stars pulsate regularly, alternately expanding and contracting, but none were known to do it so rapidly. Theoretical studies showed that such quick variations as those exhibited by the pulsars should occur only in very dense objects, denser even than white dwarfs. This led astronomers to think of neutron stars.

Enough was known from theoretical calculations, however, to rule out rapid expansion and contraction of neutron stars as the cause of the observed pulsations. The calculations showed that the vibration period of such an object would be even shorter than the observed periods of the pulsars.

A second possibility—that the rapid periods were produced by rotation of the objects—was considered. As in physical contraction and expansion, the rotation hypothesis ruled out ordinary stars and white dwarfs. To rotate several times per second, a normal star or white dwarf would have to have a surface speed in excess of the speed of light, a physical impossibility. Such an object would be torn apart by rotational forces before approaching such a rotational speed. A neutron star, on the other hand, could rotate several times per second and remain intact.

When the Crab pulsar was discovered, a great deal of additional information became available. This pulsar was found to be gradually slowing its pulsations, something that could best be explained if the pulses were linked to the rotation of the object. The rotation could slow as the pulsar gave up some of its energy to its surroundings.

This discovery cleared up another mystery. The source of energy that powers the synchrotron emission from supernova remnants had been unknown. Now it was suggested that the slowdown of the rotating neutron star could provide the necessary energy, either by magnetic forces exerted on the surrounding ionized gas or by transfer of energy from

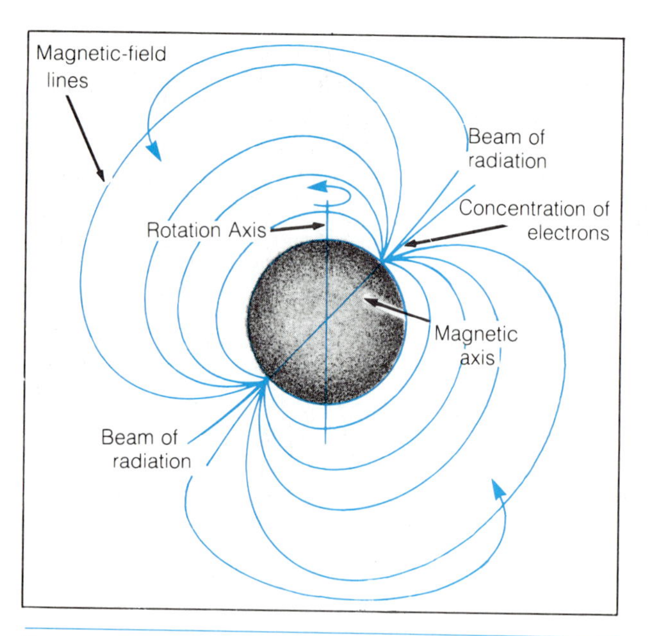

FIGURE 24.11 THE PULSAR MECHANISM. *Here we see a rapidly rotating neutron star, with its magnetic axis out of alignment with its rotation axis. Synchrotron radiation is emitted in narrow beams from above the magnetic poles, where charged particles, constrained to move along the magnetic field lines, are concentrated. These beams sweep the sky as the star rotates, and if the Earth happens to lie in a direction covered by one of the beams, we observe the star as a pulsar.*

the pulsar to the surrounding material through the emission of radio waves.

The remaining question was how the pulses were created by the rotation. Evidently a pulsar acts like a lighthouse, with a beam of radiation sweeping through space as it spins. The question was why a rotating neutron star should emit a beam from just one point on its surface.

The most probable explanation has to do with the strong magnetic fields that neutron stars are likely to have. If a neutron star has a strong field, then electrons from the surrounding gas are forced to follow the lines of the field, hitting the surface only at the magnetic poles. The result is an intense beam of electrons traveling along field lines, especially concentrated near the magnetic poles of the neutron star, where the field lines are crowded together. The rapidly moving electrons emit synchrotron radiation as they travel along the field lines, creating narrow beams of radiation from both magnetic poles of the star. If the magnetic axis of the star is not aligned with the rotation axis (Fig. 24.11), then these beams will sweep across the sky in a conical pattern as the star rotates. If the Earth happens to lie in the direction intersected by one of these beams, then we see a flash of radiation every time the beam sweeps by us.

Synchrotron radiation is normally emitted over a very broad range of wavelengths, suggesting that pulsars may "pulse" in parts of the spectrum other than the radio. Indeed this is the case: the Crab pulsar has been identified as a visible-light and X-ray pulsar (Fig. 24.12). Although the details are still somewhat vague, the rotating neutron star model seems to be the best explanation of the pulsars. Since special conditions (i.e., nonalignment of the magnetic and rotation axes, and the need for the beam to cross the direction towards the Earth) are required for a neutron star to be seen from Earth as a pulsar, it follows that there should be many neutron stars that do not manifest themselves as pulsars. In the next section we will see how some of these nonpulsating neutron stars are detected.

Neutron Stars in Binary Systems

Earlier we made a general statement that stellar remnants are often most easily observed when they are in binary systems. We have already seen that a white dwarf in a binary can flare up violently if it receives new matter from its companion star. A neutron star reacts similarly in the same circumstances.

If a neutron star is in a binary system in which the companion object is either a hot star with a rapid wind or a cool giant losing matter because of its active chromosphere and low surface gravity, some of the ejected mass will reach the surface of the neutron star. Very little of it falls directly down onto the surface; instead, much of it swirls around the neutron star, forming a disk of gas called an **accretion disk** (Fig. 24.13). The individual gas particles orbit the neutron star like microscopic planets and only fall inward as they lose energy in collisions with other particles. The accretion disk acts as a reservoir of material, slowly feeding it inwards, towards the neutron star.

The disk is very hot, because of the immense gravitational field close to the neutron star. As gas in the accretion disk slowly falls in closer to the neutron star, gravitational potential energy is converted

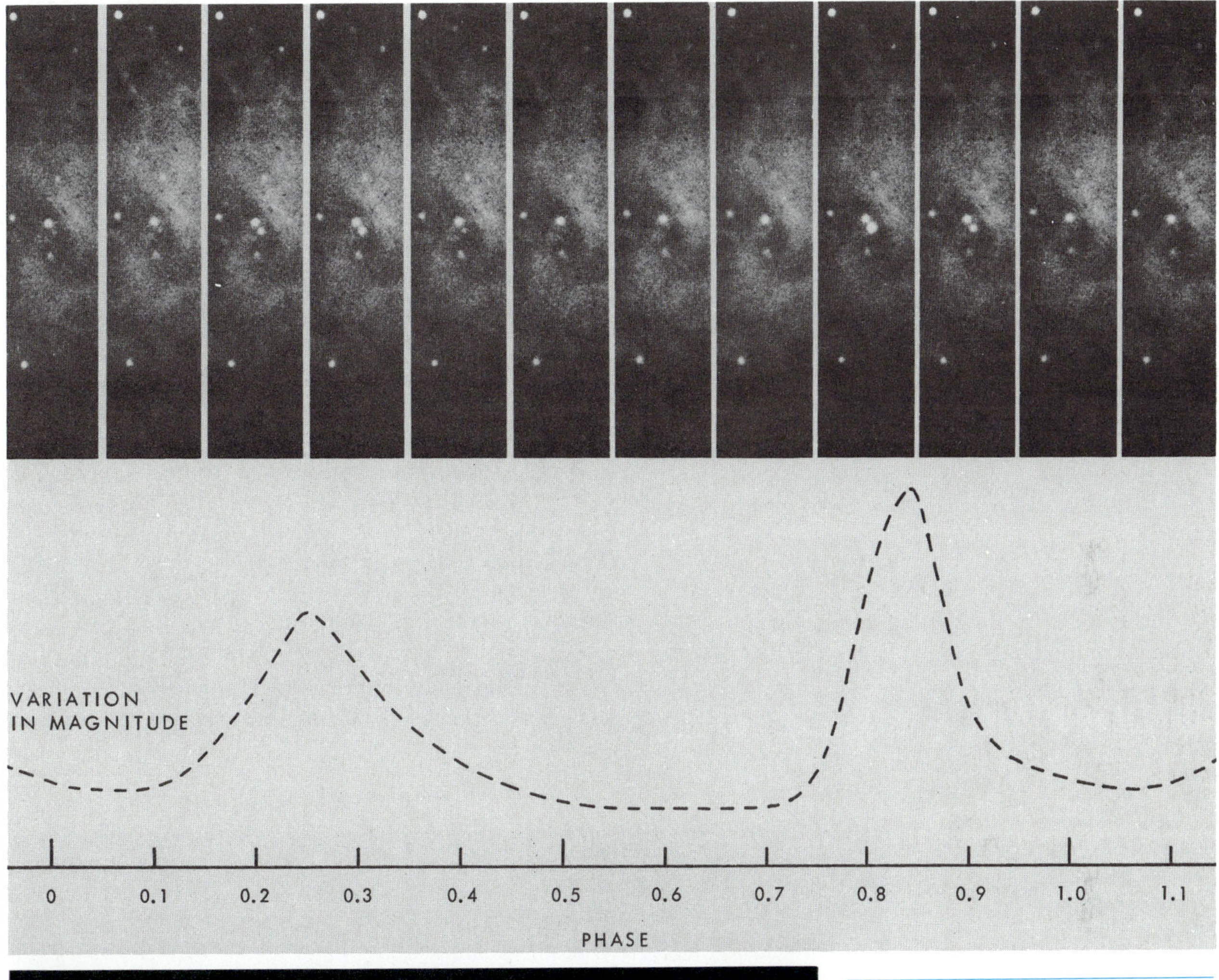

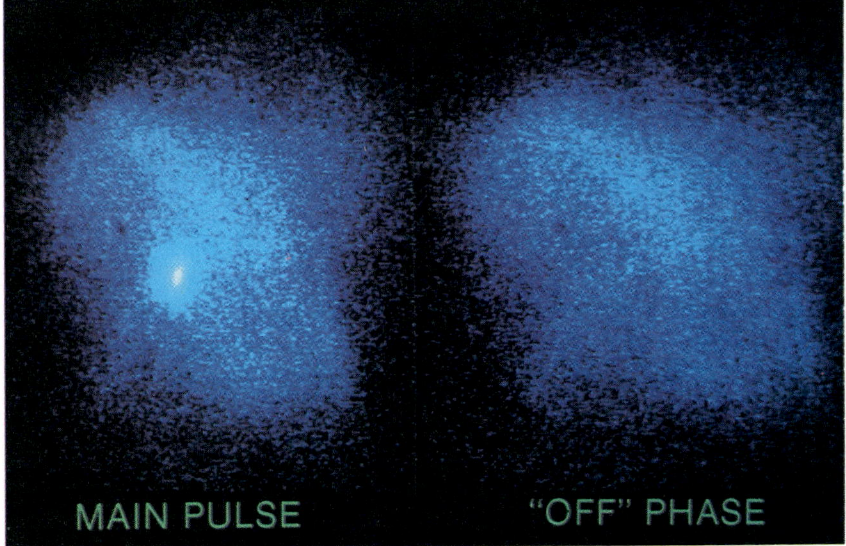

Figure 24.12 Visible and X-ray flashes from the Crab pulsar. *At top is a sequence of visible-light photographs accompanied by a light curve, showing how the pulsar in the Crab Nebula flashes on and off during its 0.03-second cycle. Below is a pair of X-ray images, showing the pulsar "on" and "off" in different parts of its cycle.* (*top:* National Optical Astronomy Observatories; *bottom:* Harvard-Smithsonian Center for Astrophysics)

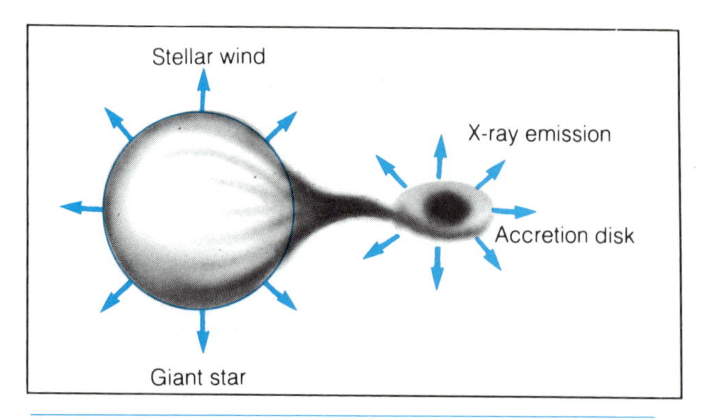

FIGURE 24.13 AN ACCRETION DISK. *Here a giant star, losing mass through a stellar wind, has a neutron star companion. Material that is trapped by the neutron star's gravitational field swirls around it in a disk, which is so hot due to compression that it glows at X-ray wavelengths. A nearly identical situation can arise when the compact companion is a black hole.*

into heat. The gas in the disk reaches temperatures of several million degrees, hot enough to emit X- rays. Hence a neutron star in a mass-exchange binary system is likely to be an X-ray source, and a number of such systems have been found in the past fifteen years (Fig. 24.14), since the advent of X-ray telescopes launched on rockets or satellites. Often it is known that an X-ray object is part of a binary system, because the X-rays are periodically eclipsed by the companion star. Eclipses are made likely by the proximity of the two stars (mass exchange would not occur unless it were a close binary) and the fact that the mass-losing companion is likely to be a large star, either an upper main-sequence star or a supergiant.

The so-called **binary X-ray sources** (a few are listed in Table 24.4) are among the strongest X-ray-emitting objects known. Many of them are, most likely, neutron stars, although the possibility also exists that some are black holes, which would similarly produce X-ray emission as material fell inwards.

There is a slightly different way in which neutron stars can emit X-rays, which also occurs in mass-exchange binaries. If the infalling material trickles down onto the neutron star in a steady fashion, then there is continuous radiation of X-rays, as we have just seen. If, however, the material falling in does so sporadically, and arrives in substantial quantities every now and then, a major nuclear outburst occurs each

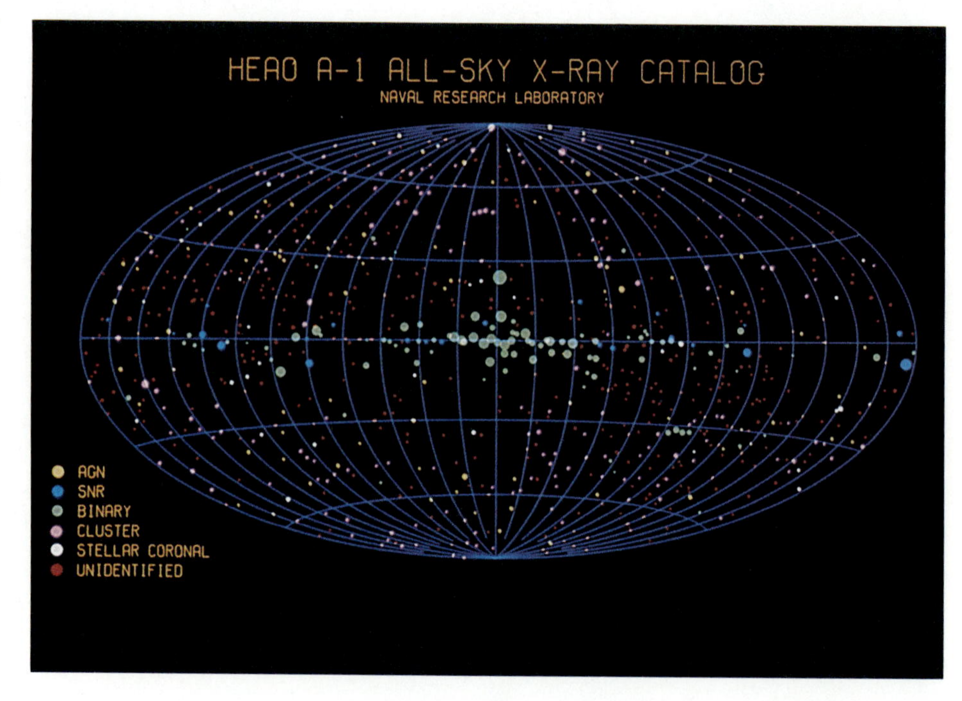

FIGURE 24.14 AN X-RAY MAP. *This shows the locations of X-ray sources cataloged by the* Einstein Observatory, *an orbiting X-ray satellite that operated in the early 1980s. Most of the sources here are binary systems in which one member is a compact stellar remnant such as a neutron star or a black hole.* (Smithsonian Astrophysical Observatory)

TABLE 24.4 SELECTED BINARY X-RAY SOURCES

Source	Period (days)	Mass of System ($M_\odot$)	Nature of Stars
Cygnus X-1	5.6 days	40	O9Ib supergiant; probable black hole
Centaurus X-3	2.09	20–25	B0Ib-III giant; pulsar (neutron star)
Small Magellanic Cloud X-1	3.89	15–25	B0Ib supergiant; pulsar (neutron star)
Vela X-1	8.95	20–30	B0.5Ib supergiant; probable neutron star
Hercules X-1	1.70	2–5	HZ Hercules (A star); pulsar (neutron star)
A 0620-00	0.32	11	K dwarf; probable black hole (companion mass $\simeq$ 10 $M_\odot$)
Large Magellanic Cloud X-3	1.70	15	B3 main sequence star; probable black hole (mass $\simeq$ 9 $M_\odot$)

time (Fig. 24.15), in close analogy to the nova process involving a white dwarf. This apparently happens in some cases, producing random but frequent X-ray outbursts. The intensity of the outburst, as in the case of a nova, depends on how much matter has fallen in and been consumed in the reactions. The neutron star binary systems where this occurs are called **bursters**. One important contrast with novae is that the flare-up of a burster occurs much more rapidly, lasting only a few seconds. Another is that most of the emission occurs only in very energetic X-rays, so these outbursts do not show up in visible light.

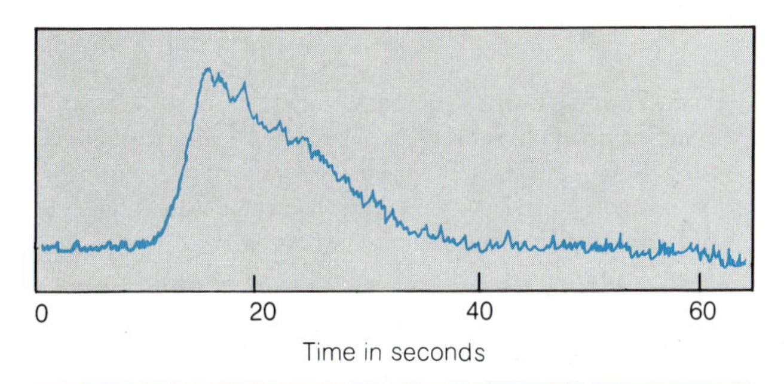

FIGURE 24.15 THE X-RAY LIGHT CURVE OF A BURSTER. *This schematic illustration of the X-ray intensity from one of these sources shows how rapidly the emission flares up and then drops off. Like the nova process, these outbursts are thought to be due to material falling onto the surface of a neutron star and igniting brief episodes of nuclear reactions.*

BLACK HOLES: GRAVITY'S FINAL VICTORY

In the last chapter we saw what happens to a massive star at the end of its life; it falls in on itself, and there is no barrier, not electron degeneracy nor neutron degeneracy, that can stop it.

The gravitational field near the surface of a collapsing star grows in strength, as the mass of the star becomes concentrated in an ever-smaller volume. This gravitational field has important effects in the near vicinity of the star, although at a distance it remains unchanged from what it was before the collapse. Close to the star, though, the structure of space itself is distorted, according to Einstein's theory of general relativity. Einstein discovered that accelerations caused by changing motion and those caused by gravitational fields are equivalent, and from this it follows that space must be curved in the presence of a gravitational field, so that moving particles follow the same path that they would follow if they were being accelerated. This applies to photons of light, just as it does to other kinds of particles.

Normally, the effects of the curvature of space are not noticeable, except under very careful observation, or when considering very large distances. Later in this text, when we discuss the universe as a whole and its overall structure, we will consider the latter situations. For now, we confine ourselves to local regions where space can be distorted by very strong gravitational fields (Fig. 24.16).

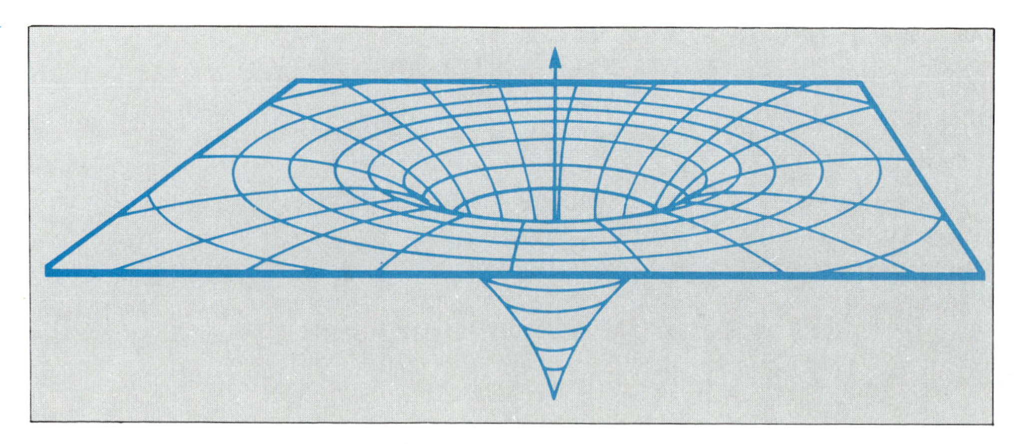

FIGURE 24.16 GEOMETRY OF SPACE NEAR A BLACK HOLE. *Einstein's theory of general relativity may be interpreted in terms of a curvature of space in the presence of a gravitational field. Here we see how this curvature varies near a black hole.*

Let us consider a photon emitted from the surface of a star as it falls in on itself (Fig. 24.17). If the photon is emitted at any angle away from the vertical, its path will be bent over further. If the gravitational field is strong enough, the photon's path may be bent over so far that it falls back onto the stellar surface. Photons emitted straight upward follow a straight path, but they lose energy to the gravitational field, causing their wavelengths to be redshifted (we discussed this phenomenon in the case of white dwarfs). When the gravitational field has become strong enough, the photon loses all of its energy and cannot escape.

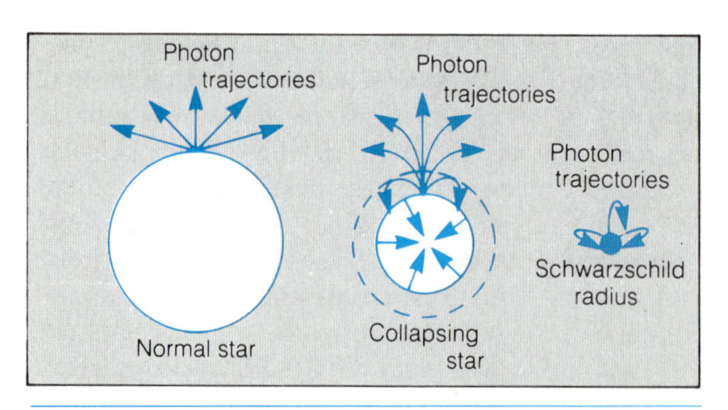

FIGURE 24.17 PHOTON TRAJECTORIES FROM A COLLAPSING STAR. *Light escapes in essentially straight lines in all directions from a normal star (left), whose gravitational field is not sufficient to cause large deflections. At an intermediate stage of collapse (center), photons emitted in a cone nearly perpendicular to the surface can escape, while others cannot. Those emitted at just the right angle go into orbit around the star, creating the ergosphere, while those emitted at greater angles fall back onto the stellar surface. After collapse has proceeded to within the Schwarzschild radius, no photons can escape.*

Another way of stating this is that the speed required for escape from the star exceeds that of light. To an outside observer, at the point when this happens the star becomes invisible, because no light from the star can reach the Earth.

The radius of the star at the time when its gravitational field becomes strong enough to trap photons is called the **Schwarzschild radius,** after the German astrophysicist who first calculated its properties some 60 years ago. The Schwarzschild radius depends only on the mass of the star that has collapsed. For a star of 10 solar masses, it is 30 km; a star of twice that mass would have twice the Schwarzschild radius, and so on (Table 24.5).

When a collapsing star has shrunk inside its Schwarzschild radius, it is said to have crossed its **event horizon,** because an outside observer cannot see it or anything that happens to it after that point. We can have no hope of ever seeing what happens inside the event horizon, but since no force is known that could stop the collapse, we assume that it continues. The mass becomes concentrated in an infinitesimally small region at the center, which is called a **singularity,** because mathematically it is a single point.

What happens to the matter that falls into a singularity is a subject for speculation, but we can be more concrete in discussing what happens just outside the event horizon. From the outsider's point of view, the time sense is distorted by the extreme gravitational field and acceleration in such a way that the collapse would seem to slow gradually, coming to a halt just as the star reached its event horizon. This slowdown, however, is most significant in the last

TABLE 24.5 SCHWARZSCHILD RADII FOR VARIOUS OBJECTS

Object	Schwarzschild Radius
Sun	3 km
Earth	0.9 cm
150-lb human	5×10^{-23} cm
Jupiter	2.9 m
Star of 50 $M_{\odot}$	150 km
Typical globular cluster	5×10^{4} km
Nucleus of a galaxy	10^{6}–10^{8} km (10^{-7}–10^{-5} pc)
Massive cluster of galaxies	10^{15} km (100 pc)
The universe	10^{26} km (10^{13} pc)

moments before the star's disappearance, and by then the star would be essentially invisible anyway, as most escaping photons are redshifted into the infrared or beyond. The star would seem to disappear rather quickly, despite the stretching of the collapse in time caused by relativistic effects. To an observer unfortunate enough to be falling into the black hole, the fall would go very quickly; the sense of time would not be distorted by relativistic effects in the same way as that of the distant observer. (The infalling observer would suffer serious discomfort from being stretched out by tidal forces.)

Mathematically, a black hole can be described completely by three quantities: its mass, its electrical charge, and its spin. The mass, of course, is determined by the amount of matter that collapsed to form the hole, plus any additional material that may have fallen in later. The electrical charge, similarly, depends on the charge of the material from which the black hole formed; if it contained more protons than electrons, for example, it would end up with a net positive charge. Because particles with opposite charges attract each other, and those with like charges repel, it is thought that electrical forces would maintain a fairly even mixture of particles during and after the formation of the black hole, so that the overall charge would be nearly zero. To illustrate this, imagine that a black hole was formed with a net negative charge. If there were ionized gas around it afterwards (as there likely would be, with some of the matter from the original star still drifting inward), the negative charge of the hole would repel additional electrons, preventing them from falling in , while protons would be accelerated inward. In time, enough protons would be gobbled up to neutralize the negative charge.

The spin of a black hole is not so easily dismissed, however. It stands to reason that if the star were spinning before its collapse, and most likely it would have been, then conservation of angular momentum would cause the rotation to speed up greatly as the star shrank. A high spin rate actually shrinks the event horizon, allowing an outside observer to see closer in to the singularity residing at the center. It is even possible mathematically—although it presents a physical dilemma—to have sufficient spin that there is no event horizon, so that whatever is in the center is exposed to view. A **naked singularity,** as this has been dubbed, would not produce the usual gravitational effects of a black hole, and it would be possible to blunder into one without any forewarning.

For the most part, it is assumed that the spin rate is never so large that a naked singularity can form, and in fact black hole properties are usually specified by mass alone, neglecting the effects of charge and spin. It is assumed that our main hope of detecting a black hole is by its gravitational effects, determined entirely by the mass.

Before turning to a discussion of how to find a black hole, we should mention that there may be other kinds of black holes, formed by processes other than the collapse of individual massive stars. The others include the so-called mini black holes, very small ones postulated to have formed under conditions of extreme density early in the history of the universe; and supermassive black holes, thought possibly to inhabit the cores of large galaxies, having formed from thousands or millions of stellar masses coalescing in the center. These other possibilities are discussed in later chapters; for now, we turn to the hunt for stellar black holes.

Do Black Holes Exist?

The mass that goes into a black hole during its formation still exists there, hiding inside its event horizon. Even though no light can escape, its gravitational effects persist. The gravitational force of the star is exactly the same as it was before the collapse, except at points so close that they would have been inside the original star.

Our best chance of detecting a black hole, then, is to look for an invisible object whose mass is too great to be anything else. Even if we do find such a thing, it is really only a circumstantial argument, because the conclusion that it is a black hole relies on the theory that says neither a white dwarf nor a neutron star can survive if the mass is sufficiently great.

The best chance of determining the mass of an object is when it is in orbit around a companion star, where Kepler's third law can be applied to find the masses. The search for black holes, then, leads us to examine binary systems, looking for invisible, but massive, objects.

This search has been facilitated by another property of black holes; if they find new material to pull into themselves, the trapped matter forms an accretion disk, just as we described in the case of a neutron star in a mass-transfer binary. We have already seen that such a disk becomes so hot that it emits X rays. Probably some of the binary X-ray sources, therefore, are due to black holes in binary systems rather than to neutron stars. The best way to distinguish between the two possible types of remnants in these systems is to determine the mass of the invisible companion, to see whether it is too great to be a neutron star.

The determination of the masses in a binary system is difficult, if not impossible, if only one of the two stars can be seen, because the information on their speeds and on the inclination angle of the orbit is incomplete. The fact that many X-ray binaries are eclipsing systems (at least the X-ray source is eclipsed; Fig. 24.18) helps to determine the tilt of the orbit in some cases. The orbital period is usually easily determined from the eclipse frequency as well, but the orbital velocities, needed in order to deduce the sizes of the orbits, are often very difficult or impossible to measure. The unseen star, of course, emits little or no light, so there is no hope of measuring Doppler shifts in its spectral lines, and the normal companion star is often a hot giant or supergiant, and these stars usually have very broad spectral lines, whose velocities cannot be measured accurately. Nevertheless, sometimes enough information can be derived to at least place limits on the mass of the invisible companion, however, and in one such case, the X-ray binary called Cygnus X-1 (Fig. 24.19), the collapsed object appears to contain at least 8 solar masses. This object is one of three leading black hole candidates, the other two being the sources LMC X-3 and A0620-00 (included in Table 24.4).

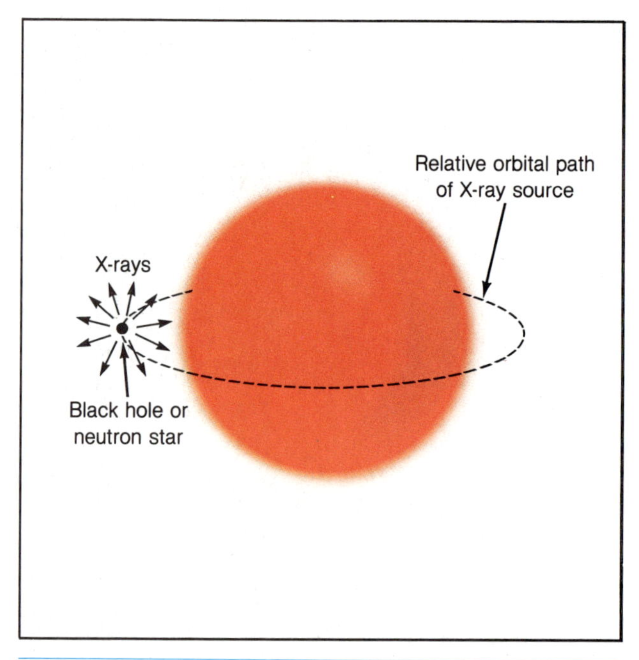

FIGURE 24.18 ECLIPSES OF AN X-RAY SOURCE IN A BINARY SYSTEM. *Because the separation of the two stars in a mass-transfer system is always very small, and because the mass-losing star is usually very large (a red giant or a hot giant or supergiant with a stellar wind), eclipses occur easily. If the mass-receiving star is a neutron star or black hole, and therefore an X-ray emitter, the X rays are likely to be eclipsed periodically by the companion. Hence many such systems are X-ray binaries.*

While the jury is still out on the question of the existence of black holes, the circumstantial evidence in favor is strong. Furthermore, it is easier to accept the existence of black holes than to find a way for a collapsing, massive star to avoid becoming one. As we have stressed, no force is known that can resist gravity in this circumstance. To believe that black

FIGURE 24.19 CYGNUS X-1. *This X-ray image obtained by the orbiting* Einstein Observatory *shows the intense source of X rays at the location of a dim, hot star which is thought to be a normal star that is losing mass to its invisible black hole companion.* (Harvard-Smithsonian Center for Astrophysics)

holes do not form, one has to invent some way of preventing collapse, and also some way of explaining the observational evidence for invisible massive objects in some binary systems. It is far less complicated to accept that black holes do form and exist in these binaries, because to do so requires the fewest unproven assumptions. The principle called **Occam's Razor,** a guiding philosophy for scientists, states that the explanation of any phenomenon that requires the fewest arbitrary assumptions is most likely to be the correct one.

Most astronomers today have adopted the concept of black holes, and not only believe that they exist, but consider them an integral part of our universe, playing numerous important roles.

PERSPECTIVE

We have hounded the stars into their graves. We have examined their corpses to see how they decay, and have come away with our minds filled with wonder at the novel forms matter can take. We have seen that the nature of stellar remnants depends entirely on how much mass is left when stars run out of nuclear fuel and collapse, the three possibilities being white dwarfs, neutron stars, and black holes. Stars that die alone, no matter what final form they take, are not likely to be detected again (except for the closest white dwarfs and the neutron stars that happen to appear to us as pulsars), while those that exist in binary systems may be reincarnated in spectacular fashion if mass exchange takes place.

Having learned all we can about stars as individuals, we are now ready to move on to larger scales in the universe, to examine galaxies, and ultimately the universe itself.

SUMMARY

1. A white dwarf gradually cools off, taking billions of years to become a cold cinder.
2. If new matter falls onto the surface of a white dwarf, for example in a binary system in which the companion star loses mass, then the white dwarf may become a cataclysmic variable, if the matter arrives in small amounts with high energy. Or it may become a nova, if matter builds up slowly to the point where a larger quantity is involved in the explosion. If it exceeds the white dwarf mass limit, it may become a neutron star or black hole, or explode as a Type I supernova.
3. Massive stars are likely to explode as supernovae (of Type II) when all possible nuclear reaction stages have ceased. The supernova explosion creates an expanding cloud of hot, chemically enriched gas known as a supernova remnant.
4. In some cases, a remnant of 2 to 3 solar masses is left behind, in the form of a neutron star, consisting of degenerate neutron gas.
5. A neutron star is too dim to be seen directly in most cases but may be observed as a pulsar (depending on the alignment of its magnetic and rotation axes, and our line of sight), and in a close binary system it may become a source of X-ray emission.
6. Some neutron stars that receive new material in clumps flare up occasionally as X-ray sources called bursters.
7. If the final mass of a star exceeds 2 or 3 solar

masses, it will become a black hole at the end of its nuclear reaction lifetime.

8. The immensely strong gravitational field near a black hole traps photons of light, rendering the black hole invisible.

9. A black hole may be detected by its gravitational influence on a binary companion, or by the X-rays it emits if new matter falls in, as in a close binary system in which mass transfer takes place. A black hole binary X-ray source can be detected only by analysis of the orbits to determine the mass of the unseen object.

REVIEW QUESTIONS

Thought Questions

1. Why is the final mass of a star, rather than the initial mass, the factor that determines what form of remnant the star will leave? How can the final mass be different from the initial mass?

2. Summarize the techniques by which white dwarfs can be detected from the Earth.

3. Explain why white dwarfs, with no source of energy, can stay hot for as long as a billion years.

4. Describe the differences between a nova, a supernova of Type I, and a supernova of Type II.

5. Compare the properties of a neutron star with those of a white dwarf. Include a description of the form of pressure force that counteracts gravity in the two types of remnants.

6. How are neutron stars detected from the Earth?

7. Explain why the pulses from pulsars were finally judged to be the result of rapid rotation of neutron stars, rather than rapid expansion and contraction or rotation of some other type of star. What was the role of the Crab Nebula and its pulsar in this discovery?

8. Both neutron stars and black holes can be X-ray sources in binary systems where mass exchange occurs. How can astronomers tell which kind of remnant is present in a given binary X-ray source?

9. Explain why some massive stars become black holes, while others become neutron stars or blow up completely, leaving no remnant.

10. Why would a neutron star or a black hole be expected to have a very rapid rotation rate?

Problems

1. Compare the average density, surface gravity, and escape velocity of a neutron star (2 solar masses, radius 10^{-4} solar radii) with the same properties of the Sun.

2. Suppose mass is added to a white dwarf and undergoes nuclear reactions, creating a nova and releasing energy according to $E = mc^2$. How much energy is released if the mass that undergoes reactions is 10^{-4} solar masses? In Type I supernova, the mass that undergoes reactions is about 1 solar mass. What is the energy released in such a supernova? In both cases, assume that only 10 percent of the mass that undergoes reactions is converted into energy.

3. If the rotation period of the Crab pulsar is 0.033 sec and its radius is 10 km, what is the speed of a point on its surface at the equator? What fraction of the speed of light is this?

4. Apart from the beams that create the pulsar effect, a neutron star glows from its surface because of its high temperature. Suppose a neutron star has a surface temperature of 1 million degrees (10^6 K). What is its wavelength of maximum emission? What kind of telescope could be used to observe it?

5. The surface area of a neutron star is so small that the luminosity is low, despite the high temperature. To show this, calculate the luminosity of a neutron star with surface temperature 10^6 K and radius 10 km, relative to the luminosity of the Sun.

6. The formula for the Schwarzschild radius of an object of mass M and radius R is $R_s = 2\,GM/c^2$. Use this formula to find the Schwarzschild radii for the following: (a) the Sun; (b) a star of 20 solar masses; and (c) the Milky Way galaxy (mass approximately 10^{45} grams).

7. Suppose an X-ray binary system is found in which the visible star is a red giant of a spectral type thought to have a mass of about 12 solar masses. The orbital period is 3.65 days, and the semimajor axis is 0.12 AU. Calculate the sum of the masses from Kepler's third law, then try to decide whether the compact companion star is a neutron star or a black hole.

ADDITIONAL READINGS

Anderson, L. 1976. X-rays from degenerate stars. *Mercury* 5(5):2.

Bethe, H. A. 1985. How a supernova explodes. *Scientific American* 252(5):60.

Black, D. L. 1976. Black holes and their astrophysical implications. *Sky and Telescope* 50(1):20 (Part I); 50(2):87 (Part 2).

Clark, David H. 1985. *The Quest for SS433*. Penguin Books.

Greenstein, G. 1985. Neutron stars and the discovery of pulsars. *Mercury* (2):34 (Part 1); (3):66 (Part 2).

Gursky, H., and E. P. J. van den Heuvel. 1975. X-ray emitting double stars. *Scientific American* 232(3):24.

Helfand, D. J. 1978. Recent observations of pulsars. *American Scientist* 66:332.

Kafatos, M., and A. G. Michalitsianos. 1984. Symbiotic stars. *Scientific American* 251(1):84.

Kaufmann, W. 1974. Black holes, worm holes, and white holes. *Mercury* 3(3):26.

Lewin, W. H. G. 1981. The sources of celestial X-ray bursts. *Scientific American* 244(5):72.

Schaefer, B. E. 1985. Gamma-ray bursters. *Scientific American* 252(2):52.

Schramm, D., and W. Arnett. 1975. Supernovae. *Mercury* 4(3):16.

Seward, F. 1986. Neutron stars in supernova remnants. *Sky and Telescope* 71(1):6.

Shaham, Jacob. 1987. The oldest pulsars in the universe. *Scientific American* 256(2):50.

Smarr, L., and W. H. Press. 1978. Spacetime: Black holes and gravitational waves. *American Scientist* 66:72.

Thorne, K. S. 1974. The search for black holes. *Scientific American* 233(6):32.

Van Horn, H. M. 1973. The physics of white dwarfs. *Physics Today* 32(1):23.

Wheeler, J. C. 1973. After the supernova, what? *American Scientist* 61:42.

P6) $R_s = 3\left(\frac{M}{M_\odot}\right)$

$R_s = \frac{2GM}{c^2}$

$G = 6.67\times10^{-8}$

3×10^{10} cm/s

$R_{s\odot} = \frac{2GM_\odot}{c^2}$

$\frac{R_s}{R_{s\odot}} = \frac{M}{M_\odot}$

$R_s = 3\left(\frac{M}{M_\odot}\right)$

a) $M = 1M_\odot$

$\frac{M}{M_\odot} = 1$

$R_s = 3$ km

b) $M = 20M_\odot$

$M/M_\odot = 20$

$R_s = 3\times20 = 60$ km

GUEST EDITORIAL

THE IMPACT OF SUPERNOVA 1987A

Stanford E. Woosley

Fascinated by explosions since his childhood in Texas, Stan Woosley went on to become one of the world's leading experts in nuclear reactions, particularly those that occur in stars. Educated at Rice University, as a graduate student he was in a unique position at the intersection of the two leading groups of theorists in the study of stellar nucleosynthesis: his dissertation advisor had been trained by William Fowler, who was later to win a Nobel Prize for his contributions to the subject, and his associate advisor was a student of A. G. W. Cameron, a pioneer in the understanding of the origin of the elements in the universe and the solar system. Woosley later worked with a group at Cambridge University headed by Fred Hoyle, another founder of modern nucleosynthesis theory, and then joined the faculty at the University of California–Santa Cruz, where he has just completed a term as Chair of the Astronomy Department. Throughout his career he has been a leader in the detailed modelling of nuclear processes in stars, and, with Thomas Weaver of the Lawrence Livermore Laboratories and others, he developed a comprehensive theory of nuclear processes in supernova explosions. The occurrence of Supernova 1987A in February of 1987 found Stan Woosley in perfect position to take the lead in determining what happened: his paper explaining the explosion of a blue supergiant star was submitted for publication at a time when many astronomers were still not convinced that indeed such a star had been involved in the supernova. In the months that have followed, Woosley has been refining his model of the event, and is regarded worldwide as the *authority on the explosion. In a way, this has brought him full circle from his childhood days of intrigue with things that blow up. In this article, Woosley discusses the significance of Supernova 1987A and describes some of his predictions for its future.*

No event in nature is more violent and powerful than the death of a massive star as a Type II supernova. For the star it is the end of a comparatively brief but brilliant life, or at least the transition to a more exotic state. For astronomers it provides not only the spectacular fireworks but a unique testing ground for theories of stellar evolution and explosion.

Throughout history the occurence of over 620 supernovae had been recorded prior to February of 1987. Almost all have occurred at vast distances because it is only at such distances that astronomers can sample a sufficient number of galaxies to compensate for the fact that each produces no more than a few supernovae per century. For example, our own galaxy is believed to produce a Type II supernova once every 40 years, but most go undetected, because, bright as they are, their optical emissions are totally obscured by interstellar dust. The last supernova clearly visible to the naked eye occurred in 1604, five years before Galileo made the first telescopic observations of the skies, and was observed extensively by Johannes Kepler.

Imagine, then, the joy and amazement of the world astronomical community when the announcement went out on February 24 of 1987 that a bright supernova, clearly visible to the unaided eye, was occurring in a galaxy "just next door," the Large Magellanic Cloud. Though just one of the many supernovae that have been detected, by virtue of its proximity and brightness this one has proved unique in many ways. It is, for example, the first Type II supernova to be observed in an irregular galaxy. Before, all were found in spiral galaxies. The light curve, showing how bright the supernova is as a function of time [see Fig. 3 in the special insert section on the supernova], has also proved unique, fainter initially by a factor of fifteen compared to other known Type II's. Perhaps the two are related—perhaps irregular galaxies produce fainter supernovae, which makes them harder

GUEST EDITORIAL

to detect. Supernova 1987A is also the first supernova for which an ordinary steller progenitor has been identified. Knowing the properties of the star that exploded gives theoreticians a big head start in understanding the behavior of the supernova.

Supernova 1987A was also unique in being observed as a neutrino source. The enormous binding energy of the neutron star produced at the middle of the star came out almost entirely in the form of neutrinos emitted during the first ten seconds of the explosion. These were detected in deep underground experiments in both the United States and Japan. The first optical detection came only three hours after the neutrinos, one hour after the shock wave from the exploding core broke through the surface of the star. Such an early detection is again unique in the study of supernovae, and severely constrains the models.

More unique properties of 1987A include a light curve powered at peak entirely by radioactive decay (a universal property of Type I supernovae, but hitherto unknown in Type II's); an accurate determination of the amount of iron produced by explosive nucleosynthesis, 0.07 solar masses; the detection, in mid-August, of X rays from the supernova, the scattered photons liberated by the radioactive dacay of ^{56}Co (radioactive cobalt) to ^{56}Fe (normal iron); and the identification of an unusual "companion" source that, at least early on, was 10 percent as bright as the supernova itself. Finally, it should not be overlooked that Supernova 1987A was the first (and perhaps the last) supernova to appear on the cover of *Time* magazine and to have its own TV show (on "NOVA," which appeared in early October, 1987).

The unique scientific results and many more that will surely follow have come about because, for the first time, astronomers have been able to study a bright supernova at all wavelengths on a frequent basis. In the southern hemisphere, radio, infrared, and optical telescopes in Chili, Australia, New Zealand, and South Africa have monitored the supernova on an almost daily basis. From an airplane in New Zealand the *Kuiper Airborne Observatory*, with its infrared telescope, has obtained spectra of the supernova. From space the *International Ultraviolet Explorer* has studied the optical brightness and the ultraviolet spectrum of the supernova; the *Solar Maximum Mission* (complemented by balloon flights in Australia) has placed upper limits on the gamma-ray emission; the Japanese satellite *Ginga*, and instruments on the Soviet space station *Mir*, have studied the X-ray emission of the supernova. And from deep underground, the only detectors of any kind that were able to "see" the supernova from the northern hemisphere have witnessed the neutrino burst as it propagated through the Earth from the other side.

We can say a few things about the future of Supernova 1987A. In the near term, high-energy radiation will begin to emerge and be observable as the expanding shell of gas spreads and gradually becomes transparent. Already X rays have been detected, as well as ultraviolet emission lines, and soon we expect to see gamma rays. The luminosity of the supernova is now being provided by the radioactive decay of ^{56}Co in the expanding cloud of debris, and the nuclei of cobalt emit gamma rays at specific energies (*i.e.*, wavelengths) during this decay process. So far the gamma ray photons have been unable to escape, but soon they will, and I expect that by early 1988 gamma-ray emission lines will be observed. [Note: As this text goes to press, the expected gamma ray lines have been detected.]

In the longer term, eventually we will be able to detect the neutron star hiding at the center of the expanding cloud. It will be a very bright source of "hard" (that is, high-energy) X rays, and might be a pulsar as well. The remnant of the supernova should turn out to be one of the very brightest X-ray sources in the sky. It will also eventually become a radio source, as the expanding gas disperses and we begin to see the emission formed by relativistic electrons. Optically and in the ultraviolet the expanding gas will produce emission lines that become stronger for several months or years, and then will eventually decline. In several decades the remnant of the explosion will be brightest in X-ray and radio wavelengths, and in centuries it will become almost exclu-

(continued on next page)

GUEST EDITORIAL

sively a radio source. We can expect that many generations of astronomers will study the supernova of 1987; perhaps some of the readers of this book will become professional astronomers and earn their doctoral degrees through the analysis of the remnant or the neutron star.

As an astronomer who has participated in this global effort, I can offer one personal observation. Especially during the first few weeks following the supernova, observers and observatories of all nations and from all continents shared data, speculations, and the sheer exhilaration of the moment. Little was held back. In the process some mistakes were made, but were quickly subjected to test and the errors freely admitted and corrected. It was science at its best. The data we now have will occupy theoreticians for at least a decade, but the memory of the shared experience will last a lot longer.

SECTION V

THE MILKY WAY

In this section we explore the great conglomeration of stars to which our Sun belongs. The Milky Way, a spiral galaxy, is itself a dynamic entity, with its own structure and evolution governed by the complex interactions of stars with each other and with the interstellar gas and dust. Although this system of some 10^{11} individual stars is infinitely more complicated than a single star, and its life story is correspondingly more difficult to unravel, great progress has nevertheless been made in piecing together the puzzle.

In the first of the three chapters of this section, we examine the overall properties of our galaxy. We learn how its size and shape, and the Sun's location within it, were deduced from observations carried out from our position within the great disk, where our view is obscured by interstellar haze. We discuss how the various parameters describing the galaxy were derived from observation, and we learn of some immense and fascinating mysteries that remain.

Chapter 26 focuses on the diffuse material between the stars. The interstellar medium is so rarefied that a superficial consideration might lead us to deem it unworthy of our time and effort, but we find, to the contrary, that the interstellar medium is a vital part of the galactic life story. Stars form from this material, and in evolving and dying they return their substance to it. The interstellar medium itself is a dynamic, active component of the galaxy, one that has in recent years provided astronomers with several major surprises. In this chapter we survey the general properties of the gas and dust that pervade space, and pay particular attention to their role in the evolution of the galaxy.

Finally, Chapter 27 ties together the disparate data gathered and examined in the preceding two chapters, and tells the story of the formation and evolution of our galaxy. We see how the major features of the Milky Way arose from a sequence of developments dictated by simple laws of physics previously learned. Although there remain some mysteries, such as the quantity of mass in the far reaches of the galactic halo and the traces of violent activity at the core of the galaxy, most of the story can be told.

Having probed the Milky Way, we will then be ready to move outward, to examine and understand the countless galaxies in the universe beyond our own.

CHAPTER 25

Structure and Organization of the Galaxy

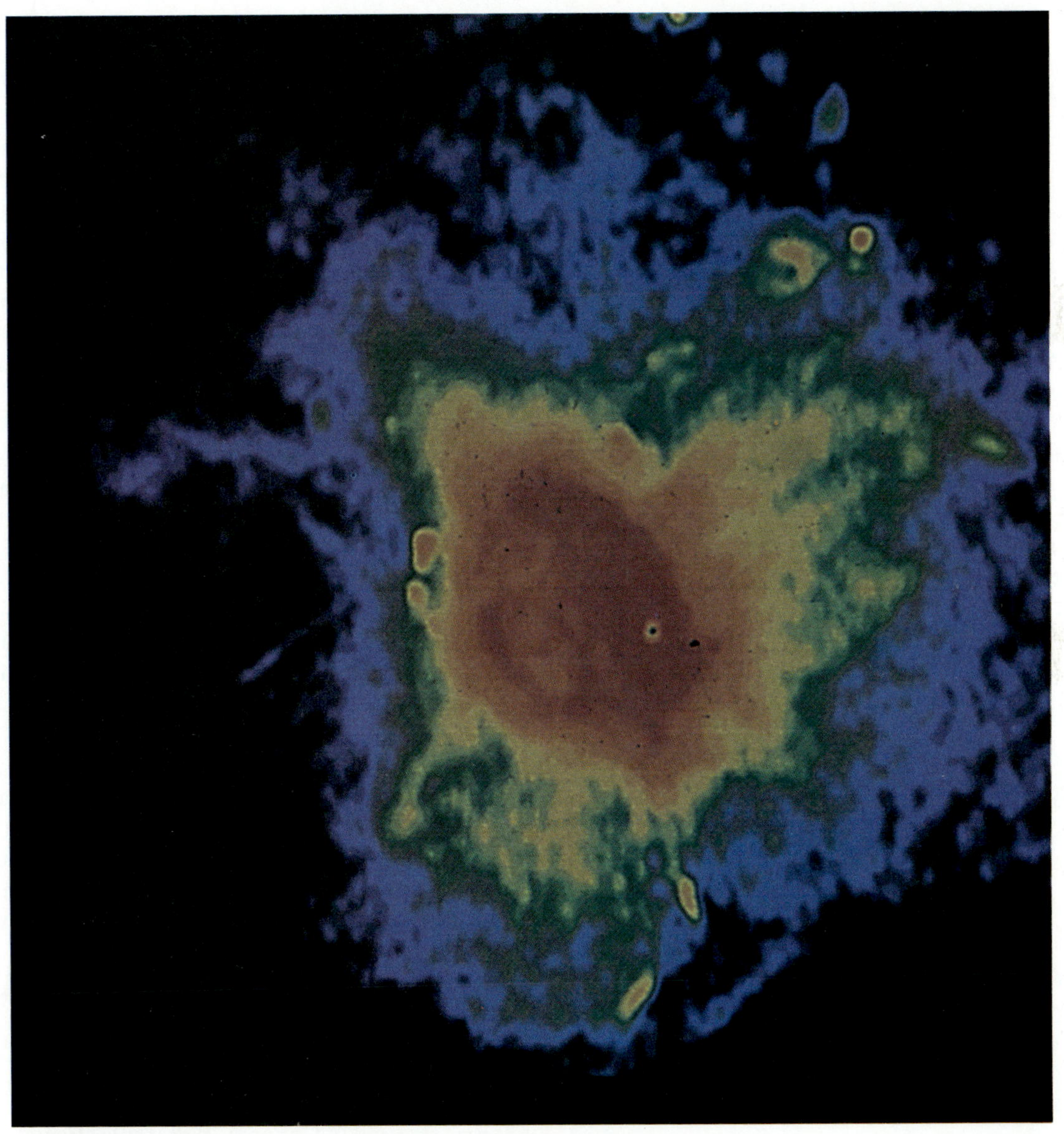

A radio map of the galactic center. (National Radio Astronomy Observatory)

CHAPTER PREVIEW

We have discussed stars as individuals, as though they existed in a vacuum, isolated from the rest of the universe. For the purpose of analyzing the structure and evolution of stars, this is a suitable approach, but if we want to understand the full stellar ecology, the properties of individuals must be discussed in the context of the larger environment.

Even a casual glance at the nighttime sky shows that the stars tend to be grouped rather than randomly distributed. The most obvious concentration is the Milky Way, the diffuse band of light stretching from horizon to horizon, clearly visible only in areas well away from city lights (Fig. 25.1). To the ancients, the Milky Way was merely a cloud; Galileo with his primitive telescope recognized it as a region of great concentration of stars. In modern times we know the Milky Way as a galaxy, the great pinwheel of billions of stars to which the Sun belongs (Table 25.1). The hazy streak across our sky is a cross-sectional, or edge-on, view of the galaxy, seen from a point within its disk.

TABLE 25.1 PROPERTIES OF THE MILKY WAY

Mass: 1.4×10^{11} $M_\odot$ (interior to Sun's position)
Diameter of disk: 30 kpc
Diameter of central bulge: 10 kpc
Sun's distance from center: 8.7 kpc
Diameter of halo: 100 kpc (very uncertain)
Thickness of disk: 1 kpc (at Sun's position)
Number of stars: 4×10^{11}
Typical density of stars: 20 stars/pc^3 (solar neighborhood)
Average density of interstellar matter: 10^{-24} g/cm^3 (roughly equal to one hydrogen atom/cm^3)
Luminosity: 2×10^{10} $L_\odot$
Absolute magnitude: -20.5
Orbital period at Sun's distance: 2.5×10^8 years

The overall structure of the Milky Way is a bit like a phonograph record, except that it has a large central bulge, the center of which is called the **nucleus** (Figs. 25.2 and 25.3). Nearly all the visible light is emitted by stars in the plane of the disk, although the galaxy also has a **halo,** a distribution of stars and star clusters centered on the nucleus but extending well above and below the disk. The most prominent objects in the halo are the **globular clusters,** very old, dense clusters of stars characterized by their distinctly spherical shape.

To envision the size of the Milky Way requires a new unit of distance. In Chapter 20 we discussed stellar distances in terms of parsecs, and we found that the nearest star is about 1.3 parsecs from the Sun. To expand to the scale of the galaxy, we speak in terms of **kiloparsecs,** or thousands of parsecs. The visible disk of the Milky Way is roughly 30 kiloparsecs (abbreviated kpc) in diameter, and the disk is a few hundred parsecs thick. Light from one edge of the galaxy takes about 100,000 years to travel across to the far edge.

Our galaxy is so large that we must discuss new methods of measuring distance, for even the main-sequence-fitting technique—the most powerful we have mentioned so far—fails when star clusters are too distant and faint to allow determination of the individual stellar magnitudes and spectral types or color indices.

VARIABLE STARS AS DISTANCE INDICATORS

It was stressed in Chapter 20 that it is always possible to determine the distance to an object if both its absolute and apparent magnitudes are known. In that

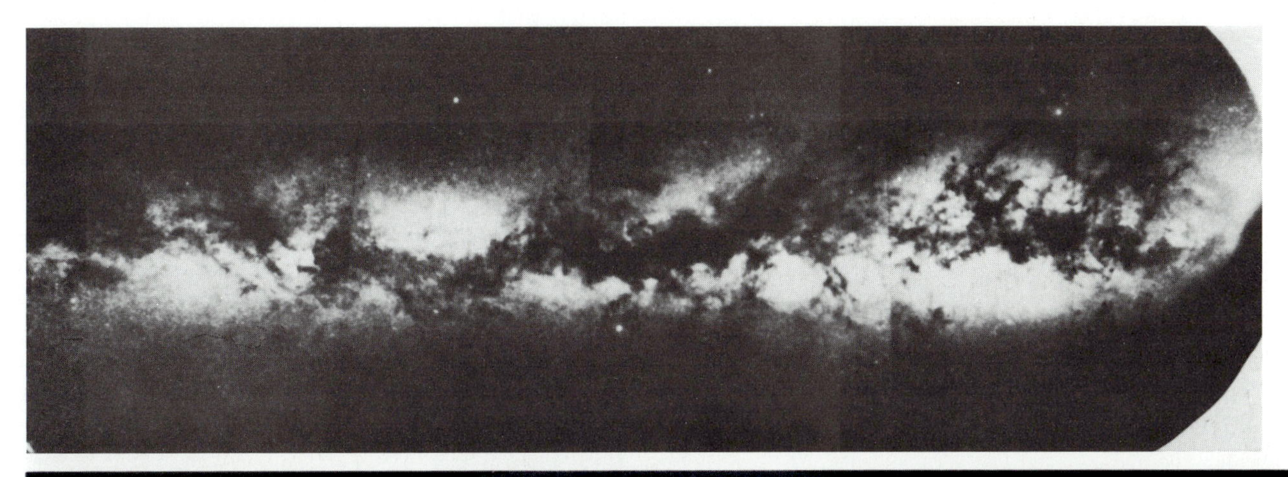

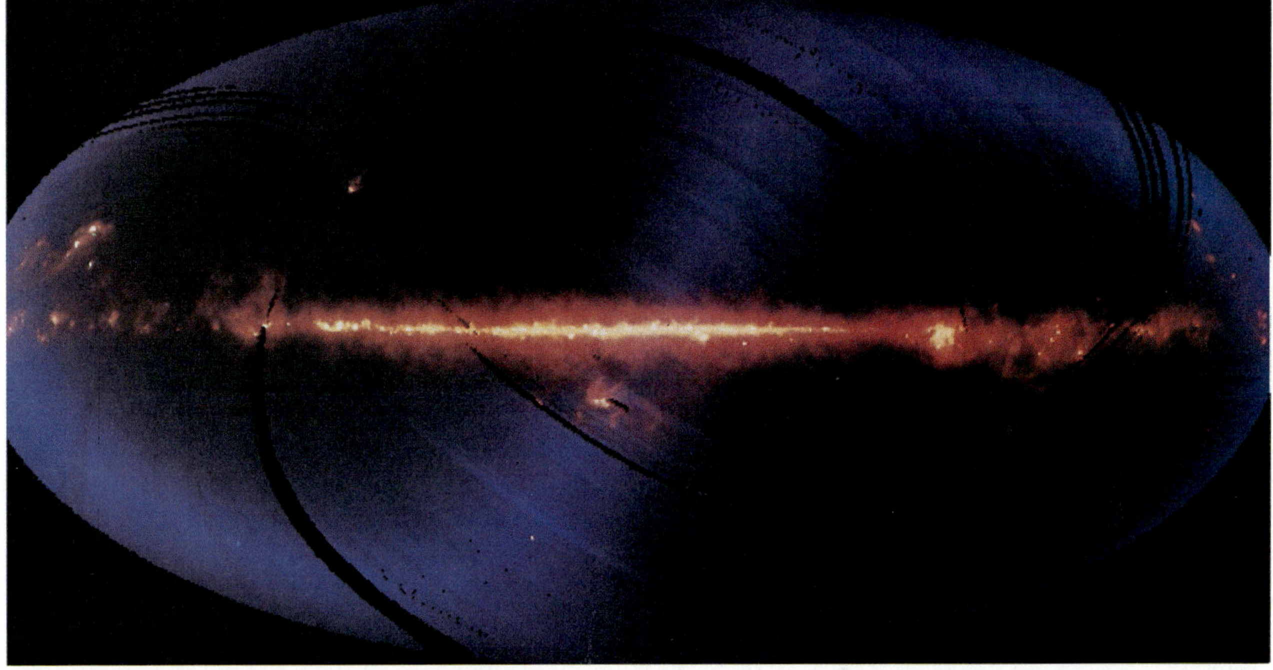

FIGURE 25.1 THE MILKY WAY. *This mosaic (top) shows a major portion of the cross-sectional view of our galaxy that we see from Earth. The bottom image is an infrared view of the Milky Way, obtained by the* IRAS *satellite. (top:* Mt. Wilson and Las Campanas Observatories, Carnegie Institution of Washington; *bottom:* NASA*)*

chapter we talked about spectroscopic parallaxes, in which the absolute magnitude of a star is derived from its spectral class, and we discussed main-sequence fitting, where a cluster of stars is observed and its main sequence determined, so that it can be fitted to the main sequence in the standard H-R diagram. Now we will learn about a special type of star whose absolute magnitude can be determined from its other properties, allowing it to be used to measure distances.

In Chapter 19 we briefly mentioned variable stars, including those which pulsate regularly. These stars physically expand and contract, changing in brightness as they do so. One of the first such stars to be discovered was δ Cephei, which is sufficiently bright

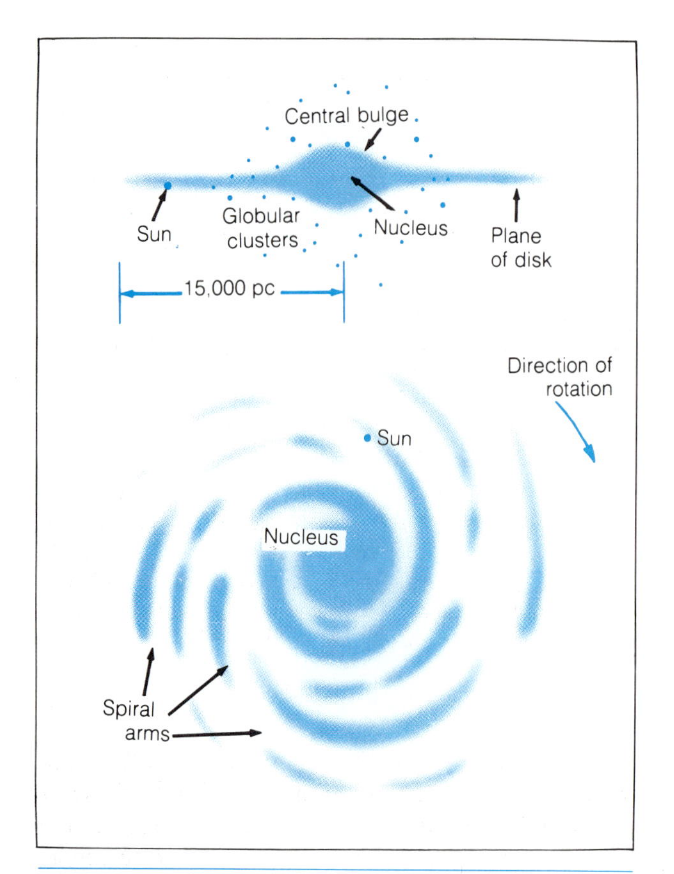

FIGURE 25.2 THE STRUCTURE OF OUR GALAXY. *These sketches illustrate the modern view of the Milky Way.*

FIGURE 25.3 A SPIRAL GALAXY SIMILAR TO THE MILKY WAY. *This galaxy (NGC 7331) probably resembles our own, as seen from afar.* (Palomar Observatory, California Institute of Technology)

to be seen with the unaided eye. Following the discovery of the variability of δ Cephei in the mid-1700s, other stars were found to have similar brightness variations, and these as a class became known as **Cepheid variables,** named after the prototype. Several other types of variable stars have subsequently been identified (Table 25.2), each with its own distinctive properties.

The usefulness of pulsating variables as distance indicators was discovered in 1912, when an astronomer named Henrietta Leavitt (Fig. 25.4) at the Harvard College Observatory carried out a study of variable stars located in the Magellanic Clouds, the two small satellite galaxies of the Milky Way. Leavitt noticed that the average brightnesses of these stars were correlated with the period of pulsation; that is, the longer the time between brightenings, the brighter the star. The Magellanic Clouds are so far away (about 55 kpc) relative to their own dimensions that all the stars are, for practical purposes, at the same distance from us. This meant that the correlation of period with apparent magnitude was actually a correlation with absolute magnitude, suggesting the possibility that the absolute magnitude could be determined from the period of pulsation.

Subsequent analysis established the numerical relationship between the period and absolute magnitude (Fig. 25.5), so that the measurement of distance to a variable star becomes a simple matter of determining the period of pulsation by counting the days between times of maximum brightness, then using the established correlation to determine the absolute magnitude. The distance is then found by comparing the absolute and apparent magnitudes, using

FIGURE 25.4 HENRIETTA LEAVITT. *Leavitt was part of the group of astronomers at Harvard who pioneered the development of spectral classification. In addition, she discovered the period-luminosity relationship for Cepheid variables in the course of studying stars in the Magellanic Clouds.* (Harvard College Observatory)

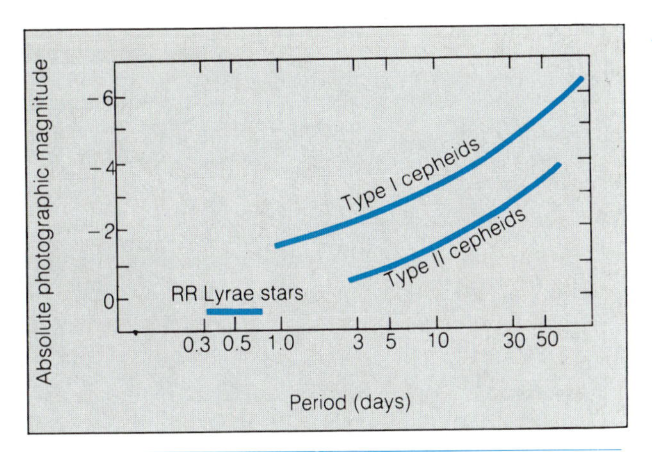

FIGURE 25.5 THE PERIOD-LUMINOSITY RELATIONSHIP FOR VARIABLE STARS. *This diagram shows how the pulsation periods for Cepheid and RR Lyrae variables are related to their absolute magnitudes. The fact that there are two types of Cepheids, with somewhat different relationships, was not recognized at first, and this led to some early confusion about distance scales.*

the standard technique discussed in Chapter 20. Because the Cepheid variables are giant stars, they are quite luminous and can be observed at great distance, even beyond the Milky Way. Therefore these stars are very useful in measuring the scale of our galaxy.

Another type of pulsating star, the **RR Lyrae variables,** also were found to be reliable distance indicators. These stars have periods of only a few hours, and they all have the same absolute magnitude. Any time we identify an RR Lyrae star by measuring its period, we immediately know its absolute magnitude and hence its distance. The RR Lyrae stars are not as luminous as the Cepheids, but they are still bright enough to be observed at large distances. These stars are often found in globular clusters and are therefore sometimes called **cluster variables.** They played

TABLE 25.2 TYPES OF PULSATING VARIABLES

Type of Variable	Spectral Type	Period (days)	Absolute Magnitude	Change of Magnitude
δ Cephei (Type I)	F,G supergiants	3–50	−2 to −6	0.1–2.0
W Virginis (Type II Cepheids)	F,G supergiants	5–30	0 to −4	0.1–2.0
RR Lyrae	A,F giants	0.4–1	0.5 to 1.2	0.6–1.3
RV Tauri	G,K giants	30–150	−2 to −3	Up to 3
δ Scuti	F giants	0.1–0.2	2	0.1
β Cephei (β Canis Majoris)	B giants	0.1–0.2	−3 to −5	0.1
Long-period variables (Mira)	M supergiants	80–600	+2 to −3	3–7

ASTRONOMICAL INSIGHT 25.1

THE INVISIBLE MILKY WAY

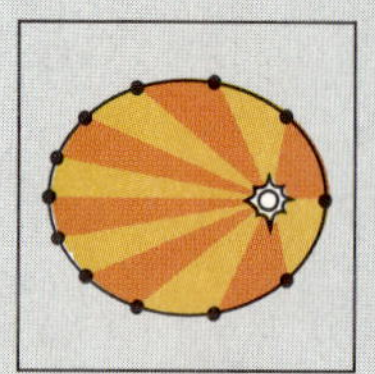

One of the most revealing facets of modern observational technology for astronomers has been the ability to see how the universe looks in wavelengths of radiation not visible to the eye. We have made frequent mention of this in earlier chapters (for example, the discussion of the Sun in Chapter 18). Our own galaxy, the Milky Way, provides another informative illustration of the power of observations in invisible wavelengths.

The series of galactic maps shown here are portraits of our galaxy in six different wavelengths, ranging from the longest (Map 1) to the shortest (Map 6). Each is organized on the same coordinate system, showing the galaxy as seen edge-on from the Earth. The direction toward the center of the galaxy is in the middle of each map, so that we are scanning along the plane of the disk (that is, in the direction of galactic longitude) when we look right or left of the center, and we are looking out of the plane (along the direction of increasing galactic latitude) when we look up or down from the center line. It is immediately apparent from all six maps that our galaxy has a disk-like form, because all six show strong concentration or symmetry about the central horizontal line.

Map 1 was made by surveying the galaxy at a wavelength of 21 cm, so that emission due to interstellar hydrogen atoms was measured. The relative intensities of this emission are color-coded, with pale violet representing the maximum intensity. We

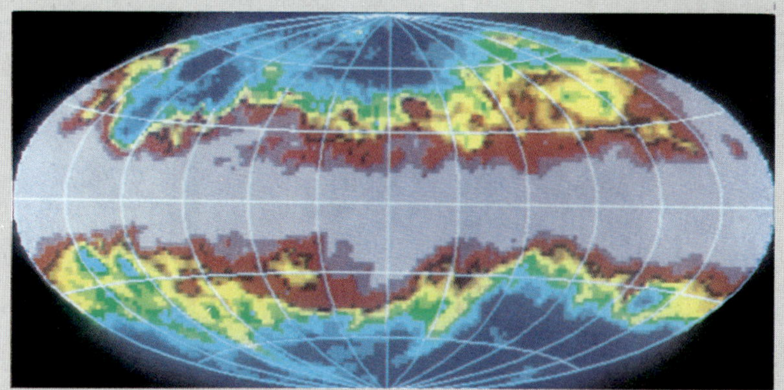

MAP 1

MAP 2

MAP 3

Top: A Stark and C. Heiles; *middle:* NASA; *bottom:* Lund Observatory

ASTRONOMICAL INSIGHT 25.1

THE INVISIBLE MILKY WAY

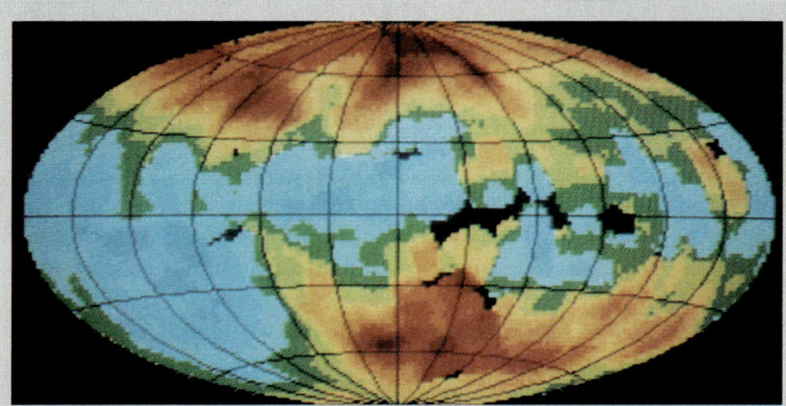

MAP 4

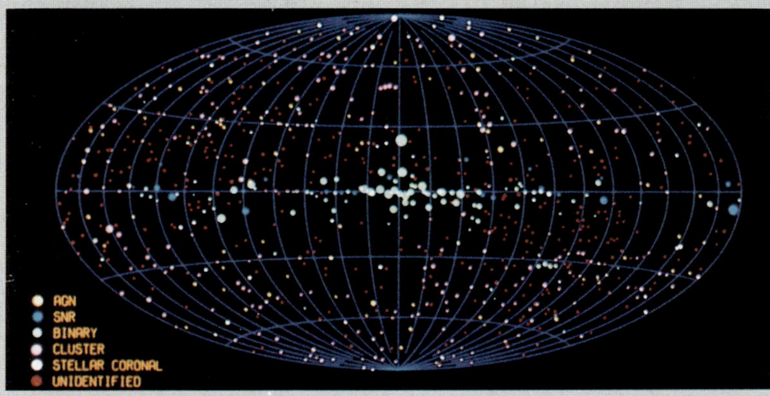

MAP 5

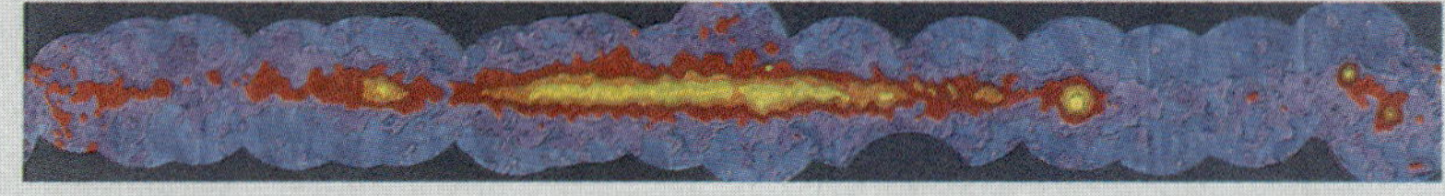

MAP 6

Top: From McCammon, *et al.* 1983, *Astrophysical Journal* 269:107, image provided by W. T. Sanders, University of Wisconsin; *middle:* Smithsonian Astrophysical Observatory; *bottom:* French National Research Council, courtesy C. Laurent

see from this that the interstellar hydrogen is concentrated in the plane of the galactic disk. Because the 21-cm emission is able to penetrate interstellar gas and dust, this map shows us material out to the farthest reaches of the galaxy. Careful analysis of the Doppler shifts of the 21-cm emission reveals the spiral structure of the disk, something not at all easy to see in a simple edge-on view such as this one.

Map 2, obtained by the *IRAS* satellite, shows the infrared emission characteristic of cold matter, particularly interstellar dust. Again we see that this material is highly concentrated in the galactic plane, even more so than the hydrogen gas represented in the first map. By tracing the distribution of cold dust in this way, we are locating the places in the galaxy where stars are forming, a very important thing to know if we are to understand the evolution of the galaxy. As in the 21-cm map, the infrared sky survey shows us material out to great distances, most of the way across the galaxy.

Map 3, representing the visible appearance of the galaxy, shows in broad scope what we have known of our galaxy through millenia. Again, we see that most of the stars, as well as the dark blotches of interstellar dust clouds, are concentrated in the plane of the disk. In this picture, we are seeing out to a limited distance, particularly in the plane, because of obscuration by gas and dust. In the plane of the Milky Way, the stars we can see lie within a few hundred parsecs from us; out of the plane, where there is less interstellar obscuration, we see stars as far as a few thousand parsecs away.

Map 4, representing low-energy X rays, is the most limited of all in the distance it reaches. The X-ray emission shown in this map arises in interstellar gas that is very close to the Sun, so we are seeing only the "local" neighborhood. The interstellar gas emits X rays because it is very hot; temperatures of nearly 1 million degrees are implied by the presence of this emission (the origin of this hot gas, in supernovae and stellar winds, is discussed in the next chapter).

Map 5 shows what the galaxy looks like when it is surveyed in high-

(continued on next page)

ASTRONOMICAL INSIGHT 25.1

THE INVISIBLE MILKY WAY

(continued from previous page)
energy X-ray wavelengths. Now instead of seeing the diffuse, hot gas in our neighborhood, we see point sources of energetic activity. Included are X-ray binary systems (discussed in Chapter 24), X-ray burst sources, and various kinds of explosive stars in our own galaxy, as well as the active nuclei of other galaxies and quasars (Chapter 31). The stellar sources of X rays in our galaxy tend to be concentrated in the plane of the disk, where most of the stars are, whereas the extragalactic sources are more readily seen at high galactic latitudes, well above or below the plane of the disk.

Finally, Map 6 shows what the galaxy looks like when observed in gamma rays. These extremely energetic photons are produced primarily in nuclear reactions. The general emission of gamma rays shown in this map is highly confined to the plane of the galactic disk, and is thought to arise from collisions between very rapid atomic nuclei (protons) with molecules of hydrogen in interstellar clouds. When these collisions take place at sufficiently high energy, nuclear reactions occur, resulting in the production of elementary particles and gamma rays. The rapidly-moving protons (called cosmic rays; see Astronomical Insight 26.2) are probably produced in supernova explosions.

There are many other wavelengths in which it would be useful to map the Milky Way. To do so is helpful for astronomers seeking to understand the distribution of matter and the physical state of the galaxy, and very important in studies of galactic evolution and for comparisons of our galaxy with others. Perhaps in future textbooks additions may be made to the series of maps shown here.

a very important role in the early measurements of the size and shape of our galaxy, as we will learn later in this chapter.

THE STRUCTURE OF THE GALAXY AND THE LOCATION OF THE SUN

Because the solar system is located within the disk of the Milky Way, we have no easy way of getting a clear view of where we are in relation to the rest of the galaxy. All we see is a band of stars across the sky, which tells us that we are in the plane of the disk. It is not so easy to determine where we are within the disk with respect to its edge and center.

The first serious attempts to solve this problem were based on determinations of the density of stars in space around the Sun. The method was to choose random regions of the sky and count the number of stars of each magnitude in these regions. The number of stars increases with decreasing brightness, because faraway stars appear faint, and in a given direction there are many more faraway stars than nearby ones. Therefore we can determine the distribution of stars in space by seeing how rapidly their number increases with decreasing brightness (that is, with increasing distance).

In the first decade of the twentieth century, the Dutch astronomer J. C. Kapteyn employed this method and got a surprising result. The density of stars appeared to fall off in all directions from the Sun, implying that our solar system is in the densest portion of the Milky Way (Fig. 25.6). The most logical interpretation was that the Sun is at the very center of the galaxy, for it seemed likely that it should be densest at the core. The thought that the Sun was at the center of the known universe made many astronomers uneasy, because in the past, every assumption that the Earth held a special place in the universe had been proved incorrect. It turns out there was indeed a flaw in the work of Kapteyn, although he could hardly have been aware of it. Before the error in Kapteyn's method was found, however, two other studies showed that the Sun was probably not at the center of the galaxy.

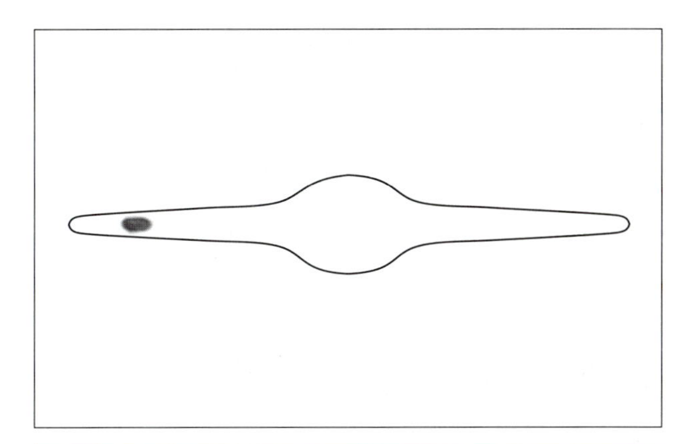

FIGURE 25.6 THE STRUCTURE OF OUR GALAXY. *This sketch illustrates the modern view of the Milky Way.*

FIGURE 25.7 HARLOW SHAPLEY. *Shapley's work on the distances to globular clusters was a key step in the determination of the size of the Milky Way. Shapley also played a prominent role in the discovery of galaxies outside our own (see Chapter 28). Much of his observational work was done before 1920, when Shapley was a staff member at the Mount Wilson Observatory. He later became director of the Harvard College Observatory.* (Harvard College Observatory)

One of these studies, carried out by Harlow Shapley (Fig. 25.7), made use of the globular clusters, the spherical star clusters found to lie outside the confines of the galactic disk. As noted earlier, these clusters tend to contain RR Lyrae variables, so it was possible for Shapley to determine their distances and hence their locations with respect to the Sun and the disk of the Milky Way (Fig. 25.8). He found that the globular clusters are arranged in a spherical volume centered on a point several thousand parsecs from the Sun, and he argued that this point must represent the center of the galaxy. It would not make physical sense for the globular clusters to be concentrated around any location other than the center of the entire galactic system.

Shapley's conclusion, first published in 1917, was not widely accepted initially, but other supporting evidence came to the fore in the 1920s. Two scientists, Jan Oort of Holland (already mentioned in connection with the origin of comets; see Chapter 16), and the Swede Bertil Lindblad, carried out careful studies of the motions of stars in the vicinity of the Sun. What they found was that these motions could best be understood if the Sun and the stars around it were assumed to be orbiting a distant point. That is, there are systematic, small-velocity differences between stars, similar to those between runners on a track who are in the inside and outside lanes (Fig. 25.9). It appeared from these studies that the Sun is following a more-or-less circular path about a point several thousand parsecs away, indicating that the center of the galaxy is located at that distant center

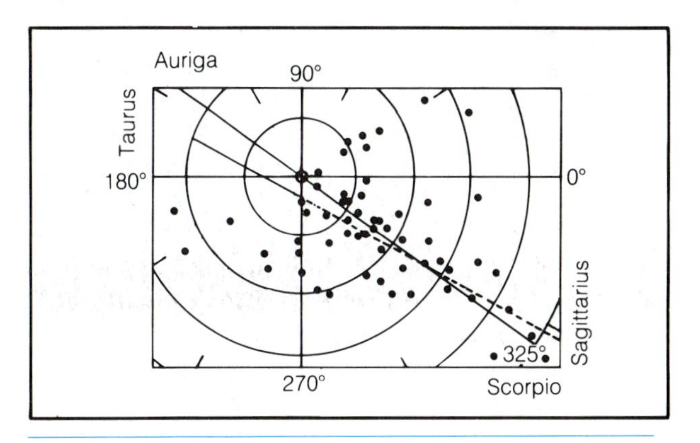

FIGURE 25.8 SHAPLEY'S MEASUREMENTS OF GLOBULAR CLUSTERS. *This is one of Shapley's original figures illustrating the distribution of globular clusters in the galaxy. The Sun is at the point where the straight lines intersect, and each circle centered on that point represents an increase in distance of 10,000 parsecs. Note that the distribution of globulars is centered some 10 to 20 kiloparsecs from the Sun.* (Estate of H. Shapley, reprinted with permission.)

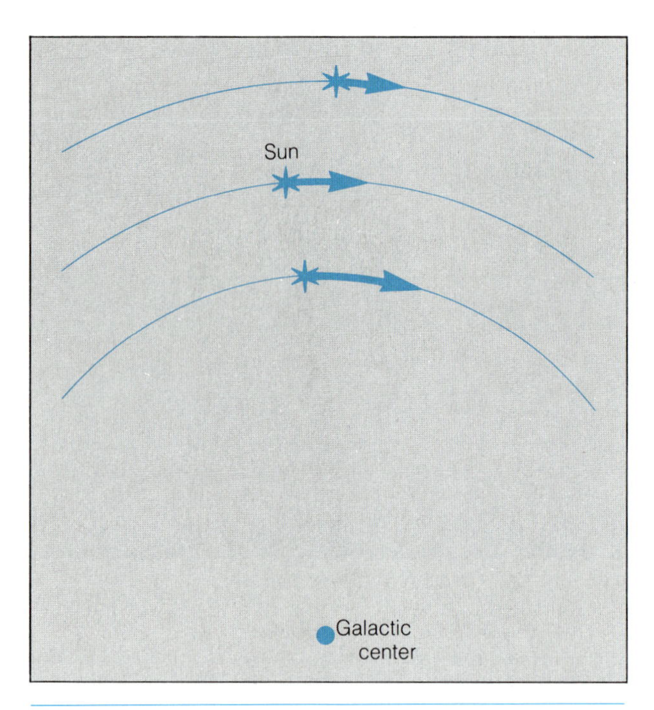

FIGURE 25.9 STELLAR MOTIONS NEAR THE SUN. *Stars just inside the Sun's orbit move faster than the Sun, whereas those farther out move more slowly. Analysis of the relative speeds (as inferred from measurements of Doppler shifts) and distances of stars like these led to the realization that the Sun and stars near it are orbiting a distant galactic center.*

of rotation. This supported Shapley's view of the galaxy, although there were still uncertainties about how far the Sun was from the center.

Let us now return to the work of Kapteyn, for the next major discovery was the one that revealed the flaw in his conclusions. In 1930 a study performed by the Swiss-American astronomer R. Trumpler showed that the galaxy is filled with an intersteller haze that makes stars appear fainter than they otherwise would. Trumpler made this discovery by examining the apparent brightnesses of star clusters whose distances he knew from their apparent sizes. What he found was that distant clusters appear much fainter than they should, and that the effect increases with increasing distance. This was the first direct evidence that there is a pervasive interstellar medium in the space between the stars; until then, it was known only that interstellar clouds and nebulae existed, as seen on photographs showing dark regions and brightly glowing gas clouds.

Trumpler's discovery explained the contradiction between Kapteyn's results based on star counts, and the picture developed by Shapley and by Lindblad and Oort. Kapteyn's star counts failed to take the interstellar haze into account, so his estimates of stellar distances were erroneous. It was the obscuration by the interstellar medium that made the density of stars appear to fall off with distance from the Sun.

Consideration of the effects of interstellar material, along with refinements in the assumed relation between pulsation period and absolute magnitude for variable stars, led eventually to a consensus on the size of the galaxy and the Sun's location within it. As mentioned earlier, the disk is some 30 kpc in diameter. The Sun is located a little less than 10 kpc from the center, or about two-thirds of the way to the edge (the most recent estimate is that the Sun is 8.7 kpc from the galactic center). As we will see in later sections, there is still some uncertainty about how far the galaxy extends beyond the Sun's position.

GALACTIC ROTATION AND STELLAR MOTIONS

An important result of the work of Lindblad and Oort was the development of an understanding of the overall motions in the galaxy. Oort's analysis especially was useful in this regard, for he showed not only that the Sun and the stars near it are orbiting the distant galactic center, but also that the rotation of the galaxy is differential, meaning that the galaxy does not spin like a rigid disk, but that each star follows its own orbit at its own speed. Thus the galaxy in the vicinity of the Sun does not act like a rigid disk but like a fluid, with each star moving as an independent particle.

This is not true of the inner portion of the galaxy. There the entire system rotates more like a rigid object; like a record on a turntable. In the region where rigid-body rotation is the rule, the speeds of individual stars increase with distance from the center, whereas in the outer portion they decrease with distance (Fig. 25.10). In the outer reaches of the galaxy, each star orbits the central portion of the galaxy with little influence from its neighbors, and the stellar orbits are approximately described by Kepler's laws. This is why the orbital speeds decrease with distance from the center in this part of the galaxy. There is an intermediate distance, just inside the Sun's orbit, where a transition between rigid-body and Keplerian orbits occurs, and it is here that stars have the greatest orbital velocities. The Sun, near this peak position, travels at about 250 kilometers per second, taking roughly 250 million years to make one complete circuit about the galaxy. In its 4- to 5-billion-year lifetime, the Sun has thus completed 15 to 20 orbits.

Individual stellar motions do not necessarily follow precise circular orbits about the galactic center. What we have described so far is the overall picture that develops from looking at the composite motions of large numbers of stars. If we look at the individual trees instead of the forest, we find that each star in the great disk has its own particular motion, which may deviate slightly from the ideal circular orbit. These individual motions are comparable to the paths of cars on a freeway, where the overall direction of motion is uniform, but a bit of lane-changing occurs here and there.

In the galaxy, a star's deviation in motion from a perfect circular orbit is called the **peculiar velocity.** The Sun's motion also deviates from perfect circular motion about the galactic center. This deviation, called the **solar motion,** is estimated from the Sun's average velocity with respect to other stars in its vicinity, along with the assumption that, on average, these nearby stars follow a circular orbit about the galaxy. The Sun has a velocity of about 20 kilometers per second with respect to the standard established by the average motion of its neighbors, that is, with respect to what it would be if it moved precisely in a circular orbit. The solar motion is in a direction about 45 degrees from the galactic center and slightly out of the plane of the disk. Most peculiar velocities of stars near the Sun are comparable, amounting to only minor departures from the overall orbital velocity of about 250 kilometers per second. In Chapter 27 we will discuss the **high-velocity** stars, which greatly deviate from the circular orbits followed by most stars in the Sun's vicinity.

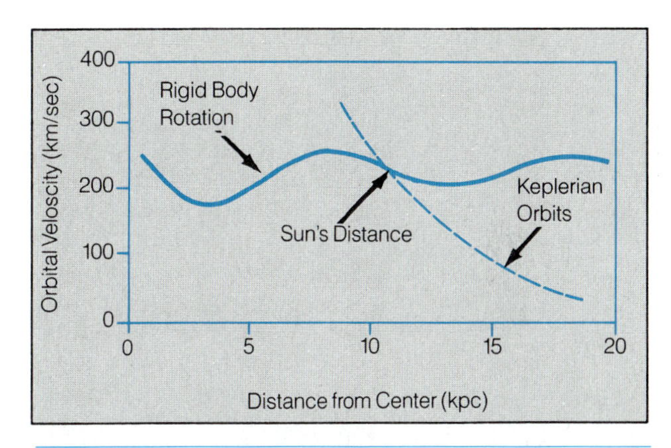

FIGURE 25.10 THE ROTATION CURVE FOR THE MILKY WAY. *This diagram shows how the stellar orbital velocities vary with distance from the center of the galaxy. The fact that the curve does not simply drop off to lower and lower velocities beyond the Sun's orbit, as it would if all the mass of the galaxy were concentrated at its center, indicates that there is a lot of mass in the outer portions of the galaxy.* (Data from M. Fich)

SPIRAL STRUCTURE AND THE 21-CM LINE

So far we have spoken of the galactic disk as though it were a uniform, featureless object, but we know that this is not a completely accurate picture. The Milky Way is a spiral galaxy, and if we could see it face-on, we would see the characteristic pinwheel shape normally found in galaxies of this type (Fig. 25.11).

There is a common misconception about the spiral structure in the Milky Way (and other spiral galaxies); namely, that there are few stars between the visible spiral arms. In reality, the density of stars between the arms is nearly the same as it is in the arms. The most luminous stars, however, the young, hot O and B stars, tend to be found almost exclusively in the arms. Because these are the brightest of stars, their residence in the arms makes the spiral structure stand out.

The fact that we live in a spiral galaxy was not easily discovered, again because of our location within it, where we see only a cross-sectional view. It was not until 1951 that investigations of the distribution of luminous stars revealed tracers of spiral structure in the Milky Way, and even those studies were limited to a small portion of the galaxy. Because of the obscuration caused by interstellar material, our view of even the brightest stars is limited to a local region, about 1000 parsecs from the Sun at most.

FIGURE 25.11 SPIRAL STRUCTURE. *This galaxy has prominent spiral arms, as ours does, because hot, luminous young stars tend to be concentrated in the arms. (The overall appearance of this galaxy is not exactly like the Milky Way, which does not have a bar-like central region as seen here.)* (© 1977 Anglo-Australian Telescope Board)

A major advance in measuring the structure of the galaxy occurred in the same year, when the 21-cm radio emission of interstellar hydrogen atoms was first detected. It had been predicted that hydrogen atoms should emit radiation at this wavelength, and the Americans E. M. Purcell and H. I. Ewen were the first to detect this emission, using a specially built radio telescope.

The great advantage of the hydrogen 21-cm emission for measuring galactic structure is its ability to penetrate the interstellar medium to great distances. Whereas the brightest stars can be detected at distances of a few hundred parsecs at most, hydrogen clouds in space can be "seen" by radio telescopes from all the way across the galaxy, at distances of several thousand parsecs. Hydrogen gas is the principal component of the interstellar medium, and it tends to be concentrated along the spiral arms, so observations of the 21-cm radiation can be used to trace out the spiral structure throughout the entire Milky Way.

When a radio telescope is pointed in a given direction in the plane of the galaxy, it receives a 21-cm emission from each segment of spiral arm in that direction (Fig. 25.12). Because of differential rotation, each arm has velocity distinct from the others that are closer to the center or farther out. Therefore what

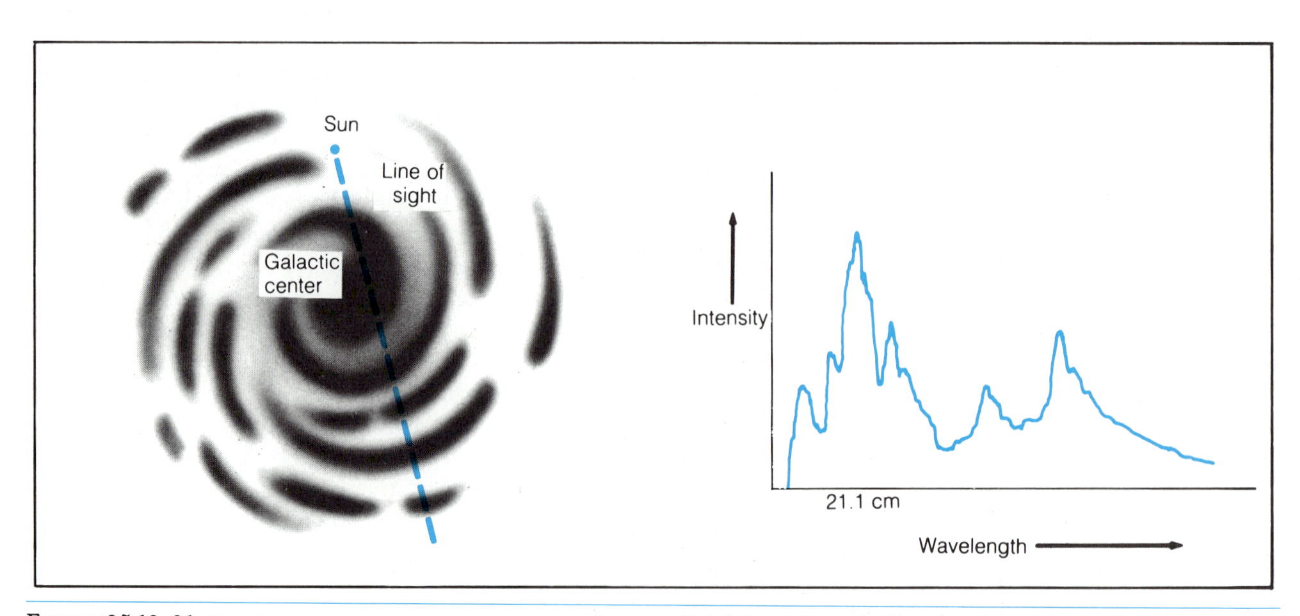

FIGURE 25.12 21-CENTIMETER OBSERVATIONS OF SPIRAL ARMS. *At left is a schematic diagram of the galaxy, showing the direction in which a radio telescope might be pointed to produce a 21-centimeter emission-line profile like the one sketched at right. There are many components of the 21-centimeter line, each corresponding to a distinct spiral arm, and each at a wavelength reflecting the Doppler shift between the velocity of that arm and the Earth's velocity.*

is seen, instead of a single emission peak at exactly 21.1-cm wavelength, is a cluster of emission lines near this wavelength, separated from each other by the Doppler effect. By combining measurements such as these with Oort's mathematical analysis of differential rotation, astronomers were able to reconstruct the spiral pattern of the entire galaxy. Radio emission from the carbon monoxide (CO) molecule has also recently been used to map the distribution and velocities of relatively dense interstellar clouds, providing further information on the galactic rotation and spiral structure (Fig. 25.13).

The pattern is quite complex, much more so than in some galaxies, where two arms are seen elegantly spiraling out from the nucleus. The Milky Way consists of bits and pieces of a large number of arms, giving it a definite overall spiral form, but not a smoothly coherent one. As we will learn in Chapter 27, the differences between the type of spiral structure seen in our galaxy and the regular appearance seen in others may reflect different origins for the arms.

THE MASS OF THE GALAXY

Once the true size of the Milky Way was determined, it became possible to estimate its total mass. This could be achieved by measuring the star density in the vicinity of the Sun and then assuming that the entire galaxy has about the same average density, but a much simpler and more accurate technique became possible with the discovery that the stars in our region of the galaxy obey Kepler's laws.

Kepler's third law, in the more complete form developed by Newton, expresses a relationship among the period, the size of the orbit, and the sum of the masses of the two objects in orbit about each other:

$$(m_1 + m_2)P^2 = a^3,$$

where m_1 and m_2 are the two masses (in units of the Sun's mass), P is the orbital period (in years), and a is the semimajor axis (in astronomical units). If we consider the Sun to be one of the two objects, and the galaxy itself to be the other, we can use this equation to determine the mass of the galaxy. As we

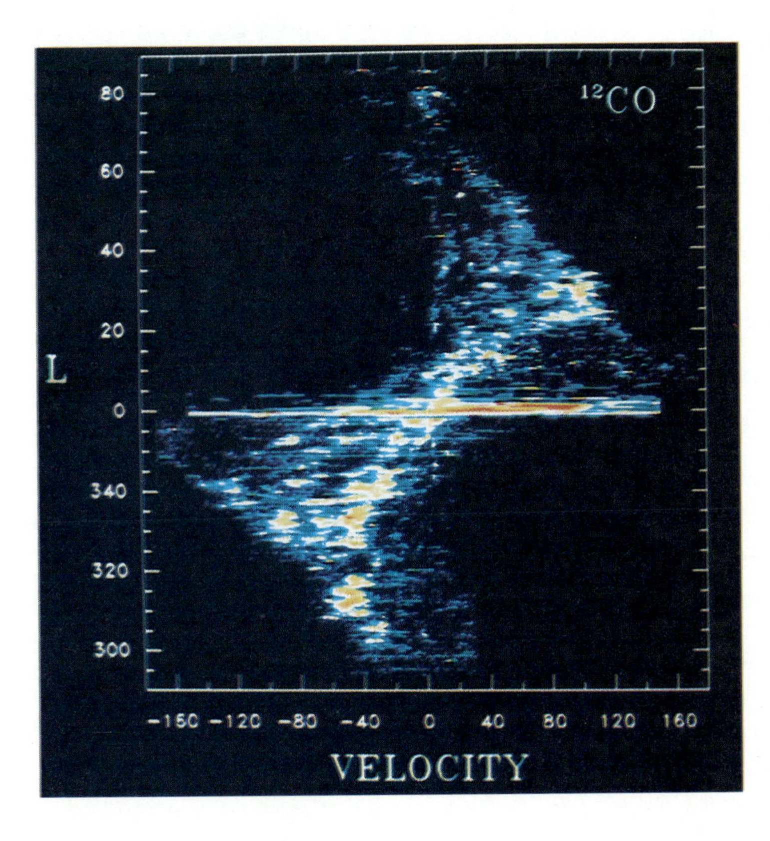

FIGURE 25.13 GALACTIC ROTATION FROM CARBON MONOXIDE OBSERVATIONS. *This map is of radial velocities of CO molecules observed with a radio telescope. It shows that interstellar gas on one side of the galactic center (which lies at galactic longitude L = 0°) is approachng us, while material on the other side of the center is receding. Data like these are used to construct the galactic rotation curve, such as in Fig. 25.10.* (From Robinson, et al. *1984. Astrophysical Journal* 282:131; image processing at the Remote Sensing Center of the University of Massachusetts)

ASTRONOMICAL INSIGHT 25.2

21-CM EMISSION FROM HYDROGEN

To understand how a hydrogen atom can emit radiation at a wavelength of 21 centimeters, we need to take a closer look at the structure of the atom. A hydrogen atom consists only of a proton, which forms the nucleus, and a single electron in orbit around it. We have already learned that the electron can occupy a variety of possible orbits, each corresponding to a different energy state. We have also learned that, if a photon of light is absorbed by the electron, the electron moves from a low energy level to a higher one, and a photon is emitted if the electron drops from a high energy level to a lower one. The wavelength of the photon is related to the energy difference between the two electron energy levels; the greater the difference, the shorter the wavelength, and vice versa.

It happens that the energy-level structure of the electron is more complicated than previously described. The electron and the proton are both spinning, and the energy of the electron depends on whether it is spinning in the same or the opposite direction as the proton. If both spin in the same (parallel) direction, then the energy state is slightly greater than when the electron and proton spin in opposite (antiparallel) directions. As before, the electron can change from one state to the other by either emitting or absorbing a photon. The energy difference is so small, however, that the wavelength of the photon is very much longer than that of visible light. It is 21.1 centimeters.

Hydrogen atoms in space tend to have the electron in the lowest possible energy state, with its spin antiparallel to that of the proton. Occasionally an atom will collide with another, however, causing the electron to jump to the higher state, and its spin is then parallel to that of the proton. Following this, the electron spontaneously reverses itself, seeking to return to the lowest energy state, and when it does so, it emits a photon of 21.1-cm wavelength. The probability that the electron will make the downward transition is very low, and the electron may remain in the upper state for as long as 10 million years before spontaneously dropping back down. There are so many hydrogen atoms in space, however, that at any given instant, many photons are being emitted. Hence radio telescopes capable of receiving this wavelength can trace the locations of hydrogen clouds throughout the galaxy and beyond.

have already seen, the orbital period of the Sun is roughly 250 million years; that is, $P = 2.5 \times 10^8$ years. The orbit is nearly circular, with the galactic nucleus at the center, so the semimajor axis is approximately equal to the orbital radius; $a = 10$ kpc $= 2 \times 10^9$ AU. Now we can solve Kepler's third law for the sum of the masses:

$$m_1 + m_2 = a^3/P^2 = 1.3 \times 10^{11} \text{ solar masses.}$$

Since the mass of the Sun (1 solar mass) is inconsequential compared with this total, we can say that the mass of the galaxy itself is about 1.3×10^{11} solar masses. The Sun is slightly above average in terms of mass, so we conclude that the total number of stars in the galaxy must be $3–4 \times 10^{11}$, that is, several hundred billion.

This method, based on Kepler's third law, refers only to the mass inside the orbit of the Sun; the matter that is farther out has no effect on the Sun's orbit. Thus when we estimate the mass of the galaxy by applying Kepler's third law in this way, we neglect all the mass that lies farther out. In order to overcome this problem, radio measurements of interstellar gas in the outer portions of the galaxy have recently been used to determine the orbital velocity (and therefore the orbital period) of material in the outer reaches of the galaxy. It was found that the velocity does not decrease as rapidly with distance as had previously been thought, and this in turn led to the conclusion that quite a bit of mass lies beyond the Sun's orbit (recall Fig. 25.10). In a later section, we will discuss the possibility that most of the mass of the galaxy lies in the halo and that the actual mass of the galaxy may exceed 10^{12} solar masses.

THE GALACTIC CENTER: WHERE THE ACTION IS

The center of our galaxy is a mysterious region, forever blocked from our view by the intervening interstellar medium (Fig. 25.14). From Shapley's work on the distribution of globular clusters, as well as Oort's analysis of stellar motions, it was known by the 1920s that the center of our great pinwheel lies in the direction of the constellation Sagittarius. There we find immense clouds of interstellar matter and a great concentration of stars. It is one of the richest regions of the sky to photograph, although the best observations can be made only from the Southern Hemisphere.

The central portion of the galaxy consists of a more-or-less spherical bulge, populated primarily by relatively cool stars. The absence of hot, young stars implies that the central region has had relatively little recent star formation, something we will have to deal with in our discussion of the history of the galaxy (Chapter 27).

Although we cannot see into the central portion of the nucleus in visible light, this region can be probed at longer wavelengths where the interstellar extinction is not such a problem (Fig. 25.15). Radio and, more recently, infrared observations have revealed some interesting features of the galactic core (Fig. 25.16). The first clue that something unusual was taking place there came from 21-cm observations of hydrogen, which revealed a turbulent mixture of clouds moving about at high speed, and which showed that one of the inner spiral arms, about 3 kpc from the center, is expanding outward at a velocity of more than 100 kilometers per second. This is suggestive of some sort of explosive activity at the center that forced matter to move outward. More recently, infrared observations of emission lines from hot gas clouds have shown that interstellar clouds near the center of the nucleus are moving quite rapidly in their orbits, which in turn implies (through the use of Kepler's third law) that they are orbiting a very massive central object.

FIGURE 25.14 THE GALACTIC CENTER. *The dark regions are dust clouds in nearby spiral arms. Because of obscuration, photographs such as this do not reveal the true galactic center, but only nearby stars and interstellar matter in the plane of the disk and the outer portions of the central bulge of the galaxy.* (© 1980 Anglo-Australian Telescope Board)

Additional evidence of violent activity is seen in another portion of the radio spectrum, at the wavelength where interstellar carbon monoxide molecules emit (see Chapter 26 for a discussion of molecules in space). Observations of CO have revealed a gigantic ring of interstellar clouds circling the galaxy at a distance of some 6 kpc from the center. It looks as though the ring could have been built up by the expansion of matter away from the galactic center, perhaps as the result of an ancient explosion. The concentration of matter at 6 kpc may be the end result of an earlier

FIGURE 25.15 AN INFRARED VIEW OF THE MILKY WAY. *The cold material in the galaxy, primarily interstellar dust, produces most of the infrared radiation seen in this composite image, obtained by the IRAS satellite. The dust is more strongly concentrated in the plane of the galactic disk than are the stars. (NASA)*

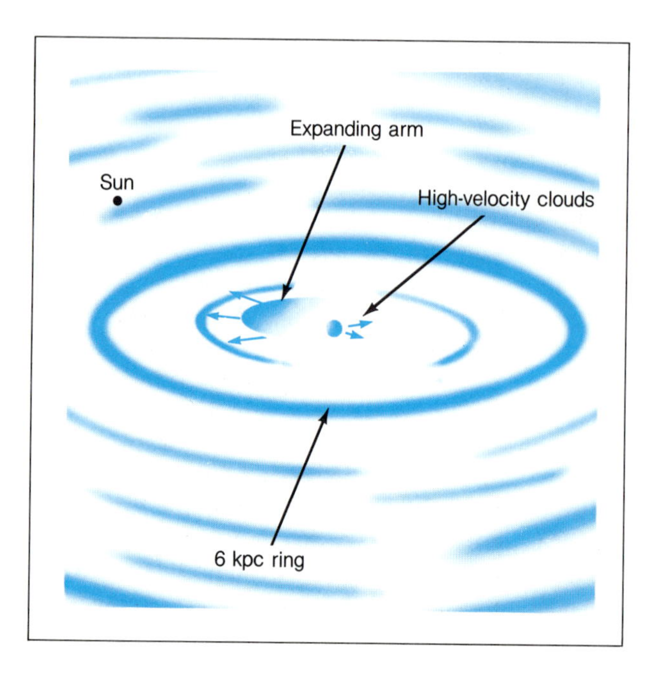

FIGURE 25.16 ACTIVITY AT THE GALACTIC CENTER. *A variety of evidence for energetic motions associated with the center of our galaxy is indicated here.*

episode of the same mysterious activity that is responsible for the current expansion of the 3-kpc arm.

Another very surprising phenomenon was recently discovered. An arc of gas has been observed at radio wavelengths (Fig. 25.17) to lie outside the plane of the galactic disk. It appears that this material is falling into the core, following a path governed by the galaxy's magnetic field.

Finally, space observations show that a very small object precisely at the center of the galaxy is emitting enormous amounts of energy in the form of X-rays. This object, whatever it is, is smaller than 1 parsec in diameter, yet it appears to be responsible for all

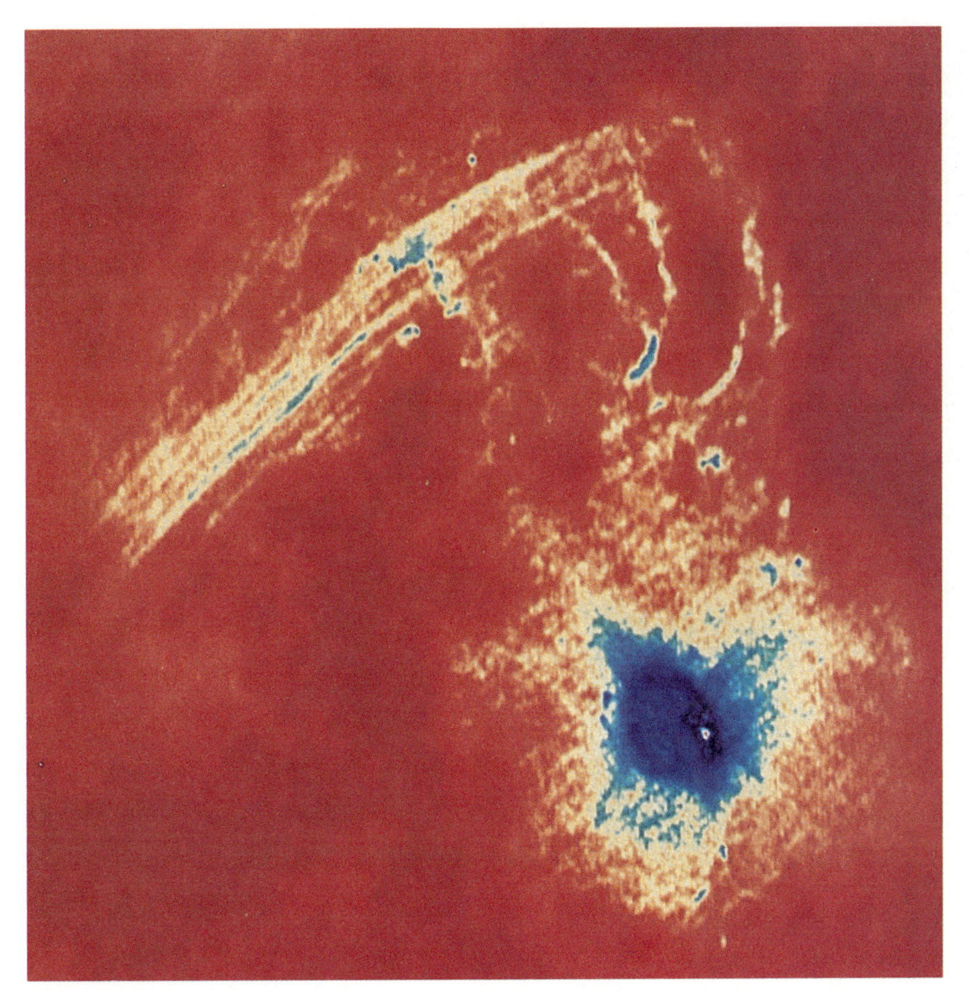

FIGURE 25.17 A RADIO IMAGE OF THE GALACTIC CENTER. *This image, made at a wavelength of 6 centimeters (where thermal continuum emission is measured, rather than a spectral feature such as the 21-centimeter line), shows an arc of gas extending some 200 parsecs above the plane of the galactic disk.* (F. Yusef-Zadeh, M. Morris, and D. Chance)

the violent and energetic activity just described. From the observed cloud velocities, astronomers have estimated the mass of the central object to be roughly 10^6 solar masses. This is only a small fraction of the total mass of the galaxy, but it is much greater than the mass of any known object within it and must therefore represent some new kind of astronomical entity. One explanation offered by astronomers is that a very massive black hole resides at the core of our galaxy (some astronomers argue that a very massive, yet compact cluster of ordinary stars could account for the observed motions, but the black hole hypothesis seems to be more widely accepted at this time). The gravitational influence of the black hole would be responsible for the rapid orbital motions of nearby objects, and matter falling in would be compressed and heated, accounting for the X-ray emission. We do not yet understand how such an object formed, but its presence would imply that at some time during the formation and evolution of the Milky Way, great quantities of matter were compressed into a very small volume at its center. Such an intense buildup of matter may have resulted from frequent collisions among stars in the early days of the galaxy, causing vast numbers of them to gradually coalesce into a single, massive object at the center. The prospect that such a beast inhabits the core of our galaxy is quite a bizarre one, yet it may tie in directly to some of the strange happenings that have been observed in the nuclei of other galaxies (see Chapter 31).

FIGURE 25.18 A GLOBULAR CLUSTER. *This cluster, designated Hodge 11, actually resides in one of the Megellanic Clouds, which are small galaxies just outside the Milky Way.* (© 1984 Anglo-Australian Telescope Board)

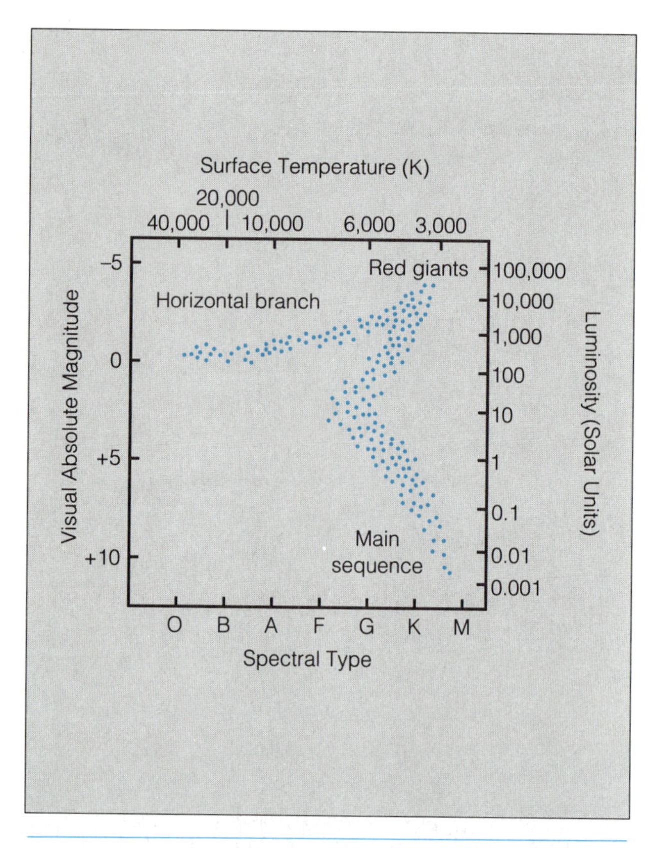

FIGURE 25.19 THE H-R DIAGRAM FOR A TYPICAL GLOBULAR CLUSTER. *The main sequence branches off at a point near the bottom, indicating the great age of the cluster.*

— GLOBULAR CLUSTERS REVISITED —

We have not yet said much about the globular clusters, the gigantic spherical conglomerations of stars that orbit the galaxy (Fig. 25.18). A large globular cluster can be an impressive sight when viewed with a small telescope, and a number of them are popular objects for astronomical photography. More than a hundred of these clusters have been catalogued; no doubt many more are obscured from our view by the disk of the Milky Way.

A single cluster may contain several hundreds of thousands of stars, and have a diameter of 10 to 20 parsecs. The mass is usually in the range of several hundred thousand to a few million solar masses. To contain so many stars in a relatively small volume, globular clusters must be very dense, compared with the galactic disk. The average distance between stars in a globular cluster is only about one-tenth of a parsec (recall that the nearest star to the Sun is more than 1 parsec away). If the Earth orbited a star in a globular cluster, the nighttime sky would be a spectacular sight, with hundreds of stars brighter than first magnitude.

An H-R diagram for a globular cluster is rather peculiar looking, compared with the familiar diagram for stars in the galactic disk (Fig. 25.19). The main sequence is almost nonexistent, having stars only on the extreme lower portion. On the other hand, there are many red giants and a number of blue stars that lie on a horizontal sequence extending from the red giant region across to the left in the diagram. These facts together point to a very great age for globular clusters. As we learned in Chapter 22, in a cluster H-R diagram, the point where the main sequence turns off toward the red giant region indicates the age of the cluster; the lower the turnoff point, the older the cluster. Using this technique to date globular clusters leads typically to age estimates of 14 to 16 billion years, comparable to the accepted age of the galaxy itself. Thus globular clusters are among the oldest objects that exist, and, as we will see, their presence in the galactic halo provides important data on the early history of the galaxy.

The sequence of stars extending across the globular cluster H-R diagram from the red giant region to the left is called the **horizontal branch** (also mentioned in Chapter 23). The evolutionary status of these stars has been the subject of a good deal of uncertainty; the theoretical models have not yet been developed that can explain how stars evolve to this region of the H-R diagram. It is not even known for certain whether these stars are moving to the left (getting hotter, perhaps following a red giant phase), or to the right. Most likely these stars have completed their red giant stages and are moving to the left on the diagram, perhaps on their way to becoming white dwarfs.

Some globular clusters have been found to contain X-ray sources in their centers (Fig. 25.20). In a few cases these are X-ray bursters, described in Chapter 24 as being neutron stars that are slowly accreting new matter on their surfaces, probably in binary systems where the companion star is losing mass. In other cases, however, the X-ray source does not have the recognizable properties of a binary system, and may instead be caused by a single object at the center of the cluster. This could be a giant black hole, formed as stars in the cluster collided and fell together in the core.

FIGURE 25.20 A GLOBULAR CLUSTER X-RAY SOURCE. *Several globular clusters have been found to contain intense X-ray sources at their centers. This X-ray image obtained with the* Einstein Observatory *clearly shows the central source of a globular cluster.* (Harvard-Smithsonian Center for Astrophysics)

A MASSIVE HALO?

The halo of the Milky Way galaxy has traditionally been envisioned as a very diffuse region, populated by a scattering of dim stars and dominated by the giant globular clusters. There was little or no evidence of any substantial amount of interstellar material, and the halo was assumed not to contain a significant fraction of the galaxy's mass.

Some of these ideas are changing as a result of very recent discoveries. We mentioned earlier that radio observations revealed an unexpectedly high amount of mass in the outer portions of the disk, indicating that the galaxy is more extensive than previously thought (Fig. 25.21). At the same time, other evidence (to be described in later chapters) led to a general expectation that many galaxies have large quantities of matter in their halos. Finally, in the early 1980s, ultraviolet observations revealed a large amount of very hot, rarefied interstellar gas in the halo of our galaxy, probably extending to distances of several

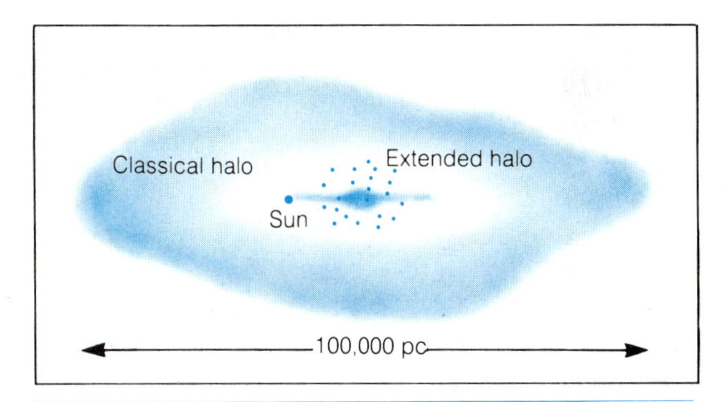

FIGURE 25.21 THE EXTENSIVE HALO OF THE MILKY WAY. *This illustrates the shape and scale of the very extended galactic halo, whose presence has been inferred from the shape of the galactic rotation curve (see Fig. 25.10) and from absorption lines formed by interstellar gas.*

thousand parsecs above and below the plane of the disk. This gas is very turbulent, with clouds traveling at speeds of several hundred kilometers per second, and is highly ionized, indicating a temperature of at least 100,000 K or more.

At the present time we cannot say much about the origin or total quantity of material in the halo of our galaxy. It is quite possible that in addition to the gas, there are enough stars, too dim to be detected, to add significantly to the total mass of the galaxy. Some astronomers believe that as much as 90 percent of the galaxy's mass may be in the halo. The *Space Telescope,* to be launched later in the 1980s, should be able to settle this question by making more sensitive measurements of both the gas and the stars in the halo.

PERSPECTIVE

We have discussed the anatomy of the galaxy, particularly its overall structure and motions. The solar system is located nearly 10 kpc from the center of a flattened, rotating disk containing a few hundred billion stars. With the help of radio and infrared observations, astronomers have unraveled the structure of the disk, which consists of spiral arms (delineated by bright, hot stars and interstellar gas), and a central nucleus that contains relatively dim, red stars along with a mysterious, energetic gremlin that stirs up the core region. Many questions remain about the structure and extent of our galaxy; some will be answered with the launch of the *Space Telescope.*

We have yet to discuss the workings of the galaxy; we have seen what it is like but have said little about why. Before we endeavor to do so, we will discuss the interstellar medium, a vital part of the galactic ecology.

SUMMARY

1. The Milky Way is a spiral galaxy consisting of a disk with a central nucleus and a spherical halo, where the globular clusters reside. The disk is about 30 kpc in diameter.

2. Distances within the Milky Way can be determined from the measurement of variable-star periods and application of the period-luminosity relation.

3. The true size of the Milky Way and the Sun's location within it were difficult to determine because the view from our location within the disk is obscured by interstellar matter.

4. Star counts were once interpreted as indicating that the Sun is in the densest part of the Milky Way, but measurements of the distribution of globular clusters and analysis of stellar motions show that the Sun is about 8.7 kpc from the center. The discrepancy was resolved when it was discovered that interstellar extinction affects the star counts.

5. The inner part of the galactic disk rotates rigidly, whereas the outer part rotates differentially, with each star following its own individual orbit, approximately described by Kepler's laws.

6. Individual stellar orbits in the Sun's vicinity are generally in the plane of the disk, and are nearly circular. Deviations from perfect circular motion by stars are called peculiar velocities, and in the case of the Sun, the solar motion.

7. The spiral structure of our galaxy is most easily and directly measured through radio observations of the 21-cm line of hydrogen atoms in space. The spiral pattern is complex, with many segments of spiral arms.

8. The mass of the galaxy, determined by the application of Kepler's third law to the Sun's orbit, is roughly 10^{11} solar masses. This technique does not take into account any mass that resides in the halo.

9. A variety of evidence indicates there is chaotic, energetic activity associated with the central core of our galaxy. The data show a compact, massive object existing there, and the best explanation is that it is a massive black hole.

10. The globular clusters that inhabit the halo of the galaxy are very old and therefore provide information about the early history of the galaxy.

11. There may be large quantities of mass in the galactic halo, in the form of interstellar gas or dim, cool stars, or both. It appears possible that as much as 90 percent of the mass of the galaxy is in the halo.

REVIEW QUESTIONS

Thought Questions

1. Explain how the use of pulsating variable stars to find distances is similar to the spectroscopic parallax method described in Chapter 20.

2. What would Kapteyn have found if there had not been any interstellar dust obscuration to modify his star-count method for determining the structure of the galaxy?

3. Compare the reasoning of Shapley in deciding that the globular clusters must be concentrated around the galactic center with that of Aristarchus who, more than 2,000 years earlier, had deduced that the Sun is at the center of the solar system.

4. Summarize the nature of stellar orbits about the galaxy. Why are the highest orbital speeds attained by stars at the position where rigid-body rotation gives way to fluid motion (Keplerian orbits)?

5. Explain why the 21-cm line of atomic hydrogen is a powerful tool for mapping the structure of the galaxy.

6. Suppose the mass of the galaxy were determined from the application of Kepler's third law to the orbit of a star 20 kpc from the galactic center. How might the result compare with that found from analysis of the Sun's orbit?

7. Summarize the evidence that a massive black hole might exist at the core of the galaxy.

8. How is it known that globular clusters are very old, relative to the rest of the galaxy?

9. Given the great age of the globular clusters, do you think their chemical composition might differ from that of younger stars?

10. What is the evidence that our galaxy might have a very massive halo, containing most of the galaxy's mass? Can you think of any observational methods for learning more about this halo?

Problems

1. Calculate the distance to a star whose parallax is $p = 0.005$ (nearly the smallest parallax that can be measured), using the method described in Chapter 19. How does your answer compare with the overall diameter of the galaxy?

2. Suppose a Cepheid variable is found to have a period of 10 days, and an average apparent magnitude of $m = +16$. From the period-luminosity relation, the average absolute magnitude for such a star is $M = -4$. What is the distance to this star? Compare your answer with the maximum distance that can be measured with the stellar parallax technique.

3. Suppose that interstellar obscuration makes a star appear 5 magnitudes fainter than it would otherwise. If the star has apparent magnitude $+17$, what would its apparent magnitude be if it were not obscured by dust? If the star's absolute magnitude is $+2$, how far away is it? Compare your answer with what you would find if you did not correct the apparent magnitude for the effects of dust.

4. Confirm the orbital period of the Sun about the galaxy by calculating how long it would take to travel the circumference of a circle whose radius is 10,000 pc at a speed of 250 km/sec. (You will have to convert to consistent units.)

5. If the rest wavelength of the hydrogen emission line is 21.1 cm, what would be the observed wavelength from an interstellar cloud whose velocity is 150 km/sec away from the Earth?

6. Suppose that in a distant spiral galaxy, the orbital speed at a distance of 5 kpc from the center is found to be 150 km/sec. What is the orbital period for stars at that point in the galaxy? What is the mass of the galaxy interior to that point?

7. If there is an object at the galactic center with a mass equal to 1 million solar masses, what is the orbital period for a star 10 parsecs from this object? If this star's orbit is circular, what is its orbital speed? Comment on the relationship of your result to observations suggesting the presence of a massive black hole at the galactic core.

8. To appreciate what the nighttime sky might be like if the Earth orbited a globular cluster star, calculate the apparent magnitudes of the first five stars in the list of the brightest stars (Appendix 8) if each star were only 0.1 pc away (the average distance between stars in a globular cluster).

ADDITIONAL READINGS

Bok, B. J. 1972. Updating galactic spiral structure. *American Scientist* 60:708.

———. 1981. The Milky Way galaxy. *Scientific American* 244(3):92.

Bok, B. J., and Bok, P. 1981. *The Milky Way*. Cambridge, Ma.: Harvard University Press.

Bok, Bart J. 1984. A bigger and better Milky Way. *Astronomy* 12(1):6.

Clark, Gail O. 1985. Ancients of the universe: Globular clusters. *Astronomy* 13(5):6.

de Boer, K. S., and B. D. Savage. The coronas of galaxies. *Scientific American* 247(2):54.

Geballe, T. R. 1979. The central parsec of the galaxy. *Scientific American* 241(1):52.

Gingerich, O., and B. Welther. 1985. Harlow Shapley and the Cepheids. *Sky and Telescope* 70(6):540.

Herbst, W. 1982. The local system of stars. *Sky and Telescope* 63(6):574.

Hodge, Paul, 1986. *Galaxies.* Cambridge Mass.: Harvard University Press.

———. 1987. The Local Group: Our galactic neighborhood. *Mercury* 6(1):2.

King, Ivan R. 1985. Globular clusters. *Scientific American* 252(6):78.

Kraft, R. P. 1959. Pulsating stars and cosmic distances. *Scientific American* 201(1):48.

Mullholland, Derral. 1985. The beast at the center of the galaxy. *Science 85* 6(7):50.

Rubin, V. 1983. Dark matter in spiral galaxies. *Scientific American* 248(6):88.

Sanders, R. H., and G. T. Wrixon. 1974. The center of the galaxy. *Scientific American* 230(4):66.

Seeley, D., and R. Berendzen. 1978. Astronomy's great debate. *Mercury* 7(3):67.

Weaver, H. 1975 and 1976. Steps towards understanding the large-scale structure of the Milky Way. *Mercury* 4(5):18, Part 1, 5(6):19.

CHAPTER 26

The Interstellar Medium

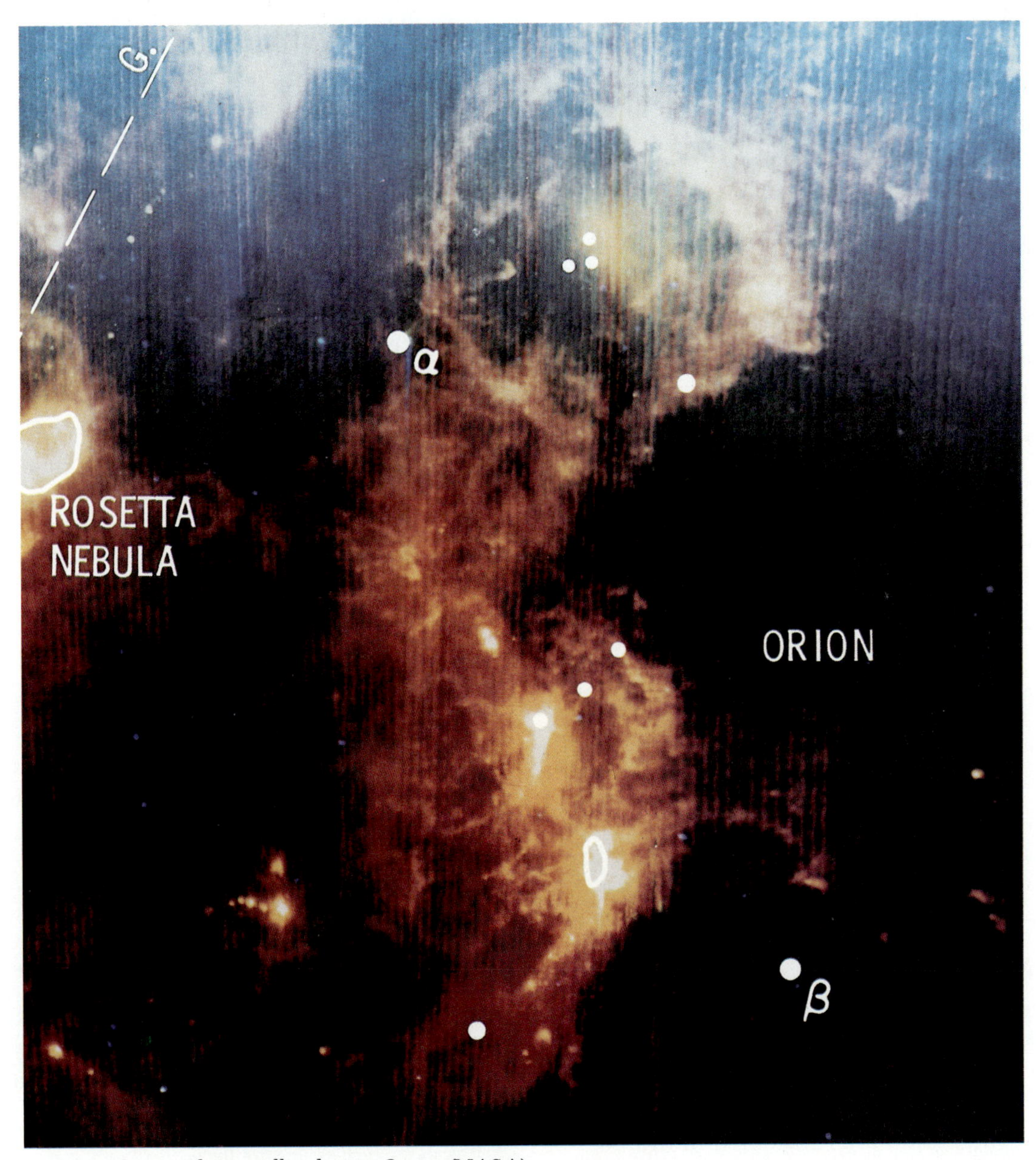

An infrared view of interstellar dust in Orion. (NASA)

CHAPTER PREVIEW

The very rarefied matter between the stars, diffuse though it is, constitutes one of the most important elements in the galactic environment. Stars form from it, and as they age and die, they return their substance to it. As we have seen, the interstellar haze limits our observations of the galactic disk to the nearest few hundred parsecs, allowing us to view only a very small fraction of the entire galactic volume.

Despite its obvious importance, the interstellar medium, by ordinary earthly standards, is so tenuous that it verges on nonexistence. Even in the densest interstellar clouds, the density of particles is less than one-trillionth of the Earth's sea-level atmospheric density. Only the most sophisticated artificial vacuum pumps can even approach the natural vacuum of space.

The medium is so pervasive, however, and the volume of space so large, that a large fraction of the total mass of the galactic disk is contained in this form. There are two distinct types of interstellar material, which are thoroughly mixed together throughout space: tiny solid particles called interstellar dust grains; and interstellar gas particles, which may be atoms, ions, or molecules, depending on the temperature and density. The grains are relatively few in number, and constitute only about 1 percent of the total mass in the interstellar medium.

THE INTERSTELLAR DUST

The fact that there are interstellar dust grains was recognized long ago. Photographs of the Milky Way show extensive, irregular dark regions (Figs. 26.1–26.4) where dense, dust-bearing clouds completely hide the stars behind them. Knowledge of the presence of these clouds is ancient, but it was not recognized until Trumpler's work in 1930 that there is a general distribution of dust grains throughout the spaces in between the obvious clouds. Today the dust is studied by a number of techniques, and a great deal has been learned about its nature. The *IRAS* satellite,

FIGURE 26.1 DUST CLOUDS ALONG THE MILKY WAY. *This is a portion of the Milky Way in Sagittarius, showing multitudes of stars as well as extensive patches of interstellar clouds.* (© 1980 Anglo-Australian Telescope Board)

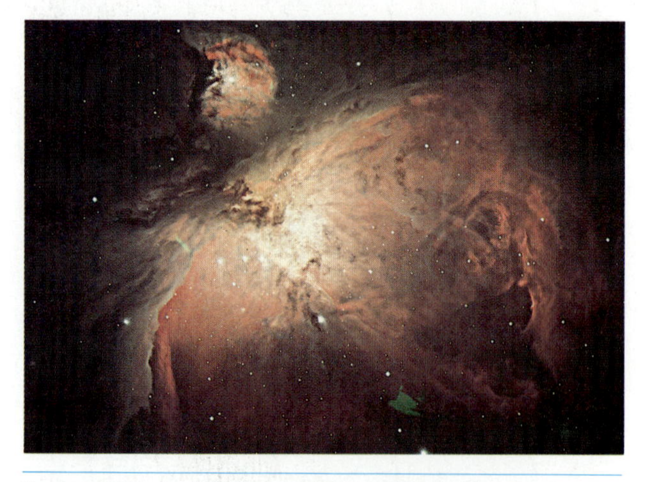

FIGURE 26.2 A GASEOUS NEBULA. *Clouds of gas and tiny dust particles such as this are the birthplaces of stars.* (© 1981 Anglo-Australian Telescope Board)

with its infrared sensors highly sensitive for detecting thermal emission from cold objects, found numerous thin, wispy dust clouds in space. It is not yet clear how these interstellar "cirrus" clouds, as they have been called, are related to other types of interstellar clouds (described later in this chapter).

Even though the dust grains constitute only a tiny fraction of the total mass in the interstellar medium, they have a very important influence on the starlight that passes through space. There are two basic effects: (1) the obscuration of light from distant stars, commonly referred to as **interstellar extinction;** and (2) the polarization of starlight. Both effects provide important information about the nature of the dust grains themselves.

FIGURE 26.3 A DARK CLOUD. *The dark streaks in the lower portion of this photograph are regions where dark clouds hide newly-forming stars. This is part of a well-studied cloud complex near the bright red supergiant star Antares.* (Fr. R. E. Royer)

The extinction of starlight by interstellar dust particles is the result of a combination of absorption and **scattering,** a process in which the light essentially bounces off of the dust grains. When a photon is scattered, its wavelength remains fixed, but its direction is altered. The scattering process is more effective for short wavelengths of light than for longer wavelengths, so a distant star appears redder than it otherwise would, because the red light that it emits reaches us more easily than the blue (Fig. 26.5). For a similar reason, the Sun appears red when it is near the horizon so that its light must traverse a long pathlength through the Earth's atmosphere; the fine particles in the air scatter the blue light but allow the red to come through relatively unhindered. Because the extinction of starlight is so much more severe at short wavelengths, ultraviolet telescopes cannot probe the great distances that visible-light telescopes can.

It is possible to determine how much dust lies in the direction of a given star by measuring how much redder the star appears than it should. Recall the color index $B-V$, the difference between the blue and visual magnitudes of a star (see Chapter 19). The redder a star is, the greater the value of $B-V$. (Do not forget: If a star is red, it is brighter in visual than in blue light, so the V magnitude is smaller than the B magnitude). The effect of interstellar extinction is to make the value of $B-V$ greater than it would otherwise be.

To estimate the amount of dust in the line of sight to a distant star, therefore, simply requires a comparison of the star's $B-V$ color index with the value it would have without extinction. The latter quantity is determined from the spectral type of the star. Remember that $B-V$ and spectral class both measure the same stellar property, namely the temperature; therefore all stars of a given spectral type have the same intrinsic value of $B-V$.

Theoretical considerations allow astronomers to determine the average size of the grains, from the variation of extinction with wavelength (known as the extinction curve; Fig. 26.6). The fact that the grains scatter blue light more effectively than red, and the way in which the scattering efficiency varies

FIGURE 26.4 AN EXTENSIVE REGION OF NEBULOSITY. *This is a well-known region of dark clouds, including one known as the Horsehead nebula. Contained within these clouds are numerous infrared-emitting objects thought to be young stars.* (© 1980 Royal Observatory, Edinburgh)

as a function of wavelength leads to the conclusion that there are two distinct types of grains, both very small: (1) those with an average diameter of about 5 $\times$ 10^{-5} cm (or about 5,000 Å, comparable in size to the wavelength of visible light); and (2) a population of even smaller grains, whose diameters are only about 10^{-6} cm (or about 100 Å), which create extremely strong ultraviolet extinction.

The polarization of starlight tells us something about the shape of the grains. In Chapter 6 we learned that when light is polarized, the electric and magnetic fields of the waves have a preferred orientation. The interstellar dust creates this effect by selectively absorbing light of one orientation, so that what gets through to us is the light with the orientation perpendicular to that (Fig. 26.7). The dust has the same

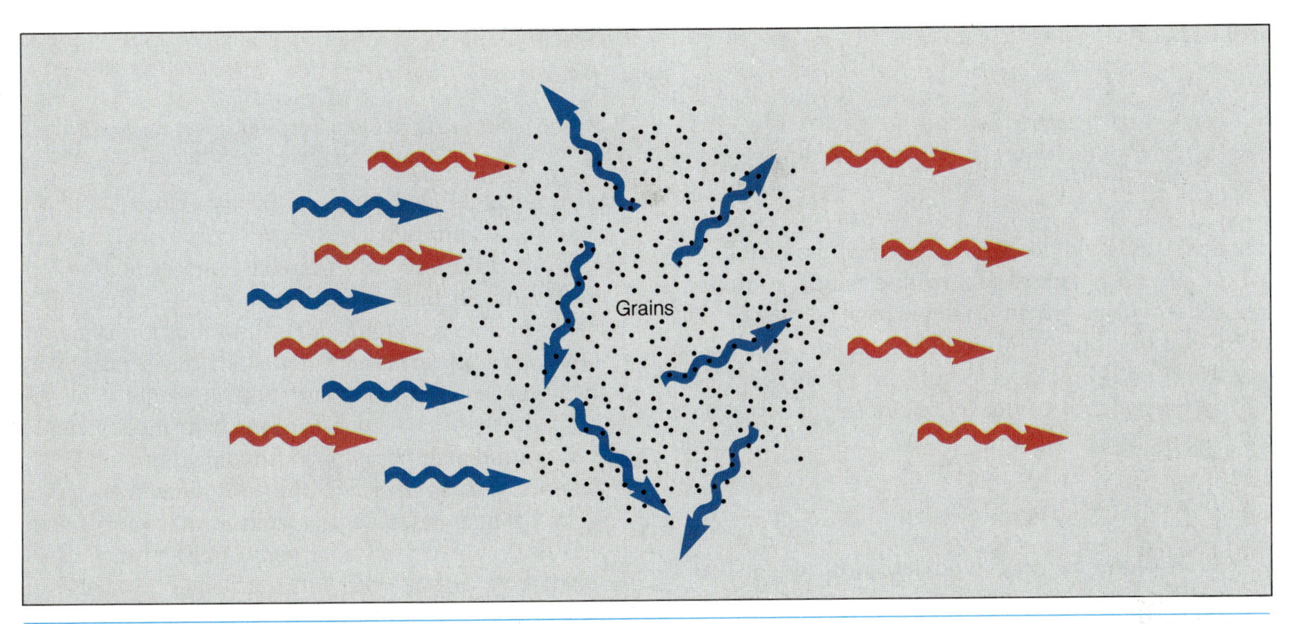

FIGURE 26.5 SCATTERING OF LIGHT BY INTERSTELLAR GRAINS. *The grains tend to absorb or deflect blue light more efficiently than red, so a star seen through interstellar material appears red.*

effect as the polarizing lenses commonly used in sunglasses.

The only way the interstellar dust grains can produce such an effect is if they are not spherical, but elongated (Fig. 26.8). They must also be aligned in some organized fashion, so that a majority of the grains between the Earth and a given star are parallel to each other. Their alignment is probably caused by an interstellar magnetic field, which has a constant orientation over large regions of space. Thus the fact that the grains produce polarization tells us something not only about their properties, but also about the galaxy—namely, that it has an overall magnetic field. It appears that the field tends to parallel the spiral arms, at least in the vicinity of the Sun.

Little has been learned about the chemical composition of the grains. Theoretical work indicates that a variety of common substances, including silicates, oxides, graphite, and iron-bearing minerals, could cause the observed extinction. There is a pronounced peak of extinction near the ultraviolet wavelength of 2,200 Å, which is best explained as being a result of absorption by a form of carbon, so it is probable that some of the grains contain this substance.

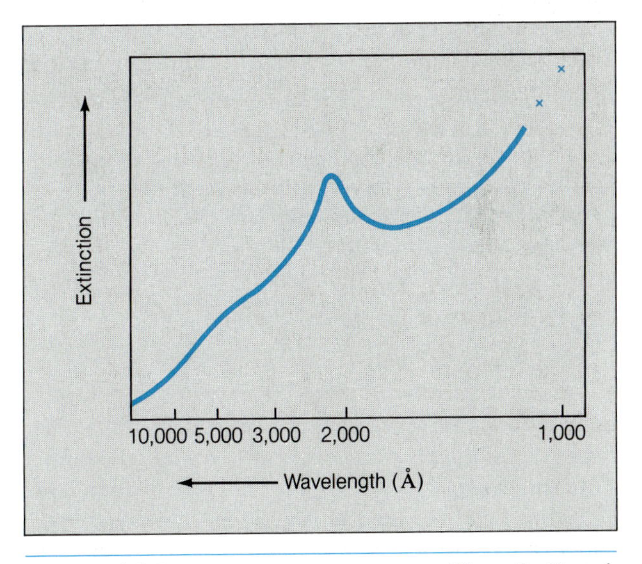

FIGURE 26.6 INTERSTELLAR EXTINCTION. *The extinction of starlight by interstellar dust is greatest at short wavelengths, especially in the ultraviolet, and decreases toward longer wavelengths, being very small in the infrared and virtually nonexistent in radio wavelengths.*

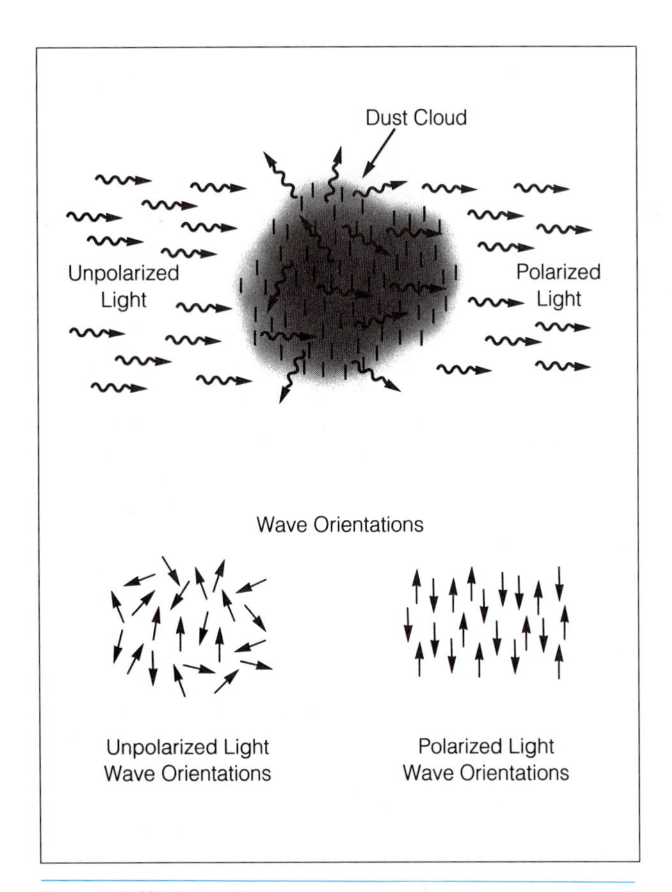

FIGURE 26.7 POLARIZATION OF STARLIGHT BY INTERSTELLAR GRAINS. *Starlight passing through the interstellar medium becomes weakly polarized, indicating that the dust grains in space are elongated in shape and tend to be aligned, probably by the galactic magnetic field.*

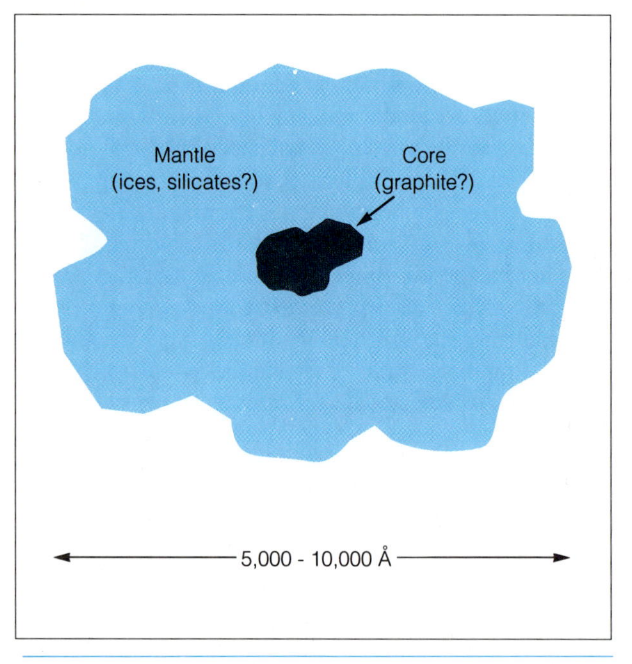

FIGURE 26.8 AN INTERSTELLAR GRAIN? *No direct information is yet available on the composition and the detailed shapes of interstellar grains. This sketch illustrates a popular model, but for a sobering look at what the grain shapes might really be like, see the photo of an interplanetary grain in Chapter 16 (Fig. 16.28).*

The origin of the interstellar grains is not well understood; there seem to be several possible processes, all of which may contribute to their existence. Tiny solid particles can form from a gas through condensation, if the proper combination of pressure and temperature is reached. As we learned in Chapter 18, this probably occurred early in the formation of the solar system, and therefore we expect that some interstellar grains are born in the vicinity of newly formed stars, then expelled into space, perhaps by a stellar wind. A more important source of grains appears to be the extended outer atmospheres of red giant and supergiant stars, where the proper conditions apparently exist, and radiation pressure from these very luminous stars expels the grains into interstellar space. In addition, infrared observations show that grains can form in the expanding material around novae and in planetary nebulae. The formation of interstellar grains is a by-product of the formation and evolution of stars, and is therefore an important part of the large-scale recycling of matter between stellar and interstellar forms.

OBSERVATION OF INTERSTELLAR GAS

Interstellar gas can be detected by several different techniques. As in the case of dust, one manifestation is in the form of rather obvious clouds that can be seen on photographs of the sky. Rather than being dark clouds, however, the visible gas regions are bright concentrations of material that is hot enough to glow.

terstellar clouds. In the line of sight to a faraway star, there are likely to be several clouds, their presence being revealed to us by the extra redness they impart to the color of the star and by the interstellar absorption lines they form in its spectrum. These most common of interstellar clouds are not usually hot enough to glow nor thick enough to show up as dark patches in the sky like the dense clouds found in some regions, and for this reason they are called **diffuse clouds.**

In comparing the properties of different types of interstellar clouds (Table 26.1), we find it convenient to think in terms of the density: the number of particles per cubic centimeter. At sea level on the Earth, the atmospheric density is about 2×10^{19} molecules per cubic centimeter. In a typical diffuse interstellar cloud, the number is more like 10 to 1000/cm^3, a factor of 10^{16} to 10^{18} less! The densest clouds, the dark ones that allow no light to pass through, have densities of about 10^6 particles/cm^3 at most, still a factor of 10^{13} less than the density of the Earth's atmosphere. The temperature of the gas in a diffuse cloud is typically in the range of 50 to 150 K and is as low as 20 to 50 K in dark clouds.

Diffuse clouds are thought to be ordinarily less than 1 parsec thick, although in many cases they may stretch for many parsecs in one dimension, as though they were thin sheets of matter, rather than being more or less spherical. As mentioned earlier in this chapter, very thin, wispy dust clouds, which may be identified with diffuse clouds, have been detected at infrared wavelengths with the *IRAS* satellite (Fig. 26.11). The total mass of a diffuse cloud is difficult to assess because of its unknown extent but is often comparable to the mass of the Sun. Thus we conclude that these clouds are not the type from which massive stars or star clusters are formed.

The composition of diffuse clouds is dominated by hydrogen, with other elements present in much smaller quantities. The hydrogen is primarily in atomic form, except in the central portions of the clouds, where the density may be sufficient to allow the formation of molecular hydrogen (H_2). The abundances of some of the elements are lower than the accepted "cosmic" values. It is as though some of the elements, among them iron, calcium, titanium, and other metallic species, are missing from the interstellar gas. The likely explanation is that these materials are preferentially in the solid dust grains, which gives us some indirect indication of what the grains are made of.

As we have pointed out, diffuse clouds are essentially transparent, but there are conditions under

TABLE 26.1 INTERSTELLAR MEDIUM CONDITIONS

Component	Temperature (K)	Density (per cm^3)	State of Gas	How Observed
Coronal gas (intercloud)	10^5–10^6	10^{-4}	Highly ionized	X-ray emission, UV absorption lines
Warm intercloud gas	1,000	0.01	Partially ionized	21-cm emission, UV absorption lines
Diffuse clouds	50–150	1–1000	Hydrogen in atomic form, others ionized	Visible, UV absorption lines, 21-cm emission
Dark clouds	20–50	10^3–10^5	Molecular	Radio emission lines, IR emission and absorption lines

FIGURE 26.11 INFRARED CIRRUS. *This image, obtained by the IRAS satellite, reveals extensive regions of diffuse, patchy clouds in interstellar space (shown here as orange against the blue background). The faint, wispy clouds seen in the upper portion here have been dubbed "infrared cirrus" because of their resemblance to cirrus clouds in the Earth's atmosphere.* (NASA)

FIGURE 26.12 AN EMISSION NEBULA. *The red color of this cloud of hot gas is due to emission by hydrogen atoms. This is a portion of the Rosette nebula.* (© 1984 Anglo-Australian Telescope Board)

which they show up in photographs. If a hot star is embedded within a cloud, its radiation ionizes and heats the gas to the point where it glows. Such a cloud is known as an **emission nebula** (Fig. 26.12), nebula being a general word for any interstellar cloud that is dense or bright enough to show up in photographs. In the case of a hot star embedded within a diffuse cloud, the emission occurs when atoms are ionized and then combine again with electrons to form new atoms. The electron usually starts out in a high energy state, and then quickly drops down to the lowest level, emitting a photon of light each time it makes a downward jump. Because most of the light from an emission nebula is from regions where hydrogen is ionized, these nebulae are called **HII regions,** the H standing for hydrogen, and the Roman numeral II indicating the ionized state (astronomers generally use this type of notation to indicate the

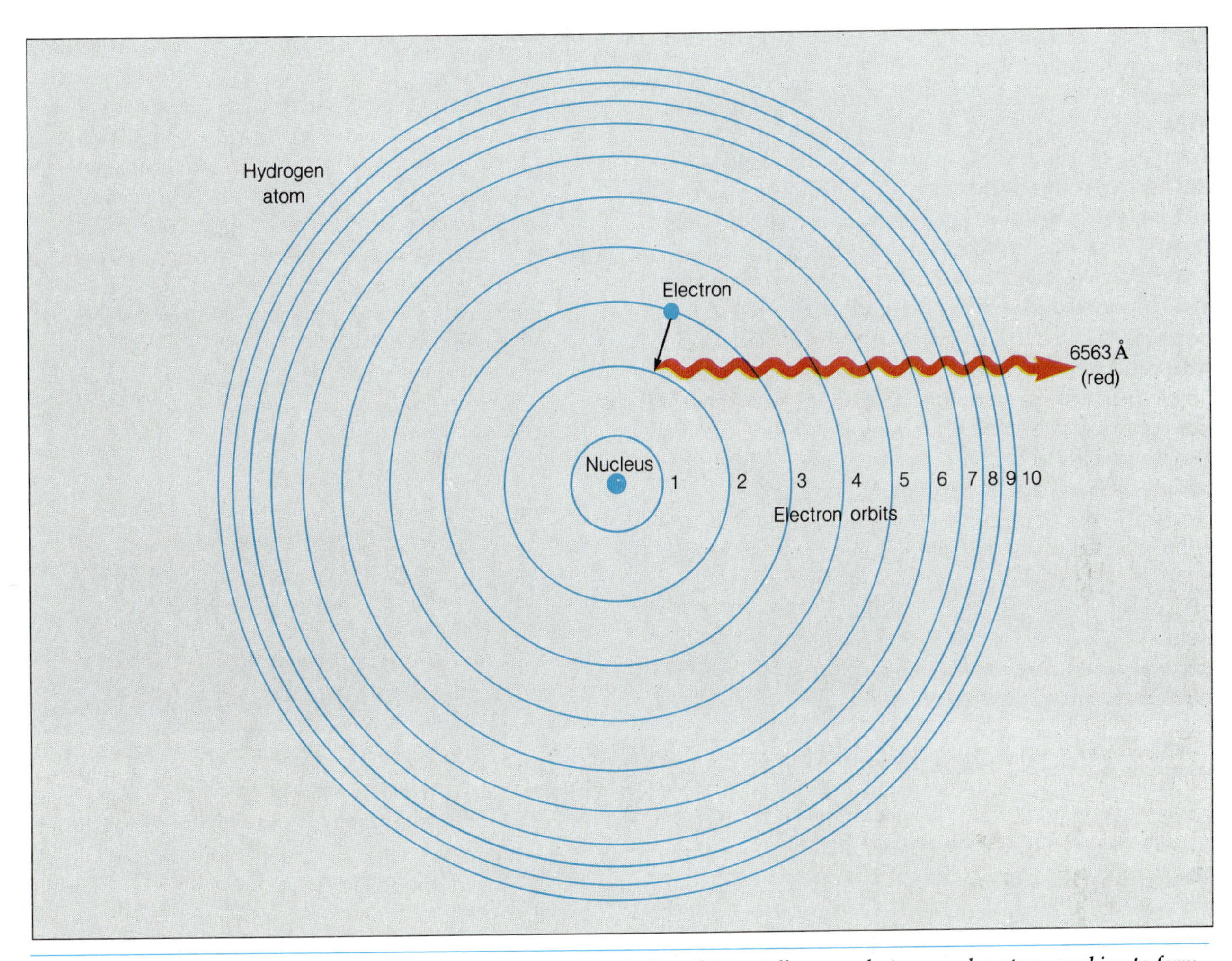

FIGURE 26.13 HYDROGEN LINES FROM AN EMISSION NEBULA. *In heated interstellar gas, electrons and protons combine to form hydrogen atoms, usually with the electron initially in an excited state. It then drops down to the lowest state, emitting photons at each step. The strongest emission in visible wavelengths corresponds to the jump from level 3 to level 2, and has a wavelength of 6563 Å. This is why emission nebulae appear red.*

degree of ionization of an element). The strongest line of hydrogen has a wavelength of 6,563 Å in the red portion of the spectrum (Fig. 26.3), so HII regions stand out especially well in photographs taken with red-sensitive film. A particularly prominent HII region surrounds the stars of the Trapezium, in the sword of Orion, and is commonly called the Orion Nebula (see Fig. 26.2).

If a diffuse cloud happens to lie just behind a hot star, it may be visible to us as a **reflection nebula** (Fig. 26.14). In this case it is the dust, rather than the gas, that causes the cloud to glow. What we see is the light that has been scattered in our direction by dust grains (Fig. 26.15). This light is always blue, for the same reason the daytime sky is blue: The tiny particles scatter blue light more effectively than longer wavelengths. The best-known reflection nebula is the one in the young star cluster called the Pleiades, and it lends photographs of this group of stars an ethereal quality (Fig. 26.16).

FIGURE 26.14 A DARK CLOUD. *The dark, patchy region here is a dense, cold interstellar cloud, the kind of region where stars can form. The interior temperature may be as low as 20 K, and the density roughly 10^5 particles per cubic centimeter, a high vacuum by earthly standards, but quite dense for the interstellar medium. Some emission nebulosity is present in this region also.* (© 1984 Anglo-Australian Telescope Board)

FIGURE 26.15 BRIGHT NEBULAE. *This shows how reflection and emission nebulae are formed. Light from a hot star is reflected off of a background cloud containing dust. The cloud has a bluish color, because short-wavelength photons are scattered most effectively. The stellar radiation also ionizes gas that is very close to the star, creating a region around it that glows at the wavelengths of certain emission lines. This type of emission nebula is also known as an H II region.*

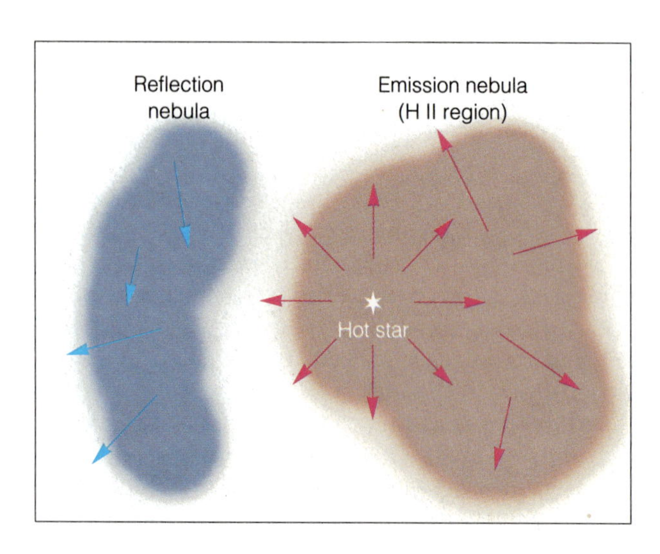

FIGURE 26.16 THE PLEIADES. *This relatively young galactic cluster is a prominent object in the northern hemisphere during the late fall and winter.* (Palomar Observatory, California Institute of Technology)

DARK CLOUDS AND THE MOLECULAR ZOO

In many ways the most fascinating of the interstellar objects are the dark clouds, whose densities and dimensions are sufficient to block out all the light of stars behind them. Dark-cloud regions are highly concentrated in the plane of the Milky Way and have a variety of amorphous sizes and shapes (see Figs. 26.1–26.4). Often bright HII regions are located within and among them, for the regions where dark clouds are common are also the sites of star formation, and therefore many young, hot stars are mixed in.

The masses of dark clouds are comparable to, or, in the cases of large cloud complexes, far in excess of a single star mass, ranging up to hundreds of thousands of solar masses in some instances. Thus it is quite possible for massive stars and star clusters to form from material in this type of cloud, and it is in dark-cloud regions that the most active star formation occurs.

As we have already noted, the temperature in the interior of a dark cloud is extremely low, perhaps only 10 to 20 K, and the density in the range of 10^3 to 10^6 particles/cm^3. Under these conditions, atoms can combine into molecular form, and a wide variety of species have been detected. The discovery of the molecular nature of the gas in dark clouds was relatively recent, the first species (the simple molecular OH) having been detected in 1963. The reason these observations were not made sooner is that sophisticated new radio astronomy techniques had to be developed.

Because these clouds are so dense, it is not possible to observe them in the same manner as the diffuse clouds. No stars can be seen through the dark clouds, so there is no way to measure absorption lines formed there. Molecules have special properties, however, that produce radio emission lines, and these lines can be observed.

A molecule is a combination of atoms in a single particle, held together by chemical bonds. These bonds may be viewed as a merging of energy states, so that some of the electrons orbit the entire combinations of atoms, acting as the glue that holds them together. The electrons still have fixed energy levels, and there is more complex structure as well. The atoms act as though they are held together with springs, in the sense that they can vibrate, and different frequencies of vibration correspond to different energy states. A molecule can also rotate on its axis, and different rotation speeds also correspond to different energy levels. The vibration and rotation states are close together, so the wavelengths of light corresponding to transitions between them are long. In most cases, the rotational levels are so closely spaced that the differences between them correspond to radio wavelengths. Thus if a molecule jumps from one rotational state to a lower one (that is, if it slows its rotation), it emits a photon at a radio wavelength. Each kind of molecule has its own characteristic set of rotational energy levels, so each has its own particular set of radio emission lines. Molecules have the same sort of fingerprints in the radio spectrum that atoms and ions do in visible and ultraviolet light. (Molecules, having such complex structure, also have spectral lines at visible and ultraviolet wavelengths, but these cannot be easily observed in dark clouds for the reasons already stated.)

In a dark cloud, the density is sufficiently high that molecules collide every now and then, and when they do they can be excited to high rotational energy levels. Once a molecule has been kicked into a rapid rotation state, it will soon slow (in a single, instantaneous jump) to a lower state, emitting a radio photon in the process. Thus a dark cloud is constantly releasing radio emission at a variety of fixed wavelengths corresponding to the types of molecules inside. As we learned in discussing the 21-cm line, radio lines are not affected by interstellar extinction, so the emission from a dark cloud escapes easily into space.

Most of the molecules that exist in dark clouds emit at wavelengths of a few millimeters or centimeters. Since the emission from a distant cloud is greatly attenuated (because of the inverse-square law) by the time it reaches the Earth, very sensitive receivers are required to detect it, and the technology for doing this has been developed only in the last two decades. Every time a new advance is made in receiver sophistication, new molecules are immediately detected.

The number of molecules discovered in space has grown to more than seventy (see Appendix 12). Most are rather simple species, the most common being diatomic molecules, which consist of only two atoms. Of these, the dominant one is molecular hydrogen which, for a subtle reason, does not emit radio waves, and has been detected in dark clouds only in active regions where it is excited to emit infrared radiation. The next most common molecule is carbon monoxide (CO), and following this the abundant species include OH (called the hydroxyl radical), water vapor (H_2O), and an assortment of other simple combinations of abundant elements.

Some rather complex large molecules have begun to be detected as well, the current grand champion being $HC_{11}N$, a species consisting of a total of thirteen atoms, eleven of them carbon. There may be even larger molecules in space, most likely other species containing long chains of carbon atoms, which form easily under the conditions that prevail inside the dark clouds.

All the molecules detected so far are made up of combinations of just a few of the most common elements, primarily hydrogen, carbon, nitrogen, oxygen, and sulfur. There probably are molecules containing other elements, but because of their lower abundances, these species are difficult to detect. The second most common element after hydrogen is helium, but this is an inert element, meaning that it cannot easily combine with other atoms to form molecules. Helium is certainly present in interstellar clouds, but only in atomic form.

Perhaps the most interesting of all the species detected in dark clouds are the **organic molecules,** those containing certain combinations of carbon atoms that are also found in living material. Their presence shows that when stars form in dark clouds, the

ASTRONOMICAL INSIGHT 26.2

THE MYSTERIOUS COSMIC RAYS

We have known for a long time that the Earth is constantly being bombarded by rapidly moving subatomic particles from space, which have been given the name **cosmic rays.** Consisting of electrons or atomic nuclei, these particles can be detected in several ways, the most common of which is to trace their tracks in special liquid-filled containers called bubble chambers. The cosmic ray particles are electrically charged, and their passage through such a container creates a trail of fine bubbles. From the length and shape of this track, we can infer the charge and speed of the particles. Cosmic ray velocities are usually near the speed of light.

The Earth's magnetic field deflects most cosmic rays, particularly the less massive and energetic ones, so that they reach the ground most easily near the poles. From the point of view of life on Earth, this is probably a good thing, for cosmic rays are suspected of causing genetic mutations and could be harmful in other ways if allowed to reach the surface unimpeded. (On the other hand, mutations caused by cosmic rays are thought to have assisted evolutionary processes in the development of life on Earth.)

Extensive observations of cosmic rays show that most are either electrons or nuclei of atoms, with all of the electrons stripped free. Most of the latter are single protons, the nuclei of hydrogen atoms, but some are nuclei of heavy elements, including metallic species known to be formed only in supernova explosions. The majority of the cosmic rays that reach the Earth originate in the Sun, but many, particularly the more massive and energetic ones, come from all directions in the plane of the galaxy and seem to fill its volume more or less uniformly. Apparently they are trapped by the galactic magnetic field and cannot escape into intergalactic space. Because their paths are curved and deflected by the magnetic field, it is impossible to tell where they come from. Therefore the source of the galactic cosmic rays is not understood, but a strong possibility is that they are released in supernova explosions. The total energy contained in cosmic rays darting about the galaxy is very large, and only supernovae seem capable of producing the necessary energy.

The distribution of cosmic-ray nuclei of various elements is generally similar to the chemical abundances found in stars and the interstellar material, although there are some differences, most of which can be attributed to nuclear reactions that can occur when the rapidly moving nuclei strike interstellar dust grains. Such reactions are responsible for the production of certain elements in space, such as the lightweight species beryllium.

Cosmic rays have little effect on our everyday lives, but they have the potential of telling us quite a bit about the energetic activities that take place in interstellar space.

ingredients for life are already present. We have seen (in Chapter 16) that amino acids have been found in meteorites, indicating that these molecules were present before the planets formed. We can speculate that perhaps these relatively complex organic molecules formed even before the first solid matter in the solar system.

INTERSTELLAR VIOLENCE AND THE ROLE OF SUPERNOVAE

Now that we have examined the principal components of the interstellar medium, let us take another look at its overall properties. We have found a diverse collection of ingredients, which can be characterized by a wide variety of temperatures and densities.

When the first spectroscopic observations of the interstellar gas were made more than 50 years ago, it was noticed that some diffuse clouds are moving through space at rather large speeds, up to 100 kilometers per second or even more (Fig. 26.17). Since then it has been found that the motions are often directed away from groups of bright stars, as though the gas were being expelled by some force originating in the stars.

When 21-cm observations of hydrogen gas became possible, similar motions were found. A variety

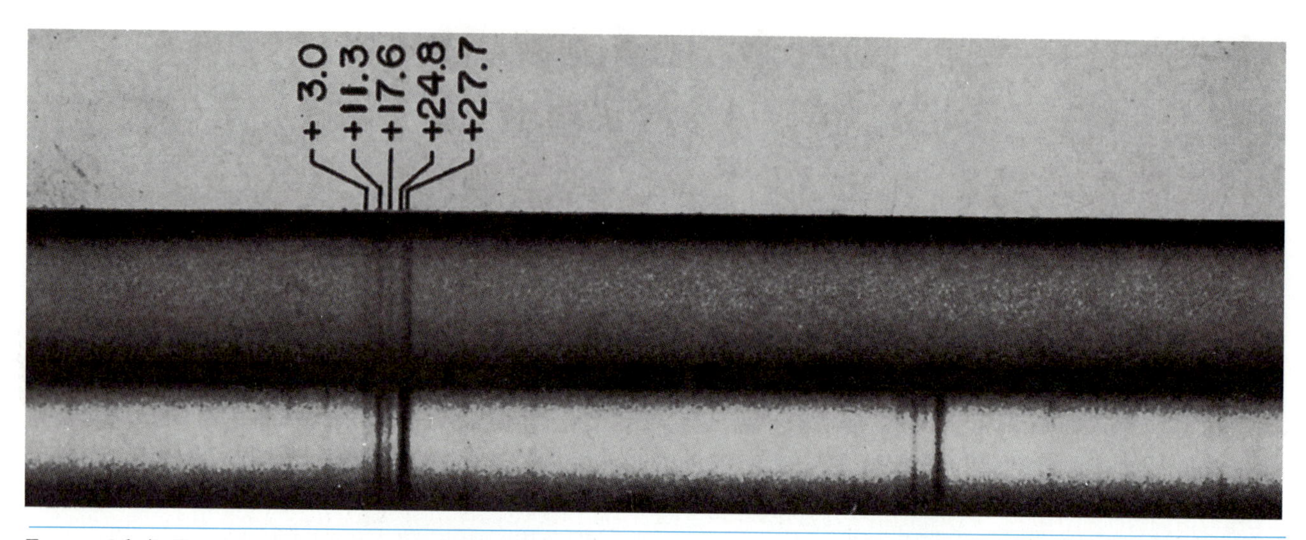

FIGURE 26.17 THE EFFECT OF CLOUD MOTIONS ON INTERSTELLAR LINES. *This is a spectrogram of a distant star that has several interstellar clouds along its line of sight. The upper spectrum shows a line due to ionized calcium; the lower spectrum shows atomic sodium (two major lines). Each line consists of several closely-spaced features representing individual clouds in the line of sight. Each cloud has a slightly different velocity, so that the lines are separated by the Doppler effect. Velocities (in km/sec) of the individual clouds with respect to the Sun are indicated.* (G. Munch)

of rapidly moving clouds was discovered, and in some regions there appeared to be expanding loops or shells of gas (Fig. 26.18). Some of these regions were associated with known supernova remnants, where it was clear that the gas was still expanding from the original explosion. In other cases, however, nothing was found at the center of expansion except perhaps a group of young stars. Whatever the cause, the picture developed by these observations was one of an interstellar medium in turmoil, with random, high-speed motions throughout.

In the 1970s new evidence of energetic phenomena in space was uncovered, in two forms. Data obtained from X-ray satellite and rocket experiments showed that the interstellar medium in the plane of the galaxy emits X-rays. At about the same time, ultraviolet spectroscopic observations revealed previously undetected ions in space that form only at very high temperatures. The combination of these X-ray and ultraviolet data led to the conclusion that the space between cool clouds is filled with a very hot gas, whose temperature is 1 million degrees or more. The density of the hot gas between the clouds is very low even by interstellar standards; it is 10^{-4} to 10^{-3} particle per cubic centimeter, the equivalent of one particle in every cubic volume 10 to 20 centimeters on a side!

We now see that the interstellar medium not only is diversified, but also is being constantly disturbed, which creates the observed high-speed motions and high temperatures. All this activity requires a substantial amount of energy input; a quiet pond does not spontaneously form ripples and eddies unless something stirs it up.

The agitation of the interstellar medium, it is now understood, is primarily the result of supernova explosions. These occur in the galaxy at a frequency of one every few decades, and each time, a vast amount of energy is released into space. Material expands outward from the site of the explosion, sweeping up a shell of gas that continues to expand, and creating a cavity of very rarefied, extremely hot gas inside the shell (Fig. 26.19). As we have seen, interstellar clouds can be compressed by the outburst, triggering star formation, and clouds can be accelerated to high speeds as well. The net effect of repeated supernova explosions in the galaxy is that a large fraction of its volume has become filled with the hot gas that is left behind

FIGURE 26.18 AN EXTENSIVE SUPERNOVA REMNANT. *This is a large portion of the Vela supernova remnant, which covers a large region of sky in the southern hemisphere.* (© 1980 Royal Observatory, Edinburgh)

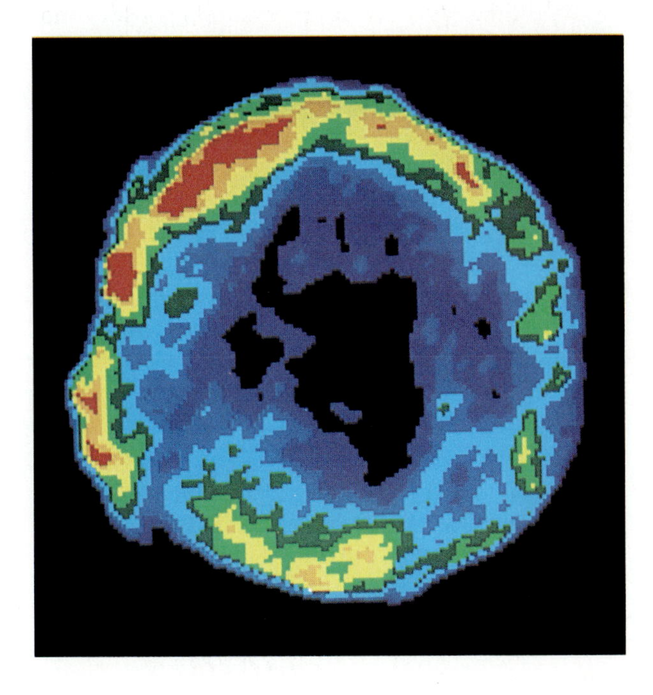

FIGURE 26.19 THE REMNANT OF TYCHO'S SUPERNOVA. *In 1572 a prominent supernova occurred, and was observed by Tycho Brahe. (The reputation his analysis earned him led to his obtaining royal funding for his observatory.) This is a radio map of the remnant of this supernova, showing how the expanding gas has created a huge, spherical bubble.* (National Radio Astronomy Observatory)

by the expanding shells. Much of the cloud material has been accelerated, so that high-velocity clouds are seen here and there, particularly in loops or fragments of spherical shells.

Another mechanism that has the same effect, and which contributes to this picture, is the expansion of stellar winds. We learned in Chapter 21 that hot, blue stars as a rule have high-speed winds; these are streams of gas flowing outward at speeds as great as 3,000 kilometers per second. Such a wind will act in some ways like a supernova explosion, sweeping up material around the star and creating an expanding shell (Figs. 26.20 and 26.21). Although a supernova releases far more energy in a brief moment, a stellar wind may last several million years, eventually injecting a comparable amount of energy into the surrounding gas. In a cluster of young stars, the winds can combine to evacuate a large bubble around the entire cluster, closely simulating the effect of a supernova explosion.

In one recently discovered case, a gigantic shell was found surrounding a vast volume in the constellation Cygnus (Fig. 26.22); the total energy required to create that structure is estimated to be comparable to that of several hundreds or even thousands of supernovae. It is not clear how this "superbubble" formed. Possibly a very large star cluster at its center experienced a number of supernovae whose expanding shells combined to form the single huge structure seen today. Several other examples of "superbubbles" are now known.

The hot, tenuous interstellar gas is known to extend well above and below the plane of the galactic disk. Perhaps the galaxy as a whole has a wind emanating outward, a composite of the expansion of all the supernova explosions and stellar winds that have occurred within it. In any case, we can no longer think of the interstellar void as a cold, desolate place where nothing ever happens.

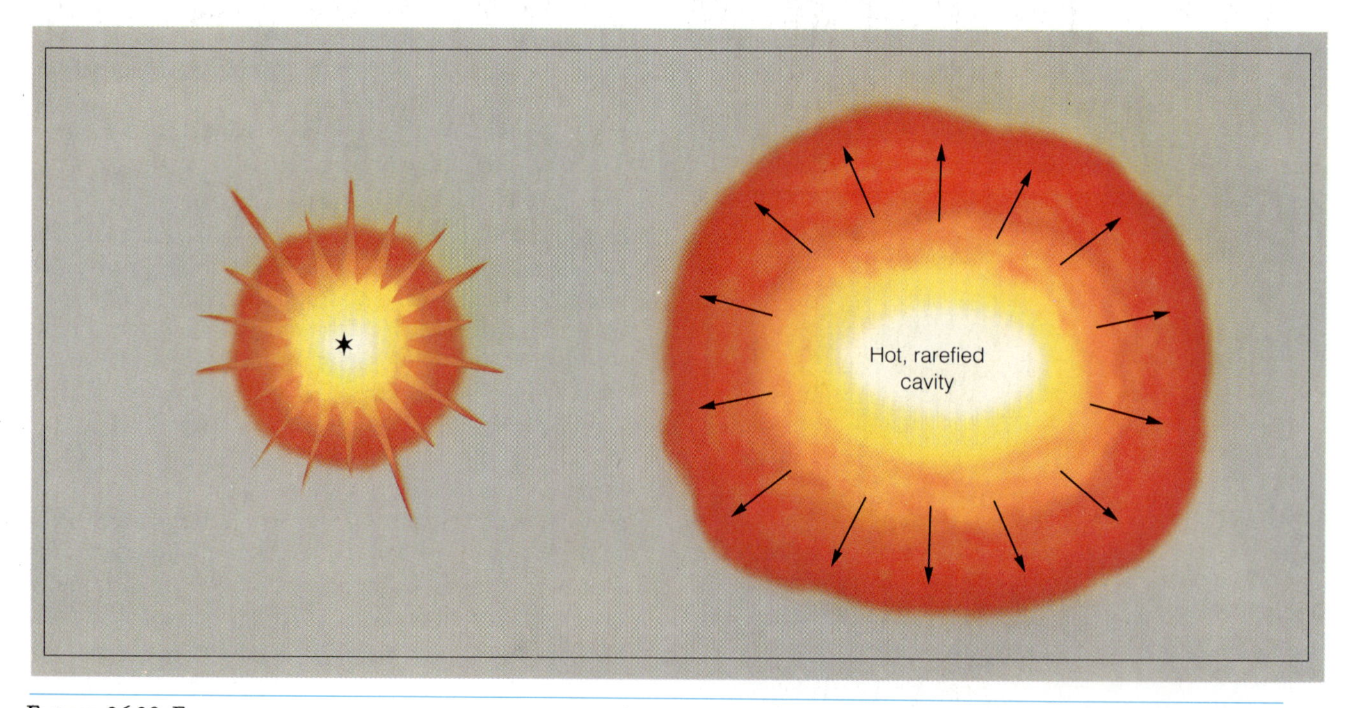

FIGURE 26.20 FORMATION OF AN INTERSTELLAR BUBBLE. *A supernova explosion or a strong stellar wind occurs in the midst of a cloudy interstellar medium (left). As the resultant shock waves expand, a spherical cavity is formed, with a concentration of swept-up gas surrounding it. The cavity, or bubble, is filled with very hot, rarefied gas. In real situations, the medium is probably not so uniform initially, and the resulting cavity is not so symmetrically shaped as shown here.*

FIGURE 26.21 A BUBBLE CREATED BY A STELLAR WIND. *This nebula, called NGC2359, contains a roughly spherical cavity that has been evacuated by the high-velocity wind from a hot, luminous star.* (© 1979 Anglo-Australian Telescope Board)

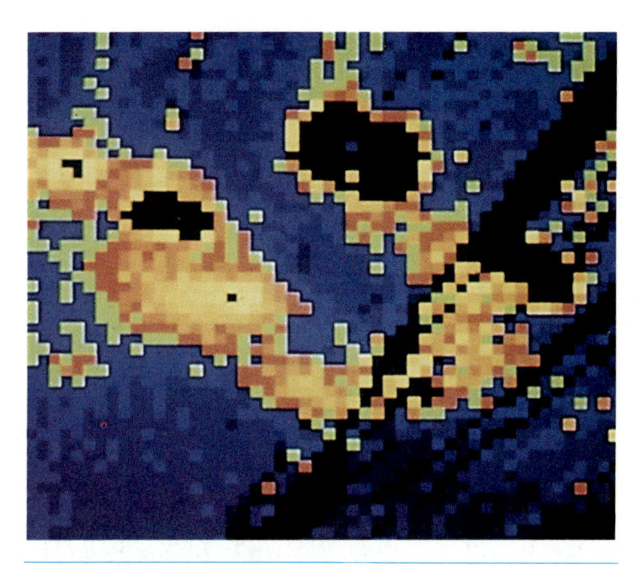

FIGURE 26.22 THE CYGNUS SUPERBUBBLE. *This diagram shows the composite X-ray emission from a huge region in the constellation Cygnus. The yellow arc-like structure outlines a gigantic bubble of superheated gas that has been swept up by a series of supernova explosions. The dimensions of the bubble are roughly 600 by 450 parsecs.* (W. C. Cash)

PERSPECTIVE

If this chapter had been written 15 years ago, it would have had a much different flavor. The interstellar medium would have been viewed as a quiescent background against which all the activities of stellar evolution are played out. Instead, we now have a picture of a dynamic role for interstellar matter in the affairs of the galaxy, with the stars providing the necessary energy as they live and die, and the interstellar material in turn creating the conditions necessary for the births of new generations. The rich interplay between stars and interstellar matter is the dominant theme of galactic ecology, and its role will be brought to the fore again in the next chapter, as we discuss the formation and evolution of the Milky Way.

SUMMARY

1. The interstellar medium, although extremely tenuous, contains a significant fraction of the mass of the galactic disk.
2. Interstellar grains cause extinction and reddening of starlight, because they scatter and absorb photons, and do so most efficiently at short wavelengths. The grains also cause polarization of starlight, indicating that they are elongated and preferentially aligned, probably by a galactic magnetic field.
3. Analysis of the interstellar extinction created by dust grains leads to the conclusion that many of the grains are about 5,000 Å in diameter, and that there is a second group of dust grains, responsible for the strong extinction of ultraviolet wavelengths, that are

much smaller in size, with diameters of about 100 Å.
4. The grains also cause polarization of starlight, implying that they are elongated in shape and that they are aligned by a galactic magnetic field.
5. The gas in space was first detected through absorption lines in the spectra of distant stars. Most of these lines lie at ultraviolet wavelengths.
6. The interstellar medium is patchy, containing cold clouds embedded in a hot, more rarefied, intercloud medium. Most of the mass in space is in the clouds, and most of the volume is filled with the hot gas.
7. Interstellar clouds have a variety of densities and temperatures. The so-called diffuse clouds have densities of about 1 to 100 particles per cubic centimeter and temperatures of 50 to 100 K. We can observe these clouds in absorption-line measurements. Also, if near a hot star, they may appear as emission or reflection nebulae.
8. The dark clouds, with temperatures of 10 to 20 K and densities up to 10^6 particles per cubic centimeter, consist of gas that is primarily molecular in form. The most common molecules are hydrogen (H_2) and carbon monoxide (CO). A large number of different species have been detected, primarily through their radio emission lines.
9. The interstellar medium displays various forms of energetic activity: There are random, often rapid, cloud motions; the gas between clouds is extremely hot; and here and there are gigantic loop and ring structures suggestive of expanding shells. All of this activity can be attributed to energy injected into the interstellar medium by supernova explosions and stellar winds.

REVIEW QUESTIONS

Thought Questions

1. Summarize the conditions in the different types of interstellar environments discussed in this chapter, and say how each type of region is observed.
2. List all the different ways in which astronomers detect and study interstellar dust.
3. Suppose a certain star lies behind a large quantity of interstellar dust. Why is it much more difficult to observe this star in ultraviolet wavelengths than in infrared wavelengths?
4. How can astronomers distinguish between a cool star that is intrinsically red and a hotter star that appears just as red because of interstellar dust?
5. How can absorption lines formed in the spectrum of a star by interstellar gas be distinguished from absorption lines formed in the atmosphere of the star itself?
6. Explain why the interstellar absorption lines for most elements lie in ultraviolet wavelengths.
7. How can most of the mass in interstellar space be in the form of clouds when the clouds fill only a small fraction of the volume of interstellar space?
8. Explain why the densest interstellar clouds are best observed at radio wavelengths.
9. What is the evidence that the interstellar medium is an active, energetic place?
10. Summarize the role of the interstellar medium in the evolution of stars.

Problems

1. If interstellar dust causes a star to appear only 1/100 as bright as it would otherwise, how many magnitudes of extinction is the dust creating?
2. Suppose a B0 star is observed to have a color index of $B-V = 0.20$. It is known from data on other B0 stars that the normal color index for a star of this type is $B-V = -0.30$. What is the color excess for this star?
3. If 15 percent of the mass of the galaxy is in the form of interstellar matter, estimate the average density of the interstellar medium by dividing the interstellar mass by the volume of the galactic disk. (Assume the mass of the galaxy is 2×10^{11} solar masses. Assume that the galactic disk has a radius of $r =$ 15,000 pc and a thickness of $z = 200$ *pc*; its volume V is equal to its area of πr^2 times its thickness, or $V = \pi r^2 z$. To get an answer in g/cm³, convert the mass to grams and all distances to centimeters before you calculate the density.)

ADDITIONAL READINGS

Blitz, L. 1982. Giant molecular cloud complexes in the galaxy. *Scientific American* 246(4):84.
Buhl, D. 1972. Light molecules and dark clouds. *Mercury* 1(5):4.

Chaisson, E. J. 1978. Gaseous nebulas. *Scientific American* 239(6):164.

Chevalier, R. A. 1978. Supernova remnants. *American Scientist* 66:712.

Eicher, David J., and David Higgins. 1987. The secret world of dark nebulae. *Astronomy* 15(9):46.

Gehrz, R. D., D. C. Black and P. M. Soloman. 1984. The formation of stellar systems from interstellar molecular clouds. *Science* 224(4651):823.

Greenberg, J. M. 1984. The structure and evolution of interstellar grains. *Scientific American* 250(6):124.

Heiles, C. 1978. The structure of the interstellar medium. *Scientific American* 238(1):74.

Herbig, G. H. 1974. Interstellar smog. *American Scientist* 62:200.

Herbst, E., and W. Klemperer. 1976. The formation of interstellar molecules. *Physics Today* 29(6):32.

Jura, M. 1977. Interstellar clouds and molecular hydrogen. *American Scientist* 65:446.

Kaler, J. B. 1982. Bubbles from dying stars. *Sky and Telescope* 63(2):129.

Knacke, R. F. 1979. Solid particles in space. *Sky and Telescope* 57(4):347.

Lewis, R. S., and E. Anders. 1983. Interstellar matter in meteorites. *Scientific American* 249(2):54.

Malin, David. 1987. In the shadow of the horsehead. *Sky and Telescope* 74(3):253.

Miller, J. S. 1974. The structure of emission nebulas. *Scientific American* 231(4):34.

Scoville, N., and J. S. Young. 1984. Molecular clouds, star formation, and galactic structure. *Scientific American* 250(4):42.

Snow, T. P. 1976. The interstellar medium: Much ado about nothing? *Griffith Observer* 40(9):2.

Verschuur, Gerritt L. 1987. Molecules between the stars. *Mercury* 16(3):66.

CHAPTER 27

THE FORMATION AND EVOLUTION OF THE GALAXY

The spiral galaxy NGC253, showing the distribution of stellar populations. (© 1980 Anglo-Australian Telescope Board)

CHAPTER PREVIEW

The galaxy that we observe today is the end product of some 10 to 20 billion years of evolution. A variety of processes have shaped it, and many have left unmistakable evidence of their action. In this chapter we will discuss the major influences on the galaxy, along with the developments that they have brought about.

We have laid the groundwork that will enable us to understand one of the most elegant concepts in the study of astronomy—the interplay between stars and interstellar matter and the long-term effects of this cosmic recycling on the evolution of the galaxy.

FIGURE 27.1 A SPIRAL GALAXY SIMILAR TO THE MILKY WAY. *This vast conglomeration of stars and interstellar matter is following its own evolutionary course, just as individual stars do. Note that the spiral arms tend to be blue in color, while the central bulge is yellow. These colors indicate the types of stars that dominate in the different parts of the galaxy.* (© 1980 Anglo-Australian Telescope Board)

STELLAR POPULATIONS AND ELEMENTAL GRADIENTS

To begin this story, we focus first on the overall structure of the galaxy (Fig. 27.1) and the types of stars found in various regions. We have already seen that most of the young stars in the galaxy tend to lie along the spiral arms. Indeed, it is the brilliance of the hot, massive O and B stars and their associated HII regions that delineates the arms, making them stand out from the rest of the galactic disk. We have also noted that in the nuclear bulge of the galaxy, as well as in the globular clusters and the halo in general, there are few young stars and relatively little interstellar material.

Careful scrutiny has revealed a number of distinctions between the stars that lie in the disk, particularly those in spiral arms, and the stars in the nucleus and halo. Analysis of the chemical compositions of the stars has shown that those in the halo and nucleus tend to have relatively low abundances of heavy elements such as metals, while in the vicinity of the Sun (and in spiral arms in general), these species are present in greater quantity. Nearly all stars are dominated by hydrogen, of course, but in the halo stars the heavy elements represent an even smaller trace than in the spiral arm stars. The relative abundance of heavy elements in a halo star may typically be a factor of 100 below that found in the Sun; if iron, for example, is 10^{-5} as abundant as hydrogen in the Sun, it may be only 10^{-7} as abundant in a halo star.

A picture has emerged in which the galaxy has two distinct groups of stars, those in the halo and nuclear bulge on one hand, and those in the spiral arms on the other (Fig. 27.2). These two groups have been designated **Population II** for the stars in the halo and **Population I** for those in spiral arms; (Table 27.1). The Sun is thought to be typical of Pop. I, and its chemical composition is usually adopted as representative of the entire group.

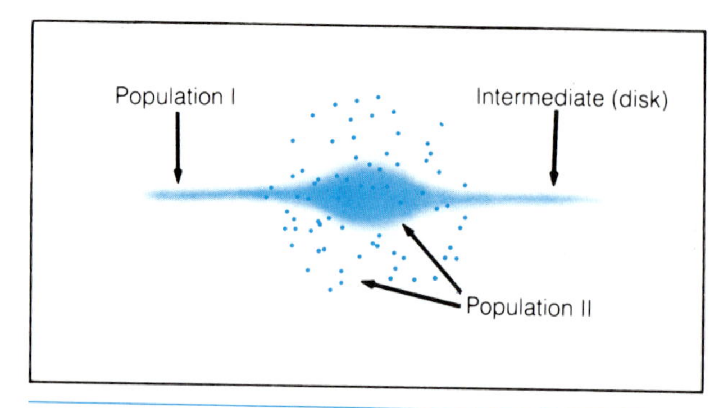

FIGURE 27.2 THE DISTRIBUTION OF POPULATIONS I AND II. *This cross-section of the galaxy shows that Population II stars lie in the halo and central bulge, whereas Population I stars inhabit the spiral arms in the plane of the disk. Intermediate population stars are distributed throughout the disk.*

Almost every time an attempt is made to classify astronomical objects into distinct groups, the boundaries between them turn out to be a little indistinct. There are usually intermediate objects whose properties fall between categories, and the stellar populations are no exception to this. For example, the stars in the disk of the galaxy that do not fall strictly within the spiral arms have intermediate properties between those of Pop. I and Pop. II, and are often called simply the **disk population.** Furthermore, there are gradations within the two principal population groups, and we speak therefore of "extreme" or "intermediate" Pop. I or Pop. II objects. It is much more accurate to view the stellar populations as a smooth sequence of stellar properties represented by very low heavy element abundances at one end, and solarlike abundances at the other.

The differences between chemical compositions from one part of the galaxy to another provide astronomers with very important clues in their efforts to piece together the history of the galaxy. Changes that occur gradually over substantial distances are referred to as **gradients,** and the variations of stellar composition from one part of the galaxy to another are therefore referred to as **abundance gradients.** There is a gradient of increasing heavy-element abundances from the halo to the disk (Fig. 27.3). Within the disk itself, there is a similar gradient from the outer to the inner portions (this does not include the stars in the nuclear bulge, which, as we have pointed out, tend to have low abundances of heavy elements).

Most of the stars near the Sun belong to Pop. I; they are disk stars orbiting the galactic center in approximately circular paths, like the Sun. There are a few nearby Pop. II stars here and there, and they are distinguished by several properties in addition to their compositions. The most easily recognized is that their motions depart drastically from those of Pop. I stars. As a rule, Pop. II stars do not follow circular paths in the plane of the disk, but instead follow highly eccentric orbits that are randomly oriented (Fig. 27.4), much like cometary orbits in the solar system. Thus, most of the Pop. II stars that are seen near the Sun are just passing through the disk from above or below it. The Sun and the other Pop. I stars in its vicinity move in their orbits at speeds around 250 km/sec, so the Sun moves very rapidly with respect to the Pop. II stars that pass through its neighborhood in a perpen-

TABLE 27.1 STELLAR POPULATIONS

	Population I	Population II
Age	Young to intermediate	Old
Distribution	Disk, spiral arms	Halo, central bulge
Composition	Normal metals	Low metals
Constituents	Disk, arm stars O,B stars Interstellar matter Type I Cepheids Type II supernovae The Sun	Low-metal stars High-velocity stars Globular clusters Type II Cepheids RR Lyrae stars Type I supernovae Planetary nebulae

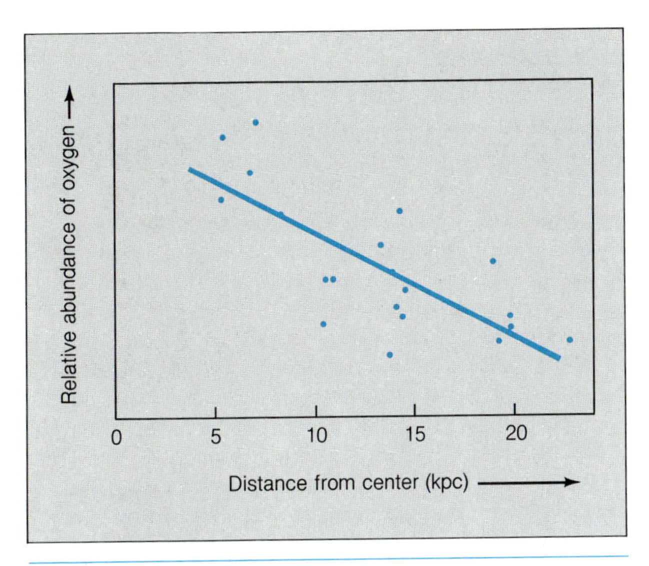

FIGURE 27.3 AN ABUNDANCE GRADIENT. *This diagram shows how the abundance of oxygen (relative to hydrogen) varies with distance from the center of the Andromeda galaxy. More stellar generations have lived and died in the dense regions near the center, so nuclear processing is further advanced in the central region than it is farther out.* (Data from Blair, W. P., Kirshner, R. P., and Chevalier, R. A. 1982, *Astrophysical Journal* 254:50)

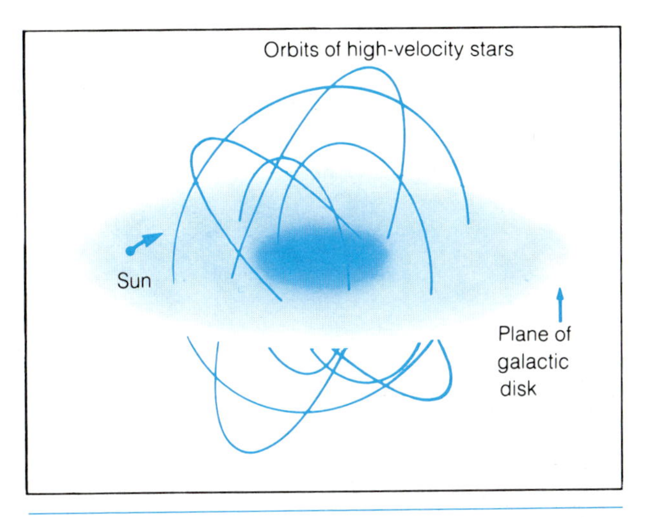

FIGURE 27.4 THE ORBITS OF HIGH-VELOCITY STARS. *These stars follow orbits that intersect the plane of the galaxy. When such a star passes near the sun, it has a high velocity relative to us, because of the sun's rapid motion along its own orbit.*

dicular direction. From our perspective here on the Earth, these Pop. II objects therefore are classified as **high-velocity stars.**

Historically, high-velocity stars were discovered and recognized as a distinct class even before the differences between Pop. I and Pop. II were enumerated. It was only later that these objects were equated with Pop. II stars following randomly oriented orbits in the galactic halo.

The systematic differences between the stellar populations arose from differing conditions at the times when they formed. In this regard it is important to recall that globular clusters, representative of Pop. II, are among the oldest objects in the galaxy, while Pop. I, on the other hand, includes young, newly formed stars. We will follow this discussion further in a later section.

STELLAR CYCLES AND CHEMICAL ENRICHMENT

The fact that stars in different parts of the Milky Way have distinctly different chemical makeups is readily explained in terms of stellar evolution. Recall (from discussions in Chapter 21) that as a star lives its life, nuclear reactions gradually convert light elements into heavier ones. The first step occurs while the star is on the main sequence, during which hydrogen nuclei deep in its interior are fused into helium. Later stages, depending on the mass of the star, may include the fusion of helium into carbon, and possibly the formation of even heavier elements by the addition of further helium nuclei. The most massive stars, which undergo the greatest number of reaction stages, also form heavy elements in the fiery instants of their deaths in supernova explosions.

All of these processes work in the same direction: They act together to gradually enrich the heavy-element abundances in the galaxy. Material is cycled back and forth between stars and the interstellar medium, and with each passing generation a greater supply of heavy elements is available. As a result, stars formed where there has been a lot of stellar cycling in the past are born with higher quantities of heavy elements than those formed where little previous cycling has occurred. Stars formed before much cycling occurred therefore tend to have low abundances of heavy elements. This explains why old stars belong to Pop. II; they formed out of material that had not yet been chemically enriched.

ASTRONOMICAL INSIGHT 27.1

A War Benefit

The discovery of stellar populations was actually based on observations of another galaxy, rather than the Milky Way, and the circumstances that led to the discovery were rather unusual.

During World War II, following the Japanese attack on Pearl Harbor in late 1941, cities on the West Coast were often kept dark at night to reduce the danger of bombardment. This was accomplished by enforced blackouts, in which all outside lights were kept turned off, and windows were covered so that no light got out. As a result of these blackouts, major cities on the West coast experienced some of the darkest nighttime skies of this century. This was a boon to local astronomers.

Many scientists were preoccupied with war-related research efforts and were not available to take advantage of the excellent observing opportunities. At the Mt. Wilson Observatory in the mountains overlooking Pasadena, however, there was a German-born astronomer named Walter Baade, a naturalized United States citizen restricted to nonsensitive research because of missing citizenship papers. Baade made extensive use of the 100-inch telescope at Mt. Wilson during the blackouts in Los Angeles, and was able to obtain extraordinarily fine photographs of galaxies neighboring the Milky Way.

One of these is the Andromeda nebula, a large spiral thought to be similar in many respects to our own galaxy. Baade's images of Andromeda were so fine and clear that he was able to make out individual stars, and to assess their properties. From colors and locations of the stars, he realized that there were systematic differences between different parts of the galaxy, and from these differences he developed the notion of two distinct stellar populations. Only later was the idea applied to the Milky Way, after careful observation revealed the same distribution of stellar properties.

The Pop. I stars, such as the Sun, are those that condensed from interstellar material that had previously been processed in stellar interiors and in supernova explosions. Therefore, as a rule, they formed at moderate to recent times in the history of the galaxy, and are not as ancient as Pop. II objects.

Besides elemental abundance variations with age, there can also be variations with location. As noted in the preceding section, there is a distinct abundance gradient within the disk of the galaxy, the stars nearer the center having higher abundances of heavy elements than those farther out. This is a reflection of enhanced stellar cycling near the center, rather than an age difference. The central portion of the disk is where the density of stars and interstellar material is highest, and therefore in this region there has been relatively active stellar processing, because star formation has proceeded at a greater rate than in the less dense outer portions of the disk. Over the lifetime of the galaxy, more generations of stars have lived and died in the central region than in the rarefied outer reaches.

The Care and Feeding of Spiral Arms

In Chapter 25 we described the structure of the galaxy and described how its spiral nature was discovered and mapped, but we said nothing about how the spiral arms formed, nor why they persist. Both questions are important, for it is clear that if the arms were simply streamers trailing along behind the galaxy as it rotates, they should wrap themselves up tightly around the nucleus (Fig. 27.5), like string being wound into a ball. Because the galaxy is old enough to have rotated at least 40 to 50 times since it formed, something must be preventing the arms from winding up, or they would have done so long ago.

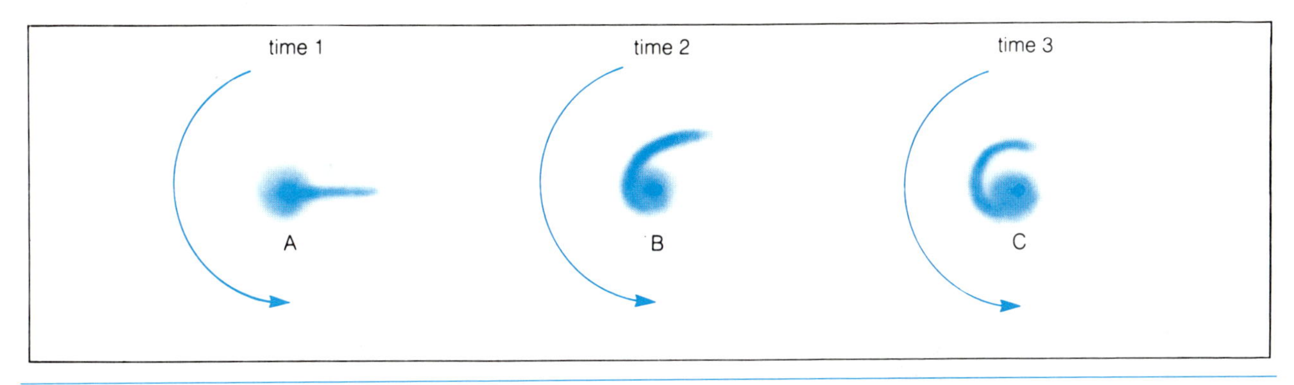

FIGURE 27.5 THE WINDUP OF SPIRAL ARMS. *If spiral arms were simple streamers of material attached to a rotating galaxy, they would, within a few hundred million years, wind tightly around the nucleus. The galaxy is much older than that, so some other explanation of the arms is needed.*

Maintaining the spiral arms, a distinct question from that of forming them in the first place, is a very complex business, and it is not yet fully understood how this is done. The first successful theory on this appeared in 1960, and is still being refined. The essence of this theory is that the large-scale organization of the galaxy is imposed on it by wave motions. We have understood waves as oscillatory motions created by disturbances, and we know that waves can be transmitted through a medium over long distances while individual particles in the medium move very little. In the case of water waves, for example, a floating object simply bobs up and down as a wave passes by, whereas the wave itself may travel a great distance. Here we are distinguishing between the wave and the medium through which it moves.

The waves that apparently govern the spiral structure in our galaxy are not transverse waves, like those in water, but are compressional waves, similar to sound waves and to certain seismic waves (the P waves; see Chapter 5 for an elaboration on the different types of waves). In this case the wave pattern consists of alternating regions of high and low density. When a compressional wave passes through a medium, the individual particles vibrate back and forth along the direction of the wave motion. There is no motion perpendicular to that direction, in contrast with water waves.

The theory of galactic spiral structure that invokes waves as a means of maintaining the spiral arms is called the **density wave theory.** This theory supposes that there is a spiral-shaped wave pattern centered on the galactic nucleus, creating a pinwheel shape of alternating dense and relatively empty regions (Fig. 27.6). The density waves have more affect on the interstellar medium than on stars, so the spiral arms are characterized primarily by concentrations of

FIGURE 27.6 THE DENSITY WAVE THEORY. *This drawing depicts the manner in which circular orbits are deformed into slightly elliptical ones by an outside gravitational force. The nested elliptical orbits are aligned in such a way that there are density enhancements in a spiral pattern. This pattern rotates at a steady rate, and does not wind up more tightly.*

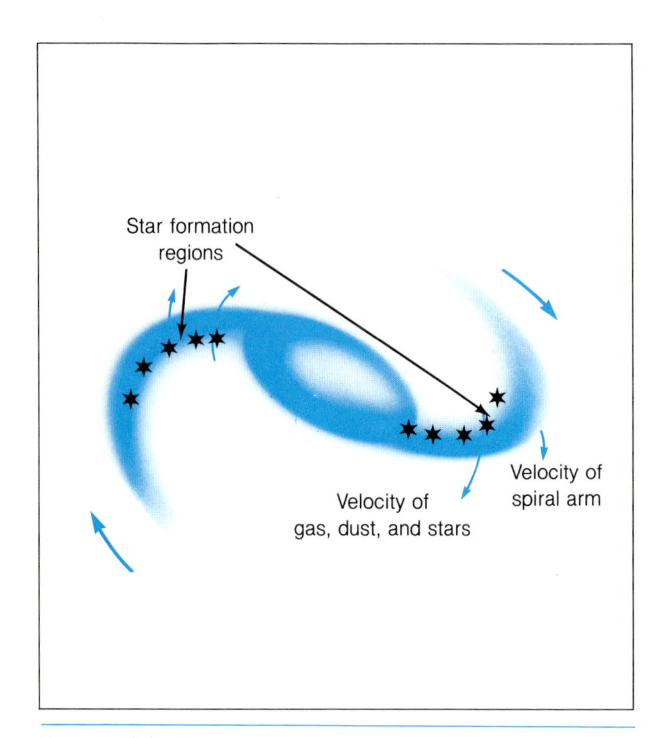

FIGURE 27.7 THE EFFECT OF A SPIRAL DENSITY WAVE ON THE INTERSTELLAR MEDIUM. *As the wave moves through the medium, material is compressed, leading to enhanced star formation. The young stars, H II regions, and nebulosity along the density wave are what we see as a spiral arm.*

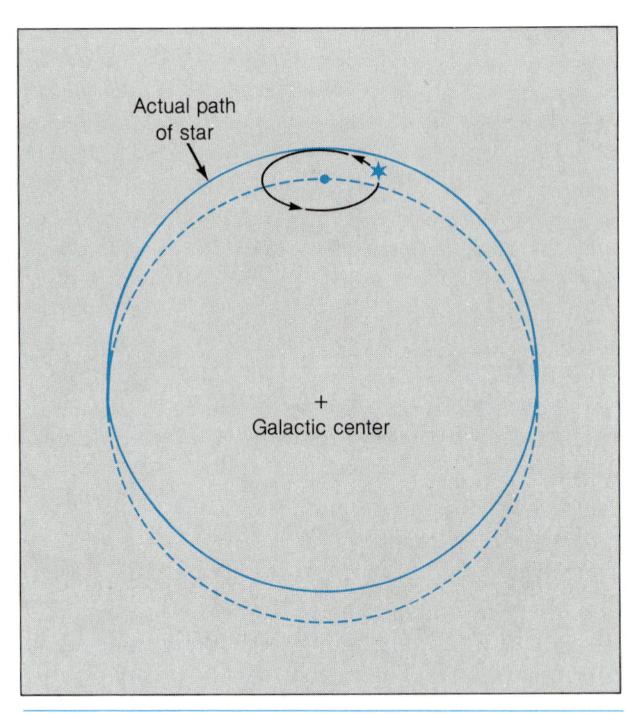

FIGURE 27.8 EPICYCLIC ORBITS. *The oscillatory motions created by spiral density waves cause individual stars to have orbits about the galaxy that are not perfect circles but oval. The shape of the orbit may be viewed as a combination of a large circle and a small circle (epicycle).* (Adapted from Shu, F. 1982, *The Physical Universe* (San Francisco: University Science Press), Fig. 12.12, p. 267.

gas and dust. These lead, in turn, to concentrations of young stars, because star formation is enhanced in regions where the interstellar material is compressed (Fig. 27.7).

In its simplest form, the wave pattern is double; that is, there are just two spiral arms emanating from the nucleus, on opposite sides. There are galaxies that have such a simple spiral structure, but many, including the Milky Way, are more complicated. The density wave theory allows the possibility of having more arms, if the waves have shorter "wavelength," or distance between them.

The waves rotate about the galaxy at a fixed rate that is constant from inner portions to the outer edge. Thus, while the outer portions of the arms appear to trail the rotation, in fact they move all the way around the galaxy in the same time period that the inner portions do. The waves are essentially rigid, in strong contrast with the motions of the stars. Just as in the case of water waves, the motion of the spiral density waves is quite distinct from the motions of individual particles (i.e., stars) in the medium through which the waves travel. In the Milky Way, individual stars orbit the galaxy at their own speed. In the inner part of the galactic disk, out to the Sun's position, the stars move faster than the density waves. This means that as a star circles the galaxy, every so often it will overtake the pass through a region of high density as it penetrates a spiral arm. Because their motion is slowed slightly when they are in a density wave, stars tend to become concentrated in the arms, just as cars on a highway become jammed together at a point where traffic flow is constricted.

The motion of an individual star about the galaxy is not precisely circular, although it is close. The spiral density waves impose slight oscillatory motions on the stars as they pass through the waves, with the result that the orbit of a star can be represented as an epicyclic motion (Fig. 27.8).

ASTRONOMICAL INSIGHT 27.2

THE MISSING POPULATION III

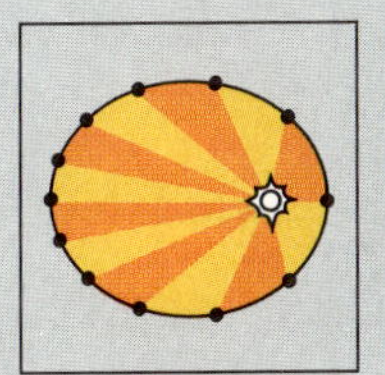

It was explained in the text that the two populations of stars should be viewed as a range of stellar characteristics such as age and metal content, rather than a strict separation of two completely distinct classes of stars. The most extreme Population II stars, those with the lowest metal abundances and the greatest ages, represent one end of the range, whereas the youngest stars in spiral arms, containing the highest metal abundances, represent the other end of the range.

The fact the extreme Population II stars contain *some* heavy elements, even though it may be only 1 percent of the metal content of Population I stars, has proven difficult to explain. In the picture of galactic formation and evolution presented in this chapter, the Population II stars are portrayed as the first stars to form from the collapsing cloud of gas that was to become the galaxy. This cloud of gas is assumed to be primordial—that is, never to have been involved in stellar formation and processing. Therefore, the cloud should have had a composition consisting almost entirely of hydrogen and helium, with *no* heavy elements. The question, then, is why the oldest stars we observe, those belonging to extreme Population II, have the metal content that they do. Even these stars must have formed from material that had been processed through stars, becoming somewhat enriched with heavy elements.

The missing population of stars formed from completely unprocessed material and having virtually zero heavy-element abundances has become known as **Population III.** Several astronomers have spent considerable time and effort in the attempt to find such stars, with no success.

No widely accepted explanation for the lack of Population III stars has been devised yet, but there have been suggestions. One is that before the pregalactic cloud collapsed, there was a brief episode of star formation and nuclear processing, in which a number of very massive, short-lived stars were formed and quickly evolved to explode in supernovae, creating heavy elements. These heavy elements then became part of the gas that collapsed to form the disk, so that stars formed during and after the collapse—the oldest stars we now observe—contained some heavy elements, and represent what we know today as Population II. None of the original Population III stars remain, because they were all too massive and short-lived to survive for long.

The chief difficulty in this and other pictures in which an early star-formation episode occurred is to explain how star formation became so active at a time when the pregalactic material was still widely dispersed in a giant, tenuous cloud. Despite our lack of understanding of *how* this occurred, it seems probable that it did, since we find no Population III stars. Furthermore, there is independent evidence that galaxies do undergo periods of intense star formation and processing; such episodes are observed in some other galaxies. Therefore, it is quite possible that our galaxy went through such a time early in its history. Several studies are under way today in an effort to understand how and why this may have occurred.

We have noted that the main reason the spiral arms stand out is that hot, luminous young stars are to be found exclusively in the arms because star formation is enhanced there, where the interstellar gas is compressed. According to the density wave theory, the most active stellar nurseries should be located on the inside edges of the arms, where stars and gas catch up with and enter the compressed region. This is difficult to check observationally for our own galaxy, but it seems to be true of others.

The most luminous stars have such short lifetimes that they evolve and die before having sufficient time to pull ahead of the arm where they were born; therefore, we find no bright, blue stars between the arms. Less luminous stars—those that live billions instead of only millions of years—can survive long enough to orbit the galaxy several times, and these stars become spread almost uniformly throughout the galactic disk, between the arms as well as in them. The Sun, for example, is old enough to have circled the galaxy some 15 to 20 times, and therefore has passed through spiral arms and the intervening gaps on several occasions. It is only coincidence that the Sun is presently in a spiral arm; it is not necessarily the one in which it formed.

The initial formation of the density wave that is the cause of the spiral arms is a separate question. A rotating disk of particles, such as the disk of a galaxy like ours, will naturally form spiral density waves if disturbed by an outside force (Fig. 27.9). This sug-

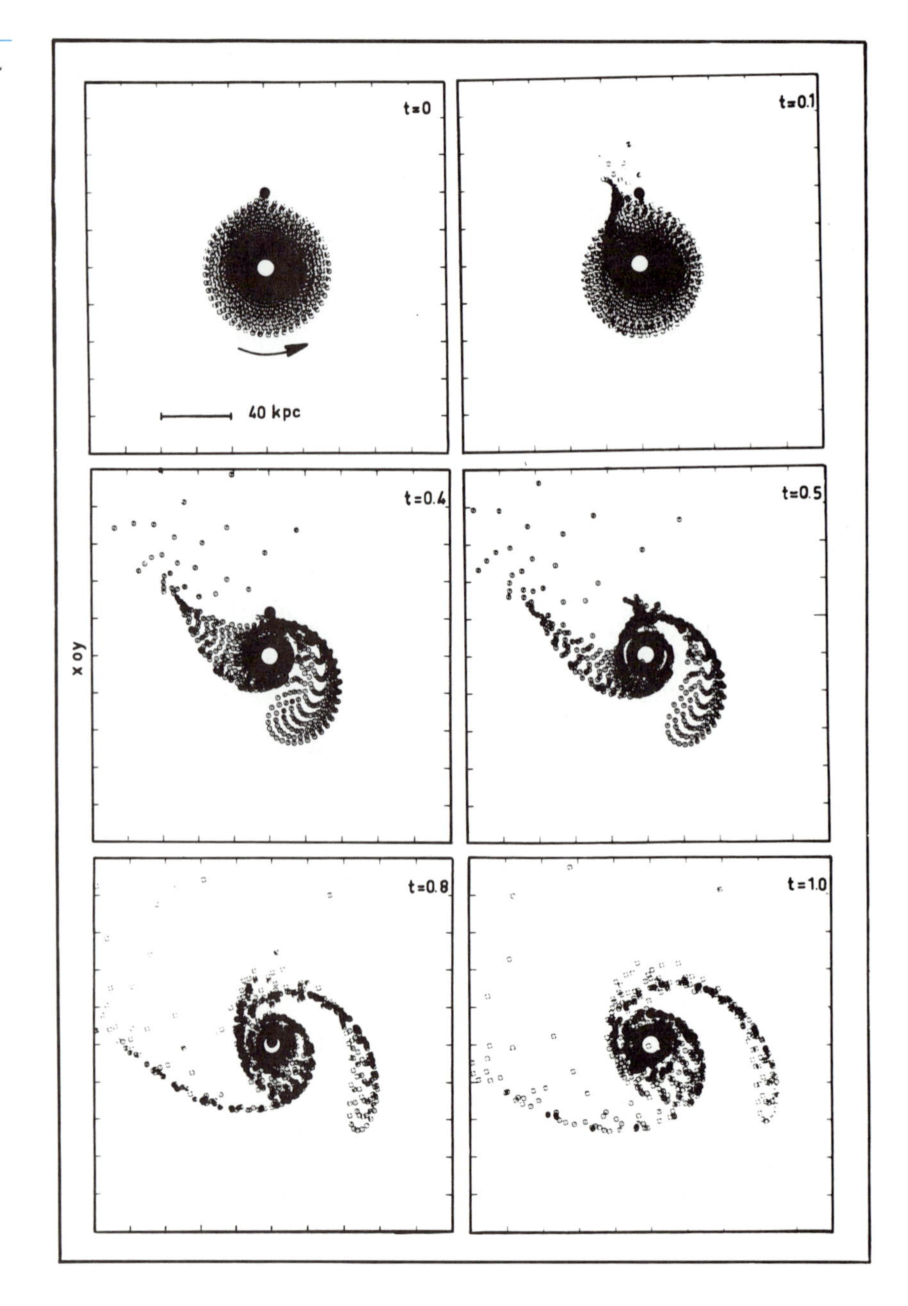

FIGURE 27.9 THE CREATION OF SPIRAL STRUCTURE. *This sequence of computer-generated models shows how a rotating disk forms spiral density waves when subjected to a gravitational force.* (After T. M. Eneev, N. H. Kuzlov, and R. A. Sunyaev, 1973, *Astronomy and Astrophysics* 22:41)

gests that the spiral density waves in our galaxy were triggered by the gravitational influence of a neighbor galaxy, most likely the Magellanic Clouds, a pair of small galaxies that orbit the Milky Way.

Another possible mechanism for creating spiral density waves, with no help from neighboring galaxies, occurs in certain galaxies that have noncircular central bulges. In these **barred spiral galaxies** (see Chapter 28), the asymmetric shape of the central region can create the gravitational disturbance needed to initiate spiral density waves. Some calculations suggest that the barred spiral structure is the more natural form for a disklike galaxy to take, and that the presence of a massive, extended halo is what prevented our own galaxy from becoming a barred spiral instead of the "normal" spiral it is. If so, this suggests that other normal spiral galaxies may also have massive halos, whereas the barred spirals may not. This will be discussed further in the next chapter.

GALACTIC HISTORY

We are now in a position to tie together all the diverse information on the nature of the Milky Way, and to develop from that a picture of its formation and evolution. There are still many important uncertainties in this picture, however, and some of these will be pointed out.

The pertinent facts that must be explained include the size and shape of the galaxy, its rotation, the distribution of interstellar material, elemental abundance gradients, and the dichotomy between Pop. I and Pop. II stars in terms of composition, distribution, and motions. The task of fitting all the pieces of the puzzle together is made easier by the fact that we can reconstruct the time sequence, knowing that stars with high abundances of heavy elements were formed more recently as a rule than those with lower abundances.

The oldest objects in the galaxy are in the halo, which is dominated by the globular clusters but contains a large (but unknown) number of isolated, dim, red stars as well. Estimates based on the main sequence turnoff in the H-R diagrams for globular clusters indicate ages of between 14 and 16 billion years, and we conclude that the age of the galaxy itself is comparable.

The spherical distribution of the halo objects about the galactic center demonstrates that when they formed, the galaxy itself was round. Evidently the progenitor of the Milky Way was a gigantic spherical gas cloud, consisting almost exclusively of hydrogen and helium. Very early, perhaps even before the cloud began to contract (Fig. 27.10), the first stars and globular clusters formed in regions where localized condensations occurred. These stars contained few heavy elements and were distributed throughout a spherical volume, with randomly oriented orbits about the galactic center. Eventually the entire cloud began to fall in on itself, and as it did so, star formation continued to occur, so that many stars were born with motions directed towards the galactic center. These stars assumed highly eccentric orbits, accounting for the motions observed today in Pop. II stars.

Apparently the pregalactic cloud was originally rotating, for we know that as the collapse proceeded, a disklike shape resulted. Rotation forced this to happen, just as it did in the contracting cloud that was to form the solar system (see Chapter 17). Rotational forces slowed the contraction in the equatorial plane, but not in the polar regions, so material continued to fall in there. The result was a highly flattened disk. Stars that had already formed before the disk took shape retained the orbits in which they were born; stars are unaffected by the fluid forces (e.g., viscosity) that caused the gas to continue collapsing to form a disk.

Stars that were formed after the disk had developed had characteristics unlike those of their predecessors in at least two respects: They contained greater abundances of heavy elements, because by this time some stellar cycling had occurred, enriching the interstellar gas; and they were born with circular orbits lying in the plane of the disk. These are the primary traits of Pop. I stars.

Since the time of the formation of the disk, there have been additional, but relatively gradual, changes. Further generations of stars have lived and died, continuing the chemical enrichment process, particularly in the inner regions of the disk, creating the chemical abundance gradient mentioned earlier. Apparently the enrichment process was once more rapid than it is today, because the present rate of star formation is too slow to have built up the quantities of heavy elements that are observed in Pop. I stars. At some time in the past, probably just when the disk was

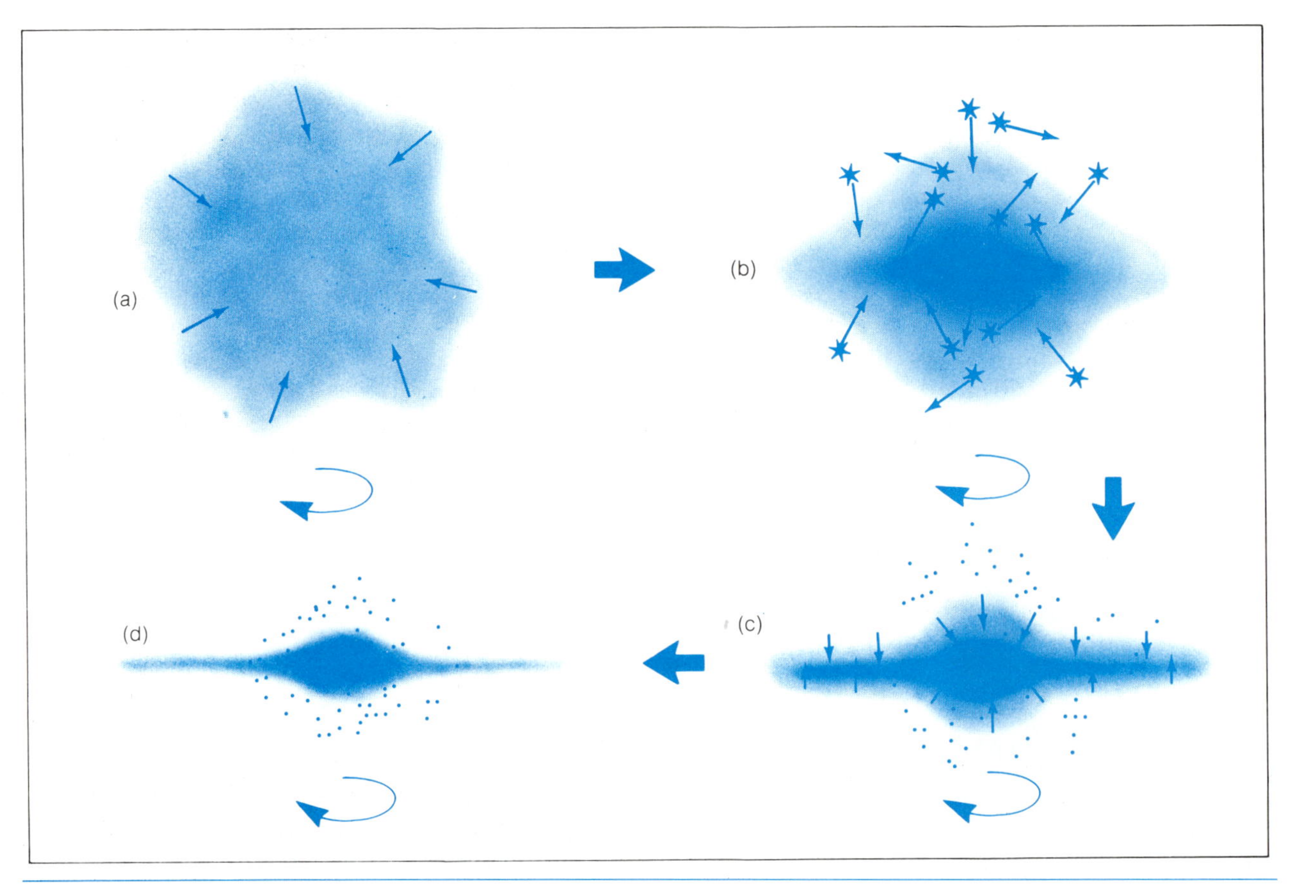

FIGURE 27.10 FORMATION OF THE GALAXY. *The pregalactic cloud, rotating slowly, begins to collapse (a). The stars formed before and during the collapse have a spherical distribution and noncircular, randomly oriented orbits (b). The collapse leads to a disk with a large central bulge, surrounded by a spherical halo of old stars and clusters (c). The disk flattens further, and eventually forms spiral arms (d). Stars in the disk have relatively high heavy element abundances because they formed from material that had been through stellar nuclear processing.*

forming, there must have been a period of intense star formation, during which the abundances of heavy elements in the galaxy jumped from almost none to nearly the present level. A large fraction of the galactic mass must have been cooked in stellar interiors and returned to space in a brief episode of stellar cycling whose intensity has not been matched since. (Today some galaxies are observed that seem to be in such a state of rapid stellar processing; these are called **starburst galaxies,** and they are discussed in Chapter 28.)

We have just about recounted all the events that led up to the present-day Milky Way, except for the formation of the spiral arms. It is not known when this took place, but it was probably soon after the formation of the disk itself. We see relatively few disk-shaped galaxies without spiral structure, which leads to the conclusion that a disk galaxy does not exist long in an armless state.

It is difficult to guess exactly what may become of the galaxy in the future. It appears that some sort of balance has been reached in the grand recycling process between stars and the interstellar material, so that the interstellar medium is being replenished by evolving and dying stars about as rapidly as it is being consumed by stellar births. Therefore we do not expect the interstellar material to gradually disappear, terminating star formation, as it apparently has in certain other types of galaxies.

PERSPECTIVE

We have now seen our galaxy in a new light, as an active, dynamic entity. Individual stars orbit its center in individual paths, yet the overall machinery is systematically organized. The majestic spiral arms rotate at a stately pace, while stars and interstellar matter pass through them. We have come to understand the active processes of stellar cycling and chemical enrichment, which still occur today. In a constant turnover of stellar generations, the deaths of the old give rise to the births of the new, while the violence of their death throes energizes a chaotic interstellar medium.

The lessons we have learned by examining our own Milky Way will be remembered as we move out into the void, probing the distant galaxies.

SUMMARY

1. Stars in the galactic halo have low abundances of heavy elements and are referred to as Population II stars, whereas those in the spiral arms of the disk have "normal" compositions and are called Population I stars.
2. There are gradients of increasing heavy-element abundance from halo to disk, and from the outer to the inner portions within the disk.
3. Population II stars have randomly oriented, highly eccentric orbits, whereas Population I stars have nearly circular orbits that lie in the plane of the disk. Population II stars, when passing through the disk near the Sun's location, are seen as high-velocity stars.
4. The variations in heavy-element abundance from place to place within the galaxy reflect variations in age: Very old stars, formed before stellar nuclear reactions had produced a significant quantity of heavy elements, have low abundances of these elements, whereas younger stars were formed after the galactic composition had been enriched by stellar evolution.
5. Spiral arms are probably density enhancements produced by spiral density waves, which rotate about the galaxy while stars and interstellar material pass through them. Because the interstellar gas is compressed in these density waves, star formation tends to occur there, and this in turn explains why young stars are found predominantly in the spiral arms.
6. The existence and characteristics of Population I and II stars can be explained in a picture of galactic evolution that begins with a spherical cloud of gas that has little or no heavy elements at first. Population II stars are those that formed while the cloud was still spherical or just beginning to collapse, but when it still had no significant quantities of heavy elements. Population I stars are those that formed later, after the galaxy had collapsed to a disk, and after stellar evolution had produced some heavy elements.
7. The spiral arms formed after the disk had been created by the collapse of the original cloud. The spiral density waves that maintain the arms probably started as a result of gravitational disturbances, possibly from nearby galaxies.

REVIEW QUESTIONS

Thought Questions

1. Summarize the properties of Population I and Population II objects in the galaxy.
2. Explain in your own words what a gradient is.
3. Explain why the Sun is thought to be a Population I object.
4. How are high-velocity stars and comets similar?
5. How are the spiral arms of the galaxy similar to the rings of Saturn?
6. Explain the distinction between the orbital periods of stars in the galaxy and the rotation period of the spiral arms.
7. Why do stars form primarily in the spiral arms of a galaxy like ours?
8. Why do the globular clusters have the lowest heavy-element abundances of any objects in the galaxy?
9. Explain why the evolution of massive stars, rather than that of the much more common low-mass stars, has been the principal contributor to the enrichment of heavy elements in the galaxy.
10. Summarize the various roles played by supernovae in the evolution of the galaxy.

Problems

1. Suppose the spectrum of a star is analyzed, and iron is found to be present in its outer layers in the

ratio of 1:40,000 compared to hydrogen. Is this a Population I or a Population II star? Explain how you decided.

2. The strong line of ionized calcium, whose rest wavelength is 3933.8 Å, is observed at a wavelength of 3932.0 Å in the spectrum of a star. To which population does this star belong? Explain how you decided.

3. The Sun moves faster in its orbit than the rotation speed of the galaxy's spiral arms, so the Sun repeatedly overtakes any given arm. If the Sun passes through the arm it is now in every 5×10^8 years, how many times has it done so in the 15-billion-year lifetime of the galaxy?

4. If the mass of the galaxy is 1.5×10^{11} solar masses, what is the orbital period of a globular cluster whose semimajor axis is 20 kpc? (Remember to convert this into astronomical units.)

ADDITIONAL READINGS

Arp, H. 1969. On the origin of arms in spiral galaxies. *Sky and Telescope,* Dec., p. 385.

Bok, B. J. 1981. Our bigger and better galaxy. *Mercury* 10(5):130.

Bok, B. J., and P. Bok. 1981. *The Milky Way.* Cambridge, Ma.: Harvard University Press.

Burbidge, G., and E. M. Burbidge. 1958. Stellar populations. *Scientific American* 199(2):44.

Hodge, Paul. 1986. *Galaxies.* Cambridge, Mass.: Harvard University Press.

Iben, I. 1970. Globular cluster stars. *Scientific American* 223(1):26.

Larson, R. B. 1979. The formation of galaxies. *Mercury* 8(3):53.

Mathewson, Don. 1985. The clouds of Magellan. *Scientific American* 252(4):106.

Scoville, N., and J. S. Young. 1984. Molecular clouds, star formation, and galactic structure. *Scientific American* 250(4):42.

Shu, F. H. 1973. Spiral structure, dust clouds, and star formation. *American Scientist* 61:524.

———. 1982. *The physical universe.* San Francisco: W. H. Freeman.

Silk, Joseph L. 1986. Formation of the galaxies. *Sky and Telescope* 72(6):582.

Weaver, H. 1975 and 1976. Steps towards understanding the spiral structure of the Milky Way. *Mercury* 4(5):18 (Pt. I); 4(6):18 (Pt. II); 5(1):19 (Pt. III).

GUEST EDITORIAL

SUPERNOVAE AND THE GALAXY

Robert Kirshner

Robert Kirshner is a Professor of Astronomy at Harvard University. His work has centered on supernovae and supernova remnants and on measuring the large-scale structure in the universe. As a graduate student at CalTech in 1972, he was working on supernova spectra when SN 1972e, the brightest supernova since 1937, was discovered. His extensive observations of that object helped set the direction of his future work, at the University of Michigan and then at Harvard. His observations of supernovae and young supernova remnants have been aimed at testing the picture of stellar evolution and nucleosynthesis in stars. His work on old supernova remnants has helped measure the properties of the interstellar gas. In between supernovae, he has helped conduct a deep survey of the three-dimensional distribution of galaxies that revealed the largest empty region: the giant void in Boötes (this is discussed in Chapter 29). In this article, Dr. Kirshner discusses some of the implications of Supernova 1987a, the spectacular event that has recently occurred in the Large Megellanic Cloud.

Human lives are so short compared to the lives of stars or the evolution of galaxies that we cannot expect to be witnesses to the violent and sudden phenomena that shape our galaxy, even if they are repeated a hundred million times. Supernovae are stellar explosions which destroy stars, hurl new elements into the interstellar gas, blast giant cavities in the gas of our galaxy, and shine with the light of a billion stars, but human life is so short that not one supernova has been visible to the naked eye in the last ten generations. Until February 23, 1987, when Oscar Duhalde, a telescope operator at Las Campanas Observatory, in Chile, noticed something odd in the Large Megallanic Cloud, a satellite galaxy of our own Milky Way, just 150,000 light years away. Duhalde's suspicions were forgotten in that night's routine until Ian Shelton, a University of Toronto astronomer, came over with the exciting news that he had photographed a new star in the Large Magellanic Cloud. Then Oscar Duhalde knew that his had been the first eyes since 1604 to glimpse the violent death of a star as a supernova.

Supernova 1987A presents a brilliant spectacle, poses a challenging set of scientific riddles, and reminds us of the central role that supernovae play in the development of a galaxy. Most stars are robustly balanced between gravitation and pressure, but supernovae result from the disaster that follows a failure of that balance. The core of a massive star collapses, emitting a giant pulse of neutrinos. The center implodes to become a neutron star; the mantle is heated and synthesizes new, heavy elements out of the light ones; and the atmosphere is blasted into the interstellar gas. The massive stars (about 19 solar masses for SN 1987A) that become supernovae in this way (the Type II supernovae) lead relatively short lives: they explode where they were formed, dying in the stellar nursery after only ten million years. The elements they produce are disgorged into the interstellar gas. The build-up of heavy elements in our own galaxy, leading to stars as rich in carbon and calcium and iron as our own Sun, is the cumulative effect of generations of stars which lived and died before the Sun and its planets condensed. Earth, and more specifically this book and its reader, are made of carbon and oxygen and calcium and iron which literally were once in the interior of a supernova. The oldest stars of our galaxy did not inherit such a rich legacy. Born too soon, they could not benefit from the enrichment of the supernovae that were to follow, and they are 100 times poorer in the elements beyond helium.

GUEST EDITORIAL

The violent disruption of a star as a supernova leads to a brilliant display of light, but 100 times more energy goes into the motion of the stellar debris. Dropping a rock in a pond makes a ripple, but detonating a supernova makes a violent shock that spreads supersonically through the interstellar gas for tens of parsecs. The shock heats the gas to temperatures of ten million degrees, after which it emits X rays and cools slowly. The shock scorches interstellar dust, vaporizing the solid matter of interstellar grains. It heats interstellar clouds so that they glow as beautiful filamentary supernova remnants. It accelerates some atoms of the interstellar gas to nearly the speed of light, generating cosmic rays and causing the supernova remnant to be a source of radio emission. Finally, after 100,000 years, the supernova shock wave spreads and weakens. But before it fragments and merges completely with the interstellar gas, a supernova's pressure wave may nudge an interstellar cloud over the edge of instability, helping it collapse and form a new cluster of stars. Most of those stars will have low masses, like that of the Sun, and long lives. But a few may be more massive, the kind that burn quickly to explode ten million years later as supernovae. And, if the interstellar dust is not too thick, and you are born at the right time, you may get to see the closing of this cosmic loop as a supernova, from a star whose formation was triggered by an earlier supernova, blazes its light. The light from hundreds of supernovae that have exploded over the last 30,000 years is traversing the galaxy right now. Perhaps, like Oscar Duhalde, you will look up one night, and your eyes, made of stuff from ancient supernovae, will see the light from the next supernova in the Milky Way.

EXTRAGALACTIC ASTRONOMY

We are ready now to move out of the confines of our own galaxy, to explore the universe beyond. The Milky Way is just one of billions of galaxies in the cosmos, and we will find that although our own system is typical of a certain class of these objects, there is a wide variety of shapes, sizes, and peculiarities associated with other galaxies.

The first chapter of this section describes both the history of the discovery of galaxies beyond the Milky Way and the observational properties of these galaxies. We rely here on what was learned in the previous section on our own galaxy, for many aspects of its structure and evolution are characteristic of other galaxies, particularly spirals. We do not discuss the evolution of individual galaxies in great detail; we assume that the forces that have shaped the Milky Way are at work in other situations as well.

Chapter 29 begins to bridge the gap between our discussion of individual objects (that is, galaxies) and the structure and organization of the universe as a whole. In this chapter we examine the distribution of galaxies, paying particular attention to groupings and clusters and the possibility that these are in turn organized into larger associations. In the process we learn that galaxies in clusters interact with each other in ways that modify their individual properties, and they also influence the nature of the clusters in which they reside. One question addressed in this chapter, which is of critical importance to our later discussions of the universe as a whole, has to do with the uniformity of the distribution of matter. The implications of this question are not discussed until Chapter 32, but the observational evidence is cited in chapters 29 and 30.

Chapter 30 describes two major aspects of the universe, and for the time being, we direct our attention away from individual objects and focus on the universe itself. The two profound observational discoveries are the expansion of the universe and the cosmic background radiation that fills it. Both are legacies of the fiery origin of the universe, and their observed properties teach us something of its early history. Here we see clearly for the first time that the universe itself is a dynamic, evolving entity. It had a beginning, has been changing since, and will continue to develop.

With the perspective gained from Chapter 30, in the next chapter we turn our attention back to individual objects, ones whose properties can best be appreciated in the context of an expanding universe. A variety of peculiar galaxies, and especially the quasistellar objects, have fantastic properties that were only discovered by astronomers because the nature of the universal expansion was already known. These objects, on the frontiers of the observable universe, can potentially tell us a great deal about its history. An important theme here is the realization that very distant objects are only seen as they were long ago, because of the light-travel time. This is useful because it allows us to probe the early history of the universe, but is also a hindrance because of the difficulty of comparing distant objects with nearby ones. The quasistellar objects are a fundamental component of the early universe, and therefore the mysteries they present for astronomers are among the most important now under study.

Finally, the last chapter of this section describes the current state of cosmology, the science of the universe as a whole. Here we tie together all the diverse information gained from the preceding chapters and assess the overall structure of the universe, and we especially emphasize the question of its future. If there is a single premier challenge to modern astronomy, it is this, and we are on the verge of knowing the answer. Having completed this section, we will be nearly finished with our introduction to astronomy.

CHAPTER 28

THE NATURE OF THE NEBULAE

A false-color image of a barred spiral galaxy. (Royal Greenwich Observatory, sponsored by the Science and Engineering Research Council, United Kingdom)

CHAPTER PREVIEW

The Hubble Classification System
Galactic Distances
The Masses of Galaxies
Luminosities, Colors, and Diameters
The Origins of Spirals and Ellipticals

We have spoken of our galaxy as one of many, a single member of a vast population that fills the universe. Given all that we have learned about our ordinary position in the cosmos, this is no surprise. Having traced our painful progression from the geocentric view to the realization that we occupy an insignificant planet orbiting an ordinary star in an obscure corner of the galaxy, we should be surprised if we found that our galaxy held any kind of unique status in the larger environment of the universe as a whole. It doesn't.

Despite the seeming inevitability of this idea, the actual proof that our galaxy is not alone was some time in coming, arriving only after considerable debate and controversy. The so-called **nebulae,** dim, fuzzy objects scattered throughout the sky, have been known since the early days of astronomical photography, but it was not until the mid-1920s that they were demonstrated to be galaxies, rather than more nearby objects such as gas clouds or star clusters. The proof of their galactic nature was announced in 1924, when the American astronomer Edwin Hubble (Fig. 28.1) reported that he had found Cepheid variables in the prominent Andromeda galaxy (until then known as the Andromeda nebula, since its true nature was not known), and used the period-luminosity relation to show that this nebula was entirely too distant to be within our own galaxy.

FIGURE 28.1 EDWIN HUBBLE. *Hubble's discovery of Cepheid variables in the Andromeda nebula led to the unambiguous conclusion that this object lies well beyond the limits of the Milky Way and must therefore be a separate galaxy. Hubble later made important discoveries about the properties of galaxies and what they tell us about the universe as a whole (see Chapter 30).* (Mt. Wilson and Las Campanas Observatories, Carnegie Institution of Washington)

FIGURE 28.2 AN ELLIPTICAL GALAXY. *These smooth, featureless galaxies are probably more common than spirals.* (National Optical Astronomy Observatories)

FIGURE 28.3 A SPIRAL GALAXY. *At left is NGC 4321, a beautiful example of a face-on spiral galaxy. During the early years of the twentieth century, there was considerable controversy among astronomers over the true nature of objects like this. The view at right shows how thin the disk of a spiral galaxy is, relative to its diameter.* (*left:* J. D. Wray, University of Texas; right: U.S. Naval Observatory)

THE HUBBLE CLASSIFICATION SYSTEM

Having established that nebulae are truly extragalactic objects, Hubble began a systematic study of their properties. The most obvious basis for establishing patterns among the various types was to categorize them according to shape. Hubble did this, designating the spheroidal nebulae **elliptical galaxies** (Fig. 28.2), as distinct from the **spiral galaxies** (Fig. 28.3). Within each of the two general types, Hubble established subcategories on the basis of less dramatic gradations in appearance (Table 28.1). The ellipticals displayed varying degrees of flattening and were sorted out ac-

TABLE 28.1 TYPES OF GALAXIES

Type	Designation	Characteristics
Elliptical	E0–E7	Spheroidal shape; subtype determined by the formula $10(1-b/a)$, where a is the long axis and b is the short axis of the galaxy.
Dwarf elliptical	dw E	Spheroidal shape; very low mass and luminosity.
S0	S0	Disklike galaxies with no spiral structure.
Spiral	Sa–Sc	Disklike galaxies with spiral arms; subtype determined by relative size of nucleus and openness of spiral structure; Sa refers to large nucleus and tightly wound arms, whereas Sc refers to small nucleus and open spiral arms.
Barred spiral	SBa–SBc	Spiral galaxies with elongated, barlike nuclei; subtypes determined the same way as in spiral galaxies.
Irregular I	Ir I	Disklike galaxies with evidence of spiral structure, but not well organized.
Irregular II	Ir II	Galaxies that do not fit into any of the other types; some Ir II galaxies have been found to be normal spirals heavily obscured by interstellar gas and dust.

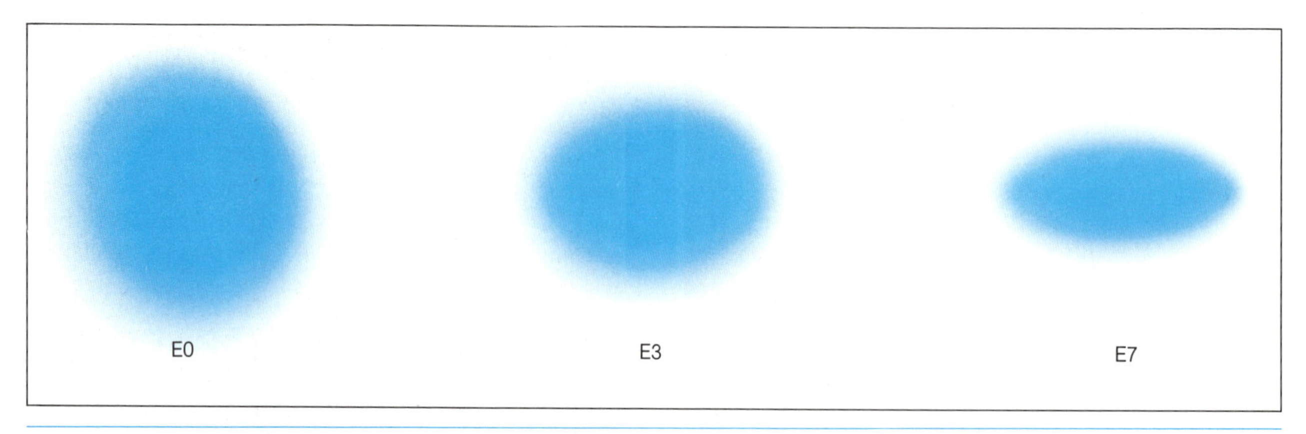

FIGURE 28.4 THE SHAPES OF ELLIPTICALS. *The numerical designation (following the letter E) is given by the formula* (1 − b/a) × 10, *where* b *is the short axis and* a *the long axis of the galaxy image. Here are three examples: the E0 galaxy has* a = b *(that is, it is circular); the E3 galaxy has* a = 1.43b; *and the E7 has* a = 3.3b. *The E7 galaxy is the most highly elongated of the ellipticals.*

FIGURE 28.5 SPIRAL GALAXIES. *Here are four typical examples of spiral galaxies. At top left is M64, top right is M66 (these designations refer to the Messier Catalog, see Appendix 15); at bottom left is NGC6946, bottom right is NGC4321 (these designations are from the* New General Catalog). (*top left, top right, and bottom left:* U.S. Naval Observatory; *bottom right:* J. D. Wray, University of Texas)

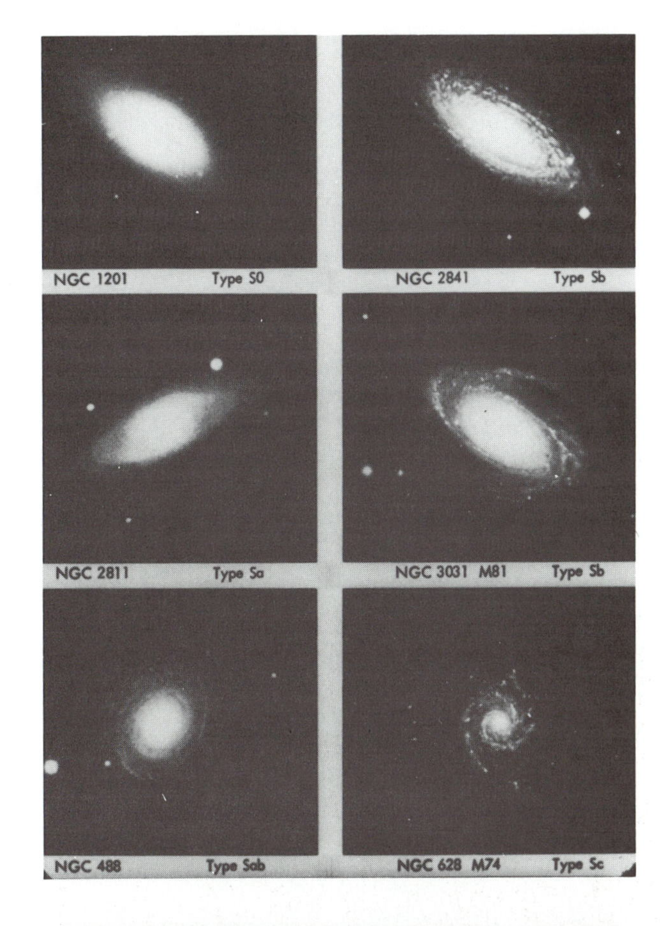

FIGURE 28.6 AN ASSORTMENT OF SPIRALS. *This sequence shows spiral galaxies of several subclasses.* (Palomar Observatory, California Institute of Technology)

cording to the ratio of the long axis to the short axis, with designations from E0 (spherical) to E7 (the most highly flattened). The number following the letter *E* is determined from the formula $10(1-b/a)$, where a is the long axis and b is the short axis, as measured on a photograph (Fig. 28.4).

Hubble's classification of the spirals was based on the tightness of the arms and the compactness of the nucleus (Figs. 28.5 and 28.6). The types ranged from Sa (tight spiral, large nucleus) to Sc (open arms, small nucleus). The Milky Way is probably an Sc in this system, although it is difficult to determine, because we cannot get an outsider's view of our galaxy. Some have classified the Milky Way as an Sb galaxy.

Hubble recognized a variation of the spiral galaxies in which the nucleus has extensions on opposing sides, with the spiral arms emanating from the ends of the extensions. These he called **barred spirals,** and assigned them the designations SBa through SBc, using the same criteria as before in establishing the a, b, and c subclasses (Figs. 28.7 and 28.8). Following

FIGURE 28.7 BARRED SPIRALS. *At left is M83, a well known example of a spiral galaxy with a barlike structure through the nucleus; at right a computer-enhanced view of another barred spiral.* (*left:* © 1980 Anglo-Australian Telescope Board; *right:* photo by H. C. Arp, image processing by J. J. Lorre, Image Processing Laboratory, JPL)

FIGURE 28.8 BARRED SPIRALS. *Roughly half of all spiral galaxies have central elongations, or bars, from which the spiral arms emanate.* (Palomar Observatory, California Institute of Technology)

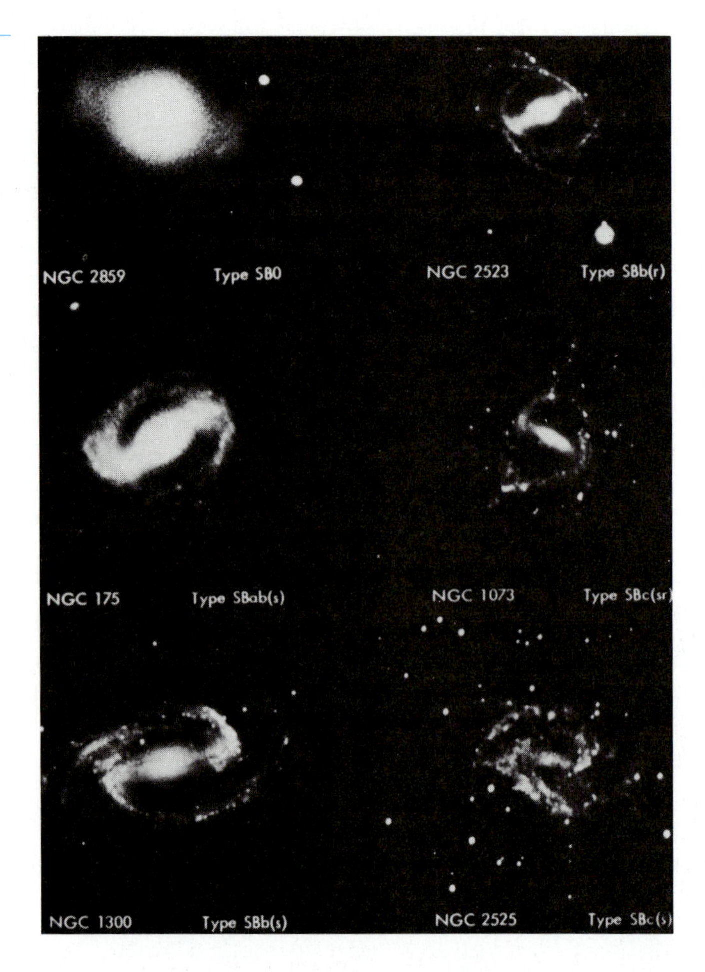

Hubble's original work on galaxy classification, astronomers recognized an intermediate class called the S0 galaxies. These appear to have a disk shape, but no trace of spiral arms.

Although Hubble himself did not draw an evolutionary connection among the various types of galaxies, he assembled an organization chart that was for a time thought by many astronomers to represent an age sequence. Because there are two types of spirals, Hubble chose not to force all the types into a single sequence, but instead split the sequence into two branches, and the result became known as the **tuning fork diagram** (Fig. 28.9). It was the prevailing belief that elliptical galaxies evolve into spirals, the originally spherical distribution of gas gradually flattening into a disk (as in S0 galaxies) and then developing spiral arms. This view was later reversed, after the discovery of stellar populations (see Chapter 27). Since elliptical galaxies consist primarily of old stars, with little interstellar matter and therefore little active star formation, and since spirals contain prominent young stars and interstellar gas and dust, it seemed reasonable to assume that spirals evolve into ellipticals. This was the accepted view as recently as the early 1960s. It was eventually found, however, that even spiral galaxies contain old populations of stars and clusters and are just as old as the ellipticals. The tuning fork diagram is not an age sequence after all, and the differences in galactic type must be explained in some other way.

Most of the galaxies listed in catalogues are spirals, which are about evenly divided between normal and barred spirals. Only about 15 percent of the listed galaxies are ellipticals, and a comparable number are

ASTRONOMICAL INSIGHT 28.1

The Shapley-Curtis Debate

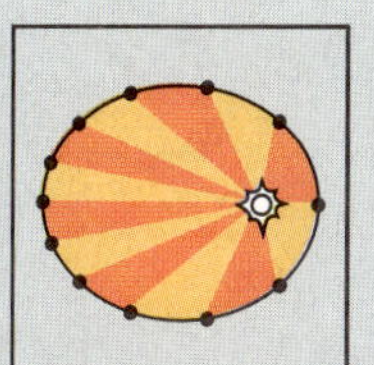

Formal debate traditionally has been a respected means of determining right in cases of dispute. Debating is a time-honored practice, undertaken by some for entertainment, and by others for more serious reasons. Our legal system, for example, is based on something akin to debates that take place in court rooms, with a judge or jury to decide who is right. Occasionally, scientific questions are argued in the format of a debate, the judge being the test of time and the acquisition of new data to test the disputed theories, the jury being the community of scientists. A famous debate on the nature of the so-called nebulae was one of the singular events in early twentieth-century astronomy.

At the opening of this century, it was not known that there were galaxies beyond our own. (In fact, as we saw in Chapter 25, it was not even known where we were located in our own galaxy, nor was there any good information on our galaxy's size and shape.) By 1920, large numbers of indistinct, hazy objects had been found sprinkled throughout the sky. Some astronomers believed these to be other galaxies; others thought they were clouds of gas within our own galaxy. Both sides marshalled evidence to support their views.

Adherents of the "island universe" hypothesis, the idea that the so-called nebulae were distant galaxies like our own, tried to show that the nebulae lay outside of our galaxy. In doing so, they argued that our galaxy was smaller than thought by some, and this in turn hinged on disagreements among astronomers over distance scales, particularly those based on pulsating variable stars in clusters. The distribution of nebulae was mapped and found to avoid the plane of the Milky Way, which indicated to the proponents of the island universe hypothesis that the nebulae lay far outside our galaxy, so that the galactic dust layer could block our view of the ones lying near the plane of the disk. Observations of novae in other galaxies (particularly our neighbor, the Andromeda galaxy) were used to show that they must be outside of our galaxy, if they were as luminous as novae seen within the Milky Way.

Proponents of the local hypothesis for the nebulae argued that the Milky Way was large enough to include them. This is where the major disagreement on distance scales came into the argument, one side believing the galaxy to be some ten times larger than did those on the other side of the issue. The problem was exacerbated by the assumption (on both sides) that there was no significant interstellar absorption, so that distances could be estimated simply from apparent magnitudes. Some "novae" were seen in other galaxies that, if comparable in luminosity to novae in our galaxy, would have to be close by, within the galaxy. (These were later identified as supernovae, far more luminous than ordinary novae.) The most telling argument favoring the local hypothesis was the observation, later shown to be incorrect, that some of the spiral nebulae were actually seen to be rotating. An object as large as the Milky Way would have to be rotating faster than the speed of light in order for the rotation to be observed over a time of only several months, so this was taken as proof that the nebulae could not be as large as galaxies, and therefore must be close by.

The chief proponents of the opposing points of view met in formal debate before the National Academy of Sciences, in Washington, D.C., on April 26, 1920. Representing the island universe hypothesis was H. D. Curtis, who had amassed most of the arguments cited above; opposing him was the formidable Harlow Shapley, who was already famous for his work on the size scale of the galaxy (see Chapter 25). The first part of the debate was devoted wholly to a discussion of the size of the Milky Way, and the second part of the debate was on the nature of the nebulae.

Shapley's arguments were considered more persuasive, and in fact he was more nearly correct on the question of the size scale of the galaxy (except for the omission of interstellar absorption). The judgment of time and the clues of new data found later by the jury ended up showing that Curtis was correct on the question of the nature of the nebulae, however. His arguments about the distribution of the nebulae, the existence of a class of "super" novae, and the invalidity of the rotation measurements for spiral nebulae were all correct, but all needed time and new information to be substantiated.

Today formal debate is rarely used as a tool for investigating scientific theories, but nevertheless the principles of debate often come into play. Usually the forum is the pages of a technical journal, where scientists publish their viewpoints and the evidence to support them. Verbal jousts often occur at scientific conferences, so the spirit of face-to-face debate is sometimes resurrected as well. Through these exchanges progress toward better understanding is made.

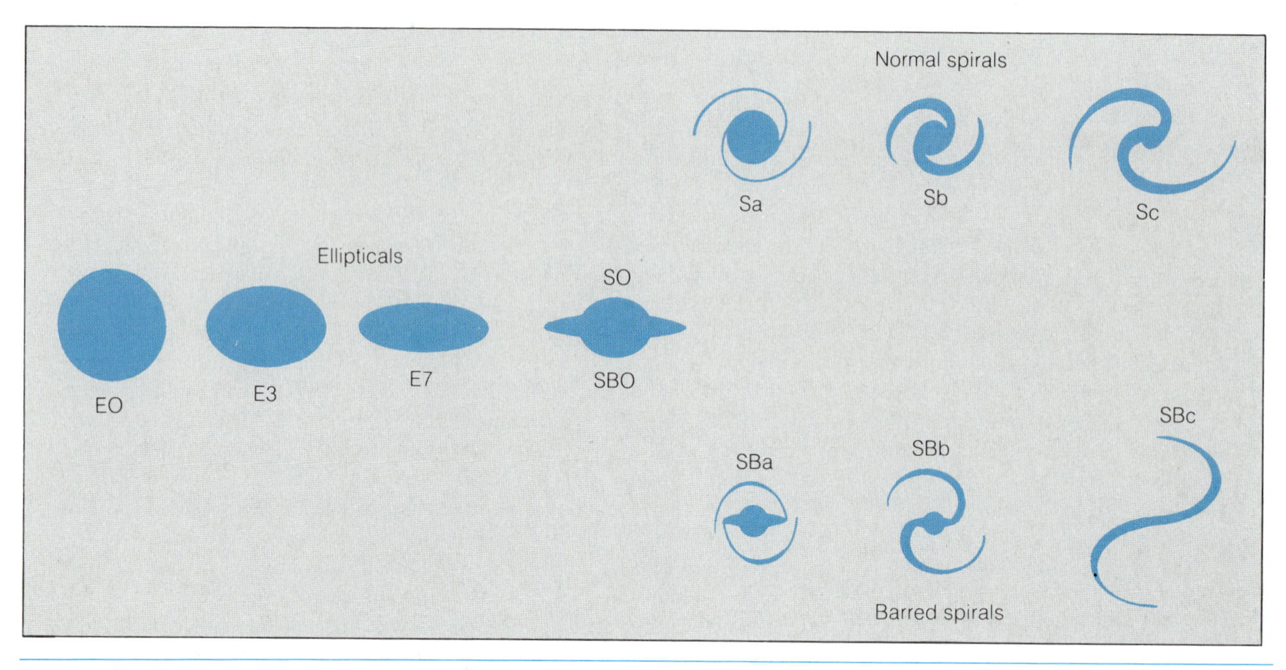

FIGURE 28.9 THE TUNING-FORK DIAGRAM. *This is the traditional manner of displaying the galaxy types originally devised by Hubble. For quite some time this was thought to be an evolutionary sequence, although the imagined direction of evolution was reversed at least once. Now it is known that in the normal course of events galaxies do not evolve from one type to another.*

S0 galaxies. The remaining few are called irregular galaxies, because they do not fit into the normal classification scheme. Apparently there are many small, dim, elliptical galaxies that are usually not sufficiently prominent to appear in catalogues, and it seems likely that ellipticals actually outnumber spirals in the universe. This certainly is the case in dense clusters of galaxies, as we will see. Table 28.2 lists a few prominent galaxies whose properties indicate the typical values for galaxies of different types.

Although the irregular galaxies are, by definition, misfits, it has proved possible to find some systematic characteristics even in their case. Most have a hint of spiral structure, although they lack a clear overall pattern, and these have been designated as **Type I Irregulars.** The rest, a small group, simply do not conform in any way to the normal standards, and are assigned the classification of **Type II Irregulars,** or peculiar, galaxies (Fig. 28.10). The Magellanic Clouds (Fig. 28.11) are irregulars of Type I.

Some of the Type II irregulars have the appearance of pairs of galaxies in collision (Fig. 28.12). In these cases, very unusual overall shapes are seen, but careful examination reveals hints of two galaxies in the process of merging, or at least passing through each other (Fig. 28.13). Since galaxies are mostly empty space, in such a collision few direct impacts of stars occur. The galaxies can pass through one another like ghosts and go their separate ways, except that their gravitational effects cause major distortions of their shapes. Even the Milky Way suffers from such an effect; the gravitational forces of the Magellanic Clouds have caused a slight warping of the galactic disk. As we will see in the next chapter, collisions among galaxies, particularly in clusters, can have very important effects on the evolution of both the individual galaxies and the clusters in which they are found.

When it became possible to observe galaxies in wavelengths other than visible light, new kinds of galaxies that do not fit the standard classification scheme were discovered. Infrared sky maps in particular revealed new types of galaxies, some not detectable in visible light. The *IRAS* satellite (Chapter 7) found many infrared galaxies that are so dim in visible wavelengths that they had not been previously noticed. These appear to be galaxies that contain very large quantities of interstellar gas and dust, so that much of the starlight is obscured by dust and reemit-

TABLE 28.2 SELECTED BRIGHT GALAXIES

Galaxy	Type	Angular Diameter (arcmin)	Diameter (kpc)	Distance (kpc)	Apparent Magnitude*	Absolute Magnitude*	Mass ($M_\odot$)
Large Magellanic Cloud	Ir I	460	7	55	0.1	−18.7	10^{10}
Small Magellanic Cloud	Ir I	150	3	63	2.4	−16.7	2×10^{9}
Andromeda (M31)	Sb	100	16	700	3.5	−21.1	3×10^{11}
M33	Sc	35	6	730	5.7	−18.8	1×10^{10}
M81	Sb	20	16	3,200	6.9	−20.9	2×10^{11}
Centaurus A	EOp	14	15	4,400	7	−20	2×10^{11}
Sculptor system	dw E	30	1	85	7	−12	3×10^{6}
Fornax system	dw E	40	2	170	7	−13	2×10^{7}
M83	SBc	10	12	3,200	7.2	−20.6	—
Pinwheel (M101)	Sc	20	23	3,800	7.5	−20.3	2×10^{11}
Sombrero (M104)	Sa	6	8	1,200	8.1	−22	5×10^{11}
M106	Sb	15	17	4,000	8.2	−20.1	1×10^{11}
M94	Sb	7	10	4,500	8.2	−20.4	1×10^{11}
M82	Ir II	8	7	3,000	8.2	−19.6	3×10^{10}
M32	E2	5	1	700	8.2	−16.3	3×10^{9}
Whirlpool (M51)	Sc	9	9	3,800	8.4	−19.7	8×10^{10}
Virgo A (M87)	E1	4	13	13,000	8.7	−21.7	4×10^{12}

*The apparent and absolute magnitudes represent the light from the entire galaxy. Because the light from a galaxy comes from a large area in the sky, a galaxy of a given apparent magnitude is not as easily seen by the eye as a star of the same magnitude. Even though several galaxies in the list are brighter than sixth magnitude, only the two Magellanic Clouds and the Andromeda galaxy are easily visible to the unaided eye.

FIGURE 28.10 A PECULIAR GALAXY. *Not all galaxies fit the standard classifications. This is M82, a well-known example of a galaxy with an unusual appearance. For a time, it wa thought that the nucleus of this galaxy was exploding, but more recent analysis has indicated that the odd appearance is due simply to a very extensive region of interstellar gas and dust surrounding the center of a spiral galaxy that is undergoing an episode of intense star formation. The image at right uses computer-enhanced colors to highlight the gas and dust.* (*left:* Lick Observatory photograph; *right:* photo by H. C. Arp, Image Processing by J. J. Lorre, Image Processing Laboratory, JPL)

FIGURE 28.11 THE MAGELLANIC CLOUDS. *These two irregularly shaped, small galaxies lie just outside the Milky Way galaxy and orbit it, taking hundreds of millions of years for each complete orbit.* (Fr. R. E. Royer)

ted in infrared wavelengths. Some of these galaxies seem to be undergoing very intense star formation and are called **starburst galaxies.** Perhaps the best-known example, although only recently recognized as a starburst galaxy, is M82, a peculiar-looking galaxy thought for a long time to be undergoing some kind of explosion (see Fig. 28.10). Starburst galaxies may be normal galaxies undergoing very intense episodes of star formation; recall from Chapter 27 that our own galaxy is thought to have had at least one early period when stars formed much more rapidly than they do now. Perhaps at some time in the past, the Milky Way was a starburst galaxy.

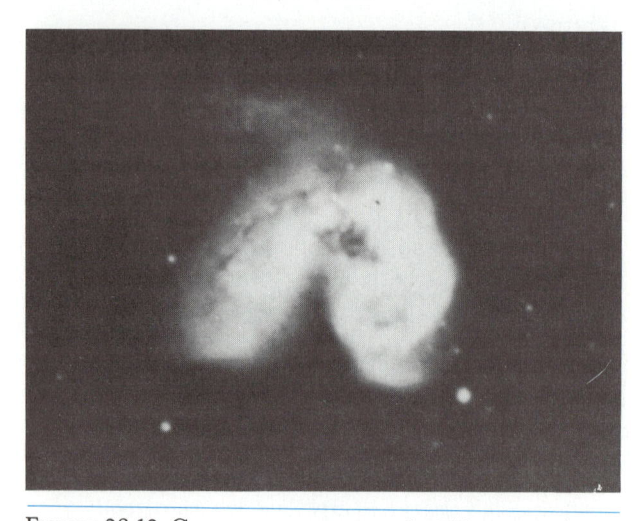

FIGURE 28.12 GALAXIES IN COLLISION? *This unusual object is thought to be a pair of galaxies running into each other.* (Palomar Observatory, California Institute of Technology)

GALACTIC DISTANCES

To probe the physical nature of the galaxies, we must first know galactic distances, because without this information we cannot deduce such fundamental parameters as masses and luminosities. We have already mentioned one technique for distance determination that can be applied to some galaxies: the use of Cepheid variables. These stars are sufficiently luminous to be identified as far away as a few million parsecs (that is, a few **megaparsecs,** abbreviated Mpc), which is sufficient to reach the Andromeda Nebula and several other neighbors of the Milky Way. This technique is not adequate, however, for probing the distances of most galaxies; other techniques had to be developed.

Recall, once again, from our discussions of stellar distance determinations (Chapter 20), that we can always find the distance to an object if we know both its apparent and absolute magnitudes. This is the basis of the spectroscopic-parallax method, the main-sequence-fitting technique, and even the Cepheid variable period-luminosity relation. A general term for any object whose absolute magnitude is known from its observed characteristics is **standard candle,** and an assortment of these are used in extending the distance scale to faraway galaxies.

The most luminous of stars are the red and blue supergiants, those which occupy the extreme upper regions of the H-R diagram. These can be seen at much greater distances than Cepheid variables, and therefore are important links to distant galaxies. The absolute magnitudes of these stars are inferred from their spectral types, just as in the spectroscopic-parallax technique, although they are so rare that there is substantial uncertainty in assuming that they conform to a standard relationship between spectral class and luminosity. In a variation on this technique, it is simply assumed that there is a fundamental limit on how luminous a star can be, and that in a collection of stars as large as a galaxy, there will always be at least one star at this limit. Then to determine the distance to a galaxy, we need only to measure the apparent magnitude of the brightest star in it, and to assume that its absolute magnitude is at the limit, which is about $M = -8$. This technique extends the distance scale by about a factor of 10 beyond what is possible with Cepheid variables—that is, to 10 or more megaparsecs.

Other standard candles come into play at greater distances. These include supernovae, which, at peak brightness, are thought to reach the same absolute magnitude. (Care must be exercised however, because the two types of supernovae have different absolute magnitudes, and because the recent supernova in the Large Magellanic Cloud has shown that some supernovae have unusual peak brightnesses.) A supernova is fantastically bright and can even outshine the entire galaxy in which it resides, reaching an absolute magnitude as bright as $M = -19$. By adopting super-

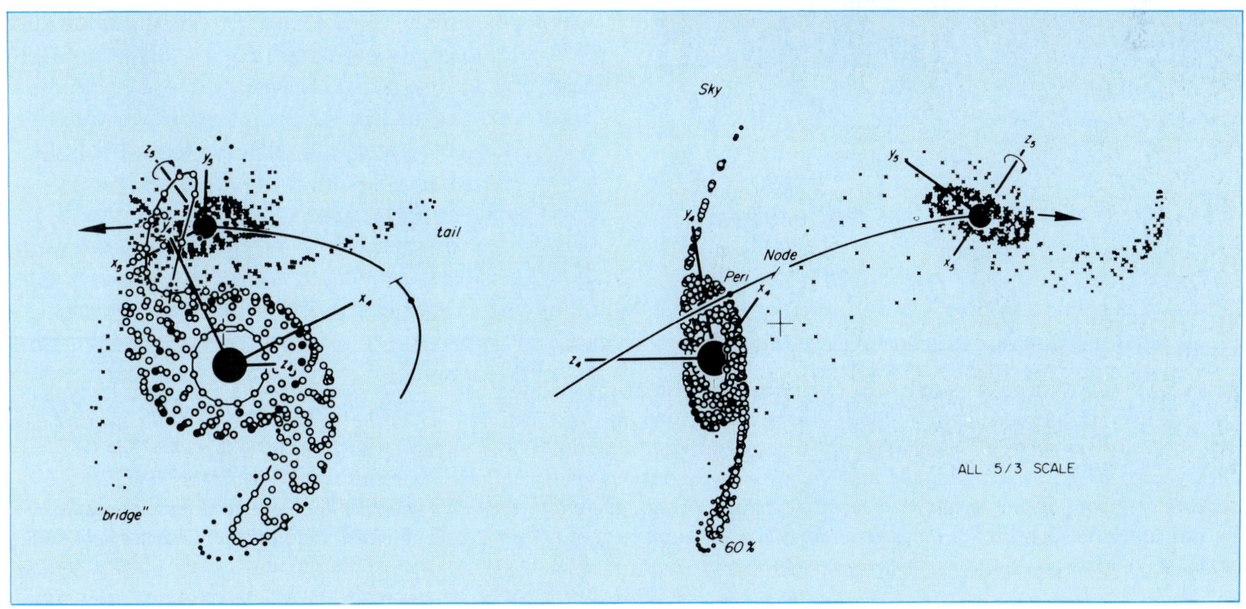

FIGURE 28.13 THE EFFECT OF A NEAR-COLLISION BETWEEN GALAXIES. *This is M51 (also known as the Whirlpool galaxy), with its smaller companion. Below is a computer simulation of the correct orientation of these two galaxies as seen from different angles. At left the orientation is as we see it; at right is a side view.* (*top:* © 1975 Royal Greenwich Observatory; *bottom:* A. Toomre and J. Toomre)

novae as standard candles, we can measure distances of hundreds of millions of parsecs. One drawback is that we can use this technique only when a supernova happens to occur in a galaxy, which may be only once every 10 to 100 years. Therefore, if we wish to estimate the distance to a specific galaxy, and we do not want to wait that long, we must use some other standard candle.

Other objects useful for this purpose include certain types of star clusters and bright HII regions. In the latter case, it is not the absolute magnitude but the diameter of the HII region that is assumed to be known. Then the apparent diameter (that is, the angular diameter) can be used to infer the distance. The ultimate standard candle is the brightest galaxy in a cluster of galaxies, in which case we adopt the assumption that this object should have the same absolute magnitude in every instance. This assumption makes it possible to measure distances to objects up to several billion parsecs away.

A new technique, called the **Tully-Fisher method,** shows great promise for spiral galaxies. This method is based on a relationship that has been discovered between the masses of spiral galaxies and their luminosities. The more massive a spiral galaxy is, the more luminous it is; hence the luminosity and therefore the distance to a galaxy can be estimated by measuring the mass. The mass is not measured directly but is inferred from the rotational speed of the galaxy, as indicated by the width of the 21-cm emission line of hydrogen (Fig. 28.14). The more rapidly the outer portions of the galaxy are rotating, the wider the 21-cm line, because of the Doppler effect. The width of the 21-cm line leads directly to an estimate of the galaxy's absolute magnitude; a comparison of this with the apparent magnitude yields a distance estimate. This technique works best when infrared magnitudes are used because infrared light is relatively unaffected by interstellar dust within the galaxy being observed. The Tully-Fisher method, along

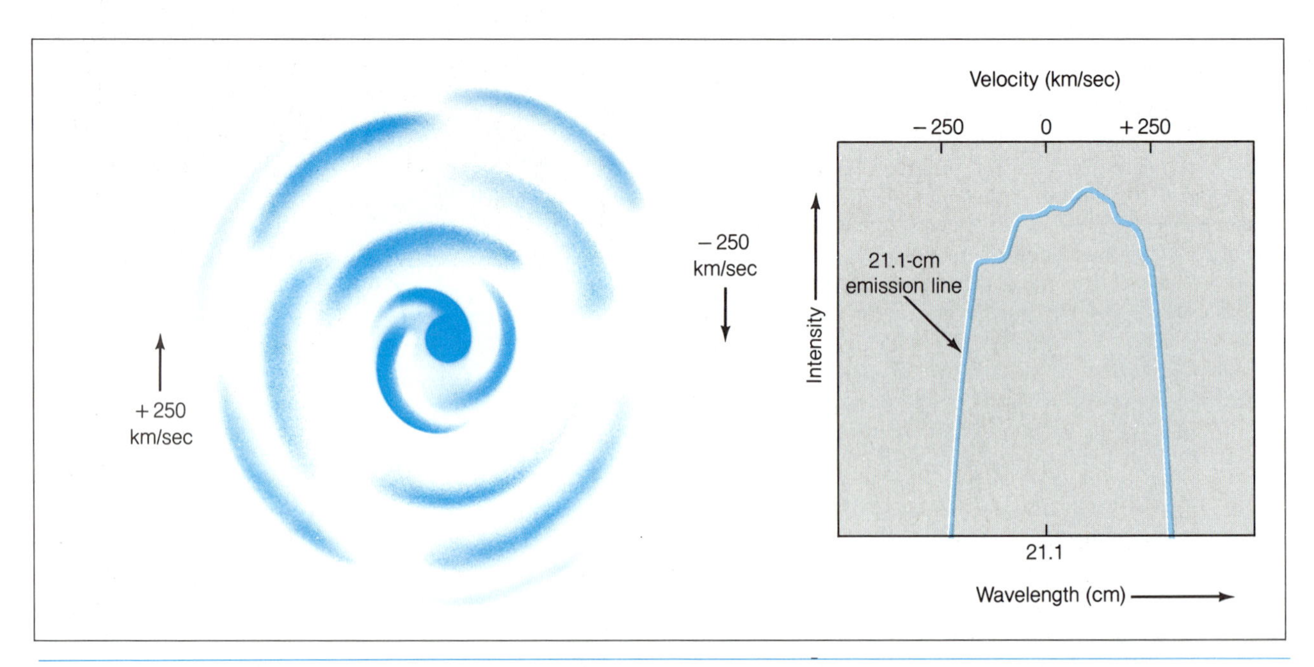

FIGURE 28.14 ROTATION VELOCITY AND THE WIDTH OF THE 21-CENTIMETER LINE. *This shows how the rotation velocity of a spiral galaxy is related to the width of the 21-centimeter emission line from hydrogen gas in the galaxy. Because of the Doppler effect, 21-centimeter photons from one side of the galaxy are shifted toward shorter wavelengths, those from the other side are shifted toward longer wavelengths, and those from the central regions are not shifted. The result is a 21-centimeter line that is broadened according to how rapidly the galaxy is rotating. Because the rotation speed is governed by the central mass of a galaxy, the width of the line is related to the mass of the galaxy, and can be used to estimate its absolute magnitude for distance determination.*

with a similar relation between luminosity and velocity dispersion for elliptical galaxies, is now being used in efforts to improve the intergalactic distance scale. Better knowledge of the distances to galaxies has important applications to our understanding of the size scale of the universe itself, as discussed in Chapters 30 and 32.

It is important to keep in mind how uncertain these techniques are. To assume that all objects in a given class are identical in basic properties such as luminosity is always risky, especially when such assumptions are applied to objects as distant as external galaxies or as rare as the brightest star in a galaxy. There is little else that can be done, however, so we must simply recognize the inherent limitations in accuracy and take them into account.

THE MASSES OF GALAXIES

The masses of nearby galaxies can be measured in the same manner as the mass of our own galaxy, by applying Kepler's third law to the orbital motions of stars or gas clouds in the outer portions. All that is required is to measure the orbital speed at some point well out from the center, and to determine how far from the center that point is (which in turn requires knowledge of the distance to the galaxy). Then Kepler's third law leads to

$$M = a^3/P^2$$

where M is the mass of the galaxy (in solar masses), and a and P are the semimajor axis (in AU) and the period (in years) of the orbiting material at the observed point (see Chapter 25).

The difficulties of this technique are that it is hard to isolate individual stars in distant galaxies and then to measure their speeds (using the Doppler shift), and that both the distance and the orientation of the galaxy must be known before the true orbital speed and semimajor axis can be determined. The orientation can usually be deduced for a spiral galaxy, since it has a disk shape whose tilt can be seen, and distance can be estimated through the use of one of the methods just outlined. By allowing light from many stars to enter the telescope and spectrograph, and then measuring the aggregate velocity of a large area of the galaxy, we can avoid the problem of isolating individual stars. This is somewhat imprecise, because there is always some internal motion within the portion of the galaxy that is observed, but it leads to useful estimates of the orbital speed in the outer regions.

Radio astronomy techniques are especially useful, because the gas that produces the radio emission (usually the 21-cm line of hydrogen is used in this case) often extends farther out in the galaxy than the visible stars (Fig. 28.15), and it is important to know the orbital velocities at the greatest possible distance from the galactic center. Furthermore, velocity measurements can be made very accurately with radio emission-line data, and interstellar extinction is not a problem.

In most cases the orbital velocities are measured at several points within a galaxy, from the center out as far as possible, and the data are plotted on a **rotation curve,** which is simply a diagram showing the variation of orbital velocity with distance from the center (Fig. 28.16). This is useful for several reasons: It provides a means of checking whether the observations go sufficiently far out in the galaxy to reach the region where the orbiting material follows Kepler's third law; it helps in determining the orientation of the galaxy; and it ensures that the velocities have been measured as far out as possible from the center.

The technique of measuring rotation curves as just described can best be applied to spiral galaxies, where there is a disk with stars and interstellar gas orbiting in a coherent fashion. In elliptical galaxies there is no such clear-cut overall motion, and a slightly different technique must be used. The individual stellar orbits are randomly oriented within an elliptical galaxy, so there is a significant range of velocities within any portion of the galaxy's volume. This range of velocities, called the **velocity dispersion** (Fig. 28.17), is greatest near the center of the galaxy, where the stars move fastest in their orbits, and is smaller in the outer regions. Furthermore, the greater the mass of the galaxy, the greater the velocity dispersion (at any distance from the center), so a measurement of this parameter can lead to an estimate of the mass of a galaxy. The velocity dispersion is deduced from the widths of spectral lines formed by groups of stars in different portions of a galaxy; the greater the internal motion within the region that is observed, the greater the widths of the spectral lines,

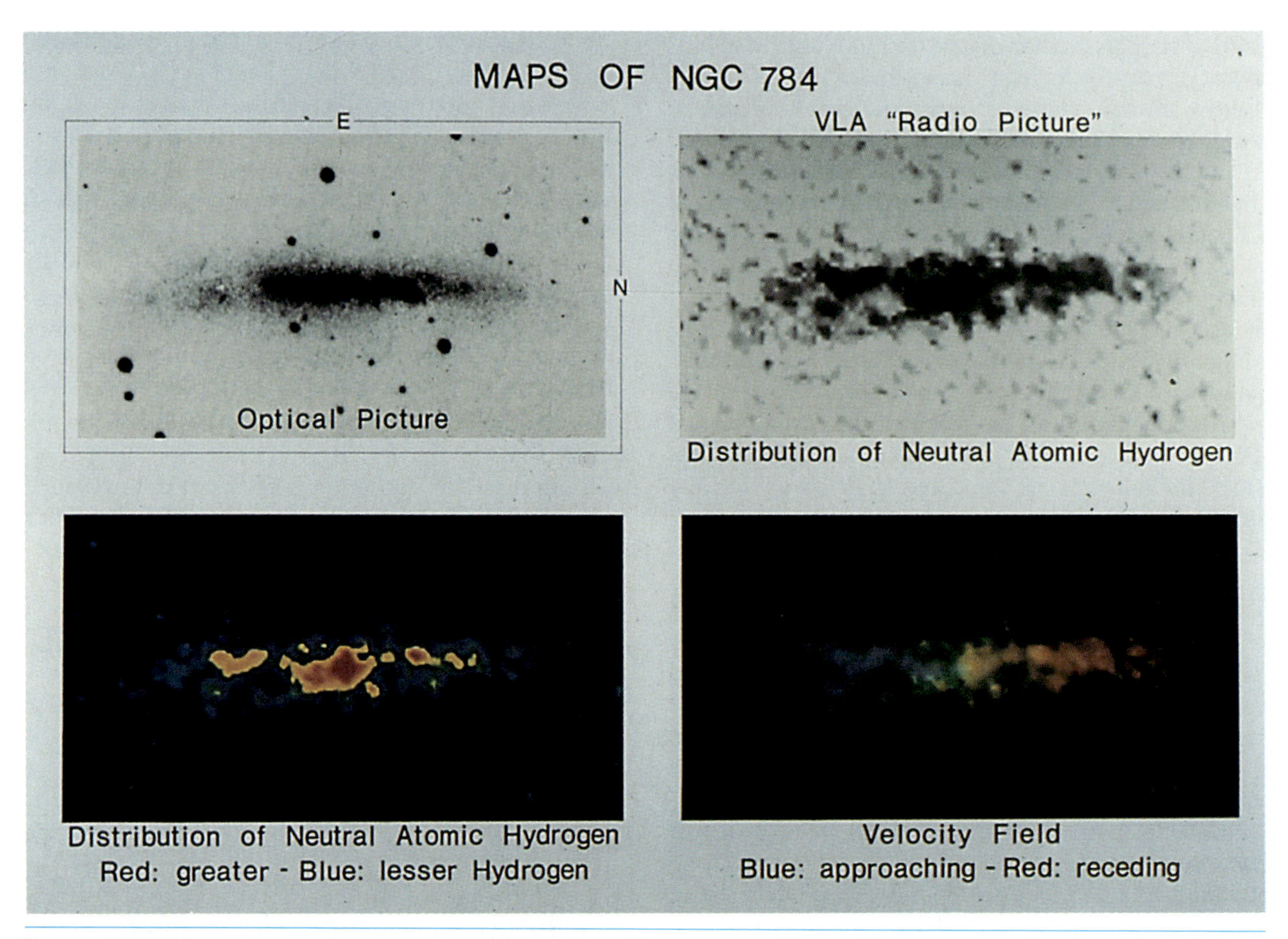

FIGURE 28.15 MEASURING A GALACTIC ROTATION CURVE. *This sequence of images illustrates how radio observations of atomic hydrogen are used to derive the rotation curve for a spiral galaxy. At upper left is a visible-light photograph of the nearly edge-on galaxy; at upper right is a radio map obtained in the 21-cm line of hydrogen. In the two lower panels the radio data have been color-coded, on the left to indicate relative intensities of the hydrogen emission; and on the right to indicate relative line-of-sight velocities. In this last panel we can see that one side of the galaxy (red) is receding and the other side (blue) is approaching. Comparison of the velocities leads to a determination of the rotational speed as a function of distance from the galactic center.* (National Radio Astronomy Observatory)

owing to the Doppler effect (caused by the fact that some stars are moving toward the Earth and others away from it).

Kepler's third law can sometimes be used in an entirely different way in determining galactic masses. There are double galaxies here and there in the cosmos, orbiting each other exactly like stars in binary systems. In these cases, Kepler's third law can be applied, leading to an estimate of the combined mass of the two galaxies. The uncertainties are even more severe than in double stars, however, because the orbital period of a pair of galaxies is measured in hundreds of millions of years. Thus the usual problems of not knowing the orbital inclination or the distance to the system are compounded by inaccuracies in estimating the orbital period, something that is usually very well known for a double star. Still, this technique is useful and has one major advantage: It takes into account all the mass of a galaxy, including whatever part of it is in the outer portions, beyond the reach of the standard velocity curve or velocity-dispersion measurements. Galactic masses estimated

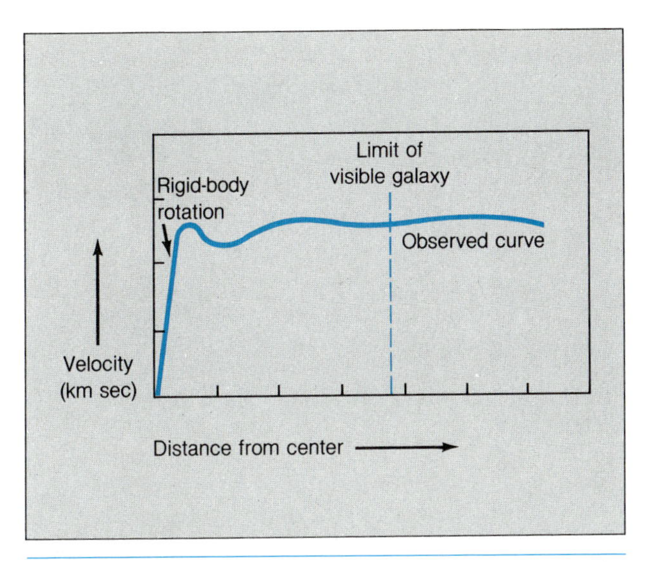

FIGURE 28.16 A SCHEMATIC ROTATION CURVE FOR A GALAXY. *This illustrates the fact that in most spiral galaxies, the rotation velocity does not drop off in the outer regions, as it would if most of the mass were concentrated at the center of the galaxy. The data on rotation speed in the outermost regions comes from radio observations.*

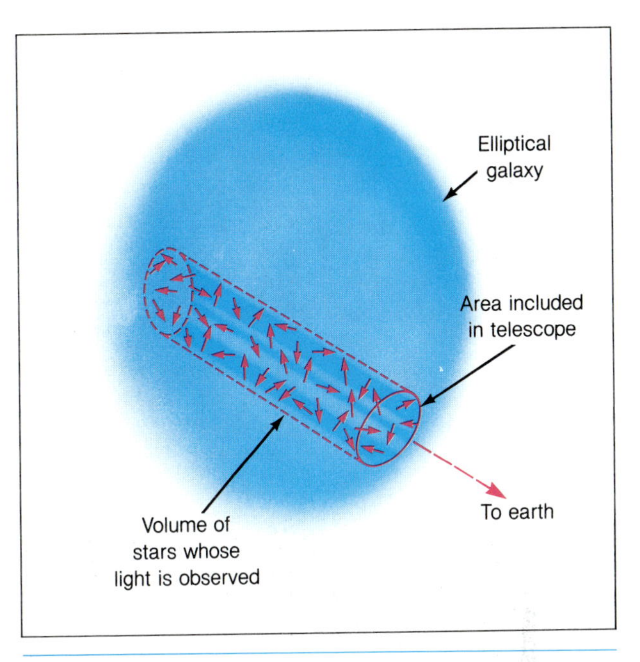

FIGURE 28.17 THE VELOCITY DISPERSION FOR AN ELLIPTICAL GALAXY. *There is no easily measured overall rotation of an elliptical galaxy, so the mass cannot be estimated from a rotation curve. Instead the average velocities of stars at a known distance from the center are used. Light from a small area of the galaxy's image is measured spectroscopically. The random motions of the stars in the observed part of the galaxy broaden the spectral lines by the Doppler effect, and the amount of broadening is a measure of the average velocity of the stars in the observed region. At a given distance from the center of the galaxy, the higher the average velocity, the greater the mass contained inside that distance.*

from double systems are generally much larger than those based on measurements of internal motions within galaxies, possibly indicating that most galaxies have extensive halos containing large quantities of matter. As noted in Chapter 25, there is independent evidence that our own galaxy has a massive halo, perhaps containing as much as 90 percent of the total mass.

LUMINOSITIES, COLORS, AND DIAMETERS

Once the distance to a galaxy has been established, its luminosity and size can be deduced directly from the apparent magnitude and apparent diameter. Both quantities are found to vary over wide ranges, with luminosities as low as 10^6 and as high as 10^{12} times that of the Sun, and diameters ranging from about 1 to 100 kpc.

Generally the elliptical galaxies display a wider range of luminosities and sizes than the spirals. The latter tend to be more uniform, with luminosities usually between 10^{10} and 10^{12} solar luminosities and diameters between 10 and 100 kpc. The smallest elliptical galaxies are called **dwarf ellipticals.** These may be very common, but because they are too dim to be seen at great distances, we can only say for sure that there are many of them near our own galaxy.

It appears that galaxies of a given type tend to be fairly uniform in other properties, just as stars of a given spectral type are the same in other ways. One quantity often used by astronomers to characterize galaxies is the **mass-to-light** (M/L) ratio, which is simply the mass of a galaxy divided by its luminosity, in solar units. Values of the mass-to-light ratio typically range from 10 to over 100. Any value larger than 1 means that the galaxy emits less light per solar mass than the Sun; that is, such a galaxy is dominated

ASTRONOMICAL INSIGHT 28.2

INFRARED GALAXIES

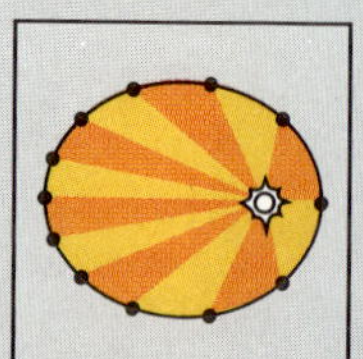

Until very recently, nearly all that we knew about the properties of galaxies and about their numbers was based on observations made in visible wavelengths of light. We now know that limiting ourselves to this small portion of the electromagnetic spectrum has severely limited our knowledge of galaxies and their evolution.

The *Infrared Astronomical Satellite (IRAS)* mapped the entire sky at four wavelengths far into the infrared portion of the spectrum. As noted in other chapters in this text, *IRAS* data have brought about major changes in our understanding of many objects in the universe, ranging from planets to newborn stars to interstellar clouds of gas and dust. It also produced new understandings of the galaxies that populate the universe.

The *IRAS* sky survey revealed huge numbers of faint, fuzzy sources of infrared emission not associated with any objects in our galaxy. These objects are now thought to be distant galaxies, and many of them are invisible in all wavelengths except the infrared. This means that most of the energy that they emit comes from very cold material. These infrared objects are now believed to be spiral galaxies so obscured by dust that they are seen only in the far infrared wavelengths where the dust emits. Most of the light from the stars in these galaxies is absorbed by dust, which is heated in the process to temperatures of several tens of degrees and then emits infrared radiation.

The discovery of the infrared galaxies has revealed a whole new class of galaxies containing far more interstellar matter than normal spirals. The interpretation of these galaxies is uncertain: Are they aging galaxies undergoing a new phase of evolution in which stars emit vast quantities of material into space; or are they newly forming galaxies undergoing rapid star formation? The second point of view, that these are young galaxies undergoing massive star formation episodes, is becoming more widely accepted. If this is correct, it has important implications for galactic evolution. It implies that intervals of very intense and rapid star formation may be normal stages in the early evolution of spiral galaxies. Visible-light observations, as well as infrared data, have already led astronomers to define a class of galaxies called **starburst galaxies,** galaxies thought to be in stages of rapid star formation. The new infrared data from *IRAS* appears to have established the widespread existence of starburst galaxies.

If a starburst stage is normal in the early evolution of spiral galaxies, then one problem confounding studies of the history of our own galaxy may have been solved. As noted in Chapter 27, the abundance of heavy elements in the Milky Way is higher than would be expected if current star formation rates were the norm throughout the history of the galaxy. This has led to speculation that our galaxy was either formed from material that had already undergone elemental enrichment by nuclear reactions in some earlier generation of stars, or that the galaxy underwent a much more intense episode of star formation sometime in the past than it is undergoing now. The discovery by *IRAS* that starburst phases of galactic evolution may be common leads astronomers to believe that our own galaxy was once a starburst galaxy. If so, the Milky Way may once have been virtually invisible in all wavelengths except the far infrared. Perhaps alien astronomers several billion light years away are at this moment discovering the Milky Way as a source of far infrared radiation, to be included in their equivalent of our *IRAS* catalogs.

in mass by stars that are dimmer than the Sun or contains substantial quantities of mass in forms other than stars (such as interstellar matter). Even the smaller values of M/L that are observed for galaxies are much larger than 1; a value of M/L = 10, for example, means that 10 solar masses are required to produce the luminosity of one Sun.

Elliptical galaxies tend to have the largest mass-to-light ratios, consistent with our earlier statement that these galaxies are relatively deficient in hot, luminous stars. Spirals, on the other hand, which contain some of these stars, have the lower M/L values. It is worth stressing again, however, that even the low values of M/L for these galaxies are much greater than 1, indicating that they too are dominated in mass by low-luminosity stars. Hence a spiral galaxy, with all its glorious bright blue stars, actually has far more dim red ones.

The colors of galaxies also can be measured, with the use of filters to determine the brightness at different wavelengths. In general, the spirals are not as red as the ellipticals, again indicating that the latter galaxies contain a higher fraction of cool, red stars. There are also color variations within galaxies; for example, in a spiral, the central bulge is usually redder than the outer portions of the disk where most of the young, hot stars reside.

Both the mass-to-light ratios and the colors of galaxies are indicators of the relative content of Population I and Population II stars. Recall that Pop. I stars tend to be younger, and included in this group are all the bright blue O and B stars. Population II, by contrast, includes only old, red, relatively dim stars. Therefore, a red overall color, along with a high mass-to-light ratio, implies that Pop. II stars are dominant, whereas a low value of M/L and a bluer color means that some Pop. I stars are mixed in. Thus, elliptical galaxies seem to consist almost entirely of Pop. II objects, whereas spirals contain a mixture of the two populations.

This dichotomy between the two types of galaxies is found also when we compare the interstellar-matter content of spirals and ellipticals. Photographs of spirals, especially edge-on views, often clearly show the presence of dark dust clouds, and face-on views typically reveal a number of bright HII regions. Neither shows up on photographs of elliptical galaxies. Radio observations of the 21-cm line of hydrogen bear this out; emission is usually strong in spirals and weak in ellipticals.

THE ORIGINS OF SPIRALS AND ELLIPTICALS

In attempting to explain the differences between spiral and elliptical galaxies, astronomers have suggested several theories, none of which is fully developed at present. All evidence shows that spiral galaxies are dynamic entities, with active star formation and recycling of material between stellar and interstellar forms, whereas elliptical galaxies have reached some sort of equilibrium in which these processes are not taking place at a significant rate. We know that there is no significant age difference between the two types of galaxies; we cannot simply conclude that the ellipticals are older and have run out of gas.

One of the earliest suggestions was that rotation is responsible for the differences between the two types. We learned in the previous chapter that our own galaxy is thought to have formed from a rotating cloud of gas that flattened into a disk as it contracted. The rotation of the cloud was the cause of the disk formation, so perhaps elliptical galaxies are the result of contracting gas clouds that did not rotate rapidly enough to form disks. It is not clear how this would account for the lack of both interstellar material and star formation, however.

One problem with this idea is that elliptical galaxies can and do rotate, yet they have not formed disks. Rotation alone, therefore, cannot be the full explanation. There must be another factor involved. Apparently the key is whether or not a disk forms before most of the gas in the contracting cloud has been consumed by star formation. If all the gas is converted into stars quickly, before the collapse has proceeded very far, the result is an elliptical galaxy (Fig. 28.18). If, on the other hand, the initial rate of star formation is not so great, then the cloud has time to form a disk while it still contains a large quantity of interstellar gas and dust. The reason timing is so important is that stars, once formed, will act as individual particles and will continue to orbit the galaxy without forming a disk. In contrast, gas acts as a fluid and will settle into a disk. (The essence of this is whether or not there is internal friction that will

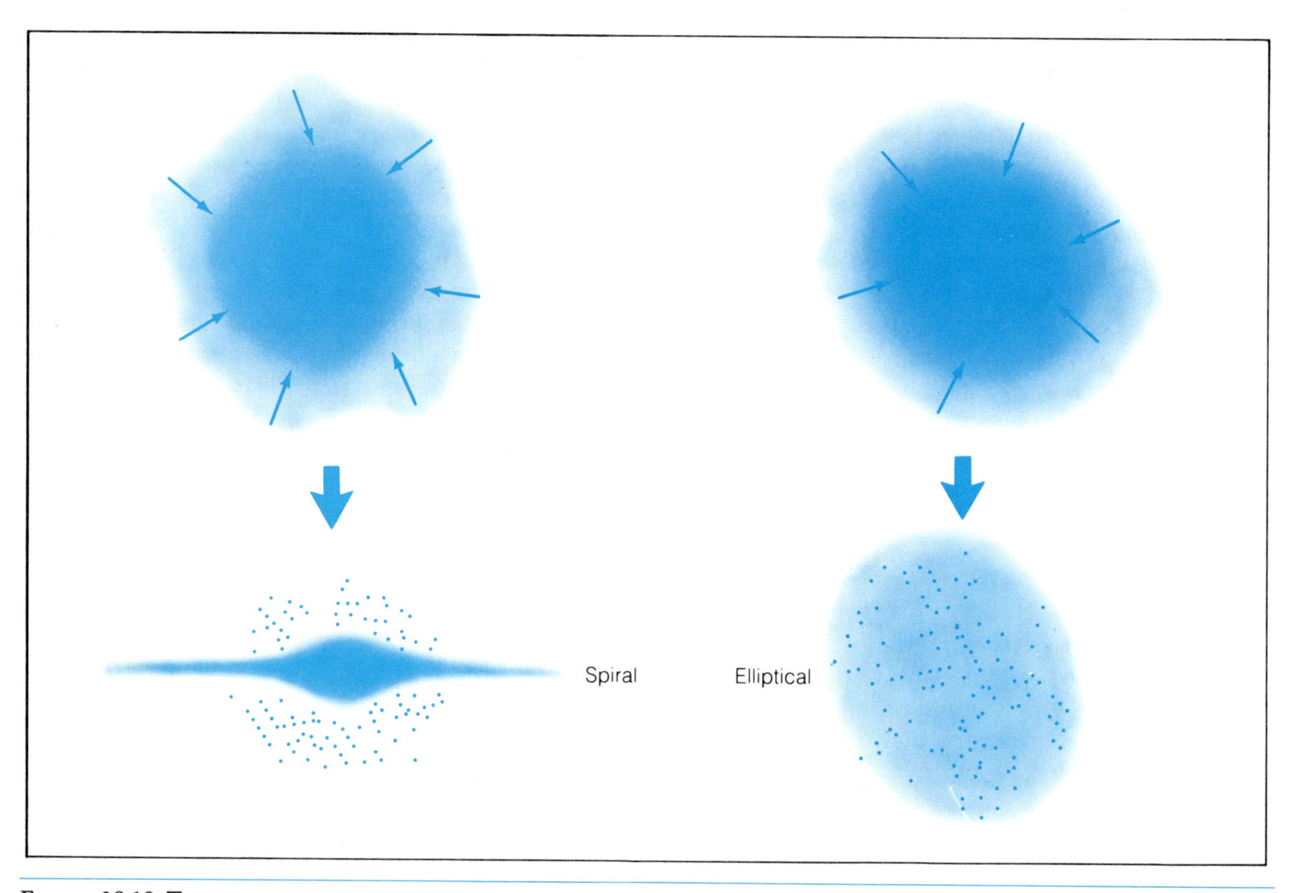

FIGURE 28.18 THE ORIGINS OF SPIRALS AND ELLIPTICALS. *Both types of galaxies begin as giant gas clouds that collapse gravitationally. If collapse to a disk occurs before all the gas is converted into stars, the result is a spiral galaxy. If the gas is entirely used up in star formation before a disk forms, the result is an elliptical. The rate of star formation relative to the rate of collapse to a disk is probably determined by the initial density of the cloud, and may be influenced by the rotation rate.*

dissipate energy and allow the material to sink into the plane of a disk; stars do not exert friction forces on each other, but gas particles do.)

Although these ideas represent progress toward understanding why some galaxies are ellipticals and others are spirals, there remain substantial questions. It is not clear why the initial rate of star formation should differ from one galaxy to another, for example, although it has been suggested that it may have to do with the density of the cloud from which the galaxy contracts. Other important factors are the rate at which the gas loses energy (that is, cools) because of the emission of radiation, and whether or not the pregalactic cloud has a magnetic field, which may determine how easily condensations leading to star formation may occur.

Other as-yet unexplained questions regarding the two types of galaxies include why ellipticals have a wider range of masses and sizes than spirals, and why many spirals have bars. Some progress has been made on the latter question; calculations show that instabilities that can develop naturally in a rotating disk may lead to the formation of a symmetric bar like those observed. It has turned out, in fact, to be more difficult to explain why all spirals do not have bars, than to account for the fact that some of them do, because the responsible instabilities should be effective in all cases. It appears that the presence of a massive halo may inhibit the formation of a bar, however, so perhaps the difference between barred spirals and those without bars is that the latter have extensive halos and the former do not.

PERSPECTIVE

We have expanded our horizons almost to the ultimate limit, by examining objects that inhabit the full volume of the universe. In this chapter we have confined ourselves to a discussion of individual galaxies and their properties, a necessary step before moving on to the large-scale organization of the universe. We are prepared now to examine groups and clusters of galaxies, to see what they can tell us about the origin and evolution of the universe itself.

SUMMARY

1. Nebulae were observed by astronomers for centuries before it was known whether they were local objects, such as gas clouds, or other galaxies, comparable to the Milky Way.
2. The controversy over the nature of nebulae peaked in 1920 and was settled in 1924, when Hubble's analysis of Cepheid variables in the Andromeda nebula proved that this object is too distant to be part of the Milky Way.
3. Galaxies are categorized by shape and fall into two general classes: spirals and ellipticals. There are also a substantial number of S0 galaxies (disk-shaped, but with no spiral structure) and irregular galaxies.
4. Distances to galaxies are measured using a variety of standard candles, such as Cepheid variables, extremely luminous stars, supernovae, and galaxies of standard types.
5. Masses of spiral galaxies are determined through the application of Kepler's third law to the outer portions, where the orbital velocity and hence the period are determined from a rotation curve. For elliptical galaxies, the internal velocity dispersion is used. Galactic masses can also be determined from the application of Kepler's third law to binary galaxies.
6. Determinations of mass based on external gravitational influences of galaxies usually yield higher values than those based on internal diagnostics such as rotation curves or velocity dispersions. This leads to the conclusion that many galaxies have massive halos.
7. Galactic luminosities and diameters vary quite widely among elliptical galaxies, but not so widely among spirals.
8. Although all galaxies are dominated by Population II stars, spirals tend to contain a greater proportion of Population I stars, have substantial quantities of interstellar matter, and generally seem to be in a state of continuous evolution and stellar cycling. Ellipticals, on the other hand, have few or no Population I stars, contain little or no interstellar matter, and generally do not seem to have active stellar cycling at present.
9. Spiral galaxies appear to originate from rotating gas clouds that flatten into disks before all the gas is used up in the star formation, whereas ellipticals seem to result when star formation consumes all of the gas before collapse to a disk occurs.

REVIEW QUESTIONS

Thought Questions

1. Summarize the properties of the different types of galaxies in the Hubble classification system.
2. How do we know that the tuning fork diagram does not represent an evolutionary sequence?
3. How can it be that elliptical galaxies are more numerous in the universe than spiral galaxies, if most of the bright galaxies are spirals?
4. Summarize the methods used to find the distances to galaxies.
5. Explain why the velocity of stars orbiting a galaxy at a given distance from the center depends on the mass contained within that distance. (This is the basis for the velocity-dispersion method and the Tully-Fisher method for measuring galactic masses.)
6. Why is it better to use radio data to determine galactic masses than it is to use visible-light data?
7. Explain why elliptical galaxies typically are redder than spiral galaxies.
8. The mass-to-light ratio for spiral galaxies is usually much larger than 1, and for elliptical galaxies it is larger yet. What does this tell us about the types of stars that emit most of the light from such galaxies?
9. Elliptical galaxies are dominated by Population II stars, whereas spiral galaxies contain the younger Population I stars. Why are spiral galaxies thought to be as old as elliptical galaxies?
10. Explain how the theories described in the text for galaxy formation can account for all the observed differences between spiral and elliptical galaxies.

Problems

1. To help demonstrate the difficulty of observing individual stars in other galaxies, calculate the apparent magnitude of a supergiant star whose absolute magnitude is $M = -8$, if such a star were observed in a galaxy that is 100,000 pc away (that is, a bit beyond the Magellanic Clouds), and in a galaxy 1,000,000 pc away (a little farther than the Andromeda galaxy).

2. Suppose an elliptical galaxy is found to be twice as long as it is wide. What type of elliptical is it?

3. A supernova of Type I is observed in a distant galaxy. The absolute magnitude of such a supernova is assumed to be -19. If the supernova has an apparent magnitude of $+24$ (near the limit that can be observed with the largest telescopes), how far away is the galaxy in which it lies?

4. Suppose that in a spiral galaxy, the rotation curve shows that at a distance of 15 kpc from the center, the orbital velocity is 150 km/sec. Assuming that the orbital motions at this distance from the center obey Kepler's laws, what is the mass of this galaxy?

Additional Readings

Bok, B. J. 1964. The large cloud of Magellan. *Scientific American* 210(1):32.

Burns, Jack O. 1986. Very large structures in the universe. *Scientific American* 255(1):38.

Geller, M. J. 1978. Large-scale structure of the universe. *American Scientist* 66:176.

Gorenstein, P., and W. Tucker. 1978. Rich clusters of galaxies. *Scientific American* 239(5):98.

Groth, E. J., P. J. E. Peebles, M. Seldner, and R. M. Soneira. 1977. The clustering of galaxies. *Scientific American* 237(5):76.

Hodge, Paul. 1981. The Andromeda galaxy. *Scientific American* 244(1):88.

———. 1986. *Galaxies.* Cambridge, Mass.: Harvard University Press.

———. 1964. The Sculptor and Fornax dwarf galaxies. *Sky and Telescope,* Dec., p. 336.

Larson, R. B. 1977. The origin of galaxies. *American Scientist* 65:188.

Mathewson, D. 1984. The clouds of Magellan. *Scientific American* 252(4):106.

Silk, J., Szaley, A. S., and Y. B. Zel-dovich. 1983. The large-scale structure of the universe. *Scientific American* 249(4):56.

Toomre, A., and J. Toomre. 1973. Violent tides between galaxies. *Scientific American* 229(6):38.

CHAPTER 29

CLUSTERS AND SUPERCLUSTERS: THE DISTRIBUTION OF GALAXIES

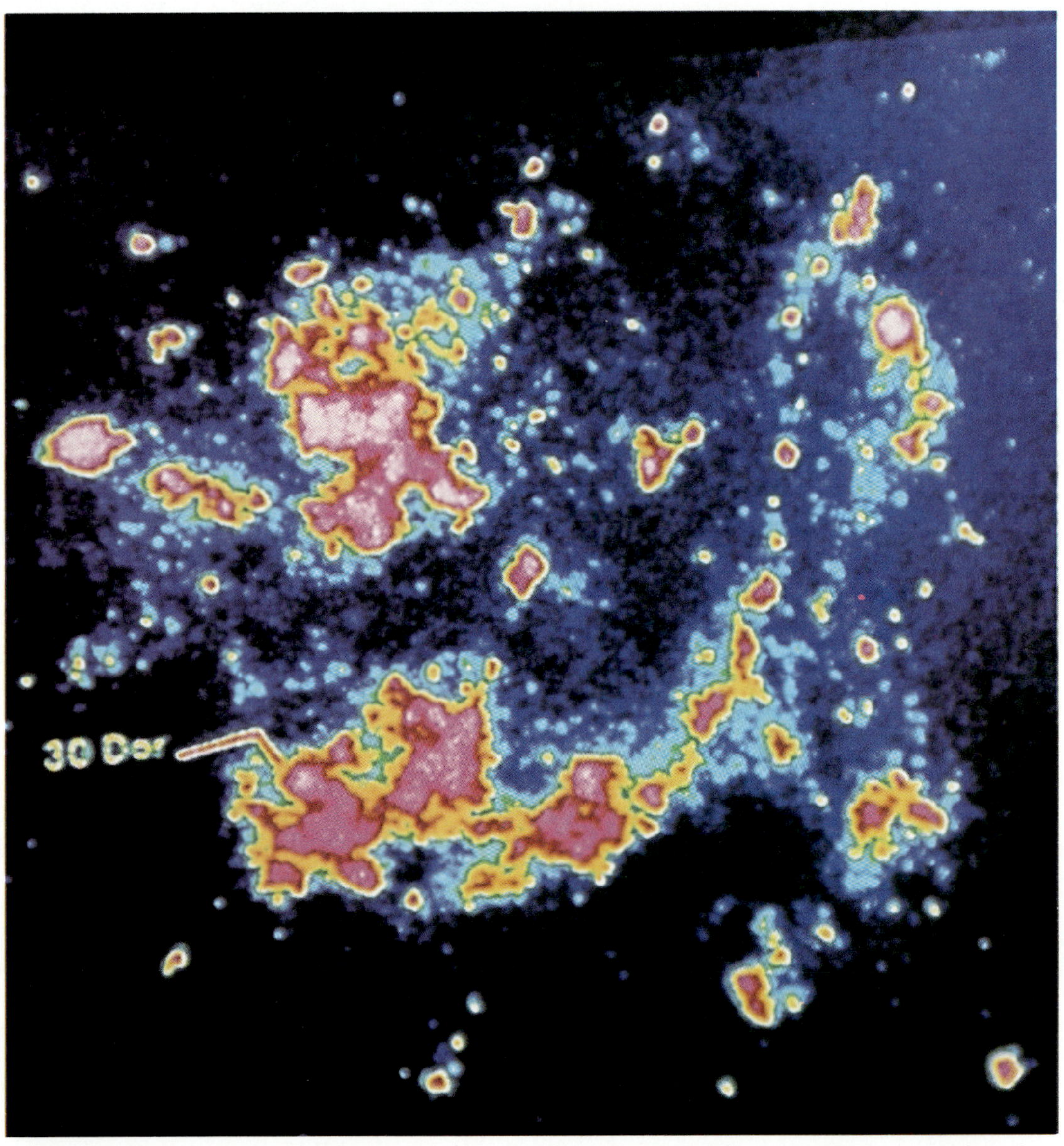

An ultraviolet view of the Large Magellanic Cloud, showing the locations of hot stars. (NASA, courtesy of A. M. Smith)

CHAPTER PREVIEW

FIGURE 29.1 A CLUSTER OF GALAXIES. *Several galaxies are visible in this photograph of a portion of a large cluster of galaxies, called the Fornax cluster.* (© 1984 Royal Observatory, Edinburgh)

Although galaxies may be considered the largest single objects in the universe (if indeed an assemblage of stars orbiting a common center can be viewed as a single object), there are yet larger scales on which matter is organized. Galaxies tend to be located in clusters (Fig. 29.1) rather than being distributed uniformly throughout the cosmos, and these in turn have an uneven distribution, with concentrations of clusters referred to as **superclusters.** It may be that superclusters are the largest things in the universe.

Clusters of galaxies range in membership from a few (perhaps half a dozen) to many hundreds or even thousands. Just as stars in a cluster orbit a common center, so are galaxies in a group or cluster gravitationally bound together, following orbital paths about a central orbit. In a large, rich cluster, the frequent encounters between galaxies have, over time, caused the members to take on a smooth, spherical distribution; in a small group like the one to which the Milky Way belongs, the arrangement of individual galaxies is more haphazard, creating an amorphous overall appearance.

THE LOCAL GROUP

We can begin by discussing the **Local Group** of galaxies, the cluster that we live in. This group consists of about thirty members arranged in a random distribution (Appendix 13 and Fig. 29.2). Despite the relative proximity of these galaxies to us, it has been difficult to ascertain their properties in some cases, because of obscuration by our own galactic disk. It is not even possible to say with certainty how many members belong to the Local Group.

Among the member galaxies are three spirals, two of which, the Andromeda galaxy and the Milky Way, are rather large and luminous. These are probably the brightest and most massive galaxies in the entire cluster. Most of the other members are ellipticals, many of them dwarf ellipticals (Fig. 29.3). There may be additional members of this type, undetected so far because of their faintness. Four irregular galaxies, two of them being the Large and Small Magellanic Clouds (Fig. 29.4), are also found within the Local Group. In addition, there are globular clusters, which are probably distant members of our galaxy, but which lie so far away from the main body of the Milky Way that they appear isolated.

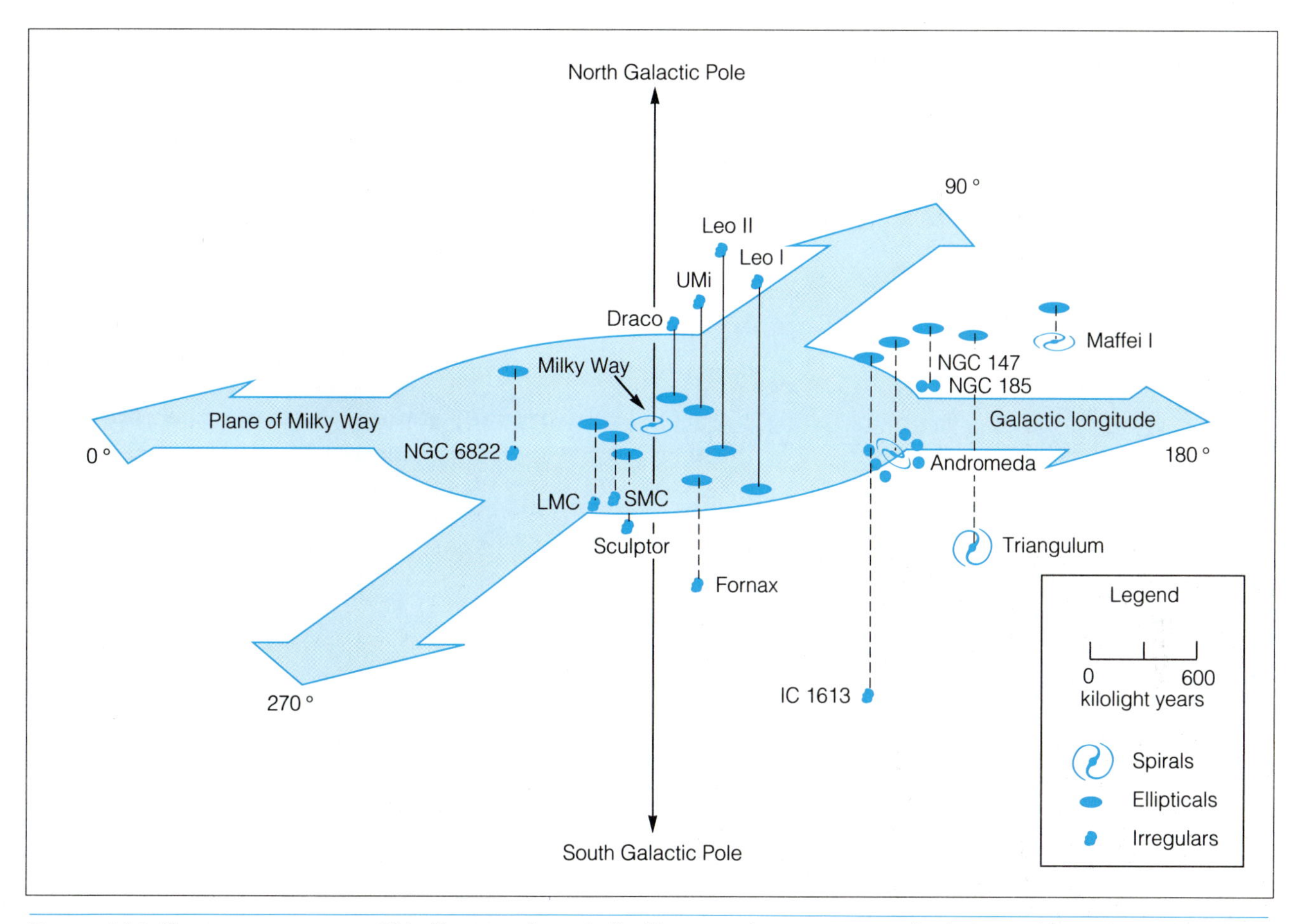

FIGURE 29.2 THE LOCAL GROUP. *This diagram schematically illustrates the arrangement of galaxies in the Local Group, with respect to the Milky Way.*

The Local Group is about 1,000 kpc in diameter and has a roughly disklike overall shape, with the Milky Way located a little off center. Beyond the outermost portions of the Local Group, there are no conspicuous external galaxies for a distance of some 1,100 kpc.

The Magellanic Clouds and the Andromeda galaxy (also commonly known as M31, its designation in the widely cited Messier Catalog; Appendix 15) have been particularly well studied, because of their proximity and prominence, and because of what they can tell us about galactic evolution and stellar processing.

The Large and Small Magellanic Clouds appear to the unaided eye as fuzzy patches, easily visible only on dark, moonless nights. They lie near the south celestial pole and can therefore be easily seen only from the Southern Hemisphere. Their name originated from the fact that the first Europeans to see them were Ferdinand Magellan and his crew, who made the first voyage around the world in the early sixteenth century.

The Magellanic Clouds are satellites of the Milky Way, having lesser masses and following orbits about it, taking several hundred million years to make each circuit. Lying between 50 and 60 kpc from the Sun, both are irregulars of Type I. They contain substantial quantities of interstellar matter and are quite obviously the sites of active star formation, with many bright nebulae and clusters of hot, young stars (Fig. 29.5). Measurements of the colors of these galaxies, and of the spectra of some of their brighter

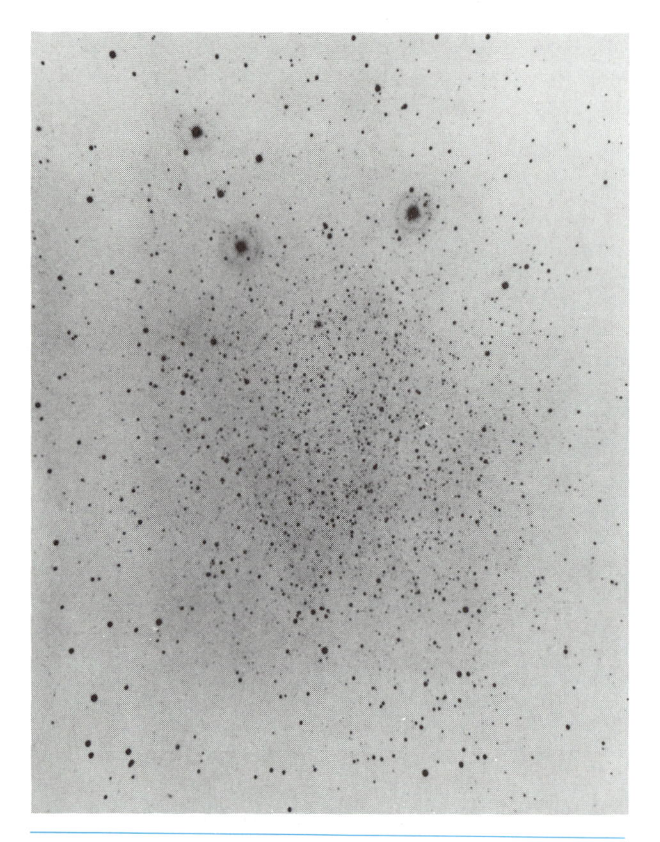

FIGURE 29.3 THE SCULPTOR DWARF GALAXY. *This loose conglomeration of stars lies about 83 kiloparsecs from the sun. Such a small, dim galaxy would not be noticed if it were much farther away, but the number of systems of this type in the Local Group leads to the assumption that they may be very common in the universe.* ((P. W. Hodge, photo taken at the Boyden Observatory)

FIGURE 29.4 THE MAGELLANIC CLOUDS. *The large cloud is shown in the upper photo, and the small cloud in the lower picture. Our nearest neighbors, these galaxies orbit as satellites of the Milky Way.* (*top:* © 1984 Royal Observatory, Edinburgh; *bottom:* R. J. Dufour, Rice University)

stars, indicate that they have somewhat lower heavy-element abundances than Pop. I stars in our galaxy. This seems to indicate that the Magellanic Clouds have not undergone as much stellar cycling and recycling as the Milky Way. These and other Type I irregular galaxies may generally be viewed as galaxies in extended adolescence, not yet having settled down into mature disks. In the Magellanic Clouds, the reason for the unrest is probably the gravitational tidal forces exerted by the Milky Way.

The great spiral of M31, the Andromeda galaxy (Fig. 29.6), is the most distant object visible to the unaided eye, lying some 700 kpc from our position in the Milky Way. All that the eye can see is a fuzzy patch of light, even if a telescope is used; but when a time-exposure photograph is taken, then the awesome disk and spiral arms stand out. The Andromeda galaxy is so large, extending over one degree across the sky, that full portraits can be obtained only by using relatively wide-angle telescope optics (most large telescopes have extremely narrow fields of view).

The Andromeda galaxy is probably very similar to our own galaxy, and thus has taught us quite a bit about the nature of the Milky Way. The two stellar populations were first discovered through studies of stars in Andromeda. In fact, the effects of stellar processing on the chemical makeup of different portions

FIGURE 29.5 THE TARANTULA NEBULA. *This spectacular region of ionized gas and hot dust is an active site of star formation in the Large Magellanic Cloud. The great supernova that occurred in early 1987 lies very near this nebula (but outside the field of view of this photograph; see the Special Insert section on the supernova).* (© 1984 Anglo-Australian Telescope Board)

of a galaxy are better determined for Andromeda then for the Milky Way. Observations of the Andromeda galaxy in invisible wavelengths (Fig. 29.7) have revealed similarities in the distribution of interstellar matter and X-ray sources as well.

Another very important by-product of observations of the Andromeda galaxy, as well as those of some other members of the Local Group, is the calibration of our distance scales for measuring objects farther away. Knowledge of distances to galaxies within the Local Group, measured through the use of the period-luminosity relationship for Cepheids, has allowed astronomers to establish the absolute magnitudes of various standard candles, which, in turn, are used to find distances to more faraway galaxies. Each

FIGURE 29.6 THE ANDROMEDA GALAXY. *At a distance of over two million light-years, this galaxy is the most distant object visible to the unaided eye. Without a telescope and a time-exposure photograph, the eye sees only an extended, fuzzy patch of light, rather than the detailed view shown here.* (Palomar Observatory, California Institute of Technology).

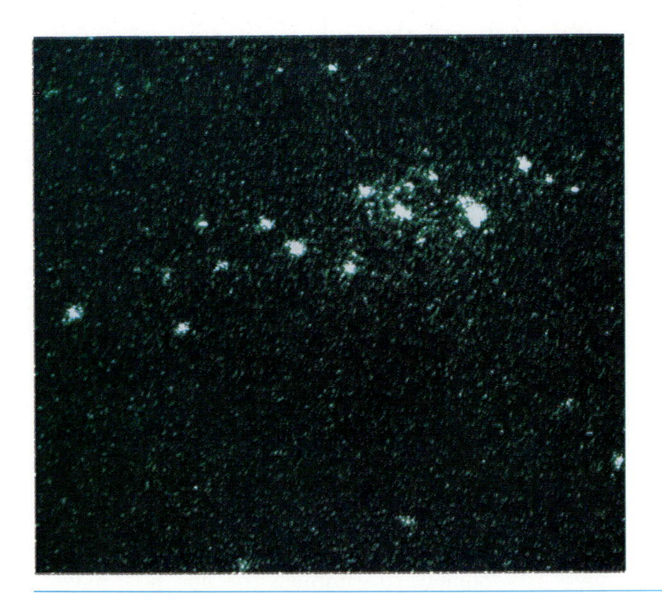

FIGURE 29.7 NEW VIEWS OF THE ANDROMEDA GALAXY. *Here are an X-ray image (left) and an infrared picture (right) of our neighbor spiral galaxy. The X-ray view reveals associations of stars with hot coronae and binary X-ray sources; the infrared image shows the locations of cold dust clouds.* (*left:* Harvard-Smithsonian Center for Astrophysics; *right:* NASA)

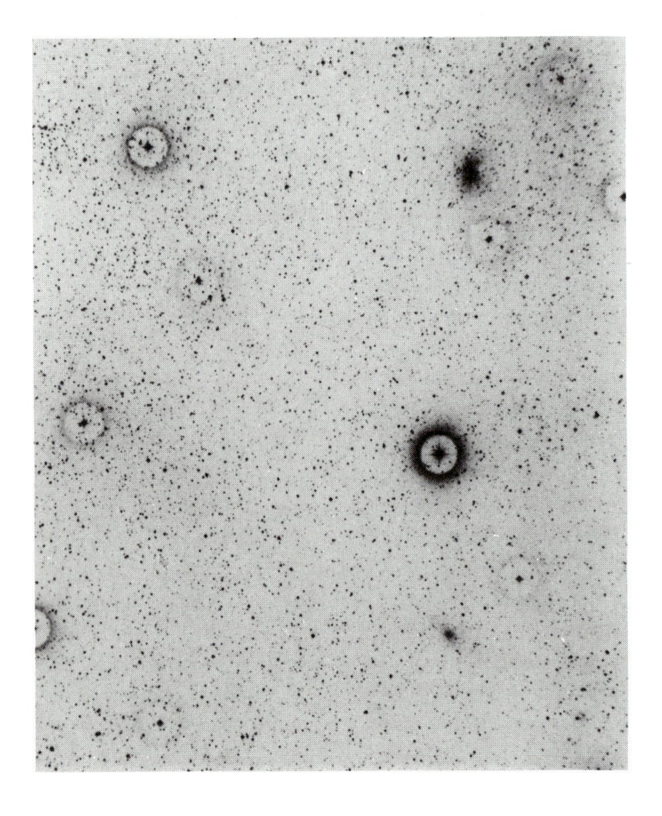

step in extending the distance scale must be built upon previous steps, and the Local Group galaxies represent an important stage in the process of reaching out to the greater distances of other clusters of galaxies. It would have been very difficult indeed to establish the distances of extragalactic objects if we did not live within a cluster of galaxies where this calibration process is possible.

Before leaving the Local Group, we return to the question of its membership, and the rank of the Andromeda galaxy and the Milky Way within it. New galaxies are occasionally discovered and found to belong to the Local Group. These are most often dwarf ellipticals. In the early 1970s, however, infrared observations revealed two new galaxies (Fig. 29.8), both lying behind the disk of the Milky

FIGURE 29.8 MAFFEI 1 AND 2. *This infrared photograph reveals two large galaxies (the fuzzy images, upper right and lower right) that were for a while considered possible members of the Local Group. More extensive analysis has shown, however, that they probably are not. (The circular marks are overexposed images of closer stars.)* (H. Spinrad)

Way and therefore obscured from our direct view. These galaxies, called Maffei 1 and Maffei 2 after their discoverer, are both rather large, one being an elliptical and the other a spiral. Their distances were not well known for a while, and it appeared possible that they were members of the Local Group. If this had been the case, they would have displaced the Andromeda galaxy and the Milky Way as the dominant members, and this would have altered our perception of the cluster as a whole. Recent evidence has shown, however, that the two galaxies are probably too distant to be members of the Local Group.

FIGURE 29.9 A PORTION OF A RICH CLUSTER OF GALAXIES. *Most of the objects in this photograph are galaxies. This cluster lies in the constellation Hercules.* (National Optical Astronomy Observatories)

RICH CLUSTERS: DOMINANT ELLIPTICALS AND GALACTIC MERGERS

In contrast with small clusters such as the Local Group, many clusters of galaxies are much larger and contain hundreds or even thousands of members (Fig. 29.9). The distribution of galaxies in one of these rich clusters tends to be smooth, so that the density of galaxies increases gradually from the outermost regions to the center, where the density of galaxies is highest. By contrast, in a small cluster like the Local Group, the galaxies are unevenly distributed, with no central dense region.

Because the density of galaxies in a rich cluster is relatively high, there are numerous close encounters between galaxies as they follow their individual orbits about the center. When two galaxies pass close by each other, they exert mutual gravitational forces that can have profound effects. Apparently these effects include the conversion of some galaxies from one type to another.

The central regions of large, rich clusters are highly dominated by elliptical and S0 galaxies, more so than in small groups or isolated galaxies. Near the center of a rich cluster, some 90 percent of the galaxies may be ellipticals or S0s (Fig. 29.10), whereas among noncluster galaxies about 60 percent are spirals. This contrast is probably partially a result of the frequent near-collisions between galaxies in dense clusters. When two galaxies have a close encounter, the tidal forces they exert on each other stretch and distort them (Fig. 29.11). Under some circumstances, any interstellar matter in them can be pulled out and

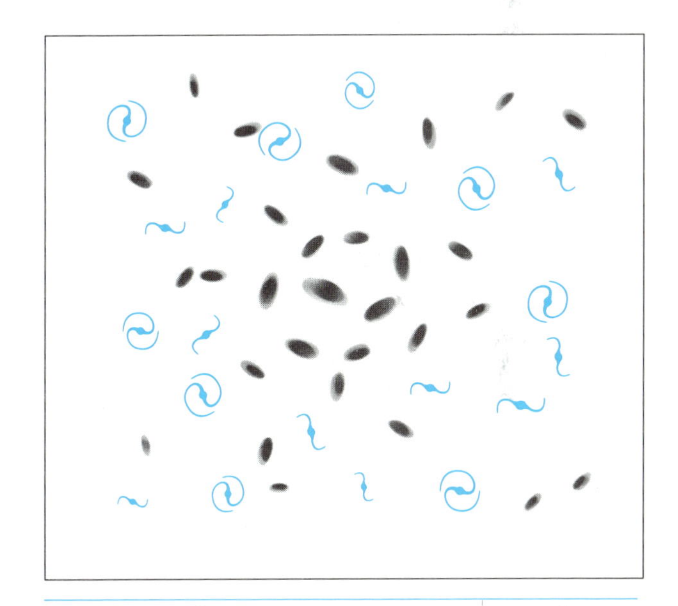

FIGURE 29.10 GALAXY TYPES IN A RICH CLUSTER. *In dense, highly populous clusters of galaxies, nearly all the members in the central portion are ellipticals, and there is often a giant, dominant elliptical at the center. Only in the outer portion are many spirals seen.*

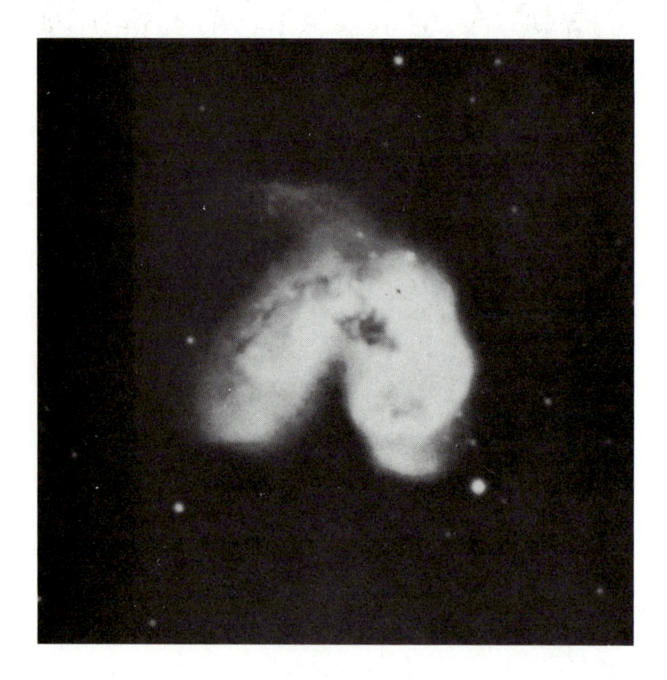

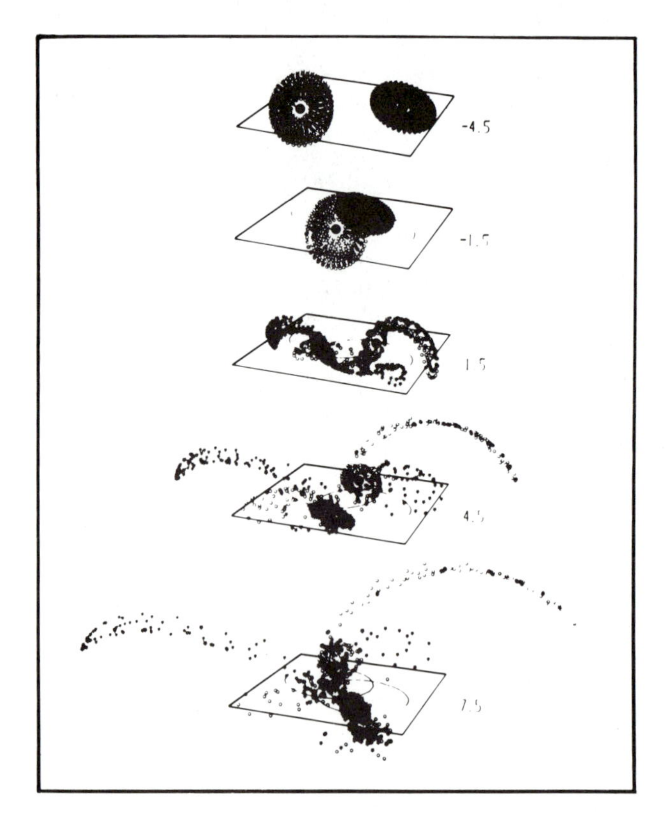

FIGURE 29.11 A COLLISION BETWEEN GALAXIES. *This shows a pair of galaxies undergoing a near-collision. At right is a computer simulation of their interaction, showing how the present appearance was created. This type of interaction between galaxies is believed responsible in some cases for the removal of interstellar matter, converting spiral galaxies to elliptical or SO galaxies.* (photograph from Palomar Observatory, California Institute of Technology; computer simulation from A. Toomre and J. Toomre)

dispersed. The outer regions, such as the halos, can also be stripped away. Intracluster gas (see next section) is even more effective at converting spirals to ellipticals. Pressure created by motion through this gas can sweep out a galaxy's interstellar matter and halt star formation. A spiral galaxy subjected to these cosmic upheavals may assume the form of an elliptical. In time, most of the spirals in an entire cluster, particularly those in the dense central region, may be converted into ellipticals and S0s.

The frequent gravitational encounters between galaxies in a rich cluster have another interesting effect: They cause a buildup of galaxies right at the center of the cluster. When two galaxies orbiting within a cluster come together, one will always gain speed and move to a larger orbit, while the other (always the more massive of the two) will lose speed and drop closer to the center. Thus a gradual sifting process, analogous to the differentiation that occurs inside some planets as the heavy elements sink toward the core, gradually builds up a dense central conglomeration of galaxies at the heart of the cluster. There these galaxies may actually merge, the end result being a single gigantic elliptical galaxy, which continues to grow larger as new galaxies fall in. In at least one case, a radio image shows a galaxy with two nuclei, apparently the result of a merger still in progress (Fig. 29.12).

The scenario just described was developed rather recently, as astronomers sought to explain why elliptical galaxies are so prevalent in many rich clusters, and why a single, giant elliptical is found at the center of many of these clusters (Fig. 29.13). These giants, designated cD galaxies, are at the extreme upper limit of known galactic sizes, masses, and luminosities, and are often peculiar in certain ways, as we will learn

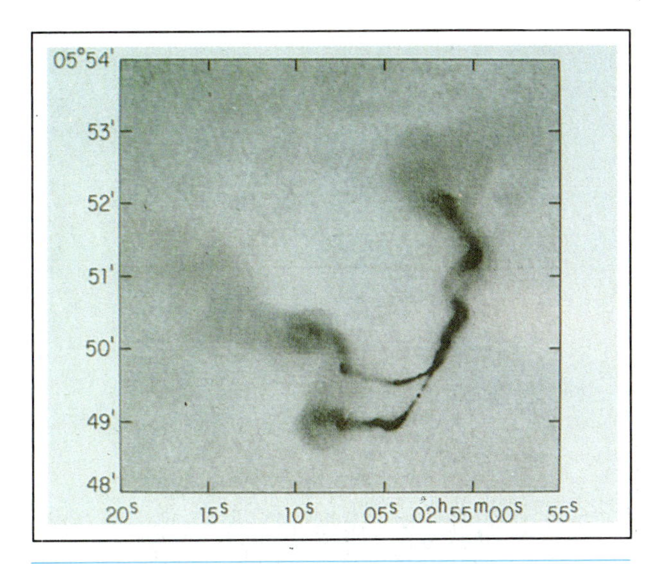

FIGURE 29.12 A RADIO IMAGE OF A GALACTIC MERGER. *This is a radio map of a giant elliptical galaxy with two nuclei (the bright points at lower center, each having a pair of wispy gaseous jets emanating from it; the jets are discussed in Chapter 22). Evidently this galaxy is in the process of forming from the merger of two galaxies whose centers have not yet completely merged.* (National Radio Astronomy Observatory)

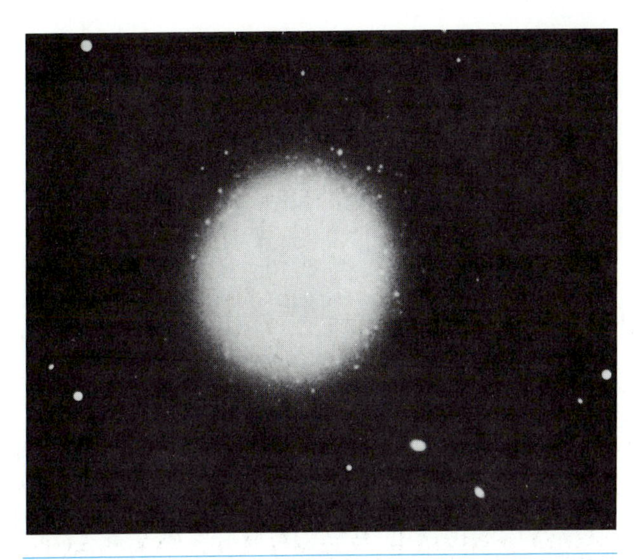

FIGURE 29.13 A GIANT ELLIPTICAL GALAXY. *This is M87, a well-known example of a dominant central galaxy in a rich cluster. This particular galaxy is discussed further in Chapter 31.* (National Optical Astronomy Observatories)

in Chapter 31. Interestingly, many galaxies outside of rich clusters also appear to have undergone mergers.

MASSES OF CLUSTERS

There are two methods for measuring the mass of a cluster of galaxies, and both are quite uncertain. This is a crucial problem, as we will see in Chapter 32, because of the importance of knowing how much mass the universe contains.

The simpler and more straightforward of the two methods is to estimate the masses of the individual galaxies in a cluster, (using techniques described in the previous chapter) and then to add up these masses. In many cases, particularly for distant clusters in which it is impossible to measure the rotation curves or velocity dispersions of individual galaxies, the only way to estimate their masses is to measure their brightnesses and then use a standard mass-to-light ratio to derive their masses. This technique is inaccurate because it depends on our knowing the distance, and because it assumes that the galaxies adhere to the usual mass-to-light ratios for their types. It also does not take into account any matter in the cluster that may lie between the galaxies.

The second method is similar to the velocity-dispersion technique used to estimate masses of elliptical galaxies (Fig. 29.14). The mass of a cluster is estimated from the orbital speeds of galaxies in its outer portions: The faster they move, the greater the mass of the cluster. This method has the advantage that it measures all the mass of the cluster, whether the mass is in galaxies or between them. But it has the disadvantage that the velocity measurements, particularly for a very distant cluster, are difficult, and it is necessary to know the cluster distance in order to determine how far from the center the observed galaxies are. Furthermore, the technique is only valid if the galaxies are in stable orbits about the cluster; if the cluster is expanding or some of the galaxies are not really gravitationally bound to it, then the results are incorrect. In rich clusters, at least, the smooth overall shape and distribution give the appearance of a bound system, and this technique is probably valid. It may not be valid for small clusters such as the Local Group, however.

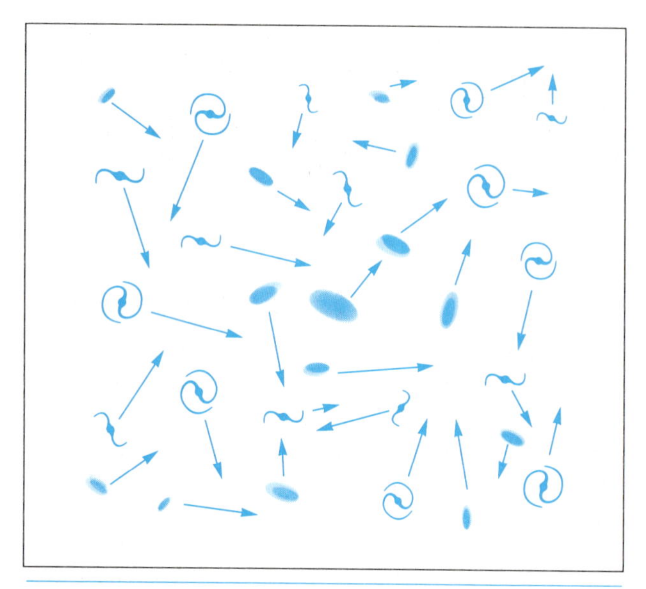

FIGURE 29.14 GALAXY MOTIONS WITHIN A CLUSTER. *Galaxies move randomly within their parent cluster. In a large cluster, where the overall distribution of galaxies is uniform, analysis of the average galaxy velocities is used to estimate total mass of the cluster, just as the velocity dispersion method is used to measure the masses of elliptical galaxies.*

FIGURE 29.15 INTRACLUSTER GAS IN THE VIRGO CLUSTER. *This is an X-ray image from the* Einstein Observatory, *showing emission from the entire central portion of this cluster of galaxies. The X rays are being emitted by very hot gas that fills the space between galaxies. The X-ray emission is shown in blue; radio images of galaxies in the cluster are shown in red.* (X-ray data from the Harvard-Smithsonian Center for Astrophysics; radio data and image processing by the National Radio Astronomy Observatory)

Whatever the reason, this method always leads to an estimated cluster mass that is much greater than that derived by adding up the masses of the visible galaxies. It is as though a large fraction of the cluster's mass is either in extensive galactic halos, which are ignored when the individual galaxy masses are estimated, or in intergalactic space.

INTERCLUSTER GAS AND X-RAY EMISSION

For a variety of reasons, astronomers have sought to know whether the space between galaxies is truly empty or whether there is an intergalactic medium analogous to the interstellar material that exists within galaxies. The quantity of intergalactic matter could play a vital role in the future evolution of the universe (see Chapter 32) and could have a profound effect on the fate of individual galaxies, in terms of their motions and their shapes. If galaxies move through a gaseous medium, drag will be created, which will help them congregate toward the centers of rich clusters and aid in stripping off their halos and interstellar gas; in this way spirals will be converted to ellipticals.

Searches for intergalactic gas that utilize techniques similar to those traditionally employed to study the interstellar matter in our own galaxy have generally yielded negative or ambiguous results. The lack of strong evidence for an intergalactic medium provides useful information, however. It tells us that there is very little gas there, and what is there must be so hot that it is essentially invisible, because a very highly ionized gas does not create visible-wavelength absorption lines.

The possibility that there is a very hot intergalactic medium appeared to be supported by recent X-ray observations of clusters of galaxies, which showed that many clusters are filled with a diffuse gas that glows at X-ray wavelengths (Fig. 29.15). The spectrum of the X-ray emission implies that this gas is indeed very hot; its temperature is estimated to be as high as 100 million degrees, much hotter than even the highly ionized gas in the interstellar medium in our galaxy. This observation raised the possibility that the general intergalactic void is filled with such gas, although if it is, the density outside the clusters must still be rather low, or the X-ray data would have revealed its presence.

A different type of X-ray measurement, however, indicated that the hot, intracluster gas is not part of a general intergalactic medium, but instead originates in the galaxies themselves. Spectroscopic measurements made at X-ray wavelengths revealed that the gas contains iron, a heavy element, in nearly the same quantity (relative to hydrogen) as in the Sun and other Pop. I stars. Such a high abundance of iron could only have been produced in nuclear reactions inside of stars. Therefore this intracluster gas must once have been involved in part of the cosmic recycling that goes on in galaxies, as stars gradually enrich matter with heavy elements before returning it to space. How the gas was then expelled into the regions between galaxies is not clear, but it may have been swept out during near-collisions between galaxies, or it may have been ejected in galactic winds created by the cumulative effect of supernova explosions and stellar winds.

SUPERCLUSTERS

We turn now to consider the overall organization of the matter in the universe. We noted at the beginning of this chapter that clusters of galaxies may represent the largest scale on which matter in the universe is organized, but that there is some evidence for a higher-order grouping. It appears that clusters of galaxies are concentrated in certain regions; these congregations of clusters are commonly referred to as **superclusters.** The Local Group appears to be part of a supercluster (Fig. 29.16).

The reality of superclusters has been difficult to establish, and for a while many astronomers were not

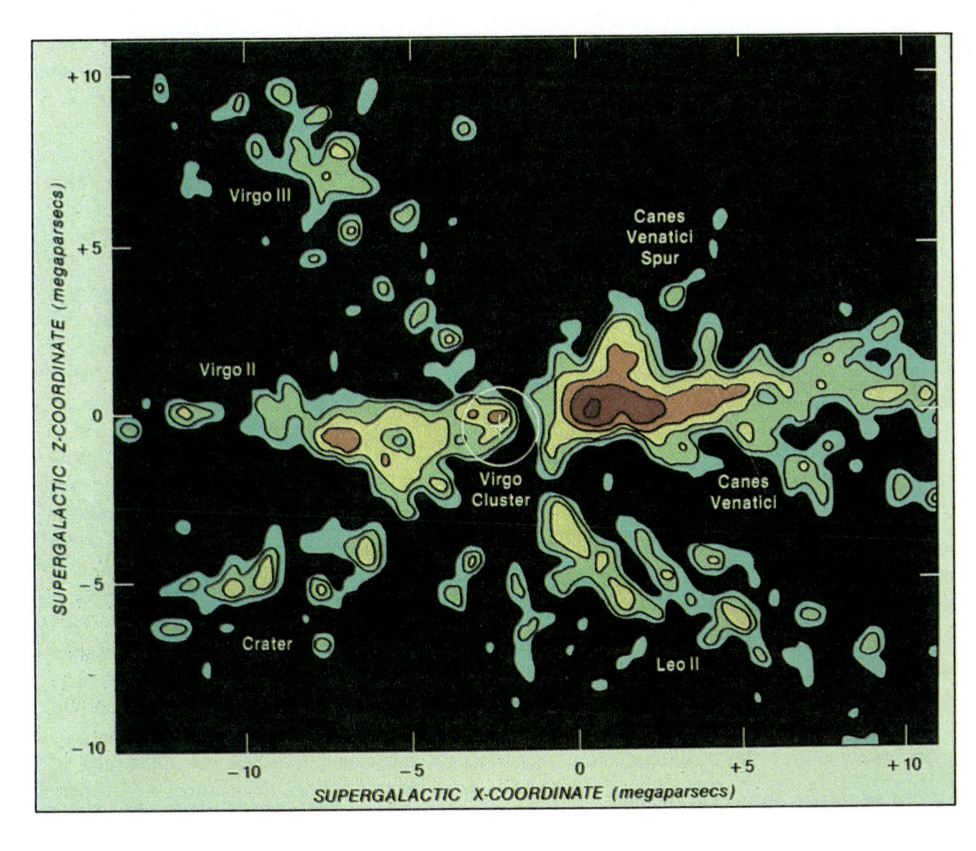

FIGURE 29.16 THE LOCAL SUPERCLUSTER. *This is an artist's concept of the cluster of galaxy clusters to which the Local Group belongs.* (Illustration from Sky and Telescope Magazine by Rob Hess, © 1982 by the Sky Publishing Corporation; also with the permission of R. B. Tully)

ASTRONOMICAL INSIGHT 29.1

THE DISTRIBUTION OF GALAXIES

A fundamental property of the universe is thought to be its homogeneity, or uniformity from one location to another. We discuss in this chapter the hierarchy of galaxies, clusters, and superclusters, but with the tacit assumption that the universe as a whole is more or less uniformly filled with these objects. As we will see in Chapter 32, the assumption of homogeneity is commonly made in the construction of models of the universe, but it is an assumption that is sometimes questioned. If the cosmos is not uniformly filled with matter, if the distribution is somehow unbalanced or uneven, it may have important implications for the formation and history of the universe.

A great deal of effort is made by astronomers to measure the distribution of matter. The technique most often used is to choose one or more specific, representative regions of the sky and carefully search for all galaxies in those regions brighter than some predetermined limit. Distances to all the detected galaxies are then estimated (usually using the Hubble expansion law described in Chapter 30), so that a three-dimensional picture can be developed of each sampled region of the sky. From this, it is possible to see what the overall distributions of galaxies is, and to discover whether any large-scale clumpiness or unevenness exists.

From a study such as that just described, the astronomical world recently received news of a large portion of space that was apparently devoid of visible matter. In the course of an extensive survey of faint galaxies, a group of astronomers found a region on the sky some 35 degrees across where there appeared to be no galaxies at distances between 240 million parsecs (Mpc) and 360 Mpc. At that distance, 35 degrees of angular diameter corresponds to a linear dimension of 180 Mpc, so it appeared that there was a void over 100 Mpc across in all dimensions, centered some 300 Mpc away.

Statistical studies show that random distributions of galaxies should produce voids, or empty zones, no larger than about 20 Mpc across. Gravitational effects, in which clumps in the distribution of matter are enhanced by the attraction of galaxies for each other, are thought to be capable of producing voids up to 35 Mpc in diameter. An empty space as large as 100 Mpc across would have profound implications, for at present there is no known way to produce such a hole in space, unless the universe is fundamentally inhomogeneous. Therefore, the announcement of the discovery of a hole in space immediately provoked a great deal of interest.

Since the original announcement of the existence of a void in the distribution of galaxies, several researchers have sought to verify the discovery. In a recent announcement, a group of astronomers have reported that the overall distribution of galaxies consists of a network of bubble-shaped voids, with the galaxies distributed in arclike sheets or filaments (see Fig. 29.17). This conclusion was based on the most complete and far-reaching sampling of galaxies yet carried out, involving some 1,100 galaxies in a large strip of sky near the north polar region of our own galaxy. (Observing in this direction avoids obscuration by interstellar gas and dust within the Milky Way.)

If the nonuniform distribution of galaxies in this recent survey is representative of the universe as a whole, this will have profound implications for theories of the early stages of the universe and of galaxy formation. The discovery of a cellular distribution of galaxies implies that the universe was somehow nonuniform from the beginning, and that galaxies simply formed where the concentrations of mass were highest.

One theory, published before the recent discovery, suggests that galaxy formation was triggered by the explosion of very massive primordial stars in supernovae. These massive stars were thought to have formed from concentrations of matter in the very young universe, before galaxies had formed. (The motivation for this theory was in part to explain how some heavy elements could have been present when galaxies formed.) A consequence of this theory was the prediction that galaxies should have formed in curved sheets as the result of expanding shock waves from the supernova explosions, creating arclike structures much like those that have now been found. This match between theory and observation appears to be a nice confirmation of the theory, except that the predicted structures are much smaller than those observed. Still, the idea that galaxies formed originally from the debris left over from primordial, gigantic explosions provides a tempting explanation for the bubblelike appearance of the distribution of galaxies in the universe that is implied by the recent observations described here.

convinced. Today, however, there are few doubts that clusters of galaxies tend to be grouped, although there is still disagreement on the significance. The best evidence lies in the distribution of rich clusters (some 1,500 of these have been catalogued), which clearly tend to congregate in certain regions, with relatively empty space in between.

The uncertainty that remains has to do with whether the groupings are random occurrences, or whether they reflect a fundamental unevenness in the distribution of matter in the universe. This is usually tested by comparing the apparent clustering of clusters with what would be expected from coincidence if they were actually distributed in a random fashion (after all, in any random scattering of objects, there will always be occasional accidental groupings). Mathematical experiments of this type appear to show that the observed superclusters could be random concentrations of clusters, and therefore may not have profound significance in terms of the universe as a whole.

Very recently, however, new surveys of galaxies have revealed what appears to be a distinct overall pattern to the distribution of superclusters, signifying that perhaps they are not the result of random accumulations of clusters. It appears that superclusters are sheetlike or stringlike, and are arranged in huge arcs spanning hundreds of millions of light-years (Fig. 29.17). Even if this pattern, which some think looks as though superclusters represented the walls of a cellular structure, is real, there is still uncertainty as to whether it indicates a fundamental structure of the universe, or merely tells us something about the formation mechanism for clusters and superclusters within a universe that is uniform on the whole. Possible causes of this distribution of clusters are discussed in the next section and in Chapter 32.

THE ORIGINS OF CLUSTERS

The question of how clusters of galaxies arose is, of course, intimately related to the mechanism for the formation of individual galaxies, already discussed briefly in Chapter 28. Many astronomers are at work on the problem, not so much for its own sake, but because of what it may tell us about the fundamental structure of the universe. As we will see, the question of how clusters and superclusters formed is related to questions of whether the universe as a whole is uniform or clumpy.

FIGURE 29.17 ARCLIKE SUPERCLUSTERS. *This shows the distribution of galaxies in a region of the sky that has recently been mapped completely, showing that superclusters are arranged in curved filamentary structures. This is evidence that the distribution of matter in the universe is not uniform, but that there is a large-scale structure.* (M. Geller, J. Huchra, M. Kurtz, and V. de Lapparent, Harvard-Smithsonian Center for Astrophysics).

There are certain similarities between the problem of galaxy formation and the problem of star formation, discussed earlier (see Chapter 22). A cloud of gas will collapse under its own self-gravity when it becomes dense enough so that gravity overcomes internal pressure. The pressure, however, depends on physical conditions such as density and temperature of the gas, so ultimately these physical conditions determine the mass of an object that will tend to form by gravitational collapse. The problem for the theorist trying to understand how galaxies and clusters of galaxies form is to discover what the physical conditions were early in the history of the universe.

There are two general classes of theories for galaxy cluster formation, corresponding to different physical conditions and therefore different initial masses of condensation. In one theory, called the "top-down" theory, the first objects to collapse had masses of around 10^{15} solar masses, corresponding to clusters of galaxies. This is called the top-down theory because clusters are thought to have formed first, and then fragmented to form individual galaxies. In the competing theory, the "bottom-up" model, individual galaxies (or even smaller objects) formed first, then aggregated through random motions and mutual gravitation to form clusters.

The key difference between the two theories is how hot the majority of the matter in the universe is, because the temperature of a gas influences the mass that is likely to become unstable and collapse under its own gravity. This may seem like an easy question to answer, but it is not, if we believe that most of the mass in the universe is dark and unobserved. We have seen that galaxies appear to have massive, invisible halos; later we will learn that many astronomers believe that the universe as a whole has about ten times more mass in it than we can see in the form of stars and galaxies. If this is true, then so-called "dark matter" in the universe dominates, and determines how galaxy formation occurred. The question, then, is whether the dark matter is hot or cold. If it is hot, then the first objects to collapse had the masses of galaxy clusters, so the top-down theory is favored. If, on the other hand, the dark matter is cold, then the first objects to collapse had masses comparable to globular clusters (about 10^5 solar masses), and the bottom-up theory must prevail.

Neither theory accounts for superclusters directly, so we must resort to other mechanisms to form them. In either model, the existence of arc-like superclusters seems to require some kind of disturbance in the early universe that would have created bubbles or ripples. In the bottom-up theory, where the dark matter is assumed to be cold and relatively small objects could form, these early disturbances could have been created by supernova explosions (or perhaps more appropriately, "super" supernova explosions), as an early generation of very massive stars formed and then exploded. The resulting blast waves could have created bubbles in the early universe that are reflected today by the structure of the superclusters.

This does not work if the dark matter in the universe is hot, however, because there would have been no early generation of massive stars. An exotic recent proposal in that case is that disturbances were created by **cosmic strings.** These can be described mathematically, but are very difficult to envision physically. Cosmic strings are described as long, thin warps in the fabric of spacetime, as though the material of the universe had been cut and overlapped. These defects in spacetime, if they exist, can have tremendous local gravitational fields, and can also have very strong emissions of electromagnetic radiation, sufficiently powerful to repel matter and form arcs and bubbles resembling today's superclusters.

The subject of cosmic strings is popular today, but the existence of such things is by no means established. Meanwhile, the question of top-down versus bottom-up models for the formation of clusters of galaxies remains open. The bottom-up scenario has a couple of advantages: (1) it more easily accounts for the arc-like structure of superclusters, as described above; and (2) it can more easily explain the shapes of galaxies. Recall that in the bottom-up model, the first objects to collapse were smaller than galaxies, more like globular clusters in mass. Many of these objects remained as they formed, and are seen today as dwarf galaxies. Others merged to form normal galaxies. In dense regions, which were on their way to becoming rich clusters of galaxies, mergers of these small pre-galactic objects would have occurred readily, using up all the interstellar gas in star formation and building up large elliptical galaxies. In less dense regions, destined to become outer parts of clusters or not to be in clusters at all, mergers would have occurred at a much slower rate, allowing individual pregalaxies to accumulate gas from their surroundings and form disks. Thus spiral galaxies tend to be found in the less dense regions of the universe, while the dense clusters of galaxies are dominated by ellipticals, consistent with observations.

Interestingly, the key to how galaxies and clusters formed ties into the possible existence and nature of unseen matter in the universe. As we will see in Chapter 32, the fate of the universe itself is dependent on the quantity and nature of dark matter, so this will be a theme not only of the rest of this text, but also of astronomy itself for the next several years.

PERSPECTIVE

We have, at last, finished our tour of the universe, having discussed nearly all the forms and types of organization that matter can take. We still must deal with a few specific kinds of objects that have not yet been described, but the overall picture of the universe in its present state is now more or less complete. We are ready to tackle questions having to do with the nature of the universe itself, its overall properties, and its dynamic nature, and we will do so before examining some of the peculiar objects that are clues to the past.

SUMMARY

1. Many galaxies are members of clusters, rather than being randomly distributed throughout the universe.
2. The Milky Way is a member of a cluster called the Local Group, which contains about thirty galaxies.
3. The nearest neighbors to the Milky Way are the Magellanic Clouds, both Type I Irregulars.
4. The Andromeda galaxy, a huge spiral of type Sb, is similar in size and general properties to the Milky Way, and the two are among the most prominent members of the Local Group.
5. Rich clusters of galaxies may contain thousands of members. In contrast to the general situation for isolated galaxies and small groups, in rich clusters elliptical and S0 galaxies are the most common, and there is a relatively small fraction of spirals.
6. The dominance of elliptical galaxies in rich clusters is probably the result of the conversion of spirals into ellipticals by tidal forces exerted by other galaxies and by drag created by intracluster gas.
7. In many rich clusters there is a giant elliptical galaxy at the center that probably formed from the merger of several galaxies that settled there as a result of collisions.
8. Masses of clusters can be determined from the sum of the masses of individual galaxies (measured by techniques described in Chapter 28), or from the internal velocity dispersion of the galaxies in the cluster.
9. Although there is no evidence for a general intergalactic medium, there is a very hot, rarefied gas filling the space between galaxies in some rich clusters. Observed best at X-ray wavelengths, this gas appears to have been created by the galaxies themselves (through mass loss in galactic winds that result from stellar winds and supernova explosions).
10. Clusters of galaxies tend to be grouped into aggregates called superclusters, but these may be random concentrations, rather than fundamental inhomogeneities of the universe. Superclusters appear to be sheetlike, with an overall distribution resembling a series of huge arclike structures, and this may tell us something about the early distribution of matter in the universe.
11. Clusters of galaxies formed because of an uneven distribution of matter at some point early in the history of the universe, but it is not known how this clumpiness originally developed. One of the two competing possibilities is that the early universe was hot, so that only very massive objects could collapse; this is the "top-down" theory, in which clusters of galaxies formed first and then fragmented to form galaxies. The other scenario, called the "bottom-up" theory, assumes that the matter in the universe was cold, so that objects as small as galaxies could condense and then aggregate into clusters. In either model, superclusters result from early disturbances in the universe, possibly caused by cosmic strings (in the top-down model) or early supernovae (in the bottom-up model).

REVIEW QUESTIONS

Thought Questions

1. Summarize the membership of the Local Group of galaxies.

2. Why does a small cluster such as the Local Group have an amorphous shape rather than a spherical, smooth shape such as is seen in rich clusters of galaxies?

3. How does our location in a cluster (the Local Group) help us to measure distances to galaxies far away, outside of our cluster?

4. Explain why the relative numbers of galaxies of different types may not be the same in a rich cluster of galaxies as it is among galaxies not belonging to such clusters.

5. Why would it be surprising if the Local Group included a giant elliptical galaxy among its members?

6. Summarize the techniques that are used for finding the masses of clusters of galaxies. For each method discussed, explain any difficulties or shortcomings.

7. Why is it thought that the intracluster gas observed in rich clusters of galaxies has been expelled from the galaxies in the cluster, rather than being truly intergalactic gas?

8. Explain why there has been some question about the significance of superclusters, even though their existence is well established.

9. What explanations have been offered by astronomers for the apparent concentration of superclusters in arclike or sheetlike structures?

10. How is the problem of explaining the formation of individual galaxies similar to the problem of explaining the formation of individual stars within a galaxy?

Problems

1. If a dwarf elliptical galaxy has an absolute magnitude of $M = -15$, and could be detected with an apparent magnitude as faint as $m = +20$, how far away can these galaxies be found? Compare this distance with the diameter of the Local Group and with the distance of the Virgo cluster of galaxies, a moderately large cluster some 15 Mpc away.

2. What is the travel time for light to reach us from: (a) the Magellanic Clouds; (b) the Andromeda galaxy; (c) the Virgo cluster; (d) a distant cluster that is 500 Mpc away? What do these times imply for our ability to understand the evolution of galaxies and of the universe?

3. Suppose the temperature of intracluster gas in a rich cluster of galaxies is 100 million degrees (10^8 K). At what wavelength does this gas emit most strongly? What kind of telescope is required to observe it?

ADDITIONAL READINGS

de Boer, K. S., and B. D. Savage. 1982. The coronas of galaxies. *Scientific American* 247(2):54.

Hirshfeld, A. 1980. Inside dwarf galaxies. *Sky and Telescope* 59(4):287.

Hodge, P. W. 1966. *Galaxies and cosmology*. New York: McGraw-Hill.

Larsen, R. B. 1977. The origin of galaxies. *American Scientist* 65:188.

Mitton, S. 1976. *Exploring the galaxies*. New York: Scribner's.

Page, T., and L. W. Page, eds. 1969. *Beyond the Milky Way*. New York: Macmillan.

Rubin, V. C. 1983. Dark matter in spiral galaxies. *Scientific American* 248(6):96.

Sandage, A., M. Sandage, and J. Kristian, eds. 1976. *Galaxies and the universe*. Chicago: University of Chicago Press.

Shapley, H., and P. W. Hodge. 1972. *Galaxies*. Cambridge, Ma.: Harvard University Press.

Strom, S. E., and K. M. Strom. 1977. The evolution of disk galaxies. *Scientific American* 240(4):56.

Talbot, R. J., E. B. Jensen, and R. J. Dufour. 1980. Anatomy of a spiral galaxy. *Sky and Telescope* 60(1):23.

Toomre, A., and J. Toomre. 1973. Violent tides between galaxies. *Scientific American* 229(6):38.

van den Bergh, S. 1976. Golden anniversary of Hubble's classification system. *Sky and Telescope* 52:410.

CHAPTER 30

UNIVERSAL EXPANSION AND THE COSMIC BACKGROUND

A microwave receiver used to measure the anisotropy of the cosmic background radiation. (George Smoot, Lawrence Berkeley Laboratory)

CHAPTER PREVIEW

A recurrent theme throughout our study of astronomy has been the dynamic nature of celestial objects. Everything we have studied in detail, from planets to stars to galaxies, has turned out to be in a constant state of change. Now that we are prepared to examine the entire universe as a single entity, we should not be surprised to see this theme maintained. That the universe itself is evolving, with a life story of its own, has been accepted by most astronomers, but there was doubt in the minds of some until recently. In this chapter we will study the evidence.

HUBBLE'S GREAT DISCOVERY

Even before it was established unambiguously that the nebulae were really galaxies, a great deal of effort went into observing them. Their shapes were studied carefully, and their spectra were analyzed in detail. Typically the spectrum of a galaxy resembles that of a moderately cool star, with many absorption lines (Fig. 30.1). Because the spectrum represents the light from a huge number of stars, each moving at its own velocity, the Doppler effect causes the spectral lines to be rather broad and indistinct. Nevertheless, it is possible to analyze these lines in some detail, and one of the early workers in this field, V. M. Slipher, discovered during the decade before the Shapley-Curtis debate in 1920 that the nebulae tend to have large velocities away from the Earth. The spectral lines nearly always were found to be shifted towards the red, and this tendency was most pronounced for the faintest of the nebulae. Slipher measured velocities as great as 1,800 km/sec.

Following his work, others continued to study spectra of nebulae, with heightened interest after Hubble's demonstration in 1924 that these objects were undoubtedly distant galaxies, comparable to the Milky Way in size and complexity. Hubble and others estimated distances to as many nebulae as possible, by use of the Cepheid variable period-luminosity relation for the nearby ones, and other standard candles such as novae and bright stars for more distant nebulae.

In 1929 Hubble made a dramatic announcement: The speed with which a galaxy moves away from the Earth is directly proportional to its distance. If one galaxy is twice as far away as another, for example, its velocity is twice as great. If it is 10 times farther away, it is moving 10 times faster.

The implications of this relationship between distance and velocity are enormous. It means that the universe itself is expanding, its contents rushing outward at a fantastic pace. All the galaxies in the cosmos are moving away from all the others (Fig. 30.2).

To envision why the velocity increases with increasing distance, it is perhaps helpful to resort to a commonly used analogy. Imagine a loaf of bread that is rising. If the dough has raisins sprinkled uniformly through it, they will move farther apart as the dough expands. Suppose that the raisins are 1 centimeter apart before the dough begins to rise, and consider what has happened 1 hour later, after it rises to the point at which adjacent raisins are 2 centimeters apart. A given raisin is now 2 cm from its nearest neighbor, but 4 cm away from the next one over, and 6 cm from the next one, and so on. The distance between any pair of raisins has doubled. From the point of view of any one of the raisins, its nearest neighbor had to move away from it at a speed of 1 cm/hr, the next raisin farther away had to move at 2 cm/hr, the next one had a speed of 3 cm/hr, and so on. If we let the dough continue to rise to the point where the adjacent raisins are 3 cm apart, then we find that all the distances between pairs have tripled over what they were to start with, and the speeds needed to accomplish this are again directly proportional to the distance between a given pair. The farther away a raisin is to begin with, the farther it must move to maintain the regular spacing during the expansion, and therefore the greater is its velocity. In the same way, the galaxies in the universe must increase their separa-

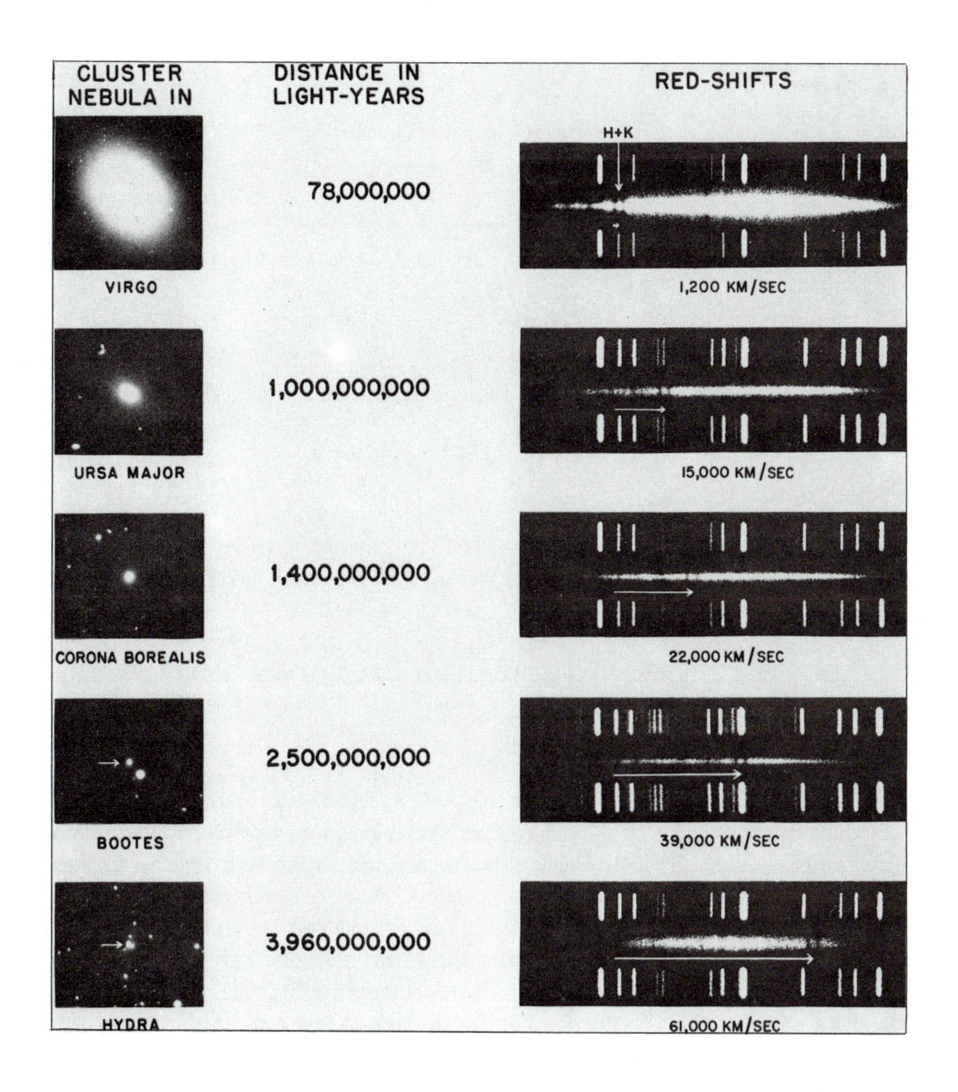

FIGURE 30.1 SPECTRA OF GALAXIES. *These examples illustrate the broad absorption lines characteristic of spectra of large groups of stars such as galaxies. It was noticed before 1920 that the spectral lines tend to be shifted toward the red in galaxy spectra, which indicates that galaxies as a rule are receding from us.* (Palometer Observatory, California Institute of Technology)

tions from each other at a rate proportional to the distance between them (Fig. 30.3), or else they would become bunched up. By observing the velocities of galaxies, astronomers are keeping watch on the raisins in order to see how the loaf of bread is coming along.

It is important to realize that it doesn't matter which raisin we choose to watch; from any point in the loaf, all other raisins appear to move away with speeds proportional to their distances from that point. Thus we do not conclude that our galaxy is at the center of the universe; from *any* galaxy, it would appear that all others are receding. It is also important to realize that the raisin-bread analogy has serious shortcomings: It suggests that the universe can be viewed as a simple three-dimensional object, with an edge and a center, which is inappropriate (this is discussed further in Chapter 32). It is better to try to envision a universe with no boundaries and no center; space itself is getting bigger. There is no location within it that can today be identified as the center, and there is no "outside" from which the universe can be viewed.

Following Hubble, several other astronomers have extended the study of galaxy motions to greater and greater distances, with the same result: As far as the telescopes can probe, the galaxies are moving away from each other at speeds proportional to their distances. Expansion is a major feature of the universe we live in, one that must be taken into account as we seek to understand its origins and its fate.

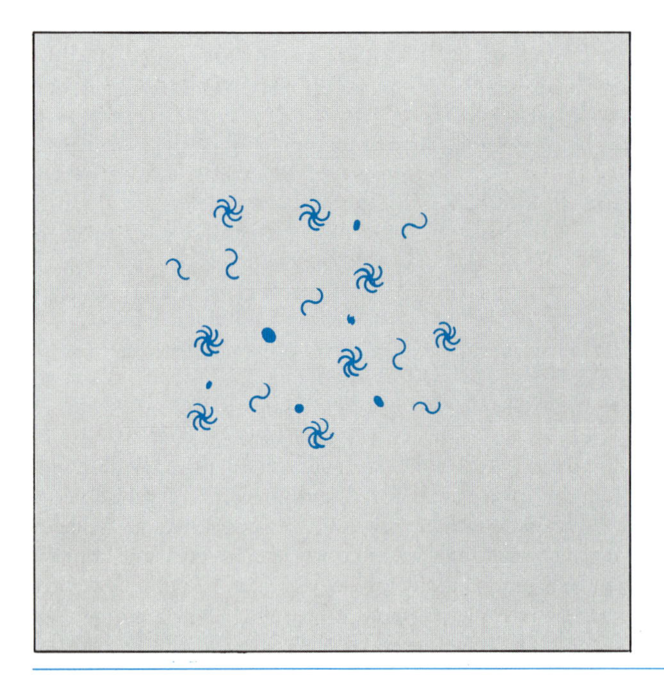

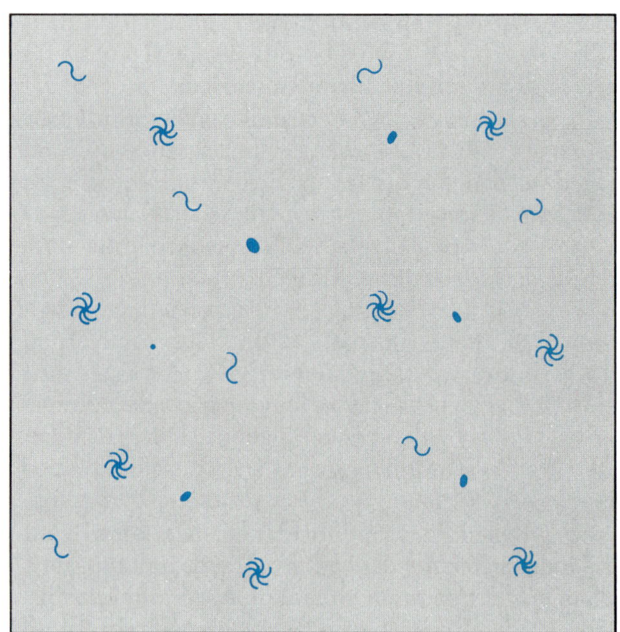

FIGURE 30.2 THE EXPANDING UNIVERSE. *These drawings show a number of galaxies at one time (left), and again at a later time (right). All of the spacings between galaxies have increased due to expansion of the universe.*

It is often convenient to display the data showing the universal expansion in a graph of velocity versus distance (as Hubble originally did; see Fig. 30.4), or on a plot of redshift versus apparent magnitude (Fig. 30.5), since increasing apparent magnitude (i.e., decreasing brightness) indicates increasing distance.

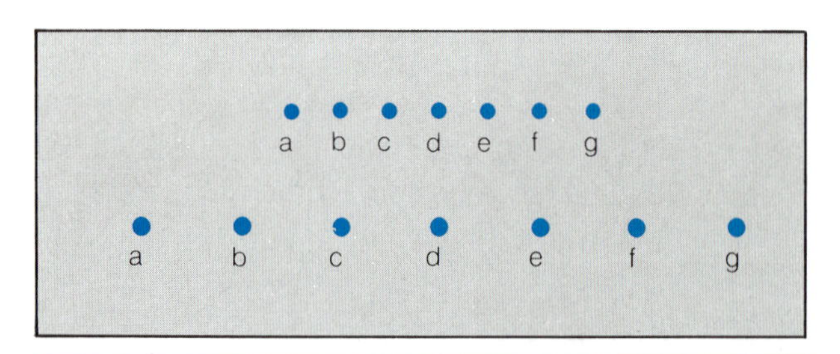

FIGURE 30.3 VELOCITY OF EXPANSION AS A FUNCTION OF DISTANCE. *At upper left is a row of galaxies, and at lower right is the same row one billion years later, when the distances between adjacent galaxies have doubled from 1 to 2 megaparsecs. From the viewpoint of an observer in* any *galaxy in the row, the recession velocity of its nearest neighbor is 1 megaparsecs per billion years; of its next nearest neighbor, 2 megaparsecs per billion years; of the next, 3 megaparsecs per billion years; and so on. The velocity is proportional to the distance in all cases, for any pair of galaxies.*

The galaxies within a cluster have their own orbital motions, which are separate from the overall expansion (Fig. 30.6). As the universe expands, the clusters move apart from each other, while the individual galaxies within a cluster do not. The motions of galaxies within a cluster are relatively unimportant for very distant galaxies that are moving away from us at speeds of thousands of km/sec, but for nearby ones, the orbital motion can rival or exceed the motion caused by the expansion of the universe. The Andromeda galaxy, for example, is actually moving *towards* the Milky Way at a speed of about 100 km/sec, whereas at its distance of 700 kpc from us, the universal expansion should give it a recession velocity of about 40 km/sec.

HUBBLE'S CONSTANT AND THE AGE OF THE UNIVERSE

The relation discovered by Hubble can be written in simple mathematical form:

$$v = Hd,$$

where v is a galaxy's velocity of recession and d is its distance, in megaparsecs. The Hubble constant, H, is

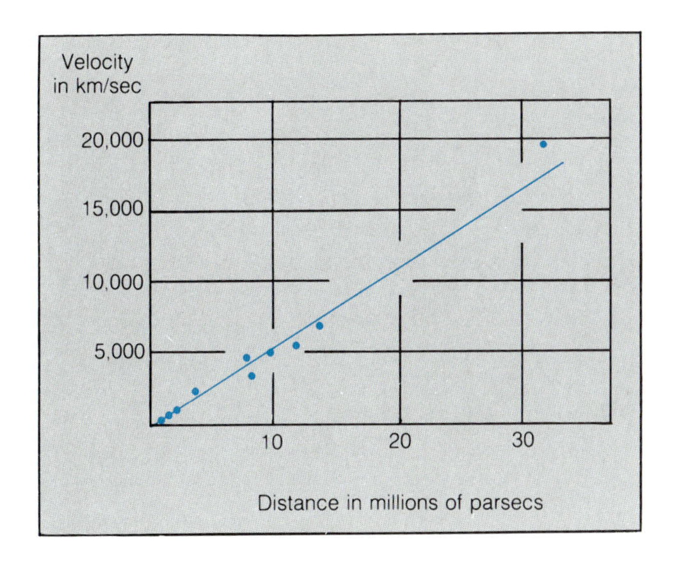

FIGURE 30.4 THE HUBBLE LAW. *This early diagram prepared by Hubble and M. Humason shows the relationship between galaxy velocity and distance. The slope of the relation has been altered since, with addition of data for more galaxies and improvements in the measurement of distances, but the general appearance of this figure is the same today.* (Estate of E. Hubble and M. Humason. Reprinted with permission)

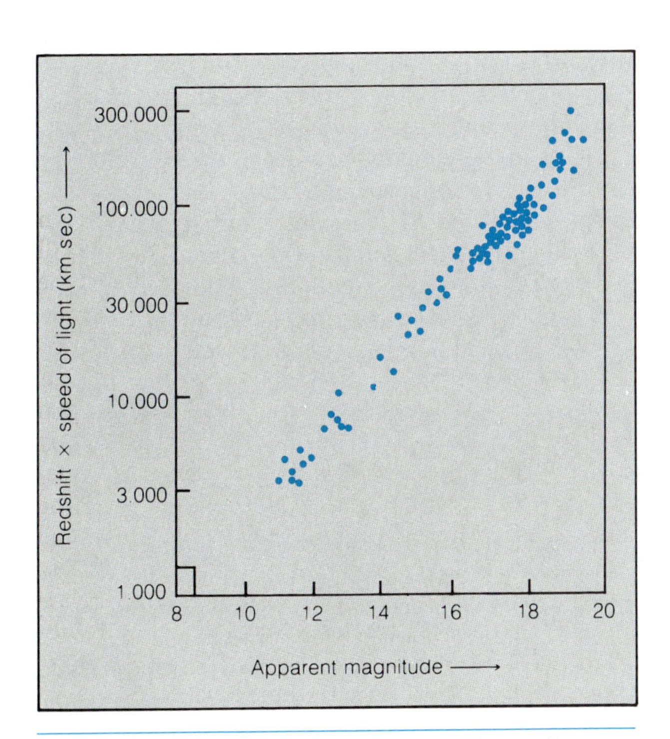

FIGURE 30.5 A MODERN VERSION OF THE HUBBLE LAW. *Often, apparent magnitude is used on the horizontal axis instead of distance, because the two quantities are related, particularly if the diagram is limited to galaxies of the same type, as this one is. At lower left the small rectangle indicates the extent of the relationship as Hubble first discovered it; today many more galaxies, much dimmer, have been included.* (Adapted from J. Silk 1980, *The Big Bang* [San Francisco: W. H. Freeman])

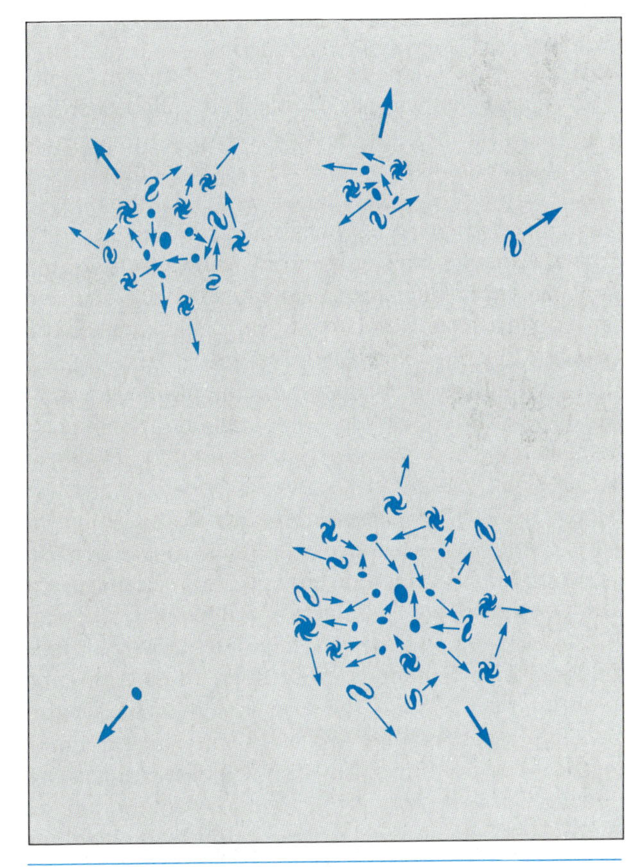

FIGURE 30.6 LOCAL MOTIONS. *Although there is a systematic overall expansion of the universe, individual galaxies within clusters, and even clusters within superclusters, have random individual motions. Hence within the Local Group the galaxies are not uniformly receding from each other.*

TABLE 30.1 EXAMPLE MEASUREMENTS OF THE HUBBLE CONSTANT

Value of H (km/sec/Mpc)	Observer	Date	Implied Age of Universe (years)
540	Hubble	1930	1.8×10^9
260	Baade	1949	3.8×10^9
180	Humason et al.	1955	5.4×10^9
75	Sandage	1956	1.3×10^{10}
56	Sandage and Tammann	1974	1.8×10^{10}
100	van den Bergh	1975	1.0×10^{10}
80	Tully and Fisher	1977	1.2×10^{10}
100	de Vaucouleurs, et al.	1979	1.0×10^{10}
95	Aaronson	1980	1.0×10^{10}
65	Mould, et al.	1980	1.5×10^{10}
50	Sandage and Tammann	1982	2.0×10^{10}
82	Aaronson and Mould	1983	1.2×10^{10}

*Information from M. Rowan-Robinson, *The Cosmological Distance Ladder*, 1985 (New York: W. H. Freeman).

then given in units of km/sec/Mpc, if v is in units of km/sec.

The value of H is difficult to establish (Table 30.1). It is done by collecting as large a body of data as possible on galactic velocities and distances, and then deducing the value of H that best represents the relationship between distance and velocity. Hubble did this first, finding that $H = 540$ km/sec/Mpc, meaning that for every megaparsec of distance, the velocity increased by 540 km/sec. The standard candles were not very well established in Hubble's day, when it was still news that the nebulae were distant galaxies, and this turned out to be an overestimate. The best modern values for H are between 50 and 100 km/sec/Mpc. Until very recently, a consensus had been developing that the best value is 55 km/sec/Mpc, but a new distance determination for a number of galaxies has now suggested a value close to 90 km/sec/Mpc. The precise value of H has extremely important implications for our understanding of the universe and its expansion, and intense research effort is being devoted to refining the estimates.

If the universe is expanding, it follows that all the matter in it used to be closer together than it is today (Fig. 30.7). If we follow this logic to its obvious conclusion, then we find that the universe was once

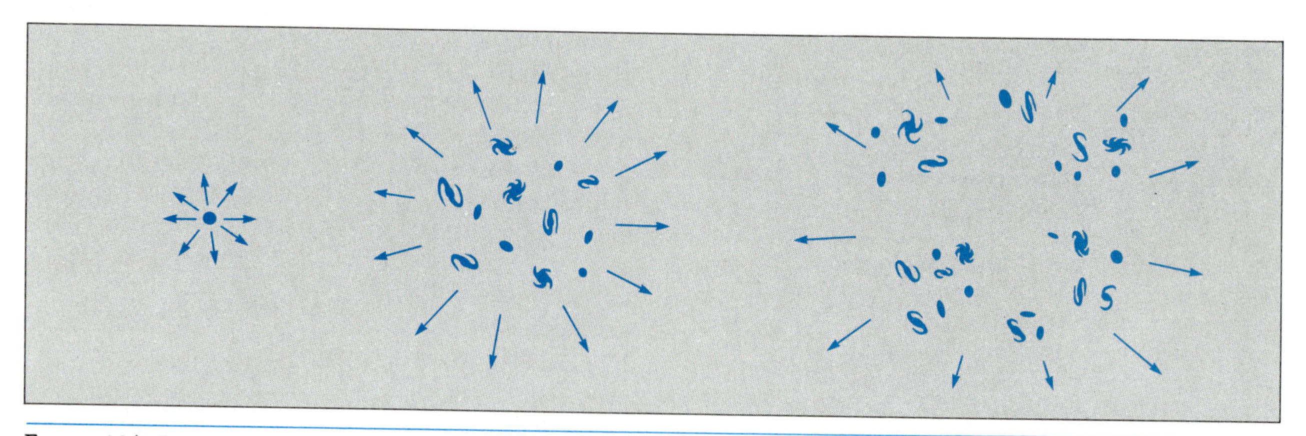

FIGURE 30.7 BACKTRACKING THE EXPANSION. *If the universe began as a single point, the present rate of expansion provides information on how long ago the expansion began. With considerable uncertainty due to the fact that the expansion rate has probably not been constant, this leads to an estimated age for the universe of between 10 and 20 billion years.*

concentrated in a single point, from which it has been expanding ever since.

From the rate of expansion, we can calculate how long ago the galaxies were all together in a single point. From the simple expression

$$\text{Time} = \text{distance/velocity},$$

we find

$$\text{age} = d/Hd = 1/H.$$

The age of the universe is equal to $1/H$, if the expansion has been proceeding at a constant rate since it began. As we will see in Chapter 32, this assumption of a constant expansion rate is not strictly true (it was more rapid in the beginning), but it gives a useful estimate of the age for now.

If we adopt, for the sake of discussion, a value for H of 75 km/sec/Mpc, then the age of the universe is 4.1×10^{17} sec = 1.3×10^{10} years. (To carry out this calculation, first you must convert the value of a megaparsec into kilometers.)

A simple calculation based on the observed expansion rate has led to a profound conclusion: Our universe is about 13 billion years old. If this value is correct, it just fits in with what we have learned about galactic ages. The Sun, for example, is thought to be about 4.5 billion years old, and the most aged globular clusters are apparently some 14 billion years old, acceptably close, within the uncertainties of the measurements. On the other hand, a much higher value of H would lead to an age for the universe that is much smaller than the estimated ages of globular clusters, and those who favor a value nearer to 100 km/sec/Mpc are facing just such a dilemma.

We can easily see what happens to our estimate of the age of the universe if we choose other values of H. For example, if Hubble's value of 540 km/sec/Mpc had been correct, then the age would have been calculated as less than 2 billion years, and if the recently suggested value of about 90 km/sec/Mpc is found to be correct, then the age is 11 billion years. In either case, there is a conflict with the estimated ages of globular clusters; it will be interesting to see how this conflict is resolved if indeed a value for H of 90 km/sec becomes widely accepted. The conflict worsens when we realize that age estimates based on $1/H$ are probably too large, because they do not take into account the fact that the expansion rate at the earliest times was somewhat more rapid than it is today. This further complicates things, if a value of 90 km/sec/Mpc is found to be best. Whatever the correct ages are, this situation serves to illustrate how uncertain some of the most important measurements in astronomy can be.

There are many other important questions about the universe that depend on the value of H, and this is the reason so much effort is being expended to define its value as accurately as possible. One of the principal reasons for development of the *Space Telescope* is to probe the most distant possible galaxies, to determine their distances and velocities so that this information can be used to help find the correct value of H.

REDSHIFTS AS YARDSTICKS

The expansion of the universe has a very practical benefit for astronomers concerned with the properties of distant galaxies: It provides a means of determining their distances (Fig. 30.8). If we know the value of

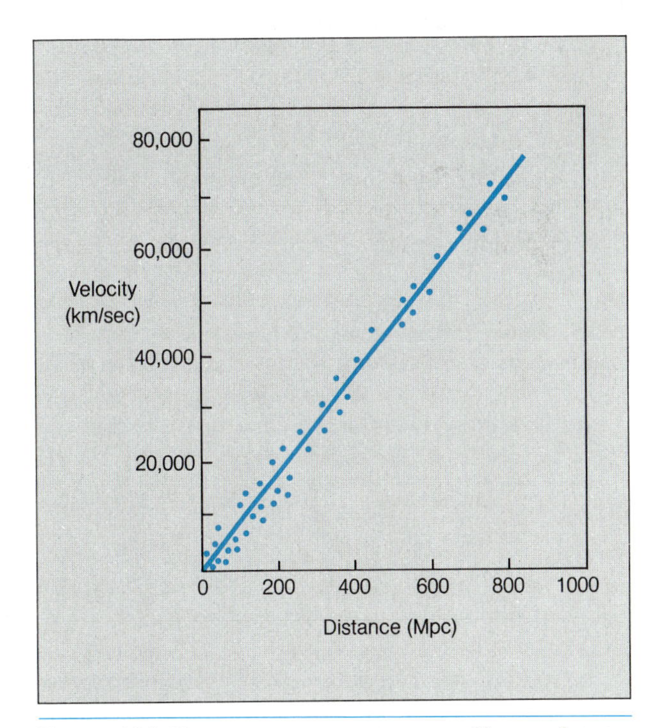

FIGURE 30.8 FINDING DISTANCE FROM THE HUBBLE LAW. *This diagram shows how the distance to a galaxy can be deduced directly from the Hubble law, if the redshift, hence the velocity, is measured.*

ASTRONOMICAL INSIGHT 30.1

THE DISTANCE PYRAMID

Having reached the ultimate distance scales, we can pause to look back at the progression of steps that got us here. As noted in the text, the use of Hubble's relation to estimate distances to the farthest galaxies depends critically on the value of H, which can only be determined from knowledge of the distances to a representative sample of galaxies. Those distances in turn depend on various standard candles, themselves calibrated through the use of other methods, such as the Cepheid variable period-luminosity relation, which are applicable only to the most nearby galaxies. Our knowledge of these close extragalactic distances rests on the known distances to objects within the Milky Way, such as star clusters containing variable stars, which are used to calibrate the period-luminosity relation. The distances to these clusters, in their turn, are known from more local techniques, such as spectroscopic parallaxes and main-sequence fitting, and these ultimately depend on the distances to the most nearby stars, derived from trigonometric parallaxes.

Distance (pc)	Key Object	Method
10^{-6}	Sun	↓ Radar, planetary motions
10^{-5}		
10^{-4}		
10^{-3}		
10^{-2}		Trigonometric parallax
10^{-1}		
1	α Centauri	
10^{1}		
10^{2}	Hyades cluster	↓ Moving cluster method
10^{3}		Main-sequence fitting
10^{4}	Limits of Milky Way	
10^{5}	Magellanic Clouds	
10^{6}	Andromeda galaxy	Cepheid variables
10^{7}		↓ Brightest stars
10^{8}	Virgo cluster	
10^{9}		
10^{10}		Brightest galaxies

Thus, the entire progression of distance scales, all the way out to the known limits of the universe, depends on our knowledge of the distance from the Earth to the Sun, the basis of trigonometric parallax measurements. At every step of the way outward, our ability to measure distances rests on the previous step. If we revise our measurement of local distances, we must accordingly alter all our estimates of the larger scales, affecting our perception of the universe as a whole. This elaborate and complex interdependence of distance determination methods is known as the **distance pyramid,** and it is sometimes depicted in graphic form (see table).

A very important step in the sequence shown above is represented by the Hyades cluster, a galactic star cluster some 43 parsecs away. The distance to this cluster is determined from the **moving cluster method,** in which a combination of Doppler shift measurements and proper motion measurements (see Chapter 19) reveal the true direction of cluster motion, so that a comparison of angular motion and true velocity then gives the distance to the cluster. Because much of the rest of the distance pyramid depends on the application of this technique to the Hyades cluster, great care has been taken in the analysis, and every so often it is redone. Whenever the distance to the Hyades is altered slightly, the impact spreads as astronomers judge how this affects other distance scales in the universe. One of the things the *Space Telescope* may accomplish is to make direct parallax measurements of stars in the Hyades, providing a new and important measurement of the distance to the cluster and thus revising, once again, our understanding of the distance scale of the universe.

H, then we can find the distance to a galaxy simply by measuring its velocity as indicated by the Doppler shift of its spectral lines. The distance is given by

$$d = v/H.$$

Thus, if we adopt $H = 75$ km/sec/Mpc, then a galaxy found to be receding at 7,500 km/sec, for example, has the distance

$$d = 7{,}500/75 = 100 \text{ Mpc}.$$

This technique can be applied to the most distant and faint galaxies, where it is nearly impossible to observe any standard candles. It is important to remember, however, that the use of Hubble's relation to find distances is only as accurate as our value of H, and the value that we use is derived in turn from distance determinations that depend on standard candles. Furthermore, because the expansion rate of the universe has not been constant, for very distant galaxies it is necessary to know what the rate was in the past to accurately determine the distance.

A Cosmic Artifact: The Microwave Background

Since Hubble's discovery of the universal expansion, many astronomers and physicists have explored its significance, both for what it may tell us about the future and for the information it provides about the past. A number of scientists, beginning in the 1940s with George Gamow and his associates, have done extensive studies of the nature of the universe in its earliest days. It was quickly realized that, if all the matter were originally compressed into a small volume, it must have been very hot, for the same reason that any gas heats up when compressed. The fiery conditions inferred for the early stages of the expansion have led to the term **big bang,** which is the commonly accepted title for all theories of the universal origin that start with an expansion from a single point.

Gamow's primary interest was in analyzing nuclear reactions and element production during the big bang (the modern understanding of this is described in Chapter 32). But in the course of his work he came to another important realization: The universe must have been filled with radiation when it was highly compressed and very hot, and this radiation should still be with us. At first, when the temperature was in the billions of degrees, it was γ-ray radiation, but later, as the universe cooled, next X-rays, and then ultraviolet light filled it. (Recall the relationship between temperature and wavelength of maximum emission, as stated in Wien's law; see Chapter 6). During the first million years or so after the expansion started, the radiation was constantly being absorbed and reemitted (i.e., scattered) by the matter in the universe, but eventually (when the temperature had dropped to around 4,000 K), protons and electrons combined to form hydrogen atoms, and from that time on there was little interaction between the matter and the radiation. The matter, in due course, organized itself into galaxies and stars, while the radiation has simply continued to cool as the universe has continued to expand (Fig. 30.9). The intensity and spectrum of the radiation is dependent only on the temperature of the universe and is described by the laws of thermal radiation (see Chapter 4).

Gamow and others made rough calculations showing that the present temperature of the radiation filling the universe should be very low indeed, about 25 K. More recently, R. H. Dicke and colleagues, quite independently, carried out new calculations, predicting an even lower temperature, about 5 K. Using Wien's law, it is easy to calculate the most intense wavelength of such radiation: If the temperature is 5 K, then the value of λ_{max} is 0.06 cm, in the microwave part of the radio spectrum.

Gamow's principal interest was in element production during the big bang, rather than in the remnant radiation. Furthermore, in the late 1940s the technology needed to detect the radiation had not yet been developed. Hence no attempt was made in Gamow's time to search for the radiation, and the idea was forgotten until Dicke's work. The remnant radiation was eventually detected in 1965 by accident, when the radio astronomers Arno Penzias and Robert Wilson found a persistent source of background noise while testing a new antenna (Fig. 30.10).

The implication soon became clear: The "noise"

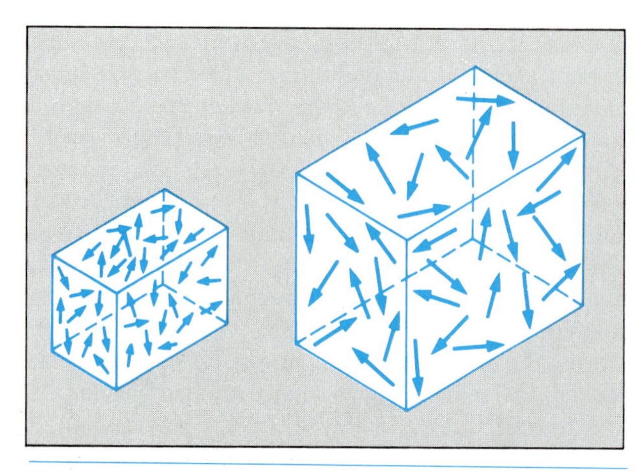

FIGURE 30.9 THE COOLING OF THE UNIVERSAL RADIATION. *At left is a box representing the early universe, filled with intense gamma-ray radiation. At right is the same box, greatly expanded. The radiation is still there, but its wavelength has increased, and the temperature represented by its thermal spectrum has decreased.*

was the universal radiation that the theorists had predicted should exist. This was one of the most significant astronomical observations ever made, for it provided a direct link to the origins of the universe. More than a piece of circumstantial evidence, this was a real artifact, the kind of hard evidence that carries weight in a court of law. Penzias and Wilson later were awarded the Nobel Prize for their discovery.

FIGURE 30.10 UNIVERSAL BACKGROUND RADIATION RECEIVER. *Robert Wilson, left, and Arno Penzias in front of the horn-reflector antenna at Bell Laboratories with which they found a persistent noise that turned out to be the remnant radiation from the big bang.* (Courtesy of AT&T)

THE CRUCIAL QUESTION OF THE SPECTRUM

There remained questions about the interpretation of the radiation, however, and doubt in the minds of some who did not accept the big bang theory of the origin of the universe. If the radiation was really the remnant of the primeval fireball, then it was expected to fulfill certain conditions, and intensive efforts were made to find out whether it did.

One important prediction was that the spectrum should be that of a simple glowing object whose emission of radiation is the result only of temperature (that is, it should be a thermal spectrum, as explained in Chapter 6). If, on the other hand, the source of the radiation was something other than the remnant of the initial universe, such as distant galaxies or intergalactic gas, it could have a different spectrum. Thus, the shape of the spectrum, its intensity as a function of wavelength, was a very important test of the interpretation of its origin.

Some of the early results were confusing, because of unforeseen complications, but now the situation seems to have been resolved. The spectrum does indeed follow the shape expected (Fig. 30.11), and the radiation is considered to be a relic of the big bang. The most difficult part of measuring the spectrum was observing the intensity at short wavelengths, around 1 millimeter or less, where the Earth's atmosphere is especially impenetrable, but this was finally accomplished with high-altitude balloon payloads.

The net result of all the effort expended in determining the spectrum is that the temperature of the radiation is 2.7 K, and its peak emission occurs at a wavelength of 1.1 mm. The radiation is commonly referred to as the **microwave background,** or the **3-degree background radiation.**

The care that went into establishing the true nature of the background radiation reflects the importance of what it has to tell us about the universe.

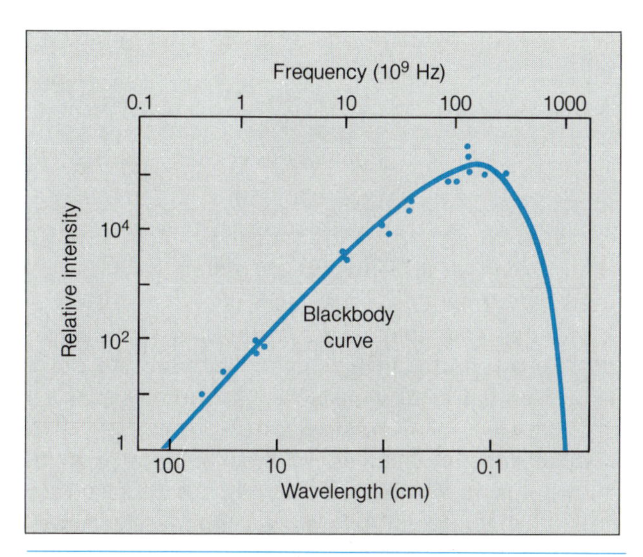

FIGURE 30.11 THE MEASURED BACKGROUND SPECTRUM. *This curve shows the spectrum of the cosmic background radiation, plotted as a function of frequency (a wavelength scale is given at the top as well). The solid line represents a thermal spectrum for a temperature of just under 3 K.*

FIGURE 30.12 A MICROWAVE RECEIVER USED TO MEASURE THE COSMIC BACKGROUND. *The pair of receiving horns are used to compare the radiation intensity in different directions.* (G. Smoot, Lawrence Berkeley Laboratory)

As we will see in Chapter 32, there are alternatives to the big bang theory, and those who support other theories have sought other explanations of the microwave radiation. Its close adherence to the spectrum expected of the radiation from the cosmic fireball envisioned in the big bang theory has presented grave difficulties for these alternative points of view.

– ISOTROPY AND DAILY VARIATIONS –

Another expectation of the background radiation, in addition to the nature of its spectrum, is that it should fill the universe uniformly, with no preference for any particular location or direction. A medium or radiation field that has no preferred orientation, but instead looks the same in all directions, is said to be **isotropic.** Some of the early observations of the 3-degree background were designed to test its isotropy, for again, failure to meet this expectation could imply some origin other than the big bang.

A test of isotropy, in simple terms, is just a measurement of the radiation's intensity in different directions, to see whether or not it varies (Fig. 30.12). From the time it was discovered, the microwave background showed a high degree of isotropy, as expected. The issue has been pressed, however, for a variety of reasons. One possible alternative explanation of the radiation was that it arises from a vast number of individual objects, such as very distant galaxies, so closely spaced in the sky that their combined emission appears to come uniformly from all over. To test this idea, astronomers have attempted to find out whether the radiation is patchy on very fine scales, as it would be if it came from a number of point sources. So far, no evidence of any clumpiness has been found, and the big-bang origin for the radiation has not been threatened.

Some clumpiness is to be expected, even if the radiation is the remnant of the early expansion of the universe. It is known that galaxies formed from density enhancements in the universe at an early time, and these condensations may have affected the radiation locally, leading today to small fluctuations in the temperature of the radiation. These fluctuations, expected to occur over angular sizes on today's sky of around 1 arcminute, have not been detected (but

ASTRONOMICAL INSIGHT 30.2

OVERLOOKING THE COSMIC BACKGROUND

Sometimes significant discoveries are delayed because no one expects them, or because few people take unconventional predictions seriously. The latter appears to have delayed the discovery of the cosmic microwave background radiation by nearly two decades.

When George Gamow predicted the existence of the cosmic background radiation in 1947, the prediction constituted a minor postscript to a major paper about the formation of the elements. In the paper Gamow and his colleagues showed that the present-day abundances of helium and other light elements were produced during the early stages of the universal expansion, while other, heavier elements must have been produced in stellar cores. No one seems to have paid much attention to the prediction about the cosmic background, despite its significance. No one made an attempt to look for the background radiation for the next 18 years.

One of the reasons for the delay was technological. Thermal radiation corresponding to a temperature of 25 K (the value predicted by Gamow and colleagues) has its peak intensity at a wavelength of 0.01 cm, or 0.1 mm, in a portion of the microwave spectrum where little work had been done by 1947. The technology to do so was just becoming available, as a result of studies of microwave emissions from the Sun, and as a response to radar developments during World War II. Thus, one reason nobody looked for the predicted radiation was that few scientists had the technology to do so.

The other reason was more subtle. Apparently few scientists were motivated to undertake a major experimental project such as the search for background radiation. The cost and time required for developing an experiment to search for the background radiation were high, and in the late 1940s there were many areas of research that appeared more promising. The basic answer to the question of why the background was not discovered before 1965 is that no one looked for it before then. At nearly the same time as Gamow's prediction, astronomers learned that observations at other radio wavelengths—particularly the 21.1 cm wavelength where atomic hydrogen should emit—would be very productive scientifically, and early radio astronomy followed these paths.

Even when the cosmic background was first detected the discovery was accidental, although there is little doubt that James Dicke and his colleagues would have found it soon if Penzias and Wilson had not. Ironically, the discovery could have been made at least 15 years earlier. The history of twentieth-century cosmology might have been quite different if astronomers had taken Gamow's prediction more seriously.

current measurement techniques are probably not sensitive enough). On a somewhat larger scale, perhaps 5 to 10 arcminutes, some theorists expect small variations in the radiation temperature because of a sort of patchiness that may have existed even earlier than the time of the first galaxy formation. During the first 100,000 years of the big bang expansion, localized regions of the universe should have expanded independently of each other, and therefore may have had slight differences in physical conditions such as temperature. If so, this would have left the cosmic radiation field slightly clumpy today, and several observers are hoping to detect the resulting irregularities in the background radiation temperature. Some preliminary evidence for such irregularities, amounting to only a few hundred-thousandths of a degree, has been found. Further work on this should reveal important information about conditions in the very early universe, before ordinary matter had formed.

In addition to any subtle localized nonuniformities expected in the big bang theory, a broad asymmetry is expected, because of the motion of the Earth with respect to the frame of reference of the radiation. Because of the Doppler shift, the observed intensity of the radiation as seen by a moving observer will vary with direction, depending on whether the observer is looking toward or away from the direction of motion. If the observer looks ahead, toward the direction of motion, then a blueshift occurs, the peak of the spectrum being shifted slightly toward shorter wavelengths (Fig. 30.13). In terms of temperature, this means that the radiation looks a little hotter when viewed in this direction. If the observer looks the other way, there is a slight redshift, and the measured temperature is cooler.

The Earth is moving in its orbit about the Sun. The Sun, in addition, is orbiting the galaxy, the galaxy is moving along its own path through the Local Group, and apparently the Local Group is moving with respect to the local supercluster. All of these motions combined represent a velocity of the Earth with respect to the frame of reference established by the background radiation, a velocity that should produce slight differences in the radiation temperature if it is viewed in different directions. Because of the Earth's rotation, if we simply point our radio telescope straight up, we should alternatively see high and low temperatures, as our telescope points toward and then (12 hours later) away from the direction of motion. This means that there should be a daily variation cycle in the radiation, referred to as the **24-hour anisotropy** or the **dipole anisotropy** (*anisotropic* is the term for a nonuniform medium).

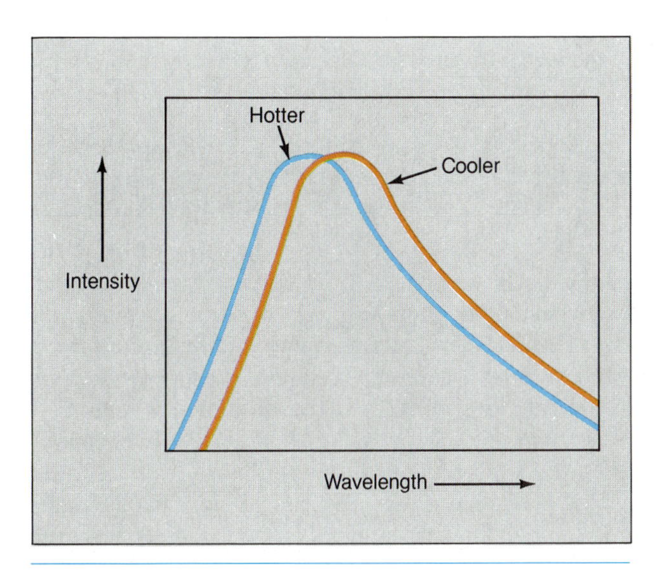

FIGURE 30.13 THE EFFECT OF A DOPPLER SHIFT ON THE COSMIC BACKGROUND RADIATION. *If the observer on the earth has a velocity with respect to the radiation, this shifts the peak of the spectrum a little, and affects the temperature that observer deduces from the measurements. Motion of the earth results in a daily cycle of tiny fluctuations in the observed temperature, known as the twenty-four-hour anisotropy.*

The actual change in temperature from one direction to the other is very small (though not as small as the localized fluctuations discussed above), and was not successfully measured until a few years ago. The observed temperature difference from the forward to the rear direction is only a few thousandths of a degree, and very sophisticated technology was required for this to be measured. The data show that the Earth is moving with respect to the background radiation at a speed of about 600 km/sec. When the Sun's known orbital velocity about the galaxy is taken into account, along with the galaxy's motion within the Local Group, this implies that the Local Group is "falling" toward the very large Virgo cluster of galaxies with a speed of about 780 km/sec. Some of this motion is caused by the gravitational influence of the Virgo cluster, and apparently some is from the effects of distant superclusters. Very recently it has been found that the motion toward the Virgo cluster is a component of a much larger motion, part of a general streaming of all the clusters in our part of the universe. This streaming motion may be caused by the gravitational attraction of something extraordinarily massive—even compared to clusters of galaxies—yet invisible. The so-called "great attractor," the subject of current excitement and controversy, may be related to the dark matter thought to contain most of the mass of the universe (see Chapter 32).

The use of the 3-degree background radiation as a tool for deducing the local motions with respect to the radiation is predicated upon the assumption that the radiation itself is fundamentally isotropic on large scales (as distinguished from the localized fluctuations discussed earlier). At present there is no reason to doubt this, but the possibility remains that it is not—that the universe itself is somehow lopsided. If it is, it will be very difficult indeed to measure its asymmetry, because of the complexity of the motions that our observing platform, the Earth, undergoes.

ASTRONOMICAL INSIGHT 30.3

THE FALL OF THE LOCAL GROUP

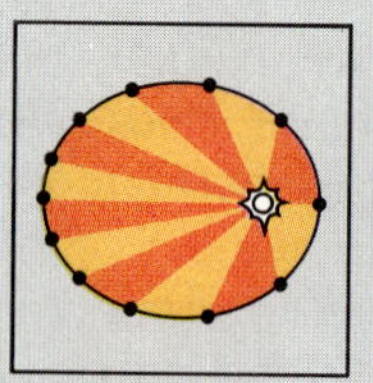

In this chapter we have learned of the expansion of the universe and the frame of reference provided by the cosmic background radiation. Galaxies are expected to have little motion in this frame of reference; that is, they should be moving apart from each other at the velocity of the universal expansion, with only minor additional motions due to gravitational forces exerted by nearby neighbors.

The dipole anisotropy in measurements of the background radiation indicates that our galaxy does move with respect to the background radiation. The standard interpretation is that this motion consists of orbital motion of the Milky Way within the Local Group, and motion of the Local Group toward the gigantic Virgo cluster of galaxies. In effect the Local Group is falling toward Virgo because of the enormous mass and gravitational force exerted by this dense cluster. The Virgo cluster is about 19 Mpc (or about 62 million light years) away, so we see that deviations from smooth motion of galaxies with the universal expansion can appear over rather large distance scales. It is as if the universe had a localized "ripple" in its expansion.

If the universe on the whole is uniform, as is normally believed, then on sufficiently large distance scales the average motions of galaxies should be in accord with the expansion. In other words, if we examine the motions of galaxies to large enough distances, we should find that on average they move with the universal expansion, and have zero velocity with respect to the reference frame of the microwave background. There should be a size scale on which all ripples in the expansion become insignificant.

Recently the size scale required to find this uniformity has had to be revised upward. A new study of galaxy distances and velocities shows that there is a general streaming motion of galaxies with respect to the background radiation out to distances as far as 100 Mpc (over 300 million light years). This size scale is still small compared to the overall size of the universe, but the streaming motion represents a far larger ripple in the expansion than had previously been expected.

The motion of the Earth with respect to the cosmic radiation, as measured through the 24-hour anisotropy, is now believed to be due mostly to this general streaming motion of galaxies, with only a small sideways component toward the Virgo cluster. The average velocity of the motion of galaxies with respect to the cosmic background is 600 to 700 km/sec, and the direction as seen from Earth is towards the Southern Cross, a prominent constellation in the southern skies (this gives a reference for the direction, but of course the galaxies are far beyond the stars that form the constellation).

The Local Group and other galaxies and clusters within 100 Mpc of it are all moving in a more or less uniform direction with respect to the universal expansion. Even the Virgo cluster is involved in this motion. The implications are quite startling: either the universe is not uniform, or there is something invisible far out in intergalactic space with enough mass to create all this motion due to its gravitational force. In other words, the Local Group is not really falling toward the Virgo cluster, but is instead falling toward something far more massive, yet invisible.

If this interpretation is correct, we do not know what the unseen massive object is. The evidence that it even exists is quite indirect, but there are other reasons for believing that the universe contains vast quantities of invisible matter (these will be discussed in Chapter 32). One of the most important questions for understanding the evolution and future of the universe depends crucially on how much unseen mass there is. The fall of the Local Group and other clusters and galaxies in its vicinity may end up having important implications for our understanding of the fate of the universe.

FIGURE 30.14 THE COSMIC BACKGROUND EXPLORER. *This satellite, to be launched in the 1990s, will make the most precise and extensive measurements yet of the microwave background radiation.* (NASA, courtesy of J. C. Mather)

Future observations are being planned to improve upon the measurements of the 3-degree background. A satellite called the *Cosmic Background Explorer* (Fig. 30.14) is scheduled to be launched in the 1990s and will provide, from its vantage point above the Earth's atmosphere, the best data yet on the spectrum and isotropy of the background radiation.

PERSPECTIVE

We have learned now that the universe, like all the subordinate objects within it, is a dynamic, evolving entity. One of the grandest stories in astronomy has been the unfolding of the concept of universal expansion. An idea forced on astronomers by the evidence of the redshifts, it has led directly to the big bang picture, which in turn tells us the age of the universe, explains the origin of some of the elements, and is verified through the presence of the microwave background radiation.

Before we finish the story by studying modern theories of the universe and their implications for its future, we must take a closer look at some of the objects that inhabit it. There is a breed of strange and bizarre entities whose very nature seems to be tied up with the evolution of the universe, and which therefore will provide us with a few additional bits of evidence for our final discussion.

SUMMARY

1. It was discovered early in this century that galaxies tend to have redshifted spectral lines, implying that they are moving away from us.
2. Hubble discovered in 1929 that the velocity of recession is proportional to distance, demonstrating that the universe is expanding.
3. The rate of expansion, expressed in terms of Hubble's constant H, is probably between 50 km/sec/Mpc and 100 km/sec/Mpc.
4. The fact that the universe is expanding implies that it originated from a single point, and the rate of expansion, if constant, tells us that it began some 10 to 20 billion years ago. This may be considered to be the age of the universe.
5. The relationship between speed and distance allows the distance to a galaxy to be estimated from its velocity, through the Doppler effect. This method of determining distance extends to the farthest limits of the observable universe.
6. The fact that the universe began in a single point implies that it was very hot and dense initially, and this in turn implies that the early universe was filled with radiation. As the expansion has proceeded, this radiation has been transformed into microwave radiation, with a thermal spectrum corresponding to a temperature of about 3 K. The cosmic background radiation was discovered in 1965.
7. To distinguish a primordial origin for the background radiation from other possible origins (such as many galaxies spread throughout the universe), it is necessary to measure the spectrum to see whether it truly is a thermal spectrum. Observations to date are consistent with the assumption that it is.

8. Another important test is to determine whether the radiation is isotropic, or whether it might be unevenly intense in different directions. Careful observations have revealed no evidence of patchiness or uneveness that might imply that the radiation arises in a large number of individual sources, or that the early universe was nonuniform in any way.
9. There is a 24-hour (or dipole) anisotropy in the radiation, however, created by the Doppler effect resulting from the Earth's motion. By comparing the radiation intensity in the forward and backward directions, it has been possible to determine the Earth's velocity with respect to the reference frame of the radiation. The Earth's motion is a combination of its orbital motion about the Sun, the Sun's orbital motion about the galaxy, the galaxy's motion within the Local Group, and the Local Group's motion within the local supercluster.

REVIEW QUESTIONS

Thought Questions

1. In your own words, explain how all galaxies can be receding from our own, yet our galaxy is not at the center of the universe.
2. Explain the distinction between the motions of galaxies within a cluster of galaxies and motions resulting from the expansion of the universe.
3. How is the value of the Hubble constant measured? Why is it difficult to obtain a precise value?
4. If the universal expansion was once more rapid than it is today, how does that affect estimates of the age of the universe that are based on the present-day expansion rate?
5. Explain how the distances we measure to the most faraway galaxies depend on how accurately we know the Sun-Earth distance.
6. Explain why the assumption that the early universe was hot and dense leads to the conclusion that it must have been filled with radiation.
7. Why was measuring the spectrum of the background radiation so important in determining whether it was truly remnant radiation from the early expansion of the universe? Why is measuring the spectrum difficult?
8. Explain why the isotropy of the background radiation is important in establishing whether it is remnant radiation from the early expansion of the universe.
9. If the background radiation is isotropic, why are there small shifts in its temperature as observed in opposite directions from the Earth?
10. What might subtle variations in the intensity of the background radiation tell us about the fundamental structure of the universe?

Problems

1. What is the speed of recession of a galaxy whose ionized calcium line is observed at a wavelength of 3,937.61 Å? (The rest wavelength of the line is 3,933.68 Å.)
2. Suppose three galaxies are situated so that the distance from galaxy A to galaxy C is twice the distance from A to B. After a period of time, the distance between galaxies A and B has doubled. How has the distance between A and C changed? How does the distance between A and C now compare with the distance between A and B?
3. The age of a supernova remnant can be determined from a technique analogous to determining the age of the universe from the Hubble constant. To see the analogy, calculate the age of a supernova remnant whose radius is 10 pc and whose outer portions are expanding at a speed of 1000 km/sec. (*Note:* You'll have to convert parsecs to kilometers.) Express your answer in years.
4. Suppose the current value of the Hubble constant is 75 km/sec/Mpc, but because the initial expansion was more rapid, the average value over the history of the universe is 80 km/sec/Mpc. What is the true age of the universe, and how does it compare with the value derived by assuming the expansion has been steady at the present rate?
5. Suppose three galaxies are observed, and in their spectra the position of the ionized calcium line is observed to be 3,936 Å for galaxy A, 4,028 Å for galaxy B, and 3,942 Å for galaxy C (the rest wavelength is 3,933 Å). Find the velocities and distances to the galaxies, assuming a Hubble constant of H = 75 km/sec/Mpc.

6. What was the wavelength of maximum emission for the universal background radiation when the temperature was 4,000 K? (This is roughly what the temperature was at the time when hydrogen atoms formed from free electrons and protons, and the radiation field became independent of the temperature of the matter in the universe.)

ADDITIONAL READINGS

Abell, G. 1978. Cosmology—the origin and evolution of the universe. *Mercury* 7(3):45.

Barrow, J. D., and J. I. Silk. 1980. The structure of the early universe. *Scientific American* 242(2):118.

Ferris, Timothy. 1984. The radio sky and the echo of creation. *Mercury* 13(1):2.

Hodge, P. 1984. The cosmic distance scale. *American Scientist* 72(5):474.

Kippenhahn, R. 1987. Light from the depths of time. *Sky and Telescope* 73(2):140.

Layzer, D. 1975. The arrow of time. *Scientific American,* 233(6):56.

Muller, R. A. 1978. The cosmic background radiation and the new aether drift. *Scientific American* 238(5):64.

Parker, Barry. 1986. The discovery of the expanding universe. *Sky and Telescope* 72(3):227.

Shu, F. H. 1982. *The physical universe.* San Francisco: W. H. Freeman.

Silk, J. 1980. *The big bang: the creation and evolution of the universe.* San Francisco: W. H. Freeman.

Webster, A. 1974. The cosmic background radiation. *Scientific American* 231(2):26.

Weinberg, S. 1977. *The first three minutes.* New York: Basic Books.

CHAPTER 31

Peculiar Galaxies, Explosive Nuclei, and Quasars

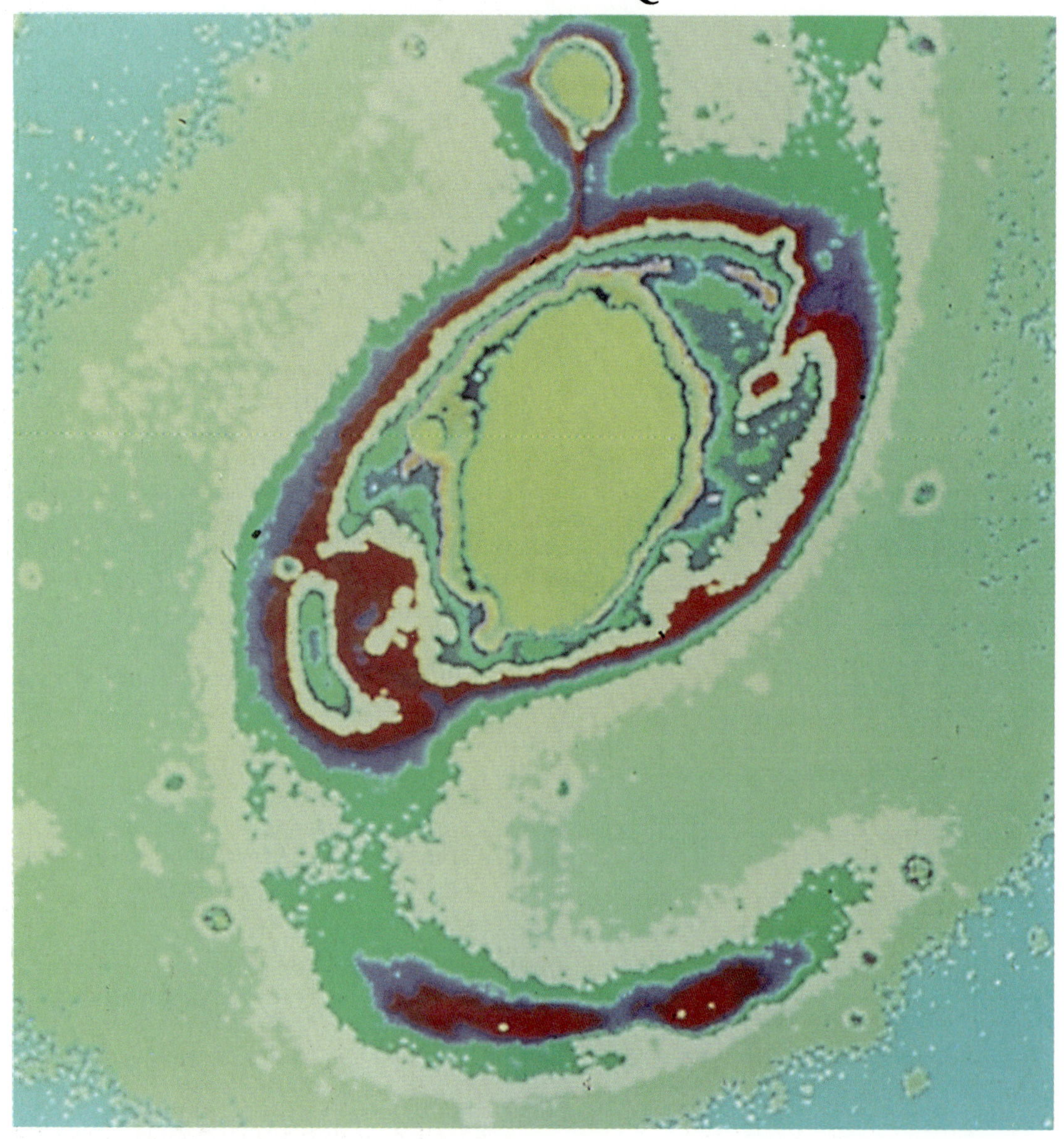

An enhanced image of a controversial connection between a galaxy and a quasar (the small round image just above the central part of the galaxy). (National Optical Astronomy Observatories)

CHAPTER PREVIEW

In discussing the characteristics of galaxies in the preceding chapters, we have overlooked a variety of objects, some of them galaxies and some possibly not, that have unusual traits. As in many other situations in astronomy, the so-called peculiar objects, once understood, have quite a bit to tell us about the more normal ones.

To fully appreciate the bizarre nature of some of these astronomical oddities, it was best to delay their introduction until this point, when we have essentially completed our survey of the universe. We know about the expansion and the big bang, and we are just in the process of tying it all together. In this chapter we will uncover a number of vital clues in the cosmic puzzle.

THE RADIO GALAXIES

Most of the first astronomical sources of radio emission to be discovered, other than the Sun, are galaxies. Most of these, when examined optically, have turned out to be large ellipticals, often with some unusual-appearing structure (Fig. 31.1). The first of these objects to be detected, and one of the brightest, is called Cygnus A (named under a preliminary cataloging system in which the ranking radio sources in constellation are listed alphabetically). This galaxy has a strange

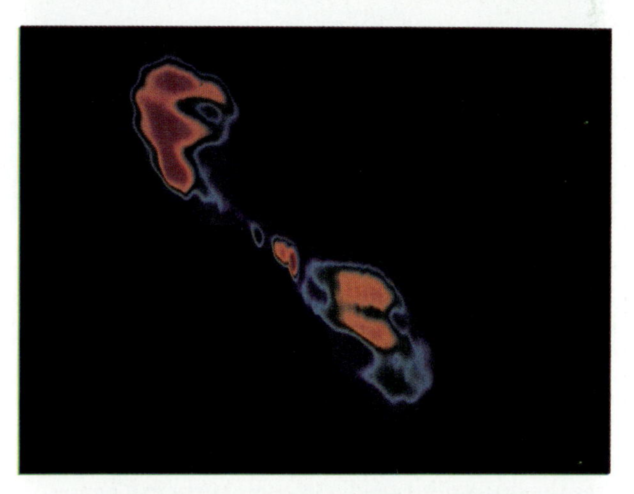

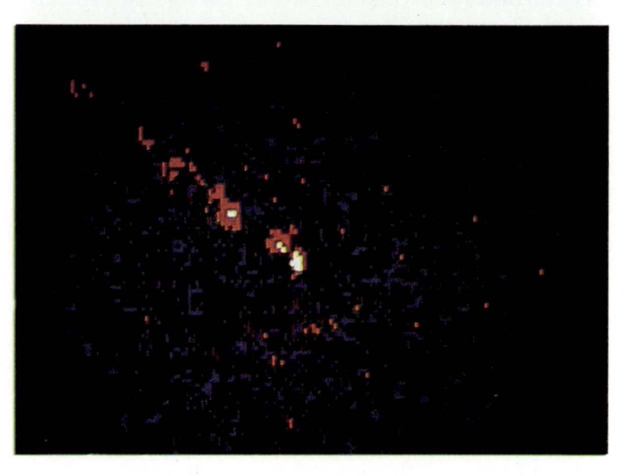

FIGURE 31.1 CENTAURUS A. *This is a giant elliptical radio galaxy, showing a dense lane of interstellar gas and dust across the central region. At top is a visible-light photograph; in the center is a radio image; and at the bottom is an X-ray image. Note that the radio and x-ray emission comes from the lobes above and below the visible galaxy.* (*top:* © 1980 Anglo-Australian Telescope Board; *middle:* National Radio Astronomy Observatory; *bottom:* Harvard-Smithsonian Center for Astrophysics)

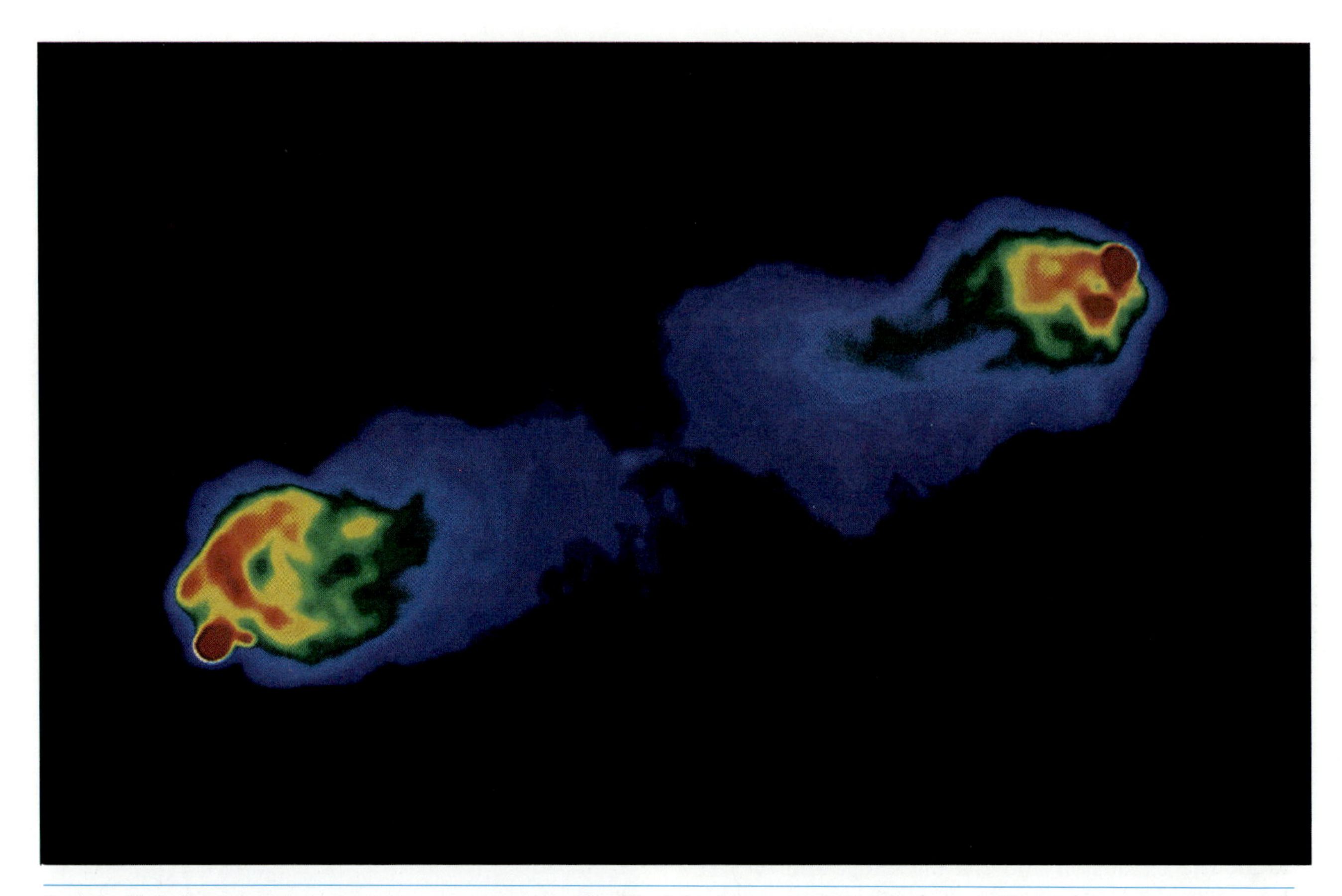

FIGURE 31.2 A RADIO IMAGE OF A RADIO GALAXY. *Here the lighter areas represent maximum radio brightness. The galaxy as seen in visible wavelengths does not even appear here; this image is completely dominated by the double-side lobes.* (National Radio Astronomy Observatory)

double appearance, and it was eventually found, after sufficient refinement of radio observing techniques (see the discussion of radio interferometry in Chapter 7) that the radio emission comes from two locations on opposite sides of the visible galaxy and well separated from it (Figs. 31.2 and 31.3). High-resolution observations with the *Very Large Array* have revealed remarkable detail in some radio galaxies (Fig. 31.4).

This double-lobed structure is a common feature of the so-called **radio galaxies,** those with unusually strong radio emission. Most, if not all, ordinary galaxies emit radio radiation, but the term "radio galaxy" is applied only to the special cases in which the radio intensity is many times greater than the norm. The core of the Milky Way is a strong radio source as viewed from the Earth, but it is not in a league with the true radio galaxies. Typical properties of radio galaxies are listed in Table 31.1.

Although the double-lobed structure is standard among elliptical radio galaxies, the visual appearance

TABLE 31.1 PROPERTIES OF RADIO GALAXIES

Galaxy type: Most often elliptical or giant elliptical.

Radio luminosity compared to optical luminosity: 0.1 to 10.

Radio source shape: Either double-lobed or compact central source; often jets are seen which emit throughout the spectrum.

Variability: Intensity variations in times as short as days may occur in the compact radio sources.

Nature of spectrum: Usually synchrotron or inverse Compton spectrum, usually polarized.

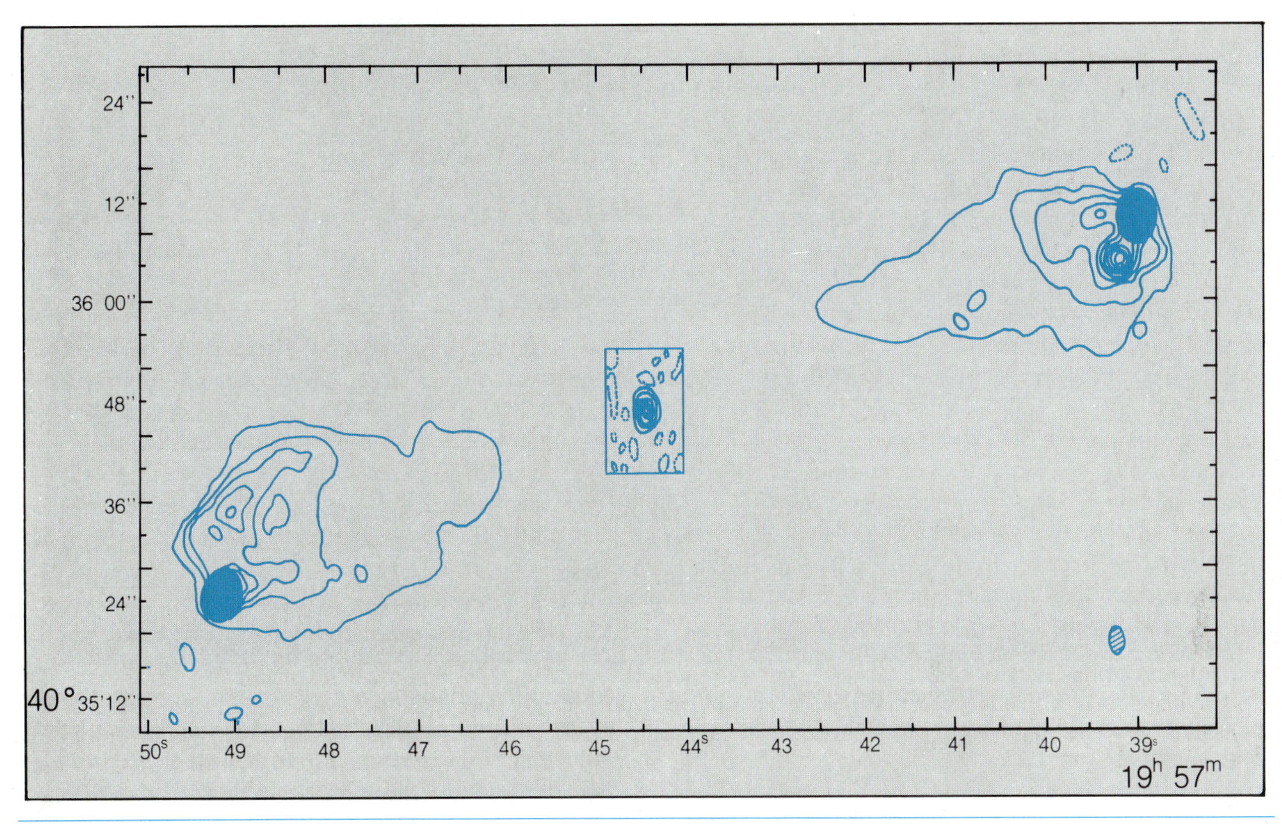

FIGURE 31.3 A MAP OF A RADIO GALAXY. *The intensity contours here represent radio emission from the two side lobes of the radio galaxy Cygnus A. The blurred image at the center illustrates the size of the visible galaxy compared to the gigantic side lobes.* (Mullard Radio Astronomy Observatory, University of Cambridge)

varies quite a bit. We have already noted the double appearance of Cygnus A; another bright source, the giant elliptical Centaurus A (the object shown in Fig. 31.1), appears to have a dense band of interstellar matter bisecting it, and it has not one, but two, pairs of radio lobes, one much farther out from the visible galaxy than the other. Other giant elliptical radio galaxies have other kinds of strange appearances. One of the most famous is M87, also known as Virgo A, which has a jet protruding from one side that is aligned with one of the radio lobes (Fig. 31.5). Careful examination shows that this jet contains a series of blobs or knots that appear to have been ejected from the core of the galaxy in sequence. Many galaxies have been discovered to have radio-emitting jets. These include galaxies already known to be peculiar, such as Centaurus A (see Fig. 31.1), and at least one spiral galaxy (Fig. 31.6). In some cases double jets are seen on opposite sides of a galaxy (Fig. 31.7).

The size scale of the radio galaxies can be enormous. In some cases, the radio lobes or jets extend as far as 5 million parsecs from the central galaxy. Recall that the diameter of a large galaxy is only one-tenth of this distance, and that the Andromeda galaxy is less than 1 million parsecs from our position in the Milky Way.

The first radio galaxies were detected in the 1940s, and by the 1950s studies of the radio spectra of these objects were in progress. It was discovered that they are emitting by the synchrotron process (already mentioned in Chapter 24), which requires a strong magnetic field and a supply of rapidly moving electrons. The electrons are forced to follow spiral paths around the magnetic field lines, and as they do so, they emit radiation over a broad range of wavelengths. The characteristic signature of synchrotron radiation, in contrast with the thermal radiation from hot objects such as stars, is a sloped spectrum with no strong

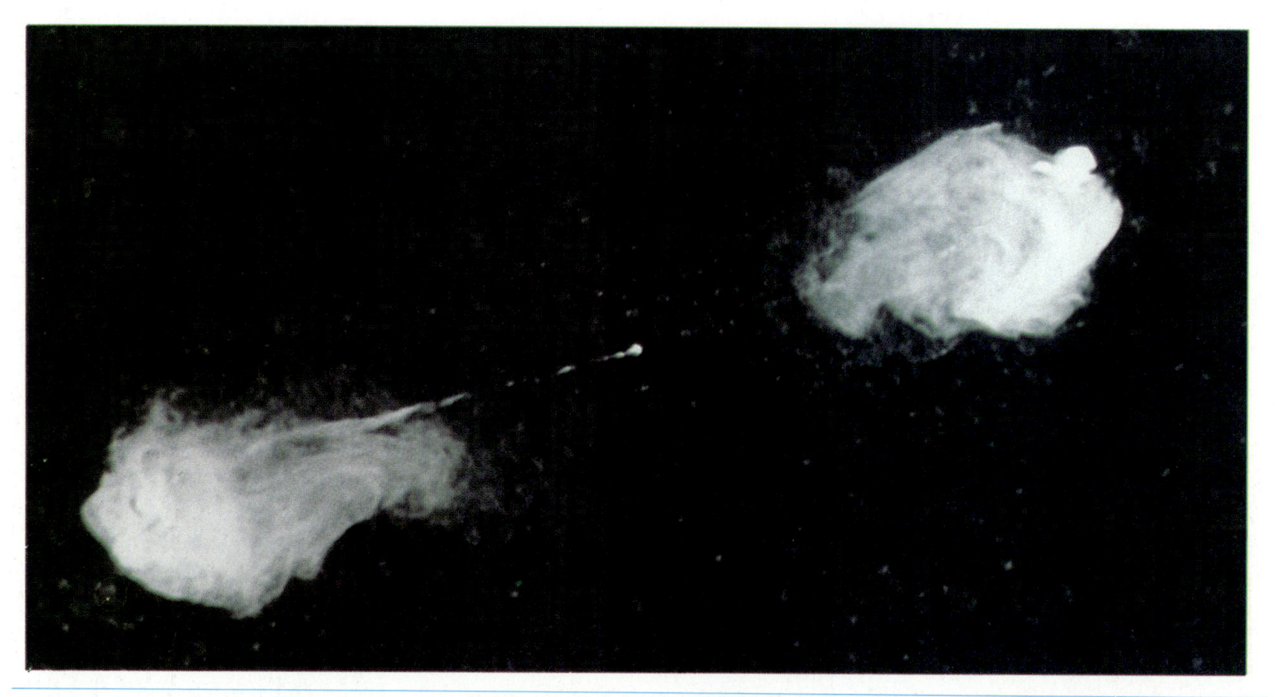

FIGURE 31.4 DETAILS OF RADIO LOBES. *This image of Cygnus A was obtained with the* Very Large Array *and shows fine detail suggesting turbulence and flows in the lobes.* (National Radio Astronomy Observatory)

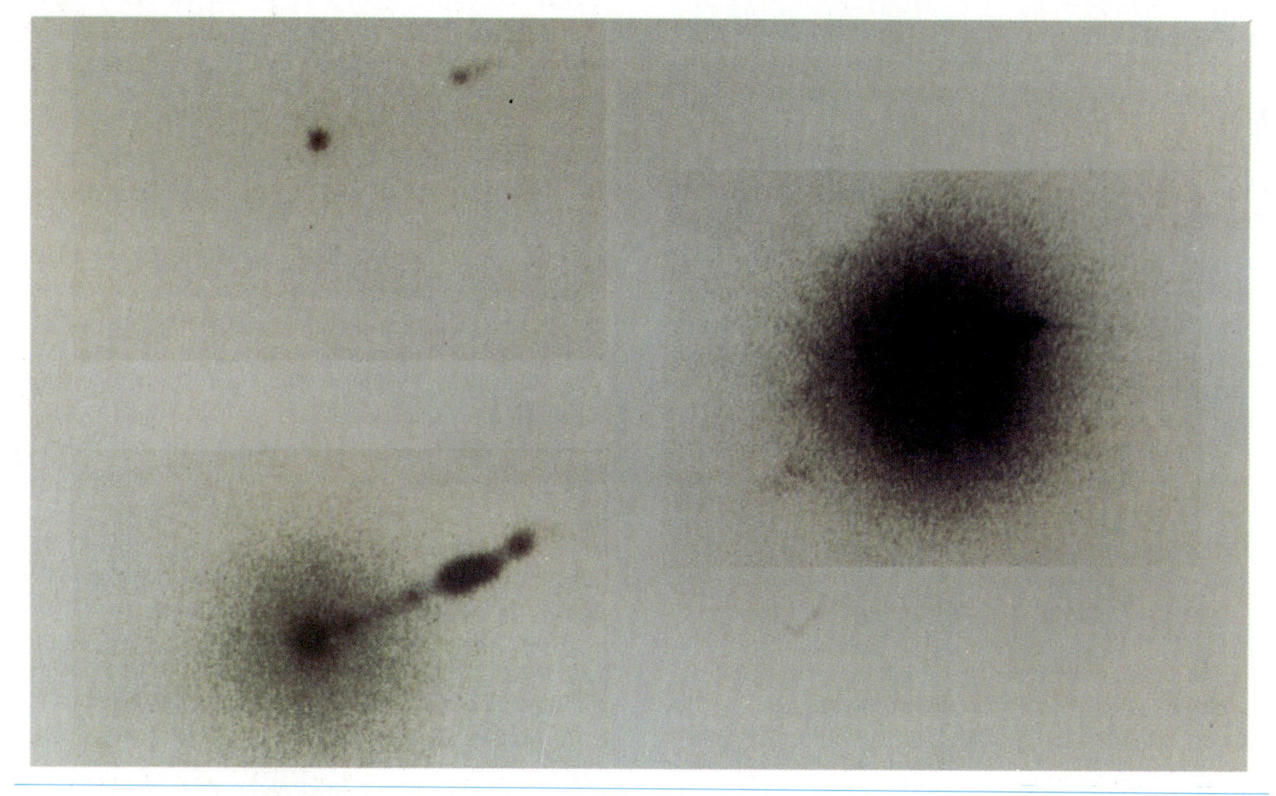

FIGURE 31.5 A GIANT ELLIPTICAL RADIO GALAXY WITH A JET. *This is M87, also known as Virgo A, showing its remarkable linear jet of hot gas. This is a short-exposure photograph, specially processed to maximize the visibility of the jet and its lumpy structure. On a longer-exposure photograph, this galaxy looks like a normal elliptical, the light from the jet being drowned out by the intensity of light from the galaxy.* (H. C. Arp)

FIGURE 31.6 JETS IN AN UNUSUAL GALAXY. *Here are three views of the barred spiral galaxy NGC1097, showing its two remarkable liner jets. At top left is a black and white photograph that has been processed to enhance the visibility of the jets. At lower left is a false-color version. An enlarged view of the inner part of the galaxy is shown at right.* (H. C. Arp)

peak at any particular wavelength. The radiation from a synchrotron source is also polarized. Whenever an object is found to be emitting by the synchrotron process, it is immediately concluded that some highly energetic activity is taking place, producing the rapid electrons that are required to create the emission.

In some cases, it appears that a different process is responsible for the radio emission, but also one that

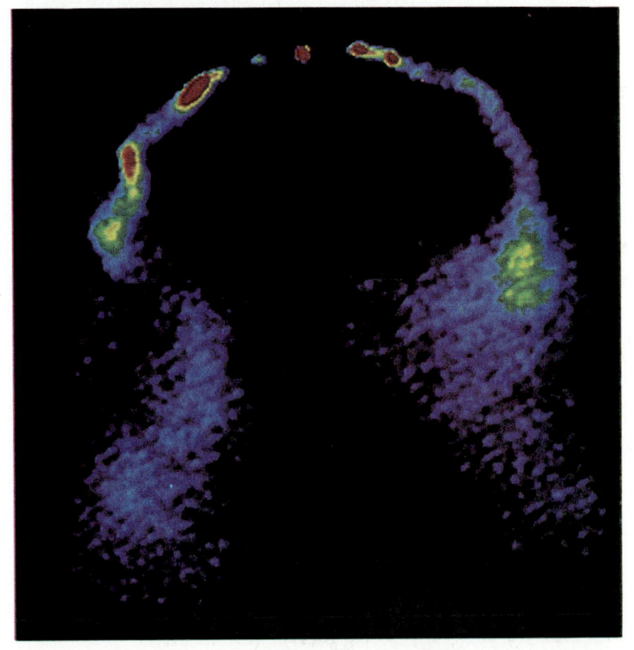

FIGURE 31.7 DUAL RADIO JETS. *While visible jets are rare, radio images often reveal jetlike structures. Here is a radio image of a galaxy with a pair of jets emanating from opposite sides, and curving away from the galaxy, apparently because the galaxy is moving through an intergalactic gas medium.* (National Radio Astronomy Observatory)

requires a supply of rapidly moving electrons. If there is ordinary thermal radio emission from a galaxy, then the presence of electrons moving near the speed of light can alter the spectrum of the radio emission, as energy is transferred from the electrons to the radiation. This process is called **inverse Compton scattering,** and it has the same implication as synchrotron radiation: There must be a source of vast amounts of energy, to produce the rapidly moving electrons required to create the radiation.

The source of this energy in radio galaxies is a mystery, although some interesting speculation has been fueled by certain characteristics of the galaxies. The double-lobed structure indicates that two clouds of fast electrons are located on either side of the visible galaxy, and this leads naturally to the impression that the clouds of hot gas may have been ejected from the galactic core, perhaps by an explosive event that expelled material symmetrically in opposite directions. This impression is heightened by the many cases in which jets have been seen protruding from the core (Figs. 31.5, 31.6, and 31.7), and by instances in which more than one pair of lobes, aligned in the same direction but with one farther out than the other, is seen. In these galaxies, it seems that superheated, highly ionized gas is continually being accelerated from the core in opposing directions.

One method envisioned by astronomers for producing prodigious amounts of energy from a small volume such as the core of a galaxy is to have a giant black hole there, surrounded by an accretion disk, and this has naturally been a suggested source for the energy in the radio galaxies. The general idea is that superheated gas that has been compressed by the immense gravitational field of the black hole escapes along the rotation axes and is channeled into two opposing jets. It is probably better to view this process not as explosive, but as a more-or-less steady expulsion of gas. It has been suggested that *all* radio galaxies have jets, and that this is the only mechanism for forming the double radio lobes. The fact that the jets are not *observed* in all radio galaxies is thought to be an effect of how the jets are aligned with respect to our line of sight. The emission from rapidly moving particles is confined to a narrow angle; therefore, an observer in the wrong direction from a jet will see little of its radiation.

The theory that massive black holes are the energy sources in radio galaxies has received some observational support. Recently it was reported that evidence for such a central, massive object has been found in the heart of M87, the radio galaxy with the well-known visible jet. Based on measurements of the velocity dispersion near the center of the galaxy, it was deduced that a large amount of mass must be confined in a very small volume at the core, and a black hole seemed the best explanation. Subsequent observations, however, have failed to confirm the reportedly high speeds of stars in the inner portions of the galaxy, and the case for a black hole in M87 has been weakened. Further efforts to find direct evidence of supermassive black holes are being made.

SEYFERT GALAXIES AND EXPLOSIVE NUCLEI

Although it is true that the strongest radio emitters among galaxies are giant ellipticals, they are by no means the only ones with evidence for explosive events in their cores. Some spiral galaxies also display such behavior. We learned in Chapter 25 that even the Milky Way is not immune; the evidence was summarized there for the existence of a compact, massive object at its center. There are other spiral galaxies with much more pronounced violence in their nuclei. These galaxies as a class are called **Seyfert galaxies** (Fig. 31.8 and Table 31.2), after the astronomer who discovered and cataloged many of them in the 1930s.

Seyfert galaxies have the appearance of ordinary spirals, except that the nucleus is unusually bright and blue, in contrast with the red color of most normal spiral galaxy nuclei. About 10 percent of them are radio emitters, with spectra indicating that the syn-

TABLE 31.2 PROPERTIES OF SEYFERT GALAXIES

Properties
Spiral galaxies
Luminosities comparable to brightest normal spirals
Bright, compact blue nuclei
Nuclei show emission lines of highly ionized gas
About 10% are radio sources
Most are variable in times of days or weeks
Emission from the nucleus is synchrotron radiation
Some have radio jets

FIGURE 31.8 A SEYFERT GALAXY. *This photograph of NGC 4151, a well-studied example, shows the enormous intensity of light from the nucleus compared to the rest of the galaxy.* (Palomar Observatory, California Institute of Technology)

chrotron process is at work. The radio emission usually comes from the nucleus, rather than from double side lobes. The spectrum of the visible light from a Seyfert nucleus typically shows emission lines, something completely out of character for normal galaxies. The emission lines, formed in an ionized gas, are always very broad, indicating speeds of several thousand km/sec in the gas that produces the emission. Evidently the cores of these galaxies are in extreme turmoil, with hot gas swirling about and tremendous amounts of energy being generated.

Few spirals show such effects, and it is not clear whether the Seyfert behavior is a phase they all pass through at some time, or whether only a few act up in this manner. Later in this chapter we will discuss evidence supporting the first of these possibilities.

THE DISCOVERY OF QUASARS

In 1960, spectra were obtained of two starlike, bluish objects that had been found to be sources of radio emissions (Fig. 31.9). No radio stars were known, and astronomers were very interested in these two objects, called 3C48 and 3C273 (their designations in a catalog of radio sources that had recently been com-

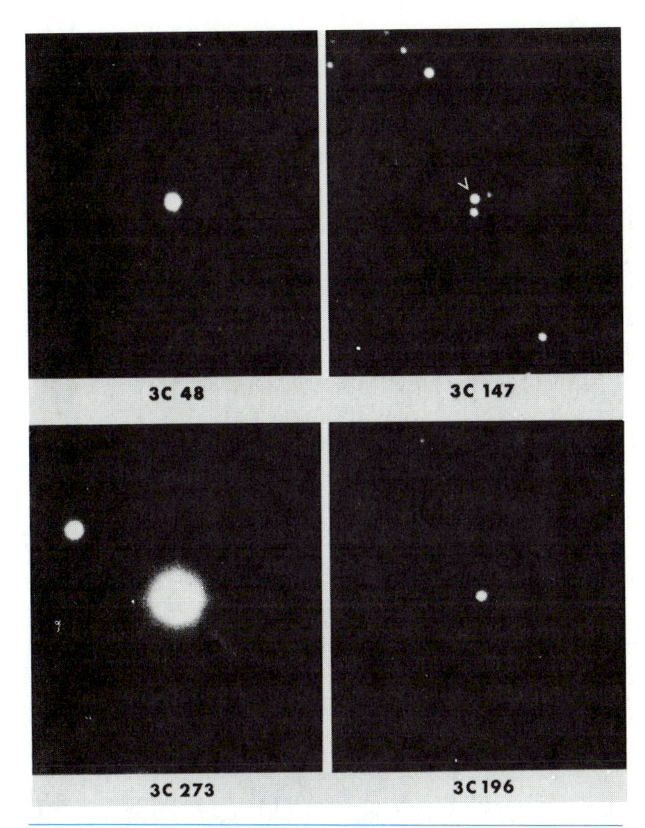

FIGURE 31.9 QUASI-STELLAR OBJECTS. *These starlike objects are quasars, whose spectra are quite unlike those of normal stars. As explained in the text, these objects are probably more distant, and therefore more luminous, than any normal galaxies.* (Palomar Observatory, California Institute of Technology)

piled by the radio observatory of Cambridge University, in England). The spectra of the two objects, which had several strong emission lines, defied understanding for some 3 years. The objects became known as **quasars** (short for "quasistellar radio sources"), or **quasistellar objects** (sometimes simply **QSOs**) because of their starlike appearance but nonstellar spectra.

It was eventually realized that the strong emission lines in the two quasars were the well-known Balmer series of hydrogen, but shifted far toward longer wavelengths than those measured in the laboratory. The shifts of the lines in 3C273 indicated a redshift of 16 percent, (that is, the object emitting these lines must be moving away from the Earth at 16 percent of the speed of light, or 48,000 km/sec!).

In 3C48, the shift was even greater, about 37 percent, corresponding to a speed of 111,000 km/sec (these speeds actually have to be corrected for relativistic effects, as described in Appendix 14; the correct values are 44,000 km/sec for 3C273 and 91,000 km/sec for 3C48).

Since the early 1960s, hundreds of additional quasars have been discovered. Only a small fraction are radio sources, and in other ways they may differ from the first two, but they invariably have highly redshifted emission lines, and in very many cases weak absorption lines are also seen. In many quasars the redshift is so huge that spectral lines with rest wavelengths in the ultraviolet portion of the spectrum are shifted all the way into the visible region. The grand champion today is a quasar with redshift of 443 percent, so that the strongest of all the hydrogen lines, with a rest wavelength of 1,216 Å, is shifted all the way to 6,602 Å. This quasar is *not* traveling away from us at 4.43 times the speed of light, however; for very large velocities, the Doppler shift formula has to be modified in accordance with relativistic effects (Appendix 14). The velocity of this quasar is 93.4 percent of the speed of light, still an enormous speed.

THE ORIGIN OF THE REDSHIFTS

To understand the physical properties of the quasars, we must first discover the reason for the high redshifts. We have already tacitly adopted the most obvious explanation, that they are a result of the Doppler effect in objects that are moving very rapidly; but we still must ascertain the nature of the motions. Furthermore, at least one alternative explanation was suggested that has nothing to do with motions at all.

One consequence of Einstein's theory of general relativity, verified by experiment, is that light can be redshifted by a gravitational field. Photons struggling to escape an intense field lose some of their energy in the process, and as this happens, their wavelengths are shifted towards the red. We have already been exposed to this concept in Chapter 24, when we discussed the behavior of light near black holes, which have such strong gravitational fields that no light can escape at all. For a while it was considered possible that quasars are stationary objects sufficiently massive and compact to have large gravitational redshifts.

This suggestion has now been largely ruled out, for two reasons. One is that there is no known way for an object to be compressed enough to produce such strong gravitational redshifts without falling in on itself and becoming a black hole. If enough matter is squeezed into such a small volume that its gravity produces redshifts as large as those in quasars, no known force could prevent this object from collapsing further. A neutron star does not have as large a gravitational redshift as those found in quasars, and we already know that the only possible object with a stronger gravitational field than that of a neutron star is a black hole.

The second objection is that, even if such a massive, yet compact, object could exist, its spectral lines would be very much broader than those observed in quasar spectra. The high pressure would distort electron energy levels so that the spectral lines would be smeared out (as in a white dwarf, but more extreme), and light emitted from slightly different levels in the object would have different gravitational redshifts, again causing spectral lines to be broadened.

For both of these reasons, we are forced to accept the Doppler shift explanation for the redshifts. The problem then is to somehow explain how such large velocities can arise. One possibility is that the quasars are relatively nearby (by intergalactic standards), and are simply moving away from us at very high speeds, perhaps as the result of some explosive event. This "local" explanation requires some care: To accept it, we must explain why no quasars have ever been found to have blueshifts. In other words, if quasars are nearby objects moving very rapidly, it is not easy to see why none happen to be approaching us but are instead all receding. It might be possible to argue that they originated in a nearby explosion (at the galactic center, perhaps) a long time ago, so any that happened to be aimed toward us have had time enough to pass by and are now receding (Fig. 31.10). There are serious difficulties with this picture, though, primarily in the amount of energy that would be required to get all these objects moving at the observed velocities, an amount that would dwarf the total light output of the galaxy over its entire lifetime.

We are left with one other alternative, one that still poses problems, but perhaps is more acceptable than the others. Let us consider the possibility that the quasars are very distant objects, moving away

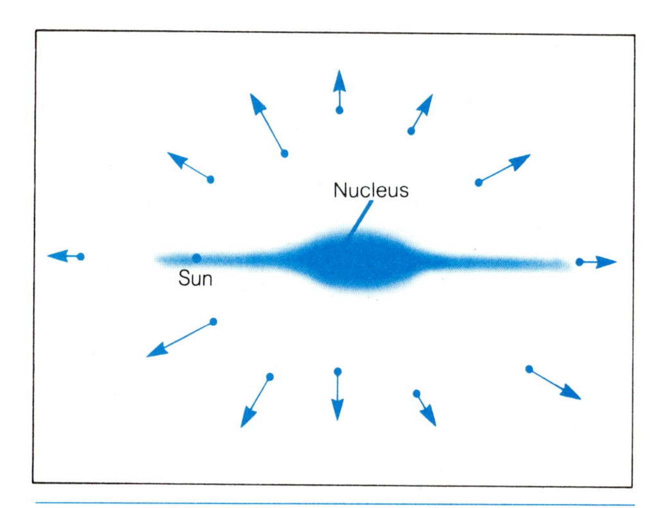

FIGURE 30.10 A "LOCAL" HYPOTHESIS FOR QUASAR REDSHIFTS. *If quasars were very rapidly moving objects ejected from our galactic nucleus some time ago, they could all be receding from our position in the disk by now. This would explain the fact that only redshifts are found, but there is no known mechanism for providing the energy required to produce such great speeds.*

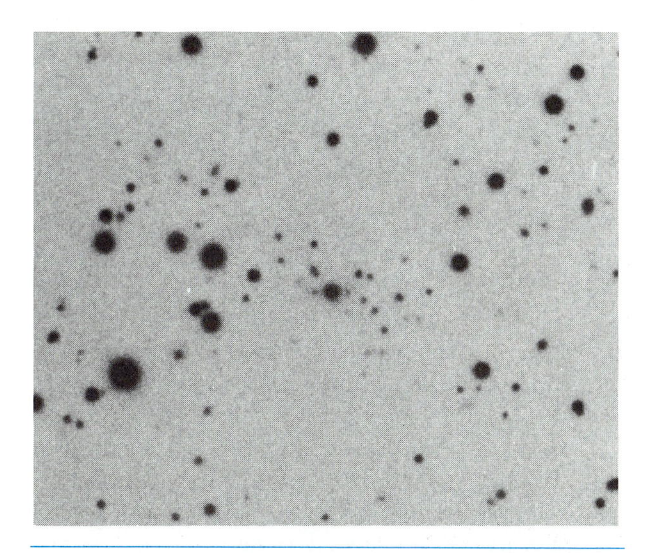

FIGURE 31.11 A QUASAR IN A CLUSTER OF GALAXIES. *The image at the center of this very long exposure photograph is a quasar. The much fainter objects around it are galaxies, forming a cluster of which the quasar is apparently a member. The galaxies and the quasar have the same redshift, indicating that they lie together at the same distance.* (H. Spinrad)

from us with the expansion of the universe. In this view they are said to be at "cosmological" distances, meaning that they obey Hubble's relation between distance and velocity, just as galaxies do. We find, if we adopt this assumption, that 3C273 is some 870 Mpc away, and 3C48 is more than 2,000 Mpc away. (These distances are based on the assumption that the rate of expansion of the universe has been constant, something that is probably not true; this will be discussed in Chapter 32.)

The best evidence that the quasars are at cosmological distances is the fact that some have been found in clusters of galaxies (Fig. 31.11), with the same redshift as the galaxies. This was an extremely difficult observation to make, because at the distances at which the quasars exist, the much fainter galaxies are very hard to see, even with the largest telescopes.

Supporting evidence for the cosmological redshift interpretation is provided by the **BL Lac objects.** Named for the prototype, an object called BL Lacertae that was once thought to be a variable star, these are elliptical galaxies with very bright central cores. The nucleus of a BL Lac object displays many of the properties of a quasar, with radio synchrotron emission, variable brightness, and enormous luminosity. The surrounding galaxy shows an almost featureless spectrum with a redshift that is consistent with its apparent distance and which is undoubtedly cosmological. Generally, the redshifts of BL Lac objects are relatively small (by quasar standards), and these apparently are nearby objects very closely related to quasars. The implications of the fact that they are embedded in galaxies will be discussed later in this chapter.

Distances of hundreds or thousands of Mpc, inferred for the quasars with very large redshifts, have enormous implications. One is that the light that we receive from them has traveled on its way to us for many billions of years. When we look at a quasar, we are looking far into the history of the universe, and we must keep this in mind as we attempt to interpret them. The fact that quasars are seen only at high redshifts, meaning great distances, indicates that they were more common in the early days of the universe than they are now. This will prove to be a very important clue to their true nature. It has been speculated that there is a limit to quasar redshifts;

ASTRONOMICAL INSIGHT 31.1

THE REDSHIFT CONTROVERSY

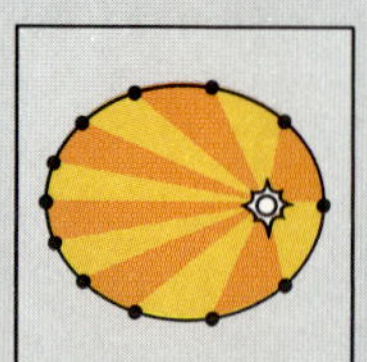

In the text we present a standard set of arguments demonstrating that quasar redshifts are cosmological, resulting from the expansion of the universe, and implying that the quasars must be very distant objects. Although it is true that most astronomers accept this viewpoint, it is not universally adopted, and there are those who favor other explanations of the redshifts.

The evidence sited by the opponents of the cosmological interpretation consists primarily of cases where quasars are found apparently associated with galaxies or clusters of galaxies, but do not have the same redshift as the galaxies. It is argued in these cases that the quasar is physically associated with the galaxy or cluster of galaxies, and is therefore at the same distance, so that its redshift cannot be cosmological.

The evidence favoring these arguments can be quite striking. When a quasar is found within a cluster of galaxies, there is a natural tendency to think of it as a physical member of the group. It can be argued from statistical grounds that the chances of accidental alignments between galaxies and quasars are so low that the associations between these objects that are observed are unlikely to be coincidental. Long-exposure photographs have even appeared to show gaseous filaments connecting a galaxy with a quasar that has a different redshift. This would seem to prove that the redshift cannot be cosmological.

This radical view of quasars leaves many important questions open. The most difficult of these is how to explain the redshifts, if they really are not cosmological, for the arguments cited in the text against gravitational and local Doppler redshifts are still valid. No satisfactory explanation of the quasar redshifts has been offered by the opponents of the cosmological viewpoint.

The counterarguments center on statistical calculations. One point is that the calculations showing that the quasar-galaxy associations have a low probability of occurring by chance are made after the fact. That is, after a few such associations were found, arguments were made that this was a very unlikely thing to occur by chance alignments of objects randomly distributed over the sky. This is somewhat akin to arguing that the chances of being dealt a certain combination of cards in a game are very low. This is true before the deal, but meaningless after the fact. Based on this and other analogies, many scientists think it incorrect to argue that chance associations of quasars and galaxies are improbable. More to the point, they argue, quasars associated with galaxies are more likely to be discovered than those that are not; therefore, it is natural to find a disporportionately high number of quasars that happen to be aligned on the sky with galaxies. This is an example of a **selection effect,** because the process by which quasars are found is not perfectly random and thorough; the attention of astronomers is naturally concentrated on regions where there are many galaxies, so those regions are more thoroughly searched for quasars.

Probably the most telling argument against the noncosmological redshifts for quasars arises from the fact (noted in the text) that a number of quasars are embedded in galaxies or within clusters of galaxies that share the same redshift. Recent observations show, in fact, that *every* quasar that is sufficiently nearby for an associated galaxy to be detected is indeed embedded within a galaxy that has the same redshift as the quasar. There has not yet been an exception to this, and the data amount to virtual proof that quasars are the cores of galaxies and that their redshifts are cosmological.

The controversy is by no means over, however. Adherents of the noncosmological interpretation of the redshifts continue to find remarkable cases of apparent galaxy-quasar connections where the redshifts do not match. Perhaps observations made in the near future, with the *Space Telescope* or other planned instruments, will provide unequivocal evidence supporting one side or the other.

that is, no quasars can be found with a redshift greater than this limit. The implication is fantastic, for it means that we are looking back to a time before the universe had organized itself into such objects as galaxies and quasars.

Another important implication of the great distances to quasars is that they are the most luminous objects known. Their apparent magnitudes (the brightest are thirteenth-magnitude objects) combined with their tremendous distances, imply that their luminosities are far greater than even the brightest galaxies, by factors of hundreds or even thousands. As we will see, explaining this astounding energy output is one of the central problems of modern astronomy.

TABLE 31.3 PROPERTIES OF QUASARS

Characterized by large redshifts.
Spectra dominated by emission lines of highly ionized gas.
Optical luminosities 100 to 1,000 times those of normal galaxies.*
About 10% are radio sources.
Appear as compact, blue objects.
Many are variable in times of days or weeks.
Emission due to synchrotron radiation.
Some have radio or optical jets.

*Assuming that quasars are at cosmological distances.

THE PROPERTIES OF QUASARS

More than 1,500 quasars have been cataloged, and more are being discovered continually. Many are being studied with care, primarily by means of spectroscopic observations. Within the past few years, it has even become possible to observe quasars at ultraviolet and X-ray wavelengths, with the use of satellite observatories such as the *International Ultraviolet Explorer* and the *Einstein Observatory*. The faintness of the quasars makes all these observations difficult, but their importance makes the effort worthwhile. The result of all the intensive work being done on quasars is that a great deal is now known about their external properties (Table 31.3), even though their origin and especially their source of energy remain mysterious.

The blue color that characterized the first quasars discovered is a general property of all quasars (except for extremely redshifted ones, where the blue light has been shifted all the way into the red part of the spectrum), but the radio emission is not. Only about 10 percent are radio sources, contrary to the early impression that they all are. This type of misunderstanding is known as a **selection effect,** because the process of selecting quasars was based at first on a certain assumption about their characteristics. For a while, the only method used to look for quasars was to search for objects with radio emission, so naturally, all those that were found were radio sources. The radio emission, and usually the continuous radiation of visible light as well, shows the characteristic synchrotron spectrum.

Apparently *all* quasars emit X-rays, again by the synchrotron process, according to the data collected by the *Einstein Observatory,* the most sensitive X-ray satellite yet launched. Hence X-ray observations may be a reliable technique for finding new quasars, particularly those of relatively low redshift, which are relatively nearby and hence have stronger X-ray emission than the more distant ones.

In some cases, photographs of quasars reveal some evidence of structure, instead of a single point of light. The most notable of these is 3C273, which played such a key role in the initial discovery of quasars. This object shows a linear jet extending from one side (Fig. 31.12), closely resembling the one emanating from the giant radio galaxy M87 (see Fig. 31.5). Perhaps this means that similar processes are occurring in these two rather different objects.

Many quasars vary in brightness, usually over times of several days to months or years (although in at least one case, variations were seen over just a few hours). This variability is very important, for it provides information on the size of the region in the quasar that is emitting the light. This region cannot be any larger than the distance light travels in the time over which the intensity varies. There is no way an entire object that might be hundreds or thousands of light-years across can vary in brightness in a time of a month or so. The light-travel time across the object would guarantee that we would see changes only over times of hundreds or thousands of years, as the light from different parts of the object reached us

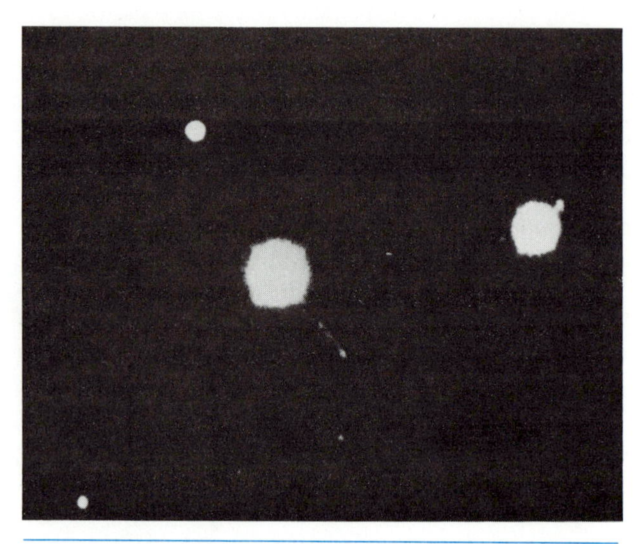

FIGURE 31.12 A QUASAR WITH A JET. *This is 3C273, one of the first two quasars discovered. This visible-light photograph reveals a linear jet very much like those seen in many radio galaxies. The radio structure of quasars is usually double-lobed, also similar to radio galaxies.* (National Optical Astronomy Observatories)

(Fig. 31.13). Therefore the observed short-term variations in quasars tell us that the fantastic energy emitted by these objects is produced within a volume no more than a light-month, or about 0.03 parsecs, in diameter. This obviously places stringent limitations on the nature of the emitting object.

The spectra of quasars have already been described in broad outline, but there is a great deal of detail as well (Fig. 31.14). All quasars have emission lines, generally of common elements such as hydrogen, helium, and often carbon, nitrogen, and oxygen (some of the latter elements only have strong lines in the ultraviolet, and therefore are best observed when the redshift is sufficently large to move these lines into the visible portion of the spectrum). In many cases, the emission lines are very broad, showing that the gas that forms them has internal motions of thousands of km/sec. The degree of ionization tells us that the gas is subjected to an intense radiation field, which continually ionizes the gas by the absorption of energetic photons.

Many quasars have absorption lines in addition to the emission features. Strangely enough, the absorption lines are usually at a different redshift (almost always smaller) than the emission lines, indicating that the gas that creates the absorption is not moving away from us as rapidly as the gas that produces the emission. Further confounding the issue is the fact that many quasars have multiple absorption redshifts; that is, they have several distinct sets of absorption lines, each with its own redshift, and each therefore representing a distinct velocity. This shows

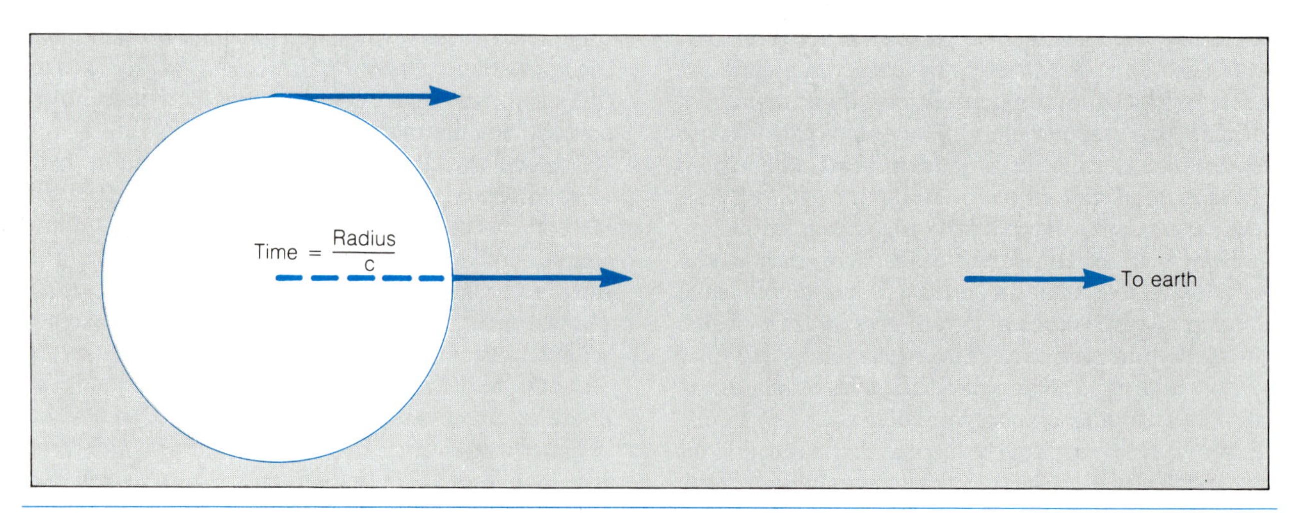

FIGURE 31.13 THE IMPLICATION OF TIME VARIABILITY FOR THE SIZE OF THE EMITTING OBJECT. *This illustrates why an object cannot appear to vary in less time than it takes for light to travel across it. As a simple case, this spherical object is assumed to change its luminosity instantaneously. On earth we observe the first hint of this change when light from the nearest part of the object reaches us, but we continue to see the brightness changing gradually as light from more distant portions reaches us.*

that there are several absorbing clouds in the line of sight. There are two possible origins for these clouds (Fig. 31.15). One is that the quasar itself, which is moving at the high speed indicated by the redshift of the emission lines, is ejecting clouds of gas, some of which happen to be aimed toward the Earth. These clouds therefore have lower velocities of recession than the quasar itself, so the redshift of the absorption lines they form is less than the redshift of the emission lines formed in the quasar.

The alternative view is that the absorption lines arise in the gaseous halos or disks of galaxies that happen to lie between us and the more distant quasar. Each galaxy has a recession velocity determined by the expansion of the universe, and because the intervening galaxies are not as far away as the quasars, they are not moving as fast and therefore have lower redshifts. If this interpretation of the absorption lines is correct, as is thought to be true in at least some cases, then their analysis should provide important information about the nature of the intervening galaxies. Of particular interest is the information quasar absorption lines might provide on the extensive halos that many galaxies are thought to have.

In some cases the absorption spectra of quasars show only lines formed by hydrogen, and none due to heavier elements. These lines apparently arise in gas that has not been enriched at all by stellar processing, and it is thought that these might be intergalactic clouds that have never formed into galaxies,

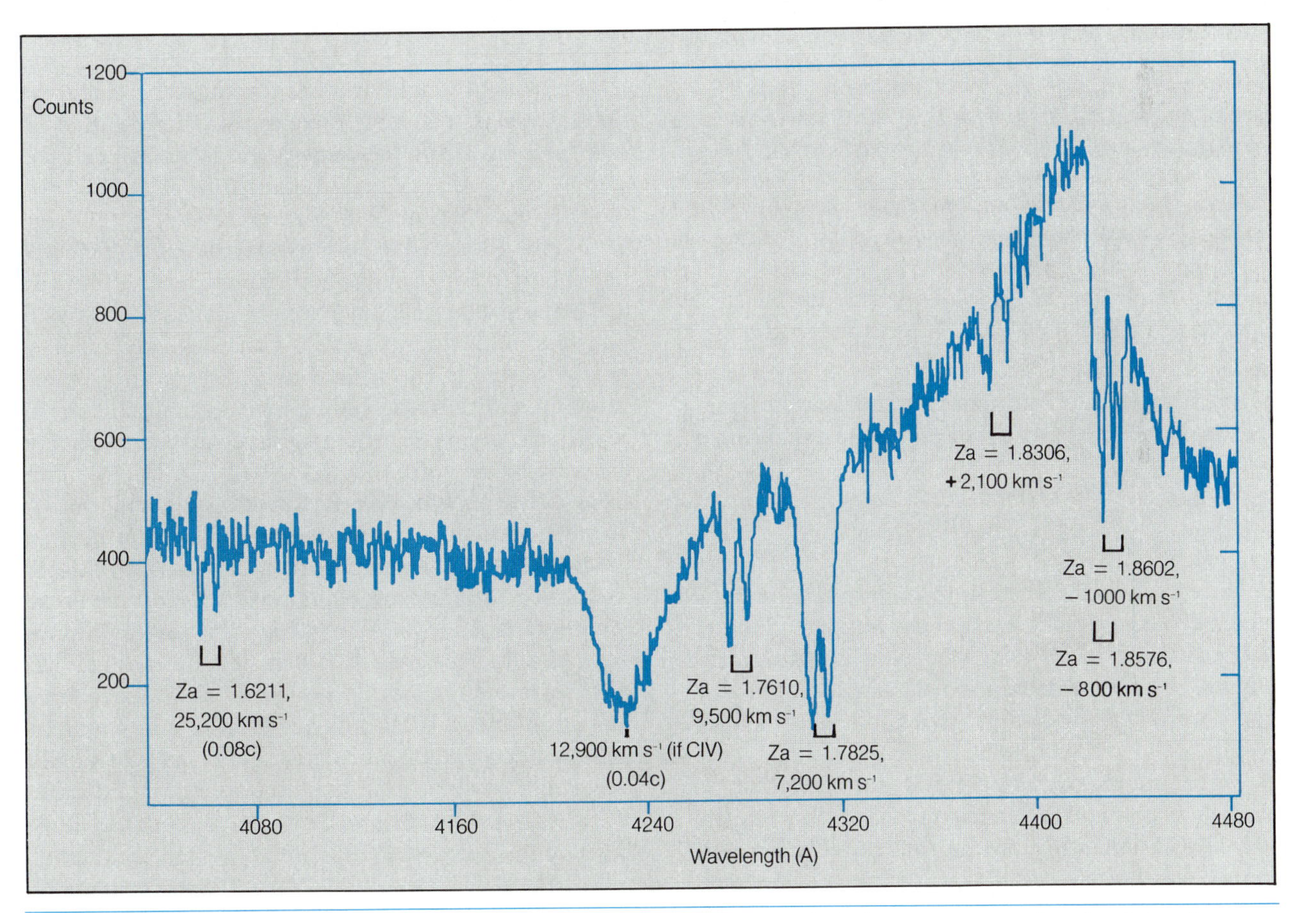

FIGURE 31.14 A QUASAR SPECTRUM. *This spectrum shows several features typical of quasar spectra: a broad emission line centered near 4,400 Å; broad absorption line near 4,240 Å; and narrow absorption line at several wavelengths. All features seen here are due to carbon that has lost three electrons (designated CIV). Redshifts are indicated.* (Data from Weymann, R. J., R. F. Carswell, and M. G. Smith 1981, *Annual Reviews of Astronomy and Astrophysics,* 19:41.)

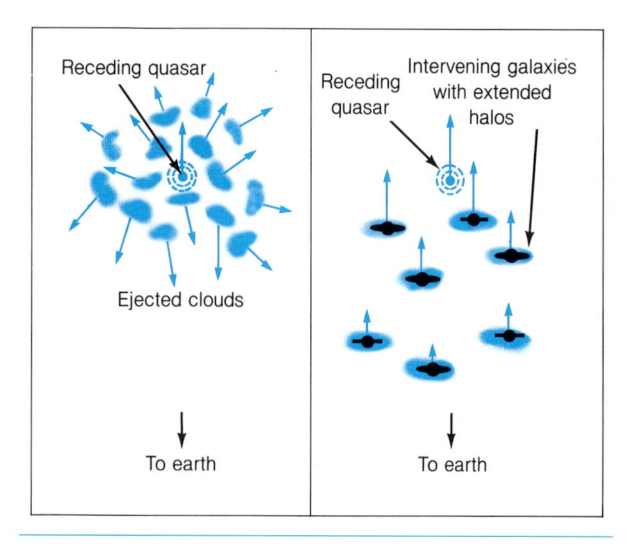

FIGURE 31.15 TWO EXPLANATIONS OF QUASAR ABSORPTION LINES. *In one scenario (left), the absorption lines are formed by clouds of gas ejected from the quasar. Clouds ejected directly toward the earth have lower velocities of recession than the quasar itself. In the other more likely view (right), the absorption lines are formed in the extended halos of galaxies that happen to lie between us and the quasar. The galaxies are closer to us than the quasar, and therefore have lower velocities of recession and smaller redshifts.*

so that no stars have formed in them. These may be material remnants of the big bang; if so, they can tell us much about the exact amount of nuclear processing that occurred in the early stages of the expansion.

GALAXIES IN INFANCY?

A variety of theories, some of them rather fanciful, have been proposed to explain the quasars. One idea has gradually become widely accepted, however, and we will restrict ourselves to discussing only that one hypothesis, keeping in mind that there are other suggestions.

The prevailing interpretation of the nature of quasars was inspired by the fact that they existed only in the long-ago past, by their association with the BL Lac objects (which clearly are galaxies), and by their resemblance to the nuclei of Seyfert galaxies. The statement that they existed only in the past is based on the fact that all are very far away, so that the light-travel time ensures that we are seeing things only as they were billions of years ago, not as they are today. The resemblance to Seyfert nuclei is striking: Both are blue; both are radio sources in about 10 percent of the cases; both vary on similar time scales; and both have very similar emission line spectra. Seyferts lack the complex absorption lines often seen in the spectra of quasars, but this would be expected if the quasar absorption lines are formed in the halos of intervening galaxies, since Seyfert galaxies are not so far away that there are likely to be many other galaxies along their lines of sight. The nuclei of Seyferts differ from quasars also in the amount of energy they emit, being considerably less luminous.

The picture that is developing is that quasars are young galaxies, with some sort of youthful activity taking place in their centers. In this view, the Seyfert galaxies are descendants of quasars, still showing activity in their nuclei, but with diminished intensity. If we carry this idea another step, we are led to the suggestion that normal galaxies like the Milky Way are later stages of the same phenomenon; recall the mildly energetic (by quasar standards) activity in the core of our galaxy.

The notion that quasars may be infant galaxies is supported by some rather direct evidence. As noted earlier, some quasars have been found associated with clusters of galaxies, showing that they can be physically located in the same region of space at the same time. In addition, there are a few examples known in which careful observation has revealed a fuzzy, dimly glowing region surrounding a quasar (Fig. 31.16). This is probably a galaxy, at the center of which a quasar is embedded. It is likely that all quasars are located in the nuclei of galaxies, which are so much fainter that they usually cannot be seen. A short-exposure photograph of a Seyfert galaxy (Fig. 31.17) shows only the nucleus, which is very much brighter than the surrounding galaxy, and it is clear that if we observed one of these galaxies from so far away that it was near the limit of detectability, all we would see would be a blue, starlike object resembling a quasar.

Although we can make a strong case that quasars are young galaxies, we are still far from understanding all of their properties. The main mystery remaining is the source of the tremendous energy of quasars and active galactic nuclei. Several possibilities have been raised, but today only one seems viable.

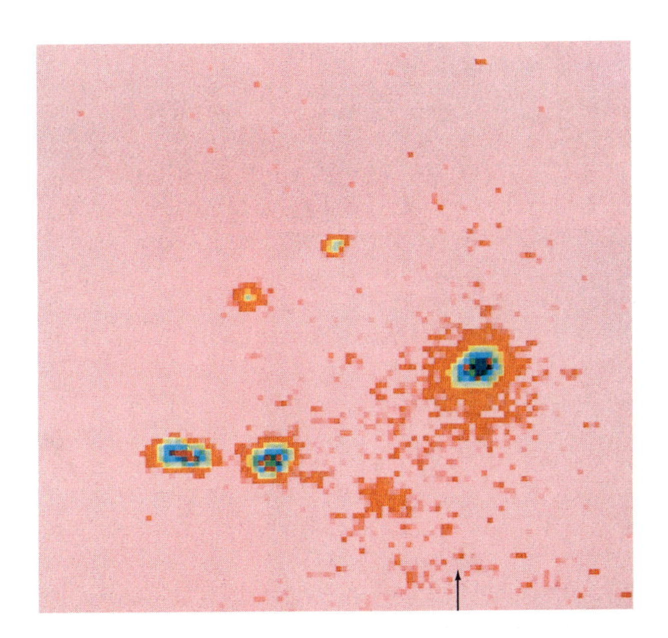

FIGURE 31.16 A QUASAR EMBEDDED IN A GALAXY. *This photograph, obtained with an electronic detector, contains several star images (top, sides), a galaxy (slightly elongated object just to the right of center), and a quasar (lower right). The quasar image is extended, just as the galaxy image is, and the fuzzy region surrounding it is indeed a galaxy. In every case where a quasar is close enough to us that a galaxy would be bright enough to be detected, one has been, indicating that all quasars lie at the centers of galaxies.* (M. Malkan)

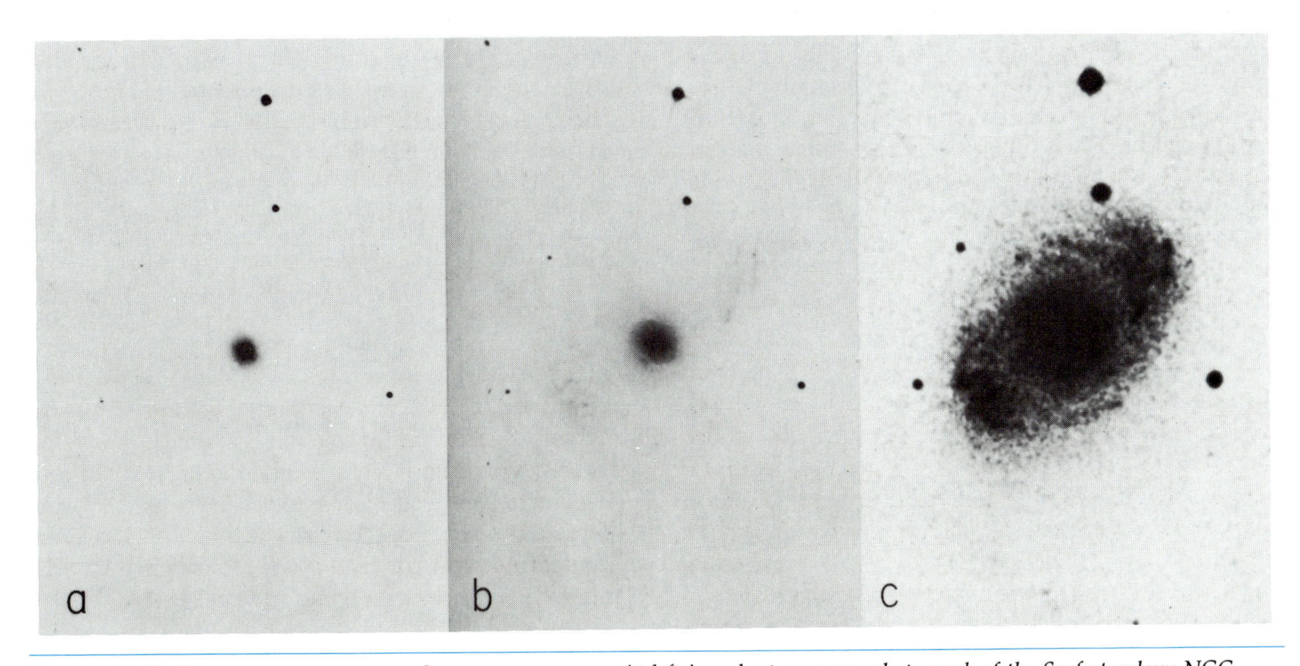

FIGURE 31.17 THREE EXPOSURES OF A SEYFERT GALAXY. *At left is a short-exposure photograph of the Seyfert galaxy NGC 4151, in which only the nucleus is seen, resembling the image of a star. The center and right-hand images are longer exposures of the same object, revealing more of its outer, fainter structure. This sequence illustrates how a quasar, which is even more luminous than a Seyfert galaxy nucleus, can appear as a starlike image even though it is embedded in a galaxy.* (Photographs from the Mt. Wilson Observatory, Carnegie Institution of Washington, composition by W. W. Morgan)

ASTRONOMICAL INSIGHT 31.2

THE DOUBLE QUASAR AND GRAVITATIONAL LENSES

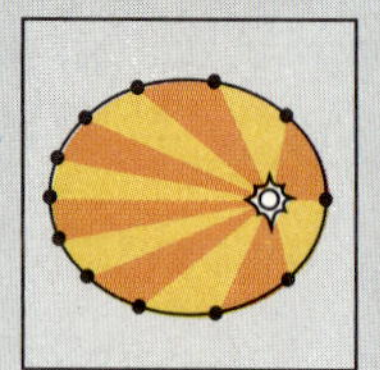

A few years ago a pair of quasars was discovered, very near each other in the sky, with properties so nearly identical that astronomers have concluded that they are in fact two images of the same quasar. If so, then astronomers have found the first example of a **gravitational lens** something predicted to be a possibility on the basis of Einstein's theory of general relativity. Since a gravitational field can bend light rays, it is possible for such a field to act as a lens. In particular, calculations have shown that the gravitational field of a galaxy can bend the light from a distant object around it. If the galaxy were perfectly aligned with the more distant object, then we would see either a single-point image, as if the galaxy were not there (this would happen only if we were precisely at the focal point of the lens), or, more likely, a circular ring-shaped image. In the more probable event that the galaxy is slightly off-center between us and the distant light source, we will see two or more distinct images, on either side of the intervening galaxy. The double quasar is thought to be such a case, the two images reaching us by coming around opposite sides of the gravitational lens formed by a galaxy between us and the quasar. Careful photographic studies have revealed what appears to be a galaxy between the two images of the quasar, helping to support this interpretation.

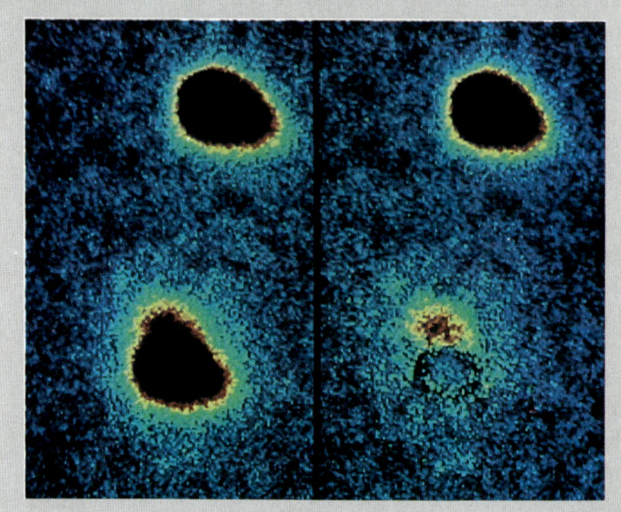

Institute for Astronomy and Planetary-Geosciences Data Processing Facility, University of Hawaii, courtesy of A. Stockton

Since the discovery of the first gravitational lens, other apparent cases have been found, all involving quasars seen behind galaxies. Sometimes more than two images of the galaxy are seen. One of the best tests of whether the multiple images really are of the same quasar would be to see whether they vary with time in the same fashion, but this is complicated by the fact that the light from the different images follows different paths to us. The length of one path may be longer than the other, so when the quasar changes

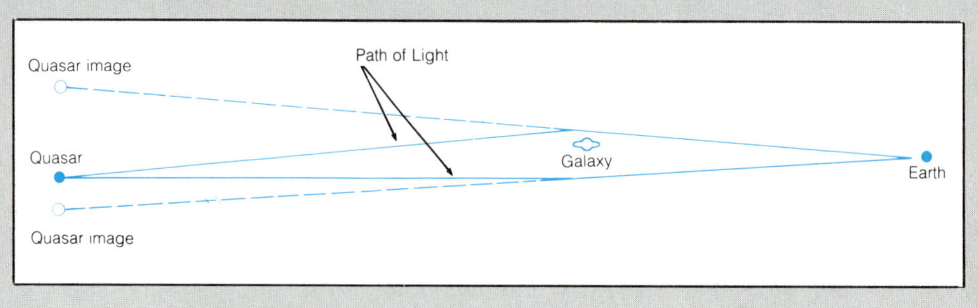

This diagram illustrates how an intervening galaxy can act as a gravitational lens, forming two images of a distant quasar.

ASTRONOMICAL INSIGHT 31.2

THE DOUBLE QUASAR AND GRAVITATIONAL LENSES

(continued from previous page)
in brightness, we see this at different times for the separate images. Therefore, we have to rely on similarities in redshift, apparent magnitude, and spectral details in determining whether the multiple quasars are images of the same object.

As we look to greater and greater distances, so that we see more and more distant objects, the sky becomes filled with images of galaxies and quasars. It stands to reason that we should find more and more cases of galaxies lying nearly in front of distant quasars, and we should expect that gravitational lenses should show up more and more. This seems to be happening, as new telescopes and detectors reveal ever-fainter objects. Eventually we may reach the point at which there are so many multiple quasar images created by gravitational lenses that we will not know how many quasars we really are seeing. It will be as though the universe is screening our view beyond a certain distance, by distorting the picture with many images of each object beyond that distance.

This possibility, one that keeps coming up in situations where large amounts of energy seem to be coming from small volumes of space, is that quasars are powered by massive black holes (Fig. 31.18). In the chaotic early days of collapse, when a galaxy was just forming out of a condensing cloud of primordial gas, a great deal of material may have collected at the center. If many stars formed there, frequent collisions could have caused them to settle in more and more tightly, until they coalesced to form a black hole. The gravitational influence of such a massive object would have then stirred up the surroundings, creating the high electron velocities required to fuel the synchrotron emission. Infalling matter would have formed an accretion disk around the black hole, from which X-rays would be emitted, in analogy with stellar black holes and neutron stars in binary systems. In this picture the source of energy in quasars is the same as in radio galaxies, described earlier in this chapter.

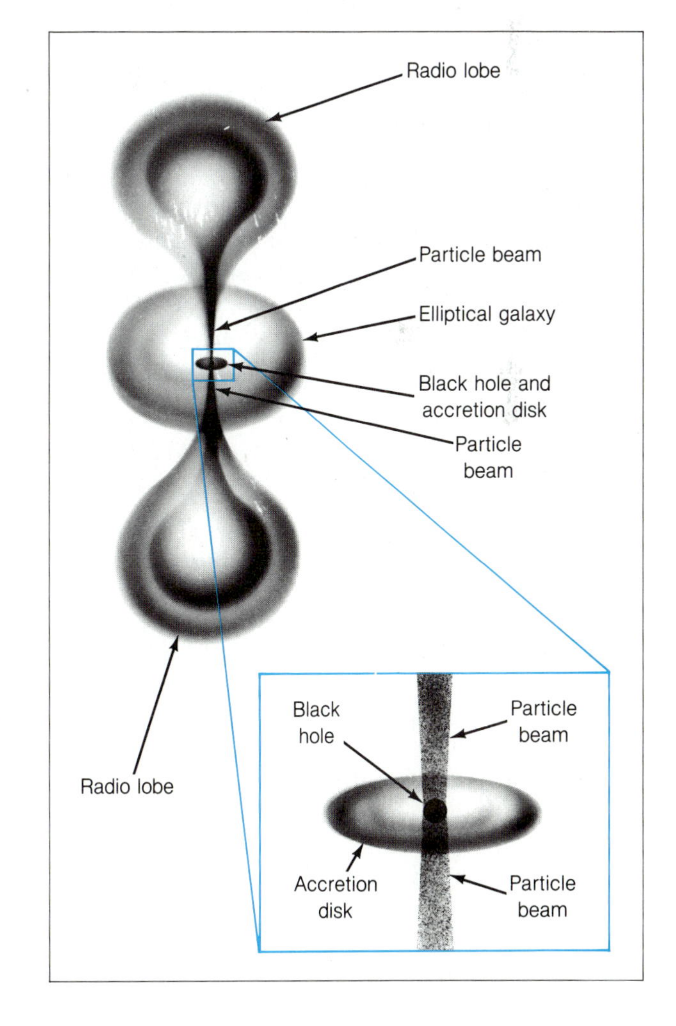

FIGURE 31.18 A GEOMETRICAL MODEL FOR A QUASAR, SEYFERT GALAXY, OR A RADIO GALAXY. *All of these objects have certain features in common that fit the picture shown here. A central object (most likely a supermassive black hole with an accretion disk) ejects opposing jets of energetic charged particles. These jets produce synchrotron radiation, and build up double radio-emitting lobes on either side of the central object. These lobes extend well beyond the confines of the galaxy in which the central object is embedded.*

Even the existence of jets of hot gas is the same; such jets have been observed emanating from several quasars.

Infrared data from the *IRAS* spacecraft have led to a suggested mechanism for building up sufficient densities in the cores of young galaxies to form supermassive black holes. The properties of some of the very luminous infrared galaxies (see Astronomical Insight 28.2) indicate that they may be pairs of galaxies that are colliding and merging. The infrared properties of quasars are very similar to those of luminous infrared galaxies, suggesting that quasars may form as the result of galactic mergers early in the history of the universe.

Whatever the origin and power source of the quasars, we will certainly learn a lot about the nature of matter and the early history of the universe when we are able to answer all the questions about them. The *Space Telescope,* with its broad wavelength coverage and great sensitivity, will provide invaluable information on these fundamentally important objects.

PERSPECTIVE

In this chapter we have learned of new and wondrous things, the mighty radio galaxies and the enigmatic quasars. Along the way we have gained several important bits of information on the universe itself. It is time to tackle the fundamental question of the origin and fate of the cosmos.

SUMMARY

1. Radio galaxies emit vast amounts of energy in the radio portion of the spectrum. These galaxies, often giant ellipticals, commonly show structural peculiarities. The radio emission is nonthermal (usually synchrotron radiation), and in most cases is produced from two lobes on opposite sides of the visible galaxy.
2. The source of energy in a radio galaxy is unknown but appears to be concentrated at the core.
3. Seyfert galaxies are spiral galaxies with compact, bright blue nuclei which produces emission lines characteristic of high temperatures and rapid motions.
4. Several point sources of radio emission were found in the early 1960s to look like blue stars, and were therefore called quasistellar objects, or quasars.
5. Quasars have emission-line spectra with very large redshifts, corresponding to speeds that are a significant fraction of the speed of light. If these redshifts are cosmological, then quasars are on the frontier of the universe, are extremely luminous, and are seen as they were billions of years ago.
6. Quasars are sometimes radio sources, usually emit X-rays, are compact and blue, and sometimes vary over times of only a few days. This implies that their tremendous energy output arises in a volume only a few light-days across.
7. Many quasars have absorption lines at a variety of redshifts (almost always smaller than the redshift of the emission lines), which may be created by matter being ejected from the quasars, or, more likely, by intervening galactic disks and halos.
8. The most satisfactory explanation of the quasars is that they are the cores of very young galaxies. This interpretation is supported by the fact that they are only seen at great distances (that is, they existed only long ago), and that they resemble the nuclei of Seyfert galaxies. In several cases, photographs have now revealed galaxies surrounding quasars, supporting this suggestion.
9. The source of the energy that powers quasars is a premier mystery of modern astronomy. The most probable explanation is that a quasar has a massive black hole at its core; this can produce enough energy (from infalling matter) to account for the great luminosity of quasars, as well as their time variability. By inference, similar objects may exist in the nuclei of galaxies such as Seyfert galaxies, radio galaxies, and the Milky Way.

REVIEW QUESTIONS

Thought Questions

1. Summarize the properties of radio galaxies, pointing out the ways in which they are distinguished from normal galaxies.
2. Why do astronomers suggest that massive black holes may be responsible for the vast amounts of energy being emitted from the cores of radio galaxies, Seyfert galaxies, and quasars?
3. Summarize the arguments for and against the hypothesis that the redshifts of quasars are cosmological,

that is, that they result from the expansion of the universe rather than from some "local" phenomenon.

4. Explain, in your own words and using your own sketch, why the time scale of variations in light from a source such as a quasar places a limit on the diameter of the emitting region.

5. Compile a list of the similarities and differences between quasars and Seyfert galaxies.

6. Why are quasar absorption lines always observed at smaller redshifts than the emission lines? Would this necessarily be true if the redshifts of the emission lines were not cosmological?

7. From what you have learned about quasars in this chapter and about the *Hubble Space Telescope* in Chapter 7, describe some important observations of quasars that might be made with the *Space Telescope*.

8. If quasars are young galaxies, how does this help demonstrate the difficulty of using faraway galaxies as standard candles in estimating large distances in the universe?

9. Explain why nearly all of the first quasars to be discovered are radio emitters, but only about 10 percent of all quasars are.

10. How can the material in intergalactic gas clouds give us information about conditions very early in the history of the universe?

Problems

1. Suppose that the red hydrogen line (rest wavelength 6,563 Å) is observed as a broad emission line from the nucleus of a Seyfert galaxy. The width of the line is 200 Å (that is, it extends for 100 Å in either direction from the rest wavelength). What is the maximum speed of the gas in the region where this line is emitted?

2. A quasar is found to have a redshift of $z = \Delta\lambda/\lambda = 2.45$. What is its recession velocity (use the relativistic Doppler shift formula given in Appendix 14)? How far away is it if the Hubble constant is $H = 75$ km/sec/Mpc? How long has the light from this quasar been on its way to us? How luminous is it if its apparent magnitude is $m = +18$?

3. If the brightness of a quasar changes in a time of one day, how large can the diameter of the light-emitting region be at most? How does your answer compare with the size of the solar system?

ADDITIONAL READINGS

Blandford, R. D., M. C. Begelman, and M. J. Rees. 1982. Cosmic jets. *Scientific American* 246(5):124.

Burns, J. O., and R. Marcus. 1983. Centaurus A: The nearest active galaxy. *Scientific American* 249(5):50.

Chaffee, F. H. 1980. The discovery of a gravitational lens. *Scientific American* 243(5):60.

Disney, M. J., and P. Veron. 1977. BL Lac objects. *Scientific American* 237(2):32.

Downes, Ann. 1986. Radio galaxies. *Mercury* 15(2):34.

Feigelson, E. D., and E. J. Schreier. 1983. The x-ray jets of Centaurus A and M87. *Sky and Telescope* 65(1):6.

Ferris, T. 1984. The radio sky and the echo of creation. *Mercury* XIII(1):2.

Gorenstein, M. V. 1983. Charting paths through gravity's lens. *Sky and Telescope* 66(5):390.

Hamilton, D., W. Keel, and J. F. Nixon. 1978. Variable galactic nuclei. *Sky and Telescope* 55(5):372.

Hazard, C., and S. Mitton, eds. 1979. *Active galactic nuclei.* Cambridge, England: Cambridge University Press.

Kaufmann, W. 1978. Exploding galaxies and supermassive black holes. *Mercury* 6(5):78.

Kellerman, K. I. 1973. Extragalactic radio sources. *Physics Today* 26(10):38.

Lawrence, J. 1980. Gravitational lenses and the double quasar. *Mercury* 9(3):66.

Margon, B. 1983. The origin of the cosmic x-ray background. *Scientific American* 248(1):104.

Osmer, P. S. 1982. Quasars as probes of the distant and early universe. *Scientific American* 246(2):126.

Overbye, Dennis. 1982. Exploring the edge of the universe. *Discover* 3(12):22.

Sandage, A., M. Sandage, and J. Kristian, eds. 1976. *Galaxies and the universe.* Chicago: University of Chicago Press.

Shipman, H. L. 1976. *Black holes, quasars, and the universe.* Boston: Houghton-Mifflin.

Silk, J. 1980. *The big bang: the creation and evolution of the universe.* San Francisco: W. H. Freeman.

Strom, R. G., G. M. Miley, and J. Oort. 1975. Giant radio galaxies. *Scientific American* 233(2):26.

Tyson, T., and M. Gorenstein. 1985. Resolving the nearest gravitational lens. *Sky and Telescope* 70(4):319.

Wyckoff, S., and P. Wehinger. 1981. Are quasars luminous nuclei of galaxies? *Sky and Telescope* 61(3):200.

CHAPTER 32

COSMOLOGY: PAST, PRESENT, AND FUTURE OF THE UNIVERSE

An artist's view of the Hubble Space Telescope *in operation, probing the structure of the universe.* (NASA)

CHAPTER PREVIEW

Given the time and opportunity, humans have, through the ages, devoted themselves to speculation on the grandest scale of all. Poets, philosophers, and theologians have approached the question of the origin and future of the universe in a countless variety of ways. So have astronomers, with the exception that they follow a somewhat restrictive set of rules; the answers they accept must not violate known laws of physics. By retaining this requirement, scientists attempt to approach problems in an objective, verifiable manner.

There are difficulties in maintaining this idealized posture, however, and we shall try to point them out. Because the universe in which we live is, as far as we know, unique, we have no opportunity to check our hypotheses by comparison with other examples. Furthermore, no matter how thoroughly and rigorously we can trace the evolution of the universe by application of known physical laws, there will always be fundamental questions that are beyond the scope of physics. As a result, even the most careful and objective scientists reach a point at which they have to make certain unverifiable assumptions, and at that point, they become philosophers or theologians. In this chapter we shall restrict ourselves to the questions that can in principle have objective, verifiable answers.

Technically, the study of the universe as it now appears is **cosmology;** that is really what the preceding thirty-one chapters have been all about. The study of its origins is **cosmogony,** and this word applies to the big bang as well as to the earlier theories on the origin of the solar system, once thought to be the entire universe. In practice, the general subject of the nature of the universe and its evolution is lumped under the heading of cosmology, and so it is on a pursuit of this subject that we now embark.

UNDERLYING ASSUMPTIONS

To even begin to study the universe as a whole, we must make certain assumptions. These can, in principle, be tested, although it is not clear whether there is a practical way to do so. One of the most important assumptions we make is that the known laws of physics can be applied to the universe as a whole. This is a bold assumption, since we are just now beginning to be able to observe significantly large distances (compared to the size of the universe), and it is possible that the laws that we know to apply to our own times and our own region are merely subsets of some large-scale set of laws, as yet undiscovered. Nevertheless, for now the only way we can proceed is to assume that our known laws can be applied.

A central rule traditionally set forth by cosmologists is that the universe must look the same at all points within it. This does not mean that the appearance of the heavens should be literally identical everywhere; it means that the general structure, the density and distribution of galaxies, and the clusters of galaxies (Fig. 32.1), should be constant. This assumption states that the universe is **homogeneous.**

A related, but slightly different assumption, has to do with the appearance of the universe when viewed in different directions. The assumed property in this case is **isotropy;** that is, the universe must look the same to an observer, no matter in which direction the telescope points. We have already encountered this concept, in connection with the cosmic microwave background radiation. Again, the assumption of isotropy does not imply identical constellations of galaxies and stars in all directions, just comparable ones.

Both of these assumptions are thought to apply only on the largest scales, larger than any obvious clumpiness in the universe, such as clusters or superclusters of galaxies. It should also be pointed out here, although it is not discussed until later in the chapter, that both assumptions have been questioned in some

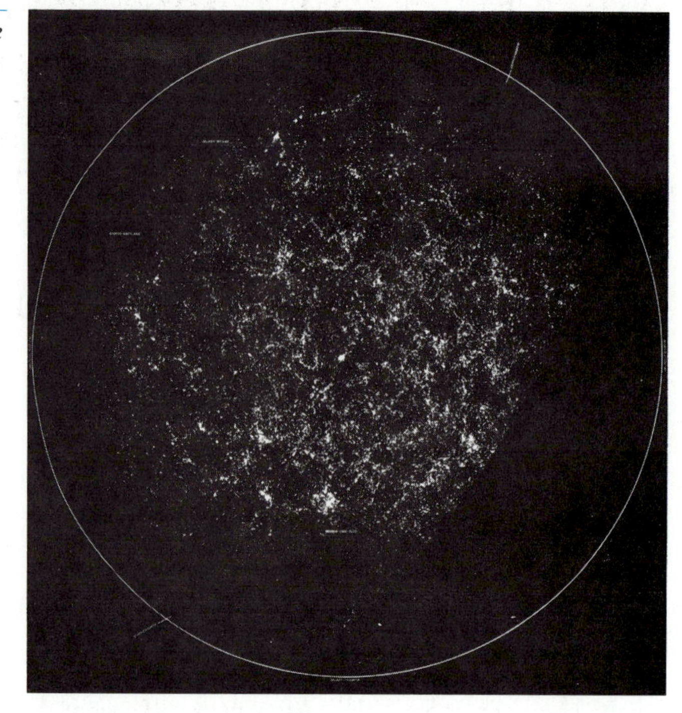

FIGURE 32.1 THE COSMOLOGICAL PRINCIPLE. *The universe is assumed to be homogeneous and isotropic, meaning that it looks the same to all observers in all directions. Here is a segment of the universe, filled with galaxies. Their distribution, while not identical from place to place, is similar throughout.* © 1978, 1979 by The Co-Evolution Quarterly, Box 428, Sausalito, California)

modern theories and observations of the structure of the universe. It is even thought surprising in some current theories that the universe should be as homogeneous as it is. Hence the assumption of homogeneity and isotropy should now be viewed more as an observation that must be explained, rather than as a postulate of cosmological theory.

The statement that the universe is homogeneous and isotropic is often referred to as the **Cosmological Principle,** and is often stated in terms of how the universe looks to observers; i.e. the universe looks the same to all observers everywhere. It is difficult to verify the Cosmological Principle. We can test the assumption of isotropy by looking in all directions from Earth, and indeed we do not find any deviations. On the other hand, we cannot test the homogeneity of the universe by traveling to various other locations to see whether things look any different. What we can do is count very dim, distant galaxies, to see whether their density appears to be any greater or less in some regions than in others, but even this is complicated by the fact that we are looking back in time to an era when the universe was more compressed and dense than it is now. The 3-degree background radiation gives us another tool for testing both homogeneity and isotropy, and it also appears to satisfy the Cosmological Principle. The best we can say for now is that the available evidence supports this principle but does not completely rule out the possibility of subtle anisotropies or inhomogeneities.

EINSTEIN'S RELATIVITY: MATHEMATICAL DESCRIPTION OF THE UNIVERSE

The simple act of making assumptions about the general nature of the universe by itself tells us very little about its past or its future. It serves to elucidate certain properties of the universe as it is today, but to tie this into a quantitative description, to develop a theory capable of making predictions that can be tested, requires a mathematical framework. In developing such a mechanism for describing the universe, the astronomer seeks to reduce the universe to a set of equations and then to solve them, much as a stellar structure theorist studies the structure and evolution of stars by constructing numerical models.

ASTRONOMICAL INSIGHT 32.1

THE MYSTERY OF THE NIGHTTIME SKY

A very simple question, asked over two centuries ago and discussed at length in the early 1800s by the German astronomer Wilhelm Olbers, leads to profound consequences. The question: Why is the sky dark at night?

Olbers argued that in an infinite universe filled with stars, every line of sight, regardless of direction, should intersect a stellar surface. The nighttime sky should therefore be uniformly bright, not dark, as the star images in the sky would literally overlap each other. The fact that this is not so presents a paradox, one worth pondering. The dilemma created by asking why the sky is dark is called **Olbers' paradox.**

One suggested solution is that interstellar extinction can diminish the light from distant stars sufficiently to make the sky appear black. Careful consideration shows, however, that this is not so. Light that is absorbed by dust in space causes the grains to heat up. If the universe were really filled with starlight, as suggested by Olbers' paradox, then the grains would become so hot that they would glow, and we would still have a uniformly bright nighttime sky.

Another suggested explanation of Olbers' paradox has to do with the expansion of the universe. The redshift due to the expansion, if we look sufficiently far away, becomes so great that the light never gets here in visible form. It becomes redshifted to extremely large wavelengths, losing nearly all of its energy. This solution to Olbers' paradox also works in an infinitely old universe, so that even if there is no beginning to the universe, there is still a horizon.

While the expansion of the universe partially explains Olbers' paradox, it does not completely account for the darkness of the sky between galaxies. Yet another solution to the paradox turns out to be more important. To understand this solution, we must realize that the universe is not really infinite, at least from a practical point of view. At sufficiently great distances we expect to see no galaxies or stars, because we will be looking back to a time before the universe began. Thus, even though the universe has no edge, there is a horizon beyond which stars and galaxies do not exist. Therefore, this argument goes, all lines of sight do not have to intersect stellar surfaces; many reach the darkness beyond the horizon of the universe, and we have a dark nighttime sky.

There is another factor that accounts for the darkness of the nighttime sky, which also has to do with the finite size of the universe. The source of energy available to produce light—that is, all the stars in the universe—is limited. If we calculate the total energy that possibly can be produced by all stars everywhere, we find that it is too little to make our sky appear bright. The universe simply does not contain enough energy.

The simple question of why the sky is dark at night, if analyzed fully, could have led to the conclusion that the universe has a finite age or a finite size a full century before this was learned from other observational and theoretical developments.

FIGURE 32.2 ALBERT EINSTEIN. *Among Einstein's great contributions was the development of a mathematical formalism to describe the interaction of gravity, matter, and energy in the universe with the nature of space-time. This framework, general relativity, has withstood all observational and experimental tests applied to it so far.* (The Granger Collection)

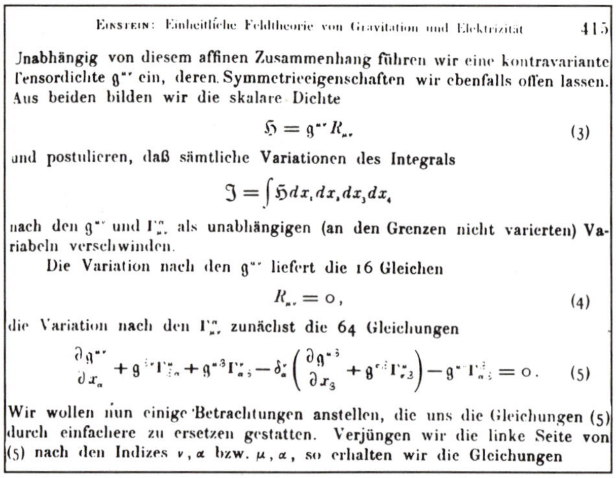

EINSTEIN: Einheitliche Feldtheorie von Gravitation und Elektrizität 415

Jnabhängig von diesem affinen Zusammenhang führen wir eine kontravariante Tensordichte $\mathfrak{g}^{\mu\nu}$ ein, deren Symmetrieeigenschaften wir ebenfalls offen lassen. Aus beiden bilden wir die skalare Dichte

$$\mathfrak{H} = \mathfrak{g}^{\mu\nu} R_{\mu\nu} \qquad (3)$$

und postulieren, daß sämtliche Variationen des Integrals

$$\mathfrak{J} = \int \mathfrak{H} dx_1 dx_2 dx_3 dx_4$$

nach den $\mathfrak{g}^{\mu\nu}$ und $\Gamma^{\alpha}_{\mu\nu}$ als unabhängigen (an den Grenzen nicht variierten) Variabeln verschwinden.

Die Variation nach den $\mathfrak{g}^{\mu\nu}$ liefert die 16 Gleichen

$$R_{\mu\nu} = 0, \qquad (4)$$

die Variation nach den $\Gamma^{\alpha}_{\mu\nu}$ zunächst die 64 Gleichungen

$$\frac{\partial \mathfrak{g}^{\mu\nu}}{\partial x_\alpha} + \mathfrak{g}^{\beta\nu}\Gamma^{\mu}_{\beta\alpha} + \mathfrak{g}^{\mu\beta}\Gamma^{\nu}_{\alpha\beta} - \delta^{\nu}_{\alpha}\left(\frac{\partial \mathfrak{g}^{\mu\beta}}{\partial x_\beta} + \mathfrak{g}^{\sigma\beta}\Gamma^{\mu}_{\sigma\beta}\right) - \mathfrak{g}^{\mu\nu}\Gamma^{\beta}_{\alpha\beta} = 0. \qquad (5)$$

Wir wollen nun einige Betrachtungen anstellen, die uns die Gleichungen (5) durch einfachere zu ersetzen gestatten. Verjüngen wir die linke Seite von (5) nach den Indizes ν, α bzw. μ, α, so erhalten wir die Gleichungen

FIGURE 32.3 THE FIELD EQUATIONS. *Here are portions of the equations that describe the physical state of the universe. To a large extent, the science of cosmology involves finding solutions to these equations.* (The Granger Collection)

The most powerful mathematical tool for describing the universe was developed by Albert Einstein (Fig. 32.2). His theory of general relativity represents the properties of matter and its relationship to gravitational fields. Within the context of his theory, Einstein developed a set of relations, called the **field equations** (Fig. 32.3), that express in mathematical terms the interaction of matter, radiation, and gravitational forces in the universe. Although alternatives to general relativity have been developed and their consequences explored, most research in cosmology today involves finding solutions to Einstein's field equations and testing these solutions with observational data.

The basic premise of general relativity is that acceleration created by a gravitational field is indistinguishable from acceleration due to a changing rate or direction of motion. One way to visualize this is to imagine that you are inside a compartment with no windows (Fig. 32.4). If this compartment is on the Earth's surface, your weight feels normal because of the Earth's gravitational attraction. If, however, you are in space and the compartment is being accelerated at a rate equivalent to 1 Earth gravity, your weight will also feel normal. There is no experimental way to tell the difference between the two situations, short of opening the door and looking out.

One consequence of the equivalence of gravity and acceleration is that an object passing near a source of gravitational field (that is, any other object having mass) undergoes acceleration and therefore follows a curved path. In a universe containing matter, this means that all trajectories of moving objects are curved, and it is often said that space itself is curved. The degree of curvature is especially high close to massive objects (Fig. 32.5), but there is also an overall curvature of the universe, a result of its total mass content. The solutions to the field equations specify, among other things, the degree and type of curvature. We will return to this point shortly.

Einstein's solution to the equations, developed in 1917, had a serious flaw, in his view: It did not allow for a static, nonexpanding universe. In what he later admitted was the biggest mistake he ever made, Einstein added an arbitrary term called the **cosmological constant** to the field equations, solely for the purpose of allowing the universe to be stationary, neither expanding nor contracting.

Others developed different solutions, always by making certain assumptions about the universe. Some

FIGURE 32.4 GENERAL RELATIVITY. *The person in the enclosed room has no experimental or intuitive means of distinguishing whether he is motionless on the surface of the earth, or in space accelerating at a rate equivalent to one earth gravity. The implication of this is that the acceleration due to motion and that due to gravity are equivalent, which in turn implies that space is curved in the presence of a gravitational field.*

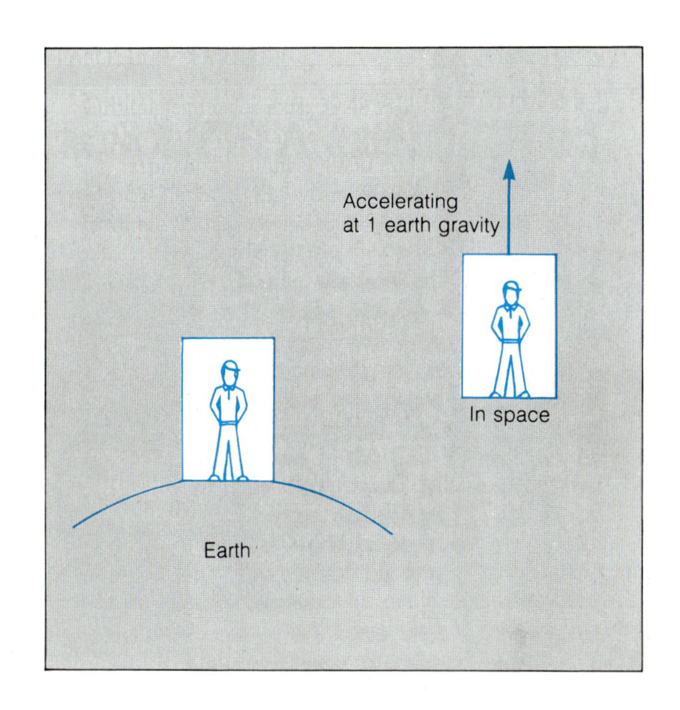

of these assumptions were necessary to simplify the field equations so that they could be solved. For example, W. de Sitter in 1917 developed a solution that corresponded to an empty universe, one with no matter in it.

By the 1920s, solutions for an expanding universe were found, primarily by the Soviet physicist A. Friedmann and later by the Belgian G. LeMaître, who went so far as to propose an origin for the universe in a hot, dense state from which it has been expanding ever since. This was the true beginning of the big bang idea, and it was developed some 3 years before Hubble's observational discovery of the expansion. Of course, once it was found that the universe is not static but is actually in a dynamic state, the original need for Einstein's cosmological constant disappeared. Nevertheless, it is usually included by modern cosmologists in the field equations, but its value is assumed to be zero.

The general relativistic field equations allow for three possibilities regarding the curvature of the universe and three possible futures (Fig. 32.6). A central question of modern studies of cosmology has to do with deciding which possibility is correct. One is referred to as "negative curvature," and an analogy to this is a saddle-shaped surface (Fig. 32.7), which is

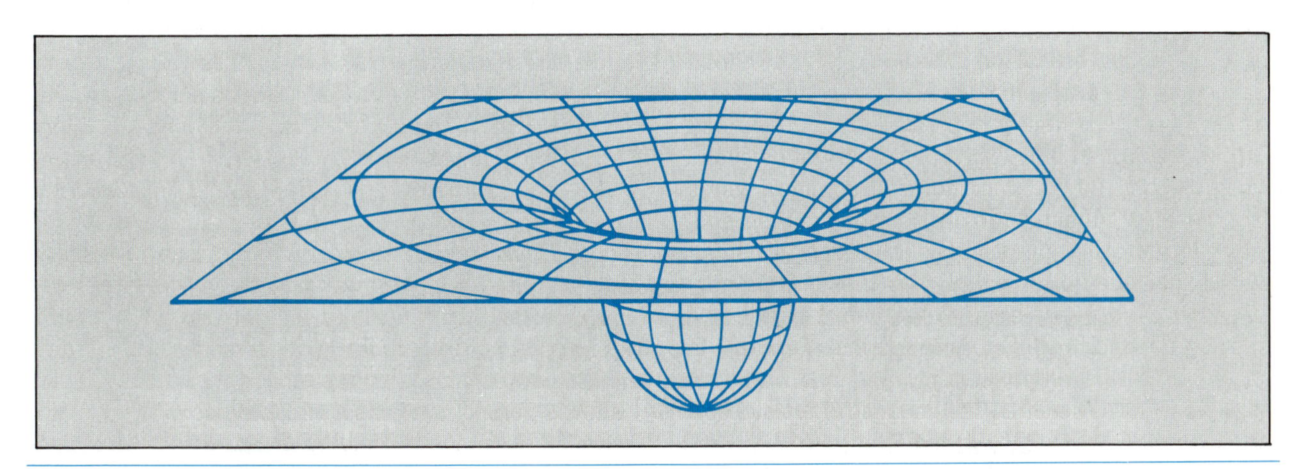

FIGURE 32.5 CURVATURE OF SPACE NEAR A MASSIVE STAR. *This curved surface represents the shape of space very close to a massive star. The best way to envision what it means to say that space is curved in this way is to imagine photons of light as marbles rolling on a surface of this shape.*

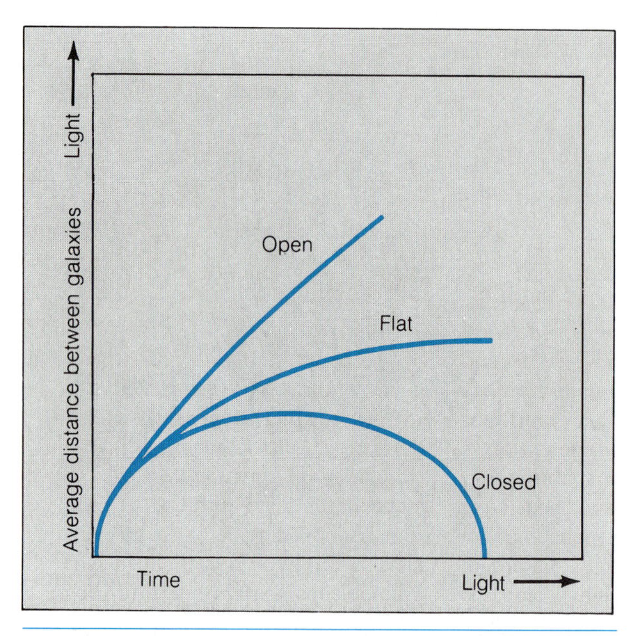

FIGURE 32.6 THE THREE POSSIBLE FATES OF THE UNIVERSE. *This diagram shows the manner in which the average distance between galaxies will change with time for the open, flat, and closed universe. In the first case, galaxies will continue to separate forever, although the rate of separation will slow. In the second case, the rate of separation will, in an infinite time, slow to a halt but will not reverse. In the third case, the galaxies eventually begin to approach each other, and the universe returns to a single point.*

curved everywhere, has no boundaries, and is infinite in extent. This type of curvature corresponds to what is called an **open universe,** a solution to the field equations in which the expansion continues forever, never stopping. (It does slow down, however, because the gravitational pull of all the matter in it tends to hold back the expansion.)

A second possibility is that the curvature is "positive," corresponding to the surface of a sphere (Fig. 32.8), which is curved everywhere, has no boundaries, but is finite in extent. This possible solution to the field equations is called the **closed universe,** and it implies that the expansion will eventually be halted by gravitational forces and will reverse itself, leading to a contraction back to a single point.

The third and last possibility is that the universe is a **flat universe,** with no curvature. This corresponds to outward expansion being precisely balanced by the inward gravitational pull of the matter in the universe, so that the expansion will eventually come to a stop but will not reverse itself. This balanced, static state requires a perfect coincidence between the energy of expansion and the inward gravitational pull of the combined total mass of the universe, and

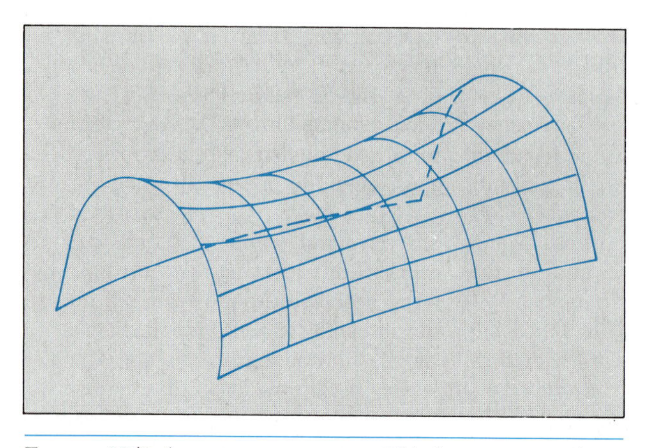

FIGURE 32.7 A SADDLE SURFACE. *This is a representation of the geometry of an open universe, which has negative curvature, is infinite in extent, has no center in space, and has no boundaries.*

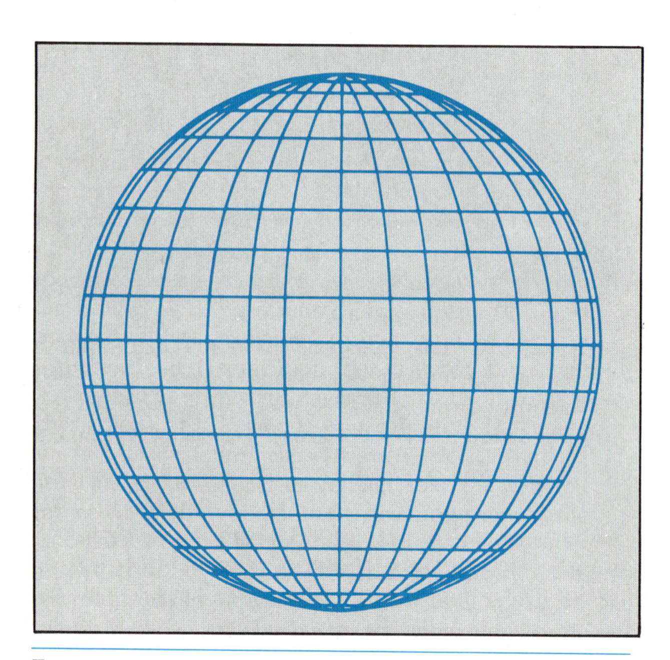

FIGURE 32.8 THE SURFACE OF A SPHERE. *This surface represents a closed universe, which has positive curvature, and a finite extent but no boundaries and no center.*

ASTRONOMICAL INSIGHT 32.2

TESTING RELATIVITY

In the framework of Einstein's theories of special and general relativity, most departures from the classical laws of physics do not become significant except under rather extreme conditions, such as velocities near the speed of light, or the presence of very intense gravitational fields, or when very large distances (compared to the size of the universe) are considered. Therefore, under normal circumstances it is difficult or impossible to tell whether the theory of relativity is correct or incorrect. There are means of testing the theory, however, and the importance of doing so has inspired many experiments.

One of the earliest tests of general relativity was attempted in 1919, when star positions near the Sun's limb were measured during a total solar eclipse, to see whether light from the stars was deflected by the Sun's gravity, as the theory predicted. The stellar positions were found to be shifted by the expected amount, scoring a resounding victory for the then-new theory. Much more recently, this experiment has been performed using radio interferometry techniques to observe the position of quasars near the Sun on the sky, and the agreement with Einstein's theory is so good that many of its competitors have been ruled out.

It has also been possible to detect the expected, very small, gravitational redshift of sunlight, and also to confirm Einstein's prediction about gradual changes in the orientation of the orbit of Mercury that would not be expected under classical physics. These and other, more complex, tests have never yet shown any measurable deviation from the expectations of general relativity theory, and there are few alternative theories still in the running. Many experiments being planned, such as those to be launched on the satellite called the *Relativity Explorer,* are designed to test further, more subtle predictions.

Long before he developed the concepts of general relativity, Einstein published his theory of **special relativity,** in which he first stated the equivalence of mass and energy and delved into the effects of velocities near that of light. One consequence of special relativity is that time should advance at a different rate for a person traveling near the speed of light than for someone who is at rest. This leads to the famous paradox, often exploited in science fiction novels, in which a space-faring traveler returns to Earth after a high-speed voyage and finds himself younger than his twin.

Although we do not yet have the ability to travel at speeds near that of light, we can develop clocks so accurate that the time-dilation effect can be detected at lower speeds. A few years ago, a pair of very high precision atomic clocks were used in an experiment; one was taken around the world by airplane at an average speed of a few hundred miles per hour, while the other remained stationary. The two clocks were synchronized at the start and were found to disagree by exactly the expected amount after the flight, providing another confirmation of Einstein's work.

One of the best laboratories for testing relativistic theory was discovered in space, in the form of a binary pulsar. This is a pair of neutron stars, one of them a pulsar, in mutual orbit. Because the gravitational fields involved are so much stronger than that of the Earth or even the Sun, many of the effects of general relativity should be much greater and easier to measure than in the experiments just described. It will still take several years of observation to answer all the questions that may be asked of the binary pulsar, but the data that are in so far are all in agreement with Einstein's predictions.

One of the most interesting consequences of general relativity is that there should be a form of radiation completely distinct from electromagnetic radiation, that is emitted whenever massive objects undergo strong acceleration. The radiation, referred to as **gravity waves,** might be detectable by means of the minute vibrations it would set off in a large object such as a rigid metal bar. Experiments have not yet reached the level of sensitivity that is thought necessary, but within a very few years it is likely that meaningful searches for gravitational radiation will be possible. The cosmic events that might produce gravity waves include the nearly instantaneous collapse of massive stars as they become supernovae, the rapid motions in close binary systems, and perhaps the mysterious energetic phenomena that occur in many galactic nuclei.

may therefore seem very unlikely. But we do not know what set of rules governed the amount of matter the universe was born with, so there is no logical reason to rule out this possibility. Most astronomers, however, generally pose the question in terms of whether the universe is open or closed, without specifically mentioning the third possibility that it is flat. As we will see in the next section, if a perfect balance has been achieved, it will be very difficult to determine this from the observational evidence, which has large uncertainties.

OPEN OR CLOSED: THE OBSERVATIONAL EVIDENCE

Substantial intellectual and technological resources are devoted today to the question of determining the fate of the universe. Theorists are at work developing and refining the solutions to the field equations, or seeking alternatives to general relativity that might provide equally valid or more valid representations. Observers are busily attempting to test the theoretical possibilities by finding situations in which competing theories should lead to different observational consequences. This is a difficult job, because most of the differences will show up only on the largest scales, and therefore to detect them requires observations of the farthest reaches of the universe.

Within the context of general relativity, the premier question is whether the universe is open or closed. Historically, there have been two general observational approaches to answering this: Determine whether there is enough matter in the universe to produce sufficient gravitational attraction to close it; or measure the rate of deceleration of the expansion to see whether it is slowing rapidly enough to eventually stop and reverse itself. A third approach, tried recently, shows great promise. In this method, the number of galaxies at different redshifts is counted, directly measuring the rate of expansion at different epochs in the past.

Total Mass Content

The total mass in the universe is the quantity that determines whether or not gravity will halt the expansion. The field equations express the mass content of the universe in terms of the density, the amount of mass per cubic centimeter. This is convenient for observers, because it is obviously simpler to measure the density in our vicinity than to try to observe the total mass everywhere in the universe. The field equations can be solved for the value of the density that would produce an exact balance between expansion and gravitational attraction (the density corresponding to a flat universe). If the actual density is greater than this critical density, then there is sufficient mass to close the universe, and the expansion will stop. The critical density depends on the value of Hubble's constant H; for a value near 55 km/sec/Mpc, it is calculated to be roughly 6×10^{-29} g/cm^3, or about 3 protons per cubic meter, a very low value by any earthly standards.

The most straightforward way to measure the density of the universe is to simply count galaxies in some randomly selected volume of space, add up their masses, and divide the total by the volume. Care must be taken to choose a very large sample volume, so that clumpiness from clusters of galaxies is not important. This technique yields very low values for the density, only a small percentage of the critical density. It may seem that we have already answered the question, and that the universe is open, but this method may overlook substantial quantities of mass.

One clue to this arises from determinations of cluster masses based on the velocity dispersion of the galaxies in them (see discussion in Chapter 20). These mass measurements, for reasons still not entirely clear, always yield much higher values than the estimated total mass of the visible galaxies in the cluster. The disparity can be as great as a factor of 10 or more, so if the larger values are correct, the average density of the universe comes closer to the critical value. Even the larger values based on velocity-dispersion measurements, however, fall short of the amount needed to close the universe. If the mass density is to exceed the critical value, there must be large quantities of matter in some form that has not yet been detected (Fig. 32.9). The hidden matter cannot be concentrated inside clusters of galaxies, because its presence would have been detected by the velocity-dispersion measurements.

If there is a lot of dark matter in the universe, it could take several different forms. One possibility that is obvious to us by now is that there may be many black holes in the space between clusters. The idea that there may be black holes in intergalactic

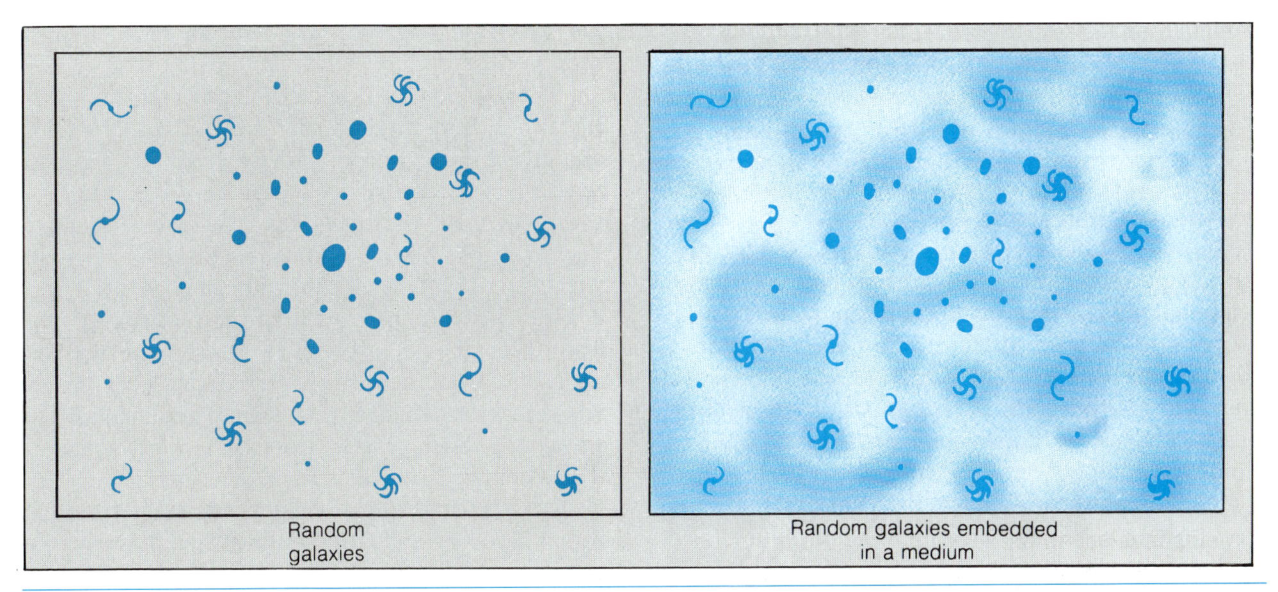

FIGURE 32.9 DARK MATTER IN THE UNIVERSE. *Are the isolated galaxies that we see all that there is in the universe? Or are they merely points that happen to glow, embedded in a universal sea of unknown substance, whose mass density overwhelms that represented by the galaxies?*

space has been seriously suggested by the astrophysicist Stephen Hawking, who hypothesized the existence of countless numbers of "mini" black holes, formed during the early stages of the big bang. These would have very small masses, much smaller even than the mass of the Earth, but would be so numerous that they could easily exceed the critical density. Current observational evidence argues against the existence of such objects in great numbers, however.

Another possibility is that neutrinos, the elusive subatomic particles produced in nuclear reactions, have mass. Recall from Chapter 18 that these particles permeate space, freely traveling through matter and vacuum alike, but that standard theory says they are massless. A recent controversial experiment indicates that this last assertion may not be correct, that they may contain minuscule quantities of mass after all. If so, they are sufficiently plentiful to provide more than the critical density, thereby closing the universe. Further experiments designed to determine whether the neutrino has mass have been performed, all of them so far being more consistent with the standard notion that these particles are massless.

Significant information on the possibility that neutrinos have mass came from the recent supernova (Supernova 1987a) in the Large Magellanic Cloud. Neutrinos from reactions that took place during the explosion were detected at the Earth, and the time of arrival and the energies of these neutrinos allowed scientists to place a limit on how massive they can be. It was found that neutrinos probably cannot be massive enough to bring the average density of the universe up to the point where it is closed. Therefore, if the universe is closed, the dark, unseen matter must be in some form other than neutrinos.

There are other kinds of elementary particles that have been postulated to exist, on the basis of elementary particle theory. These particles have not been detected experimentally (the energies required far exceed those possible in any existing particle accelerators), but it is possible that they are sufficiently numerous and massive to close the universe. We must await new generations of particle accelerators before we will have an answer to the question of the significance of such particles.

If we base our current thinking only on the density of matter that has actually been observed, we are forced to conclude from the available observational evidence that the mass density of the universe is less than the critical density, and therefore that the universe is open. It is worth stressing, however, that many scientists believe there are substantial quantities of dark matter in the universe, and that it might be closed.

The Deceleration of the Expansion

Another approach to answering the question of whether the universe is open or closed is perhaps being pursued more vigorously today. The objective is to compare the present expansion rate with what it was early in the history of the big bang, to see how much slowing, or deceleration, has occurred (Fig. 32.10). The expansion has certainly slowed in any case; the question is how much. If it has only decelerated a little, then we infer that is not going to slow down enough to stop and reverse itself; but if it has decelerated quite a bit, then we conclude that the expansion is coming to a halt, and that the universe is closed.

Establishing the deceleration rate requires knowing both the present expansion rate and the expansion rate at a time long ago, shortly after the big bang began. Neither quantity is easy to determine: We already learned that the value of the Hubble constant H, which tells us the present expansion rate, is uncertain to begin with. To measure the expansion rate at early times in the history of the universe is even more difficult, for it involves determining the distances and velocities of very faraway galaxies, so distant that we see them as they were when the universe was young. Such objects are now at the very limits of detectability, and it is hoped that the *Space Telescope* will push the frontier back far enough to permit observations of velocities in the early days of the expansion.

We still must rely on standard candles to establish the distance scale, and for such distant objects this procedure becomes even more uncertain than usual. When we look so far back in time, we are seeing galaxies that are much younger than those near us and therefore may have different properties than mature galaxies. For example, their content of stellar populations may be rather different than those in nearby galaxies, their brightest stars and nebulae may be different, and even their total galactic luminosities may be different. It may not be valid to assume that the brightest galaxy in a cluster of galaxies has the same absolute magnitude as in more nearby clusters.

The present evidence on deceleration is that the universe has not slowed very much and therefore is not on its way to stopping and beginning to contract. As we have seen, however, the observational uncertainties are great, and this conclusion is not very firm.

An indirect technique for measuring the deceleration has been recently employed and avoids many of the difficulties of observing extremely distant galaxies. Some light elements were created in the early stages of the expansion, and the amounts that were produced depend on the fraction of the universal energy that was in the form of mass rather than radiation at the time of element formation. The primordial mass density is in turn related to the present-day deceleration of the expansion, so if the primordial density is deduced from the abundances of elements that were created, we can infer the deceleration rate. The strat-

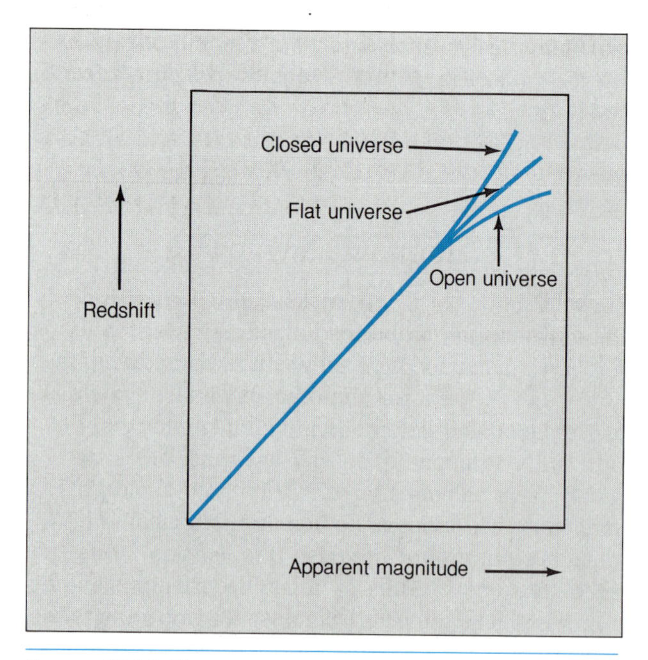

FIGURE 32.10 THE EFFECT OF DECELERATION IN THE HUBBLE DIAGRAM. *This shows the relationship between velocity of recession and distance for the three possible cases: closed (upper curve), flat (middle), or open (lower). Present data seem to favor the open universe, although there is substantial uncertainty, largely because of problems in applying standard candle techniques to galaxies so far away that they are being seen as they were at a young age. The distinction is also made difficult by the subtlety of the contrast between the shapes of the curves.*

egy, therefore, is to measure the abundance of some element that was produced in the big bang, and from that to derive the deceleration.

The most abundant element (other than hydrogen) produced in the big bang is helium, so if we could measure how much helium was created, this would be a good indicator of the deceleration. One problem with this element, however, is that it is also produced in stellar interiors, so it is difficult to determine how much of what we see in the universe today is really left over from the big bang. Another problem with using helium as a probe of the early universal expansion rate is that the quantity produced is not expected to vary much with different expansion rates; that is, the primordial helium abundance, even if we knew it well, is not a very sensitive indicator of the early expansion rate.

Better candidates are ^{7}Li, an isotope of the element lithium, and deuterium, the form of hydrogen that has one proton and one neutron in the nucleus. These species were also produced in the big bang, and, as far as is presently known, are not made in any other way. To date, most efforts have been concentrated on measuring the present-day abundance of deuterium, although in principle ^{7}Li can be measured as well. While deuterium can be destroyed by nuclear processing in stars, it is not produced in that way (even if it were, it would not survive the high temperatures of stellar interiors without undergoing further reactions). The present deuterium abundance in the universe should therefore represent an upper limit on the quantity created in the big bang. Direct measurements of the amount of deuterium in space became possible in the 1970s, with the launch of the *Copernicus* satellite, which made ultraviolet spectroscopic measurements and was able to observe absorption lines of interstellar deuterium atoms (Fig. 32.11). The abundance of deuterium that was found is sufficiently high that it implies a low density, in turn pointing to a small amount of deceleration. Thus this test, in agreement with others, indicates that the universe is open. There is some uncertainty, though, because the *Copernicus* data seem to indicate an uneven distribution of interstellar deuterium throughout the galaxy, and this would not be expected if the deuterium formed exclusively in the big bang. A clumpy distribution of deuterium in the galaxy could indicate that some of it is somehow produced by stars, in which case its abundance is not a strict test of deceleration. A recent measurement of the ^{7}Li abundance in the galaxy is consistent with the deuterium results; that is, the observed quantity of ^{7}Li is also large enough to imply that the universe is open.

The new technique alluded to earlier, the direct measurement of the expansion rate at different times, based on counting galaxies with different redshifts, may be viewed as a modification of the attempts to infer the deceleration of the expansion. The result of the first application of this technique is that the universe appears to be flat; that is, the expansion appears to have just the rate needed to eventually stop (in an infinite time) but never to contract again. Certainly this technique will be further exploited, and this result further scrutinized in the coming years.

It is worth noting that all the observational techniques tried so far to discover whether the universe is open or closed have pointed towards a fairly close balance. That is, the data do not point to a universe that is closed or open by a wide margin. Whatever the correct answer is, we can say that the universe is close to being flat. This is very significant, for such a close balance is not necessarily expected in the big bang theory, and must be regarded in this theory as

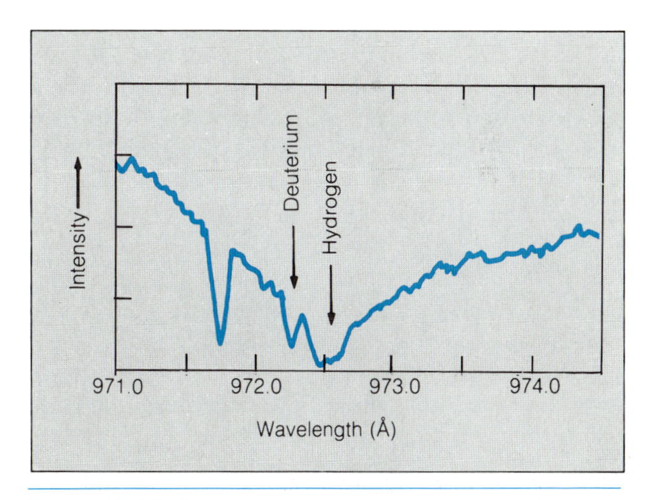

FIGURE 32.11 AN ULTRAVIOLET ABSORPTION LINE OF INTERSTELLAR DEUTERIUM. *The abundance of deuterium in space is determined from the analysis of absorption lines it forms in the spectra of background stars. Here we see a weak deuterium line close to a strong absorption feature due to normal hydrogen. The spectrum shown here was obtained with the* Copernicus *satellite.* (J. B. Rogerson and D. G. York)

a coincidence. There are other models, however, in which a flat or nearly flat universe is expected; one of these is described in the next section.

The Inflationary Universe

The big bang cosmology is widely accepted, but there remain some nagging problems. One is the improbability for a universe starting from a singularity as envisioned in the big bang to be as symmetric as the observed universe. It is unlikely that the universe would appear as homogeneous and isotropic as observations tell us it is. A great deal of effort has been put into testing this property of the universe, because any departures from homogeneity would provide information on imbalances or asymmetries in the early epochs of the expansion. It is also thought surprising that the universe should be so closely balanced between being open and closed.

A new kind of expansion model has been developed in which the close balance and the homogeneity of the universe are easily understood. Calculations have shown that, at a certain point very early in the history of the universe, when the temperature was around 10^{27} K, conditions were right for small regions to separate themselves from the rest of the universe and then to expand very rapidly to much larger sizes. A reasonable analogy would be the creation of bubbles in a liquid, with the bubbles then growing much larger almost instantaneously. In this cosmological model, each "bubble" becomes a universe in its own right, with no possibility of communication with other bubbles.

The major advantage of this so-called **inflationary universe** model is that a universe born of a tiny cell in the early expansion would be expected to remain symmetric and homogeneous as it grew. The reasons for this are somewhat abstract but have to do with the fact that the various portions of the tiny region that was to expand into the present universe were so close together that they were in a form of equilibrium, with little or no variation in physical conditions being possible. In a larger region, such as in the standard big bang cosmology, no such equilibrium would have existed, and there would be no reason to expect such uniformity throughout the resulting universe.

Another advantage of the inflationary model is that in such a universe, the expansion would be expected to go on forever but would continue to slow, approaching a stationary state; that is, a flat universe is the natural result of rapid expansion from a tiny bubble as envisioned in this model. As we have seen, observations tell us that our universe is nearly flat, and may be precisely flat. If it were not so closely balanced, it would not be so difficult to determine whether it is open or closed.

A third promising feature of the inflationary model is its agreement with the observational data on another kind of particle, one that has yet to be unambiguously detected. Some big bang theorists predict the formation in the early universe of a large number of **magnetic monopoles,** subatomic particles with single magnetic poles, instead of the usual pair of opposing poles. Such particles would be very difficult to detect, the best technique being to look for tiny electrical currents they would briefly create in wires they passed near. So far, only one possible detection of a magnetic monopole has been reported, in contradiction with the big bang theory, which leads to the expectation that these particles should not be so rare. The inflationary model, however, leads to the formation of only a very few magnetic monopoles in each bubble that formed before expanding into a universe. Therefore, the lack of many observed monopoles is more consistent with the inflationary model than the big bang theory.

The homogeneity of a bubble in the inflationary universe is an advantage in explaining some observations, as we have seen, but it is also a disadvantage. In such a uniform universe, it is very difficult to understand how the matter ever became lumpy enough to form galaxies and clusters of galaxies. We have already spoken of possible turbulence in the early universe and other mechanisms having to do with concentrations of matter, but in the inflationary model it is very difficult to see how such concentrations would ever have formed.

The inflationary model, first suggested in 1981, is currently under active discussion among cosmologists. There will continue to be refinements to it, as more of the observational constraints are confronted. It will be very interesting to see whether it withstands the test of continued scrutiny.

THE HISTORY OF EVERYTHING

We are now at the forefront of modern cosmology, having outlined the present state of knowledge, and having pointed out the basis for current and planned observational tests. At this point it is useful to review the development of the universe, highlighting some of the significant events (Figs. 32.12 and 32.13). The very fact that we can do this, that we can say with any degree of certainty what conditions were like at the beginning and at points along the way, is a triumph of modern science.

It is impossible to physically describe the universe at the precise moment that the expansion (or the inflation of our "bubble") began; it is physical nonsense to deal with infinitely high temperature and density. It is possible, however, to calculate the conditions immediately after the expansion started, and at any later time. Many of the most interesting events in the early history of the universe, and the ones currently under the most active investigation, occurred very early, before even a ten-thousandth of a second had passed. Under the conditions of density and temperature that existed then, matter and the forces that act on it were quite different from anything we can experience, even in the most advanced laboratory experiments. Even the familiar subatomic particles such as protons, neutrons, and electrons could not exist, but were replaced by *their* constituent particles.

Current particle physics theory holds that the most fundamental particles are **leptons** and **quarks.** Modern theory also provides a basis for believing that all four of the fundamental forces in nature (see Astronomical Insight 5.1) may really be manifestations of the same phenomenon. So far it has been possible to show that three of the fundamental forces (the electromagnetic force and the weak and strong nu-

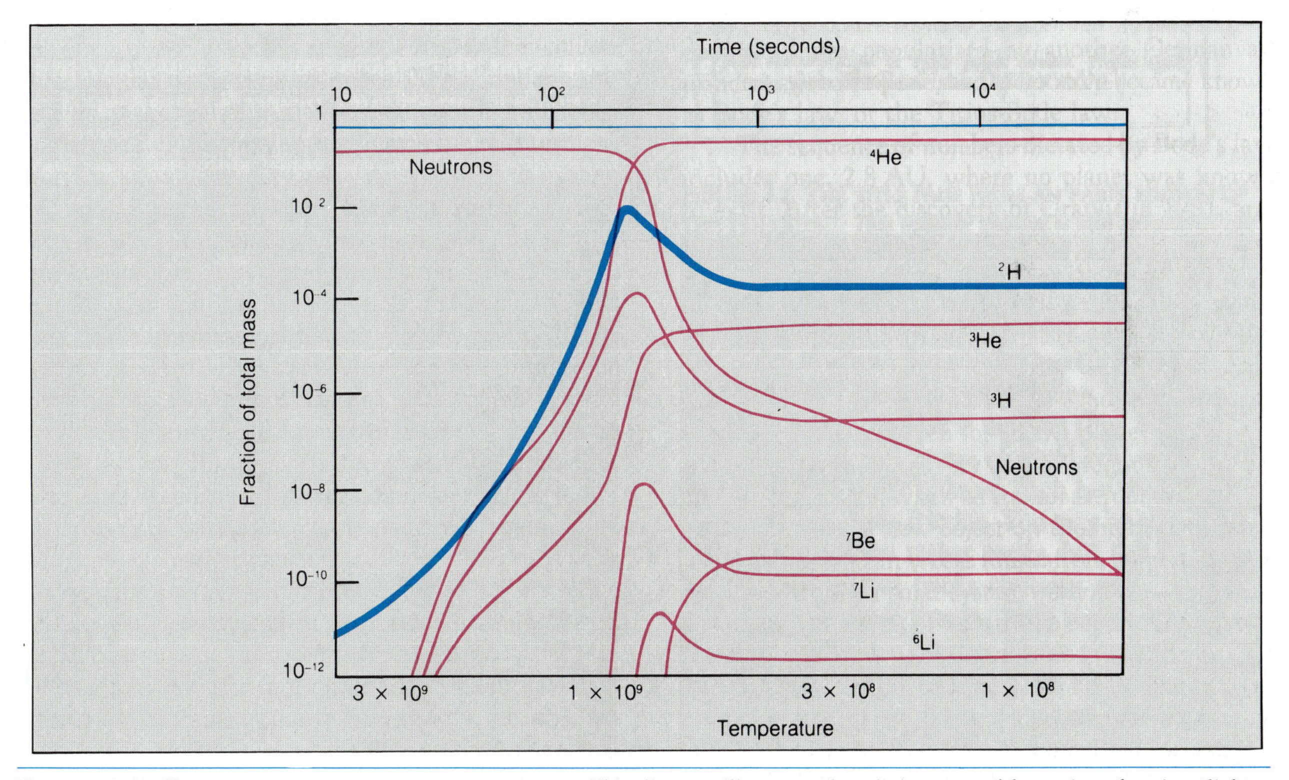

FIGURE 32.12 ELEMENT FORMATION IN THE BIG BANG. *This diagram illustrates the relative rates of formation of various light elements by nuclear reactions during the early stages of the big bang expansion. The production rates are shown as functions of time and temperature.* (Data from Wagner, R. V. 1973, *The Astrophysical Journal* 179:343)

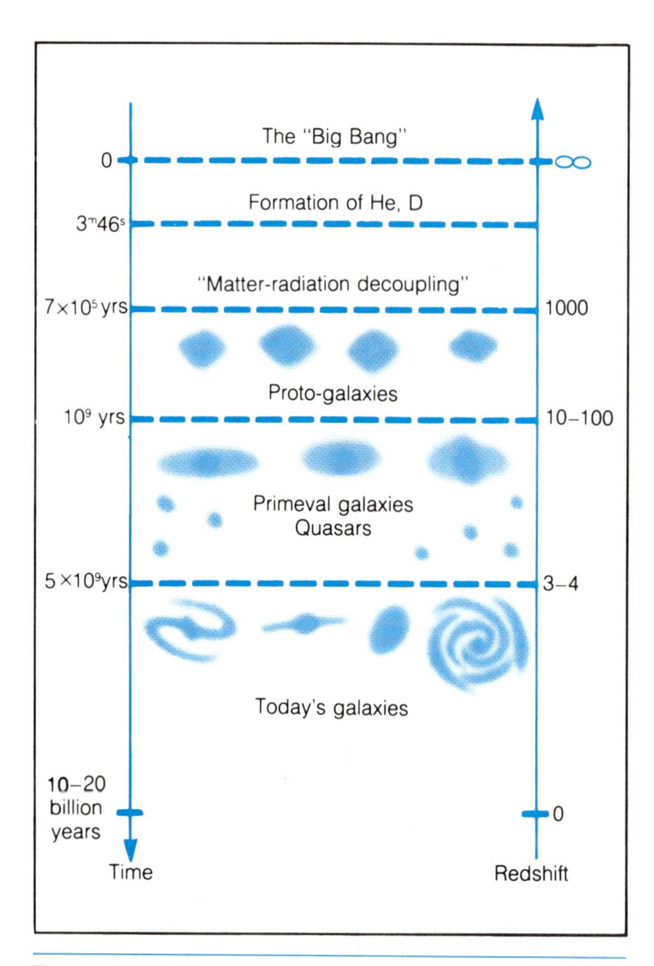

FIGURE 32.13 THE EVOLUTION OF THE UNIVERSE. *This diagram shows, as a function of time since the start of the big bang, significant stages in the evolution of the universe. It also shows the types of objects that existed at each stage, and the redshift at which they are (or would be) observed at the present time.* (J. M. Shull)

clear forces) are manifestations of the same basic interaction. Under physical conditions that we are used to, these forces behave very differently, but when the density and temperature are very high, as they were early in the expansion of the universe, the forces are indistinguishable. Some aspects of the theory that shows this have been confirmed by laboratory experiments. The theoretical framework connecting the three forces is called **Grand Unified Theory** (**GUT,** for short). This name may be somewhat exaggerated, because the fourth force, gravity, has not yet been shown to be unified with the other three, and some theories of gravity indicate that it may not be.

The unification of the other three forces implies that, in the very early moments following the beginning of the expansion, until 10^{-35} seconds had passed, the universe contained only leptons and quarks and related particles and radiation, and the only forces operating in it were gravity and the unified force that was later to become recognizable as the electromagnetic, strong, and weak forces.

Let us now jump forward in time to an epoch in the early universe when matter was beginning to take on more familiar forms, and the four fundamental forces were already acting as four distinct forces. At 0.01 seconds after the beginning of the expansion, the temperature was perhaps 100 billion degrees (10^{11} K), and electrons and positrons began to appear. The temperature dropped to 10 billion degrees (10^{10} K) by 1.09 seconds after the start, and by then protons and neutrons were appearing. At this point, most of the energy of the universe was still in the form of radiation. Within a few minutes, conditions became better suited for nuclear reactions to take place efficiently, and the most active stage of element creation began. The principal products, in addition to the helium and deuterium already mentioned, were tritium (another form of hydrogen, with one proton and two neutrons in the nucleus) and the elements lithium (three protons and four neutrons) and to a minor extent beryllium (four protons and five neutrons). Nearly all the available neutrons combined with protons to form helium nuclei, a process that was complete within 4 minutes of the beginning of the expansion. At this point, with some 22 to 28 percent of the mass in the form of helium, the reactions were essentially over, except for some production of lithium and beryllium over the next half hour.

The expansion and cooling continued, but nothing significant happened for a long time after the nuclear reactions stopped. Eventually the density of the radiation had reduced sufficiently that its energy was less than that contained in the mass (that is, the energy derived from $E = mc^2$ became greater than that contained in the radiation). At that point, it is said, the universe became matter-dominated, rather than radiation-dominated.

The matter and radiation continued to interact, however, because free electrons scatter photons of light very efficiently. The strong interplay between matter and radiation finally ended nearly a million years after the start of the big bang, when the tem-

ASTRONOMICAL INSIGHT 32.3

PARTICLE PHYSICS AND COSMOLOGY

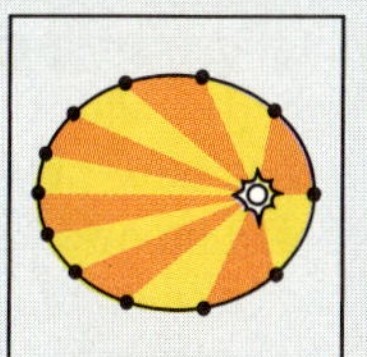

There is some irony in the fact that the science of the smallest scales, particle physics, is finding its greatest opportunities in cosmology, the science of the largest scales. One of the most rapidly-advancing areas in modern particle physics coincides with one of the brightest fields in modern astrophysics, bringing full circle one of the themes of astronomy, the relationship between the science of the atom and that of the cosmos.

In particle physics, scientists are concerned with discovering the nature of matter and energy at its most fundamental level. We now know that the atom, and the major subatomic particles such as protons and neutrons, are actually composed of more basic particles, known in general as elementary particles. The most modern theories of elementary particles postulate that all matter is made up of two types of basic particles. One type, which includes electrons, are the **leptons;** the others, which can have several different combinations of properties, are called **quarks.** Quarks and other particles known only from theory are sought experimentally in particle accelerators, enormous machines that can accelerate particles to such high kinetic energies that when they collide they are broken down into their constituent elementary particles. Because there is an equivalence between matter and energy (as expressed in Einstein's theory of relativity), elementary particles can be created from pure energy in sufficient quantities. Thus, the objective in building particle accelerators to probe elementary particles is to inject enough energy into the accelerators to create the particles being sought. Progress is made every time new levels of energy are reached.

The planned Superconducting Super Collider (SSC), having recently been approved for construction by the United States government, will provide higher energies than any existing accelerators, and will give scientists new capabilities for producing elementary particles. This instrument will consist of a huge, circular tunnel (53 miles in circumference) in which particle beams will be accelerated by gigantic magnets and then allowed to collide. The collision energies will be comparable to particle energies in the early universe, only a small fraction of a second after the expansion began.

Even the SSC, however, will not answer questions about the very earliest times. Some of the particles that are predicted by theory will require energies so high that it may never be practical to create them in man-made accelerators. This is where astrophysics, specifically cosmology, may come to the rescue. Energies high enough to create *all* possible particles certainly existed during the earliest stages of the universal expansion. Thus, elementary particle physicists may find some of the answers they seek by studying what is known about the big bang.

We cannot actually observe the composition of the universe at the time of the early big bang, but we know that today's universe is the product of that era. Thus, by examining the detailed atomic composition of the universe today, particle physicists hope to deduce the mixture of particles that inhabited the universe during its initial moments. A remarkable degree of success has already been achieved in just this way, as particle theorists have been able to derive, with considerable certainty, the physical conditions that must have prevailed in early times. We are confident in our knowledge about the formation of the elements in the early universe because the theory of elementary particles is capable of reproducing the observed abundances of the elements believed to be present a fraction of a second after the start of the expansion. Now the hope is that the process can be reversed; that particle physicists can learn about elementary particle processes by observing the state of today's universe.

perature became low enough to allow electrons and protons to combine into hydrogen atoms. The atoms still absorbed and reemitted light, but much less effectively, because they could do so only at a few specific wavelengths. From this time on, the matter and the radiation went separate ways. The radiation simply continued to cool as the universe expanded, reaching its present temperature of 2.7 K some 10 to 20 billion years after the expansion began.

Sometime in the first billion years or so, the matter in the expanding universe became clumpy and fragmented into clouds and groups of clouds that eventually collapsed to form galaxies and clusters of galaxies. As we noted in Chapter 20, the cause of the clumpiness and fragmentation is as yet unknown. Once galaxies began to form, the subsequent evolution of matter followed steps already described in the preceding chapters.

Because there is so much uncertainty about how the initial stages of fragmentation took place, there is a great deal of interest in probing into this process. To do so will be another of the goals of the *Space Telescope*.

WHAT NEXT?

To discuss the future of the universe is obviously a speculative venture. We cannot even answer, with absolute certainty, the basic question whether it is open or closed. There have been attempts to calculate future conditions in the universe, in analogy to the theoretical work on the early stages discussed in the last section. In the case of the future of the universe, however, the uncertainties are much greater, and the following descriptions should be regarded as *very* speculative.

If the universe is open, then it will have no definite end; it will just gradually run down. The radiation background will continue to decline in temperature, approaching absolute zero. As stellar processing continues in galaxies, the fraction of matter that is in the form of heavy elements will continue to grow, and the supply of hydrogen, the basic nuclear fuel, will diminish. It is predicted that all the hydrogen will be gone by about 10^{14} years after the birth of the universe, so the universe in which stars dominate has now lived approximately one ten-thousandth of its lifetime. The recycling process between stars and the interstellar medium will continue until this time, but gradually the matter will become locked up in black holes, neutron stars, white dwarfs, and black dwarfs. Dead and dying stars will continue to interact gravitationally, eventually colliding often enough in their wanderings that all planets will be lost (by about 10^{17} years) and galaxies will dissipate as their constituent stars are lost to intergalactic space (by about 10^{18} years). Further speculation shows that new physical processes will take over at later times. The new GUT theory tells us that the proton, a basically stable particle, may disintegrate in a very low-probability process that occurs, on average, once in 10^{32} years for a given proton. When the universe reaches an age of about 10^{20} years, enough protons will begin to evaporate here and there that the energy produced will keep the remnant stars heated, although only to the modest temperature of perhaps 100 K. At 10^{32} years, most protons will have decayed, and the universe will consist largely of free positrons, electrons, black holes, and radiation (the extremely cold remnant of the big bang). The final stage that has been foreseen occurs at an age of 10^{100} years, when sufficient time has passed for all black holes to evaporate, and nothing is left but a sea of positrons, electrons, and radiation (theory says that black holes can disintegrate with a very low probability, meaning that if we wait long enough, eventually they will do so).

If the universe is closed, then someday, perhaps some 50 billion years from now, the expansion will stop and will be replaced by contraction. The deterioration described above will still take place, until the final moments when the universe once again becomes hot and compressed, entering a new singularity. In some views, purely conjectural and without possibility of verification, such a contraction would be followed by a new big bang, and the universe would be reborn. This concept of an oscillating universe, pleasing to the minds of many, will not be fulfilled unless the present weight of the evidence favoring an open universe is found to be in error.

PERSPECTIVE

Our story is essentially complete. We have now discussed the universe and all of the various objects and structures in it, and we have described what is known of its evolution and its fate.

We have, however, omitted consideration of what is perhaps the most important ingredient of all: life. Although much of what we might say about this is beyond the scope of astronomy, it is appropriate to assess what can be learned from objective scientific consideration. We do so in the next chapter.

SUMMARY

1. In cosmological studies, astronomers usually adopt the Cosmological Principle, which states that the universe is both homogeneous and isotropic. Existing observational data tend to support this assumption.

2. Einstein's theory of general relativity, which describes gravitation and its equivalence to acceleration, is used to mathematically describe the universe as a whole. In the context of this theory, field equations are written that represent the interaction of matter, radiation, and energy in the universe. Solutions of the field equations define the properties of the universe.

3. There are three general solutions to the field equations that are considered by modern cosmologists. They correspond to a closed universe (positive curvature), an open universe (negative curvature), and a flat universe (no curvature).

4. A major question in modern astrophysics is whether the universe is closed (expansion eventually to be reversed) or open (expansion never to stop).

5. There are two general types of tests of whether the universe is open or closed: (1) to ascertain whether there is enough mass in the universe to gravitationally halt the expansion; or (2) to determine the rate of deceleration of the expansion, to see whether it is slowing enough to eventually stop.

6. The total mass content is measured in terms of the average density of the universe, and the matter that is visible in the form of galaxies is not sufficient to close the universe. Various suggestions have been made concerning other forms in which the necessary mass could exist.

7. The deceleration is measured in two ways: (1) by comparing past and present expansion rates, through observations of very distant galaxies; and (2) by inferring the early expansion rate from the present-day abundances of elements that formed only during the big bang.

8. Both the total mass content that is observed and the inferred deceleration of the expansion indicate that the universe is open. This conclusion is not universally accepted, and observational tests are continuing.

9. The early stages of the universal expansion, up to the time when matter and radiation decoupled, can be described quite precisely and certainly by modern physics. Following the initial amount of infinite density and temperature, the first atomic nuclei formed just over 1 second later, and all of the early element production was finished within a few minutes. Matter and radiation decoupled almost a million years later; it is not so well understood how the universe subsequently organized itself into stars and galaxies.

10. The future of the universe appears to have two possibilities. If it is open, it will gradually become cold and disorganized. If it is closed, it will eventually contract, perhaps to a new beginning in another big bang.

REVIEW QUESTIONS

Thought Questions

1. Explain how the universe can be homogeneous even though galaxies are concentrated in clusters and superclusters.

2. Recall from the previous chapter that there is a daily anisotropy in the cosmic background radiation. Does this violate the assumption that the universe is isotropic? Explain.

3. Explain in your own words why general relativity implies that space is curved near massive objects.

4. Summarize the future of the universe under the three possibilities discussed in the text; that is, what will happen if the universe is open, closed, or flat?

5. Explain why it is difficult to determine the average mass density of the universe from observations.

6. It has been suggested (as discussed in Chapter 31) that most or perhaps all galaxies have very massive black holes in the cores. If so, would this change observational estimates of the mass density of the universe, and therefore help determine whether the universe is open or closed? Explain.

7. Why is the deuterium abundance a better indicator of the early expansion rate of the universe than helium?

8. What are the advantages of the inflationary universe theory over the conventional big bang model?
9. Why did element production occur only during the first few minutes of the universal expansion?
10. Summarize the ways in which the *Space Telescope* will help to answer questions about the future of the universe.

Problems

1. The interstellar medium in the disk of our galaxy has an average density of about one hydrogen atom for every 10 cm^3 of volume. Convert this density into units of g/cm^3, assuming that the mass of a hydrogen atom is 1.7×10^{-24} g. How does your answer compare with the critical density for closing the universe? What implications does this have for whether the universe is open or closed?
2. Suppose that, on average, the universe contained one galaxy like ours (mass about 10^{11} solar masses) for every 10 Mpc^3. What is this density in terms of $grams/cm^3$, and how does it compare with the critical density for closing the universe? (*Hint*: A volume of 10 Mpc^3 corresponds to a cube that is 2.2 Mpc on each side; convert this to centimeters, and then compute the volume of the cube in cubic centimeters.)

ADDITIONAL READINGS

Barrow, J. D., and J. Silk. 1980. The structure of the early universe. *Scientific American* 242(4):118.
Bartusiak, Marcia. 1985. Sensing the ripples in space-time. *Science 85* 6(3):58.
Bartusiak, Marcia. 1986. *Thursday's Universe,* New York: Times Books.
Boslough, John. 1981. The unfettered mind: Stephen Hawking. *Science 81* 2(9):66.
Dicus, D. A., J. R. Letaw, D. C. Teplitz, and V. L. Teplitz. 1983. The future of the universe. *Scientific American* 248(3):90.
Ferris, T. 1977. *The red limit: The search for the edge of the universe.* New York: William Morrow.
———. 1984. The radio sky and the echo of creation. *Mercury* 13(1):2.
Field, G. B. 1982. The hidden mass in galaxies. *Mercury* 11(3):74.
Gaillard, M. K. 1982. Toward a unified picture of elementary particle interactions. *American Scientist* 70(5):506.
Gott, J. R., III, J. E. Gunn, D. N. Schramm, and B. Tinsley. 1976. Will the universe expand forever? *Scientific American* 234(3):62.
Gribbin, John. 1986. *In search of the big band.* Bantam Books.
Guth, A. H., and P. J. Steinhardt. 1984. The inflationary universe. *Scientific American* 250(5):90.
Hawking, S. W. 1984. The edge of spacetime. *American Scientist* 72(4):355.
Overbye, Dennis. 1985. The shadow universe: Dark matter. *Discover* 6(5):12.
———. 1983. "The universe according the Guth," *Discover* 4(6):92.
Page, D. N., and M. R. McKee, 1983. The future of the universe. *Mercury* 12(1):17.
Penzias, A. A. 1978. The riddle of cosmic deuterium. *American Scientist* 66:291.
Schramm, D. N. 1974. The age of the elements. *Scientific American* 230(1):69.
Schwinger, Julian. 1986. *Einstein's legacy.* New York: W. H. Freeman.
Shu, F. 1983. The expanding universe and the large-scale geometry of spacetime. *Mercury* 12(6):162.
Silbar, M. L. 1982. Neutrinos: Rulers of the universe? *Griffith Observer* 46(1):9.
Silk, J., A. S. Szalay, and Y. B. Zel'dovich. The large-scale structure of the universe. *Scientific American* 249(4):56.
Trefil, J. S. 1978. Einstein's theory of general relativity is put to the test. *Smithsonian* 11(1):74.
———. 1983. How the universe began. *Smithsonian* 14(2):32.
———. 1983. How the universe will end. *Smithsonian* 14(3):72.
Trefil, James S. 1983. *The moment of creation.* New York: Scribners.
Tucker, W., and K. Tuckere, 1982. A question of galaxies. *Mercury* 11(5):151.
Weinberg, S. 1977. *The first three minutes.* New York: Basic Books.
Will, C. M. 1983. Testing general relativity: 20 years at progress. *Sky and Telescope* 66(4):294.

GUEST EDITORIAL

PATTERNS IN THE UNIVERSE

Margaret J. Geller

Dr. Geller is one of the world's foremost authorities on the structure of the universe. Starting with her graduate work at Princeton, her research has consistently focused on questions of the largest scope and the grandest scale imaginable as she has probed the overall distribution and motions of galaxies and what they tell us about the shape of the universe. Since receiving her doctorate at Princeton in 1975, Dr. Geller has been associated during most of her career with the astronomy programs at Harvard and the Smithsonian Astrophysical Observatory, where she is now a research staff member. Besides being a leader in cosmological research, Dr. Geller has also served the national astronomical community in several capacities, including her present position as a Councilor of the American Astronomical Society and her activities on several advisory committees to both NASA and the National Science Foundation. In this article, she shares some of her thoughts on recent discoveries about the distribution of galaxies in the universe.

For more than 4,500 years we humans have been making maps. We have mapped the surface of the Earth in excruciating detail. The concept of map-making extends to regimes we cannot directly explore—we make maps of the positions of atoms in solid materials and in DNA, the key to life itself. From the smallest to the largest physical systems maps reflect the human drive to know. Today one of the grand frontiers is the exploration of the universe over distances of hundreds of millions—even billions—of light years. With these maps, we hope to come to understand the formation and evolution of very large patterns in the distribution of galaxies.

Science and map-making are closely related. The development of the theory of continental drift is a historical example. During the early twentieth century, clear knowledge of the large-scale geographic features of the Earth (particularly the shapes of the continents) led to the idea. Before 1912, Alfred Wegener noticed that the continents can be fitted together like pieces of a jigsaw puzzle and made the creative leap to suggest they once were a single giant land mass which later fragmented. The continents drifted apart to their present locations. The idea of a dynamic Earth was greeted with more than a bit of skepticism—it was even ridiculed. Now plate tectonics, which explains continental drift, is a full-fledged physical theory with abundant observational support.

The daunting enormity of the universe presents challenges to cosmic map-makers which differ from those faced by map-makers on the Earth. When we talk about "mapping the universe" we mean that we chart the positions of galaxies like our own Milky Way in a three-dimensional space we call "redshift space." Practical limitations dictate that a feasible study of the distribution of galaxies could include measurements for 10,000 or perhaps 100,000 galaxies. Current surveys include a few thousand galaxies. In other words, we have mapped out a fraction of the universe comparable with the fraction of the Earth covered by the state of Rhode Island—about 1/100,000.

Although we are limited to mapping out a small fraction of the volume of the universe we would like to try to answer a well-defined, perhaps deceptively simple, question: How big are the structures, the patterns (if any), in the distribution of galaxies? Once we focus on the question, the challenge is to obtain the best possible limits on the properties of big structures (if they exist).

The first step in making a map is the construction of a catalog of the positions of galaxies on the sky. Fortunately, during the 1960's Fritz Zwicky and his collaborators catalogued the positions of more than 300,000 galaxies brighter than a specific limiting magnitude.

(continued on next page)

GUEST EDITORIAL

Until the 1930's, extragalactic astonomers were limited to the "flatland" of maps of positions in the sky. In 1929 Hubble's discovery of the universal expansion opened the way for exploration of the three-dimensional structure of the universe. We get the third dimension by measuring the redshifts of galaxies which, in turn, give us estimates of their distances from us.

Modern detector technology makes it possible to use a 1.5-m telescope (small by today's standards) to measure the redshifts for galaxies within about 500 million light years of the Earth. The Smithsonian Astrophysical Observatory operates a 1.5-m telescope on Mt. Hopkins, just south of Tucson, Arizona. Every clear moonless night, John Huchra and I use this telescope to measure redshifts of galaxies. To measure a typical redshift in our survey, we collect photons from each galaxy for about half an hour. Within the next five to ten years (depending on the weather in Tucson!) we plan to measure the redshifts for some 12,000 galaxies. During the spring of 1985, Valérie de Lapparent, John Huchra, and I completed the first sample in this project. We chose the geometry of the sample for sensitivity to large patterns in the distribution of galaxies—patterns which persist over distances of a hundred million light years or more. We measured redshifts for the approximately 1100 galaxies in a 6° wide strip stretching for 120° across the sky. The strip extends across the entire right ascension range [see the map in Fig. 29.17] with a declination range of 26.5° to 32.5°. Why this particular strip? The number of galaxies included is about what we can measure in one spring of observing time on the 1.5-m telescope. The strip is roughly overhead in Tucson and seemed a convenient starting place.

In this map we are at the vertex of a cone. The map shows only two of the three deminsions of the survey; the declination coordinate is perpendicular to the page. In the azimuthal direction we plot the right ascension—the coordinate on the sky which runs along the long dimension of the strip. Along the radial direction we plot the velocity derived from the redshift. Each point is a galaxy comparable with the Milky Way. Perhaps the map has a subliminal appeal because the pattern of galaxies looks like a human figure. In any case, the pattern is striking!

We designed a survey to look for big structures and we found some. Nearly all the galaxies lie in thin, sharp structures which loop across the map. If you treat the pattern of galaxies as a connect-the-dot exercise, your pencil outlines the vast dark regions we call voids. These regions contain few if any bright galaxies. The largest of these voids (centered at 15^h and 7500 km/sec) has a diameter of 5000 km/sec or 150 million light years.

The map of recession velocities is not exactly a map of distances. There are distortions because of the motions of galaxies internal to systems like the Coma cluster. The torso of the "human figure" is the Coma cluster. It is stretched out along a line pointing toward us at the origin. The length of this "finger" is a measure of the typical motions of galaxies within the cluster. From the width and length of this finger we can estimate the amount of matter in the cluster by applying Newton's laws. [See the discussion of velocity dispersion measures of cluster masses, in Chapter 29.] When we perform this exercise we are faced with one of the profound problems in extragalactic astronomy—the dark matter problem. If we measure the amount of light from the galaxies in the cluster and figure out the mass of stars responsible for the light, the total falls far short of the estimate based on Newton's laws. We come up short by a factor of ten! In other words, we see only about 10 percent of the mass; the rest is dark.

Other distortions of the velocity map may occur if all the galaxies in some region have a motion relative to the expansion of the universe. In general, we expect these velocities to be small compared with the extent of the voids we see. Except for the fingers associated with the rich clusters we expect that the big patterns in our velocity map are a reasonable representation of what we would see if we could measure distance directly. From here on we treat the map for the most part as though the "velocity" label read "distance."

Although our map is only a thin slice, we can use it to figure out the most likely three-dimensional arrangement of bright galaxies. We have two guides in searching for the

GUEST EDITORIAL

answer to this puzzle: pure geometric intuition and theoretical models for the way the distribution of galaxies might look.

What sort of geometric structure would yield the pattern we see in almost any slice? There are several familiar things which fit the bill. Imagine, for example, taking a slice through the soapsuds in your kitchen sink. Make the slice thin compared with the diameter of a typical bubble. The surfaces of the bubbles—thin films of soap—would outline a pattern very much like the one defined by the galaxies. The interiors of the bubbles correspond to voids. Of course a soap bubble is about twenty-seven orders of magnitude smaller than a void in the galaxy distribution and the physics of soapsuds is very different from the physics of large-scale structures in the universe. Only the geometries are similar. Other familiar structures like honeycombs and sponges also have similar structure; the distinctions among these are subtle and currently beyond the discriminatory power of the data. We do know that the distribution of galaxies is made up of thin sheets in which the galaxies lie. These sheets surround or nearly surround vast voids. The sheets which appear to run all the way across the survey are the surfaces of several adjacent "bubbles." Since the completion of the first slice we have finished surveying three more slices which support this picture.

The pattern we observe in the distribution of galaxies has profound implications. One of the simplest and perhaps most disconcerting is that the largest voids we can see are the largest we can detect within the limits set by the dimensions of our survey. There are, at the moment, no direct limits on even larger structures. The enormous dark, but not necessarily empty, regions are common. Every survey big enough to contain one does so. These results make it clear that if we could sample the universe in volumes 100 million light years on a side, the distribution of galaxies would differ substantially from one volume to the next. Some volumes ("patches") might be devoid of bright galaxies; others might contain lots of galaxies in one or several thin sheet-like structures.

How big a region do we have to survey in order that one region looks essentially the same as another? The diameters of the largest voids we have seen so far are about one percent of the radius of the visible universe—still small. If we find similarly striking voids with much larger dimensions we might have to abandon one of the fundamental ideas in cosmology, the idea that the universe is homogeneous on some sufficiently large scale. We have to make deeper surveys to answer this question.

The development of the large voids and sheets in the universe depends on many poorly known or even unknown properties of the universe and its contents. However, the action of gravity on large scales must be an integral component of any explanation of the formation of structure we see. Galaxies and clusters of galaxies are held or "bound" together by the force of gravity. These large systems develop from smaller concentrations in the early universe. The origin and properties of the small lumps and bumps are one of the puzzles we have to solve.

We also have to understand the role of the ubiquitous dark stuff. Because we do not see the dark matter we don't know what it is or where it is. If we could look directly at the way matter, rather than light-emitting matter, is distributed in the universe, we might see that it is actually very uniform and that the galaxies form preferentially where the density of matter is somewhat higher than the average. Because the higher-density regions are rare, the galaxy formation process automatically leaves large holes without having to move galaxies across enormous distances. Although computer simulations based on these ideas look much like the data, it is worrisome that we do not know the physics which determines where galaxies form.

If these models are right, the striking contrast between the voids and sheets in our maps is misleading. Galaxies are then the merest flotsam and jetsam in a uniform sea of dark matter. Although the phenomonology of galaxies might be interesting in itself, they would be poor guides to the dynamics of the universe as a whole. The dark voids would be full of stuff—dark stuff and perhaps faint galaxies which would not be included in our surveys.

(continued on next page)

GUEST EDITORIAL

Physics other than the action of gravity may be important in making the structure we see. Perhaps explosions occurred when the age of the universe was less than a billion years and then drove the hydrogen into thin shells which fragmented to form galaxies. This suggestion is appealing because it is easy to imagine how it produces a bubble-like pattern like the one we observe. Here, too, the voids could be full of dark stuff. But these models have their drawbacks. Although there are many explosive, energetic phenomena in the universe ranging from supernovae to the sources of energy in the nuclei of galaxies, there are none which release sufficient energy over a short enough period to make the largest observed shells. Even worse, if there were such enormous explosions we should probably have detected their effect as irregularities in the very smooth microwave radiation background.

The unanswered questions in cosmology are profound. I often feel that we are missing some fundamental element in our attempts to understand the large-scale structure of the universe. Advances in particle physics may eventually lead to the identification of the mysterious dark matter which appears to dominate the large-scale dynamics of the universe. Detection of one of these particles would be a revolution in both particle physics and cosmology.

It is both sobering and exciting to think once again about the analogy between mapping the Earth and mapping the universe. It took centuries to learn the shapes of continents well enough to prompt the intuitive suggestion of continental drift. Then it took another half century to substantiate the theory.

Perhaps our approach to mapping the universe is more sophisticated than the early approaches to mapping the Earth. But the universe is immense; we are encumbered because we can only admire it from afar. The techiques required to uncover the fundamental physics governing the largest realm we can explore may be well beyond our current technology.

Discovering the universe will undoubtedly keep us busy, awed, and fascinated for a long time to come. I often ask myself what we will learn about large-scale structure during my lifetime. There will be surprises, answers to old questions, and uncovering of new questions. At every stage we will think we understand, but at every stage we will have nagging doubts.

SECTION VII

Life in the Universe

We finish our survey of the dynamic universe with a discussion of living organisms and the prospects for finding that we are not alone in the cosmos. To many, this is the most central question we can ask. In the sole chapter of this section, we will examine the astronomer's attempt to answer it.

In discussing the question of life elsewhere in the universe, we will begin by looking into the origins of life here on Earth. In doing so, we will learn of biological evolution and the evidence that has been unearthed to tell us of our own beginnings. To do this, we will step outside the normal arena of astronomy, but we will then return to it in assessing whether the conditions that led to life on Earth might exist on other planets in our solar system and in the galaxy at large. We will discuss the probability that technological civilizations might be thriving here and there in the Milky Way.

We will complete our brief treatment of life in the universe by discussing the strategy for searching for other civilizations. It seems most likely that radio communications will be our initial means of contact, and there is an interesting exercise in logic involved with choosing the wavelength at which to search for or to send signals.

CHAPTER 33

THE CHANCES OF COMPANIONSHIP

Artist's conception of a permanently manned space station. (NASA)

CHAPTER PREVIEW

We have attempted to answer all of the fundamental questions about the physical universe that can be treated scientifically. Having done this, we know our place in the cosmos: We know something of its scale and age, and we realize how insignificant our habitat is.

In this chapter we contemplate whether we as living creatures are unique in the universe, or whether even that distinction must be shared. Nothing that we have learned so far leads us to rule out the possibility that other life forms, some of them intelligent, may exist. We believe that the Earth and the other planets in our solar system are a natural by-product of the formation of the Sun, and we have evidence that some of the essential ingredients for life were present on the Earth from the time it formed. Similar conditions must have been met countless times in the history of the universe, and will occur countless more times in the future.

FIGURE 33.1 CHARLES DARWIN. *Scientific inquiries by Darwin led to an understanding of evolution, one of the most profound concepts of human intellectual development.* (The Granger Collection)

Science cannot yet tell the full story of how life began, however, and we have not found any evidence that it actually does exist elsewhere. The mystery remains.

LIFE ON EARTH

We can start our discussion of possible extraterrestrial life by discussing the origins of the only life we know. Besides giving us some insight into the processes thought to have been at work on the Earth, this will help us later, when we are speculating about whether the same processes have occurred elsewhere.

Before the time of Charles Darwin (Fig. 33.1), in the mid-1800s, the view was widely held that life could arise spontaneously from nonliving matter. Darwin's work in the study of evolution, showing how species develop gradually as a result of environmental pressures, made such an idea seem improbable.

An alternative to the spontaneous formation of life was proposed in 1907 by the Swedish chemist S. Arrhenius, who suggested that life on the Earth was introduced billions of years ago from space, originally in the form of microscopic spores that float through the cosmos, landing here and there to act as seeds for new biological systems. This idea, called the **panspermia** hypothesis, cannot be ruled out, but several arguments make it seem unlikely. It would take a very long time for such spores to permeate the galaxy, and there would have to be a very high density of them in space for one or more of them to reach the Earth by chance. More importantly, it seems very unlikely that the spores could survive the hazards of space, such as ultraviolet light and cosmic rays. Even if the panspermia concept is correct, the question of the ultimate origin of life remains, although it is transferred to some other location. In view of what is known today about the evolution of life and the early conditions on the Earth, scientists generally agree that life arose through natural processes occurring here, and was not introduced from elsewhere.

It is believed that the early atmosphere of the Earth was composed chiefly of hydrogen and hydrogen-bearing molecules such as ammonia (NH_3) and methane (CH_4), as well as water (H_2O). Therefore, the first organisms must have developed in the presence of these ingredients.

In the 1950s, scientists began to perform experiments in which they attempted to reproduce the conditions of the early Earth. The starting point of these experiments was to place water in containers filled with the type of atmosphere just described. Water was introduced because it is apparent that the Earth had oceans from very early times, and because it is thought that life started in the oceans, where the liquid environment provided a medium in which complex chemical reactions could take place. Reactions occur much more slowly in solids and in gases—in solids because the atoms are not free to move about easily and interact, and in gases because the density of gas is low, and particles are relatively unlikely to encounter each other. Water is the most stable and abundant liquid that can form from the common elements thought to be present when the earth was young.

The first of these experiments (Fig. 33.2) was performed in 1953 by the American scientists H. Urey and S. Miller, who concocted a mixture of methane, ammonia, water, and hydrogen, and exposed it to electrical discharges, a possible source of energy on the primitive Earth. (Ultraviolet light from the Sun is another, but it was more difficult to work with in the lab.) After a week, the mixture turned dark brown, and Urey and Miller analyzed its composition. What they found were large quantities of amino acids, complex molecules that form the basis of proteins, which are the fundamental substance of living matter. Other experimenters later showed that exposure to ultraviolet light produced the same results. These experiments demonstrated that at least some of the precursors of life probably existed in the primitive oceans almost immediately after the Earth had cooled enough to support liquid water. Other similar experiments

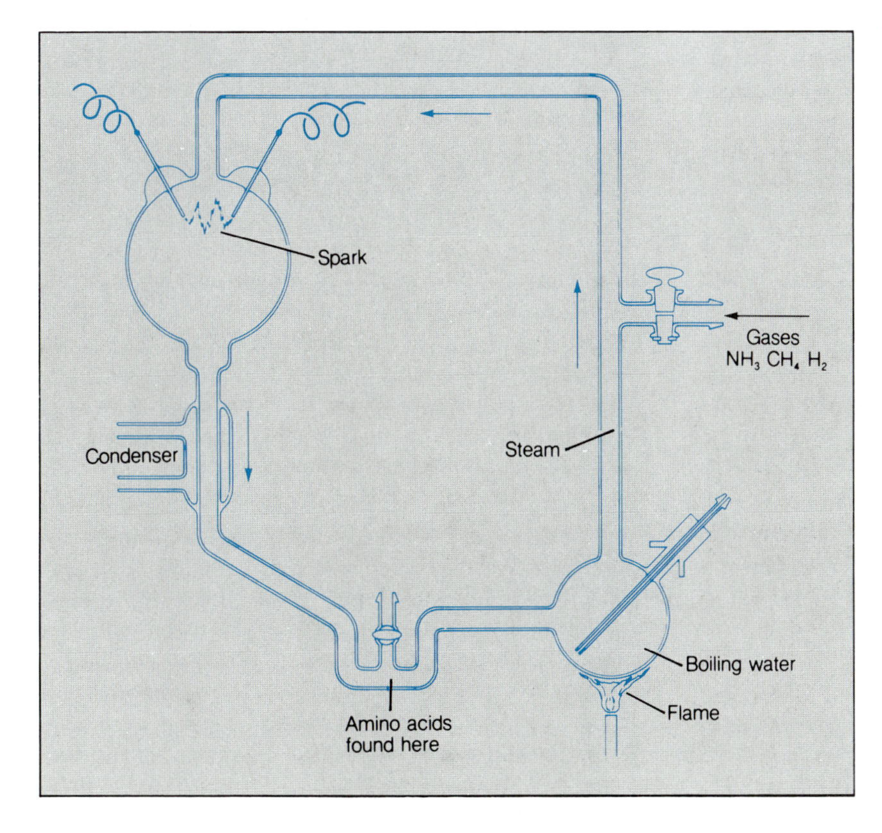

FIGURE 33.2 SIMULATING THE EARLY EARTH. *This is an apparatus constructed by Urey and Miller to reproduce conditions on the primitive earth, in hopes of learning how life forms could have developed.* (© West Publishing Co.)

FIGURE 33.3 PRIMORDIAL AMINO ACIDS. *This is a section of the Murchison meteorite, which fell in Australia. Amino acids found in this carbonaceous chondrite were apparently present in it when it fell.* (Photo by John Fields, the Trustees, the Australian Museum.)

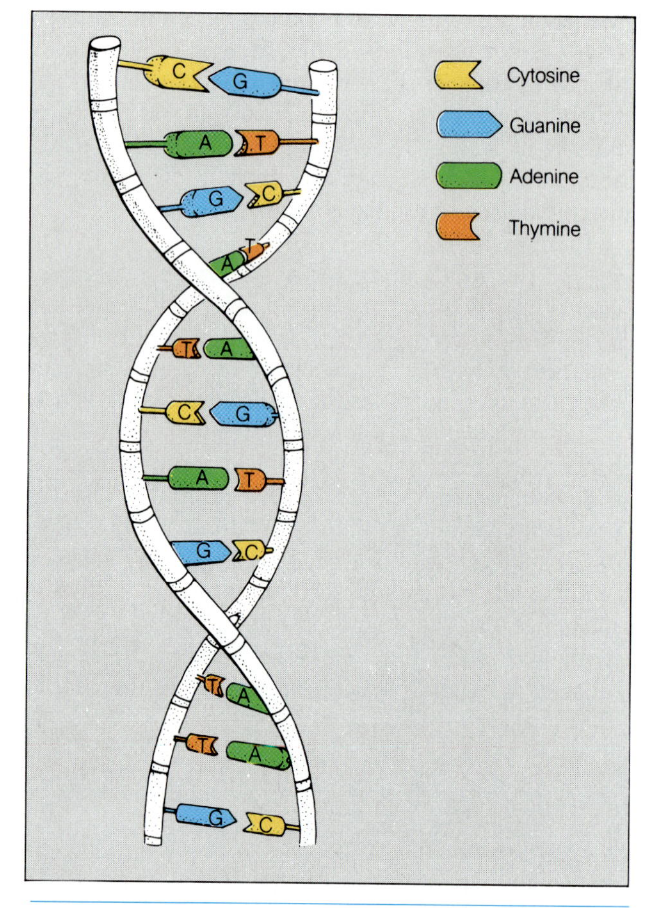

FIGURE 33.4 THE DNA MOLECULE. *This is a schematic diagram of a section of a DNA molecule. DNA carries the genetic code that allows organisms to reproduce themselves. A critical question in understanding the development of life on earth is to learn how DNA arose.* (© West Publishing Co.)

have produced more complex molecules, including sugars and larger fragments of proteins.

As noted in Chapter 16, it may be that amino acids were present in the solar system even before the Earth formed, because traces of them have been found in some meteorites (Fig. 33.3), and we know that meteorites are very old, representing the first solid material in the solar system. We also know that several kinds of complex molecules, including organic (carbon-bearing) molecules, exist inside dense interstellar clouds (see Chapter 26), and we may speculate that perhaps amino acids may also have formed in these regions. To have survived on the Earth, these primordial amino acids would have to reach our planet sometime *after* its molten period.

It is not so clear what direction things took once amino acids and other organic molecules existed. Somehow these building blocks had to combine to form **ribonucleic acid (RNA)** and **deoxyribonucleic acid (DNA;** Fig. 33.4). These very complex molecules carry the genetic codes that allow living creatures to reproduce. Experiments have successfully produced molecules that are fragments of RNA and DNA from conditions like those that prevailed on the early Earth, but not the complete forms required. Maybe it is a simple matter of time; if such experiments could be performed for years or millennia, perhaps the vital forms of these proteins would appear. This is one of the areas of greatest uncertainty in our present knowledge of how life began.

Fossil records (Fig. 33.5) tell us that the first microorganisms appeared some 3 to 3.5 billion years ago, when the Earth was barely 1 billion years old. Following their appearance, the evolution of increasingly complex species seems to have followed naturally (Table 33.1). At first the development was very slow, only reaching the level of simple plants such as algae 1 billion years later. Increasingly elaborate multicellular plant forms followed, and gradually altered the Earth's atmosphere by introducing free oxygen.

Meanwhile, the gases hydrogen and helium, light and fast-moving enough to escape the Earth's gravity, essentially disappeared. Nitrogen, always present from outgassing and volcanic activity, became more predominant through the decay of dead organisms. By about 1 billion years ago, the Earth's atmosphere had reached its present composition.

The first broad proliferation of animal life occurred about 600 million years ago, and the great reptiles arose some 350 million years later. The dinosaurs died out after about 200 million years, and mammals came to dominance about 65 million years ago. Our primitive ancestors appeared only in the last 3 or 4 million years. Once the development of intel-

TABLE 33.1 STEPS IN THE EVOLUTION OF LIFE ON EARTH

Era	Period	Age (yr)	Biological Developments
Archeozoic		3.5×10^9	
	Archean		No life forms
Proteozoic		$1.5–3.5 \times 10^9$	
	Algonkin		Radiolaria; marine algae
Paleozoic		$0.25–1.5 \times 10^9$	
	Cambrian		Marine faunas, primitive vegetation
	Ordovician		Fishlike vertebrates
	Silurian		Air-breathing invertebrates, first land plants
	Devonian		Fishes and amphibians; primitive ferns
	Mississippian		Ancient sharks, mosses, ferns
	Pennsylvanian		Amphibians, insects
	Permian		Primitive reptiles, mosses, ferns
Mesozoic		$70–250 \times 10^6$	
	Triassic		Reptiles, dinosaurs, ferns
	Jurassic		Toothed birds, primitive mammals, palms
	Cretaceous		Decline of dinosaurs, modern insects, birds, snakes
Cenozoic		$1.5–70 \times 10^6$	
	Paleocene		Large land mammals
	Eocene		Primates, first horse, modern plants
	Oligocene		Larger horse, modern plants
	Miocene		Proliferation of mammals, larger horse, modern plants
	Pliocene		Grassland mammals, earliest hominids
Modern		1.5×10^6 to present	
	Pleistocene		Hominids, modern-size horse, modern plants
	Holocene		*Homo sapiens,* present-day vegetation

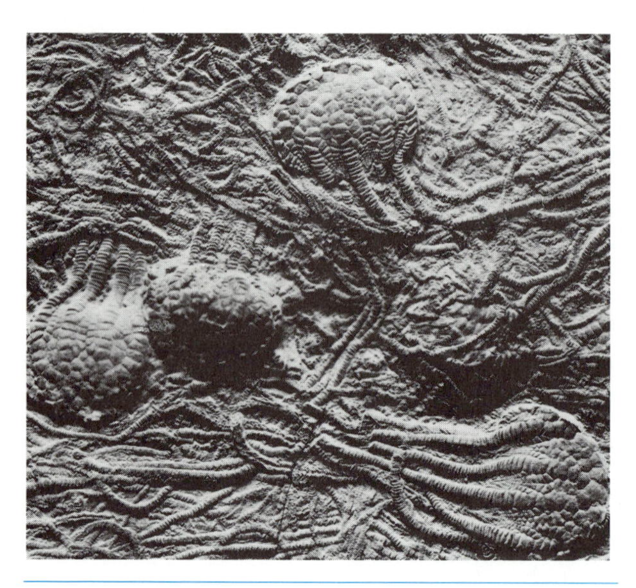

FIGURE 33.5 A FOSSIL MICROORGANISM. *This is evidence that primitive life forms existed on the earth billions of years ago.* (Photograph by Division of Photography, Field Museum of Natural History)

ligence had provided the ability to control the environment, the entire world became our ecological home, and our physical evolution essentially stopped. It remains to be seen whether future ecological pressures will create future evolution of the human species.

COULD LIFE DEVELOP ELSEWHERE?

The scenario just described, if at all accurate, seemingly should occur almost inevitably, given the proper conditions. If this is so, then the question of whether life exists elsewhere amounts to asking whether the conditions that existed on the primitive Earth could have arisen elsewhere. (For a dissenting view—that life was still improbable—see Astronomical Insight 33.1.)

It is clear that no other planet in the solar system could have provided an environment exactly like that on the early Earth. It therefore seems unlikely that Earth-like organisms will be found anywhere else in the solar system. (But recall the argument in Chapter 13 that the atmosphere of Jupiter may come close to reproducing the conditions of the early Earth.) Mars is the only planet where life forms have been sought so far (Fig. 33.6), but even there the conditions are quite unlike those on the Earth. Given the billions of stars in the galaxy, however, and the vast number that are very similar to the Sun, it seems highly probable that the proper conditions must have been reproduced many times in the history of the galaxy.

So far we have worked from the tacit assumption that life on other planets, if it exists, must be similar to life on the Earth, and we have considered only the question of whether other Earth-like environments may exist. We may question the premise that life could have developed only in the form that we are familiar with, however. Here, obviously, we must indulge in speculation, having no examples of other types of life at hand for examination.

Life as we know it is based on carbon-bearing molecules, and it has been argued that only carbon has the capability of combining chemically in a sufficiently wide variety of ways with other common elements to produce the complexity of molecules thought to be necessary. After all, it is clear that the basic elements available to begin with must be the same everywhere, given the homogeneous composition of the universe. This may seem to rule out life forms based on anything other than carbon. However, it has been pointed out that another common element, silicon, also has a very complex chemistry and therefore might provide a basis for a radically different type of life. If so, we cannot begin to speculate on the conditions necessary for such life forms to arise.

Another assumption that might be subject to question is that water is a necessary medium. As noted earlier, it is the most abundant liquid that can form under the temperature and pressure conditions of the early Earth, and it is thought that only a liquid medium could support the required level of chemical activity. Other liquids can exist under other conditions, however, and it is interesting to consider whether life forms of a wholly different type than we are familiar with might arise in oceans of strange composition. It is interesting to speculate, for example, about what goes on in the lakes of liquid methane or liquid nitrogen that are thought to exist on Titan, the mysterious giant satellite of Saturn.

If we are satisfied that life probably has formed naturally in many places in the universe, we can address a related, and to many, a more important, question: Given the existence of life, how likely is it that intelligence will follow? Here we have no means of

FIGURE 33.6 SEARCHING FOR LIFE IN THE SOLAR SYSTEM. *Mars was long thought to be the most likely home in the solar system for extraterrestrial life. Here we see a* Viking *lander in a simulated Martian environment. The* Viking *missions reached Mars in 1976 but found no evidence for life forms.* (NASA)

answering, except to reiterate that as far as we know, the evolution of our species on Earth was the natural product of environmental pressures.

THE PROBABILITY OF DETECTION

In view of the limitations that prohibit faster-than-light travel, it is exceedingly unlikely that we will ever be able to visit other solar systems, seeking out life forms that may live there. We will continue to explore our own system, so there is a reasonable chance that if life exists on any of the other planets of our Sun, we will someday discover it. It seems, however, that our best hope of finding other intelligent races in the galaxy will be to make long-range contact with them through radio or light signals. Since this requires both a transmitter and a receiver, we can only hope to contact other civilizations as advanced as ours is, with the capability of constructing the necessary devices for interstellar communication.

A mathematical exercise in probabilities has been used for some years as a means of assessing, as objectively as possible, the chances for making contact with an extraterrestrial civilization. The aim is to separate the question into several distinct steps, each of which could then be treated independently. The underlying assumption is that the number of technological civilizations in our galaxy today with the capability for interstellar communication is the product of the number of planets that exist with appropriate conditions, the probability that life developed on those planets, the probability that such life has developed intelligence that gave rise to a technological civilization, and finally, the likelihood that the civilization has not killed itself off, through evolution or catastrophe.

Mathematically, the so-called Drake equation (after Frank Drake, who has been its best-known advocate) is written

$$N = R_* f_p n_e f_l f_i f_c L,$$

where N is the number of technological civilizations presently in existence, R_* is the number of stars of appropriate spectral type formed per year in the galaxy, f_p is the fraction of these that have planets, n_e is the number of Earth-like planets per star, f_l is the fraction of these on which life arises, f_i is the fraction of those planets on which intelligence has developed, f_c is the fraction of planets with intelligence on which a technological civilization has evolved to the point at which interstellar communications would be pos-

ASTRONOMICAL INSIGHT 33.1

The Case for a Small Value of $\mathcal{N}$

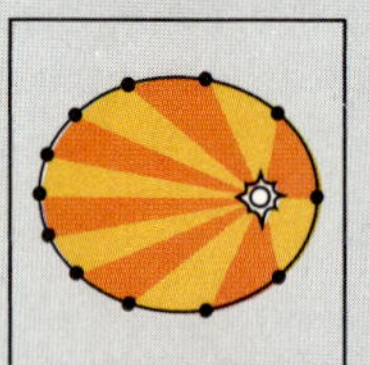

Although it is an interesting exercise to write down an equation for $\mathcal{N}$, the number of technological civilizations in the galaxy, no two astronomers can agree on the values of the various terms.

There is one very pessimistic argument, developed recently, that concludes that we are alone in the galaxy. This is based on the likely fact that if there were other civilizations, a significant fraction of them would have arisen long before ours. This is a straightforward consequence of the great age of the galaxy, some 5 to 10 billion years greater than the age of the Earth. This argument postulates further that if technological civilizations, once begun, live a long time, then the galaxy must be inhabited by a number of very advanced races, much older and more mature than our own. Up to this point, there is little controversy about any of these assertions.

The next logical step in the argument, however, is debatable; it says that any advanced civilization that has survived for millions or even billions of years will necessarily have done so by perfecting some form of interstellar travel and colonizing other planets. Once this had started, the argument goes, the spread of a given civilization throughout the galaxy would accelerate rapidly, and by this time in galactic history, no habitable planet such as the Earth would remain uncolonized. The first successful galactic civilization, in this view, would quickly rise to complete dominance.

The only logical corollary, if the argument is accepted up to this point, is that no other civilizations exist, because if they did, the Earth would have been colonized long ago. The absence of interstellar visitors on the Earth is taken as proof that there are no other civilizations in the galaxy.

Those who adhere to this line of reasoning believe that $\mathcal{N} = 1$. Therefore, at least one of the terms in Drake's equation must be very much smaller than the more optimistic values discussed in the text. One suggestion is that the temperate zone around a Sun-like star where planets can have moderate conditions is really very much narrower than usually supposed, perhaps because of the more ready development of a Venus-like greenhouse effect than is normally thought to be the case. If the temperate zone is very small, then the term n_e, representing the number of habitable, Earth-like planets per star, would be very much smaller than the value of 0.1 adopted in the text.

Another term that has recently been singled out is f_l, the fraction of Earth-like planets on which life begins. As we learned in the text, somehow the amino acids that were present on the primitive Earth had to arrange themselves into special combinations to produce the long, chainlike protein molecules RNA and DNA. From a purely statistical point of view, the probability that the proper combination would happen to come together by chance is very small. This apparently happened on the Earth, but it may be so unlikely that it has not occurred anywhere else in the galaxy.

Whatever the correct values of the terms in the equation, the debate will rage on until our civilization reaches its life expectancy L and dies, until we discover another civilization, or until we ourselves expand to colonize the galaxy and find no one else out there.

---

sible, and, finally, L is the average lifetime of such a civilization.

By expressing the number of civilizations in this way, we can isolate the factors about which we can make educated guesses from those about which we are more ignorant. It is an interesting exercise to go through the terms in the equation one by one, to see what conclusions we reach under various assumptions. People who do this have to make sheer guesses for some of the terms, and the result is a variety of answers ranging from very optimistic to the very pessimistic. In the following, we will adopt middle-of-the-road numbers for most of the unknown terms.

The first two factors, R_* and f_p, are, in principle, quantities that can be known with some certainty from observations. The rate of star formation in the equation refers to stars similar to the Sun. A much cooler star would not have a temperate zone around it where a planet could have the moderate temperatures needed for life to begin (we are, throughout this exercise, limiting ourselves to life forms similar to our own), and a very hot star would be short-lived, so that there would be insufficient time for life to develop on its planets, before the parent star blew up in a supernova explosion. Taking these considerations into account, it is estimated by some that up to ten suitable stars form in our galaxy per year; for the sake of discussion, we will be more cautious and adopt $R_* = 1/\text{year}$. This may even be a little high for the present epoch in galactic history, but the star-formation rate was surely much higher early in the lifetime of the galaxy, and low-mass stars of a solar type formed that long ago are still in their prime. Thus, the adoption of an average formation rate of one Sun-like star per year is probably reasonable.

From what we know of the formation of our solar system, it seems that the formation of planets is almost inevitable, except perhaps in double- or multiple-star systems. For the sake of discussion, let us assume that $f_p = 1$, that is, that all stars of solar type have planets. Observations made with future instruments such as the *Space Telescope* may soon provide real information on this term.

The number of Earth-like planets, n_e, is highly uncertain and depends on how wide the zone is where the appropriate temperature conditions could exist. Recent studies show that this zone might be rather small, that is, that the Earth would not have been able to support life if its distance from the Sun were only about 5 percent closer or farther than it is. Estimates of n_e vary from 10^{-6} to 1. Let us be moderate and assume $n_e = 0.1$; that is, that in one out of ten planetary systems around solar type stars, there is a planet within the temperate zone where life can arise.

Now we get to the *really* speculative terms in the equation. We have no way of estimating how likely it is that life should begin, given the right conditions. From the seeming naturalness of its development on Earth, it can be argued that life would always begin if given the chance. Let us be optimistic here and agree with this, adopting $f_l = 1$.

Again, the chance of this life developing intelligence and furthermore advancing to the point of being capable of interstellar communication is completely unknown. All we know is that in the only example that has been observed, both things happened. For the sake of argument, we therefore set both f_i and f_c equal to 1.

At this point, it is instructive to put the values adopted so far into the equation. We find

$$\begin{aligned} N &= (1/\text{year})(1)(0.1)(1)(1)(1)L \\ &= 0.1L \end{aligned}$$

Having taken our chances and guessed at the values for all the other terms, we now face a critical question: How long can a technological civilization last? Ours has been sufficiently advanced to send or receive interstellar radio signals for only about 50 years, and there is sufficient instability in our society to lead some pessimists to think that we will not last many more decades. If we take this viewpoint and adopt 50 years as the average lifetime, then we find:

$$N = 5,$$

meaning that we should expect the total number of technological civilizations present in the galaxy at one moment to be very small, about 5. If this is correct, then the average distance between these outposts of civilization is nearly 20,000 light-years (Fig. 33.7). The time it would take for communications to travel between civilizations would therefore be very much longer than their lifetimes, and there would be no hope of establishing a dialogue with anyone out there. If this estimate is correct, it is no surprise that we haven't heard from anybody yet.

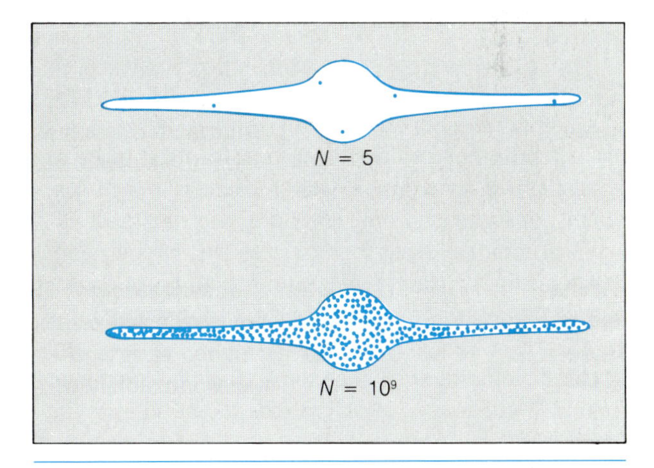

FIGURE 33.7 POSSIBLE VALUES OF N. *The number of technological civilizations in the galaxy (N) may be very small (upper), in which case the average distance between them is very large; or N may be large, so that the distance between civilizations is relatively small. The chances for communication with alien races are much higher in the latter case, where the distances are only a few tens of light-years.*

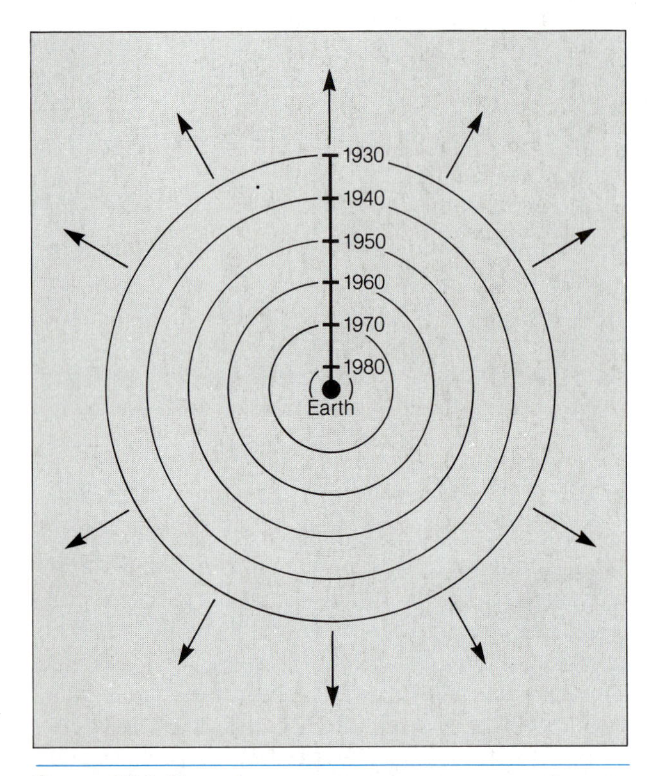

FIGURE 33.8 EARTH'S MESSAGE TO THE COSMOS. *As our entertainment and communications broadcasts travel out into space, they provide a history of our culture for anyone who may be receiving the signals. At the present time, the growing sphere that is filled with our broadcasts has a radius of over fifty light-years.*

We can be more optimistic, though, and assume that a technological civilization solves its internal problems and lives much longer than 50 years. Extremely optimistic people would argue that a civilization is immortal; that it colonizes star systems other than its own, so that it is immune to any local crises such as planetary wars or suns expanding to become red giants. In that case, allowing a few billion years for the development of such civilizations, we can set $L = 10^{10}$ years (i.e., nearly equal to the age of the galaxy), and we find

$$\mathcal{N} = 10^9,$$

where the average distance between civilizations is only about 30 light-years (see Fig. 33.7), coincidentally just a little less than the distance our own radio signals have traveled (Fig. 33.8) since the early days of radio and television. If this estimate of $\mathcal{N}$ is correct, we should be hearing from somebody very soon.

We have presented two extreme views of the likelihood that other civilizations exist in our part of the galaxy. As we mentioned earlier, opinions among the scientists who seriously study this question vary throughout this large range. Those who favor the optimistic viewpoint advocate the idea that we should make deliberate attempts to seek out other civilizations.

— THE STRATEGY FOR SEARCHING —

The probability arguments outlined in the previous section are amusing and perhaps somewhat instructive, but obviously not very accurate. There are entirely too many unknowns in the equation for us to develop a reliable estimate of the chances for galactic companionship. Perhaps we will not know for certain what the answer is until we make contact with another civilization.

The problem of developing an experiment to search for or to send interstellar signals is that we do not know the ground rules. There is an infinite number of ways in which a distant civilization might choose to try to communicate, and it is impossible to search for them all. We must try to guess what the most probable technique would be.

In view of the power that is transmitted by radio signals and the relative lack of natural "noise" in the galaxy in that part of the spectrum, it has often been assumed that radio communications are most likely to succeed, although other techniques have been tried (Fig. 33.9). In the early 1960s, the U.S. National Radio Astronomy Observatory (Fig. 33.10) "listened" for transmissions from two nearby Sun-like stars, tau Ceti and epsilon Eridani, without success. More recently, larger surveys of neighboring stars have been done in the United States and in the Soviet Union, with similar results. About forty different searches for extraterrestrial civilizations have been attempted so far.

One of the problems encountered by anyone wanting to search for extraterrestrial radio signals is obtaining observing time on a radio telescope. Most major radio telescopes are already in high demand by astronomers carrying out more standard kinds of research, and it is very difficult to have much time assigned for such a low-probability program as a search for extraterrestrial life. At most, one may obtain a few days a year on a given telescope. Since 1983, however, one radio telescope has been devoted entirely to the search for life. A radio telescope belonging to Harvard University has been dedicated to a program in which a sophisticated computer is used to automatically scan a wide range of wavelengths in as many stars as possible. This search is being funded primarily by private donations (the major donor was movie producer Steven Spielberg).

With the exception of the Harvard program, most searches so far have been confined to a single radio wavelength. The wavelength chosen in most cases was 21 cm, partly for practical reasons (radio telescopes designed for observing at this wavelength already existed), and partly because it was hypothesized that if others out there deliberately wanted to send signals, they might choose to do so at a wavelength that other astronomers around the galaxy would be observing anyway, even if they weren't trying to find distant civilizations. Another wavelength that has been considered, but tried only in a very limited search, lies in the microwave region, between the emission lines of interstellar water vapor and OH. Dubbed the "water hole," this region has the advantages that it, too, might be a target of extraterrestrial astronomers, and it is in the quietest part of the radio spectrum (various objects, ranging from supernova remnants to radio galaxies, create radio noise that fills the galaxy, and the "water hole" lies in a region of the spectrum where this noise is minimized; see Fig. 33.11). As in the Harvard search, new programs being planned today will cover wide ranges of wavelengths, avoiding the need to try to outguess the extraterrestrials in trying to decide which specific wavelength to observe.

One pessimistic viewpoint is that we are all listening, and no one is sending. In that case, we can only hope to pick up accidental emissions, such as entertainment broadcasts on radio or television, and these would be much weaker and more difficult to detect. Deliberately sent signals can be detected over far greater distances than accidental transmissions, because more power would be put into a deliberate signal, and it could be directed specifically at candidate stars, whereas our accidental radio and television

FIGURE 33.9 ANOTHER MESSAGE FROM EARTH. *This recording of a message from earth is traveling beyond the solar system aboard the* Voyager *spacecraft. Only an advanced race of beings would have the technological skills necessary to learn how to listen to and decode the message of peace that it contains.* (NASA)

FIGURE 33.10 A RADIO TELESCOPE USED IN THE SEARCH FOR LIFE. *This large dish, at the U. S. National Radio Astronomy Observatory, in Green Bank, West Virginia, has been used in efforts to detect signals from alien civilizations. Most of these attempts have been made at the 21-centimeter wavelength of atomic hydrogen emission.* (National Radio Astronomy Observatory)

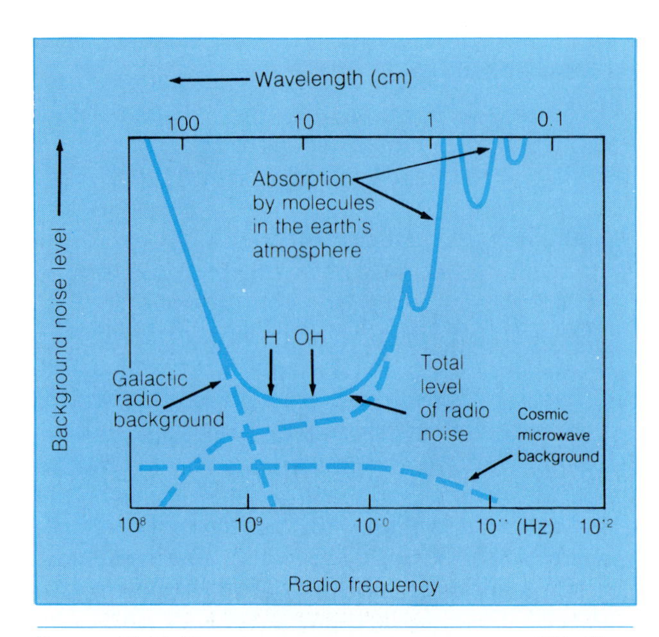

FIGURE 33.11 THE GALACTIC NOISE SPECTRUM. *This diagram shows the relative intensity of the background noise in the galaxy as a function of wavelength. The quietest portion of the spectrum is in the vicinity of the "water hole" between the natural emission wavelengths of H_2O and OH molecules.* (Data from D. Goldsmith and T. Owen 1980, *The Search for Life in the Universe* [Menlo Park: Benjamin Cummings])

signals are broadcast indiscrimately in all directions. Therefore, it makes a big difference whether someone out there is trying to send a message or not.

In 1974, a message from Earth was sent from the great Arecibo radio dish (see Fig. 7.24) toward the globular cluster M13, about 25,000 light-years away. The reason for choosing a globular cluster was that there are hundreds of thousands of stars in it, many of which are similar to the Sun in spectral type, and all of them would be within the beam of the transmission. The message consists of a stream of numbers which, when arranged in a two-dimensional array, form a pattern that illustrates schematically such things as the structure of DNA, the form of the human body, Earth's population, and the location of the Earth in the solar system. If this message is received and understood by anyone in M13, it will be some 25,000 years from now, and any answer they may send back will arrive here about 50,000 years from now.

It is not clear what the future will bring, in terms of deliberate searches for extraterrestrial intelligence. The most ambitious project yet studied seriously is called *Project Cyclops* (Fig. 33.12), but its cost will probably prevent it from ever being built. The plan is to build a veritable forest of large radio antennae, along with a very sophisticated data processing sys-

FIGURE 33.12 PROJECT CYCLOPS. *The grandest plan seriously considered for the purpose of communicating with extraterrestrial civilizations,* Project Cyclops *would consist of 1000 to 2500 radio antennae, each 100 feet in diameter. This huge array, some 16 kilometers in extent, would be capable of detecting signals over a large portion of the radio spectrum, from planetary systems 1000 or more parsecs away.* (NASA)

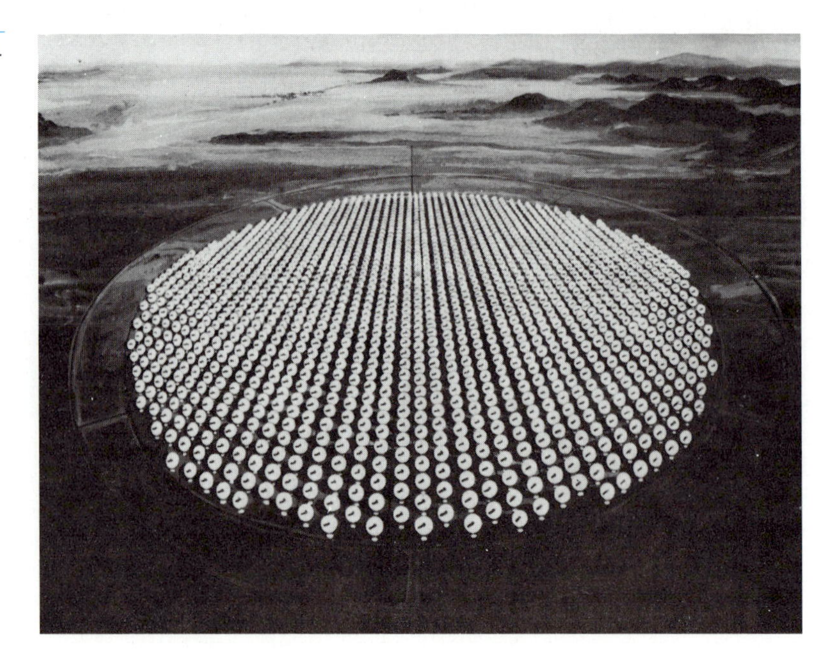

ASTRONOMICAL INSIGHT 33.2

THE COMPLEXITY OF UFO'S

In this chapter we have discussed the strategies for detecting extraterrestrial civilizations using radio communications, with little or no mention of the possibility of direct physical contact. Yet there is a forty-year history of "sightings" of objects thought to be alien spacecraft visiting Earth that suggests that the first contact might be in the form of face-to-face meetings. Since 1947, there have been countless reports of strange objects or lights in the sky, and some have attributed these "Unidentified Flying Objects" or "UFO's" to alien visitors.

To help decide what to think about UFO's, it is very useful to apply Occam's Razor, the principle that the simplest explanation for an observation is most likely the correct one. By simplest we mean the explanation that requires the fewest unsupported assumptions. Let's see how the alien spacecraft interpretation of UFO's fares under this criterion.

The laws of physics dictate that a spacecraft cannot travel faster than the speed of light. Therefore, allowing for acceleration time, it must take decades or centuries for a ship to travel the distance from even the nearest star to the Earth (recall that the nearest star is over 4 light-years away; calculations based on known propulsion systems indicates that it would take roughly 40 years to get there from here, and changing propulsion system to some as-yet uninvented type would still require some acceleration time). Furthermore, the energy required for such a journey is immense; it has been estimated that the energy needed to boost a colonizing spaceship from Earth to a nearby star is enough to power the entire United States at its present rate of consumption for one hundred years! Is it likely that Congress would ever approve such an expenditure? This enormous energy requirement would be faced by any civilization planning interstellar travel, and cannot be lessened by advanced technology (unless there are ways of violating the known laws of physics, but there is no evidence that this is possible, and there is plentiful evidence that it is not).

Given the time and energy constraints imposed on any civilization thinking about interstellar travel, it is very difficult indeed to accept the notion that UFO's are sightseeing or spying spacecraft from another star system; that an alien civilization would expend the required time and energy routinely and often. It is far simpler to accept the alternative theory of UFO's, that they are phenomena that occur in the Earth's atmosphere, and are not spacecraft. This explanation is made all the more likely by the fact that virtually every UFO sighting that has been carefully studied has been found to be due to something normal such as an airplane, a balloon of some sort, or a bright planet seen near the horizon.

Many people are tempted to believe that UFO's represent visiting aliens, but objective scientific analysis makes this belief very difficult to accept. Occam's Razor tells us that a much more mundane explanation of UFO's is likely to be correct, and that our best hope of contacting alien civilizations lies in long-range radio communications, which travel at the speed of light and require very little energy. This conclusion is supported by more discussion in the Guest Editorial following this chapter. Of course, the fact that no one really *knows* what alien technology and psychology might be like makes the speculation all the more interesting.

tem, so that up to 1 million stars could be scanned for artificial emissions over a broad wavelength region, in a few years' time. By searching over a wide range of wavelengths, the scientists planning this effort hope to eliminate much of the guesswork about which is the most likely portion of the spectrum to be used by alien civilizations. This also enhances the chances of detecting accidental emissions, which, if they are anything like our own radio and television signals, may be spread over much of the spectrum. The *Project Cyclops* antennae could also be used to send powerful radio signals, allowing the Earth to broadcast its own beacon for others to find, and providing for the possibility of two-way communications.

The principal problem in implementing *Cyclops*, or any other large-scale effort, is the enormous cost. The version of *Project Cyclops* outlined above is currently estimated to require some 10 billion dollars in expenditures. Such sums are rarely devoted to astronomy, and if they were, there would still be considerable debate over whether to spend it on this or on some other research programs that might appear to have greater certainty of success. Presumably, alien astronomers in other solar systems would have to face the same questions, although we can hope that sufficiently advanced civilizations will have overcome the limitations of their planetary resources, and will be willing and able to tackle such a massive task.

A less ambitious project, called the *Search for Extraterrestrial Life (SETI)*, has been funded by NASA. The plan is to use existing radio dishes and specially developed, sophisticated computers to carry out a limited search of several stars. Because of advances in computer technology since the time when *Project Cyclops* was first planned, the far less expensive *SETI* program will be capable of a search nearly as thorough. *SETI* will scan some 8 million wavelengths for possible signals and is capable of probing about 1 million nearby Sun-like stars over several years' time. This is a large number of stars, but still only a very small fraction of the total in our galaxy. For *SETI* to succeed in detecting another civilization, the value of $\mathcal{N}$ will have to be quite high. While the scope of *SETI* is not as grand as that envisioned for *Cyclops*, it is nevertheless noteworthy that a federal government agency has taken the task seriously enough to fund a search for extraterrestrial life. The *SETI* computer system is now undergoing its first tests and will soon be in full operation.

PERSPECTIVE

In this chapter we have introduced a bit of speculation, well founded perhaps in the discussion of life on Earth, but less so in the sections on the possibilities of life elsewhere.

We are prepared now to understand and appreciate new advances in astronomy; having completed our study of the present knowledge of the universe, we will be able to put into perspective the major new discoveries that are sure to come as technology and theory continue to improve. The *Space Telescope*, the large optical telescopes now being designed, the comparable advances in radio, infrared, and X-ray techniques—all of these will inevitably lead to novel and unforeseen breakthroughs in our view of the universe. The next few decades will be exciting times for astronomy, and we can anticipate their coming with a sense of excitement and wonder.

SUMMARY

1. While there were early beliefs that life started on Earth spontaneously or by primitive spores from space, most scientists today accept the theory that life began through natural, evolutionary processes.

2. Amino acids, fundamental components of living organisms, are formed readily in experiments designed to simulate early conditions on the Earth.

3. The steps that led to the development of the necessary forms of RNA and DNA are not yet fully understood, and probably occurred over a very long period of time.

4. Fossil evidence provides a record of the evolutionary steps leading from the first primitive life forms to modern life.

5. The conditions that prevailed on the early Earth have probably been duplicated on other planets in the galaxy, though probably not on other planets in the solar system.

6. It is often assumed that only Earth-like life could develop, because carbon is nearly unique in the complexity of its chemistry, but there have been suggestions that at least one other element (silicon) may have the necessary properties.

7. Estimates of the number $\mathcal{N}$ of technological civilizations now in the galaxy can be made, with great uncertainty, based on what is known of the formation rate of Sun-like stars, and what is guessed for the probability of such stars' having planets with the proper conditions, and the probability that life, leading to intelligence and technology, will develop on these planets. Estimates range from $N = 1$ to $N = 10^9$.

8. Some attempts have been made to search for radio signals at 21 cm and other wavelengths, but with no success.

9. The chances for detecting or being detected by other civilizations depend strongly on whether or not deliberate attempts are made to send signals.

10. Future projects to search for and to send signals have been planned, and one, the *Search for Extraterrestrial Intelligence (SETI)*, has been funded by NASA and is now undergoing preliminary testing. *SETI* will search about 1 million stars in 8 million radio wavelengths and should result in a detection if the number of technological civilizations in the galaxy is large.

REVIEW QUESTIONS

Thought Questions

1. Why is the natural evolution of life on Earth considered by scientists to be the most likely explanation?

2. To what extent has life on Earth modified its atmosphere? Does this suggest a method for determining, from remote spectroscopic observations, whether a distant planet might have life?

3. Why would a planet only slightly closer to the Sun than the Earth probably not be able to support life? (*Hint:* Recall what you learned in Chapter 10 about the cause of the contrasting conditions on Venus and the Earth.)

4. Summarize the arguments that lead some scientists to expect life elsewhere to be similar in chemistry to life on Earth.

5. Based on what is assumed about how life started on the Earth, do you think it likely that life has developed in the atmosphere of Jupiter? Do you think it could develop in an interstellar cloud?

6. Recall, from Chapter 27, the distinction between Population I and Population II stars, and where these stars are located in the galaxy. Do you think planets orbiting Population II stars are as likely to have life forms as planets orbiting Population I stars like the Sun? Explain.

7. Explain in your own words the meaning of the various terms in the Drake equation, and how this equation represents the number of technological civilizations in the galaxy at any one time.

8. Why are radio communications thought to be the best method for detecting evidence for other civilizations?

9. What are the factors that govern the choice of wavelength at which to search for signals?

10. Explain why signals from a distant civilization would be easier to detect if they were sent deliberately rather than being accidental transmissions (such as our own entertainment broadcasts).

Problems

1. Under the panspermia hypothesis, life propagates through the galaxy in the form of spores that travel randomly about, occasionally landing on planets. If such a spore travels at the speed of a typical interstellar cloud (say 10 km/sec), how long would it take to reach the Earth from the nearest star, 1.3 pc away? How long would it take to travel the distance from the galactic center to the solar system (10,000 pc)? How does your answer compare with the age of the galaxy?

2. Suppose a planet was orbiting a star exactly like the Sun, and the planet's semimajor axis was 0.9 AU. How would the intensity of sunlight on that planet compare with the intensity of sunlight on the Earth? What effect might this have on the possibility that life could develop on this planet?

3. Repeat the calculation of N, the number of technological civilizations in the galaxy, if the probability of life starting on an Earth-like planet (the term f_l in the Drake equation) is only 10^{-4} instead of 1, as assumed in the text. Assume that the lifetime L is 10^{10} years, then redo the calculation for $L = 10^5$ years.

4. If *I Love Lucy* broadcasts started in 1953, in what year did they first reach alpha Centauri, 1.3 pc away? When will they reach Canopus, 30 pc away? When might we expect an answer from Canopus, if anyone is there to receive our signals?

5. Using Appendix 8, list the stars among the fifty brightest that might have planets on which life could develop.

ADDITIONAL READINGS

Abt, H. A. 1979. The companions of sun-like stars. *Scientific American* 236(4):96.

Ball, John A. 1980. Extraterrestrial intelligence: where is everybody? *American Scientist* 68:656.

Barber, V. 1974. Theories of the chemical origin of life on earth. Mercury 3(5):20.

Goldsmith, D., ed. 1980. *The quest for extraterrestrial life*. Mill Valley, Cal.: University Science Books.

Guin, Joel. 1980. In the Beginning. *Science 80* 1(5):45. (the development of life on Earth)

Horowitz, Norman. 1986. *To utopia and back: The search for life in the solar system*. New York: W.H. Freeman.

Olson, Edward C. 1985. Intelligent life in space. *Astronomy* 13(7):6.

O'Neill, G. K. 1974. The colonization of space. *Physics Today* 27(9):32.

Overbye, Dennis. 1982. Is anyone out there? *Discover* 3(3):20.

Papagiannies, M. D. 1982. The search for extraterrestrial civilizations—a new approach. *Griffith Observer*. 11(1):112.

———. 1984. Bioastronomy: The search for extraterrestrial life. *Sky and Telescope* 67(6):508.

Pelligrino, C. R. 1979. Organic clues in carbonaceous meteorites. *Sky and Telescope* 57(4):330.

Pollard, W. G. 1979. The prevalence of earth-like planets. *American Scientist* 67:653.

Rood, R. T., and J. S. Trefil. 1981. *Are we alone?* New York: Scribner.

Sagan, Carl. 1982. The search for who we are. *Discover* 3(3):30.

Sagan, C., and F. Drake. 1975. The search for extraterrestrial intelligence. *Scientific American* 232(5):80.

Sagan, C., and I. S. Shklovskii. *Intelligent life in the universe*. San Francisco: Holden-Day.

Sheaffer, Robert. 1986. *The UFO Verdict*. Buffalo, N.Y.: Prometheus Books.

Special Report on Life in Space. 1983. *Discover* 4(3).

Tipler, F. J. 1982. The most advanced civilization in the galaxy is ours. *Mercury* 11(1):5

Wetherill, C., and W. T. Sullivan. 1979. Eavesdropping on the earth. *Mercury* 8(2):23.

GUEST EDITORIAL

TRYING TO PSYCHE OUT EXTRATERRESTRIAL CIVILIZATIONS

Frank D. Drake

Dr. Drake, a radio astronomer with research interests in solar system and galactic astronomy, is Dean of Natural Sciences at the University of California–Santa Cruz. He moved there following some 20 years as a member of the faculty at Cornell University, which followed his tenure in senior posts at the United States National Radio Astronomy Observatory and at the Jet Propulsion Laboratory of the California Institute of Technology. Despite his impressive record in solar system research and science administration, Dr. Drake is probably best known for his work on the question of extraterrestrial civilizations. He pioneered the modern methods of treating the subject, instigated many of the searches that have been attempted, and is a central figure in all organized studies of the question, now serving as president of the SETI Institute, a NASA-funded center for the search for extraterrestrial intelligence. His equation describing the number of civilizations in the galaxy as a function of various probabilities that can be independently assessed has set the stage for most serious discussions on the subject.

The chances of success and the cost and effort of a search for radio signals from other worlds depend on N, the number of detectable civilizations. You have read here of the thinking which leads to estimates of N, and as you have seen, more than one scenario for the development of civilizations has been invoked to predict N. On the one hand there are rather reserved people who feel that we should only assume that what has happened on our Earth and in our civilization is typical, and that vast projects in cosmic engineering are not likely. Such people conclude that there are indeed many civilizations like ours, populating planets like ours and perhaps a few space colonies near their planets. These civilizations can be found most easily and perhaps only by searching diligently with radio telescopes.

There is another school of thought which says that civilizations like ours will accede to a compulsion to sustain ever larger populations of their creatures. If so, they must eventually establish living room beyond their own planet, first in suitable places in their planetary system, and later, in a heroic enterprise, on planets of other stars. Even at a slow rate of colonization of one star after another, the rate at which populations increase, if they are anything like us, will lead to the colonization of every star in the galaxy in a time of a hundred million years or so, or about 1 percent of the age of the galaxy. Clearly, if even one such civilization had arisen in the past, their colonists would be here by now and we would know of them. This might be called the "cancer" scenario: all it takes is one "malignant" civilization, and it quickly spreads to every star in the galaxy. And yet they have not come to us. If one accepts this picture as realistic, it implies that there are, for some reason, very few civilizations and that we may be alone or at least the most advanced.

Which of these basic scenarios is right? It is important to know, for one scenario says that a treasure of discoveries awaits us if we mount a respectable search for extraterrestrial intelligence. The other says we would be wasting our money and talents. There is nothing to find: just wait awhile and *we* will be the extraterrestrials. Thinking about it, you will see that the choice between these scenarios depends not, as we would very much wish, on the laws of physics or the arrangement of the universe. Rather, it depends on the actual priorities, needs, values, and even social systems and governments of advanced civilizations.

The price to be paid by a civilization to adopt various modes of

(continued on next page)

GUEST EDITORIAL

life is enormous, and choices will not be made casually. Let's look at the interstellar colonization scenarios. Despite what we have been led to believe by science fiction, movies, and television, the colonization of the planets, and especially the stars, is an unbelievably costly undertaking, no matter how advanced you are. It is easy to compute that the cost in energy alone required to send a single colonist to a nearby star system in a time of 100 years, which already seems long, is enough energy to provide a good life to 200,000 people, if they are like us. Take into account the inefficiencies in rockets and fuel production, and the price goes up to the cost of supporting 2,000,000 people luxuriously throughout their lives. Send 100 colonists to start a colony, and the cost is equal to the cost of supporting roughly the whole population of the United States for all their lives. In other words, to launch a minimal colony at reasonable speed would require, in effect, the shutting down of America for 100 years. Would we do that?

Now you may say, "But they may have access to cheap energy sources." True, but energy will never be free; it must not only be generated, it must be stored and transmitted. There will always be an economic question as to how to get the most benefit from a given amount of energy. For a small fraction of the cost of sending a small colony to the stars, you could create glorious living space by enclosing the arctic in heated domes, or building floating cities on the oceans, or building a multitude of large colonies in space near the home planet.

What will they do? The decisions are not at all obvious. My own feeling is that they will choose to remain near home. They may embark on marvelous enterprises. They may build great cities in the sky. But they will be within sight of home, and not at distant alpha Centauri or their equivalent. A truly intelligent civilization will use its resources to the greatest benefit for the most creatures; and that is done by acknowledging the monstrous penalty in costs which would be paid if one tried to flaunt the laws of physics and relativity. Only do what can be done "slowly," that is, "slowly" compared to the speed of light!

Of course, others may think otherwise. Perhaps it is worth the cost to experience the adventure of starting anew on another star. Perhaps you decide to support only a small elite population, allotting enormous resources to each individual. The personal ration of materials and energy may be enough to provide for interstellar travel.

Who can say? And that is exactly my point. The laws of physics and the great distances which separate the stars have made it difficult for civilizations to choose what is good for them. It will depend on their goals and philosophies. It will be hard for *them* to decide, and surely impossible for *us* to decide what they may do. Until we learn the histories of many civilizations, we will be in no position to psyche out the extraterrestrials.

Thus the search for extraterrestrial intelligence for now must be an experimental one. Mind alone will not work to give us answers. Only the lengthy scanning of huge telescope mirrors and giant radio antennae can. We should not be disappointed; science has always worked best when we experimented with open minds. Now we should get on with the job.

APPENDIX 1 Symbols commonly used in this text

Symbol	Meaning	Symbol	Meaning
Å	Angstrom, a unit of length often used to measure wavelength of light; 1 Å = 10^{-8} cm	$\Delta\lambda$	The Greek letters delta lambda, used to designate a shift in wavelength, as in the Doppler effect
c	Standard symbol for the speed of light	γ	The Greek letter gamma, sometimes used to designate a gamma-ray photon
G	Standard symbol for the gravitational constant	λ	The Greek letter lambda, usually used to designate wavelength
H	Symbol for the constant in the Hubble expansion law	ν	The Greek letter nu, the standard symbol for frequency, also used to designate a neutrino in nuclear reactions
h	Standard symbol for the Planck constant	π	The Greek letter pi, usually used to designate the parallax angle; also used for the ratio of the circumference of a circle to its diameter
K	Kelvin, the unit of temperature in the absolute scale		
z	Symbol commonly used to designate the Doppler shift ($z = \Delta\lambda/\lambda = vc$ for velocities much less than the speed of light)		
α	The Greek letter alpha, sometimes used to designate an alpha particle		

APPENDIX 2 Physical and mathematical constants

Constant	Symbol	Value
Speed of light	c	2.9979250×10^{10} cm/sec
Gravitation constant	G	6.670×10^{-8} dyn cm²/gm
Planck constant	h	6.62620×10^{-27} erg/sec
Electron mass	m_e	9.10953×10^{-28} gm
Proton mass	m_p	1.672648×10^{-24} gm
Stefan-Boltzmann constant	σ	5.67032×10^{-5} erg/cm²deg⁴sec
Wien constant	W	0.289789 cm/deg
Boltzmann constant	k	1.38066×10^{-16} erg/deg
Astronomical unit	AU	1.49599×10^{8} km
Parsec	pc	3.085678×10^{13} km = 3.261633 light-years
Light-year	ly	9.460530×10^{12} km
Solar mass	$M_\odot$	1.9891×10^{33} gm
Solar radius	$R_\odot$	6.9600×10^{10} cm
Solar luminosity	$L_\odot$	3.827×10^{33} erg/sec
Earth mass	$M_\oplus$	5.9742×10^{27} gm
Earth radius	$R_\oplus$	6378.140 km
Tropical year (equinox to equinox)		365.241219878 days
Sidereal year (with respect to stars)		365.256366 days = 3.155815×10^{7} sec

Appendix 3 The elements and their abundances

Element	Symbol	Atomic No.	Atomic Weight[1]	Abundance[2]
Hydrogen	H	1	1.0080	1.00
Helium	He	2	4.0026	0.085
Lithium	Li	3	6.941	1.55×10^{-9}
Beryllium	Be	4	9.0122	1.41×10^{-11}
Boron	B	5	10.811	2.00×10^{-10}
Carbon	C	6	12.0111	0.000372
Nitrogen	N	7	14.0067	0.000115
Oxygen	O	8	15.9994	0.000676
Fluorine	F	9	18.9984	3.63×10^{-8}
Neon	Ne	10	20.179	3.72×10^{-5}
Sodium	Na	11	22.9898	1.74×10^{-6}
Magnesium	Mg	12	24.305	3.47×10^{-5}
Aluminum	Al	13	26.9815	2.51×10^{-6}
Silicon	Si	14	28.086	3.55×10^{-5}
Phosphorus	P	15	30.9738	3.16×10^{-7}
Sulfur	S	16	32.06	1.62×10^{-5}
Chlorine	Cl	17	35.453	2×10^{-7}
Argon	Ar	18	39.948	4.47×10^{-6}
Potassium	K	19	39.102	1.12×10^{-7}
Calcium	Ca	20	40.08	2.14×10^{-6}
Scandium	Sc	21	44.956	1.17×10^{-9}
Titanium	Ti	22	47.90	5.50×10^{-8}
Vanadium	V	23	50.9414	1.26×10^{-8}
Chromium	Cr	24	51.996	5.01×10^{-7}
Manganese	Mn	25	54.9380	2.63×10^{-7}
Iron	Fe	26	55.847	2.51×10^{-5}
Cobalt	Co	27	58.9332	3.16×10^{-8}
Nickel	Ni	28	58.71	1.91×10^{-6}
Copper	Cu	29	63.546	2.82×10^{-8}
Zinc	Zn	30	65.37	2.63×10^{-8}
Gallium	Ga	31	69.72	6.92×10^{-10}
Germanium	Ge	32	72.59	2.09×10^{-9}
Arsenic	As	33	74.9216	2×10^{-10}
Selenium	Se	34	78.96	3.16×10^{-9}
Bromine	Br	35	79.904	6.03×10^{-10}
Krypton	Kr	36	83.80	1.6×10^{-9}
Rubidium	Rb	37	85.4678	4.27×10^{-10}
Strontium	Sr	38	87.62	6.61×10^{-10}
Yttrium	Y	39	88.9059	4.17×10^{-11}
Zirconium	Zr	40	91.22	2.63×10^{-10}
Niobium	Nb	41	92.906	2.0×10^{-10}
Molybdenum	Mo	42	95.94	7.94×10^{-11}
Technetium	Tc	43	98.906	—
Ruthenium	Ru	44	101.07	3.72×10^{-11}
Rhodium	Rh	45	102.905	3.55×10^{-11}
Palladium	Pd	46	106.4	3.72×10^{-11}
Silver	Ag	47	107.868	4.68×10^{-12}
Cadmium	Cd	48	112.40	9.33×10^{-12}
Indium	In	49	114.82	5.13×10^{-11}
Tin	Sn	50	118.69	5.13×10^{-11}
Antimony	Sb	51	121.75	5.62×10^{-12}
Tellurium	Te	52	127.60	1×10^{-10}
Iodine	I	53	126.9045	4.07×10^{-11}
Xenon	Xe	54	131.30	1×10^{-10}
Cesium	Cs	55	132.905	1.26×10^{-11}
Barium	Ba	56	137.34	6.31×10^{-11}
Lanthanum	La	57	138.906	6.46×10^{-11}
Cerium	Ce	58	140.12	4.37×10^{-11}
Praseodymium	Pr	59	140.908	4.27×10^{-11}
Neodymium	Nd	60	144.24	6.61×10^{11}
Promethium	Pm	61	146	–
Samarium	Sm	62	150.4	4.57×10^{-11}
Europium	Eu	63	151.96	3.09×10^{-12}
Gadolinium	Gd	64	157.25	1.32×10^{-11}
Terbium	Tb	65	158.925	2.63×10's^{-12}
Dysprosium	Dy	66	162.50	1.29×10^{-11}
Holmium	Ho	67	164.930	3.1×10^{-12}
Erbium	Er	68	167.26	5.75×10^{-12}
Thulium	Tm	69	168.934	2.69×10^{-11}
Ytterbium	Yb	70	170.04	6.46×10^{-12}
Lutetium	Lu	71	174.97	6.92×10^{-12}
Hafnium	Hf	72	178.49	6.3×10^{-12}
Tantalum	Ta	73	180.948	2×10^{-12}
Tungsten	W	74	183.85	3.72×10^{-10}
Rhenium	Re	75	186.2	1.8×10^{-12}
Osmium	Os	76	190.2	5.62×10^{-12}
Iridium	Ir	77	192.2	1.62×10^{-10}
Platinum	Pt	78	195.09	5.62×10^{-11}
Gold	Au	79	196.967	2.09×10^{-12}
Mercury	Hg	80	200.59	1×10^{-9}
Thallium	Tl	81	204.37	1.6×10^{-12}
Lead	Pb	82	207.19	7.41×10^{-11}
Bismuth	Bi	83	208.981	6.3×10^{-12}
Polonium	Po	84	210	—
Astatine	At	85	210	—
Radon	Rn	86	222	—
Francium	Fr	87	223	—
Radium	Ra	88	226.025	—
Actinium	Ac	89	227	—
Thorium	Th	90	232.038	6.61×10^{-12}
Protactinium	Pa	91	230.040	—
Uranium	U	92	238.029	4.0×10^{-12}
Neptunium	Np	93	237.048	—
Plutonium	Pu	94	242	—
Americium	Am	95	242	—
Curium	Cm	96	245	—
Berkelium	Bk	97	248	—
Californium	Cf	98	252	—
Einsteinium	Es	99	253	—
Fermium	Fm	100	257	—
Mendelevium	Md	101	257	—
Nobelium	No	102	255	—
Lawrencium	Lr	103	256	—

[1]The atomic weight of an element is its mass in *atomic mass units.* An atomic mass unit is defined as one-twelfth of the mass of the most common isotope of carbon, and has the value 1.660531×10^{-24} grams. In general, the atomic weight of an element is approximately equal to the total number of protons and neutrons in its nucleus.

[2]The abundances are given in terms of the number of atoms of each element compared to hydrogen, and are based on the composition of the Sun. For very rare elements, particularly those toward the end of the list, the abundances can be quite uncertain, and the values given should not be considered exact.

APPENDIX 4 Temperature scales

At the most basic level, temperature can be defined in terms of the motion of particles in a gas (or a solid, or a liquid). We all have an intuitive idea of what heat is, and we are all familiar with at least one scale for measuring temperature.

The most commonly used scales are somewhat arbitrarily defined, with zero points not representing any truly fundamental physical basis. The popular *Fahrenheit* scale, for example, has water freezing at a temperature of 32°F and boiling at 212°F. On this scale absolute zero, the lowest possible temperature (where all molecular motions cease), is −459°F.

The centigrade (or Celsius) scale is perhaps better founded, although it is based on the freezing and boiling points of water, rather than the more fundamental absolute zero. In this system, the freezing point is defined as 0°C, and 100°C is the boiling point. This scale has the advantage over the Fahrenheit scale that there are exactly 100° between the freezing and boiling points, rather than 180°, as in the Fahrenheit system. To convert from Fahrenheit to centigrade, we must subtract 32° first, and then multiply the remainder by 100/180, or 5/9. For example, 50°F is equal to 5/9 × (50 − 32) = 10°C. To convert from centigrade to Fahrenheit, first multiply by 9/5 and then add 32°. Thus, −10°C = 9/5 × (−10 + 32 = 14°F. On the centigrade scale, absolute zero occurs at −273°C.

The temperature scale preferred by scientists is a modification of the centigrade system. In this system, named after its founder, the British physicist Lord Kelvin, the same degree is used as in the centigrade scale; that is, one degree is equal to one one-hundredth of the difference between the freezing and boiling points of water. The zero point is different from the zero point on the centigrade scale, however; it is equal to absolute zero. Hence on this scale water freezes at 273°K and boils at 373°K. Comfortable room temperature is around 300°K. In modern usage the degree symbol (°) is dropped, and we speak simply of temperatures in units of Kelvins (273 K, for example).

APPENDIX 5 Radiation laws

Several laws described in Chapter 6 apply to continuous radiation from hot objects such as stars. In the text these laws were discussed in general terms, and a few simple applications were explained. Here the same laws are given in more precise mathematical form, and their use in that form is illustrated.

WIEN'S LAW

In general terms, Wien's law says that the wavelength of maximum emission from a glowing object is inversely proportional to its temperature. Mathematically this can be written as:

$$\lambda_{max} \propto 1/T,$$

where λ_{max} is the wavelength of strongest emission, T is the surface temperature of the object, and $\propto$ is a special mathematical symbol meaning "is proportional to."

Experimentation can determine the *proportionality constant*, specifying the exact relationship between λ_{max} and T, and Wien did this, finding

$$\lambda_{max} = W/T,$$

where W has the value 0.29 if λ_{max} is measured in centimeters and T in degrees absolute. With this equation it is possible to calculate λ_{max}, given T, or vice versa. Thus, if we measure the spectrum of a star and find that it emits most strongly at a wavelength of 2000 Å = 2×10^{-5} cm, then we can solve for the temperature:

$$T = W/\lambda_{max} = 0.29/2 \times 10^{-5} = 14{,}500 \text{ K}.$$

This is a relatively hot star, and it would appear blue in the eye. Note that it was necessary to measure the spectrum in ultraviolet wavelengths in order to find λ_{max}.

When solving problems using Wien's law, it is always possible to use the equation form, as we have just done in this example. Often, however, it is more convenient to compare the properties of two objects by considering the ratio of the temperatures or of the wavelengths of maximum emission. In effect this is what we did in the text when we compared two objects of different temperatures in order to determine how their λ_{max} values compared. For example, we said that if one object is twice as hot as another, its value of λ_{max} is half that of the other.

We can see how this ratio technique works by writing the equation for Wien's law separately for object 1 and object 2:

$$\lambda_{max1} = W/T_1, \quad \text{and} \quad \lambda_{max2} = W/T_2.$$

Now we can divide one equation by the other:

$$\frac{\lambda_{max1}}{\lambda_{max2}} = \frac{W/T_1}{W/T_2}$$

or

$$\frac{\lambda_{max1}}{\lambda_{max2}} = \frac{T_2}{T_1}.$$

The numerical factor W has canceled out, and we are left with a simple expression relating the values of λ_{max} and T for the two objects. Now we see that if $T_1 = 2T_2$ (object 1 is twice as hot as object 2), then

$$\frac{\lambda_{max1}}{\lambda_{max2}} = \frac{T_2}{2T_2} = \frac{1}{2},$$

or λ_{max} for object 1 is one-half that for object 2. In this extremely simple example, it probably would have been easier just to work it out in our heads, but what if we have a case where one object is 3.368 times hotter than the other, for example?

A great deal can be learned from making comparisons in this way. Astronomers often use the sun as the standard for comparison, expressing various quantities in terms of the solar values. Another reason for using comparisons occurs when the numerical constants (such as Wien's constant in the foregoing examples) are not known. If the trick of comparing is kept in mind, it is often possible to work out answers to astronomical questions simply by carrying around in one's head a few numbers describing the sun.

The Stefan-Boltzmann Law

As discussed in the text, the energy emitted by a glowing object is proportional to T^4, where T is the surface temperature. This can be written mathematically as

$$E = \sigma T^4,$$

where σ stands for a proportionality constant that has the value 5.7×10^{-5}, if the centimeter-gram-second metric units are used.

If we now consider how much surface area a star has, then the total energy it emits, called its *luminosity* and usually denoted L, is

$$\begin{aligned} L &= \text{surface area} \times E \\ &= 4\pi R^2 \sigma T^4, \end{aligned}$$

where R is the radius of the star. This equation is called the Stefan-Boltzmann law.

As in other cases, the law can be used directly in this form, or we can choose to compare properties of stars by writing the equation separately for two stars and then dividing. If we do this, we find

$$\frac{L_1}{L_2} = \frac{R_1^2 T_1^4}{R_2^2 T_2^4} = \left(\frac{R_1}{R_2}\right)^2 \left(\frac{T_1}{T_2}\right)^4$$

The constant factors 4π and σ have canceled out.

As an example of how to use this expression, suppose we determine that a particular star has twice the temperature of the sun but only one-half the radius, and we wish to know how this star's luminosity compares to that of the sun. If we designate the star as object 1 and the sun as object 2, then

$$\frac{L_1}{L_2} = \left(\frac{1}{2}\right)^2 (2)^4 = \frac{1}{4} \times 16 = 4$$

This star has four times the luminosity of the sun.

The Stefan-Boltzmann law is particularly useful because it relates three of the most important properties of stars to each other.

The Planck Function

The radiation laws described above and in the text are actually specific forms of a much more general law, discovered by the great German physicist Max Planck. Wien's law, the Stefan-Boltzmann law, and some others not mentioned in the text bear the same kind of relation to Planck's law as the laws of planetary motion discovered by Kepler do to Newton's mechanics. Kepler's laws were first discovered by observation, but with Newton's laws of motion it is possible to derive Kepler's laws theoretically. In the same fashion the radiation laws discussed so far were found experimentally but can be derived mathematically from the much more general and powerful Planck's law.

Planck's law is usually referred to as the Planck function, a mathematical relationship between intensity and wavelength (or frequency) that describes the spectrum of any glowing object at a given temperature. The Planck function specifically applies only to objects that radiate solely because of their temperature and do not reflect any light or have any spectral lines. The popular term for such an object is *blackbody,* and we often refer to radiation from such an object as *blackbody radiation* or, more commonly, *thermal radiation,* the term used in the text. Stars are not perfect radiators, but to a good approximation can be treated as such; hence in the text we apply Wien's law and the Stefan-Boltzmann law to stars (and even planets in certain circumstances) without pointing out the fact that to do so is only approximately correct.

The form of the Planck function for the radiation intensity B as a function of wavelength is

$$B = \frac{2hc^2}{\lambda^5} \frac{1}{e^{hc/\lambda kt} - 1},$$

where h is the Planck constant, c is the speed of light, and k is the Boltzmann constant (the values of all three are tabulated in Appendix 2), λ is the wavelength (in cm), and T is the temperature of the object (on the absolute scale). The symbol e represents the base of the natural logarithm, something not used elsewhere in this text; for the present purpose, this may be regarded simply as a mathematical constant with the value 2.718.

In terms of frequency ν rather than wavelength, the expression is

$$B = \frac{2h\nu^3}{c^2} \frac{1}{e^{h\nu/kT} - 1}.$$

Either expression may be used to calculate the spectrum of continuous radiation from a glowing object at a specific temperature. In practice the Planck function is used in a wide assortment of theoretical calculations that call for knowledge of the intensity of radiation so that its effects can be assessed on physical conditions such as ionization.

Appendix 6 Major telescopes of the world (2 meters or larger)

Observatory	Location*	Telescope
Keck Observatory (California Institute of Technology and the University of California)	(Mauna Kea, Hawaii)	10.0-m Keck Telescope
Special Astrophysical Observatory	Mount Pastukhov, USSR	6-m Bol'shoi Teleskop Azimutal'nyi
Hale Observatories	Palomar Mountain, California	5.08-m George Ellery Hale Telescope
Smithsonian Observatory	Mount Hopkins, Arizona	4.5-m Multiple Mirror Telescope
Royal Greenwich Observatory	La Palma, Canary Islands	4.2-m William Herschel Telescope
Cerro Tololo Observatory	Cerro Tololo, Chile	4.0-m
Anglo-Australian Observatory	Siding Spring Mountain, Australia	3.9-m Anglo-Australian Telescope
Kitt Peak National Observatory	Kitt Peak, Arizona	3.8-m Nicholas U. Mayall Telescope
Royal Observatory Edinburgh	Mauna Kea, Hawaii	3.8-m United Kingdom Infrared Telescope
Canada-France-Hawaii Observatory	Mauna Kea, Hawaii	3.6-m Canada-France-Hawaii Telescope
European Southern Observatory	Cerro La Silla, Chile	3.57-m
Max Planck Institute (Bonn)	Calar Alto, Spain	3.5-m
Astronomical Research Corporation	(Sacremento Peak, New Mexico)	3.5-m
Lick Observatory	Mount Hamilton, California	3.05-m C. Donald Shane Telescope
Mauna Kea Observatory	Mauna Kea, Hawaii	3.0-m NASA Infrared Telescope
McDonald Observatory	Mount Locke, Texas	2.7-m
Haute Provence Observatory	Saint Michele, France	2.6-m
Crimean Astrophysical Observatory	Simferopol, USSR	2.6-m
Byurakan Astrophysical Observatory	Byurakan, USSR	2.6-m
Mount Wilson and Las Campanas Observatories	Mount Wilson, California	2.5-m Hooker Telescope
Mount Wilson and Las Campanas Observatories	Las Campanas, Chile	2.5-m Irenée du Pont Telescope
Royal Greenwich Observatory	Canary Islands	2.5-m Isaac Newton Telescope
Dartmouth College and University of Michigan	Kitt Peak, Arizona	2.4-m McGraw-Hill Telescope
Wyoming Infrared Observatory	Mount Jelm, Wyoming	2.3-m Wyoming Infrared Telescope
Steward Observatory	Kitt Peak, Arizona	2.3-m Wyoming Infrared Telescope
Mauna Kea Observatory	Mauna Kea, Hawaii	2.2-m
Max Planck Institute (Bonn)	Calar Alto, Spain	2.2-m
Max Planck Institute	(South West Africa)	2.2-m
University of Mexico Observatory	(Mexico)	2.16-m
La Plata Observatory	La Plata Argentina	2.15-m
Kitt Peak National Observatory	Kitt Peak, Arizona	2.1-m
McDonald Observatory	Mount Locke, Texas	2.1-m Otto Struve Telescope
Karl Schwarzschild Observatory	Tautenberg, East Germany	2.0-m (largest Schmidt telescope)

*Telescopes under construction have their locations indicated in parentheses.

APPENDIX 7 Planetary and satellite data

Orbital Data for the Planets

Planet	Sidereal Period	Semimajor Axis	Orbital Eccentricity*	Inclination of Orbital Plane	Rotation Period	Tilt of Axis
Mercury	0.241 yr	0.387 AU	0.2056	7°0′15″	$58^{d}65$	28°
Venus	0.615	0.723	0.068	3°23′40″	243	3°
Earth	1.000	1.000	0.0167	0°0′0″	$23^{h}56^{m}$	23°27′
Mars	1.881	1.524	0.0934	1°51′0″	$24^{h}37^{m}$	23°59′
Jupiter	11.86	5.203	0.0485	1°18′17″	$9^{h}56^{m}$	3°5′
Saturn	29.46	9.555	0.0556	2°29′33″	$10^{h}39^{m}$	26°44′
Uranus	84.01	19.22	0.0472	0°46′23″	$16^{h}10^{m}$	97°55′
Neptune	164.79	30.11	0.0086	1°46′22″	$18^{h}12^{m}$	28°48′
Pluto	248.5	39.44	0.250	17°10′12″	$6^{d}24^{m}$	125°

*The eccentricity of an orbit is defined as the ratio of the distance between foci to the semimajor axis. In practice it is related to the perihelion distance P and the semimajor axis a by $P = a(1 - e)$, where e is the eccentricity; and to the aphelion distance A by $A = a(1 + e)$.

Physical data for the planets

Planet	Mass*	Diameter*	Density	Surface Gravity	Escape Speed	Temperature	Albedo
Mercury	0.0558	0.382	5.50 g/cc	0.38 g	4.3 km/sec	100–700 K	0.06
Venus	0.815	0.951	5.3	0.90	10.3	730	0.75
Earth	1.000	1.000	5.517	1.00	11.2	200–300	0.31
Mars	0.107	0.531	3.96	0.38	5.0	130–290	0.15
Jupiter	318.1	10.79	1.33	2.64	60.0	130	0.34
Saturn	95.12	8.91	0.68	1.13	36.0	95	0.34
Uranus	14.54	4.05	1.20	0.89	21.2	52	0.34
Neptune	17.2	3.91	1.58	1.13	23.5	50	0.29
Pluto	0.002	0.18	1.7	0.06	1.2	40	0.4

*The masses and diameters are given in units of the earth's mass and diameter, which are 5.974×10^{27} grams and 12,734 km, respectively.

Satellites

Planet	Satellite	Semimajor Axis	Period	Diameter	Mass	Density
Earth	Moon	3.84×10^{5} km	$27^{d}322$	3476 km	7.35×10^{25}	3.34 g/cm³
Mars	Phobos	9.38×10^{3}	0.3189	27 × 22 × 19	9.6×10^{18}	2:
	Deimos	2.35×10^{4}	1.2624	15 × 12 × 11	1.9×10^{18}	2:
Jupiter	Metis	1.280×10^{5}	0.295	20:	9.5×10^{19}	
	Adrastea	1.290×10^{5}	0.298	20 × 20 × 15	1.9×19^{19}	
	Amalthea	1.81×10^{5}	0.498	270 × 170 × 150	7.2×10^{21}	
	Thebe	2.22×10^{5}	0.675	110 × 90	7.6×10^{20}	
	Io	4.22×10^{5}	1.769	3630	8.92×10^{25}	3.53
	Europa	6.71×10^{5}	3.551	3138	4.87×10^{25}	3.03
	Ganymede	1.07×10^{6}	7.155	5262	1.49×10^{26}	1.93
	Callisto	1.88×10^{6}	16.689	4800	1.08×10^{26}	1.70
	Leda	1.11×10^{7}	238.7	16:	5.7×10^{18}	
	Himalia	1.15×10^{7}	250.6	186:	9.5×10^{21}	

APPENDIX 7 Planetary and satellite data, continued

Planet	Satellite	Semimajor Axis	Period	Diameter	Mass	Density
	Lysithea	1.17×10^7	259.2	36:	7.6×10^{19}	
	Elara	1.18×10^7	259.7	76:	7.6×10^{20}	
	Ananke	2.12×10^7	631	30:	3.8×10^{19}	
	Carme	2.26×10^7	692	40:	9.5×10^{19}	
	Pasiphae	2.35×10^7	735	50:	1.9×10^{20}	
	Sinope	2.37×10^7	758	36:	7.6×10^{19}	
Saturn	Atlas	1.377×10^5	0.602	40 × 20		
	Prometheus	1.394×10^5	0.613	140 × 100 × 80		
	Pandora	1.417×10^5	0.629	110 × 90 × 80		
	Epimetheus	1.514×10^5	0.694	140 × 120 × 100		
	Janus	1.514×10^5	0.695	220 × 200 × 160		
	Mimas	1.855×10^5	0.942	392	4.5×10^{22}	1.43
	Enceladus	2.381×10^5	1.370	500	7.4×10^{22}	1.13
	Telesto	2.947×10^5	1.888	34 × 28 × 26		
	Calypso	2.947×10^5	1.888	34 × 22 × 22		
	Tethys	2.947×10^5	1.888	1060	7.4×10^{23}	1.19
	No Name I	$3.3{:} \times 10^5$	?	15–20		
	Dione	3.774×10^5	2.737	1120	1.05×10^{24}	1.43
	Helene	3.781×10^5	2.737	36 × 32 × 30		
	No Name II	$3.8{:} \times 10^5$	?	15–20		
	No Name III	$4.7{:} \times 10^5$	?	15–20		
	Rhea	5.271×10^5	4.518	1530	2.50×10^{24}	1.33
	Titan	1.222×10^6	15.945	5150	1.35×10^{26}	1.89
	Hyperion	1.481×10^6	21.277	410 × 260 × 220	1.71×10^{22}	
	Iapetus	3.561×10^6	79.330	1460	1.88×10^{24}	1.15
	Phoebe	1.295×10^7	550.5	220		
Uranus	1986U7	4.97×10^4	0.33	40		
	1986U8	5.38×10^4	0.38	50		
	1986U9	5.92×10^4	0.43	50		
	1986U3	6.18×10^4	0.46	60		
	1986U6	6.27×10^4	0.48	60		
	1986U2	6.46×10^4	0.49	80		
	1986U10	6.61×10^4	0.51	80		
	1986U4	6.99×10^4	0.56	60		
	1986U5	7.53×10^4	0.62	60		
	1986U1	8.60×10^4	0.76	170		
	Miranda	1.294×10^5	1.413	480	7.5×10^{22}	1.26
	Ariel	1.910×10^5	2.520	1330	1.4×10^{24}	1.14
	Umbriel	2.663×10^5	4.144	1110	1.3×10^{24}	1.82
	Titania	4.359×10^5	8.706	1600	3.5×10^{24}	1.60
	Oberon	5.835×10^5	13.463	1630	2.9×10^{24}	1.28
Neptune	Triton	3.543×10^5	5.877	3800	1.3×10^{26}	
	Nereid	5.512×10^6	360.2	300	2.1×10^{22}	
Pluto	Charon	1.97×10^4	6.387			

APPENDIX 8 Stellar data

The Fifty Brightest Stars

Star		Spectral Type	Apparent Magnitude	Distance	Position (1980) Right Ascension	Declination
α Eri	Achernar	B3 V	0.51	36 pc	$01^h37^m.0$	−57°20′
α UMi	Polaris	F8 Ib	1.99	208	02 12.5	+89 11
α Per	Mirfak	F5 Ib	1.80	175	03 22.9	+49 47
α Tau	Aldebaran	K5 III	0.86	21	04 34.8	+16 28
β Ori	Rigel	B8 Ia	0.14	276	05 13.6	−08 13
α Aur	Capella	G8 III	0.05	14	05 15.2	+45 59
γ Ori	Bellatrix	B2 III	1.64	144	05 24.0	+06 20
β Tau	Elnath	B7 III	1.65	92	05 25.0	+28 36
ε Ori	Alnilam	B0 Ia	1.70	490	05 35.2	−01 13
ζ Ori	Alnitak	09.5 Ib	1.79	490	05 39.7	−01 57
α Ori	Betelgeuse	M2 Iab	0.41	159	05 54.0	+07 24
β Aur	Menkalinan	A2 V	1.86	27	05 58.0	+44 57
β CMa		B1 II-III	1.96	230	06 21.8	−17 56
α Car	Canopus	F0 Ib-II	−0.72	30	06 23.5	−52 41
γ Gem	Alhena	A0 IV	1.93	32	06 36.6	+16 25
α CMa	Sirius	A1 V	−1.47	2.7	06 44.2	−16 42
ε CMa	Adhara	B2 II	1.48	209	06 57.8	−28 57
δ CMa		F8 Ia	1.85	644	07 07.6	−26 22
α Gem	Castor	A1 V	1.97	14	07 33.3	+31 56
α CMi	Procyon	F5 IV-V	0.37	3.5	07 38.2	+05 17
β Gem	Pollux	K0 III	1.16	11	07 44.1	+28 05
γ Vel		WC8	1.83	160	08 08.9	−47 18
ε Car	Avior	K3 III?	1.90	104	08 22.1	−59 26
ζ Vel		A2 V	1.95	23	08 44.2	−54 38
β Car	Miaplacidus	A1 III	1.67	26	09 13.0	−69 38
α Hya	Alphard	K4 III	1.98	29	09 26.6	−08 35
α Leo	Regulus	B7 V	1.36	26	10 07.3	+12 04
γ Leo		K0 III	1.99	28	10 18.8	+19 57
α UMa	Dubhe	K0 III	1.81	32	11 02.5	+61 52
α Cru A	Acrux	B0.5 IV	1.39	114	12 25.4	−62 59
α Cru B	Acrux	B1 V	1.86	114	12 15.4	−62 59
γ Cru	Gacrux	M4 III	1.69	67	12 30.1	−57 00
β Cru		B0.5 III	1.28	150	12 46.6	−59 35
ε UMa	Alioth	A0p	1.79	21	12 53.2	+56 04
α Vir	Spica	B1 V	0.91	67	13 24.1	−11 03
η UMa	Alkaid	B3 V	1.87	64	13 46.8	+49 25
β Cen	Hadar	B1 III	0.63	150	14 02.4	−60 16
α Boo	Arcturus	K2 III	−0.06	11	14 14.8	+19 17
α Cen A	Rigil Kentaurus	G2 V	0.01	1.3	14 38.4	−60 46
α Cen B	Rigil Kentaurus	K4 V	1.40	1.3	14 38.4	−60 46
α Sco	Antares	M1 Ib	0.92	160	16 28.2	−26 23
α TrA	Atria	K2 Ib	1.93	25	16 46.5	−68 60
λ Sco	Shaula	B1 V	1.60	95	17 32.3	−37 05
θ Sco		F0 Ib	1.86	199	17 35.9	−42 59
ε Sgr	Kaus Australis	B9.5 III	1.81	38	18 22.9	−34 24
α Lyr	Vega	A0 V	0.04	8	18 36.2	+38 46
α Aql	Altair	A7 IV-V	0.77	5	19 49.8	+08 49
α Pav	Peacock	B2.5 V	1.95	95	20 24.1	−56 48
α Cyg	Deneb	A2 Ia	1.26	491	20 40.7	+45 12
α Gru	Al Na'ir	B7 IV	1.76	20	22 06.9	−47 04
α PsA	Fomalhaut	A3 V	1.15	7	22 56.5	−29 44

APPENDIX 8 Stellar data, continued

Nearby stars (within 5 parsecs of the Sun)*

Star	Spectral Type	Visual Apparent Magnitude	Visual Absolute Magnitude	Parallax	Distance	Proper Motion
α Centarui	G2V	−0.1	4.8	0.753″	1.33 pc	3.68″/yr
Barnard's star	M5V	9.5	13.2	0.544	1.84	10.31
Wolf 359	M8V3	13.5	16.7	0.432	2.31	4.71
BD + 36°2147	M2V	7.5	10.5	0.400	2.50	4.78
Luyten 726-8	M6Ve	12.5	15.4	0.385	2.60	3.36
Sirius	A1V	−1.5	1.4	0.377	2.65	1.33
Ross 154	M5Ve	10.6	13.3	0.345	2.90	0.72
Ross 248	M6Ve	12.3	14.8	0.319	3.13	1.58
ε Eridani	K2V	3.7	6.1	0.305	3.28	0.98
Ross 128	M5V	11.1	13.5	0.302	3.31	1.37
Luyten 789-6	M6V	12.2	14.6	0.302	3.31	3.26
61 Cygni	K5Ve	5.2	7.5	0.292	3.42	5.22
ε Indi	K5Ve	4.7	7.0	0.291	3.44	4.69
τ Ceti	G8V	3.5	5.9	0.289	3.46	1.92
Procyon	F5V	0.4	2.7	0.285	3.51	1.25
Σ 2398	M4V	8.9	11.2	0.284	3.52	2.28
BD + 43°44	M1Ve	8.1	10.4	0.282	3.55	2.89
CD − 36°15693	M2Ve	7.4	9.6	0.279	3.58	6.90
G51-15		14.8	17.0	0.273	3.66	1.26
L725-32	M5Ve	11.5	13.6	0.264	3.79	1.22
BD + 5°1668	M4V	9.8	12.0	0.264	3.79	3.73
CD − 39°14192	M0Ve	6.7	8.8	0.260	3.85	3.46
Kapteyn's star	M0V	8.8	10.8	0.256	3.91	8.89
Kruger 60	M4V	9.7	11.7	0.254	3.94	0.86
Ross 614	M5Ve	11.3	13.3	0.243	4.12	0.99
BD − 12°4523	M5V	10.0	11.9	0.238	4.20	1.18
Wolf 424	M6Ve	13.2	15.0	0.234	4.27	1.75
van Maanen's star	W.D.	12.4	14.2	0.232	4.31	2.95
CD − 37°15492	M3V	8.6	10.4	0.225	4.44	6.08
Luyten 1159-16	M8V	12.3	14.0	0.221	4.52	2.08
BD + 50°1725	K7V	6.6	8.3	0.217	4.61	1.45
CD − 46°11540	M4V	9.4	11.1	0.216	4.63	1.13
CD − 49°13515	M3V	8.7	10.4	0.214	4.67	0.81
CD − 44°11909	M5V	11.2	12.8	0.213	4.69	1.16
BD + 68°946	M3.5V	9.1	10.8	0.213	4.69	1.33
G158-27		13.7	15.5	0.212	4.72	2.06
G208-44/45		13.4	15.0	0.210	4.76	0.75
BD − 15°6290	M5V	10.2	11.8	0.209	4.78	1.16
40 Eridani	K0V	4.4	6.0	0.207	4.83	4.08
L145-141	W.D.	11.4	12.6	0.206	4.85	2.68
BD + 20°2465	M4.5V	9.4	10.9	0.203	4.93	0.49
70 Ophiuchi	K1V	4.2	5.7	0.203	4.93	1.13
BD + 43°4305	M4.5Ve	10.2	11.7	0.200	5.00	0.83

*Data from Lippencott, L. S., 1978, *Space Science Reviews* 22:153. Many of the stars listed here are multiple systems; in these cases, the data in the table refer only to the brightest member of the system.

APPENDIX 9 The constellations

Name	Genitive	Abbreviation	Position	
			Right Ascension	Declination
Andromeda	Andromedae	And	01^h	$+40°$
Antlia	Antliae	Ant	10	−35
Apus	Apodis	Aps	16	−75
Aquarius	Aquarii	Aqr	23	−15
Aquila	Aquilae	Aql	20	+05
Ara	Arae	Ara	17	−55
Aries	Arietis	Ari	03	+20
Auriga	Aurigae	Aur	06	+40
Bootes	Bootis	Boo	15	+30
Caelum	Caeli	Cae	05	−40
Camelopardalis	Camelopardalis	Cam	06	−70
Cancer	Cancri	Cnc	09	+20
Canes Venatici	Canum Venaticorum	CVn	13	+40
Canis Major	Canis Majoris	CMa	07	−20
Canis Minor	Canis Minoris	CMi	08	+05
Capricornus	Capricorni	Cap	21	−20
Carina	Carinae	Car	09	−60
Cassiopeia	Cassiopeiae	Cas	01	+60
Centaurus	Centauri	Cen	13	−50
Cepheus	Cephei	Cep	22	+70
Cetus	Ceti	Cet	02	−10
Chamaeleon	Chamaeleonis	Cha	11	−80
Circinis	Circini	Cir	15	−60
Columba	Columbae	Col	06	−35
Coma Berenices	Comae Berenices	Com	13	+20
Corona Australis	Coronae Australis	CrA	19	−40
Coronoa Borealis	Coronae Borealis	CrB	16	+30
Corvus	Corvi	Crv	12	−20
Crater	Crateris	Crt	11	−15
Crux	Crucis	Cru	12	−60
Cygnus	Cygni	Cyg	21	+40
Delphinus	Delphini	Del	21	+10
Dorado	Doradus	Dor	05	−65
Draco	Draconis	Dra	17	+65
Equuleus	Equulei	Equ	21	+10
Eridanus	Eridani	Eri	03	−20
Fornax	Fornacis	For	03	−30
Gemini	Geminorum	Gem	07	+20
Grus	Gruis	Gru	22	−45
Hercules	Herculis	Her	17	+30
Horologium	Horologii	Hor	03	−60
Hydra	Hydrae	Hya	10	−20
Hydrus	Hydri	Hyi	02	−75
Indus	Indi	Ind	21	−55
Lacerta	Lacertae	Lac	22	+45
Leo	Leonis	Leo	11	+15
Leo Minor	Leonis Minoris	LMi	10	+35

APPENDIX 9 The constellations, continued

Name	Genitive	Abbreviation	Position	
			Right Ascension	Declination
Lepus	Leporis	Lep	06	−20
Libra	Librae	Lib	15	−15
Lupus	Lupi	Lup	15	−45
Lynx	Lincis	Lyn	08	+45
Lyra	Lyrae	Lyr	19	+40
Mensa	Mensae	Men	05	−80
Microscopium	Microscopii	Mic	21	−35
Monoceros	Monocerotis	Mon	07	−05
Musca	Muscae	Mus	12	−70
Norma	Normae	Nor	16	−50
Octans	Octantis	Oct	22	−85
Ophiuchus	Ophiuchi	Oph	17	00
Orion	Orionis	Ori	05	+05
Pavo	Pavonis	Pav	20	−65
Pegasus	Pegasi	Peg	22	+20
Perseus	Persei	Per	03	+45
Phoenix	Phoenicis	Phe	01	−50
Pictor	Pictoris	Pic	06	−55
Pisces	Piscium	Psc	01	+15
Piscis Austrinus	Piscis Austrini	PsA	22	−30
Puppis	Puppis	Pup	08	−40
Pyxis	Pyxidis	Pyx	09	−30
Reticulum	Reticuli	Ret	04	−60
Sagitta	Sagittae	Sge	20	+10
Sagittarius	Sagittarii	Sgr	19	−25
Scorpius	Scorpii	Sco	17	−40
Sculptor	Sculptoris	Scl	00	−30
Scutum	Scuti	Sct	19	−10
Serpens	Serpentis	Ser	17	00
Sextans	Sextantis	Sex	10	00
Taurus	Tauri	Tau	04	+15
Telescopium	Telescopii	Tel	19	−50
Triangulum	Trianguli	Tri	02	+30
Triangulum Australe	Trianguli Australi	TrA	16	−65
Tucana	Tucanae	Tuc	00	−65
Ursa Major	Ursae Majoris	UMa	11	+50
Ursa Minor	Ursae Minoris	UMi	15	+70
Vela	Velorum	Vel	09	−50
Virgo	Virginis	Vir	13	00
Volans	Volantis	Vol	08	−70
Vulpecula	Vulpeculae	Vul	20	+25

APPENDIX 10 Mathematical treatment of stellar magnitudes

LOGARITHMIC REPRESENTATION

In the text magnitudes are discussed in terms of the brightness ratios between stars of different magnitudes. We generally avoided discussion of cases where two stars differ by a fraction of a magnitude, because in such cases it is no longer simple to calculate the brightness ratio corresponding to the magnitude difference. If star 1 is 0.5 magnitudes brighter than star 2, for example, what is the brightness ratio? Or if star 1 is a factor of 48.76 fainter than star 2, what is the difference in magnitudes?

Astronomers use an exact mathematical relationship between magnitude differences and brightness ratios, written as

$$m_1 - m_2 = 2.5 \log (b_2/b_1),$$

where m_1 and m_2 are the magnitudes of two stars, and b_2/b_1 is the ratio of their brightnesses. The notation "log (b_2/b_1)" means the **logarithm** of this ratio; a logarithm is the power to which 10 must be raised to give this ratio. Hence, for example, if $b_2/b_1 = 100$, $\log (b_2/b_1) = \log (10^2) = 2$, because 10 must be raised to the second power to give 100. The magnitude difference is $2.5 \log (100) = 2.5 \times 2 = 5$. Similarly, if $b_2/b_1 = 0.001$, then $\log (b_2/b_1) = \log (0.001) = \log (10^{-3}) = -3$, and in this case the magnitude difference is $2.5 \times -3 = -7.5$ (the minus sign indicates that star 1 is brighter than star 2 in this example).

The method works equally well in cases where the power of 10 is not a whole number, as in the example where $b_2/b_1 = 48.76$. Here $\log (b_2/b_1) = \log (48.76) = 1.69$ (this is usually found by consulting tables of logarithms or by using a scientific calculator). In this example, the magnitude difference is $2.5 \log (b_2/b_1) = 2.5 \log(48.76) = 2.5 \times 1.69 = 4.23$, so star 1 is 4.23 magnitudes fainter than star 2..

The equation can be used in other ways as well; solving for b_2/b_1 yields

$$b_2/b_1 = 10^{(m_1 - m_2)/2.5} = 10^{0.4(m_1 - m_2)}.$$

Thus, if we know that the magnitudes of two stars differ by $m_1 - m_2$, we multiply this difference by 0.4 and raise 10 to the power $0.4(m_1 - m_2)$, again using a calculator or tables, to get the brightness ratio b_2/b_1. As a simple example, suppose $m_1 - m_2 = 5$; then $0.4(m_1 - m_2) = 0.4 \times 5 = 2$, and $10^{0.4(m_1 - m_2)} = 10^2 = 100$, as we knew it should. As a more complex example, consider the stars Betelgeuse (magnitude +0.41) and Deneb (magnitude +1.26). From the equation above, we see that Betelgeuse is $10^{0.4(1.26 - 0.41)} = 10^{0.34} = 2.19$ times brighter than Deneb.

While in most cases it is possible to follow the discussions of magnitudes and brightness ratios in the text without using this exact mathematical technique, it is still useful to be familiar with it.

THE DISTANCE MODULUS

Whenever we know both the apparent and absolute magnitudes of a star, a comparison of the two will give its distance. In the text we have seen how to make this calculation by the following several steps:

1. Convert the difference $m - M$ between apparent and absolute magnitude into a brightness ratio, that is, a numerical factor indicating how much brighter or fainter the star would appear at 10 parsecs distance than at its actual distance.
2. Using the inverse square law, determine the change in distance required to produce this change in brightness.
3. Multiply this distance factor by 10 parsecs to find the distance to the star.

To do calculations mentally in this way can be laborious, especially in cases where the magnitude difference does not correspond neatly to a simple numerical factor, as it did in the examples given in the text. Hence astronomers use a mathematical equation expressing the relationship between distance and the distance modulus $m - M$. This equation, which works equally well for all cases, is

$$d = 10^{1 + .2(m - M)},$$

where d is the distance in parsecs to a star whose apparent magnitude is m and whose absolute magnitude is M.

In a simple example, where $m = 9$ and $M = -6$, we have

$$d = 10^{1 + .2(15)} = 10^{1 + 3} = 10^4 \text{ parsecs}.$$

Now let's try a more complex case. Suppose the star is an M2 main sequence star, so that $M = 13$, as found from the H-R diagram. The apparent magnitude is $m = 16$. Our equation tells us that

$$\begin{aligned} d &= 10^{1+.2(16-13)} \\ &= 10^{1+.6} \\ &= 10^{1.6} \\ &= 39.8 \text{ parsecs.} \end{aligned}$$

It is necessary to use a slide rule, calculator, or mathematical table to do this of course, but it is still relatively straightforward compared to following the mental steps outlined above.

The Effect of Extinction on the Distance Modulus

When we discussed distance determination techniques (in Chapter 20), we ignored the effects of interstellar extinction. In any method that depends on the apparent brightness of a star, however, extinction can be important, particularly for very distant stars. Because the effect of extinction is to make a star appear fainter than it otherwise would, the tendency is to overestimate distances if no allowance is made for it.

Recall that in the spectroscopic parallax technique, the distance modulus $m - M$ is used to find the distance to a star from the equation given above:

$$d = 10^{1+.2(m-M)},$$

where m is the apparent magnitude, M is the absolute magnitude, and d is the distance to the star in parsecs.

To correct this equation for extinction, we add a term, A_v, which refers to the extinction (in magnitudes) in visual light. Thus if the extinction toward a particular star makes that star appear 2 magnitudes fainter than it otherwise would, $A_v = 2$. If we insert this into the equation, we find:

$$d = 10^{1+.2(m-M-A_v)}$$

Let us consider a simple example, a star whose apparent magnitude is $m = 12.4$ and whose absolute magnitude is $M = 2.4$. First let us consider its distance if extinction is ignored:

$$d = 10^{1+.2(12.4-2.4)} = 10^3 = 1000 \text{ parsecs.}$$

Now suppose it is determined that the extinction in the direction of this star amounts to one magnitude. Now the distance is

$$d = 10^{1+.2(12.4-2.4-1)} = 10^{2.8} = 631 \text{ parsecs.}$$

One magnitude is only a modest amount of extinction, yet by neglecting it, we overestimated the distance to this star by almost 60 percent. We can see from this example that extinction can have a drastic effect on distance estimates.

It is worthwhile to add a note about how the extinction A_v is determined. It is possible to determine how much redder, in terms of the $B-V$ color index, a star appears because of extinction. To carry that a step further, astronomers define a *color excess* called $E(B-V)$, which is the difference between the observed and intrinsic values of $B-V$:

$$E(B-V) = (B-V)_{\text{observed}} - (B-V)_{\text{intrinsic}}.$$

Studies of the variation of interstellar extinction with wavelength show that the extinction at the visual wavelength is approximately three times the color excess; that is:

$$A_v = 3E(B-V).$$

Hence determination of excess reddening leads to an estimate of A_v, and this in turn can be used in the modified equation for the distance.

Appendix 11 Nuclear reactions in stars

In the text we did not spell out the details of the reactions that occur in stellar cores, although they are quite simple. To do so here, we will use the notation of nuclear physics. This is basically a shorthand in symbols. For example, a helium nucleus, containing two protons and two neutrons, is designated ^4_2He, the subscript indicating the **atomic number** (the number of protons) and the superscript the **atomic weight** (the total number of protons and neutrons). Similarly a hydrogen nucleus is ^1_1H, and deuterium, a form of hydrogen with an extra neutron in the nucleus, is ^2_1H. A special symbol (ν) is used for the **neutrino,** a massless subatomic particle emitted in some reactions, and e^+ indicates a **positron,** which is equivalent to an electron but has a positive electrical charge. The symbol γ indicates a gamma ray, a very short wavelength photon of light emitted in some reactions.

The Proton-Proton Chain

Using this notation system, we can now spell out the proton-proton chain:

$$^{1}_{1}H + ^{1}_{1}H \rightarrow ^{2}_{1}H + e^{+} + \nu$$
$$^{2}_{1}H + ^{1}_{1}H \rightarrow ^{3}_{2}He + \gamma.$$

The $^{3}_{2}He$ particle is a form of helium, but not the common form. Once we have this particle it will combine with another:

$$^{3}_{2}He + ^{3}_{2}He \rightarrow ^{4}_{2}He + ^{1}_{1}H + ^{1}_{1}H.$$

We end up with a normal helium nucleus. A total of six hydrogen nuclei went into the reaction (remember, the first two steps had to occur twice, in order to produce two $^{3}_{2}He$ particles for the final reaction), while there were two left at the end, so the net result is the conversion of four hydrogen nuclei into one helium nucleus.

The CNO Cycle

The CNO cycle, which dominates at higher temperatures, is more complex, involving not only carbon but also nitrogen and oxygen. Each of these elements has more than one form, with differing numbers of neutrons. Some of these **isotopes** are unstable and spontaneously emit positrons, decaying into other species in the process. Here is the CNO cycle:

$$^{12}_{6}C + ^{2}_{1}H \rightarrow ^{13}_{7}N + \gamma$$
$$^{13}_{7}N \rightarrow ^{13}_{6}C + e^{+} + \nu$$
$$^{13}_{6}C + ^{1}_{1}H \rightarrow ^{14}_{7}N + \gamma$$
$$^{14}_{7}N + ^{1}_{1}H \rightarrow ^{15}_{8}O + \gamma$$
$$^{15}_{8}O \rightarrow ^{15}_{7}N + e^{+} + \nu$$
$$^{15}_{7}N + ^{1}_{1}H \rightarrow ^{12}_{6}C + ^{4}_{2}He.$$

Here we end up with a helium nucleus and a carbon nucleus, while the particles going into the reaction were four hydrogen nuclei and a carbon nucleus. Along the way three isotopes of nitrogen and one of oxygen were created and then converted into something else, leaving neither element at the end. As in the proton-proton chain, the net result is the conversion of four hydrogen nuclei into one helium nucleus.

The Triple-Alpha Reaction

Stars that have used up all the available hydrogen in their cores may become hot enough to undergo a new reaction in which helium is converted into carbon. Helium nuclei, consisting of two protons and two neutrons, are called alpha particles, and the reaction is called the triple-alpha reaction since three of these particles are involved:

$$^{4}_{2}He + ^{4}_{2}He \rightarrow ^{8}_{4}Be$$
$$^{8}_{4}Be + ^{4}_{2}He \rightarrow ^{12}_{6}C + \gamma.$$

In this reaction sequence, $^{8}_{4}Be$ is a form of the element beryllium, and the end product, $^{12}_{6}C$, is the most common form of carbon. The second step must follow very quickly after the first because $^{8}_{4}Be$ is unstable and will break apart in two $^{4}_{2}He$ particles in a very short time. Therefore the third $^{4}_{2}He$ particle must react with the $^{8}_{4}Be$ particle almost immediately so that this reaction sequence can be viewed as a three-particle reaction.

Other Reactions

Following helium burning in the triple-alpha reaction, other reactions can take place if the stellar core becomes hot enough. The first of these reactions are **alpha-capture reactions,** in which one form of nucleus adds an alpha particle to become a new form with two more protons and two additional neutrons. One example is the carbon alpha capture:

$$^{12}_{6}C + ^{4}_{2}He \rightarrow ^{16}_{8}O + \gamma,$$

in which $^{16}_{8}O$, the most common form of oxygen, is produced. In another sequence of alpha captures, nitrogen is converted into another isotope of oxygen, which can then be converted into neon:

$$^{14}_{7}N + ^{4}_{2}He \rightarrow ^{18}_{8}O + e^{+} + \nu,$$

and

$$^{18}_{8}O + ^{4}_{2}He \rightarrow ^{22}_{10}Ne + \gamma.$$

Additional alpha-capture reactions can occur, but at high enough temperatures other, more complex, reactions also take place. For example, two carbon nuclei can react, forming a number of different products, including sodium, neon, and magnesium. Two oxygen nuclei can also react, creating such species as sulfur, phosphorus, silicon, and magnesium. As a massive star evolves and its core contracts and heats following each reaction stage, a wide variety of elements is created with generally increasing atomic numbers. As explained in the text, the heaviest and most complex element produced in stable reactions in stellar cores is iron ($^{52}_{26}Fe$). Once a star has an iron core, it cannot undergo any further reaction stages without major disruption by a supernova explosion or collapse to a neutron star or black hole.

APPENDIX 12 Detected interstellar molecules*

Number of Atoms	Symbol	Name	Number of Atoms	Symbol	Name
2	H_2	Molecular hydrogen	4	$NCNH^+$	Protonated hydrogen cyanide
	C_2	Diatomic carbon		HNCS	Isothiocyanic acid
	CH	Methylidyne		C_3N	Cyanoethynyl
	CH^+	Methylidyne ion		C_3O	Tricarbon monoxide
	CN	Cyanogen		H_2CS	Thioformaldehyde
	CO	Carbon monoxide	5	C_4H	Butadiynyl
	CS	Carbon monosulfide		C_3H_2	Cyclopropenylidene
	OH	Hydroxyl		HCO_2H	Formic acid
	NO	Nitric oxide		CH_2CO	Ketene
	NS	Nitrogen sulfide		HC_3N	Cyanoacetylene
	SiO	Silicon monoxide		NH_2CN	Cyanamide
	SiS	Silicon sulfide		CH_2NH	Methanimine
	SO	Sulfur monoxide		CH_4	Methane
3	H_2D^+	Protonated hydrogen deuteride		SiH_4	Silane‡
	C_2H	Ethynyl	6	C_5H	Pentynylidyne‡
	HCN	Hydrogen cyanide		CH_3OH	Methanol
	HNC	Hydrogen isocyanide		CH_3CN	Methyl cyanide
	HCO	Formyl		CH_3SH	Methyl mercaptan
	HCO^+	Formyl ion		NH_2CHO	Formamide
	N_2H^+	Protonated nitrogen	7	CH_2CHCN	Vinyl cyanide
	HNO	Nitroxyl		CH_3C_2H	Methylacetylene
	H_2O	Water		CH_3CHO	Acetaldehyde
	HCS^+	Thioformyl ion		CH_3NH_2	Methylamine
	H_2S	Hydrogen sulfide		HC_5N	Cyanodiacetylene
	OCS	Carbonyl sulfide	8	$HCOOCH_3$	Methyl formate
	SO_2	Sulfur dioxide		CH_3C_3N	Methylcyanoacetylene
	SiC_2	Silicon dicarbide‡	9	CH_3C_4H	Methyldiacetylene
4	C_2H_2	Acetylene‡		CH_3CH_3O	Dimethyl ether
	C_3H	Propynylidyne		CH_3CH_2CN	Ethyl cyanide
	H_2CO	Formaldehyde		CH_3CH_2OH	Ethanol
	NH_3	Ammonia		HC_7N	Cyano-hexa-tri-yne
	HNCO	Isocyanic acid	11	HC_9N	Cyano-octa-tetra-yne
	$HOCO^+$	Protonated carbon monoxide	13	$HC_{11}N$	Cyano-deca-penta-yne

*This list does not include isotopic variations (identical molecules except that one or more atoms are in rare isotopic forms, such as deuterium in place of hydrogen, or ^{13}C instead of the much more common ^{12}C).

‡These species have been detected only in the dense circumstellar clouds surrounding red giant or supergiant stars.

APPENDIX 13 Clusters of galaxies

GALAXIES OF THE LOCAL GROUP

Galaxy*	Type‡	Absolute Magnitude	Position: Right Ascension	Position: Declination
M31 (Andromeda)	Sb	−21.1	00ʰ40.ᵐ0	+41° 00′
Milky Way	Sbc	−20.5	17 42.5	−28 59
M33 = NGC 598	Sc	−18.9	01 31.1	−30 24
Large Magellanic Cloud	Irr	−18.5	05 24	−69 50
IC 10	Irr	−17.6	00 17.6	+59 02
Small Magellanic Cloud	Irr	−16.8	00 51	−73 10
M32 = NGC 221	E2	−16.4	00 40.0	+40 36
NGC 205	E6	−16.4	00 37.6	+41 25
NCG 6822	Irr	−15.7	19 42.1	−14 53
NCG 185	Dwarf E	−15.2	00 36.1	+48 04
NGC 147	Dwarf E	−14.9	00 30.4	+48 14
IC 1613	Irr	−14.8	01 02.3	+01 51
WLM	Irr	−14.7	23 59.4	−15 44
Fornax	Dwarf sph	−13.6	02 37.5	−34 44
Leo A	Irr	−13.6	09 56.5	+30 59
IC 5152	Irr	−13.5	21 59.6	−51 32
Pegasus	Irr	−13.4	23 26.1	+14 28
Sculptor	Dwarf sph	−11.7	00 57.5	−33 58
And I	Dwarf sph	−11	00 42.8	+37 46
And II	Dwarf sph	−11	01 13.6	+33 11
And III	Dwarf sph	−11	00 32.7	+36 14
Aquarius	Irr	−11	20 44.1	−13 02
Leo I	Dwarf sph	−11	10 0.58	+12 33
Sagittarius	Irr	−10	19 27.1	−17 47
Leo II	Dwarf sph	− 9.4	11 10.8	+22 26
Ursa Minor	Dwarf sph	− 8.8	15 08.2	+67 18
Draco	Dwarf sph	− 8.6	17 19.4	+57 58
Carina	Dwarf sph		06 40.4	−50 55
Pisces	Irr	− 8.5	00 01.2	+21 37

*Galaxy names are derived from a variety of sources, including several catalogs (such as those designated M, NGC, and IC) and colloquial names bestowed by discoverers. Many in this list are simply named after the constellation where they are found.

‡The galaxy types listed here are described in the text, except for the *Dwarf sph* designation which stands for "dwarf spheroidal" and refers to dwarf galaxies that do not fit easily into the designation of dwarf ellipticals. Note that the absolute magnitudes of some of these are comparable to those of the brightest individual stars in our galaxy.

APPENDIX 13 Clusters of galaxies

Clusters of Galaxies within 1000 Mpc*

Cluster	Distance	Radial Velocity	Diameter	Number of Galaxies	Density of Galaxies
Virgo	19 Mpc	1180 km/sec	4 Mpc	2500	500/Mpc³
Pegasus I	65	3700	1	100	1100
Pisces	66	250	12	100	250
Cancer	80	4800	4	150	500
Perseus	97	5400	7	500	300
Coma	113	6700	8	800	40
UMa III	132		2	90	200
Hercules	175	10,300	0.3	300	
Cluster A	240	15,800	4	400	200
Centaurus	250		9	300	10
UMa I	270	15,400	3	300	100
Leo	310	19,500	3	300	200
Cluster B	330		4	300	200
Gemini	350	23,300	3	200	100
CrB	350	21,600	3	400	250
Bootes	650	39,400	3	150	100
UMa II	680	41,000	2		400
Hydra	1000	60,600			

*Data from Allen C. W., 1973, *Astrophysical quantites*, 3d ed. (London: Athlone Press).

APPENDIX 14 The relativistic Doppler effect

The simple formula given in the text relating the Doppler shift of an object's spectrum to its velocity is accurate only when the velocity is much less than the speed of light. That simple relation is

$$v = (\Delta\lambda/\lambda)c,$$

where v is the object's velocity, $\Delta\lambda$ is the shift in wavelength of a spectral line whose rest wavelength is λ, and c is the speed of light.

When v is a significant fraction of c, the correct equation must be used. The error in determining v caused by using the incorrect formula is 1 percent of v when $\Delta\lambda/\lambda$ is only 0.02; and the error becomes 5 percent when $\Delta\lambda/\lambda$ is 0.1. Thus failure to use the correct, relativistic formula becomes important for speeds of only a few percent of the speed of light.

For simplicity of notation, astronomers usually use the symbol z to represent the Doppler shift; that is,

$$z = \Delta\lambda/\lambda.$$

From Einstein's theory of special relativity, it can be shown that the correct relationship between the shift in wavelength and the speed of the emitting object is

$$z = \Delta\lambda/\lambda = \sqrt{\frac{1 + v/c}{1 - v/c}} - 1,$$

which leads to the following solution for the velocity:

$$v = c\left[\frac{(Z + 1)^2 - 1}{(Z + 1)^2 + 1}\right]$$

Let us apply this to the redshifts of quasars. One of the first quasars discovered, 3C273, has $z = 0.16$, which would imply a velocity of $0.16c = 48{,}000$ km/sec, if we used the simple, nonrelativistic equation. Use of the relativistic formula leads instead to $v = 0.147c = 44{,}200$ km/sec. In this case use of the wrong formula leads to an error of almost 9 percent. The quasars with the highest known redshifts have z greater than 4.0, which would lead to the nonsensical conclusion that their speeds are more than four times that of light, if the nonrelativistic equation were used. Use of the correct equation for $z = 4.0$ gives $v = 0.923c = 277{,}000$ km/sec, still an enormous velocity.

APPENDIX 15 The Messier catalog

Number	Right Ascension	Declination	Magnitude	Description
M1	$05^h33.^m3$	+22°01′	11.3	Crab nebula in Taurus
M2	21 32.4	−00 54	6.3	Globular cluster in Aquarius
M3	13 41.3	+28 29	6.2	Globular cluster in Canes Venatici
M4	16 22.4	−26 27	6.1	Globular cluster in Scorpio
M5	15 17.5	+02 07	6	Globular cluster in Serpens
M6	17 38.9	−32 11	6	Open cluster in Scorpio
M7	17 52.6	−34 48	5	Open cluster in Scorpio
M8	18 02.4	−24 23		Lagoon nebula in Sagittarius
M9	17 18.1	−18 30	7.6	Globular cluster in Ophiuchus
M10	16 56.0	−04 05	6.4	Globular cluster in Ophiuchus
M11	18 50.0	−06 18	7	Open cluster in Scutum
M12	16 46.1	−01 55	6.7	Globular cluster in Ophiuchus
M13	16 41.0	+36 30	5.8	Globular cluster in Hercules
M14	17 36.5	−03 14	7.8	Globular cluster in Ophiuchus
M15	21 29.1	+12 05	6.3	Globular cluster in Pegasus
M16	18 17.8	−13 48	7	Open cluster in Serpens
M17	18 19.7	−16 12	7	Omega nebula in Sagittarius
M18	18 18.8	−17 09	7	Open cluster in Sagittarius
M19	17 01.3	−26 14	6.9	Globular cluster in Ophiuchus
M20	18 01.2	−23 02		Triffid nebula in Sagittarius
M21	18 03.4	−22 30	7	Open cluster in Sagittarius
M22	18 35.2	−23 55	5.2	Globular cluster in Sagittarius
M23	17 55.7	−19 00	6	Open cluster in Sagittarius
M24	18 17.3	−18 27	6	Open cluster in Sagittarius
M25	18 30.5	−19 16	6	Open cluster in Sagittarius
M26	18 44.1	−09 25	9	Open cluster in Scutum
M27	19 58.8	+22 40	8.2	Dumbbell nebula; planetary nebula in Vulpecula
M28	18 23.2	−24 52	7.1	Globular cluster in Sagittarius
M29	20 23.3	+38 27	8	Open cluster in Cygnus
M30	21 39.2	−23 15	7.6	Globular cluster in Capricornus
M31	00 41.6	+41 09	3.7	Andromeda galaxy
M32	00 41.6	+40 45	8.5	Elliptical galaxy, companion of M31
M33	01 32.8	+30 33	5.9	Spiral galaxy in Triangulum
M34	02 40.7	+42 43	6	Open cluster in Perseus
M35	06 07.6	+24 21	6	Open cluster in Gemini
M36	05 35.0	+34 05	6	Open cluster in Auriga
M37	05 51.5	+32 33	6	Open cluster in Auriga
M38	05 27.3	+35 48	6	Open cluster in Auriga
M39	21 31.5	+48 21	6	Open cluster in Cygnus
M40	12 20	+59		Double star cluster in Ursa Major
M41	06 46.2	−20 43	6	Open cluster in Canis Major
M42	05 34.4	−05 24		Orion nebula
M43	05 34.6	−05 18		Small extension of the Orion nebula
M44	08 38.8	+20 04	4	Praesepe; open cluster in Cancer
M45	03 46.3	+24 03	2	The Pleiades; open cluster in Taurus

APPENDIX 15 The Messier Catalog, continued

Number	Right Ascension	Declination	Magnitude	Description
M46	07 40.9	−14 46	7	Open cluster in Puppis
M47	07 35.6	−14 27	5	Open cluster in Puppis
M48	08 12.5	−05 43	6	Open cluster in Hydra
M49	12 28.8	+08 07	8.9	Elliptical galaxy in Virgo
M50	07 02.0	−08 19	7	Open cluster in Monocerotis
M51	13 29.0	+47 18	8.4	Whirlpool galaxy; spiral galaxy in Canes Venatici
M52	23 23.3	+61 29	7	Open cluster in Cassiopeia
M53	13 12.0	+18 17	7.7	Globular cluster in Coma Berenices
M54	18 53.8	−30 30	7.7	Globular cluster in Sagittarius
M55	19 38.7	−31 00	6.1	Globular cluster in Sagittarius
M56	19 15.8	+30 08	8.3	Globular cluster in Lyra
M57	18 52.9	+33 01	9.0	Ring nebula; planetary nebula in Lyra
M58	12 36.7	+11 56	9.9	Spiral galaxy in Virgo
M59	12 41.0	+11 47	10.3	Elliptical galaxy in Virgo
M60	12 42.6	+11 41	9.3	Elliptical galaxy in Virgo
M61	12 20.8	+04 36	9.7	Spiral galaxy in Virgo
M62	16 59.9	−30 05	7.2	Globular cluster in Scorpio
M63	13 14.8	+42 08	8.8	Spiral galaxy in Canes Venatici
M64	12 55.7	+21 48	8.7	Spiral galaxy in Coma Berenices
M65	11 17.8	+13 13	9.6	Spiral galaxy in Leo
M66	11 19.1	+13 07	9.2	Spiral galaxy, companion of M65
M67	08 50.0	+11 54	7	Open cluster in Cancer
M68	12 38.3	−26 38	8	Globulalr cluster in Hydra
M69	18 30.1	−32 23	7.7	Globular cluster in Sagittarius
M70	18 42.0	−32 18	8.2	Globular cluster in Sagittarius
M71	19 52.8	+18 44	6.9	Globular cluster in Sagitta
M72	20 52.3	−12 39	9.2	Globular cluster in Aquarius
M73	20 57.8	−12 44		Open cluster in Aquarius
M74	01 35.6	+15 41	9.5	Spiral galaxy in Pisces
M75	20 04.9	−21 59	8.3	Globular cluster in Sagittarius
M76	01 40.9	+51 28	11.4	Planetary nebula in Perseus
M77	02 41.6	−00 04	9.1	Spiral galaxy in Cetus
M78	05 45.8	+00 02		Emission nebula in Orion
M79	05 23.3	−24 32	7.3	Globular cluster in Lepus
M80	16 15.8	−22 56	7.2	Globular cluster in Scorpio
M81	09 54.2	+69 09	6.9	Spiral galaxy in Ursa Major
M82	09 54.4	+69 47	8.7	Irregular galaxy in Ursa Major
M83	13 35.9	−29 46	7.5	Spiral galaxy in Hydra
M84	12 24.1	+13 00	9.8	Elliptical galaxy in Virgo
M85	12 24.3	+18 18	9.5	Elliptical galaxy in Coma Berenices
M86	12 25.1	+13 03	9.8	Elliptical galaxy in Virgo
M87	12 29.7	+12 30	9.3	Giant elliptical galaxy in Virgo
M88	12 30.9	+14 32	9.7	Spiral galaxy in Coma Berenices
M89	12 34.6	+12 40	10.3	Elliptical galaxy in Virgo
M90	12 35.8	+13 16	9.7	Spiral galaxy in Virgo

APPENDIX 15 The Messier Catalog, continued

Number	Right Ascension	Declination	Magnitude	Description
M91				Not identified; possibly M58
M92	17 16.5	+43 10	6.3	Globular cluster in Hercules
M93	07 43.6	−23 49	6	Open cluster in Puppis
M94	12 50.1	+41 14	8.1	Spiral galaxy in Canes Venatici
M95	10 42.8	+11 49	9.9	Barred spiral galaxy in Leo
M96	10 45.6	+11 56	9.4	Spiral galaxy in Leo
M97	11 13.7	+55 08	11.1	Owl nebula; planetary nebula in Ursa Major
M98	12 12.7	+15 01	10.4	Spiral galaxy in Coma Berenices
M99	12 17.8	+14 32	9.9	Spiral galaxy in Coma Berenices
M100	12 21.9	+15 56	9.6	Spiral galaxy in Coma Berenices
M101	14 02.5	+54 27	8.1	Spiral galaxy in Ursa Major
M102				Not identified; possibly M101
M103	01 31.9	+60 35	7	Open cluster in Cassiopeia
M104	12 39.0	−11 35	8	Sombrero galaxy; spiral galaxy in Virgo
M105	10 46.8	+12 51	9.5	Elliptical galaxy in Leo
M106	12 18.0	+47 25	9	Spiral galaxy in Canes Venatici
M107	16 31.8	−13 01	9	Globular cluster in Ophiuchus
M108	11 10.5	+55 47	10.5	Spiral galaxy in Ursa Major
M109	11 56.6	+53 29	10.6	Barred spiral galaxy in Ursa Major

GLOSSARY

Absolute magnitude The magnitude a star would have if it were precisely 10 parsecs away from the sun.
Absolute zero The temperature where all molecular or atomic motion stops, equal to $-273°C$ or $-459°F$.
Absorption line A wavelength at which light is absorbed, producing a dark feature in the spectrum.
Acceleration Any change—either of speed or direction—in the state of rest or motion of a body.
Accretion disk A rotating disk of gas surrounding a compact object (such as a neutron star or black hole), formed by material falling in.
Albedo The fraction of incident light that is reflected from a surface, such as that of a planet.
Alpha-capture reaction A nuclear fusion reaction in which an alpha particle merges with an atomic nucleus. A typical example is the formation of ^{16}O by the fusion of an alpha particle with ^{12}C.
Alpha particle A nucleus of ordinary helium containing two protons and two neutrons.
Amino acid A complex organic molecule of the type that forms proteins. Amino acids are fundamental constituents of all living matter.
Andromeda galaxy The large spiral galaxy located some 700,000 parsecs from the sun; the most distant object visible to the unaided eye.
Angstrom The unit normally used in measuring wavelengths of visible and ultraviolet light; one angstrom is equal to 10^{-8} centimeter.
Angular diameter The diameter of an object as seen on the sky, measured in units of angle.
Angular momentum A measure of the mass, radius, and rotational velocity of a rotating or orbiting body. In the simple case of an object in circular orbit, the angular momentum is equal to the mass of the object, times its distance from the center of the orbit, times its orbital speed.
Annual motions Motions in the sky caused by the earth's orbital motion about the sun. These include the seasonal variations of the sun's latitude and the sun's motion through the zodiac.
Annular eclipse A solar eclipse that occurs when the moon is near its greatest distance from the earth, so that its angular diameter is slightly smaller than that of the sun, and a ring, or annulus, of the sun's disk is visible surrounding the disk of the moon.
Anticyclone A rotating wind system around a high-pressure area. On the earth, an anticyclone rotates clockwise in the Northern Hemisphere and counterclockwise in the Southern Hemisphere.
Antimatter Matter composed of the antiparticles of ordinary matter. For each subatomic particle, there is an antiparticle that is its opposite in such properties as electrical charge, but its equivalent in mass. Matter and antimatter, if combined, annihilate each other, producing energy in the form of gamma rays according to the formula $E = mc^2$.
Aphelion The point in the orbit of a solar system object where it is farthest from the sun.
Apollo asteroid An asteroid whose orbit brings it closer to the sun than 1 astronomical unit (AU).
Asteroid Any of the thousands of small, irregular bodies orbiting the sun, primarily in the asteroid belt between the orbits of Mars and Jupiter.
Asthenosphere The deep portions of the earth's mantle, below the zone (the lithosphere) where convection currents are thought to operate. The term is also applied to similar zones in the interiors of the moon and other planets.
Astrology The ancient belief that earthly affairs and human lives are influenced by the positions of the sun, moon, and planets with respect to the zodiac.
Astrometric binary A double star recognized as such because the visible star or stars undergo periodic motion that is detected by astrometric measurements.
Astrometry The science of accurately measuring stellar positions.
Astronomical unit (AU) A unit of distance used in astronomy, equal to the average distance between the sun and the earth; 1 AU is equal to 1.4959787×10^8 kilometer.
Atomic number The number of protons in the nucleus of an element. The atomic number defines the identity of an element.
Atomic weight The mass of an atomic nucleus in atomic mass units [one atomic mass unit (amu) is defined as the average mass of the protons and neutrons in a nucleus of ordinary carbon, ^{12}C]. For most atoms, the atomic weight is approximately equal to the total number of protons and neutrons in the nucleus.
Autumnal equinox The point where the sun crosses the celestial equator from north to south, around September 21. See also **equinox** and **vernal equinox.**

Barred spiral galaxy A spiral galaxy whose nucleus has linear extensions on opposing sides, giving it a barlike shape. The spiral arms usually appear to emanate from the ends of the bar.
Basalt An igneous silicate rock common in regions formed by lava flows on the earth, the moon, and probably on the other terrestrial planets.
Beta decay A spontaneous nuclear reaction in which a neutron decays into a proton and an electron, with a neutrino being emitted also. The term has been generalized to mean

any spontaneous reaction in which an electron and a neutrino (or their antimatter equivalents) are emitted.

Big bang A term referring to any theory of cosmogony in which the universe began at a single point, was very hot initially, and has been expanding from that state since.

Binary star A double star system in which the two stars orbit a common center of mass.

Black hole An object that has collapsed under its own gravitation to such a small radius that its gravitational force traps photons of light.

Bode's law Also known as the Titius-Bode relation, a simple numerical sequence that approximately represents the relative distances of the inner seven planets from the sun. The relation is derived by adding 4 to each number in the sequence 0, 3, 6, 12, . . ., and then dividing each sum by 10, resulting in the sequence 0.4, 0.7, 1.0, 1.6, . . ., which is approximately the sequence of planetary distances from the sun in astronomical units.

Bolide An extremely bright meteor that explodes in the upper atmosphere.

Bolometric magnitude A magnitude in which all wavelengths of light are included.

Breccia Lunar rocks consisting of pebbles and soil fused together by meteorite impacts.

Burster A sporadic source of intense X rays, probably consisting of a neutron star onto which new matter falls at irregular intervals.

Carbonaceous chondrite A meteorite containing chondrules having a high abundance of carbon and other volatile elements. Carbonaceous chondrites, thought to be very old, have apparently been unaltered since the formation of the solar system.

Cassegrain focus A focal arrangement for a reflecting telescope in which a convex secondary mirror reflects the image through a hole in the primary mirror to a focus as the bottom of the telescope tube. This arrangement is commonly used in situations where relatively lightweight instruments for analyzing the light are attached directly to the telescope.

Catastrophic theory Any theory in which observed phenomena are attributed to sudden changes in conditions or to the intervention of an outside force or body.

Celestial equator The imaginary circle formed by the intersection of the earth's equatorial plane with the celestial sphere. The celestial equator is the reference line for north-south (declination) measurements in the standard equatorial coordinate system.

Celestial pole The point on the celestial sphere directly overhead at either of the earth's poles.

Celestial sphere The imaginary sphere formed by the sky. It is a convenient device for discussing and measuring the positions of astronomical objects.

Center of mass In a binary star system, or any system consisting of several objects, the point about which the mass is "balanced;" that is, the point that moves with a constant velocity through space while the individual bodies in the system move about it.

Cepheid variable A pulsating variable star, of a class named after the prototype δ Cephei. Cepheid variables obey a period-luminosity relationship and are therefore useful as distance indicators. There are two classes of Cepheid variables, the so-called classical Cepheids, which belong to Population I, and the Population II Cepheids, also known as W Virginis stars.

Chondrite A stony meteorite containing chondrules.

Chondrule A spherical inclusion in certain meteorites, usually composed of silicates and always of very great age.

Chromosphere A thin layer of hot gas just outside the photosphere in the sun and other cool stars. The temperature in the chromosphere rises from about 4000 K at its inner edge to 10,000 or 20,000 K at its outer boundary. The chromosphere is characterized by the strong red emission line of hydrogen.

Closed universe A possible state of the universe in which the expansion will eventually be reversed and which is characterized by positive curvature, being finite in extent but having no boundaries.

CNO cycle A nuclear fusion reaction sequence in which hydrogen nuclei are combined to form helium nuclei, and other nuclei, such as isotopes of carbon, oxygen, and nitrogen, appear as catalysts or by-products. The CNO cycle is dominant in the cores of stars on the upper main sequence.

Color index The difference B—V between the blue (B) and visual (V) magnitudes of a star. If B is less than V (that is, if the star is brighter in blue than in visual light), the star has a negative color index and is a relatively hot star. If B is greater than V, the color index is positive, and the star is relatively cool.

Coma The extended glowing region that surrounds the nucleus of a comet.

Comet An interplanetary body, composed of loosely bound rocky and icy material, which forms a glowing head and extended tail when it enters the inner solar system.

Configuration The position of a planet or the moon relative to the sun-earth line.

Confusion limit A natural, fundamental limitation on the astronomer's ability to probe the universe that cannot be overcome by technological improvement.

Conjunction The alignment of two celestial bodies on the sky. In connection with the planets, a conjunction is the alignment of a planet with the sun, an inferior conjunction being the occasion when an inferior planet is directly between the sun and the earth, and a superior conjunction being the occasion when any planet is directly behind the sun as seen from the earth.

Constellation A prominent pattern of bright stars, historically associated with mythological figures. In modern usage each constellation incorporates a precisely defined region of the sky.

Continental drift The slow motion of the continental masses over the surface of the earth, caused by the motions of the earth's tectonic plates, which in turn are probably caused by convection in the underlying asthenosphere.
Continuous radiation Electromagnetic radiation that is emitted in a smooth distribution with wavelength, without spectral features such as emission and absorption lines.
Convection The transport of energy by fluid motions occurring in gases, liquids, or semirigid material such as the earth's mantle. These motions are usually driven by the buoyancy of heated material, which tends to rise while cooler material descends.
Corona The very hot, extended outer atmosphere of the sun and other cool main sequence stars. The high temperature in the corona ($1-2 \times 10^6$ K) is probably caused by the dissipation of mechanical energy from the convective zone just below the photosphere.
Cosmic background radiation The primordial radiation field that fills the universe, having been created in the form of gamma rays at the time of the big bang, but having cooled since so that today its temperature is 3 K, and its peak wavelength is near 1.1 millimeters, in the microwave portion of the spectrum. Also known as the 3° background radiation.
Cosmic ray A rapidly moving atomic nucleus from space. Some cosmic rays are produced in the sun, while others come from interstellar space and probably originate in supernova explosions.
Cosmogony The study of the origins of the universe.
Cosmological constant A term added to the field equations by Einstein to allow solutions in which the universe was static (neither expanding nor contracting). Although the need for the term disappeared when it was discovered that the universe is expanding, the cosmological constant is retained in the field equations by modern cosmologists but is usually assigned the value zero.
Cosmological principle The postulate, made by most cosmologists, that the universe is both homogeneous and isotropic. It is sometimes stated as, "The universe looks the same to all observers everywhere."
Cosmological redshift A Doppler shift toward longer wavelengths that is caused by a galaxy's motion of recession due to the expansion of the universe.
Cosmology The study of the universe as a whole.
Coudé focus A focal arrangement for a reflecting telescope, in which the image is reflected by a series of mirrors to a remote, fixed location where a massive, immovable instrument can be used to analyze it.
Cyclone A rotating wind system about a low-pressure center, often associated with storms on the earth. On the earth, cyclones rotate counterclockwise in the Northern Hemisphere and clockwise in the Southern Hemisphere.

Declination The coordinate in the equatorial system that measures positions in the north-south direction, with the celestial equator as the reference line. Declinations are measured in units of degrees, minutes, and seconds of arc.
Deferent The large circle centered on or near the earth on which the epicycle for a given planet moved, in the geocentric theory of the solar system developed by ancient Greek astronomers such as Hipparchus and Ptolemy.
Degenerate gas A gas in which either free electrons or free neutrons are as densely spaced as allowed by laws of quantum mechanics. Such a gas has extraordinarily high density, and its pressure is not dependent on temperature as it is in an ordinary gas. Degenerate electron gas provides the pressure that supports white dwarfs against collapse, and degenerate neutron gas similarly supports neutron stars.
Deuterium An isotope of hydrogen containing in its nucleus one proton and one neutron.
Differential gravitational force A gravitational force acting on an extended object, so that the portions of the object closer to the source of gravitation feel a stronger force than the portions that are farther away. Such a force, also known as a tidal force, acts to deform or disrupt the object and is responsible for many phenomena, ranging from synchronous rotation of moons or double stars to planetary ring systems and the disruption of galaxies in clusters.
Differentiation The sinking of relatively heavy elements into the core of a planet or other body. Differentiation can only occur in fluid bodies, so any planet that has undergone this process must once have been at least partially molten.
Distance modulus The difference $m - M$ between the apparent and absolute magnitudes for a given star. This difference, which must be corrected for the effects of interstellar extinction, is a direct measure of the distance to the star.
Diurnal motion Any motion related to the rotation of the earth. Diurnal motions include the daily risings and settings of all celestial objects.
Doppler effect The shift in wavelength of light caused by relative motion between the source of light and the observer. The Doppler shift, $\Delta\lambda$, is defined as the difference between the observed and rest (laboratory) wavelengths for a given spectral line.
Dwarf elliptical galaxy A member of a class of small spheroidal galaxies similar to standard elliptical galaxies except for their small size and low luminosity. Dwarf galaxies are probably the most common in the universe but cannot be detected at distances beyond the Local Group of galaxies.
Dwarf nova A close binary system containing a white dwarf in which material from the companion star falls onto the other at sporadic intervals creating brief nuclear outbursts.

Eclipse An occurrence in which one object is partially or totally blocked from view by another or passes through the shadow of another.
Eclipsing binary A double star system in which one or both stars are periodically eclipsed by the other as seen from earth. This situation can occur only when the orbital plane of the binary is viewed edge-on from the earth.

Ecliptic The plane of the earth's orbit about the sun, which is approximately the plane of the solar system as a whole. The apparent path of the sun across the sky is the projection of the ecliptic onto the celestial sphere.
Electromagnetic force The force created by the interaction of electric and magnetic fields. The electromagnetic force can be either attractive or repulsive and is important in countless situations in astrophysics.
Electromagnetic radiation Waves consisting of alternating electric and magnetic fields. Depending on the wavelength, these waves may be known as gamma rays, X rays, ultraviolet radiation, visible light, infrared radiation, or radio radiation.
Electromagnetic spectrum The entire array of electromagnetic radiation arranged according to wavelength.
Electron A tiny, negatively charged particle that orbits the nucleus of an atom. The charge is equal and opposite to that of a proton in the nucleus, and in a normal atom the number of electrons and protons is equal so that the overall electrical charge is zero. The electrons emit and absorb electromagnetic radiation by making transitions between fixed energy levels.
Ellipse A geometrical shape such that the sum of the distances from any point on it to two fixed points called foci is constant. In any bound system where two objects orbit a common center of mass, their orbits are ellipses with the center of mass at one focus.
Elliptical galaxy One of a class of galaxies characterized by smooth spheroidal forms, few young stars, and little interstellar matter.
Emission line A wavelength at which radiation is emitted, creating a bright line in the spectrum.
Emission nebula A cloud of interstellar gas that glows by the light of emission lines. The source of excitation that causes the gas to emit may be radiation from a nearby star or heating by any of a variety of mechanisms.
Endothermic reaction Any nuclear or chemical reaction that requires more energy to occur than it produces.
Energy The ability to do work. Energy can be in either kinetic form, when it is a measure of the motion of an object, or potential form, when it is stored but capable of being released into kinetic form.
Epicycle A small circle on which a planet revolves, which in turn orbits another, distant body. Epicycles were used in ancient theories of the solar system to devise a cosmology that placed the earth at the center but accurately accounted for the observed planetary motions.
Equatorial coordinates The astronomical coordinate system in which positions are measured with respect to the celestial equator (in the north-south direction) and a fixed direction (in the east-west dimension). The coordinates used are declination (north-south, in units of angle) and right ascension (east-west, in units of time).
Equinox Either of two points on the sky where the planes of the ecliptic and the earth's equator intersect. When the sun is at one of these two points, the lengths of night and day on the earth are equal. See also **autumnal equinox** and **vernal equinox.**
Erg A unit of energy equal to the kinetic energy of an object of 2 grams mass moving at a speed of 1 centimeter per second, but defined technically as the work required to accelerate at 1 cm/sec^2 a mass of 1 gram through a distance of 1 centimeter.
Escape velocity The velocity required for an object to escape the gravitational field of a body such as a planet. In a more technical sense, the escape velocity is the velocity at which the kinetic energy of the object equals its gravitational potential energy; if the object moves any faster, its kinetic energy exceeds its potential energy, and it can escape the gravitational field.
Event horizon The "surface" of a black hole; the boundary of the region from within which no light can escape.
Evolutionary theory Any theory in which observed phenomena are thought to have arisen as a result of natural processes requiring no outside intervention or sudden changes.
Excitation A state in which one or more electrons of an atom or ion are in energy levels above the lowest possible one.

Fluorescence The emission of light at a particular wavelength following excitation of the electron by absorption of light at another, shorter wavelength.
Focus (1) The point at which light collected by a telescope is brought together to form an image; (2) one of two fixed points that define an ellipse (see also **ellipse**).
Force Any agent or phenomenon that produces acceleration of a mass.
Frequency The rate (in units of hertz, or cycles per second) at which electromagnetic waves pass a fixed point. The frequency, usually designated ν, is related to the wavelength λ and the speed of light c by $\nu = c/\lambda$.
Fusion reaction A nuclear reaction in which atomic nuclei combine to form more massive nuclei.

Galactic cluster A loose cluster of stars located in the disk or spiral arms of the galaxy.
Gamma ray A photon of electromagnetic radiation whose wavelength is very short and whose energy is very high. Radiation whose wavelength is less than one angstrom is usually considered to be gamma-ray radiation.
Gegenschein The diffuse glowing spot, seen on the ecliptic opposite the sun's direction, created by sunlight reflected off interplanetary dust.
Globular cluster A large spherical cluster of stars located in the halo of the galaxy. These clusters, containing up to several hundred thousand members, are thought to be among the oldest objects in the galaxy.
Gram A unit of mass, equal to the quantity of mass contained in 1 cubic centimeter of water.
Granulation The spotty appearance of the solar surface (the photosphere) caused by convection in the layers just below.

Gravitational redshift A Doppler shift toward long wavelengths caused by the effect of a gravitational field on photons of light. Photons escaping a gravitational field lose energy to the field, which results in the redshift.
Greatest elongation The greatest angular distance from the sun that an inferior planet can reach, as seen from earth.
Greenhouse effect The trapping of heat near the surface of a planet by atmospheric molecules (such as carbon dioxide) that absorb infrared radiation emitted by the surface.

H II region A volume of ionized gas surrounding a hot star (see also **emission nebula**).
H-R diagram See **Hertzsprung-Russell diagram.**
Half-life The time required for half of the nuclei of an unstable (radioactive) isotope to decay.
Halo (1) The extended outer portions of a galaxy (thought to contain a large fraction of the total mass of the galaxy, mostly in the form of dim stars and interstellar gas); (2) the extensive cloud of gas surrounding the head of a comet.
Helium flash A rapid burst of nuclear reactions in the degenerate core of a moderate mass star in the hydrogen shell-burning phase. The flash occurs when the core temperature reaches a sufficiently high temperature to trigger the triple-alpha reaction.
Hertz A unit of frequency used in describing electromagnetic radiation; 1 hertz (1 Hz) is equal to one cycle or wave per second.
Hertzsprung-Russell diagram A diagram on which stars are represented according to their absolute magnitudes (on the vertical axis) and spectral types (on the horizontal axis). Because the physical properties of stars are interrelated, they do not fall randomly on such a diagram but lie in well-defined regions according to their state of evolution. Very similar diagrams can be constructed using luminosity instead of absolute magnitude and temperature or color index in place of spectral type.
High-velocity star A star whose velocity relative to the solar system is large. As a rule high-velocity stars are Population II objects following orbital paths that are highly inclined to the plane of the galactic disk.
Homogeneous Having the quality of being uniform in properties throughout. In astronomy this term is often applied to the universe as a whole, which is postulated to be homogeneous.
Horizontal branch A sequence of stars in the H-R diagram of a globular cluster, extending horizontally across the diagram to the left from the red giant region. These are probably stars undergoing helium burning in their cores by the triple-alpha reaction.
Hubble constant The numerical factor, usually denoted H, which describes the rate of expansion of the universe. It is the proportionality constant in the Hubble law $v = Hd$, which relates the speed of recession of a galaxy (v) to its distance (d). The present value of H is not well known, with estimates ranging between 55 and 90 kilometers per second per megaparsec.
Hydrostatic equilibrium The state of balance between gravitational and pressure forces that exists at all points inside any stable object such as a star or planet.

Igneous rock A rock that was formed by cooling and hardening from a molten state.
Impact crater A crater formed on the surface of a terrestrial planet or a satellite by the impact of a meteoroid or planetesimal.
Inertia The tendency of an object to remain in its state of rest or of uniform motion. This tendency is directly related to the mass of the object.
Inferior planet One of the planets whose orbits lie closer to the sun than that of the earth (Mercury or Venus).
Infrared radiation Electromagnetic radiation in the wavelength region just longer than that of visible light; that is, radiation whose wavelength lies roughly between 7000 and 0.01 centimeter.
Interferometry The use of interference phenomena in electromagnetic waves to measure positions precisely or to achieve gains in resolution. Interferometry in radio astronomy entails the use of two or more antennae to overcome the normally very coarse resolution of a single radio telescope; in visible-light observations, the object is to eliminate the distorting effects of the earth's atmosphere.
Interstellar cloud A region of relatively high density in the interstellar medium. Interstellar clouds have densities ranging between 1 and 10^6 particles per cubic centimeter and, in aggregate, contain most of the mass in interstellar space.
Interstellar extinction The obscuration of starlight by interstellar dust. Light is scattered off dust grains, so that a distant star appears dimmer than it otherwise would. The scattering process is most effective at short (blue) wavelengths, so that stars seen through interstellar dust are reddened and dimmed.
Inverse square law In general any law describing a force or other phenomenon that decreases in strength as the square of the distance from some central reference point. In particular the term *inverse square law* is often used by itself to mean the law stating that the intensity of light emitted by a source such as a star diminishes as the square of the distance from the source.
Ion Any subatomic particle with a nonzero electrical charge. In standard practice, the term *ion* is usually applied only to positively charged particles, such as atoms missing one or more electrons.
Ionization Any process by which an electron or electrons are freed from an atom or ion. Ionization generally occurs in two ways: by the absorption of a photon with sufficient energy or by collision with another particle.

Ionosphere The zone of the earth's upper atmosphere, between 80 and 500 kilometers altitude, where charged subatomic particles (chiefly protons and electrons) are trapped by the earth's magnetic field. See also **Van Allen belts.**
Isotope Any form of a given chemical element. Different isotopes of the same element have the same number of protons in their nuclei, but different numbers of neutrons.
Isotropic Having the property of appearing the same in all directions. In astronomy this term is often postulated to apply to the universe as a whole.

Kelvin A unit of temperature, equal to 0.01 of the difference between the freezing and boiling points of water, used in a scale whose zero point if absolute zero. A Kelvin is usually denoted simply by K.
Kiloparsec A unit of distance, equal to 1000 parsecs.
Kinetic energy The energy of motion. The kinetic energy of a moving object is equal to one-half times its mass times the square of its velocity.
Kirkwood's gaps Narrow gaps in the asteroid belt created by orbital resonance with Jupiter.

Light-gathering power The ability of a telescope to collect light from an astronomical source; the light-gathering power is directly related to the area of the primary mirror or lens.
Limb darkening The dark region around the edge of the visible disk of the sun or of a planet caused by a decrease in temperature with height in the atmosphere.
Liquid metallic hydrogen Hydrogen in a state of semi-rigidity that can exist only under conditions of extremely high pressure, as in the interiors of Jupiter and Saturn.
Lithosphere The layer in the earth, moon, and terrestrial planets that includes the crust and the outer part of the mantle.
Local Group The cluster of about thirty galaxies to which the Milky Way belongs.
Logarithm The logarithm of a number is the power to which 10 must be raised to equal that number. For example, $100 = 10^2$; so the logarithm of 100 is 2.
Luminosity The total energy emitted by an object per second (the power of the object). For stars the luminosity is usually measured in units of ergs per second.
Luminosity class One of several classes to which a star can be assigned on the basis of certain luminosity indicators in its spectrum. The classes range from I for supergiants to V for main sequence stars (also known as dwarfs).
Lunar month The synodic period of the moon, equal to 27 days, 7 hours, 43 minutes, 11.5 seconds.
L wave A type of seismic wave that travels only over the surface of the earth.

Magellanic Clouds The two irregular galaxies that are the nearest neighbors to the Milky Way and are visible to the unaided eye in the Southern Hemisphere.
Magma Molten rock from the earth's interior.
Magnetic braking The slowing of the spin of a young star such as the early sun by magnetic forces exerted on the surrounding ionized gas.
Magnetic dynamo A rotating internal zone inside the sun or a planet, thought to carry the electrical currents that create the solar or planetary magnetic field.
Magnetosphere The region surrounding a star or planet that is permeated by the magnetic field of that body.
Magnitude A measure of the brightness of a star, based on a system established by Hipparchus in which stars were ranked according to how bright they appeared to the unaided eye. In the modern system a difference of five magnitudes corresponds exactly to a brightness ratio of 100, so that a star of a given magnitude has a brightness that is $100^{1/5} = 2.512$ times that of a star one magnitude fainter.
Main sequence The strip in the H-R diagram, running from upper left to lower right, where most stars that are converting hydrogen to helium by nuclear reactions in their cores are found.
Main sequence fitting A distance-determination technique in which an H-R diagram for a cluster of stars is compared with a standard H-R diagram to establish the absolute magnitude scale for the cluster H-R diagram.
Main sequence turnoff In an H-R diagram for a cluster of stars the point where the main sequence turns off toward the upper right. The main sequence turnoff, showing which stars in the cluster have evolved to become red giants, is an indicator of the age of the cluster.
Mantle The semirigid outer portion of the earth's interior extending from roughly the midway point nearly to the surface and consisting of the mesosphere (the lower portion) and the asthenosphere.
Mare (*pl.* maria) Any of several extensive, smooth lowland areas on the surface of the moon or Mercury that were created by extensive lava flows early in the history of the solar system.
Mascon A contraction of "mass concentration," referring to any of several locations on the near side of the moon where regions of enhanced density are found.
Mass-to-light ratio The mass of a galaxy, in units of solar masses, divided by its luminosity, in units of the sun's luminosity. The mass-to-light ratio is an indicator of the relative quantities of Population I and Population II stars in a galaxy.
Mean solar day The average length of the solar day as measured throughout the year; the mean solar day is precisely 24 hours.
Megaparsec (Mpc) A unit of distance, equal to 10^6 parsecs.
Meridian The great circle on the celestial sphere that passes through both poles and directly overhead; that is, the north-south line directly overhead.
Mesosphere (1) The layer of the earth's atmosphere between roughly 50 and 80 kilometers in altitude, where the temperature decreases with height; (2) the layer below the asthenosphere in the earth's mantle.

Metamorphic rock A rock formed by heat and pressure in the earth's interior.
Meteor A bright streak or flash of light created when a meteroid enters the earth's atmosphere from space.
Meteorite The remnant of a meteroid that survives a fall through the earth's atmosphere and reaches the ground.
Meteoroid A small interplanetary body.
Meteor shower A period during which meteors are seen with high frequency, occurring when the earth passes through a swarm of meteoroids.
Micrometeorite A microscopically small meteorite.
Microwave background See **cosmic background radiation.**
Milky Way Historically the diffuse band of light stretching across the sky; our cross-sectional view of the disk of our galaxy. In modern usage the term *Milky Way* refers to our galaxy as a whole.

Neutrino A subatomic particle without mass or electrical charge that is emitted in certain nuclear reactions.
Neutron A subatomic particle with no electrical charge and a mass nearly equal to that of the proton. Neutrons and protons are the chief components of the atomic nucleus.
Neutron star A very compact, dense stellar remnant whose interior consists entirely of neutrons, and which is supported against collapse by degenerate neutron gas pressure.
Newtonian focus A focal arrangement for reflecting telescopes in which a flat mirror is used to reflect the image through a hole in the side of the telescope tube.
Nonthermal radiation Radiation not due only to the temperature of an object. The term is most often applied to sources of continuous radiation such as synchrotron radiation.
Nova A star that temporarily flares up in brightness, most likely as the result of nuclear reactions caused by the deposition of new nuclear fuel on the surface of a white dwarf in a binary system. See also **recurrent nova.**
Nucleus The central, dense concentration in an atom, a comet, or a galaxy.

OB association A group of young stars whose luminosity is dominated by O and B stars.
Occam's razor The principle that the simplest explanation of any natural phenomenom is most likely the correct one. *Simple* in this case usually means requiring few assumptions or unverifiable postulates.
Oort cloud The cloud of bodies, hypothesized to be orbiting the sun at a great distance, from which comets originate.
Open universe A possible state of the universe in which its expansion will never stop, and it is characterized by negative curvature, being infinite in extent and having no boundaries.
Opposition A planetary configuration in which a superior planet is positioned exactly in the opposite direction from the sun as seen from earth.
Optical binary A pair of stars that happen to appear near each other on the sky but are not in orbit; not a true binary.
Orbital resonance A situation in which the periods of two orbiting bodies are simple multiples of each other so that they are frequently aligned and gravitational forces due to the outer body may move the inner body out of its original orbit. This is one mechanism thought responsible for creating the gaps in the rings of Saturn and Kirkwood's gaps in the asteroid belt.
Organic molecule Any of a large class of carbon-bearing molecules found in living matter.

Paleomagnetism Veestigial traces or artifacts of ancient magnetic fields.
Parallax Any apparent shift in position caused by an actual motion or displacement of the observer. See also **stellar parallax.**
Parsec A unit of distance, equal to the distance to a star whose stellar parallax is 1 arcsecond. A parsec is equal to 206, 265 AU, 3.03×10^{13} kilometers, or 3.26 light-years.
Peculiar velocity The deviation in a star's velocity from perfect circular motion about the galactic center.
Penumbra (1) The light, outer portion of a shadow, such as the portion of the earth's shadow where the moon is not totally obscured during a lunar eclipse; (2) the light, outer portion of a sunspot.
Perfect cosmological principle The postulate, adopted by advocates of the so-called steady state theory of cosmology, stating that the universe is homogeneous and isotropic with respect to space and time. It is commonly stated as, "The universe looks the same to all observers everywhere, in all directions, and at all times."
Perihelion The point in the orbit of any sun-orbiting body where it most closely approaches the sun.
Permafrost A permanent layer of ice just below the surface of certain regions on the earth and probably on Mars.
Photometer A device, usually using a photoelectric cell, for measuring the brightnesses of astronomical objects.
Photon A particle of light having wave properties but also acting as a discrete unit.
Photosphere The visible surface layer of the sun and stars; the layer from which continuous radiation escapes and where absorption lines form.
Planck constant The numerical factor h relating the frequency v of a photon to its energy E in the expression $E = hv$. The Planck constant has the value $h = 6.62620 \times 10^{-27}$ erg second.
Planck function (also known as the Planck law) The mathematical expression describing the continuous thermal spectrum of a glowing object. For a given temperature, the Planck function specifies the intensity of radiation as a function of either frequency or wavelength.
Planetary nebula A cloud of glowing, ionized gas, usually taking the form of a hollow sphere or shell, ejected by a star in the late stages of its evolution.

Planetesimal A small (diameter up to several hundred kilometers) solar system body of the type that first condensed from the solar nebula. Planetesimals are thought to have been the principal bodies that combined to form the planets.
Plate tectonics A general term referring to the motions of lithospheric plates over the surface of the earth or other terrestrial planets. See also **continental drift.**
Polarization The preferential alignment of the magnetic and electric fields in electromagnetic radiation.
Population I The class of stars in a galaxy with relatively high abundances of heavy elements. These stars are generally found in the disk and spiral arms of spiral galaxies and are relatively young. The term *Population I* is also commonly applied to other components of galaxies associated with the star formation, such as the interstellar material.
Population II The class of stars in a galaxy with relatively low abundances of heavy elements. These stars are generally found in a spheroidal distribution about the galactic center and throughout the halo and are relatively old.
Positron A subatomic particle with the same mass as the electron but with a positive electrical charge; the antiparticle of the electron.
Potential energy Energy that is stored and may be converted into kinetic energy under certain circumstances. In astronomy the most common form of potential energy is gravitational potential energy.
Precession The slow shifting of star positions on the celestial sphere caused by the 26,000-year periodic wobble of the earth's rotational axis.
Primary mirror The principal light-gathering mirror in a reflecting telescope.
Prime focus The focal arrangement in a reflecting telescope in which the image is allowed for form inside the telescope structure at the focal point of the primary mirror, so that no secondary mirror is needed.
Prograde motion Orbital or spin motion in the "normal" direction; in the solar system, this is counterclockwise as viewed from above the North Pole.
Proper motion The motion of a star across the sky, usually measured in units of arcseconds per year.
Proton-proton chain The sequence of nuclear reactions in which four hydrogen nuclei combine, through intermediate steps involving deuterium and ^{3}He, to form one helium nucleus. The proton-proton chain is responsible for energy production in the cores of stars on the lower main sequence.
Protostar A star in the process of formation, specifically one that has entered the slow gravitational contraction phase.
Pulsar A rapidly rotating neutron star that emits periodic pulses of electromagnetic radiation, probably by the emission of beams of radiation from the magnetic poles, which sweep across the sky as the star rotates.
P wave A seismic wave that is a compressional, or density, wave. P waves can travel through both solid and liquid portions of the earth and are the first to reach any remote location from an earthquake site.

QSO See **quasi-stellar object.**
Quadrature The configuration where a superior planet or the moon is 90 degrees away from the sun, as seen from the earth.
Quantum The amount of energy associated with a photon, equal to h, where h is the Planck constant, and v is the frequency. The quantum is the smallest amount of energy that can exist at a given frequency.
Quantum mechanics The physics of atomic structure and the behavior of subatomic particles, based on the principle of the quantum.
Quasar See **quasi-stellar object.**
Quasi-stellar object Any of a class of extragalactic objects characterized by emission lines with very large redshifts. The quasi-stellar objects are thought to lie at great distances, in which case they existed only at earlier times in the history of the universe; they may be young galaxies.

Radiation pressure Pressure created by the forces exerted by photons of light when they are absorbed or reflected.
Radiative transport The transport of energy, inside a star or in other situations, by radiation.
Radioactive dating A technique for estimating the age of material such as rock, based on the known initial isotopic composition and the known rate of radioactive decay for unstable isotopes initially present.
Radio galaxy Any of a class of galaxies whose luminosity is greatest in radio wavelengths. Radio galaxies are usually large elliptical galaxies, with synchrotron radiation emitted from one or more pairs of lobes located on opposite sides of the visible galaxy.
Ray A bright streak of ejecta emanating from an impact crater, especially on the moon or on Mercury.
Recurrent nova A star known to flare up in nova outbursts more than once. A recurrent nova is thought to be a binary system containing a white dwarf and a mass-losing star, in which the white dwarf sporadically flares up when material falls onto it from the companion.

Red giant A star that has completed its core hydrogen-burning stage and has begun hydrogen shell-burning, which causes its outer layers to become very extended and cool.
Reflecting telescope A telescope that brings light to a focus by using mirrors.
Reflection nebula An interstellar cloud containing dust that shines by light reflected from a nearby star.
Refracting telescope A telescope that uses lenses to bring light to a focus.
Refractory The property of being able to exist in solid form under conditions of very high temperature. A refractory element is one that is characterized by a high temperature of

vaporization; refractory elements are the first to condense into solid form when a gas cools, as in the solar nebula.
Regolith The layer of debris on the surface of the moon created by the impact of meteorites; the lunar surface layer.
Resolution In an image the ability to separate closely spaced features, that is, the clarity or fineness of the image. In a spectrum the ability to separate features that are close together in wavelength.
Retrograde motion Orbital or spin motion in the opposite direction from prograde motion; in the solar system retrograde motions are clockwise as seen from above the North Pole.
Right ascension The east-west coordinate in the equatorial coordinate system. The right ascension is measured in units of hours, minutes, and seconds to the east from a fixed direction on the sky, which itself is defined as the line of intersection of the ecliptic and the celestial equator.
Rille A type of winding, sinuous valley commonly found on the moon.
Roche limit The point near a massive body such as a planet or star inside which the tidal forces acting on an orbiting body exceed the gravitational force holding it together. The location of the Roche limit depends on the size of the orbiting body.
RR Lyrae variable A member of a class of pulsational variable stars named after the prototype star, RR Lyrae. These stars are blue-white giants with pulsational periods of less than one day and are Population II objects found primarily in globular clusters.

Saros cycle An eighteen-year, eleven-day repeating pattern of solar and lunar eclipses caused by a combination of the tilt of the lunar orbit with respect to the ecliptic and the precession of the plane of the moon's orbit.
Scattering The random reflection of photons by particles such as atoms or ions in a gas or dust particles in interstellar space.
Schwarzschild radius The radius within which an object has collapsed at the point when light can no longer escape the gravitational field as the object becomes a black hole.
Secondary mirror The second mirror in a reflecting telescope (after the primary mirror), usually either convex in shape, to reflect the image out a hole in the bottom of the telescope to the cassegrain focus, or flat, to reflect the image out of the side of the telescope to the Newtonian focus or along the telescope mount axis to the coudé focus.
Sedimentary rock A rock formed by the deposition and hardening of layers of sediment, usually either underwater or in an area subject to flooding.
Seeing The blurring and distortion of point sources of light, such as stars, caused by turbulent motions in the earth's atmosphere.
Seismic wave A wave created in a planetary or satellite interior, usually caused by an earthquake.
Selection effect The tendency for a conclusion based on observations to be influenced by the method used to select the objects for observation. An example was the early belief that all quasars are radio sources, when the principal method used to discover quasars was to look for radio sources and then see if they had other properties associated with quasars.
Semimajor axis One-half of the major, or long, axis of an ellipse.
Seyfert galaxy Any of a class of spiral galaxies, first recognized by Carl Seyfert, with unusually bright blue nuclei.
Shear wave A wave that consists of transverse motions, that is, motions perpendicular to the direction of wave travel.
Sidereal day The rotation period of the earth with respect to the stars, or as seen by a distant observer, equal to 23 hours, 56 minutes, 4.091 seconds.
Sidereal period The orbital or rotational period of any object with respect to the fixed stars or as seen by a distant observer.
Solar day The synodic rotation period of the earth with respect to the sun; that is, the length of time from one local noon, when the sun is on the meridian, to the next local noon.
Solar flare An explosive outburst of ionized gas from the sun, usually accompanied by X-ray emission and the injection of large quantities of charged particles into the solar wind.
Solar motion The deviation of the sun's velocity from perfect circular motion about the center of the galaxy; that is, the sun's peculiar velocity.
Solar nebula The primordial gas and dust cloud from which the sun and the planets condensed.
Solar wind The stream of charged subatomic particles flowing steadily outward from the sun.
Solstice The occasion when the sun, as viewed from the earth, reaches its farthest northern (summer solstice) or southern point (winter solstice).
Spectrogram A photograph of a spectrum.
Spectrograph An instrument for recording the spectra of astronomical bodies or other sources of light.
Spectroscope An instrument allowing an observer to view the spectrum of a source of light.
Spectroscopic binary A binary system recognized as a binary because its spectral lines undergo periodic Doppler shifts as the orbital motions of the two stars cause them to move toward and away from the earth. If lines of only one star are seen, it is a single-lined spectroscopic binary; if lines of both stars are seen, it is a double-lined spectroscopic binary.
Spectroscopic parallax The technique of distance determination for stars in which the absolute magnitude is inferred from the H-R diagram and then compared with the observed apparent magnitude to yield the distance.
Spectroscopy The science of analyzing the spectra of stars or other sources of light.
Spectrum An arrangement of electromagnetic radiation according to wavelength.
Spectrum binary A binary system recognized as a binary because its spectrum contains lines of two stars of different spectral types.

Spin-orbit coupling A simple relationship between the orbital and spin periods of a satellite or planet, caused by tidal forces that have slowed the rate of rotation of the orbiting body. Synchronous rotation is the simplest and most common form of spin-orbit coupling.
Spiral density wave A spiral wave pattern in a rotating, thin disk, such as the rings of Saturn or the plane of a spiral galaxy like the Milky Way.
Spiral galaxy Any of a large class of galaxies exhibiting a disk with spiral arms.
Standard candle A general term for any astronomical object, the absolute magnitude of which can be inferred from its other observed characteristics, and which is therefore useful as a distance indicator.
Steady state theory The theory of cosmology in which the universe is thought to have had no beginning and is postulated not to change with time.
Stefan-Boltzmann law The law of continuous radiation stating that for a spherical glowing object such as a star the luminosity is proportional to the square of the radius and the fourth power of the temperature.
Stefan's law An experimentally derived law of continuous radiation stating that the energy emitted by a glowing body per square centimeter of surface area is proportional to the fourth power of the absolute temperature.
Stellar parallax The apparent annual shifting of position of a nearby star with respect to more distant background stars. The term *stellar parallax* is often assumed to mean the parallax angle, which is one-half of the total angular motion a star undergoes. See also **parallax** and **parsec.**
Stellar wind Any stream of gas flowing outward from a star, including the very rapid winds from hot, luminous stars; the intermediate-velocity, rarefied winds from stars like the sun; and the slow, dense winds from cool supergiant stars.
Stratosphere The layer of the earth's atmosphere, between 10 and 50 kilometers in altitude, where the temperature increases with height.
Subduction The process in which one tectonic plate is submerged below another along a line where two plates collide. A subduction zone is usually characterized by a deep trench and an adjoining mountain range.
Supercluster A cluster of clusters of galaxies.
Supergiant A star in its late stages of evolution that is undergoing shell burning and is therefore very extended in size and very cool; an extremely large giant.
Supergranulation The pattern of large cells seen in the sun's chromosphere when viewed in the light of the strong emission line of ionized hydrogen.
Superior planet Any planet whose orbit lies beyond the earth's orbit around the sun.
Supernova The explosive destruction of a massive star that occurs when all sources of nuclear fuel have been consumed, and the star collapses catastrophically.
S wave A type of seismic wave that is a transverse, or shear, wave and can travel only through rigid materials.
Synchronous rotation A situation in which the rotational and orbital periods of an orbiting body are equal so that the same side is always facing the companion object.
Synchrotron radiation Continuous radiation produced by rapidly moving electrons traveling along lines of magnetic force.
Synodic period The orbital or rotational period of an object as seen by an observer on the earth; for the moon or a planet the synodic period is the interval between repetitions of the same phase or configuration.

Tectonic activity Geophysical processes involving motions of tectonic plates and associated volcanic and earthquake activity. See also **plate tectonics** and **continental drift.**
Temperature A measure of the internal energy of a substance. At the atomic level, temperature is related to the motions of atoms and molecules that make up the substance.
Thermal radiation Continuous radiation emitted by any body whose temperature is above absolute zero.
Thermal spectrum The spectrum of continuous radiation from a body that glows because it has a temperature above absolute zero. A thermal spectrum is one that is described mathematically by the Planck function.
Thermosphere The layer of the earth's atmosphere above the mesosphere, extending upward from a height of 80 kilometers, where the temperature rises with altitude.
3° background The radiation field filling the universe, left over from the big bang. See also **cosmic background radiation.**
Tidal force A gravitational force that tends to stretch or distort an extended object. See **differential gravitational force.**
Total eclipse Any eclipse in which the eclipsed body is totally blocked from view or totally immersed in shadow.
Transit telescope A telescope designed to point straight overhead and accurately measure the times at which stars cross the meridian.
Transverse wave See **shear wave.**
Triple-alpha reaction A nuclear fusion reaction in which three helium nuclei (or alpha particles) combine to form a carbon nucleus.
Trojan asteroid Any of several asteroids orbiting the sun at stable positions in the orbit of Jupiter, either 60 degrees ahead of the planet or 60 degrees behind it.
Troposphere The lowest temperature zone in the earth's atmosphere, extending from the surface to a height of about 10 kilometers, in which the temperature decreases with altitude.
T Tauri star A young star still associated with the interstellar material from which it formed, typically exhibiting brightness variations and a stellar wind.
24-hour anisotrophy The daily fluctuation in the observed temperature of the cosmic background radiation caused by a

combination of the earth's motion with respect to the background and its rotation.
21-centimeter line The emission line of atomic hydrogen, whose wavelength is 21.11 centimeters, emitted when the spin of the electron reverses itself with respect to that of the proton. The 21-centimeter line is the most widely used and effective means of tracing the distribution of interstellar gas in the Milky Way and other galaxies.

Ultraviolet The portion of the electromagnetic spectrum between roughly 100 and 300 angstroms.
Umbra (1) The dark inner portion of a shadow, such as the part of the earth's shadow where the moon is in total eclipse during a lunar eclipse; (2) the dark central portion of a sunspot.

Van Allen belts Zones in the earth's magnetosphere where charged particles are confined by the earth's magnetic field. There are two main belts, one centered at an altitude of roughly 1.5 times the earth's radius, and the other between 4.5 and 6.0 times the earth's radius.
Velocity curve A plot showing the orbital velocity of stars in a spiral galaxy versus distance from the galactic center.
Velocity dispersion A measure of the average velocity of particles or stars in a group or cluster with random internal motions. In globular clusters and elliptical galaxies, the velocity dispersion can be used to infer the central mass.
Vernal equinox The occasion when the sun crosses the celestial equator from south to north, usually occurring around March 21. See also **autumnal equinox** and **equinox.**
Visual binary Any binary system in which both stars can be seen through a telescope or on photographs.
Volatile The property of being easily vaporized. Volatile elements stay in gaseous form except at very low temperatures and did not condense into solid form during the formation of the solar system.

Wavelength The distance between wavecrests in any type of wave.
White dwarf The compact remnant of a low-mass star, supported against further gravitational collapse by the pressure of the degenerate electron gas that fills its interior.
Wien's law An experimentally discovered law applicable to thermal continuum radiation, which states that the wavelength of maximum emission intensity is inversely proportional to the absolute temperature.

X ray A photon of electromagnetic radiation in the wavelength interval between about 1 and 100 angstroms.

Zeeman effect The broadening or splitting of spectral lines due to the presence of a magnetic field in the gas where the lines are formed.
Zenith The point directly overhead.
Zero-age main sequence The main sequence in the H-R diagram formed by stars that have just begun their hydrogen-burning lifetimes and have not yet converted any significant fraction of their core mass into helium. The zero-age main sequence forms the lower left boundary of the broader band representing the general main sequence.
Zodiac A band circling the celestial sphere along the ecliptic that is broad enough to encompass the paths of all the planets visible to the naked eye; in some usages the sequence of constellations lying along the ecliptic.
Zodiacal light A diffuse band of light visible along the ecliptic near sunrise and sunset, created by sunlight scattered off interplanetary dust.

INDEX

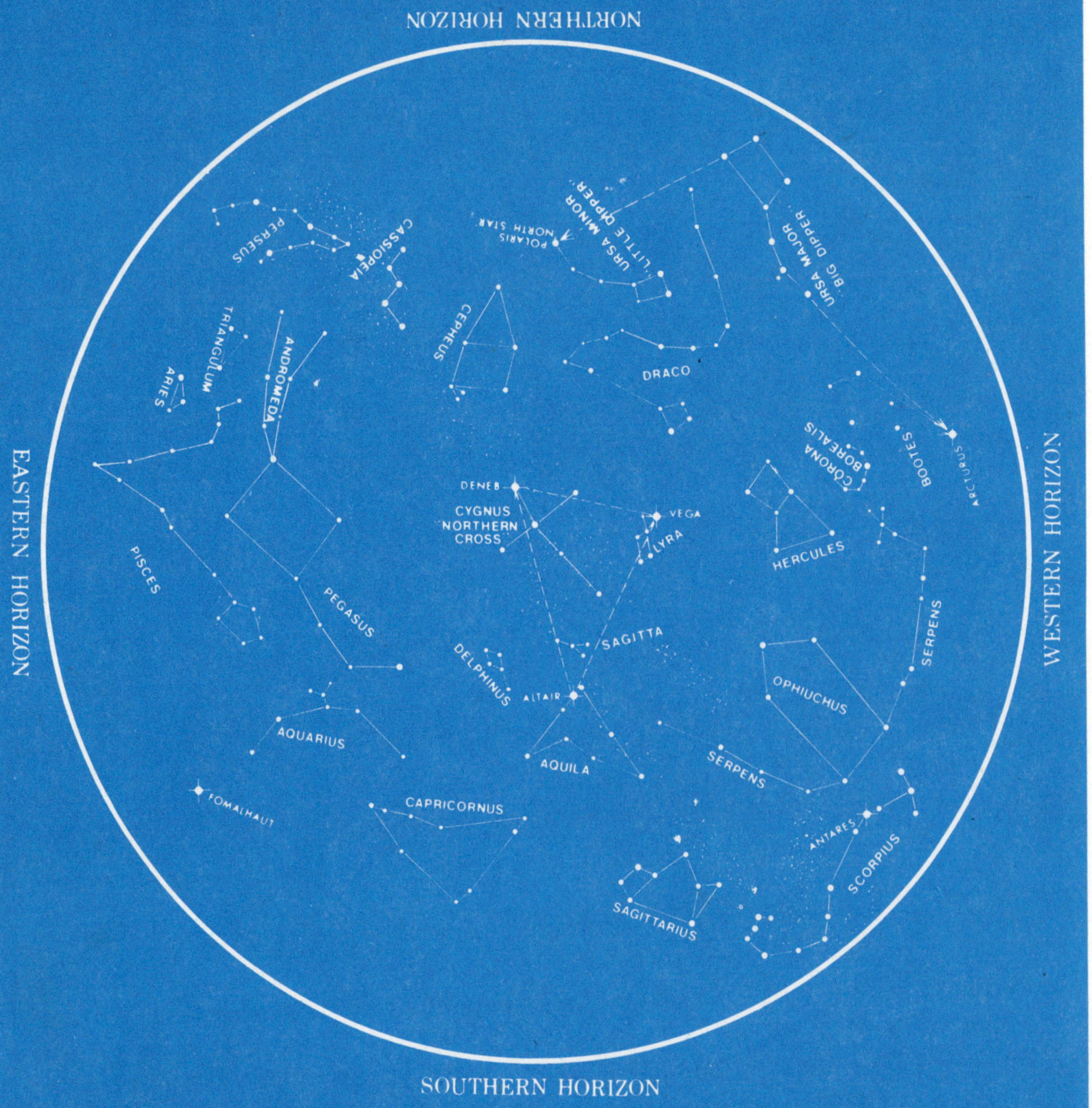

THE NIGHT SKY IN SEPTEMBER

Latitude of chart is 34°N, but it is practical throughout the continental United States.

To use: Hold chart vertically and turn it so the direction you are facing shows at the bottom.

Chart time (Local Standard):

10 p.m. First of month
9 p.m. Middle of month
8 p.m. Last of month